TRACE METALS IN THE ENVIRONMENT

VOLUME 6—Cobalt

An Appraisal of Environmental Exposure

TRACE METALS IN THE ENVIRONMENT

VOLUME 6—Cobalt

An Appraisal of Environmental Exposure

Edited by

Ivan C. Smith
Bonnie L. Carson

With Contributions by

Bonnie L. Carson
Christopher J. Cole
Betty L. Herndon
Robert A. Jacob
Joy L. McCann
Glenn M. Trischan
Ralph R. Wilkinson

Midwest Research Institute
Kansas City, Missouri

Published 1981 by Ann Arbor Science Publishers, Inc.
230 Collingwood, P.O. Box 1425, Ann Arbor, Michigan 48106

Library of Congress Catalog Card Number 77-88486
ISBN 0-250-40362-5
Manufactured in the United States of America

Technical Report No. 7, December 29, 1979
Contract No. N01-ES-8-2153
MRI Project No. 4643-L

for

National Institute of Environmental Health Sciences
P.O. Box 12233, Research Triangle Park, North Carolina 27709
Attn: Warren T. Piver

The work on which this publication is based was performed pursuant to Contract No. N01-ES-8-2153 with the National Institute of Environmental Health Sciences, National Institutes of Health, Department of Health and Human Services

PREFACE

In 1972 the National Institute of Environmental Health Sciences (NIEHS) initiated a program with Midwest Research Institute (MRI) to appraise environmental exposure to specific heavy metals in all their chemical forms and to evaluate the potential environmental and human health effects of the resulting exposures.

This book on cobalt is one of a series of comprehensive documents that resulted from this program. It appraises sources of environmental cobalt in all its chemical forms, and evaluates the potential environmental and human health effects of resulting exposure. Specifically, this book identifies natural and anthropogenic sources of cobalt in the environment; evaluates the chemical forms, quantities and modes of transport of cobalt moving through the environment; assesses human exposure to cobalt through inhalation, drinking water, the food chain and dermal contacts; and, finally, assesses the human health hazards expected from these environmental exposures.

Chapters of this book were written by senior- and mid-level staff at MRI with expertise in environmental chemistry (Bonnie L. Carson, Senior Chemist; Dr. Ralph R. Wilkinson, Senior Technology Scientist; and Christopher J. Cole, Assistant Environmental Scientist); pharmacology (Dr. Betty L. Herndon, Senior Physiologist); biochemistry (Dr. Robert A. Jacob, Senior Chemist); analytical chemistry [Dr. Glenn M. Trischan, Associate Analytical Chemist (currently at the Globe Union Division, Johnson Control Corporation, Milwaukee, Wisconsin)]; and agronomy (Joy L. McCann, Assistant Scientist). The technical editors, Bonnie L. Carson and Dr. Ivan C. Smith, have each served as project leader and/or principal investigator on the NIEHS program since its inception. The editors with to express their appreciation to Dr. Ernest Angino of the University of Kansas for technical editing of Chapters IV and V. Particular thanks are due to Dr. Edward W. Lawless, Head of the Technology Assessment Section, and Ms. Doris Nagel for editorial comments; to the MRI library staff, who helped procure the more than 2000 references cited herein; and to the MRI word processing staff, which is headed by Ms. Audene Cook, especially Ms. Elaine J. Kent and Ms. Barbara Malson.

Finally, we thank Dr. Warren T. Piver (Project Officer) and Dr. Hans Falk of NIEHS for their support, encouragement and patience throughout the many years of our association.

Ivan C. Smith is Director of the Saudi Arabian-United States Program for Cooperation in the Field of Solar Energy at the Solar Energy Research Institute, a division of Midwest Research Institute. Prior to this assignment he was Senior Advisor for Environmental Science at Midwest Research Institute, where he was actively involved in studies of the relationship between the geochemical environment and health and disease for over 15 years. His areas of expertise include identification and control of chemical emissions to the environment, their environmental persistence, routes to the human population, and impact on human health and the environment.

Dr. Smith was co-chairman (1976-1979) of the National Academy of Sciences Subcommittee, Geochemical Environment in Relation to Health and Disease, a charter member and president (1977-1978) of the Society of Environmental Geochemistry and Health, and is a past chairman of the Kansas City Section of the American Chemical Society. He is co-author of seven books, a book chapter and many technical publications and reports. He has served on numerous national panels relating to this field. He is currently Associate Editor of *Science of the Total Environment*.

Dr. Smith received his PhD in Physical Chemistry from Kansas State University in 1961 and his BS from Emporia State University in 1956. He is a member of Sigma Xi, Phi Kappa Phi, Lambda Delta Lambda, Phi Lambda Upsilon and the Scientific Research Society of America.

Bonnie L. Carson is an Associate Chemist at Midwest Research Institute in Kansas City, Missouri. She has been the principal investigator for four years on the project that resulted in documents on thallium, silver and zirconium.

Ms. Carson graduated *summa cum laude* from the University of New Hampshire in 1963 with a BA in Chemistry. She received her MS in Organic Chemistry from Oregon State University in 1966. She has been an organic chemistry laboratory instructor at the University of Waterloo in Ontario; an assistant abstractor at Chemical Abstracts Service; and a free-lance Russian translator. She is a member of the American Chemical Society, Iota Sigma Pi, Phi Kappa Phi, the American Institute of Chemists, the Society for Environmental Geochemistry and Health, and the American Association for the Advancement of Science. She has co-authored with Dr. Smith a journal article on osmium and a book chapter on trace metals in new and used petroleum products.

TABLE OF CONTENTS

TABLE OF CONTENTS (continued)

TABLE OF CONTENTS (continued)

TABLE OF CONTENTS (continued)

List of Figures

List of Tables

SUMMARY

A. Uses

The three major industrial uses of cobalt in the 1970's have been in superalloys, magnet alloys, and salts and paint driers. Each category accounts for approximately 20 to 30% of total U.S. cobalt consumption. These and other uses are summarized below.

1. Superalloys: Superalloys are alloys based on iron, nickel, or cobalt that maintain their strength at high temperatures approaching their melting points. They were developed for, and still find their biggest use in, vanes, blades, and disks of gas turbine jet aircraft engines, which operate at temperatures up to 1650°C. Cobalt-base alloys are currently used for static parts such as vanes and combustion cans of turbines, and significant amounts of cobalt are used in the nickel-base alloys required for rotating turbine blades. Up to 23 kg cobalt is used in each large-thrust jet engine. However, larger amounts of cobalt are probably used per engine in marine and industrial gas turbines; examples of each currently in use weigh 19 and 150 MT, respectively. Other uses for superalloys include space vehicles, rocket engines, experimental aircraft, nuclear reactors, steam power plants, and petrochemical apparatus.

Conventional machining of wrought superalloys removes up to 90% or more of the material as chips. New machining and net shape forging processes allow great materials savings. Various processes produce dispersion-free superalloy powders. The controlled and/or vacuum atmospheres required preclude or minimize occupational exposures. Dispersion-strengthened superalloy powders require a chemical precipitation step of a metal carbonate from solution onto particles such as ThO_2 followed by hydrogenation. The powders themselves must be handled in an inert gas atmosphere. Sintering in a hydrogen atmosphere or compaction while encased in steel cans substantially reduces the potential for occupational exposure.

Directionally solidified cobalt-containing eutectic castings are thought to be the best near-term (through the late

1980's) method to improve jet engine efficiency by significantly increasing higher engine operating temperatures. Ultra-high-temperature ceramics are not expected to replace superalloys completely in the hot section of jet engines until the late 1980's.

Although the principal superalloys market in recent years has been for retrofitting existing aircraft engines, the largest aircraft orders in history have been placed in the last few months. Cobalt superalloy demand in 1979 is about twice what it was in 1977. From 4,000 to 6,000 airliners are expected to be produced through the 1980's worldwide, but factors other than the current high price of cobalt could deter this growth, e.g., fears of a new oil crisis, higher fuel prices, inability of the airline industry to sustain expansion, and supply problems with other elements (e.g., titanium, molybdenum, and columbium). However, the jet engine market is expected to remain strong for 3 to 5 years, with an estimated 1981 production of 14,600 jet engines.

Nonaerospace uses of superalloys are expected to grow faster in the next 20 years than the jet engine market. Superalloys may find use for apparatus in critical corrosion environments such as geothermal and sour gas wells and the chemical processing industry as well as in land-based electrical power generation systems. The industrial gas-turbine and marine gas-turbine fields are expected to grow 7 to 8%/year until the late 1980's.

By the 1990's, a significant growth in the nuclear industry could increase further the use of superalloys. Most nuclear plants use cobalt-base alloys for high-temperature, sliding, and rotating wear, and corrosive operating environments. Since the exposed alloys become radioactive, maintenance and disposal are troublesome, and substitution by cobalt-free radiation-resistant alloys is probable.

Although some trucks and prototype buses have been produced with gas turbines, the lack of mass production methods has hindered expansion into this market to date.

The high price of cobalt in 1978-1979 has reduced the use of cobalt-base alloys by major jet engine and superalloy manufacturers. However, development and certification of cobalt-free substitute alloys may take 3 to 8 years. Besides substitution by nickel-base alloys, Pratt & Whitney Aircraft Group of United Technologies Corporation intends to reduce its gross cobalt consumption by 25% in 1980 by recycling machining chips and increasing the use of advanced powder metallurgical techniques, such as hot isostatic pressing, which reduce waste.

2. Magnetic alloys: The major use of cobalt in magnetic alloys is for permanent magnet compositions. Most of these (until the price of cobalt reached $5/lb) were Alnico alloys, the chief of which is Alnico V (24% cobalt). The Alnico alloys have been used widely for electrical motors and loudspeakers for automobile radios, television receivers, and phonographs. Electromagnets led to the decline in use of permanent magnets in electrical motors. Ceramic (barium ferrite) magnets are rapidly encroaching on the use of Alnico magnets in loudspeakers. Alnico magnets used exclusively in automobile radio loudspeakers in 1976 were not used at all in 1980 models. Among possible substitutes, the rare earth-cobalt magnets are the most likely substitutes for Alnico V. $Sm-Co_5$ is the material easiest to alloy. The samarium-cobalt magnets have a maximum energy product five to six times greater than that of Alnico V. Other rare earths and natural mixtures of rare earths (misch metal) are being used by General Electric Company's Hitachi Magnetics. Currently, Alnicos account for 40% or less of the permanent magnet market and Alnicos may decline to 20% of this market if the price of cobalt remains at $25/lb. Substitution of rare earth-cobalt magnets and ceramics for all Alnico applications could reduce cobalt consumption in magnets by 50% or 1.9 million lb/year (860 MT/year). However, a more moderate decline of only 15 to 30% is expected. Thus, in 1980, the permanent magnet market will be comprised of 25% each Alnico magnets and ceramic magnets with the rare earth-cobalt magnets making up the remaining 50%.

Present and potential uses of the rare earth-cobalt magnets include miniaturization of specialized electrical motors, electronic components, products requiring large repulsive forces, and permanent magnet suspension systems for high-speed interurban transport or materials handling. Consumer products containing the magnets include magnetic earrings and flexible magnetic signs.

Little information was found on efforts to identify substitutes for cobalt in soft magnets such as Permendur, 2 V-Permendur, Supermendur, Perminvars, and Hiperco, which contain 25 to 50% cobalt. These alloys are used in transformers, amplifiers, switching devices, receivers, diaphragms, sonar applications, and core material for motors, generators, and transformers.

3. Cemented carbides: Cemented carbide (hard metal) tools are produced primarily from tungsten carbide (WC) and cobalt powders. Their major use is for metal-cutting operations. The second largest use is for mining and quarrying tools, a use expected to grow because of the urgent need for fossil fuels. The last year for which data on consumption of cobalt in cemented carbides (299 MT) were available was 1969.

Cobalt may be removed in wear particles from these tools as cobalt metal or oxides (CoO or $CoWO_4$). Coatings of TiN or NiC have greatly reduced diffusion and abrasion wear. Many of the operations used in forming the tool expose the worker to cobalt-containing dusts. The majority of aerosols, however, do not exceed the threshold limit value (TLV) (0.1 mg/m^3) for cobalt.

Nickel has been little used as a binder in the past 25 years for cemented carbides. The industry feels that substitution of nickel for cobalt may give products that are inferior to the cobalt-bonded grades found acceptable today. Substitution will be swift in merely abrasion-resistant applications such as sand blast nozzles, but percussive tools require greater care in finding materials for substitution. Since the current cobalt-containing material is vastly superior to any known substitutes, customers may continue to pay the price. The tungsten market is also sensitive to political and economic pressures, but cemented carbides are not likely to be supplanted altogether by materials such as ceramics, cermets, and titanium carbide.

4. Hard-facing and wear-resistant alloys: Hard-facing is the application by welding of a specific alloy to a part, primarily to deter wear. About 180 to 450 MT cobalt has been used annually in hard-facing alloys since the Korean War. Cobalt-containing hard-facing alloys, however, comprise only 1 to 2% of the world market for hard-facing materials. Hard-facing alloys

include the Stellite®-type alloys (Co-Cr alloys with W and/or Mo), tungsten carbides, and tungsten-cobalt alloys. In addition to hard-facing, the Stellite®-type alloys are cast into metal-cutting, turning, and forming tools. The wear-resistant alloys are described as Co-, Fe-, or Ni-base alloys containing > 1% C with W and Cr to form carbides.

Industrial uses of hard-faced materials include automotive, 30 to 35%; petrochemical, 20%; chain saws, 10 to 15%; steel, 8%; and miscellaneous, 12%.

Where cobalt alloy use has been essential, there will be no quick substitutions. Nickel alloys are not a good substitute where resistance to moderate impact, abrasion, corrosion, high temperature, and nonlubricated metal-to-metal contacts are required. However, substitutions are expected to be swift for applications where cobalt alloys have been preferred but not essential. Ironically, within the past 10 years, the automotive industry had switched to using cobalt hard-facing alloys for coating internal combustion and diesel engine valves and valve seats. Now they are searching seriously for substitutes. An iron-base alloy has been developed that can replace cobalt alloys for moderate- and heavy-duty applications. Valve manufacturers, however, do not believe that cobalt alloys can be eliminated altogether, because their unique combination of toughness, corrosion resistance, and abrasion resistance is not met fully by any potential substitutes.

Although cobalt use in hard-facing alloys increased about 3%/year up through 1978, the price increase has caused a substantial decrease in its use in 1979.

Cobalt-bearing wear-resistant alloys are used in many food and drug industry cutting and scraping operations. The possibility of the constituent metals' contaminating such products is probably negligible. Food contamination by cobalt has been noted after certain grinding operations, prior to analysis.

5. Control alloys: Low-expansion cobalt alloys have been used in instruments and for glass-to-metal seals in electronics. ASTM F-15 (Kovar) and other cobalt-containing alloys were used extensively in semiconductor enclosures until 1973.

For example, ASTM F-15 was used for leads and headers (chip bases) in metal cans where the leads were glass-sealed into the headers. Because ASTM F-15 has the same coefficient of thermal expansion as borosilicate glasses, cracking was minimized during cooling. Cobalt alloys were also used for lead frames, which are metal alloy skeleton structures providing support and the exterior electrical connections that will be soldered to the circuit board. In 1973, the semiconductor industry switched from predominantly ASTM F-15 for seals and lead frames to copper and 42% Ni-58% Fe alloy. By 1990, both nickel and cobalt are expected to be phased out of the semiconductor industry. Before the substitution of cheaper materials, about 848 MT cobalt was used annually by the semiconductor industry, but only about 270 MT/year will be used in the next decade.

The price of cobalt would probably have to reach $60 to $80/lb before cobalt would be replaced in control alloys for thermostatic systems such as thermostatically controlled circuit breakers and automotive and appliance switches. Less than 23 MT of such cobalt alloys is used annually.

6. Steels: Since 1967, between 260 and 550 MT cobalt has been used annually in high-speed tool steels (5 to 12% Co), hot-work tool steel ($\leq$ 4.25% Co), cold-work tool steel ($\leq$ 3.25% Co), maraging steels, high-temperature stainless steels (0.2% Co), and miscellaneous alloys such as quench-hardened steels (4% Co). High-speed tool steels are used principally for cutting tools in high-speed machining. They have historically constituted the biggest use of cobalt in steels. Hot-work tool steels are used for punches, forging dies, and dies for the hot extrusion of brass. Cold-working steels are used for cold forming and cutting dies and tools. Maraging steels have been used since 1962 for tool and die applications.

Demand for high-speed steels in 1979 has been high; the automotive and appliance markets were the strongest. However, because of the "cobalt crisis," substitutes for cobalt alloys are being sought. Several producers are reported to have ceased production of cobalt-containing tool steels; yet, as recently as May 1979, the scarcity of cobalt had not yet caused a reduction in its content in the product mix at plants producing vacuum remelted metals because the customers' specifications had not changed.

7. Other alloys: Corrosion-resistant cobalt-base alloys are used in the chemical and pulp and paper industries. Low cobalt (1 to 2.5%) alloys are used in chemical corrosion applications such as power plant and municipal incinerator stacks.

Alloys containing about 30 to 70% cobalt have been widely used in dental and joint prostheses as well as in nails, screws, plates, etc., in bone repair surgery. Vitallium (62.5 to 65% Co) is one of the most widely known of these alloys. Cobalt alloys are not likely to be replaced in these markets.

Spring alloys (for watch mainsprings, instrument springs, electrical contacts, etc.) include Elgiloy and Cobenium (40% Co) and beryllium-copper alloys (0.2 to 2.7% Co, 0.1 to 2.85% Be, the rest Cu).

Little recent information was found for cobalt alloys used for heating elements, filaments, and cathodes. They contain $<$ 2 to 60% cobalt.

Miscellaneous cobalt alloys are used in jewelry and other ornamental applications. They range from alloys of cheap metals to alloys with gold and palladium. Cobalt is also used in brasses and bronzes for electrical instruments.

8. Metal coatings: Cobalt salts (the chloride, sulfamate, ammonium sulfate, and sulfate) used for electroplating are included in the major use category salts and driers. Although cobalt has been substituted for nickel during periods of nickel shortages, unalloyed cobalt is not widely used for electroplating or electroforming pure cobalt. Cobalt alloys are often electroplated to form thin films on fast-switching memory devices, magnetic drums, and magnetic disks. The major use of unalloyed cobalt electrodeposits is electrodeposited composite coatings for aircraft parts. In these coatings, an insoluble compound such as chromium carbide is dispersed in the plating bath and finally in the deposited matrix.

The major electrodeposited cobalt alloy is cobalt-nickel. These coatings have been used for decorative and corrosion resistance, wear resistance, electroforming, electrojoining, and magnetic applications. Electroformed cobalt-nickel alloys are used as molds and dies for metals and plastics and for aerospace applications. An electrodeposited tin-cobalt alloy has been developed as a decorative plating substitute for chromium plating in less stringent service conditions.

Electroless coatings of Co-P are produced by autocatalytic reduction of cobalt ions _in situ_ by hypophosphite ions on a catalytic surface. Elemental phosphorus, which is produced by reduction hydrolysis of $H_2PO_2^-$, alloys with the cobalt. The electroless Co-P coatings have magnetic properties applicable to high-density magnetic storage devices associated with peripheral file memories for computers (magnetic tapes, drums, cards, and primarily, disks).

Metal coatings are also produced by spraying melted and atomized cobalt metal or cobalt alloys. Although metal is "lost" as overspray, it is generally trapped in efficient collection equipment.

9. _Driers_: Cobalt salts and soaps are the most active and generally useful paint driers. They are used not only in oil-based paints but also in the newer latex paints that have been oil- or alkyd-modified and in unmodified butadiene-resin latex paint systems. Up to 0.25% cobalt is added to the vehicle. The soaps themselves are provided as formulations with 6 to 12% cobalt. Cobalt driers catalyze oxidation at the film surface. They are generally used in combination with other metals to slow down the oxidation, thereby preventing wrinkling and promoting deep drying. Cobalt soaps are also used as catalysts to promote curing or hardening of unsaturated polyester resins in building glass fiber laminates, foundry core oil binders, composition board binders, and silicone resins. In the period 1975 to 1978, U.S. consumption of cobalt in driers fluctuated from 350 to 1,200 MT/year.

10. _Catalysts_: Cobalt compounds are used commercially as catalysts for hydroformylations of olefins (oxo synthesis), petroleum refining, and oxidation of organic compounds. They show promise in emission control devices, although currently commercial automobile emission catalytic converters do not use cobalt catalysts. They are possible choices as catalysts for future synthetic fuels production from coal.

The oxo synthesis or oxonation converts monoolefins to aldehydes and alcohols in the presence of dicobalt octacarbonyl, $Co_2(CO)_8$, and cobalt hydrocarbonyl, $HCo(CO)_4$, or a cobalt carbonyl-phosphine complex. 2-Ethylhexanol is a major product produced from n-butyraldehyde, which is manufactured by the cobalt-catalyzed oxo process from propylene. The principal use of 2-ethylhexanol is in the manufacture of bis(2-ethylhexyl) phthalate, a plasticizer for flexible poly(vinyl chloride). An improved process has been commercially adopted for butyraldehyde synthesis that substitutes a rhodium-based catalyst for a cobalt-based one. Other major products that are produced by the oxo synthesis are higher fatty alcohols (C_{11+}), which are used in the production of plasticizers and detergents.

Principal catalysts for hydrodesulfurization of petroleum distillates and residues are CMA and tungsten-nickel sulfide. CMA comprises 3% cobaltous oxide, 10% molybdenum trioxide, and 87% alumina. The service lifetime of the CMA catalyst is at least 2 years. It is regenerated *in situ* by treating with air and steam at $\leq$ 600°C 10 to 50 times before replacement.

Liquid hydrocarbon fuels and chemicals can be manufactured from carbon monoxide and hydrogen in the presence of cobalt or iron catalysts. This Fischer-Tropsch synthesis can be adapted to coal conversion to synthetic fuels. An effective cobalt catalyst comprises cobalt-thoria-magnesia on diatomaceous earth (silica). The only operating commercial synthetic fuels plant is the coal-liquefaction plant in Sasolburg, South Africa, which uses an iron Fischer-Tropsch catalyst. The technology may be adopted in the United States. Other coal conversion catalysts being studied include Pt-Co; Co-Ni (a Raney alloy); and nonmetallic oxides, sulfides, and oxysulfides.

Major products produced in the presence of cobalt catalysts include terephthalic acid (by oxidation of p-xylene), which is used in the production of polyester fibers and phenol (via benzoic acid produced by the cobalt-catalyzed oxidation of toluene [2% of U.S. phenol capacity in 1975] or via cumene peroxidation [88% of U.S. phenol capacity]). Interesting uses of cobalt catalysts in consumer products are in enamels for self-cleaning ovens and in biodegradable and environmentally degradable plastic products.

11. Decolorizers, pigments, and frits: Addition of < 0.02% blue cobalt compounds to pottery, glazes, and enamels masks the yellowish tinge imparted by impurities such as iron oxide. Several oxide, silicate, and aluminate combinations of cobalt with other metal oxides are used as pigments for porcelain decoration. Cobalt pigments may be used for the body stain, the underglaze stain, and the overglaze color. Kitchen utensils are given the characteristic mottled white and blue appearance by adding 0.001 to 0.002% cobalt sulfate and/or nickel sulfate to the enamel slip just before application.

Cobalt pigments are used in artists' oil paints and in printing inks for fabric and paper. Cobalt-bearing ultramarine blue is one of the most frequently used pigments and extenders in printing ink manufacture. Bank notes are printed with cobalt-pigmented inks. Cobalt blue (cobalt silicate or aluminate) is also used as a pigment for hydraulic cement. Several organometallic dyes used in the United States contain parts per million levels of cobalt. Cobalt is not used in hair dyes on the American market although it has been fairly common in foreign metallic hair dyes.

Up to 3% cobalt (usually 0.5 to 0.6%) as a salt or oxide is used in the ground coat or frit necessary for porcelain enameling of steel for bathroom fixtures, large appliances, and kitchenware. The blue-black ground coat, which is 0.0025 to 0.0075 mm thick, becomes visible if the cover coat of porcelain is chipped away.

12. Miscellaneous uses of cobalt compounds: Cobalt compounds are also used in the preparation of vitamin B_{12}, feed supplements for ruminants, plant fertilizers, moisture indicators, and the preparation of other cobalt compounds.

B. Chemistry

1. Physical properties and descriptive inorganic chemistry: Cobalt belongs to Group VIII of the periodic classification of elements comprising the triad of iron, cobalt, and nickel and the six platinum-group metals. Its electronic configuration is $[Ar]3\underline{d}^7 4\underline{s}^2$. The uncomplexed cobaltous ion ($3\underline{d}^7$) is stable in

aqueous solution, but the cobaltic ion ($3\underline{d}^6$) is a powerful oxidizing agent unless it is present in an anhydrous crystal lattice or is complexed with various ligands. These stabilized complexes are formed when cobalt(II) is oxidized in the presence of complexing agents; even air is sufficient for complexes such as the ammino chlorides.

Cobalt is a blue-white hard metal that exists in two allotropic forms, the hexagonal-close-packed (below 417°C) form and the face-centered-cubic form.

Cobalt melts at 1493°C, boils at 3100°C, and has a density of 8.90 g/cm^3. It is essentially unaffected in the bulk state at temperatures up to 300°C by air, oxygen, and water vapor, but it is oxidized by all of these at higher temperatures. Reduction of cobalt oxides by hydrogen gas gives finely divided cobalt metal, which is pyrophoric in air.

Cobalt forms commercial alloys with several elements and these alloys are described in some detail in Chapter II.

Cobalt is electrochemically more active than nickel but less so than iron; it has a standard electrode potential of -0.277 V. Cobalt is attacked by dilute hydrochloric or sulfuric acid and concentrated nitric acid with release of hydrogen gas; however, it is passivated upon attack by concentrated nitric acid at -10°C. Cobalt is stable to dilute alkaline solutions.

The reaction conditions giving the cobaltous and/or cobaltic acetates, carbides, carbonyls,* carbonates, halides, nitrates, phosphates, sulfates, and silicides are described in Chapter III.A.

Cobaltous oxide, CoO, is formed upon the controlled oxidation of cobalt metal by air or oxygen; by thermal decomposition of the hydroxide, the carbonate, or a higher cobalt oxide;

---

* The carbonyls are less hazardous than those of most metals because of instability in air and low vapor pressure, e.g., the vapor pressure of dicobalt octacarbonyl at 15°C is 0.072 mm Hg (9.6 Pa).

or by treating cobalt metal with steam. Cobalto-cobaltic oxide, Co_3O_4, is probably a spinel-type compound consisting of $CoO \cdot Co_2O_3$. It is formed above 100°C. Further oxidation of the lower oxides gives cobaltic oxide Co_2O_3, which decomposes to CoO and oxygen above 300°C.

Numerous cobalt sulfides are known. Thermal decomposition of the sulfides CoS_2, Co_3S_4, and CoS usually gives the sulfide C_9S_8. Above 900°C, Co_9S_8 decomposes to metallic cobalt and sulfur; but under roasting conditions (elevated temperatures in the presence of oxygen), the sulfate, monoxide, and cobalto-cobaltic oxide are formed. The sulfate decomposes to the oxides at 730 to 950°C. Since the cobalto-cobaltic oxide also decomposes in this range, the major cobalt species produced by metallurgical roasting of sulfide ores or burning of coal is probably cobaltous oxide.

Cobaltous oxide is probably the ultimate thermal decomposition product of any cobalt species in oxygen or air prior to its own decomposition at 1800°C to metallic cobalt. It has a very low vapor pressure at elevated temperatures. At 1471°C (1744°K), the partial pressures above a Co-O system are 3.32×10^{-6} atm for cobalt metal, 3.48×10^{-8} atm for CoO, and 4.86×10^{-7} atm for oxygen gas.

At temperatures below 300°C, Co_3O_4 and CoO are formed from ultrafine cobalt aerosols in the air. The cobalto-cobaltic oxide is the predominant product in oxygen or air, but only CoO forms in water vapor.

2. <u>Corrosion</u>: Chapter III.B. details the oxide products formed on the surface of cobalt metal and alloys when exposed to oxygen- and sulfur-containing corrosive media. In almost all cases, CoO is the cobalt compound that would be present in any scale spalled off the metal surface. For metals in service below 900°C, both CoO and Co_3O_4 may be present in the scale. The presence of $CoWO_4$ is likely in scales spalled from tungsten-bearing alloys. $CoCr_2O_4$ is less likely to be present in spalled-off scales of cobalt alloys. Sulfides are formed from cobalt and its alloys in the presence of sulfur and hydrogen sulfide, but these are transformed into the oxides in the presence of air. Cobaltous oxide is therefore the major cobalt species

that can be lost in particulate form from jet engine or industrial gas turbines. Since spalling leads to metal failure, it is probably not an important pathway by which cobalt from in-service metals reaches the environment. However, catastrophic oxidation of cobalt-bearing alloys, especially in marine atmospheres, by the process of sulfidation, or "hot corrosion," is not unknown. Hot corrosion occurs when a gas turbine is operated on fuels containing $\sim$ 0.25% sulfur in a saltwater environment. The Na_2SO_4 that forms under these conditions causes severe corrosion. The sulfides that form on the metal surface are rapidly converted to oxides.

Pipes made of alloys containing cobalt as a constituent or as an impurity will give corrosion products on exposure to steam or hot water at high pressures that contain small amounts of cobalt oxide. Most of the cobalt oxide produced will probably deposit in the system itself. For example, the solubility of cobalt oxides in superheated steam at 4 to 24 MPa and 400 to 550°C is in the lower parts per billion range. Such concentrations are typical of environmental waters.

When metals are discarded in landfills, corrosion by soil biota should also be considered. Cobalt is more resistant to biocorrosion by _Penicillium expansum_ at pH 6.8 and 25°C than zinc, tin, aluminum, copper, and cadmium. The lowest removal rate for cobalt may be due to its toxicity (copper and cadmium also supported far less microbial growth than did the other metals).

3. Chemistry of selected cobalt complexes: The following equilibria among complex hydroxocobalt species occur in basic cobalt solutions:

$$Co^{2+} \underset{\text{pH 8}}{\rightleftharpoons} Co(OH)^{+} \rightleftharpoons Co_4(OH)_4^{4+} \underset{\text{pH 10}}{\rightleftharpoons}$$

$$Co(OH)_2 \rightleftharpoons Co(OH)_3^{-} \underset{\text{pH 12}}{\rightleftharpoons} Co(OH)_4^{2-}$$

Fresh solutions of pentacyanocobaltous ion, $[Co(CN)_5H_2O]^{3-}$, readily absorb hydrogen gas to form the complex hydride $[HCo(CN)_5]^{3-}$. This complex is useful for homogeneous catalytic hydrogenation of unsaturated compounds and reductive aminations of α-oxo acids and esters.

Hydridocobalt tetracarbonyl, $HCo(CO)_4$, decomposes reversibly to give hydrogen and dicobalt octacarbonyl, $Co_2(CO)_8$. It is a strong acid in aqueous solution. The compound forms cobalt-carbon bonds by addition to epoxides, carbon monoxide, and olefins in synthetic organic reactions.

Dicobalt octacarbonyl is also used in synthetic organic reactions including catalytic hydroformylation of olefins (the oxo reaction), hydrogenation of olefins and aldehydes, and hydrosilation of olefins.

Cobalt forms very stable complexes with nitrilotriacetic acid (NTA), ethylenediaminetetraacetic acid (EDTA), and acetylacetonate (acac), e.g., $NaCoNTA \cdot H_2O$, $Co_2Co(NTA)_2 \cdot 3H_2O$, $Co(acac)_2 \cdot 2H_2O$, and Na[Co-EDTA]. Chelation of cobalt by such multidentate ligands as EDTA strongly enhances its mobility in water and soil.

The cobalt ammines are historically important in that they were among the compounds studied by Werner in his development of the theory of coordination compounds. Their chemistry is applied in several commercial cobalt recovery and refining processes. For example, when a vigorous stream of air is drawn through a solution of a cobalt(II) salt containing ammonia, the corresponding ammonium salt, and activated carbon, the hexammine salts $[Co(NH_3)_6]X_3$ where X = Cl, Br, or NO_3 are obtained. The reaction is involved in a commercial cobalt refining process. The chemistry of the very stable cobalt(III) complexes is extensive, but there are only two simple cobalt(III) compounds, the sulfate and the fluoride.

In 3 $\underline{M}$ HCl solutions, the hexaaquocobalt ion is converted to $Co(H_2O)_5Cl^+$ and $Co(H_2O)_2Cl_2$. In 8 $\underline{M}$ HCl solutions, species formed are $Co(H_2O)Cl_3^-$ and $CoCl_4^{2-}$.

4. Compounds of biological and environmental interest: Certain cobalt coordination compounds such as porphyrin complexes and Schiff base complexes in the presence of a coordinating solvent have been studied because they can bind oxygen reversibly like hemoglobin.

Cyanocobalamin or vitamin B_{12} is a cobalt coordination compound found in soil, water, sewage sludge, manure, and dried estuarine mud. The unique features of its structure are the cobalt-carbon bond and the corrinoid ligand system (similar to porphyrin), which includes four pyrrole nuclei joined in a ring.

Several cobalamins are known in which the cyanide group is replaced by another group. In cobamamide, this group is 5'-deoxyadenosine. Other cobalamins include those with cobalt bonds to hydroxide ion and to a methyl group. In these cobalamins, cobalt has a formal oxidation state of three. Chemical reduction produces reduced forms whose detailed chemical nature is unknown.

Cyanocobalamin, B_{12}-CN, reacts with many acylating or alkylating agents to give the corresponding derivatives. Methylcobalamin in aqueous solution and in the presence of excess oxygen, photochemically yields B_{12}-OH and formaldehyde; but in the absence of oxygen, methyl- and ethylcobalamin are photostable compounds. Methylcobalamin is capable of transferring a methyl group to ions of certain metals and metalloids such as Hg, Sn, As, Se, and Te. Such methylation reactions of methylcobalamin are suspected of occurring in the environment and several examples have been confirmed in the laboratory. Environmental occurrences of cobalamins are due to their synthesis by bacteria. Bioalkylations in the environment appear to involve the bacteria *Pseudomonas*, *Aeromonas*, *Flavobacterium*, *Escherichia coli*, and *Methanobacterium*.

The question of trace metal flux in the environment may not depend entirely upon corrinoid systems. Methylation of metalloids, for example, may also involve coenzymes containing sulfur atoms such as *S*-adenosylmethionine.

5. *Analytical chemistry*: Dry ashing at 450°C often with the aid of nitric acid or magnesium nitrate is generally satisfactory for cobalt determinations in organic matrices. Use of sodium carbonate for dry ashing of plant material results in cobalt losses. Losses also occur during rigorous dry ashing procedures. Low-temperature ashing, however, is inadequate for the preparation of coal for DC arc emission analysis purposes because of incomplete destruction of the matrix. Wet ashing does not cause obvious losses.

Cobalt can be separated from interfering materials by precipitation, chelation solvent exchange, chromatography, electrodeposition, and ion-exchange methods. Specific chelating materials include dithizone, ammonium pyrrolidine dithiocarbamate (APDC), sodium diethyldithiocarbamate (NDDC), oxine (8-hydroxyquinoline), trifluoroacetylacetone, chitosan, and Chelex 100 chelating resin. Dithizone is one of the oldest and most widely used extracting agents for separating cobalt. Thiocyanate complexes can be extracted with organic solvents such as amyl alcohol and/or diethyl ether.

During separation steps, variable cobalt recoveries may arise from equilibrium shifts and the significant dependency of extraction efficiency on pH.

Relatively few organic standard reference materials (SRM's) are available that have certified cobalt concentrations. However, many certified cobalt concentrations in inorganic standards are available from the U.S. Geological Survey (USGS) and the National Bureau of Standards (NBS). The International Atomic Energy Agency (IAEA) also has a series of organic standards, almost all of whose analyses for interlaboratory comparison have been performed by neutron activation analysis (NAA). The NBS Certificates of Analysis do not identify the specific analytical techniques by which the certified or "information only" values were determined. There has been good overall agreement between the NBS informational value for cobalt in coal and fly ash SRM's, but the data for orchard leaves and bovine liver reported by numerous laboratories vary considerably from the NBS informational value. For example, reported values for NBS SRM-1577 bovine liver range from 0.12 to 0.41 mg Co/kg dry weight compared to the informational value of 0.18 mg Co/kg. When the same laboratory analyzed SRM-1577 and SRM-1571 orchard leaves by both atomic absorption (AAS) and NAA, NAA gave consistently higher cobalt concentrations. However, in another IAEA interlaboratory comparison, the average cobalt concentration in an oyster homogenate was 1.9 ± 0.8 mg Co/kg by AAS and only 0.41 ± 0.02 mg Co/kg dry weight by NAA. Possible sources of systematic error among the laboratories using atomic absorption techniques may be due to low-level contamination or interference problems. In data reported in Chapter VII, we have noted that cobalt values determined by NAA in foods and aquatic organisms are significantly lower values than values determined by AAS. Of course, in these cases, the specific samples were not identical as in the case of the round robins with SRM's.

Cobalt contamination of biological tissues has been reported from disposable syringes (blood), technical grades of anticoagulants (blood), Menghini needles (liver biopsy), tungsten carbide and alumina-ceramic mortars and pestles or grinding vials, and a jar mill with flint balls (oats). Other sources of contamination may be collection devices, storage containers, or preparation reagents. Blanks should always be analyzed.

Cobalt in acidic solutions (e.g., pH 1.5) is not measurably adsorbed to the walls of borosilicate or polyethylene storage containers in 1 year. At pH 8.1, cobalt losses range up to 7% in borosilicate glass and 15% in polyethylene containers within a 20-day storage period. Contamination from high-density linear polyethylene containers has been documented for 10% nitric acid solutions stored in the frozen state for 4 years.

The most useful and still frequently used classical analytical technique for cobalt determinations employs colorimetric/photometric methods.* Single-element instrumental techniques most frequently used for cobalt determinations are AAS and electrochemical methods, usually polarography. Multielement techniques that have been used successfully for cobalt determinations include emission spectroscopy (ES), plasma emission spectroscopy (PES), NAA, and x-ray techniques, primarily fluorescence (XRF). Spark source mass spectrometry (SSMS) and catalytic reactions are among additional methods that are finding developing uses. However, AAS and NAA continue to be the preferred methods for cobalt analyses. The sensitivities of the methods are described in Chapter III. Obviously a method sensitive enough for the expected cobalt concentrations must be used. In many multielement surveys of environmental samples, the method sensitivity has not been low enough for cobalt.

Colorimetric determinations of an element require extensive separations. They have been widely used for both inorganic and organic samples and are still being developed. During the period 1975-1978, more than 30 reports of colorimetric determinations of cobalt were cited in Analytical Abstracts. When comparing cobalt concentrations in materials of the same class by

---

* Colorimetry in various modifications is known as spectrophotometry, photometry, or absorptrometry.

various methods, the older colorimetric values often fall within the range of values determined by NAA. Iron, copper, and nickel are the most frequently reported analytical interferences. Other elements also cause problems. However, the general trend is away from colorimetric/photometric methods towards instrumental techniques with less stringent requirements for sample preparation to minimize interferences.

The electrochemical method of analysis has been used primarily for aqueous matrices. Interferences from surfactants, zinc, tellurium, iron, chromium, and manganese have been reported in cobalt determinations. The need for constant electrolyte properties, interelement and matrix interferences, and the length of time per analysis all limit the appeal of electrochemical methods.

XRF techniques offer a rare opportunity for rapid nondestructive simultaneous multielement sample analysis. Only other x-ray methods and possibly NAA offer this advantage. Major limitations of XRF methods involve matrix effects and insufficient understanding of x-ray line characteristics with respect to analyte concentration. Analysis is essentially limited to the sample surface so that the homogeneous distribution in the bulk sample and the sample surface is necessary to assure accurate quantitative analysis. Although the method is capable of low parts per million (ppm) sensitivity, competing analytical techniques are available having equal or greater sensitivity with shorter analysis times. Generally, use of XRF techniques for the determination of cobalt has been limited to geological, metallurgical, and fuel matrices. When used for environmental samples and SRM's, the cobalt concentrations have sometimes been below the detection limits. The applicability of XRF techniques for the determination of cobalt concentrations less than 1 mg/kg is limited.

Arc and spark ES techniques are widely used for rapid semiquantitative analyses of solids and liquids. Both organic and inorganic samples have been analyzed by ES. PES is used more frequently for quantitative analyses and is generally restricted to liquefied samples. The sensitivity of the method for cobalt depends on the spectral emission line chosen for the analysis. The high excitation energy available in either arc or spark discharges minimizes the dependency of analyte response on its original oxidation state or chemical speciation. Matrix effects can often be compensated by buffer systems. However, the semi-quantitative nature of the data, particularly when internal standards are not used, limits the attractiveness of this method.

Unlike other ES methods, PES requires the nebulization of sample solutions into a plasma discharge so that the possibilities of direct analysis of solid material is limited. Emission excitation is generally accomplished using an electronic plasma discharge in either a DC discharge configuration or a radio frequency inductively coupled plasma (ICP). The ICP/PES system appears to be suitable for analyzing biological tissues and inorganic samples, with suitable sensitivity (e.g., at the 3453.5 Å emission line, 0.029 mg Co/kg in environmental materials dissolved in perchloric acid and 5 μg/liter in diluted blood serum). The use of PES techniques, especially those employing ICP sources, for rapid simultaneous multielement analyses of environmental and biological matrices, is increasing. The major analytical limitations are the requirement for liquid samples and the possibility of unknown matrix and interelement effects in uncharacterized systems.

AAS has been used for the determination of cobalt in a multitude of organic and inorganic matrices as exemplified by the compiled data in the tables of Chapters IV and VII. Flame atomization techniques are generally capable of determining cobalt at the low ppm level, whereas electrothermal (nonflame or flameless) methods of atomization, including carbon rod and graphite furnace methods, have sensitivities in the low parts per billion (ppb) concentration range. The detection limits attained in specific situations depend largely on the sample matrix and preparative methods used. The sensitivity of the flameless methods allows analyses of low elemental concentrations previously unattainable in biological, medicinal, and toxicological applications.

Problems encountered in AAS analysis of samples of cobalt may be either chemical or spectroscopic in nature. Chemical interferences may be the result of either sample matrix or sample preparation factors. For example, the results for cobalt determination in silicate rock were significantly elevated above actual concentrations unless background correction methods were employed to compensate for spurious matrix absorption. Matrix effects problems are more pronounced where electrothermal atomization techniques are used. For example, cobalt was found to co-volatilize with the major salts in saline waters to an extent that was dramatically dependent upon the salt content. The use of narrow band widths and high resolution monochromators is necessary when the 2407 Å cobalt line is used due to the abundance of spectral lines in that region. Analyses using overlapping spectral lines are subject to interferences from matrix components and have very limited analytical response linearity range.

In activation analysis, the decay of each radionuclide produced by irradiating the sample emits a characteristic radiation having an equally characteristic decay rate. The reaction $^{59}Co(n,\gamma)^{60}Co$ has been most widely used for cobalt activation analyses. The detection limit for the reaction is 0.0005 μg. The classical method of NAA involves isolation of an analyte from matrix inferences, after irradiation, through a series of radiochemical separations and usually involves rather extensive sample handling procedures. However, small quantities of an analyte may be effectively separated and quantitatively analyzed in otherwise intractable matrices. Instrumental neutron activation analysis (INAA) dispenses with preanalysis radiochemical separations. Identification and quantitative analysis in INAA relies upon the measurement of characteristic photoemissions separated through a high-resolution gamma-ray spectrometer. INAA permits more rapid sample analysis than is possible using radiochemical separations but may be subject to interferences from other matrix components.

Cobalt determination in urine has been a specific analytical problem as can be seen by comparing literature values (see Chapter VII.C.). NAA results for urine specimens in one study were generally low compared to previously reported values. The authors concluded that inadequate procedures or inaccurate sampling had occurred in some of the previous studies.

Numerous examples of the use of NAA and INAA for environmental and biological samples are cited in Chapters III.E., IV, and VII. In a multielement analysis of freshwater by INAA, blank corrections for cobalt were not employed for the sample container or acid used, but a correction of 0.04 µg/liter was needed to compensate for contributions from the quartz irradiation ampule. Cobalt impurities in filter materials must usually be corrected for. Interferences in NAA of petroleums, the reactions $^{60}Ni(n,p)^{60}Co$ and $^{63}Cu(n,\alpha)^{60}Co$, should be anticipated.

The former is the more serious source of interference due to the abundance of nickel in petroleum.

Miscellaneous techniques that have been used for successful cobalt determinations are also described in some depth in Chapter III. These include catalytic reactions involving the Co(II)-catalyzed oxidation of an organic material by hydrogen peroxide under basic conditions, SSMS, chromatography, molecular fluorescence, and laser microprobe ES.

SSMS is an instrument-intensive technique which offers high sensitivity and selectivity with minimum sample treatment. Direct standard comparison and isotope dilution techniques are the two general methods of SSMS analysis used. In isotope dilution, a known quantity of one analyte isotope is enriched in the sample and analyzed. The intensities of the analyte ion and the enriched ion are compared to allow for quantitative elemental analysis. Analyte elements, therefore, must have at least two isotopes to use the technique. Direct standard comparison SSMS is generally limited to semiquantitative applications.

C. Geochemistry and Occurrence

The crustal abundance of cobalt is about 25 mg/kg. Most of the cobalt presumably substitutes for Fe(II) and Mg(II) ions in easily weathered silicate minerals such as olivine, hornblende, augite, and biotite. The most important cobalt minerals are linnaeite, Co_3S_4; carrollite, $CuCo_2S_4$; safflorite, $CoAs_2$; skutterudite, $(Co,Fe)As_3$; erythrite, $Co_3(AsO_4)_2 \cdot 8H_2O$; and glaucodot, (Co,Fe)AsS. Most of these minerals have been mined or concentrated for their cobalt content as have pentlandite, $(Fe,Ni,Co)_9S_8$; cobaltite, (Co,Fe)AsS; heterogenite, CoOOH; asbolite, which are manganese oxides plus cobalt; gersdorffite, (Ni,Co)AsS; pyrrhotite $(Fe,Ni,Co)_{\underline{x}-1}S_{\underline{x}}$; pyrite, $(Fe,Ni,Co)S_2$; and sphalerite, Zn(Co)S.

The concentrations of cobalt (an average unless stated otherwise) in crustal rocks are 150 or 200 mg/kg in ultramafic (ultrabasic) rocks; 45 or 48 mg/kg in basaltic rocks; 0.09 to 16.9 in granitic rocks; 7 to 118 mg/kg in metamorphic rocks; 4 to 147 mg/kg in schists; 19 or 20 mg/kg in clays and shales; 4 mg/kg in sands; and 1.2 mg/kg in sandstone and limestone. The cobalt concentration usually increases with increasing iron concentration. During weathering, cobalt tends to behave like iron, except when separation of iron and manganese occurs, whereupon cobalt follows manganese. Chemical weathering of ultrabasic rocks into laterite deposits is one of the chief ways that cobalt has concentrated in mineral deposits.

Among the better known cobalt deposits according to geological classification are the nickel-copper sulfide ores of the Sudbury district, Ontario, and the Duluth gabbro complex southeast of Ely, Minnesota (hypogene deposits associated with mafic intrusive igneous rocks); Cornwall and Morgantown deposits in Pennsylvania (contact metamorphic deposits associated with mafic intrusive igneous rocks); Riddle, Oregon, and Moa Bay, Cuba (lateritic or weathered deposits); Ducktown, Tennessee (massive sulfide deposits in metamorphic rocks, chiefly of volcano-sedimentary origins); the Fredericktown, Missouri, and the Bou Azzer area in Morocco (hydrothermal deposits); the Blackbird district, Idaho, and the copper-cobalt deposits of Zaire and Zambia (strata-bound deposits*); and deep-sea manganese nodules (deposits formed as chemical precipitates). The USGS has estimated that each deposit of the U.S.-identified cobalt resources contains 3 to 150 million lb (1,400 to 68,000 MT) cobalt, except for the Duluth gabbro complex, which contains at least 1 billion lb (450,000 MT) cobalt. The Missouri deposits contain 1,600 to 1,800 mg Co/kg; the Riddle, Oregon, deposits, 500 mg Co/kg. One of the richest deposits with 6,000 to 7,300 mg Co/kg is the Blackbird district.

The Blackbird deposit has been mined in the period 1917-1920, 1938-1941, 1951-1960, and 1966-1969. The only nearby population was the company town, Cobalt, Idaho, which housed the miners. Noranda Mines Ltd. has discovered more cobalt-copper deposits in the district and plans to reopen the Blackbird mine.

Cobalt is associated with arsenic and antimony in polymetallic sulfide ores and in lead, zinc, and copper sulfide ores. However, the average cobalt concentration in most base metal sulfide minerals is probably not as high as it is in average

---

* The Blackbird district has been traditionally classified with hydrothermal deposits.

ultrabasic rocks. The average cobalt concentration in Coeur d'Alene sphalerites is ~ 55 mg/kg.

Cobalt also accompanies nearly all pitchblende and urainite (uranium) ores, but little information was found on this topic (the little that was is discussed in Part E of this summary).

Cobalt may be enriched in the quartz-pebble conglomerate type of gold deposit. The few cobalt concentrations found in U.S. gold ores are not high; however, cobalt concentrations reported for U.S. iron oxide minerals and ores (< 4 to 143 mg/kg) are low and reports linking cobalt environmental contamination to mining them were not found.

All aspects of coal mining technology are examined in some depth in Chapter IV from the viewpoint of cobalt occurrence in the wastes. However, cobalt in acid coal mine drainage and cobalt in emissions from coal burning are described in Chapter V on environmental transport (and thus, in the next section of this summary).

Representative averages for cobalt in U.S. coals and fly ashes are 5 and 25 mg/kg, respectively. Cobalt shows a slight tendency to become enriched on the finer (uncollected) fly ash particles. Bottom ash and boiler slag at a few U.S. coal-burning plants have been reported to contain ~ 7 to 30 mg Co/kg.

Flue gases contain considerable quantities of cobalt. The effluent from a flue gas scrubber of a coal-burning power plant contains cobalt concentrations of 100 to 700 μg/liter. Coal cleaning wastes, stored in the open, can be a source of acid drainage. A typical waste from the Illinois Coal Basin contains 30 mg Co/kg, 60% of which is removed in the laboratory by leaching for 4 days with an equal weight of water.

Cobalt is expected in new coal technology wastes, possibly as the carbonyl. The latter may be emitted from fixed-bed catalyst regeneration in a coal liquefaction plant (this is also possible in a petroleum refinery), from quenching and cooling of coal gasifier off-gas, from sulfur recovery operations during coal liquefaction or gasification, and from reactor off-gas in coal liquefaction. One estimate is that 5,450 MT cobalt would be volatilized annually during gasification of 1 billion short tons of coal containing 6 mg Co/kg. (By comparison, Bureau of Mines data give < 600 million short tons of coal consumed annually in recent years in the United States.)

Coal gasification pilot plant aqueous effluents contain 2 μg Co/liter; the process waters themselves contain up to at least 10 μg Co/liter.

Solvent-refined coal contains only ~ 3 to 5% of its original cobalt concentration and 73% of this is in organic forms. (Only 17% is in the organic form in the original coal.) A residue from evaporation of the solvent contains ~ 97% of the original cobalt.

Examination of the potential wastes of a coal liquefaction facility using 30,000 short tons coal/day and a cobalt-, nickel-, and iron-catalyzed Fischer-Tropsch process shows that emissions of cobalt carbonyl are unlikely during normal operation but are possible for short periods during shutdown operations.

Cobalt is consistently present as a minor constituent of crude oils. The cobalt was either derived from marine organisms, the primary source of the petroleum, or was removed from seawater solution and incorporated in the asphalt sheet structure. Cobalt correlates well with the nitrogen content but may not be associated with the porphyrins in the oils. Cobalt concentrations in crude oils range from 0.001 to 10 mg Co/kg. Oil shales contain an average 10 mg Co/kg; medium- to extra-heavy fuel oils, 0.03 to 1.4 mg Co/kg,* and gasoline, < 0.1 mg Co/kg. Asphalts, asphaltenes, asphaltites, and rock bitumens contain up to 122 mg Co/kg.

Upon retorting of raw oil shale, simulating a potential commercial process, only 0.4 to 0.5% of the cobalt in the raw shale is transferred to the oil; the remainder is left in the spent shale. Process water contains 2 to 5 μg/liter. Other oil-shale processing waters from pilot plants contain up to 30 μg/

---

* New data are presented in the summary of Chapter V that raise the maximum reported value for cobalt in fuel oils tabulated in Chapter IV.

liter. Cobalt has also been found at the microgram per liter level in some U.S. oil-field waters (brines).

Inorganic fertilizer materials including those derived from phosphate rocks generally contain cobalt concentrations (< 10 mg/kg) lower than those of ordinary soils. Manures have been reported to contain ~ 0.001 to ~ 0.2 mg Co/kg fresh weight and about 8 mg/kg dry weight. Common concentrations in sewage sludge are ≤ 15 mg Co/kg although values as high as 100 mg Co/kg have been reported. Miscellaneous organic fertilizer materials include up to 6.25 mg Co/kg dry weight. If a fertilizer is to claim cobalt as a nutritional additive, it should contain ≥ 5 mg Co/kg. It has been estimated that a common agricultural fertilizer containing ~ 10 mg Co/kg and applied at the usual rate of 500 lb/acre makes little contribution to the cobalt concentration of the soil. The amount applied is probably less than that removed by the annual crop.

Cobalt-deficient soils are generally either treated with phosphate fertilizers to which cobaltous sulfate has been added or are sprayed directly by a cobalt sulfate solution to remedy the deficiency. Fertilizing with cobalt can have a dramatic effect on the cobalt content of pasture plants.

Peats are used more for soil improvements than as fuels. The few cobalt concentrations found for peat are in the range < 8 to 12 mg Co/kg.

Because it is used as a catalyst in terephthalic acid production, cobalt is found as an impurity in some polyester fibers (~ 30 to 40 mg Co/kg). Manufactured papers contain 0.01 to 0.36 mg Co/kg. Conventional (heat-dried) printing inks and coatings for cans may contain 5 to 60% cobalt blue pigment or 0 to 0.2% cobalt naphthenate drier. Inks that can be dried by ultraviolet light are becoming more important than heat-dried inks and coatings. They may contain up to 50% pigment. Cobalt emissions are probably negligible.

Cosmetic talcum powders formulated prior to 1973 contained 4 to 88 mg Co/kg. Cobalt either substitutes for magnesium in the talc (a sheet silicate) or for the iron in the brucite layer of the talc.

Cobalt is also present in dust-free chrysotile asbestos fibers (40 to 120 mg Co/kg) and the associated dusts (120 to 310 mg Co/kg).

Most soils contain cobalt concentrations in the range 1 to 40 mg/kg; the U.S. average is 7.2 mg Co/kg. Almost 75% of U.S. soil and other surficial materials sampled to a depth of 20 cm contains cobalt in the range 3 to 15 mg/kg; 15% contains < 3 mg Co/kg; and ~ 10% contains 15 to 80 mg Co/kg. Soils containing < 0.5 to 3 mg Co/kg are deemed deficient because plants growing on them have insufficient cobalt (< 0.08 to 0.1 mg Co/kg) to supply the dietary requirements of cattle and sheep. The largest world areas of cobalt-deficient soils are from sandstones, especially if calcareous. Cobalt-deficient soils include the humus groundwater podzols of the southeastern U.S. and the podzols, brown podzolic soils, and humus groundwater podzols in the northeastern section of the United States. Because of leaching, presumably during formation of podzol morphology, humus groundwater podzols (low clay soils subjected to a fluctuating water table) appear to be nearly stripped of reactive forms of cobalt (as well as of iron and manganese, which have been reduced to soluble forms).

Soils sometimes appear to be contaminated by airport and highway traffic. In Grand Rapids, Michigan, the soil in the airport area contained more than 3 times the cobalt concentration found in residential soils. Cobalt was nearly twice as high in acid extracts of soil samples within 5 km of two heavily traveled Japanese highways as in those from beyond 5 km. Freeway dust sampled near the Los Angeles International Airport gave saturation extracts at pH 5.1 containing 2.0 mg Co/liter. However, saturation extracts of most California soils, representing 30 soil series, revealed cobalt in only two (at < 0.01 and 0.14 mg Co/liter).

Chicago street dusts contain 5.6 to 12.6 mg Co/kg. In the small town of Lancaster, England, the street dusts contain 9.6 mg Co/kg; the average for a rural English road is 7.1 mg Co/kg. These street dusts would appear to be comprised of natural soil dusts.

Soils and stream sediments are strongly enriched in nickel and cobalt from nearby ore deposits. Of 960 soil samples taken at the Blackbird mine ca. 1975, only eight had concentrations > 100 mg Co/kg, and most values that were > 30 mg Co/kg were in the Blackbird mine area itself. Stream sediments contained averages (depending on rock types) up to 45.88 mg Co/kg.

Soils near smelter processing cobalt-bearing ores are also contaminated by cobalt. At Sudbury, Ontario, where ~ 71 MT cobalt was released annually from 1961 to 1971, cobalt concentrations in the surface soil within 0.8 to 7.4 km of the smelters was 42 to 154 mg/kg compared with 17 to 19 mg/kg at 50 km. If the latter values are near-background concentrations, soil contamination is evident up to at least 19 km (31 to 48 mg Co/kg).

Values of up to 107 μg Co/liter in U.S. drinking waters have been reported, but values ≤ 2 μg Co/liter are more common. The National Community Water Supply Study has found 0.0 to 19 μg Co/liter (average 2.2 μg/liter) with 62% of the samples containing > 1 μg/liter. There is no recommended U.S. standard for cobalt in drinking water. Little information was found on cobalt in groundwaters. Up to 5 μg Co/liter has been reported in the spring and well waters surrounding the Salton Sea.

Finished drinking water plants do not generally greatly change the concentration of cobalt in the raw waters they treat.

Concentrations of 1 to 10 μg Co/liter are reported in streams close to populated areas. Streams running through mining districts and areas of extensive agricultural land usage generally have cobalt concentrations in the range 11 to 50 μg/liter. Highest values reported were 4,500 μg/liter in Mineral Creek near Big Dome, 3 miles south of Ray, Arizona, which is near the site of a copper smelter and copper-molybdenum mine, and 6,500 μg Co/liter as the maximum concentration in the Little St. Francis River, which has received effluents from the Fredericktown cobalt mining and milling operations.

Rainwater concentrations range from 0.045 μg Co/liter at Quillayute, Washington, to 2.9 μg Co/liter in the Swansea Valley, Wales, where the highest average ambient air cobalt concentrations have been reported.

Vitamin B_{12} cobalt concentrations* in natural waters are considerably lower than total cobalt concentrations, with

* The concentration of cobalt in vitamin B_{12} is 4.3%.

values of 0.001 to 0.150 μg Co/liter reported in world lakes. In the Cayuga Lake Drainage Basin, one of the Finger Lakes in New York state, for example, the percent of total soluble cobalt represented by cobalt in vitamin B_{12} rose from ~ 0.5 to 0.6% in late June to 11.00% in early August. Cobalt is most efficiently converted to vitamin B_{12} in August by blue-green algae.

Except where there is gross heavy metal contamination by industrial or sewage effluents, the concentration of cobalt in freshwater sediments is generally < 20 mg/kg. Freshwater (lake) manganese nodules contain higher concentrations (70 to 650 mg/kg).

In the highly industrialized Maumee River Basin in Ohio, "extractable" cobalt was detected in 99.1% of the sediment samples (0.006 to 83.3 mg/kg; average 0.9 mg/kg) but in only 26.5% of the water samples (at 13 to 49 μg/liter). Even at Toledo, Ohio, where the river is very contaminated with heavy metals, sediment concentration was only 8 mg Co/kg.

The sediments in the southern Great Lakes, near industrialized cities, are more enriched in cobalt than are the sediments in the northern Great Lakes. In Buffalo Harbor, for example, the cobalt enrichment factor for the concentration at the surface relative to the background concentration at depth has been reported as 5.3.

Recent reported values for cobalt concentrations in seawater range from 0.0097 μg/liter in the North Central Pacific to 0.046 μg/liter in the Northeast Pacific off the Oregon coast. Reported vitamin B_{12} concentrations in seawater range up to 0.2 μg/liter (which is not compatible with the maximum total cobalt concentration stated above). Most reported B_{12} values are < 0.025 μg/liter. The high values are reported in waters from Vineyard Sound, Woods Hole, Massachusetts (highs attributed to length of storage of unfiltered water), and from the northwest shore of Florida.

The mean cobalt concentration in uncontaminated marine sediments is 24.5 mg/kg dry weight. Cobalt has very similar concentrations in freshwater, estuarine, and coastal marine sediments. Some reports indicate that estuary sediments contain lower average cobalt concentrations than are in the freshwater sediments of the rivers draining into the estuary. There is little variation in cobalt concentration with depth even in San Francisco Bay sediments.

Interstitial brines trapped in marine sediments have high cobalt concentrations. Those from San Francisco Bay contain up to 60 μg/liter; those from the Red Sea sediments (with 4 to 400 mg Co/kg) contain up to 4,100 μg/liter. East Pacific cores contain interstitial waters with up to 20 μg Co/liter. Pore waters of sediments in the estuarine water of the River Conway (0.05 μg/liter) in north Wales contain 490 μg Co/liter.

Cobalt is concentrated (up to 7,400 mg/kg) in the iron-rich amorphous oxides of Pacific Ocean manganese nodules. The copper and nickel are concentrated in the manganese-rich crystalline oxides.

Elevated cobalt concentrations in marine sediments resulting from anthropogenic contamination have been reported in Piraeus Harbor (14 mg Co/kg), receiving untreated industrial and domestic wastes from Athens, Greece; Narragansett Bay (up to 10 mg Co/kg, not much different from background levels), receiving metal plating and other wastes from the Naval Air Rework Facility; the mouth of Althaus Creek (27 to 37 mg Co/kg), receiving tailings pond effluents (130 mg Co/liter laboratory supernatant) from a nickel refinery in North Queensland, Australia; Kaohsiung Harbor, southwest of Taiwan Island (10 to 50 mg Co/kg), receiving mixed industrial wastes where the highest concentrations have been observed; Humber Estuary (6 to 30 mg Co/kg), receiving sewage effluent from 20% of the English population as well as industrial effluents; Jamaica Bay, New York City (average 5.98 mg Co/kg but 21.4 mg Co/kg in the southwestern portion, receiving petroliferous pollution from fuel terminals and heavy industry), having highest cobalt concentrations in the vicinity of Kennedy Airport runways (which jut out into the Bay); and Foundry Cove, Cold Spring, New York (< 6 to 552 mg Co/kg), receiving wastes from a nickel-cadmium battery plant that at one time contained cobalt in amounts representing 10% of the nickel used.

Most reported average concentrations in U.S. dry sewage sludges range from 3 to 24 mg Co/kg. Canadian values are similar. A semiquantitative spectrochemical analysis of sewage sludges in 12 Oklahoma cities gave an average of 67 mg Co/kg. A survey of

10 Indiana cities gave an average of 16.4 mg Co/kg with a range of 5.85 to 59.1 mg Co/kg. The highest value found was at Kokomo, Indiana, which is, incidentally, the site of the Stellite Division of Cabot Corporation, a major producer of cobalt alloys.

Canadian studies indicated that up to 60 kg Co/day (0.02 MT/year) is discharged by 10 southern Ontario treatment plants into Lake Ontario. Raw sewage contained 2 to 40 μg Co/liter; primary effluents and secondary effluents, 1 to 30 μg Co/liter; final effluent, 1 μg Co/liter; and raw sludge, 7 to 30 mg Co/kg. The cobalt removal efficiency could not be calculated. A Soviet study of sewage from two Ukrainian cities indicated that mechanical purification removed about 50% of the cobalt and that subsequent biological treatment permitted a total removal of 83 to 89% of the cobalt. In storage ponds designed for land disposal of sewage effluents for the Great Lakes region, cobalt was distributed 12 to 17% in the water, 0.5 to 10% in the plants, and the rest in the pond sediments, which indicates a removal efficiency of 83 to 88%. Cobalt removal varied from 0 to 85% from the sewage at Binghamton/Johnson City and Endicott, New York, sewage treatment plants. The effluents, however, generally contained < 10 μg Co/liter.

About 35% of the U.S. sewage sludges are incinerated. Most U.S. incinerator ashes contained 11 to 57 mg Co/kg except for Palo Alto, California, whose sewage sludge ashes contained 290 mg Co/kg. The other metals content of this sludge indicated that the cobalt source was probably the electronics industry.

Many U.S. cities dump sewage sludge into the oceans, but ocean dumping must cease by the end of 1981 according to a recent Environmental Protection Agency (EPA) regulation. Some enrichment is seen in marine sediments receiving such sewage sludge and effluents. At the Los Angeles sewage effluent outfall, the cobalt concentration in the effluent is diluted from 10 μg/liter to 0.2 μg/liter in the seawater. About 50% of the cobalt is in particulates. In oxidized waters, the insoluble cobalt sulfides in the discharges are oxidized to soluble compounds, but reduced sediments occur for wide areas around such outfalls. Only 0.5% of the sewage particulate, however, settles in the reduced area. About 3 MT cobalt is contributed to the southern California Bight from sewage treatment plants (≤ 6 μg Co/liter in sewage effluent and 0.03 μg Co/liter from sludge effluent), about 5.3 MT is contributed by surface runoff, and about 6 MT is contributed by all other possible sources for ocean dumping.

Discharge of activated sludge and sewage may represent the largest source of cobalamins in natural waters other than their production by benthic bacteria. Sewage sludges contain < 2 to 27 mg cobalamin/kg dry weight (the highest concentration would represent 1.1 mg Co/kg). Highest concentrations are observed in late July.

A few reports were found of cobalt concentrations in industrial wastewaters (other than primary metallurgical), process waters, or sludges: < 0.1 to 6.7 μg/liter in an electroplating process (nitric acid rack-stripping) water; 0.5 and 3 μg Co/liter in final effluents from aircraft metal finishing plants; < 10 to 90 μg Co/liter in dyebath effluent from an Ontario carpet mill that spins and dyes synthetic fibers; < 0.2 mg Co/liter in spray and quench waters from municipal incinerators (at Alexandria, Virginia; Washington, D.C.; and Montgomery County, Maryland); 4 to 6 μg Co/liter in California food processing plant wastewaters; 2 to 11 μg Co/liter in wastewaters from metal and chemical processing plants in California; 10 to 10,000 mg Co/liter in aqueous process streams when used as a catalyst; and 1,600 μg Co/liter in deep well injected wastes (and 20 μg Co/liter in groundwater contaminated by the well) from production of terephthalic acid by oxidation of *p*-xylene at the Hercules Inc., plant near Wilmington, North Carolina.

Cobalt has been removed from aqueous industrial wastes by precipitation by soda ash, uptake by algae (with a concentration factor of 342, removal efficiency is 53.5%) or biota and sediments in biological ponds (96% removal within 4 days), coagulation and filtration (40 to 75%), and cobalt ion-specific ion-exchange resins ($\sim$ 99% removal).

The bromide, formate, and sulfamate cobaltous salts are EPA-designated hazardous substances to be regulated in discharges into navigable waters or adjoining shorelines from vessels or facilities. These salts are apparently the only ones EPA considers having potential for significant discharges, several others having been deleted from an earlier listing. Discharges of these cobalt salts must be reported when equal to 454 kg or more. Discharges identified under various permits or permit applications, however, are exempt from these regulations.

Cobalt concentrations reported for ambient air range from 0.128×10^{-6} ng/m^3 over the Inland Sea of Japan to 610 ng/m^3 in Cleveland, Ohio (a transient maximum). Urban air concentrations usually average < 4 ng/m^3. Average values around 10 ng/m^3 at the National Air Surveillance Network sites are unusual. The highest average cobalt concentration reported in ambient air is 48 ng/m^3 at Clydach, Wales, where nickel and cobalt are refined. Heavy industry and traffic appear to be the source of some of the cobalt in urban locales. Cobalt usually has a low enrichment factor in air particulates compared to soil, however, and is usually concentrated on soil-size particles. The enrichment factor, EF, is defined as:

$$EF = \frac{[Co] / [n]_{atmosphere}}{[Co] / [n]_{crust}}$$

where $\underline{n}$ is an appropriate normalizing element presumed not to have any anthropogenic source. EF's higher than 7 have been reported at Chadron, Nebraska, summer 1973 (22 normalized to manganese); at Kellogg, Idaho, in the vicinity of lead and zinc smelting activities (14 normalized to scandium); at industrial and residential areas in Ghent, Belgium (17 and 12, respectively, normalized to aluminum); at Liege, Belgium (an industrial town) and Mechelen, Belgium (a small town) (14 and 23 normalized to aluminum); at Jungfraujoch, Switzerland (15 normalized to aluminum); at Laksely, Norway (17 normalized to aluminum); and at Clydach, Wales (240 normalized to scandium). It is not clear why EF's should be high at isolated places like Jungfraujoch, a mountain pass which is 3.4 km above sea level.

D. Environmental Transport

1. Soils: Soils tend to hold more cobalt with increasing pH and higher contents of clay, natural organic substances (if well drained), and hydrous manganese and iron oxides. Conversely, the mobility of cobalt is increased by lowering these factors. Very strong organic complexing agents such as EDTA, which is probably a widespread contaminant of U.S. waters, greatly enhance cobalt mobility. Weaker organic complexing agents (such

as those present in plant extracts) also increase cobalt extractability, but both types of complexes decrease cobalt plant uptake.

When a high-organic soil is not well-drained, the redox potential (Eh) and the pH of the soil decrease due to exclusion of oxygen by waterlogging and buildup of carbon dioxide from decomposing plant substances. These factors favor valence reduction and the dissolution of manganese and iron with the subsequent release of adsorbed species such as cobalt.

Cobalt is adsorbed by clays, apparently both as the metallic ion (by ion exchange) and as hydrolyzed species such as $CoOH^{+}$. Principal sites of cobalt sorption may be external clay surfaces where chemical weathering and physical abrasion may have introduced defect structures favorable to chemical bonding.

On manganese oxides in soils, cobalt is present due to precipitation, to sorption, and/or to incorporation in the crystal lattice. When cobalt salt solutions are added to high-manganese soils, most of the original adsorbed cobalt becomes nonextractable (e.g., by 2.50% acetic acid) upon aging for several months and largely unavailable for plant uptake. It has been suggested that Co^{2+} may interchange with both Mn^{2+} and Mn^{3+} in the hydrous δ-MnO_2 crystal structure. Co^{2+} may be oxidized to Co^{3+} on the manganese oxide at the oxide-water interface.

In organic soils, cobalt salt solutions are held in the upper horizons (the biologically active zones). However, in the presence of EDTA, a widely used decontaminating agent at nuclear facilities, cobalt becomes much more soluble, more mobile, and less concentrated in upper soil horizons.

2. Water: Many factors complicate the transport and speciation of cobalt in natural waters and sediments so that few generalities can be made. Anthropogenic pollution appears to enhance the solubility of cobalt in freshwater, sometimes at least, by complexing it with sewage-derived organics.

Although suspended solids are frequently thought to be the major mode of aqueous transport for heavy metals, the percentage of cobalt carried in this way is highly variable: Danube

River (1961-1970), 27.4 to 85.9%; Rio Puerco (New Mexico), 86%; the North Sea, 34%; Main River (West Germany), 33.4 to 42.2%; Lake Washington (Washington state), 0%; Strait of Juan de Fuca (between Puget Sound and the Pacific Ocean), 11 to 15%; natural seawaters, 11.1%; seawater enriched in marine algal organic matter, 10.5%; Columbia River (as ^{60}Co downstream from the Hanford Works, which is near Richland, Washington), 95 to 98%; Glomma River (at Fetsund, Norway), 13%; Lake Trehorningen (at Baerum, Norway), 58 to 71%; Amazon and Yukon Rivers, > 98%; samples from contaminated and uncontaminated sites in the Haw and New Hope Rivers (North Carolina), 88% and 92%, respectively; and Susquehanna River (near its source in New York), 9%.

The Susquehanna River in Pennsylvania has been reported to carry 90 to 150 short tons cobalt per year to the sea on suspended particulates. The concentration of suspended cobalt in this highly polluted river is 27 μg/liter (500 mg Co/kg in the suspended material). More commonly, the cobalt concentration in the suspended material of rivers is 7 to 94 mg/kg.

Some authors have described the speciation of cobalt in the suspended material and bottom sediments. The large unpolluted Amazon and Yukon Rivers carried cobalt in the following forms: 1.6 to 1.7% dissolved; 4.7 to 8.0% adsorbed; 27.3 to 29.2% precipitated and coprecipitated in metallic (i.e., iron and manganese oxide) coatings; 12.9 to 19.3% in inorganic solids; and 43.9 to 51.4% in crystalline sediments. In the Haw and New Hope Rivers, the average water values reported for cobalt at control sites were 8.0% dissolved; 31.4% adsorbed; 20.6% in oxide coatings; 10.7% in solid organics; and 29.1% in crystalline minerals. The corresponding values for contaminated sites were 12.2, 27.4, 19.2, 14.6, and 26.5%, respectively. At one site downstream from a sewage treatment plant and points of textile discharge on the Haw River, the cobalt was only 2% dissolved. In the lower reaches of the Yamaska and Saint Francois Rivers collected at Saint Marcel and Pierreville, Quebec, Canada, ~ 40% of the cobalt in bottom sediment samples were bound to iron and manganese oxides. More than 50% was in detrital silicate minerals, resistant sulfides, and a small amount of organic substances resistant to acidic peroxide treatment. Reported sediment speciation of cobalt in the lower Rhine River is ~ 60% rock dust, ~ 20% carbonate, ~ 10% humate, ~ 7% sorbed, and ~ 3% hydroxide.

Often the cobalt concentration in bottom sediments is approximately the same at various depths. Rather than suppose that this indicates that little contamination by cobalt must have occurred, it is more logical to conclude that the concentrations have been rendered homogeneous by resuspension, mining, and redeposition by dredging, shipping activity, the action of the current, storm activity, and burrowing organisms. Radiocobalt released by a nuclear research center into Lake Maggiore, Italy, contaminated an area of about 0.03 km^2. The average ^{60}Co activities in the first five layers of 1.5 cm each were approximately the same, and their total represented 90% of the total from 0 to 9.0 cm beyond which depths ^{60}Co was generally absent.

A few inventories have been reported of cobalt sources to water bodies. At least 1% of the cobalt in Chesapeake Bay has anthropogenic sources. No estimates have been made of input by rain and atmospheric fallout, but northern Chesapeake Bay, excluding Baltimore Harbor, is estimated to receive annually 150 MT cobalt from rivers, 30 MT from salt water advection, 4 MT from shore erosion, and only 0.5 MT from municipal wastewater. Baltimore Harbor receives an estimated 1 MT Co/year from rivers and storm drainage, 1 MT Co/year from municipal wastewater, and 0.7 MT Co/year from excess sediment. Lakes Huron and Superior are estimated to receive 30 and 35 MT Co/year from anthropogenic sources and 95 and 205 MT Co/year from natural sources, respectively.

An elaborate model for the speciation of cobalt in seawater gives the following rank order for concentrations of species representing free Co^{2+} or Co^{2+} complexed with the following ligands: $Cl^- >$ free $Co^{2+} > CO_3^{2-} > SO_4^{2-} > OH^- > Br^- >> NH_3$. Adsorbed species concentrations fall between those of Br^- and NH_3. The order for freshwater is free $Co^{2+} \gtrsim CO_3^{2-} > OH^- > SO_4^{2-} > Cl^- >> PO_4^{3-} \gtrsim NH_3 > Br^-$. Adsorbed species concentrations are much more important in freshwater, ranking third between those of CO_3^{2-} and OH^-.

In a laboratory experiment, the probable speciation of ^{60}Co added to natural seawater as the chloride after 3 hr was 55.3% $CoCl^+$, 5.5% Co^{2+}, 1.0% $CoCl_2$ plus $CoCl_3^-$ plus $CoCl_4^{2-}$, 9.6%. solid CoOOH on MnO_2 or FeOOH, 27.0% unassigned soluble cobalt, and 1.50% unassigned insoluble cobalt.

In another seawater model, the total Co^{2+} concentration is estimated to be 10^{-8} $\underline{M}$, which is greater than that of $CoCl^+$ and $CoSO_4$ (about 10^{-9} $\underline{M}$, each). Organic complexes represent only 0.1% of the cobalt. The $CoCl^+$ and $CoSO_4$ each represent 10% of the cobalt.

Addition of humic acid to natural waters may merely increase the colloidally dispersed metals rather than form truly soluble humic acid complexes. These insoluble complexes may represent a significant fraction of cobalt in natural waters. For example, a recent report stated that the colloidally bound fraction of cobalt in Chesapeake Bay near Howell Point represented > 10% of the cobalt in the water. However, Co^{2+} is among the divalent cations having the least affinity for humic and fulvic acids. Thus, as salinity increases, the slight flocculation of cobalt observed is attributed to flocculation of humic acids and hydrous iron oxides which is counteracted by the formation of the soluble strong chloride complex.

Natural organic compounds in the water frequently increase desorption and solubilization of cobalt from the inorganic fractions of sediments or suspended material. The organic complexing capacity for cobalt is higher in streams receiving waste effluents. For example, when the chemical oxygen demand in the Rhone River in France attained 15 mg/liter, cobalt was almost 100% complexed. About 2 to 3% of the material in sediments is organic matter, and humic compounds are the major function of the organic matter. Up to 20 mg organic matter per liter may be suspended in the water. (Cobalt is complexed primarily by the amino acids.)

EDTA is widely used in agriculture, food and drug processing, photography, and textile and paper manufacturing. Thus, EDTA is likely to be a constituent of industrial discharges. The Co-EDTA chelate is very stable environmentally. At near neutral pH, the chelate is destroyed only slightly by photodegradation (0.4% at pH 6.9 under 4,000 footcandles of artificial sunlight).

In suspended particulates and sediments, cobalt may be intercalated as cations between the layers of clay lattices and/ or adsorbed on ion-exchange positions; it may be complexed by natural organic matter (more so in freshwater than in seawater); it may be precipitated as cobaltic hydroxide after oxidation of coagulated cobaltous hydroxide at pH 7.7, and it may be adsorbed on hydrous iron and manganese oxides. Adsorption-desorption equilibria of sediment and suspended particulates depend on time, pH, concentration of particulate and cobalt in water, the mineralogy of the sediment, and flocculation and desorption effects caused by increasing salinity.

In bottom sediments, the cobalt distribution is frequently related to that of fine-grained material, which is usually dominated by clay minerals. Cobalt is often more concentrated in fine-grained sediment fractions.

Cobalt adsorption by oxide minerals, whether silicates or metal oxides and hydroxides, is low at low pH, but when the pH is increased (pH 7 to 10) so that hydrolysis of Co^{2+} becomes significant, the percentage adsorption rises to 100%.

The scavenging and coprecipitation effects of manganese oxides may be partly attributed to oxidation-reduction processes. As the pH of the water increases, Mn^{2+} precipitates as Mn_3O_4 or MnOOH, which disproportionates to MnO_2 and Mn^{2+}. Both Co^{2+} and Co^{3+} and other heavy metal ions with approximately the same dimensions as Mn^{2+} and Mn^{3+} in manganese oxides enter the lattice during disproportionation of Mn_3O_4.

As mentioned above, increasing the content of soluble organic matter in natural waters increases the concentration of dissolved cobalt. Thus, in an artificial system at a suspended load of 1.0 g/liter, cobalt was 93% adsorbed by the suspended material ($<$ 1% organic) in Var River (Monaco) water at pH 8.3; but addition of 25% sewage decreased adsorption to 87%. Sewage, however, did not affect the degree of cobalt desorption in the presence of seawater, which was about 20% when 0 to 100% sewage was present.

The fate of radiocobalt released to natural waters has been studied at several sites. Radiocobalt concentrations in the sediments of New London Harbor, Thames River, Connecticut (where nuclear-powered ships have been serviced) decreased by an average factor of 33 between 1966 and 1972. The decrease was due not only to radioactive decay but also to dilution by uncontaminated sediment and reduction in the radioactive discharges. The radioactive sediment sites, however, had not shifted during that period. In Humboldt Bay, which received irregular discharges from a 65-Mw nuclear power reactor at Eureka, California, however, lateral transport of ^{60}Co was observed in the sediments and may have been caused by leaching/readsorption and/or scouring/sedimentation.

Radiocobalt after its release from the reactors at the Hanford Works between January 1964 and January 1965 showed $\geq$ 50% depletion from the river between Pasco and Vancouver by sediment uptake except during May through July due to scouring of the river bottoms. At McNary Dam, a large fraction of the ^{60}Co was in the fine sediments (which comprised only 0.096% of the whole sediment sample) and was largely associated with the organic matter.

Only 9% of the ^{60}Co activity released to Whiteoak Creek in Tennessee (as high-nitrate solutions) was retained in the bottom sediment within the 21-mile reach extending from the mouth of the Clinch River to the mouth of Whiteoak Creek. Cobalt sorption in Whiteoak Creek sediments is due both to ion exchange with silicate materials and adsorption by manganese and iron oxides. Only 20% of ^{60}Co released to Whiteoak Creek is associated with the suspended sediment. Perhaps more ^{60}Co is present as dissolved ions. Although highest sediment concentrations were immediately downstream from the mouth of Whiteoak Creek, secondary peaks occurred at Clinch River miles 17.5 and 15.0. The sedimentation pattern was presumably affected by diffusion, flow distribution, and turbulence. The variation in ^{60}Co activity in Clinch River sediment cores corresponded to that of the major radionuclide released, ^{137}Cs. Presumably, both were incorporated into the bottom sediments by sedimentation of suspended solids that had acquired their radionuclide content before entry into the Clinch River.

Industrial sludge containing stable cobalt discharged to the Saar River in the vicinity of Mettlach, West Germany, produced sediments on the right bank containing 72 mg Co/kg at 0 to 4 cm depth, 92 mg/kg at 4 to 8 cm, and 133 mg/kg at 8 to 12 cm. However, in the sediments of the opposite bank, the surface concentration of cobalt (10 mg/kg) was not much higher than at nearby points on that side of the river. For at least 400 m downstream from the outfall on the right bank, cobalt concentrations were still elevated (31 to 70 mg/kg at the surface).

The aforegoing cases are for cobalt in wastes intentionally released into natural water bodies. Other studies touch on the fate of cobalt in water after leaching and/or mechanical mobilization from land. The appearance of cobalt in the South Fork of the Coeur d'Alene River may be due, however, to the former dumping of tailings from lead-zinc-silver mining and milling operations. There was 27 μg Co/liter in water of the polluted South Fork but only 0.4 μg/liter in the unpolluted North Fork. The surface sediments (0 to 3 cm) of the river delta contained 21.6 mg Co/kg (ca. 1974) compared with only 0.17 mg Co/kg in the sediments of the north part of Coeur d'Alene Lake. In the lake sediments near the delta, cobalt concentrations were 5.32 mg/kg at 0 cm depth, 15.9 mg/kg at 10 cm, 22.9 mg/kg at 20 cm, 7.85 to 10.4 mg/kg at 30 to 70 cm, and 0.16 mg/kg at 80 cm.

Cobalt concentrations are sometimes elevated in sediments of streams draining unmined areas where ore deposits containing cobalt minerals are located. Sediments of streams draining black shales may also show anomalous cobalt concentrations (e.g., 30 mg Co/kg in central Equador).

Besides point sewage discharges to natural water bodies, cobalt may be contributed in urban runoff. In Durham, North Carolina, the urban runoff in the Third Fork Creek Drainage Basin (which drains into the Neuse River and the Cape Fear system) contained 160 μg Co/liter.

The cobalt in continental runoff could be derived from both natural and anthropogenic sources. Shelf waters may contain significantly higher amounts of cobalt than offshore waters.

A 42.5-fold difference was found in cobalt concentrations between water contaminated by agricultural runoff in Collier County, Florida (the Barren River Canal System of Choskolosku Bay) (34.0 μg/liter) and water of an undeveloped area (0.8 μg/liter). In the agricultural area, cobalt is applied to the soil in fertilizer or to the plants in nutritional sprays.

Cobalt has sometimes been found in the leachates of sanitary landfills. It has been reported at 4 to 70 μg Co/liter in leachates from five to six landfills in Norway in August during dry weather. The highest cobalt concentration was associated with the lowest pH and the highest BOD, COD, and contents of organics, phosphorus, iron, zinc, and calcium. A summary of leachates observed in one large study of land disposal sites indicates that spills and industrial wastes and treated electronics wastes gave leachates with highest cobalt concentrations (220 and 170 µg Co/liter, respectively). Municipal and pickling wastes gave leachates containing 80 µg Co/liter; metal finishing wastes, 50 µg Co/liter; wastes from pharmaceutical chemical and pesticides plants, 40 µg Co/liter; and mixes of industrial and municipal wastes and/or fly ash, 10 µg Co/liter. These values were found in monitoring wells in Massachusetts, Michigan, New Jersey, New York, Pennsylvania, and Wisconsin.

Cobalt exists in sulfides associated with coal and ore bodies and appears in the acid mine drainage that results from weathering of these sulfides when sulfuric acid is produced and ferrous and other heavy metal ions are released. Elevated cobalt concentrations have been observed in waters and sediments in regions where acid mine drainage occurs. For example, Maryland coal mine drainage contains 8 to 1,400 µg Co/liter (median 210 µg Co/liter for 21 sites). Mine drainage may be treated by aeration (to oxidize Fe^{2+} to Fe^{3+}) and lime neutralization to precipitate ferric and other heavy metal ions as a sludge (which is disposed of in lagoons, abandoned mines, and underground workings) but not all of the cobalt is removed by this treatment. For example, neutralizing drainage water containing 1,100 µg Co/liter to pH 6.0 removed 75% of the cobalt (0.64 mg Co/kg in the sludge), leaving a solution containing 270 µg Co/liter. An average 25 µg Co/liter was found in 38% of Ohio streams receiving acid mine drainage. Sediments in Lake Anna in Virginia contained 8.12 to 34.95 mg Co/kg. The concentrations of cobalt and other heavy metals were higher in Lake Anna sediments for a distance 3 miles below a source of acid mine drainage from abandoned pyrite mines. Colorado surface streams receiving manganese-enriched mine drainage contained 1 to 49 µg Co/liter.

Tailings piles containing sulfide minerals can be a source of leached cobalt as well as acid mine drainage. The mechanism is the same. Seepage water at pH 2 from a high-pyrite tailings pile in the Eliot Lake, Ontario, uranium district contained 3,800 µg Co/liter. Coal preparation wastes represent another source from which cobalt is highly leachable. The wastes are discharged in a damp condition and are wet by rain for long periods before they are covered.

Ore-forming processes, erosion and weathering, and mechanical and chemical mobilization account for the presence of cobalt in water bodies near operations mining and milling cobalt-containing ores. At Fredericktown, Missouri, several thousand tons of cobalt has been produced from the sulfide ores. Surface and underground mining as well as milling have left large dumps and mill tailings piles and ponds. These are sources of cobalt by both chemical and mechanical mobilization. Upon comparing southeastern Missouri stream sediment concentrations, it appears that the cobalt enrichment is largely restricted to the Fredericktown area although streams associated with lead-zinc mining and milling do show somewhat elevated cobalt concentration (36 to 38 mg Co/kg) in their bottom sediments. The sediments of the Little St. Francis River, which receives the Fredericktown drainage, contain 6 to 1,744 mg Co/kg (mean, 215.4 mg Co/kg); and the river water contains 1 to 6,500 μg Co/liter (mean, 23.6 μg Co/liter). When a tailings pond was breached during a major spring flood in 1977, tailings were dispersed to about 16 km downstream in the Little St. Francis River. Apparently, cobalt and nickel were transported together in closely associated mineral phases.

In the Bathurst District in New Brunswick, Canada, the massive lead-zinc-copper sulfide deposits contain 100 to 3,500 mg Co/kg. Cobalt is apparently released from the sulfide deposits as soluble cobalt sulfate (as in acid mine drainage). Manganese hydroxides and oxides in the stream sediments appear to be the most efficient collectors of cobalt. No information was found regarding cobalt transport from the cobalt deposits in Tennessee and Pennsylvania.

3. Air: Little has been reported regarding atmospheric transport of cobalt compared with the systematic studies of soil and water systems. Information compiled herein indicates that metallurgical processing plants can be significant point sources of cobalt although estimates of emissions from these sources were not found. Some workers have concluded that the combustion of coal and petroleum products are the major anthropogenic sources of cobalt to the atmosphere.

Ambient air concentrations of cobalt at Kellogg, Idaho, were in the high end of the range of ordinary average urban concentrations and were 14 times above background. The base metal ores of the Coeur d'Alene district are considered a cobalt resource, but the specific process (from lead smelting or zinc roasting) releasing cobalt to the atmosphere has not been identified.

The cobalt concentration (1.9 ng/m^3) in the ambient air of Tucson, Arizona, which is located within a 193-km radius of half of U.S. copper smelting activity, is not attributed to copper smelting. A pilot plant roasting and otherwise treating cobalt-bearing laterites is also in the Tucson vicinity. In Cleveland, Ohio, a maximum 610 ng Co/m^3 was observed at a site near (but not the nearest to) the location of a plant producing cobalt-containing beryllium-copper alloys. Since there are many metallurgical plants in the area, the association is only tenuous. However, most of the high average values were found downwind from the beryllium-copper plant. The Mond process nickel refinery in Clydach, Wales, appears to be the major local cobalt pollution source in the Swansea Valley. Average cobalt concentrations in the air were 48 ng Co/m^3 at Clydach and 15 to 16 ng Co/m^3 in nearby Llansamlet and Trebanos. Lower concentrations were found at towns farther away. A weaker source of cobalt was recognized in the vicinity of Kidwelly and Burryport but has not been identified. Cobalt emissions may be related to former copper smelting or to zinc metallurgy in Burryport. The water solubility of cobalt is higher in the atmospheric particulates (78 to 86% water-soluble cobalt) at the four most contaminated sites than in those (53 to 63%) collected at sites farther from the source of contamination. At Clydach, there was a total deposition of atmospheric cobalt of 4.5 μg/cm^2/year; 3.0 μg/cm^2/year was soluble. The range of ambient air cobalt concentrations at Clydach was 3 to 300 ng Co/m^3, and cobalt was strongly correlated with iron, not with nickel.

Mention was made in the preceding subsection that cobalt concentrations in soils decreased markedly with increasing distance from the smelters at Sudbury, Ontario, due to atmospheric emissions; however, associated ambient air concentrations have not been reported.

Inside a metallurgical plant in Ghent, Belgium, the cobalt concentration in the air was highest at a steel wire drawing operation where a tungsten carbide die was probably used. The concentration was 107 times higher than in the outside air (absolute values were not reported). Cobalt is used in relatively high amounts in cemented carbides so this result is not unexpected; however, presumably cobalt impurities in the steel, zinc, or copper gave elevated cobalt concentrations relative to outdoor air at the bolt factory, 10.3; zinc plating station, 1.6; galvanizing station, 18; copper plating station, 62; and welding wire packaging station, 4.9. Thus, it may be naive to attribute cobalt emissions in urban environments that have many metallurgical industries to sites that are known to be using or processing high-cobalt materials. These questions will probably not be resolved until cobalt is routinely included in the metals determined directly in the monitored emissions of the suspect plants.

Several studies report the extent to which cobalt contaminates the environs of a coal burning plant. The variability in those estimates may arise from the degree of control on the emissions and the plant size. A coal burning plant in the USSR burning 14,000 tons brown coal daily would emit 1.8 $kg/km^2/year$ within a 1,017 km^2 area (18 km radius) if the lowest value for the cobalt concentration in the fly ash is used (30 to 200 mg/kg). A coal-fired plant in Brno, Czechoslovakia, emits 2.96 kg $Co/km^2/year$ within a radius of 1.5 km. A power plant burning Ohio coal containing 5 mg Co/kg on the eastern shore of Lake Michigan emits 10 kg $Co/km^2/year$ within a radius of $\sim$ 17 km. A model 1,400 Mw power plant burning coal containing 0.5 mg Co/kg would be expected to deposit a maximum of 1.2 $kg/km^2/year$ within a 20 km radius.

The trace metals in emissions from the coal-fired Allen Steam Plant are the subject of many reports. Highest soil concentrations (in the top 2.5 cm) of cobalt are at 8 km north (23.6 mg Co/kg) and 13 km south (21.0 mg Co/kg). Minimum values (within the 32 km transect north and south) are at 30.5 km north (11.3 mg/kg) and 1.4 km south (7.6 mg/kg). There is little difference in cobalt concentrations at 2.5 cm- and 15-cm depths, whether close to or far from the plant. Fly ash from the plant contains 25 to 70 mg Co/kg.

In the model coal combustion plant mentioned above, the cobalt concentration is expected to rise by a factor of 1.3 in green plants after 40 years of operation. The plant concentration factor used was 87, a very unrealistic value for most plants. Emissions from a coal gasification plant are estimated to raise the cobalt concentration in plants by a factor of 1.1 after 40 years. Neither process is expected to volatilize as much as 10% of the cobalt in the coal. A mass balance study for cobalt at coal-fired power plants indicates that only 2% of the cobalt contained in the coal is discharged when emissions are controlled by a scrubber or an electrostatic precipitator. Almost 30% of the cobalt, however, is discharged on the uncollected fly ash from a cyclone-fed boiler. Fly ash appears to be the most enriched solid stream produced from coal combustion compared with bottom ash, economizer ash, or scrubber solids. In 1972-1973, the Allen Steam Plant emitted 0.4 to 0.8% of the cobalt in the coal, about 50% was concentrated in the slag (bottom ash), and about 55% was trapped in the electrostatically precipitated fly ash. Total emissions were 21 kg Co/year. Laboratory-scale experiments indicated there are essentially no cobalt emissions during pressurized fluidized-bed combustion of coal.

Investigators at the University of Maryland estimated that about 11 to 15% of the cobalt in particulates emitted from a stack controlled by a electrostatic precipitator is < 2 μm in diameter.

Investigators at the University of California, Livermore, stated that cobalt is more concentrated in smaller particles from a scrubber than from an electrostatic precipitator (ESP), and that since more "respirable" cobalt will be emitted from a scrubber than from an ESP, the potential alveolar deposition of cobalt would be at least twice as much for scrubber emissions as for ESP emissions. However, the mass median diameters estimated for the particulates by these workers were sometimes much higher than < 2 μm (up to 12 μm). Another group pointed out that 20 to 30% of all fly ash particles with physical diameters < 5 μm are encapsulated or agglomerated, thereby reducing their health hazard. The cobalt was largely on the surface layer of particles of bulk fly ash, and 18% of it dissolved in a dimethyl sulfoxide-water mixture after 48 hr at 25°C. Another report stated that 25% of the cobalt in 50 mg of a fly ash fraction of volume median diameter 2.2 μm was dissolved after 5 min in a synthetic fluid presumably meant to represent lung fluid.

Literature values for the enrichment factor of cobalt in coal-combustion particulates range from <2 to ~ 5. Since the enrichment factor for cobalt in the ambient air particulates in six U.S. cities ($\sim$ 4 to 9) overlaps this range, this fact has been taken as support for the view that coal burning is a significant contributor to urban aerosol cobalt. However, the same reasoning applied by other authors to particulates emitted from burning urban refuse (EF's of $\sim$ 3 to 7) led to the conclusion that municipal incineration is a minor source of atmospheric cobalt.

Only one value was found for the net plume sample at ground level, 3.86 $\pm$ 0.55 ng Co/m^3 for the Chalk Point Generating Station in Maryland. This value is more than nine times higher than the average concentration in the ambient air, 0.41 $\pm$ 0.11 ng Co/m^3.

At least four studies compare cobalt emissions from burning coal and oil. A 1971 report (using 1967 and 1969 statistics) estimates that 8% of the total cobalt in coal annually consumed worldwide is emitted to the air. At 5 mg Co/kg in coal and 0.2 mg Co/kg in crude oil, 700 MT cobalt would be emitted annually by burning coal and 30 MT would be emitted by burning oil. Another 1971 report and recent studies estimate that oil burning produces half to almost the same magnitude of cobalt emissions. The second 1971 report estimates that 50 MT cobalt is emitted annually in the Chicago, Milwaukee, and northwest Indiana area. The total is comprised of 22.7 MT from coal burning, 1.8 MT from coking, and 25.4 MT from fuel oil. There are assumed to be no cobalt emissions from metallurgy, cement plants, and transportation. In the calculations, cobalt concentrations are assumed to be 90 mg Co/kg in coal-burning particulate emissions and 1,500 mg Co/kg in fuel oil particulate emissions.

A 1977 report estimates that residual oil burning for generation of electricity in the United States in 1974 emitted 145 MT cobalt (based on 2.24 mg Co/kg in the residual oil). The same report estimates that coal-burning power plants in 1974 emitted 145 MT (127 MT from plants burning bituminous coal containing 4.0 mg Co/kg, 5.0 MT from plants burning lignite containing 4.0 mg Co/kg, and 12 MT from burning anthracite containing 90.0 mg Co/kg).

Since electricity generation was presumed in this 1977 report to account for 63% of particulate generated by fossil fuel combustion, an estimate of 290/0.63 or 460 MT total cobalt emission from all fossil fuel combustion was made although the authors stated that coal burning alone accounts for 95% of the particulates emitted from electricity generation. Thus, the extrapolation is only valid for coal burning: $145/0.63 \times 0.95 \simeq 240$ MT.

In a 1979 inventory made of ambient air sources of cobalt in Washington, D.C., in the summer of 1974, the cobalt contributions by various sources were estimated to be coal combustion, 35%; oil combustion, 36%; refuse combustion, 0.7%; marine aerosols, < 0.1%; and soil, 32%. These estimates give a predicted total atmospheric concentration of 1.0 ng Co/m^3 compared with the observed atmospheric concentration of 1.1 ± 1.0 ng Co/m^3. Since Washington, D.C., has little heavy industry, the inventory seems logical, but it is not clear what concentrations were used to derive the estimated values. Only relative cobalt concentrations in the "source" are given. If this means the coal, oil, refuse, and soil, for example, and if coal is assumed to contain about 5 mg Co/kg, then the cobalt concentrations in the oil, refuse, and soil would be 72.5, 0.6, and 2.8 mg Co/kg, respectively. Surely, the oil concentration would be too high at 14.5 times the concentration in coal. However, this is about the same ratio for cobalt in the fuel oil/coal emissions described for the Chicago, Milwaukee, and northwest Indiana area.

In Chapter V, reported emissions of cobalt from fuel oil combustion are included rather uncritically because of the 1971 EPA report that $\sim$ 0.3 to 12 mg Co/kg may be added to oil-fired furnaces in the form of proprietary compounds to suppress particulate emissions. Upon investigation, such use could not be verified.*

If cobalt concentrations in particulates generated by fuel oil combustion are indeed as high as 1,500 mg/kg, as much as

---

* According to a personal communication from Dick Husta, National Account Manager, Ethyl Corporation, New York, New York, in late October 1979, cobalt-containing formulations are not sold for this purpose. (Ethyl Corporation markets a manganese-containing product for this purpose called CI-2.)

860 MT total cobalt may be emitted from all fuel oil combustion in the United States. If most of the cobalt is in residual fuel oils, the value should be halved. Even 430 MT cobalt from fuel oils seems rather high based on the few fuel oil values given in Table IV-9, especially since the use of cobalt as soot suppressants is questionable. One study gives some support to such a high value: cobalt emissions from three commercial boilers fired with Nos. 4, 5, and 6 fuel oils were < 0.066 to 2.0 kg/million liters burned. Two residential units emitted 0.11 to 0.12 kg/million liters burned. The maximum emission factor gives 320 MT Co/year applied to 1.6×10^{11} liters residual fuel oil/year.

Another questionable concentration mentioned above is 2.24 mg Co/kg in residual fuel oils, which led to the estimate of a 145 MT cobalt emission from electricity generation by burning fuel oils. The highest value given in Table IV-9 is 0.3 mg/kg. Robert Jungers* of the U.S. EPA in Research Triangle Park, North Carolina, reported the following data on cobalt concentrations in fuel oils.

No. 2 fuel oil (a distillate fraction)	<0.000007 to 0.00017 μg/ml (average of 6 samples, 0.00015 mg/kg if the density is 845 g/liter)
No. 5 fuel oil (residual)	0.0072 to 0.012 μg/ml (three samples)
No. 6 fuel oil (residual)	0.16 to 1.3 μg/ml (average of 5 samples, 0.71 mg/kg if the density is 944 g/liter)

Approximately half of the 300 million MT fuel oil used annually is distillate. The cobalt emissions from burning distillate fuel oil would be negligible (0.02 MT/year) if No. 2 fuel oil typifies distillate fuel oils and these few analyses are typical of cobalt concentrations. Mr. Jungers cautioned that _he_

---

* Personal communication, October 1979. The source of the cobalt concentration is cited in the 1977 report as Darrel von Lehmden, with whom Mr. Jungers formerly worked.

would not consider the arithmetic means of his analyses as necessarily being typical of the fuel oil fractions.

Data for residual fuel oils usually include only Nos. 5 and 6 (and sometimes No. 4). Only heating oil sales data are broken out by type,* and sales data for 1977 and 1974** indicate that there is about 5 times as much No. 6 as No. 5 used for heating. For purposes of estimation one can reasonably assume that all residual fuel oil consumed has a substantially higher cobalt content than that of distillate fuel oils. In 1974, 475 million barrels or 71 million MT residual fuel oil was used for electricity generation and only 85 million barrels of distillate fuel oil. At 0.71 mg Co/kg, 71 million MT No. 6 fuel oil would contain 50 MT cobalt, not the 145 MT estimated for 1974 in the 1977 report described above. Since electricity generation accounted for almost exactly 50% of residual fuel oil consumption, total cobalt emission from burning fuel oil would have been only 100 MT in 1974. Thus, if the total coal burning estimate of 240 MT developed above is reasonable, the total Midwest Research Institute (MRI) estimate for cobalt emissions from fossil fuel combustion is 340 MT. And the inventories that give approximately equal weight to coal burning and oil burning as atmospheric sources of cobalt are called into question.

Gasoline contains < 0.1 mg Co/kg, and gasoline additives contain < 0.0002 to 0.0360 mg/liter. An isolated report of 391.0 mg Co/kg in an automobile exhaust pipe deposit might be attributed more to cobalt in the hard-facing alloy used on the exhaust valves and valve seats of internal combustion engines than to its content in the motor fuel.

Average concentrations for cobalt in the combustible fraction of urban refuse range from < 3 to 7 mg/kg, values that are compatible with cobalt concentration in paper (1 to 4 mg/kg; 13 mg Co/kg in the colored comics sections of newspapers) and handpicked plastics (< 8 mg/kg) from urban refuse. One study reports that the fine bottom ash, fly ash, and atmospheric particles

---

* Personal communication in October 1979 from Mary Zitomer, Petroleum Specialist, Office of Energy Data and Interpretation, Department of Energy, Washington, D.C.

** Personal communication in October 1979 from the American Petroleum Institute library in Washington, D.C.

from urban refuse burning contain 70, 100, and 12 mg Co/kg, respectively. About 1% of the cobalt is in the atmospheric particles. Fly ash scrubber water contains up to 500 μg Co/liter. In other studies, reported cobalt concentrations in the fly ash are 27 and 35 mg Co/kg; and the suspended particles that are emitted contain 1.8 to 12 mg Co/kg (average 6.6 mg Co/kg). Of the suspended particles, 34 to 41% may be in the $\leq$ 2 μm size range.

In a mass balance study for 1 week's operation, the weekly incinerator input was 2 kg in the combustibles and twice as much in the noncombustibles. There was 4 kg in the fine bottom ash and 2 kg in the collected fly ash. About 0.05 kg/week or 2.6 kg/year was emitted in the atmospheric particles.

Refuse incineration is judged not to be a major source of atmospheric cobalt although the EF (normalized to aluminum) for cobalt emitted from the Washington, D.C., area incinerators was $\sim$ 3 to 7 compared with EF's in ambient air of $\sim$ 5 to 8. One group estimates the contribution of urban refuse burning to cobalt ambient air concentrations to be 0.7%.

Another source of cobalt to the air may be cement works. A Czechoslovakian study indicates that depositions of cobalt emitted from two cement plants (1.0 to 1.5 kg Co/km^2 within a 2.5 km radius) are comparable to about 0.3 of those from a coal-fired plant. Whether this concentration would arise from similar raw materials used in other countries is not known. Perhaps, contaminated metallurgical materials are being used as a source of iron.*

---

* Duisburger Kupferhütte iron oxide wastes have been used in cement works in Lengerich, West Germany, for years. A recent story in The Christian Science Monitor, August 31, 1979, reported that the thallium pollution (attributed to the iron oxide material) of the environs is so bad that wild and domestic animals are losing their hair. Seventeen cement, glass, and other factories have been using this iron oxide waste product for many years.

E. Processing

The processing of cobalt metal, alloys, cemented carbides, and compounds is described in detail in Chapter II. Chapter VI, summarized here, describes primary and secondary processing of cobalt-bearing materials.

Minor amounts (< 1,000 MT/year) of cobalt were produced in the United States as a by-product of magnetite iron ores mined in Pennyslvania. A separate concentrate of cobaltous pyrite was roasted in Sparrows Point, Maryland, and a copper-bearing solution produced upon water quenching of the roasted concentrate was refined in Wilmington, Delaware. Losses of cobalt at the latter plant were probably in the form of cobalt sulfate and cobalt oxide or hydroxide in the manganese oxide precipitate.

Cobalt was mined at Fredericktown, Missouri, in 1944-1946 and again in 1955-1961. Anschutz Uranium Company plans to resume mining the ores, which contain 2.5 to 2.75% lead, 0.5 to 0.75% copper, 0.25% each cobalt and nickel, and 5 to 6% iron. Annual production of cobalt will be 680 to 910 MT. In the past, the cobalt concentrates were partially roasted and pressure leached. Copper was removed from the leaching solution by cementation on iron; iron was removed by oxidation and precipitation; and nickel was precipitated as the nickel ammonium sulfate, leaving cobaltic ammine sulfate in solution. Hydrogenation then precipitated cobalt as the metal powder.

Cobalt has been produced sporadically from the Blackbird district in Lemhi County, Idaho, since 1918. Peak production was 1,390 MT cobalt in 1958. About 70% of the cobalt in the concentrate was recovered. About 4,000 MT cobalt at a concentration of 0.23% (2,300 mg/kg) is in the accumulated tailings. The Bureau of Mines recently developed a process to recover 60 to 90% of the cobalt in these tailings. Cobalt was recovered (following pressure leaching) in the 1950's by hydrogenation, and later, by electrolysis. The cobalt was purchased by the U.S. Government to fulfill stockpile requirements. When the government contracts ended, cobalt recovery was no longer economical. Because of the recent dramatic rise in the price of cobalt, Noranda Mining Company plans to expand mining activities in the Blackbird district. New techniques for the separation of cobaltite from pyrite and pyrrhotite have been developed. High-pressure leaching and precipitation of arsenic as ferric arsenate, which will be buried at the mine, will give an impure cobalt sulfate solution, which will be toll refined by others until solution purification and electrowinning circuits are constructed.

Cobalt is present in the nickel magnesium silicate ores mined and smelted near Riddle, Oregon, by Hanna Nickel Smelting Company. Presumably, the major outlet for cobalt in these ores (0.05%) is in the ferronickel product, which contains 0.5% cobalt. Cobalt concentration in the smelter air averages 0.002 mg Co/m^3, well below the Occupational Safety and Health Administration (OSHA) standard or threshold limit value of 0.1 mg Co/m^3. EPA has found < 50 μg Co/liter in process water from the first settling pond but reports no cobalt concentration in the 460 m^3 water discharged daily at pH 8.7 from the second pond.

Lateritic Cuban ores were pressure leached at Port Nickel, Louisiana, for 6 months just prior to the 1959 Cuban revolution. Since November 1974, Amax Nickel Company has operated the plant and has recovered cobalt and nickel from imported copper-nickel mattes by hydrometallurgical methods including atmospheric leaching and hydrogenation to precipitate cobalt metal from solution. Amax has produced 270 to 385 MT/year and will reach full capacity (907 MT cobalt/year) by the end of 1979.

Lateritic resources of nickel and cobalt similar to the ores mined in the Philippines occur in southwestern Oregon and northwestern California. These domestic laterites contain 0.06 to 0.25% cobalt. Suitable smelting and refining processes must be developed before large-scale mining of these laterites can begin. The Bureau of Mines has described a potential process comprising roasting, selective reduction, and electrowinning. Cobalt recoveries are 80 to 90%. A pilot plant in Tucson, Arizona, is being converted by Universal Oil Products to treat domestic laterites by this process.

The Bureau of Mines has experimentally produced a copper-nickel concentrate from the low-grade nickel resource near Ely, Minnesota (the Duluth gabbro) containing 0.1 to 0.4% cobalt. Although commercialization is not imminent, it has been estimated that 410 to 910 MT cobalt could eventually be produced annually from the Duluth gabbro complex.

Deep-sea manganese nodules contain an average 0.3% cobalt, but their full-scale exploration will probably not be realized until 1990. The nodules would be collected and concentrated at sea and metal values recovered at a land-based plant. National Academy of Sciences and National Oceanic and Atmospheric Administration studies have stated that marine disturbances will be minimal, but little has been reported on potential environmental disturbances on land. Impacts will probably be similar to those of other hydrometallurgical and pyrometallurgical processes used for nickel and cobalt recovery. Forecasts for cobalt production from manganese nodules range from about 2,000 to 14,000 MT/year. No internationally acceptable treaty has been negotiated at the United Nations Law of the Sea Conference sessions (first held in 1968), and private companies are reluctant to proceed with mining deep-sea nodules until legalities are clarified.

Cobalt occurs in base metal ores, concentrates, tailings, and smelter products at concentrations that are seldom much higher than those found in ultrabasic rocks ($\sim$ 200 mg/kg). Smelter emissions and tailings are composed of smaller particles than background dusts arising from rocks and soils, and therefore are more leachable and more absorbable via the lungs. Because of the lack of significant cobalt enrichment in base metal ores, smelter products, etc., the amounts of cobalt in various base metal process streams were not estimated as had been done for the rarer metals in previous MRI studies of this series. Even if the average cobalt concentration in iron ores is not much more than the crustal abundance (25 mg/kg), the vast quantities of ore mined in the United States ($\sim$ 80 million MT) annually might be expected to mobilize an amount of cobalt ($\sim$ 2,000 MT) representing a large fraction of the cobalt consumed annually in this country.

Fish (sculpin) foraging in benthic sediments contaminated by taconite tailings (dumped at Silver Bay, Minnesota) in Lake Superior have been found to contain about the same concentration of cobalt as fish localized in noncontaminated areas.

Zinc concentrates are often somewhat enriched in cobalt, averaging about 9 to 10 times more cobalt than the associated lead concentrate or the crustal abundance. If pyrite is associated with the lead and zinc minerals, cobalt will be preferentially concentrated in the pyrite.

In the lead smelting circuit, the major removal of cobalt is in the speiss obtained after reverberatory smelting of the copper dross. Speiss and matte from copper dross smelting are treated at copper smelters in Tacoma, Washington, and El Paso, Texas. Roasting, digestion, and electrowinning recover metal values (cobalt as the oxide) from speiss. A minor transfer of cobalt is via the lead desilverization crusts. In the zinc pyrometallurgical circuit, cobalt apparently is found in retorting residues that are recycled or discarded to the sinter bank.

Cobalt is frequently mentioned as an impurity in the zinc electrolyte and the copper-cadmium cake at U.S. electrolytic zinc plants. Small amounts of cobalt have been produced annually by the Bunker Hill electrolytic zinc plant in Idaho; e.g., the plant at Kellogg, Idaho, has frequently isolated purification residues containing cobalt (e.g., 3.0 MT in 1965), but their fate has not been reported. Losses of cobalt in iron precipitates from zinc sulfate leach solutions are only $\sim$ 1%. Cobalt has been found in concentrations of 27 to 30 mg/liter in purified cadmium electrolytes at two U.S. electrolytic cadmium plants. Spent zinc electrolyte, which typically contains $<$ 0.1 mg Co/liter, must be periodically or continuously bled for purification or discharge to waste to remove cobalt and other impurities. The remaining spent electrolyte is recycled to leaching operations.

One report implicated an electrolytic zinc refinery in Hobart, Tasmania, as a source of atmospheric pollution in residential areas by cobalt and other heavy metals (via dust blowing from exposed stockpiles). However, the site having the highest ambient concentrations of zinc, cadmium, copper, iron, etc., did not have the highest concentration of cobalt (1,490 mg/kg). Presumably, some other metallurgical process was the source of the cobalt.

Copper ores, concentrates, and smelter products contain highly variable concentrations of cobalt ranging from undetectable to $\geq$ 1,000 mg/kg. Probably, $\geq$ 30% of the cobalt in a copper ore remains in the tailings. A mass balance for the smelting and refining processes is difficult to make because of the variability in reported transfers in such processes. We have estimated that, for each 100 kg cobalt in U.S. copper ores, 70 kg cobalt is transferred to the concentrate. The final distribution of the 70 kg cobalt in the concentrate may be as follows: 3.5 kg in the waste reverberatory slag, 13.5 kg in the recycled converter slag, 49 kg (70% of the cobalt in the concentrate) in the by-product nickel sulfate, and 3.1 kg in the cathode copper. If fire-refined (a minor process in the United States), almost all of the cobalt is transferred from the blister copper to slag.

The nickel sulfate is sold for chemical uses, including electroplating. Crude nickel sulfate from Asarco Inc. electrolytic copper refineries in Tacoma, Washington; Baltimore, Maryland; and Perth Amboy, New Jersey, was refined in Perth Amboy; and much of the cobalt was probably precipitated from the solution with the copper and zinc sulfides.

Although we have found no data specifically on cobalt in U.S. copper smelter emissions, the smelting of cobalt-containing nickel-copper sulfide ores in Sudbury, Ontario, is known to have released about 71 MT cobalt annually between 1961-1971. Nearby soils and rainwater are highly contaminated with cobalt.

One U.S. copper mine discharges mine water containing 320 µg Co/liter at a rate of 2,300 kg/year, but $< 5\%$ of the copper industry has to dispose of acid mine waters.

Sulfuric acid or ammonia leaching used to recover copper also dissolve cobalt from tailings and oxidized ores. Sulfuric acid leach solutions at five of six U.S. copper mines monitored contained 3.30 to 72.0 mg Co/liter. Bleed solutions to reduce such buildup are evaporated in lagoons. Any cobalt remaining after cementation or electrowinning of copper probably remains in the barren leach solution. After electrowinning, one vat-leach acid solution is reported to have contained 51.0 mg Co/liter. Such wastes are dried in an impoundment.

In the Southwest where the majority of U.S. copper production occurs, mill wastewaters are fully recycled. One mill wastewater contained 1.68 mg Co/liter, which was equivalent to 11 MT Co/year at that mill. Of six operations not in the Southwest, three precipitate heavy metals by lime and settling procedures.

Environmental samples and human hair have been found to be contaminated with cobalt in the vicinity ($\leq$ 4.3 km) of the smelters of the gold-producing mines at Yellowknife, Northwest Territories, Canada. Since the cobalt concentrations in ores were not reported, it is difficult to draw analogies with U.S. practices. Cobalt has been found in concentrations up to 0.2% in U.S. gold concentrates. Significant fractions of cobalt are dissolved in the cyanide solutions used for leaching gold.

Cobalt is apparently not a problem in U.S. uranium milling wastes. One of five uranium mills was reported to have mill wastewater containing detectable cobalt, 1.7 mg/liter. However, its tailings pond decant contained < 50 μg Co/liter.

Cobalt concentrates have been reported in the literature from diverse ores, but they were apparently not of any commercial significance. These have included cobalt concentrates produced during flotation of talc in Vermont, bauxite in Arkansas, and low-grade manganese ores in Alabama. Cobalt (stainerite) ores have been concentrated in Goodsprings, Nevada.

Little environmental information has been reported in the literature that corresponds to emissions from primary metallurgy of cobalt-bearing materials in other countries besides the instances of environmental contamination in Canada already mentioned. International Nickel Company sends its matte to Clydach, Wales, for nickel recovery by the Mond carbonyl process. Since only nickel is volatilized in the process and cobalt is leached from the residue for its recovery, it is difficult to pinpoint the reason for the extraordinarily high average cobalt concentration (48 ng/m^3) in the ambient air of Clydach and surrounding towns in the Swansea area of Wales. Perhaps cobalt is released when the residue is roasted prior to leaching. The environmental contamination of this area is discussed in detail in Chapter V. In Chapter VII, it is noted that the lung tissues of older residents of Duisburg, West Germany, contain 7.5 times more cobalt than do lung tissues of residents of nearby Cologne (Köln). In the recent past, the Duisburger Kupferhütte Refinery has recovered cobalt from pyrite residues containing up to 6,000 mg Co/kg. Annual production has fallen from 1,100 MT cobalt in 1966 to 50 MT in 1978.

Within the next 5 years, world cobalt production may attain nearly 40,000 MT annually; and the United States will be producing 4,000 MT. Environmental contamination by cobalt in the U.S. may well be on the rise, but serious pollution problems were not seen as a result of high mining activity for cobalt in the 1950's.

Secondary recovery of cobalt in the United States is quite small ($\leq$ 1% of the total cobalt market) according to government sources. In the past, most of the cobalt-bearing scrap that was recycled in the United States was new and clean. Off-grade contaminated metal or residues, grindings, and mixed scrap were reportedly shipped to Japan, West Germany, and Belgium for recycling. It now appears, however, that recycling of cobalt has been significantly more extensive than previously believed. Furthermore, recycling will increase greatly because of the present high price of primary cobalt.

The processes actually in use for secondary recovery, despite the copious patent literature, are difficult to identify. Clean material is probably only remelted; the remainder must be sorted and graded. More and more cobalt users are segregating their cobalt scrap, and U.S. recovery is being encouraged. The U.S. Air Force is retaining all of its reclaimed scrap for military use. This is decreasing the low supply of U.S. high-temperature alloy scrap, however. It is not clear whether condemned aircraft engine parts have been or will be recycled.

Cobalt in worn-out cutting tool inserts and wear parts may be reclaimed as pure cobalt powder by the zinc embrittlement process, by the Coldstream process, or by chemical extraction. Leaching and mechanical crushing methods for reclaiming scrap carbides are slow and expensive. In the zinc embrittlement process, the cemented carbide mass is treated with molten zinc at 800°C to form a Zn-Co alloy, disintegrating the mass. The zinc is subsequently recovered from the alloy by distillation. Currently, about 20% of the carbide powder that is used is recycled. Zinc embrittlement is, or soon will be, the most common method used for such recovery. The Coldstream process employs a cold, dry atmosphere and pulverizes the zinc-embrittled carbides to allow classification of the particles by grain size.

Recycling of spent cobalt-containing catalysts is increasing in the United States. One New York City firm has been recovering cobalt from Co-Mo-Cu catalysts for 3 to 4 years by a Japanese patented process. The firm is also stockpiling Co-Mo/Al_2O_3 catalysts, but does not know of a suitable recovery process. Another plant came on stream in 1978 to recover cobalt as a mixed nickel cobalt carbonate product from unidentified spent catalysts.

Iron and steel scrap consumed in the United States between 1960 and 1969 contained an estimated 59 to 86 MT cobalt. If cobalt was uniformly diluted in the iron and steel, its average concentration would have been 1.1 mg Co/kg. However, certain steels intentionally contain significantly higher concentrations ($\leq$ 0.4% Co). Scrap from high-speed tool steels is presently being segregated and returned to domestic producers for cobalt recovery. In the past, the low percentages of cobalt in recycled high-speed steels were probably lost by dilution when recycled with noncobalt-bearing steels.

F. Cobalt in the Food Chain

1. The aquatic food chain: Cobalt attains high concentration factors (CF's)* in lower aquatic organisms (algae and invertebrates); but in any one food chain, the CF's generally decrease dramatically as the higher trophic levels are reached.

The literature data are difficult to generalize because some authors report on a dry weight basis, others on a wet weight basis, and some do not state the basis. Assuming that wet weight concentrations when multiplied by 10 give values comparable to dry weight concentrations, one can generalize the following CF's for aquatic organisms on a dry weight basis:

---

* Concentration in the organism divided by its concentration in the medium, or sometimes, in the food ingested.

Marine Algae--3,000 to 20,000
Freshwater Algae--400 to 2,000,000

Marine Mollusks--$\leq$ 20,000
Freshwater Mollusks--100 to 14,000 (1 to 300 in soft parts)

Freshwater Insect Larvae--100,000

Other Marine Invertebrates--100 to 40,000 (2,000 in crustacean muscle)
Other Freshwater Invertebrates--1 to 11,000 (500 to 3,500 in crustacean muscle)

Marine Fish--100 to 4,000 (5 to 500 in muscle)
Freshwater Fish--40 to 1,000 ($\leq$ 24,000 in one report)

Chapter VII contains an extensive discussion of the variables that influence concentration factors in aquatic organisms. These variables include population density, feeding mechanisms, age, speciation of cobalt, duration of exposure, seasonal phenomena, and size. In general, the high surface-to-volume ratio of plankton species favors uptake by adsorption. Absorption of cobalt by invertebrates is predominantly via ingestion of food particles, whereas absorption by the vertebrates is primarily via the gills from the water. Organisms that are filter feeders and come in contact with the sediment are also likely to take up more cobalt because the cobalt concentrations in sediments are much higher than in water. The edible parts of the organism generally have lower concentrations than the shell and other tissues.

In the most contaminated environments, cobalt uptake is apparently under homeostatic control, especially in the vertebrates. Thus, as the concentration of cobalt in the water increases, the concentration in the tissues does not rise proportionally.

In southeastern Missouri, where nickel and cobalt have been mined, the concentration of cobalt in the algae in the contaminated Little St. Francis River (224 µg Co/liter), which received milling effluents, was 307 mg/kg dry weight. Thus, the CF for cobalt was 1,374. Algae in nearby rivers (1 to 1.5 µg Co/liter) not receiving milling effluents had more typical CF's

(12,200 to 25,000) and dry weight concentrations of 12.2 to 32.3 mg Co/kg. Thus, in water containing about 200 times as much cobalt, the cobalt concentration in the algae was only 10 to 15 times higher while the CF was about 0.1 to 0.5 as large as values for uncontaminated sites.

2. The terrestrial food chain: Although data are plentiful for concentrations of stable cobalt in terrestrial plants, few of these data have been correlated with other trophic levels. With ^{60}Co, at least one study finds cobalt CF's increasing up a food chain, culminating in predaceous insects. In feeding studies, concentration factors from food to animal are generally < 1. One study finds identical cobalt concentrations in the milk from two groups of cows grazing on two soils with very different cobalt concentrations and proportionally different cobalt concentrations in the plants growing on them (i.e., the plants show almost identical concentration factors).

The most comprehensive comparison of cobalt concentrations in components of an ecosystem is of a tropical moist forest system in Panama. The study reveals some very different patterns from those observed in temperate climates. In the tropical ecosystem, most of the cobalt is held in the vegetation rather than in the soil. Animal tissue concentrations of cobalt (up to 1,700 mg/kg dry weight) are much higher than in their temperate system counterparts, and cobalt appears to concentrate more in bone than in soft tissues. The mammals, birds, and reptiles have significantly higher cobalt concentrations than do insects. Although the herbivores show CF's of about 10 based on the cobalt concentrations in the vegetation ($\sim$ 40 mg/kg dry weight), the carnivores represented by reptiles and amphibians show CF's of about 100 based on the cobalt concentrations of the insects (13 mg/kg). The CF from soil to plant was 26.

In temperate systems, plant CF's generally range from 0.1 to 10, but some plants are known to have CF's from 10 to 100.

Cobalt in decaying plant material is more available to soil fauna than inorganic cobalt in soils of basic pH. Plant uptake is also lower on high pH soils and from soils high in manganese dioxide.

Even when growing on soils to which cobalt has been added (e.g., in fertilizers, sewage sludge, fly ash, mineral wastes), plants show cobalt CF's generally < 1. For example, in soils treated with cobalt sulfate to give very high soil concentrations of 100, 200, and 400 mg Co/kg, the cobalt concentrations in cotton plant leaves were $\sim$ 30 to 40, $\sim$ 120 to 150, and $\sim$ 220 to 240 mg/kg, respectively. Thus the CF's were $\sim$ 0.3 to 0.4, $\sim$ 0.6 to 0.7, and $\sim$ 0.6, respectively.

3. Cobalt in the human diet: Numerous studies have reported daily cobalt intake from food. Unfortunately, the data fall into two widely separated ranges, 2 to 96 μg/day and 160 to 1,005 μg/day. Four studies using NAA give the best agreement at 9 to $>$ 40 μg/day. Five studies using other analytical methods report intakes of 10 to 96 μg Co/day.

The food group contributing the most cobalt to the diet also varies among studies. Three studies report that grains and cereals are highest, two, potatoes, and one each, fish, sugar, and coffee. Several studies have found high cobalt concentrations in coffee, and one report states that unfiltered coffee effusions from dry coffee (containing an average of 0.93 mg Co/kg) contained $\sim$ 5 to 10 μg Co/cup. If this is true, coffee drinkers may derive as much cobalt from this source as from the rest of their diet.

Cooking and canning may cause significant losses of cobalt from the fresh food. Roasting and grilling meat result in 0 to 10% cobalt losses, but stewing and boiling cause 20 to 90% losses. About 70% of the cobalt content of vegetables is lost upon canning.

Reports are contradictory as to whether milling grain causes a loss (70 to 90%) or gain (13%) in cobalt content in the flour or meal. One investigator reported at different times that samples of refined sugar had both more and less cobalt than raw sugar. Since grinding equipment may be made with cobalt alloys, it is likely that contamination is the cause when a gain is observed.

Food stored in containers made of isophthalic acid-based polyester resins and glazed kitchen earthenware may extract cobalt.

For example, 1 liter of 50% alcohol or 8% vinegar might extract up to 350 μg cobalt from a polyester resin bottle.

When cobalt salts were used as foam-stabilizing additives for beer, the maximum authorized dose was 1 mg Co/liter. A synergistic action between alcohol and cobalt apparently produced the fatal cases of cardiac myopathy associated with the use of cobalt foam-stabilizing agents. Beers naturally contain 2 to 50 μg Co/liter.

Although vitamin B_{12} (4.3% Co) is essential to mammals, it does not comprise the bulk of cobalt in the diet. Rich dietary sources are mammalian liver, kidney, and heart and marine bivalves. Boiling fish and goat meat causes 24 to 95% losses of vitamin B_{12}.

4. Cobalt in human tissues and the body burden: A large variability exists in reported tissue concentrations of cobalt (especially in whole blood and urine: 0.17 to 262 μg Co/liter). The following values (in mg Co/kg dry weight) have been selected as being reasonably representative: lungs, 0.2; hair, 0.5; kidneys, 1; liver, 1; brain, 0.1; heart, 0.2; intestine, 0.8; muscle, 1; spleen, 1; skin, 0.03; and fat, 0.1. The best data for human tissue concentrations lead to the calculation of a human body burden of 1.5 mg cobalt.* Despite the discrepancies in the few reported bone concentrations of cobalt, we estimate that this value is not larger than 15 mg. If the daily diet of Reference Man contains ~ 0.01 to 0.02 mg Co/kg fresh weight (based on ~ 20 to 40 μg Co/day intake), it would appear that humans do not bioconcentrate cobalt based on the value of 0.02 mg Co/kg (1.5 mg/70 kg) in their tissues.

Hair does not appear to be a good indicator of ambient exposure to cobalt, but cobalt concentrations were dramatically higher in the hair of a worker at a cemented tungsten carbide plant. The concentration of cobalt in the lungs of 50- to

---

* In fact, the International Commission on Radiological Protection gives a total body burden of < 1.5 mg cobalt for 70-kg Reference Man, but some of the major tissue values given do not appear to fall in line with the data compiled herein.

68-year-old residents of Duisburg, West Germany, where cobalt-containing pyrites are smelted, was increased by 7.5 times the concentration in lungs of 50- to 68-year-old residents of nearby but less industrialized Cologne. Even in Cologne, the cobalt concentrations of the lungs of residents increase with age. Several studies show infant lung concentrations of 0.03 to 0.04 mg Co/kg dry weight and average adult values of 0.1 to 0.2 mg/kg. Liver and heart concentrations also increase with age.

G. Physiological Effects

1. Pharmacokinetics

a. Absorption: Cobalt and iron differ from most other trace metals in that they are the only ones that are absorbed better by the intestine in an organically bound form than in the free ionic state. Human absorption of cobalt from the diet is variable, appearing to average about 80% in some studies but being as low as 1 to 44% in other studies (especially when administered as an inorganic salt). The body apparently absorbs Co^{2+} by the same mechanism by which it absorbs iron. Absorption of the two is mutually inhibiting. Because cobalt absorption varies substantially by species, no single best animal model for human absorption has been described. An overall observation on experimental oral dosing of cobalt indicates that the actual percentage absorption depends on the size of the dose, i.e., whether the cobalt was given to experimental animals in trace amounts (fractions of μg) or carrier amounts (> 100 μg).

Oral absorption is decreased by administering cobalt after a meal, pretagging cobalt to a protein, and ingestion of an antacid. Absorption by intact skin and from the lungs has been shown when workers have been contaminated by cobalt radioisotopes. Most of an absorbed dose soon appears in the urine. Tests with guinea pigs show that intact skin absorption of cobalt is low: only 1.72% of a dose of radioactive cobalt appeared in the urine and 0.9% in the liver when the skin was intact. Absorption of CoO from the lungs of dogs was greater than that of Co_3O_4; after 8 days, only 10% of CoO but 85% of Co_3O_4 remained in the lungs.

Cobalt absorption is slower in ruminants, presumably due to binding by ruminal microflora before incorporation into vitamin B_{12}. A similar binding by cecal flora also occurs in chickens.

All mammals including man require for hemoglobin synthesis a cobalt-containing tetrapyrrolic ring, cyanocobalamin (vitamin B_{12}). Dietary standards recommend a daily intake of 5 μg vitamin B_{12} (containing 0.2 μg Co). Its absorption from the ileum is dependent upon the presence of hydrochloric acid and the intrinsic factor, a mucopolysaccharide secreted by the parietal cells of the gastric mucosa. Without the intrinsic factor, B_{12} absorption is $< 2\%$ compared to $\sim 70\%$ in its presence.

As soon as vitamin B_{12} is passed through the ileum cell wall, it is bound to one of three globulins, the transcobalamins, as it enters the portal system.

b. Distribution: Cobalt occurs at a high level in the liver, presumably because of its portal circulation. Muscle, spleen, kidney, thymus, adrenals, pancreas, lungs, stomach wall, and bone have each also been reported to be the organ with the highest concentration or total content of cobalt in various distribution studies. Generally, cobalt given either parenterally or orally does not appear to accumulate in a specific target organ in many species. The concentrations of cobalt carried in the blood appear to be highly variable. Perhaps some of the analytical methods used have distinct limitations for cobalt detection in this matrix or perhaps exposure to cobalt from sources other than diet and drinking water also are highly variable. However, in the latter instance, one would expect much greater variation in reported cobalt concentrations in other tissues than is observed (see Chapter VII). If the range of cobalt concentrations in human diets is as broad as those reported in Table VII-18, the variation in tissue distribution might be explained by the different-sized doses of an element. Variations in animal distribution studies may be partly explained by the time that had elapsed between dosing and organ analysis for cobalt.

In rats fed a diet containing 1.3 mg Co/kg dry weight for a lifetime, the normal cobalt concentrations (determined by NAA) were (in mg/kg dry weight): kidney, 0.96; heart,

0.30; liver, 0.37; and muscle, 0.04. Generally, cobalt concentrations were independent of age. The values agree remarkably well with those given for one human being in Table VII-21.

Rats given trace amounts of cobalt in the diet for 5 days attained equilibrium between ingested and excreted cobalt within 48 hr. Retention was 0.6%. In another short-term rat study, retention at a 10-μg oral dose of cobalt was 2.7%. In longer term oral dosing (40 days), the equilibrium body burden in rats was $\sim$ 20%. Retention is probably inversely related to the size of the dose.

Transplacental distribution of cobalt to fetal calf liver, kidney, and intestine was observed after repeated injection or oral dosing of the pregnant cow with cobalt. Some reports remark that, despite wide variations in pasture content of cobalt, the cobalt concentration of the milk is constant; others claim a wide variation in cobalt concentrations in milk. Passive transport of cobalt from the maternal circulation to the fetus occurs in rabbits. In birds, about 10% of an absorbed dose of cobalt is incorporated into bird eggs over a period of 2 weeks.

Vitamin B_{12} is transported to the body tissues as a complex with transcobalamin II, a β-globulin. Transcobalamin I is an α-globulin with minor transport functions whose major role is as a storage depot for B_{12} in the form of methylcobalamin. In the tissues, vitamin B_{12} is bound to a variety of protein receptors. In plasma (only 0.1% of total body content), B_{12} is present as methyl-, 5'-deoxyadenosyl-, or hydroxocobalamin bound to proteins called the transcorrins. About 50 to 90% of the total body store of B_{12} is found in the liver. Cobalt ingested as vitamin B_{12} is retained more strongly than when in an inorganic form.

c. Excretion: Cobalt is excreted in both urine and feces of humans. The cobalt that is absorbed is largely excreted by the urine with some (perhaps $\sim$ 5% of that absorbed) excretion via the bile and feces. The large fraction of ingested cobalt eliminated via the feces indicates that much of it is probably unabsorbed. Cobalt balance studies in both humans and animals show great variability. Whether this reflects biological variability in cobalt absorption and excretion or analytical problems is not clear.

Once cobalt reaches the tissues, elimination is slow. Some body pools of radiocobalt have been reported to have biological half-lives ranging up to 69 years. The longest biological half-lives have been observed after inhalation exposure. The very long half-lives may be due to uptake by pulmonary lymph nodes. A study of long-term excretion of cobalt in rats showed that the bone ultimately contained the largest stores of the intravenously dosed cobalt.

The main route of vitamin B_{12} excretion is by way of the bile. About 40 μg/day passes into the jejunum, but most is reabsorbed in the ileum, utilizing the intrinsic factor mechanism. The small amount that remains unabsorbed ($\sim$ 3 to 6 μg/day) leaves the body in the feces. This amount of B_{12} includes the amount synthesized in the colon by bacteria. Urinary excretion of vitamin B_{12} is confined to the vitamin that is not protein bound and amounts to 0 to 0.25 μg/day. The biological half-life of vitamin B_{12} is about 400 days.

2. Biochemistry

a. Enzymatic effects: The mode of action of cobalt is multilocular and is not fully understood. A large part of its effects are produced by enzymatic impairment which leads to depressed tissue respiration and further depressive effects on energy metabolism. For example, Co^{2+} blocks the Krebs citric acid cycle and cellular respiration. It also depresses some drug-metabolizing enzymes of the liver and stimulates others.

The cobaltous ion can substitute for other divalent cations in several enzymes. For example, Co^{2+} can substitute for Zn^{2+} and activate alcohol dehydrogenase, lactate dehydrogenase, carboxypeptidase A, carbonic anhydrase, and alkaline phosphatase.

Besides reducing drug-oxidizing activity, cobalt affects other liver enzymes. Cobalt is the most efficient ion for inducing heme oxygenase in the liver. It also affects hepatic catalase and the mitochondrial enzyme aminolevulinic acid synthetase. Animals dosed with cobalt show elevated serum concentrations of lactate dehydrogenase, serum glutamic oxaloacetic transaminase (SGOT), and serum glutamic pyruvic transaminase (SGPT).

Since most studies of the biochemistry of cobalt involve doses of 30 to 60 mg $CoCl_2$/kg daily to rats, the relationship is probably not significant when considering environmental exposure.

The important action of cobalt on increasing hemoglobin and the apparent synergism or sparing effect on iron may be due to the enzymatic effects of cobalt. Cobalt appears to regulate by repression the synthesis of δ-aminolevulinic acid (ALA) synthetase. This mitochondrial enzyme is rate-limiting for the sequence of enzymatic steps necessary for heme production. Cobalt also regulates the rate-limiting enzyme for heme degradation in a negative feedback manner analogous to the control exerted on this enzyme by heme itself. In animal experiments, δ-ALA synthetase initially falls; but within a few hours, the depression is followed by a dramatic rise in enzyme activity. The contradictions in the published research have not yet been resolved. The decreased synthesis of heme produced by cobalt and the erythropoietic response to cobalt are opposing effects which need much further clarification.

Vitamin B_{12} is a coenzyme for three enzyme systems in animals and bacteria. These systems play important roles in metabolizing proprionate, methylating homocysteine to give methionine, and reducing ribonucleotides. In bacteria, the latter reaction involves the conversion of the ribose moiety in a ribonucleotide to deoxyribose when DNA is to be formed.

Ruminal microflora of multigastric animals convert dietary cobalt to vitamin B_{12}. Man and other monogastric species are dependent on these animals and bacteria for the production of vitamin B_{12}.

b. Metabolism: Cobalt, by reacting with the thiol groups of amino acids, proteins, and other coenzymes or cofactors, inactivates several biochemical pathways, especially those associated with the production of cellular energy. Treating animals with $\sim$ 10 mg $CoCl_2$/kg for 11 to 12 days increases serum or plasma triglycerides, cholesterol, and lipoproteins. The mechanism of the lipemic response to cobalt is not understood. Some evidence points to inhibition by cobalt of the enzyme responsible for the catabolism of circulating lipoproteins.

The demand for vitamin B_{12} in the body for the function of hemopoiesis exceeds that for any other clinically recognizable physiological function. Thus the most characteristic sign in humans of a deficiency of vitamin B_{12} is development of a macrocytic anemia or characteristic lesions of the nervous system.

The metabolic roles of vitamin B_{12} and folic acid (vitamin B_c) are intimately combined. By methyl transfers from 5-methyltetrahydrofolate to methylcobalamin and from methylcobalamin to homocysteine, tetrahydrofolic acid can be recycled back into the folate pool and methionine is synthesized. The effects of B_{12} deficiency on the peripheral nervous system might be explained by the effect of cobalt on lipoprotein synthesis in myelin sheaths. The isomerization of methylmalonate to succinate (catalyzed by the cobamide coenzyme) could be the rate-limiting reaction for the production of the lipoprotein.

c. <u>Antagonistic/synergistic effects</u>: Administration of cobalt salts to animals and humans increases the formation of red blood cells, but this may not be a normal physiologic stimulus of erythrocyte synthesis but rather may be a toxic manifestation of cobalt activity. When cobalt inhibits tissue respiration and oxidative phosphorylation, tissue hypoxia occurs. This hypoxia is the probable stimulus to the secretion of erythropoeitin, the hormone that stimulates bone marrow stem cells to produce erythrocytes.

Cobalt appears to have a synergistic toxic effect with ethanol on the heart and on general growth rate.

The lung toxicities of cobalt and other ingredients of cemented tungsten carbide are greater together than when administered separately.

Selenite and vitamin E protect against the myocardial toxicity of cobalt. Cobalt protects against the testicular necrosis caused by cadmium. Animals ingesting ≥ 4 mg Co/kg show an anemia indicative of depression of iron absorption.

Cobalt deficiencies render sheep more susceptible to selenium toxicity and children more sensitive to the toxicity of vitamin D.

d. Nutrition: Vitamin B_{12} concentrations of 0.2 μg/liter are sufficient to prevent any signs of clinical, hematological, or biochemical deficiencies in humans. Malabsorption is the primary cause of the B_{12} deficiency anemias rather than deficient dietary intakes except where animal protein intake is low. Fish tapeworm may contribute to some human deficiencies.

Deficiency of vitamin B_{12} or folate results in defective synthesis of DNA which is manifested early in hemopoietic tissue where megaloblastic changes are seen in developing red and white cells and in the megakaryocytes. Vitamin B_{12} deficiency also causes demyelinating neurological lesions of unknown biochemical etiology. B_{12} absorption is decreased by ethanol intake in the presence of adequate protein and vitamins in healthy adults. Humans suffering from pernicious anemia tolerated doses of 500 to 1,000 μg B_{12} daily for 10 to 70 months without unfavorable reactions.

The minimum daily dose of cobalt for most animals is 1 to 10 μg except for ruminants whose daily requirement is 100 μg. The diet of pigs and chicks should contain (dry weight basis) 0.01 to 0.02 mg Co/kg; that of ruminants, 0.07 to 0.10 mg Co/kg. Cobalt deficiency diseases have been observed wherever the herbage is very low in cobalt due to inadequate soil concentrations. Ruminants require cobalt so that their ruminal microflora can synthesize the vitamin B_{12} which is required for propionate metabolism, etc. Propionate metabolism is more important to energy metabolism in ruminants than in nonruminants. Deficiency symptoms in ruminants include wasting with anemia and anorexia.

Cobalt compounds (acetate, carbonate, chloride, oxide, and sulfate) are commonly added to animal feeds, to the salt lick, or to fertilizers. Cobalt pellets placed into the rumen are also effective in preventing cobalt deficiency in ruminants.

Maximum tolerated concentrations of cobalt in rations include 200 mg/kg for pigs, 20 mg/kg for cattle, 50 mg/kg for sheep, and 4 mg/kg for poultry.

Although cobalt is only known to be required as a constituent of vitamin B_{12}, some evidence points to a cobalt requirement for the optimal utilization of low doses of iodine by the thyroid.

3. Effects on humans and animals

a. Toxicity

(1) Human toxicity: Cobalt has recognized toxic effects in humans on the thyroid, the heart, and possibly, the kidney, in addition to the occupational lung disease seen in the cemented carbide industry and allergic manifestations.

Epidemics of severe congestive heart failure in heavy beer drinkers from the addition of $\sim$ 1 mg Co/liter to beer as a foam stabilizer were documented in Louvain, Belgium; Omaha, Nebraska; Minneapolis, Minnesota; and Quebec City, Canada. In most cases, a cyanosis limited to the face and trunk was diagnostic. Progressive signs of heart failure appeared (dyspnea, tachycardia, low blood pressure, and small volume arterial pulses) and SGOT levels were strikingly elevated. Mortality was $> 40\%$ in 100 patients in these four localities. The autopsied victims (20 deaths in 50 cases) showed diffuse vacuolar degeneration and myofibrilysis in the heart and thyroid changes. Histological findings were quite similar to those found in animal experiments with cobalt alone.

A long-standing high level intake of alcohol appears to be associated in a causal way with myocardial disease. Because only a few chronic heavy drinkers developed cardiomyopathy, individual susceptibility must be important. The addition to beer of low concentrations of cobalt, with its own enzyme-inhibitory and cardiac toxic effect, apparently exacerbated the condition of the heart brought about by chronic alcoholism. Perhaps nutritional deficiencies of alcoholism also played a part.

Myocardial toxicity due to industrial exposure to cobalt is known, but rare.

Patients receiving 0.17 to 3.9 mg Co/kg/day for 6 days to 8 months, usually for the treatment of anemia, showed 20 to $> 90\%$ depression in iodine uptake which was typical of disease states producing an inhibition of the protein binding of inorganic iodide. Doses as high as 200 mg $CoCl_2$/day were given. Goiters and classic signs of hypothyroidism occurred as early as 6 weeks after initiation of cobalt therapy. By 1970, all anti-anemia preparations containing cobalt had been removed from the U.S. market. Other symptoms that were seen in patients receiving cobalt anti-anemia preparations included anorexia, nausea, vomiting, diarrhea, tinnitus and neurogenic deafness, erythema and hot sensations, skin rashes, and substernal aches.

Studies in rats indicate that either excessive or deficient intake of cobalt renders a decrease in the absolute content and concentration of iodine in the thyroid gland. Relationships between cobalt concentrations in the environment and the disease rate in areas of endemic goiter have not been shown.

The erythropoietic effect of cobalt is probably a toxic action. For example, the polycythemia induced by cobalt in laboratory mammals is inhibited by common nutrients. Dosing in rats with 4 and 0.25 mg $CoCl_2$ for 4 weeks was accompanied by a generalized stress response as evidenced by a dose-response agreement in the size of the adrenal. Cobalt-dosed rats excreted the hemoglobin precursor coproporphyrin in the urine at more than twice the rate of controls. Increases in urinary excretion of coproporphyrin are indicative of toxicity to the blood-forming tissues and are seen after other heavy metal poisoning.

Besides the peripheral neuritis and auditory nerve symptoms mentioned above, other manifestations of cobalt neurotoxicity include headaches, weakness and irritability, changes in the electrical activity of the brain, and changes in reflexes, especially visual. (These transient symptoms were seen in 118 workers suffering from acute industrial exposure to cobalt carbonyl vapor.) Optic nerve atrophy has been described in at least two patients dosed with totals of 0.9 and 32 g $CoCl_2$.

Patients with chronic renal failure are commonly anemic and, outside the United States, are often still treated with cobalt chloride. Fourteen such patients on maintenance dialysis were treated with 25 mg $CoCl_2$/day for 4 weeks followed by 50 mg $CoCl_2$/day for four more weeks in 11 of the patients to increase their hematocrit 23% by the end of 8 weeks. Three of the patients experienced a hearing loss and one, nausea. In another group of 23 renal failure patients receiving 50 mg $CoCl_2$/day for 3 months, one died of congestive heart failure 3 months after receiving the cobalt dosage. Blood cobalt levels were significantly higher in dialysis patients who had been treated 13 to 20 months previously with cobalt than in dialysis patients who had not received cobalt. Prolonged retention of cobalt in the blood of patients with severe renal failure could lead to toxicity.

The kidney and erythropoietin appear to be involved in the mechanisms of hemoglobin increase by cobalt. Perhaps cobalt acts by producing local ischemia of kidney receptors controlling erythropoietin production. This could be done by suppression of cellular energy metabolism by inhibiting sulfhydryl groups of some oxidative enzymes, resulting in diminished oxygen utilization by the tissues. However, in anephric patients, cobalt would have to act by extrarenal sources of erythropoietin.

Immunologic changes were described in four workers from the cemented carbide industry; two had developed respiratory signs after 2 months' inhalation exposure, and two had developed a bronchiogenic irritation after a year's exposure. One of the four had a greatly elevated IgE (immunoglobulin E); two of the four had antinuclear antibodies and positive tuberculin reactions; but none reacted positively to intradermal challenge by cobalt or the other metals.

Among $>$ 4,200 workers surveyed in six studies of the cemented tungsten carbide industry, 1.7 to 9.4% of the populations showed dermal hypersensitivity or respiratory allergic symptoms. An erythematous, papular type of dermatitis, most marked in friction areas, and positive patch tests to metallic cobalt powder were often reported. Cutting fluids used in grinding hard metal contained up to 200 mg Co/kg* and produced eczema in some workers. Besides occupational exposures in the manufacture of cobalt-containing alloys, coatings, pigmented glass and ceramics, paint, and drugs, farm workers may show an allergic dermatitis from handling the cobalt preparations used in agriculture. In 10 to 60% of the reported cases of cement eczema, cobalt sensitivity occurs together with chromium sensitivity.

---

* One study of 60 cobalt-allergic patients determined that the mean threshold was 2,700 mg/liter in water and 3,100 mg/kg in petrolatum, but 14 of the patients were exquisitely sensitive to cobalt at all dilutions, showing no threshold.

Isolated cobalt sensitivity in cement eczema is seldom seen. Although the free cobalt oxides present in cement are water insoluble, they are probably dissolved in the presence of amino acids that may be present on eczematous skin.

Dermal hypersensitivity to cobalt was observed in one child and was attributed to the cobalt catalyst impurity in his plastic pens and eyeglass frames and in the cobalt-impurity in his nickel-plated wristwatch band.

A large number of orthopedic studies on the rejection of prosthetic implants made of cobalt alloys have suggested that a cobalt allergy contributes to this rejection. However, allergy to nickel is most often responsible for tissue reactivity of prostheses.

Some patients treated with therapeutic doses of vitamin B_{12} have also exhibited allergic reactions.

Among its therapeutic applications, cobalt is used as an antidote to cyanide poisoning. Use of Co-EDTA should be reserved for acute cases of cyanide exposure that are truly life-threathening because the effects of intravenous injections of Co-EDTA include atrial fibrillation and ventricular rhythm disturbances.

Cobalt exposure via cigarette smoking may be at least one order of magnitude lower than cobalt in diet (even if the magnitude in diet is taken as $\sim$ 40 μg/day). At 1 μg Co/cigarette, 0.1 μg Co was transferred to the smoke. In other experiments, only 4.2% of the cobalt in the tobacco was transferred to the smoke condensate from an unfiltered cigarette, and 0.87% from a filtered cigarette.

Common complaints among workers occupationally exposed to cobalt powders include partial or complete loss of the sense of smell, dyspnea, gastrointestinal distress, and weight loss. Lung fibrotic changes are attributed to the action of cobalt even when cobalt is associated with tungsten and titanium carbides.

Long-term exposures in the cemented tungsten carbide industry give rise to extrinsic asthma, diffuse interstitial pneumonitis, fibrosis, and/or pneumoconiosis. About 5% of the complaints are allergic in nature. Of $\sim$ 5,000 workers exposed to cemented tungsten carbides described in the literature, $\sim$ 500 showed various symptoms of respiratory toxicity. Cobalt concentrations in the workplace air have often exceeded the threshold limit value (TLV) of 0.1 mg/m^3. Concentrations as high as 79 mg Co/m^3 have been reported in the older literature. (Hundreds of mg cobalt as the oxide was noted in workplace air around a vibrating sieve. Cobalt powders no longer need to be manufactured at the cemented carbide plant via the oxide as they were decades ago.) Long-term exposure to cobalt metal and fume at even 0.1 mg Co/m^3 produces diffuse, interstitial lung disease with a 50% reduction in vital capacity. Workers with allergy-like symptom patterns to the respiratory involvement show improvement away from the work environment.

Although cobalt metal presumably represents the bulk of occupational cobalt inhalation exposures, the oxides and salts may also produce intoxication. A hydrometallurgical cobalt plant may produce aerosols containing, for example, CoNiS, $Co(NH_3)_6CO_3$, $[Co(NH_3)_5SO]_2SO_3 \cdot 2H_2O$, $Co(OH)_2$, $CoSO_4$, and Co metal. At one such plant, workers absorbed at least 694 μg/day since this was the average urinary excretion. Air concentrations at this plant ranged from 0.063 to 1.605 mg Co/m^3. Blood values of erythropoietin, red blood cell count, etc., were not significantly different from those of unexposed controls, but the albumin and lactic dehydrogenase isoenzyme were highly significantly greater in the exposed group. Lung disease was not reported.

The TLV in the United States for cobalt metal and fume as adopted by the Occupational Safety and Health Administration (OSHA) is 0.1 mg/m^3 which is a time-weighted average (TWA) for a conventional workday. Several countries have set the same limit, but the USSR and the Federal Republic of Germany have set a standard of 0.5 mg Co/m^3. In the USSR, TLV's have also been set for cobalt oxide (0.5 mg/m^3), dicobalt octacarbonyl (0.01 mg/m^3) and cobalt hydrocarbonyl and its decomposition products (0.01 mg/m^3).

Epidemiological studies other than occupational ones and dermal sensitivity studies usually do not associate cobalt alone with specific disorders. Thus, although a 1978 report states that the well water in an industrialized city in India contains 2.00 mg Co/liter, the high incidence of "mental deformation and defective speech" among the inhabitants was probably ascribable to high levels of other heavy metals in the water, including lead and cadmium.

Periodontal disease in New Zealand sheep was associated with the high concentrations of cobalt, nickel, chromium, copper, and titanium in the topsoil.

In only one of three studies was a correlation (an inverse one) found between cobalt in drinking water and the incidence of caries in USSR schoolchildren.

In other studies of diseases related to the geochemical environment in various parts of the USSR, elevated concentrations or "imbalanced ratios" of cobalt (and several other elements) in the environment were related to alterations in osteogenesis, stomach and lung cancer, and lung cancer in coal miners. Yet other epidemiological studies found no correlation between cobalt and cancer. Another study related esophageal cancer to cobalt-deficient soil. Some studies claim a relationship between environmental cobalt deficiencies and endemic goiter; others find no relation. Perhaps cobalt has some protective effect against atheriosclerosis, since 70-year-olds (but not 30-year-olds) dying from coronary atheriosclerosis had depressed concentrations of cobalt in their blood and aorta compared to concentrations in controls of the same age.

Cobalt along with several other elements may have a protective effect against cardiovascular disease as their occurrences in drinking water correlate negatively with the disease.

(2) Animal toxicity: The lethal doses of cobalt and its compounds in acute toxicity tests depend on the route, the form, and the animal tested. For purposes of comparison, the rat is most useful since most compounds and most routes have been tested in this species. As expected, cobalt is usually

more toxic when given by injection or intratracheally than when administered orally or intragastically. The LD_{50} or lowest lethal oral dose (LD_{LO}) in rats for cobalt metal, the carbonate, cobaltous oxide, Na_2[Co-EDTA], and the stearate ranged from 1,000 to 2,000 mg/kg. Co_2O_3 is even less toxic having an intraperitoneal LD_{50} of 5,000 mg/kg.

Cobalt compounds having LD_{50}'s or LD_{LO}'s when given orally to rats in the range 100 to 1,000 mg/kg include the tetracarbonyl, the nitrate, the chloride, the fluoborate, and the sulfate. The lactate is more toxic when given intraperitoneally (7.5 mg/kg) than is the chloride given intravenously (20 mg/kg).

Co^{2+} given as the sulfate to white male Swiss mice intraperitoneally had an LD_{50} of 21 mg Co/kg, making it 19th in toxicity among 42 metals tested. It was 127.2 times more toxic than Na^+. (Pb^{2+} was 120.3 times more toxic and Hg^{2+} was 2,283.0 times more toxic than Na^+.)

Numerous animal studies on the role of cobalt in cardiomyopathy are reviewed in Chapter VIII. The mechanism of cobalt toxicity to the rat heart is based on a cobalt-induced reduction in the oxidation of pyruvate. Animal pathology is more pronounced if thiamine and protein deficiency is present.

Blood pressure rise seen in cats following cobalt chloride administration intravenously at $\sim$ 0.016 mg Co/kg is attributed to stimulation of the medullary circulatory center. Blood pressure drop after 8 mg Co/kg is ascribed to a direct action on the blood vessels.

An erythropoietic effect was seen after 6 to 7 months in rats drinking water containing $\geq \sim$ 2 mg Co/liter as mentioned above. Rats exposed to cobalt metal aerosols of 0.05 mg/m^3 showed initially depressed but ultimately elevated hemoglobin and erythrocyte count after 2.5 months. Rats exposed to 0.005 mg Co/m^3 showed an increase only in hemoglobin, but pathomorphological changes were seen in lungs and other organs even at this level. Toxic effects included decreased activity of lymphatic tissue in spleen, decreased reactivity of bronchial cell membranes and lymphatic tissue in lungs, and protein dystrophy in kidneys. Protein and carbohydrate metabolisms were also disturbed. Rats exposed to 0.001 mg/m^3 for 3 months showed no significant changes in these parameters. The authors of the aerosol study recommend 0.001 mg Co/m^3 as the maximum permissible average daily concentrations for air in populated areas. (This no-effect level is equivalent to 1,000 ng/m^3, > 20 times higher than the average cobalt concentration in the air at Clydach, Wales.)

The mechanisms of the erythropoietic effect of cobalt appear to involve early decomposition of both old and newly formed erythrocytes under the action of hypoxia at higher levels of cobalt exposure, which elicits the hemopoietic effect on bone marrow. Injected doses of cobalt chloride in rats at 10 times the recommended human dose for anemia produce hyperplasia of bone marrow, particularly in the erythropoietic elements.

Cobalt chloride produces severe toxicity to the α cells of the pancreatic isles of Langerhans; e.g., ∼ 10 mg $CoCl_2$/kg intravenously in dogs cause a complete loss of α cells in many of the pancreatic islets within 24 hr of dosing due to autolysis. Regeneration of the α cells begins within 5 days. Since hyperglycemia can be elicited in pancreatectomized animals by cobalt, this response may be due to cobalt's toxic effect on the exocrine part of the pancreas rather than its effect on the α cells. Degenerative changes and hyperglycemia are also seen in teleost fish when injected intramuscularly at a dose of 100 mg Co salt/kg. Pigeons given 1 mg $CoCl_2$/kg intraperitoneally showed damage of both the endocrine (α) cells of the pancreas and the exocrine pancreatic epithelium. Apparently, cobalt accumulates primarily in the exocrine pancreas, in the zymogen granule lipids.

The mechanisms through which cobalt salts induce changes in the pancreas and inhibition of insulin release are hypothesized as follows: cobalt stabilizes insulin in the β cells and the β cells concentrate cobalt from the blood, leading to toxicity in these cells and subsequent metabolic derangements. The initial hyperglycemia may be due to some extrapancreatic factor such as epinephrine, but subsequent increments in blood glucose levels are ascribed to the degranulation of the β cells.

Most laboratory animals do not serve as a satisfactory model for studying the effects of cobalt on the thyroid. Cobalt administration, however, has produced goiter in guinea pigs and thyroid effects in chicks.

Although brain implants of cobalt metal or salts are commonly used to study epilepsy in animals, the epileptogenic etiology of cobalt is unknown. Cobalt implants cause extensive neuronal death and other severe local effects. Perhaps its action is due to the interference of ionized Co^{2+} with transport ATPase activity in whole brain. In addition, Co^{2+} competes with Ca^{2+} in neurons of several lower life forms. In human epilepsy, there is usually no definite focus; but when there is, the histochemistry resembles cobalt epilepsy with abnormal neurons that have few dendrites. In the focal area, there are reduced levels of Na^+- and K^+-ATPase activity and of transmitter amino acids. There is also an impaired ability to synthesize acetylcholine in both human and cobalt-induced epilepsy.

Behavioral effects have been reported in rats drinking water containing $CoCl_2$ for 6 days/week for 7 months at doses of 0.5 mg/kg. This dose corresponds to a level of $\sim$ 2,000 μg Co/liter in the water. The effect on reflex learning (lengthening of the latent period and increase in the number of extinctions of conditioned reflexes) seen in rats drinking $\sim$ 2,000 μg Co/liter was not seen in rats receiving water containing $\sim$ 200 μg Co/liter. These rats showed no behavioral differences from those of controls.

Experiments with rats breathing 0.5 mg Co/m^3 showed metabolic and tissue changes that were described above. Pigs breathing 0.1 mg Co metal/m^3 for 3 months showed a lung compliance only 66% that of the controls, but compliance in the exposed group returned to control levels within 1 month after cessation of cobalt exposure. Blood chemistry studies of the exposed pigs revealed an increase in α-, β-, and γ-globulins over those of controls, a net increase in total protein, and inversion of the albumin/globulin ratio. Such changes may be interpreted as early indicators of lung cell damage. Pigs exposed to 0.1 mg Co/m^3 and 1.0 mg Co/m^3 excreted 29 and 220 μg Co/liter, respectively, compared to 18 μg/liter by the controls. This study, showing serious effects with relatively low exposure, is cited in the 1977 printing of the 1971 edition of Documentation of Threshold Limit Values for Workroom Air as the reason for suggesting a lowering of the adopted TLV for cobalt metal dust and fume from 0.1 to 0.05 mg/m^3.

Other serious effects were seen in rats who were exposed to 0.48 mg Co powder/m^3 for 4 months. Circulatory and dystrophic changes were seen in the mucosal tissue with destruction of epithelium cellular elements and some epithelial atrophy. Such levels also caused a drop in all respiratory mucosal enzymes within 2 months. Respiratory tract mucosal structure 1 month after cessation of the 4 months' exposure had not yet returned to normal.

Cobalt metal when introduced into lung tissue has a more pronounced effect than does cobalt oxide. A single 50-mg intratracheal dose of cobalt oxide dust in guinea pigs produced an acute response, but the lungs were normal within 1 month. Twelve months after a single 50-mg intratracheal dose of cobalt metal particles the lungs showed a diffuse central fibrocellular infiltration. In another experiment, only one of six guinea pigs survived an intratracheal dose of 10 mg cobalt metal dust.

Liquid cobalt carbonyl hydride is more toxic than solid dicobalt octacarbonyl. The former, $Co(CO)_4H$, has a 30-min LD_{50} of 165 mg/m^3 in rats. Chronic experiments with unstated levels of $Co(CO)_4H$ produced some deaths and fibrotic phenomena in the lungs of the dead rats. In these chronic experiments, 12 to 15% of the quantity of inhaled $Co(CO)_4H$ was excreted daily in the urine.

Although Chapter VIII summarizes reports of studies testing the effects of cobalt ion on aquatic animals at concentrations up to more than 1,800 mg/liter, few are pertinent to the environmental concentrations found in U.S. waters summarized in Chapter IV. For waters that are not especially contaminated ($<$ 10 µg Co^{2+}/liter), there is one report of beneficial growth effects in the ciliated protozoan, _Tetrahymena pyriformis_. In this range, reproductive and teratogenic effects in _Daphnia magna_ and _Limnea_ are also noted in this summary in G.3.(4) and G.3.(5). In rivers generally contaminated by urban effluents, mine drainage, or agricultural runoff, a typical concentration of 45 µg Co^{2+}/liter retards the movement toward light of the larval form of the crab _Carcinus maenas_ within 2 to 3 days. Between concentrations of 330 and 990 µg Co^{2+}/liter, the first lethal effects are reported in zooplankton (48-hr LC_{50}'s) and snails (30-day LD_{50}'s). Teratogenic effects in snails are also noted. See

G.3.b.(4). Several freshwater fish species including rainbow trout show no effects at 1,000 μg Co/liter (56 hr for the trout in tapwater). At 1,300 μg Co/liter, the zooplankton D. magna is immobilized, and T. pyriformis has a 96-hr LC_{50} of 1,550 μg Co/liter in distilled water. Cobalt concentrations higher than these are rarely reported, usually being the concentration in an effluent or in acid mine drainage. However, two surveys have reported concentrations of 4,500 and 6,500 μg Co/liter in water bodies receiving effluents from a copper-molybdenum mill near Ray, Arizona, and the cobalt-nickel-copper mill at Fredericktown, Missouri, respectively. Algae grew in the latter waters (the Little St. Francis River) so that an appraisal of cobalt effects on life forms at these highest environmental water levels is appropriate. A concentration of 3,840 μg Co/liter is the 48-hr LC_{50} for the freshwater zooplankton Cyclops abyssorum prealpinus; 4,500 μg Co/liter retards within 1 day the movement toward light of C. maenas larvae; and concentrations of 7,300 to 8,800 μg Co/liter are fatal to about half or more of the population of Fundulus fish and Austropotamobius pallipes snails within about 4 to 5 days. A concentration of about 10,000 μg Co/liter (10 mg/liter) is the 10-day "lethal concentration limit"* for large sticklebacks (Gasterosteus aculeatus) and the 96-hr LC_{50} for the snail, Orconectes limosus. Higher cobalt concentrations are invariably fatal to the fish tested within 1 day to 1 week.

b. Special toxicities

(1) Mutagenicity: Mutagenic actions or mitotic aberrations due to cobalt salts have been reported in studies on rat tumor (rhabdomyofibrosarcomata) cells, root tips of Vicia fabais and Allium cepa (onion), Saccharomyces cerevisiae (yeast), and Pisum (pea) rootlets and seeds. Cultured human lymphocytes exposed to 0.06 to 0.6 μg Co/liter showed an increased frequency of diploidy formation. In studies looking for mutagenicity of cobalt, none was found in cultures of human fibroblasts (although the chromosomes were often over-contracted), Bacillus subtilis strains H17 Rec^{+} and M4S Rec^{-}, leukocyte cultures, bacteriophage T4 strains, cobalt-poisoned cattle, and Escherichia coli. Some authors claim that the mutagenic ions of Pb, Cd, Co,

* This concentration was no more lethal to the test fish than was tapwater to control fish.

Cu, and Mn stimulate initiation of RNA synthesis at concentrations that inhibit overall RNA synthesis, whereas nonmutagenic metals do not. Cobaltous chloride at a concentration of 44 mg Co/liter stimulated chain initiation but inhibited overall RNA synthesis.

(2) Organ, tissue, or *in vitro* effects: Dosed at toxic levels, cobalt at the cellular level produces histological anoxia, is a mitochondrial toxin, and inactivates cytochrome oxidase and a variety of sulfhydryl-group-containing enzymes. Feeding amino acids containing sulfhydryl groups can offer some protection against excess cobalt.

(3) Carcinogenicity: Animal studies indicate that the carcinogenicity of cobalt metal and its salts is low. Most of the tumors observed are sarcomas at the injection sites. Cobalt given to rats in intraperitoneal doses of 10 μg twice a week for 8 weeks potentiated the development of the skin tumors induced by 20-methylcholanthene. Cobalt oxide dust injected intratracheally into hamsters appeared to act as a synergist in diethylnitrosamine carcinogenesis of the respiratory tract, but the carcinogenic effects of ethylnitrosourea in rats were not altered by administration of cobalt chloride.

In some studies, its compounds have an antineoplastic effect; in others, cobalt compounds are without effect against tumors.

In vitro, cobalt chloride (590 μg Co/liter) induced 3.7% transformed colonies of fetal hamster cells, indicating carcinogenic character; and cobalt metal powder after dissolution in horse serum produced cytological changes in cultured rat myoblasts similar to those seen in rhabdomyosarcomas induced *in vivo* by cobalt powder. Presumably, a cobalt-protein complex is the carcinogenic agent *in vivo* rather than Co^{2+}.

Although present evidence is not conclusive, cobalt metal and cobalt salts are not significant causative agents of human cancer. Industrial cases where malignancy has been associated with metals have usually involved exposure to a mixture, thereby preventing assignment of causation to one specific metal such as cobalt. Epidemiologic studies in Zaire, Canada, the United States, and other countries where cobalt is mined do not support an association between cobalt and neoplasm. No carcinogenic action of cobalt was observed in more than 20 years by the staff physician at the Olen, Belgium, cobalt plant. Tumor formation in anemia patients treated with cobalt chloride has been reported in only two patients (one had a reticulum cell sarcoma and the other, a giant follicular lymphoma).

Although nickel carbonyl is carcinogenic, no studies have reported carcinogenic effects of various cobalt carbonyl compounds used in industry.

Vitamin B_{12} is generally recognized as safe (GRAS) as a food additive and no reports of direct carcinogenic effects due to its exposure are known. It does appear to enhance tumor growth, however. Dosing with other organic compounds or salts has not produced neoplasms.

(4) Teratogenic effects: A slight increase in malformed adult offspring was observed when mother wasps were given a one-time dose of cobalt sulfate. Cobaltous chloride and nitrite salt solutions induced fetal cleft palates when injected alone into mouse dams.

Cobalt inhibited cleft formation caused by cortisone injections in one study, but in another study cobalt compounds had no effect on the cortisone-induced anomalies. In the latter study, however, cobalt compound prevented cleft palate induced by the anticonvulsant phenytoin.

In vitro, $CoCl_2$ at 5.9 mg Co/liter reduced collagen synthesis in the embryonic rat skullcap. Growth of embryonic chick tissue *in vitro* was totally inhibited by 10 mg Co/liter. At doses of 0.75 mg $CoCl_2$/kg chick embryo, the development of the endocrine pancreas was disturbed. Treatment of chick embryos with a 0.5% solution of $CoCl_2$ resulted in delayed organization patterns, overdevelopment of neural and mesodermal structures, and neural tube and notochord position abnormalities. Injection of 0.4 to 0.5 mg $CoCl_2$ into 4-day-old chicken eggs produced a teratogenicity rate of 2.8%, gross abnormalities being confined to the lower extremities and the eyes. Histological examination revealed hyperplasia, necrosis, and edema. At levels fairly common to environmental waters (0.16 to 8.1 μg Co/liter as $CoCl_2$), freshwater snails (*Limnaea* sp.) showed development abnormalities of shells, feet, and gastrulae. *Helisoma* snail eggs showed decreased survival and deformities of the shell and gut when exposed to $CoCl_2$ at 1 to 100 mg/liter. Concentrations of 0.59 to 5.9 mg Co/liter affected mainly the skeletal structure of sea urchin embryos exposed to $CoCl_2$ in artificial seawater.

Cobalt deficiencies in Missouri soils were correlated with swine birth defects in a survey of 40,000 hog producers.

Supplementing the diets of female rats with 10 to 1,000 μg vitamin B_{12} per kilogram of diet showed no adverse effects on offspring over five generations.

Cobalt has not been shown to cause significant teratogenic effects in humans even when pregnant women have been dosed with 75 to 100 mg $CoCl_2$/day for treatment of anemia.

There is a concentration gradient of cobalt between mother and fetus and passive transport of cobalt across the placenta. Study of maternal/fetal cobalamin transport in rats also shows the transfer occurring against a concentration gradient in rats (cobalt levels in maternal plasma are lower than those in fetal plasma in late pregnancy). In cows, $\sim$ 0.01% of a total oral dose of cobalt transfers across the placenta to the fetus during the last 2 months of pregnancy.

Human placental concentrations of cobalt reflect environmental contamination, being highest in industrialized cities.

(5) Other reproductive effects: Cobalt as an intrauterine wire inhibited pregnancy in rats and hamsters at the time the fertilized egg reached the uterus but before implantation. Exposing wasps to cobalt sulfate (770 mg Co/liter) lowered total egg production and decreased hatchability. Cobalt sulfate and chloride at $\leq$ 100 μg/liter had negligible effects on the reproduction of *Aeolosoma headleyi* (annelid worms). A concentration of 10 μg Co/liter caused 16% reproductive impairment in *D. magna*. In germination tests with *Nicotiana tabacum* seeds, cobalt was one of the most inhibitory metals tested. Concentrations of $\sim$ 30 and 570 μg Co^{2+}/liter resulted in 10 and 50% inhibition, respectively. For seeds of most species, however, soaking in cobalt-treated solutions enhances germination.

Generally, studies of the effects of cobalt or B_{12} on reproduction reflect a beneficial effect presumably related to optimum nutrition, especially in ruminants. Rats and mice show longer survival and better growth in the litters when the dams' diets have been supplemented with B_{12} than when the dams are fed a nonsupplemented diet. In other studies, B_{12}-depleted rats lost their fertility. Sows fed a B_{12}-deficient diet suffered a high incidence of abortion.

4. Effects on plants: Most of the information developed on effects of cobalt on plants is based on plant growth in nutrient solutions containing dissolved cobalt.

Cobalt concentrations as low as 2 to 6 μg/liter have been tested on a few plant species. The rubber plant *Hevea brasiliensis* showed increased growth when grown for 1 year in a nutrient solution containing 5 μg Co/liter. The growth of the alga *Chlorella* and of microbe *Phytophthora* species was not affected in this concentration range. Concentrations in the range 10 to 100 μg Co/liter were favorable for the growth of tomatoes, sweet peppers, legumes, and *Chlorella*; but 50 μg Co/liter has been proposed as an upper limit for cobalt in irrigation water, and since 10 to 50 μg Co/liter may well be found in municipal sewage effluent or contaminated river waters, more investigation of the effects of such cobalt concentrations on additional commercially important plant species are advisable. However, soil-grown plants would be exposed to less cobalt from irrigation water than hydroponically grown plants even though the solution concentrations were the same because of the binding effects of soil components on cobalt, making it less available to the plants.

Concentrations of about 200 to 250 μg Co/liter were still favorable to the growth of tomato, sweet pepper, *H. brasiliensis*, and algal species *Chlorella* and *Crucigina*. However, 200 to 350 μg Co/liter lowered the respiration rates of *Chlorella fusca*. At 500 to 590 μg Co/liter, growth of *Chrysanthemum*, *Chlorella*, and *Haematococcus* (alga) were reduced 10 to 50%. A concentration of 600 μg Co/liter reduced by 36% the ability of microflora to convert glucose to L-malate. Thus, cobalt concentrations in the range 200 to 600 μg/liter have proved detrimental to life processes of various plant species but are not directly fatal. At 1,000 μg Co/liter (1 mg/liter), however, some plants such as algae begin to be severely affected (e.g., Chlorella growth is inhibited 93%). Yet legumes may still be benefited by such concentrations. For example, Rhizobium-inoculated soybeans showed increased uptake of nitrogen from air and total plant growth at 1 mg Co/liter--12 times that of plants grown in the absence of cobalt. However, in another soybean study, concentrations of only 60 to 300 μg Co/liter produced a slight reduction in growth and turgidity; in fact, reduction in turgidity was noted at concentrations as low as 6 to 60 μg Co/liter.

Contact of plants with water containing concentrations of cobalt above 1 mg/liter will probably occur only where soils or holding ponds directly receive cobalt-containing industrial wastes or where cobalt sprays are used as fertilizer (an uncommon practice). Dilution and/or soil uptake will reduce the effective concentration of cobalt to which any plants present are exposed. Few industrial wastewaters were identified that contained cobalt in concentrations > 1 mg/liter.

Seed treatments with concentrations in the range 1 to 5 mg Co/liter were favorable to the growth of oats and oak shoots. Generally, concentrations as high as 450 mg Co/liter were beneficial to the growth of a variety of plants when their seeds were soaked in cobalt-containing solution. Algal growth is inhibited 80 to 100% and microflora are generally inhibited at 1 to 5 mg Co/liter. By 5.9 to 10 mg Co/liter, the growth of most kinds of plants is moderately to severely retarded.

Even 10 mg Co/kg in soil or sand reduced the growth of oat and timothy plants, with chlorosis appearing in the oat plants (grown in sand and nutrient solution). Concentrations of 100 mg Co/kg soil and 500 mg Co/kg soil decreased the dry weight of timothy grown on the soils by 80% and 100%, respectively. Very severe chlorosis was produced in oats growing in soils containing 778 mg Co/kg. Unfortunately, these studies did not determine plant-available cobalt concentrations in soil solution so that some comparison could be made with soil studies and hydroponic studies.

Concentrations of 115 to 227 mg Co/liter are fatal to yeast strains, yet some higher plants have survived higher concentrations. The growth of pea plants was reduced by 70% by a concentration of 2,000 mg Co/liter, and bush beans showed 75% reduced growth and severe leaf chlorosis by growth in solutions containing 5,900 mg Co/liter. _Chlorella pyrenoidosa_, however, was killed within 36 hr when exposed to 2,480 mg Co/liter.

H. Environmental Losses and Hazard Assessment

Probable major sources of cobalt in the atmosphere are coal and residual fuel oil burning and attrition in use of superalloys, hard-facing alloys, and cemented tungsten carbides. The chief form of cobalt entering the environment from atmospheric emissions is probably CoO. Major contributors of cobalt to water include the electronics industry (electrolytic and electroless coating liquid and solid wastes as well as leachates from landfills of the latter); sewage effluents; urban runoff; mining and milling base metal and gold sulfidic ores, especially cobalt ores; and agricultural runoff where cobalt has been purposely added especially as nutrient sprays. The probable speciation of cobalt in these environmental losses is given in Chapter IX. In freshwater, cobalt ultimately assumes the following forms: Co^{2+}, $CoCO_3$, $Co(OH)_2$, and adsorbed species. In sewage effluents, organic complexation appears to make cobalt somewhat more water-soluble.

Present maximum ambient air and water concentrations of cobalt are not likely hazards to human health. They are 1 to 5% of the concentrations known to cause adverse effects in experimental animals. Although cobalt causes tumors at the injection site in animals and promotes the activity of certain chemicals causing lung and skin cancers, it has been only weakly linked to human cancers. Some epidemiological studies correlated cobalt with skin and lung cancers while others could not make such a correlation. No association of cobalt with neoplasms has been made in countries where cobalt is mined.

Cobalt salts are mutagenic to some species but not to others. Cobalt salts are teratogenic _in vivo_ or _in vitro_ to rats, mice, chicks, wasps, snails, and sea urchins; but dosing pregnant women with 75 to 100 mg $CoCl_2$/day causes no birth defects in the fetuses they carried. Neither diet, coffee or drinking water, or cigarettes is likely to elevate one's daily intake of cobalt to detrimental levels. There appears to be no human health hazard from known levels of cobalt in the environment.

On the other hand, concentrations of cobalt frequently found in ambient waters have produced developmental and reproductive impairment effects in snails and *Daphnia*, reduced growth and turgidity in hydroponically grown soybeans, and inhibited respiration in algae. Fish are not adversely affected until exposed to much higher levels. Animals eating plants growing on cobalt-contaminated soils probably will not be adversely affected since various domestic animals are known to tolerate up to 200 mg Co/kg feed.

Because of the known adverse effects of cobalt in humans and other lower life forms, cobalt should continue to be included in multielement surveys of environmental media, human diets, and human tissues. The possible effect of increasing environmental cobalt levels on the methylation of heavy metals by methylcobalamin should also be studied.

I. INTRODUCTION

Cobalt is a significant element in the earth's crust (crustal abundance 25 mg Co/kg); it is used in substantial amounts by industry ($\sim$ 6,000 to 9,000 metric tons [MT] annually) and has important biological properties. The selection of cobalt for this study was particularly appropriate because recent political problems in Zaire (the source of 60% of the free world's cobalt) have caused the price to rise almost fourfold within the last 2 years. Thus, trade literature reports of substitutions, future markets, and new primary sources have been an exceptionally fruitful source for the kinds of information that usually required direct industrial contacts in our previous studies of less abundant heavy metals. In some ways, then, the recent "cobalt crisis" greatly facilitated this study.

On the other hand, because cobalt is not rare and because it is nutritionally required by ruminants and (as a component of vitamin B_{12}) by other mammals, there was far more literature to assemble and evaluate, e.g., on its occurrence in the environment; its physiological effects, both from deficiency and excess viewpoints; and its uptake by plants and animals. The literature on the latter topic is unusually extensive because ^{60}Co is one of the major radioisotopes released from nuclear reactors and nuclear bomb explosions.

The literature on certain topics has not been treated as exhaustively for cobalt as in our reports on other metals with less voluminous data. For other topics, the present reviews are much more comprehensive (e.g., cobalt in the food chain, Chapter VII) and the analytical chemistry of cobalt, Chapter III.E. And we have devoted an entire chapter to environmental transport (Chapter V).

The aim of this study was to identify and quantify, to the extent possible, the potential and actual releases of cobalt from known and plausible sources and to assess the associated hazards. Information was gathered from published literature. In past studies, we have extensively utilized mass balance data for industrial processes to estimate the magnitude of environmental releases of each heavy metal. Seldom did we have the related information on environmental concentrations. The cobalt literature, on the other hand, is replete with information on environmental concentrations, much of which can be related to discharges from known sources. This is really the information needed to assess human health hazards and effects on wildlife. Two factors, however, made quantification of the magnitude of environmental losses of cobalt from specific processes very difficult. One is its high natural background concentration in rocks (up to 200 mg Co/kg in ultrabasic rocks), soil (average 7 mg Co/kg), and aquatic sediments (average $\sim$ 25 mg Co/kg). The other is that most of the available literature on the primary metallurgy of cobalt and on cobalt as an impurity in base metal ores describes processes used in foreign countries and is not particularly applicable to U.S. processes.

The findings of this study are presented in the following eight chapters:

- Uses of Cobalt, Its Alloys, and Compounds;
- Chemistry;
- Geochemistry and Occurrence;
- Environmental Transport;
- Primary and Secondary Processing of Cobalt-Bearing Materials;
- Cobalt in the Food Chain;
- Physiological Effects; and
- Environmental Cobalt Losses and Assessment of Health Hazards to Humans and Other Life Forms.

Each chapter has its own bibliography; references are cited by author(s) and publication date. Tables and figures are placed in the text where first mentioned. Because of the extreme length and detail of the chapters, the most pertinent information has been repeated (without references) in the foregoing Summary.

II. USES OF COBALT, ITS ALLOYS, AND COMPOUNDS

Bonnie L. Carson,* Ralph R. Wilkinson,** and
Christopher J. Cole***

Cobalt consumption in the United States between 1967 and 1978 varied from 11.5 to 18.9 million lb/year or 5,670 to 8,555 MT/year. Major uses are superalloys (16 to 26% of total consumption in 1971-1978), magnet alloys (16 to 30% of the total), and salts and driers (19 to 30% of the total). Bureau of Mines statistics for U.S. cobalt consumption are given in Table II-1.

This chapter on cobalt uses is divided into the following subsections: Superalloys; Magnet Alloys; Cemented Carbides; Hard-Facing and Wear Resistant Alloys; Control Alloys; Steels; Other Alloys; Metal Coatings; Driers in Paints and Related Uses; Catalysts; Decolorizers, Pigments, and Frits; and Miscellaneous Uses of Cobalt Compounds as well as a discussion of the Cobalt Crisis of 1978-1979. To avoid redundant subsections in the processing chapter, brief processing details are given in this chapter. For several applications, such as superalloys, processing information is necessary for a better understanding of the use itself and of future trends. The extraordinary rise in prices and the short supply of cobalt in 1978-1979 are alluded to throughout the chapter. They are the reasons that substitutes for cobalt-containing materials are being sought so intensively at this writing. It will also be alluded to in discussions of revised and expanded processing of primary and secondary cobalt-bearing materials.

Because of the small amounts used and because the hazard is one of ionizing radiation not metal toxicity, the use of ^{60}Co in radiation therapy and tracer applications in research and industry will not be further described.

A. Superalloys

1. Compositions, properties, and applications: Between 1947 and 1967, consumption of cobalt in superalloys comprised 31% of total U.S. cobalt consumption (Charles River Associates, 1969); between 1971 and 1976, it comprised 16 to 26% of the total (see Table II-1). Superalloy consumption of cobalt in 1977 and 1978

---

* Sections A, C through I, L, M.

** Section J.

*** Section B.

TABLE II-1

U.S. COBALT CONSUMPTION, 1967 - 1979, THOUSAND POUNDS
(Metric tons)a/

Use	1967	1968	1969	1970	1971	1972	1973	1974	1975	1976	1977	1978	1979 (3 mo)
Superalloys	3,447 (1,564)	(see nonferrous alloys	3,675 (1,667)	2,322 (1,053)	1,983 (900)	3,012 (1,366)	3,282 (1,489)	4,090 (1,855)	2,255 (1,023)	2,777 (1,260)	3,097 (1,405)	4,251 (1,928)	1,300 (590)
Magnetic alloys					2,278 (1,033)	3,441 (1,561)	4,302 (1,951)	3,457 (1,568)	2,033 (922)	3,525 (1,599)	3,466 (1,572)	3,752 (1,702)	977 (443)
Permanent	2,486 (1,128)	2,700 (1,225)	2,560 (1,161)	2,374 (1,077)	-	-	-	-	-	-	-	-	-
Cutting and wear-resistant materials (including cemented carbides)	916 (415)	707 (321)	1,747 (792)	1,395 (633)	1,230 (558)	1,273 (577)	2,511 (1,139)	2,578 (1,169)	1,403 (636)	1,586 (719)	1,256 (570)	1,654 (750)	467 (213)
Cemented carbides	486 (220)	516 (234)	660 (299)		-	-	-	-	-	-	-	-	-
Welding and alloy hard-facing rods and materials	864 (392)	495 (225)	302 (137)	181 (82)	246 (112)	199 (90)	391 (177)	423 (192)	475 (215)	518 (235)	376 (171)	668 (303)	91 (42)
Steel													
Carbon			5 (2.3)		1 (0.45)	3 (1.36)	2 (0.91)	5 (2.27)	-	-	-	690 (313)	317 (144)
Stainless and heat-resisting	136 (62)	145 (66)	73 (33)	114 (52)	50 (22.7)	39 (17.9)	32 (14.5)	39 (17.7)	37 (16.8)	41 (18.6)	72 (32.7)		
Full alloy	-	-	282 (128)	136 (62)	196 (88.9)	217 (98.4)	226 (103)	249 (113)	185 (83.9)	236 (102)	345 (161)		
High strength, low-alloy	-	-		-	-	7 (3.18)	45 (20.4)	8 (3.63)	3 (1.26)	W	W		
Tool or high-speed steel	514 (233)	553 (251)	570 (259)	534 (242)	318 (144)	361 (164)	518 (235)	690 (313)	291 (132)	223 (101)	231 (105)		
Other steel	652 (296)	615 (279)		250 (113)	-	-	-	-	-	-	-	-	-

TABLE II-1 (concluded)

Use	1967	1968	1969	1970	1971	1972	1973	1974	1975	1976	1977	1978	1979 (3 mo)
Mill products made from metal powder	-	-	39 (18)	With-held	-	-	-	-	-	-	-	-	-
Nonferrous alloys	-	3,061 (1,388)	660 (299)	549 (249)	532 (241)	651 (295)	789 (358)	780 (353)	695 (315)	812 (368)	624 (283)	543 (246)	115 (52)
Electrical materials	759 (344)	954 (433)	-	-	-	-	-	-	-	-	-	-	-
Other alloys	-	-	1,108 (503)	981 (249)	470 (213)	676 (307)	755 (342)	639 (290)	395 (179)	654 (297)	231 (105)	703 (319) (includes miscel-laneous)	118 (54) (in-cludes miscel-laneous)
Total metallic (including not specified)	11,205 (5,083)	9,950 (4,513)	-	10,751 (4,877)	-	-	-	-	-	-	10,179 (4,617)		
Catalysts	626 (284)	721 (327)	286 (130)	402 (182)	474 (215)	702 (318)	1,150 (522)	1,378 (625)	1,112 (504)	1,446 (656)	199 (90)	405 (184)	486 (220)
Salts and driers: lacquers, varnishes, paints, ink, pig-ments, enamels, glazes, feed, electroplating, etc.	1,592 (722)	1,826 (828)	2,577 (1,169)	2,616 (1,187)	2,744 (1,245)	2,691 (1,221)	3,569 (1,619)	3,635 (1,649)	2,871 (1,302)	3,985 (1,808)	3,778 (1,714)	5,399 (2,449)	541 (245)
Ground coat frit	286 (130)	201 (91)	133 (60)	129 (59)	137 (62)	144 (65)	165 (75)	133 (60)	84 (38)	90 (41)	97 (44)		
Pigments (ceramic)	134 (61)	211 (96)	191 (87)	155 (70)	146 (66)	165 (75)	217 (98)	192 (87)	129 (59)	207 (94)	124 (56)		
Glass decolorizer	70 (32)	67 (30)	74 (34)	69 (31)	60 (27)	61 (28)	64 (29)	51 (23)	41 (19)	30 (14)	18 (8)	231 (105)	320 (145)
Other chemical and ceramic uses	63 (29)	29 (13)	5 (2.3)	7 (3)	102 (46)	173 (78)	197 (89)	151 (68)	212 (96)	7 (3)	-		
Miscellaneous and unspecified (in-cludes electric steels, cast irons, mill products from metal powder)	1,390 (631)	863 (391)	1,104 (501)	1,403 (636)	1,532 (695)	315 (143)	526 (239)	363 (165)	566 (257)	345 (156)	472 (214)	see "Other alloys"	
GRAND TOTAL (excluding cast irons through 1972)	13,976 (6,340)	13,000 (5,897)	15,390 (6,981)	13,367 (6,063)	12,500 (5,670)	11,498 (5,215)	18,741 (8,501)	18,861 (8,555)	12,787 (5,800)	16,482 (7,476)	14,395 (6,530)	18,140 (8,228)	4,732 (2,146)

a/ Sources: Reno (1969); DeHuff (1969, 1971); Corrick (1972); Anonymous (1974d); Cere (1978); and Gorham International, Inc. (1979b, 1979c).

was about 21 to 23%. Table II-2 shows that the absolute magnitude of cobalt used in superalloys has been about 900 to 1,800 MT annually since 1965.

Superalloys are alloys based on iron, nickel, or cobalt that maintain their strength at high temperatures approaching their melting points. They contain > 10% chromium and < 50% iron with high nickel or cobalt concentrations plus other elements such as molybdenum, tungsten, columbium, titanium, aluminum, zirconium, and boron. They were developed for, and still find their biggest use in, vanes, blades, and disks of gas turbine jet aircraft engines.* In 1971, about 90% of the dollar value of superalloy shipments was used in civilian and military aircraft (National Materials Advisory Board [NMAB], 1971).

The high-temperature strength of cobalt-base superalloys is due to solid solution strengthening** of the basic Co-Cr-C alloy with refractory metals such as tungsten, molybdenum, columbium, and tantalum and to the formation and dispersion of complex stable carbides throughout the matrix (<u>Cobalt Monograph</u>, 1960).***

---

* Early in World War II, the investment cast cobalt-base dental alloy Vitallium was found to have mechanical and microstructural properties suitable for service at 1000°F (538°C) in turbosupercharger buckets (blades) in jet aircraft. HS-21 was the first modification of Vitallium developed especially for such high-temperature use.

** Solid solution strengthening is achieved because the added elements, having different atomic sizes from those of the matrix, create lattice distortion that hinders dislocation movement and thereby increases creep strength (Molloy and Green, 1976).

*** Cobalt-base superalloys are characterized by a face-centered cubic (austenite) solid solution matrix-containing complex carbides (Bradley, 1978). Carbides may be present in superalloys as coarse primary particles or as a finer precipitate formed intragranularly (mainly titanium carbide) or intergranularly (carbides of chromium, molybdenum, and tungsten). Discrete particles of carbides restrict grain boundary migration, thus improving creep strength (Molloy and Green, 1976). Since carbide second phases are the principal means by which cobalt alloys are strengthened, cobalt-base superalloys contain 0.25 to 1.0% carbon compared with ≤ 0.20% carbon in austenitic stainless steels and cast nickel-base superalloys (Sims, 1972).

TABLE II-2

U.S. COBALT CONSUMPTION IN SUPERALLOYS[a/]

Year	Cobalt, lb	Superalloys, lb
1947	941,000	-
1950	2,226,000	-
1952	6,408,000	-
1953	5,322,000	-
1954	2,754,000	-
1965	3,675,000	-
1966	4,001,000	-
1967	3,771,000	-
1968	1,838,000	90,000,000
1971	1,983,000	-
1972	3,012,000	-
1973	3,282,000	90,000,000
1974	4,090,000	-
1975	2,255,000	90,000,000
1976	2,777,000	-
1977	3,097,000	-
1978	4,251,000	-
1980	-	115,000,000
1985	-	146,000,000

a/ Sources: Charles River Associates (1969) for cobalt in 1947-1967, Metal Statistics 1974 (Anonymous, 1974d), and Metal Statistics 1978 (Cere, 1978) for cobalt in 1971-1977; Gorham International, Inc. (1979b) for cobalt in 1978; and the National Materials Advisory Board (1971) for superalloys in years indicated and for cobalt in 1968.

Nickel stabilizes the face-centered cubic (high-temperature) form of cobalt (Grant et al., 1961).*

At first, cobalt-base superalloys were more important, but the technology of nickel-base alloys surpassed that of cobalt for several years. Vacuum melting allowed the addition of aluminum and titanium to nickel-base alloys without significant losses. Investment casting was modernized in the late 1950's. It allowed for production of increasingly complex component configurations and for air cooling technology, which increased turbine operating temperatures and, thereby, engine efficiency (Hulsizer, 1978).

By 1969, the limits of nickel systems had been reached. Interest returned to cobalt alloys, which can be used at temperatures a few hundred degrees higher. Cobalt alloys had problems of limited low-temperature (200-700°C) short time yield and limited oxidation resistance above 1800°F (980°C). Their reduced structural stability is related to the phase transformation of cobalt. Following elevated temperature exposure, cobalt alloys may show low room temperature ductility and unacceptably low tensile and rupture ductility at intermediate temperatures.

Cobalt alloys may be superior in strength to nickel alloys at low stresses, but they are weaker at low and intermediate temperatures (< 800°C) in rupture and tension tests (Sims, 1969).

---

* In some superalloys, e.g., MAR-M-302, the face-centered cubic β-cobalt lattice tends to transform to the hexagonal α-cobalt lattice during solidification. The transformation does not occur in alloys whose nickel content is higher (Nafziger and Lincoln, 1976).

Addition of sufficient chromium to nickel or cobalt alloys improves the oxidation resistance of alloys on which only the poorly adhering low oxide NiO or CoO forms. The formation of protective layers of $NiCr_2O_4$, Cr_2O_4, or Al_2O_3 occurs at 15 to 20% chromium; but at least 30% chromium must be added to prevent the formation of CoO and to form the protective $CoCr_2O_4$. Only fair oxidation resistance is achieved by adding 20% chromium to cobalt-base alloys (Sims, 1969).

Gas turbines operate at temperatures up to 1650°C; the combustion gas is cooled to 1100-1315°C before it enters the turbine section. Turbine components operating above 540°C must be constructed from heat-resisting, oxidation resistant materials such as superalloys. The components include turbine blades (buckets),* vanes, disks, combustion cans, ducts, cases, and liners. Nickel-base alloys are generally used for turbine blades;** nickel- or nickel-iron-base alloys for turbine wheels; and nickel- or cobalt-base alloys for vanes*** and combustion cans. Modern turbine lifespans are 100,000 hr (or about 5 year operating at 8 hr/day). The limiting factor is often the surface stability to oxidation and hot corrosion of the superalloys. As mentioned above, cobalt-base alloys are superior to nickel-base alloys in hot corrosion resistance; but they are far inferior to nickel-base alloys ($<$ 20% Co) in oxidation resistance (Fawley, 1972).

---

* Blades is the term used for aircraft; buckets is the industrial term.

** Cobalt-base alloys would fail under most stresses and temperatures experienced by most rotating blades in current engines. Wrought cobalt and cast cobalt-base alloys have been used in turbine blades in the past.

*** Almost all industrial hot static parts are cast cobalt-base alloys.

As the high-temperature properties of superalloys have improved, they have been used increasingly in gas turbine engines. In the 1940's the average percent superalloy as a fraction of gas turbine engine weight was only 10% (Desforges, 1977); by 1966, the average was 25% (Morrow, 1976). Danesi and Semchyshen (1972) estimated that about 40% of the engine weight was superalloy with 50 to 69% superalloy in the newer engines of jumbo jets (Pratt and Whitney's JT9-D, General Electric's CF-6, Rolls Royce's RB-211) and of military jet engines (General Electric's F-101 and Pratt and Whitney's F-401). In 1976, Morrow stated that commercial engines contain 46% superalloy compared with 58% in military engines. By 1977, up to 70% superalloy was used in gas turbines (Desforges, 1977). Perhaps as much as 50 lb (23 kg) cobalt is used in each large-thrust (20,000 to 50,000 lb) jet engine (Anonymous, 1978d).* Table II-3 lists the specific uses for certain superalloys. These are not given in as much detail as the extensive list published by Morral (1969) (Table II-4). Table II-5 lists the aircraft powered by several commercial jet engines.

Fawley (1972) reviewed progress in superalloys used in the United States. Although the jet engine gas turbine industry is still the largest high-temperature application for superalloys, other uses include space vehicles, rocket engines, experimental aircraft, nuclear reactors, steam power plants, and petrochemical apparatus.

Besides cast and wrought alloys used in gas turbines, coating alloys having compositions MCrAlY, where M = Fe, Ni, and/or Co, have just begun to be used (Beale, 1978). Turbine components are usually coated by aluminum-base coatings (Alpak) deposited by the pack diffusion process (Beale, 1978; Siegel, 1979).

---

* Probably much more is used in marine and industrial gas turbines; examples of each currently in use weigh 19 MT (21 short tons) and 150 MT (165 short tons), respectively (Morrow, 1976).

TABLE II-3

COBALT IN SUPERALLOYS AND OTHER HIGH-TEMPERATURE ALLOYS

Base	Name	Percent[a/] Co	Ni	Cr	Mo	Other Elements	Condition	Applications	Current and Future Trends	Competing Materials and Other Remarks
Ni	Astroloy (General Electric Company)	17.0	55	15.0	5.3	Al, Ti, C, B	Wrought, conventional powder metallurgy (PM), or cast	Major use in powder forging of jet engine disks (by Wyman Gordon and Pratt & Whitney) (Gessinger and Bomford, 1974).[b/]	In general, superalloy demand by aerospace will be high for several years. Land-based gas turbine and other heat- and corrosion-resistant industrial uses of superalloys are growing.	Co-base alloys are critical for hot rotating parts, but substantial reductions in their use are planned by major aerospace firms.
Ni	B-1900	10.0	64.5	8.0	6.0	C, Ta, Ti, Al, B, Zr	Cast	Turbine blades[c/]		
Co	FSX-414	52.0	10.0	29.0	-	W, Fe, C, B	Cast			
Ni	Hastelloy® alloy S (Cabot Corporation)	2.0	-	15.5	14.5	Fe, Si, Mn, W, C, La, B		Gas turbine seal rings (Cabot Corporation, 1979).		
Ni	Hastelloy® alloy X	1.5	47.6	21.8	9.0	C, Mn, Si, 18.5% Fe, W	Wrought or cast	Nearly all gas turbine engines use Hastelloy X or Haynes 188 in one or more combustion zone components in gas turbine engines (Cabot Corporation, 1976). Furnace applications (Grant et al., 1961).		60 short tons Co was used in Hastelloy X in 1972 when it was by far the most popular alloy for combustors (Morrow, 1976).
Co	Haynes® alloy No. 25 (Cabot Corp.)	49	10	20	-	15% W, Fe, Si, Mn, C	Wrought	Widely used for jet engine components in 1975 (Haflich, 1975).		
Co	Haynes® alloy No. 31	52.0	10.5	25.5	-	7.5% W, C, Mn, Si, Fe	Cast	Commonly used high-Co alloy in jet engines (Haflich, 1978). Also used in nozzles for sampling hot gases from burning coal (Morral, 1966).		

TABLE II-3 (continued)

Base	Name	Percent[a/] Co	Ni	Cr	Mo	Other Elements	Condition	Applications	Current and Future Trends	Competing Materials and Other Remarks
Co	Haynes® alloy No. 188	42.3	22.0	22.0	-	14% W, Fe, C, La	Wrought	Combustion zone components in gas turbine engines. Each Saturn V booster which launched the moon capsule in the Apollo program used ~14 tons Stellite Div. alloys (Cabot Corp., 1976). Used in the preburner liner of the space shuttle (Blair, 1975).	The alloy is expected to replace Hastelloy X in advanced aircraft engines (Wessling, 1975). Availability in small tubing opens potential uses in thermocouple sheaths and nuclear components (Superior Tube Company, 1978).	Haynes 188 has a 38°C temperature advantage over Hastelloy X (Cabot Corp., 1976). Established use by 1976 after development in the early 1960's shows the long times needed for testing before commercial production (Wessling, 1975).
Co	Haynes® developmental alloy No. 1002	Base	16.0	22.0	7.0	Ta, Fe, Mn, Al, Zr, La, Si, C	Cast	New in 1974. Gas turbines. Not mentioned in 1979 Cabot Corp. brochure.		Adding Al and rare earths to some Co-base alloys has produced alloys of oxidation resistance equal to that of contemporary Ni-base alloys (Wasielewski and Rapp, 1972). More use of rare earths was expected in 1976 (Davidson, 1976).
Co	Haynes Stellite® alloy No. 21 (Cabot Corp.)	~60	2.8	27	5.0	C, Mn, Si, Fe, W	Cast	Investment cast turbo supercharger blades for the B-17 Flying Fortress aircraft in World War II (Meetham, 1976). Turbine nozzle diaphragms of jet engines (Grant et al., 1961).		Currently not described as a high-temperature alloy by Cabot Corp. (1979).
Ni	IN-100	15.0	61.5	10.0	3.0	Al, Ti, C, B, Zr, V	Cast, conventional PM	Blades (Morrow, 1976c). By May 1979, used only in the F-100 engine that powers the F-15 and F-16 fighter planes (Furst, 1979b). In 1975, one major jet engine superalloys (Haflich, 1975). Major use in power forging of jet engine disks (by Pratt & Whitney) (Gessinger and Bomford, 1974).		

TABLE II-3 (continued)

Base	Name	Percent[a/] Co	Ni	Cr	Mo	Other Elements	Condition	Applications	Current and Future Trends	Competing Materials and Other Remarks
Ni	IN-587	20.0	47.2	28.5	-	Cb, Al, Ti, C, B, Zr	Wrought			
Ni	IN-591	12.0	~57	3.0	-	19% W, Ta, Al, C, B, Zr	Cast			
Ni	IN-597	20.0	48.4	24.5	1.5	Cb, Al, Ti, C, B, Zr, Mg	Wrought			
Ni	IN-738	8.5	62.3	16.0	1.8	C, Ti, Al, B Zr, W, Cb, Ta	Cast	Turbine blades. Both IN-738 and IN-792 were "under intensive development by turbine builders" in 1974. Turbine stators in the larger auxiliary power unit engine of the Boeing 747 jetliner (Haflich, 1975).	There was a European trend for the use of IN-738 in industrial and marine gas turbines (Meetham, 1976). IN-738 and IN-792 are considered the next generation of alloys beyond those such as Inco 713 and IN-100 (Haflich, 1975).	
Ni	IN-792	9.0	61.2	12.5	2.0	Ti, Al, B, Zr, W, Ta				
Ni	Incoloy 903 (Huntington Alloys, Div. International Nickel Co.)	15	38	-	-	Fe, Al, Cb, Ti	Cast or wrought. Also PM. (de Bord, 1976).	New in 1974. A new use for Co alloys was in gas turbines for natural gas pipelines (Anonymous, 1974c). Widely used by Rocketdyne in the space shuttle main engine: transition rings, turbine inlet-housing support-strut rings, heat-exhanger lines, structural rings, and coolant manifolds (Tillack, 1976).		Alloy 903 has a low coefficient of thermal expansion and a relatively low modulus of elasticity, giving it resistence to thermal fatigue and shock better than that of leading superalloys.
Ni	Inconel alloy No. 617 (International Nickel Co.)	12.5	55.4	22.0	9.0	C, Al	Wrought	As support for Pt gauze catalyst in nitric acid plants (1600 to 1700°F); as inner housings in a 55-Mw land-based turbine engine; for high-temperature heat-treating, the lighter weight baskets carry bigger loads and the mufflers have thinner walls (Anonymous, 1976b). Combustion cans, transition liners, turbine seals in gas turbine engines (Tillack, 1976).		Inconels 625 and 718, containing no Co, were the leading Ni-base superalloys in the aerospace industry in 1976 (de Bord, 1976). They could replace aerospace fastener uses of Co alloys (Mari, 1978d).

TABLE II-3 (continued)

Base	Name	Percent[a/] Co	Ni	Cr	Mo	Other Elements	Condition	Applications	Current and Future Trends	Competing Materials and Other Remarks
	Inconel alloy No. 700	30	45	15	3	Si, Mn, C, Fe, Ti, Al				Inconels 713 and 718 and Mar-M-247 (10% Co) might substitute for high-Co alloys in less-hot areas: vanes, combustors, transition lines (Haflich, 1978).
Ni	Inconel MA 754	Low	-	20	-	0.6% Y as the oxide	Mechanically alloyed, wrought (de Bord, 1976).[b/]	In 1976, General Electric began its use for turbine vanes and shrouds in the F-404 engine for the F-18 fighter plane and the F-101 for the B-1 bomber (Blum, 1976).	In 1976, this oxide dispersion-strengthened alloy was being evaluated in other advanced engines (de Bord, 1976).	The B-1 bomber program was cancelled in 1977 (Harkins, 1979). Some high-Co alloys might be replaced by low-Co mechanically alloyed powders such as Inconels MA 754 and 956, but they are very expensive (Mari, 1978d).
Ni	M-52	10.0	56.2	20.0	10	C, Ti, Al, B		Turbine blades.		
Ni	M-252	10	50.5	19.0	10.0	Ti, Al, B, Fe, C	Wrought	One of the first U.S. Ni alloys for turbine blades (Meetham, 1976).		
Ni	Mar-M-002 (Martin Marietta Corporation)	10.0	58.3	9.0	-	Ti, C, Al, B Zr, 10.0% W, Hf, Ta		Turbine blades.		Directionally solidified eutectic alloy, e.g., 70.5% Ni, 20.5% Cb, 6% Cr, 2.5% Al, C (Larsen, 1978).
Ni	Mar-M-200	10.0	60	9.0	-	12.5% W, Al, Ti, C, B, Zr	Cast	Turbine blades.		
Ni	Mar-M-246	10	59.8	9.0	2.5	W, Ti, Al, B, Ta, Zr	Cast			
Co	Mar-M-322	61	-	21.5	-	W, Ti, Zr, Ta	Cast			

TABLE II-3 (continued)

Base	Name	Percent[a/] Co	Ni	Cr	Mo	Other Elements	Condition	Applications	Current and Future Trends	Competing Materials and Other Remarks
Ni	Mar-M-421	9.0	61	16	2.0	Al, Ti, W, Cb, B, Zr, C				
Ni	Mar-M-432	20	50.4	15.5	-	Al, Ti, W, Ta, Cb				
Co	Mar-M-509	54.7	10	23.5	-	Ti, W, Ta, C, Zr	Cast	Hot stage sections of the JT9D engine for the Boeing 747 jet and the CF-6 engine for the McDonnell Douglas DC-10 airliner (Mari, 1979e). Alloy among the most commonly used high-Co superalloys (Haflich, 1978).		
-	MP35N	35.0	35.0	20.0	10.0	-		Aerospace fasteners (Morral, 1969).	Use in sutures, surgical needles, and prosthetics explored in 1969 (Anonymous, 1969).	Multiphase® alloys were developed by Du Pont and licensed to Standard Pressed Steel Co. Melted by Latrobe Steel Co. (Anonymous, 1969).
Co	MP20N	50	20	20	10	-		Extrusion tooling material for brass (Anonymous, 1969).		
Co	MP159	35.7	25.5	19.0	7.0	Cb, Ti, Al				
-	Multimet® alloy	20	20	21	3	W, Fe, Si, Mn, C, Cb, Ta, N		Blast panels (Cabot Corporation, 1979).		
-	N-155	20	20	21	3	C, Mn, Si, W, Cb, N	Cast or wrought	Sheet form--tail cones, afterburners (Harris et al., 1961).		
-	Nicrotung	10.0	x	x	x	Si, Mn, C, Ti, Al, B				

TABLE II-3 (continued)

Base	Name	Percent[a/] Co	Ni	Cr	Mo	Other Elements	Condition	Applications	Current and Future Trends	Competing Materials and Other Remarks
Ni	Nimonic 80A (Henry Wiggin and Co., Ltd.)	1.1	75.5	19.5	-	Al, Ti, C	Wrought	Turbine blades. Some Nimonic alloys used in current jet engines in U.K. (Meetham, 1976).		Nimonic alloys, except Nimonic 75, require a very pure H_2 atmosphere or a very high vacuum for brazing. Usually use a flux and preplate with Ni (Rhys and Betteridge, 1962).
Ni	Nimonic 90	16.5	59.0	19.5	-	Al, Ti, C, Mn, B, Zr	Wrought			
Ni	Nimonic 95	20.0	54.4	20.0	-	C, Ti, Al	Wrought	Turbine blades.		
Ni	Nimonic 105	20	53	15	5	Al, Ti, Mn, Si, C, B, Zr	Wrought			
Ni	Nimonic 115	13.2	60.0	14.3	3.3	Al, Ti, C, B, Zr	Wrought			
Ni	Nimonic 263	20	51	20	5.9	Al, Ti, C, B, Si, Zr	Wrought			
Ni	Pyromet CTX-1 (Carpenter Technology)	16	38	-	-	Al, Ti, Cb		Rocket thrust chambers, casings, seals, hot-work dies, extrusion dies, punches, mandrels, ordnance hardware, steam turbine blades, gauge blocks, springs, die-casting dies (Anonymous, 1976b).		The alloy exhibits low thermal expansion (Anonymous, 1976b).
Ni	Refractaloy 26	20	x	x	-	Si, Mn, C, Fe, Ti, Al				
Ni	Refractaloy 80	30				W, Cb				
Ni	René 41 (Allvac Metals Co.)	11.0	37.3	19.0	10.0	Ti, Al, B, 18% Fe, C	Wrought			
Ni	René 80	9.5	60	14.0	4.0	W, Al, Ti, C, B, Zr	Cast	Turbine blades.[b/]		

TABLE II-3 (continued)

Base	Name	Percent[a/] Co	Ni	Cr	Mo	Other Elements	Condition	Applications	Current and Future Trends	Competing Materials and Other Remarks
Ni	René 95	8.0	61.3	14.0	3.5	W, C, Ti, Al, B, Zr, Cb	Conventional PM. Wrought	Major use in powder forgings of jet engine disks (by General Electric Company) (Gessinger and Bomford, 1974).		
Co	S-816	42	20.0	20.0	4.0	W, Cb, Fe	Wrought	Wheels and blades (Harris et al., 1961). One of the early wrought superalloys along with L-605 and J-1650 (Sims, 1969).		
Co	TD Co	60.0	20.0	18.0	-	2.0% ThO_2	Wrought			
Ni	Thetaloy	12.5	50.0	25.0	3.0	7% W, 2.5% Mn	Cast			
Ni	TRW-NASA	7.5	51.5	6.1	2.0	W, Ta, Cb, Al, Ti, C, B, Zr, Re, Hf	Cast	Turbine airfoils. Strongest available experimental alloy for turbine blades in 1969 (Wile, 1969).		
Ni	Udimet 500 (Special Metals Co., Div. of Allegheny Ludlum Industries)	18	53.2	19	4	Al, Ti, C, B, Zr	Wrought or cast	One of first Ni-base alloys for turbine blades (Morrow, 1976).		Udimet 500 competes with IN-738 (8.5% Co). They have hot-corrosion resistance comparable to that of Co-base alloys (Fawley, 1972).
Ni	Udimet 700	18.5	53.6	15	5.2	Al, Ti	Wrought, PM, or cast			Special Metals is working with customers to develop substitutes for Co alloys such as the newly developed Udimet 720 for gas turbines, a Ni-base alloy with W, Mo, and Ti (Mari, 1976).
Ni	Udimet 710	15	55	18	3	Al, Ti, W				
Co or Ni	Udimet ODS series, e.g.	-	-	-	-	-				
Co	-	53	25	16	-	5% Al, 1% Y_2O_3		Potential sheet material for jet engine burner cans and afterburner components.		

TABLE II-3 (continued)

Base	Name	Percent[a/] Co	Ni	Cr	Mo	Other Elements	Condition	Applications	Current and Future Trends	Competing Materials and Other Remarks
-	UMCo-50 UMCo-51	48 to 52	-	27 to 29	-	-		Developed by a large Cu producer ca. 1950. UMCo-50 used for components of thermal metallurgical process apparatus and for gas and pulverized-coal burners (Morral, 1966).		
-	Unitemp AF	7.2	x	x	x	C, W, Fe, Al, B, Zr				
	VM alloys (developed by NASA)	70 to 74	-	3	-	17.5 to 25% W		Primarily for space power systems (Morral, 1969).		
Ni	Waspaloy (United Aircraft Corp.)	13.5	56.2	19.5	4.3	Al, Ti	Wrought or cast	Welded fabrications and rotor parts (Wile, 1969). Disks. One of first U.S. Ni-base superalloys for blades. Currently used in turbine and compressor disks for the RB-211 engine (powers Lockheed Tri-Star aircraft) and the Olympus 593 engine (powers the Concorde) (Meetham, 1976) and in the turbine disks and shafts of the space shuttle (Blair, 1975).		Incoloy 901 (no Co) is also used for compressor disks of modern engines.
Co	WI-52	63	-	21.0	-	W, Cb, Fe, Mn, Si, C	Cast	Turbine vanes. One of major jet aircraft superalloys in 1968 (Wile, 1969).		
Co	X-40	56.5	10	25.5	-	W, C, Mn	Cast	Blades for B-17 Flying Fortress in World War II (Meetham, 1976).		
Co Co	Dispersion-strengthened Co-base alloys.	- -	20 20	18 30	- -	2% ThO_2 2% ThO_2		In engine testing, these alloys were superior to established inlet guide vane materials in resistances to bowing, oxidation, corrosion, and thermal fatigue at leading edges (Gangler, 1973).		See also FSX 414, Udimet ODS, and Inconel MA 754.

TABLE II-3 (concluded)

Base	Name	Percent a/ Co	Ni	Cr	Mo	Elements	Condition	Applications	Current and Future Trends	Competing Materials and Other Remarks
-	Co/TaC eutectic system		-	15	-	13% TaC as fiber		One of the two eutectic systems (the other was Ni_3Al/Ni_3Cb) that, in 1973, had the greatest potential for hot engine components (Gangler, 1973).		
Ni	Boride casting alloys	8.4 or 10	Bal	9 to 16	1 to 3	~0.1% B, W, Ta, Cb, Al, Ti, C, Zr	Cast	Blades.		Grain control features of carbide replaced by borides.

a/ Sources for the nominal compositions of the alloys were: Anonymous (1969), Anonymous (1974b), Anonymous 1974c), Anonymous (1976b), Anonymous (1977a), Cabot Corporation (1979), Cobalt Monograph (1960), Davidson (1976), Decker (1969), Gangler (1973), Gessinger and Bomford (1974), Grant et al. (1961), Mari (1978a), Meetham (1976), Morral (1966), Morral (1969), Morrow (1976), and Tillack (1976).

b/ Use of conventional powder metallugy (PM) superalloys in turbine blades and vanes is limited because of the high cost of inert-gas-atomized powders and preforms. Coarse-grained PM Ni-base alloys cannot compete for high-temperature blade use until their cost can be made less than that of cast materials (Gessinger and Bomford, 1974).

c/ About 5,000 tons of these alloys were produced in 1972 (Meetham, 1976).

TABLE II-4

PRESENT AND FORMER USES OF COMMERCIAL COBALT-BASE SUPERALLOYS
(Morral, 1969)

Components Used in Hot Sections of Engines	
Support rings	Nozzle diaphragm valves
Burner liners	High-temperature valves and springs
Exhaust cone assemblies	Gas-turbine rotors and buckets
Stressed and Unstressed Parts Operable at $\leq$ 1150-1200°C	
Turbine blades	Manifolds
Disks	Turboprop engine parts
Combustion chambers	Pneumatic controls
Jet stacks	Temperature- and pressure-sensing systems
Jet combs	Closed-loop turbine systems
Aircraft afterburner parts	Rotor-type louvers
Fasteners	Post-entry lips for jet engines
Tail cones	Nozzle rings
Rocket chambers	Rocket nozzles
Shingles for space capsules and boost glider vehicles	Jigs to hold jet blades for finishing operations
Thrust reversers	Shafts
Nozzle guide vanes	

TABLE II-5

COMMERCIAL JET ENGINES AND THE AIRCRAFT THEY POWER
(Harkins, 1979)

Type	Manufacturer	No. Produced	Thrust, lb	Aircraft Powered by
CF-6	General Electric	1,300	$\leq$ 54,000	Boeing 747 Jumbo Jets
CF-34	General Electric	-	8,000	Craft for feeder airlines
CFM-56	General Electric	-	22 - 27,000	Re-engine McDonnell-Douglas DC-8 aircraft
JT8D	Pratt and Whitney Aircraft	> 9,000	-	Most widely used commercial engine in the world, e.g., Boeing's 727's and 737's and McDonnell-Douglas DC-9's
JT8D-209	Pratt and Whitney Aircraft	18,500	-	New Super 80 version of the DC-9. Is quieter, is more fuel-conserving, and emits less pollution
JT9-D	Pratt and Whitney Aircraft	-	$\leq$ 53,000	Boeing 747 and McDonnell-Douglas DC-10 Series 40
RB-211 Family (e.g., Dash 524 and Dash 535)	Rolls Royce	660	30 - 55,000	Lockheed Tri-Star, Boeing 747, Boeing 757 (short- to medium-range airliner)
RB-432 (under development)	Rolls Royce	-	16 - 18,000	Aimed at short-haul airliners

2. Processing

a. Melting and remelting: Schlatter (1971) of Latrobe Steel Company described the vacuum-induction melting (VIM) technology of high-temperature alloys, which gives the best control over the entire alloy chemistry of known melting methods. Vacuum melting has been used increasingly for high-temperature cobalt alloys and Co-Cr-Ni steels although their melting is more common in electric-arc* or open induction furnaces. Subsequent melting is by vacuum-arc remelting (VAR), electroflux or electroslag remelting (EFR or ESR),** electron-beam melting (EBM), or plasma-arc remelting (PAR). The VIM charge must be especially clean, usually 100% virgin alloy or 100% vacuum-melted scrap. Impurities (e.g., Pb, Bi, Sn, and Cu) and some alloying elements (e.g., Cr and Mn) are lost during vacuum melting by volatilization. To avoid volatilization loss of manganese, its alloys are melted under a partial inert gas pressure. Most superalloys are melted in air rather than vacuum unless they contain reactive aluminum, titanium, or boron (Cobalt Monograph, 1960).

Nafziger and Lincoln (1976) reviewed the electroslag remelting of superalloys, which was becoming more common. The most common secondary melting techniques for superalloys had been vacuum-arc and vacuum-induction processes. ESR has the advantages of giving an improved yield, workable ingot surfaces, improved hot workability, and improvement in some mechanical properties. It also allows control of solidification to minimize segregation and close control of ingot chemistry.

ESR has been used primarily for nickel-base alloys. In the early 1970's, Hastalloy X accounted for more than half of Stellite's electroslag production. Less attention was being

---

* Carbon pickup and lack of homogeneity are drawbacks to using an arc furnace for melting (Cobalt Monograph, 1960).

** In EFR, a consumable electrode of the final alloy composition is submerged in molten, chemically reactive flux. The flux is resistance heated by electrical current between the electrode and the molten pool beneath the flux. The droplets from the melting electrode are refined as they pass through the flux (Desforges, 1977).

directed toward ESR of cobalt-base alloys because property advantages of the process did not differ much from those of other melting techniques. Yet, because of the presence of volatile elements in cobalt-base alloys, ESR should be preferred to vacuum processes, which leave rough surfaces. Flux compositions for cobalt superalloys contain CaF_2, CaO, and Al_2O_3. Chemical refinement was not one of the advantages found by the Bureau of Mines in ESR of the cobalt-base superalloys MAR-M-302, MAR-M-509, and X-45 in furnace atmospheres of air or helium.

b. Forming and fabrication:* Early production of superalloy gas turbine jet engine components was by investment casting (the "lost wax" process). In 1969, 25% of superalloy production was cast directly into jet engine parts (major manufacturers were Howmet, TRW, and Precision Cast Parts); and 75% was sold in the form of fabricated parts to turbine producers or of unwrought alloys to specialty shops (Charles River Associates, 1969).

The EFR and VAR processes produce billets that are forged into components. The VAR process lends itself to casting in vacuum by the lost wax process to give the products. (Vaughn, 1978). Jackman (1976) reviewed forming and fabrication of superalloys. Operations include primary hot working, whereby cast ingots are forged, extruded, or rolled into wrought billet or bar. Secondary working (i.e., rolling, open die forging, closed-die forging, upset forging, extruding, roll forging, and ring rolling) converts the billet or bar into sheets, shaped bars, rings, etc. Cold-working techniques similar to those used on stainless steel are also used occasionally. Superplastic forming, a recent advance, allows close tolerances on complex shapes with materials savings. Hot or cold working controls the grain size and modifies the internal structure of the alloy (Cobalt Monograph, 1960).

Heat treatments are less complex for cobalt-base superalloys than for nickel-base alloys. Wrought alloys may be heat treated to produce recrystallization or stress relief.

---

* See also next subsection on powder metallurgy.

Solution treatment at 870 to 980°C (sometimes to 1425°C) followed by aging at 730 to 800°C to reprecipitate the carbides in a finer form is sometimes done to increase rupture strength of cast alloys (Planinsek and Newkirk, 1979; Desforges, 1977; and Sims, 1972).

Joining superalloys may be done by fusion welding, brazing, diffusion bonding, bolting, and riveting.

Nickel- and cobalt-base superalloys are difficult to machine by conventional machining methods. Special techniques such as electrodischarge, electrochemical, chemical, electron beam, and laser machining must be used (Jackman, 1976). Grinding removes more superalloy than other machining processes: 200 cm^3/min compared with 49 cm^3/min each for turning and milling,* 16 cm^3/min for electrochemical machining, and only 1.6 cm^3/min for electrodischarge machining (Desforges, 1977).

The processing steps are the biggest cost in producing jet engine superalloy components, e.g., a 1-kg turbine blade has a material cost 0.5% the finished component cost (Desforges, 1977).

c. Powder metallurgy: Friedman and Ansell (1972) and Gessinger and Bomford (1974) have reviewed powder metallurgical techniques for superalloys. Thermomechanical treatment to produce strengthening mechanisms by dislocation substructure strengthening, control of shape, control of grain size, and orientation of the grains is possible with superalloys produced by powder metallurgy because of their minimum macrosegregation. In conventional powder metallurgy, oxidation occurred when the elemental powders were processed, and homogenization was incomplete by liquid- or solid-phase sintering. The major breakthrough was the development of producing, collecting, and densifying prealloyed powders. Another advance due to powder metallurgy has been dispersion strengthening to increase strength up to temperatures near the melting point. Both dispersion-free and dispersion-strengthened alloy powders are discussed here with respect to their production, consolidation and working, and thermomechanical processing.

---

* Conventional machining, for example, of wrought MA-754 (an oxide dispersion-strengthened alloy) from bars or shapes removes 92% of the raw material as chips. Developments in hot-working to near net shapes by General Electric are envisioned as shortening contour milling time by 30% and saving 50% of the material (Anonymous, 1979a).

Dispersion-free powder metallurgy superalloys can be produced by: (a) inert gas atomization, (b) the rotating electrode process (REP), (c) the soluble gas process, (d) ultrasonic disintegration, or (e) the Coldstream process.

The first three processes involve atomization from the melt. Inert gas atomization is done on a large scale commercial basis by Kelsey-Hayes, Federal-Mogul, and Universal Cyclops, as well as in laboratories.* The melt is poured through a refractory orifice and atomized into coarse particles by a high-pressure gas stream (argon, helium, even steam for cobalt-base alloys) and then sieved. In the REP process, the end of the electrode made of the alloy is melted (by an arc from a non-rotating-tungsten electrode or plasma); and as the electrode rotates, drops of molten metal fly off, freeze in flight, and drop to the floor of the helium- or argon-filled tank. The soluble gas process is performed by transferring the vacuum induction-melted alloy from a hydrogen-pressurized chamber into an expansion chamber via a ceramic tube. The fine spray of molten droplets is cooled, and the powder is collected under vacuum and stored under a nonreactive gas. Homogeneous Metals produces superalloy powder by this method using from 23- to 230-kg charges.

In the Coldstream process, a relatively coarse argon-atomized prealloyed powder is introduced into a high velocity airstream, accelerated, passed through a venturi nozzle, and shattered by impact on a hard metal surface. The purity of the powders was not good in 1974, but the powders have very good compressibility by conventional powder metallurgical techniques.

Ordinarily, superalloy powders cannot be consolidated by conventional techniques, such as cold pressing or sintering. They must generally be compacted hot by extrusion, forging, or hot pressing in a protective atmosphere or in a vacuum. Vacuum hot pressing is done in dies made of refractory metal foil-lined graphite or molybdenum-base alloys with presses of the same material. Hot isostatic pressing (HIP) is done by

* Carpenter Technology plans to produce gas-atomized high temperature alloys for aircraft, and atomized powder forming was expected in November 1978 to come on stream soon (Mari, 1978d).

isostatically pressing powder-filled preshaped cans of steel or glass at > 10,000 psi and 1010 to 1150°C for 1 to 4 hr. The process can produce a fully dense object such as a turbine disk or a forging preform. Spark sintering is a highly specialized method for difficult to process alloys. It involves rapid heating of the compacted powder in a cylindrical die by passing current through graphite punches.

Dense preforms or canned powders may be extruded or forged. Extruded bars can be forged, hot rolled, etc. Hot isostatically pressed bar stock can also be forged.

Thermomechanical processing (TMP) of dispersion free powder metallurgy superalloys can be done using both hot- and cold-working techniques to obtain a microstructure capable of superplasticity. Superplasticity, allowing low-energy deformation of the billet, is characterized by a fine equiaxed grain structure (1 to 10 μm) that is stable at temperatures above half the melting point. Adiabatic heating during extrusion or low-pressure forging (e.g., the Gatorizing process of Pratt and Whitney) are suitable techniques. As mentioned above, TMP is also used to strengthen the powder metallurgy superalloys.

Dispersion-strengthened powders can be produced by (a) selective reduction processes, (b) mechanical alloying, and (c) partial oxidation of prealloy and powders. In the selective reduction processes such as that originated by Du Pont for making TD Nickel powders or the DS Nickel production method of Sherritt-Gordon, a basic nickel carbonate is chemically precipitated from solution onto particles (such as ThO_2) followed by reduction by hydrogen. Cobalt-base and iron-base dispersion-strengthened materials have been made by similar methods. The cobalt- and nickel-base powders are very fine and require inert gas handling to avoid contamination. Alloying of metals such as aluminum that form oxides not readily reduced is difficult. The mechanical alloying process (patented by J. S. Benjamin of International Nickel Company, Inc., in 1971) involves alloying of component powders in a high-energy mill. The powder is then canned in steel and hot extruded, which allows recrystallization during the postextrusion heat treatment at about 1320°C (Tillack, 1976).

Mechanical alloying allows products of alloys combining oxide dispersion strengthening and γ' precipitation hardening. Laboratory studies of dispersion strengthening by internal or partial oxidation revealed several problems that had prevented commercial exploitation as of 1974.

Procedures for compaction and working of dispersion-strengthened alloys must be chosen carefully since the final mechanical properties are dependent upon the final structure of the alloys. Hot or cold working are necessary. Sintering is done under hydrogen below the incipient melting point. Sintering or hot pressing can only be considered a preliminary step. Extrusion into bar, tube, etc., is the most widely used process for compaction and working of dispersion-strengthened alloys, which may then be rolled or swaged.

Thermomechanical processing of dispersion-strengthened alloys include cold swaging, cold rolling, and recrystallization heat treatments.

The cobalt-containing alloys René 95, Astroloy, and IN-100 are poorly machineable. Much work has been aimed at net shape forging of the expensive alloys by using special dies heated hotter than in conventional forging. Net shape forging produces a shape close to that of the final configuration, reducing alloy consumption. For example, a conventionally forged turbine disk of René 95 weighed 930 lb compared with 537 lb by net shape forging with hot dies. Pratt and Whitney Aircraft's Gatorizing Process net shape forges nickel-base alloys isothermally, usually with IN-100 powder (deBord, 1976).

Lupi (1978) (manager of technical sales for Crucible Compaction Metals Operation) reported the first use of rotating jet engine parts from powdered René 95 (compacted by hot isostatic pressing) for a production model engine.* HIP is increasingly used in the jet engine and premium cutting tool industry to produce near net shapes. The ratio of material weight to end product weight in conventional forging is about 6:1, whereas the ratio is only 2.5:1 with HIP. By 1982, Pratt and Whitney Aircraft hopes to have reduced the ratio to 2:1 (Furst, 1979b). The HIP process is of interest not only to manufacturers of gas turbine engine components but also for production of surgical implants, hot work tooling dies, and materials for oil well production (Lupi, 1978).

---

* The T700 jet engine produced by the Aircraft Engine Group of General Electric Company.

Powdered cobalt alloys are used for coating gas turbine engine components. Airco Temescal, Berkeley, California, coats superalloy turbine blades and vanes (200,000 to 300,000 pieces annually) to halt corrosion, erosion, oxidation, and abrasion with families of alloys called "cocralys" (Co, Cr, Al, Y) and "nicralys" (Ni, Cr, Al, Y) by a physical vapor deposition (PVD) method utilizing an electron beam to ionize the metals. The coatings are about 6 mils (0.0254 mm) thick. Presently, only two other plants produce PVD coatings--Pratt and Whitney Aircraft and Chromalloy Corporation. (PVD coatings also have a potential for use by the automotive industry.)

Metco, Inc., Westbury, Long Island, New York, a unit of Perkin-Elmer Corporation, has used a competing process for about a year. The soft vacuum plasma spraying process avoids the oxidation of the cocralys and nicrolys that would occur with plasma systems operating in the atmosphere and the oxidation of the blade material at the high temperatures used (927°C) (Thornton, 1979).

Rapid solidification rate (RSR) is a promising new development in superalloy powder metallurgy. Pouring a cobalt-free superalloy powder against a spinning disk in a helium gas (ultracold) environment produces a superalloy having the desirable characteristics of a cobalt-base alloy. The unique microstructure obtained gives hot workability and heat resistance. The process is being examined by Pratt and Whitney for potential production of turbine vanes and blades (Turk, 1979).

d. Processing trends: Processing advances rather than composition changes were expected in 1969 to lead to greater improvements in superalloy performance. These methods included use of: directionally solidified eutectics, dispersed compounds, and prealloyed powders; slip casting; instant sintering; plating of Co-W-type alloys from aqueous solutions; and electrophoresis to codeposit particulate insoluble compounds (e.g., oxides or solid lubricants) with the metal (Morral, 1969). Some of these methods are being used today, as described above.

Investment cast nickel-base superalloy blades can be significantly improved by the directional solidification process (Hulsizer, 1978). Directional solidification involves controlled grain growth so that all crystals are aligned in the longitudinal direction (Bradley, 1978). In one technique (the Battelle-Geneva Process), a molten charge is supercooled by perhaps as much as 200°C and then progressively cooled from one end within 5 min (Anonymous, 1978a). Directionally solidified alloys will be extended from military-to-commercial aircraft and industrial turbines (Danesi and Semchyshen, 1972).

Directionally solidified eutectic castings are thought to be the best near term (through the late 1980's) method to improve jet engine efficiency by significantly increasing higher engine operating temperatures. (Ultrahigh-temperature ceramics are expected to replace superalloys completely in the hot section of jet engines by the late 1980's.) Pratt and Whitney Aircraft, which pioneered in directionally solidified casting technology, has produced conventional directionally solidified cast blades and vanes since about 1975. In 1978, the firm tested directionally solidified eutectic blades for small helicopter turbine engines. The production alloy used was cobalt-containing, nickel-base Mar-M-200 + Hafnium, but the directionally solidified eutectic alloy studied contained no cobalt. However, the former alloy for single crystal directionally solidified casting was being studied for possible use in blades and vanes. Eliminating grain boundaries from conventional directionally solidified castings would raise engine operating temperatures by 10°C* and make the components competitive with those made of directionally solidified eutectics (Larsen, 1978).

In general, superalloy progress depends on the ability of components to be manufactured by high-volume production processes. Processing innovations will extend temperature and/or strength capabilities (Danesi and Semchyshen, 1972).

3. Present and future trends

a. Aerospace use of superalloys and competing materials: During the recent lull in the aircraft business, the principal superalloys market was for retrofitting existing aircraft engines. Currently, the aerospace market is expanding.

---

* Larsen (1976) stated that directionally solidified casting work by Pratt and Whitney Aircraft added 94°C to the operating temperature.

Aerospace markets were up 10 to 15% in 1978 over 1977 due to a modernizing program in jet air fleets (Mari, 1976).*

In the past few months, the largest aircraft orders in history have been placed so that the cobalt superalloy demand is about twice what it was in 1977; yet the producers are receiving only 70% of their 1977 consumption** (Mari, 1979a). Firm orders for more than 700 jet airliners were made in 1978. From 4,000 to 6,000 airliners are expected to be produced through the 1980's worldwide (Harkins, 1979). Other factors enumerated by Harkins besides the cobalt crisis that could deter this growth are fears of a new oil crisis, higher fuel prices, lack of the ability of the airline industry to sustain expansion, and availability problems with other elements (Ti, Mo, Cr, Cb, Ta, etc.).

Prizinsky (1978) reported that the jet engine market was expected to remain strong for 3 to 5 years. A Universal Cyclops spokesman estimated production of 13,000 jet engines in 1978 and 14,600 engines in 1981. General Electric Company predicted $90 billion worth of commercial aircraft construction in the next decade.

---

* Simmons, president of Allegheny Ludlum Steel Corporation (quoted by Howard, 1976b), had projected a growth rate of 13.7% in durable equipment purchases by the airlines from 1976 to 1985, and even more would have to be spent to replace and update existing equipment, achieve fuel economies, and meet noise-abatement standards. On the other hand, the National Materials Advisory Board (NMAB) in 1971 had predicted that the civilian and military aircraft use of superalloys would decline (but remain major until at least 1980) due to the greater thrust output per pound being achieved by new engines. The NMAB predicted declines in use through 1970-1973, growth to reach the 1970 level in 1975, and growth at a rate of 5%/year until 1985. The growth was predicted upon an increased use of gas turbine aircraft engines and greatly increased industrial use.

** All consumers were placed on allocation in 1978 (70% of 1977 purchases), and the situation is unchanged in mid-1979 (Anonymous, 1979c).

b. Other industrial uses of superalloys: In 1978, the nonaerospace markets for superalloys were in a lull (Prizinsky, 1978). Yet Vaughn (1978) of Teledyne Allvac was predicting that other uses for superalloys will become significant in the next 10 to 15 years. The industrial gas turbine and marine gas turbine fields were expected to grow 7 to 8%/year for the next 10 years, and use of superalloys would be augmented by significant growth in the nuclear industry in the 1990's. The nonaerospace uses are expected to grow faster in the next 20 years than the jet engine market. Their longer lifetime and/ or reduced downtime for maintenance would favor superalloys over cheaper stainless or other steels for use in noncritical applications. Among potential users are the chemical processing, petrochemical, agrichemical, pulp and paper, and high-temperature processing industries. The heaviest service requirements are for small diameter, thin wall heat exchangers. An emerging market for the nickel-base and titanium alloys is for valves and pumps, e.g., in municipal and sludge treatment plants.

Superalloys may find use in critical corrosion environments such as in apparatus for geothermal and geopressurized wells, fusion power, ocean thermal energy conversion, and coal liquefaction (Vaughn, 1978). New markets for superalloys in the petrochemical industry include turbines to take advantage of waste heat (Prizinsky, 1978). In the oil field industry, "sour" gas wells are drilled to such depths that ordinary alloy steel cannot withstand the heat. Superalloys should be able to fill the need for pipe and tubing materials. Another potential market is for working parts of gas compressors, which operate in highly corrosive environments at 1000°F (538°C) (Howard, 1976b). Potential growth in water desalination applications was expected in the 1970's by the National Materials Advisory Board (1971).

Energy conservation is possible using superalloy materials in land-based electrical power generation systems so that operating temperatures can be increased to the levels of aircraft turbines. The efficiency of converting fuels to energy would be increased from present levels of $\sim$ 40% to between 55 and 60% (thermal fatigue and oxidation would be less of a problem so that longer operating times are possible) (Gangler, 1973).

Utilities were expected to begin to increase their investment in durable equipment in 1980 (Howard, 1976b). Curtiss Wright, Woodridge, New Jersey, the former aircraft engine producer, is using nickel-base and cobalt-base alloys in land turbine engines for electric power generation (Blum, 1976).

Most nuclear plants use cobalt-base materials for high-temperature, sliding and rotating wear, and corrosive operating environments. They become radioactive, however; and maintenance and disposal of such exposed parts is troublesome. Du Pont developed a new nickel-base alloy, Tribaloy T-700, which is radiation resistant (Anonymous, 1977b). The nuclear market is probably a small one for cobalt-containing superalloys. The main structural materials are heat resistant stainless steels and iron-based superalloys (Howard, 1976b).

In the automotive industry, superalloys have been used as valves in diesel trucks, buses, and off-the-road equipment. The automotive superchargers cast from superalloys was developed originally for trucks (Prizinsky, 1978a). The NMAB noted a trend in 1971 to use gas turbines in ships, buses, trucks, and construction equipment. However, land use turbines were expected to need 15% less superalloys because output power per/pound and fail safe performance are less critical. Gas turbines were used in some General Motors and Ford trucks in 1971-1972. Superalloys could provide advantages in passenger automobiles since the newer, less-polluting automobile gasoline engine requires higher operating temperatures and more severe engine environments (Howard, 1976b). In 1972, the commercial introduction of the automotive turbine engine was believed possible within 10 years, but mass production methods were needed. Vacuum investment casting is not satisfactory for mass production (Danesi and Semchyshen, 1972).

c. Substitutes for cobalt-containing superalloys and other measures: Strong demand for superalloys for jet aircraft gas turbines should spur consumption of cobalt; however, the tight cobalt supply in 1978 and 1979 has caused many superalloy producers and consumers to look for ways to reduce or avoid the use of cobalt. Some of the approaches are mentioned in Table II-3. Although advanced engine development programs are looking at nickel-base alloys as substitutes for cobalt, cobalt still seems critical in the hot section of a turbine for hardening the blades (Mari, 1978b).

Development and certification of cobalt-free substitute alloys for jet engines may take 3 to 8 years according to Slack of Pratt and Whitney Aircraft Group of United Technologies Corporation, but nickel-base alloys are being used whereever possible. Snapp of Certified Alloys Products, Inc., did not believe in April 1979 that engine manufacturers would rather pay the high price for cobalt-base alloys than resort to substitution. Cobalt-free Incos 713 and 718 are being substituted for cobalt-base alloys by Pratt and Whitney. Inco 718 can be substituted for the 2 million lb Waspalloy used annually for engine disks (containing $\sim$ 122 MT cobalt). The 700,000 tons of cobalt alloys used annually in the engines for F-15 and F-16 fighters are being replaced by other substitutes (Mari, 1979f, reporting on the Critical Metals Forum held in New York City on April 27, 1979).

Nigro of the Pratt and Whitney Aircraft Group told Furst (1979b) on May 7, 1979, that his company intended to reduce its gross cobalt consumption by 25% in 1980 by substituting nickel-base superalloys,* recycling cobalt-containing chips produced in machining, and increasing the use of advanced powder metallurgical techniques such as HIP to produce some rotating engine parts. Blades that are directionally cast of single crystal noncobalt materials are another development that could reduce cobalt consumption. These blades will be used in the JT9D-R4 engine for the Boeing 767 airliner.

Cabot Corporation is collaborating with both Pratt and Whitney Aircraft and General Electric on nickel-base superalloy substitutes for cobalt-base alloys in jet engines (Mari, 1978d).

---

* For example, nickel-base alloys are now being used for the high-volume vane replacement market, e.g., in the J57, TF33, and J75 military engines and the JT3, JT4, and JT3D civilian engines.

Interim methods for handling the cobalt crisis at Special Metals Corporation include use of merchant (higher-priced free market) cobalt, direct recycling of scrap into the melt (limited in the past, especially where rotating parts are involved), study of lower cobalt alloys both within and outside of specifications, and study of substitutions of noncobalt for cobalt-containing superalloys (Harkins, 1979).

Other reports of cobalt substitution in superalloys have been featured in the press recently. Stellite Division, Cabot Corporation, is developing a low-cobalt or cobalt-free nickel-base alloy for combustor use; but its production is a year or so away (Mari, 1978d). AiResearch Manufacturing Company, Phoenix, Arizona, a division of the Garrett Corporation, has a research program to identify nickel-base substitutes for high-cobalt ($\geq$ 50%) alloys in jet engines (Haflich, 1978). Avco Lycoming, Stratford, Connecticut, Division of Avco Corporation, has already changed some of the noncritical sheet metal parts in jet engines from cobalt-base to nickel-base (Anonymous, 1978d). Substitutions for specific alloys are mentioned in Table II-3.

Harkins (1979), Vice President - Sales and Marketing, Special Metals Corporation, an Allegheny Ludlum Industries company, emphasized at the Cobalt Crisis Conference in April-May 1979 that the mechanism that he favored for ending the "cobalt crisis," at least in the United States where much of the use is in defense, would be to release material from the U.S. government stockpile. Corder (1979) of the Federal Preparedness Agency implied that stockpile release* was unlikely because the amounts held are much less than the recently reassessed goals

---

* Harkins and others present at the conference had concluded that the present shortage and high price of cobalt is due less to problems of production and shipment in Zaire than to the large amounts of cobalt released from the stockpile in the early 1970's. Harkins pointed out that the superalloys business is crucial to the economic health of the country, being second only to agricultural products in their positive effect on the balance of trade.

for aquisition.* Economic reasons have been rejected heretofore in applying the "common defense" criterion of the Strategic and Critical Materials Stockpiling Act (Harkins, 1979).

One of the most promising substitutes for nickel- and cobalt-base superalloys appears to be titanium aluminides. The Air Force is hoping to cut jet engine weight 3 to 16% by the use of titanium aluminides to replace nickel-base superalloys below 1200°F (649°C). Titanium alloys are sufficient at < 1000°F (< 538°C). Adding columbium increased the ductility and strength of titanium aluminide and retarded oxidation. Parts made of Ti_3Al are scheduled for production within 3 years. It is being considered for use in nozzle flaps, casings, and inner ducts (Weintraub, 1979). Pratt and Whitney Aircraft is working on Ti-3Al, which can be used to fabricate large turbine case and frame components (Thornton, 1978). Hopefully, coatings can increase the serviceability of Ti_3Al to 1500°F (816°C). Ti_3Al parts are easier to produce by conventional melting and forging than TiAl, which must be produced from powder. Since the aluminides weigh half what nickel superalloys do, the weight of engine support structures can also be much less (Weintraub, 1979).

Another Air Force manufacturing development contract may ultimately ease the tight cobalt supply for jet engines. The Aircraft Engine Group of General Electric Company will fabricate small turbine blades from Nitac, a nickel-base directionally solidified eutectic alloy containing TaC fibers, for its advanced Technology Engine Gas Generator (under development for early and middle 1980's) (Thornton, 1978).

In 1976, it was believed that the use of ceramics in gas turbine engines was years away (Larsen, 1976).

---

* The goal of 85,415,000 lb (38,744 MT) cobalt was established in October 1976 to meet a 3-year requirement in case of military emergency. Presently only 40,803,000 lb (18,508 MT) cobalt is in the stockpile (Strauss, 1979).

B. Magnetic Alloys

Consumption of cobalt in magnet alloys is generally the second or third largest single use of cobalt in the United States. Salts and driers, superalloys, and magnet alloys sales have been of the same magnitude in recent years. Values for cobalt consumption in magnet alloys in the last decade are given in Table II-1.

1. Permanent magnets: Cobalt-containing permanent magnet alloys are listed in Table II-6 along with an overview of their unique characteristics and important uses. In 1917, introduction of 36% cobalt increased the coercivity of quench-hardened steels from values of 40 to 70 oersteds to as high as 250 oersteds. Magnet alloys developed after 1930 as substitutes for magnet steels included Remalloy, Cunife, and Cunico.

The Alnicos quickly followed. These alloys produced more energy per unit volume than most other commercial magnets until the advent of the rare earth-cobalt magnets.

Cobalt was recognized early as contributing substantially to the magnetic properties of the aluminum-nickel-iron alloys. Its hexagonal close-packed crystal structure gives its alloys a high coercivity value. Cobalt's use in Alnico alloys began when the advantages of heat-treating magnetic material were recognized and cobalt's contribution to increasing the Curie temperature in the alloy was discovered. Also, cobalt increases the saturation magnetization of iron.

A major improvement in Alnico alloys came in 1938 when heat treating the alloy was first done in a magnetic field to orient the strongly ferromagnetic precipitate in the weakly ferromagnetic matrix. This led to the development of the Alnico V alloy (Bilbrey, 1962).

Further development of Alnico alloys has led to more control over the orientation of the matrix crystals as well as of the precipitating particles during nucleation. This has led to the discovery of several alloys with extremely high energy products (Cobalt Monograph, 1960).

TABLE II-6

COBALT-CONTAINING PERMANENT MAGNETS

Classification and Specific Alloys (year introduced)	% Co	Other Elements	Treatment	Applications	Competing Materials	Future Trends	Other Remarks	Reference
Co-containing quench-hardened steels (magnet steels) (1917)	2 to 40	Fe, Cr, W, C, Mn	Heat-treated cast--must be machined.	Synchronous hysteresis motors. U-shaped magnets.	Remalloy, chromium steel, Alnico, Vicalloy	Obsolete.	Invented by Honda and Tagaki.	Cobalt Monograph (1960)
Alnicos (1931)	5 to 35	Fe, Ni, Al, Cu	Usually cast to size and shape; only Alnico II can be sintered.	Magnet rotors for generators and electrical motors. Magnetos in aircraft, truck, tractor, outboard motors. Flowmeter, temperature control, loudspeaker.	Ceramic magnets	Will continue to be main alloy composition for magnets due to low cost.	The magnet is mechanically weak and very brittle. Four major manufacturers: Arnold Engineering (Boston); Crucible Steel; General Electric; Indiana General, plus 4 or 5 smaller firms.	Cobalt Monograph (1960), Strnat et al. (1976), Ruder et al. (1961), Charles River Associates (1969)
Primarily Alnico V	24	Al, Ni, Cu, Fe						
Alnico I	5	Al, Ni, Fe						
Alnico II	12.5	Al, Ni, Cu, Fe	Bonded.					
Alnico IV	5	Al, Ni, Fe						
Alnico VI	24	Al, Ni, Cu, Ti, Fe						
Alnico XII	35	Al, Ni, Ti						
Pt-Co (1936)	25	Pt		Battery-powered wristwatch. Traveling wave tubes. Finite biasing magnets in Faraday effect devices for wave guides and antennas.	$R-Co_5$	Very expensive; very limited use.	Highest maximum energy product, ordered structure, stable under long-term exposure to air.	Cobalt Monograph (1960), Craik (1972)

TABLE II-6 (continued)

Classification and Specific Alloys (year introduced)	% Co	Other Elements	Treatment	Applications	Competing Materials	Future Trends	Other Remarks	Reference
R-Co_5 (1965-1966)	60	Rare earth elements Sm, Yb, and Y; misch metal (mixture of rare earths)	Power metallurgy methods; bonding.	Microwave tubes and filters, frictionless bearings, stepping motors, torque couplers, focusing magnets in traveling tubes, miniaturized electronics. Watt-hour meter, rotary or linear magnetic bearings, flexible magnetic sign, windshield wipers motor, klystrons, magnetrons, miniature electric motors, electrical switches and relays, mechanical actuators or valves, automated ignition systems (starter motor), fractional horsepower motor, accelerometers, gyros, spectrometers, tachometers, computer printer, alternator.	Alnico in small magnetic fields; hard ferrites; Supermendur.	Products requiring repulsive forces. Magnetic levitation and linear motor propulsion of tracked vehicles. Medical uses in seals of colostomy bag, heart pumps, catheters, surgical clamps. The high price of cobalt is expected to seriously affect the growth of the R-Co_5 magnet. Five years earlier the samarium and other rare earths were thought to be scarce and were high priced. However, magnet usage is seen on the rise. Jewelry, e.g., magnetic earrings, clasps for necklaces and stickless stickpins. Permanent magnets suspension systems for high-speed interurban transport or materials handling. Shift from rotating coil to rotating magnet. Use of plastic-bonded alloy in place of sintering where magnet under stress.	R = rare earth element. High magnetocrystalline anisotropy. Brittle. Electronics industry cannot substitute cobalt in parts but can miniaturize. Manufacturers: Colt-Industries Crucible Magnets Division (1974).	Strnat (1978), Ruder et al. (1961), Chynoweth (1976), Anonymous (1978), Strnat et al. (1976), Thornton (1974), Gorham International, Inc. (1979a), Haflich (1979)
R_2Co_{17}	40	Fe, Cu, Zr, Sm, other rare earths	Sintering.					

TABLE II-6 (continued)

Classification and Specific Alloys (year introduced)	% Co	Other Elements	Treatment	Applications	Competing Materials	Future Trends	Other Remarks	Reference
Chromindur	15	Al, Zr, Cr, Fe	Cold-formed.	Telephone receiver.	Remalloy.	Bell Telephone will be replacing Remalloy with this alloy in the 10 million receivers produced annually.	Ductile. Use in receivers and telephone station apparatus will give $\geq$ 12.5% savings in Co.	Tredwell (1978), Lowe (1979)
Remalloy (early 1930's) or Comol	12	Fe, Mo	Quenched, heat-aged, hot-stamped.	Telephone receiver.	Chromindur.			Cobalt Monograph (1960)
Alcomax III and IV		Addition of columbium to Alnico-type alloy	Sintered.	Large-scale production of small magnets.	Alnico.		Increased coercivity without lowering remanence or energy product.	Cobalt Monograph (1960)
Vicalloy	52	Fe, V	Cold-worked, low-temperature aging.	Magnetic recordings. Automobiles (unspecified). High precision synchronous motors in washing machines. Compass needle, clocks. Hysteresis motors, storage of magnetic impulses in the telecommunication industry.			Cast in large billets and rolled into thin sheets.	Cobalt Monograph (1960), Charles River Associates (1969), Bilbrey (1962)
Cunico	29	50% Cu, 21% Ni	Cast in sand or precision casting.	Maximum size 1 in.			Ductile, malleable, highest coercivity.	Cobalt Monograph (1960),
Cunife I and II	2.5	Fe, Ni, Cu	Hot short ingot < 1-1/4 in. Cold-worked.	Magnetic recordings, airborne instruments, magnetic sheets and wires, television accessories.			Highest residual induction, malleable, ductile, machinable.	Cobalt Monograph (1960), Charles River Associates (1969)

TABLE II-6 (concluded)

Classification and Specific Alloys (year introduced)	% Co	Other Elements	Treatment	Applications	Competing Materials	Future Trends	Other Remarks	Reference
Cobalt ferrites (ceramic)		Fe, O, Mo	Iron powders with insulating binders granulated, aligned in magnetic field, sintered.	Loudspeaker magnets. DC motor stator segments and rings. Cordless appliances; small motors in automobiles, e.g., windshield wiper motors. Small engine magnetos. Latches.			Oxides-electrical resistivities are on the order of a million times greater than other alloys. Properties due to anisotropy and small size.	Cobalt Monograph (1960), Economos (1965)
Vectolite	26% Co_2O_3	30% Fe_2O_3, 44% Fe_3O_4		Rotors, tachometers, moving magnet instruments.			Not produced in 1969.	Charles River Associates (1969), Cobalt Monograph (1960), Ruder et al. (1961)

Some of the main disadvantages of Alnico magnets are their extreme brittleness, mechanical weakness, and unmachinability except by grinding. As a consequence, they have to be cast to desired size and shape (Nesbitt, 1967).

The Alnico alloys have been used extensively for electrical motors and loudspeakers for automobile radios, television receivers, and phonographs. The use of permanent magnets in motors has declined due to the substitution of electromagnets. In 1961, Alnicos accounted for $> 50\%$ of the total amounts of all permanent magnet materials. Alnico V, at 90% of the Alnicos total, was by far the most important alloy (Ruder, 1961). Loudspeaker use of Alnicos was variously estimated at about 30 to 40% or 75 to 80% in the 1960's (Charles River Associates, 1969). Competition from the ceramic (barium ferrite) magnets for loudspeakers had reduced the share of Alnicos in the permanent magnet market to about 40% by the time the price of cobalt reached \$5/lb. The automobile radio speakers will contain only ceramic materials in the 1980 models, yet they were totally Alnico magnets in 1976 (Wholey, 1979).

If the price of cobalt remains at the current \$25/lb, the use of Alnicos may decline to 20% of the permanent magnet market. Wholey (1979) of Hitachi Magnetics further speculated that at \$100/lb, consumption of cobalt magnets would drop to $< 10\%$. He felt the future for cobalt-containing magnets would hinge on cobalt prices no higher than \$14 to 17/lb.

Among choices of ceramic magnets, hybrid magnet circuits, electromagnets, and the rare earth-cobalt magnets, the latter with their high-energy density are the likeliest substitutes for Alnico V. Hitachi is also examining materials such as Mn-Al-C, Fe-Cr-Co, and Alnico 9 and giving new emphasis to processing techniques (Wholey, 1979).

The rare earth-cobalt alloys, introduced in 1965, were the culmination of the quest for good permanent magnets with a large energy product, a high remanence, high coercivity, and thermal stability in a small inexpensive package. The rare earth alloys offer high-saturation magnetism, high-Curie temperatures, and large uniaxial magnetic anisotropy (Strnat, 1970). The rare earth-cobalt alloy material easiest to alloy has been

$SmCo_5$, which has two to three times the energy per unit volume and 5 to 20 times greater resistance to demagnetization than Pt-Co and Alnico 8. The price of samarium-cobalt magnets has fallen dramatically in the last decade as new sources of the rare earths have been found; alloy composition has been improved; and, perhaps most significantly, better production techniques have been developed (Strnat et al., 1976).

The rare earth-cobalt alloy type magnet can be manufactured using powder metallurgy techniques. Component materials are ground to 10 to 20 μm, mixed in exact proportions in the absence of oxygen, and are sintered or bound together by resin or soft metal (solder). Sintering can be done in a vacuum or hydrogen atmosphere to minimize oxidation of the rare earth element on the surface of the powder grains. An alternative to this step is precipitation with copper or some other nonmagnetic element and then casting the crystallites, which are approximately 100 mm and already magnetized. These crystallites can be aligned in a magnet field and bound with resins, soft metals, or heat.

If sintering is done, the alloy is then densified by pressing. If isostatic pressing is used, the powder is aligned in a magnetic field and then removed from the field before annealing or binding. If it is pressed in a die, then densification and cementing occur under the influence of a magnetic field.

Machining to specifications can be done by grinding, spark erosion, or ultrasonic abrasion. One of the advantages of binding by resin or rubber is that the material does not need machining to smooth its surface (Weihrauch et al., 1972).

By the early 1970's, rare earth-cobalt alloys ($R\text{-}Co_5$) were becoming commercially available. The samarium-cobalt magnets, which have a maximum energy product five to six times greater than that of Alnico V alloy, are useful for miniaturization of specialized electrical motors and other rotating machines, electronics components, and products requiring large repulsive forces. A few new uses are the development of flexible magnetic signs, a permanent magnet suspension system for high-speed interurban transport or materials handling, and a shift from rotating coil to rotating magnet in electrical motors. $R\text{-}Co_5$ magnets are economically competitive with the 50% cobalt-50% platinum alloy which has been the magnet of choice in special design needs such as miniaturization and focusing magnets in traveling bore tubes. The use of rare earth magnets in loud-speakers will probably grow at the expense of Alnicos, as described below. Hitachi Magnetics is switching from the samarium-cobalt alloys to alloys of praseodymium or praseodymium and samarium with cobalt. Natural mixtures of rare earths (misch metal) are also being considered (Wholey, 1979).

Through the first half of the 1970's, $R\text{-}Co_5$ usage was limited by the availability of rare earth elements, in particular samarium. Lately, new sources of rare earth elements have been found and new combinations of rare earth alloys have been developed. Currently, the major limiting factor on the industry's growth appears to be a limited cobalt supply. In 1977, the value of the rare earth magnet industry was estimated at $6 million; the industry was predicted to grow to $100 million in the 1980's (Hazlett, 1977).

At the Critical Metals Forum, April 27, 1979, Friedman, President of Metallurgical Industries, predicted that if all Alnico applications were to shift in the next 5 years to ferrite and rare earth-cobalt magnets, cobalt consumption in magnets would decline 50% (1.9 million lb), which is equal to about 10% of the total U.S. consumption of cobalt in 1978 (Koflowitz, 1979b).

Wholey (1979), on the other hand, predicted a decline of only 15 to 30% in the use of cobalt in magnets.

Wholey stated in a press session at the Cobalt Crisis Conference that substitution of rare earth-cobalt magnets for Alnicos is presently preferred to substitution by ceramic or hard ferrite magnets or to the continued use of Alnicos at higher prices. Use of Alnico magnets will be only 25% of the total in 1980 compared with 25% ceramic magnets and 50% rare earth-cobalt magnets (Saville, 1979b).

2. Soft magnets: The magnetically soft cobalt-containing alloys include Permendur, 2V-Permendur, Supermendur, Perminvars, and Hiperco--alloys containing 25 to 50% cobalt. These alloys are used in transformers, amplifiers, switching devices, receiver diaphragms, sonar applications, and core material for motors, generators, and transformers (Cobalt Monograph, 1960). Substitution for cobalt in soft magnets is presently thought to be improbable (Mari, 1978e).

The cobalt ferrites are intermediate in properties between permanent and soft magnets. Their applications have been shown in Table II-6.

C. Cemented Carbides

1. Applications: Cemented carbide (hard metal) tools produced from tungsten carbide (WC) and cobalt powders (with titanium, columbium, and/or tantalum carbide additions for metal-cutting grades) are used primarily for metal-cutting operations. The second largest use is for mining and quarrying tools, a use that is expected to grow because of the urgent need for fossil fuels (Lardner, 1976; Mari, 1979b). Hard metal inserts in chisel, wedge, or hemispherical button shape for percussive, rotary, or rotary percussive rock drilling are almost exclusively WC-Co alloys, containing 3 to 20% cobalt (Larsen-Basse, 1973). The older cemented carbide tips brazed to steel shanks have been largely replaced by the clamped indexable inserts having three to eight cutting edges so that new ones can be brought into operation as previous ones are worn without any need for grinding or tool setting.* The brazed tips were commonly reground 10 to 20 times during their service life (Spriggs and Bettle, 1975).

Table II-7 reviews specific applications of cemented carbides.

---

* Recyclers of tungsten noted for decades that the indexable cemented carbide tool inserts (unfortunately termed "throw-away inserts") had been used on only half of their cutting edges. Stricter shop rules could immediately extend the life of these indexable inserts. In addition, coatings of TiC, Ti nitride, or Al_2O_3 can lengthen the service life of cobalt-containing tools (Anderson, 1979).

TABLE II-7

COBALT IN CEMENTED CARBIDES

Material	Tungsten Carbide Grain Size, μm	Percent Cobalt	Other Ingredients	Treatments	Applications	Competing Materials	Future Trends	Other Remarks
Cemented tungsten carbide		3-25	WC	• Cold pressing in dies, presintering at 815°C, machining, and sintering • Hot isostatic pressing (HIP) for large articles, e.g., projectile cores • Cold extrusion and final sintering to produce rods and bars • Slip casting and final sintering for complicated shapes	Cutting tool material for steels, nonferrous metals, and nonmetallics (see metal-cutting grades below); blanking and forming dies; wire-drawing dies; rolls; shafts; valves; gages; rock bits for drilling. Mounting diamonds in cutting tools (Bilbrey, 1962).	Ni as a binder has performance drawbacks in most applications, but Co can be diluted with Ni to avoid performance problems in some uses (Saville, 1978). Ceramics, arc hardfacing, flame and plasma coatings, and numerous cast and wrought alloys where WC's advantages are marginal (Clark, 1979).	Cemented carbide sales for mining and metal cutting tools should continue high in 1979 due to the continuing need for fossil fuels exploration and production (Mari, 1979b). The use by the Chinese for WC drill bits for drilling and exploring China's large crude oil reserves will increase demand for U.S. tools made of Co and W (Anonymous, 1979f).	
For cold rolls	fine	e.g., 15			Burnished cemented carbide rolls are "almost universally" used to produce thin high-quality steel sheet. They are responsible for the ultrathinness of modern razor blades and earlier ones made of 1% carbon steel. They are also used to produce tape-form Cu of accurately controlled thickness for electrical conductors. WC-Co also suitable for hot rod mills (Spriggs and Bettle, 1975).			

TABLE II-7 (continued)

Material	Tungsten Carbide Grain Size, μm	Percent Cobalt	Other Ingredients	Treatment	Applications	Competing Materials	Future Trends	Other Remarks
For mining softer and more friable rocks such as coal, potash, gypsum	3	7-9			Chain saws, shearer disks fitted with coal picks (tips brazed to shanks) rotary tools. (Less Co in coal picks, higher WC grain size; coal picks are subject to severe thermal shock) (Spriggs and Bettle, 1975). In 1975, ~40% of cemented carbide use was in mining tools (Lardner, 1975).	Substitutes for cemented carbides in many mining tool applications are less likely to be found than for cutting tool uses (Lardner, 1975).		Mining and quarrying is the second widest application of cemented carbides after metal-cutting (Lardner, 1976).
For "hard rock" mining								
For rock blasting	fine	7-9						
For percussive drills	2.5-4	7-12		HIP to reduce porosity	Tunnelling machine for boring large sewer tunnels, etc.			
For high-pressure applications dies anvils	2.5-3.0 1.0	9-11 5.5			Production of synthetic diamond, cubic boron nitride	Cemented carbides have a relatively short life in this application, but no other material is suitable (Spriggs and Bettle, 1975).		The anvils and dies are supported in steel belts (Spriggs and Bettle, 1975).
For abrasion-resistant applications	fine	low		For parts that cannot be made of solid cemented carbide, the surface subject to excessive wear can be protected by thin cemented carbide plates brazed, soldered or bonded onto them (Spriggs and Bettle, 1975).	Tire studs, horseshoe nails, ball-point pen balls, carpet loopers, and knives, textile thread guides, brick mold plates, ceramic forming jigs, work-rest blades. Surfacing applications: fan blades for conveying powdered coal in electrical generating stations (Spriggs and Bettle, 1975). About 15% of cemented carbide use in 1975 was for wear use (Lardner, 1975).			

TABLE II-7 (continued)

Material	Tungsten Carbide Grain Size, μm	Percent Cobalt	Other Ingredients	Treatment	Applications	Competing Materials	Future Trends	Other Remarks
For corrosion resistance	1.0	5-7			Where severe abrasion conditions remove the protective film from traditional stainless materials, cemented carbide alloys can provide the only practical solution, e.g., for the low-quality gas-turbine engine fuels, valve and pump components for carrying abrasive slurries such as seawater and sand, and the nozzle and target plate in mild homogenization (Spriggs and Bettle, 1975).			
Metal-cutting grades			WC, TiC, minor amounts of TaC, and/or CbC		In 1975, ~45% of cemented carbide use was for cutting applications (Lardner, 1975).			Adding TiC to simple WC-Co materials sharply increased the steel-cutting efficiency.
For wire drawing	1.5	6		Die reground to a larger size when it becomes enlarged due to abrasion.	Cheap substitute for diamond dies for wire drawing, especially for the larger bore sizes (Spriggs and Bettle, 1975).			
For extrusion punches	1.4	9-16						
For continuous cutting of steel	~1.5	11	≤20% TiC, 15% TaC					

TABLE II-7 (concluded)

Material	Tungsten Carbide Grain Size, μm	Percent Cobalt	Other Ingredients	Treatment	Applications	Competing Materials	Future Trends	Other Remarks
General Motors steel-cutting grades								
CU-1 to CU-5	1-3	5.8-25	No Ti		Cutting tools, dies, punches, crushing hammers.			
CW-1 to CW-5	2-7	6.0-10.7	0.30-25.33% Ta		Cutting tools, steel finishing			
CX-1 to CX-8	1-5	3.5-25.0	No Ti or Cb		Cutting tools, dies, heavy impact applications.			
CY-1 to CY-12	1-4	4.0-13.0	No Ta or Cb, 3.2-25.6% Ti		Precision boring of steel, cutting tools, dies			
CZ-1 to CZ-6		3.0-10.0	No Cb		Light finishing of cast iron and nonferrous metals, precision boring, cutting tools.			

2. Wear in service: The cemented carbide tools wear during service. Dies for wire drawing become enlarged due to abrasion and are reused several times by continually resizing to larger diameter dies (Spriggs and Bettle, 1975). Attrition wear in extrusion punches removes cobalt, leaving grooves and exposed tungsten carbide grains (Cobalt Monograph, 1960).

Larsen-Basse (1973) reviewed the literature on the wear of hard metals during rock drilling. Drill bit temperatures greater than 320°C are common; under the severest conditions, the temperatures may attain 490°C. In abrasion tests with calcite, wear of cemented carbide rotary drill bits occurred by three mechanisms: uniform removal of both carbide and cobalt, initial removal of cobalt followed by uprooting of carbide grains, and pieces torn out due to cracking of especially large carbide grains.

Cobalt may oxidize (to CoO and complex oxides such as $CoWO_4$) at the high temperatures attained in the cutting edge of cemented carbides. Temperatures exceeding 1000°C may be reached in the contact area between rake face and chip. Temperatures exceeding 700°C are possible near the end cutting edge. Oxidized areas can be dislodged by scratching, which leaves pits. Under certain operating conditions, the tool fails; and pieces of varying size break away from the cutting edge. Some oxidation resistance can be achieved by coating the tool with titanium carbide (Tuininga, 1967).

Cemented carbide tip coatings of gold-colored TiN or dark gray TiC, first used in 1970, are the greatest single advance in cemented carbide tools since they were introduced. Alumina coatings can also be used, but intermediate layers of mutually compatible materials must be used to overcome bonding problems.

Diffusion wear occurs on the rake face of the carbide tool while cutting steels at conventional speeds. The tip surface diffuses into or alloys with the steel and is carried away with the swarf (the fine metallic particles removed by the cutting tool). Additions of titanium, tantalum, and columbium carbides at $\leq$ 30% can substantially reduce diffusion wear and the rake face cratering it creates; but since the additions reduce the tool's strength and resistance to thermal shock, only grades used for light high-speed finishing work can use such high-percentage amounts of the additional carbides. Abrasion wear as well as diffusion wear is observed on the flank surface of the carbides.

Coating the tips with titanium carbide has been the best remedy for diffusion and abrasion wear, although in 1975 they suffered from a tendency to spall and crack in thicknesses $> 8\ \mu m$.

By 1975, the coated tips were very successful in turning operations but much less so in the smaller milling market. However, improvements and expansion in their use in milling were expected (Lardner, 1975).

3. Processing techniques and worker exposure: Worker exposure to cobalt-containing airborne dusts during tool use has not been examined as closely as exposure during cemented carbide production. McDermott (1971) found that the loading in the mills in which powders of tungsten carbide and cobalt (with tantalum carbide or titanium carbide for special grades) are ground and mixed in the presence of a volatile solvent (water is avoided to reduce oxidation) does not present dust problems. Most loading areas are equipped with local ventilation. After the slurry is dried by heat *in vacuo* to recover the solvent, the powder is compressed into desired shapes, presintered at 1000°C, usually in a hydrogen atmosphere, and cut and shaped by grinding, etc. The fairly soft product, having a consistency like that of chalk, would present dust problems without adequate ventilation during the machining operations. The machined carbide shape is then sintered at 1500°C in horizontal furnaces similar to those used for presintering or in vertical vacuum induction furnaces.* Some tools are manufactured by sand blasting the tips or blanks and brazing them into drills, lathe tools, reamers, saws, and other

* More than 50% of cemented carbide cutting tools produced in the United States are vacuum sintered; vacuum sintering is expected to increase over the next several years since it allows better quality control and is a cheaper operation than the hydrogen process (Kahaner, 1978a).

holders. Brazing, usually with induction heating equipment and with fluoride type fluxes, also requires ventilation controls. Final grinding may be done either wet or dry.

Dust particles from forming operations collected from the ventilation ductwork had a geometric mean particle size of 1.2 to 1.9 μ (the range was $<$ 1 to 14 μ with 96 to 97.5% smaller than 10 μ). About 70% of the 171 samples of workplace air collected by McDermott (1971) were less than the threshold limit value for cobalt (0.1 mg/m^3); see Table II-8. Exposures have been reduced in this industry by improved ventilation and the availability of ready-to-use cobalt powder since the late 1950's. In the 1930's and early 1940's, cobalt rondelles were dissolved in nitric acid, the salt calcined and oxidized, and the oxide reduced to powder. From the late 1940's to the late 1950's, only the last step was performed.

McDermott (1971) cautioned that workers may be exposed to carbide dust during any operation where carbide tools are sharpened by means of diamond wheels. As of 1971, as many as 3,000 workers were at risk in Michigan alone.

4. Substitution and future trends: News reports in recent months have often mentioned substitution of nickel or other materials for cobalt as a binder for cemented carbides (for example, Saville, 1978). In the past 25 years, the use of nickel, nickel-chromium, and nickel-molybdenum binders has been small compared with cobalt binders. A little iron-bonded material has had special applications (Clark, 1979).

There has been little research on nickel-bonded carbides in the last 25 years. Substitution of nickel for cobalt may well give products that are inferior to those found acceptable today. Research on ball milling, sintering, and composition modification will probably be needed to produce nickel-bonded materials comparable to the cobalt-bonded grades. The cemented carbide industry is presently studying the advisability of investing in these research programs or of continued reliance on high-priced cobalt, assuming the supply remains adequate.

TABLE II-8

CEMENTED CARBIDE WORKPLACE AIR CONCENTRATIONS OF COBALT, mg/m^3, AT SEVEN PLANTS[a]

(McDermott, 1971)

Operation	Cobalt Concentration, mg/m^3 Mean	Range
Metal preparation	0.04	0.0 - 0.14
Forming	0.03	0.01 - 0.15
Sintering	0.03	0.005 - 0.14
Press operations		
Slug presses	0.39	0.01 - 0.90
Blank presses	0.07	0.0 - 0.40
Machine tool operations		
Lathes	0.20	0.01 - 0.80
Surface grinders	0.09	0.0 - 0.80
Milling	0.01	0.01 - 0.01
Drilling	0.04	0.0 - 0.10
Saws	0.19	0.01 - 0.40
Slicers	0.94[b]	0.01 - 4.60
Pelletizers	0.47[c]	0.03 - 2.16
Packing	0.13	0.01 - 0.25
Equipment cleaning	0.54	0.04 - 0.82
Ball-mill loading	0.03	
Screening	0.25	0.06 - 0.44
Powder mixing	0.29	0.01 - 0.80
Miscellaneous	2.40	0.01 - 6.70

a/ Not all plants performed all operations. The number of samples collected for each operation ranged from 1 to 80. Sample collection was by an electrostatic precipitator from the breathing zone except for the first three groups.

b/ The average is 0.34 mg/m^3 when the high value is excluded.

c/ The average is 0.13 mg/m^3 when the high value is excluded.

In wear resistant applications such as sand blast nozzles and wear plates, substitution of nickel for cobalt will be swift. In percussive tools, the material requirements are such that substitution must be made with care. For some applications, the material cost is modest compared with labor, inventory, and customer technical service costs. Here, substitution will meet with customer resistance. However, the cost of mining and wear parts is > 70% due to the material content of the carbide product. Yet customers may well continue to pay the price because the current cobalt-containing cemented carbide material is vastly superior to any known substitutes. In other applications where carbide is only marginally advantageous, likely alternate choices are arc hard facing, flame and plasma coatings, ceramics,* and numerous cast and wrought alloys (Clark, 1979).

The sensitivities of the tungsten as well as the cobalt supply to political and economic pressures made forecasts of future cemented carbide markets difficult. Lardner (1975) did not foresee that ceramics, cermets, or titanium carbide would supplant cemented carbides altogether.

At present, there apparently has not been as much substitution for cobalt as reports would indicate (Teplitz, 1979). Strong demand for cemented carbide mining tools during expanded energy exploration would indicate continued cobalt use in this application.

Pertinent to trends in the cemented carbide industry are the two new U.S. cobalt-powder plants. GTE Sylvania's Chemical and Metallurgical Division started up a fine cobalt

---

* The work for which ceramic tool materials are best suited is rare (Lardner, 1976). GTE Sylvania has apparently lost interest in its ceramic silicon nitride as a possible substitute for cemented carbide in jigs and fixtures for brazing and welding (Teplitz, 1979). At least one firm, V/R Wesson Company, Division of Fansteel, Inc., Waukegan, Illinois, is reported to be trying to use cheaper ceramics and cermets more often as a result of the high prices of cobalt-containing cemented carbides (Saville, 1979).

powder plant at Towanda, Pennsylvania, in 1978. It reportedly uses recycled cobalt materials for feedstocks. Metallurgie Hoboken Overpelt SA, Antwerp, Belgium, is the owner of an extra-fine cobalt powder plant in Laurinburg, North Carolina, which is slated to go on-stream in 1980 and to produce ≥ 5,000 tons (≥ 4,500 MT) cobalt powder annually (Mari, 1978c).

D. Hard-Facing and Wear-Resistant Alloys

Hard facing is the application by welding of a specific alloy to a part, primarily to deter wear. Since the Korean War, when hard-facing rods were used to weld military jet engine components, the demand for cobalt in hard-facing alloys has been about 400,000 to 1,000,000 lb/year (180 to 450 MT/year). Consumption data for specific years are given in Table II-9. Prior to the appearance of the superalloys category in the Bureau of Mines data, data were collected under the heading "Cast Cobalt-Chromium-Tungsten-Molybdenum Alloys" (See Table II-1). These are the wear-resistant Stellite®-type* alloys in Table II-9, whose softer and tougher grades are used in gas turbine vanes and blades (Fritzlen and Elbaum, 1961).

The wear resistant applications of the Stellite®-type alloys include vehicular components and machinery for industrial and metallurgical processing. These applications are detailed in Table II-10 along with the hard-facing applications.

Hard-facing alloys are applied from surfacing welding rods, probably most often by the oxyacetylene process. Other welding processes include use of metallic arc, argon arc, heliarc, and submerged welding arc. They may also be applied as Stellite® powders containing 42 to 52% cobalt by spray fuse, powder welding, and sprinkle fuse processes (Morral, 1966).

---

* Cobalt-chromium alloys with tungsten and/or molybdenum were first developed and patented by Elwood Haynes (Fritzlen and Elbaum, 1961). Haynes Stellite Company, Division of Union Carbide Corporation, has become the Stellite Division of Cabot Corporation. Only this company can use the name Stellite® to describe its products.

TABLE II-9

U.S. CONSUMPTION OF COBALT IN HARD-FACING AND WELDING ALLOYS

Year	Cobalt (lb)	Reference[a/]
1946-1949	54,000-116,000	CRA
1950	260,000	CRA
1951	575,000	CRA
1952	505,000	CRA
1953-1960	350,000-625,000	CRA
1965	1,055,000	MS 72
1966	953,000	MS 72
1967	864,000	MS 72
1968	495,000	MS 72
1969	302,000	MS 72
1970	181,000	MS 72
1971	246,000	MS 76
1972	199,000	MS 76
1973	391,000	MS 78
1974	423,000	MS 78
1975	475,000	MS 78
1976	518,000	MS 78
1977	376,000	MS 78
1978	668,000	GI 79

a/ CRA = Charles River Associates, 1969
MS 72 = Metal Statistics 1972 (Mustero, 1972)
MS 76 = Metal Statistics 1976 (Roditi, 1976)
MS 78 = Metal Statistics 1978 (Cere, 1978)
GI 79 = Gorham International, Inc. (1979b)

TABLE II-10

COBALT IN HARD-FACING AND WEAR-RESISTANT (CAST) ALLOYS[a/]

(Cobalt Monograph, 1960)

Classification and Specific Alloys	% Cobalt	Other Elements	Treatment	Applications	Competing Materials	Trends	Other Remarks
Hard-facing and wear-resistant alloys, ASTM classification:			Readily weldable by gas- or arc-welding (see text). Large pieces must be preheated before welding. Oil quenching after welding if toughening needed. No aging treatment usually needed. Compositions intended to give required hardness in the as-deposited or as-cast condition. Most not amenable to rolling or forging. Shapes are finished by grinding if necessary.	Hard-facing: To reclaim worn parts or extend service life of new equipment. Alloys can be applied from welding rods onto a variety of parts. There are hard-facing applications in agriculture, the automotive industry, brick and cement plants, coke and gas plants, concrete mixing equipment, earth moving and mining machinery (drill bits, power-shovel teeth, bulldozer and scraper blades), iron and steel manufacture, metalworking, oil fields and refineries, and power-generating plants. Specific parts that are hard-faced include: valves and valve seats for handling abrasive materials and in internal combustion engines; lathe and grinder centers; burnishing tools; hot and cold reduction rolls, guide and feed rolls in steel mills; screw conveyors, etc., in brick manufacture; pokers, clinker arms, coke pusher shoes, etc., in gas and coke producers; rabble disks in ore roasters; knives and blades for cutting hot or cold materials, e.g., hay, sugar cane, and hot steel; dies and punches.	Fe, Mo, or Ni for Co (and Mo for W). In 1960 austenitic Mn steel was popular for wear-resistant applications in Europe. Low-cobalt Haynes Alloy No. 716, uses 20 to 25% less Co than Stellite® No. 6 and does not have the problems of Stellite® 711 (Furst, 1979c). Metallurgical International recently developed a Co-based plasma-arc hard-facing powder, S-32, with half the Co of conventional powders (63% in S-1, S-12, S-156, S-6). S-32 could not be used in nuclear valves because of their more stringent specifications (Anonymous, 1978c).	High prices for Co will preclude the use of Co in alloys for hard-facing valves according to Bhansali of Cabot Corp. Stellite Div. (Stern, 1978). Ni alloys are being increasingly substituted for Co in hard-facing alloys (Mari, 1978c).	
R CoCr-A (Haynes Stellite® Alloy No. 6)	Balance	0.9 to 1.4% C, 1.00% Mn, 3.0 to 6.0% W, 3.0% Ni, 26.0 to 32.0% Cr, 1.0% Mo, 3.0% Fe, 0.4 to 2.0% Si, 0.50% others					
R CoCr-B (Haynes Stellite® Alloy No. 12)	Balance	1.2 to 1.7% C, 1.00% Mn, 7.0 to 9.5% W, 3.0% Ni, 26.0 to 32.0% Cr, 1.0% Mo, 3.0% Fe, 0.4 to 2.0% Si, 0.50% others					

TABLE II-10 (continued)

Classification and Specific Alloys	% Cobalt	Other Elements	Treatment	Applications	Competing Materials	Trends	Other Remarks
R CoCr-C (Haynes Stellite® Alloy No. 1)	Balance	2.0 to 3.0% C, 1.00% Mn, 11.0 to 14.0% W, 3.0% Ni, 26.0 to 33.0% Cr, 1.0% Mo, 3.0% Fe, 0.4 to 2.0% Si, 0.5% others		Castings: Metal-cutting tools (discussed under "Machine-tool alloys" below). Burnishing tools, especially for railroad car and locomotive axles, piston rods, crank pins, etc., to produce a surface highly fatigue-resistant. Sleeves and bushings for handling abrasive materials, e.g., screw conveyors, rock crushers, ball mills, mixing equipment, pump seals, handling equipment in cement and steel mills, bearings where little or no lubricant can be used, e.g., in food handling. Exhaust valves for aircraft engines, diesel-engine valve-seat inserts, exhaust ports for marine diesel engines, slippers for rocket slides, inserts in machine gun breeches during World War II.			Cobalt wear-resistant alloys have better toughness, impact resistance, and oxidation resistance than competing wear-resistant alloys (Charles River Associates, 1969).
Haynes Stellite® Alloy No. 711 (Stellite Division, Cabot Corporation, and Mitsubishi Metal Corporation)	25 to 30	7 to 12% Ni, 27% Cr, 10% Mo + W, 23% Fe, 2.7% C		New in 1978. Will be used in valves, augurs, extrusion screws, chain saws, and other industrial products (Anonymous, 1979e; Isleib, 1978).	Was to be a substitute for Haynes Stellite® Alloy No. 1 (Furst, 1979a), but it proved to be too hard and brittle and has not been well accepted (Furst, 1979c).		HS-711 performs better at higher temperatures than conventional Stellites® (containing an average 65% Co) and is ~25% cheaper (Anonymous, 1979e).
Other Haynes Stellite® alloys	1.3 to 67.0	Cr, Fe, W, Mo, C, Si, and sometimes Ni, B, or V		Hard-facing: Bearings, bushings, sleeves for shafting, and shafting itself, e.g., in marine propeller shafts and automotive water- or fuel-pump shafting. Bearings for rock bits	Haynes Stellite® Alloy No. 721, to be introduced by July 1, 1979, is intended as a substitute for Haynes Stellite® Alloy No. 21. (Furst, 1979c).		

TABLE II-10 (continued)

Classification and Specific Alloys	% Cobalt	Other Elements	Treatment	Applications	Competing Materials	Trends	Other Remarks
				in oil-well drilling; for screw conveyors, elevators, rock crushers, and mixing equipment. Food-processing machinery.			
Tungsten carbides				Hard-facing in a petroleum refinery using these materials was described by Locke (1974). Union Carbide's Linde Division has a growing hard-facing business, coating steel mill rolls with WC by the plasma-arc or other proprietary processes (Mari, 1979b).			Carbides are the hardest and most wear-resistant hard-facing materials (Gilder et al., 1961). Haystellite® tungsten carbides containing 40% Hastelloy® Alloy B or Haynes Stellite Alloy No. 6 are described by the Cabot Corporation (1979).
Tungsten-cobalt alloys	9	91% W					
Machine-tool alloys (cast nonferrous metals)		1 to 3% C, 4 to 25% W		Principal use for turning and forming tools and other metal-cutting operations. High-temperature applications. Also used for low coefficient of friction in cutting common metals at lower temperatures. Uses include milling cutters, trimming dies, gauges, burnishing rollers, counter bores, reamers, and cutoff blades.	Ordinary high-carbon or high-speed steels for operations <540°C.		Co-Cr-W alloys show excellent recovery of hardness after heating, even to >760°C.
Tough or soft (typical alloy)	64.00	1.00% C, 4.00% W, 30.00% Cr, 0.50% Mn, 0.50% Si	Alloys can be cast. Cutting tips are brazed or welded to steel shanks, or the alloy is used as a removable tool in a holder. Alloys sometimes heat-treated to enhance hardness.	Gas-turbine vanes and blades.			Since the metal-cutting grades become brittle in use, they must be supported or used in heavy sections to prevent fracture.
Medium (typical alloy)	58.50	1.50% C, 9.00% W, 30.00%	Simple machining with carbide tools (Fritzlen and Elbaum, 1961).				

TABLE II-10 (concluded)

Classification and Specific Alloys	% Cobalt	Other Elements	Treatment	Applications	Competing Materials	Trends	Other Remarks
		Cr, 0.25% Mn, 0.75% Si					
Hard (typical alloy)	46.50	2.50% C 18.00% W, 32.00% Cr, 0.60% Mn, 0.40% Si	Finished by grinding only (Fritzlen and Elbaum, 1961).				
Surgical-type alloy	65	20% W, 15% Cr		Surgical instruments unaffected by organic acids or ordinary antiseptics.			

a/ Wear-resistant alloys are Co-, Fe-, or Ni-base alloys containing >1% C with W and Cr to form carbides (Anonymous, 1976a).

In the flame-plating process, WC-Co is sprayed onto the base metal at $\leq$ 400°F (204°C). Kitchen and hunting knives produced by the process were first marketed in 1959 (Bilbrey, 1962). Although the original welding process for hard facing was still used in 1974, spray welding, plasma, and detonation gun methods had become popular. The latter process is done by remote control. The detonation is carried out in a soundproof room (Locke, 1974). Terminology for hard-facing procedures used by Cabot Corporation (1975) included spray gun (flame spray) hard facing, manual torch hard facing, plasma spray surfacing, and plasma arc weld surfacing.

The variety of hard-facing alloys include: overlays of machinery steels; high-alloy, homogeneous deposits for highly abrasive conditions; heterogeneous deposits of tungsten carbide or vanadium carbide for extreme abrasion and cutting; and nonferrous alloys for service where abrasion is accompanied by high temperature and/or corrosion. Cobalt-containing weld rods and electrodes comprise the largest fraction of the nonferrous hard-facing alloys. The most popular alloys are Stellites® 1, 6, and 12. Yet, cobalt-containing hard-facing alloys comprise only 1 to 2% of the world market for hard-facing materials. In the United States, however, cobalt consumption for hard facing in 1978 (1.6 million lb; 460 MT) was about 8.5% of the total U.S. cobalt consumption. Table II-11 details specific industrial users of cobalt hard-facing alloys (Modic, 1979).

TABLE II-11

USERS OF COBALT-CONTAINING HARD-FACING MATERIALS
(Modic, 1979)

Industry	Percent of Total
Automotive	30-35%
Petrochemical	20%
Chain saws	10-15%
Steel	8%
Miscellaneous	12%

Manufacturers are testing low-cobalt, nickel-base, and iron-base alloys in accelerated programs to find substitutes for the high-cobalt hard-facing alloys. Where cobalt alloy use has been essential, there will be no quick substitutions. In uses requiring resistance to moderate impact, abrasion, corrosion, and high temperature as well as those involving nonlubricated metal-to-metal contact, cobalt alloys are best; nickel alloys are not a good substitute (Modic, 1979). In applications where cobalt alloys have not been essential (just foolproof) change will be faster.

The automotive industry is looking for substitutes for cobalt hard-facing alloys* for coating internal combustion and diesel engine valves and valve seats (Modic, 1979). Ironically, the automotive industry had switched to cobalt alloys over the past 10 years because of their availability (Saville, 1979b). Formerly, valves of engines for tractors and 4-cylinder passenger cars required cobalt-based hard-facing alloys, but TRW Valve Division has developed an iron-base hard-facing alloy containing no cobalt that meets all requirements for moderate- and heavy duty applications. Its oxidation resistance is not as good as that of cobalt alloys, but its sulfidation resistance is superior. A May 14, 1979, news story (Furst 1979c) entitled "Taking the Cobalt Out of Valves" detailed other substitute alloys coming on the market for valves in other industries as well. (This information is incorporated into Table II-10). The current task for the manufacturers of the new alloys is to educate the valve makers away from specifying high-cobalt alloys. Yet panelists at the Valve Manufacturers Association meeting in

---

* In addition to hard-facing alloys, internal combustion engine valve applications of cobalt alloys included the diverter valve part made of Du Pont Triballoy 480 (12% Co) for automobile engine emissions. About 20 g cobalt in this alloy was used per auto in 1975 (20,000 to 25,000 lb of alloy). Triballoy 480 was produced by mixing one part Triballoy 400 (62 Co-28 Mo-8 Cr-2 Si) with four parts AISI 316 stainless steel powder (Thornton, 1974).

mid-May 1979 pointed out that cobalt alloys cannot be eliminated altogether because their unique combination of toughness, corrosion resistance, and abrasion resistance is not met fully by any potential substitutes. They are especially critical in the aerospace applications (Koflowitz, 1979c).

Although cobalt use in hard-facing alloys increased about 3%/yr up through 1978, the price increase has caused a substantial decrease in its use in 1979. Part of the shortfall in supply to the rod maker has been made up by recycling rod stub ends, obsolete materials, and spent castings (Modic, 1979).

The wear resistant cobalt-base alloys have many machine tool applications in which they come into contact with food, drugs, eating utensils, and fabrics. A Cabot Corporation (1977) brochure described such uses of wrought Haynes® alloys Nos. 6B and 6K (containing $\sim$ 70% cobalt): cutting knives for carpet cushions; scale pans, hoppers, and gates to control the flow of soybeans; earthenware dish-trimming knives; doctor blades to scrape corn flakes from pressing rolls ("Metallic contamination is held to an acceptable minimum"); slicker blades for leather cutting; hunting knives; knives for stripping corn kernels from the cob at canneries; doctor blades for stripping antibiotics from filter drums; knives for cutting reclaimed wool; and apple-paring knives. Other uses for wear resistant Haynes® and Haynes-Stellite® alloys include steam turbine erosion shields, ball bearings, surgical scissors, fluid valve seats, textile thread guides, variable vane bushings, diesel valve seat inserts, and diesel fuel pump parts (Cabot Corporation, 1979).

E. Control Alloys

Low-expansion cobalt alloys have been used in instruments and for glass-to-metal seals in electronics. Several alloys and their applications are described in Table II-12.

The use of cobalt-containing alloys in semiconductor enclosures (packages) was extensive up until 1973. In metal cans, where the leads were glass sealed into the headers (chip bases), ASTM F-15 was used both for the leads and headers because it has the same coefficient of thermal expansion as borosilicate glasses. This minimized cracking during cooling.

TABLE II-12

COBALT IN SPECIAL-EXPANSION AND CONSTANT-MODULUS ALLOYS
(Cobalt Monograph, 1960)

Classification and Specific Alloys	% Cobalt	Other Elements	Treatment	Applications	Competing Materials	Future Trends and	Other Remarks
Low-expansion alloys				Primarily for instruments where constant dimensions must be maintained, e.g., tapes and wires for geodesic measurements and length standards, resistors, variable condensers, jet-engine mounts, thermostatic controls. Glass-to-metal seals in electronics.			The first low-expansion alloy was Invar (64% Fe - 36% Ni).
Fe-Ni-Co alloys		Fe, Ni	Annealing. Kovar can be welded, soldered, or brazed. Most commonly brazed under H_2 with pure Cu or Cu-Ni as filler. May need to be electrolytically plated by Cu or Ni before brazing (Rhys and Betteridge, 1962). Pretinning before soft soldering.	Sealing to borosilicate glass. These alloys give the most satisfactory match with the thermal expansion of glass.		Eventually, steel and Cu will replace both Co and Ni in integrated circuits. (Saville, 1979b).	
Super-Invar	3.5 to 6	62.5 to 64% Fe 30.5 to 34% Ni					
Kovar, ASTM F-15	17	29% Ni, 54% Fe			N-42 (42% Ni, rest Fe) is a potential substitute for Kovar alloy (Mari, 1978e).		
Fernico, Therlo, Rodar	Similar to Kovar	Similar to Kovar					
Super Nilvar	4 to 6	31% Ni, balance Fe					
NI-SPAN-C	Ni + Co 42.10	Fe, Cr, Ti, Al, Mn	Annealing, aging, cold-working (Anonymous, 1968).				
Corning G-5	23	37% Fe, 30% Ni, 8% Cr		Suitable for sealing to lead-lime glasses. Not widely used in 1960.			

TABLE II-12 (continued)

Classification and Specific Alloys	% Cobalt	Other Elements	Treatment	Applications	Competing Materials	Future Trends	Other Remarks
Co-Fe-Cr-Ni alloys, e.g.,	7.8 to ~40	Fe, Ni, Cr					
high-cobalt Fe-Ni-Co alloy	42.2	39.5% Fe, 18.3% Ni	Powder metallurgical techniques.	Suitable for sealing to soft glasses (developed in Poland).			
Co-Fe-Cr alloys, e.g., Stainless Invar, Japanese Invar	53.14 to 54.32	36.22 to 36.92% Fe, 9.09 to 9.87% Cr, plus C, Mn, and/or Si		Used over a limited temperature range.			
Ni-Co alloys	20 to 95	5 to 80%					
Other cobalt-bearing alloys							
EMK	19	71% Fe, 10% Mo, 0.8% Si		Glass-sealing (developed in Europe).			
Wonico	5	80% W, 13% Ni		Glass-sealing (developed in Europe).			

TABLE II-12 (concluded)

Classification and Specific Alloys	% Cobalt	Other Elements	Treatment	Applications	Competing Materials	Future Trends	Other Remarks
Constant-modulus alloys				Major use as hairsprings and balance wheels in clocks and watches. Also used in precision instruments--seismographs, spring balances, bourdon tubes, tuning forks for radio synchronization.			
Co-Elinvar	51.5 to 60	25 to 38.5% Fe, 8 to 15% Cr, 0.18 to 0.19% Ni, 0.46 to 0.68% Si, 0.07% C, 0.5% Mn, 0.1% Al					
Velinvar	Balance	28.4 to 34.5% Fe, 7.0 to 10.2% V, 0.5% Mn, 0.1% Al					

During metal can assembly, the surface-oxidized heads and leaders were heated in an inert furnace atmosphere to $\sim 1000°C$ with borosilicate glass powder. The surface oxide was removed by etching; then the assembly was barrel plated with gold followed by bonding of the silicon transistor chip to the header at 400°C in an inert atmosphere, connecting of the gold wires, and resistance welding of a 99.5% pure nickel can to the header (Anonymous, 1972b; Jordan, 1976).

In ceramic-metallized integrated circuit packages where the silicon chip is fired into an alumina ceramic base, ASTM F-15 was used less than the 42% Ni-58% Fe alloy for lead frames.* Temperatures involved in the manufacture of the metal oxide semiconductor are $\leq$ 1650°C. These high-reliability packages are used to protect large scale semiconductor chips containing thousands of individual circuit elements (Anonymous, 1970; 1972b).

For plastic-encapsulated semiconductors used in radios, televisions, and other entertainment systems, the semiconductor chip is bonded to a lead frame, connected electrically to its leads, and finally encapsulated in plastic. The standard lead frame for these uses was 99.5% nickel. However, the flat pack semiconductor package, which was developed for use in the space program, used ASTM F-15 for the covers and lead frames. Processes involved wet hydrogen and nitrogen atmospheres and temperatures up to 1050°C (Anonymous, 1972b).

The amount of cobalt used in lead frames can be reduced by using only a thin strip of Kovar bonded to copper. This has been done to some extent for the past 20 years and is becoming more widespread (Kahaner, 1978b).

---

* Lead frames are metal alloy skeleton structures providing support and the exterior electrical connections that will be soldered to the circuit board (Anonymous, 1970). Polymetallurgical Corporation, North Attleboro, Massachusetts, sells the raw materials for lead frames to Western Electric Company, IBM Corporation, Texas Instruments, Inc., National Cash Register Company, and other semiconductor manufacturers (Kahaner, 1978b).

In 1973, the semiconductor industry switched from predominantly ASTM F-15 for seals and lead frames to copper and 42% Ni-58% Fe alloy (alloy 42) because of the cost advantage of the latter materials. Iron-nickel and copper materials have also replaced F-15 in eyelets and copper-cored F-15 leads. (Both nickel and cobalt will have been phased out of semiconductor materials by 1990.)

About 935 tons (848 MT) cobalt was used annually in alloys by the semiconductor industry for lead frames. Since the substitution of cheaper materials, only about 300 tons/year (270 MT/year) will be used in the next decade.

Cobalt in control alloys for thermostatic systems, (e.g., thermostatically controlled circuit breakers and automotive and appliance switches) probably would not be replaced until the price of cobalt reaches \$60 to \$80/lb (all other factors being equal). Less than 50,000 lb (23 MT) of such cobalt alloys are used annually; one of these is Trueflex 1513, which contains 16% cobalt (Moffat, 1979).

F. Cobalt in Steels

Since 1967, about 260 to 550 MT cobalt has been used in steel annually (see Table II-1). The types of steel that contain cobalt are classified in Table II-13 as high-speed tool steels (tungsten- and molybdenum-types) (5 to 12% Co), hot work tool steel ($\leq$ 4.25% Co), cold work tool steel ($\leq$ 3.25% Co), maraging steels, high-temperature stainless steels (0.20% Co), and miscellaneous alloys such as quench-hardened steels (4% Co). (The latter are no longer produced commercially.) High-speed tool steels are used principally for cutting tools in high-speed machining (Morral, 1966; *Cobalt Monograph*, 1960); the tungsten type alloys have been largely supplanted by the molybdenum type (Prizinsky, 1979). High-speed steels have historically been the biggest use of cobalt in steels. Hot work tool steels are used for punches, forging dies, and dies for hot extrusion of brass. The cold-working alloys are used for dies for severe cold-forming operations and cutting tools for cold trimming and shearing.

TABLE II-13

COBALT IN STEELS
(Cobalt Monograph, 1960)

Classification and Specific Alloys	% Cobalt	Other Elements	Treatment	Applications	Competing Materials	Future Trends	Other Remarks
High-speed tool steels	5 to 12	0.8 to 1.55% C, 1.5 to 20% W, 3.75 to 4.5% Cr, 1 to 2.25% V, 0 to 9.5% Mo, Fe (Morrall, 1966).	Conventional powder metallurgy techniques: melting mixtures of low-alloy powders, sintering, and pressing. Particle metallurgy process patented by Crucible Stainless Steel in use since 1970. Melted alloy atomized to powder which undergoes hot isostatic compaction. Billets are forged or fabricated into bars and coils (Vicini, 1976). Casting.	Preferred cutting tool material for many applications (Farren et al., 1961). Principally as cutting tools for high-speed machining, e.g., cutting cast iron, chilled rolls, heat-treated alloy steels, stainless steels, and Ti-base alloys; grinding; turning; and broaching.	The alloys compete with each other, Stellites®, WC--all contain Co. After Korean War, WC rapidly increased its market share at expense of high-speed steels (Charles River Associates, 1969). WC especially strong for turning, boring, and milling, but high-speed steel still dominant for drilling (Lardner, 1975).	Use in cutting tools has been erratic; but a slow, steady growth was predicted by W. Haswell, vice-president of technology, Crucible Stainless Steel Div., Colt Industries, Inc. (Vicini, 1976). In 1976, sales of particle metallurgy products from high-speed steels expected to triple within 5 yr at Crucible Stainless Steel. The Syracuse, N.Y., plant was undergoing a 25% expansion.	Co-containing steels are more brittle than Co-free high-speed steels. Subject to checking and cracking upon rapid heating or cooling.
Tungsten-type	5 to 12	12 to 20% W, Cr, V, C, Fe		Cast tool steel is used only where compressive strength and wear-resistance are needed such as in forming dies for auto bumpers and for making rolls for tube mills (Anonymous, 1978b).			Tungsten alloys now account for only 7% of tool steels (Prizinsky, 1979). Stellites® (Co-Cr-W-C) are also used for machine tools (Morral, 1966).
T4	5			Heat, shock			
T5	8			Heat, abrasion			
T6	12			High heat			
T8	5			Heat, shock			

TABLE II-13 (continued)

Classification and Specific Alloys	% Cobalt	Other Elements	Treatment	Applications	Competing Materials	Future Trends	Other Remarks
T15	5			Heat, high abrasion. Grinding completely shaped tools. Tools for aerospace production; lamination dies for electric motors, transformers, generators, steel rolling mills $\leq$ 12 in. in diameter. Special-purpose cutting tools (Anonymous, 1979e).	No close counterpart (Farren et al., 1961).		
Rex 76 (Crucible Steel)	9	C, 30% Mn, 30% Si, Cr, V, W, Mo, Fe		Special-purpose cutting tools (Anonymous, 1979e).			Superior to T15.
Molybdenum-type		5 to 8% Mo, 2 to 6% W, Fe, Cr, W, V, C	Co steels more difficult to machine, heat treat, and grind than T_1 and the leading Mo-types that contain no Co (Farren et al., 1961).				T1 (18% W) was the standard high-speed tool steel until World War II. The W shortage led to the use of Mo-base types (Prizinsky, 1979).
M6	12			High heat, high abrasion.	M47 (5% Co) super high-speed steel was revived in 1978 and promoted by one company to substitute for high-speed steels containing 8 or 12% Co (Mari, 1978e).		
M30	5			Heat, abrasion.			
M33	8			Applications requiring high red hardness (Anonymous, 1979e).			
M34	8			Heat, abrasion.			
M35	5			Heat, abrasion.	No close counterpart (Farren et al., 1961).		
M36	8			Heat, abrasion.	No close counterpart (Farren et al., 1961).		

TABLE II-13 (continued)

Classification and Specific Alloys	% Cobalt	Other Elements	Treatment	Applications	Competing Materials	Future Trends	Other Remarks
M41	5			Machining heat-treated high-strength steels, superalloys, and Ti; special purpose cutting tools (Anonymous, 1979).			
M42					M11 (a V-Cr steel).		
M43							
M44							
M47	5						
Hot-work tool steel	Generally 0.5 to 0.6 (Charles River Associates, 1969).						
Tungsten-chromium-cobalt, H19 (Allegheny-Ludlum Steel Corp.)	4.25	W, Cr, 2% V, Mo, C, Fe		Severe hot-working applications. Tools that encounter repeated shocks such as punches, inserts in forging dies, and intrusion dummy blocks. Dies for hot extrusion of brass. Mandrels for shell forging. Extension dies for Al backer blocks (Cobalt Monograph, 1960; Charles River Associates, 1969).	W, Cr, W-Cr hot-work tool steels. H19 competes with H14 (more W and Mo, no Co). High-W grades such as H21 or H26 are only slightly inferior to H19.		
Cold-work tool steel (high carbon, high-chromium die steels)	Generally 1 to 2 (Charles River Associates, 1969).			Dies for severe cold-forming operations, some hot-working. Both D2 and D5 steels are suitable for blanking, cold trimming, warm trimming of thin flash, cold shearing under abrasive conditions, drawing heavy gages of sheet metal, extrusion punches, cold-extrusion dies, thread-rolling dies, complex master gages.			Cobalt addition improves wear and abrasion-resistance.
D2	0.75						
D5	3.25	Fe, Cr, C, Mo			D7 (4% instead of 3% C) (Charles River Associates, 1969). D2, D4, or another steel in D or A series of AISI grades (Mari, 1978e).		

TABLE II-13 (concluded)

Classification and Specific Alloys	% Cobalt	Other Elements	Treatment	Applications	Competing Materials	Future Trends	Other Remarks
Maraging steels		Ni, Mo, Ti, Al, Fe	Simple heat treatment, 3 hr at 480°C. Easily fabricated (Charles River Associates, 1969).	First developed for the aerospace industry. Used in experimental missile cases in 1963 - 1964. Aircraft landing gear components, dies for Al extrusion, high-strength bolts, gimbal bearings, torque transmission shafts, high-pressure polyethylene units. They lack the abrasion and wear-resistance of carbon steels but good for tool and die use because of dimensional stability (Charles River Associates, 1969).		In 1969, future demand seemed dependent on space and missile programs because of the high cost of maraging steels (Charles River Associates, 1969).	Introduced in 1962 by International Nickel Co. Several thousand pounds Co was being used in maraging steels by 1964 (Charles River Associates, 1969.
Grade 200 ksi	8.5						
Grade 250 ksi	7.5						
Grade 300 ksi	9.0						
Miscellaneous alloy steels							
Quench-hardened steel (Republic Steel)	4	9% Ni, 25 to 30% C, Fe		Not commercial in 1969. The 25% C type was used in propellant cases. The 30% C type was used in the Boeing 747 for engine hangar mounts, parts of tail sections, and in torsion links of the landing gear (Charles River Associates, 1969).			
High-temperature stainless steels							
AISI 348	0.20						

The maraging steels were introduced in 1962 for aerospace applications. They are good for tool and die use because of their high dimensional stability (Cobalt Monograph, 1960; Charles River Associates, 1969).

Data for cobalt consumption in steels since 1946 have been compiled in Table II-14. The National Materials Advisory Board (1971) grouped such data approximately the same way for the years 1966 to 1969, but there were large discrepancies for the "high speed and tool steel" value in 1966 (586,000 lb versus 411,000 lb in the table; yet the totals match) and all the "other steel alloys except stainless" category (thousand pounds): 846 in 1966, 516 in 1967, 470 in 1968, and 282 in 1969 (making the totals about 140,000 lb less in 1967 and 1968 than those in Table II-14).

Specialty steel melt shops may use either ESR and VAR, depending on the customers' specifications, metallurgical considerations, or costs. ESR is a cheaper process that can produce a better steel quality with less ingot discard. ESR is essential for alloys with volatile components such as nitrogen and manganese (Howard, 1976).

Current demand for high-speed steels is high; the automotive and appliance markets are the strongest (Saville, 1979a). At least one tool steel producer is encouraging its customers to reduce the cobalt in the tools they buy, especially those containing M30 or M40 series steels, whenever the demands on the tool are not great. National Twist Drill Division of Lear Siegler, Inc., is also investigating the use of substitutes in high-speed steel formulations and asking its customers to segregate their cobalt-containing scrap for recycling by domestic mills (Wrigley, 1979b). Up to 10 to 15% savings in cobalt consumption can be made, according to Andrews of Allegheny-Ludlum Industries, Inc., Pittsburgh, by reducing the cobalt charge slightly, carefully monitoring the metal, and collecting materials for recycling (Mari, 1979a). Armco's Advanced Materials Division sees PH13-8 alloy as a potential substitute for cobalt-containing steels. By May 1979, the scarcity of cobalt had apparently not yet caused a reduction in its content in the product mix at plants producing vacuum remelted metals because the

TABLE II-14

U.S. CONSUMPTION OF COBALT IN STEELS
(Metric tons)[a]

Year	High-Speed Tool Steel	All Others Besides High-Speed[b]	Maraging Steel	Carbon	Stainless and Heat-Resisting	Full Alloy	High-Strength Low-Alloy	Tool	Total
1946-1948									Avg 175
1949									Avg 429
1950-1958									Avg 160
1959-1961									378-385
1962-1968									Avg 565
1963			18[c]						
1964			178[c]						
1965	138	417	109[c]						555
1966	186	464	117[c]						650
1967	233	296	128[c]						529
1968	251	279							530
1969	259	163							422
1970	242	113							356
1971				-	23	89	-	144	256
1972				1	18	98	3	164	284
1973				1	14	103	20	235	373
1974				2	18	113	4	313	450
1975				-	17	84	1	132	234
1976				-	19	107	W	101	>227
1977				-	33	161	W	105	>298
1978				-	-	-	-	-	313

a/ Sources: Charles River Associates (1969); Metal Statistics 1972 (Mustero, 1972); Metal Statistics 1976 (Roditi, 1976); Metal Statistics 1978 (Cere, 1978); and Gorham International, Inc. (1979b).

b/ Other tool steels in 1969 used about 91 MT cobalt per year (Charles River Associates, 1969).

c/ Not included in totals since they would have been included under "all other" statistics.

customers' specifications had not yet changed (Howard, 1979). Another recent news story (Nordberg, 1979a), however, stated that several producers had ceased production of cobalt-containing tool steels between April 1978 and early 1979.

G. Other Alloys

1. Corrosion resistant, dental, and prosthetic alloys: Corrosion resistant alloys are nickel-, cobalt-, and iron-base alloys containing large percentages of chromium, molybdenum, or other alloying elements. Table II-15 mentions some of the cobalt-base alloys used in the chemical and pulp and paper industries. Low-cobalt alloys in the Hastelloy® alloys B- and C-series are used in chemical corrosion applications such as power plant and municipal incinerator stacks. New Hastelloy® alloys Bp2 and C-276 (2.5% cobalt) are described as the "workhorses" in these applications (Thornton, 1977). Hastelloy® alloy B (2.5% Co), C-4 (2.0% Co), and G (2.5% Co) are also of commercial interest with stainless steels and other corrosion resistant alloys for use in sulfur dioxide scrubbers (Michels and Hoxie, 1978). Hastelloy® and Haynes® alloys are also used for shell and tube heat exchangers in chemical process equipment. Fasteners for extremely corrosive environments may also be made of these alloys (Cabot Corporation, undated).

The uses of Vitallium and other cobalt-chromium alloys for dental applications and prosthetics are described more fully in Table II-16.

2. Spring alloys: Applications of cobalt-containing alloys for springs are described in Table II-17. Little statistical information was found. Cobalt has been used as a grain refiner in certain beryllium-copper alloys, but Kawecki Berylco Industries has begun to substitute nickel for cobalt in some of the beryllium-coppers (Furst, 1979a).

TABLE II-15

COBALT IN CORROSION-RESISTANT ALLOYS[a/]
(*Cobalt Monograph*, 1960; Cabot Corporation, 1979; Cabot Corporation, undated; Anonymous, 1976a)

Classification and Specific Alloys	% Cobalt	Other Elements	Condition or Treatment	Applications
Potentially useful alloys (1960)				
Fe-Co	Optimum 60 atom % versus 50% H_2SO_4			
Ni-Co	30 to 80			Resistance to brine, HCl, KOH, and 5% H_2SO_4.
Mn-Co				
Cr-Co	75 to 80 atom %			Better than Co or Cr alone in resistance to nitric, hydrochloric, and sulfuric solutions.
Haynes Stellite® alloy No. 6B	53	30% Cr, 4.5% W, 1.2% C; Ni, Fe, Si, Mn	Wrought	Used in the pulp and paper industry where wear resistance is also needed: digesters, machines for repulping scrap paper, in caustic measuring tanks and lime and caustic tanks. Alloy No. 6B sleeves and bushings support agitator shafts; sheet used for impeller paddles in the lime mud tank plate welded to the leading edge of rotor vanes in the repulping machine (Dyer et al., 1976). Bearings in saltwater pumps for S mines and as wear strips in massive gate valves for handling salts and brine (Morral, 1966).
Haynes Stellite® alloy No. 25			Wrought (easily fabricated)	Chemical industry, e.g., handling fuming nitric acid, chlorine-handling valve stems used where stainless steels are inadequate or where product purity must be maintained.
Hastelloy® alloy B	2.5	Ni, Mo, Cr, Fe, Si, Mn, C, V	Wrought	Furnace components. Carburizing equipment (Grant et al., 1961). Isomerization contactors. Ethylbenzene alkylation.
Hastelloy® alloy B-2	1.0	Ni, Mo, Cr, Fe, Si, Mn, C	Wrought	Isomerization contactors. Ethylbenzene alkylation.
Hastelloy® alloy C-4	2.0	Ni, Mo, Cr, Fe, Si, Mn, C, Ti	Wrought	Acetic acid and other organic synthesis apparatus components.
Hastelloy® alloy C-276	2.5	Ni, Mo, Cr, Fe, W, Si, Mn, C, V	Wrought	Process environments involving strong oxidizing agents, brines, chlorine gas and solutions.
Hastelloy® alloy G	2.5	Ni, Mo, Cr, Fe, W, Si, Mn, C, Cb, Ta, Cu	Wrought	SO_2 and SO_3 scrubbers. Resists both acid and alkaline solutions.

a/ Corrosion-resistant alloys are Ni-, Co-, or Fe-base alloys containing large fractions of Cr and Mo. Cobalt alloys are generally too expensive and too hard to work with for corrosion-resistant applications unless resistance to high temperature and abrasion is also needed. See also Vitallium for dental applications and bone surgery.

TABLE II-16

COBALT IN DENTAL ALLOYS AND PROSTHETIC DEVICES
(*Cobalt Monograph*, 1960)

Classification and Specific Alloys	% Cobalt	Other Elements	Treatment	Applications	Competing Materials	Other Remarks
Dental alloys						
Vitallium	62.5	30.8% Cr, Mo, C, Si, Mn, Fe	Cast (investment casting)	Cast alloys - cast denture bases, complex partial dentures, and certain types of bridgework. Widely used in U.S., Germany, and Great Britain. Wrought alloys - orthodontic appliances.	Au alloys, but Co-base alloy dentures are considerably lighter than gold dentures for equivalent size and rigidity.	In 1949, five times as much Co alloys was used in cast dentures as were precious metals. These alloys have good castability, resist tarnish and abrasion, are compatible with mouth tissues, and have low density.
Nobilium	65	28.0% Cr, Ni, Mo, C, Si, Mn, Fe, V	Cast			
Wisil	66.2	27.0% Cr, Mo, C, Si, Mn, Fe	Cast			
Croform	60.0	30.0% Cr, Mo, C, Si, Mn, Fe	Cast			
Virillium	67.9	24.1% Cr, Ni, Mo, C, Si, Mn, Fe	Cast			
Ticonium	28.7	27.4% Cr, 37.5% Ni, Mo, C, Si, Al, Be	Wrought. Hot- and cold-worked into wire and strip. Can be age-hardened.			
Wiptam	45.5	28.3% Cr, 24.4% Ni, C, Si, Mn	Wrought. Strengthened only by cold-working.			

TABLE II-16 (concluded)

Classification and Specific Alloys	% Cobalt	Other Elements	Treatment	Applications	Competing Materials	Other Remarks
Alloys used in bone surgery						
Vitallium	65	30% Cr, 5% Mo	Cast; powder metallurgy a recent advance (Saville, 1979c).	Nails, screws, plates, etc., in bone repair surgery. Major surgical implant use in replacements for arthritic or fractured hip, knee, ankle joints, etc. Co-Cr implants preferred except in young patients where the 20-year lifetime would be insufficient. Annual use of Ti-based and Co-Cr surgical implants: 80,000 hip, 60,000 knee, and 10,000 ankle joints (Fox, 1978).	Co alloys not likely to be replaced in this market significantly. Hip Ti-V-Al implants may be an improvement over Co-alloy implants (Saville, 1979c).	Surgical implant market is growing at $\geq$ 20% per year (Saville, 1979c). Cobalt must be bought a year ahead of final prosthetic sale. High alloy costs have doubled costs of implants (Saville, 1979c).
Stellite®-type			Precision castings.			
Variant of Ticonium	29	28% Cr, 6% Mo, balance Ni		Used where malleability is desired.		Ticonium showed toxicity as a surgical prosthesis due to the presence of Be.
Orthochrome		Cr, Mo	Cast	Total hip replacements.		
Protosul		Ni, Cr, Mo Ti	Wrought	Total hip replacements.		

TABLE II-17

COBALT IN SPRING ALLOYS
(Cobalt Monograph, 1960)

Classification and Specific Alloys	% Cobalt	Other Elements	Treatment	Applications	Competing Materials	Other Remarks
Elgiloy, Cobenium	40	20% Cr, 15% Ni, 7% Mo, 2% Mn, 0.04% Be, 0.15% C, balance Fe		Mainsprings in watches; springs for sewing machines, electric contacts, optical equipment, movie cameras, firing-pin retractors; Belleville-type springs; fountain-pen nibs; balls for ball bearings; reed valves; drawing instruments; wristwatch bands; garment stays; traveler rings (in weaving equipment); weighing-scale draft bands; toaster and snap-switch components; medical and dental equipment; metal drive bands in electronic fire-control systems for night aircraft; ammunition components that operate only once, perhaps after long storage; (one case prior to 1960) spring to assist functioning of a human heart valve.		Elgiloy developed ca. 1950 by Battelle Memorial Institute for the Elgin National Watch Company. Elgiloy's set resistance is > 3 times that of carbon steel, and its fatigue strength is nearly double. It is extremely resistant to atmospheric corrosion, eliminating mainspring failures in watches. One of a few spring materials unaffected by atomic radiation.
Beryllium-copper alloys, e.g., Berylcos, Beraloys, Ampcoloy 95-20 55-C, 200-C, 250-C, Alloy W5, Oakes Trodaloy No. 1, Tuffalloy 55	0.2 to 2.7	0.1 to 2.85% Be, rest Cu	Easily fabricated in the soft condition. Mechanical properties attained by age-hardening.	Springs and clips requiring high electrical conductivity--contact springs, clips, and blades; brush springs; spring clamps. Also satisfactory for connectors, circuit-breaker components, clutch plates, switches, electronic attachments, spline shafts, poppet valves, welder tips, diaphragms, and dies for forging and extruding Al, Mg, and Ti. Excellent for nonsparking tools for hazardous environments. Applications requiring wear resistance--solenoid plunger guides, switch ratchets, bearing cages.	Co used as a grain refiner; but Kawecki Berylco Industries, Reading, Pa., is substituting Ni for Co in some of its Cu-Be alloys (Furst, 1979a). Bell Laboratories has substituted Spinodal (Cu-Ni-Sn) alloys for the Cu-Be alloys used for years in telephone equipment relay springs and connectors (Berman, 1979).	High thermal and electrical conductivities, high tensile and fatigue strengths, excellent hardness, wear resistance, corrosion resistance. Most withstand temperatures 200 to 260°C without losing spring properties.
Haynes® [iron-base?] alloy No. 5			Cold reduction gives optimum hardness and improves strength and ductility.	Springs, diaphragms, bearings, and rupture disks.		The only Haynes® alloy designated by "5" in a 1979 Cabot Corporation brochure is an iron-base alloy. No Co content is indicated.

3. Heating element, filament, and cathodic alloys: Table II-18 describes cobalt alloys used for heating elements, filaments, and cathodes in 1960. Recent information on these applications has not been discovered.

4. Miscellaneous alloys: Most of the information in Table II-19 was taken from the Cobalt Monograph (1960). More recent information described potential substitute alloys for tin brasses, cartridge brasses, and phosphor bronzes (Anonymous, 1974a; Butt, 1970); cobalt-platinum electrodes for electrolysis of aqueous metal chloride solutions to produce chlorine, hypochlorite, and chlorates (Krohn and Bohn, 1972-1973); and lead-cobalt battery alloy for an electric automobile to go into commercial production in 1980 (Wrigley, 1979a).

H. Metal Coatings

1. Electrodeposited coatings: Morral and Safranek (1974) reviewed the electrodeposition of cobalt and cobalt alloys. Early commercial applications included electroformed cobalt record stampers and bright nickel alloy electroplate containing 17 to 18% cobalt for plating appliances and automotive parts. Cobalt has also been added to commercial bright nickel alloys during periods of nickel shortages.* The aerospace industry has investigated electroformed parts made of cobalt superalloys and electrodeposited Co-Mo, Co-W, and Co-Ni alloys as well as dispersion-strengthened cobalt alloy electrodeposits where oxides, borides, or carbides are very finely dispersed in the matrix.

---

* Cobalt was used as a temporary nickel-plating substitute on steel during the 1969 nickel strike in Canada. At that time, cobalt was cheaper than nickel. A half nickel-half cobalt process was developed as a substitute for 100% nickel plating. It was not suitable, however, for automobiles because of poorer corrosion resistance (Charles River Associates, 1969).

TABLE II-18

COBALT IN HEATING-ELEMENT, FILAMENT, AND CATHODE ALLOYS
(Cobalt Monograph, 1960)

Specific Alloys	% Cobalt	Other Elements	Treatment	Applications	Other Remarks
Cobanic	45	55% Ni, 0.1% C		Filament, cathode cores.	Generally, cobalt alloys for cathode filaments and cores have higher electrical resistivity and tensile strengths at elevated temperatures than non-cobalt-bearing alloys.
Cochrome	60	24% Fe, 12% Cr, 2% Mn		Heating elements, filament. Cobalt-containing heating elements are used in household appliances and delicate research instruments.	
Filnie M	45	55% Ni		Filament.	
Filnie F				Filament.	
Hilo	18	75% Ni, 5.0% Fe, 2.0% Ti, 0.25% Mn		Filament, cathode cores.	
Hilo-modified	20	78% Ni, 0.5% Fe, 0.3% Ti, 0.25% Mn, 1.0% Si		Filament, cathode cores.	
Hilamic	18	79% Ni, 2% Fe, 1% Si		Cathode filaments.	
Kanthal A	2	69% Fe, 5.7% Al, 23% Cr, 0.06% C		Heating elements. Kanthal alloys are recommended for all types of industrial heating equipment.	Kanthal heating elements resist oxidation up to 1300°C.
Kanthal D	2	Fe, Al, Cr, C		Heating elements.	
Kanthal DS		Fe, Cr, Al, Ca		Appliance elements.	
Konel	1.72	73% Ni, 8.8% Ti		Lamp filaments.	
Kovar	18	28% Ni, 54% Fe		Filament.	
RCA N 91	21.6	78.1% Ni; Mn, C, Mg		Cathode filaments.	
RCA N 97	40.0	59.5% Ni; Mn, C, Si		Cathode filaments.	

TABLE II-19

COBALT IN MISCELLANEOUS COMMERCIAL ALLOYS
(Cobalt Monograph, 1960)

Classification and Specific Alloys	% Cobalt	Other Elements	Treatment	Applications	Future Trends and Other Remarks
Jewelry alloys					
Gold alloys	> 0.9		Can be electrodeposited (Krohn and Bohn, 1972 - 1973).	To increase hardness, wear resistance, and ease of polishing gold. Au-Co alloys can also be plated onto electrical terminals (Krohn and Bohn, 1972 - 1973).	
Permium 205				Cheap jewelry with good corrosion resistance.	
Lemarquand's alloy	8	39% Cu, 7% Ni, 37% Zn, 9% Sn			
Le Mats Paris metal	1	5% Fe, 71 to 81% Cu, 5% Zn, 6 to 16% Ni, 2% Sn			
Argental B	5	85% Cu, 10% Sn			
Platinum substitute					
Coopers	5	70% Ag, 25% Pd			
Warne's metal	11	37% Sn, 26% Ni, 26% Bi		Silver substitute in ornamental articles.	
Several alloys	Low Co	High Cu, Zn, Ni		Ornamental uses, not necessarily classified as jewelry.	

TABLE II-19 (continued)

Classification and Specific Alloys	% Cobalt	Other Elements	Treatment	Applications	Future Trends and Other Remarks
Cobalt-aluminum alloys					
Engine piston alloys	0.3 to 1.2	17 to 25% Si, 1.4 to 5.3% Cu, 0.5 to 4.2% Ni; Mg, Fe, Ti, Mn, Zn, Cr		In Germany for pistons in motorcycle and motor-scooter engines.	
	0.7	4.0% Cu, 14.5% Si, 0.8% Mg, 0.03% Ti, 4.0% Ni, 0.3% Fe, balance Al			
Alcan M116				Cobalt added to Al-Si alloys to hinder $FeSiAl_5$ formation, which causes embrittlement.	
Alpax H au cobalt (Canadian and French commercial alloys)					
Al-Mg alloys				Patented compositions claim that Co improves ductility, increases strength and high-temperature property stability.	
Alloy W	0.23 to 0.28	0.57 to 0.90% Cu, 0.28 to 0.36% Ni, 0.14 to 0.17% Mg; V, W, Cr, Be, balance Al		Die castings to be anodized (avoids the pin-point porosity and streaking seen with Si-containing alloys).	

TABLE II-19 (continued)

Classification and Specific Alloys	% Cobalt	Other Elements	Treatment	Applications	Future Trends and Other Remarks
Antimonides of aluminum (indium and germanium)	0.001 to 5 atom %			Co additions increase the carrier mobility and magnetoresistance of these semiconductors.	
Cobalt-copper alloys[a]					
-	3 to 5	Cu		Thermoelectric couples.	
Aluminum bronzes, e.g.,	1	8% Al, balance Cu		Sometimes added to increase hardness and tensile strength.	
Zinc-copper brasses, e.g.,	0.5	{80% Cu, 20% Zn		Co may be added to increase the strength, but the alloys are hard and brittle.	
-	2.5	97% Cu, 0.5% Si	Easier to cast than Be-Cu alloys.		
Cobron (CDA alloy 664) (Olin Corp.)		Cu, Fe, Zn		Substitute where tin brasses and cartridge brasses have performed marginally and phosphor bronzes are too expensive. Potential uses are electrical springs, fuse clips, terminals, bearings, bushings, thrust washers, weather strip, clutch plates, spring washers, and relay switches (Anonymous, 1974). Approved last year by the Rural Electrification Administration for use as a protective shield for underground distribution wire. Its low magnetic permeability reduces the loss of telephone signals that occurs with alternate materials (Berman, 1979).	

TABLE II-19 (continued)

Classification and Specific Alloys	% Cobalt	Other Elements	Treatment	Applications	Future Trends and Other Remarks
Sun Bronze	40 to 60	30 to 50% Cu, 10% Al		High-temperature fittings.	
Alcoloy Alloy 688, Coronze Alloy 638 (Olin Corporation)		Al, Si, Cu		Substitutes for phosphor bronze and nickel silver in telephone instruments, switchgear, electronic components, radios and TV, automotive electrical systems, appliance switches and controls, jewelry springs (Butt, 1970). The use of Coronze in water meters gives good protection against freezing, Berman (1979).	In 1970, these alloys provided cost savings as substitutes for phosphor bronze and nickel silver due to low density (more parts per pound) and better processability (Butt, 1970).
-		> 50% Cu, usually substantial amounts of Ni and Zn		Corrosion-resistant parts in chemical equipment.	
Cobalt-silicon alloys, e.g.,	Balance	13% Si, 5% Mn, 7% Cr		Some use as insoluble anodes in electrowinning of Cu.	
Miscellaneous alloys					
-		3 to 12% Fe, balance Zn		Sliding shoe insert for collecting current from trolley wires.	
-	0.7 - 1	7% Sn; Sb, As, B		Patented bearing material.	
Co-Pt			Plating from HCl solutions onto Ti or Ti-alloy cathodes (Krohn and Bohn, 1972 - 1973).	Electrodes for electrolysis of aqueous metal chloride electrolyte to produce Cl_2, hypochlorite, and chlorates (Krohn and Bohn, 1972 - 1973).	See also magnetic alloys.

TABLE II-19 (concluded)

Classification and Specific Alloys	% Cobalt	Other Elements	Treatment	Applications	Future Trends and Other Remarks
Pb-Co		Pb		Electric vehicle battery.	Automobile containing the 0.8 MT Pb-Co battery as its main energy source to go into production by Electric Auto Corp. in 1980. Battery requires 90 min to recharge to 100% capacity. Car has maximum speed >70 mi/hr (112 km/hr) and a range of 112 to 160 km between battery charges (Wrigley, 1979).

a/ See also "Jewelry alloys" above and beryllium coppers under "Spring alloys" in Table II-17.

The concentrations of cobalt salts in representative baths for electroplating cobalt and cobalt alloys are given in Table II-20. Control processes include filtration, anode bagging, and electrolytic purification. (Morral and Safranek, 1974).

The properties of electrodeposited cobalt are listed in Table II-21. In this table, it is pointed out that the coercivity of cobalt coatings depends on the electrodeposition conditions: thus, a pH > 4.5 and temperatures of 40 to 90°C in the plating bath give high-coercivity coatings; but a pH < 2.0 and room temperature give low-coercivity coatings suitable for fast-switching memory devices. The high-coercivity coatings are used on magnetic drums and disks.

Walker and Cruise (1978a) state that the electrodeposited composite coating (ECC) appears to be the major use of unalloyed cobalt electrodeposits. In these coatings, an insoluble compound such as chromium carbide is dispersed in the plating bath and finally in the deposited matrix. The hexagonal-close-packed structure of cobalt gives the ECC good wear resistance and frictional characteristics. Use of the cobalt-chromium carbide system in aircraft air conditioning and brake cooling equipment, compressor blade roots, rotary engine casings, piston rings (chiefly for diesel engines), and helicopter rotor shafts has advantages over the competing plasma or flame-spraying technique. The advantages include no distortion due to heating, production of smooth coatings of a designated thickness, and better geometric flexibility of the coated parts. The ECC is limited, however, in the size of the component being coated. Nonetheless, an ECC coating comprising a cobalt matrix with 30% chromium carbide powder has been used on thousands of aircraft engine production parts at Bristol Aerojet and Rolls Royce plants in England (Anonymous, 1979b).

The major electrodeposited cobalt alloy is cobalt-nickel. These coatings are used for decoration and corrosion resistance, wear resistance, electroforming, electrojoining, and magnetic applications (Walker and Cruise, 1978a). Ni-18% Co alloys have also been used as undercoats in plating silver;

TABLE II-20

COBALT SALTS IN PLATING BATHS
(Morral and Safranek, 1974;
Walker and Cruise, 1978a)

Type	Cobalt Salt	g/liter	pH	Temperature, °C	Comments
For cobalt deposits					
Chloride	$CoCl_2.6H_2O$	90-105	2.5-3.5	49-54	Boric acid is usually added to Co baths as a buffer.
Sulfamate[a/]	$Co(SO_3NH_2)_2$	450	-	20-50	
Ammonium Sulfate	$Co(NH_4)_2(SO_4)_2.6H_2O$	175-200	5.0-5.2	25	30 ml/liter formamide. High plating rates make bath suitable for electroforming.
Sulfate	$CoSO_4.7H_2O$ $CoCl_2.6H_2O$ (optional)	330-565 45	3.0-5.0	35-38	Popular bath. Easy to control.
For alloy deposits with					
Nickel, 50%	$CoSO_4.7H_2O$	29	3.7-4.0	66	300 g/liter $NiSO_4.7H_2O$ and 50 g/liter $NiCl_2.6H_2O$; Co and Ni anodes
Nickel	$Co(BF_4)_2$	60-100	-	25	
Nickel	$CoCl_2.6H_2O$	28.3	9.3	60	
Molybdenum, 40%	$CoSO_4.7H_2O$	85	10.5	25	174 g/liter $K_4P_2O_7$ 48 g/liter $Na_2MoO_4.2H_2O$; Co anodes.
Phosphorus, 5%	$CoCl_2.6H_2O$ $CoCO_3$ to adjust pH	180	0.05-2.0	75-95	50 g/liter H_3PO_4, 15 g/liter H_3PO_3; Co anodes.
Tungsten, 30%	$CoCl_2.6H_2O$	100	8.5-8.7	82-90	Co-30% W anode
Tungsten	$CoSO_4.7H_2O$	150	1.5-2.0	55	31.5 g/liter $Na_2WO_4.2H_2O$
Chromium[b/]	$CoSO_4.7H_2O$	430-470	4.7	50	500 g/liter Cr carbide

a/ Sulfamate baths containing 2 to 16.5 g Co/liter have been used industrially since about 1950 to give 12-60% Co-Ni deposits (Hammond, 1970).
b/ Walker and Cruise (1978b) described these baths and their operating conditions in more detail.

TABLE II-21

PROPERTIES OF ELECTRODEPOSITED COBALT
(Safranek, 1972)

Property	Electrodeposit	Wrought Metal	Application of Electrodeposit
Density, g/cm^3	8.8	8.85	
Electrical resistivity	8.4-10	6.24	
Coefficient of thermal expansion	14.5	13.8	
Coercivity, oersteds	8-25	-	Magnetic films and shields. Soft magnetic coatings with a low coercivity (≤150 Oe) are used for fast-switching memory devices.
Co a/			
Co-1% P b/	1-11	-	Thin films on nonmagnetic substrates for recording information in computers.
Ni-Co-P c/	0.1-4.5	-	
Co d/	600-1,200	-	Thin films with high coercivity (>200 Oe) used for high-density information storage on drums or disks.
Co-3% P	1,200-1,400	-	
Co-25% Ni	100-500	-	
Co-Ni-P	1,400-2,000	-	
Tensile strength, psi			
Co	172,000	37,000	
Co-20% W e/	155,000		
Ni-45% Co	200,000-275,000		
Elongation %			
Co	12-19	50-55	
Co-20% W	< 1	-	
Ni-4% C	4-5	-	
Microhardness, kg/mm^2			
Co	~190 to ~450	-	For some uses, Co alloys surpass Cr electrodeposits for wear and abrasion resistance.
Co-P	~400 to ~1,200 (if heat treated)	-	
Co-W	~450 to ~750	-	
Ni-Co	~400 to ~500	-	
Stress range (for highly stressed deposits) in kg/m^3 at 0.012 to 0.080 psi	~15 to ~55	-	

a/ Low-coercivity cobalt is electrodeposited from baths at pH <2.0 and at 20-25°C. The structure of the cobalt produced is both hexagonal-close-packed and face-centered cubic.
b/ Cobalt-phosphorus alloys are electrodeposited or are applied by electroless plating in solutions containing sodium hypophosphite.
c/ Ni-Co-P alloys may be deposited from electroless baths or as supersaturated solid solutions.
d/ High-coercivity cobalt is electrodeposited from baths at pH >4.5 and at 40-90°C, which favors the hexagonal-close-packed structure.
e/ Cobalt-tungsten alloys may be deposited as supersaturated solid solutions.

gold; and chromium, e.g., in automobile bumpers, jewelry, and printed electrical circuits (de Bie and Doyen, 1962b). Molds of electroformed cobalt-nickel alloys are suitable for zinc-base die casting, plastic injection molding, typemetal casting, and pressing gold and silver coins and medallions. Electroformed nickel-cobalt has several aerospace applications: waveguides, jet engine ducts, reflectors, venturi nozzles, and rocket engine thrust chambers and vanes. Magnetic wire coated with electrodeposited Fe-Co-Ni alloy is used widely in the storage system of crash recorders, compulsory equipment on all civilian airliners. Ni-Fe-Co alloy deposits are used as magnetic shields and in high-coercivity magnets, magnetic recording, and thin-film memory devices (Walker and Cruise, 1978a).

Adding 0.2% cobalt to electroplated gold electrical contacts (cyanide citrate bath) has improved the adhesion of the coating and wear resistance (Walker and Cruise, 1978a).

At a symposium on Tin and Its Alloys in the United Kingdom, April 1978, J. O. Grady of Oxy Metal Industries (G. B.) Ltd. discussed decorative tin-cobalt alloy electrodeposits on nickel as a substitute for chromium plate. Drawbacks of chromium plating include the disposal problems of the concentrated electrolyte containing up to 400 g/liter chromic acid. Tin-nickel and tin-zinc electrodeposits are also being used for decorative plating in the move away from chromium. Both acid fluoride and slightly alkaline sulfate baths* are used commercially for tin-cobalt electrodeposition. The latter (Acrolyte) was developed by Udylite Research and was introduced by Oxy Metal Industries (G.B.) Ltd. (Anonymous, 1978e). The optimum deposit comprises 78% tin and 22% cobalt. Although not as hard or as wear resistant as chromium deposits, the deposits can be soldered and are suitable for chromium substitute use in certain service conditions (Grady, 1978).

* In these nearly neutral solutions, the salts are solubilized by an organic stabilizer (Grady, 1978).

Maki and Tanaka (1975) reviewed the electrochemistry of cobalt, but practical applications were only briefly mentioned. Besides electrodeposition of cobalt alloys, cobalt electrochemistry is applied to electropolishing and toward the electrowinning and electrorefining of cobalt. They did not give any information pertinent to environmental pollution or occupational exposure associated with electroplating of cobalt.

2. Electroless coatings: The electroless deposition of Co-P involves an autocatalytic reduction of cobalt ions *in situ* by hypophosphite ions on a catalytic surface.

Reactions include: $Co^{2+} + H_2PO_2^- + OH^- \xrightarrow{catalyst} Co^0 + 2H^+ + HPO_3^{2-}$ and $H_2PO_2^- + OH^- \longrightarrow HPO_3^{2-} + H_2$. Elemental phosphorus, produced by reduction hydrolysis of $H_2PO_2^-$, alloys immediately with cobalt (Gutzeit et al., 1971).

Pearlstein (1969) gave the composition of a bath for electroless deposition of a Co-P alloy. The solution, containing 30 g/liter of $CoCl_2 \cdot 6H_2O$, was used at pH 9-10 and 90°C. An alkaline solution for electroless deposition of nickel on plastics was reported to contain 5 g cobalt chloride per liter (International Nickel Company, 1973).

Electroless cobalt or Co-P coatings have magnetic properties applicable to high-density magnetic storage devices associated with peripheral file memories for computers (magnetic tapes, drums, cards, or primarily, disks). (γ-Fe_2O_3 is one of the more conventional media.) The magnetic and other physical properties of electroless cobalt deposits are discussed in depth by Fisher (1972).

3. Sprayed coatings: Simon et al. (1973) described the various techniques of metal deposition by spraying metal (pure or alloyed) that had been melted in a flame or arc and atomized by a blast of compressed gas. (See also Anonymous, 1967). Since the process does not heat the parts to cause warpage, spraying is commonly used to manufacture machine elements, build up worn parts, or salvage mismachined parts. It is also used to provide corrosion and wear-resistant coatings on iron and steel (see Table II-22). Metallizing uses the metal in wire

TABLE II-22

SPRAYED WEAR-RESISTANT COATINGS
(Anonymous, 1972a)

Coating	Method of Application[a/] M	T	P	Applications and Other Comments
Iron-, nickel-, and cobalt-base	X	X	X	The group is used for hard-bearing coatings; coatings resistant to abrasive grains, hard surfaces, fretting (with or without intended motion), cavitation, and particle erosion.
Tungsten carbide-cobalt composite in a fused Ni-Cr matrix		X		Percentage WC-Co composite 30-50%. Used for coatings resistant to abrasive grains, hard surfaces, fretting (intended motion), and particle erosion.
Tungsten carbide-cobalt composites			X	Coatings for above-described applications except cavitation-resistant coatings.
Tungsten carbide-cobalt composite unique blends with Ni-Cr alloy and Ni-Al composite powders		X	X	Often used as less costly substitute for Cr plating. Same applications as above.

a/ M = Metallizing process
T = Thermo spray process
P = Plasma flame spray process

form. Thermal spraying uses powdered materials such as tungsten carbide. Cobalt alloys, cobalt-zirconia blend, and tungsten carbide-cobalt have been deposited by plasma arc (flame-spraying) processes.

Simon et al. (1973) also discussed the air pollution control equipment needed for metal-spraying operations. For example, in plasma arc spraying, cobalt powder alloy and tungsten carbide-cobalt show deposition efficiencies of 60 to 70%; i.e., 30 to 40% of the powder is lost as overspray and must be trapped by equipment such as water wash scrubbers ($\leq$ 96% collection efficiency), absolute filters (> 99% efficiency), baghouses (for metal oxides, not metals) (> 99% efficiency), and mechanical scrubbers ($\geq$ 95% efficiency). Spraying is done in a fume hood with 37 to 61 m/min in-draft velocities.

In addition to wire and powder, cobalt compounds can be used to produce metal coatings. For example, cobalt acetylacetonate, $Co(C_5H_7O_2)_3$, is used in vapor plating of cobalt-nickel and cobalt-base alloys (Morral, 1979).

I. Driers in Paints and Related Uses

Approximately 350 to 1,200 MT/year cobalt driers were produced in the United States in the period 1975-1978 (Chase, 1979b). Cobalt driers are the most active and generally useful paint driers. Table II-23 lists the cobalt salts and soaps used as driers. Cobalt driers are known as "top" driers because they catalyze oxidation at the film surface to give a tack free state in a short time (cobalt driers are 1.5 to 6 times more rapid than lead* or manganese soaps). Because the catalyzed oxidation is so rapid, it may prevent deep drying and cause wrinkling. For this reason, cobalt driers are generally used in combination with other metals that slow down the oxidation (de Bie and Doyen, 1962b; Anonymous, undated b).

* Lead is almost always used with cobalt and/or manganese driers. It is the best auxiliary drier, giving good through-dry (Stewart, 1967).

TABLE II-23

COBALT SOAPS AND SALTS USED AS DRIERS
(Morral, 1979; Anonymous, undated a)

Compound[a/]	Applications: Paints	Varnishes	Enamels	Lacquers	Printing Inks
$Co(C_2H_3O_2)_2 \cdot 4H_2O$		X		X	X
Co borate					X
Co linoleate $Co(C_{18}H_{31}O_2)_2$	X	X			X
Co naphthenate	X	X			X
Co octoate $Co(C_{18}H_{15}O_2)_2$					X
Co oleate $Co(C_{18}H_{33}O_2)_2$	X	X			X
CoO	X	X			
Co hydrate $Co(OH)_2$					X
Co resinate $Co(C_{44}H_{52}O_4)_2$	X	X	X		X
Co_2SnO_4	X	X			
Co stearate $Co(C_{17}H_{35}CO_2)_2$					
$CoSO_4 \cdot 7H_2O$ or $CoSO_4 \cdot H_2O$					X
Co soyate					X
Co tallate	X	X			X
Co(II) tungstate $CoWO_4$	X	X	X		X

a/ All of these compounds contain divalent cobalt. However, cobalt is at least transiently in the cobaltic form during drying. The normal color is red-violet-blue, but the oxidation products that form in a vehicle during prolonged storage give the green color of cobalt(III).

Since only 0.005 to 0.2% cobalt is added to the oil, cobalt soaps can be used in coatings of any color despite the strong color of cobalt. However, since cobalt driers cause discoloration in baked finishes, manganese driers are preferred (de Bie and Doyen, 1962b; Anonymous, undated b).

Although generally used with lead or manganese, cobalt is used alone in white baking alkyds and in vehicles for aluminum paints (Anonymous, undated b).

Cobalt driers allowed the use of cheap oils, e.g., soya oil and fish oils, previously thought unfit for drying. In addition to oil-base paints, cobalt driers (0.01 to 0.02% Co) can be added to latex paints that have been oil- or alkyd-modified. Cobalt also accelerates hardening of films of unmodified butadiene-styrene resin latex paint systems (de Bie and Doyen, 1962b; Anonymous, undated b).* Cobalt driers alone or with calcium, manganese, or lead driers are used for epoxy resin esters, polyurethane, and polysulfide resins (Anonymous, undated b).

Cobalt naphthenate or octoate is used with peroxide catalysts to accelerate curing of unsaturated polyester resins in building glass fiber laminates (Anonymous, undated b).

Cobalt driers may also be used in putty and caulking compounds (Berkowitz et al., 1973).

Cobalt naphthenate, octoate, and tallate are used as catalysts in hardening foundry core oil binders and composition board binders (Anonymous, undated b). For example, 0.06 to 0.6% cobalt as cobalt naphthenate is used to cure molding sand bonded with linseed oil or resins. Other driers have been found to be unsuitable for drying core sand (de Bie and Doyen, 1962b).

Cobalt and zinc octoates (0.05 to 0.2% metal) are used for low-temperature curing of coatings of silicone organic

---

* Stewart (1969) recommended $\leq$ 0.1% cobalt additions to all other vehicles but up to 0.25% in butadiene-styrene vehicles.

polymers or silicone-modified resins. These coatings for metal, however, are more often cured with heat (de Bie and Doyen, 1962b).

Cobalt soaps are generally supplied in formulations containing 6 to 12% cobalt. They are prepared by fusion or precipitation. In fusion, the cobalt hydroxide, oxide, or salt of a weak acid is heated with the organic acid. Precipitation is performed by treating an aqueous sodium soap solution with an aqueous solution of a soluble cobalt salt (Meinstein and Spielman, 1968).

J. Catalysts

Cobalt compounds are utilized as catalysts for the production of industrial chemicals, and the refining of petroleum products. Future synthetic fuels production and/or processing of coal may require cobalt catalysts. They are also potentially useful in automobile emission control devices.

1. Industrial chemicals production: The oxonation (oxo synthesis) or hydroformylation reaction catalytically converts monoolefins to aldehydes and alcohols. In the two step, high-pressure variation of the oxo reaction, a mixture of olefin, carbon monoxide, and hydrogen gas react in the presence of dicobalt octacarbonyl, $Co_2(CO)_8$, to form an aldehyde which is subsquently reduced to an alcohol. The cobalt is introduced into the reactor as cobalt naphthenate which is readily available as a drying agent for paints, inks, varnishes, etc. Subsequently, cobalt naphthenate forms dicobalt octacarbonyl, $Co_2(CO)_8$, and cobalt hydrocarbonyl, $HCo(CO)_4$, both of which are thought to be the catalytic species responsible for homogeneous reaction.

The intermediate aldehyde may be straight- or branched-chain depending on the particular olefin and the experimental conditions. Straight-chain olefins yield about 60% normal and 40% α-branched aldehydes or alcohols. Branched olefins yield only branched products.

In the one-step, low pressure variation of the oxo reaction, a mixture of olefin, carbon monoxide, and hydrogen gas react in the presence of a cobalt carbonyl-phosphine complex, $Co_2(CO)_6[(C_4H_9)_3P]_2$, to form the alcohol directly. An undesirable side reaction to form paraffins also occurs. Table II-24 presents details of the two oxo processes (Gautreaux et al., 1978).

Feedstocks include propylene, C_2-C_{20} linear α-olefins, and C_8-C_{15}-branched olefins depending on availability and the alcohol desired. Principal products manufactured by the oxo process include several fatty acids, butyraldehydes, butanols, and higher aliphatic alcohols up to C_{20}. 2-Ethylhexanol is a major product prepared from _n_-butyraldehyde produced by the cobalt-catalyzed oxo process from propylene. A newer commercial process produces _n_-butyraldehyde by rhodium-catalyzed hydrocarbonylation of propylene. Union Carbide, Davy Powergas, and Johnson Matthey developed this low-pressure rhodium-based catalytic oxo process, which produces more than twice the _n_-butyraldehyde at lower pressure and temperature than the cobalt process (Matthey Bishop, Inc., 1978). The principal use of 2-ethylhexanol is in the manufacture of bis(2-ethylhexyl) phthalate, a plasticizer for flexible poly(vinyl chloride) (Lowenheim and Moran, 1975).

Cobalt naphthenate is prepared:

- By reacting a soluble cobaltous salt (e.g., cobaltous sulfate) with the sodium salt of the organic chosen (e.g., sodium naphthenate).

- By reacting cobalt powder with naphthenic acid in the presence of oxygen.

Cobalt carbonyl is prepared by heating cobaltous acetate in cyclohexane and in the presence of a mixture of Co and H_2 (de Bie and Doyen, 1962b).

Harshaw Chemical Company (undated, ca. 1971) markets a series of cobalt metal or oxide catalysts on silica supports; and cyclopentadienylcobalt dicarbonyl is marketed by Pressure Chemical Company (1972).

TABLE II-24

INDUSTRIAL CATALYTIC USES OF COBALT

Process Designation	Reaction	Reaction Conditions	Facilities Utilizing Process	Remarks
Oxonation (two step)	$R\text{-}CH{=}CH_2 + CO + H_2 \xrightarrow{\text{Co cat.}}$	200-300 atm.; 130-190°C	Exxon Chemical, Company, Baton Rouge, Louisiana	Cat. concentration 0.1-1.0%. Product separated by fractional distillation.
	$R\text{-}CH_2\text{-}CH_2\text{-}CHO + H_2 \xrightarrow{\text{Ni cat.}} R\text{-}CH_2\text{-}CH_2\text{-}CH_2OH$	50-200 atm.; 150-200°C		
(one step)	$R\text{-}CH{=}CH_2 + CO + 2\ H_2 \xrightarrow{\text{Co cat.}} R\text{-}CH_2\text{-}CH_2\text{-}CH_2OH$	50-100 atm.; 170-200°C	Shell Development Company, Geismar, Louisiana	Cat. concentration 0.5%. Product separated by degassing and distillation.
Fischer-Tropsch	$CO + 2\ H_2 \xrightarrow{\text{Co cat.}}$ Aldehydes, alcohols, ketones, gasolines, diesel fuels, oils, paraffins	1-10 atm.; 180-200°C		Has been too expensive to compete with petroleum as a source of fuel.
Hydrodesulfurization	$C_2H_5SH + H_2 \xrightarrow[\text{Cat.}]{\text{Co}} C_2H_6 + H_2S$ $CH_3SCH_3 + 2\ H_2 \xrightarrow[\text{Cat.}]{\text{Co}} 2\ CH_4 + H_2S$	10-50 atm.; 260-430°C	All major petroleum refineries	Higher temperatures favor hydrodesulfurization but promote hydrocracking and coke formation on catalyst surface.
Hydrodenitrification	$C_5H_5N + 5\ H_2 \xrightarrow[\text{Cat.}]{\text{Co}} C_5H_{12} + NH_3$	As above.		
Hydrodeoxygenation	$C_6H_5OH + H_2 \xrightarrow[\text{Cat.}]{\text{Co}} C_6H_6 + H_2O$	As above.		

Cobalt content in catalysts can be recovered by thermal decomposition of the carbonyl, extraction with dilute acetic acid, or precipitation with oxalic acid.

The Fischer-Tropsch synthesis in which liquid hydrocarbon fuels and chemicals are manufactured from CO and H_2 has been improved considerably since its discovery in 1923 and during World War II served to supply Germany with various gaseous, liquid, and solid hydrocarbons, including lubrication oil, soaps, edible fats, etc. (Mills and Cusumono, 1979).

The process is deceptively simple and involves combining carbon monoxide and hydrogen gas at moderate temperatures and pressures in the presence of cobalt or iron catalysts. The reaction is highly exothermic and the heat generated must be dissipated via heat exchanger tubes to prevent decomposition/disproportionation of the products formed. Sulfur compounds in the feed gas must be removed to prevent catalyst poisoning. Typical reaction conditions are 180 to 200°C and 1 to 10 atm.

An effective catalyst is composed of cobalt-thoria-magnesia on diatomaceous earth (kieselguhr or Celite™). It is easily prepared by mixing boiling solutions of cobalt and thorium nitrates and sodium carbonate which yields precipitates of cobalt and thorium bicarbonates (de Bie and Doyen 1962b). These substances are isolated, re-suspended in water, and added to a slurry mixture of magnesium oxide and kieselguhr. After rapid mixing and filtration, the filter cake is washed free of extraneous salts, dried, and reduced at 350 to 450°C in a current of dry hydrogen gas. The final catalyst formulation is $100Co:5ThO_2:8MgO_2:200SiO_2$. The reduced catalyst is best maintained in storage under a carbon dioxide atmosphere (<u>Cobalt Monograph</u>, 1960).

Iron catalysts have also been utilized in the Fischer-Tropsch reaction because of lower cost. However, cobalt catalysts have greater long term stability at elevated temperatures (Young, 1960). Cobalt catalysts have been recommended for the hydrogenation of CO-containing blast furnace gases. (Kölbel, 1957). Table II-24 presents details of the Fischer-Tropsch process. Its use in coal conversion processes is described in Subsection 3.

2. Petroleum refining: In the last 20 years, hydrodesulfurization of crude oil has become increasingly important due to rising demands of low and middle boiling range distillates and gasolines. Contributing economic factors included:

- Increasing sulfur content of available crude oils.

- Mandatory requirements for sulfur-free liquid fuels for industrial and consumer use.

- Availability of relatively low cost, by-product hydrogen gas.

- Protection of expensive platinum reforming catalysts which would be poisoned by sulfur (Kay, 1956, cited in the Cobalt Monograph, 1960).

Hydrodesulfurization refers to a catalytic process for removal of sulfur (and also oxygen and nitrogen) from petroleum distillates and residues. Typical hydrodesulfurization, denitrification, and deoxygenation reactions and operating conditions are given in Table II-24. Additional reactions which may occur during hydrodesulfurization are hydrogenation and hydrocracking:

$$C_3H_7CH{=}CH_2 + H_2 \xrightarrow[\text{cat.}]{\text{Co}} C_5H_{12} \quad \text{(hydrogenation)}$$

$$C_{10}H_{22} + H_2 \xrightarrow[\text{cat.}]{\text{Co}} 2\ C_5H_{12} \quad \text{(hydrocracking)}$$

With proper control of operating conditions a petroleum refiner can reduce sulfur content to $\leq 0.1\%$ in distillates.

The principal catalysts employed for hydrodesulfurization are (3%) cobaltous oxide: (10%) molybdenum trioxide: (87%) aluminum oxide (CMA), and tungsten-nickel sulfide. CMA catalyst can be prepared in several ways, e.g.:

- By immersing freshly prepared γ-alumina in separate or combined solutions of ammoniacal cobalt salts and ammonium molybdate, filtering, and calcining at $\leq 600°C$.

- By coprecipitating cobalt and molybdenum as oxides on γ-alumina and calcining (Cobalt Monograph, 1960).

A small amount (5%) of silica may be added to stabilize the CMA catalyst. Iron oxide may be added to conventional CMA catalyst when hydrotreating lubricating oils and greases.

Commercial CMA catalysts are available as granules, tablets, or extruded small diameter (2.5 mm O.D.) rods of high porosity (de Bie and Doyen, 1962b).

In general, the CMA catalyst life may be 2 years or longer depending on operating conditions and extent of use. It is customary to regenerate the catalyst *in situ* by passing an air and steam mixture containing 1 to 2% oxygen over it at $\leq$ 600°C for 1 to 3 days. This process removes any carbonaceous material (coke) on the catalyst surface and 10 to 50 separate regenerations are common before replacement.

3. Cobalt catalysts for coal conversion: As yet, catalytic coal conversion to liquid or gaseous fuels is not economically attractive because of technical problems. The following discussion is based on *Catalysis in Coal Conversion* by Cusumano et al. (1978). Cobalt compounds being studied are listed in Table II-25.

Catalytic systems being studied include Pt-Co, as well as Ag-Au and Cu-Au to upgrade or refine low-sulfur coal liquids and to catalyze CO/H_2 synthesis reactions. Bimetallic nickel, ruthenium, or cobalt catalysts favor lower methane and coke formation, which is caused by hydrogenolysis of C-C bonds. Pt-Co is prepared as a stabilized supported catalyst.

Bimetallic catalysts such as Co-Ni and Ni-Cu are prepared as Raney alloys, e.g., by leaching a finely powdered aluminum-cobalt-nickel alloy with caustic soda at 15 to 20°C.

In addition, cobalt compounds are used in the non-metallic catalysts: oxides, sulfides, and oxysulfides. Unsupported oxide or mixed oxide catalysts are produced, e.g., by flame decomposition, freeze drying, gel precipitation, or sol-gel formation. Sulfides and oxysulfides may be obtainable by sulfidation. Supported nonmetallic catalysts are commonly prepared by impregnation, adsorption from solution, and gel precipitation. Cobalt molybdate-impregnated stabilized alumina is a widely used supported nonmetallic catalyst.

Not much attention has been given to regenerating coal conversion catalysts. Coked catalysts in the petroleum industry are commonly regenerated by controlled slow oxidation. However, oxidation procedures can only be applied to catalysts used with low-ash coal.

TABLE II-25

COBALT COMPOUNDS PERTINENT TO COAL CONVERSION CATALYSIS
(Cusumano et al., 1978)

Compound	Comment
$PtCoO_2$	Precursor for hydrodesulfurization catalyst.
Co_3N_2	Nitrides are more stable than carbides in low H_2S concentrations.
$Co_{21}Hf_2B_6$ (perovskite structure) Co_2P	Borides and phosphides have high thermal stability. Resistant to high H_2S concentrations. May resist carbiding.
Co-Al alloy	Metals are unstable in high H_2S.

There is only one operating commercial Fischer-Tropsch coal liquefaction plant--in Sasolburg, South Africa. The Sasol-1 facility has produced liquid and solid hydrocarbons from coal since 1956. Here, the promoted powdered iron catalyst is replaced approximately every 50 days. Cobalt catalysts slowly deteriorate with time, due to sulfide and sulfate formation, carbon deposition (coke formation), and sintering. A much larger facility, Sasol-2, is scheduled to go on stream in 1980 (Worthy, 1979; Cusumano et al., 1978; Rousseau, 1966).

Currently, there are no U.S. facilities employing the Fischer-Tropsch reaction based on coal conversion. However, Fluor Engineers and Constructors, a subsidiary of the Fluor Corporation, and Sasol have agreed in principle to market this technology in the United States in the future (Anonymous, 1979h). The major emphasis in the U.S. coal conversion process would be production of liquid fuels, including gasolines, jet fuel, diesel, and home heating fuel oil.

4. Automotive emissions control: Current automotive emission standards designate legal limits (grams/kilometers) for carbon monoxide (CO), unburned hydrocarbons (UHC), and nitrogen oxides (NO_x) emissions under the 1970 Clean Air Act.

Several catalysts have been developed to reduce the quantities of these pollutants in automotive exhaust. The most successful systems utilize rugged, extruded ceramic honeycomb supports (monoliths) composed of mullite ($3Al_2O_3 \cdot 2SiO_2$) or cordierite ($2MgO \cdot 5SiO_2 \cdot 2Al_2O_3$). The monoliths are coated with thin layers of substrate (e.g., alumina or zirconia) and catalyst (e.g., a Group VIII metal or metal oxide).

Table II-26 presents various catalysts for the oxidation of a number of components in automotive exhaust (Mills and Cusumano, 1979).

Present Pt-Pd catalysts used since 1975 for automotive emissions control have limited lifetimes (25,000 to 50,000 miles of vehicle operation) and are poisoned by sulfur and lead compounds. Newer three-way catalysts contain platinum and rhodium. Nothing in the trade literature at this time suggests that cobalt will be used commercially in catalytic converters for automobiles.

5. Miscellaneous cobalt catalysts and research: Table II-27 briefly presents several miscellaneous industrial uses and research involving cobalt catalysts. Applications include:

- Synthesis of nitriles
- Synthesis of alcohols

TABLE II-26

SUITABLE CATALYSTS FOR REDUCING POLLUTANTS IN AUTOMOTIVE EXHAUST

Pollutant	Catalyst	Remarks	References
H_2 and UHC as CH_4 and CH_3OH	$Pt\text{-}Pd\text{-}CeO_2$	1-3 wt % Pt-Pd	Mills and Cusumano (1979)
CO	$(R.E.)_{1-x}Pb_xMnO_3$ where $0.4 \leq x \leq 0.6$ and $R.E.CoO_3$	R.E. = La, Pr, Nd perovskite structure	Voorhoeve et al. (1972)
CO and NO	Ru-BaO $Ru\text{-}La_2O$	Forms volatile oxide, RuO_4	McEvoy (1975) Shelef and Gandhi (1974)
NO_x, CO, and UHC	$Pt\text{-}Rh\text{-}Al_2O_3$ (3-way catalyst)	Very narrow air-fuel ratio operating range, 14.7-14.8 wt/wt	Mills and Cusumano (1979)
	$CuO\text{-}CoO\text{-}MnO_2\text{-}Al_2O_3\text{-}NiO$	No_x reduction in first-stage; CO and UHC oxidation in second stage	Warshaw et al. (1976)

TABLE II-27

MISCELLANEOUS USES OF COBALT CATALYSTS

Category	Reaction	Form of the Cobalt Catalyst	Remarks	Reference
Synthesis of nitriles	$CH_3\text{-}CH{=}CH_2 + NH_3$ (and C_3H_8) $\xrightarrow[100\text{ atm.}]{370^\circ C}$ $CH_3CN + C_2H_5CN$	$Co\text{-}MgO\text{-}SiO_2$	Mixed nitrile products including $n\text{-}C_3H_7CN$ and iso-C_3H_7CN	Teter and Olson (1953)
Synthesis of alcohols	$H_2 + CO \longrightarrow C_2H_5OH$ plus higher alcohols	3ZnO:CoO	Up to 39% C_2H_5OH in condensate products at 320-330°C and 120 atm.	Klyukvin et al. (1932; cited by Natta et al. 1957)
Synthesis of aromatic carboxylic acids	$CH_3C_6H_4CH_3$ + air $\xrightarrow[15\text{-}35\text{ atm.}]{140\text{-}155^\circ C}$ mixture of aromatic carboxylic acids	Cobalt toluate	Produces phthalic anhydride, isophthalic acid, terephthalic acid, and benzoic acid. Other catalysts used include Mn salts.	Burney et al. (1959); Himel (1954); Lowenheim and Moran (1975)
Fluorination of hydrocarbons	$C_7H_{16} + 32\ CoF_3 \xrightarrow[N_2\text{ as diluent}]{250\text{-}350^\circ C} C_7F_{16} + 32\ CoF_2 + 16\ HF$	$CoF_2 + F_2 \xrightarrow{150\text{-}200^\circ C} CoF_3$	Fluorination of hydrocarbons is highly exothermic	Fowler et al. (1947)
Polymerization of olefins	$CH_2{=}CH_2 \xrightarrow[70\text{ atm.}]{100\text{-}120^\circ C}$ 1-butene, 2-butene, hexenes, and higher hydrocarbons	18% Co on activated charcoal	Forms polyethylene-type solids, greases	Peters and Evering (1953)
Oxygen generation from water	$H_2O + h\nu$ (visible) $\xrightarrow[\text{catalysts}]{\text{Co and Ru}} O_2 + 2H^+$	$Co[(NH_3)_5Cl]Cl_2$, $Ru(2,2'\text{-bipyridine})_3^{2+}$, and RuO_2	pH 4.05 in acetate buffer	O'Sullivan (1979)
Self-cleaning oven surfaces	Catalytic oxidation of food spatter	$MnO.3Mn_2O_3.SiO_2$ (40-80%) $B_2O_3.SiO_2$ (10-15%) CoO (0-20%)	Oven operating temperature 200-500°C. Cobalt oxides may also be included as a common constituent of porcelain enamels and frits rather than as a catalyst.	Chay (1974). See also Lee (1970), Eisen and Feighan (1970), and Ott (1973)
Air oxidation of waste sulfate liquors	{Phenols, Cresols, β-Naphthol, Na_2SO_3 in solution} + air $\xrightarrow[60^\circ C,\ pH\ 7\text{-}9]{CoCl_2}$ oxidation products	$CoCl_2$		Hurwitz et al. (1959)

TABLE II-27 (concluded)

Category	Reaction	Form of the Cobalt catalyst	Remarks	Reference
Phenol synthesis	$C_6H_5CH_3 + 1.5O_2 \xrightarrow[cat.]{Co} C_6H_5CO_2H + H_2O$ $C_6H_5CO_2H + 1/2O_2 \xrightarrow[cat.]{Cu} C_6H_5OH + CO_2$ $C_6H_5CH(CH_3)_2 + O_2 \longrightarrow C_6H_5C(CH_3)_2OOH$ $C_6H_5C(CH_3)_2OOH \longrightarrow C_6H_5OH + (CH_3)_2CO$		The toluene oxidation and cumene peroxidation processes are the most recently adopted commercial processes for manufacturing phenol. The competitiveness of these with other processes are based on local availability of raw materials. Only one plant used the toluene oxidation process ca. 1975 (2% of total phenol capacity in the U.S.). At that time 88% of capacity was based on the cumene peroxidation process. Morral (1979) stated that cobalt is used as a catalyst for this process, but Lowenheim and Moran (1975) do not identify the peroxidation catalyst.	Lowenheim and Moran (1975)
Environmental degradation and biodegradation of blends of polyethylene and an oxyalkanoyl polymer or a dialkanoyl polymer and biodegradation of polyethylene materials		Co octanoate and Co naphthenate	Compositions contain ~0.01 to 0.1% Co or Fe. Suggested uses include mulch film, transplanter containers, disposable containers, and shipping boxes and crates.	Union Carbide Corporation patents by Clendenning et al. (1975a, 1975b) and Potts et al. (1973)

- Synthesis of aromatic carboxylic acids
- Fluorination of hydrocarbons
- Polymerization of olefins
- Oxygen generation from water
- Self-cleaning oven surfaces
- Air oxidation of toxic waste liquors

K. Decolorizers, Pigments, and Frits

1. Decolorizers: Cobalt compounds, due to their blue shade, are used to decolorize pottery, glazes and enamels, and glasses. In pottery bodies, $< 0.02\%$ cobalt masks the yellowish tinge from impurities such as iron oxide. Cobalt may be added as a mixture of cobalt oxide with kaolin, as an aqueous solution of a soluble salt, or as a freshly precipitated gel of cobalt carbonate. Cobalt (0.002 to 0.01%) is used as a decolorizer in white enamels and transparent glasses. As a decolorizer in the glass industry, 1 to 2 mg Co oxide/kg is used together with 15 mg Se/kg, 0.05% arsenic as arsenious oxide, and 0.15% sodium nitrate (de Bie and Doyen, 1962a).

2. Pigments and dyes: Table II-28 lists numerous cobalt compounds used as pigments for porcelain decoration. There are two cobalt blue pigments: cobalt silicate or smalt, which gives a deep violet blue, and cobalt aluminate or Thenard's blue, which gives a turquoise color. Mixing the oxide, silicate, or aluminate with other metal oxides gives numerous other shades: Canton blue ($BaCo_3$), Rinman green (ZnO), jade green (Cr oxide), olive green (Al_2O_3, ZnO, Cr oxide), violet and purple (Cr and Sn oxides), mulberry (Mn oxide), pink (MgO), Cerulean blue (Sn oxide), black (oxides of Fe, Ni, Cr, and Mn), and gray (Ni and Sn or Zr oxides) (de Bie and Doyen, 1962a; Morral, 1979).

Cobalt compounds may be used as pigments in the body stain (stained glazes or baked clay), the underglaze stain, and the overglaze color in the surface decoration of porcelain. For example, for painting untreated ceramic ware, the coating contains 0.4 to 0.5% cobalt oxide.

TABLE II-28

COBALT COMPOUNDS USED AS PIGMENTS, DYES, AND DECOLORIZERS
(Morral, 1979)

Compound	Application
$Co_3(AsO_4)_2.8H_2O$	Light blue color for painting on glass and porcelain, coloring glass
$CoO.Al_2O_3$	Painting porcelain
CoB	Ceramels
$CoCO_3$	Pigments, ceramics
$CoCl_2.6H_2O$	Dye mordant
$CoCrO_4$	Green tint in ceramics
$CoSiF_6.6H_2O$	Ceramics
Co(III) hydroxide	Paints
$LiCoO_2$	Ceramics
$Co(NO_3)_2.6H_2O$	Pigments, decoration for stoneware and porcelain, hair dyes
CoO	Glass decorating, coloring and whitener
Co(II,III) oxide [Co_3O_4]	Enamels
$Co_3(PO_4)_2.8H_2O$	Glazes, enamels, pigments
$K_3Co(NO_2)_6.1.5H_2O$	Oil and water-color pigment, paint for glass and porcelain, rubber colorant, Fisher's yellow
Co resinate [$Co(C_{44}H_{52}O_4)_2$]	Lustrous coating for chinaware, pottery, textiles
$CoSO_4$	Ceramics
$CoSO_4.7H_2O$	Pigments for porcelain, glazes
$NH_4CoPO_4.H_2O$	Ceramic pigment
Co silicate	Painting porcelain

Cobalt oxide (0.5 to 5%) may be added to the molten bath during preparation of enamel for ceramics and metal sheets. For the latter, cobalt is generally added to the frit (see the next subsection). Kitchen utensils are given the characteristic mottled white and blue appearance by adding 0.001 to 0.002% cobalt sulfate and/or nickel sulfate to the enamel slip just before application (de Bie and Doyen, 1962a).

Cobalt is used in concentrations up to 0.5% for coloring blue glass and crystal. The highest concentration is for goggles for furnace workers and welders. Only a few milligrams per kilogram are required in tinted automobile windshields and blued camera lenses (de Bie and Doyen, 1962a).

Cobalt pigments are also used as pigments for oil paints* and printing inks for fabric and paper. Cobalt stannate (Cerulean blue), $K_3Co(NO_2)_6$ (Aureolin yellow; turns blue if baked), and cobalt phosphate (cobalt violet) are used in oil paints. Cobalt oxide is used in black silicone enamels where service temperatures exceed 315°C. Cobalt-bearing ultramarine blue is one of the most frequently used pigments and extenders in printing ink manufacture. Cobalt-pigmented inks are used for printing bank notes (Morral, 1979; de Bie and Doyen, 1962a). Cobalt blue is also used as a pigment for hydraulic cement (Berkowitz et al., 1973).

Shaver et al. (1975) stated that cobalt-containing streams are recycled in the production of cobalt black (see Figure II-1), Co_2O_3, and that wastes are small.

Cobalt has been fairly common in foreign metallic hair dyes. In the 1950's, lead, silver, and copper were the predominant heavy metals in hair dyes on the American market. With a developer (pyrogallol or hydrosulfide), cobalt salts produce light-brown shades. For example, Broux of Paris used cobalt nitrate alone or mixed with silver and ammonium compounds to produce 20 shades from blonde to black (Wall, 1957). The only cobalt compound listed in the 1977 CTFA (Cosmetic, Toiletry and Fragrance Association) Cosmetic Ingredient Directory (Estrin et al., 1977) is the naphthenate.

3. Frits and ceramic coatings: In porcelain enameling of steel, a ground coat or frit containing cobalt oxides or salts

* Cobalt blues are generally too expensive for paints (Shaver et al., 1975).

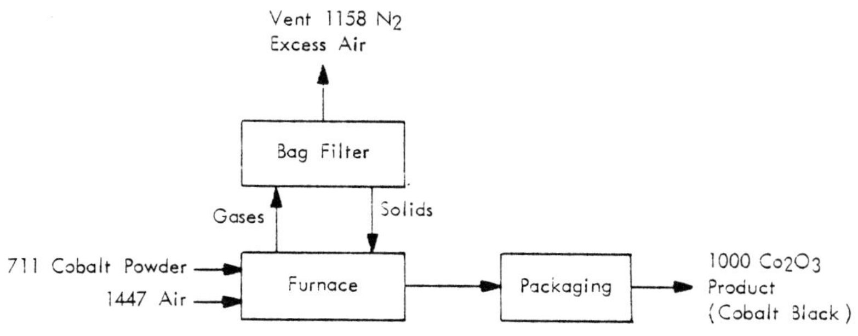

Figure II-1. Production of Cobalt Black.
(Shaver et al., 1975)

(Numbers are based on units of mass required to give 1,000 mass units of product.)

provides adherence and inhibits reactions between the enamel finish and the base metal. In 1947, more than 15% of total cobalt consumption was for frit (276 MT). Consumption of cobalt in this use was 136 MT in 1952, 270 MT in 1964, and 91.2 MT* in 1968 (Charles River Associates, 1969).

The usual cobalt concentration in the 0.0025 to 0.0075 mm-thick ground coat or frit is 0.5 to 0.6% although the concentration may range from 0.2 to 3%. A final enamel is used to mask the blue-black ground coat, but the frit on bathtubs and kitchen ware often becomes visible when the cover coat is chipped (Morral, 1979; de Bie and Doyen, 1962a).

Cobalt compounds may also be used in ground coats with alumina for low-carbon and low-alloy steels for service

---

* No reason was given for the decrease in consumption. It may be connected to the decline in use of porcelain coatings for kitchen appliances (except for ranges) with the increasing popularity of nonporcelain colored coatings.

up to 750°C in industrial flue pipes and gas turbine exhaust pipes. Cobalt-containing ceramic coatings are also used on superalloys (de Bie and Doyen, 1962a).

L. Miscellaneous Uses of Cobalt Compounds

The cobalt compounds used for vitamin B_{12} preparation, feed supplements for ruminants, plant fertilizers, moisture indicators, and preparation of other cobalt compounds are listed in Table II-29. Other miscellaneous uses of cobalt compounds are described in Table II-30.

M. The Cobalt Crisis of 1978-1979

1. Factors leading to the crisis: Cobalt and nickel were selling at depressed prices in early 1978 due to oversupply so that production had decreased.* Total cobalt consumption in the United States was 25% more in 1978 than it had been in 1977. Consumer inventories, therefore, were drastically reduced.

In May 1978, Katangan rebels invaded Shaba Province, Zaire, about the same time that there was a strike at the Belgian refinery. U.S. customers of Zairian cobalt were placed on allocation: 70% of 1977 consumption. The apprehension over further political problems in Zaire has further depressed the market.

The producers' price for cobalt rose from $6.40/lb in December 1977 to $25/lb in February 1979 (Sozacom, the marketing arm of Gecamines, Zaire, set the pace for all price increases) (Gorham International, Inc., 1979a). The free market or dealer price was almost $50/lb at the end of 1978; by January 16, 1979, it had fallen to between $41 and $43/lb (Nordberg, 1979a). The $25/lb producers' price is still in effect as of this writing (mid-August 1979) (Gorham International, Inc., 1979d).

---

* Lowe (1979) pointed out that the shortage of cobalt is due to the "insufficient investment in cobalt production during the very lean years experienced by copper and nickel producers between 1974 and 1979."

TABLE II-29

COBALT COMPOUNDS USED IN NUTRITION, MOISTURE INDICATORS, AND PREPARATION OF OTHER COMPOUNDS
(Morral, 1979)

Compound	Vitamin Preparation, Feed Supplement, Etc.	Hydrometers, Moisture Indicators	Preparation of Cobalt Compounds
Co(II) acetate tetrahydrate	X		
$CoBr_2$		X	
$CoBr_3$		X	
$CoCO_3$	X		X
Co(III) basic carbonate $2CoCO_3 \cdot Co(OH)_2 \cdot H_2O$			X
Co(II) citrate dihydrate $Co_3(C_6H_5O_7)_2 \cdot 2H_2O$	X		
Co(III) hydroxide			X
CoI_2		X	
$Co(NO_3)_2 \cdot 6H_2O$	X		
Co(II) oxalate (Hydrous form)			X (catalysts)
Co(II) succinate tetrahydrate $Co(C_4H_4O_4).4H_2O$	X		
$CoSO_4 \cdot 7H_2O$	X		
$Co(SCN)_2 \cdot 3H_2O$		X	
Cobalamine (cyanocobalamin, vitamin B_{12}) $C_{63}H_{88}CoN_{14}O_{14}P$	X		
$NH_4CoPO_4 \cdot H_2O$	X[a/]		

a/ Plant fertilizer.

TABLE II-30

MISCELLANEOUS USES OF COBALT COMPOUNDS
(Morral, 1979)

Compound	Application
Co(II) acetate tetrahydrate	Sealing in aluminum anodizing solutions.[a/]
$CoCO_3$	Temperature indicator.
$CoCl_2 \cdot 6H_2O$	Barometers, absorb poison gas and NH_3, flux for Mg refining, solid lubricant.
Cobalt disodium ethylenediaminetetracetate $CoNa_2(C_{10}H_{12}N_2O_8) \cdot H_2O$	Support of other chelating compounds in medicinal and tree-spray preparation.
Co(III) hydroxide	Storage batteries.[b/]
Co(II) oxalate	Temperature indicator.
CoO	Semiconductor, cobalt powder preparation for sintered Co_6W_6C, thermistors.[c/]
Co(II,III) oxide Co_3O_4	Semiconductors, grinding wheels.
$Co(ClO_4)_2$	Chemical reagent.
$K_3Co(NO_2)_6 \cdot 1.5H_2O$	Cobalt analysis.
$CoWO_4$	Antiknock agents.
$NH_4CoPO_4 \cdot H_2O$	Cobalt analysis.

a/ Shades from bronze to black in the oxide layer produced on anodized aluminum and its alloys are imparted by immersing the anodized part in 4 to 20% cobalt acetate solutions followed by potassium permanganate or ammonium sulfide. Immersing in cobalt acetate solution also improves the light fastness of dyed anodized aluminum (de Bie and Doyen, 1962b).

b/ Silver and cobalt sulfates (0.005 to 0.5%) have been suggested for use in the lead storage battery positive plates from which the cations migrate to protect the grids from oxygen and acids. The precipitated hydroxides of cobalt and nickel on the anodes in another lead storage battery increased the charge capacity and decreased the tendency to evolve gases during charging (de Bie and Doyen, 1962b).

c/ Thermistors are used for voltage regulation and temperature measurement (de Bie and Doyen, 1962b).

A major factor in the current shortage of cobalt has been the lack of U.S. government stockpile releases, to which the industry had become accustomed. Substantial amounts of cobalt had been released from the stockpile from the early 1970's until September 1976.

2. African producers: Zaire accounts for more than 60% of free world cobalt production; Zambia, its neighbor, produces another 15%.* Sozacom (Société Zairoise de Commercialisation des Minerais) is the marketing arm of the Zairian metals producers (Stundza, 1979).

In 1978, mining constituted 30% of Zaire's gross national product (in 1975, cobalt revenue was only 8% of Zaire's GNP). In 1978, mining products constituted 60% of its exports, which provided almost 80% of external revenue of the Zairian monetary system. Mining provided 15% of Zairian employment and 20% of salaries (Matho, 1979).

In April 1978, Katangan secessionists took over the Kolwezi mining district, but Zairian forces aided by France, Belgium, and Morocco soon drove them out (Nordberg, 1979c).

William Moffat, vice-president of African Metals Corporation, New York, U.S. agent for Sozacom, described the technician problem in Zaire in late 1978. In May over 600 Belgian technicians working at the copper-cobalt mines and refineries evacuated the Kolwezi area of Shaba province, anticipating return when the region's safety was guaranteed. In the meantime, Zairian workers took their places. When only some of the key Belgian technicians were asked to return, the Belgians demanded that all of their positions be restored. Zaire then terminated all the Belgians. Subsequently, a small group of Belgian and French technicians have commuted from secure areas, and a few French nationals have been recruited. There was hope for a while that the Zairians would eventually restore the Belgians to their jobs (Mari, 1978c).

---

* Slightly more than half of the 8 million MT of world land-based cobalt reserves are in the copper sulfide belt lying in Zaire and Zambia--about 4.3 million MT are in Zaire and about 1.2 million MT are in Zambia. The cobalt concentration in the ores is about 1,000 mg/kg (Lowe, 1979).

Current Zairian cobalt capacity is 16,000 MT/year (Matho, 1979). Plans for increasing production to 21,000 MT/year by 1981 have been delayed (Nordberg, 1979a). Recent cobalt production in Zaire has fallen 17 to 33% below the monthly target production of 1,200 MT (Nordberg, 1979c). The fall-off in Zairian production since 1974 and 1975 implied to Reddy of Charles River Associates that the mining and refining complex had been deteriorating even before the 1978 rebel invasion (Mari, 1979c).

A Belgian source (Scheidweiler, 1979), however, claims that reports of Zaire's production problems as the aftermath of the Shaba province invasion have been exaggerated. Production in 1978 was about 13,000 MT--more than in the two preceding years. However, a large Peko copper-cobalt project that would produce 6,000 MT/year of cobalt was delayed because of the invasion (Chase, 1979a). It is possible that cobalt production, which has increased with respect to copper, is from stocks; but Gecamines (La Generale des Carriers et des Mines) may have chosen to process copper ores richer in cobalt at this time. In recent years, the decline in the Co-Cu ratio was due to refinery problems, not lack of feed. There is admittedly a maintenance problem due to lack of capital for spare parts and a huge transport problem because of political conditions in Central Africa (Scheidweiler, 1979). The Benguela Railway in Angola, the fastest way to move exports from Zaire, was only recently reopened, having been closed in 1975 (Matho, 1979). By August 1979, shipments had been shifted completely from rail to air (Gorham International, Inc., 1979d).

A bright note in the Zaire situation was an interest free loan approved in mid-May 1979 by the International Development Association (IDA), a World Bank affiliate, to improve the country's railroads between Shaba province and the port cities. The loan of $20 million, far less than the $157.8 million total project cost, will go toward rolling stock and track improvements (Anonymous, 1979g).

Zambia intends to become a bigger producer of cobalt due to discovery of new cobalt mineralization containing 150,000 MT cobalt. Production is expected to rise to 9,500 MT/year by 1982 (9,000 MT by 1985 according to Nordberg, 1979b). An iron-copper-cobalt alloy produced by electrowinning would represent the bulk of the new cobalt production (Nordberg, 1979a). Market observers do not believe that Zambia can raise its cobalt output this year to 2,300-3,600 MT (Mari, 1979c) from 1,600 MT in 1978 (Nordberg, 1979b).

The Zambian sales agent reportedly recognized in late December 1978 that cobalt prices would have to be stabilized to reduce the incentive for substitution of other materials for cobalt (Anonymous, 1979d).

Janke (1979) of the Institute for the Study of Conflict, London, analyzed the political situation in Africa at the "Cobalt Crisis" conference in April-May 1979. He foresaw more Marxist military government in Africa. Zaire and Zambia are likely candidates once their current leaders lose control. Fighting still goes on in Angola, which depends on Cubans to maintain internal control. A foreign contingency remained in Zaire to keep order after the Shaba invasion but were to withdraw in June. The extremely poor population provides revolutionary potential in Zaire. Despite all the current problems, Janke did not feel that Zaire was "under any greater threat today than in the past." Conditions are probably better in some respects since relations with Angola have been repaired.

A less sanguine viewpoint was offered by Lepkowski (1979) in a June issue of Chemical Engineering News, who cited the London Economist in stating, "The country's agricultural system...has broken down, and the government could collapse anytime because much of the country's meager revenue is hoarded by the corrupt leadership."

3. Outlook: U.S. cobalt consumption was projected by the Bureau of Mines in 1977 (Sibley, 1977) to increase 2.9%/year until the year 2000. Rest-of-world demand was forecast at a 3.3% annual increase. Even in February 1979, Gorham International, Inc. (1979a) believed that this forecast was reasonable, perhaps even conservative.

Another pre-crisis forecast was that of Fischman and Landsberg (1972) in Population Resources and the Environment, who predicted that U.S. cobalt demand in the United States in the year 2000 could range from 8,900 MT for a low demand from a low population to 16,100 MT for a high demand from the larger population projected by the Population Commission. Their complete projections are given in Table II-31. Rest-of-world projections for the year 2000 were taken from Minerals Facts and Problems (Reno, 1970).

TABLE II-31

PROJECTED COBALT DEMAND TO THE YEAR 2020
(thousand metric tons)[a/] (Fischman and Landsberg, 1972)

United States	Year 2000	Year 2020	Total 1968-2000	Total 1968-2020
High population model				
High demand	16.1	31.5	326	798
Low demand	9.9	13.5	254	490
Low population model				
High demand	14.3	26.0	308	707
Low demand	8.9	11.7	245	454
Rest of world				
High demand	25.4	37.6	627	1,257
Low demand	18.6	22.7	529	942
Total				
High demand	41.5	69.1	954	2,055
Low demand	27.6	34.4	774	1,396

a/ Projections for the U.S. were derived from a combined macroeconomic/input-output projection model of the U.S. economy. Relative recovery from secondary materials was presumed to remain unchanged from the value in 1970.

On the other hand, Reddy of Charles River Associates*, predicted that cobalt substitution in key markets such as magnets, hard facing, and cutting tools could lead to a 30% reduction in U.S. cobalt consumption in the next 5 years (Mari, 1979f). Reddy, addressing the annual meeting of the National Association of Recycling Industries, in early April 1979, foresaw a state of flux in the cobalt market for the next few years. When Zairian production returns to the 1974 levels of 17,500 MT, which is unlikely within the next 2 years, the price should drop substantially.** Prices could continue to rise if demand remains strong and producers cannot increase production quickly enough, but they may fall if demand slackens due to a recession, and producers (including recyclers) have been able to increase their cobalt production as planned (Mari, 1979d).

The Department of the Interior also foresees a high probability of continued cobalt market disruptions through the 1980's (Stundza, 1979).

The Department of Commerce put into effect in January 1979 a system for reporting exports of materials containing $\geq$ 10% cobalt*** to determine if such exports were adversely affecting U.S. supplies and prices. Under the Export Administration Act of 1969, Commerce could eventually formally monitor exports. The materials falling under the reporting system are unwrought and wrought forms of nickel, nickel waste and scrap, unwrought and waste scrap from cobalt and cobalt alloys, cobalt oxides and hydroxides, and cobalt molybdate catalysts (Wesson, 1979).

---

* In February 1979, Charles River Associates began a 6-month multiclient study to assess potential trends in the availability and price of cobalt. Substitution and conservation possibilities, political and military developments in Zaire, and the potential for new and expanded supplies will be examined. (MRI did not acquire this study; the subscription fee was $15,000) (Mari, 1979a.)

** CRA estimated that if Zairian production rose to 1974 levels, the free market price would fall to about $10/lb.

*** Current statistics do not show the amounts of cobalt being exported in materials containing $<$ 50% cobalt.

Hewitt of Gorham International, Inc. (as told to Haflich, 1979) believed in January 1979 that the U.S., which has recently taken steps to monitor exports containing $\geq$ 10% cobalt, would soon put export controls on cobalt although no defense-related producers have been unable to obtain cobalt. He foresaw no stockpile releases by the government. Supplies will remain tight for another 2 years even though Zairian cobalt production is rumored to have returned to pre-insurrection levels.

If export controls were to be applied to cobalt exports, Reddy predicted that the domestic cobalt supply would increase, thereby decreasing the price of merchant and scrap cobalt in the United States.* The merchant price would tend to rise again as the flow of merchant cobalt into the U.S. would slow down. Exports would have to exceed imports sold at the merchant price before there would be a net price reduction (Koflowitz, 1979a).

A panel discussion at the conclusion of the Cobalt Crisis Conference in April-May 1979 of cobalt producers and users brought out the following predictions:

- There will be a slow growth in the production rate of cobalt with disruptions but not shortages (McPhail, International Nickel Company).

- By 1982-1983, world production of cobalt will be 35,000 MT (Scheidweiler, Société Générals des Minerais, Belgium).

- For the next 2 years, there will be no change in price**, even though the supply of cobalt will be increased (Modic, Stoody Company).

---

* The price of scrap cobalt (followed by merchant prices) might rise, however, if scrap processing costs were to rise.

** The price increase in cobalt by Zaire is due to the lack of capital by Gecamines, which is financing the entire country (Phiri, Roan Consolidated Mines, Zambia). Conference participants were very interested in how much money it would take to get cobalt production in Zaire back to normal. Funds are needed for importation of raw materials and replacement parts for the mining and refining operations.

BIBLIOGRAPHY. CHAPTER II

Anderson, J. W., "Cobalt Reclamation Attracts Broad Industry Spectrum," Am. Metal Market, **87**(18), 16 (1979).

Anonymous, Toxicity and Printing Ink IV, Reprint from the Am. Ink Maker, Technical Committee of the National Association of Printing Ink Makers, Elmsford, New York [undated].

Anonymous, Witco Metallic Carboxylates as Catalysts, Fungicides, Lubricant Additives, Paint Driers, Witco Chemical Organics Division, New York, New York [undated].

Anonymous, The Metco Flame Spraying Processes, Metco, Inc., Westbury, Long Island, New York, 1967.

Anonymous, Properties of Some Metals and Alloys, The International Nickel Company, Inc., New York, New York, 1968.

Anonymous, "Multiphase Alloys--Strength and Corrosion Resistance," Moly Pays Off, **2**(1), 14 (1969).

Anonymous, "Nickel Alloys for Electronics; Metal to Ceramic Seals Improve Reliability," Nickel Topics, **23**(4) [unpaginated reprint] (1970).

Anonymous, Wear Resistant Coatings Applied by Flame Spraying (Product Bulletin), Metco, Inc., Westbury, Long Island, New York, 1972a.

Anonymous, "Nickel Alloys Assure Reliability," Nickel Topics, **25**(2) [unpaginated reprint] (1972b).

Anonymous, "Substitute New Copper Alloy for Tin Brasses," Mater. Eng., **80**(6), 73 (1974a).

Anonymous, "Metal Tales: One of Plenty, One of Want. Inco's Page Says Supply Adequate," Am. Metal Market/Metalworking News, **81**(53), 4 (1974b).

Anonymous, "Expect Demand for Hi-Temp Cobalt to Remain Healthy," Am. Metal Market, **81**(129), 6A (1974c).

Anonymous, Metal Statistics 1974, The Purchasing Guide of the Metal Industries, 67th ed., American Metal Market, Fairchild Publications, Inc., New York, New York, 1974d.

Anonymous, "Names Tell the Story," Am. Metal Market, **83**(229), 12 (1976a).

Anonymous, "Superalloys Coming Down to Earth with Application in Critical Hot Spots," Am. Metal Market, **83**(229), 24-26 (1976b).

Anonymous, High Temperature, High Stength, Nickel Base Alloys, 3rd ed., The International Nickel Company, Inc., New York, New York, July 1977a, 52 pp.

Anonymous, "New Tribaloy T-700 Suited for Nuclear Plants: Du Pont," Am. Metal Market/Metalworking News, **85**(118), 27 (1977b).

Anonymous, "Superalloys. Directional Solidification of Mar-M-200," Current Awareness Bull. (Metals and Ceramics Information Center, Battelle, Columbus), No. 47, 3 (1978a).

Anonymous, "Cast Tool Steel Demand Termed Steady," Am. Metal Market, **86**(151), 22A (1978b).

Anonymous, "New Hard Facing Powder Uses Half Usual Cobalt," Am. Metal Market, **86**(156), 25 (1978c).

Anonymous, "Engine Makers Anxious about Cobalt," Am. Metal Market, **86**(214), 26 (1978d).

Anonymous (R.R.D.), "Symposium on Tin and Tin Alloy Coatings," Tin and Its Uses, No. 117, 13-15 (1978e).

Anonymous, "Metals. Metals Fabrication. Turbine Components," Current Awareness Bull. (Metals and Ceramics Information Center, Battelle, Columbus), No. 55, 4-5 (1979a).

Anonymous, "Electrodeposited Composite Coatings for Aircraft Engine Parts," Current Awareness Bull. (Metals and Ceramics Information Center, Battelle, Columbus), No. 56, 6 (1979b).

Anonymous, "Search for Substitutes Frustrates Alloy Users," Am. Metal Market, **87**(18), 13 (1979c).

Anonymous, "Sherritt Gordon 'S' Cobalt Powder Price Withdrawn," Am. Metal Market, **87**(29), 2 (1979d).

Anonymous, "U.S., Japan Firms Offer New Alloy," Am. Metal Market, **87**(46), 3 (1979e).

Anonymous, "Most Favored Nation Status May Pave Way for Chinese Tungsten," Am. Metal Market, **87**(48), 9 (1979f).

Anonymous, "Zaire Getting $20-Million Loan," Am. Metal Market, **87**(96), 24 (1979g).

Anonymous, "Coal: Closing Cost Gap with Oil," Chem. Week, **125**(11), 45-47 (1979h).

Beale, H. A., "New Uses Found for Vacuum Coating," Am. Metal Market, **86**(140), 14 (1978).

Berkowitz, J. B., G. R. Schimke, and V. R. Valeri, Water Pollution Potential of Manufactured Products. Catalog Section II - Product Listing, EPA-R2-73-179c, Office of Research and Monitoring, U.S. Environmental Protection Agency, U.S. Government Printing Office, Washington, D.C., 1973.

Berman, M., "Alloy Research Efforts Oriented to Meet Customer Specifications," Am. Metal Market, **87**(116), 20 (1979).

Bilbrey, J. H., Jr., Cobalt. A Materials Survey, Bureau of Mines Information Circular, PB 192 859, National Technical Information Service, U.S. Department of Commerce, Springfield, Virginia, 1962.

Blair, M., "High Temps Face Stiff Fight in Space Shuttle Applications," Am. Metal Market, 82(249), 14 (1975).

Blum, A., "Aircraft Buying in Holding Pattern for '77," Am. Metal Market, 82(229), 15 (1976).

Bradley, E. F., "Superalloys: Tough Enough to Take the Heat," Am. Metal Market, 86(214), 24-25, 33 (1978).

Burney, D. E., G. H. Weisemann, and N. Fragen, "A New Process for Oxidation of Aromatics," 5th World Petroleum Congress, New York, New York, 1959.

Butt, S. H., "Cobalt Makes Possible Better Copper Alloys," Am. Metal Market, 78, 8A (August 3, 1970).

Cabot Corporation, High-Performance Alloys for Chemical Process Equipment, Bulletin F-30,532, Stellite Division, Cabot Corporation, Kokomo, Indiana, [undated], 4 pp.

Cabot Corporation, Stellite® Hard-Facing Products, F-30,472B, Cabot Corporation, Kokomo, Indiana, 1975, 23 pp.

Cabot Corporation, Stellite Division Advertisement, Am. Metal Market, 83(229), 15 (1976).

Cabot Corporation, Haynes Wrought Wear-Resistant Alloys, F-30,542B, Stellite Division, Cabot Corporation, Kokomo, Indiana, 1977, 19 pp.

Cabot Corporation, Stellite Division - Alloys at a Glance, F-30,214D, Stellite Division, Cabot Corporation, Kokomo, Indiana, 1979.

Cere, P., Ed., Metal Statistics 1978, Purchasing Guide of the Metal Industries, 71st ed., American Metal Market, Fairchild Publications, New York, New York, 1978.

Charles River Associates, Economic Analysis of the Cobalt Industry, PB 195 821, National Technical Information Service, U.S. Department of Commerce, Springfield, Virginia, December 1969.

Chase, M., "Zaire Cobalt, Copper - Schwedweiler [sic] Optimistic on Output Boost," Am. Metal Market, **87**(85), 1, 24 (1979a).

Chase, M. M., Ed., Metal Statistics 1979, The Purchasing Guide of the Metal Industries, 72nd ed., American Metal Market, Fairchild Publications, a Division of Capital Cities Media, Inc., 1979b.

Chay, D. M., "Braunite-Glass Catalytic Particles and Method of Preparation," U.S. Patent 3,791,995, February 12, 1974; Chem. Abstr., **81**, 6658q (1974).

Chynoweth, A. G., "Electronic Materials: Functional Substitutions," Science, **191**, 725-732 (1976).

Clark, E. B., "Hard Metal," Paper presented at the Cobalt Crisis Conference, April 29 - May 1, 1979, Oak Brook, Illinois, Sponsored by Gorham International, Inc., Gorham, Maine.

Clendinning, R. A., J. E. Potts, and S. W. Cornell, "Environmentally Degradable-Biodegradable Blend of an Oxyalkanoyl Polymer and an Environmentally Degradable Polymer," U.S. Patent 3,867,324, February 18, 1975a.

Clendinning, R. A., J. E. Potts, and S. W. Cornell, "Environmentally Degradable Biodegradable Blends of a Dialkanoyl Polymer and an Environmentally Degradable Ethylene Polymer," U.S. Patent 3,901,838, August 26, 1975b.

Cobalt Monograph, Prepared in Collaboration with the Staff of Battelle Memorial Institute, Edited by Centre d'Information du Cobalt, Brussels, Belgium, 1960.

Corder, R. E., "The Strategic and Critical Materials Stockpile," Paper presented at the Cobalt Crisis Conference, April 29 - May 1, 1979, Oak Brook, Illinois, Sponsored by Gorham International, Inc., Gorham, Maine.

Corrick, J. D., "Cobalt" in Minerals Yearbook, 1970, Vol. I., Metals, Minerals, Fuels, Bureau of Mines, U.S. Department of the Interior, U.S. Government Printing Office, Washington, D.C., 1972, pp. 415-421.

Craik, D. J., "Cobalt-Platinum Permanent Magnets," Platinum Metals Rev., **16**(4), 129-137 (1972).

Cusumano, J. A., R. A. Dalla Betta, and R. B. Levy, Catalysis in Coal Conversion, Academic Press, New York, New York, 1978, 272 pp.

Danesi, W. P., and M. Semchyshen, "The Future in Superalloys" in Superalloys, S. C. Thomas, Ed., Wiley, New York, 1972, pp. 565-574.

Davidson, J. H., "High Temperature Materials Requirements of the Metallurgical Industries," Rev. Int. Hautes Temp. Refract., **13**(3), 163-171 (1976).

de Bie, E., and P. Doyen, "Cobalt Oxides and Salts," Cobalt, **11**(2), 3-13 (1962a).

de Bie, E., and P. Doyen, "Cobalt Oxides and Salts. Cobalt Catalysts," Cobalt, **11**(3), 3-15 (1962b).

DeBord, J. H., "Oxide Dispersion Adds Some Muscle to Alloy Products. Aerospace Market Tested by New High Nickel Types," Am. Metal Market, **83**(120), 25A, 27A (1976).

Decker, R. F., "Strengthening Mechanisms in Nickel-Base Superalloys," Paper presented at the Steel Strengthening Mechanisms Symposium, May 5-6, 1969, Zurich, Switzerland, Sponsored by Climax Molybdenum Company, Available from International Nickel Company, New York, New York, 1970, pp. 1-24.

DeHuff, G. L., "Cobalt" in Minerals Yearbook, 1968, Vol. I-II, Bureau of Mines, U.S. Department of the Interior, U.S. Government Printing Office, Washington, D.C., 1969, pp. 405-410.

DeHuff, G., "Cobalt" in Minerals Yearbook 1969, Vol. I-II, Bureau of Mines, U.S. Department of the Interior, U.S. Government Printing Office, Washington, D.C., 1971.

Desforges, C. D., "Metals and Alloys for High Temperature Applications. Current Status and Future Prospects," Rev. Int. Hautes Temp. Refract., **24**(1), 28-45 (1977).

Economos, G., "Ferrites" in Kirk-Othmer Encyclopedia of Chemical Technology, 2nd ed., Vol. 8, Electron Tube Materials to Ferrites, A. Standen, Executive Ed., Interscience Publishers, John Wiley and Sons, Inc., New York, New York, 1965, pp. 881-901.

Eisen, H., and J. A. Feighan, "Oven Cleaning Compositions Containing an Oxidation Catalyst," S. African Patent 68 05,139, February 9, 1970; Chem. Abstr., **73**, 67910x (1970).

Estrin, N. F., J. P. Ferguson, W. C. Coale, and L. C. Rhoads, Eds., CTFA Cosmetic Ingredient Dictionary, 2nd ed., The Cosmetic, Toiletry, and Fragrance Association, Inc., Washington, D.C., 1977.

Farren, P. C. (ASM Committee on Cutting Tools Chairman), et al., "The Selection of Material for Cutting Tools" in Metals Handbook, Vol. I, Properties and Selections of Metals, 8th ed., T. Lyman, H. E. Boyer, P. M. Unterweiser, J. E. Foster, J. P. Hontas, and H. Lawton, Eds., American Society for Metals, Novelty, Ohio, 1961, pp. 671-682.

Fawley, R. W., "Superalloy Progress" in Superalloys, C. T. Sims, Ed., Wiley, New York, 1972, pp. 3-29.

Fischman, L. L., and H. H. Landsberg, "Chapter 4. Adequacy of Nonfuel Minerals and Forest Resources" in Population, Resources and the Environment, Vol. III of U.S. Commission on Population Growth and the American Future Research Reports, R. G. Ridker, Ed., U.S. Government Printing Office, Washington, D.C., 1972, pp. 77-101.

Fisher, R. D., "Properties of Electroless Cobalt and Recording Technology" in Symposium on Electrodeposited Metals as Materials for Selected Applications, November 3-4, 1971, Metals and Ceramics Information Center, Battelle Columbus Laboratories, Columbus, Ohio, 1972, pp. 38-52.

Fowler, R. D., W. B. Burford III, J. M. Hamilton, Jr., R. G. Sweet, C. E. Weber, J. S. Kasper, and I. Litant, "Synthesis of Fluorocarbons," Ind. Eng. Chem., **39**, 292-298 (1947).

Fox, J., "Improved Processes Strengthen Alloys for Surgical Implants," Am. Metal Market, **86**(140), 19 (1978).

Friedman, G. I., and G. S. Ansell, "Powder Metallurgy" in Superalloys, S C. Thomas, Ed., Wiley, New York, 1972, pp. 427-450.

Fritzlen, G. A., and J. K. Elbaum, "Cobalt-Chromium-Tungsten-Molybdenum Wear-Resistant Alloys" in Metals Handbook, Vol. I, Properties and Selection of Metals, 8th ed., T. Lyman, H. E. Boyer, P. M. Unterweiser, J. E. Foster, J. P. Hontas, and H. Lawton, Eds., American Society for Metals, Novelty, Ohio, 1961, pp. 669-671.

Furst, A., "Cabot Trying to Stretch Cobalt and Moly Supplies," Am. Metal Market/Metalworking News, **87**(59), 31 (1979a).

Furst, A., "Pratt & Whitney to Reduce Cobalt Use 25%," Am. Metal Market/Metalworking News, **87**(89), 10 (1979b).

Furst, A., "Taking the Cobalt Out of Valves," Am. Metal Market, **87**(94), 25 (1979c).

Gangler, J. J., "New Materials in the Aerospace Industries," Monogr. Rep. Ser., Inst. Metals, London, **36** (Microstruct. Des. Alloys, Proc. Int. Conf. Strength Metal Alloys, 3rd, Vol.2), 237-269 (1973).

Gautreaux, M. F., W. T. Davis, and E. D. Travis, "Alcohols, Higher Aliphatic (Synthetic)" in Kirk-Othmer Encyclopedia of Chemical Technology, 3rd ed., M. Grayson, Ed., Vol. 1, a Wiley-Interscience Publication, John Wiley and Sons, New York, New York, 1978, pp. 747-754.

Gessinger, G. H., and M. J. Bomford, "Powder Metallurgy of Superalloys," Int. Metall. Rev., **19** (June), 51-76 (1974).

Gorham International, Inc., The Cobalt Communiqué, No. 1, 1-6 (1979a).

Gorham International, Inc., The Cobalt Communiqué, No. 2, 1-5 (1979b).

Gorham International, Inc., The Cobalt Communiqué, No. 3, 1-5 (1979c).

Gorham International, Inc., The Cobalt Communiqué, No. 4, 1-6 (1979d).

Grady, J. O., "An Alternative to Chromium Plating," Finish. Ind. (UK), **2**(7), 31 (1978).

Grant, N. J. (ASM Committee on Heat-Resistant Castings Chairman), et al., "Heat-Resistant Alloy Castings" in Metals Handbook, Vol. I, Properties and Selections of Metals, 8th ed., T. Lyman, H. E. Boyer, P. M. Unterweiser, J. E. Foster, J. P. Hontas, and H. Lawton, Eds., American Society for Metals, Novelty, Ohio, 1961, pp. 443-453.

Gutzeit, G., E. B. Saubestre, and D. R. Turner, "Nonelectrolytic Metal Coating Processes. C. Catalytic Chemical Methods" in Electroplating Engineering Handbook, 3rd ed., A. K. Graham, Ed., Van Nostrand, New York, New York, 1971, pp. 486-507.

Haflich, F., "Jet Engine Producer Boosts Nickel-Based Alloys, Ceramics," Am. Metal Market, **82**(249), 20 (1975).

Haflich, F., "AiResearch Studying Alloying Substitutes, Cobalt Market Turmoil Triggers Look at Nickel-Based Materials for Critical Jet Engine Parts," Am. Metal Market/Metalworking News, **86**(156), 12 (1978).

Haflich, F., "Analyst Predicts Increasing Producer Cobalt Prices," Am. Metal Market/Metalworking News, **87**(15), 2 (1979).

Hammond, R. A. F., "Nickel Plating from Sulfamate Solutions. V. Cobalt and Nickel-Cobalt Alloy Deposits from Sulfamate Baths," Metal Finish. J., **16**(189), 279-282, 284-285 (1970).

Harkins, T. R., "Cobalt Crisis--As Seen by a Superalloy Manufacturer," Paper presented at the Cobalt Crisis Conference, April 29 - May 1, 1979, Oak Brook, Illinois, Sponsored by Gorham International, Inc., Gorham, Maine.

Harris, W. J., Jr. (ASM Committee on Wrought Heat-Resisting Alloys Chairman), et al., "Wrought Heat-Resisting Alloys" in Metals Handbook, Vol. I, Properties and Selection of Metals, 8th ed., T. Lyman, H. E. Boyer, P. M. Unterweiser, J. E. Foster, J. P. Hontas, and H. Lawton, Eds., American Society for Metals, Novelty, Ohio, 1961, pp. 466-488.

Harshaw Chemical Company, Harshaw Catalysts for Industry, The Harshaw Chemical Company, Division of Kewanee Oil Company, Cleveland, Ohio, [undated, ca. 1971].

Hazlett, C., "Rare Earth, Cobalt Magnet Mart May Top $100-Million in 1980s," Am. Metal Market, **85**(191), 30 (1977).

Himel, C. M., "Preparation of Toluic Acids," U.S. Patent 2,696,499, December 7, 1954; Chem. Abstr., **49**, 3257d (1955).

Howard, H., "Electroslag Versus Vacuum Arc. Economy and Efficiency of Furnaces Weighed by Specialty Steel Producers in Selection," Am. Metal Market, **83**(144), 12 (1976a).

Howard, H., "New Year, Improved Market Projected by Superalloy Markets," Am. Metal Market, 83(229), 18-19 (1976b).

Howard, H., "Call for Steelmaking Capacity. The Boom in Aerospace Alloys and Concerns over Product Liability Have Placed New Demands on the Steel Industry for Vacuum Processed Materials," Am. Metal Market, 87(91), 12, 18, 19 (1979).

Hulsizer, W., "Gas Turbine Technology," Nickel Topics, 31(1), 3 (1978).

Hurwitz, E., E. Ciabettari, R. A. Wolff, and I. Bernstein, "Air Oxidation of Waste Sulfite Liquors," Ind. Eng. Chem., 51(10), 1301-1304 (1959).

International Nickel Company, Inc., Plating on Plastics. Extended Abstracts (1965-1971), The International Nickel Company, Inc., New York, New York, 1973.

Isleib, C. R., "New and Established Commercial Alloys for the Power Industry," Paper presented at the 16th Power Conference, August 15-17, 1978, Wrightsville Beach, North Carolina, Sponsored by (and available from) the International Nickel Company, Inc., New York.

Jackman, L. A., "Forming and Fabrication of Superalloys," Electrochem. Soc., 77-1 (Proc. Symp. Prop. High Temp. Alloys, 1976), 42-58 (1976).

Janke, P., "Political Stability for African Metals," Paper presented at the Cobalt Crisis Conference (Banquet Address), April 29 - May 1, 1979, Oak Brook, Illinois, Sponsored by Gorham International, Inc., Gorham, Maine.

Jordan, R. W., "Pressing Need for Nickel. Circuit Miniaturization Creates Revolution in the World of Electronics," Am. Metal Market, 83(120), 24A, 27A (1976).

Kahaner, L., "Sintering's Popularity Gains in Cemented Carbide Production," Am. Metal Market, 86(140), 16 (1978a).

Kahaner, L., "Clad System Can Replace Alloys for Lead Frames," Am. Metal Market/Metalworking News, **86**(166), 43 (1978b).

Koflowitz, K., "Cobalt Export Curbs, No Price Help: Reddy," Am. Metal Market, **87**(86), 7 (1979a).

Koflowitz, L., "Friedman: Cobalt Substitution Trend Will Continue," Am. Metal Market/Metalworking News, **87**(89), 31 (1979b).

Koflowitz, L., "Cobalt Still a Necessity in Value Making," Am. Metal Market/Metalworking News, **87**(99), 25 (1979c).

Koflowitz, L., "Cobalt Profile" in Metal Statistics 1979, Purchasing Guide of the Metal Industries, 72nd ed., American Metal Market, Fairchild Publications, New York, New York, 1979d, p. 49.

Kölbel, H., "Verwertungsmöglichkeiten kohlenoxydhaltiger Abgase zur Synthese von Kohlenwasserstoffen" ["Possibilities for Utilization of Exhaust Gases Containing Carbon and Oxygen for Synthesis of Hydrocarbons"], Chem. Ing. Techn. (Weinheim), **29**(8), 505-511 (1957).

Krohn, A., and C. W. Bohn, "Electrodeposition of Alloys: Present State of the Art," Electrodeposition and Surface Treatment, **1**(3), 199-211 (1973).

Lardner, E., "Current Developments in Cemented Carbide Cutting Tools," New Tool Mater. Cutting Tech., Int. Conf., **1**, 1-19 (1975).

Lardner, E., "Alloy Systems and Processes in Cemented Carbide Production," Metals Technol. (London), **3**(Pt. 5-6), 237-245 (1976).

Larsen, R., "Research Points to Growth," Am. Metal Market, **82**(229), 10 (1976).

Larsen, R., "As Temperatures Rise Eastern Engine Builders Counting on Directionally Solidified Castings to Bridge Performance Gap between Now and the Late 1980's," Am. Metal Market/Metalworking News, 86(54), 4A (1978).

Larsen-Basse, J., "Wear of Hard-Metals in Rock Drilling: a Survey of the Literature," Powder Metall., 16(31), 1-32 (1973).

Lee, W. S., "Porcelain Enamels for Self-Cleaning Cooking Ovens," U.S. Patent 3,547,098, December 15, 1970; Chem. Abstr., 75, 52361m (1970).

Lepkowski, W., "Politics and the World's Raw Materials," Chem. Eng. News, 57(23), 14-19 (1979).

Locke, J. J., "Cobalt Alloy Overlays in a Petro-Chemical Refinery," Cobalt (Engl. Ed.), No. 2, 25-31 (1974).

Loewe, W., "Bell Labs New Magnetic Alloys Represent 12.5% Cobalt Savings," Am. Metal Market/Metalworking News, 87(142), 57 (1979).

Lowe, S. C., "Cobalt in Zambia," Paper presented at the Cobalt Crisis Conference, April 29 - May 1, 1979, Oak Brook, Illinois, Sponsored by Gorham International, Inc., Gorham, Maine.

Lowenheim, F. A., and M. K. Moran, Faith, Keyes, and Clark's Industrial Chemicals, 4th ed., Wiley-Interscience, John Wiley and Sons, New York, New York, 1975.

Lupi, R. D., "Superalloy Jet Parts Cut Waste, Costs," Am. Metal Market, 86(80), 18 (1978).

Maki, N., and N. Tanaka, "Cobalt (Electrochemistry)," Encycl. Electrochem. Elem., 3, 43-210 (1975).

Mari, A., "Nickel Alloys Step Toward Tomorrow," Am. Metal Market/Metalworking News, 86(54), 10A (1978a).

Mari, A., "Users of Superalloys Scouting Substitutes for Cobalt," Am. Metal Market, 86(104), 25 (1978b).

Mari, A., "Belgians Choose N.C. Site for Cobalt Powder Plant," Am. Metal Market, **86**(183), 1, 7 (1978c).

Mari, A., "Cobalt in Crisis," Am. Metal Market, **86**(214), 14, 16, 32 (1978d).

Mari, A., "Panel Says Cobalt Crunch Spurring Substitution Moves," Am. Metal Market, **86**(228), 29 (1978e).

Mari, A., "Analysts Slate World Cobalt Market Study," Am. Metal Market, **87**(2), 9 (1979a).

Mari, A., "Consumption Pace Expected to Hold Up," Am. Metal Market, **87**(3), 9-11 (1979b).

Mari, A., "S.O.S.! The Squeeze on Supplies--A Signal Heard 'Round the World'--Shows Little Indication of Abating, as Competition for Cobalt Intensifies and Output Increases Remain in Doubt," Am. Metal Market, **87**(18), 10 (1979c).

Mari, A., "Events in Zaire Seen Key to Cobalt Prices," Am. Metal Market, **87**(71), 10 (1979d).

Mari, A., "Material Problems Squeezing Alloy Producers: Snapp," Am. Metal Market, **87**(84), 39 (1979e).

Mari, A., "P&WA Doing Extensive R&D on Cobalt-Free Engine Alloys," Am. Metal Market, **87**(88), 7 (1979).

Matho Ngoma, "Zairian Cobalt Production and Supply," Paper presented at the Cobalt Crisis Conference, April 29 - May 1, 1979, Oak Brook, Illinois, Sponsored by Gorham International, Inc., Gorham, Maine.

Matthey Bishop, Inc., Advertisement, Chem. Eng. News, **56**(45), 41 (1978).

McDermott, F. T., "Dust in the Cemented Carbide Industry," J. Am. Ind. Hyg. Assoc., **32**, 188-193 (1971).

McEvoy, J. E., Ed., Catalysts for the Control of Automotive Pollutants, Advances in Chemistry Series No. 143, American Chemical Society, Washington, D.C., 1975.

Meetham, G. W., "The Contribution of Materials to the Development of the Gas Turbine Engine," Metall. Mater. Technol., **8**(11), 589-593, 602 (1976).

Meinstein, S., and H. R. Spielman, "Chapter 22. Metallic Soaps: Driers, Suspending Agents, Flow Modifiers, Flatting Agents and Sanding Aids" in Technology of Paints, Varnishes and Lacquers, C. R. Martens, Ed., Reinhold Book Corporation, a Subsidiary of Chapman-Reinhold, Inc., New York, 1968, pp. 390-421.

Michels, H. T., and E. C. Hoxie, "Some Insights into Corrosion in SO_2 Exhaust Scrubbers," presented at the ASM Conference on Materials Reliability Problems in Fossil Fired Power Plants," digested as "Corrosion in SO_2 Exhaust Scrubbers--Nickel-Containing Alloys to the Rescue," Nickel Topics, **31**(2), 7-10 (1978).

Mills, G. A., and J. A. Cusumano, "Catalysis" in Kirk-Othmer Encyclopedia of Chemical Technology, 3rd ed., M. Grayson, Ed., Vol. 5, Castor Oil to Chlorosulfuric Acid, a Wiley-Interscience Publication, John Wiley and Sons, New York, New York, 1979, pp. 16-61.

Modic, F. A., "Cobalt in Hardfacing Products," Paper presented at the Cobalt Crisis Conference, April 29 - May 1, 1979, Oak Brook, Illinois, Sponsored by Gorham International, Inc., Gorham, Maine.

Moffat, P. K., "Cobalt in Control Alloys," Paper presented at the Cobalt Crisis Conference, April 29 - May 1, 1979, Oak Brook, Illinois, Sponsored by Gorham International, Inc., Gorham, Maine.

Molloy, W. J., and D. R. Green, "Development, Production and Application," Metall. Metal Form., **43**(7), 215, 218-221 (1976).

Morral, F. R., "Cobalt Alloys in Process Metallurgy Operations," J. Metals, **18**(10), 1115-1118 (1966).

Morral, F. R., "Cobalt-Base Alloys in Aerospace," Aircr. Struct. Mater. Appl., Natl. SAMPE (Soc. Aerosp. Mater. Process Eng.) Tech. Conf., 111-121 (1969).

Morral, F. R., "Cobalt Compounds" in Kirk-Othmer Encyclopedia of Chemical Technology, 3rd ed., Vol. 6, Chocolate and Cocoa to Copper, M. Grayson, Executive Ed., a Wiley-Interscience Publication, John Wiley and Sons, Inc., New York, New York, 1979, pp. 495-510.

Morral, F. R., and W. H. Safranek, "Cobalt and Cobalt Alloys" in Modern Electroplating, 3rd ed., F. A. Lowenheim, Ed., Wiley, New York, 1974, pp. 152-164.

Morrow, H., III, "The Use of Molybdenum in Superalloys," Mosaic (Greenwich, Connecticut), 1(4), 3-7 (1976).

Mustero, R. J., Ed., Metal Statistics 1972, The Purchasing Guide of the Metal Industries, 65th ed., American Metal Market, Fairchild Publications, Inc., New York, New York, 1972.

Nafziger, R. H., and R. L. Lincoln, "Electroslag Remelting of Superalloys," U.S. Bur. Mines, Bull., 669, 154-178 (1976).

National Materials Advisory Board, Trends in the Use of Ferroalloys by the Steel Industry of the United States, Publication NMAB-276, National Academy of Sciences--National Academy of Engineering, Washington, D.C., 1971.

Natta, G., U. Colombo, and I. Pasquon, "Direct Catalytic Synthesis of Higher Alcohols from Carbon Monoxide and Hydrogen" in Catalysis, Vol. 5, Reinhold Publishing Corporation, New York, New York, 1957, pp. 131-174.

Nesbitt, E. A., "Magnetic Materials" in Kirk-Othmer Encyclopedia of Chemical Technology, Vol. 12, Iron to Manganese, 2nd ed., A. Standen, Executive Ed., Interscience, a Division of John Wiley and Sons, New York, 1967, pp. 737-772.

Nordberg, D., "Zaire Pricing Hits Zambia Advance Cobalt Agreements," Am. Metal Market, 87(11), 1, 8 (1979a).

Nordberg, D., "European Efforts Underway to Guard Against Shortfall," Am. Metal Market, 87(18), 11 (1979b).

Nordberg, D., "U.S., IMF Concerned over Zaire, Report Cobalt Sale at Free Mart Prices," Am. Metals Market, 87(47), 1, 7 (1979c).

O'Sullivan, D., "Oxygen Generated from Water Photochemically," Chem. Eng. News, 57(30), 24-25 (1979).

Ott, R. E., "Cooking Device Coated with a Vesicular Material Having Porous, Sorptive Surface Structure," U.S. Patent 3,742,930, July 3, 1973; Chem. Abstr., 79, 117988a (1973).

Pearlstein, F., "Electroless Deposition of Metals--Principles and Applications," Conf., Natl. Assoc. Corros. Eng., Proc., 24th 1968, 499-507 (1969).

Peters, E. F., and B. L. Evering, "Vapor-Phase Ethylene Polymerization," U.S. Patent 2,658,059, November 3, 1953; Chem. Abstr., 48, 3022b (1954).

Planinsek, F., and J. B. Newkirk, "Cobalt and Cobalt Alloys" in Kirk-Othmer Encyclopedia of Chemical Technology, 3rd ed., Vol. 6, Chocolate and Cocoa to Copper, M. Grayson, Executive Ed., a Wiley-Interscience Publication, John Wiley and Sons, Inc., New York, New York, 1979, pp. 481-494.

Potts, J. E., S. W. Cornell, and A. M. Sracic, "Biodegradable Polyethylene Materials," Ger. Offen. 2,247,170, April 1973.

Pressure Chemical Company, Materials and Services for Research, Pressure Chemical Company, Pittsburgh, Pennsylvania, 1972.

Prizinsky, D., "Upbeat Demand Laid to Aircraft Boom," Am. Metal Market, 86(214), 18 (1978).

Prizinsky, D., "Tool Steel Makers Foresee Ample '79 Supply," Am. Metal Market, 87(3), 12 (1979).

Reno, H. T., "Cobalt" in Minerals Yearbook, 1967, Vol. I-II, Bureau of Mines, U.S. Department of the Interior, U.S. Government Printing Office, Washington, D.C., 1969, pp. 381-385.

Rhys, D. W., and W. Betteridge, "Brazing for Elevated Temperature Service," Metal Ind., **101**(2), 2-4, 27-30, 45-46 (1962).

Roditi, E., Ed., Metal Statistics 1976, The Purchasing Guide of the Metal Industries, 69th ed., American Metal Market, Fairchild Publications, Inc., New York, New York, 1976.

Rousseau, P. E., "The Production of Gas, Synthetic Oil and Chemicals from Low-Grade Coal in South Africa" in 15th Sectional Meeting of the World Power Conference, Vol. II, Tokyo, Japan, October 1966, Japanese National Committee, World Power Conference, Tokyo, Japan, 1966, pp. 912-924.

Ruder, W. E. (ASM Committee on Permanent Magnet Materials Chairman), et al., "Permanent Magnet Materials" in Metals Handbook, Vol. I, Properties and Selections of Metals, 8th ed., T. Lyman, H. E. Boyer, P. M. Unterweiser, J. E. Foster, J. P. Hontas, and H. Lawton, Eds., American Society for Metals, Novelty, Ohio, 1961a, pp. 779-785.

Ruder, W. E. (ASM Committee on Magnetically Soft Materials Chairman), et al., "Magnetically Soft Materials" in Metals Handbook, Vol. I, Properties and Selection of Metals, 8th ed., T. Lyman, H. E. Boyer, P. M. Unterweiser, J. E. Foster, J. P. Hontas, and H. Lawton, Eds., American Society for Metals, Novelty, Ohio, 1961b, pp. 785-797.

Saville, S., "Tool Makers Look to Alternate Materials," Am. Metal Market, **86**(205), 8 (1978).

Saville, S., "Tight Cobalt, Moly Supply Upsets Timetable: Boeltger," Am. Metal Market, **87**(83), 1, 9 (1979a).

Saville, S., "Success in Hunt for Cobalt Substitutes Could Turn Current Crisis into Surplus," Am. Metal Market, **87**(88), 1, 10 (1979b).

Saville, S., "Implant Market Shows Vitality," Am. Metal Market, 87(152), 8A, 11A (1979c).

Scheidweiler, P., Remarks on African and European Cobalt Production, presented at the Cobalt Crisis Conference, April 29 - May 1, 1979, Oak Brook, Illinois, Sponsored by Gorham International, Inc., Gorham, Maine.

Schlatter, R., "Vacuum Induction Melting Technology of High-Temperature Alloys," Elec. Furn. Conf., Proc., 29, 2-11 (1971).

Shaver, R. G., L. C. Parker, E. F. Rissman, K. M. Slimak, and R. C. Smith, Assessment of Industrial Hazardous Waste Practices, Inorganic Chemicals Industry, PB 244 832, EPA-SW-104C, National Technical Information Service, U.S. Department of Commerce, Springfield, Virginia, March 1975.

Shelef, M., and H. S. Gandhi, "Reduction of Nitric Oxide in Automobile Emissions. Stabilization of Catalysts Containing Ruthenium," Platinum Metals Rev., 18(1), 2-14 (1974).

Sibley, S., Cobalt-1977, Mineral Commodity Profiles, MCP-5, Publications Distribution Branch, Bureau of Mines, U.S. Department of the Interior, Pittsburgh, Pennsylvania, July 1977.

Siegel, M., Development Engineer, Detroit Diesel Allison, Division of General Motors Corporation, Indianapolis, Indiana, personal communication (Carson), May 1979.

Simon, H., C. R. Anderson, and A. J. Wilson, "Chapter 7. Mechanical Equipment. Ceramic Spraying and Metal Deposition Equipment" in Air Pollution Engineering Manual, 2nd ed., J. A. Danielson, Ed., U.S. Environmental Protection Agency, Office of Water and Air Programs, Office of Air Quality Planning and Standards, Research Triangle Park, North Carolina, May 1973, pp. 421-433.

Sims, C. T., "A Contemporary View of Cobalt-Base Alloys," J. Metals, 21(12), 27-42 (1969).

Sims, C. T., "Cobalt-Base Alloys" in Superalloys, C. T. Sims, Ed., Wiley, New York, 1972, pp. 145-174.

Spriggs, G. E., and D. J. Bettle, "Properties and End Uses of Cemented Carbides," Powder Metall., **18**(35), 53-70 (1975).

Stern, L., "Nickel-Based Hardfacing Alloys Can Reduce Costs," Am. Metal Market, **86**(205), 8 (1978).

Stewart, W. J., "Driers" in Treatise on Coatings, Vol. 1, Film-Forming Compositions, Part 1, Marcel Dekker, Inc., New York, New York, 1967, pp. 211-223.

Stewart, W. J., Paint Driers and Additives, Federation Series on Coatings Technology, Unit Eleven, W. H. Madson, Ed., Federation of Societies for Paint Technology, Philadelphia, Pennsylvania, 1969.

Strauss, S. D., "Stockpile Strategies Shifting," Am. Metal Market, **87**(18), 15 (1979).

Strnat, K. J., "The Recent Development of Permanent Magnet Materials Containing Rare Earth Metals," IEEE Trans. Magn., **6**(2), 182-190 (1970).

Strnat, K. J., K. M. D. Wong, and H. Blaettner, "Bonded Rare Earth-Cobalt Permanent Magnets," Proc. Rare Earth Res. Conf., 12th, **1**, 31-40 (1976).

Stundza, T., "AIME Told Interior Dept. Designs Disruption Index," Am. Metal Market, **87**(37), 1, 20 (1979).

Superior Tube Company, "Alloy 188 (Seamless and WELDRAWN®)," Data Memorandum No. 51, Superior Tube Company, Norristown, Pennsylvania, November 1978, 4 pp.

Teplitz, B., "Recycled Tungsten Finds Ready Market," Am. Metal Market, **87**(3), 15, 16 (1979).

Teter, J. W., and L. E. Olson, "Cobalt-Magnesia-Kieselguhr Catalyst for Nitrile Production," U.S. Patent 2,658,041, November 3, 1953; Chem. Abstr., **48**, 2291c (1954).

Thornton, J., "Cobalt Demand at Record Height. Long-Term Outlook: Great Consumption," Am. Metal Market, 81(148), 1, 9 (1974).

Thornton, J., "High-Temp Alloys Gain," Am. Metal Market, 85(119), 18A (1977).

Thornton, J., "Air Force Contracts Provide Debut for Two New Alloys," Am. Metal Market, 86(214), 29-30 (1978).

Thornton, J., "Coatings Cover Market Territory. Use of Vacuum Processes in Coating Aerospace Parts Is Growing, and High Volume Metallizing of Auto Parts Is Catching On," Am. Metal Market, 87(91), 14, 15 (1979).

Tillack, D. J., "Recent Advances in Nickel Alloys for Aerospace Applications," Natl. SAMPE Tech. Conf., 8 (Bicenten. Mater.), 169-181 (1976).

Tredwell, D., "New Alloy May Greatly Reduce Production Cost of Magnets Used in Telephone Receivers," Am. Metal Market/ Metalworking News, 86(25), 15 (1978).

Tuininga, E. J., "Wear and Oxidation Phenomena on the End Cutting Edge of Carbide Turning Tools," CIRP, Ann. Int. Inst. Prod. Eng. Res., 14(4), 449-463 (1967).

Turk, P., "Cobalt Reduction Efforts Underway at Aerospace Firms," Am. Metal Market, 87(152), 10A, 18A (1979).

Vaughn, V. A., "Challenge of 'Other Markets'," Am. Metal Market, 86(214), 27-28 (1978).

Vicini, J., "Crucible Stainless Steel Says Particle Metallurgy Products May Triple in Sales in 5 Years," Am. Metal Market/Metalworking News, 83(160), 30 (1976).

Voorhoeve, R. J., J. P. Remeika, P. E. Freeland, and B. T. Matthias, "Rare-Earth Oxides of Manganese and Cobalt Rival Platinum for the Treatment of Carbon Monoxide in Auto Exhaust," Science, 177(4046), 353-354 (1972).

Walker, R., and B. Cruise, "The Use and Production of Electrodeposited Cobalt," Metal Finish., 76(5), 25-29 (1978a).

Walker, R., and B. Cruise, "The Use and Production of Electrodeposited Cobalt," Metal Finish., 76(6), 45-48 (1978b).

Wall, F. E., "Chapter 21. Bleaches, Hair Colorings, and Dye Removers" in Cosmetics: Science and Technology, E. Sagarin, Ed., Interscience Publishers, Inc., New York, New York, 1957, pp. 479-530.

Warshaw, A., J. S. Negra, and J. F. Tourtellotte, "Catalyst for Treating Combustion Exhaust Gas," U.S. Patent 3,956,189, May 11, 1976; Chem. Abstr., 85, 148338s (1976).

Wasielewski, G. E., and R. A. Rapp, "High-Temperature Oxidation" in Superalloys, C. T. Sims, Ed., Wiley, New York, New York, 1972, pp. 287-316.

Weihrauch, P. F., A. E. Paladino, D. K. Das, W. R. Reid, L. Lesensky, E. C. Wettstein, and A. A. Gale, "Manufacture of Rare Earth-Cobalt Magnets," Am. Inst. Phys (AIP) Conf. Proc., 1972, No. 10 (Pt. 1), 638-642 (1973).

Weintraub, P., "New Material to Cut Weight of Jet Engines," Am. Metal Market/Metalworking News, 87(44), 12, 18 (1979).

Wessling, J., "Superalloy Development: No Overnight Sensation," Am. Metal Market, 82(249), 19 (1975).

Wesson, S., "Gov't. Plans Export Reporting System for Items Bearing at Least 10% Cobalt," Am. Metal Market, 87(3), 1, 4 (1978).

Wholey, L. C., "Magnets," Paper presented at the Cobalt Crisis Conference, April 29 - May 1, 1979, Oak Brook, Illinois, Sponsored by Gorham International, Inc., Gorham, Maine.

Wile, G. J., "Materials and Processes for Improved Gas Turbines," Moly Pays Off, 2(1), 1-3 (1969).

Worthy, W., "Synfuels: Uncertain and Costly Fuel Option," Chem. Eng. News, **57**(35), 20-28 (1979).

Wrigley, A., "Auto Town's Focus Is Still Mainly on Smaller, Lighter, More Maintenance-Free Batteries," Am. Metal Market, **87**(55), 20 (1979a).

Wrigley, A., "Detroit Feels Cobalt, Moly Supply Pinch," Am. Metal Market, **87**(84), 23 (1979b).

III. CHEMISTRY

Glenn M. Trischan,* Ralph R. Wilkinson,** and Bonnie L. Carson***

This chapter summarizes the inorganic, organic, and analytical chemistry of cobalt appropriate to the objectives of this study. Several topical reviews are available. Geochemistry is discussed in Chapter IV.

A. Physical Properties and Descriptive Inorganic Chemistry

Reviews of cobalt inorganic and organic chemistry include *Cobalt* by R. S. Young (1948); *Cobalt: Its Chemistry, Metallurgy, and Uses* by R. S. Young (1960); *Cobalt Monograph* (1960) edited by Centre d'Information du Cobalt; and *The Chemistry of Iron, Cobalt and Nickel* by D. Nicholls (1973). Other excellent source materials include "Cobalt Oxides and Salts" by E. de Bie and P. Doyen (1962a, 1962b); *Pure Cobalt and Its Properties* by H. Winterhager and J. Krüger (1965 and 1966); *Advanced Inorganic Chemistry* by F. A. Cotton and G. Wilkinson (1972); *Cobalt* by Vu Quang Kinh (1972); *Cobalt* by P. R. Mitchell (1976); "Cobalt and Cobalt Alloys" by F. Planinsek and J. B. Newkirk (1979); and *Cobalt Compounds* by F. R. Morral (1979). Young (1960) also gives a good review of the radiochemistry of cobalt.

The electronic configuration of cobalt is [Ar] $3d^7\ 4s^2$, placing it in Group VIII of the periodic classification of elements with iron, nickel, and the platinum-group metals. Cobalt heads a vertical subgroup followed by rhodium and iridium characterized by the same number of electrons in the outermost shells. (We note in passing that the chemistries of iron, cobalt, and nickel are quite similar; and for this reason, these elements are placed in Group VIII as a triad and often discussed together.)

Uncomplexed cobalt(II), a $3d^7$ ion, is stable in aqueous solution, whereas under similar conditions, cobalt(III), a $3d^6$ ion, is a powerful oxidizing agent unless it is present in an anhydrous crystal lattice or is complexed with various ligands including ammonia, ethylenediamine, cyanide, dimethylglyoxime, etc.

---

* Section E.

** Sections A, C, and D.

*** Section B.

The chemistry of cobalt(III) complexes is diverse and the cobalamins (vitamin B_{12} group) are of importance in biological systems and are discussed in a later subsection of this chapter.

Cobalt is a blue-white hard metal which can be machined although it tends to be brittle. (If cobalt contains $\sim$ 1% carbon, the alloy is more readily machinable and can be swaged or rolled when hot.) Cobalt exists in two allotropic forms, hexagonal close-packed (below 417°C) and face-centered-cubic; and metal-working (deformation) processes, even at room temperature, may cause allotropic transformations to occur.

Cobalt melts at 1493°C, boils at 3100°C, and has a density of 8.90 g/cm^3. Numerous physical and mechanical properties are tabulated in the Cobalt Monograph (1960), Young (1960), Winterhager and Krüger (1965 and 1966), Vu Quang Kinh (1972), Nicholls (1975), and Planinsek and Newkirk (1979).

Cobalt alloys with a number of elements as binary, ternary, or quaternary systems. Commercially important alloys have been developed for their superior permanent magnetic qualities, extreme wear-resistance, high-temperature/high-stress properties, or corrosion resistance. Cobalt alloys have been discussed earlier in Chapter II.

In the bulk state at ambient temperatures up to 300°C, cobalt is essentially unaffected by air, oxygen, water vapor, etc. Cobalt is oxidized by steam, air, or oxygen forming layers of CoO and Co_3O_4 depending on the particular conditions. (The latter oxide is probably a spinel type consisting of $CoO{\cdot}Co_2O_3$ in which Co(II) ions occupy tetrahedral and Co(III) ions occupy octahedral interstitial lattice positions.) Reduction of cobalt oxides with hydrogen yields finely divided cobalt metal, which is pyrophoric in air.

The metal reacts with a number of elements and compounds including the halogens, boron, oxygen, phosphorus, sulfur, arsenic, antimony, ammonia, carbon monoxide, nitric oxide, etc.

Cobalt is electrochemically more active than nickel but less so than iron based on reversible electrode potential measurements:*

$$Fe^{2+} + 2e = Fe \qquad E° = -0.441\ V$$

$$Co^{2+} + 2e = Co \qquad E° = -0.277\ V$$

$$Ni^{2+} + 2e = Ni \qquad E° = -0.250\ V$$

Cobalt is attacked by dilute hydrochloric or sulfuric acid with release of hydrogen gas. Concentrated nitric acid also attacks cobalt; but if the reaction is carried out near -10°C, a protective oxide barrier is formed and the metal is said to be "passive" under this condition. Cobalt is not attacked by dilute alkaline solutions.

The cobaltic/cobaltous electrochemical couple favors the lower valence state. Thus:

$$[Co(H_2O)_6]^{3+} + e = [Co(H_2O)_6]^{2+} \qquad E° = 1.84\ V$$

In the presence of complexing agents, Co(III) is stabilized, which accounts for the great preponderance of complex ions containing trivalent cobalt. Thus:

$$[Co(NH_3)_6]^{3+} + e = [Co(NH_3)_6]^{2+} \qquad E° = 0.1\ V$$

Following are descriptions of some important inorganic cobalt compounds and complexes. Tables III-1 and III-2 summarize information regarding the physical properties and equilibrium constants of selected cobalt compounds and ionic species.

Cobaltous acetate is prepared by treating the carbonate with acetic acid followed by crystallization. The acetate is used

---

* Standard electrode potentials taken at an activity of one molal at 24°C. The sign of the electrode potential is the same as the electrostatic charge on the elemental electrode relative to the standard hydrogen electrode.

TABLE III-1

PHYSICAL PROPERTIES OF SELECTED COBALT COMPOUNDS[a/]

Compound	Formula	Melting Point	Boiling Point	Color, Crystalline Form	Solubility g/100 ml Water and Other Solvents
Cobalt acetate tetrahydrate	$Co(C_2H_3O_2)_2 \cdot 4H_2O$	Dehydrates 140°C	-	Red-violet, monoclinic	Soluble in hot and cold water, alcohol
Cobalt carbonyl	$Co_2(CO)_8$	51°C	Dec. > 52°C	Orange crystals	Insoluble in hot and cold water; slightly soluble in alcohol, CS_2
	$Co_4(CO)_2$	-	Dec. > 60°C	Black crystals	Slightly soluble in hot water, acids, benzene
Cobalt carbonate	$CoCO_3$	Decomposes (Dec.)	-	Red, trigonal	Insoluble in hot and cold water, NH_3; soluble in acids
Cobalt fluoride	CoF_2	~ 1200°C	-	Pink, monoclinic	Soluble in water, acids
	CoF_3	Dec. to $Co(OH)_3$	-	Brown, hexagonal	Insoluble in alcohol, ether, benzene
Cobalt chloride hexahydrate	$CoCl_2 \cdot 6H_2O$	86°C	-	Red, monoclinic	499 g/liter (20°C); very soluble in alcohol
Cobalt bromide hexahydrate	$CoBr_2 \cdot 6H_2O$	678°C (in N_2)	-	Green, hexahydrate is deliquescent	66.7 (50°C); 68.1 (97°C); 77.1 alcohol
Cobalt hydroxide	$Co(OH)_2$	Dec. 168°C to form CoO	-	Blue-green or rose-red powder	Very slightly soluble in water; soluble in acids
	$Co(OH)_3$	Dec. 150°C to form Co_3O_4	-	Dark brown or black powder	Insoluble in water; soluble in HCl
Cobalt nitrate hexahydrate	$Co(NO_3)_2 \cdot 6H_2O$	55-56°C	133.8°C	Red, monoclinic	Soluble in alcohol; slightly soluble in water
Cobalt oxide	CoO	Dec. 1800°C	-	Dark gray	Insoluble hot or cold H_2O, NH_4OH, alcohol
Cobalto-cobaltic oxide	Co_3O_4	Dec. 900-950°C to form CoO	-	Black	Insoluble hot or cold H_2O, HNO_4 aqua regia; soluble H_2SO_4
Cobaltic oxide	Co_2O_3	Dec. 895°C to form CoO	-	Black	Insoluble hot or cold H_2O, alcohol
Cobalt phosphate octahydrate	$Co_3(PO_4)_2 \cdot 8H_2O$	Dehydrates 200°C	-	Red powder	Slightly soluble in water; soluble in acid
Cobalt sulfate heptahydrate	$CoSO_4 \cdot 7H_2O$	96.8°C	-	Red-pink monoclinic	Soluble in water, alcohol
Cobalt silicide	$CoSi_2$	1277°C	-	-	Insoluble in water, H_2SO_4; soluble in hot HCl
Cobalt sulfide	CoS	> 1100°C	-	Red octahedral	Insoluble in water; slightly soluble in acid

a/ Sources: Stokinger (1963), Wender et al. (1957b), Merck Index, 9th ed. (Windholz et al., 1976), and Morral (1979).

TABLE III-2

SOLUBILITY OR STABILITY CONSTANTS OF SELECTED COBALT COMPOUNDS AND SPECIES[a/]

Compound or Species	Log Solubility or Stability Constant at 25°C	Chemical Equilibria
$Co(OH)_2(s)$	Ksp = -15.7 (pink, aged) -14.8 (pink) -14.2 (blue)	$Co^{2+} + 2H_2O \rightleftharpoons Co(OH)_2 + 2H^+$
$Co(OH)^+$	-9.65	$Co^{2+} + H_2O \rightleftharpoons Co(OH)^+ + H^+$
$Co(OH)_3^-$	-31.5	$Co^{2+} + 3H_2O \rightleftharpoons Co(OH)_3^- + 3H^+$
$CoCl^+$	-2.4	$Co^{2+} + Cl^- \rightleftharpoons CoCl^-$
$CoSO_4$	2.47	$Co^{2+} + SO_4^{2-} \rightleftharpoons CoSO_4$
$Co(NH_3)_6^{3+}$	β_6 = 35.2	$Co^{3+} + 6NH_3 \rightleftharpoons Co(NH_3)_6^{3+}$
$[Co(NH_3)_5H_2O]^{3+}$	β_5 = 30.8	$Co^{3+} + 5NH_3 \rightleftharpoons Co(NH_3)_5^{3+}$
$Co(CN)_6^{3-}$	β_6 = 19 (in 5 M $CaCl_2$)	$Co^{3+} + 6CN^- \rightleftharpoons Co(CN)_6^{3-}$
$CoSCN^+$	1.72	$Co^{2+} + SCN^- \rightleftharpoons CoSCN^+$
$CoCO_3(s)$	Ksp = -12.8	$Co^{2+} + CO_3^{2-} \rightleftharpoons CoCO_3$
$Co_3(PO_4)_2(s)$	Ksp = -34.7	$3Co^{2+} + 2PO_4^{3-} \rightleftharpoons Co_3(PO_4)_2$
$CoNTA^-$	10.6	$Co^{2+} + NTA^{3-} \rightleftharpoons CoNTA^-$
$CoEDTA^{2-}$	16.2	$Co^{2+} + EDTA^{4-} \rightleftharpoons CoEDTA^{2-}$
$CoEDTA^-$	36	$Co^{3+} + EDTA^{4-} \rightleftharpoons CoEDTA^-$
Co acetyl-acetonate	5.40 4.14	$Co^{2+} + acac^- \rightleftharpoons Co\ acac^+$ $Co\ acac^+ + acac^- \rightleftharpoons Co(acac)_2$

Sources: Sillén and Martell (1964)
Mesmer and Baes (1975)
Lange's Handbook of Chemistry (Dean, 1973)
McCrary and Howard (1979)

a/ 25°C and zero ionic strength

as a bleaching agent and drier for varnishes and lacquers, in aluminum anodizing electrolytic solutions, and as an animal dietary supplement. When a normally pink solution of cobaltous acetate is heated, it turns blue, which forms the basis of some sympathetic (invisible) inks. Cobaltic acetate is obtained either by electrolytic oxidation or ozonation of cobaltous acetate in solution.

Cobalt forms three carbides, Co_3C, Co_2C, and CoC_2. Carbides are prepared either by equilibrating cobalt and carbon at elevated temperatures (1000 to 1300°C) or by passing carbon monoxide over finely divided cobalt near 225°C.

At least three cobalt carbonyls are known: (a) dicobalt octacarbonyl, $Co_2(CO)_8$; (b) tetracobalt dodecacarbonyl, $Co_4(CO)_{12}$; and (c) hexacobalt hexadecacarbonyl, $Co_6(CO)_{16}$.

Dicobalt octacarbonyl is conveniently prepared by reducing cobaltous carbonate in ligroin with hydrogen gas in the presence of carbon monoxide (Wender et al., 1957a). Alternatively, the octacarbonyl is formed from finely divided cobalt and carbon monoxide at 100 atm and 200°C.

Tetracobalt dodecacarbonyl may be prepared by thermally decomposing the octacarbonyl at 50°C in an inert atmosphere (Hileman, 1964).

Hexacobalt hexadecacarbonyl is obtained from tetracobalt dodecacarbonyl by reaction with an alkali metal in tetrahydrofuran to form an intermediate, $[Co_6(CO)_{14}]^{4-}$. This anion is oxidized in solution by ferric chloride, and upon extraction with n-pentane, hexacobalt hexadecacarbonyl is obtained (Chini, 1967).

Cobalt carbonyls are not particularly hazardous because of their low vapor pressure and their instability toward air. These materials are useful in preparation of metallic cobalt powder and in catalyzing the oxo synthesis (hydroformylation of olefins) (Wender et al., 1957b).

Cobaltous carbonate occurs as cobalt spar (spherocobaltite) and is isomorphous with magnesium and ferrous carbonates. Treatment of cobaltous ion with alkali carbonates yields a precipitate of variable composition, $xCoCO_3 \cdot yCo(OH)_2 \cdot zH_2O$ (cobalt content 45 to 47%). To prepare the normal carbonate, the precipitation is carried out in an atmosphere of carbon dioxide followed by dehydration (Sidgwick, 1950). Applications include ceramic pigments and trace-mineral feed supplements for cattle and sheep. Cobaltic carbonate is formed by air or chemical oxidants.

Several cobalt halides are known, including: CoF_2, $CoF_3 \cdot 7H_2O$, $CoCl_2 \cdot 6H_2O$, $CoBr_2 \cdot 6H_2O$, and CoI_2. Cobaltous fluoride is easily prepared by neutralization of $Co(OH)_2$ with HF. Cobaltic fluoride may then be prepared from CoF_2 by (a) electrolytic oxidation in HF or (b) further reaction with fluorine gas at 200°C. Cobaltic fluoride is very useful as a fluorinating agent for hydrocarbons (Fowler et al., 1947a and 1947b).

Cobaltous chloride is obtained by heating cobalt in the presence of chlorine gas. The hexahydrate readily dissolves in water to yield a pink solution or in alcohol to yield a blue solution. $CoCl_2$ is widely used as an indicator in silica gel and other desiccants; in the anhydrous condition, the salt is blue-violet whereas the hexahydrate is pink.

Red cobaltous bromide hexahydrate is easily prepared by treating cobaltous hydroxide or carbonate with hydrobromic acid. Upon dehydration the red salt turns green, and this property is utilized in hygrometers. Cobaltous iodide exists in two modifications: an α-form, black hexagonal crystals which are hygroscopic, and a β-form, yellow crystals of undetermined structure.

Cobaltous hydroxide is obtained as a pink or blue precipitate upon alkaline treatment of a soluble cobaltous salt. The solubility of the hydroxide is $\sim$ 2 x 10^{-5} g/liter. The hydroxide is slightly amphoteric (amphoteric point pH 10.9) and redissolves in excess alkali to form a deep blue solution, probably containing $Co(OH)_4^{2-}$ ions. Plots of solubility versus pH and Eh show minima at approximately pH 11 (Paces, 1973).

Cobaltic hydroxide readily forms when air is passed through a suspension of cobaltous hydroxide and allows separation of cobalt from nickel. Other useful oxidants include chlorine, hypochlorite, or neutral hydrogen peroxide.

Cobaltous nitrate is obtained by nitric acid treatment of cobalt metal, oxide, hydroxide, or carbonate as the hexahydrate at room temperature. The nitrate is commercially important for manufacture of cobalt pigments, sympathetic inks, ceramics, porcelain, stoneware, catalysts, and the preparation of vitamin B_{12} supplements.

Cobaltous phosphate may be prepared by reacting soluble cobaltous salts with orthophosphoric acid or soluble phosphates. It is relatively insoluble in water, and the hydrated salt, $Co_3(PO_4)_2 \cdot 8H_2O$, is used as a pigment. A pyrophosphate, $Co_2P_2O_7 \cdot 8H_2O$, is also known and is prepared in a similar manner.

Cobaltous sulfate occurs in nature as bieberite, $CoSO_4 \cdot 7H_2O$, and is isomorphous with $FeSO_4 \cdot 7H_2O$. It is a commercially important cobalt salt and does not lose or absorb water as easily or as rapidly as cobalt chloride or nitrate. The principal uses of cobaltous sulfate include: a diet supplement or drench for sheep and cattle; a fertilizer additive for pasture land; a pigment and glaze for porcelain; a component of electroplating baths; and a drying agent in paints, varnishes, and inks.

Double sulfates and basic sulfates containing cobaltous ion and alkali metals or ammonium ion have been characterized. Examples include:

$$(NH_4)_2SO_4 \cdot CoSO_4 \cdot 6H_2O,$$

$$CoSO_4 \cdot 3Co(OH)_2, \text{ and}$$

$$2CoSO_4 \cdot 3Co(OH)_2 \cdot 5H_2O.$$

Cobaltic sulfate, $Co_2(SO_4)_3 \cdot 18H_2O$, is prepared by electrolytic oxidation of cobaltous sulfate in a solution of 8 _N_ sulfuric acid. It is unstable in water and evolves oxygen gas.

Cobalt forms several compounds with silicon, including:

Compound	Co_2Si	Co_3Si_2	CoSi	$CoSi_2$	$CoSi_3$
% Co	80.77	75.91	67.75	51.23	41.18

These are obtained by melting the elements together or by combining cobaltous oxide and silicon dioxide in an electric furnace. CoSi forms when silicon carbide (SiC) is heated with cobaltous oxide (CoO).

Cobalt silicides are generally inert toward acids or bases, depending on the composition of the particular silicides. Thus, cobalt-rich compounds (Co_2Si and Co_3Si_2) are highly resistant to concentrated alkaline solutions, whereas silicon-rich compounds (CoSi, $CoSi_2$, and $CoSi_3$) are generally resistant to concentrated mineral acids. Fluorides and hydrofluoric acid attack all of the above cobalt silicides.

Cobalt silicides are used as insoluble anodes in the electrolysis of copper solutions and may be useful as cobalt-silicon alloy additions to molten iron or steel (Climax Molybdenum Company, 1946).

Cobalt forms three oxides: cobaltous oxide (CoO), cobalto-cobaltic oxide (Co_3O_4 or $CoO \cdot Co_2O_3$), and cobaltic oxide (Co_2O_3). The commercial products are often mixtures of these oxides and are useful as pigments in ceramics and enamels, as decolorizers in pottery and glass, for surface decoration of porcelain, and as an adherent under coatings for vitreous enamels (de Bie and Doyen, 1962a).

Cobaltous oxide may be prepared by controlled oxidation of the metal in air (or oxygen atmosphere); by thermal decomposition of the hydroxide, the carbonate, or a higher cobalt oxide; or by treating cobalt metal with steam, i.e.,

$$Co + H_2O \longrightarrow CoO + H_2.$$

Heating cobaltous oxide in air (above 100°C) causes formation of cobalto-cobaltic oxide, Co_3O_4. The latter is isomorphous with Fe_3O_4 (magnetite).

Cobaltic oxide is obtained by further oxidation of cobaltous oxide or cobalto-cobaltic oxide and is generally obtained as the monohydrate. This compound is unstable when heated above 300°C and evolves oxygen.

$$Co_2O_3 \longrightarrow CoO + O_2$$

According to Table III-1, most inorganic compounds of cobalt thermally decompose to cobaltous oxide at sufficiently high temperatures. Cobaltous oxide does not volatilize appreciably *per se*. It decomposes to cobalt to give primarily cobalt metal and oxygen as vapor species. At 1578°K, the partial pressures of species in equilibrium with the Co-O system are 1.71 x 10^{-7} atm for cobalt, 1.92 x 10^{-9} atm for CoO, and 4.01 x 10^{-8} atm for O_2. Even at 1744°K, these partial pressures are only 3.32 x 10^{-6}, 3.48 x 10^{-8}, and 4.86 x 10^{-7} atm, respectively (Grimley, 1966). These temperatures are even higher than those prevailing in roasting and coal-burning operations.

Gen et al. (1970) heated aerosol cobalt metal particles of diameter 0.016 μm to 0.083 μm (160 to 830 Å) in oxygen, air, or water vapor at 20 to 300°C for 1 hr. Only Co_3O_4 formed in oxygen. A slight amount of CoO was also formed in air along with the major product Co_3O_4. However, in water vapor, CoO was the only product.

Several cobalt sulfides are known: CoS, CoS_2, Co_2S_3, Co_3S_4, and Co_9S_8. The common form, CoS, readily forms when cobaltous ion is treated with sulfide ion at pH < 5, which is a familiar procedure in inorganic qualitative analysis schemes. Amorphous α-CoS is soluble in acids when freshly precipitated,* but on standing, it converts to acid-insoluble β-CoS (crystalline) and small amounts of Co_9S_8. Cobalt sulfides may also be formed by directly combining the elements.

Meunier et al. (1955) studied the mechanism of roasting cobalt sulfides. Thermal decomposition of the sulfides usually gives the sulfide Co_9S_8 according to the following reactions:

$$CoS_2 \xrightarrow{400\text{-}700^\circ C} 1/2\ Co_9S_8 + 5/9\ S_2$$

$$Co_3S_4 \xrightarrow{500\text{-}600^\circ C} 1/3\ Co_9S_8 + 2/3\ S_2$$

$$CoS \xrightarrow[N_2]{750^\circ C} Co_9S_8$$

$$CoS \xrightarrow{970^\circ C} Co_9S_8 + CoS$$

* The actual composition of α-CoS is more nearly Co(SH)OH.

Above 900°C, Co_9S_8 decomposes to metallic cobalt and sulfur; but under roasting conditions, the following reactions may occur:

$$Co_9S_8 \xrightarrow[air]{240°C} Co_9S_8O_4 \text{ (probably } 2Co_3S_4 \cdot Co_3O_4)$$

$$Co_9S_8 + 33/2\ O_2 \overset{400°C}{\rightleftharpoons} 8CoSO_4 + CoO$$

$$Co_9S_8O_4 + 19/2\ O_2 \overset{400-650°C}{\rightleftharpoons} 8CoSO_4 + CoO$$

$$3CoO + 1/2\ O_2 \overset{650°C}{\rightleftharpoons} Co_3O_4$$

$$Co_3O_4 \overset{700-1000°C}{\rightleftharpoons} 3CoO + 1/2\ O_2$$

$$CoS + 3CoSO_4 \overset{550°C}{\rightleftharpoons} 4CoO + SO_2$$

In addition, thermal decomposition reactions of cobalt sulfate may occur to produce CoO at 730 to 900°C and Co_3O_4 at 940 to 950°C in oxygen. In air, the decomposition of Co_3O_4 into CoO and oxygen occurs at 864 to 920°C.

B. Corrosion

Morral (1970) reviewed the oxidation, sulfidation, and hot corrosion of cobalt and cobalt alloys. The amount of literature published in the field is large, and the scope of the present discussion will be largely limited by considerations of possible release of corrosion products of cobalt alloys into the environment as well as the chemical forms and of corrosion reducing the service time of a component. This discussion does not describe the corrosion products to be expected when cobalt-containing alloys are used with corrosive industrial chemicals. When such corrosion occurs, it is the product itself that is contaminated. A threat of direct environmental contamination is not posed.

Table III-3 includes most of the studies mentioned by Morral (1970) and Wasielewski and Rapp (1972) that characterized the chemical composition of the scales on cobalt and cobalt alloys during exposure to oxygen- and sulfur-containing corrosive media. In almost all cases, the cobalt compound that would be present in any scale spalled off the metal surface is CoO. For metals in service below 900°C, both CoO and Co_3O_4 may be present.

TABLE III-3

OXIDATION AND OTHER CORROSION PRODUCTS OF COBALT AND COBALT-BEARING ALLOYS

Cobalt Metal or Alloy	Atmosphere/Solution	Temperature and Time	Oxidation and Corrosion Products: Inner Layer	Intermediate Layer	Outer or Only Layer Mentioned	Comments	Reference
Co powder	O_2 or air	-78, -22, 26°C	Multiple thin oxide layers at each temperature.				Yung-Fang Yu et al. (1957; cited by Morral, 1970)
Co metal	Aerated NaCl solution	30 days			CoO		Pollard (1964; cited by Morral, 1970)
Co granules packed in plastic (from Africa)	Air				Brownish adherent scale of CoOOH	CoOOH produced by air oxidation of $Co(OH)_2$	Morral (1970)
Anodic Co	Electrolyte solution				$Co(OH)_2$	Co_3O_4 coats the anode between 0.3 and 0.6 V. Co_3O_4 is reduced to $Co(OH)_2$ or CoO above 0.6 V.	Goehr (1966; cited by Morral, 1970)
Co metal		~200°C short periods			CoO		Studies of Gulbranson and associates cited by Morral (1970)
		~200 °C longer periods	CoO		Co_3O_4	Film becomes very thick	
		500°C for 1 hr			CoO		
Co metal	O_2 or air	750-1000°C	Void	Gray CoO scale	Compact CoO scale		Foster and Reynik (1968; cited by Morral, 1970)
Single crystal Co	Air	400°C 0.5 hr			Polycrystalline Co_3O_4		Newkirk and Martin (1963; cited by Morral, 1970)
Co metal	Air, 1 atm to 0.0001 torr				Co and Co_3O_4		Goswami (1965; cited by Morral, 1970)

Cobalt Metal or Alloy	Atmosphere/Solution	Temperature and Time	Oxidation and Corrosion Products: Inner Layer	Intermediate Layer	Outer or Only Layer Mentioned	Comments	Reference
Co metal	CO_2	< 900°C			Nonadherent CoO scale	Only CoO is stable above 900°C. Although a very thin outer Co_3O_4 layer forms at > 800°C, the oxidation involves CoO formation.	Nizhelskii and Vladimirova (1966; cited by Morral, 1970)
Co metal	O_2 or air	1200°C, 12 hr			900-μ-thick oxide scales	Several studies cited by Morral (1970) indicate that Co_3O_4 forms on the surface by nucleation at random locations or subgrain boundaries.	Wood et al. (1967; cited by Morral, 1970)
80% Co-Ni alloy	O_2 or air	1200°C, 12 hr			500-μ-thick oxide scales		
Co metal	O_2, 0.1-1 atm				CoO	Activation energy 35 kcal/mole	
					CoO + Co_3O_4 scales	Activation energy 58 kcal/mole. Krüger et al. (1966; cited by Morral, 1970) concluded that the diffusion of Co^{2+} over the cation vacancies of the oxide layer controls the oxidation rate. At lower pressures O_2 diffuses inward; but after the outer Co_3O_4 layer forms, the oxidation rate is independent of the O_2 pressure.	Kofstad et al. studies cited by Morral (1970)

TABLE III-3 (continued)

Cobalt Metal or Alloy	Atmosphere/Solution	Temperature and Time	Oxidation and Corrosion Products: Inner Layer	Intermediate Layer	Outer or Only Layer Mentioned	Comments	Reference
Co-4% Al alloys	O_2, 100 torr		CoO plus α Al_2O_3		Co_3O_4		Hagel (1965; cited by Morral, 1970)
Co-Cb alloys	O_2 or air	800-900°C	Intense scaling produces Cb_2O_5, Co_3O_4, CoO, and sometimes Co metal in the oxidized layer.				Voitovitch (1966; cited by Morral, 1970)
Co-rich Cu alloys	O_2 or air	> 700°C			CoO, Cu_2O, and Cu		Voitovich (1965; cited by Morral, 1970)
Co-Fe alloys	O_2 or air				CoO plus FeO; possibly $CoFe_2O_4$		Voitovich (1966; cited by Morral, 1970)
73% Co-Ni alloys	O_2 or air				Solid solution of CoO and NiO	Scales were richer in Co than were the alloys.	Voitovich (1965; cited by Morral, 1970)
Co-Ta alloys	O_2 or air	≤ 800°C			CoO, Co_3O_4, Co, and a little α-Ta_2I_5	Observed for a Co-rich alloy.	Voitovich (1965; cited by Morral, 1970)
Co-15% W alloys	O_2 or air				$CoWO_4$ and CoO	No catastrophic oxidation observed. Frantsevich and Voitovich (1964), however, noted significant oxidation at 900°C although corrosion resistance was apparently improved below 800°C.	Preece and Lucas (1952-1953; cited by Morral, 1970)
Electrodeposits of W-Co on steel	O_2 or air	< 300°C			Scarcely any oxide film.		Yamamoto (1960; cited by Morral, 1970)
		> 400°C			Co_3O_4 plus a deposit that had been carburized.		
		> 900°C	$CoWO_4$		Only CoO on the carburized deposit.		

TABLE III-3 (continued)

Cobalt Metal or Alloy	Atmosphere/Solution	Temperature and Time	Oxidation and Corrosion Products: Inner Layer	Intermediate Layer	Outer or Only Layer Mentioned	Comments	Reference
Low Cr-Co alloys	O_2 or air		CoO, Cr_2O_3, $CoCr_2O_4$		CoO	Co^{2+} diffuses faster than Cr^{3+}. Buildup of Cr_2O_3 in the scale inhibits Co^{2+} diffusion and reduces the oxidation rate.	Phalnikar et al. (1956; cited by Morral, 1970)
25% Cr-Co alloy	O_2 or air				Mixtures of CoO plus Cr oxides	This is the alloy with optimum oxidation resistance.	Phalnikar et al. (1956; cited by Morral, 1970)
> 25% Cr-Co alloys	O_2 or air				Cr oxides, essentially Cr_2O_3		Phalnikar et al. (1956; cited by Morral, 1970)
10-35% Cr-Co alloys	O_2, 760 mm Hg to 0.1 torr	800-1300°C	CoO, Cr_2O_3, and $CoCr_2O_4$		CoO	Oxidation was approximately parabolic and dependent on O_2 pressure. The two-layered oxide scale was porous. In addition, there was an internally oxidized zone containing Cr_2O_3.	Studies of Kofstad et al., cited by Morral (1970)
Co-Cr alloys with a a ternary alloying element	O_2 or air		Ternary-element-rich layer (if W, Al, etc.)	Cr-rich layer	Co-rich layer	Al, B, Ce, Zr, and Ta may improve oxidation resistance, depending on the composition Cr content.	Davin et al. (1967; cited by Morral, 1970)
Co-20% Cr- 1-3% Mn or Si	O_2 or air	900-1000°C 5 hr			CoO, Cr_2O_3, $CoCr_2O_4$	Severe oxide spalling occurred in the Si-doped alloys at 900-1200°C	Douglas and Armijo (1967; cited by Morral, 1970)

TABLE III-3 (continued)

Cobalt Metal or Alloy	Atmosphere/Solution	Temperature and Time	Oxidation and Corrosion Products: Inner Layer	Intermediate Layer	Outer or Only Layer Mentioned	Comments	Reference
L-605 and X-40	O_2 or air	1200°C	SiO_2, Cr_2O_3, $CoCr_2O_4$		Low melting scale of $CoWO_4$, Co_3O_4, CoO, and/or $CoCr_2O_4$	Catastrophic attack due to presence of $CoWO_4$ not identified in scale of X-40, but X-40 scales exfoliated on cooling.	Wlodec (1964; cited by Morral, 1970, and Wasielewski and Rapp, 1972)
WI-52	Air	870-1095°C, ≤ 100 hr			Mostly CoO in scales that spalled on cooling. Mostly $CoCr_2O_4$.	Observed on samples that had been ground and lapped through 14 μ diamond. $CoCb_2O_4$ incompatible with bulk $CoCr_2O_4$ spinel. Observed on samples that had been metallographically polished. Oxidation slower. No spalling	Lowell and Drell (1968; cited by Morral, 1970. Lowell (1969; cited by Wasielewski and Rapp, 1972)
Mar-M-302	O_2 or air	850-1200°C			CoO, Cr_2O_3, $CoCr_2O_4$, $CoWO_3$, $CrTaO_4$, $CoTa_2O_6$	Oxidation diffusion is controlled up to 1200°C.	Morral (1970) cited both Felton and Gregs (1964) and Kosak and Lombard (1967).
ATS-12, a German Co-Cr-Fe alloy	O_2 or air	1000°C prolonged exposure	Cr- and Fe-rich $(Co,Ni)_3O_2$·-(Co,Ni)O		Co- and Mn-rich NiO·CoO solid solutions. Layer spalled easily on cooling.	The alloy composition was 0.28% C, 1.4% Mn, 20% Cr, 16% Fe, 11% Ni, 2% Mo, 1.65% Cb, 3% V, and 46% Co.	Studies of Bollenrath et al. cited by Morral (1970)
Sintered TiC-Co alloy	O_2 or air				Mixtures of CoO and rutile TiO_2		Yasuo Sasaki (1961; cited by Morral, 1970)
WC-Cb-25% Co (v/v) alloys	O_2 or air	500-800°C			WO_3, Co_3O_4 (CoO), Cb_2O_5, plus an unidentified phase.	Film non-protective because of its low mechanical strength.	Meerson and Panov (1967; cited by Morral, 1970)

TABLE III-3 (continued)

Cobalt Metal or Alloy	Atmosphere/Solution	Temperature and Time	Oxidation and Corrosion Products: Inner Layer	Intermediate Layer	Outer or Only Layer Mentioned	Comments	Reference
Very thin Co metal layers	S at very low pressure	< 300°C			Co_9S_8, Co_3S_4, and hexagonal CoS or face-centered-cubic CoS_2	CoS and CoS_2 are destroyed at 300-500°C.	Studies of Lafourcade and coworkers cited by Morral (1970)
	S	500-900°C			Co_3S_4 disappears while Co_9S_8 remains		
	S	300-700°C			Co_2O_3 and Co_3O_4	Co_2O_3 destroyed at higher temperatures.	
Bulk	S	< 650°C			Co_9S_8	Layers very porous, almost powdery.	Davin (1966; cited by Morral, 1970)
Co metal	S, 120 torr	700°C			Co_9S_8 and Co_3S_4		
99.77% Co metal	$H_2S + 35\%\ H_2$	600°C	Porous		Dense, large grain size	Two layers.	Davin and Coutsouradis (1967; cited by Morral, 1970)
	$H_2S + 35\%\ H_2$	650°C, 5 hr			Co_9S_8	Two layers.	
	$H_2S + 35\%\ H_2$	700°C, 1 hr				Only one layer. Second layer formed on longer expo-sure.	
Co powder	H_2S, 14 torr	275-350°C, ≤ 20 hr			Co_9S_8	At 275-345°C, the reaction suggested was: $9\ Co + 8H_2S \longrightarrow Co_9S_8 + H_2O$. At 180-300°C and 14-760 torr, the reaction suggested was: $3CoO + 4H_2S \longrightarrow Co_3S_4 + 3H_2O + H_2$.	Colson (1964; cited by Morral, 1970).

TABLE III-3 (continued)

Cobalt Metal or Alloy	Atmosphere/Solution	Temperature and Time	Oxidation and Corrosion Products: Inner Layer	Intermediate Layer	Outer or Only Layer Mentioned	Comments	Reference
Bulk Co	SO_2 SO_2 10^{-5} torr	> 360°C 809°C			CoO, CoS, Co		Pannetier and Davignon(1964; cited by Morral, 1970)
Co metal	Air + SO_2	≤ 1200°C	CoS		CoO	Two layers.	Arkharov and Blankova (1959; cited by Morral, 1970)
Co metal	SO_2	500-1200°C	Co_9S_8	Co_9S_8 inclusions	CoO		Konev et al. (1966; cited by Morral, 1970)
Co metal	Na_2SO_4 in air	950°C, 2 hr			CoO		Beltran and Seybolt (1966; cited by Morral, 1970)
Co metal	Na_2SO_4	950°C, 2 hr			CoO scale plus Co-Co_4S_3 (sic) eutectic		Beltran and Seybolt (1966; cited by Morral, 1970)
80% Co-20% Cr alloy	H_2S	900°C, 2 hr	CrS		Co_4S_3 (sic) and Co_9S_8		Davin and Coutsourandis (1962, 1965; cited by Morral, 1970)
80% Co-20% Cr alloy	H_2S	900°C, 15 hr	40% S 35% Cr 20% Co	45% S 37% Cr 14% Co	58% Co 37% S		
Co-10% Cr alloy	Combustion gas contaminated with S and NaCl	1000°C 56 hr	Cr_2O_3	Cr_2O_3 plus CrS as well as a Co sulfide layer	CoO	Four layers	Fuller and Suorez (1961; cited by Morral, 1970, but possibly incorrectly attributed)

TABLE III-3 (concluded)

Cobalt Metal or Alloy	Atmosphere/Solution	Temperature and Time	Oxidation and Corrosion Products: Inner Layer	Intermediate Layer	Outer or Only Layer Mentioned	Comments	Reference
Co alloys	Hot corrosion testing apparatus	910-1040°C			NiO, NiO·CoO, a spinel (M_3O_4), Cr_2O_4, Na_2SO_4, etc. (depending on the other element). α-Co in the Y-containing alloy. $Cb_2Co_4O_3$ in the Cb alloy.	Cobalt-base alloy composition was 0.45% C, 10% Ni, 25% Cr plus one of the following: 8-15% Ta or W, 4% Ti or Cb, 1% Ce, 0.15% Y or La, or 0.5% Zr.	Bergman and Kaufman (1967; cited by Morral 1970)
IN-100	O_2 or air	~ 870°C	Al_2O_3, TiO_2, TiN		(Ni,Co)O, $(Ni,Co)Cr_2O_4$, $NiTiO_3$		Wasielewski (1967; cited by Wasielewski and Rapp, 1972)
60% Co + 20% Ni + 20% Cr alloy	Air	1100-1200°C 5 hr	Cr_2O_3, other M_2O_3	NiO, CoO, Co_3O_4	Several μ - thick layer (~ 0.5 of total thickness) NiO, CoO, Co_3O_4 (forms on cooling)	Inner layer harder and more adherent in La-containing than La-free alloys.	Haftka and Terao (1973)
Co-10 to 22% Cr-2.6 to 13.2% Al alloys	Cyclic oxidation apparatus	1100°C and 1200°C, 1-hr cycles	Al_2O_3, $CoAl_2O_4$	CoO, $CoCr_2O_4$, $CoAl_2O_4$	CoO massive spalling and catastrophic failure		Barrett and Lowell (1978)

The presence of $CoWO_4$ is likely in scales spalled from tungsten-bearing alloys. $CoCr_2O_4$ is less likely to occur in spalled-off scales. Sulfides formed from cobalt and its alloys in the presence of sulfur and hydrogen sulfide include Co_3S_4 and Co_9S_8, but these are transformed into the oxides in the presence of air. Thus, the major cobalt species that can be lost in particulate form from jet engine or industrial gas turbines during operation is probably CoO.

Morral (1979) summarized the high-temperature oxidation behavior of cobalt metal. Co_2O_3 forms at temperatures below 300°C. At 300 to 900°C, a double-layered oxide scale is formed when cobalt metal is exposed to air or oxygen. The scale comprises an outer Co_3O_4 layer and an inner CoO layer. Both layers are comprised of CoO above 900°C because Co_3O_4 decomposes. Scales formed in the range 600 to 750°C crack on cooling and flake off the surface.

When superalloys oxidize, four processes may occur. In surface scaling, the metal is converted directly to the oxide. Internal oxidation can produce stress and reduce fatigue resistance. During thermal cycling, heterogeneous scale growth results in incompatible oxide formation, exfoliation, and subsequent faster oxidation of unprotected metal. This third process is designated oxide spalling. Protective oxides, especially Cr_2O_3, may vaporize. Loss of this protective oxide promotes faster oxidation rates (Wasielewski and Rapp, 1972).

Both internal and external stress may lead to failure of the protective films. For example, corrosive attack on grain boundaries or intermetallic phases may reduce the alloy's service life. Carbides and precipitates at grain boundaries may promote intercrystalline oxidation. Thermal shock and static or dynamic stress during oxidation can lead to cracking and spalling of protective layers (Morral, 1970).

Nickel-base superalloys have superior oxidation resistance over cobalt-base alloys in dynamic burner-rig testing. The protective oxide scales on nickel-base alloys are α-Al_2O_3; on conventional cobalt-base alloys, α-Cr_2O_3. Additions of aluminum and rare earths to cobalt-base alloys have produced some alloy compositions with oxidation resistance comparable to that of contemporary nickel-base alloys (Wasielewski and Rapp, 1972).

During oxidation of superalloys, the fast-growing oxide nuclei NiO and CoO overgrow the slower-growing nuclei of Cr_2O_3, Al_2O_3, and other protective oxides. The latter grow laterally and down grain boundaries, finally halting the rapid development of the overgrowth scale. However, the protective effect may be reduced upon cracking and spalling or blistering of the scale, formation of low-stability products in the overgrowth layer, formation of spinels (M_3O_4) or other ternary oxides, depletion of the protective element in the alloy (Cr, Al, etc.), evaporation of the protective scale, etc. (Wasielewski and Rapp, 1972).

Barrett et al. (1978) classified 12 types of common nickel-, cobalt-, and iron-base high-temperature alloys as to the type of scales they form in cyclic oxidation tests (1-hr cycles at 1038 to 1149°C or 0.05-hr cycles at 1093°C) in both a dynamic burner rig and in a static air furnace for up to 100 hr. The burner rig test was more severe; it showed an increased tendency for oxide spalling, which was caused by the vaporization of the chromium oxide in the protective scale and the more drastic cooling procedure (air-blast quench). The types of scales were I, NiO or CoO; II, Cr_2O_3-chromite spinel; III, α-Al_2O_3-aluminate spinel; and IV, ThO_2-blocked Cr_2O_3. When the protective scale broke down under the continuous oxidation, each was ultimately transformed to the Class I type. Both Class I and Class II scales will spall. Classes III and IV are generally the most protective types of scales. Cobalt-containing alloys that showed control of oxidation by scales of Cr_2O_3-chromite spinel were those containing $>$ 16% Cr and $\lesssim$ 3% Al included HA-188, L-605, X-40, and WI-52 (all were cobalt-base alloys). IN-100 (15% Co) and B-1900 (10% Co) formed Class IV protective scales. Failure in the nickel-base alloys was usually by NiO formation after aluminum was depleted. Cobalt-base alloys are more spall prone than nickel-base or iron-base alloys because the Class II scale spalls more drastically when CoO is formed.

Sulfidation is the effect on metals of atmospheres containing a sulfur vapor, sulfur dioxide, or hydrogen sulfide. Hot corrosion combines the effects of oxidation and sulfidation and occurs when superalloys are used in jet engines and gas turbines, especially in marine atmospheres. Hot corrosion has been defined as the phenomenon that occurs when a gas turbine is operated on fuels containing $\sim$ 0.25% sulfur in a salt water environment. Severe corrosion is caused by the Na_2SO_4, which forms under these conditions. The sulfides that form are rapidly converted to oxides. Reactive and, usually, low-melting combustion residues that may originate solely from the fuel or may result from reaction with atmospheric constitutents promote catastrophic oxidation (Morral, 1970).

All cobalt-base superalloys have hot corrosion resistance superior to that of nickel-base alloys. Alloys such as X-40/X-45 have been used for tens of thousands of hours in industrial and aircraft turbines burning lower grade fuels that were high in sulfur in atmospheres where salt was present. Cobalt alloys are used "where the toughest sulfidation conditions are expected." The reactions involved in hot corrosion may be (Sims, 1969):

$$2NaCl + S + 2O_2 \longrightarrow Na_2SO_4 + 2HCl$$

$$Na_2SO_4 + 4Co + 1/2\ Cr_2O_3 + 3/4\ O_2 \longrightarrow Na_2CrO_4 + 3CoO + CoS$$

Stringer (1977) recounted some early examples of hot corrosion in high-temperature cobalt- and nickel-base alloys that shortened their service life in marine, aircraft, and industrial engines exposed to both sulfur and salt (e.g., in coastal environments). Since gas turbine manufacturers are probably reluctant to describe hot corrosion engine failures, extensive information is not available in the literature. Such failures are undoubtedly exceptions rather than the rule.

Cobalt-base alloys have better resistance to sulfidation because of their higher chromium content. Formation of chromium-containing spinels (M_3O_4) gives some protection, but reaction with sulfur can deplete the chromium content. Later the sulfur attacks the nickel to form a nickel-nickel sulfide eutectic. Finally, the sulfides are rapidly oxidized.

Cobalt-containing spinels formed during conventional oxidation probably include Co_3O_4, $CoAl_2O_4$, $NiCo_2O_4$, and $CoCr_2O_4$. Cobalt may appear in sulfide phases also: $(Cr, Co, Ni, Mo,\text{-}Al)_2S_3$ and $(Cr, Co, Ni)_2S_3$ (Bergman, 1967).

Cobalt-base alloys may also be more resistant to hot corrosion than are nickel-base alloys because sulfur diffuses at a slower rate through cobalt alloys or because sulfur is less soluble in the cobalt alloys (Morral, 1970).

Radiocobalt apparently is a frequent major contaminant of cooling water released from nuclear reactors. Copious literature references exist on the environmental behavior of ^{60}Co in sediments and biota of water bodies receiving such effluents. In some cases, cobalt alloys are purposely used in reactor environments. However, Cordovi (1972) explained that cobalt was present in corrosion products from austenitic metal surfaces in a shim-controlled pressurized water (nuclear) reactor (PWR) plant because it is an impurity in the base metals. At one such plant, the corrosion film on the austenitic metal surface comprised (weight percent on an oxide basis): iron, 24; nickel, 11; chromium, 36; and cobalt, 0.17. The compositions of the circulating (in cooling water)* and deposited corrosion products were significantly different, although the cobalt oxide concentration was about the same: iron, 58 to 59; nickel, 8; chromium, 0.6 (circulating) to 2 (deposit); manganese, 0.3 to 0.4, and cobalt, 0.2% (oxide basis). The reactor radiation and high heat fluxes at fuel element surfaces favor the transition from dissolved metal ion in the cooling water to agglomerated deposits of relatively insoluble corrosion product oxide.

Tskhvirashvili et al. (1970) studied the solubility of construction material metal oxides in superheated steam at 4 to 24 MPa. The maximum solubility of cobalt oxides in the temperature range studied was 5 μg/kg at 550°C at the lowest pressure. At the higher pressures, the solubility fell from $\sim$ 12 to 28 μg/kg at 400°C to < 10 g/kg at 550°C. Such concentrations do not appear to be alarmingly contaminated compared with common environmental water concentrations ($\sim$ 2 μg/liter).

---

* The cooling water is mildly acidic to alkaline and contains lithium and boric acid. Alloy 600 or Alloy 800 steam generator tubes represent 85 to 95% of the total structural surface area that is contacted by the coolant in a PWR plant. Most of the rest of the components contacted by the coolant are made of austenitic stainless steel. Stainless steel components are used in boiling water reactors.

Figure III-1 shows the theoretical conditions of pH and electrode potential that lead to corrosion, immunity, or passivation of cobalt at 25°C.

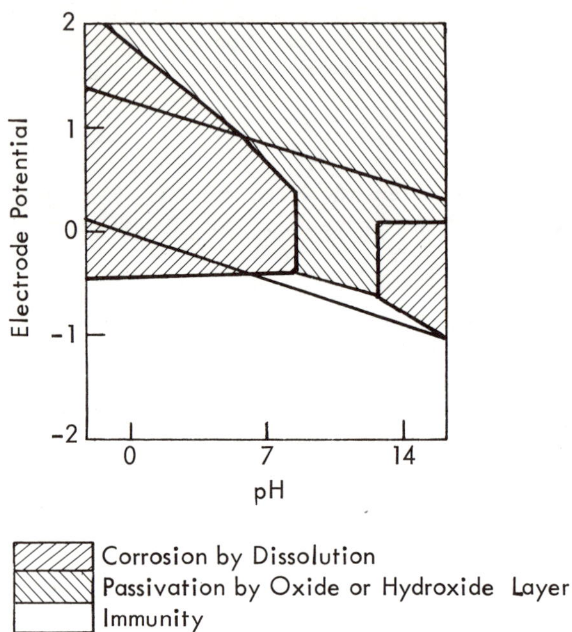

Figure III-1. Theoretical Conditions of Corrosion, Immunity, and Passivation of Cobalt at 25°C (Obrecht and Pourbaix, 1969)

In addition to corrosion by the atmosphere, steam or water, and industrial solutions, metals can be corroded by biological action. Cobalt is no exception, although it appears to be quite resistant. Siegel and Siegel (1979) studied the biocorrosion of pure metal foils by Penicillium expansum at pH 6.80 and 25°C. All of the 0.0025-cm-thick foils were perforated by a localized colony, but the lowest percentage removal (0.014% after 30 days) was observed for cobalt. The order observed was Zn > Sn > Al > Cu = Cd > Er > Co. Copper, cadmium, and cobalt, because of their toxicity, supported far less growth than did the other metals. For example, for cobalt the density was 0.8 ± 0.4 colonies/cm^2 compared to 155 ± 85 colonies/cm^2 for tin.

C. Chemistry of Selected Cobalt Complexes and Coordination Compounds

The physical, chemical, and structural properties of many inorganic and organic cobalt complexes and coordination compounds have been characterized.

The cobalt ammines, e.g., $[Co(NH_3)_5Cl]Cl_2$, are of historical interest. Alfred Werner devoted nearly 20 years (1893-1911) to researching the chemistry of cobalt ammines and other coordination compounds and in the process developed the essentials of the modern theory of coordination compounds and complex ions (Moeller, 1952). He developed the concepts of primary valency (oxidation state), secondary valency (coordination number), and spatially directed secondary valency (structural isomerism).

The original coordination theory has been extended by Lowry (1923) and Sidgwick (1927) (two-electron covalent bond), Van Vleck (1935) (molecular orbitals), Pauling (1960) (hybrid orbitals), and Orgel (1952 and 1961) (electrostatic-crystal field). A full discussion of these concepts is beyond the scope of this report.

The literature on the properties and structure of cobalt complex ions and coordination compounds is vast, and the reader is referred to standard texts for further information (Cotton and Wilkinson, 1972; Nicholls, 1973).

Table III-4 presents data on cobalt complex ions and coordination compounds selected on the bases of historical importance, usefulness for inorganic and organic syntheses, and environmental concern.

Several hexammine cobalt complexes are known and the determination of their properties and characteristics was a keystone in Werner's work. When aqueous cobaltous chloride solutions are treated with excess ammonia and oxidized by air or an oxidizing agent, the compounds described in Table III-5 can be isolated.

TABLE III-4

SELECTED COBALT COMPLEXES AND COORDINATION COMPOUNDS

Complex	Preparation	Color	Structure	Melting Point Boiling Point
Dicobalt octacarbonyl	$2CoI_2 + 8CO + 4Cu \xrightarrow[100\ atm]{200°C} Co_2(CO)_8 + 4CuI$	Orange-brown solid	Co-Co bond with two bridging CO groups	51°C (m.p.)
Cobalt carbonyl hydride	$2Co + 8CO + H_2 \xrightarrow[250\ atm]{180°C} 2Co(CO)_4H$	Yellow liquid	Trigonal bipyramid with hydrogen bonded to cobalt	-26°C (m.p.)
Hexa-aquo cobaltous ion	$Co(OH)_2 + dil.\ HNO_3 \rightarrow Co(NO_3)_2 \cdot 6H_2O$ $Co(NO_3)_2 + H_2O \rightarrow [Co(H_2O)_6]^{2+}$	Pink solution	Octahedral	-
Tetrachlorocobaltate(II) ion	$[Co(H_2O)_6]^{2+} \underset{H_2O}{\overset{Cl^-}{\rightleftharpoons}} [CoCl_4]^{2-}$	Blue solution	Tetrahedral	-
Hexammine cobaltic chloride	$CoCl_2 + 6NH_3 \rightarrow [Co(NH_3)_6]Cl_2$	(Unstable)	Octahedral	-
	$[Co(NH_3)_6]Cl_2 + O_2 \xrightarrow{activated\ C} [Co(NH_3)_6]Cl_3$	Orange solution	Octahedral	
Pentacyanocobaltate(II) ion	$Co^{2+} + 5CN^- \longrightarrow [Co(CN)_5]^{3-}$	Dark green solution	Square pyramidal	-
Tetrathiocyanatocobaltate(II) ion	$Co^{2+} + 4SCN^- \longrightarrow [Co(SCN)_4]^{2-}$	Blue solution	Tetrahedral	-

TABLE III-5

COBALTIC AMMINE CHLORIDE COMPLEXES

Compound	Composition	No. of Ions
Luteocobaltic chloride (orange-yellow crystals)	$[Co(NH_3)_6]Cl_3$	4
Roseocobaltic chloride (pink crystals)	$[Co(NH_3)_5(H_2O)]Cl_3$	4
Purpureocobaltic chloride (violet crystals)	$[Co(NH_3)_5Cl]Cl_2$	3
Violeocobaltic chloride (violet crystals, cis-form)	$[Co(NH_3)_4Cl_2]Cl$	2
Praseocobaltic chloride (green crystals, trans-form)	$[Co(NH_3)_4Cl_2]Cl$	2

It is now understood that the ammonia and water molecules and some of the chloride ions are bonded to the cobalt ion within the coordination sphere and are unavailable to participate in normal chemical substitution reactions. The nature of the bonding within the coordination sphere involves sharing of atomic and molecular electronic orbitals.

Cobalt also forms a series of chlorocomplexes in hydrochloric acid including:

$$[Co(H_2O)_6]^{2+} + Cl^- \xrightleftharpoons{3\ \underline{M}\ HCl} [Co(H_2O)_5Cl]^+ + H_2O$$

$$[Co(H_2O)_5Cl]^+ + Cl^- \rightleftharpoons [Co(H_2O)_2Cl_2] + 3H_2O$$

$$[Co(H_2O)_2Cl_2] + Cl^- \xrightleftharpoons{8\ \underline{M}\ HCl} [Co(H_2O)Cl_3]^- + H_2O$$

$$[Co(H_2O)Cl_3]^- + Cl^- \rightleftharpoons [CoCl_4]^{2-} + H_2O$$

Thus, the ionic and solvent media play an important role in determining the cobalt species that predominates. A similar set of equilibria occurs for the species present in basic cobalt solutions:

$$Co^{2+} \underset{pH\ 8}{\rightleftharpoons} Co(OH)^{+} \rightleftharpoons Co_4(OH)_4^{4+} \underset{pH\ 10}{\rightleftharpoons} Co(OH)_2 \rightleftharpoons Co(OH)_3 \underset{pH\ 12}{\rightleftharpoons} Co(OH)_4^{2-}$$

The tetrathiocyanatocobaltate ion, $[Co(SCN)_4]^{2-}$, forms the basis of a colorimetric determination of cobalt (Furman, 1962). The mercury salt is precipitated in the gravimetric estimation of cobalt (Vogel, 1962) and is also used as a calibrating salt for Gouy magnetic susceptibility measurements (Figgis and Lewis, 1964).

Fresh solutions of pentacyanocobaltate ion $[Co(CN)_5\text{-}H_2O]^{3-}$ readily absorb hydrogen gas at 0°C to form a complex hydride $[HCo(CN)_5]^{3-}$, which is useful for homogeneous catalytic hydrogenation of unsaturated compounds and reductive amination of α-oxo acids and esters (Jones, 1968; Murakami and Kang, 1963). Thus, phenylalanine can be prepared from phenylpyruvic acid or its oxime:

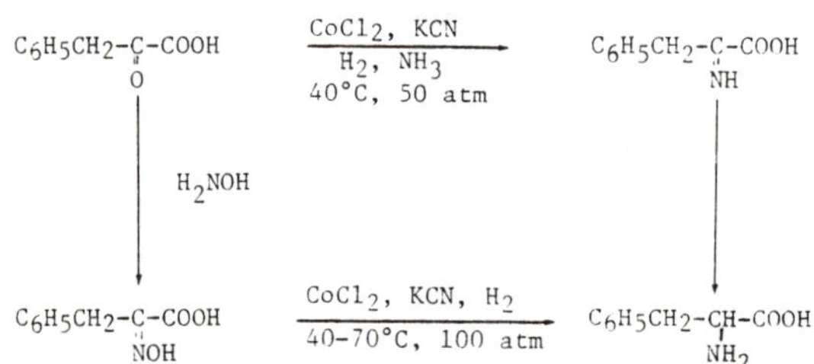

Unsaturated hydrocarbons are reduced by hydrogen gas and $[HCo(CN)_5]^{3-}$ only when they are capable of coordination to the cobalt (Murakami and Kang, 1963). When cinnamic acid is reduced, the primary product is dihydrocinnamic acid; sorbic acid gives γ,δ-hexenoic acid. Nitrobenzene is reduced to aniline (Murakami et al., 1963).

Cobalt forms several carbonyls and nitrosyls including carbonyl hydride and halides; nitrosyl halides, cyanides, and sulfur compounds; cyanides, etc.

Cobalt carbonyl hydride (the polarity of the cobalt-hydrogen bond is in doubt, but reactions are formulated as if the cobalt tetracarbonyl is the negative species) combines with olefins and epoxides to form acylcobalt tetracarbonyls. The compound attacks the double bond of olefins forming alkylcobalt carbonyls wherein the cobalt atom is bonded directly to a carbon atom:

$$CH_3CH_2CH_2CH = CH_2 + HCo(CO)_4 \longrightarrow CH_3CH_2CH_2\underset{|\atop Co(CO)_4}{C}HCH_3 + CH_3CH_2CH_2CH_2CH_2Co(CO)_4$$

When cobalt carbonyl hydride is added to ethylene epoxide in the presence of carbon monoxide, β-hydroxyethylcobalt tetracarbonyl is formed:

$$HCo(CO)_4 + \overset{O}{\overbrace{CH_2CH_2}} \xrightarrow[Co]{20°C} (CO)_4CoCH_2CH_2OH \xrightarrow{Co} (CO)_4CoCOCH_2CH_2OH$$

Acylcobalt tetracarbonyls react with alcohols to form esters (Heck, 1968). Thus:

$$CH_3I + CO + NaCo(CO)_4 \xrightarrow[1\ atm]{50°C} CH_3COCo(CO)_4 + NaI$$

$$CH_3COCo(CO)_4 + C_2H_5OH \xrightarrow{50°C} CH_3COOC_2H_5 + HCo(CO)_4$$

The alcohol can also serve as the reaction solvent. A hindered tertiary amine is usually added to complex cobalt hydrocarbonyl, which can reduce the intermediate acyltetracarbonyl:

$$(C_6H_{11})_2NC_2H_5 + HCo(CO)_4 \longrightarrow [(C_6H_{11})_2NHC_2H_5]^+ Co(CO)_4^-$$

Dicobalt octacarbonyl is used in a variety of synthetic organic chemistry reactions including: (a) catalytic hydroformylation of olefins (oxo reaction); (b) hydrogenation of olefins and aldehydes; (c) hydrosilation of olefins; (d) carboxylation via acylcobalt tetracarbonyls; and (e) carbonylation. These reactions are dealt with at length in Chalk and Harrod (1968), and the important industrial catalytic applications have been noted earlier in this chapter.

Several chelating agents (e.g., NTA, EDTA, acac)* form inner complexes with cobalt. If the resulting metal complex has a neutral charge and simultaneously satisfies both the oxidation number and the coordination number of the central metal ion, the metal can be extracted into organic solvents and volatilized at elevated temperatures without decomposition. Cobalt acetylacetonate ($Co(acac)_2$) is a case in point. Its structure is given below.

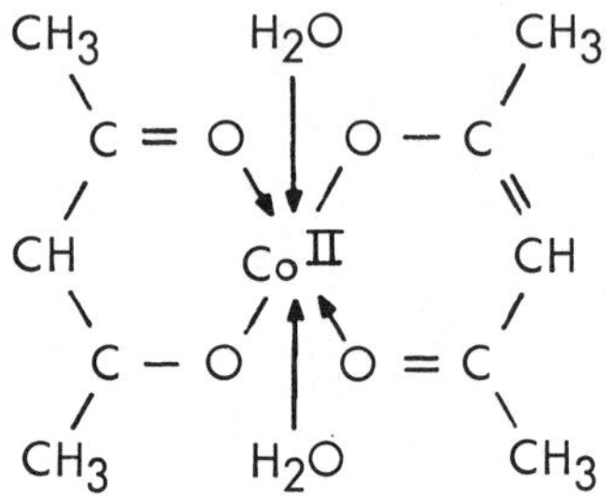

```
CH3        H2O        CH3
  \         |         /
   C = O    |    O - C
  /      \  ↓  /      \\
CH          CoII        CH      (trans-form)
  \      /  ↑  \      /
   C - O    |    O = C
  /         |         \
CH3        H2O        CH3
```

Bis(acetylacetonato)cobalt(II)

The value of the formation or stability constant (log K = 10.6) indicates a very stable species (McCrary and Howard, 1979). The diaquo compound can be dehydrated at 30 mm Hg pressure and 60°C and can be purified by sublimation at reduced pressure. It is soluble in organic solvents (e.g., benzene and chloroform). In the solid state, the anhydrous compound is tetrameric and has an octahedral structure (Cotton and Elder, 1965; Ellern and Ragsdale, 1968).

Cobalt also forms a very stable complex with EDTA (log K = 16.3). The structure of the chelate formed between cobaltous ion and EDTA is given:

---

* NTA is nitrilotriacetic acid and its salts. EDTA is ethylenediaminetetraacetic acid and its salts. The acac is 2,4-pentanedione or acetylacetone.

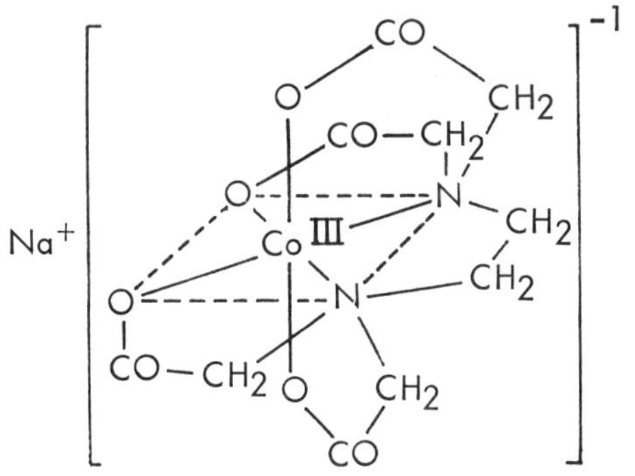

Cobalt forms two chelates with NTA, $NaCoNTA \cdot H_2O$ and $Co_2Co(NTA)_2 \cdot 3H_2O$. The synthesis, structural properties, and infrared spectra of cobalt NTA chelates have been described by Rajabalee (1974).

D. Cobalt Compounds of Biological and Environmental Interest

This subsection discusses compounds of biological and environmental interest in which cobalt is bonded to oxygen, cyanide, methyl, and other alkyl groups. Cobalt compounds may serve as simple models of complex biological and environmental systems.

Cobalt-oxygen reversible compounds are rare, but several have been characterized according to formation conditions, stoichiometry, and structure. Thus, cobalt-Schiff base complexes absorb molecular oxygen in the presence of DMF, DMSO, pyridine, or substituted pyridines. Crumbliss and Basolo (1970) investigated the reaction of Co(II)(acacen) dissolved in dimethylformamide (DMF) with molecular oxygen. These authors found that oxygen is reversibly taken up by the cobalt complex and that a stable black crystalline compound is formed at $\leq$ -10°C and with log K formation = 2.1.

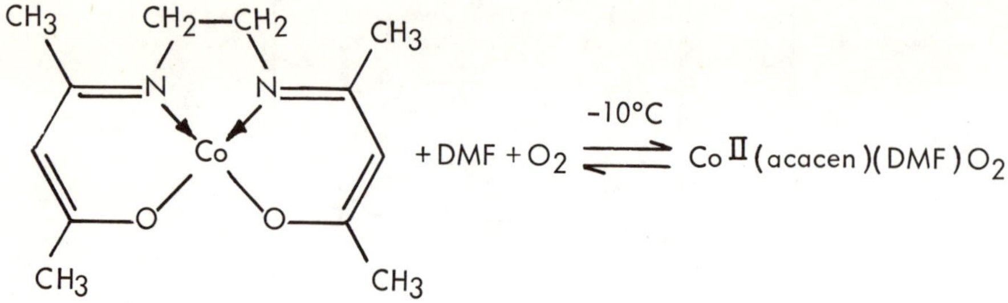

$$\text{Co(acacen)} + DMF + O_2 \underset{}{\overset{-10^\circ C}{\rightleftharpoons}} Co^{II}(\text{acacen})(DMF)O_2$$

The structure of the oxygen adduct is given below:

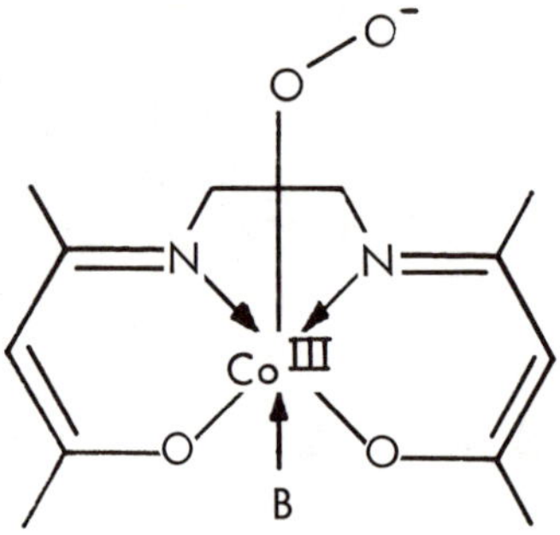

where Co(III) is complexed by a Schiff base and a coordinating base solvent, B. Known adducts of oxygen-cobalt-Schiff base-solvent base are given in Table III-6.

TABLE III-6

MONOMERIC OXYGEN ADDUCTS OF COBALT(II) COMPLEXES

Specific Adduct	Coordinating Solvent	Adduct Color Solution	Crystal
$Co(acacen)(DMF)O_2$	Dimethylformamide	Opaque black	Black
$Co(acacen)(py)O_2$	Pyridine	Opaque black	Red-brown
$Co(acacen)(CH_3\text{-}py)O_2$	4-methylpyridine	Opaque black	Carmine red
$Co(acacen)(CN\text{-}py)O_2$	4-cyanopyridine	Opaque black	Reddish black
$Co(acacen)(H_2N\text{-}py)O_2$	4-aminopyridine	Opaque reddish black	Brown

Other examples of cobalt-containing Schiff bases are N,N'-ethylenebis(salicylideneimine)Co^{II} and N,N'-dipropylenamine-bis(salicylideneimine)Co^{II}.

Oxygenation of Co(acacen) is rapid and complete in 3 min at 0.5 atm P_{O_2} in DMF at -10°C, and oxygen release under vacuum is complete in a matter of minutes. The oxygenation-deoxygenation cycle is repeatable at least over 10 cycles according to Crumbliss and Basolo (1970).

By analogy to oxyhemoglobin and on the basis of electron spin resonance and infrared spectroscopy, it is believed the oxygen is bonded to the cobalt as a superoxide ion, Co(III)-O_2^- (Hoffman et al., 1970). The iron-oxygen linkage in oxyhemoglobin is postulated to be similar in nature, Fe(II)-O_2^- (Pauling, 1964).

Other oxygen adducts, e.g., bis(dimethylglyoximato)-cobalt(II), N,N'-ethylenebis(3-methoxysalicyliminato)cobalt(III), and N,N'-dipropyleneaminebis(salicylideneiminato)cobalt(II), contain oxygen bridges with no unpaired electrons. The schematic structure of 1:2 oxygen adducts of N,N'-ethylenebis(salicylidene-iminato)cobalt(II) with anionic ligands is postulated to contain an oxygen bridge and is given below:

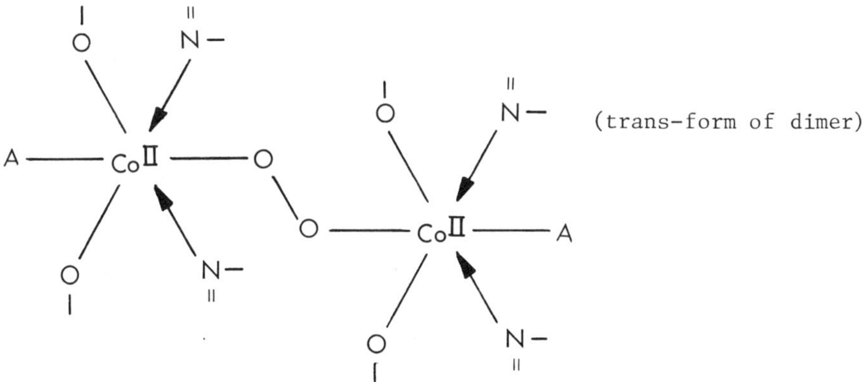

where A is a coordinating anionic ligand. The anionic ligand (e.g., pyridine oxide, thiocyanate, azide, or acetate) may compensate the charge transferred from cobalt to oxygen (Floriani and Calderazzo, 1969).

Certain porphyrin complexes of cobalt in organic solvents reversibly form oxygen adducts in the presence of coordinating nitrogen base (Hoffman and Petering, 1970). Thus, the dimethyl ester of Co(II)-protoporphyrin IX (designated Co(II)P) in 10% pyridine-water media reacts with molecular oxygen at 15°C as follows:

$$\text{Co(II)P} + O_2 + \text{py} \longrightarrow \text{Co(III)P-(py)-}O_2$$

The proposed structure of the adduct is given below:

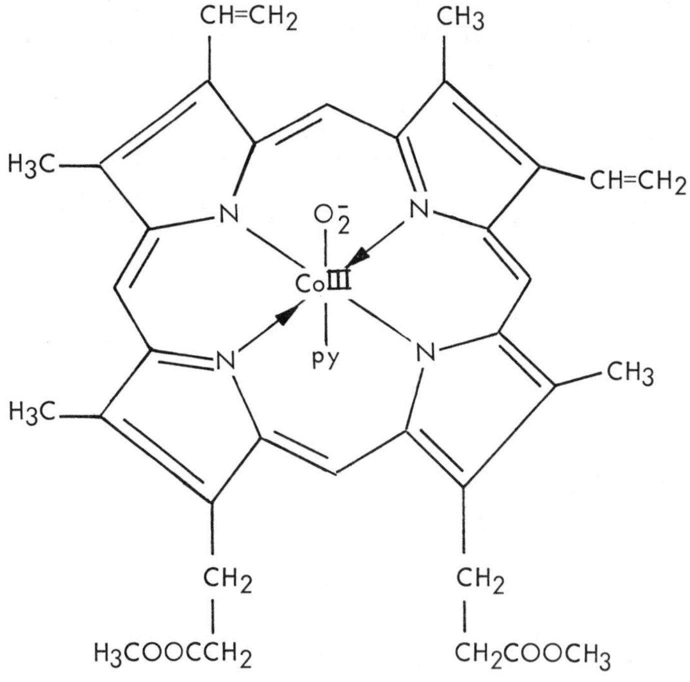

Cyanocobalamin or vitamin B_{12} is a cobalt coordination compound found in soil, water, sewage sludge, manure, and dried estuarine mud (Windholz et al., 1976). It is produced by intestinal microorganisms in the rumen of cattle and sheep (Sauchelli, 1969). The commercial source of vitamin B_{12} is fermentation residue from antibiotics production.

The unique features of cyanocobalamin (CN-B_{12}) are the corrinoid ligand system and the cobalt-carbon bond. The corrinoid ligand system is similar to a porphyrin and contains four pyrrole nuclei joined in a ring with three bridging carbon atoms and six conjugated double bonds. The cobalt atom is centrally located within the corrinoid ligand system and is bonded by three nitrogen atoms via coordinate covalent bonds and one nitrogen atom via a covalent bond. The two remaining octahedral positions are occupied by a cyanide group (cobalt-carbon bond) and a nitrogen atom from a 5,6-dimethylbenzimidazole group.

The structure of cyanocobalamin is given below.

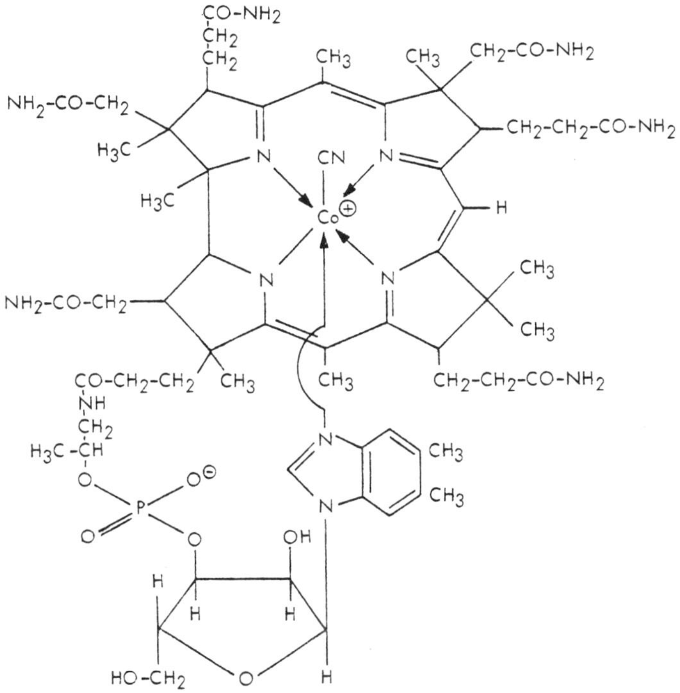

Several cobalamins are known, including cobamamide, in which the cyanide group is replaced with 5'-deoxyadenosine. This cobalamin acts as a coenzyme and in the presence of dioldehydrases causes conversion of ethylene glycol to acetaldehyde or 1,2-propanediol to propionaldehyde (Frey et al., 1970; Essenberg et al., 1971).

$$CH_3\underset{OH}{C}HCH_2OH \xrightarrow[\text{cobamamide}]{\text{Enzyme}} CH_3CH_2CHO + H_2O$$

Other cobalamins include those with cobalt bonds to hydroxide ion and the methyl group. In each case, the cobalt atom is considered to have a formal oxidation state of three.

Cobalamins have been chemically reduced in neutral or alkaline solutions with hydrogen gas using a catalyst, borohydrides, zinc dust in acid solution, chromous acetate, etc. (Wagner, 1966). Reduction of cyanocobalamin (Co^{III}-CN) yields B_{12r} (Co^{II}), an orange-brown product, and subsequently B_{12s} (Co^{I}), a gray-green product. Both materials are oxidized to B_{12}-OH (Co^{III}) by air. B_{12s} decomposes water to yield hydrogen gas and B_{12r}, i.e.:

$$B_{12s}\ (Co^{I}) + H_2O \longrightarrow B_{12r}\ (Co^{II}) + 1/2\ H_2 + OH^-$$

The detailed chemical nature of B_{12r} and B_{12s} is unknown at present.

Several interesting reactions with alkyl halides, dialkyl sulfates, trialkyl phosphates, acylating agents, acetylenes, olefins, ethylene oxide, and tetrahydrofuran have been noted. Table III-7 illustrates some of these reactions (Johnson et al., 1963; Wagner, 1966).

Similarly, B_{12}-CN or B_{12}-OH yields cobalamin sulfonates, B_{12}-SO_3^-, when treated with bisulfite, sulfurous acid, or dithionite. When B_{12}-OH is treated with glutathione (GSH) in aqueous solution (pH 2 to 8), a violet product is obtained which is then transformed to methylcobalamin (B_{12}-CH_3) in the presence of methylating agents (Wagner and Bernhauer, 1964; Dubnoff, 1964). Thus:

$$B_{12}\text{-}OH \xrightarrow{GSH} B_{12}\text{-}SG \xrightarrow{CH_2N_2} B_{12}\text{-}CH_3$$

Cobalamins may be sensitive or insensitive to photochemical reactions depending on the particular compound and the experimental conditions. Methylcobalamin in aqueous solution and in the presence of excess oxygen photochemically yields B_{12}-OH and formaldehyde (Wagner and Bernhauer, 1964). However, in the

TABLE III-7

SELECTED REACTIONS OF $B_{12}s$

Reactant	Cobalamin Product (cobalt-carbon bonds)
$CH{\equiv}CH$	$B_{12}-CH{=}CH_2$
$CH{\equiv}CBr$	$B_{12}-C{\equiv}CH$
_n_-C_4H_9I	$B_{12}-C_4H_9$
tert-C_4H_9I	No reaction
BrCN	B_{12}-CN (cyanocobalamin)
CH_3COCl	$B_{12}-COCH_3$
CH_2N_2	$B_{12}-CH_3$ (methylcobalamin)
$(C_2H_5)_3PO_4$	$B_{12}-C_2H_5$
$CH_2{-}O{-}CH_2$ (epoxide)	$B_{12}-CH_2CH_2OH$
SO_2Cl_2, H_2O	$B_{12}-SO_3^-$

absence of oxygen, methyl- and ethylcobalamin are photostable compounds. It is generally believed that photolytic cleavage of the cobalt-carbon bond is the primary degradation pathway.

Other reactions of vitamin B_{12} and its analogues include: (a) hydrolytic reactions in acid solution in the dark involving electrophilic displacement at the cobalt atom; (b) alkaline oxidation resulting in substitution or cyclization reactions; and (c) transfers of alkyl or acyl groups from alkyl-B_{12} or acyl-B_{12} derivatives under anaerobic or aerobic conditions in the dark or as a photolysis, e.g.:

$$B_{12}-CH_3 \xrightarrow[2)\quad O_2]{1)\ h\nu\ +\ R'SH} CH_3SR' + B_{12}-OH$$

$$B_{12}\text{-}COCH_3 \xrightarrow[2)\ CH_3I]{1)\ NH_3} B_{12}\text{-}CH_3 + CH_3CONH_2$$

(Wagner, 1966).

Methylcobalamin is capable of transferring a methyl group to certain metal ions, including Hg(II), Tl(III), Pt(II), and Au(I) (Agnes et al., 1971). This is but one example of a growing body of literature involving bioalkylation of metals in the environment (Wood, 1975; Ridley et al., 1977; Anonymous, 1977a; and Wood et al., 1978). Methylcobalamin methylates thallium(III) ions and other heavy metal ions at 25°C in aqueous solutions buffered between $0 \leq pH \leq 5$. In the reaction with thallium(III), the reaction products are B_{12}-OH, i.e., B_{12} with cyanide replaced by hydroxide, and methylthallium(III):

$$Tl^{3+} + B_{12}\text{-}CH_3 \xrightarrow{H_2O} B_{12}\text{-}OH + CH_3Tl^{2+} + H^+$$

In the presence of excess chloride ion, methyl chloride is formed.

$$CH_3Tl^{2+} + Cl^- \longrightarrow CH_3Cl + Tl^+$$

The environmentally stable thallous salts do not react with methylcobalamin (Agnes, 1971).

In the case of methylation of platinum by methylcobalamin, both $PtCl_4{}^{2-}$ and $PtCl_6{}^{2-}$ are required. The postulated mechanism is given below and involves a two-electron redox "switch."

$$Pt^*Cl_6{}^{2-} + \underset{\text{(complex)}}{PtCl_4{}^{2-}\ldots B_{12}\text{-}CH_3} \xrightarrow[pH\ 5.0]{H_2O}$$

$$Pt^*Cl_4{}^{2-} + CH_3PtCl_5^- + B_{12}(H_2O) + Cl^-$$

At pH 1.0 and in moderate to high chloride ion concentrations (0.01 to 1.0 M), methyl chloride is the final methylated product.

---

* Here the asterisks denote a specific platinum atom for the purpose of illustrating changes in the oxidation state.

Four types of *in vitro* mechanisms have been proposed for vitamin B_{12}-dependent methyl transfer reactions: (a) cleavage of the Co-CH_3 bond with the transfer of the methyl carbanion to an electrophilic attacking metal ion in one of its higher oxidation states with cobalt remaining as Co(III); (b) cleavage of the Co-CH_3 bond with the transfer of the methyl carbonium ion to an attacking nucleophile with cobalt(III) being reduced to cobalt(I); (c) cleavage of the Co-CH_3 bond with the transfer of a methyl radical to an attacking free radical with cobalt(III) being reduced to cobalt (II); and (d) a "redox switch" mechanism where metal ions complex with the corrinoid system to weaken the Co-CH_3 bond toward attack by weak electrophiles. Of these mechanisms, the transfers of carbanions and carbon free radicals appear to be the most prominent routes (Fanchiang et al., 1978).

The reactions between Tl^{3+}, Hg^{2+}, Pb^{4+}, or Pd^{2+}, and methylcobalamin are examples of methyl group transfer via a carbanion.

$$H_2O + Hg^{2+} + \underset{(Co^{III})}{B_{12}\{\cdot\cdot CH_3} \longrightarrow CH_3Hg^+ + \underset{(Co^{III})}{B_{12}\text{-}OH} + H^+$$

Under aerobic conditions rate constants generally are in the millisecond range.

Free radical attack on the Co-CH_3 bond by a transient Sn^{III} ion is postulated to occur under anaerobic conditions. Thus:

$$Sn^{2+} + Fe^{3+} \rightleftharpoons [Sn^{3+}] + Fe^{2+}$$

$$\underset{Sn(III)}{[Sn^{3+}]} + \underset{(Co^{III})}{B_{12}\cdot\{\cdot CH_3} \xrightarrow{H_2O} \underset{Sn(IV)}{CH_3Sn^{3+}} + \underset{(Co^{II})}{B_{12}(H_2O)}$$

$B_{12}(H_2O)$ is rapidly oxidized to a cobalamin with Co(III) in the presence of excess Fe(III). Similar reactions occur with Au^{3+} or Cr^{2+} and methylcobalamin.

Based on methylation reactions between several metal ions and methylcobalamin, Dizikes et al. (1978) and Ridley et al. (1977) proposed a relationship between the standard reduction potential (E°) and the methylation mechanism. For those metal redox couples above +0.85 V, reactions with methylcobalamin occur via carbanion formation; and for redox couples below +0.50 V, the methylation reaction involves free radical formation. For those metals with standard reduction potentials in the range +0.50 V to +0.80 V, the mechanism is described in terms of a "redox switch" because both oxidation states of the metal are required for methylation to occur. Metals having standard reduction potentials in the redox switch range include As(V)/As(III), Pt(IV)/Pt(II), and Au(III)/Au(II). Thus, $PtCl_4{}^{2-}$ forms complexes with methylcobalamin and thereby weakens the Co-CH_3 bond and makes it susceptible to electrophilic attack:

$$PtCl_4^{2-} \cdot B_{12}\text{-}CH_3 + PtCl_6^{2-} + H_2O \longrightarrow B_{12}(H_2O) + PtCl_4^{2-} + CH_3PtCl_5^{-} + Cl^{-}.$$

The mechanism is identical to that proposed by Agnes et al. (1971). The structure of the complex formed by platinum tetrachloroplatinate ion and methylcobalamin is not known in detail at this time although it is under study by nuclear magnetic resonance spectroscopy (Fanchiang et al., 1978).

Various metals (Hg, Sn, Pd, Pt, Au, and Tl) and metalloids (As, Se, and Te) are suspected of being methylated in the environment by methylcobalamin, and several of these examples have been confirmed in the laboratory (Chau and Wong, 1978). Other metals (Pb, Cd, and Zn) may not be as susceptible to methylation because of weak alkyl-metal bonds and extensive hydration or hydrolysis of these metal ions in solution.

Bioalkylation of certain metals and metalloids in the environment appears to involve *Pseudomonas*, *Aeromonas*, *E. coli*, *Flavobacterium*, and *Methanobacterium* bacteria (Chau and Wong, 1978; Huey et al., 1974). Many microorganisms produce B_{12} naturally. Thus, the production of alkylated metals and metalloids in the environment may involve cobalamin which is constantly being regenerated by bacteria.

Mechanisms for methylation of metalloids may also involve coenzymes containing sulfur atoms, e.g., S-adenosylmethionine. The question of trace metal flux in the environment and its significance to higher forms of life may or may not depend heavily upon corrinoid systems, but at least these systems serve as convenient laboratory models for the extremely complex biological environment.

E. Analytical Chemistry

The determination of cobalt concentration in a variety of matrices has resulted in the development of numerous sample preparation and analysis procedures. From the plentiful data available on the determination of cobalt, the pertinent literature has been reviewed and is cited. A compendium of all the published literature has not been attempted.

1. Reviews

a. General reviews: Several reviews and authoritative treatises on the analytical chemistry of cobalt are available. The text by Young (1960) emphasized classical analytical methods including gravimetric, titrimetric, colorimetric, potentiometric, polarographic, and spectrographic techniques. Pyatnitskii (1966) addressed both separation techniques, which included precipitation, chelation, chromatography, and electrochemical isolation, and the analytical techniques of radioactivation and catalytic reactions, as well as gravimetric, titrimetric, photometric, and polarographic methods. More recently, Young (1976) reviewed atomic absorption, catalytic, colorimetric, electrometric, gravimetric, spectrographic, titrimetric, and x-ray fluorescence techniques. The cited reviews included analytical methods for cobalt determination in organic matrices such as plant and animal tissues and inorganic matrices including minerals, ores, alloys, soil, and water.

b. Sample preparation: Sample preparation methods generally are classified with respect to their ability to release an analyte from inorganic and organic matrices. Frequently an effective decomposition method for an inorganic matrix is ineffective when organic matrices are encountered. The inorganic matrices requiring preparative measures are generally solids, aqueous liquids, and atmospheric particulates. During sample preparation it is often necessary to consider the limitations of the analytical techniques to be employed to minimize possible matrix interferences in the prepared sample.

The general analyses of organic and inorganic matrices are presented biennially by a number of authors in the April review issue of Analytical Chemistry in odd-numbered years.

Topics reviewed in the April 1979 issue included the determination of cobalt in air, soil, water, fuels, foods, and tissues and liquids analyzed in clinical chemistry.

(1) Preparation of inorganic matrices: A thorough general review of procedures for the decomposition of inorganic matrices was recently published by Sulcek et al. (1977). The use of open and closed (pressurized) systems involving inorganic acids and fusion methods using alkaline, acid, and redox fluxes were discussed in detail. The potential analyte losses and the compatibility of preparative matrices with various analytical instrumental methods were presented.

Another excellent review of general decomposition methods for inorganic and organic matrices was by O'Haver (1976). Dry ashing with and without the use of ashing aids, wet ashing with various acids, alkali salt fusions, and low-temperature plasma ashing were discussed, and there were a number of supplementary references.

A review of general preparative methods for geological materials has been published by Reeves and Brooks (1978). The review outlines a variety of preparative techniques having potential and documented application to most inorganic matrices including atmospheric particulates, glasses, and ashes. Sample preparations by wet-acid dissolution and fusion techniques as well as separation methods were described briefly along with the appropriate analytical technique used.

(2) Preparation of organic matrices: General wet- and dry-ashing methods for the destruction of organic matrices were presented by Gorsuch (1970), who discussed $HClO_3$, HNO_2, H_2SO_4, acid combinations, dry ashing in closed and open systems, the use of ashing aids, oxidative fusion, and low-temperature oxygen plasma ashing. Cobalt was among the specific elements discussed. The results of several recovery studies were presented along with a limited review of cobalt determinations in several matrices by various techniques. The author noted that, in many cases, dry ashing at 450°C should be satisfactory with the aid of HNO_3 or $Mg(NO_3)_2$. Wet-ashing methods were not found to be sources of serious loss. It was suggested that wet-ashing procedures be selected on the basis of convenience and the ability to completely destroy the matrix of interest.

There is an older, but similarly general, text by Thiers (1957) who discussed contamination control in trace element analyses. Dry- and wet-ashing techniques were discussed briefly. The work included references to earlier reviews on sample preparation. Limited data on the recovery of known concentrations of various metals from organic matrices were presented. Recoveries of 105 to 112% were reported for cobalt in the range of 1 to 10 mg/kg for protein after ashing at 450°C. Information regarding potential contamination sources and possible preventive measures were presented for a number of metals; there were several specific references to cobalt.

Dry ashing of high-organic matrices such as coal in silica crucibles at 500°C for 2 hr was reported (Gluskoter et al., 1977). Barium nitrate was employed as an ashing aid. Satisfactory results for several metals determined by emission spectrography were found.

Wet ashing of refuse using a combination of $HNO_3/HF/HClO_4$ was reported by Campbell et al. (1977). Subsequent analyses by flame atomic absorption spectrophotometry and energy dispersive x-ray spectrography were performed.

A comparison of low-temperature oxygen plasma ashing and muffle furnace ashing at 500°C by radiotracer studies using ^{60}Co in blood was reported (Oikawa, 1977). Recovies of 103% were observed for low-temperature oxygen plasma ashing, and 98% for muffle furnace ashing. During a study of preparative methods for coal samples, low-temperature oxygen plasma ashing was found not to be satisfactory for subsequent arc emission spectrographic analysis (Ruch et al., 1974).

The utility of wet ashing with $HNO_3/HClO_4$ or HNO_3/H_2O_2 mixtures with heating in a microwave oven was studied (Abu-Samra et al., 1975). A potential advantage of the technique may be in the reduced hazard of drying $HClO_4$ in the presence of organic material.

The preparation of food samples using a V_2O_5-catalyzed decomposition employing $HNO_3/H_2SO_4/H_2O_2$ was reported (Baetz and Kenner, 1975). Sample cleanup consisted of column separation using Chelex 100 resin. The method was not recommended for cobalt determinations at the 100-ppb (μg/kg) level due to reported variable recoveries.

c. Methods of separation: In many analytical schemes, it is often desirable or necessary to separate cobalt from interfering materials. Isolation methods also afford the opportunity to concentrate the analyte and to minimize background interferences.

There were several general reviews of separation techniques. Precipitation, chelation, chromatography, and electrochemical methods for the separation of cobalt were discussed by Pyatnitskii (1966). More recently Reeves and Brooks (1978) reviewed precipitation, chelation/solvent exchange, electrodeposition, and ion-exchange methods for isolation of interfering species.

Details of several chelation/extraction techniques were summarized by O'Haver (1976). Specific chelating materials including dithizone, ammonium pyrrolidine dithiocarbamate, sodium diethyldithiocarbamate, oxine (8-hydroxyquinoline), and Chelex 100 chelating resin were discussed.

Lakanen (1962) classified extractants on the basis of their chemical properties: neutral salts, alkalis and alkaline salts, acids, acid salts, chelating agents, reducing agents, water and CO_2-saturated water, and combinations. The results for six different extraction methods for 13 trace elements in soil were presented.

Many useful separation schemes and specific procedural details were found in classical colorimetric/photometric texts, such as Sandell (1959), Sandell and Onishi (1978), and Snell and Snell (1949 and 1959).

The use of dithizone as an extracting agent is one of the oldest and most prevalent separation techniques. Baudot et al. (1976) reported the use of a dithizone-carbon tetrachloride system at controlled pH with subsequent separation by thin-layer chromatography on silica gel. Sensitivities on the order of 10^{-7} g/liter were reported. The method was reported to be directly applicable in hydrology. After the destruction of organic matter, biological samples such as blood, urine, and excrement were similarly extracted.

The use of a dithizone-chloroform system for the extraction of sediments treated with $HF/H_2SO_4/HClO_4$ was described (Armannsson, 1977). After extraction, the organic phase was oxidized using $HClO_4$ and finally taken up in an HCl solution for analysis by atomic absorption methods. Recoveries on the order of 95% were reported. A similar method was used by Tewari and Bhatt (1973) for the direct extraction of metals from biological tissues. They reported the extraction of the cobalt complex at pH 6 to 8. The method was found to be superior to dry-ashing and wet digestion techniques.

Another frequently encountered class of extracting agents is the dithiocarbamates. Early work was performed by Pohl (1953) using sodium diethyldithiocarbamate and by Stetter and Exler (1955) using sodium pyrrolidine dithiocarbamate. Lakanen (1962) employed pyrrolidine dithiocarbamic acid in chloroform for the quantitative extraction and concentration of 18 metals at a single pH from soil extracts. The influence of pH and the presence of other complexing agents were studied. After combustion of the organic complex, analyses were performed by direct current (DC) arc-emission techniques. Chau (1971) used ammonium pyrrolidine dithiocarbamate for the analysis of natural waters by atomic spectrometric methods.

Another useful extraction system employed thiocyanate complexes. Sandell (1959) reported the use of solvents such as amyl alcohol, ethyl ether, methyl isobutyl ketone, or amyl alcohol/ethyl ether for extraction of the cobalt complex. With the use of ammonium thiocyanate, a recovery of 86% was reported for a single extraction step. The thiocyanate complexation procedure may be used advantageously to separate cobalt from iron and/or nickel interference.

The use of trifluoroacetylacetone to chelate ^{60}Co in seawater for subsequent extraction with toluene was reported (Lee and Burrell, 1972).

Chelating resins were used to effect the separation and concentration of cobalt in solution mixtures. Chelex 100 was employed to separate alkali and alkaline earth elements from estuarine water and seawater prior to electrothermal atomic absorption analyses (Kingston et al., 1978). Recoveries of 99.5 ± 0.3%

and 97.3 ± 0.2% were observed for 100- and 1-liter volumes of seawater, respectively. Similarly, Amberlite CG400 resin in the thiocyanate and/or chloride forms was used for the preparation of seawater matrices prior to spectrophotometric analyses (Kiriyama and Kuroda, 1977). Elution of the analyte was accomplished with 2 M $HClO_4$. The use of chitosan, a chelating, polymeric, chromatographic support for the separation of trace elements from seawater was described by R. Muzzarelli et al. (1970). Experiments with chitosan and Chelex 100 resins using zinc and mercury found comparable capacities for both resins. Complete collection from salt solutions and seawater of 0.0104 μg of ^{60}Co was reported. Subsequently, 89% of the collected material was eluted using 60 ml of 0.1 N EDTA. Collection efficiency dropped to 80% for ^{60}Co in a brine solution of twofold-concentrated seawater. Chitosan was the subject of a subsequent patent by N. W. Muzzarelli (1972).

Numerous other preparation and extraction techniques and modifications for a variety of matrices were reported. These additional methods are highlighted in subsequent sections with the specific analytical techniques for which they were employed.

2. Analytical quality assurance: The accurate analysis of various atmospheric, liquid, and solid materials that may be organic and/or inorganic in composition requires careful sample preparation and analysis. Frequently, procedures for the analysis of a selected material may be nonexistent, inapplicable, or invalidated for specific analytes or matrices. The emphasis of this subsection is placed on procedures that should be followed and precautions that should be taken during the development of new analytical methods. Documented problems encountered in the preparation of sample materials for cobalt analysis are discussed.

a. Standard reference materials: The establishment of the accuracy and precision of an analytical protocol is necessary to validate data generated from real-world samples. To establish analytical accuracy and precision, the reproducible recovery of a known analyte quantity from the sample matrix of interest must be documented. For evaluating analytical protocols, several organizations have undertaken development of SRM's for both inorganic and organic matrices.

(1) Inorganic reference materials: A large number of metals and metal alloys (> 70) have been prepared by the National Bureau of Standards (NBS) for use as SRM's. Cobalt concentrations in the range of 0.008 to 51.6% by weight are available in numerous ferrous alloys, specialty alloys, and nonferrous alloys having copper or nickel bases. Glass reference materials with nominal concentrations of 1, 50, and 500 mg Co/kg contain (informational values in parentheses) (390), (35.5), and 0.71 mg Co/kg. The reader should consult the NBS catalog of SRM's for details on currently available standards (Office of Standard Reference Materials, 1979).

Some care should be taken in selection of the SRM since some have been prepared specifically for microanalyses such as electron microprobe or spark source mass spectral analyses rather than spectroscopic methods which generally consume larger sample quantities.

The U.S. Geological Survey makes available reference glass (Table III-8) and rock (Table III-9) samples. A detailed list of over 140 standard rock materials, their sources, and applicable references was published recently by Reeves and Brooks (1978). Unfortunately, information on the certified contents of the various rock materials was not provided. In lieu of duplicating the complete list, the sources cited are presented in Table III-10.

(2) Organic reference materials: In contrast to the variety of inorganic SRM's available, only a very limited number of reference materials which have certified cobalt concentrations are available. But a number of additional materials are available which have noncertified cobalt concentration values reported for information purposes only. Two sources of a variety of materials are the U.S. NBS and the International Atomic Energy Agency (IAEA). A compilation of materials available through these two organizations is presented in Table III-11. The analytical techniques used by NBS during certification are summarized for all elements analyzed in each SRM on the Certificate of Analysis; thus, the specific technique(s) employed for cobalt analyses cannot be directly identified. Interlaboratory analyses of the cobalt content of reference standards prepared by IAEA were almost entirely performed by neutron activation techniques (Górski et al., 1974). The reported values are the overall mean of acceptable laboratory averages.

TABLE III-8

U.S. GEOLOGICAL SURVEY REFERENCE GLASS STANDARDS
(Myers et al., 1976)

Designation	Approximate Cobalt Concentration (mg/kg)	Actual Cobalt Concentration (mg/kg)	Evaluation of Quantitative Assay
GSB	0.5	2	Reservations
GSC	5	6	Suggested
GSD	50	35	Accepted
GSE	500	450	Accepted

TABLE III-9

REFERENCE ROCK MATERIALS AVAILABLE FROM THE U.S. GEOLOGICAL SURVEY

Designation	Matrix	Cobalt Concentration (mg/kg)
AGV-1	Andesite	18.7
BCR-1	Basalt	41.6
BHVO-1	Basalt	44.8 - 45.5
DTS-1	Dunite	> 215
G-2	Granite	4.14
GSP-1	Granodiorite	6.61
MAG-1	Marine mud	17.6
PCC-1	Peridotite	105 - > 215
QLO-1	Quartz latite	6.9
RGM-1	Rhyolite	2.3
SCO-1	Cody shale	10[a]
SOC-1	Mica schist	20[a]
STM-1	Nepheline syenite	< 1.00

a/ Semiquantitative estimate

TABLE III-10

SOURCES OF STANDARD ROCK MATERIALS
(Reeves and Brooks, 1978)

Mineral Resources Division, Box 903, Dodoma, Tanzania.
E. Aleksiev, Geological Institute, Sofia, Bulgaria.
Bureau of Analysed Samples, Newham Hall, Middlesborough, U.K.
A. B. Poole, Department of Geology, Queen Mary College, London.
H. de la Rocher, Centre de Recherches Petrographiques et Geochimiques, 15 Rue N.-D. des Pauvres, Nancy, France.
K. Schmidt, Zentrales Geologisches Institut. Invalidenstrasse 44, 104 Berlin, Germany (D.D.R.).
A. A. Kukharenko, Department of Mineralogy, Leningrad State University.
A. H. Gillieson, Mines Branch, 555 Booth St., Ottawa.
G. F. Smith Chemical Co., Box 23344, Columbus, Ohio 43223.
National Bureau of Standards, Washington, D.C. 20234.
F. J. Flanagan, U.S. Geological Survey, Reston, Virginia 22092.
N. H. Suhr, Mineral Constitution Laboratories, Pennsylvania State University, University Park, Pennsylvania 16802.
Atsushi Ando, Geological Survey of Japan, Misamoto-cho, Kawasaki, Japan.
H. P. Beyers, South African Bureau of Standards, Pvt. Bag 191, Pretoria, South Africa.
D. M. Moore, Department of Geology, Knox College, Galesburg, Illinois 61401.
New Brunswick Laboratory, U.S. Atomic Energy Commission, Box 150, New Brunswick, New Jersey 08903.
H. Schwander, Mineralogisches-Petrographisches Institut der Universität, CH-4000 Basel, Switzerland.
E. Jaeger, Mineralogisches-Petrographisches Institut der Universität, CH-3000 Bern, Switzerland.
Engineered Materials, Box 363, Church St. Station, New York 10008.
S. Y. Taha, Laboratories Division, Natural Resources Authority, Box 2220, Amman, Jordan.
W. D. Johns, Source Clay Repository, Department of Geology, University of Missouri, Columbia, Missouri 65201.
Branch of Exploration Research, U.S. Geological Survey, Denver, Colorado 80225.
R. I. Barnhisel, Department of Agronomy, University of Kentucky, Lexington, Kentucky 40506.
G. Holmgren, Soil Conservation Service, 1325 North St., Lincoln, Nebraska 68508.
International Atomic Energy Agency, Analytical Quality Control Services, Laboratory Seibersdorf, Box 590, A-1011 Vienna, Austria.
Laboratory of Marine Radioactivity, Monaco.
O. H. J. Christie, Mass Spectrometric Laboratory, University of Oslo, Box 1048, Blindern, Oslo 3, Norway.
H. J. Rosler, Bergakademie Freiberg, Sektion Geowissenschaften, 92 Freiberg, Germany (D.D.R.).
Václav Zyka, Institute of Mineral Raw Materials, 28403, Kutna Hora, Czechoslovakia.
E. Jarosewich, Department of Mineral Sciences, Smithsonian Institute, Washington, D.C. 20560.

TABLE III-11

ORGANIC MATRIX REFERENCE MATERIALS HAVING REPORTED COBALT CONTENT
(Office of Standard Reference Materials, 1978; Gorski et al., 1974)

Material	Supplier	Catalog Number	Reference Concentration (mg/kg)
Rice flour	NBS	1568	0.02
Spinach	NBS	1570	(1.5)
Orchard leaves	NBS	1571	(0.2)
Tomato leaves	NBS	1573	(0.6)
Pine needles	NBS	1575	(0.1)
Bovine liver	NBS	1577	(0.18)
Cobalt cyclohexanebutyrate	NBS	1055b	14.8%
Dried whole animal blood	IAEA	A-2	0.419
Calcinated animal bone	IAEA	A-3/1	0.463
Milk powder	IAEA	A-8	0.0166
Wheat flour	IAEA	V-2/1	0.0368
Dried potatoes	IAEA	V-4	0.0205

Several other sources of reference materials are available. To aid in the evaluation of waste-to-energy systems which consume urban refuse, Haynes et al. (1977; see also Campbell et al., 1976) developed a ground refuse material and determined a number of elemental concentrations in it by atomic absorption and x-ray fluorescence methods. Additional information including elemental analyses of this reference material by spark source mass spectrography and atomic absorption spectrophotometry has been published (Gorman et al., 1978; Golembiewski et al., 1978).

An additional series of organometallic stock standards is available through the Conostan Division, Continental Oil Company. Organocobalt standards are available individually or in mixtures of up to 23 elements. This standard was designed specifically for the determination of wear metals in lubricating oils by atomic absorption or emission spectrochemical methods.

Another standard material consisting of a mixture of several metals, including cobalt, in gelatin is available through the Eastman Kodak Company, Rochester, New York. This material is useful for the simulation of proteinaceous matrices. Although the determination of cobalt concentrations has been of continuing interest in a variety of materials and its analysis by current analytical capabilities is possible at levels below 1 ppm (1 mg/kg), low-concentration certified reference materials for many matrices of current interest have been lacking. Furthermore, relatively subtle differences between the SRM matrix and the sample matrix of interest may result in significantly different behavior by the two materials when treated by a specific analytical protocol. Fortunately, alternative supplementary methods are available for cases in which SRM's are not available or the sample matrix is unknown or unpredictable.

Many analytical data about cobalt concentrations in NBS SRM 1571 Orchard Leaves and SRM 1577 Bovine Liver have been generated. Some of the results reported are summarized in Table III-12. There is a similar compilation for SRM 1632 Coal and SRM 1633 Coal Fly Ash; this is given in Table III-13. It is interesting to note that the overall agreement is quite good between the NBS informational value and those values reported for the coal and fly ash SRM's. In contrast, the data for orchard leaves and

TABLE III-12

COMPARISON OF REPORTED COBALT ANALYSES OF NBS ORCHARD LEAVES AND BOVINE LIVER

SRM 1571 Orchard Leaves (mg/kg dry weight)	SRM 1577 Bovine Liver (mg/kg dry weight)	Reference
(0.2)	(0.18)	NBS--information only
0.138 ± 0.010	0.23 ± 0.03	Goldberg (1972)
< 0.3	< 0.3	Goldberg (1972)
0.12	0.15	Goldberg (1972)
0.18 ± 0.02	0.29 ± 0.01	Goldberg (1972)
0.13 ± 0.02	0.41 ± 0.12	Kusaka et al. (1977)
0.17	0.21	Becker et al. (1974)
0.29 ± 0.1	0.23 ± 0.1	Zikovsky and Schweikert (1977)
0.1 - 0.18	0.15 - 0.29	Becker (1977)
0.1	-	Morrison and Potter (1972)
-	0.17	Nadkarni and Morrison (1973)
-	0.24	Henzler et al. (1974)
-	0.34	Masironi (1974)
-	0.225	Masironi (1974)
-	0.26	Masironi (1974)
-	0.39	Masironi (1974)
-	0.12	Masironi (1974)
-	0.174	Masironi (1974)

TABLE III-13

COMPARISON OF REPORTED COBALT ANALYSES OF NBS COAL AND COAL FLY ASH

SRM 1632 Coal (mg/kg dry weight)	SRM 1633 Coal Fly Ash (mg/kg dry weight)	Reference
(6)	(38)	NBS--information only
5.70 ± 0.12	40.3 ± 0.4	Rowe and Steinnes (1977)
5.13 - 6.2	35.4 - 46	Literature value cited by Rowe and Steinnes (1977)
5.7 ± 0.4	41.5 ± 1.2	Ondov et al. (1975)
6.04 ± 0.29[a/] 6.4 ± 0.28[a/]	42.0 ± 1.9	Ragaini et al. (1975)
5.6 ± 0.3	41.2 ± 1.0	Zoller, cited by Ragaini et al. (1975)
5.2 ± 0.4	40 ± 2	Abel and Rancitelli (1975)
-	41.5 ± 1.2	Bambynek et al. (1972; cited by Giauque et al., 1977)
-	< 130	Giauque et al. (1977)
5.9	46	Klein et al. (1975)
-	38 ± 4	Sheibley (1975)

a/ Presumably one of these values is a typographical error.

bovine liver vary considerably from the NBS informational values. It is possible that significantly different preparation and handling techniques were used by the analysts working with orchard leaves and bovine liver. The possibilities of analyte losses to container walls, low-level contamination, and/or matrix interferences during analysis exist. If one assumes that all reported analyses are of similar quality, the nonhomogeneous distribution of cobalt in orchard leaves and bovine liver is possible, given the much lower concentration of cobalt in these materials compared to coal and fly ash SRM's. The importance of the sample quantity analyzed and the analyte species and concentration present with respect to securing accurate quantitative information from the analysis of reference materials was pointed out by Lister (1977).

Additional discussion of interferences or problem areas is presented with the specific analytical techniques described below.

b. Interlaboratory comparisons: The verification of analyses or the development of a specific reference material may be undertaken through interlaboratory analyses. A sufficiently large quantity of homogeneously mixed materials must be prepared to allow verification analyses and, in the case of reference materials, to provide a reserve quantity of material for future use. The preparation step frequently entails the pooling of materials from many similar sources as is the case with blood/serum matrices. In the case of liquids, thorough mixing may be the only preliminary processing step. In the case of solids, a drying and/or comminuting process may be necessary to assure thorough mixing and the availability of representative aliquots from the final mixture.

The establishment of a minimum handling protocol is necessary. This is especially important for the identification of the aliquot to be analyzed and the long-term stability of the reference material. Standard reference materials provided by NBS specify SRM storage conditions; SRM aliquot preparation, such as drying prior to weighing; and the minimum aliquot to be analyzed to assure representative sampling of the SRM. For example NBS SRM 1571 Orchard Leaves should be stored in the dark at 10 to 30°C, dried a minimum of 4 hr at 85°C prior to weighing, and a minimum aliquot of 250 mg taken (Cali, 1977). Establishment of such storage and sampling procedures will help assure that a properly prepared reference material is sampled in a similar manner by all individual users of the material.

Validation of the analyte content of a reference material consists of two phases: (a) establishing the actual analyte content and (b) identifying the reproducibility of analyses. To establish the analyte content, at least two independent quantitative analytical methods must be employed. For example, atomic absorption spectrophotometry and neutron activation analysis are often complementary techniques. In addition to providing selective quantitative measurements by two different routes, the sample preparation is often significantly different for each technique. The minimized possibility of matrix interferences or failure to completely release the analyte from the sample matrix is the primary benefit of employing two or more independent preparation and analysis schemes. In the course of this preliminary testing a sufficient number of aliquots should be analyzed to allow statistical analysis of the results for each technique employed. Interlaboratory comparisons allow a variety of analytical techniques to be used and minimize possible operator bias. Precision values beyond the expected ranges of the analytical technique employed may be indicative of homogeneity or stability problems of the proposed reference material. Failure of two independent techniques to provide comparable quantitative data also may be indicative of technique associated deficiencies with the matrix or preparation of interest.

The results of several interlaboratory comparisons for the preparation of new reference materials or the extension of SRM's to additional analytes have been published.

The analysis of dried animal whole blood, calcinated animal bone, powdered milk, wheat flour, and dried potatoes was published by the IAEA (Górkski et al, 1974). A total of 31 laboratories using at least six independent analytical methods participated, which produced 453 data sets for 10 elements in the five different materials tested.

The thorough procedures and data generated during the preparation of reference rock materials by the USGS were described in detail in the Geological Survey Professional Paper 840

(Flanagan, 1976). The use of multiple analytical preparations and techniques as well as statistical analysis to identify acceptable, qualified, and unacceptable analyte data were presented.

A classic study for the extended use of available SRM's was published by Ondov et al. (1975). Using neutron activation techniques, four laboratories generated concentration and precision data for certified and uncertified elements in NBS SRM 1632 Bituminous Coal and SRM 1633 Coal Fly Ash. This work has greatly extended the utility of the two SRM's.

More recently Ondov et al. (1977) reported an interlaboratory comparison of neutron activation and atomic absorption analyses of size-classified fly ashes. A comparison of quantitative and statistical data for the 12 elements analyzed in NBS SRM-1633 and size-classified fly ashes by both techniques was presented.

A recent 52-laboratory survey using 20 multielement ores and mining concentrates was reported by Lister (1977). Of this group, five ores were analyzed for cobalt by neutron activation analysis. It was reported that, unfortunately, very little sample material remained for use as reference standards.

Dybczynski et al. (1978) reported a 35-laboratory intercomparison of trace elements in water. Cobalt was analyzed using neutron activation analysis, atomic absorption spectrophotometry, and x-ray fluorescence. After statistical treatment of all reported data, a mean of 11.9 ± 0.8 μg/liter for all accepted results by all methods was found; the actual value was 12.1 μg/liter.

Goldberg (1972) reported a three-laboratory intercomparison of cobalt in NBS SRM 1571 Orchard Leaves and SRM 1577 Bovine Liver. The data for SRM 1577 were somewhat variable; they ranged from 0.15 to 0.29 mg Co/kg. The cobalt concentration found in Orchard Leaves (0.12 to 0.13 mg/kg) was slightly less than the NBS uncertified value of 0.2 mg/kg. A comparison of atomic absorption and neutron activation analyses of both materials by a single laboratory produced consistently higher results using neutron activation.

A round robin analysis of cobalt concentration in SRM 1577 Bovine Liver was sponsored by WHO-IAEA (Masironi, 1974). Neutron activation and some atomic absorption analyses were performed. A range of 0.12 to 0.39 mg/kg was reported.

The concentrations of 12 elements were determined in an oyster homogenate by 87 laboratories in an IAEA study reported by Fukai et al. (1978). More than 60% of the results reported for cobalt concentrations were within an acceptable range. A comparison of data generated by atomic absorption and neutron activation methods resulted in average concentrations of 1.9 ± 0.8 mg Co/kg and 0.41 ± 0.02 mg Co/kg dry weight by each method, respectively. The authors concluded that atomic absorption analyses tend to produce high cobalt results. Possible sources of systematic error among the laboratories using atomic absorption techniques may reflect low level contamination or interference problems; however, the specific individual methods of analysis are not reported, and identification of problem areas is open to conjecture.

c. Standard additions method: Interlaboratory comparisons generate a large body of quantitative and statistical data, but such a major undertaking is often beyond the scope of the individual analyst who has only a limited number of samples. A more limited approach is the use of a standard additions technique. Details of the technique, which involve the addition of known quantities of analyte to the sample matrix of interest, can be found in general analytical chemistry texts (Ewing, 1969, and Willard et al., 1974).

Initially, the sample is analyzed to establish an observed analyte concentration. Subsequently, a known quantity of the analyte is added to the unknown sample which should produce a proportional increase in the analyte signal. The analyte fortification should be added in a minimal volume to prevent unnecessary dilution of the sample matrix. Single or multiple additions may be performed. Problems with the matrix interference or the analytical operating conditions may be identified with this technique. An excellent presentation on the uses and limitations of the standard additions method was presented by Klein and Hach (1977). The standard additions method is compatible with the gamut of analytical techniques in electrochemistry and spectroscopy.

d. Sampling, preservation, and preparation problems: Accurate quantitative sample analysis begins with the collection and preservation of high-integrity samples. Maienthal and Becker (1976) published a survey of sampling, sample handling, and long-term storage procedures for environmental materials. One citation from Speecke et al. recorded the contamination of four successive blood aliquots with trace elements from a disposable syringe. Cobalt contamination was 0.9 μg/liter for the first 20-ml aliquot and had dropped to 0.2 μg/liter for the fourth aliquot.

Anand et al. (1975) discussed some aspects of sample collection and stability of body fluids for trace element analysis. Cobalt contamination of blood samples by technical grades of anticoagulants was reported. A citation from Versieck et al. documented significant cobalt and other trace elemental contamination of liver tissue from Menghini needles used for the collection of biopsy samples.

The use of extremely sensitive analytical methods capable of highly accurate analyses require the use of collection, preservation, and preparation procedures that will maintain high-sample integrity by minimizing sources of contamination or analyte losses. A number of observations have been made in this respect. Zief and Mitchell (1976) recently published a general text on contamination control in elemental analysis. They reported on basic considerations in quantitative ultratrace analysis, sources of contamination, selection of suitable collection and storage materials, purification of reagents, and analysis of reagent impurities. The introduction of cobalt contamination from tungsten carbide and alumina-ceramic mortars and pestles or grinding vials was documented. The successful recovery (99 to 102% range) of radioactive cobalt from blood using low-temperature ashing was noted.

An older general text by Thiers (1957) addressed both losses and contamination during the preparation of organic matrices. Contamination during sample manipulation, contamination measurement, cleaning techniques, and reagent purification was discussed. Use of a hammermill, a Wiley mill, and a hand mortar and pestle was not found to produce cobalt contamination during oat-grinding operations; however, a jar mill with flint balls introduced the equivalent of 0.02 mg Co/kg of oat grain.

The loss of cobalt during dry ashing of plant material with Na_2CO_3 was referenced. In contrast, the recoveries of cobalt added to protein and subsequently dry ashed at 450°C ranged from 105 to 112% for 2.5 to 25-μg quantities. It was also recorded that ^{60}Co at 13.3 μg/liter in acid solutions could be stored in borosilicate glass bottles without measurable loss for at least 500 days. Similarly, a 3 mg Co/liter aqueous acidic solution was stored in polyethylene containers with no detectable change over 1 year.

Many specific instances of losses or contamination have been published. As the sensitivity of analytical methods improves, these problems are more frequently encountered. A number of studies have been conducted to identify preparation problems. A rather extensive study of cobalt losses from coal samples during up to 20 hr of dry ashing verified the relative stability of cobalt (Ruch et al., 1974). Another dry-ashing study revealed retention of cobalt on crucible walls unless an ashing aid or fusion salt was present (O'Haver, 1976).

Iyengar et al. (1978) documented the loss of cobalt from rat tissue under vigorous dry-ashing conditions. Alternatively, cobalt was found to be quantitatively retained with the use of moderate ashing procedures.

Although low-temperature ashing has been performed in many instances, its use for the preparation of coal for DC arc emission purposes is not recommended due to incomplete destruction of the sample matrix (Ruch et al., 1974).

The success of any preparation requires the complete release of the analyte from the crystalline matrix or the compound species of interest. The formation of insoluble or slightly soluble cobalt sulfates during the preparation of sulfoarsenide or sulfur-containing materials was reported (Sulcek et al., 1977). The use of alkaline fusion techniques, sometimes called Freiberg decomposition, was recommended. A strong affinity for carbon was found for Co^{2+} in HCl solutions (Oikawa, 1977).

A study of analyte losses to storage containers made of borosilicate glass or polyethylene has been conducted (Robertson, 1968). At pH 8.1, cobalt losses over a 20-day storage period were 0 to 7% in borosilicate glass compared to 15% in polyethylene containers. At pH 1.5 negligible losses of cobalt were encountered over a 60-day storage period.

During separation steps, variable recoveries of cobalt may be encountered. This is frequently the result of equilibrium shifts and the significant dependency of extraction efficiency on pH (Chau, 1971). In some cases the removal of interfering species may be accomplished by neutralization methods (Young, 1960).

Positive errors or contamination can originate from a number of sources, collection devices, storage containers, or preparation reagents (Oikawa, 1977). The need to analyze blanks from collection devices, containers, and reagents for neutron activation analyses was emphasized (Erasmus et al., 1977). Contamination from high-density linear polyethylene has been documented for 10% HNO_3 solutions frozen for 4 years (Marchant and Klopper, 1978). Similarly, the use of high-purity reagents, such as Ultrex® from Baker or Suprapure® from Merck, is mandatory for quantitative analysis at current detection limits.

3. Methods of detection and analysis

a. Background remarks: The determination of cobalt in numerous organic and inorganic matrices has been of continuing long term interest. As a result, analytical protocols have been developed which employ an assortment of different detection methods. The most useful and still frequently used classical analytical technique employs colorimetric/photometric methods. A variety of instrumental techniques for single and multielement determinations are available for the sensitive selective measurement of cobalt. The single element techniques most frequently encountered are atomic absorption spectrophotometry and electrochemical methods, usually polarography. Most multielement analytical techniques have been successfully employed for cobalt determinations and include emission spectrography, plasma emission spectroscopy, neutron activation analysis, and x-ray techniques, primarily fluoresence. Methods which are being used more and more include spark source mass spectrometry and catalytic reactions. Of all the techniques, atomic absorption spectrophotometry and neutron activation analysis currently hold equal portions of all the cobalt analyses reported. Several reviews compared the utility and limitations of a number of techniques. Young (1960) reviewed some classical methods which are summarized in Table III-14. Pyatnitskii (1966) presented information on a combination of classical and instrumental methods including gravimetry, titrimetry, photometry, polarography, catalytic reactions, and radioactivation analysis.

TABLE III-14

CLASSICAL METHODS OF COBALT ANALYSIS
(Young, 1960)

Method	Interferences	Cobalt Concentration Range
Gravimetry	Bi, Cr, Cu, Fe, Ag, Sn, Ti, V, Zr	Wide
Electrolysis	Fe, PO_4^{3-}	Percent
Titrimetry	Mn, Cu, Ni	-
Colorimetric	Fe, Ni, Cu (most frequent)	< 1 μg
Polarographic	Bi, Cu, Cr, Fe, Pb, Mn, Mo, W, V	50 - 400 mg/liter

A thorough discussion of modern methods of elemental analysis was published by Pinta (1978). Topics presented included principles, methods, and applications of emission spectroscopy, flame and flameless atomic absorption spectrometry, activation analysis, x-ray fluorescence spectrometry, and molecular fluorescence. The author made extensive compilations of the analytical applications of instrumental techniques to organic and inorganic matrices reported during the 1971-1974 period.

An excellent review of current general spectroscopic methods was edited by Winefordner (1976). Specific methods discussed included atomic emission, absorption, and fluorescence spectrometry, molecular photometry and fluorescence, nuclear methods/neutron activation analysis, x-ray methods, and spark source mass spectrometry.

A review of general instrumental methods including emission spectroscopy, neutron activation analysis, x-ray fluorescence spectrometry, and spark source mass spectrometry appeared in the literature (Anonymous, 1977b). A limited comparison of emission spectrographic, atomic absorption spectrometric, and neutron activation methods was made (Oliver and Cox, 1967).

Although their work is primarily directed to the analysis of inorganic/geological materials, Reeves and Brooks (1978) presented an excellent overview of gravimetric and titrimetric methods, absorptrometry/photometry, molecular fluorimetry, emission spectroscopy, atomic emission, fluorescence and absorption spectrometry, x-ray emission spectrometry, radiometric/radioactivation methods, electrochemical methods, and spark source mass spectrometric methods.

Biennial general reviews on the basis of techniques are presented in the April review issues of Analytical Chemistry during even-numbered years. The most recent review, issued in 1978, included cobalt determination by ultraviolet-visible absorption spectrometry, emission spectroscopy, flame emission/atomic absorption/atomic fluorescence spectrometry, nuclear techniques (including neutron activation analysis), x-ray diffraction and spectrometry, and electrochemistry.

Although the complexity of analysis and the actual detection limit attained is directly dependent upon the matrix analyzed and the analytical technique employed, some general comparisons can be made. A compilation and comparison of various instrumental techniques is presented in Table III-15. It must be realized that a number of cited detection limits may be based on the analysis of standard or defined matrix materials which reflect an optimum situation. Deteriorated or variable sensitivity may be encountered during the analysis of samples having unknown or inconsistent matrices.

While some freedom is available in the selection of analytical methods, one is frequently limited to accessible equipment or faced with the necessity of subcontracting when costly specialized equipment is required, as is usually the case for neutron activation analysis or spark source mass spectrometry.

TABLE III-15

INSTRUMENTAL METHODS FOR COBALT ANALYSIS

Method	Remarks	Cobalt Detection Limit	Reference
DC arc emission			
Spectrometry	Geological material, λ = 3453.5 Å	2 - 5 μg/g	Pinta (1978)
Spectrography	Graphite matrix, λ = 3454 Å	20 μg/g	Reeves and Brooks (1978)
Stallwood jet source	Graphite matrix, λ = 3454 Å	5 μg/g	Reeves and Brooks (1978)
DC plasma emission spectrometry	Liquid matrix, λ = 3405 Å	0.001 μg/ml	Veillon (1976)
Spark emission spectrometry	Liquid matrix, λ = 3454 Å	2 μg/ml	Pinta (1978)
Radiofrequency plasma emission spectrometry	Liquid matrix, λ = 3454 Å	0.003 μg/ml	Pinta (1978)
Flame atomic emission spectrometry	$N_2O-C_2H_2$ flame, λ = 3454 Å	0.03 μg/ml	Pinta (1978)
Flame atomic absorption spectrometry	Air-C_2H_2 flame, λ = 2407 Å	0.002 μg/ml	Veillon (1976)
Electrothermal atomic absorption spectrometry			
Carbon rod atomizer	λ = 2407 Å	5 x 10^{-12} g	Culver (1975)
Graphite furnace atomizer	Air purge, λ = 2407 Å	80 x 10^{-12} g	Anonymous (1974)
	Air purge interrupt, λ = 2407 Å	25 x 10^{-12} g	Anonymous (1974)
Flame atomic fluorescence	Air-C_2H_2 flame, λ = 2407 Å	1.5 μg/ml	Pinta (1978)
Neutron activation analysis	^{60}Co γ-ray spectrometry	0.0002 μg	Reeves and Brooks (1978)
	^{60}Co β counting	0.04 μg	Reeves and Brooks (1978)
X-ray fluorescence	-	100 - 1,000 μg/liter	Anonymous (1977b)
Spark source mass spectrography	-	0.1 μg/liter	Anonymous (1977b)
Polarography	-	50 - 400 μg/g	Reeves and Brooks (1978)
Molecular fluorescence spectrometry	Salicylfluorene/H_2O_2	0.0001 μg/ml	St. John (1976)

b. <u>Colorimetric methods of analysis</u>: Colorimetry, otherwise known as spectrophotometry, photometry, or absorptrometry, for the analysis of cobalt has utilized a variety of absorptive cobalt complexes.

Snell and Snell (1949 and 1959) prepared two volumes describing general colorimetric analyses. The earlier volume presented methods for the analysis of cobalt in various ferrous and nonferrous alloys, ores, silicates, soil, biological tissue, and plant matter using a number of complexing agents. The later volume expanded the analytical applications to dust, minerals, feed, and fertilizer. The use of ion-exchange procedures was also presented.

In his classic text on the general subject, Sandell (1959) noted, "The colorimetric trace determination of an element requires more or less extensive separations. This requirement is both a source of weakness and strength." A chapter specifically devoted to cobalt analyses and applications discusses the use of dithizone, nitrosonaphthol, and thiocyanate separation procedures. Applications to silicates, soils, ferrous and nonferrous alloys and biological materials are presented. The sensitivities of a variety of colorimetric methods for cobalt analyses, as tabulated by Sandell, are listed in Table III-16.

TABLE III-16

SENSITIVITIES OF SOME COLORIMETRIC METHODS FOR COBALT
(Sandell, 1959)

	Sensitivity $\mu g\ Co/cm^2 \equiv \log I_o/I = 0.001$
Nitroso-R salt	0.0019 (420 nm); 0.0042 (520 nm)
2-Nitroso-1-naphthol-4-sulfonic acid	0.004 (525 nm)
2-Nitroso-1-naphthol ($CHCl_3$)	0.0042 (530 nm)
o-Nitrosoresorcinol	0.0025 (430 nm)
Thiocyanate (isoamyl alcohol)	0.055 (620 nm); 0.009 (310 nm)
Tetraphenylarsonium thiocyanate ($CHCl_3$)	0.034 (620 nm)
Hydrochloric acid (concentrated)	0.2 (625 nm)

Recently, a revised edition of the general aspects of trace metal photometric analysis was published by Sandell and Onishi (1978).

A brief review of general colorimetric principles and methods was presented by St. John (1976). From the work of Cheng and Bray (1955), a sensitivity of 0.0016 $\mu g/cm^2$ was cited for 1-(2-pyridylazo)-2-naphthol (PAP).

The review of ultraviolet and light absorption spectrometry for the 1975-1977 period, prepared by Howell and Hargis (1978), identified over 20 methods for the analysis of cobalt in metals, soils, silicates, and concentrated NaCl solutions.

During the 1975-1978 period, more than 30 reports of colorimetric methods for the determination of cobalt in a variety of matrices were tabulated in Analytical Abstracts.

Detection limits in the range of 0.0019 to 0.2 $\mu g/ml$ for cobalt were reported for a variety of analytical reagents (Pyatnitskii, 1966).

A review of colorimetric methods for various elements, including cobalt in plants, soils, and water (which included 59 citations) was prepared by Verigina (1975). An earlier review of photometric and spectrometric methods for the analysis of soils was published by Verigina and Dobritskaya (1974). Both reviews appear to be fairly recent and extensive, but being in Russian, they present a language barrier.

A review of analytical methods for ferrous metallurgy by Straub and Hurwitz (1979) cited 10 recently reported colorimetric methods.

In general the presence of iron, copper, and/or nickel have produced the most frequently reported analytical interferences. A number of other elements have also been found to cause problems. The sample matrix, preparation or analysis technique, and concentrations of analyte and interferent are all important considerations for the selection of a colorimetric analytical protocol having minimal interferences in the desired analysis.

A number of colorimetric methods have been reported recently which primarily deal with the analyses of aqueous matrices such as seawater or other natural waters.

After a 2-liter water sample was concentrated using Amberlite CG400 and the column stripped with 80 ml of 2 M $HClO_4$, cobalt was determined at a wavelength of 510 nm using 4-(2-pyridylazo)resorcinol in the presence of EDTA and KCN (Kiriyama and Kuroda, 1977). A concentration of 0.1 to 0.2 μg/liter could be observed with a 10% coefficient of variation.

The concentration, separation, and determination of Co^{2+} in seawater using 1-nitroso-2-naphthol and carbon tetrachloride was reported by Honjo et al. (1978). A working curve extended as low as 10^{-6} M Co at an absorption wavelength of 410 nm. Recoveries of 97 to 99% were reported for the method.

Korkisch and Goedl (1974) discussed the determination of cobalt in natural waters. After concentration and separation of the complex on Dowex 1-X 8 anion exchange resin, elution of the Co-thiocyanate complex was accomplished with 6 M HCl in tetrahydrofuran and 2-methoxyethanol. Concentrations of 0.04 to 1.9 μg Co/liter in a number of water sources were reported.

Motomizu (1973) presented a method for the analysis of cobalt in seawater using 2-nitroso-5-diethylaminophenol and back extraction. The method was found useful over a range of 0 to 0.24 μg Co/liter with a relative standard deviation of 4% for 0.15 μg/liter. Absorbance was measured at 462 nm.

A colorimetric method for the direct measurement of absorption by an ion-exchange resin which sorbs a sample complex species was reported by Yoshimura et al. (1976). Cobalt as the thiocyanate complex was absorbed on Dowex 1-X2. An approximately 10-fold increase in sensitivity compared to solution colorimetric was reported. The presence of Fe(III) and NaOH in molar ratios of 1 and 10^4 with respect to cobalt, respectively, produced the most serious interferences.

The determination of Co(II) in pharmaceutical materials is a second area for which several colorimetric methods have been reported. Eldawy et al. (1976) employed Na 7-nitroso-8-hydroxyquinoline-5-sulfonate as the chromogenic reagent. A selective method using McIlvaine's pH 8 citric acid phosphate buffer was developed. The absence of interferences from Fe(II), Cu(II), Zn(II), Ca(II), Mg(II), as well as from hormones, vitamins, and additives was reported. A sensitivity of 0.37 mg Co/liter of sample solution was reported.

Sikorsha-Tomicka (1975) discussed the use of pyrazinecarboxylic acid amide with Co(II) in alkali media for the analysis of Co(II) in pharmaceutical preparations. Cobalt in the range of 0.5 to 15 mg/liter was determined at an absorbance maximum of 380 nm.

The analysis of cobalt in air samples has been described by Kharkover and Kharkover (1973). A minimum measurable concentration of 2 μg/10 ml of extract was reported using 8-mercaptoquinoline (thiooxin) for the analysis of air samples collected near a nonferrous metallurgy plant.

The determination of cobalt by the catalytic oxygenation/decoloration of gallocyanin (GAL) by H_2O_2 was reported by Hirayama and Unohara (1978). Cobalt was determined in the range of 0.2 to 2.4 mg/liter at an absorbance maximum of 520 nm. Reagents were 1.6×10^{-3} $\underline{M}$ GAL, 2.4×10^{-1} $\underline{M}$ H_2O_2, pH 9.4 (borate buffer), and 2 to 4 μg Mg(II)/ml, which were allowed to react for 10 min at 25°C prior to analysis.

A limited number of additional recent applications of colorimetric/photometric methods for cobalt analysis to organic and inorganic matrices are summarized in Table III-17.

Although a number of recent applications of colormetric/photometric methods are evident, the general trend appears to favor more instrument intensive techniques such as atomic absorption or neutron activation analysis. This trend appears to be a result of less stringent sample preparation requirements to minimize interferences.

c. Electrochemical methods of analysis: In surveying the literature reporting the analysis of cobalt concentrations by various electrochemical methods, the use of polarographic

TABLE III-17

ADDITIONAL COLORIMETRIC METHODS OF COBALT ANALYSIS

Matrix	Reagent	Reference
Liquid fertilizers	nitroso-R salt	Gonchakova et al. (1977)
Plants and soils	nitroso-P salt	Tsap et al. (1977)
Plants	p-nitrosophenylhydrazone of diacetylmono-oxime	Pearson and Seim (1977)
Fertilizer, plants, soil	furoin thiosemicarbazone	Bhaskare and Deshmukh (1975)
Cyanocobalamin	-	Vasilikiotis et al. (1977)
U, U_3O_8, $UO_2(NO_3)_2$	2-nitroso-5-dimethylaminophenol	Motomizu (1975)
Pharmaceuticals	1-(5-mercapto-1,3,4-thiadiazol-2-ylazo)-2-naphthol; and 4-(1,3,4-thiadiazol-2-ylazo)resorcinol	Domagaline and Zareba (1977)
Organocobalt	tiron-type compounds	Debol et al. (1977)
Biological materials	4-nitrophenylhydrazone of diacetylmono-oxine	Peirson and Seim (1977)
Plants, soil	1,3-diphenyl-3-thiosopropan-1-one	Hoppe and Uhkmann (1977)
NaCl	1-(2-pyridylazo)-2-napthol	Gürtler (1977)
Water	Dowex 1-X2/NH_4SCN	Yoshimura et al. (1976)
Pharmaceuticals	Na 8-hydroxy-7-nitrosoquinoline-5-sulfonate	Eldawy et al. (1976)
Water	Chromofoam	Braun and Farag (1974)

techniques is most frequently encountered. Electrochemical techniques which have been reported for the determination of cobalt much less frequently include amperometry, anodic stripping voltammetry, coulometry, and potentiometry.

The basic differences in electrochemical methods of analysis are frequently subtle. Voltammetry generally consists of monitoring current voltage relations in dilute electrolytic solutions by applying a voltage to a small polarizable electrode with respect to a reference electrode. The voltage is scanned in a negative potential direction, and the resultant current changes are measured. Polarography is a term usually reserved for voltammetric analyses using a dropping mercury electrode. Amperometry is also related to voltammetry although the relationship monitored is one of current analyte concentration at a controlled constant potential. Coulometry refers to the analytical application of Faraday's hour of electrolysis, which relates the quantity of chemical change to a quantity of electricity. Both current and true measurements are made. Potentiometric methods employ Nernstian potential responses as measured through nonpolarized electrodes which are in zero current condition (Willard et al., 1974, and Ewing, 1969). This discussion of electrochemical methods is limited to polarographic techniques, which have been most frequently used for cobalt analyses.

Several general reviews of polarographic analyses have been published. The polarographic behavior of divalent cobalt, nickel, and iron with respect to pH, complex formation, organic solvents, absorption, and double-layer formation reported in 93 references was reviewed by Pavlov and Bondar (1973). A more recent and extensive review of polarography in biochemistry, pharmacology, and toxicology was prepared by B'rezina and Volke (1975). In excess of 240 references were cited; however, a number of additional analytes also were addressed.

A general presentation by Nelson (1970) reviewed the use of polarography for environmental studies. The use of anodic stripping techniques as well as a variety of specific procedures for metals, anion, pesticides, polymer effluents, and surfactants in water was described. Sensitivities of 10^{-6} $\underline{M}$ for most ions were noted.

An older general review of 57 references by Maienthal and Taylor (1967) considered the use of polarography and anodic stripping and linear sweep voltammetry for the analysis of trace inorganics in water. The analysis of 12 elements including cobalt in reactor water at the 1 ppb (1 μg/liter) level was cited.

The bulk of the electrochemical work reported has involved the analysis of aqueous matrices. Piccardi and Legittimo-Cellini (1974) reported the qualitative and quantitative polarographic determination of cobalt and 12 other elements in wastewaters. Organic substances present were oxidized using HNO_3 and $HClO_4$. Analyses at pH 4.6, 9.2, and 12 were performed with all detection limits at < 0.1 mg/liter.

Heigl (1978) more recently described the determination by polarography of nine metals including cobalt in wastewaters. Preparation with H_2SO_4, HNO_3, and Na_2SO_3 with heating was described. The analysis of cobalt by differential pulse polarography at a detection limit of 0.03 mg/liter was indicated. Analysis of a synthetic wastewater matrix containing 0.50 mg Co/liter was indicated. Analysis of a synthetic wastewater matrix containing 0.50 mg Co/liter produced a result of 0.50 mg/liter with a standard deviation of ± 0.02. Possible interferences between Co^{2+} and Zn^{2+} were demonstrated.

Poole (1970) described the application of phase-selective alternating current (AC) polarography and square wave polarography to the analysis of various metals in water. A comparison of these two techniques with conventional DC polarography was made. For the AC polarographic technique, a sensitivity of 10^{-6} formula weights per liter was reported. Discussion of common interferences, such as surfactants, and their elimination was presented.

The determination of cobalt in nonferrous metal ore or products was described by Vasil'eva et al. (1974). An electrolyte of 1 $\underline{M}$ NH_4Cl, 0.02 $\underline{M}$ dimethylglyoxime, and NH_4OH produced the most suitable polarograms at pH 9. Interferences from Te, Zn, and to a lesser degree, Fe, Cr, and Mn were problems.

Harvey and Dutton (1973) reported the use of photooxidation prior to the pulse polarographic determinations of cobalt as the dimethylglyoximate. The detection limit of the method was reported to be 0.6 ng Co/liter. Recoveries of 97 to 81% were observed for cobalt fortifications of 3.4 to 2,076 ng, respectively, from seawater. Some additional electrochemical procedures are summarized in Table III-18.

TABLE III-18

ADDITIONAL ELECTROCHEMICAL ANALYSES OF COBALT

Matrix	Comments	Range	Reference
Glazed ceramics	4% acetic acid extraction	0.5 - 500 mg/liter	Buldini (1977)
Polyester resin	Dry ashing	0.01 - 0.06%	Radovici et al. (1976)
Fruit	Dry ashing at 450°C	0.1 - 1.5 mg/kg	Lapitskaya et al. (1975)
Seawater	Concentration on MnO_2	0.6 ng/liter	Harvey and Dutton (1973)
Cyanocobalamin (vitamin B_{12})	*N*-bromosuccinimide-arsenite potentiometric titration	5.9 - 35.4 mg	Aboul Kheir et al. (1974)

Although electrochemical methods have been successfully applied to the analysis of small quantities of cobalt in several matrices, the various techniques as a whole have seen limited application. The need for a liquid matrix of constant electrolyte properties, interelement and matrix interferences, and the length of time per analysis have all contributed to the limited appeal of electrochemical methods.

d. X-ray techniques: X-ray fluorescence techniques generally are capable of determining a number of elements (atomic number, $Z \geq 9$) at concentrations ranging from parts per million to virtually 100%. The technique offers a rare opportunity for rapid nondestructive simultaneous multielement sample analysis, which is not possible with other elemental analysis techniques except other x-ray methods and possibly neutron activation analysis. As a result, sample preparation is often minimal. Additionally, analysis of nonmetals such as the halogens can be accomplished; this is not directly possible by other spectroscopic techniques except by neutron activation analysis. Solids and liquids may be analyzed.

Wavelength dispersive x-ray fluorescence equipment consists of an excitation source, sample holder, collimators, analyzing crystal, and x-ray detector. A more simplified configuration is found in the energy dispersive x-ray fluorescence spectrometer, which consists of an x-ray source, sample holder, and an x-ray detector. The energy dispersive configuration allows the use of maximized solid angle viewing which, when coupled with large semiconductor detectors may result in improved sensitivity and detection limits (Parsons, 1976).

Sample excitation may be accomplished in several ways. The sample may be subjected to cathodic emissions having energies greater than the analyte excitation threshold (direct emission x-ray spectrometry). More frequently, the sample is bathed in an x-ray beam sufficiently energetic to excite the analyte x-ray spectrum (fluorescence spectrometry). The use of charged-particle-induced x-ray emission, which involves sample bombardment with protons or other massive ions, has been reported. The advantage of the last technique is in the relatively small sample sizes ($\sim$ 20 mg) and the resultant decrease in background levels compared to the other excitation technique. Unfortunately, the penetrating power of the bombarding particle ion is significantly less than that of the x-ray used for excitation purposes. Although gamma-source x-ray instrumentation is available, the detection limits are not generally amenable to trace analysis. Applications are limited to ores and semiquantitative field analyses. EKCO Instruments Limited (Anonymous, 1969) reported the use of a ^{238}Pu gamma source in conjunction with a Mn/Fe filter pair for the analysis of cobalt at a detection limit of 1 g/kg.

The major limitations of x-ray fluorescence methods involve matrix effects and insufficient understanding of x-ray line characteristics with respect to analyte concentration. Since fluorescence excitation is produced by bombardment of the sample with x-rays having limited penetrating potential (0.01 to 0.1 mm), analysis essentially is limited to the sample surface. As a result, the homogeneous distribution in the bulk sample and the sample surface is necessary to assure accurate quantitative analysis. Finally, although the method is capable of operation in the low (0.01 to 1 μg detection limits for $Z > 15$) parts per million analyses, competing analytical techniques can operate with equal or greater sensitivity with shorter analysis times.

Parsons (1976) has presented a general review of x-ray methods including fluorescence techniques for chemical analysis. The topics discussed are limited to basic instrumental considerations. A presentation of both x-ray fluorescence equipment and documented analytical applications in a variety of matrices was published by Pinta (1978). In addition to discussion of general instrument construction, an extensive discussion of sample preparation, including grinding and fluxing procedures, and a review of documented analysis in organic and inorganic matrices were presented.

Reeves and Brooks (1978) discussed the application of x-ray techniques to the analysis of geological materials. The use of sample grinding procedures to produce relatively uniform matrices and homogeneously distributed analyte was presented. The authors reported a number of reviews and texts on the advances in x-ray techniques and the existence of an abstracting service specifically for the x-ray emission field.

Duggan et al. (1972) and Garcia et al. (1973) published reviews of specific applications of charged-particle-induced x-ray emission.

While the primary technique addressed in this section is x-ray fluorescence, other x-ray techniques such as charged-particle-induced x-ray emission will be included where appropriate. Generally, analytical applications of x-ray fluorescence techniques for the assay of cobalt have involved geological, metallurgical, and fuel matrices; however, its application to the analysis of environmental matrices is of primary interest.

Valkovic (1977) has reviewed the applications of proton-induced x-ray fluorescence in medicine for the analysis of a variety of trace elements. The work of Jones et al. (1972) in the analysis of eight elements including cobalt in normal bone compared to human osteogenic sarcoma was cited. Early work establishing elemental concentrations in hair ranging from 0.1 to 100 mg/kg for 15 elements was also mentioned.

Rhodes et al. (1974) determined by x-ray fluorescence 17 elements in urban air particulate from 12 Texas cities. Extensive sampling and analysis considerations were presented.

The estimated detection limit for cobalt was 18 ng/m^3 for a 24-hr sampling period, which is poor since ambient air concentrations of cobalt in most samples are considerably lower.

The use of x-ray fluorescence for the analysis of 10 trace metal concentrations in plants has been reported by Beitz (1973). Sample preparation consisted of homogenization with an organic binder and subsequent pelletizing.

Marcie (1968) employed a rapid x-ray fluorescence procedure for the analysis of trace elements in water. Water samples were extracted three times with ammonium pyrrolidine dithiocarbamate and $CHCl_3$. After evaporation of the organic phase to 0.25 to 0.5 ml, the extract was spotted on Whatman No. 40 or No. 41 filter paper and analyzed.

Giauque et al. (1977) reported the use of x-ray fluorescence for the determination of 40 elements in USGS and NBS reference materials. Cobalt was extracted using diluted HNO_3; however, the concentrations in the three USGS materials and NBS coal fly ash (SRM 1633) were below the technique detection limit of 108 to 130 mg/kg.

Bassos et al. (1973) reported on the concentration of reducible metal ions from diluted solution by electrodeposition on pyrolytic graphite electrodes prior to x-ray fluorescence analysis. Known concentrations of cobalt in the range of 5 to 40 μg/15 ml total volume were quantitatively recovered with relative standard deviations $< 12\%$.

The use of proton-induced x-ray fluorescence for the analysis of cardiac tissue was reported by Mandler and Semmler (1974). A cobalt concentration of 70 mg/kg dry weight was observed in an aliquot of < 1 mg. Cobalt was not found in dried cardiac tissue from another source.

Luo et al. (1978) have described the use of proton-induced x-ray fluorescence for the analysis of anode slimes from a copper refinery. The sensitivity for cobalt was 10^{-9} g. Cobalt was observed at $\sim$ 20 mg/kg with a relative standard deviation of 4.2%.

The use of radioisotope sources to induce x-ray fluorescence has been reported by Molinari (1976) for the on-stream analysis of slurries. Eight elements were determined using either ^{224}Cm or ^{238}Pu.

A number of additional x-ray fluorescence applications or techniques are summarized in Table III-19.

TABLE III-19

ADDITIONAL X-RAY FLUORESCENCE ANALYSES OF COBALT

Matrix	Remarks	Reference
Cellulose	Hypan chelating ion exchanger	Burba and Lieser (1977)
Catalysts	Hydrodesulfurization catalysts	Labrecque (1977)
Water	Ion-exchange concentration	Cesareo et al. (1976)
Organic extracts	Diethyldithiocarbamate complexes	Iwasaki et al. (1974)

Although x-ray fluorescence methods have occasionally been applied to the determination of trace concentrations of cobalt, the use of concentration steps or the more sensitive ion-induced x-ray fluorescence methods have been necessary. The applicability of x-ray fluorescence techniques for the determination of trace concentrations ($\leq$ 1 mg Co/kg) is limited.

e. Emission spectrometry: Emission spectrometric techniques generally employ an electrical discharge which may consist of arc, spark, or plasma discharges to produce an excited, emitting analyte species. The analyte emission is sorted by monochromator or polychromator systems for subsequent detection photographically (spectrography) or electrometrically (direct reading). The method is generally operated in a simultaneous multielement mode at parts per million concentrations. Arc and spark emission methods have found wide application for rapid semiquantitative analyses of solids and liquids. Plasma emission techniques more frequently are used for quantitative analyses and are generally restricted to liquefied samples. Photographic and electronic detectors have both found wide application in arc/spark emission work. In contrast, plasma emission techniques usually employ electronic detection and computerized signal processing.

A text on the history and development of emission spectrography including emission sources, optics, spectrometers and spectrographs, and applications was edited by Barnes (1976). Barnes (1978) has also presented a general review of recent developments and applications of emission spectroscopy.

The sensitivity of the method for cobalt is dependent on the spectral emission line chosen for analysis. The three analytical lines of interest reported by Pinta (1978) are 4289 Å, 3409 Å, and 3367 Å, which provide analytical ranges of 1 to 30 mg/kg, 30 to 300 mg/kg, and 300 to 3,000 mg/kg, respectively. Spectral buffers, separations, and internal standard methods as well as practical applications also are discussed. A more extensive list of reported sensitivities for cobalt using a variety of emission sources and systems is presented at the top of Table III-15.

Delespaul et al. (1976) described the analysis of airborne particulates and gaseous pollutants using ES techniques. Material was collected on cellulose filters for 24 hr. Subsequently, the filters were sectioned and wetted with a NaCl buffer, and an internal standard of Ir and Pd was added prior to dry ashing at 440°C for 2 hr. The ash was mixed with graphite prior to analysis. Intercomparisons with neutron activation and x-ray fluorescence analyses were reported.

The spectrographic determination of 36 elements in industrial wastewater was described by Leroy and Lincoln (1974). Water samples were concentrated by evaporation at 120°C and mixed with powdered graphite (1:6). Germanium as the dioxide was added to the graphite and used as an internal standard. Cobalt was analyzed in the range of 0.001 to 0.01 mg/100 ml and 0.01 to 1.0 mg/100 ml using emission lines at 3453.50 Å and 3449.44 Å, respectively. The relative standard deviation was 11.6% for analyses using the 3453.5 Å line.

A much older work by Kopp and Kroner (1965) described the direct-reading emission analysis of 19 minor elements in natural waters. Samples were filtered, HNO_3 added, and the solution concentrated by evaporation to produce a 5-μl aliquot containing 100 mg of dissolved solids. Sample excitation was accomplished using a rotating disk high-voltage spark technique. Cobalt was determined using the emission line at 3453.5 Å having a reported analytical range of 0.5 to 20 mg/liter. Recoveries of known quantities of cobalt from various river water samples ranged from 95 to 138%. A relative standard deviation of 3.82% for the rotating disk spark technique was compared to a 5.04% relative standard deviation for a uni-arc method.

Kuz'min et al. (1968) reported concentrations of 13 trace metals in water. Extraction of analyte complexes was at pH 5.5 to 6.0 with 8-hydroxyquinoline and sodium diethyldithiocarbamate using a 3:1 mixture of chloroform and isoamyl alcohol. An AC arc was employed as the excitation source. A sensitivity for cobalt of 0.02 ppb (μg/liter) was reported.

The analysis of sediment leachates was reported by Moselhy et al. (1977) using a rotating disk high-voltage spark technique. Interferences with the cobalt analysis at 3453.5 Å were not observed. An intercomparison of ES and atomic absorption results was included.

Direct reading spectrometry was imployed by Watson and Russell (1978) for the analysis of geological samples for 11 trace elements. Samples were mixed (1:3) with a buffer consisting of 20% LiF, 0.03% GeO_2 (internal standard), and the balance graphite. A DC arc source using an Ar-O_2 Stallwood jet stabilization system was employed. Results and comparison of photographic and spectrometric (electronic) detection was included for a variety of NIM (South African Bureau of Standards) and USGS rocks. Generally good agreement with reported values was observed. Electronic detection consistently produced slightly elevated results, and photographic detection produced frequently slightly lower results than reported values.

A rather old work by Myers et al. (1958) discussed in detail the development of a DC arc spectrographic procedure for the analysis of rocks, minerals, and ores. Semiquantitative results were reported for selected elements from a 68-element survey. The preparation of standard reference plates for the variety of elements was described. Six different cobalt emission lines were used during analysis, namely those at 3405.1, 3412.3, 3412.6, 3449.2, 3453.5, and 3462.8 Å. Generally, samples were ground to ≤ 100 mesh and mixed with graphite prior to analysis. An approximate detection limit of 5 mg/kg was reported for cobalt. The results of cobalt analysis for 30 ore and rock samples and for alternate analytical techniques were compared. Generally, ES results were slightly greater than determined previously. A statistical data analysis was also presented.

The use of spectrographic procedures for the analysis of impurities in high-purity manganese and copper as reported by Job et al. (1974) and Publicover (1965). Cobalt in manganese was analyzed, after buffering with 85% (w/w) BaF_2, in the 10 to 100 ppm (mg/kg) range using the 2424.93 Å analysis lines. A relative standard deviation of ± 10% was reported. Cobalt in electrolytically pure copper was determined using the 3453.50 Å analytical emission line in a concentration range of 1 to 10 mg/kg. The problem of preventing sample contamination when sampling high-purity material was discussed in detail.

Additional recent ES methods employing spectrometric and spectrographic procedures are summarized in Table III-20.

Emission spectroscopy for rapid simultaneous semiquantitative multielement analysis has been used widely. The high excitation energy available in either arc or spark discharges is of great utility in minimizing the dependency of the analyte response on its original oxidation state or chemical species. Matrix effects can often be compensated by buffer systems. The capability to analyze solids directly with little sample preparation by ES is another significant advantage. Given its many advantages, the general semiquantitative quality of analysis when internal standards are not used is often insufficient when high-accuracy data are required.

TABLE III-20

SOME RECENT EMISSION SPECTROCHEMICAL ANALYSES FOR COBALT

Matrix	Comments	Reference
Plants	550°C dry ashing; 2 - 200 ppm ± 15% analysis range	Dixit et al. (1974)
Plants, soils, soil extracts	450°C dry ashing; detection limits of < 1 ppm of 0.1 μl of solution or 10 ng for 10 mg of solid	Gabriels and Cottenie (1976)
Metalloenzymes	Helium microwave plasma excitation; 5 μl sample aliquots, ~ 10 ng/ml detection limit	Kawaguchi and Vallee (1975)
Organic extracts	Metal chelates, 0.1 μg detection limit	Zheleznova and Zarasevich (1975)
Formic acid	Preconcentration by electrodeposition	Krasil'shchik et al. (1977)
Polyorganosiloxane varnish	Detection limit of 5 ppm	Trokhachenkova et al. (1977)
Crude oil	Sensitivity, precision, and line pair correlations deteriorate as crude oil concentration in $CHCl_3$ solution increases	Lakatos (1974)
Petroleum and products	Fe_2O_3 internal standard used in petroleum ashes	Nasser (1974)
Ion-exchange resins	Dry ashing at 450 to 480°C for 1 hr	Raginskaya et al. (1973)
Whole blood, serum, steel	Inductively coupled plasma emission	Kniseley (1974)
Air particulate	Hydrofluoric acid dissolution of fiberglass filters	Sugimae (1975)
Ocean sediments	Comparison of ES and NAA results	Kurinov et al. (1974)

f. Plasma source emission spectroscopy: Plasma source emission spectroscopy is a relatively recent development that is similar to other ES methods; however, the ease of quantitative analysis and several significant instrumental differences merit its discussion as a separate technique. Emission excitation generally is accomplished using an electronic plasma discharge in either a DC discharge configuration, pioneered by Valente and Schrenk (1970), or the radiofrequency inductively coupled plasma, championed by Fassel and co-workers (Wendt and Fassel, 1965, and Fassel and Kniseley, 1974). Unlike other ES methods, sample solutions are nebulized into the plasma discharge while the analysis of solid materials can be accomplished in only a few limited cases at present. Emission analysis is accomplished using either monochromator or, more often, polychromator systems. Frequently these optical systems are identical to ES systems. Although photographic detection can be employed, the full utility of the system can only be realized using electronic detection coupled with computerized data processing. Following the introduction of the first commercial PES units in 1975, the development and use of the PES technique has been of increasing interest.

Barnes (1978) has included references to PES developments in his recent review of developments and applications of ES.

Considerable discussion of the general analytical applications of plasma emission and absorption techniques, as well as discussions of flame, arc, and spark methods, was presented in a recent review by Fike et al. (1977).

An older presentation of PES capabilities when using an ICP source was published by Fassel and Kniseley (1974). A detection limit of 0.003 mg/liter was reported for cobalt. More recently Winge et al. (1979) reported the existence of 13 cobalt emission lines between 2388.92 Å and 2286.16 Å which were suitable for quantitative analysis at detection limits in the range of 0.0060 to 0.023 mg/liter. The data are summarized in Table III-21.

The use of ICP/PES for the analysis of 14 elements in waste streams subject to the National Pollutants Discharge Elimination System (NPDES) was presented by Olson and Bour (1979).

TABLE III-21

COBALT EMISSION WAVELENGTHS REPORTED BY WINGE ET AL. (1979)

Wave-length (nm)	Estimated Detection Limit (μg/ml)
238.892	0.0060
228.616	0.0070
237.862	0.0097
230.786	0.0097
236.379	0.011
231.160	0.013
238.346	0.014
231.405	0.016
235.342	0.017
238.636	0.021
234.426	0.021
231.498	0.023
234.739	0.023

McQuaker (1979) recently described the calibration of an ICP/PES system for the analysis of environmental materials in a solution matrix containing 3.5% $HClO_4$. The analysis of water, soil, airborne material, and biological tissue was considered. A detection limit of 0.029 mg Co/kg was reported at the 3453.5 Å emission line although analyses were made over a calibrated concentration range of 0.10 to 20 mg/kg. The absence of interelement interferences was noted at the analytical wavelength employed. This finding was in contrast to the severe iron interference with cobalt determinations at 2388.9 Å reported by Winge et al. (1977).

Kniseley et al. (1973) reported the use of ICP/PES for the analysis of nine elements in whole blood and serum. A detection limit for cobalt in serum was given as 0.005 mg/liter for analysis at the 3453 Å emission wavelength. Samples of serum were diluted twofold and blood diluted tenfold prior to analysis.

Ronan and Kunselman (1976) determined cobalt in water using ICP/PES at the 2389 Å emission wavelength. A cobalt detection limit of 4 μg/liter with a relative standard deviation of 1% was reported. Analysis of a known water sample containing 100 μg Co/liter produced an anlytical result of 97 $\pm$ 3 μg/liter.

The analysis of 17 major and trace elements for routine water quality testing was described by Garbarino and Taylor (1979). Cobalt analysis in the range of 3 μg/liter to 10 mg/liter using the emission line at 2389 Å was reported. The analyses of four reference water standards prepared by the USGS were presented. The relative standard deviation for 10 replicate analyses ranged from 8.4 to 21% for cobalt concentrations of 4.8 to 14.8 μg/liter. The analytical stability over 5 days was found to fall within 9.9 and 18% RSD for the same 4.8 to 14.8 μg Co/liter concentrations.

The use of PES techniques, especially those employing ICP sources, for rapid simultaneous multielement analyses of environmental and biological matrices is increasing. The major analytical limitations are that liquid samples are generally required and the possibility of unknown matrix and interelement effects exists in uncharacterized systems.

g. Atomic absorption spectrophotometry methods: The use of atomic absorption techniques has found widespread application in the determination of metallic elements including cobalt at the parts per million and subparts per million levels. Following the pioneering work of Walsh in the mid-1950's, considerable development of methods and instrumentation has taken place. As noted by Gelman (1972), the analysis of cobalt in a plethora of matrices using a variety of analytical schemes has been reported in the literature. Analyses of both organic and inorganic matrices including hair (Lau and Ashmead, 1975), serum (Welz and Wiedeking, 1970), tissue (Johnson, 1976), plant materials (Gelman, 1972; Allen and Parkinson, 1969; Tsap et al., 1977; and Gehrke et al., 1977), water (Gläser, 1978 and Strain et al., 1976), soil (Selivisova et al., 1975), rocks and sediments (Warren and Carter, 1975; Hannaker and Hughes, 1978; Armansson, 1977; Hansen and Hall, 1977; and Allcott and Lakin, 1975), and air particulates (Ranweiler and Moyers, 1974; Beyer, 1969; and Thompson et al., 1970) have been published. Since the sheer volume of published literature precludes an all inclusive source list, a brief overview of some typical analytical applications and methods follows.

Numerous texts describe the operating principles and analytical applications of atomic absorption techniques (Ramirez-Munoz, 1968; Slavin, 1968; L'vov, 1970; and Price, 1972). A thorough discussion of spectroscopic methods including atomic absorption techniques was presented by Dean and Rains (1969). A more recent overview of atomic absorption theory and application was included in the text by Pinta (1978). The use of atomic absorption spectrophotometric (AAS) methods for the analysis of geological materials was presented by Angino and Billings (1972); and more recently by Reeves and Brooks (1978). The analysis of heavy metal pollutants in water by AAS methods was presented by Burrell (1974).

The general application of AAS techniques to the analysis of agricultural, biological, and medicinal problems was presented in a somewhat older text by Christian and Feldman (1970). More recently Christian (1972) reviewed AAS applications to the analysis of various elements in blood, urine, tissues, organs, plants, and soils, although the majority of cobalt analyses cited were in soil and plant materials. Willis (1973) addressed the analysis of biological materials by flame and nonflame AAS techniques.

At least one periodical, Atomic Absorption Newsletter, is devoted entirely to developments in the field of AAS and includes a semiannual bibliography of literature on new or improved AAS applications and techniques. Similarly, fundamental and application reviews in Analytical Chemistry present a large number of analytical considerations and applications.

Atomic absorption techniques as applied to cobalt analysis may be divided into two general categories on the basis of the atomizing device employed. Flame atomization techniques are generally capable of low parts per million analyses, while electrothermal methods of atomization including carbon rod and graphite furnace methods are capable of performing successful analyses at the low parts per billion concentration range. The detection limits attained in specific situations are in large part dependent on the sample matrix and the preparative methods employed. The following discussion considers both flame and nonflame atomization analyses for cobalt determination.

Rains (1975) reported a number of analytical absorption lines for the AAS analysis of cobalt using a fuel-lean air-acetylene flame. The absorbing wavelength and the reported sensitivity in micrograms per milliliter needed to produce a 1% absorbance signal are summarized in Table III-22. The most sensitive cobalt absorption wavelength reported to date was at 2407 Å with a reported sensitivity of 0.19 μg/ml/1% absorbance or a corresponding calculated detection limit of 0.002 μg/ml. Peterson and Manning (1971) reported actual cobalt detection limits in water of 10 μg/liter by flame and 0.7 μg/liter by graphite furnace at the 2407 Å wavelength. A 4.9% coefficient of variation was reported for concentrations of 2 μg Co/liter using the furnace technique.

TABLE III-22

COBALT ABSORPTION WAVELENGTHS AND CORRESPONDING SENSITIVITIES
(Dean and Rains, 1975, and Anonymous, 1973)

Wavelength (Å)	Relative Sensitivity	Sensitivity (μg/ml/1% absorbance)
2407	1.0	0.19
2425	1.2	0.23
2412	1.8	-
2521	2.0	0.38
2436	2.9	1.3
3044	12.	2.3
3527	22.	4.1
3466	30.	-
2474	62.	-
3018	110.	-
3413	-	0.35

In a comparison of flame atomic emission and absorption techniques, Christian and Feldman (1971) reported flame atomic absorption techniques to offer an approximately 15-fold improvement in sensitivity over flame atomic emission methods.

Precision in the flame AAS measurement of a variety of elements was reported by Bower and Ingle (1979). Cobalt analyses using an air-acetylene flame were reported to have an optimum precision of 0.14% relative standard deviation for an absorbance of 1.1. The effect of analysis wavelength, flame type, hollow cathode lamp current, and bandpass on the analytical precision in the analysis of cobalt was highly dependent on slit width/bandpass. At low cobalt concentrations, flame transmission flicker and lamp flicker are dominant factors in poor analytical precision.

Lau and Ashmead (1975) described the use of a nitric-perchloric acid digestion scheme for the flame AAS analysis of 13 trace elements in human hair. A linear response for cobalt over the concentration range of 0.25 to 1.25 mg/kg using the 2407 Å absorption wavelength was reported. Recoveries of known quantities of cobalt were found to average 96.0%.

The analysis of 11 elements in human liver was described by Johnson (1976). Tissues were digested using a nitric-hydrochloric acid mixture and heating to 100°C. Possible metals losses to the lipid residue were found to be negligible. An average recovery of 100% was observed for 0.5 mg/liter concentrations of cobalt using flame AAS; however, specific analytical conditions were not cited.

Gelman (1972) reported the air-acetylene flame analysis of cobalt in plant material using a hydrochloric acid dissolution step followed by chelation with ammonium pyrrolidine dithiocarbamate (APDC) and extraction with methyl isobutyl ketone (MIBK). Cobalt concentrations ranging from 0 to 0.5 mg/kg were determined with an estimated extraction efficiency of 60%.

The use of AAS for trace elemental analyses of 17 elements in toxicological materials was described by Berman (1969).

The analysis of cobalt in refuse material was described by Campbell et al. (1976). Samples were prepared by digestion in a mixture of nitric/hydrofluoric acids with the addition of perchloric acid as necessary.

A comparison of flame and nonflame atomization methods for the analysis of cobalt in No. 6 residual fuel oil was cited by Hofstader et al. (1976). A measured concentration of 272 μg/kg was observed by flame AAS after wet ashing of the sample compared to 240 μg/kg after wet ashing by nonflame atomization. Using direct analysis of the fuel by nonflame atomization AAS, the observed cobalt concentration was 258 μg/kg.

The use of the APDC-MIBK extraction system for the determination of cobalt in saline water was described by Brooks et al. (1967). Using an air-acetylene flame, an approximate detection limit of 0.10 μg/liter was reported. The coefficient of variation for replicate analyses was found to be 15%.

More recently Gläser (1978) reported the use of a perchloric acid digestion followed by complexation with 1,10-phenanthroline and extraction into nitrobenzene for the flame analysis of six elements in waters. Aliquots of 100 μl were analyzed for each element.

Ranweiler and Moyers (1974) have published a method for the analysis of air particulates for 22 elements. Samples were collected on polystyrene filters which were subsequently ashed at 400 to 425°C. The residue was dissolved in a mixture of HF, HCl, and HNO_3 in Teflon®* acid digestion bombs. An average recovery of 100.5% was reported for 5-μg quantities of cobalt with a standard deviation of 16.9%. The rather large standard deviation observed in the recovery studies was attributed to the small quantities of cobalt actually analyzed which approached the detection limit of the method.

Analysis of particulate matter collected on fiberglass filters was described by Beyer (1969). A simple extraction using HCl with heating followed by water washings was employed. Cobalt was analyzed using an oxidizing air-acetylene flame and monitoring of the 2407 Å absorption wavelength.

---

* Registered trademark of E. I. du Pont de Nemours and Company, Inc., Wilmington, Delaware.

Preparation of fiberglass and membrane filters by low-temperature plasma ashing with subsequent extraction using an HCl/HNO_3 mixture has been documented by Thompson et al. (1970). The recoveries of cobalt from low-temperature ashing and muffle furnace ($\sim$550°C) procedures were to be 96 and 97%, respectively.

Armannsson (1977) described the dithizone extraction of cobalt from rocks and sediments for direct flame AAS analysis. Extraction below pH 9 was necessary to assure the quantitative extraction of cobalt.

Hannaker and Hughes (1978) employed a multistep extraction scheme for the analysis of 14 elements in geological materials. Rock samples were treated with HF, HNO_3, H_2SO_4, $HClO_4$, and HCl prior to extraction. Cobalt was determined in the organic phase following an oxine/MIBK extraction by flame AAS methods. A detection limit of 0.04 mg/kg was reported for cobalt using a 2-g sample aliquot at a precision $\leq$ 10% relative deviation.

Warren and Carter (1974) employed a mixture of HNO_3, $HClO_4$, and HF in a pressurized bomb to dissolve powdered rock samples. The dissolved samples were treated with HCl and solid boric acid prior to direct flame AAS analysis. Cobalt was analyzed using an air-acetylene flame and the 2407 Å absorption wavelength with deuterium arc background correction. A linear working range of 0 to 4 μg/ml was reported. The results of analyses of USGS reference materials were reported.

Although flame atomization techniques have been used widely in AAS, the development of electrothermal (nonflame or flameless) atomization apparatus and methodology has resulted in a concomitant improvement in analytical sensitivity. The improved analytical sensitivity has allowed analyses of low elemental concentrations that were previously unattainable, which has been of particular benefit in biological, medicinal, and toxicological applications.

Although absolute detection limits and sample aliquot size vary among different electrothermal atomizers such as carbon rod versus graphite furnace atomizer units, electrothermal devices generally offer several orders of magnitude improvement in sensitivity. Typically a flame atomization system is capable of detecting 0.19 μg/ml/1% absorbance (Dean and Rains, 1975). In contrast, manufacturers' data indicate detection limits of approximately 0.0027 μg/ml (20-μl aliquot) and approximately 0.001 μg/ml (5-μl aliquot) for the graphite furnace and carbon rod atomizers, respectively (Anonymous, 1974, and Culver, 1975). Generally, the same absorption wavelength at 2407 Å is used for both flame and electrothermal AAS analyses; however, potential background interference problems may be more serious with electrothermal techniques.

Alder et al. (1976) reported the analysis of cobalt in hair by the direct ashing and analysis of sample segments in the atomizer. Segments of hair ~ 1 cm long or 100 μg in weight were ashed for 30 sec at 1100°C prior to atomization at 2600°C for cobalt analysis. The detection limit was reported to be 0.1 mg/kg for 100-μg sample weights using the 2407 Å absorption wavelength. A comparison of the dry-ashing technique to wet ashing with tetramethylammonium hydroxide (TMA) for alternate hair sections produced results of 0.38 and 0.31 mg/kg, respectively.

Slavin et al. (1975) described the analysis of nine elements in meats by AAS following a simple acid digestion using HNO_3. Cobalt was determined using a graphite furnace program of 30 sec drying at 150°C, 30 sec charring at 800°C, and atomization for 8 sec at 2700°C. Cobalt data for four types of meat tissue were reported.

Oyamada et al. (1977) described an extraction-furnace AAS procedure for the analysis of biological materials. Samples were ashed with 10 N HCl and subsequently extracted with 5% tri-n-octylamine in carbon tetrachloride. A water reextraction was performed prior to analysis. Recoveries > 97% having relative standard deviations < 10% were reported for 0.1 μg/kg concentrations. Interferences from Sb, As, Cd, Cr, Se, Pb, Ni, Zn, and Fe were reported to be absent.

A comparison of direct-injection/selective volatilization methods with the use of APDC/MIBK extractions for the analysis of cobalt in saline water was reported by Segar (1971).

Co-volatilization of cobalt with the bulk of seawater salts during the temperature program was noted. Cobalt responses observed for seawater samples were approximately 15% of the corresponding values obtained from distilled water using the same temperature programs. It was concluded that acceptable precision and accuracy could be obtained using standard additions techniques at a detection limit of 50 μg/liter of cobalt. Alternative approaches using APDC/MIBK extractions were investigated. Direct atomization of the organic extract allowed a detection limit of 400 ng Co/liter. In contrast, ashing of the organic extract and redissolution of the ash in HNO_3 produced a detection limit of 10 ng/liter.

Simmons (1975) reported the determination of cobalt in plant material by electrothermal atomization AAS. Samples were digested in a mixture of HNO_3 and $HClO_4$ and extracted using 2-nitroso-1-naphthol in isoamyl acetate. Mean recoveries of 98.2 to 108.5% for added cobalt quantities of 9.7 to 48.5 ng were reported. Possible interferences and a comparison with values obtained using flame atomization methods were included.

The analysis of metals in medicinal preparations was described by Rousselet et al. (1975). The cobalt content of B_{12} was determined to confirm the required dosage by flame AAS methods. A furnace technique employing a dithizone/MIBK extraction method was also presented.

Foltz et al. (1977) reviewed the chemical analysis of food and reported the use of chelating ion-exchange methods for the AAS analysis of foods. Electrothermal atomization AAS methods were cited for use after wet digestion of the sample and subsequent extraction. A laser atomization technique for the analysis of plants and liver tissue was also cited.

Casula and Devoto (1975) applied graphite furnace techniques to the analysis of atmospheric particulates in the work environment. Sampling and analysis methods and their advantages were discussed.

A sequential extraction procedure for the speciation of trace metals in particulate material was presented by Tessier et al. (1979). Five binding categories were described:

(a) exchangeable, (b) carbonate bound, (c) iron and manganese oxide bound, (d) organically bound, and (e) residual material held in the crystal structure of primary and secondary minerals.

The reviews of natural waters analyses by Fishman and Erdmann (1975 and 1977) have presented a number of articles concerning the analysis of metals. Frequent use has been made of ion-exchange techniques prior to AAS analysis.

A comparison of extracting agents for use with carbon furnace AAS procedures has been made by Mizoguchi and Iskii (1977). Extraction pH ranges and optimum ashing programs are discussed for oxine, 1-nitroso-2-napththol, APDC, DDTC, and PAN.

Additional reports of AAS analyses of cobalt in organic matrices are summarized in Table III-23.

Problems encountered in the AAS analysis of samples for cobalt may be either chemical or spectroscopic in nature. Chemical interferences may be the result of either sample matrix or sample preparation factors. Christian (1972) has noted the negative influence of HCl in combination with a number of metals on the absorption of cobalt as reported by Kabanova and Andreeva. The addition of acetic acid to soil extracts was reported to compensate the suppression. Statistical analysis of interlaboratory analyses of marine organisms was reported by Fukai et al. (1978). Although possible analytical explanations were not offered, the average value determined by AAS (1.9 ± 0.8 μg/g) was significantly at variance with the results of neutron activation analysis (0.41 ± 0.02 μg/g). Positive error in the analysis of silicate rock was noted by Warren and Carter (1974). They reported that the results for cobalt analyses were significantly elevated above actual concentrations if background correction methods were not employed to compensate for spurious matrix absorption.

Problems with matrix effects become more pronounced with the use of electrothermal atomization techniques. Cobalt was found to co-volatize with the major salts in saline waters (Hansen and Hall, 1977). The extent of co-volatilization was found to vary drastically with salt content. As the salt content was increased, a rapid decrease in the analyte absorption signal was observed. The problem of co-volatilization has also been noted by Willis (1973).

TABLE III-23

RECENT ATOMIC ABSORPTION ANALYSES OF ORGANIC MATRICES FOR COBALT

Matrix	Comments	Reference
Fish tissues	Pressure decomposition, APDC extraction, > 95% recovery; electrothermal analysis, 6 pg detection limit.	Harms and Kunze (1977)
Biological Materials	Wet ashing with extraction with trifluoroacetylacetone; analysis by GC-flame AA.	Wolf (1977)
Brewing materials	Electrothermal analysis; elemental mass balances through the brewing process reported.	Helin and Slaughter (1977)
Plankton	Mixture of $HNO_3/H_2SO_4/H_2O_2/HF$ most effective treatment; 0.37 to 0.70 μg/g detected.	Machiroux and Dupont (1976)
Multivitamins	Extraction with HCl/MIBK/IPA, 1:5:4, separation with Dowex 1-X8 ion-exchange resin; 0.22 ppm Co produced 1% absorbance.	Korkisch and Huebner (1976)
Lung	Plasma ashing with HNO_3; > 90% recovery by electrothermal AA; coefficient of variation < 8%; detection limit = 0.05 ng.	Kawaraya et al. (1976)
Liver	Dry ashing, extraction with APDC/MIBK.	Gelman (1976)
Hair	Direct analysis by graphite furnace.	Alder et al. (1976a, 1976b)
Liver	Wet digestion with HNO_3/HCl.	Johnson (1976)
Blood serum	-	Alt and Massmann (1976)
Biological material	Dry ashing at 450 to 500°C for vegetable matter, manure, compost, sludge.	Andersson (1976)
Plants	Acid digestion, complexation extraction.	Simmons (1975)
Meat	Acid digestion.	Slavin et al. (1975)
Feed	Dry ashing at 450°C; complexation extraction.	Hageman et al. (1975)
Blood serum	Electrothermal AAS.	Muzzarelli and Toccheti (1975)

Due to the abundance of spectral lines in the cobalt analysis region of 2407 Å, the use of narrow band widths and high-resolution monochromators is necessary. Analyses conducted employing overlapping spectral lines will be subject to interferences from matrix components and will have a very limited analytical response linearity range.

h. Activation analysis: Activation analysis techniques are a class of extremely sensitive elemental analysis methods in which samples are made radioactive by bombardment with suitable nuclear species. Neutron activation analysis is the most common variety of activation analysis. Generally, samples are bombarded with thermal (slow) neutrons in a nuclear reactor for NAA purposes, although fast neutrons produced in an accelerator have been used in some cases. Neutron activation analysis methods have been the focus of recent intensive interest due to their potential to perform simultaneous multielement analyses with high sensitivity.

The basic principle of activation analysis involves the irradiation of a sample with a flux of particles and the resultant conversion of the sample matrix element into various radionuclides. The decay of each radionuclide present will produce a characteristic radiation having an equally characteristic decay rate. Thus, elemental components of a sample can be unambiguously identified by a unique combination of energy emission and decay rate.

Two general approaches can be employed in activation analyses. The first more classical method involves the isolation of an analyte from matrix interferences after irradiation through a series of radiochemical separations. This approach generally requires rather extensive sample-handling procedures; however, small quantities of an analyte may be effectively separated and quantitatively analyzed in otherwise intractable matrices. The alternate approach, referred to as instrumental neutron activation analysis dispenses with preanalysis radiochemical separations. Identification and quantitative analysis in INAA relies upon the measurement of characteristic photoemissions separated through a high-resolution gamma-ray spectrometer. The use of INAA allows for a more rapid sample analysis than is possible using radiochemical separations but may be subject to interferences from other matrix components. Generally, the reaction $^{59}Co(n,\gamma)^{60}Co$ has been most widely used for cobalt activation analyses.

A number of texts dealing with basic general principles and applications of NAA have been published (Lenihan and Thomson, 1965), DeSoete et al., 1972; Kruger, 1971; Lyon, 1964; Koch, 1960; and Nargolwalla and Przybylowicz, 1973).

An excellent presentation of NAA equipment and methods was written by DeSoete et al. (1972). Considerable attention was given to applications including a bibliographical survey of NAA literature through the late 1960's tabulated by element which outlined some 15 different group separations systems for use in radiochemical sample treatment and the design of effective radiochemical protocols.

An excellent text was prepared by Kruger (1971) on activation analysis. Detailed discussions of activation sources, sample handling, detectors, and general principles were given. Extensive presentations of applications and practices as well as limitations to the technique were included.

Lenihan and Thomson (1965) published a good fundamental text on activation analysis. Basic operating concepts, sources of error, interferences and other factors affecting accuracy were discussed.

Pinta (1978) included a concise discussion of activation analysis in his text on methods for trace element analysis. A brief basic discussion of practical operating principles as well as a relatively current review of general applications was presented.

In his general presentation, Parsons (1976) estimated the detection limit of the $^{59}Co(n,\gamma)^{60}Co$ to be 0.0005 μg. Potential interferences from ^{60}Ni and ^{63}Cu were noted. The estimated detection limit using fast neutron activation was 0.02 mg for the reaction $^{59}Co(n,2n)^{58}Co$. In the case of (n, particle) reactions from thermal neutron activation, a detection limit of 1.3 mg was estimated for the photopeak at 0.846 Mev for the reaction $^{59}Co(n,\alpha)^{56}Mn$.

Nargolwalla and Przybylowicz (1973) addressed rather extensively the topic of neutron generators and their applications to activation analysis.

Ganesan et al. (1978) investigated the use of 1-phenyl-3-methyl-4-benzoylpyrazol-5-one (PMBP) as an extractive concentration reagent for cobalt in biological materials prior to NAA. Equilibration times of 3 min were reported compared to the usual 30-min equilibration times needed for more commonly used reagents such as 1-nitroso-2-naphthol or 2-nitroso-1-naphthol. As little as 0.11 μg cobalt was determined in oyster homogenate at a precision of $\pm$ 10%. The technique was applied to animal muscle and plant matter.

Morrison and Potter (1972) employed a combination of radiochemical separations and INAA techniques for the multi-element analysis of plant and animal materials. Although only qualitative analyses of freeze-dried kidney and heart tissues were reported, values of 0.1 mg/kg cobalt were reported for duplicate analysis of NBS SRM 1571 Orchard Leaves, compared to the reported but uncertified value of 0.2 mg Co/kg.

Cornelis et al. (1975) described the use of NAA for the analysis of bulk and trace elements in urine. A survey of literature values for a number of elements in urine was compiled and presented. Due to a wide spread of reported values, cobalt as well as manganese and chromium was identified as being a specific problem area. The author concluded that "some results have been obtained by inadequate procedures or after inaccurate sampling." A study of possible adsorptive losses in polyethylene containers and/or storage stability using ^{60}Co was conducted. A number of elements including cobalt were found to be adequately stored in polyethylene at room temperature and in room light over a 3-day period. Results of NAA of several urine specimens were also reported; they were generally low (< 10 μg/24 hr) compared to previously reported values. The necessarily careful use of metal-free manipulations during sample handling was cited for successful low-concentration analysis.

The use of both NAA and INAA for the determination of trace elements in food grain standard reference materials was reported by Gills et al. (1978). Cobalt in 300-mg sample aliquots was analyzed by INAA techniques. A value of 0.14 ± 0.01 mg/kg cobalt in NBS SRM 1571 Orchard Leaves was reported. Values for wheat flour and rice flour were also determined.

Yule (1966) employed NAA methods to determine a number of elements in whole blood, urine, milk, tapwater, "pure" water, and polyethylene vials. Cobalt was found at the 1 mg/liter level in whole blood but was not observed in the other matrices. Sensitivities of 0.02 to 0.2 mg Co/kg were reported for cobalt in the remaining sample matrices.

Schuster et al. (1973) reported the use of NAA in the study of metallosis. Seven metallic elements including cobalt were studied.

Yang et al. (1977) employed dinonylnaphthalenesulfonic acid (HD) as a preconcentrating agent for the analysis of trace metals in natural waters. Exclusion of interfering elements such as sodium, potassium, chlorine, and bromine was noted. Cobalt was extracted at efficiencies of 32.6 and 94.4% for samples acidified with 0.180 M and 0.018 M $HClO_4$, respectively. A precision of 3.0% was observed for quantitative recoveries. Rice samples were also analyzed using the extraction procedure.

The use of the trifluoroacetylacetone-toluene system for the extraction of selected elements from seawater for subsequent NAA was reported by Lee and Burrell (1972). Cobalt could not be extracted using the simple two-component system. With the addition of a critical amount of isobutylamine, the quantitative extraction of cobalt could be accomplished.

Cyanide complexation of heavy metals and separation with anion-exchange filter papers were used by Kolaczkowski and Jester (1973) to analyze surface waters by NAA. Relatively large sample aliquots ($\sim$ 55 ml) were irradiated for analysis. Interferences from ^{24}Na, ^{82}Br, and I_2 were minimized.

Lenvik et al. (1978) analyzed six elements in fresh water without preconcentration using NAA. Aliquots of 5 ml were heated with HNO_3/H_2O_2 prior to the addition of HCl. The solutions were transferred to a Dowex 1-X8, 100/200 mesh, chloride-form column. Cobalt was eluted using 4 M HCl. A recovery of

96.4% and a standard deviation of 1.9% were reported for cobalt. Detection limits of 0.004 and 0.03 μg/liter were reported for NaI (Tl) and Ge (Li) detectors, respectively.

Similarly, Hirose et al. (1978) employed Chelex 100 and Dowex A-1 resins for the preconcentration of rare earth elements and heavy metals. Unlike previous workers, they also employed the loaded resin as the support material during activation. After irradiation, cobalt was eluted using 2 M HNO_3. Recoveries of 99.4 and 98.0% were observed in the preconcentration step for 4 μg of cobalt and in the overall analysis recovery of 80 μg of cobalt, respectively.

The multielement INAA of fresh water was described by Salbu et al. (1975). Unfiltered aliquots of 5 ml were irradiated. Blank corrections for cobalt were not employed for the sample container or acid used; however, a correction of 0.04 μg/liter was necessary to compensate for contributions from the quartz irradiation ampule. Based on a 5-ml aliquot, the detection and determination limits for cobalt were 0.007 and 0.05 μg/liter, respectively.

Erasmus et al. (1976) compared the use of thermal and epithermal irradiation methods for the analysis of aerosols, underground water, slimes, and sediments. Corrections for filter blanks and container blanks were necessary. Cobalt was best analyzed using thermal irradiation.

INAA was used by Schelenz (1977) for the analysis of river water. Water was freeze-dried prior to analysis such that 80 to 100 mg of material was irradiated. Cobalt concentrations of 0.21 to 0.43 μg/liter were found. Another article (Merritt, 1971) also reported the use of freeze-drying for the analysis of lake, river, and ground waters. Internal standards consisting of ^{198}Au and ^{54}Mn were employed. A problem with the superimposition of lower energy peaks on a relatively high Compton background was noted.

Estuarine suspended particulates were analyzed by Ellis and Chattopadhyay (1979) using INAA. Increasing salinity appeared to produce a negative interference in the cobalt analysis. As salinity increased from 2.23 to 29.9 ppTh the observed cobalt concentration decreased from 22 mg/kg to below the detection limit. What was probably observed, however, was the dissolution of cobalt from the suspended particulate as soluble $CoCl^+$.

Gordon (1971) employed INAA for the nondestructive analysis of metropolitan particulate matter collected on filters and cascade impactors. Samples were analyzed without removal of the collection substrate. Cobalt concentrations of 0.001 to 0.025 ng/m^3 were reported. Alternatively, particulate collected on polystyrene filters has been prepared for NAA by dry ashing at 400 to 425°C (Duce et al., 1974). The sample then is treated with HCl/HNO_3 and HF/HCl mixtures prior to analysis.

Dams et al., (1972) have evaluated a number of filter materials and impaction substrates for the collection of aerosols for trace elemental analyses. Ten different filter materials were considered which included both cellulose and polystyrene compositions. Cobalt impurities were found in the range of 0.1 to 0.8 ng/cm^2.

Seven elements were analyzed in a series of inorganic standard reference materials by Ravnik et al. (1976). Destructive radiochemical separations with thermal NAA were employed for the analysis due to the low sensitivity and poor resolution of the detector available. Calcium carbonate (NBS Nos. 606 and 915) and magnesite, glass sand, and kaolin from the Institute of Mineral Raw Materials, Kutna Hora, Czechoslovakia were analyzed. Good agreement with reported values of cobalt in all matrices analyzed was found.

Laul and Rancitelli (1977) employed a combination of INAA and radiochemical separations for the analysis of six USGS geological samples and NBS Orchard Leaves. Gamma-ray spectra for a number of samples were presented. Agreement between observed values and reference values of cobalt was good. Errors due to counting statistics were in the range of ± 1 to 3%.

INAA was used by Plant et al. (1976) for multielement geochemical mapping. A total of 32 elements were determined satisfactorily by this method. Samples were irradiated for 12 hr in a thermal neutron flux of 4×10^{12} n cm^{-2} sec^{-1}. A detection limit of 0.05 mg/kg was reported for cobalt. Precision was found to be ± 5%. Accuracy based on the analysis of selected USGS geological reference materials was good. The application of the techniques to geochemical mapping was demonstrated.

Morrison et al. (1969) employed chemical separation and INAA methods for the analysis of 45 elements in USGS basalt (BCR-1), andesite (AGV-1), and granite (G-2). Nuclides were separated on the basis of volatilization, extraction with hydrated antimony pentoxide, ion exchange, and solvent extraction with tri-n-butyl phosphate. Generally good agreement with reference values was found for cobalt.

Filby and Shah (1975) discussed sources of error in the NAA of petroleums. Interferences from the reactions $^{60}Ni(n,p)^{60}Co$ and $^{63}Cu(n,\alpha)^{60}Co$ were anticipated. Although the former reaction is the more serious source of interference due to the abundance of nickel in petroleum, an error of only 6.7% was found for oil containing Ni:Co ratio of 570. For crude oil having high cobalt concentrations (12.7 mg/kg), the relative standard deviation based on five replicate analyses was 2.1%.

Additional reports of cobalt analysis in organic/biological matrices by NAA and INAA are summarized in Table III-24.

Randa (1976) investigated the use of epithermal neutron activation analysis (ENAA) in routine INAA of minerals materials. It was possible to analyze 36 elements including cobalt nondestructively after 20 hr of irradiation. Results for six geological materials were reported.

An alternate less-studied activation method, photon activation analysis (PAA), involved the use of gamma-ray bombardment to initiate photonuclear reactions for nuclide production. An excellent paper by Chattopadhyay and Jervis (1975) included an overview of the pertinent literature. Instrumental PAA was applied to the multielement analysis of crop-growing soils. The reaction $^{59}Co(\gamma,n)^{58}Co$ was employed for cobalt analyses. Potential interference by the reaction $^{107}Ag(\gamma,n)^{106m}Ag$ was noted. A detection limit of 3 μg ^{58}Co was reported. A summary of various elemental concentrations in soil as a function of depth was presented.

TABLE III-24

RECENT DETERMINATIONS OF COBALT IN ORGANIC MATRICES BY NEUTRON ACTIVATION ANALYSIS

Matrix	Remarks	Reference
Tea	Direct analysis	Kasrai et al. (1977)
Kidney stones	Direct analysis	Donev et al. (1977)
Muscle tissue	Wet digestion, column chromatography clean-up	D'Hondt et al. (1977
Human concretions	Direct analysis	Jacimovic et al. (1977)
Foods	Direct analysis	Tanner and Friedman (1977)
Blood	Filtered and washed prior to analysis	Zdankiewicz and Fasching (1976)
Tissue	Direct analysis	Kostic et al. (1977)
Biochemical materials	Direct analysis	Laul and Rancitelli (1977)
Human tissue and blood	Radiochemical separations	Tjioe et al. (1977)
Milk powder	-	Lo and Yeh (1976)
Tobacco	Radiochemical separations	Wyttenbach et al. (1976)
Marine organisms	Direct analysis of freeze dried material	Grieg and Jones (1976)
Beach asphalt	-	May and Presley (1975)
Urine	-	Cornelis et al. (1975)
Biological material	-	Steinnes (1975)

Recently, Yamashita and Suzuki (1978) employed PAA for the analysis of six elements in bovine liver, orchard leaves, and Bowen's kale. Samples were pelletized and activated. Subsequently the samples were ashed using HNO_3-$HClO_4$, dried, and redissolved with water. The solution was treated with tropolone sulfate, a chelating agent, and mixed with ion-exchange resin prior to gamma-ray spectrometry analysis. The observed cobalt concentrations were comparable to values obtained using other conventional methods.

i. Miscellaneous methods: A number of techniques have been used successfully for the analyses of cobalt in various matrices which have not been subjected to extensive study. Methods employed have ranged from wet chemical techniques to instrument intensive techniques. Some of these methods are outlined below.

Montano and Ingle (1979a) investigated the use of the lucigenin chemiluminescence reaction in alkaline H_2O_2 for the quantitative analysis of cobalt. The species Co^{2+} was identified as the most effective reaction activator with a resultant linear analytical range of 20 ppt to 100 ppm. The reaction was found to be dependent on time as well as cobalt concentration. In addition to a 72-element interference study, including 11 anions, a review of pertinent references on lucigenin chemiluminescence was presented. Extending their work, Montano and Ingle (1979b) reported the application of the lucigenin chemiluminescence reaction for the analysis of cobalt in plant extracts and tapwater. Iron was found to produce an enhancement interference. In contrast, magnesium was observed to inhibit the cobalt chemiluminescence reaction. Cobalt was determined in Orchard Leaves at a concentration of 0.12 mg/kg $\pm$ 18% relative standard deviation. The chelating agent 1-nitroso-2-naphthol was found to have an extraction efficiency of 65 to 105%. Tapwater having a concentration of $\sim$ 0.10 μg Co/liter was analyzed. A review of alternate useful chemiluminescence reactions including those with luminol and gallic acid was presented.

A number of other cobalt-catalyzed reactions are presented in Table III-25. The reaction characteristics in common among all reports are the cobalt(II)-catalyzed oxidation of an organic material by hydrogen peroxide under basic conditions.

TABLE III-25

APPLICATIONS OF CATALYTIC REACTIONS FOR COBALT DETERMINATION

System	Remarks	Reference
Luminol/H_2O_2	Chemiluminescence detection, detection limit = 0.2 ng/ml	Angelova et al. (1978)
Catechol/H_2O_2	Colorimetric detection at 365 nm	Kreingol'd et al. (1976)
1,10-phenanthroline/ H_2O_2	Temperature change detection, detection limit = 10 μg/liter	Pantaler et al. (1975b)
Catechol/H_2O_2	Semiquantitative for 5 to 50 ng/ml	Weisz et al. (1975)
2-aminopyridine/H_2O_2	Detection limit of 0.01 to 0.06 ng /ml for various solution matrices	Pantaler et al. (1975a)
-diantipyrinyl-3,4-dimethoxytoluene/ H_2O_2	Colorimetric detection at 467 nm, detection limit = 0.16 pg/ml	Trofimov et al. (1974)
Tartaric acid /H_2O_2	Temperature change detection, 0.05 to 1 μg Co(II)	Kreingol'd et al. (1974)

One instrument intensive technique which offers high sensitivity and selectivity with minimum sample treatment is spark source mass spectrometry (SSMS). Two general methods of SSMS analysis may be employed, direct standard comparison or isotope dilution techniques. In the first case, comparison of ion intensities from standard and sample are performed for quantitative estimation. In the case of isotope dilution SSMS analysis, a known quantity of one analyte isotope is enriched in the sample and analyzed. Quantitative analysis is accomplished by comparison of analyte ion intensity and the enriched ion intensity. Isotope dilution SSMS generally allows for quantitative elemental analysis; however, analyte elements must have at least two isotopes to allow use of the technique. In contrast, direct standard comparison SSMS is generally limited to semiquantitative applications. An absolute detection limit of 0.05 ng of cobalt was reported (Elser, 1976). The same source reported applied detection limits of 0.007 $\mu g/m^3$ and 0.35 $\mu g/ml$ for cobalt in air particulate and water samples, respectively. A brief description of instrumental design and equipment as well as a brief review of applications in biological, environmental, and industrial analyses was published by Elser (1976). A somewhat briefer review of equipment and applications to geochemical analysis was presented by Reeves and Brooks (1978). A table of atomic and molecular ions for SSMS analysis of complex sample mixtures has been compiled by Burdo and Morrison (undated) which is useful for the identification of analytes and possible interfering species. A number of analytical applications employing SSMS methods are summarized in Table III-26.

TABLE III-26

RECENT APPLICATIONS OF MASS SPECTRAL TECHNIQUES FOR COBALT DETERMINATION

Matrix	Remarks	Reference
Lunar rock	Stabilized spark, detection of 1 ppm with < 10% coefficient of variation	Ramendik (1977)
Coal, food	Low temperature plasma ashing preparation	Donahue et al. (1977)
Steel and alloys	Detection sensitivity of 17 elements in metallic matrices, analysis of reference materials	Van Hoye et al. (1976)
Organometallics	Spectra and integrated ion current for tetraphenylporphyrin complexes of 11 elements	Davis et al. (1976)
Mussels	Tabulation of 10 elements in various mussels after dry ashing at 250°C for 3 hr	Ball et al. (1975)

Several chromatography systems were reported for cobalt analyses. Poonia et al. (1972) reported the use of paper chromatography with a solvent system consisting of chloroform, acetone, isoamyl alcohol, and HCl (1:1:1:0.5) for analysis of a variety of cations. Methods for individual elements and/or mixtures were presented. Johri and Mehra (1970) determined microgram quantities of Co(II) and seven other cations from mixed solutions by ascending thin-layer chromatography on a silica gel support. Quantitative analysis was accomplished using ring oven techniques. An ion chromatography system was developed by Girard (1979) for the analysis of Co(II). Separation of as little as 0.5 mg Co(II)/kg was accomplished using Aminex A-4 and/or Aminex A-9 columns. The analyte was detected by secondary controlled potential coulometry using the reaction

$$[CuDTPA]^{3-} + Co^{2+} + 2e^{-} — [CoDTPA]^{3-} + Cu^{\circ}$$

where CuDTPA = copper diethylenetriaminepentaacetic acid complex. Tavlarides and Neeb (1978) investigated the gas chromatographic analysis of metals as dialkyldithiocarbamate chelates. Although an extensive compilation of thermal data was presented for various metal complexes, only limited data on cobalt complexes were available.

In a general discussion of molecular fluorescence methods of analysis, St. John (1976) noted several systems for fluorimetric analysis of cobalt. Detection limits of 0.0001 and 0.06 mg/liter were reported for the salicylfluorene/H_2O_2 and 1-(2-pyridylazo)-2-naphthol systems, respectively. Four other possible systems for application in cobalt analysis which utilized carminic acid, PAN, Rhodamine B, and Superchrom Blue Black Ectra were mentioned; however, detection limits were not reported.

Giovannie et al. (1976) investigated the use of laser microprobe emission spectrometry for the analysis of metals in animal organs and tissue material. Standards were prepared by the addition of saline standard solutions to a rabbit muscle homogenate. The mixture was subsequently dried, ashed at 450°C for 24 hr, and pelletized prior to analysis. Linear working ranges were identified; however, specific applications were not reported.

The movement of radioactive cobalt (^{60}Co) in hydrologic systems was evaluated by Claasen (1970). The method, which separated cobalt first as the hydroxide and then as the thiocyanate complex with MIBK, was capable of detecting as little as 0.5 pCi/sample. With modification the method could be used for the analysis of fluvial sediments and soils.

Van de Wiel et al. (1974) reported three methods for the saturation analysis of cyanocobalamin (vitamin B_{12}). A binding protein, either pooled human serum, bacterial binding proteins, or antibodies, was employed. Studies were conducted using 30 pg of ^{57}Co-labeled vitamin B_{12}. Free and bound species were subsequently separated.

BIBLIOGRAPHY. CHAPTER III

Abel, K. H. and L. A. Rancitelli, "Major, Minor, and Trace Element Composition of Coal and Fly Ash, as Determined by Instrumental Neutron Activation Analysis" in Trace Elements in Fuel, Advances in Chemistry Series No. 144, American Chemical Society, Washington, D.C., 1975, pp. 118-138.

Aboul Kheir, A., M. Ayad, and M. M. Amer, "Semimicro Determination of Cobalt(2+) Using N-Bromosuccinimide," Bull. Fac. Pharm., Cairo University, 13(1), 137-143 (1974).

Abu-Samra, A., J. S. Morris, and S. R. Koirtyohann, "Wet Ashing of Some Biological Samples in a Microwave Oven" in Trace Substances in Environmental Health-IX, D. D. Hemphill, Ed., Proceedings of University of Missouri's 9th Annual Conference on Trace Substances in Environmental Health, Columbia, Missouri, June 10-12, 1975, University of Missouri, Columbia, Missouri, 1976, pp. 297-301.

Agnes, G., S. Bendle, H. A. O. Hill, F. R. Williams, and R. J. P. Williams, "Methylation by Methylvitamin B_{12}," Chem. Commun., No. 15, 850-581 (1971).

Alder, J., D. Alger, A. Samuel, and T. West, "The Design and Development of a Multichannel Atomic Spectrometer for the Simultaneous Determination of Trace Metals in Hair," Anal. Chim. Acta, 87(2), 301-311 (1976a).

Alder, J. F., A. J. Samuel, and T. S. West, "The Single Element Determination of Trace Metals in Hair by Carbon-Furnace Atomic Absorption Spectrometry," Anal. Chim. Acta, 87, 313-321 (1976b).

Allcott, G. H., and H. W. Lakin, "The Homogeneity of Six Geochemical Exploration Reference Samples" in Geochemical Exploration 1974, Proceedings of the Fifth International Geochemical Exploration Symposium, Vancouver, British Columbia, Canada, April 1-4, 1974, sponsored and organized by the Association of Exploration Geochemists, I. L. Elliott and W. K. Fletcher, Eds., Elsevier Scientific Publishing Company, New York, New York, 1975, pp. 659-681.

Allen, S. E., and J. A. Parkinson, "The Application of Atomic Absorption in the Analysis of Ecological Materials," Spectrovision, **22**, 2-4 (1969).

Alt, F., and H. Massmann, "Determination of Trace Elements (Manganese, Cobalt) in Sera by Atomic Absorption Spectrometry," Fresenius' Z. Anal. Chem., **279**(2), 100-101 (1976).

Anand, V. D., J. M. White, and H. V. Nino, "Some Aspects of Specimen Collection and Stability in Trace Element Analysis of Body Fluids," Clin. Chem., **21**(4), 595-602 (1975).

Andersson, A., Swedish J. Agric. Res., **6**(2), 145-150 (1976); Anal. Abstr., **31**, 5D6.

Angelova, G., A. Panova, and D. Bakurdzhieva, "New Chemiluminescence Reaction for Cobalt Determination," Metalurgiya (Sofia), **33**(6), 22-24 (1978); Chem. Abstr., **89**, 225517m (1978).

Angino, E. E., and G. K. Billings, Atomic Absorption Spectrometry in Geology. (Methods of Geochemistry and Geophysics, No. 7), 2nd ed., Elsevier Scientific Publishing Company, New York, New York, 1972.

Anonymous, Catalog L 3182-12-69, EKCO Instruments, Ltd., London, 1969.

Anonymous, Analytical Methods for Atomic Absorption Spectrophotometry, Perkin-Elmer Corporation, Norwalk, Connecticut, March 1973.

Anonymous, Analytical Methods for Atomic Absorption Spectroscopy Using the HGA Graphite Furnace, Perkin-Elmer Corporation, Norwalk, Connecticut, April 1974.

Anonymous, "Tracking Trace Metals in the Biosphere," Chem. Eng. News, **55**(17), 23 (1977a).

Anonymous, "Simultaneous Multi-Element Analysis of Aqueous Solutions," Notes Water Res. (United Kingdom), No. 8, 1-4 (1977b).

Armannsson, H., "The Use of Dithizone Extraction and Atomic Absorption Spectrometry for the Determination of Cadmium, Zinc, Copper, Nickel, and Cobalt in Rocks and Sediments," Anal. Chim. Acta, **88**(1), 89-95 (1977).

Baetz, R. A., and C. T. Kenner, "Determination of Trace Metals in Foods Using Chelating Ion Exchange Concentration," J. Agri. Food Chem., **23**(1), 41-45 (1975).

Ball, D. F., M. Barber, and P. G. T. Vossen, "Application of High Resolution Spark Source Mass Spectroscopy for the Determination of Trace Elements in Mussels," Sci. Total Environ., **4**(2), 193-200 (1975).

Barnes, R. M., Ed., Emission Spectroscopy, Dowden, Hutchinson, and Ross, Stroudsburg, Pennsylvania, 1976.

Barnes, R. M., "Emission Spectroscopy," Anal. Chem., **50**(5), 100R-120R (1978).

Barrett, C. A., and C. E. Lowell, "The Cyclic Oxidation Resistance of Cobalt-Chromium-Aluminum Alloys at 100 and 1200°C and a Comparison with the Nickel-Chromium-Aluminum Alloy System," Oxid. Met., **12**(4), 293-311 (1978).

Barrett, C. A., J. R. Johnston, and W. A. Sanders, "Static and Dynamic Cyclic Oxidation of 12 Nickel-, Cobalt-, and Iron-Base High-Temperature Alloys," Oxid. Met., **12**(4), 343-377 (1978).

Baudot, Ph., J. L. Monal, M. H. Livertoux, and R. Truhaut, "Identification de métaux toxiques après extraction et chromatographie sur couches minces de leurs dithizonates. Applications toxicologiques," ["Identification of Toxic Metals after Extraction and Thin-Layer Chromatography of their Dithizonates. Toxicological Applications"], J. Chromatogr. **128**, 141-147 (1976).

Becker, D. A., "Environmental Sample Banking Research and Methodology" in Trace Substances in Environmental Health-X, D. D. Hemphill, Ed., Proceedings of University of Missouri's 10th Annual Conference on Trace Substances in Environmental Health, Columbia, Missouri, June 8-10, 1976, University of Missouri, Columbia, Missouri, 1977, pp. 353-359.

Becker, R. R., A. Veglia, and E. R. Schmid, "Instrumental Neutron Activation Analysis of Standard Biological Materials," Radiochem. Radioanal. Lett., **19**, 343-354 (1974).

Beitz, L., "Analyse von Schadstoffen in Pflanzen mit dem Röntgenfluoreszenzverfahren" ["Analysis of Toxic Elements in Plants by the X-Ray Fluorescence Method"], Chemiker-Zeitung, **97**(8), 424-426 (1973).

Berman, E., "Applications of Atomic Absorption Spectrometry to Trace Metal Analyses of Toxicological Materials," Prog. Chem. Toxicol., **4**, 155-178 (1969).

Beyer, M., "Preparation of Metallic Air Pollutants for Atomic Absorption Analysis," At. Absorpt. Newsl., **8**(1), 23 (1969).

Bhaskare, C. K., and S. K. Deshmukh, "Spectrophotometric Determination of Magnesium(II) with 2,4-Dinitrophenylazoresorcinol", Sci. J. Shivaji Univ., 15, 171-175 (1975); Chem. Abstr., **88**, 31622a (1978).

Bower, N. W., and J. D. Ingle, "Precision of Flame Atomic-Absorption Spectrometric Measurements of Aluminum, Chromium, Cobalt, Europium, Lead, Manganese, Nickel, Potassium, Selenium, Silicon, Titanium, and Vanadium," Anal. Chem., **51**(1), 72-76 (1979).

Braun, T., and A. B. Farag, "Chromofoams [Patent Pending] Qualitative and Semiquantitative Tests with Chromogenic Organic Reagents Imobilized in Plasticized Open-Cell Polyurethane Foams," Anal. Chim. Acta, **73**(2), 301-309 (1974).

B'rezina, M., and J. Volke, "Polarography in Biochemistry, Pharmacology and Toxicology," Prog. Med. Chem., **12**, 247-292 (1975).

Brooks, R. R., B. J. Presley, and I. R. Kaplan, "APDC-MIBK Extraction System for the Determination of Trace Elements in Saline Waters by Atomic-Absorption Spectrophotometry," Talanta, **14**, 809-816 (1967).

Buldini, P. G., "Differential-Pulse Polarographic Determination of Heavy Metals Released from Ceramic Glazes," Analyst (London), **102**(1221), 921-928 (1977).

Burba, P., and K. H. Lieser, "Separation of Trace Elements on a Cellulose Exchanger with 1-(2-Hydroxyphenylazo)-2-naphthol (Hyphan) as Functional Group and Subsequent Determination by X-ray Fluorescence Analysis," Fresenius' Z. Anal. Chem., 286(3-4), 191-197 (1977).

Burdo, R. A., and G. H. Morrison, Table of Atomic and Molecular Lines for Spark Source Mass Spectrometry of Complex Sample-Graphite Mixes, Report No. 1670, Cornell University, Ithaca, New York (undated).

Burrell, D. C., Atomic Spectrometric Analysis of Heavy-Metal Pollutants in Water, Ann Arbor Science Publishers, Inc., Ann Arbor, Michigan, 1974.

Cali, J. P., National Bureau of Standards Certificate of Analysis. Standard Reference Material 1571. Orchard Leaves, Office of Standard Reference Materials, National Bureau of Standards, U.S. Department of Commerce, Washington, D.C., 1977 revision.

Campbell, W. J., H. E. Marr, III, and S. L. Law, "Determination of Trace and Minor Elements in the Combustible Fraction of Urban Refuse" in Methods and Standards for Environmental Measurement. Proceedings of the 8th Materials Research Symposium, September 20-24, 1976, W. H. Kirchhoff, Ed., Special Publication 464, U.S. National Bureau of Standards, Washington, D.C., 1977, pp. 157-164.

Casula, D., and G. Devoto, "Sui Vantaggi Presentati dall'Uso della Fornace a Frafite Come Nuova Technica di Assorbimento Atomico nell'Analisi dell'Inquinamento dell'Ambiente di Lavoro" ["Present Advantages on the Use of a Graphite Furnace as a New Technique for the Atomic Absorption Analysis of Working Environment Pollutants"], Proc. Int. Symp. Recent Adv. Assess. Health Eff. Environ. Pollut., 3, 1693-1696 (1975).

Cesareo, P., S. Sciati, and G. E. Gigante, Int. J. Appl. Radiat. Isot., 27(1), 58-60 (1976); Anal. Abstr., 31, 1H32.

Chalk, A. J., and J. F. Harrod, "Catalysis by Cobalt Carbonyls" in Advances in Organometallic Chemistry, Vol. 6, F. G. A. Stone and R. West, Eds., Academic Press, New York, New York, 1968, pp. 119-170.

Chattopadhyay, A., and R. E. Jervis, "Multielement Determination in Market-Garden Soils by Instrumental Photon Activation Analysis," Anal. Chem., 46(12), 1630-1639 (1974).

Chau, Y. K., "Determination of Trace Metals in Natural Waters" in International Symposium on Identification and Measurement of Environmental Pollution, [Proc.], B. Westley, Ed., National Research Council of Canada, Ottawa, Ontario, 1971, pp. 354-357.

Chau, Y. K., and P. T. S. Wong, "Occurrence of Biological Methylation of Elements in the Environment" in Organometals and Organometalloids: Occurrence and fate in the Environment, E. E. Brinckman and J. M. Bellama, Eds., Society Symposium Series No. 83, American Chemical Society, Washington, D.C., 1978, pp. 39-53.

Cheng, K. L., and R. H. Bray, "1-(2-Pyridylazo)-2-Naphthol as a Possible Analytical Reagent," Anal. Chem., 27, 782-785 (1955).

Chini, P., "Hexacobalthexadecacarbonyl and Its Derivatives," Chem. Commun., 9, 440-441 (1967); Chem. Abstr., 67, 54251j (1967).

Christian, G. D., "Atomic Absorption Spectroscopy for Determination of Elements in Medical Biological Studies," Top. Curr. Chem., 26, 78-112 (1972).

Christian, G. D., and F. J. Feldman, Atomic Absorption Spectrophotometry Applications in Agriculture, Biology, and Medicine, Wiley-Interscience, New York, New York, 1970.

Christian, G. D., and F. J. Feldman, "Comparison Study of Detection Limits Using Flame-Emission Spectroscopy with the Nitrous Oxide-Acetylene Flame and Atomic-Absorption Spectroscopy," Appl. Spectosc., 25(6), 660-663 (1971).

Claasen, H. C., "The Determination of Low Levels of Cobalt-60 in Environmental Waters by Liquid Scintillation Counting," Anal. Chim. Acta, 52, 229-235 (1970).

Climax Molybdenum Company, "Ferrous Alloys Containing Cobalt and Silicon," Brit. Pat. No. 579,924, August 21, 1946; Chem. Abstr., 41, 1200a (1947).

Cobalt Monograph, Centre d'Information du Cobalt, M. Weissenbruch, Ltd., S.A., Brussels, Belgium, 1960, 515 pp.

Cordovi, M. A., "Corrosion Considerations in Light Water-Cooled Nuclear Power Plants," Presented at the International Nickel Power Conference, Kyoto, Japan, 1972, available from the International Nickel Company, Inc., New York, New York.

Cornelis, R., A. Speeke, and J. Hoste, "Neutron Activation Analysis for Bulk and Trace Elements in Urine," Anal. Chim. Acta, **78**, 317-327 (1975).

Cotton, F. A., and R. C. Elder, "Crystal Structure of Tetrameric Cobalt(II) Acetylacetonate," Inorg. Chem., **4**(8), 1145-1151 (1965).

Cotton, F. A., and G. Wilkinson, Advanced Inorganic Chemistry, 3rd ed., Wiley-Interscience Publishers, New York, New York, 1972, pp. 874-890.

Crumbliss, A. L., and F. Basolo, "Monomeric Oxygen Adducts of *N*,*N*'-Ethylenebis(acetylacetoniminato) Ligand Cobalt(II). Preparation and Properties," J. Am. Chem. Soc., **92**(1), 55-60 (1970).

Culver, B. R., Analytical Methods for Carbon Rod Atmizers, Varian Techtron Pty., Springvale, Victoria, Australia, 1975.

Dams, R., K. A. Rahn, and J. W. Winchester, "Evaluation of Filter Materials and Impaction Surfaces for Nondestructive Neutron Activation Analysis of Aerosols," Environ. Sci. Technol., **6**(5), 441-448 (1972).

Davis, B. A., K. S. Hui, D. A. Durden, and A. A. Boulton, "Qualitative and Quantitative Mass Spectrometry of Some Biologically Important Trace Metals as Their Tetraphenylporphyrin Chelates," Biomed. Mass Spectrom., **3**(2), 71-76 (1976).

Dean, J. A., Ed., Langes' Handbook of Chemistry, 11th ed., McGraw-Hill Book Company, New York, New York, 1973, pp. 5-45 to 5-65.

Dean, J. A., and T. C. Rains, Flame Emission and Atomic Absorption Spectrometry, Vols 1-3, Dekker, New York, 1969.

Debal, E., R. Chassin, S. Peynot, and O. Poliakoff, "Applications de la colorimetric de precision a la microanalyse elementaire" ["Applications of Precision Colorimetry to Elementary Trace Analysis"], Talanta, 24(8), 491495 (1977).

di Bie, E., and P. Doyen, "Cobalt Oxides and Salts," Cobalt, 11(2), 3-13 (1962a).

di Bie, E., and P. Doyen, "Cobalt Oxides and Salts. Cobalt Catalysts," Cobalt, 11(3), 3-15 (1962b).

Delespaul, I., H. Peperstraete, and T. Rymen, "Analysis and Calibration Techniques for Measuring of Airborne Particulates and Gaseous Pollutants" in Methods and Standards for Environmental Measurement. Proceedings of the 8th Materials Research Symposium, September 20-24, 1976, W. H. Kirchhoff, Ed., Special Publication 464, U.S. National Bureau of Standards, Washington, D.C., 1977, pp. 565-572.

De Soete, D., R. Gijbels, and J. Hoste, Neutron Activation Analysis, Wiley-Interscience Publishers, New York, New York, 1972.

D'Hondt, P., P. Lievens, J. Versieck, and J. Hoste, "Determination of Trace Elements in Animal and Human Muscle by Semi-Automated Radiochemical Neutron Activation Analysis," Radiochem. Radioanal. Lett., 3(4-5), 231-240 (1977).

Dixit, M. G. Bhale, and A. Thoms, "Emission Spectrographic Determination of Trace Elements in Plant Materials," Indian J. Pure Appl. Phys., 14(6), 485-487 (1974).

Dizikes, L. J., W. P. Ridley, and J. M. Wood, "A Mechanism for the Biomethylation of Tin by Reductive Co-C Bond Cleavage of Alkylcobalamins," J. Am. Chem. Soc., 100(3), 1010-1012 (1978).

Domagalina, E., and S. Zareba, Farm. Pol., 33(4), 207-209 (1977); Anal. Abstr., 34, 1E9.

Donev, I., S. Mashkarov, L. Marichkova, and G. Gotseo, "Quantitative Investigation of Some Trace Elements in Renal Stones by Neutron Activation Analysis," J. Radioanal. Chem., 37(1), 441-449 (1977).

Donohue, D. L., J. A. Carter, and J. C. Franklin, "Separated Isotopes as Internal Standards in Spark Source Mass Spectrometry," Anal. Lett., **10**(5), 371-379 (1977).

Dubnoff, J. W., "A Cobalamin Glutathione Complex," Biochem. Biophys. Res. Commun., **16**(5), 484-488 (1964).

Duce, R. A., G. L. Hoffman, J. L. Fasching, and J. L. Moyers, "The Collection and Analysis of Trace Elements in Atmospheric Pariculate Matter over the North Atlantic Ocean" in Special Environmental Report No. 3, Secretariat of the World Meteorological Organization, Geneva, Switzerland, 1974, pp. 370-379.

Duggan, J. L., W. L. Beck, L. Albrecht, L. Munz, and J. D. Spaulding, "X-Rays Induced by Charged Particles," Adv. X-ray Anal., **15**, 407-423 (1972).

Dybczynski, R., A. Tugsavul, and O. Suschny, "Problems of Accuracy and Precision in the Determination of Trace Elements in Water as Shown by Recent International Atomic Energy Agency Intercomparison Tests," Analyst (London), **103** (1228), 734-744 (1978).

Eldawy, M. A., A. S. Tawfik, and S. R. Elshabouri, "Rapid and Sensitive Colorimetric Determination of Cobalt(II)," J. Pharm. Sci., **65**(5), 664-666 (1976).

Ellern, J. B., and R. O. Ragsdale, Hexacoordinate Complexes of Bis(2,4-pentanedionato)cobalt(II). Inorganic Syntheses, Vol XI, W. L. Jolly, Ed., McGraw-Hill Book Company, New York, New York, 1968, pp. 82-89.

Ellis, K. M., and A. Chattopadhyay, "Multielement Determination in Estuarine Suspended Particulate Matter by Instrumental Neutron Activation Analysis," Anal. Chem., **51**(7), 942-947 (1979).

Elser, R. C., "Spark Source Mass Spectrometry" in Trace Analysis Spectroscopic Methods for Elements, J. D. Winefordner, Ed., John Wiley and Sons, New York, New York, 1976, pp. 383-418.

Erasmus, C. S., J. Sargeant, J. P. F. Sellschop, and J. I. W. Watterson, "Multielement Analysis of Air and Water Pollutants in Gold Mines by Thermal and Epithermal Neutron Activation," in Methods and Standards for Environmental Measurement. Proceedings of the 8th Materials Research Symposium, Sept. 20-24, 1976, W. H. Kirchhoff, Ed., Special Publication 464, U.S. Bureau of Standards, Washington, D.C., 1977, pp. 129-136.

Essenberg, M. K., P. A. Frey, and R. H. Abeles, "Studies on the Mechanism of Hydrogen Transfer in the Coenzymes B_{12}-Dependent Dioldehydrase Reaction II," J. Am. Chem. Soc., **93**(5), 1242-1251 (1971).

Ewing, G. W., Instrumental Methods of Chemical Analysis, 3rd ed., McGraw-Hill, New York, 1969.

Fanchiang, Y.-T., W. P. Ridley, and J. M. Wood, "Kinetic and Mechanistic Studies on B_{12}-Dependent Methyl Transfer to Certain Toxic Metal Ions" in Organometals and Organometalloids: Occurrence and Fate in the Environment, F. E. Brinckman and J. M. Bellama, Eds., American Chemical Society Symposium Series No. 82, American Chemical Society, Washington, D.C., 1978, pp. 54-64.

Fassel, V. A., and R. N. Kniseley, "Inductively Coupled Plasma-Optical Emission Spectroscopy," Anal. Chem., **46**(13) 1110A-1120A (1974).

Figgis, B. N., and J. Lewis, "The Magnetic Properties of Transition Metal Complexes" in Progress in Inorganic Chemistry, Vol. 6, F. A. Cotton, Ed., 1964, p. 187.

Fike, R. S., C. W. Frank, and G. B. Dreher, "High Temperature Atomic and Molecular Spectroscopy," CRC Critical Reviews in Analytical Chemistry, December, 1977, pp. 37-87.

Filby, R. H., and K. R. Shah, "Neutron Activation Methods for Trace Elements in Crude Oils" in The Role of Trace Metals in Petroleum, T. F. Yen, Ed., Ann Arbor Science Publishers Inc., Ann Arbor, Michigan, 1975, pp. 89-110.

Fishman, M. J., and D. E. Erdmann, "Water Analysis," Anal. Chem., 47(5), 334R-361R (1975).

Fishman, M. J., and D. E. Erdmann, "Water Analysis," Anal. Chem., 49(5), 139R-158R (1977).

Flanagan, F. J., Ed., Descriptions and Analyses of Eight New USGS Rock Standards, Geological Survey Professional Paper 840, U.S. Department of the Interior, U.S. Government Printing Office, Washington, D.C., 1976.

Floriani, C., and F. Calderazzo, "Oxygen Adducts of Schiff's Base Complexes of Cobalt Prepared in Solution," J. Chem. Soc. (A), 946-953 (1969).

Foltz, A. K., J. A. Yeransian, and K. G. Sloman, "Food," Anal. Chem., 49(5), 194R-220R (1977).

Fowler, R. D., W. B. Burford, III, J. M. Hamilton, Jr., R. G. Sweet, C. E. Weber, J. S. Kasper, and I. Litant, "Synthesis of Fluorocarbons," Ind. Eng. Chem., 39(3), 292-298 (1947a).

Fowler, R. D., H. C. Anderson, J. M. Hamilton, Jr., W. B. Burford, III, A. Spadetti, S. B. Bitterlich, and I. Litant, "Metallic Fluorides in Fluorocarbon Synthesis," Ind. Eng. Chem., 39(3), 343-345 (1947b).

Frey, P. A., M. K. Essenberg, R. H. Abeles, and S. S. Kerwar, "Comments on a Proposed Mechanism of Action on B_{12} Coenzymes," J. Am. Chem. Soc., 92(14), 4488-4489 (1970).

Fukai, R., B. Oregioni, and D. Vas, "Interlaboratory Comparability of Measurements of Trace Elements in Marine Organisms: Results of Intercalibration Exercise on Oyster Homogenate," Oceanol. Acta, 1(3), 391-396 (1978).

Furman, N. H., Standard Methods of Analysis, Vol. 1, Van Nostrand Company, Inc., Princeton, New Jersey, 1962, pp. 389-390.

Gabriels, R., and A. Cottenie, "Direct Reading Emission Spectrometric Determination of Major and Trace Elements in Plants, Soils, and Soil Extracts," Lab. Pract., 25(12), 835-838 (1976).

Ganesan, R., A. K. Ghose, and B. C. Haldar, "Substoichiometric Estimation of Cobalt in Biological Materials by Neutron Activation Analysis Using 1-Phenyl-3-Methyl-4-Benzoylpyrazol-5-one," Radiochem. Radioanal. Lett., 35(3), 153-159 (1978).

Garbarino, J. R., and H. E. Taylor, "An Inductive-Coupled Plasma Atomic-Emission Spectrometric Method for Routine Water Quality Testing," Appl. Spectrosc. 33(3), 220 (1979).

Garcia, J. D., R. J. Fortner, and T. M. Kavanagh, "Inner-Shell Vacancy Problem in Ion-Atom Collisions," Rev. Mod. Phys. 45, 111-177 (1973).

Gehrke, C. W., P. R. Rexrood, and R. A. Sweeney, "Fertilizers," Anal. Chem., 49(5), 186R-194R (1977).

Gelman, A. L., "Determination of Cobalt in Plant Material by Atomic Absorption," J. Sci. Food Agric., 23(3), 299-305 (1972).

Gelman, A., "A Note on the Determination of Cobalt in Animal Liver," J. Sci. Food Agric., 27(6), 520 (1976).

Gen, M. Ya., E. V. Shtol'ts, I. V. Plate, and E. A. Fedorova, "Effect of Particle Size and Surface State on the Structure and Magnetic Properties of Cobalt Aerosols," Fiz. Metal. Metalloved., 30(3), 645-648 (1970), Chem. Abstr., 74, 92710k (1971).

Giauque, R. D., R. B. Garrett, and L. Y. Goda, "Determination of Forty Elements in Geochemical Samples and Coal Fly Ash by X-Ray Fluorescence Spectrometry," Anal. Chem., 49(7), 1012-1017 (1977).

Gills, T. E., M. Gallorini, and H. L. Rook, "The Determination of Trace Elements in New Food Grain SRM's Using Neutron Activation Analysis," J. Radioanal. Chem., 46(1), 21-25 (1978).

Giovannini, E., G. B. Principato, and F. Rondelli, "Standards for Iron, Cobalt, Nickel, Copper, and Zinc in Laser Microprobe Emission Spectrometry of Biological Material," Anal. Chem., 48(11), 1517-1519 (1976).

Girard, J. E., "Ion Chromatography with Coulometric Detection for the Determination of Inorganic Ions," Anal. Chem., 51(7), 836-839 (1979).

Gläser, E., "Atomabsorptionsspektroskopische Bestimmung von Eisen, Kupfer, Zink, Cadmium, Mangan und Kobalt in Natur- und Brauchwässern mit der, Injektionsmethode nach vorhergehender extraktiver Anreicherung," ["Atom-Absorption Spectroscopic Determination of Iron, Copper, Zinc, Cadmium, Manganese and Cobalt in Waters by Means of Injection Method After an Extractive Concentration"], Acta Hydrochim. Hydrobiol., 6(1), 83-86 (1978).

Gluskoter, H. J., R. R. Ruch, W. G. Miller, R. A. Cahill, G. B. Dreher, and J. K. Kuhn, Trace Elements in Coal: Occurrence and Distribution, Circular 499, Illinois State Geological Survey, Urbana, Illinois, 1977.

Goldberg, E. D., Baseline Studies of Pollutants in the Marine Environment and Research Recommendations, Deliberations of the International Decade of Ocean Exploration (IDOE) Bseline Conference, New York, New York, May 24 to 26, 1972, PB 233-959, National Technical Information Service, U.S. Department of Commerce, Springfield, Virginia, 1972.

Golembiewski, M., K. Ananth, G. Trischan, and E. Baladi, Environmental Assessment of a Waste-to-Energy Process, Braintree Municipal Incinerator, Braintree, Massachusetts, Draft Final Report, Contract No. 68-02-2166, Midwest Research Institute, for the Industrial Environmental Research Laboratory, Office of Research and Development, U.S. Environmental Protection Agency, Cincinnati, Ohio, August 1978.

Gonchakova, N. N., T. N. Vladimirskaya, and Z. S. Kozlovskaya, "Determination of Cobalt in Liquid Complex Fertilizers," Metody Anal. Kontrolya Proizvod. Khim. Promsti., 8, 30-31 (1977); Chem. Abstr., 88, 103762 (1978).

Gordon, G. E., "Instrumental Activation Analysis of Atmospheric Pollutant and Pollution Source Material" in International Symposium on Identification and Measurement of Environmental Pollution, [Proc.], B. Westley, Ed., National Research Council of Canada, Ottawa, Ontario, 1971, pp. 138-143.

Gorman, P., M. Marcus, K. Ananth, and M Golembiewski, Environmental Assessment of a Waste-to-Energy Process, Union Carbide Purox® System, Contract No. 68-02-2166, Midwest Research Institute for Industrial Environmental Research Laboratory, Office of Research and Development, U.S. Environmental Protection Agency, Cincinnati, Ohio, 1978.

Górski, L., J. Heinonen, and O. Suschny, Final Report on the Intercomparison of Trace Multielement Analysis in Dried Animal Whole Blood (A-2), Calcinated Animal Bone (A-3/1), Powdered Milk (A-8), Wheat Flour (V-2/1), Dried Potatoes (V-4), IAEA/RL/25, International Atomic Energy Agency, Vienna, Austria, July 1974.

Gorsuch, T. T., The Destruction of Organic Matter, Pergamon Press, New York, New York, 1970.

Grieg, R., and J. Jones, Arch. Environ. Contam. Toxicol., **4**(4), 420-434 (1976); Anal. Abstr., **32**, 6D7.

Grimley, R. T., "Mass-Spectrometric Study of the Vaporization of Cobalt Oxide," J. Chem. Phys., **45**(11), 4158-4162 (1966).

Gürtler, O., "Extraktionsspektralphotometrische Spurenbestimmung von Kobalt und Nickel in Konz. Natriumchlorid-Lösung," ["Extraction-Spectrophotometric Determination of Trace Amounts of Cobalt and Nickel in Concentrated Sodium Chloride Solutions"], Fresenius' Z. Anal. Chem., **284**(3), 206-207 (1977).

Haftka, F. J., and N. Terao, "Variation of Chemical Composition on the Surface of Cobalt-Base Alloys by Oxidation in Air," Jap. J. Appl. Phys., **12**(3), 329-336 (1973).

Hageman, L., L. Torma, and B. Ginther, "Analysis of Feed Grains and Forages for Traces of Cobalt by Flameless Atomic Absorption Spectroscopy," J. Assoc. Off. Anal. Chem., **58**(5), 990-994 (1975).

Hannaker, P., and T. C. Hughes, "Improved Stepwise Solvent Extraction Scheme for Geochemical Trace Analysis Using Flame Atomic Absorption," J. Geochem. Explor. **10**, 169-180 (1978).

Hansen, R. K., and R. H. Hall, "An Improved General Method for the Reduction of Analytical Errors in Flame Emission and Atomic Absorption Spectrometry," Anal. Chim. Acta, 92, 307-320 (1977).

Harms, V., and J. Kunze, Z. Lebensm.-Unters. Forsch, 164(3), 204-207 (1977); Anal. Abstr. 34, 3D46.

Harvey, B. R., and J. W. R. Dutton, "The Application of Photo-Oxidation to the Determination of Stable Cobalt in Sea Water," Anal. Chim. Acta, 67(2), 377-385 (1973).

Haynes, B. W., S. L. Law, and W. J. Campbell, Metals in the Combustible Fraction of Municipal Solid Waste, Report of Investigations 8244, U.S. Bureau of Mines, Washington, D.C., 1977.

Heck, R. F., "Organic Syntheses via Alkyl- and Acylcobalt Tetracarbonyls" in Organic Syntheses via Metal Carbonyls, Vol. 1, I. Wender and P. Pino, Eds., Interscience Publishers, New York, New York, 1968, pp. 373-403.

Heigl, A., "Polarographische Bestimmung von Kupfer, Blei, Zinn, Cadmium, Nickel, Zink, Eisen, Kobalt und Chrom in Abwasser" ["Polarographic Determination of Copper, Lead, Tin, Cadmium, Nickel, Zinc, Iron, Cobalt and Chromium in Wastewater"], Chimia, 32(9), 339-344 (1978).

Helin, T. R. M., and J. C. Slaughter, "Determination of Metals in Brewing Materials by Flameless Atomic Absorption Spectroscopy," J. Inst. Brew., 83(1), 15-16 (1977).

Henzler, T. E., R. J. Korda, P. A. Helmke, M. R. Anderson, M. M. Jimenez, and L. A. Haskin, "Accurate Procedure for Multielement Neutron Activation Analysis of Trace Elements in Biological Materials," J. Radioanal. Chem., 20, 649-663 (1974).

Hileman, J. C., Preparative Inorg. Reactions, Vol. 1, W. L. Jolly, Ed., Interscience Publishers, New York, New York, 1964, p. 101.

Hirayama, K., and N. Unohara, ["Studies on Spectrophotometric-Catalytic Determination of Trace Amounts of Cobalt Based on the Catalytic Oxidation of Gallocyanin by Hydrogen Peroxide"], Nippon Kagaku Kaishi, 11, 1498-1502 (1978); Chem. Abstr., 90, 33493z (1979).

Hirose, A., K. Kobori, and D. Ishii, "Determination of Rare Earth Elements and Heavy Metals in River Water by Preconcentration on Chelex 100 and NAE," Anal. Chim. Acta, 97(2), 303-310 (1978).

Hoffman, B. M., and D. H. Petering, "Coboglobins: Oxygen-Carrying Cobalt-Reconstituted Hemoglobin and Myoglobin," Proc. Natl. Acad. Sci., 67(2), 637-643 (1970).

Hoffman, B. M., D. L. Diemente, and F. Basolo, "Electron Paramagnetic Resonance Studies of Some Cobalt(II) Schiff Base Compounds and Their Monomeric Oxygen Adducts," J. Am. Chem. Soc., 92(1), 61-65 (1970).

Hofstader, R. A., O. I. Milner, and J. H. Runnels, Eds., Analysis of Petroleum for Trace Metals, Advances in Chemistry Series, No. 156, American Chemical Society, Washington, D.C., 1976.

Honjo, T., A. Shuhei, and K. Toshiyasu, "Concentration and Separation of a Trace Amount of Cobalt(II) in Sea-water as Its 1-Nitroso-2-Naphtholate by Extraction Chromatography," Bull. Chem. Soc. Japan, 51(4), 1239-1240 (1978).

Hoppe, J., and E. Uhlemann, "Extraktionphotometrische Bestimmung von Cobalt mit Thiodibenzoylmethan in Eisenlegierungen und organischen Material" ["Extraction Photometric Determination of Cobalt with Thiodibenzoylmethane in Iron Alloys and Organic Materials"], Z. Chem., 17(2), 72 (1977).

Howell, J. A., and L. G. Hargis, "Ultraviolet and Light Absorption Spectrometry," Anal. Chem., 50(5), 243R-261R (1978).

Hübel, W., 'Hydrido Metal Carbonyls and Related Compounds' in "Metal Carbonyls: Preparation, Structure, and Properties" in Organic Syntheses via Metal Carbonyls, Vol. 1, I. Wender and P. Pino, Eds., Interscience Publishers, a division of John Wiley & Sons, New York, New York, pp. 188-200.

Huey, C., F. E. Brinckman, S. Grim, and W. P. Iverson, "Role of Tin in Bacterial Methylation of Mercury" in Proc. Int. Conf. Transp. Persistent Chem. Aquat. Ecosyst., Part II, National Research Council, Ontario, Canada, May 1-3, 1974, Chem. Abstr., **82**, 167336u (1975).

Iwasaki, K., K. Tanaka, and N. Takagi, Japan. Analyst, 23(10), 1179-1184 (1974); Anal. Abstr., **29**, 2B7.

Iyengar, G. V., K. Kasperek, and L. E. Feinendegen, "Retention of the Metabolized Trace Elements in Biological Tissues Following Different Drying Procedures. I. Antimony, Cobalt, Iodine, Mercury, Selenium and Zinc in Rat Tissues," Sci. Total Environ., **10**(1), 1-16 (1978).

Jacimovic, L., R. Draskovic, and B. Ostojic, "Determination of Some Trace Elements in Some Human Concretions," J. Radioanal. Chem., **37**(1), 415-419 (1977).

Job, V. A., S. B. Kartha, G. Krishnamurty, and S. Gopal, "Spectrographic Analysis of High Purity Manganese Dioxide," Z. Anal. Chem., **271**, 26-28 (1974); Chem. Abstr., **81**, 145222f (1974).

Johnson, A. W., L. Mervyn, N. Shaw, and E. L. Smith, "A Partial Synthesis of the Vitamin B_{12} Coenzyme and Some of Its Analogues," J. Chem. Soc., 4146-4156 (1963).

Johnson, C. A., "The Determination of Some Toxic Metals in Human Liver as a Guide to Normal Levels in New Zealand. Part I. Determination of Bi, Cd, Cr, Co, Cu, Pb, Mn, Ni, Ag, Tl and Zn," Anal. Chim. Acta, **81**, 69-74 (1976).

Johri, K. N., and H. C. Mehra, "Trace Metal Analysis by Combined Thin-Layer Chromatography Incorporating Fluorescent Support and Ring Oven Colorimetry," Microchem. J., **15**(4), 642-648 (1970).

Jones, J. M., J. T. McCall, and L. M. Elveback, "Trace Metals in Human Osteogenic Sarcoma," Mayo Clin. Proc., **47**, 476-478 (1972).

Jones, M. M., "Chapter IV" in Ligand Reactivity and Catalysis, Academic Press, New York, New York, 1968, pp. 195-198.

Kasrai, M., M. J. Shoushtarian, M. H. Bozorgzadeh, "Determination of Trace Elements in Tea Leaves by Neutron Activation Analysis," J. Radioanal. Chem., **41**(1-2), 73-79 (1977).

Kawaguchi, H., and B. L. Vallee, "Microwave Excitation Emission Spectrometry Determination of Picogram Quantities of Metals in Metalloenzymes," Anal. Chem., **47**(7), 1029-1034 (1975).

Kawaraya, T., M. Kawasaki, K. Haruki, K. Tomito, M. Oka, and S. Horiguchi, Bunseki Kagaku, **25**(7), 464-68 (1976); Anal. Abstr., **32**, 2D10.

Kharkover, S. V., and M. Z. Kharkover, ["Successive Extraction-Photometric Determination of Cobalt, Manganese and Copper in the Air"], Gig. Sanit., **38**(11), 107-108 (1973); Biol. Abstr., HEEP/74110447.

Kingston, H. M., I. L. Barnes, T. J. Brady, T. C. Rains, and M. A. Champ, "Separation of Eight Transition Elements from Alkali and Alkaline Earth Elements in Estuarine and Seawater with Chelating Resin and Their Determination by Graphite Furnace Atomic Absorption Spectrometry," Anal. Chem., **50**(14), 2064-2070 (1978).

Kiriyama, T., and R. Kuroda, "Anion-Exchange Separation and Spectrophotometric Determination of Cobalt in Sea-Water," Z. Anal. Chem., **288**(5), 354-356 (1977).

Klein, D. H., W. Andren, J. A. Carter, J. F. Emery, C. Feldman, W. Fulkerson, W. S. Lyon, J. C. Ogle, Y. Talmi, R. I. Van Hook, and N. Bolton, "Pathways of Thirty-Seven Trace Elements Through Coal-Fired Power Plant," Environ. Sci. Technol., **9**(10), 973-979 (1975).

Klein, R., Jr., and C. Hach, "Standard Addition Uses and Limitations in Spectrophotometric Analysis" in Methods and Standards for Environmental Measurement. Proceedings of the 8th Materials Research Symposium, Sept. 20-24, 1976, W. H. Kirchhoff, Ed., Special Publication 464, U.S. National Bureau of Standards, Washington, D.C., 1977, pp. 61-68.

Kniseley, R. N., Analytical Applications of Inductively Coupled Plasma-Optical Emission Spectroscopy, Rep. Atom. Energy Comm. U.S., IS-T-626, 1974.

Kniseley, R. N., V. A. Fassel, and C. C. Butler, "Application of Inductively-Coupled Plasma Excitation Sources to the Determination of Trace Metals in Microliter Volumes of Biological Fluids," Clin. Chem., 19(8), 807-812 (1973).

Koch, R. C., Activation Analysis Handbook, Academic Press, New York, New York, 1960.

Kolaczkowski, A., and W. A. Jester, "Activation Analysis of Heavy Metals in Surface Waters Using Ion Exchange Filter Papers and Cyanide Complexing," J. Radioanal. Chem., 16(1), 21-29 (1974).

Kopp, J. F., and R. C. Kroner, "A Direct-Reading Spectrochemical Procedure for the Measurement of Nineteen Minor Elements in Natural Water," Appl. Spectrosc., 19(5), 155-159 (1965).

Korkisch, J., and L. Gödl, "Anwendung von Ionenaustauschverfahren zur Bestimmung bon Spurenelementen in naturlichen Wassern - IV. Uran, Kobalt und Kadmium" ["The Use of Ion Exchange Methods to Determine Trace Elements in Natural Waters. IV. Uranium, Cobalt and Cadmium"], Talanta, 21(10), 1035-1046 (1974).

Korkisch, J., and H. Huebner, ["Atomic Absorption Spectrophometric Determination of Cobalt, Copper, Manganese, and Zinc in Multivitamin Preparations After Separation by Means of Anion Exchange"], Mikrochim. Acta, 2(3-4), 311-316 (1976).

Kostic, K, R. J. Draskovic, M. Ratkovic, D. Kostic, and R. S. Droskovic, "Determination of Some Trace Elements in Different Organs of Normal Rats," J. Radioanal. Chem., 37(1) 405-413 (1977).

Krasil'shchik, V., N. Kuznetsova, and T. Manova, Zh. Anal. Khim., 32(4), 837-840 (1977); Anal. Abstr., 33, 5C14.

Kreingol'd, S. U., V. K. Komarova, V. I. Maitsokha, and E. M. Yutal', Zh. Anal. Khim., 29(12), 2324-2328 (1974). Anal. Abstr., 29, 6B123.

Kreingol'd, S. U., L. I. Sosenkova, and I. F. Vzorova, Metody Anal. Kontrol. Proizvod. Khim. Prom., 2, 38-41 (1976); Anal. Abstr., 32, 5B22.

Kruger, P., Principles of Activation Analysis, Wiley-Interscience Publishers, New York, New York, 1971.

Kurinov, A. D., A. P. Lisitsyn, V. N. Lukashin, and L. A. Smakhtin, Okeanologiya, **14**(3), 555-563 (1974); Anal. Abstr., **30**, 2H17.

Kusaka, Y., H. Tsuji, Y. Tamari, T. Sagawa, S. Ohmori, S. Imai, and T. Ozaki, "Neutron Activation Analysis of Biologically Essential Trace Elements in Environmental Specimens Using Pyrrolidinedithiocarbamate Extraction," J. Radioanal. Chem., **37**, 917-926 (1977).

Kuz'min, N. M., L. M. Yakimenko, and V. R. Kalinachenko, "Opredelenie Primesei v Vode" ["Determination of Impurities in Water"], Metody Anal. Khim. Reaktivov Prep., No. 15, 84-85 (1968); Chem. Abstr., **69**, 69624y (1968).

Labrecque, J. J., "Determination of Molybdenum and Cobalt in Hydrodesulfurization Catalysts Using a Computerized Radioisotope X-Ray Fluorescence System," J. Radioanal. Chem., **41**(1-2), 127-135 (1977).

Lakanen, E., "On the Analysis of Soluble Trace Elements," Ann. Agric. Fenniae, **2**, 109-117 (1962).

Lakatos, I., Acta Chim. Hung., **82**(4), 391-402 (1974); Anal. Abstr., **29**, 4C55.

Lapitskaya, S. K., V. G. Sviridenko, and A. S. Palamarchuk, Zh. Anal. Khim., **30**(3), 629-631 (1975); Anal. Abstr., **30**, 4F2.

Lau, H. K. Y., and H. Ashmead, "Trace Minerals in Human Hair by Atomic Absorption Spectrophotometry-Recovery Study," Anal. Lett., **8**(11), 815 (1975).

Laul, J. C., and L. A. Rancitelli, "Multielement Analysis by Sequential Instrumental and Radiochemical Neutron Activation," J. Radioanal. Chem., **38**, 461-475 (1977).

Lee, M.-L., and D. C. Burrell, "Extraction of Cobalt, Iron, Indium and Zinc from Seawater by Means of the Trifluoroacetylacetone Toluene System," Anal. Chim. Acta, **62**(1), 153-161 (1972).

Lenihan, J. M. A., and S. J. Thomson, Eds., Activation Analysis Principles and Application, Academic Press, New York, New York, 1965.

Lenvik, K., E. Steinnes, and A. C. Pappas, "The Simultaneous Determination of Arsenic, Cadmium, Cobalt, Mercury, Molybdenum, and Zinc in Fresh Water by Neutron Activation Analysis," Anal. Chim. Acta, **97**(2), 295-301 (1978).

LeRoy, V. M., and A. J. Lincoln, "Spectrochemical Method for the Determination of 36 Elements in Industrial Effluent," Anal. Chem., **46**(3), 369-373 (1974).

Lister, B., "Second Inter-Laboratory Survey of the Accuracy of Ore Analysis," Trans. Inst. Min. Metall. Sect. B, **86**, 133-148 (1977).

Lo, J., and J. Yeh, "Determination of Trace Elements in Powdered Milk by Neutron Activation Analysis," Radioisotopes, **25**(4), 206-209 (1976).

Lowry, T. M., "Factors Influencing Coordination," J. Soc. Chem. Ind., **42**, 316-319 (1923); Chem. Abstr., **17**, 2240 (1923).

Luo, C. S., M. H. Chang, C. T. Chang, J. C. Chou, H. H. Hsu, and H. T. Tsai, "Quantitative Trace Element Analysis by Proton-Induced X-Rays," Ho Tzu K'o Hsueh, **15**(1), 29-35 (1978); Chem. Abstr., **89**, 225553v (1979).

L'vov, B. V., Atomic Absorption Spectrochemical Analysis, Translated by J. H. Dixon, Adam Hilger Ltd., London, 1970.

Lyon, W. S., Jr., Ed., Guide to Activation Analysis, Van Nostrand, New York, New York, 1964.

Machiroux, R., and J. C. DuPont, "Dosage des metaux lourds dans le plancton par spectrometrie d'absorption atomique sans flamme" ["Determination of Heavy Metals in Plankton by Flameless Atomic Absorption Spectrometry"], Anal. Chim. Acta, **85**(2), 231-239 (1976).

Maienthal, E. J., and D. A. Becker, "A Survey on Current Literature on Sampling, Sample Handling, for Environmental Materials and Long Term Storage," Interface, 5(4), 49-61 (1976).

Maienthal, E. J., and J. K. Taylor, "Polarographic Methods in Determination of Trace Constituents of Water," Am. Chem. Soc., Div. Water Waste Chem., Preprints, 7(1), 92-102 (1967).

Mandler, J. W., and R. A. Semmler, "Trace Element Analysis Using Charged Particle Techniques" in Proceedings of the Second International Conference on Nuclear Methods in Environmental Research, J. R. Vogt, and W. Meyer, Ed., CONF-740701, National Technical Information Service, U.S. Department of Commerce, Springfield, Virginia, 1974, pp. 130-137.

Marchant, J. W., and B. C. Klopper, "Microgram Metal Contamination of Water/Nitric Acid After Four Years in Linear Polyethylene Containers," J. Geochem. Explor., 9, 103-107 (1978).

Marcie, F. J., "X-Ray Fluorescent Analysis of Trace Toxic Elements in Water," Norelco Reports, 15, 3-5, 25 (1968).

Masironi, R., Ed., Trace Elements in Relation to Cardiovascular Diseases (Status of the Joint WHO/IAEA Research Programme), World Health Organization, Geneva, Switzerland, 1974.

May, L. A., and B. J. Presley, "Determination of Trace Elements in Beach Asphalts by Neutron Activation Analysis," J. Radioanal. Chem., 27(2), 439-445 (1975).

McCrary, A. L., and W. L. Howard, "Chelating Agents" in Kirk-Othmer Encyclopedia of Chemical Technology, 3rd ed. Vol. 5, John Wiley and Sons, New York, New York, 1979, pp. 339-368.

McQuaker, N. R., P. D. Kluckner, and G. N. Change, "Calibration of an Inductively Coupled Plasma-Atomic Emission Spectrometer for the Analysis of Environmental Materials," Anal. Chem., 51(7), 888-895 (1979).

Merritt, W. F., "Identification and Measurement of Trace Elements in Fresh Water by Neutron Activation" in International Symposium on Identification and Measurement of Environmental Pollution [Proc.], B. Westley, Ed., National Research Council of Canada, Ottawa, Ontario, 1971, pp. 358-362.

Mesmer, R. E., and C. F. Baes, Jr., The Hydrolysis of Cations. Part II., Oak Ridge National Laboratory, Oak Ridge, Tennessee, 1975, pp. 110-113.

Meunier, F., L. Meunier, and H. Vanderpoorten, "Sur le mécanisme du grillage des sulfures de cobalt" ["The Mechanism of Roasting of Cobalt Sulfide"], Industrie chim. belge, **20**(Special No.), 494-499 (1955).

Montano, L. A., and J. D. Ingle, Jr., "Investigation of the Lucigenin Chemiluminescence Reaction," Anal. Chem., **51**(7), 919-926 (1979a).

Montano, L. A., and J. D. Ingle, Jr., "Determination of Cobalt by Lucigenin Chemiluminescence," Anal. Chem., **51**(7), 926-930 (1979b).

Mitchell, P. R., "Cobalt," Method. Chim., **8**, 307-344 (1976).

Mizoguchi, T., and H. Iskii, "Study of Extracting Agents for the Determination of Cobalt and Nickel by Solvent Extraction-Carbon-Furnace Atomic Absorption Spectrometry," Bunseki Kagaku, **26**(12), 839-843 (1977); Chem. Abstr., **89**, 16158f (1978).

Moeller, T., Inorganic Chemistry: An Advanced Textbook, John Wiley & Sons Inc., New York, New York, 1952, p. 229.

Molinari, A., "Radioisotope Probe for On-Stream Slurry Analysis," Instrum. Min. Metall. Ind., **4**, 61-68 (1976); Chem. Abstr. **86**, 19928f (1977).

Mor[r]al, F. R., "Corrosion of Cobalt and Cobalt Alloys" in Proc., Conf., Natl. Assoc. Corros. Eng., 25th, **1969**, 260-275 (1970).

Morral, F. R., "Cobalt Compounds" in Kirk-Othmer Encyclopedia of Chemical Technology, 3rd ed., Vol. 6, Chocolate and Cocoa to Copper, M. Grayson, Executive Ed., A Wiley-Interscience Publication, John Wiley & Sons, Inc., New York, New York, 1979, pp. 495-510.

Morrison, G. H., J. T. Gerard, A. Travesi, R. L. Currie, S. F. Peterson, and N. M. Potter, "Multielement Neutron Activation Analysis of Rock Using Chemical Group Separations and High Resolution Gamma Spectrometry," Anal. Chem., **41**, 1633-1637 (1969).

Morrison, G. H., and N. M Potter, "Multielement Neutron Activation Analysis of Biological Material Using Chemical Group Separations and High Resolution Gamma Spectrometry," Anal. Chem., **44**(4), 839-842 (1972).

Moselhy, M., D. W. Boomer, J. N. Bishop, and P. L. Diosady, "A Multielement-Direct Reading Method for the Spectral Analysis of Sediment Leachates" in Methods and Standards for Environmental Measurement, W. H. Kirchhoff, Ed., Proceedings of the 8th Materials Research Symposium, Sept. 20-24, 1976, Special Publication 464, U.S. National Bureau of Standards, Washington, D.C., 1977, pp. 137-150.

Motomizu, S., "The New Method for the Determination of Cobalt in Sea Water by Solvent Extraction with 2-Nitroso-5-Diethylaminophenol," Anal. Chim. Acta, **64**(2), 217-224 (1973).

Motomizu, S., "Spectrophotometric Determination of Micro-Amounts of Cobalt in Uranium Oxide, Pure Uranium, and a Uranyl Salt with 2-Nitroso-5-Dimethylaminophenol," Analyst, (London), **100** (1186), 39-45 (1975).

Murakami, M., and J. W. Kang, "Homogeneous Hydrogenation of Organic Compounds with Cobalt Ions. Part VII. Homogeneous Hydrogenations of Organic Compounds with a Cyano Cobalt Complex as a Catalyst," Bull. Chem. Soc. Jap., **36**(7), 763-769 (1963).

Murakami, M., R. Kawai, and K. Suzuki, "Catalytic Reduction of Unsaturated Bonds with Cyanocobalt Complex Ions and Side Reactions," Nippon Kagaku Zasshi, **84**(8), 669-673 (1963); Chem. Abstr., **60**, 4053b (1964).

Muzzarelli, N. W., "Chitin and Chitosan as Chromatographic Supports and Adsorbents for Collection of Metal Ions from Organic and Aqueous Solutions and Sea Water," U.S. Patent 3,635,818, January 18, 1972.

Muzzarelli, R. A. A., G. Raith, and O. Tubertini, "Separations of Trace Elements from Sea Water, Brine and Sodium and Magnesium Salt Solutions by Chromatography on Chitosan," J. Chromatogr., **47**, 414-420 (1970).

Muzzarelli, R., and R. Rocchetti, "Atomic-Absorption Determination of Manganese, Cobalt and Copper in Whole Blood and Serum, with a Graphite Atomizer," Talanta, **22**(8), 683-685 (1975).

Myers, A. T., R. G. Havens, and P. J. Dunton, "A Spectrochemical Method for the Semiquantatative Analysis of Rocks, Minerals, and Ores," U.S. Geol. Surv. Bull., **1084**, 207-229 (1958).

Myers, A. T., R. G. Havens, J. J. Connor, N. M. Conklin, and H. J. Rose, Jr., Glass Reference Standards for the Trace Element Analysis of Geological Materials - Compilations of Interlaboratory Data, Geological Survey Professional Paper 1013, U.S. Department of the Interior, U.S. Government Printing Office, Washington, D.C., 1976.

Nadkarni, R. R., and G. H. Morrison, "Multielement Instrumental Neutron Activation Analysis of Biological Materials," Anal. Chem., **45**, 1957-1960 (1973).

Nargolwalla, S. S., and E. P. Przybylowicz, Activation Analysis with Neutron Generators, John Wiley & Sons, New York, New York, 1973.

Nasser, M. I., "The Modulated Variable Internal Standard Method," Appl. Spectrosc., **28**(4), 356-357 (1974).

Nelson, J. E., Pollution and Polarography, Chemtrix, Inc., Beaverton, Oregon, 1970, 12 pp.; abstr. in Cadmium in Water. A Bibliography, PB 218-829, National Technical Information Service, U.S. Department of Commerce, Springfield, Virginia, March 1973, p. 61.

Nicholls, D., The Chemistry of Iron, Cobalt and Nickel, Pergamon Press, New York, New York, 1973.

Obrecht, M. F., and M Pourbaix, "Corrosion of Metals in Potable Water Systems," Proc. 3rd Int. Congr. Met. Corros, Moscow 1966, **4**, 228-245 (1969).

Office of Standard Reference Materials, Institute for Materials Research, NBS, Catalog of NBS Standard Reference Materials, 1979-80 Edition, National Bureau of Standards, U.S. Department of Commerce, U.S. Government Printing Office, Washington, D.C., 1979.

O'Haver, T. C., "Chemical Aspects of Elemental Analysis" in Trace Analysis Spectroscopic Methods for Elements, J. D. Winefordner, Ed., John Wiley & Sons, New York, New York, 1976, pp. 63-78.

Oikawa, K., Trace Analysis of Atmospheric Samples, John Wiley & Sons, New York, New York, 1977.

Oliver, R. T., and E. P. Cox, "Nonferrous Metallurgy. I. Light Metals," Anal. Chem., **39**(5), 102R-110R (1967).

Olsen, K. W., and E. L. Bour, "Inductively Coupled Plasma-Atomic Emission Spectrometry - A Tool for the Analysis of National Pollutants Discharge Elimination System Wastes," Paper No. 193 presented at the Meeting of the American Chemical Society, Honolulu, Hawaii, April 1-6, 1979.

Ondov, J. M., W. H. Zoller, I. Olmez, N. K. Aras, G. E. Gordon, L. A. Rancitelli, K. H. Abel, R. H. Filby, K. R. Shah, and R. C. Ragaini, "Elemental Concentrations in the National Bureau of Standards' Environmental Coal and Fly Ash Standard Reference Materials," Anal. Chem., **47**(7), 1102-1109 (1975).

Ondov, J. M., R. C. Ragaini, R. E. Heft, G. L. Fisher, D. Silberman, and B. A. Prentice, "Interlaboratory Comparison of Neutron Activation and Atomic Absorption Analyses of Size-Classified Stack Fly Ash" in Methods and Standards for Environmental Measurement. Proceedings of the 8th Materials Research Symposium, Sept. 20-24, 1976, W. H. Kirchhoff, Ed., Special Publication 464, U.S. National Bureau of Standards, U.S. Department of Commerce, Washington, D.C., 1977, pp. 157-164.

Orgel, L. E., "The Effects of Crystal Fields on the Properties of Transition-Metal Ions," J. Chem. Soc., 4857-4761 (1952).

Orgel, L.E., An Introduction to Transition-Metal Chemistry: Ligand Field Theory, John Wiley & Sons, Inc., New York, New York, 1961.

Oyamada, N., M. Ishaizaki, S. Ueno, F. Katuoka, R. Murakami, K. Dubota, and K. Katsumura, ["Determination of Cobalt in Biological Samples by Flameless Atomic Absorption Spectrometry Using A Carbon Tube Atomizer"], Ibaraki-Ken Eisei Kenkyusho Nempo, 15, 59-64 (1977); Chem. Abstr., 89, 125515 (1978).

Paces, T., "Active Mineral Surfaces: Origin and Possible Effects on Trace Elements in Natural Water Systems" in Trace Substances in Environmental Health-VI, D. D. Hemphill, Ed., Proceedings of University of Missouri's 6th Annual Conference on Trace Substances in Environmental Health, Columbia, Missouri, June 13-15, 1972, University of Missouri, Columbia, Missouri, 1973, pp. 361-368.

Pantaler, R. P., L. D. Alfimova, and A. M. Bulgakova, Zh. Anal. Khim., 30(5), 946-950 (1975a); Anal. Abstr., 30, 5B203.

Pantaler, R. P., L. D. Alfimova, and A. M. Bulgakova, Zh. Anal. Khim., 30(9), 1834-1836 (1975b); Anal. Abstr., 31, 2B185.

Parsons, M. L., "Nuclear Methods" in Trace Analysis Spectroscopic Methods for Elements, J. D. Winefordner, Ed., John Wiley & Sons, New York, New York, 1976a, pp. 279-344.

Parsons, M. L., "X-Ray Methods" in Trace Analysis Spectroscopic Methods for Elements, J. D. Winefordner, Ed., John Wiley & Sons, New York, New York, 1976b, pp. 345-382.

Pauling, L., Chapter III in The Nature of the Chemical Bond, 3rd ed., Cornell University Press, Ithaca, New York, 1960.

Pavlov, V. N., and V. V. Bondar, ["Polarographic Behavior of Divalent Cobalt, Nickel, and Iron"], Usp. Khim., 42(6), 987-1008 (1973); Chem. Abstr., 79, 72770e (1973).

Pearson, R. M., and J. J. Seim, "Spectrophotometric Determination of Cobalt Complexed with the p-Nitrophenylhydrazone of Diacetylmonoxime and Ethylenediamine," Anal. Chem., 49(4), 580-582 (1977).

Peterson, G. E., and D. C. Manning, "Atomic Absorption Analysis of Metal Pollutants in Water Using a Heated Graphite Atomizer" in International Symposium on Identification and Measurement of Environmental Pollution, [Proc.], B. Westley, Ed., National Research Council of Canada, Ottawa, Ontario, 1971, pp. 380-385.

Piccardi, G., and P. Legittimo Cellini, "Applicazione del Metado Polarografico alla Determinazione di Metalli in Acque di Scarico" ["Application of Polarographic Methods to the Determination of Metals in Waste Water"], Ann. Chim. (Rome), **64**(1-2), 1-6 (1974).

Pinta, M., Modern Methods for Trace Element Analysis, translated from the French by STS, Incorporated, Ann Arbor Science Publishers Inc., Ann Arbor, Michigan, 1978.

Planinsek, F., and J. B. Newkirk, "Cobalt and Cobalt Alloys" in Kirk-Othmer Encyclopedia of Chemical Technology, 3rd ed., Vol. 6, M. Grayson, Executive Ed., Wiley-Interscience Publication, John Wiley & Sons, Inc., New York, New York, 1979, pp. 481-494.

Plant, J., G. C. Goode, and J. Herrington, "An Instrumental Neutron Activation Method for Multi-Element Geochemical Mapping," J. Geochem. Explor., **6**(3), 299-319 (1976).

Pohl, F. A., "Methods for the Spectrochemical Determination of Traces. I. Analysis of Waters," Z. Anal. Chem., **139**, 241-247 (1953).

Poole, R. L., Jr., The Application of Phase Selective Alternating Current Polarography to the Analysis of Heavy Metals in Water, Ph.D. Dissertation, Georgia Institute of Technology, Atlanta, Georgia, August 1973; abstr. in Cadmium in Water. A Bibliography, PB 218-829, National Technical Information Service, U.S. Department of Commerce, Springfield, Virginia, March 1973, p. 30.

Poonia, N. S., V. Bhagwat, and H. S. Singh, "A Versatile Solvent System for the Paper Chromatography of Inorganic Cations," Mikrochim. Acta, **1**, 36-41 (1972).

Price, W. J., Analytical Atomic Absorption Spectrometry, Heyden, London, 1972.

Publicover, W. E., "Spectrochemical Analysis of Oxygen-Free Electrolytically Pure Copper by a Globule Arc Procedure," Anal. Chem., 37(13), 1680-1684 (1965).

Pyatnitskii, I. V., "Analytical Chemistry of Cobalt," Oldbourne Press, London, 1966; translated from Russian, Israel Program for Scientific Translations, Jerusalem, Israel, 1966.

Radovici, A., R. Radovici, and M. Vultur, Revista Chim., 27(6), 538-539 (1976); Anal. Abstr., 32, 2C77.

Ragaini, R. C., H. R. Ralston, D. Garvis, and R. Kaifer, Trace Elements in California Aerosols. Part 1. Instrumental Neutron Activation Analysis Techniques, UCRL-51850, Lawrence Livermore Laboratory, National Technical Information Service, U.S. Department of Commerce, Springfield, Virginia, 1975.

Raginskaya, L. K., T. G. Manova, and V. I. Efremova, Zh. Anal. Khim., 28(9), 1843-1845 (1973); Anal. Abstr., 28, 1J29.

Rains, T. C., "Iron, Cobalt, and Nickel," Flame Emiss. At. Absorpt. Spectrom., 3, 216-246 (1975).

Rajabalee, F. J. M., "The Chelates of Divalent Cu, Ni, Zn, Pb, Hb, Co and Mn with Nitrilotriacetic Acid," J. Inorg. Nucl. Chem., 36, 557-564 (1974).

Ramendik, G. I., Zh. Anal. Khim., 32(10), 1990-1998 (1977); Anal. Abstr., 33, 6B18.

Ramirez-Munoz, J., Atomic Absorption Spectroscopy, Elsevier, Amsterdam, The Netherlands, 1968.

Randa, Z., "Analytical Possibilities of Epithermal Neutron Activation in Routine INAA of Mineral Materials," Radiochem. Radioanal. Lett., 24(3), 157-168 (1976).

Ranweiler, L. E., and J. L. Moyers, "Atomic Absorption Procedure for Analysis of Metals in Atmospheric Particulate Matter," Environ. Sci. Technol., 8(2), 152-156 (1974).

Ravnik, V., M. Dermelj, and L. Kosta, "Determination of Some Trace Elements (Fe, Co, Cr, Zn, Cu, Mn, and In) in Different Series of Standard Reference Samples by Neuton-Activation Analysis," Mikrochim. Acta, 1976(1), 153-164 (1976).

Reeves, R. D., and R. R. Brooks, Trace Element Analysis of Geological Materials, John Wiley & Sons, New York, New York, 1978.

Rhodes, J. R., A. H. Pradzynski, and R. D. Sieberg, "Energy Dispersive X-Ray Emission Spectrometry for Multielement Analysis of Air Particulates," Air Qual. Instrum., 2, 1-16 (1974).

Ridley, W. P., L. J. Dizikes, and J. M. Wood, "Biomethylation of Toxic Elements in the Environment," Science, 197, 329-332 (1977).

Robertson, D. E., "The Adsorption of Trace Elements in Sea Water on Various Container Surfaces," Anal. Chim. Acta, 42, 533-536 (1968); Chem. Abstr., 69, 69613u (1968).

R. J. Ronan, and G. Kunselman, "Thousands of Metal Analyses per Man Day--A Reality in U.S. EPA's Central Regional Laboratory: Multielement (23) Analysis by an Inductively Coupled Argon Plasma Atomic Emission System (ICAP-AES)" in Methods and Standards for Environmental Measurement. Proceedings of the 8th Materials Research Symposium, Sept. 20-24, 1976, W. H. Kirchhoff, Ed., Special Publication 464, U.S. Bureau of Standards, Washington, D.C., 1977, pp. 107-112.

Rousselet, F., V. Courtois, and M. L. Girard, "Applications de la spectroscopie d'absorption atomique à l'analyse des éléments métalliques dans les médicaments" [" Applications of Atomic Absorption Spectroscopy to the Analysis of Metallic Elements in Medicinal Preparations"], Analusis, 3(3), 132-138 (1975).

Rowe, J. J., and E. Steinnes, "Instrumental Activation Analysis of Coal and Fly Ash with Thermal and Epithermal Neutrons," J. Radioanal. Chem., 37, 849-856 (1977).

Ruch, R. R., H. J. Gluskoter, and N. F. Shimp, "Occurrence and Distribution of Potentially Volatile Trace Elements in Coal," Environmental Geology Notes, No. 72, Illinois State Geological Survey, Urbana, Illinois, August 1974.

St. John, P. A., "Fluorometric Methods for Traces of Elements" in Trace Analysis Spectroscopic Methods for Elements, J. D. Winefordner, Ed., John Wiley & Sons, New York, New York, 1976, pp. 213-278.

Salbu, B., E. Steinnes, and A. C. Pappas, "Multielement Neutron Activation Analysis of Fresh Water Using Ge(Li) Gamma Spectrometry," Anal. Chem., **47**(7), 1011-1016 (1975).

Sandell, E. ., Colorimetric Determination of Traces of Metals, 3rd ed., Interscience Publishers, New York, New York, 1959.

Sandell, E. B., and H. Onishi, Photometric Determination of Traces of Metals: General Aspects, John Wiley & Sons, New York, New York, 1978.

Sauchelli, V., Trace Elements in Agriculture, Van Nostrand Reinhold Company, New York, New York, 1969, pp. 200-201.

Schelenz, R., "Multielement Analysis of River Water" in Methods and Standards for Environmental Measurement. Proceedings of the 8th Materials Research Symposium, Sept. 20-24, 1976, W. H. Kirchhoff, Ed., Special Publication 464, U. S. National Bureau of Standards, Washington, D.C., 1977, pp. 113-120.

Schuster, J., F. Lux, and R. Zeisler, ["Study of Metallosis Using Neutron Activation Analysis"], Monatsschr. Unfallheilkd., **76**(12), 537-548 (1973).

Segar, D. A., "The Determination of Trace Metals in Saline Waters and Biological Tissues Using the Heated Graphite Atomizer" Paper No. 71-1051 presented at the American Institute of Aeronautics and Astronautics' Joint Conference on Sensing of Environmental Pollutants, Palo Alto, California, November 8-10, 1971.

Sheibley, D. W., "Trace Elements by Instrumental Neutron Activation Analysis for Pollution Monitoring" in Trace Elements in Fuel, Advances in Chemistry Series No. 141, American Chemical Society, Washington, D.C., 1975, pp. 98-117.

Sidgwick, N. V., "Coordination Compounds and the Bohr Atom," J. Chem. Soc., **123**, 725-730 (1923); Chem. Abstr., **17**, 2668 (1923).

Sidgwick, N. V., The Chemical Elements and Their Compounds, Vol. II, Oxford Press, London, England, 1950, pp. 1375-1425.

Siegel, S. M., and B. Z. Siegel, "Bio-Corrosion by Fungi," ACS Div. Environ. Chem. Preprints, **19**(1), 65-67 (1979).

Sikorska-Tomicka, H., "Spektrofotometryczne Oznaczanie Kobaltu Pirazynoamidem" ["Spectrophotometric Determination of Cobalt with Pyrazinecarboxylic Acid Amide"], Chem. Anal. (Warsaw), **20** (5), 1025-1030 (1975).

Sillén, L. G., and A. E. Martell, Stability Constants, Special Publication No. 17, The Chemical Society, London, England, 1964.

Simmons, W. J., "Determination of Low Concentrations of Cobalt in Small Samples of Plant Material by Flameless Atomic Absorption Spectrophotometry," Anal. Chem., **47**(12), 2015-2018 (1975).

Sims, C. T., "A Contemporary View of Cobalt-Base Alloys," J. Metals, **21**(12), 27-42 (1969).

Slavin, S., G. E. Peterson, and P. C. Lindahl, "Determination of Heavy Metals in Meats by Atomic Absorption Spectroscopy," At. Absorpt. Newsl., **14**(3), 57-59 (1975).

Slavin, W., Atomic Absorption Spectroscopy, Interscience Publishers, New York, New York, 1968.

Snell, F. D., and C. T. Snell, Colorimetric Methods of Analysis, Vol. 2, 3rd ed., D. Van Nostrand, Princeton, New Jersey, 1949.

Snell, F. D., and C. T. Snell, Colorimetric Methods of Analysis, Vol. 2A, 3rd ed., D. Van Nostrand, Princeton, New Jersey, 1959.

Steinnes, E., "A Two-Group Separation Scheme for the Determination of Eleven Trace Elements in Biological Material by Neutron Activation Analysis," Anal. Chim. Acta, 78(2), 307-315 (1975).

Stetter, A., and H. Exler. "Eine Schnellmethode zur Anreicherung von Schwermetallspuren mittels Na-Pyrrolidin-Dithiocarbamate (Na-t-carbat)" ["A Rapid Method for Determination of Heavy Metal Traces by Means of Sodium Pyrrolidine Diothiocarbamate"], Naturwissenschaften, 42, 45 (1955).

Stokinger, H. E., "The Metals (Excluding Lead)" in Industrial Hygiene and Toxicology, F. A. Patty, Ed.; Vol. II, Toxicology, 2nd ed., D. W. Fassett and D. D. Irish, Eds., Interscience Publishers, Division of John Wiley & Sons, Inc., New York, New York, 1963, pp. 987-1194.

Strain, W. H., A. Flynn, E. G. Mansour, F. R. Plecha, W. J. Pories, and O. A. Hill, Jr., "Trace Element Content of Household Water" in Trace Substances in Environmental Health-IX, D. D. Hemphill, Ed., Proceedings of University of Missouri's 9th Annual Conference on Trace Substances in Environmental Health, Columbia, Missouri, June 10-12, 1975, University of Missouri, Columbia, Missouri, 1976, pp. 41-46.

Straub, W. A., and J. K. Hurwitz, "Ferrous Metallurgy," Anal. Chem., 51(5), 155R-170R (1979).

Stringer, J., "Hot Corrosion of High-Temperature Alloys," Annu. Rev. Mater. Sci., 7, 477-509 (1977).

Sugimae, A., "Sensitive Emission Spectrometric Method for the Analysis of Airborne Particulate Matter," Anal. Chem., 47(11), 1840-1843 (1975).

Sulcek, Z., P. Povondra, and J. Dolezal, "Decomposition Procedures in Inorganic Analyses," CRC Crit. Rev. Anal. Chem., 6(3), 255-323 (1977).

Tanner, J. T., and M. H. Friedman, "Neutron Activation Analysis for Trace Elements in Foods," J. Radioanal. Chem., 37(2), 529-538 (1977).

Tavlaridis, A., and R. Neeb, "Dialkydithiocarbamate als Reagentien zur Gas-Chromatographischen Bestimmung von Metallen" ["Dialkyldithiocarbamates as Reagents for the Gas-Chromatographic Determination of Metals. I. Thermal Behavior, Vapor Pressures and Gas Chromatography of Some Dialkyldithiocarbamato Chelates"], Fresenius' Z. Anal. Chem., 293(3), 211-219 (1978).

Tessier, A., P. G. C. Campbell, and M. Bisson, "Sequential Extraction Procedure for the Speciation of Particulate Trace Metals," Anal. Chem., 51(7), 844-851 (1979).

Tewari, S. N., and N. Bhatt, "Separation and Identification of Metal Dithizonates by Thin-Layer Chromatography and Its Application in Toxicological Analysis," Mikrochemica Acta, 1973, 337-340 (1973).

Thiers, R. E., "Contamination in Trace Element Analysis and Its Control" in Methods of Biochemical Analysis, Vol. 5, D. Gluck, Ed., Interscience Publishers, New York, New York, 1957.

Thompson. R. J., G. B. Morgan, and L. J. Purdue, "Analysis of Selected Elements in Atmospheric Particulate Matter by Atomic Absorption," At. Absorpt. Newsl., 9(3), 53 (1970).

Tjioe, P., J. de Goeij, and J. Houtman, "Extended Automated Separation Techniques in Destructive Neutron Activation Analysis; Application to Various Biological Materials, Including Human Tissue and Blood," J. Radioanal. Chem., 37(2), 511-522 (1977).

Trofimov, N. V., N. A. Kanaev, and A. I. Busev, Zh. Anal. Khim., 29(10), 2001-2007 (1974); Anal. Abstr., 29, 5B245.

Trokhachenkova, P., N. Gradskova, D. Zhinkin, and N. Makulov, Zh. Anal Khim., 32(7), 1308-1311 (1977); Anal. Abstr., 34, 3C111.

Tsap, M. L., Z. V. Proskura, N. S. Gedz, and S. P. Sheredeko, ["Spectrophotometric Determination of Trace Quantities of Cobalt in Agrochemical Samples Without Using Organic Extracting Agents"], Agrokhimiya, 120-125 (1977); Chem. Abstr., 87, 163691 (1977).

Tskhvirashvili, D. G., L. E. Vasadze, and V. D. Gotsiridze, "Issledovanie Rasprodeleniya mezhdu Vodoi i Parom Nekotorykh Okislov Tyashlykh Metallov i ikh Rastvorimosti v Vodyanom Pare" ["Study of the Distribution of Some Heavy Metal Oxides Between the Water and Vapor and Their Solubility in Water Vapor"] in Voprosy Konvektivnogo Teploobmena i Chistoty Vodyanogo Para [The Problem of Convective Heat Exchange and Steam Purity], Metsniereba, Tbilisi, Georgian S.S.R., 1970.

Valente, S. E., and W. B. Schrenk, "Design and Some Emission Characterisitics of an Economical D. C. Arc Plasma Jet Excitation Source for Solution Analysis," Appl. Spectrosc., **24**, 197-205 (1970).

Valkovic, V., "Proton-Induced X-Ray Emission: Applications in Medicine," Nucl. Instr. Methods, **142**, 151-158 (1977).

Van de Wiel, D. F. M., L. Koster-Otte, W. T. Goedemans, and M. G. Woldring, ["Saturation Analysis of Vitamin B_{12} (Cyanobobalamin)"] Pharm. Weekbl., **109**, 651-659 (1974).

Van Hoye, E., R. Gijbels, and F. Adams, "Relative Sensitivity Coefficients for the Analysis of Steel by Spark-Source Mass-Spectrometry," Talanta, **23**(5), 369-375 (1976).

Van Vleck, J. H., "The Group Relation Between the Mullikin and Slater-Pauling Theories of Valence," J. Chem. Phys., **3**, 803-806 (1935); Chem. Abstr. **30**, 662[5] (1936).

Vasil'eva, L. N., Z. L. Yustus, and N. B. Kogan, ["Luminescence Determination of Holmium and Erbium Trace Impurities in Pure Cerium Dioxide"], Zh. Anal. Khim., **29**(9), 1754-1757 (1974); Chem. Abstr., **82**, 67838r (1975).

Vasilikiotis, G. S., Th. A. Kouimtzis, and A. Voulgaropoulos, "Spectrophotometric Determination of Cyanocabalamin (as Cobalt)," Microchem. J., **22**, 479-483 (1977).

Vassos, B. H., R. F. Hirsch, and H. Letterman, "X-Ray Microdetermination of Chromium, Cobalt, Copper, Mercury, Nickel, and Zinc in Water Using Electro-Chemical Preconcentration," Anal. Chem., **45**, 792-794 (1973).

Veillon, C., "Optical Atomic Spectroscopic Methods" in Trace Analysis Spectroscopic Methods for Elements, J. D. Winefordner, Ed., John Wiley & Sons, New York, New York, 1976, pp. 123-182.

Verigina, K. V., ["Evaluation of Current Chemical Methods for Determining Trace Elements Used in Scientific Research and Industrial Laboratories"] in Fiziol. Rol. Prakt. Primen. Mikroelem., Obz. Dokl. Vses. Soveshch. Mikroelem, 7th 1975, J. Pewe, Ed., Zinatne, Riga, USSR, 1976, pp. 247-262; Chem. Abstr., **8**(70), 116920e (1977).

Verigina, K. V., and Yu. I. Dobritskaya, "Determination of Trace Elements in Soils and Plants by Chemical Methods" in Metody Opred. Mikroelem. Po'chvakh, Rast. Vodakh," I. G. Vazhenin, Ed., Kolos, Moscow, USSR, 1974, pp. 25-145; Chem. Abstr., **84**, 120275 (1976).

Vogel, A. J., A Textbook of Quantitative Inorganic Analysis, 3rd ed., Longmans Green and Company, Ltd., London, England, 1961, pp. 529-531.

Vu Quang Kinh, "Cobalt," Monogr. Met. Haute Purete, **1**, 368-387, (1972); Chem. Abstr., **82**, 128165t (1975).

Wagner, F., "Vitamin B_{12} and Related Compounds" in Annual Review of Biochemistry, P. D. Boyer, Ed., Vol. 35, Part 1, Annual Rev. Inc., Palo Alto, California, 1966, pp. 405-434.

Wagner, F., and K. Bernhauer, "New Aspects of the Structure of Corrinoid Coenzymes," Ann. N. Y. Acad. Sci., **112**, 580-589 (1964).

Warren, J., and D. Carter, "Determination of Trace Amounts of Copper, Vanadium, Chromium, Nickel, Cobalt and Barium, in Silicate Rock Using Flame Atomic Absorption Spectrometry," Can. J. Spectroscopy, **20**(1), 1-5 (1975).

Wasielewski, G. E., and R. A. Rapp, "High Temperature Oxidation" in Superalloys, C. T. Sims, Ed., John Wiley & Sons, New York, New York, 1972, pp. 287-316.

Watson, A. E., and G. M. Russell, "The Analysis of Geological Samples for Trace Elements by Direct-Reading Emission Spectrometry," Spectrochim. Acta, Part B, **33B**(5), 143-152 (1978).

Weisz, H., S. Pantel, and I. Vereno, "Anwendung katalysierter Reaktionen zur halbquantitativen Bestimmung mit Holfe der Ringofenmethode" ["Use of Catalyzed Reactions for Semiquantitative Determination with the Aid of the Ring Oven Methods"], Mikrochim. Acta, **2**(3), 287-296 (1975).

Welz, B., and E. Wiedeking, "Bestimmung von Spurenelementen in Serum und Urin mit Flammenloser Atomisierung" ["Determination of Trace Elements in Serum and Urine with Flameless Atomization"], Z. Anal. Chem., **252**, 111-117 (1970); NASA Technical Translation F-13,983, National Aeronautics and Space Administration, Washington, D.C., October 1971.

Wender, I., H. W. Sternberg, S. Metlin, and M. Orchin, "Dicobalt Octacarbonyl," Inorg. Syn., **5**, 190-192 (1957a).

Wender, I., N. W. Sternberg, and M. Orchin, "The Oxo Reaction" in Catalysis, Vol. V, P. H. Emmett, Ed., Reinhold Publishing Corp., New York, New York, 1957b, pp. 73-130.

Wendt, R. H., and V. A. Fassel, "Induction-Coupled Plasma Spectrometric Excitation Source," Anal. Chim. **37**, 920-922 (1967).

Willard, H. H., L. L. Merritt, Jr., and J. A. Dean, Instrumental Methods of Analysis, 5th ed., D. Van Nostrand, New York, New York, 1974.

Willis, J. B., "Recent Advances in the Analysis of Biological Materials by Atomic Absorption Techniques," Endeavor, **32**(117), 106-111 (1973).

Windholz, M., S. Budavari, L. Y. Stroumtsos, and M. N. Fertig, Eds., The Merck Index, 9th ed., Merck and Company, Inc., Rahway, New Jersey, 1976.

Winefordner, J. D., Ed., Trace Analysis Spectroscopic Methods for Elements, John Wiley & Sons, New York, New York, 1976.

Winge, R. K., V. A. Fassel, R. N. Kniseley, E. DeKalb, and W. J. Haas, Jr., "Determination of Trace Elements in Soft, Hard, and Saline Waters by the Inductively Coupled Plasma, Multi-Element Atomic Emission Spectroscopic (ICP-MAES) Technique," Spectrochim. Acta, Part B, 32, 327 (1977).

Winge, R. K., V. J. Peterson, and V. A. Fassel, "Inductively Coupled Plasma-Atomic Emission Spectroscopy: Prominent Lines," Appl. Spectrosc., 33(3), 206 (1979).

Winterhager, H., and J. Krüger, "Pure Cobalt and Its Properties," Cobalt, 29, 185-195 (1965).

Winterhager, H., and J. Krüger, "Pure Cobalt and Its Properties," Cobalt, 30, 27-38 (1966); Chem. Abstr., 67, 35505d (1967).

Wolt, W., "Coupled Gas Chromatography-Atomic Absorption Spectrometry," J. Chromatogr., 134(1), 159-165 (1977).

Wood, J. M., "Metabolic Cycles for Toxic Elements" in International Conference on Heavy Metals in the Environment, October 27-31, 1975, Abstracts, Institute for Environmental Studies, University of Toronto, Toronto, Ontario, Canada, 1975.

Wood, J. M., A. Cheh, L. J. Dizikes, W. P. Ridley, S. Rakow, and J. R. Lakowicz, "Mechanisms for the Biomethylation of Metals and Metalloids," Fed. Proc., 37(1), 16-21 (1978).

Wyttenbach, A., S. Bajo, and A. Haekkinen, Betr. Tabakforsch., 8(5), 247-249 (1976); Anal. Abstr., 32, 1D3.

Yamashita, M., and N. Suzuki, ["Photon Activation Analysis of Trace Metals in Biological Samples by Using Ion Exchange Absorption of Metal Complexes"], Kakuriken Kenkyu Hokoku (Tohoko Daigaku), 11(1), 95-104 (1978); Chem. Abstr., 90, 99496J (1979).

Yang, M. H., P. Y. Chen, C. L. Tseng, S. J. Yeh, and P. S. Weng, "Determination of Trace Elements by Neutron Activation Analysis Using Dinonylnaphthalene Sulfonic Acid as a Preconcentrating Agent," J. Radioanal. Chem., 37, 801-811 (1977).

Yoshimura, K., H. Waki, and S. Ohashi, "Ion-Exchanger Colorimetry: I. Micro Determination of Chromium, Iron, Copper and Cobalt in Water," Talanta, **23**(6), 449-454 (1976).

Young, R. S., Cobalt, American Chemical Society Monograph No. 108, Reinhold Publishing Corporation, New York, New York, 1948.

Young R. S., Cobalt: Its Chemistry, Metallurgy, and Uses, Reinhold Publishing Corporation, New York, New York, 1960.

Young, R. S., "The Analytical Chemistry of Cobalt," Chemsa, **2**(11), 196-197 (1976).

Yule, H. P., "Reactor Neutron Activation Analysis. Instrumental Sensitivities in Six Matrix Materials," Anal. Chem., **38**(7), 818-821 (1966).

Zdankiewicz, and J. Fasching, "Analysis of Whole Blood by Neutron Activation: A Search for a Biochemical Indicator of Neoplasia," Clin. Chem., **22**(8), 1361-1365 (1976).

Zheleznova, A., and N. Zarasevich, Vest. Mosk. Yos. Univ., Ser. Khim., **16**(6), 718-721 (1975); Anal. Abstr., **31**, 1B10.

Zief, M., and J. W. Mitchell, Contamination Control in Trace Element Analysis, John Wiley & Sons, New York, New York, 1976.

Zikovsky, L., and E. A. Schweikert, "Comparison of Nondestructive Proton and Neutron Activation: the Case of Biological Samples," J. Radioanal. Chem., **37**, 571-580 (1977).

IV. GEOCHEMISTRY AND OCCURRENCE

Bonnie L. Carson and Christopher J. Cole

A. Geochemistry

The crustal abundance of cobalt has been calculated largely on the basis of concentrations in igneous and sometimes metamorphic rocks. Values reported (in mg Co/kg) include: 10 (Clark and Washington, 1924; cited by Parker, 1967); 18 (Vinogradov, 1962); 23 (Mason, 1958, and Rankama, 1954; cited by Parker, 1967); 25 (Taylor, 1964); 27 (Carr and Turekian, 1961); 30 (Clark and Washington, 1922); and 40 (Goldschmidt, 1937). According to Mitchell (1955), the bulk of cobalt is in the more easily weathered minerals such as olivine, hornblende, augite, and biotite. Ilmenite and magnetite are moderately stable toward weathering action. The cobalt in the crystal lattice of rock-forming minerals is unavailable to other environmental media until released by weathering.

The following discussion of the geochemistry of cobalt was taken from Wedepohl (1974 to 1978) and Carr and Turekian (1961). The occurrence in the air, soil, and ore minerals are described in more detail in subsequent subsections. Behavior of cobalt in these media are described in Chapter V.

The Co^{2+} ion has the configuration [Ar] $3d^7$ and is stabilized in octahedral and tetrahedral coordinations. In octahedral coordinations, the ionic radius of Co^{2+} is 0.735Å, falling between those of Mg^{2+} and Fe^{2+}. Thus, Co^{2+} can substitute for Mg^{2+} and Fe^{2+} in silicate minerals. Cobalt can also exist as Co^{3+} ions in phases such as CoOOH, formed in highly oxidizing environments.

In addition to the minerals containing cobalt listed in Table IV-1, cobalt occurs in manganese oxide minerals in wads (asbolane, cobaltomelane) and in deep-sea manganese nodules. Table IV-2 summarizes the occurrence of cobalt in noncobalt minerals and rocks.

TABLE IV-1

COBALT MINERALS
(Wedepohl, 1974-1978)

Mineral	Formula	Structure
Wairauite	CoFe	CsCl (tentative)
Jaipurite	CoS	NiAs
Freboldite	CoSe	NiAs
Langisite	(Co,Ni)As	NiAs
Modderite	CoAs	Distorted NiAs
Pentlandite[a]	$(Fe,Ni,Co)_9S_8$	Unique[b]
Linnaeite[a,c]	Co_3S_4	Spinel[d]
Carrolite[a,c]	$CuCo_2S_4$	Spinel[d]
Bornhardtite	Co_3Se_4	Spinel[d]
Tyrrellite	$(Cu,Co,Ni)_3Se_4$	Spinel[d]
Cattierite	CoS_2	Pyrite
Bravoite	$(Fe,Ni,Co)S_2$	Pyrite
Trogtalite	$CoSe_2$	(disordered cations)
Penroseite	$(Cu,Co,Ni)Se_2$	(disordered cations)
Villamaninite	$(Cu,Ni,Co,Fe)(S,Se)_2$	(disordered cations)
Cobaltite[a]	(Co,Fe)AsS	Ordered, distorted pyrite
Cobaltian ullmannite	(Ni,Co)SbS	Ordered pyrite
Willyamite	(Co,Ni)SbS	Pyrite
Hastite	$CoSe_2$	Marcasite
Safflorite[c]	$CoAs_2$	Marcasite
Loellingite	$(Fe,Co,Ni)As_2$	Marcasite
Costibite	CoSbS	Marcasite
Clinosafflorite	$CoAs_2$	Monoclinic
Paracostibite	CoSbS	Pyrite-marcasite
Cobaltian arsenopyrite	(Fe,Co)AsS	Arsenopyrite
Skutterudite[a,c]	$(Co,Fe)As_3$	Related to pyrite
Chathamite	$(Fe,Co,Ni)As_{3-2}$	Related to pyrite
Chloanthite	$(Ni,Co)As_{3-2}$	Related to pyrite
Aplowite	$(Co,Mn,Ni)SO_4 \cdot 4H_2O$	-
Moorhouseite	$(Co,Ni,Mn)SO_4 \cdot 6H_2O$	-
Bieberite	$CoSO_4 \cdot 7H_2O$	-
Cobaltomenite	$CoSeO_3 \cdot 2H_2O$	-
Erythrite[a,b]	$Co_3(AsO_4)_2 \cdot 8H_2O$	-
α-Roselite	$\alpha\text{-}Ca_2Co(AsO_4)_2 \cdot 2H_2O$	-
β-Roselite	$\beta\text{-}Ca_2(Co,Mg)(AsO_4)_2 \cdot H_2O$	-
Forbesite	$(Co,Ni)_2[AsO_3(OH)]_2 \cdot 7H_2O$	-
Kirchheimerite	$Co(UO_2)(AsO_4)_2 \cdot nH_2O$	-
meta-Kirchheimerite	$Co(Uo_2)(AsO_4)_2 \cdot 8H_2O$	-
Sphaerocobaltite[e]	$CoCO_3$	Calcite
Julienite	$Na_2Co(SCN)_4 \cdot 8H_2O$	-
Lusakite	$(Co,Mg,Fe)_4Al_9O_6(SiO_4)_8(OH)_2$	Staurolite
Glaucodot[a,b]	(Co,Fe)AsS	-
Heterogenite[a,c]	CoO(OH)	-
Asbolite[a,c]	Manganese oxides plus cobalt	-
Gersdorffite[a]	(Ni,Co)AsS	-
Pyrrhotite[a]	$(Fe,Ni,Co)_{x-1}S_x$	-
Pyrite[a]	$(Fe,Ni,Co)S_2$	-
Sphalerite[a]	Zn(Co)S	-

a/ Minerals high in cobalt that have been mined or concentrated for their cobalt content (Vhay et al., 1973).

b/ Cube clusters of Co in alternate octants of the unit cell; S atoms form a cubic closest packed sublattice.

c/ Primary, most important minerals (Boyle, 1969).

d/ Transforms to a defect NiAs structure at elevated pressures and temperature.

e/ Secondary or supergene minerals (Boyle, 1969).

TABLE IV-2

COBALT IN MINERALS AND ROCKS
(Wedepohl, 1974-1978; Carr and Turekian, 1961)

	Cobalt Concentration (mg/kg)		
	Mean	Range	Comments
Iron meteorites and stony irons, iron phase	~ 5,000	-	Small correlation with Ni.
Chondrites	~ 600	-	-
Achondrites	-	2-80	-
Lunar materials	-	27-47	-
Tektites	-	8-15	-
Ultramafic rocks	150 or 200 (Parker, 1967)[a/]	-	-
Granitic rocks	-	0.09-16.9	High-Ca granites generally contain higher Fe, Mg, and Co than low-Ca granites.
Coexisting phases in granitic and metamorphic rocks	-	0.11-88	Lows in potassium feldspar, quartzite, and muscovite. Highs in biotite, garnet, and hornblende.
Coexisting phases in basalts and eclogites	-	10-256	Lows in plagioclase and garnet. Highs in spinel, olivine, and chlorite.
Basaltic rocks	45 or 48[a/]	4-110	Most basalts have 30-80 mg Co/kg.
Coexisting minerals of gabbroic rocks in the Stillwater complex, Montana.			
Plagioclase	-	1.0-2.5	-
Clinopyroxene	-	47-89	-
Orthopyroxene	-	65-107	-
Sedimentary rocks			
Clays and shales	19 or 20 (Parker, 1967)[a/]	-	-
Shales	-	4-115	High in Sulfur Spring shale, St. Genevieve County, Missouri.
Pellites	-	16-22	-
Argillaceous rocks	15.9	-	-
Limestones	1.2	-	-
Quartzite	0.27	-	-
Sandstone	1.2	-	-
Sands	4	-	-
Greywackes	-	15-22	-
Recent shells	-	0.019-0.24	-
Metamorphic rocks	-	7-118	Low in the felsic zone of banded gneiss. Highs in garnets, staurolites, and kyanites.
Schists	-	4-147	-
Silicate minerals	-	< 4-256	-

a/ Parker (1967) cited average values given by Vinogradov or Turekian and Wedepohl.

The cobalt concentration in mafic and ultramafic (ultrabasic) rocks is strongly correlated with the concentrations of iron and magnesium; but in granitic rocks, cobalt is primarily correlated with magnesium. Since magnesium concentrations tend to follow those of calcium, cobalt levels are higher in high-calcium granitic rocks. Cobalt is not markedly differentiated between crystalline octahedral sites and quasi-octahedral sites during the fractional crystallization of basaltic magna. Cobalt does not form residual silicate minerals during weathering because it is so easily solubilized. As discussed in more detail in Chapter V, the distribution of cobalt in the soil closely follows that of iron oxides and manganese oxides.

In sediments of natural water bodies, the cobalt concentrations reported range from 9 mg/kg (sediments in the Sea of Japan) to 160 mg/kg (carbonate-free sediments deep in the Indian and Pacific Oceans). The concentration of cobalt in sedimentary rocks ranges from 0.27 to 22. The cobalt concentration generally increases with increasing iron concentration. Vine and Tourtelot (1970) reported a median value of 10 mg Co/kg in 779 black shales from North America. In shales, cobalt correlates poorly with magnesium. During weathering, cobalt tends to behave like iron except when separation of iron and manganese occurs, whereupon cobalt follows manganese.

There is no major redistribution of cobalt in the formation of metamorphic rocks; however, both iron and cobalt increase in schists relative to shales because quartz and feldspar tend to be segregated, thereby increasing the concentration of mafic minerals. Cobalt is as closely associated with magnesium in gneisses as in granites.

Phosphate, carbonate, and hydroxyl are the anions regulating the solubility of cobalt in aerated waters; in addition to these anions, sulfide is important in anoxic, sulfate-rich waters.

Cobalt has been concentrated in mineral deposits by magmatic segregation with nickel, chromium, and other elements; by hydrothermal activity (deposited by replacement in rocks or in underground openings with other minerals); and by chemical weathering of ultrabasic rocks into laterite deposits (Reno, 1970).

The occurrence of cobalt in mineral deposits is discussed in greater detail in the next section.

B. Occurrence

1. Cobalt deposits: Geological classifications of cobalt deposits according to their genetic environment or occurrence are (Vhay et al., 1973): (a) hypogene deposits associated with mafic intrusive igneous rocks; (b) contact metamorphic deposits also associated with mafic instrusive igneous rocks; (c) lateritic (weathered deposits); (d) massive sulfide deposits in metamorphic rocks, chiefly of volcano-sedimentary origin; (e) hydrothermal deposits; (f) strata-bound deposits; and (g) deposits formed as chemical precipitates.

Among the better known of each type are: (a) the nickel-copper sulfide ores of the Sudbury district, Ontario; the Duluth gabbro complex southeast of Ely, Minnesota; the Stillwater complex in Montana; and the Lynn Lake area in Manitoba; (b) Cornwall, Lebanon County, Pennsylvania; Morgantown (the Grace Mine), Berks County, Pennsylvania; (c) Riddle, Oregon; Puerto Rico; eastern Cuba; New Caledonia; and several other countries; (d) Ducktown, Tennessee; (e) the Blackbird district, Idaho; Bou Azzer area, Morocco; Outokumpu, Finland; Mississippi Valley ores, especially Fredericktown and Mine La Motte areas of southeast Missouri; (f) copper-cobalt deposits of Zaire and Zambia; the copper-rich shales of the Kupferschiefer of central Europe; and (g) deep-sea manganese nodules.

The identified cobalt resources of the United States contain more than 1 billion lb (450 million kg) cobalt; those of the world, more than 9.9 billion lb (4.5 billion kg). Table IV-3 lists the identified U.S. resources. Vhay et al. (1973) described in much greater detail the identified U.S. deposits as well as hypothetical and speculative resources.

Hypothetical resources include mostly various sulfide ore bodies. The gabbro deposit near Ely, Minnesota, has been only partly evaluated. If the cobalt concentration is as high as 0.038%, 12 billion lb (5.5 billion kg) cobalt may be present in

TABLE IV-3

IDENTIFIED RESOURCES OF COBALT IN THE UNITED STATES
(Vhay et al., 1973)

Deposit	State	Cobalt Concentration, %	Contained Cobalt, millions of lb	Deposit Type
Brady Glacier and others	Alaska	not given	125	-
Batesville manganese district	Arkansas	not given	(some)	lateritic
Chiefly laterites	California	not given	25	lateritic
Blackbird and Coeur d'Alene districts	Idaho	0.6 some in base metals	50 (15-225)	hydrothermal
Union and others, Katahdin Iron Works	Maine	0.05-0.1	130	hypogene
Sulfide deposits in gabbro, Ely area	Minnesota	≥ 0.01	1,000 (conservative estimate)	hypogene
Southeast Missouri district, Viburnum trend	Missouri	0.16-0.18	150	hydrothermal
Manganese deposits at the contact of the Missouri Eminence and Gasconade Dolmites a/	Missouri	0.5	10	lateritic
Stillwater complex	Montana	not given	(some)	hypogene
Goodsprings district	Nevada	0.3-0.5	(some)	hydrothermal
Blackhawk and other districts with U, Zn, Ag, and Ni; in manganese ores	New Mexico	 0.09 0.01-0.5	6	hydrothermal
Chiefly laterites in the Webster district	North Carolina	0.16	3	lateritic
Chiefly laterite, Riddle	Oregon	0.05	34	lateritic
Contact metamorphic deposits in southeast region, Cornwall and Morgantown	Pennsylvania	0.02-0.056	66	contact metamorphic
Ducktown district	Tennessee	0.4-0.5	40	massive sulfide
By-product of talc operations	Vermont		(some)	
Piedmont region; gossan lead (also includes Sykesville district, Carroll County, Maryland)	Virginia	0.007-0.05 0.12-2.24 0.0.-0.07	25	hydrothermal laterites massive sulfide
Chiefly laterite	Washington	not given	20	lateritic
Upper Mississippi Valley district, includes adjacent parts of Illinois and Iowa	Wisconsin	not given	10	hydrothermal

a/ Described in the text but not specifically mentioned under identified resources.

these deposits. More deposits like those of the Gap mine may be found in the Coatsville area in Pennsylvania. There is an excellent probability of finding additional deposits of cobalt in Alaska. New England states may have other nickel-cobalt-copper deposits near presently known ones. Southeastern Pennsylvania may contain more cobalt minerals associated with iron ore of contact metamorphic origin. The southern Piedmont and the Blue Ridge provinces in the Appalachian region are especially likely to contain more massive sulfide bodies of the Ducktown, Tennessee, type. Extensions of known districts in the mid-continental United States may reveal more Mississippi Valley-type ores containing cobalt and nickel. The lead deposits in Arkansas were mentioned as possibly containing cobalt.

Speculative resources in the United States are mafic and ultramafic rocks in the Appalachian and Cordilleran regions; more massive sulfide bodies in the Great Lakes region; arsenide minerals in the copper region of the Upper Michigan Peninsula; and Cornwall-type deposits along the edges of the diabase intrusions in the Triassic troughs of Virginia, Maryland, and New Jersey. Replacement deposits might be found along the diabase contact of the Triassic trough of the Connecticut Valley.

According to Bennett (1977), the Blackbird mine in Idaho was the "only readily available domestic source of cobalt in the United States." He reported on a reconnaissance geologic and geochemical survey conducted in the summer of 1976 in Lemhi County. Copper-cobalt mineralization appears to extend at least 12 miles (19.2 km) from the mine north to Garden Creek as judged from soil and stream sediment analyses. The only town in the area, which was built to house miners for the currently inactive Blackbird mine, is Cobalt, Idaho. It is now occupied only by caretakers and maintenance personnel.

The presence of cobalt was first noted in the Blackbird ores soon after development began circa 1893. From 1917 to 1920, 55 short tons (49.5 MT) cobalt concentrate (average 17.74% cobalt) was separated from 4,000 short tons (3,600 MT) ore (by Union Carbide at the Haynes-Stellite property [McIntyre, 1979]). The mine was closed due to high transportation and refining costs. From 1938 to 1941, Uncle Sam Mining and Milling Company mined 3,657 tons (3,291 MT) ore and concentrates containing copper, gold, and silver. Cobalt drew penalties when the concentrates were purchased by the smelter (Bennett, 1977).

Renewed exploration of the area occurred during World War II by several companies, the Bureau of Mines, and the U.S. Geological Survey. Calera Mining Company, formed by Howe Sound Company, began production in 1951. A cobalt refinery was built in Garfield, Utah, that used a high-pressure leach process. The Blackbird mine produced 10,732,256-kg cobalt concentrate and 2,995-MT copper in 1957, its peak year. Underground work ceased in 1958. The open pit thereafter produced 1,340,090-kg cobalt concentrate. The mine closed altogether in 1960 due to loss of government contracts, low copper prices, and South African cobalt imports. Machinery Center, Inc., reopened the mine in 1966; but it was closed again in 1969 (Bennett, 1977; McIntyre, 1979).

Noranda Mines, Ltd. recently purchased a 75% interest in the Blackbird deposit and has subsequently explored and claimed cobalt-copper showings over a 30-mile strike, generally outside the Blackbird mine claims. Current reserves in these properties indicate that 2,000 short tons cobalt could be recovered annually for more than 10 years. There are 4 million tons of ore containing 0.73% cobalt based on a very conservative 0.6% cut off (McIntyre, 1979).

Reserves in the Blackbird mine itself, according to Bennett (1977) are 32,204 MT cobalt. The Blackbird ore reserves, stated Bennett, average about 0.64% cobalt underground and 0.43% cobalt in the open pits.

Noranda's geologists concluded that the Blackbird occurrence is a strata-bound deposit rather than vein-type, as had been previously thought. Folds, breccias, boudinage, dewatering defoliations, shear foliations, and load structures in the hanging wall and footwall are the principal signs of soft sedimentary deformation. Cobalt and copper were probably an integral part of the system during sedimentation, compaction, and diagenesis because the mineralized stratigraphy shows deformations analogous to those of the footwall and hanging wall sequences (McIntyre, 1979).

In the traditional workings of the Blackbird ore body, the bulk of the ore comprises coarse-grained cobaltite in quartz and biotite associated with chalcopyrite, pyrite, pyrrhotite, arsenopyrite, tellurides, and native gold. The Noranda geologist also found fine-grained cobaltite in biotite-quartz schist with the same associated suite of minerals and in tourmalinized compaction breccias, where cobaltite and gold are the principal minerals (McIntyre, 1979). Bennett (1977) reported that erythrite (cobalt bloom) forms red, pink, and gray incrustations and stains on the outcrops at prospects on the West Fork of Blackbird Creek and at the upper workings of the Sweet Repose mine on Panther Creek.

Besides ores whose cobalt concentration is elevated, cobalt occurs as a minor impurity in various sulfide minerals, many of which are mined in vast amounts. Representative data for cobalt levels in U.S. lead-zinc and copper ores and smelter products were insufficient to calculate detailed mass balances for cobalt in the appropriate circuits. One might assume from the compilation of information in Table IV-4 that the average cobalt concentration in most of these ores is within the range of soil and slag-forming rock concentrations. If so, then cobalt emissions from mining and possibly smelting are not likely to be any more hazardous than natural dusts.

Cobalt is associated with arsenic and antimony in polymetallic sulfide ores. It is found as cobaltite with bismuth in cosalite deposits and with lead, zinc, and copper sulfides. Cobaltite is also found in molybdenum-bearing skarn deposits and in all types of nickel deposits including sulfides and arsenides as well as lateritic nickel-cobalt deposits from the weathering of basic and ultrabasic rocks (Boyle, 1969).

In addition to sulfide ores, cobalt accompanies nearly all uranium-bearing (pitchblende and uraninite) ores. There are relatively large amounts of cobalt in the pitchblende districts of southeastern Utah (Burnham, 1959). According to Boyle (1969), of 11 types of uranium deposits, cobalt is associated with three: black carbonaceous shales; veins, lodes, pipes, and disseminations; and sandstone deposits. Cobalt is found in vanadium-bearing uranium deposits in sandstone that are widespread in the United States.

TABLE IV-4

COBALT IN SULFIDE MINERALS OF THE UNITED STATES AND CANADA

Mineral	Location	Number of Samples[a]	Cobalt Concentration (mg/kg) Mean	Range[b]	Remarks	Reference
Pyrite	Tintic district, Utah	28	-	< 4-84	Spectrographic analyses	Carr and Turekian (1961)
Pyrite (Pb-Zn-Cu mineralization)	Central City Mining district, Colorado	9	-	12-540		
Sphalerite	Central City Mining district, Colorado	7/14	-	< 5-100	Cobalt can substitute for Zn in lattice	Sims et al. (1961)
Pyrite	Copper Hill, Tennessee	4	-	115-1,650		Carr and Turekian (1961)
Chalcopyrite	Arizona	42/75	11.5	N.D.-200	Spectrochemical analyses. Samples were removed by carbide dental burrs. Co, In, and Ni has "essentially the same distribution pattern" as silver and tin in the southwestern states except that cobalt's appearance was weak in the central belt, especially in Arizona where Cu is rich.	Burnham (1959)
Sphalerite	Arizona	40/46	125	N.D.-800		Burnham (1959)
Chalcopyrite	California	10/13	11	N.D.-100		Burnham (1959)
Sphalerite	California	6/9	11	N.D.-50		Burnham (1959)
Chalcopyrite	Colorado	10/18	5	N.D.-50		Burnham (1959)
Sphalerite	Colorado	14/26	80	N.D.-200		Burnham (1959)
Chalcopyrite	New Mexico	18/36	4.9	N.D.-70		Burnham (1959)
Sphalerite	New Mexico	38/42	69	N.D.-300		Burnham (1959)
Chalcopyrite	Nevada	25/38	34	N.D.-300		Burnham (1959)
Sphalerite	Nevada	19/44	26	N.D.-400		Burnham (1959)
Chalcopyrite	Utah	7/14	241	N.D.-3,000	The highest value was at the Lucky Strike mine; there was 300 Co mg/kg at the Happy Jack mine. The Co concentration was low at Bingham Canyon, where much Cu is mined annually.	Burnham (1959)
Sphalerite	Utah	6/18	9	N.D.-100		Burnham (1959)
Chalcopyrite	Montana	0/2	-	-		Burnham (1959)
(Sulfide concentrates that were mainly galena)	United States	39/129	-	6-30 when detected	Semiquantitative	Mosier et al. (1975)
Sphalerite	Coeur d'Alene district, Idaho	62/64	~ 55	20-390	Spectrographic analysis. Cobalt was not found in Coeur d'Alene district galena.	Frykland (1964)

TABLE IV-4 (concluded)

Mineral	Location	Number of Samples[a]	Cobalt Concentration (mg/kg) Mean	Range	Remarks	Reference
Pyrrhotite	Highland-Surprise mine, Surprise vein, Coeur d'Alene district, Idaho	8/10	-	32-100 when detected		Frykland (1964)
Sphalerite	Adirondacks	1	200	-	Strock found $\leq$ 200 mg Co/kg in zinc concentrates. In most, the concentration range was 10-100 mg Co/kg.	Strock (1945)
Sphalerite	Joplin, Missouri	1	not detected	-		Strock (1945)
Sphalerite	Low-temperature deposits in Missouri, Kansas, Arkansas, Kentucky (except central Kentucky), Massachusetts, New York, and Colorado	-	detected	-	Spectrographic analyses	Stoiber (1940)
Sphalerite	Intermediate-temperature deposits in Maine, Colorado, Texas, and New Mexico	-	detected	-		Stoiber (1940)
Sphalerite	High-temperature:					Stoiber (1940)
	Vermont	-	detected	-		
	Balmat, New York	-	not detected	-		

a/ When given as a fraction, the value represents the detection frequency.
b/ N.D. = Not detected. The detection limit was at least 3 mg Co/kg.

Arsenide deposits of native silver with nickel and cobalt and the quartz-pebble conglomerate type of gold deposit may be enriched in cobalt. Cobalt also occurs with platinum-group metals in the Sudbury, Ontario, nickel-copper ores; in the Bushveld complex of South Africa; and in skarn deposits associated with basic and ultrabasic rocks intruding into carbonate rocks (Boyle, 1969).

Cobalt concentrations in iron ores are discussed in the following subsection.

2. Minerals used in high volume

a. Iron ores: Cobalt has been found in concentrations up to 143 mg/kg in iron oxide minerals (see Table IV-5). Grigor'ev (1966) found cobalt in 13 of 20 types of iron deposits. Cobalt was not found in:

- Ilmenite-magnetite ores in gabbro and amphibolites;
- Apatite-magnetite in ultrabasic rocks of the alkaline series;
- Fluorite-barite-siderite associated with alkaline syenites;
- Skarn-magnomagnetite ores in sedimentary-volcanic rocks (Paleozoic);
- Hematite ores in sandy clay;
- Magnetite-hematite ores in sedimentary and volcanic rocks; and
- Hydrogoethite-leptochlorite-siderite ores of oolite texture.

TABLE IV-5

COBALT IN IRON OXIDE MINERALS AND ORES
(Carr and Turekian, 1961)

Sample Description	Location	Cobalt Concentration (mg/kg)
Clay ironstone	Cedar Point, Maryland	< 4
Magnetite, Sanford ore	Moriah, Essex County, New York	12
Magnetite	Magnet Cove, Arkansas	64
Limonite	Buchersburg, Harz, Germany	< 4
Magnetite	Furneau, Quebec, Canada	143
Magnetite	Lyon Mountain, New York	29
Limonite	Salisbury, Connecticut	68
Limonite	Wayne County, Missouri	10
Bog iron ore	Lake Saltonstall, East Haven, Connecticut	4
Hematite, Fayal mines	Eveleth, Mesabi District, Minnesota	42
Hematite	Marquette, Michigan	5

b. Fossil fuels

(1) Coal and coal-related wastes

(a) Coals and fly ashes: The cobalt concentration in U.S. coals averages about 5 mg/kg; in fly ashes, about 25 mg/kg. Representative averages and ranges are given in Table IV-6. After reviewing the literature, Gluskoter (1975) concluded that cobalt was among the elements in coal more closely associated with the inorganic substances. Zubovic (1966; cited by Gluskoter, 1975) had estimated, however, that the organic affinity of cobalt was 53%.

TABLE IV-6

COBALT IN COALS AND COAL ASHES

Sample Description	Cobalt in Coal, mg/kg		Cobalt in Coal Ash mg/kg		Cobalt in Fly Ash mg/kg		Method	Remarks[a]	Author (year)
	Mean	Range	Mean	Range	Mean	Range			
United States Northern Great Plains	2.7			< 5 to 90			Spectrographic		Zubovic et al. (1961; cited by Abernethy and Gibson, 1963) Zubovic et al. (1960)
Penn., anthracite				10 to 90			Spectrographic		Nunn et al. (1953; cited by Abernethy and Gibson, 1963)
Tex., Colo., N.D., S.D.				100 to 10,000					Deull and Annell (1956; cited by Abernethy and Gibson, 1963)
W.V.				50 to 180					Headlee and Hunter (1955; cited by Abernethy and Gibson 1963)
Europe, Nova Scotia				< 10 to 2,000					Five foreign references cited by Abernethy and Gibson (1963)
United States Eastern Province			180				Spectrochemical	Approximate lower limit of detection, 20 mg Co/kg; D.F. 100%.	Abernethy et al. (1969)
Interior Province			193					D.F. 98%	Abernethy et al. (1969)
Western States			97					D.F. 98 %	Abernethy et al. (1969)
Pittsburgh seam mine float dust		2 to 5					Optical emission spectroscopy	No differentiation of Co in coal fractions by particle size.	Kessler et al. (1971)

TABLE IV-6 (continued)

Sample Description	Cobalt in Coal, mg/kg Mean	Range	Cobalt in Coal Ash mg/kg Mean	Range	Cobalt in Fly Ash mg/kg Mean	Range	Method	Remarks	Author (year)
Upper Freeport Coal		5 to 10 in fractions by particle size					Optical emission spectroscopy	Particles < 1.0 μ comprised 14.56% of the Co in the respirable dust (≤ 3.3 μ).	Kessler et al. (1971)
Mo. coals		3 to 6.6					Optical emission spectroscopy	Weathered (oxidized) coal contains more Co and Ni than unweathered coal but less Zn, Ge, and Sn. For this reason, samples should be taken from fresh exposures.	Zubovic et al. (1967, cited by Ebens and Connor, 1971)
United States Appalachian	5.1	0.5 to 23		140 to 210			Spectrographic		Magee et al. (1973) (A literature review); see, for example, Zubovic et al. (1960)
Interior-Eastern (Ill., Ind., western Ky., Mich.)	3.8			130 to 230			Spectrographic		Magee et al. (1973)
Interior-Western (Iowa, Mo., Neb., Kan., Okla., Ark., Tex.) } Southwestern (bracketed together)	4.4 to 5.4	0.4 to 16[b/]							
Interior-Western (Iowa, Mo., Neb., Kan., Okla., Ark., Tex.)				230 to 550					Magee et al. (1973)
Southwestern				10 to 70					Magee et al. (1973)
Northern Great Plains	2.7	0.4 to 12	60				Spectrographic		Magee et al. (1973)

TABLE IV-6 (continued)

Sample Description	Cobalt in Coal, mg/kg		Cobalt in Coal Ash mg/kg		Cobalt in Fly Ash mg/kg		Method	Remarks	Author (year)
	Mean	Range	Mean	Range	Mean	Range			
U.S. EPA coal samples of unknown origin		3.84 to 5.98					Neutron activation analysis		Sheibley (1973)
U.S. EPA bottom ash sample			4.72						
U.S. EPA fly ash samples					6.6, 6.8				
Twenty-five (mostly Illinois Basin) coal samples							Optical emission photographic	D.L. 20 mg Co/kg. Often not very good agreement between the two optical emission methods.	Ruch et al. (1973)
	6	0.5 to 23					Optical emission--spectrometric	D.L. 3 mg Co/kg. Also analyzed a few whole coals by neutron activation.	Ruch et al. (1973)
United States Whole coals (all but 19 of 101 from Illinois Basin), moisture free.	9.57	1-43					Optical emission spectrometric and neutron activation analysis	Correlation coefficients for 101 coals: Co-Ni 0.7, Co-Cu 0.5 (only 0.4 in Illinois Basin coals), other chalcophile elements ~ 0.4. Of the whole coals, 62 contained 2 to 10 mg Co/kg and 24 contained 10 to 18 mg Co/kg.	Ruch et al. (1974)
Illinois Basin	9.15	2.00-34.00							
Illinois coal, laboratory fractionation by washing to remove mineral matter, fractions having specific gravities > 1.60		8 to 22							
United Kingdom Barnsley vitrain seam			217						Hamilton (1974)

TABLE IV-6 (continued)

Sample Description	Cobalt in Coal mg/kg		Cobalt in Coal Ash mg/kg		Cobalt in Fly Ash mg/kg		Method	Remarks	Author (year)
	Mean	Range	Mean	Range	Mean	Range			
New Zealand							Spectrographic		Sim and Lewin (1975)
Coal	< 3	N.D. to 71.1		N.D. to 250					
Wet ash (~ 7% of coal)				100 to 250					
Dry ash				< 5 to 250					Sim and Lewin (1975)
Grate ash (average 5.1% of coal)				< 5 to 100					Sim and Lewin (1975)
Coking fly ashes					30	10 to 50			Sim and Lewin (1975)
United States									
Illinois		1.8 to 49					Optical emission spectroscopy--spectrographic method	D.L. in ash, 10 mg Co/kg. Average relative standard deviation, 25%.	Dreher and Schleicher (1975)
Illinois		1.0 to 42					Optical emission--spectrometric method	D.L. in ash, 2 mg Co/kg. Average relative standard deviation, 4.1%.	Dreher and Schleicher (1975)
United States							Instrumental neutron activation analysis	Co-Fe ratio in NBS-EPA coal standard 0.00064, same as Co-Fe ratio in diabase rock. This is a little higher than their ratio in the earth's crust (values of Mason, 1966)--0.00050.	Abel and Rancitelli (1975)
Fly ash					25				
Magnetic fraction (10% of total)					50				
Nonmagnetic fraction					16				

TABLE IV-6 (concluded)

Sample Description	Cobalt in Coal mg/kg		Cobalt in Coal Ash mg/kg		Cobalt in Fly Ash mg/kg		Method	Remarks	Author (year)
	Mean	Range	Mean	Range	Mean	Range			
Fly Ash							Instrumental neutron activation analysis and x-ray fluorescence		Natusch et al. (1977)
Midwestern					30	10 to 100			
Western					20				
Australian									Swaine (1977)
Bituminous		10 to 20				10 to 20			
Hard Brown		20 to 25							
Brown		25							
World									Swaine (1977)
90% of coal ashes				5 to 300					
Australian								Values in fly ash comparable to those in soils--1 to 40 mg Co/kg	Swaine (1978)
New South Wales and Queensland	4	< 0.6 to 30				10 to 30			
90% of above coals		0.6 to 20							

a/ D.F. = detection frequency.
D.L. = detection limit.
b/ A weathered sample from Arkansas, containing 44 mg Co/kg, was excluded.

Many elements are enriched in the smaller particle fraction of airborne fly ashes. However, with the elements Be, Ba, Co, Cr, Cu, Ni, Sr, U, and V, the enrichment with decreasing particle size is not so pronounced. For a coal containing 2.1 ± 0.2 mg Co/kg, Coles et al. (1979) found 1.0 ± 0.1 mg Co/kg fly ash particles of mass median diameter (MMD) 18.5 μm; 1.9 ± 0.2 mg Co/kg, MMD 6.0 μm; 2.1 ± 0.3 mg Co/kg, MMD 3.7 μm; and 2.4 ± 0.3 mg Co/kg, MMD 2.4 μm. Straughan et al. (1979) found only 10 1 mg Co/kg in western coal fly ash fractions > 250 μm, but 13.6 to 15 mg Co/kg in fractions < 53 to 250 μm. Fisher et al. (1979) reported 8.9 mg Co/kg in a fly ash fraction of 20 μm; 16.3 mg Co/kg in the 6.3-μm fraction; 19.0 mg Co/kg in the 3.2-μm fraction; and 21.0 mg Co/kg in the 2.2-μm fraction.

(b) Current technology wastes: Boiler deposits generated from burning an Australian bituminous coal containing 4 mg Co/kg (air-dried; 20 mg/kg in the ash) in a spreader-stoker-fired boiler contained 25 to 80 mg Co/kg. The minimum concentration was found in the economizer dust; the maximum, in the off-white deposit on top of the superheater tubes. Variability with depth was shown for cobalt concentrations in superheater deposits from a chain-grate stoker-fired boiler burning bituminous coal: 15 and 25 mg Co/kg in the innermost layers, 150 mg Co/kg in the interior, and 100 mg Co/kg in the outer deposit. Here, the original coal contained 10 mg Co/kg (80 mg/kg in the ash) (Brown and Swaine, 1964).

Trends in power plant ash production from 1966 to 1985 are shown in Figure IV-1. Power plant ash as of the autumn of 1978 was the sixth most plentiful mineral resource in the United States. In another 10 years, it may be fourth. The 1977 power plant ash collection comprised 71.5% fly ash (48.5 million tons or 44 Tg), 20.8% bottom ash (14.1 million tons or 12.8 Tg), and 7.7% boiler slag (5.2 million tons or 4.7 Tg). About 21% of the collected ash is recycled; i.e., it is used as filler material for concrete blocks, roofing products, ice control, and especially for asphalt roads, construction sites, land reclamation, and ecology dikes (Anonymous, 1978b).*

---

* This total was derived from 5.7 MT fly ash, 4.2 MT bottom ash, and 2.8 MT boiler slag that were "recycled."

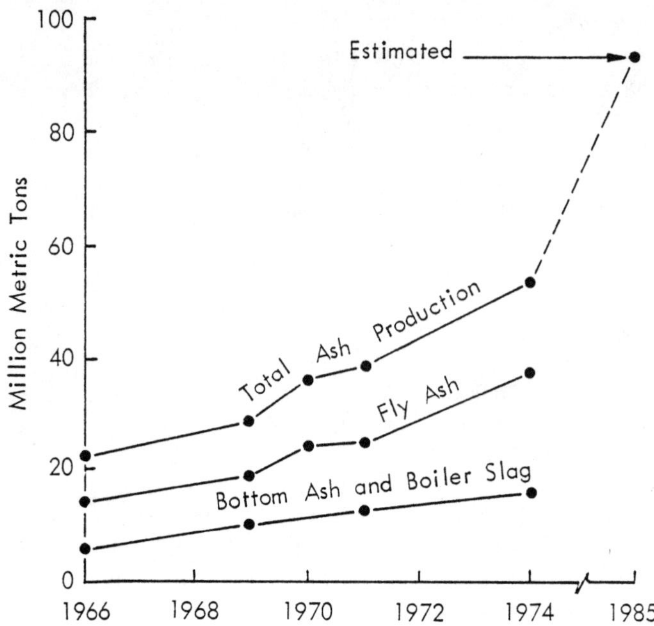

Figure IV-1. Production of Coal Ash from Power Generation in the United States (Straughan et al., 1979).

formed of fly ash should have less tendency for cobalt leaching and/or runoff than ashes disposed of in landfills or waste ponds.

Guthrie and Cherry (1976) (see also Cherry and Guthrie, 1977) studied for 2 years the ash-basin effluent in a swamp-drainage system at the Savannah River Project's coal-fired power plant, Aiken, South Carolina. The overall mean concentration of cobalt in the drainage system was 2.92 mg/liter with approximately 0.1 mg/liter in the water (approximately 1% of the cobalt), 10.6 mg/liter in the sediment (approximately 70% of the cobalt), and the rest in the biota. The overall cobalt concentration ranked 24th among the 40 elements determined, being higher, for example, than the concentrations of iodine, cadmium, molybdenum, antimony, mercury, and indium. Before dredging the system, the cobalt concentration in the water was 0.027 to 0.23 (average 0.07) mg/liter, after dredging, 0.03 to 0.07 (average 0.05) mg/liter. Dredging also reduced the total amount of cobalt in the biota. This work is also discussed in the section of Chapter VII on the aquatic food chain.

Aqueous effluents from coal-burning power plants are relatively high in cobalt. Cobalt in concentrations ranging from 100 to 700 μg/liter were found in the effluent from a closed-loop wet lime and limestone scrubber system for flue gases (Chu et al., 1977).

Typical Illinois-basin coal preparation (cleaning) wastes contain 30 mg Co/kg. After 4 days of aqueous leaching in the laboratory with an equal weight of water, 60% of the cobalt was removed from the waste. Nickel and cobalt were rated as highly leachable elements. Only calcium was leached to a greater extent (79%) by this treatment (Williams et al., 1977). These waste piles are stored in the open and can be a source of acid drainage.

(c) New technology wastes: Lebowitz et al. (1975) identified potential sources of atmospheric cobalt emissions from the extraction and processing of coal and oil. Fixed-bed catalyst regeneration in a petroleum refinery or coal liquefaction plant may produce hazardous cobalt carbonyl gas. Cobalt, possibly as the carbonyl, may be present in emissions from quenching and direct-cooling of gasifier off-gas during coal gasification, from sulfur recovery operations during coal liquefaction or gasification, and from reactor off-gas in coal liquefaction.

Rhodes (1973) described the potential water and air pollution that may be associated with coal gasification plants. These plants will be large, handling 10,000 to 25,000 tons coal/day (9,100 to 23,000 MT/day). Rhodes expected that at operating temperatures ($\leq$ 1760°C), cobalt would be among the elements volatilized from the coal. If 1 billion short tons coal is used annually for gasification (to produce 15 to 20 x 10^{12} SCF/year or 4.2 to 5.7 x 10^{11} m^3/year and it contains an average 6 mg Co/kg, then 6,000 tons (5,450 MT) cobalt would be volatilized annually.

Somerville and Elder (1978) leached laboratory-prepared ashes of North Dakota lignite meant to simulate Lurgi gasifier ash. The ash produced daily would contain 81 kg Co/day. At 1.6% leached by refluxing for 16 to 24 hr in boiling demineralized water, the maximum leachate rate would be

1.3 kg Co/day. However, an actual gasifier lignite ash was only 0.77% leachable. (The two ashes contained 13 and 6 mg Co/kg, respectively.)

Coal gasification processes produce aqueous wastes containing heavy metals and organics. In the scrubber-decanter wastewater, Forney et al. (1974; cited by Klemetson, 1977) found only 2 μg Co/liter compared with 6 to 130 μg/liter for other trace metals. The same data may have been referred to by Alford (1974), who mentioned that Synthane coal-gasification pilot plant effluent contained 2 μg Co/liter, which was the lowest concentration of 26 elements determined by spark source mass spectrometry. Other process waters from *in situ* coal gasification are reported to contain up to at least 10 μg Co/liter (see Table IV-7).

Aqueous coal conversion effluents have cobalt concentrations sufficiently high to have a medium potential for toxicity and bioaccumulation hazards, concluded Hildebrand et al. (1977). The conclusion was based on the aquatic toxicity value of 10 μg Co/liter reported by Biesinger and Christensen (1972) and the average fish bioaccumulation factor (320)* reported by Vanderploeg et al. (1975). Two aqueous streams from the Char-Oil-Energy Development, FMC Corporation, Princeton, New Jersey, were sampled. The dryer stage liquor had an average 5 μg Co/liter, and the product separator liquor had an average 500 μg/liter.

In 1977 there were four coal liquefaction processes being actively studied at the pilot plant stage. Filby et al. (1978) studied the pollution potential of the Solvent Refined coal process developed by Pittsburgh and Midway Coal Mining Company under a U.S. Energy Research and Development Administration (U.S. ERDA) contract. A 50-ton/day pilot plant was operating at Fort Lewis, Washington. The solvent-refined coal (SRC) contained only 4 to 5% of its original cobalt concentration. The highest cobalt concentration in a process stream was 22.7 mg/kg (see Table IV-8). Of the cobalt in the SRC, 73% was in organic form. Only 17% of the original coal cobalt was in organic form.

---

* This is the average based on calculations using fresh weight concentrations of cobalt in fish living in mesotrophic lakes.

TABLE IV-7

COBALT IN ENERGY-RELATED PROCESS STREAMS
(Pellizzari, 1978)[a]

Sample	Cobalt Concentration (mg/kg) Mean	Range
Liquid from a 150-ton oil-shale retort processing experiment[b]	0.01	-
Discarded cavern oil shale from *in situ* oil-shale processing[c]	11 ± 4	-
Boiler blow-down water from *in situ* oil-shale processing[c]	0.01	-
Liquid condensate from Rosebud coal during low Btu-gasification[d]	< 0.008	≤ 0.009 ± 0.009 - 0.015 ± 0.005
Liquid samples from dewatering wells prior to *in situ* coal gasification[e]	-	0.001 - 0.002
Process water from *in situ* coal gasification[e]	0.003	-
Process tar from *in situ* coal gasification[e]	0.08 ± 0.02	-
Well and seam waters from *in situ* coal gasification[b,e]	< 0.02	≤ 0.004 - 0.03
Tar from production well after *in situ* gasification[e]	0.85 ± 0.15	-
Product waters from well linkages during *in situ* coal gasification[b]	-	0.005 - < 0.1
Product tars from wells and well linkages during *in situ* coal gasification[b]	-	0.007 - 0.04
Groundwater from *in situ* coal gasification[b]	0.001	-
Post-gasification water samples after *in situ* coal gasification[b]	< 0.006	0.001 - 0.01

a/ Analysis by flame atomic absorption spectroscopy.
b/ Laramie Energy Research Center, Hanna, Wyoming.
c/ Occidental Oil Company.
d/ Morgantown Energy Research Center, West Virginia.
e/ Lawrence Livermore Laboratory, California.

TABLE IV-8

COBALT IN SOLVENT REFINED COAL (SRC) PROCESS FRACTIONS
(Filby et al., 1978)

	mg/kg	% of Total Cobalt
Coal	3.7	-
SRC[a]	0.31	5.0
Wet filter cake (solids filtered from the coal solution after hydrogenation)	12.8	-
Pyridine insolubles (residue after treating wet filter cake with pyridine)[a]	22.7	86.0
Recycled process solvent (distilled at 290 to 450°C from coal solution)	0.031	-
Light ends (an oil boiling at 60 to 230°C)	0.035	-
Process water (recycled)	0.012	-

a/ SRC + pyridine insolubles ≃ 75% by weight of the original feed coal. Pyridine insolubles = 13.6% in this run.

Other data for the SRC process, presumably for the same plant, indicated that the SRC (0.26 mg Co/kg) contained 3% of the original amount of cobalt (5.2 mg/kg). The residue (32 mg Co/kg) contained almost 97% of the original cobalt since the amounts in the process solvent (0.0026 mg Co/kg) and process naphtha (0.0044 mg Co/kg) were negligible (Fruchter et al., 1978).

Loran and O'Hara (1978) studied the feasibility of placing a Fisher-Tropsch facility producing liquid hydrocarbons and substitute natural gas by indirect coal liquefaction adjacent to a coal mine in the Eastern Region of the Interior Coal Province of the United States. The plant would use 30,000 tons/day cleaned bituminous coal containing 3.4% sulfur. Modeling showed that the emissions could meet ambient air quality standards after atmospheric dispersion. The process water would be recycled or discharged after treatment. Cobalt, nickel, and iron would be used as catalysts or be present in the low-carbon steel structural equipment. At low pressures and high prevailing temperatures, no carbonyls would be expected to form during normal operation of the facility. However, conditions might be favorable for carbonyl formation for short periods during shutdown operations. Loran and O'Hara stated, "The normal safe practise of flaring vent streams, along with operation of all contaminant removal systems, will prevent release of carbonyls to the atmosphere." They recommended that plant personnel who handle the catalysts or enter reaction vessels must be trained in proper procedures and be supplied with protective apparatus.

(2) Petroleum products: Crude oils consistently contain nickel and vanadium as major metallic constituents and cobalt, copper, chromium, manganese, iron, lead, and zinc as minor constituents (Bell, 1960).

Metallic components in petroleum are currently classified as follows:

- Metalloporphyrin chelates
  - a. Vanadyl and nickel porphyrins
  - b. Chlorophyll _a_ and other hydroporphins

c. Highly aromatic porphin* chelates
d. Porphyrin decomposition ligands

- Complexes of tetradentate mixed ligands with transitional metals such as V, Ni, Fe, Cu, Co, and Cr.
  a. Simple complexes from resin molecules
  b. Chelates from asphaltene sheets
  c. Organometallics (e.g., of Hg, Sb, and As)
  d. Carboxylic acid salts of the polar functional groups of resins (e.g., salts of Mo and Ge)
  e. Colloidal minerals

"At present it is without doubt that petroleums originate from marine organisms (Yen, 1975)." Most authors have concluded that the metals in petroleum have been derived primarily from the biological material (Filby, 1975). Gulyaeva and Punanova (1973; cited by Filby, 1975) showed that many elements have much lower concentrations in petroleum than in marine organisms and that vanadium is much higher than in organisms. The concentrations of cobalt, boron, chromium, iodine, lead, nickel, and silver, however, in both marine organisms and petroleum are generally similar. Filby also suggested that cobalt, chromium, copper, iron, mercury, and zinc could have been removed from seawater solution rather than being deposited in the bodies of dead organisms. The metals are apparently incorporated into the asphalt sheet structure through complexing at "holes" bordered by atoms of N, S, or O. Almost all of the metals present in the crude oil are found in the refinery residues.

Although cobalt has correlated well with the nitrogen content of western crude oils (Hyden, 1961), Filby (1975) concluded that cobalt is not associated with the porphyrins in the oils.

Values reported for cobalt concentrations in petroleum and shale are given in Table IV-9. They range from 0.001 to 10 mg Co/kg in crude oils. Fewer analyses for oil shales were found; cobalt values were usually < 30 mg/kg. Medium to extra heavy fuel oils contain 0.03 to 0.3 mg Co/kg, and gasoline, < 0.1 mg Co/kg.

---

* The unsubstituted nucleus.

TABLE IV-9

COBALT IN PETROLEUM AND SHALE

Sample Description	No. of Samples[a]	Cobalt Concentration in the Whole Sample, µg/kg: Average	Cobalt Concentration in the Whole Sample, µg/kg: Range	Cobalt Concentration in the Ash, %: Average	Cobalt Concentration in the Ash, %: Range	Analytical Method	Remarks on Analytical Method	Other Remarks	Authors (year)
U.S. crude oils	29			0.092	0.00X-0.X 0.9 maximum	Spectrographic and chemical methods	Samples ashed at 550°C		Erickson et al. (1954)
Solid asphalts (15 from Utah)	22			0.3	0.000X-X 9 maximum				Erickson et al. (1954)
Petroliferous rocks (17 from Utah, 8 from Colorado, many sandstones and asphalts)	27				0.000X-0.X			Athabasca tar sands (Alberta, Canada) contained 0.00X% Co in the ash.	Erickson et al.
U.S. petroleums	13/24				≤0.01	Semiquantitative spectrography	Dry ashing		Ball et al. (1960)
	7/24				0.01-0.1				Ball et al. (1960)
	4/24				0.1-1.0				Ball et al. (1960)
Wilmington, California crude oil				0.022				Values for asphaltenes and deasphaltened Wilmington crude and ash percent do not allow estimation of concentration in the nitrogen concentrate. Presumably, the magnitude of at least one value given is in error. If the nitrogen concentrate contained as much as 1% Co, the 0.0512 g ash of the concentrate would contain 510 g--more than reported for the crude oil	Ball et al. (1960)
Wilmington deasphaltened crude				0.X					
Wilmington asphaltene (7% of crude oil)				0.X					
Nitrogen concentrate of deasphaltened Wilmington crude (adsorbed on Florisil)				X					

TABLE IV-9 (continued)

Sample Description	No. of Samples	Cobalt Concentration in the Whole Sample, μg/kg: Average	Range	Cobalt Concentration in the Ash, %: Average	Range	Analytical Method	Remarks on Analytical Method	Other Remarks	Authors (year)
Rock-asphalt bitumens	55% ~ 31%				≤ 0.00X ≤ 0.0X				Bell (1960)
Western U.S. oils (mostly Wyoming, Oklahoma, California, and Kansas)	16				0.0005-0.005	Emission spectrographic	Detection limit for Co in ash ~ 5 mg/kg	Co correlation coefficient with crude oil characteristics: N content, 0.64; gravity of residue, 0.61; and residual C content, 0.54. (The correlation coefficient for other metals with N content were 0.69, Ni; 0.60, V; and 0.64, Ga. The latter was not as reliable as those for Ni and V.)	Hyden (1961)
Asphaltite, Seven Rivers Formation, Carlsbad, New Mexico			110,000-120,000		0.11-0.14	Quantitative spectrochemical analysis	Results based on four determinations for each sample	The ash sample also contained 33-43% U, 8.5-9.0% As, 0.74-0.80% Ni, etc.	Horr et al. (1961)
North American black shales	779			10				Median values	Vine and Tourtelot (1970)

TABLE IV-9 (continued)

Sample Description	No. of Samples	Cobalt Concentration in the Whole Sample, μg/kg: Average	Cobalt Concentration in the Whole Sample, μg/kg: Range	Cobalt Concentration in the Ash, %: Average	Cobalt Concentration in the Ash, %: Range	Analytical Method	Remarks on Analytical Method	Other Remarks	Authors (year)
Asphaltic crude oil, Santa Maria Valley, California			200–230		0.0220–0.0260				Horr et al. (1961)
Tarry, asphaltic crude oil, Herric Oil Pool, Albany County, New York			54–59		0.0160–0.0180				Horr et al. (1961)
Refinery residues	5				0.0005–0.005			Refinery residues contain 99% of the metals in the crude oils	Hyden (1961)
	11				0.005–0.05				
Italian crude oils	2		< 1.5			Neutron activation analysis	Spurious contribution of the reaction $^{60}Ni(n, p)^{60}Co$ was $\leq$ 0.1% of the Co concentration. The reaction $^{63}Cu(n,\alpha)^{60}Co$ contributed $\leq$ 0.03%.		Colombo et al. (1964)
	2	1.5 3							
Italian asphalts	4	106	55–140			Neutron activation analysis			Colombo et al. (1964)

TABLE IV-9 (continued)

Sample Description	No. of Samples	Cobalt Concentration in the Whole Sample, μg/kg		Cobalt Concentration in the Ash, %		Analytical Method	Remarks on Analytical Method	Other Remarks	Authors (year)
		Average	Range	Average	Range				
High-purity lubricant mineral oils. Found only in a paraffinic neutral produced by conventional refining	1/10	1,700				Neutron activation analysis		The presence of Co, As, Cu, Hg, and Mn in the oils was attributed to contamination from fluid processing and/or equipment corrosion and erosion.	Jester and Klaus (1967)
Crude oils:	10	1,710 ± 110				Neutron activation analysis	Detection limit 0.6 μg Co/kg. Co concentrations were corrected for the interference from ^{60}Ni (n,p)^{60}Co. At 241 μg/kg, the contribution of the interference was 0.34%. At 31 μg/kg, 6.67%.		Shah et al. (1970)
California	5	3,340	241-12,751						
Libya	3	350	32-877						
Louisiana	1	29							
Wyoming	1	9							
Crude oils						Neutron activation analysis			Bryan et al. (1970)
Wilmington, California		2,200					No significant change in concentration from sea-water attack.		
Duri Field, Sumatra	6/40	1,900					No change from attack by sea-water or bacteria.		

TABLE IV-9 (continued)

Sample Description	No. of Samples	Cobalt Concentration in the Whole Sample, μg/kg		Cobalt Concentration in the Ash, %		Analytical Method	Remarks on Analytical Method	Other Remarks	Authors (year)
		Average	Range	Average	Range				
Residual oils						Neutron activation analysis			Bryan et al. (1970)
Minas		900							
Minas (after 24-hr contact with seawater)		< 470							
Santa Fe No. 6		2,600							
Nigeria		1,500							
Sumatra residuum		500 ± 100							
Sumatra residuum (after 24-hr vigorous contact with seawater)		1,600 ± 400							
Oil samples, slicks, and spills	63/349 (18.1%)	Detected				Neutron activation analysis			Lukens et al. (1971)
	24/185 (13.0%)		~ < 214- ~ 10,000						
	15/185		~ 1,000- ~ 1,470						
California crude oil						Thermal neutron activation analysis		There were large variations in trace element concentration "even within a given oil well." Negligible amounts of Co in RF-1 were extracted by distilled water.	Filby and Shah (1971)
Sample RF-1		12,700							
Sample RF-1 (filtered)		23,100							
Sample RF-2		13,900							
Sample RF-2 (filtered)		14,400							
Sample RF-3		14,500							

TABLE IV-9 (continued)

Sample Description	No. of Samples	Cobalt Concentration in the Whole Sample, μg/kg		Cobalt Concentration in the Ash, %		Analytical Method	Remarks on Analytical Method	Other Remarks	Authors (year)
		Average	Range	Average	Range				
Venezuela crude oil (filtered)		198				Thermal neutron activation analysis			Filby and Shah (1971)
Venezuela crude oil (filtered)		178				Thermal neutron activation analysis			Filby and Shah (1971)
Louisiana crude oil		1				Thermal neutron activation analysis			Filby and Shah (1971)
Libya crude oil		35.4				Thermal neutron activation analysis			Filby and Shah (1971)
Fractions of California crude oil RF-1						Thermal neutron activation analysis		Resins and asphaltenes exist in colloidal form in oil. Co, Fe, Cr, and Zn were concentrated to the same extent as Ni. All (except perhaps Fe) might exist as complexes with deoxyphylloerythroetioporphyrin (DPEP) as does Ni.	Filby and Shah (1971)
Methanol soluble		800							
Resins		10,700							
Asphaltenes		122,000							

TABLE IV-9 (continued)

Sample Description	No. of Samples	Cobalt Concentration in the Whole Sample, μg/kg		Cobalt Concentration in the Ash, %		Analytical Method	Remarks on Analytical Method	Other Remarks	Authors (year)
		Average	Range	Average	Range				
Molecular-weight range fractions of RF-1 asphaltene						Thermal neutron activation analysis			Filby and Shah (1971)
300-1,000 (11%)		2,670						Only the first fraction contains Ni as the DPEP complex.	
1,000-4,000 (23.2%)		30,000							
4,000-8,000 (50.6%)		167,000							
8,000-22,000 (15.2%)		176,000							
Molecular-weight range fractions of RF-1 resins						Thermal neutron activation analysis		Concluded that Co not in porphyrins in the oils	Filby (1975)
300-1,000 (29.4%)		4,370							
1,000-4,000 (21.2%)		10,000							
4,000-8,000 (49.4%)		24,900							
8,000-22,000 (~ 0%)									
Spiked crude and residual oils		100 (μg/ℓ)				Neutron activation analysis	Found to within ± 30%		Gulf Radiation Laboratory (1973)
		1,000 (μg/ℓ)				Neutron activation analysis	Found to within ± 20%		
							Method useful down to 10 μg Co/ℓ. In actual samples, the sensitivity range was 100-1,000 μg/ℓ.		

TABLE IV-9 (continued)

Authors (year)

Sample Description	No. of Samples	Cobalt Concentration in the Whole Sample, µg/kg		Cobalt Concentration in the Ash, %		Analytical Method	Remarks on Analytical Method	Other Remarks	Authors (year)
		Average	Range	Average	Range				
Fuel oil		301				Neutron activation analysis	Standard deviation 14% in nine determinations		Filby and Shah (1975)
Crude oils	84/88	53.7	$\leq$ 2,000			Neutron activation analysis	Detection limit for Co 0.2 µg/kg	Neither Co or Mn was related to Fe in these crudes. Factor analysis and correlation coefficients seemed to rule out incorporation of these metals by corrosion of well casing, tubing, separators, treaters, or storage tanks.	Hitchon, Filby, and Shah (1975)
Oil shales of Parachute Creek member of Eocene Green River Formation, Piceance Creek Basin, Colorado, and Uinta Basin, Utah						Semiquantitative 6-step spectrography	The samples were ashed at 525°C	The oil shales contained 13-27% extractable oil. The mineral fractions comprised quartz, dolomite, calcite, albite, and K feldspar.	Desborough et al. (1976)
Mahogany bed	6	10,000	5,000-20,000						
Bed in R-4 zone	4	10,000	7,000-15,000						

TABLE IV-9 (continued)

Sample Description	No. of Samples	Cobalt Concentration in the Whole Sample, μg/kg		Cobalt Concentration in the Ash, %		Analytical Method	Remarks on Analytical Method	Other Remarks	Authors (year)
		Average	Range	Average	Range				
U.S. gasolines			5-100 (μg/ℓ)			Isotope dilution spark source mass spectrometry			Carter et al. (1976)
Copper-rich shales, Greer County, Oklahoma							Lockwood cites his Ph.D. dissertation for the description of the analytical method	These shales are already being developed or mined	Lockwood (1976)
Magnum Deposit			17,700-24,000					Co significantly correlated with quartz, illite, V, Ag, Pb, Ni, Fe_2O_3, K_2O, ZnO, organic matter, in $\geq$ 1 Magnum Deposit shale and with quartz, illite, Ag, Pb, V (1.0 in red shales), Fe_2O_3, K_2O, ZnO, and U in Creta Deposit. Cobalt may be in interlayer or structural positions in illite. The correlation was thought not to be due to adsorption because there was no correlation with chlorite, whose properties are similar to those of illite.	
Creta Deposit			19,500-30,700						

TABLE IV-9 (concluded)

Sample Description	No. of Samples	Cobalt Concentration in the Whole Sample, µg/kg		Cobalt Concentration in the Ash, %		Analytical Method	Remarks on Analytical Method	Other Remarks	Authors (year)
		Average	Range	Average	Range				
Fuel oils (Belgium)						Neutron activation analysis		The crude oils contained 4-60 µg Co/kg.	Block and Dams (1978)
"Light" (corresponds to fuel oil No. 4 on U.S. market)			1.3-1.4						
"Medium Heavy"			30-70						
"Heavy" (Fuel oil No. 5 falls between "medium heavy" and "heavy")			20-90						
"Extra Heavy" (corresponds to fuel oil No. 6)			70-200						
Oil shale, Green River formation, western Colorado			5,200-6,500			Atomic absorption	Samples prepared by low-temperature ashing for Co determination		Shendrikar and Faudell (1978)

a/ Where a fraction or a percent is given, the value denotes the frequency of detection of cobalt.

One lubricant oil was reported to have 1.7 mg Co/kg, possibly due to contamination. The asphalts, asphaltenes, asphaltites, and rock-bitumens contain up to 122 mg Co/kg.

Hyden (1961) classified oil fields according to geographic distribution, geologic age, and type of reservoir rock. The lowest cobalt concentrations were in the ashes of crude oils of the Rocky Mountain region from sandstones of Cambrian (0.0015%) and Triassic (0.0007%) ages. In this region high values of $\geq$ 0.3% were found in the ash of petroleums from sandstone of Cretaceous Age, sandstone and claystone of Tertiary Age, and sandstone and limestone of Jurassic Age. Such high values were also found in the ash of petroleums of the west coast region from sandstones of Tertiary Age.

Shendrikar and Faudel (1978) retorted three samples of raw oil shale from the Green River Formation of western Colorado under conditions simulating a potential commercial process.* Only 0.4 to 0.5% of the cobalt in the raw shale (5.2 to 6.5 mg Co/kg) was transferred to the oil (0.2 to 0.26 mg Co/kg). The remainder was in the spent shale. However, the mass balance was not perfect (100.3 to 111.7%); some cobalt was found in the process water (22 to 5 μg/liter). Fruchter et al. (1978) observed a similar mass balance for cobalt during oil shale retorting but higher concentrations in the process water (23 and 650 μg/liter). Their data are presented in Table IV-10.

Other values for cobalt in oil shale retorting process waters may be found in Table IV-7 in the preceding subsection.

Cobalt was found at the μg/liter level in some of the 823 oil field waters (brines) from the United States and Canada analyzed spectrochemically by Rittenhouse (1969). Matusevich and Popov (1978) found up to 11.0 μg Co/liter in U.S.S.R. petroleum and gas field stratal waters.

c. Fertilizers and peat

(1) Concentrations in fertilizers: Swaine (1962) compiled literature values for numerous determinations of

* The Fischer assay process for retorting oil shale is done at $\leq$ 932°F ($\leq$ 500°C).

TABLE IV-10

COBALT MASS BALANCE DURING IN-SITU RETORTING OF OIL SHALE
(Fruchter et al., 1978)

Material	Material Weight, LLL[a] Simulation	Cobalt Concentration (mg/kg) LLL Simulation[b]	Laramie Simulation[c]	Percent of Cobalt in Product Material, LLL Simulation
Raw oil shale	125.2 kg	8.6 ± 0.1	11	-
Spent oil shale	85.6%[d]	11.8	21.1	99.6
Crude shale oil	10.54%[d]	0.35 ± 0.8	0.37 ± 0.02	0.4
Process water	3.77%[d]	0.023 ± 0.002	0.65 ± 0.004	-

a/ LLL = Lawrence Livermore Laboratory.
b/ Determined by flameless atomic absorption analysis.
c/ Determined by neutron activation analysis.
d/ Given in the reference as kg; however there was no indication of other products or losses.

cobalt in fertilizer materials. Most of the common fertilizer materials generally had cobalt contents lower than those of ordinary soils (i.e., $\lesssim$ 10 mg/kg). These fertilizers included ammonium sulfate, cyanamide, calcium and sodium nitrate, urea, phosphates, and superphosphates (except serpentine-superphosphate where the serpentine contributes most of the cobalt), bird guano, bone meal, basic slags, and potassium-bearing salts and minerals (except one U.S. potassium carbonate sample with ~ 30 mg Co/kg). Manures contained from ~ 0.001 to ~ 0.2 mg Co/kg fresh weight or from 0.18 to 6 mg/kg* dry weight basis. Sewage sludges contained up to 100 mg Co/kg dry weights although values reported were more commonly ≤ 15 mg/kg. Municipal compost in Great Britain has 40 to 60 mg/kg cobalt, but only 1 to 1.5 mg/kg was available. Miscellaneous organic materials including tankage, fish scrap, blood, and seaweeds had up to 6.25 mg/kg cobalt, dry weight. A few limestones were reported with high values (e.g., < 5 to 35 mg/kg in 275 Scottish limestones, but 221 had < 5 mg/kg; and 130 mg/kg in one Norwegian sample); but most had less than ~ 16 mg/kg with the average probably < 5 mg/kg.

Among miscellaneous fertilizer materials with very high cobalt concentrations were 300 to 19,000 mg/kg in copper slags (Germany and Holland); lignite ash, 400 mg/kg (North Dakota); burnt pyrites, 145 mg/kg (USSR); manganese sludge, 27,

* More recently determined values for cobalt in manure (dry weight) are not very much different. Mazur (1972) found 0.11 to 6.76 mg Co/kg in 58 farmyard manures with an average of 1.71 mg/kg.

38 mg/kg (USSR); and an asbestos industry by-product (probably hydrated magnesium silicate), 300 mg/kg (Yugoslavia).

Clark and Hill (1958) found 0.8 to 11.6 mg Co/kg in commercial phosphate rocks from Florida, South Carolina, Tennessee, Idaho, Wyoming, and Montana by a colorimetric method. They determined 0.8 to 3.0 mg Co/kg in ordinary superphosphate, 1.3 to 4.8 mg Co/kg in triple superphosphate, and 1.6 to 3.1 mg/kg in ammonium phosphate. The cobalt content of the processed phosphates was not any higher when sulfide ores were the source of sulfuric acid than when sulfur was the source.

Contamination by milling and overburden causes the trace element content of agricultural limestone to be higher than in the unquarried rock. Although the average cobalt concentration in U.S. limestones found by Chichilo and Whittaker (1961) was < 1 mg/kg, limestones from all west north central states had higher maximum cobalt concentrations (3 to 6 mg/kg) and an average concentration up to 3 mg/kg. Limestone produced while mining lead-zinc ores had the highest values: up to 54 mg Co/kg, average 18 mg/kg.

The Association of American Fertilizer Control Officials has proposed that the minimum amount of cobalt in a fertilizer that claims it as a nutritional additive be ≥ 5 mg/kg (Sparr et al., 1968).

(2) Addition to soil from fertilizers: Andersson (1977) analyzed 77 dry manures collected in Sweden for heavy metals by atomic absorption, finding 2.1 to 24 mg Co/kg (average 7.8 mg/kg)* in 74 of them (three high-cobalt values excluded). Commercial feedstuffs contained 0.15 to 49 mg Co/kg (dry weight). At a manuring rate of 5 tons dry weight/hectare (ha), the maximum amount of cobalt added to the plow layer (0 to 20 cm) would be 48 g/ha ($<$ 0.4% nitric acid-soluble) compared with 13 kg nitric acid-soluble of naturally occurring cobalt per ha in the soil.

---

* Hung (1978) found values very close to this by flame and nonflame atomic absorption spectrometry by dry ashing at $\leq$ 600°C (7.66 and 8.46 mg Co/kg) and by wet digestion with nitric acid (7.53 and 6.41 mg Co/kg) in dry cattle and horse feces, respectively. Feeds contained 1.66 to 6.35 mg Co/kg.

The common agricultural fertilizer materials containing about 10 mg Co/kg that are applied at a usual rate of 500 lb/acre will result in little increase in the cobalt content of soil since the amount applied is probably less than that removed by an annual crop (Stojkolska and Cooke, 1958). For example, Chichilo and Whittaker (1961) calculated that only 19% of U.S. agricultural limestones contained sufficient cobalt to supply the 0.0014 lb Co/acre removed by crops in 5 years of crop rotation (2 years alfalfa; 1 year each corn, soybeans, and oats) when applied at the rate of 4 tons/acre.

Thus, cobalt-deficient soils are generally treated with fertilizers containing cobalt salts. In Australia and New Zealand, for example, either a small amount of cobaltous sulfate* is added to superphosphate fertilizer or a spray of cobaltous sulfate is applied. Sauchelli (1969) described common practice for treating cobalt-deficient soils in New Zealand. On a pumice soil, 79 g Co/ha (applied in superphosphate fertilizer) will prevent deficiency for 3 years; 20 g/ha applied each autumn is sufficient for 12 months.

Barufke (1962) found that treating lowland moor (meadow soils) of the Brandenburg district (0.3 to 1.25 mg Co/kg) with 8 kg $CoSO_4$ (3 kg Co)/ha along with 100 kg $CuSO_4$/ha increased the cobalt concentration of hay from 0.03 to 1.33 mg /kg (dry weight).

Sauchelli (1969) cautioned that lime or other alkaline treatment should be avoided when using cobalt fertilizers on high-molybdenum soils. Plants growing on cobalt-treated soils take up abnormally high amounts of molybdenum. When the molybdenum content of the soil is already high, the molybdenum concentrations in the plants after cobalt fertilization may be as high as those on teart land, where grazing cattle exhibit scouring due to excess molybdenum in their diet.

(3) Peat: Coal origins involve biochemical and metamorphic stages in the sequence: peat $\longrightarrow$ lignite $\longrightarrow$ bituminous coal $\longrightarrow$ anthracite coal. The major use of peat in

* $CoSO_4$ is dissolved in the acid used to produce normal superphosphate.

the United States, however, is for soil improvement rather than fuels. Casagran and Erchull (1976) studied the peat-forming environments in Okefenokee Swamp, Georgia. At Minnie's Lake, < 8 mg Co/kg was found in peat layers at 0 to 109 cm depth; 8 or 12 mg Co/kg, at 150 to 335 cm. At Chesser Prairie, 8 mg Co/kg was found at 13 to 51 cm depth, with lower values at greater depths. No other U.S. values were found in the literature. Tarakanova (1971) found average values up to 11 mg Co/kg in peats of different genetic types. Her results are given in Table IV-11.

TABLE IV-11

COBALT IN PEATS OF DIFFERENT GENETIC TYPES
(Tarakanova, 1971)

	Extractability	Cobalt Concentration Dry Weight (mg/kg) Average	Maximum
Reed	93%	11	60 (6/60: > 30)
Cedar	61%	2.5	18
Pine	55%	2	14

3. Commercial products: Cobalt is used as a catalyst in terephthalic acid production. It is not surprising then that Vogt and Hashii (1973) found cobalt in three of eight polyester fibers analyzed by neutron activation but not in acetate, acrylic, modacrylic, nylon, or rayon fibers. Two polyester fibers (by Celanese) contained 43.8 and 38.9 mg Co/kg. The other polyester fiber (by FMC Corporation) contained 29.5 mg Co/kg.

The occurrence of cobalt in paper products (as described in Chapter V, subsection C.3., on urban refuse burning) is largely due to the use of cobalt in printing inks. Lieser et al. (1977) found only 0.06 mg Co/kg in cellulose. And Simon et al. (1977) found only 0.01 to 0.36 mg Co/kg in 11 manufactured papers. Other potential sources of heavy metals in paper include clay, alumina, titanium dioxide, zinc oxide, talc, the wood pulp, the process water, or the process apparatus.

The major user of inks printed on metal is the can-manufacturing industry (~ 6,800 MT annually compared with < 450 MT for all other uses). As of 1976, about 90% of the cans were printed on the sheet metal (outside surface only) before the cans were formed. About 85% of the inks used in 1976 were cured by heating to 166 to 177°C for 6 min either by the dry offset printing process or by the wet lithographic process. These conventional (heat-cured) metal-decorating inks may contain cobalt as cobalt blue pigment (5 to 60%) or cobalt naphthenate drier (0 to 0.2%). The application rate may be as low as 45-g ink per hour. Ordinarily there are no emissions from the conventional process (Sallee, 1976).

In 1976, the remaining 15% of the inks used for printing cans was cured by ultraviolet light. Because of substantial thermal energy savings, American Can Company planned to use the ultraviolet process in 50% of its metal coating and all of its metal printing operations by 1980. In the ultraviolet process, exhaust gases containing solvents are incinerated, since other emission controls are unsatisfactory (Sallee, 1976).

Cralley et al. (1968) detected cobalt in 20 of 22 cosmetic talcum powders (body, bath, and all-purpose) by atomic absorption analysis. The range of concentrations was < 9 to 67 mg Co/kg. Ten samples contained ≥ 10 mg Co/kg (nine of these 10 to 25 mg Co/kg). Cobalt may be a substitute for magnesium in talc, $Mg_3Si_4O_{10}(OH)_2$, a sheet silicate, or for the iron in the brucite layer of the talc (Rohl et al., 1976). Some talc deposits contain tremolite, chrysotile, pyrophyllite, serpentine, and other basic minerals from which talc may be derived. Cobalt may also be an impurity in the zinc, titanium, manganese, or iron compounds sometimes added to talcs as pigments and emulsifiers.

Rohl et al. (1976) found cobalt in 5 of 20 commercial cosmetic talcs at 4 to 88 mg Co/kg (average 47 mg/kg; detection limit 3 mg/kg). Four samples containing 21 to 88 mg Co/kg were also high in nickel, chromium, and ferrous oxide. Consumer-grade talc had not been used in these products, which were formulated prior to June 1973. Although cosmetic and pharmaceutical talcs should contain no asbestiform minerals, as late as 1968 some consumer talcum products did contain asbestiform minerals.

Jacobs (1975) found 30 mg Co/kg in a bulk chrysotile asbestos by spark source mass spectrometry; 3.0 mg Co/kg, in bulk asbestos; and 0.89 mg Co/kg in bulk crocidolite.

Monkman (1975) reported that the average concentration of cobaltous oxide in dust-free chrysotile asbestos fibers from 18 mines was 90 mg/kg (40 to 120 mg/kg). The dust fraction contained 120 to 310 mg Co/kg. The chrysotile fraction of the dust contained 30 to 130 mg Co/kg. The nonchrysotile fraction of the dust contained 30 to 240 mg Co/kg (average 190 mg Co/kg). Note that the nonchrysotile dust had an average cobalt concentration greater than that in the dust-free fiber. Other species present in the dust were Fe_2O_3, 6.64%; Al_2O_3, 1.77%; FeO, 1.90%; MnO, 0.151%; Cr_2O_3, 0.235%; NiO, 0.246%; CuO, 50 mg/kg; and ZnO, 150 mg/kg.

4. Soils, water, and air

a. Soils: The cobalt concentration of most soils lies in the range 1 to 40 mg/kg. Hawaiian soils containing as much as 156 mg Co/kg have been reported, however. The average cobalt concentration in lunar soils collected during the Apollo mission is $\sim$ 50 mg/kg (Morrison and Nadkarni, 1973). Cobalt was first detected in plants in 1841, but it was not until 1935 that it was found to be essential for ruminants (Hemphill, 1972)

Shacklette (1971) found cobalt in 855 of 863 conterminous U.S. soils and other surficial materials sampled to a depth of 20 cm. The geometric mean cobalt concentration was 7.2 mg/kg. Figure IV-2 shows the relative abundance of cobalt in these surficial materials. Figure IV-3 shows that almost 75% of U.S. surficial materials contain cobalt in the range 3 to 15 mg/kg; $\sim$ 15% contain $<$ 3 mg Co/kg; and $\sim$ 10% contain 15 to 80 mg/kg.

Soils containing $<$ 0.5 to 3 mg Co/kg are deemed deficient because plants growing on them have insufficient cobalt ($<$ 0.08 to 0.1 mg Co/kg) to supply the dietary requirements of cattle and sheep (Sauchelli, 1969).

(1) Cobalt-deficient soils: Ruminant deficiencies of cobalt have been noted on very different soils in Australia, New Zealand, Great Britain, Norway, Canada, Kenya, the

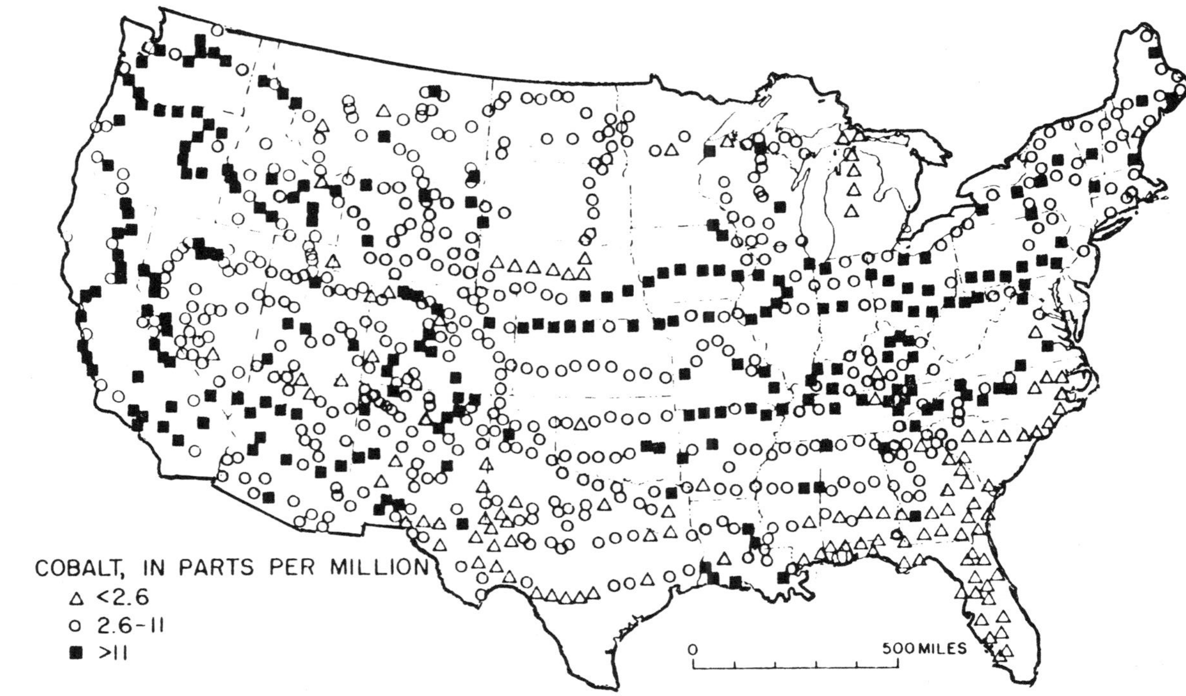

Figure IV-2. Abundance of Cobalt in Soils and Other Surficial Materials of the Conterminous United States (Shacklette, 1971).

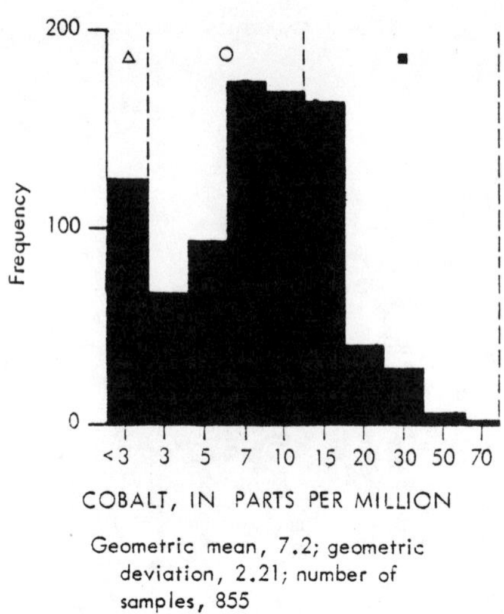

Figure IV-3. Histogram Showing Frequency Classes of Analytical Values for Cobalt in Samples of Surficial Materials and the Symbols Used to Indicate These Classes on Figure IV-2 (Shacklette, 1971).

United States, etc. The largest areas of deficient soils are from sandstones, especially if calcareous. Deficiencies are also likely on soils developed from limestone but not residual soils derived from limestones. Loess soils are often markedly deficient. On the other hand, alluvial deposits and soils from shales are seldom deficient in cobalt (Mitchell, 1955). As stated above, soils containing < 3 mg Co/kg are associated with cobalt deficiency diseases in ruminants. Cobalt-deficient soils include the humus groundwater podzols of the southeastern United States and the podzols, brown podzolic soils, and humus groundwater podzols in the northeastern section of the United States. Podzols are generally coarse-textured soils which develop under coniferous or mixed forest or under heath vegetation in a temperate to cold, moist climate, with low amounts of cobalt (Kubota, 1965). They are rather thoroughly decomposed aluminosilicates with formation of free iron and aluminum hydroxides and silica. (Iron and aluminum are carried from the A horizon into the deeper soil horizons and often form hardpan (Vinogradov, 1959).) Development of podzol morphology lowers the small amounts of cobalt inherited by sandy soils from their parent material. Some sandy podzols, however, do permit the growth of plants with sufficient cobalt for ruminants. Because of leaching, presumably during formation of podzol morphology, humus groundwater podzols appear to be nearly stripped of reactive forms of cobalt (Kubota, 1965).

Groundwater podzols are soils low in clay that are subjected to a fluctuating water table. When the water level is high, iron and manganese are reduced to soluble forms and much is leached away. In the absence of clay to adsorb it, cobalt is mostly leached away (Hodgson, 1970).

Kubota (1972), in a review of cobalt-deficient soils, described the sandy soils of the eastern U.S. as "Aquods" (Aeric Haplaquod). As the texture of the A_p horizon of soils varies from sand to fine sandy loam, then to loam and silt loam, and finally to silty clay loam or clay, the cobalt concentration increases with increasing iron content. The strongly weathered acid Aquods tend to develop more in coarse-textured soil parent materials than in fine-textured deposits. Table IV-12 shows that pronounced leaching of cobalt has occurred to beyond the rooting depths of common plants in an Aquod soil.

TABLE IV-12

COBALT CONCENTRATIONS IN HORIZONS OF AN AERIC HAPLAQUOD (HIGHLY WEATHERED SANDY SOIL) OF NEW YORK STATE (Kubota, 1972)

Depth, cm	Horizon	Cobalt, mg/kg
0 to 10	A_1	1.1
10 to 30	A_2	0.9
30 to 36	Bh	1.8
36 to 46	Birm	2.3
46 to 61	Bir	4.0
61 to 102	B_3	4.0
102 to 147	C_1	2.8
147 to 165	C_2	3.1

Soil horizon sampling is not necessary in the sampling of a soil like that of Nevada. A Nevada aquic xerofluvent soil formed in recent alluvium showed the following little-varying cobalt distribution by horizon: A_p, 8.5 mg/kg; C_1, 10.6 mg/kg; C_2, 10.6 mg/kg; C_3, 10.4 mg/kg; C_4, 9.2 mg/kg (Kubota, 1965).

In the sandy coastal plains deposits, the principal kinds of soils are those with distinct horizonation (Haplaquods) containing 0.39 mg Co/kg and those without distinct horizonation (Quartzipsamments) with 1.71 mg Co/kg. In the sandy glacial drift deposits of the Northeast,* the haplorthods are soils with distinct horizonation due to soil formation. These contain 2.13 mg Co/kg. The Udipsamments, soils without distinct horizonation, contain 5.97 mg Co/kg (Kubota, 1965).

Table IV-13 shows the cobalt contents of the A_1 and A_2 horizons of various soil groups and the dithionite extractability of the cobalt. Table IV-14 shows the cobalt concentrations and dithionite extractability of cobalt in the A and B horizons of deficient soils.

The leaves of swamp black gum plants are cobalt concentrators (up to 900 mg/kg). Yet Kubota (1965) found that they did not significantly affect the cobalt distribution and dithionite extractability in humus groundwater podzols (see Table IV-15).

Kubota (1968) compiled information from all published and unpublished data on cobalt developed and available at the U.S. Plant, Soil, and Nutrition Laboratory: 1,811 legume samples and 1,149 grass samples representing 366 counties and 42 states. All of the cobalt analyses were by the *o*-nitrocresol method. The median cobalt concentration in the grasses was 0.04 mg/kg. Grass concentrations do not clearly reflect differences in soil cobalt concentrations; legumes more clearly show these differences. The soils of cobalt adequate areas produce legumes,

---

* The poorly developed (young) soils and interrupted drainage characterizing glaciated areas result in peat bogs that may be deficient in Mn, Cu, Mo, and I, as well as cobalt. The lakes that were scoured out by the glaciers and impounded by glacial debris gradually filled with clays, marl, and finally, peat (Cannon, 1970).

TABLE IV-13

COBALT IN A_1 AND A_2 HORIZONS OF SOUTHEASTERN SOILS AND ITS DITHIONITE EXTRACTABILITY
(Kubota, 1965)

A_1 and A_2 Horizons of Soil Groups	Cobalt	
	Total, mg/kg	Dithionite Extractable/Total
Podzols and groundwater podzols	0.63	15%
Low-humic and humic gley[a] soils	1.86	11%
Well-drained, red-yellow podzolic and gray-brown podzolic soils	6.95	46%

a/ Gleying, a phenomenon that develops under excess moisture, is one of the effects noted in soil profiles that results in mobilization of cobalt and many other trace elements. Gleying produces a blue-gray soil horizon that is compact, sticky, and often structureless (Mitchell, 1955).

TABLE IV-14

COBALT IN THE A AND B HORIZONS OF DEFICIENT SOILS AND ITS DITHIONITE EXTRACTABILITY
(Kubota, 1965)

	Horizon	Total Co (mg/kg)	% Extractable by Dithionite
		Northeastern U.S.	
Podzols and brown podzolic soils	A	0.50	14
	B_2	1.50	15
Humus-iron groundwater podzols	A	1.09	14
	B_2	2.66	8
Low-humic gley soils	A	3.07	13
	B	3.29	19
		Southeastern U.S.	
Humus groundwater podzols	A	0.37	16
	B	0.38	13
Low-humic and humic gley soils	A	0.91	8
	B	1.07	5

TABLE IV-15

COBALT DISTRIBUTION IN TWO HUMUS GROUNDWATER PODZOLS
(Kubota, 1965)

Horizon	Depth (in.)	Cobalt Total (mg/kg)	Cobalt Dithionite Extractability (mg/kg)
		Under Black Gum Plants	
A_1	0-3	0.32	0.05
A_2	3-6	0.18	0.02
B_{hm}	6-12	0.32	0.02
B_h	12-18	0.38	0.01
B_2	18-26	0.34	0.01
		Not Under Black Gum Plants	
A	0-10	0.28	0.15
B_h	10-17	0.28	0.06
C	17-28	0.25	0.06

$\leq$ 90% of which contain $\geq$ 0.07 mg Co/kg. Moderately low-cobalt soils produce legumes, $\sim$ 75% of which contain $\geq$ 0.07 mg Co/kg. But only $\sim$ 20% of legumes grown on low-cobalt soils contain $\leq$ 0.07 mg Co/kg.

Western U.S. soils and the north central U.S. soils generally produce legumes whose median cobalt concentration is 0.17 mg/kg. Legumes grown on soils of the Piedmont and Upper Coastal Plain contain a median concentration of 0.19 mg Co/kg. Also, soils of the Ridge and Valley region of the eastern United States produce legumes of median cobalt concentration 0.14 mg/kg.

Legumes grown in moderately low-cobalt areas include those of the Central Loess and Drift Plain of Iowa, southern Minnesota, and adjacent states with a median concentration 0.08 to 0.09 mg Co/kg (no cobalt deficiency in cattle), those of northern Michigan and Door County, Wisconsin, with 0.09 mg Co/kg (cobalt deficiency reported as "Grand Traverse" disease), and those of New England and northern New York with 0.11 mg Co/kg. In the latter areas, most of the soils producing legumes with > 0.1 mg Co/kg are from valleys and broad glaciated drift plains. The spodosols of loam and silt loam contain 8 to 12 mg Co/kg. The orthods of sand to sandy-loam textures have 5 to 8 mg Co/kg.

But there are small areas of sandy soils where even legumes do not have enough cobalt. Parts of New York and New Hampshire have reported a cobalt-deficiency problem. In northern New York state, the problem occurs in broad alluvial fans and ridges that interfinger with heavy clay soils (Kubota, 1968).

The New England areas of low-cobalt soils include the Merrimac River-Saco River Drainage Basin and Cape Cod.* The median 0.05 mg Co/kg in legumes grown on these soils is scarcely more than that found in grasses. The early New Hampshire settlers called the cobalt-deficiency problem in their ruminants "Chocuruas Curse" and "Albany Ail." They fed balls of Saco clay to their animals to alleviate the problem. On necks of land projecting into bays, Massachusetts "neck ail" prevailed (Kubota, 1968). In cobalt-deficient soils (haplorthods) based on White Mountain granite, the total soil concentration is 2.5 to 7.2 mg Co/kg and the red clover cobalt concentration is 0.04 to 0.16 mg/kg. Soils from other rocks contain 5.0 to 11.9 mg Co/kg; the red clover, 0.07 to 0.15 mg/kg (Kubota, 1972).

The other major low-cobalt area is the Lower Atlantic Coastal Plain that extends from North Carolina to Florida. The cobalt deficiency occurs mostly in seasonally wet lowlands. The median concentration of cobalt in the legumes is 0.05 mg/kg; in the grasses, 0.06 mg/kg; and in the black gum leaves, < 5 mg/kg. These low-cobalt soils are the strongly acid Aquods of sand and loamy sand textures. The associated Aquults have more cobalt. The Haplaquods contain 1 to 2 mg/kg total cobalt, only small amounts of which are extractable. There are also small areas of clay and silty clay soils, but they are fine-textured and poorly drained. They do have more cobalt than the Haplaquods and produce plants with higher cobalt concentrations (Kubota, 1968).

In the Carolina coastal plain, less than one-third of the vegetation meets the minimum requirements of 0.04 to

---

* The glacial drift plain in granitic rock materials of the White Mountains extends from the Saco River in southern Maine to the Merrimac River in New Hampshire. A smaller area in southeast Massachusetts includes all of Cape Cod.

0.07 mg Co/kg dry weight for ruminants (Cannon, 1970). In Florida, not only are ruminants afflicted by "salt-sickness," a combined deficiency of iron, copper, and cobalt, but children suffer from anemia due to these deficiencies (Harris, 1967; cited by Cannon, 1970).

Kubota and co-workers have published papers on the relation of the cobalt concentration in soils to its concentration in swamp black gum (*Nyssa sylvatica* var. *biflora*) since an average 5 mg Co/kg in black gum leaves may serve as an indicator of cobalt deficiency in forage. Kubota and Lazar (1958) analyzed the leaves of black gum and broomsedge growing on the soils of the mid-humid Gulf and Atlantic Coastal Plains states wherein the cobalt content increases with increasing clay content of the soil. Kubota et al. (1960) studied black gum uptake again to determine possible cobalt deficiences in southern Arkansas and northern Louisiana. The results of these two studies are presented in Table VII-15 in Chapter VII. As would be expected, the lowest cobalt concentrations in black gum leaves were found in plants growing on the coastal sands of the Gulf and Atlantic Coastal Plains states and the excessively drained Regosols of southern Arkansas and northern Louisiana.

(2) *Soils contaminated by human activities*: The average soil cobalt concentrations in a 30-km by 26-km rectangle enclosing the City of Grand Rapids, Michigan, were 2.3 mg/kg in the residential area, 2.7 mg/kg in the agricultural area, 2.8 mg/kg in the industrial area, and 7.9 mg/kg in the airport area (atomic absorption analyses). The enrichment of cobalt in the airport area soils compared with its concentration in the residential area soils was 3.43. In the area studied, only about 6.3 MT of the daily total (91 MT) particulate air pollution comprised metals and metal oxides; 75% of the total air pollutants (910 MT) was due to motor vehicles; and 14%, to industrial emissions (Klein, 1972).

Minami and Araki (1975) extracted with 0.1 *N* hydrochloric acid the arable soils near two national Japanese roads that had "comparatively heavy" traffic. As can be seen from Table IV-16, cobalt was elevated in extracts of soil samples taken within 5 m of the roads. The correlation coefficients for

TABLE IV-16

COBALT IN SOILS NEAR WELL TRAVELED ROADS IN JAPAN, mg/liter
(Minami and Araki, 1975)

Distance from Road	Range	Mean
< 5 m	0.3 - 1.6	0.9
5 - 20 m	0.1 - 1.0	0.5
> 20 m	trace to 0.8	0.5
All samples	-	0.6

cobalt with the concentrations of other heavy metals [p (significance) < 0.01] in the < 5-m group of samples was Pb, 0.589; Cu, 0.671; Ni, 0.624; and Zn, 0.496.

Another study implicating air or highway traffic as a source of elevated soil cobalt was that of Bradford et al. (1971), who examined the soil saturation extracts of 68 California soil samples representing 30 soil series. Cobalt was found in only two soils at < 0.01 and 0.14 mg/liter. Saturation extracts of Lassen adobe clay at pH ≥ 3.9 contained 0.015 to 0.290 mg Co/liter (increasing concentration with decreasing pH). Freeway dust sampled near the Los Angeles International Airport gave saturation extracts with 2.0 mg Co/liter at pH 5.1 or 1.1 mg Co/liter at pH 5.9. Connor et al. (1971), however, did not detect any difference in cobalt concentrations in a subsurface Missouri soil under a cedar tree within ~ 15 m of a hard road surface or beyond (spectrographic analysis, cobalt detection limit 5 mg/kg).

Rahn and Harrison (1976) had reported an average 8.00 mg Co/kg (5.6 to 12.6 mg/kg) in Chicago street dusts, values quite comparable to those for Lancaster, England, a small town having no metal-emitting industries. Lancaster urban street dusts contained an average 9.6 mg Co/kg (6.1 to 14.4 mg/kg). The average for rural roads was a little lower, 7.1 mg Co/kg (Harrison, 1979) (see Table IV-17).

Hopke et al. (1977) found a strong relationship of cobalt in Urbana street dusts with the automobile, but concentrations were not reported.

TABLE IV-17

COBALT IN DUST SAMPLES, LANCASTER, ENGLAND, mg/kg
(Harrison, 1979)

	Mean	Range
Car Parks		
Total	10.0	6.6 - 12.0
Extractable[a/]	4.7	2.0 - 7.9
Urban Streets		
Total	9.1	6.1 - 14.4
Extractable	3.6	-
Households Distant from Major Roads		
Total	8.5	6.4 - 12
Extractable	3.3	2.0 - 5.7
Rural Roads		
Total	7.1	5.0 - 8.4
Extractable	3.4	2.5 - 3.8

a/ Extracted with 0.07 N hydrochloric acid.

There has been little published information on environmental pollution by geothermal power plants that rely on liquid-dominated systems. One report was found (Crockett, 1976) on background cobalt concentrations of 8.2 to 10.0 mg/kg (average 8.9) (analysis by neutron activation) in the soil at the San Diego Gas and Electric Company/U.S. ERDA experimental geothermal power plant just before it started operating. All fluids were to be reinjected except the noncondensible gases. Information from the subsequent semiannual monitoring activities has not been found.

Soils are often strongly enriched in nickel and cobalt from nearby ore deposits. Although the Ni-Co ratio in average rocks and average soils is 4.0, in soils near or on nickel ores the Ni-Co ratio is 20 (Vinogradov, 1959). Anomalous concentrations of cobalt in soils and stream sediments have been used to locate cobalt ore deposits such as those in Uganda and northern Rhodesia. Not only will analysis of residual soils help locate bedrock deposits of cobalt ores, cobalt may also have migrated into and may have been precipitated in the transported overburden. Hawkes and Webb (1962) have summarized the geochemical results for cobalt by numerous investigators.

Cater et al. (1975; cited by Bennett, 1977) found as much as 150 mg/kg (average 14 mg/kg) cobalt in stream sediments in the vicinity of the Blackbird mine in Lemhi County, Idaho. In Bennett's study of soils and sediments in the Blackbird mine vicinity, sediment values > 105 mg Co/kg* and soil values > 84 mg Co/kg were deemed anomalous. The average cobalt concentrations in stream sediments according to rock types ranged from a minimum of 9.46 (quartz) to 45.88 mg Co/kg (basic rocks). Soils according to rock types contained an average 13.13 to 84.85 mg Co/kg. Of 960 soil samples, only 8 had cobalt concentrations > 100 mg/kg. Most values that were > 30 mg Co/kg were in the Blackbird mine area itself. Cobalt correlated best with copper in soils (0.74) and in sediments (0.79).

Lakanen (1965) analyzed soils and stream sediments that had been sampled in the vicinity of the Outokumpu copper mine in Finland. Values of 2 to 32 mg Co/kg found were not much different from cobalt concentrations in other Finnish soils. (The concentration of cobalt in ammonium acetate-acetic acid extracts of the soil ranged from 0.06 to 0.73 mg/liter of soil.) Cobalt concentrations decreased with depth except in a coarser fine sand soil in which there was 7 mg Co/kg at 0 to 20 cm depth and 13 to 17 mg Co/kg at 20 to 80 cm depth.

Not only do mining and milling activities or merely undisturbed cobalt mineralization enrich cobalt concentrations in local soils and stream sediments, but also smelting activities are known to increase cobalt concentrations in the smelter vicinity. A 1972 EPA report (Environmental Protection Agency, 1972) on elevated blood lead levels in children in the vicinity of the ASARCO Inc. lead smelter in East Helena, Montana, mentioned that, besides lead and zinc, other heavy metals including cobalt "appear to be highly concentrated in soils from near the smelter" (spectrographic analysis), but that "further study would be required to verify this, however."

---

* Only 14 of 169 stream sediment samples had > 100 mg Co/kg.

The major North American cobalt-containing deposits currently being worked are the copper-nickel sulfide ores in the Sudbury area in Ontario. Costecu and Hutchinson (1972) have reported on the contamination of the area by cobalt and other associated minor metals. (See also Hutchinson and Whitby, 1974a, 1974b.) The Sudbury area has three major copper-nickel sulfide ore smelters: Copper Cliff is 3.2 km west of Sudbury; Coniston is 13 km east of Sudbury; and Falconbridge, 7.2 km northeast. In addition, there are 13 mines, 1 open pit, 6 concentrators, 3 iron ore recovery plants, and 1 copper refinery. Fifty kilometers from the Sudbury area, soil concentrations of cobalt at the surface (surface concentration underlined) to 10 to 13 cm in depth were 17 to <u>19</u> mg/kg. At 19 km, 31 to <u>48</u>; 13 km, 6 to <u>56</u>; 10 km, 24 to <u>33</u>. Within distances 0.8 to 1.3 km from the area, surface concentration ranged irregularly from 42 to 154 mg/kg. The cobalt concentration generally decreased irregularly with depth. The highest concentration at the 10 to 13 cm depth was 82 mg/kg (127 mg/kg surface) closest to the smelter area.

The concentrations of cobalt in the soil water extracts are as high as 4.6 mg/liter (see Table IV-18). Other data on soil contamination by cobalt emissions are discussed in section V.C.2. on fossil fuels burning.

TABLE IV-18

COBALT IN SOIL WATER EXTRACTS FROM THE CONISTON SMELTER VICINITY
(pH 2.92 to 3.86 when observed)
(Costecu and Hutchinson, 1972)

Distance from Coniston Smelter, km	At Surface mg/liter	mg/liter at Depths of 5-7.6 cm	10-13 cm
0.5	4.6	0.2	ND[a/]
0.95	3.4	1.5	0.4
1.0	2.2	0.5	0.4
1.2	4.4	1.0	0.2
2.4	0.2	0.1	0.1
4.6-31.0	ND	ND	ND

a/ ND = not detected.

b. Water

(1) Drinking water and effects of finishing treatments: The few studies of cobalt concentrations in drinking water reveal that values around ≤ 2 μg Co/liter are rather common. However, values of up to 107 μg Co/liter have been reported.

Durfor and Becker (1962) in their survey of the public water supplies of the 100 largest cities in the United States did not report cobalt concentrations, but stated that most public water supplies contained cobalt. (The detection limit for cobalt by spectrographic analysis was 10 mg/kg in the residue from the acidulated dissolved solids of the water.)

McCabe et al. (1970) described the 1969 Public Health Service survey of community water supply systems--969 distribution systems representing nine geographic locations in the United States. Craun and McCable (1975) reported that the average cobalt concentration found in this study was 2.2 μg/liter. More than 1 μg/liter was found in 62% of the samples. According to Taylor (1971), 0.0 to 19 μg/liter was found in this National Community Water Supply Study (2,500 samples).

Greathouse and Craun (1979) found cobalt in only 9.81% of 3,834 tapwater (grab) samples at concentrations above the detection limit of 0.1 μg/liter (analyses by proton-induced x-ray emission). The tapwaters were from 35 geographical areas representative of the United States. When found, the average cobalt concentration was 21.02 μg/liter (2.6 to 106.9 μg per liter).

Strain et al. (1975) analyzed tapwater samples from 50 homes in Albany, New York; Cleveland, Ohio; Rochester, New York; and Wilkesbarre-Scranton, Pennsylvania by atomic absorption spectrometry. The samples, which had been collected from taps that had not been used overnight, contained ≤ 3.3 μg Co/liter.

McCabe and Vaughn (1969) reported the following frequency distribution for cobalt in 550 Chicago, Illinois, drinking water samples collected at the industrial plant faucets: 0 to 1 μg/liter, 54%; 2 to 3 μg/liter, 29%; 4 to 5 μg/liter, 11%; 6 to 7 μg/liter, 4.7%; and 8 to 9 μg/liter, 0.05%. Kopp (1970) reported that cobalt had been found in only 0.5% of 380 U.S. samples (22 and 29 μg/liter at two sites).

Schroeder et al. (1967) reported 0.90 to 5.28 μg Co/liter in Vermont drinking waters. A spring water at its source contained 2.50 μg Co/liter; from a copper pipe, however, the tapwater contained 5.28 μg Co/liter; from an iron pipe, 5.04 μg Co/liter.

Bradford (1971) found < 0.5 μg Co/liter in the well waters from the San Joaquin Valley, California, but found up to 5 μg Co/liter in spring and well waters surrounding the Salton Sea. Some of the drinking waters analyzed by McCabe et al. (1970) may have been groundwaters; but cobalt concentrations were not associated with specific water types in the review paper by Craun and McCabe (1975) that reported the average cobalt concentration.

It is interesting to note that Borneff (1974) found 10 μg/liter in the Duisburg drinking water in the Federal Republic of Germany. (Duisburg smelted high-cobalt pyrites for many years.) Elevated lung tissue concentrations of cobalt have also been found in Duisburg residents (Persigehl et al., 1976).

Agrawal et al. (1978) found 2.00 mg Co/liter in the drinking water of the four wells supplying Cambay City, India (population 75,000). The concentrations of lead (0.40 mg/liter) and cadmium (0.05 mg/liter) exceeded the World Health Organization (WHO) drinking water limits by 8 and 5 times, respectively. The authors state that there was a high incidence of "mental deformation and defective speech among the inhabitants."

Neither the WHO water quality limits in international or European standards nor the National Academy of Sciences (1977) has recommended any standard for cobalt in drinking water since "the maximum no-observed-adverse-health-effect concentration is more than an order of magnitude greater than the occurrence in natural or drinking water supply."

Nilsson (1971) studied the effect of water treatment on an artificial solution containing 12 mg Co/liter in tapwater. First, phosphate (6 mg P/liter) was added and then 100 mg aluminum sulfate per liter was added at pH 6.5 to 7.0. The cobalt concentration rose to 14 mg/liter. Treating the phosphate solution with calcium hydroxide at pH 9.5 (instead of aluminum sulfate) reduced the cobalt concentration to 0.1 mg/liter.

Boström and Wester (1967; cited by Andelman and Shapiro, 1973, and by Andelman, 1974) reported that the ratio of the cobalt concentration in tapwater to its concentration in raw water (0.1 μg/liter) was 0.7 at Malmo; 1 at Stockholm; and 3.0 at Goteburg, indicating that the cobalt concentration is not seriously altered by purification measures at these Swedish water supply plants. Andelman and Shapiro (1973) cited another study showing little effect on cobalt concentrations due to finished drinking water treatment. Zacharious, Brent, and Andelman in 1971 (cited by Andelman and Shapiro, 1973) had reported that cobalt concentrations in the raw Thames River water, England, in the winter of 1971 were little different from those in the finished water. Their data are presented in Table IV-19.

TABLE IV-19

THE EFFECT OF FINISHED DRINKING WATER TREATMENTS ON THE COBALT CONCENTRATIONS IN WATER

(Zacharious et al., 1971; cited by Andelman and Shapiro, 1973)

Water Type	Cobalt Concentration (μg/liter)	
	Midwinter 1971	Late Winter 1971
Raw Thames River water		
Total	6.1	5.1
Filtrate	5.2	4.6
Post-sedimentation		
Total	4.4	5.2
Filtrate	-	5.1
Post-filtration		
Total	5.2	4.6
Filtrate	4.4	4.4
Post-chlorination		
Total	-	5.2
Filtrate	4.8	4.4

Jackson (1972, 1973) reported that the U.S. Geological Survey had surveyed public water supplies in North Carolina serving ≥ 500 customers from the early 1960's to the early 1970's. In the northern Piedmont, cobalt was not found in 41 samples of raw or finished water sampled. For example, on July 9, 1970, raw Neuse River water contained < 2 μg Co/liter and the finished water from it contained < 3 μg Co/liter. However, on March 9, 1971, a sample containing 10 μg Co/liter was withdrawn from a well in Fuquay-Varina, Wake County, cased to 40 ft (12 m) with a water level at 30 ft (9 m). There was no information in the report that would indicate why this well water contained cobalt while similar wells in the vicinity did not. Cobalt was found more frequently in the southern Piedmont waters sampled: up to 54 μg Co/liter in raw waters and up to 8 μg Co/liter in finished waters. Table IV-20 presents the cobalt data from the 1973 study. The variations in cobalt concentrations in these waters are not often explicable on the basis of underlying rocks, pH, industrial users, or finished water treatment.

(2) Freshwater and sediments

(a) Freshwater bodies and precipitation

(i) Total cobalt: Total cobalt concentrations in natural freshwaters of the United States fall below the detection limit of 1 μg/liter with spectometric techniques. The level of 1 to 10 μg/liter of Co represents the second largest group and is found in streams close to populated areas. The next most frequently detected concentration range is 11 to 50 μg/liter, which is associated with streams running through mining districts and areas of extensive agricultural land usage and in some cases compounded by urban inputs. Highest values were 6,500 μg Co/liter in the Little St. Francis River near the Fredericktown, Missouri, cobalt mine and 4,500 μg/liter in Mineral Creek near Big Dome, 3 miles south of Ray, Arizona, which is near the site of a copper smelter and copper-molydenum mines. Table IV-21 is a compilation of literature values for cobalt in U.S. and Canadian surface waters.

In Quillayute, Washington, up to 0.045 μg Co/liter was found in rainwater by Rancitelli et al. (1970; cited by Billings, 1974).

TABLE IV-20

COBALT IN SOUTHERN PIEDMONT, NORTH CAROLINA, RAW AND FINISHED WATERS
(Jackson, 1973)

Raw Water Source	Date Sampled	Cobalt Concentration in Raw Water (μg/liter)	Finished Water Treatment Plant Location	Water Treatment Methods	Cobalt Concentration in Finished Water (μg/liter)	Principal Industrial Users
Lake Hickory, impoundment at the Catawba River	3/23/72	3	Longview, Catawba County (cobalt not found at Hickory)	Prechlorination, alum coagulation, sedimentation, microfloc filtration, addition of phosphates for corrosion control (soft water), lime pH adjustment, postchlorination when needed	1	Textile mills and a soft drink bottling company
Maiden Creek		3	Maiden, Catawba County	Included carbon adsorption, rapid sand filtration, caustic soda for pH control, and coagulation	0	Textile mills
Catawba River	4/12/72	7	Belmont, Gaston County[a]	$KMnO_4$, prechlorination, coagulation with alum and soda ash, carbon, rapid sand filtration, phosphates, soda ash, postchlorination if needed, fluoridation	7	Piedmont Processing Company, Belmont Converting Company, Dixie Yard, Stowe Thread
Catawba River	4/12/72	3	Mount Holly		3	Textile mills and dyers
Hoyles Creek		7	Stanley, Gaston County[a]		7	Textile mills and dyers
South Yadkin River	3/22/72	8	Statesville, Iredell County[b]		1	Textile mills, metalworking plants, Carnation Company
Long Creek plus City Lake	4/11/72	2	Bessemer City, Gaston County[a]	Similar to that at Belmont. No $KMnO_4$	2	Textile mill, dyeing and finishing plant, Gaston Industries, Central Industries
Indian Creek	4/11/72	6	Cherryville, Gaston County[a]	Carbon or $KMnO_4$, etc.	7	
Rankin Lake (pumped from South Fork of Catawba River and Long Creek)		2	Gastonia, Gaston County[a]		3	Textile and dyeing mills
South Fork of Catawba River	4/10/72	4	Lincolnton, Lincoln County		4	Textile mills

TABLE IV-20 (concluded)

Raw Water Source	Date Sampled	Cobalt Concentration in Raw Water (μg/liter)	Finished Water Treatment Plant Location	Water Treatment Methods	Cobalt Concentration in Finished Water (μg/liter)	Principal Industrial Users
Catawba River		15	Charlotte-Mecklenberg Utility Department, Mecklenberg County c/		8	Barnhardt Manufacturing, Celanese, Lance, Inc., poultry and dairy plants
Kannapolis Lake, impoundment of Irish Buffalo Creek	5/9/72	54	Kannapolis, Rowen County d/		0	No information given
Yadkin River, headwaters of Badin Lake	6/7/72	0	Albemarle, Stanly County e/	Treatment included anthracite filtration	6	
Badin Lake		6	Badin		0	

a/ Volcanic slate rocks underlie Gaston County and some other counties in the vicinity.
b/ County underlain by mica-schist, gneiss, granite.
c/ Diorite and granite are predominant rocks underlying the county.
d/ County underlain by granite. Soft water.
e/ County underlain by tuffaceous argillite. Hard water with moderate amounts of iron.

TABLE IV-21

THE OCCURRENCE OF COBALT IN SURFACE FRESHWATERS

Location	Cobalt Concentration (μg/liter) Mean	Range	No. of Samples	Total No. of Samples	Date of Sampling	Method of Analysis	Detection Limit (μg/liter)	Remarks	Authors
United States									
Nationwide		< D.L. to 48		1,577	10/1/62-9/30/67	Spectrophotometry	20	Suspended trace metals were not measured. Number of positive occurrences=44. D.F.=2.8%. Weekly samples. D.L. based on a total dissolved solids of 400 mg/liter.	Kopp and Kroner (1970)
Nationwide		< D.L.	~454	> 720	10/1-15/70	Flame spectrophotometry	1	D.F.=37%. Samples were filtered.	Durum et al. (1971)
		1 to 5	217						
		> 5	67						
Alabama									
Statewide		< D.L.	7	18		Flame spectrophotometry	1		Durum et al. (1971)
		1 to 5	1						
		6 to 10	2						
Tallapoosa River near Montgomery	45		1						
Alabama River at Claiborne	22		1						
Black Warrior River near Tuscaloosa	22		1						
Tombigbee River near Carlton	12		1						
Mobile River at Mobile	12		1						
Big Creek Reservoir, Mobile	17		1						
Tennessee River at Whitesburg	12		1						
Tennessee River at Smithsonia	25		1						
Mobile River at Mt. Vernon Landing	0.89		1	4	1958 and 1959	Emission spectrographic			Durum (1960)

TABLE IV-21 (continued)

Location	Cobalt Concentration (μg/liter) Mean	Range	No. of Samples	Total No. of Samples	Date of Sampling	Method of Analysis	Detection Limit (μg/liter)	Remarks	Authors
Alaska									
Statewide		< D.L.	8	9		Flame spectrophotometry	1		Durum et al. (1971)
		1 to 5	1						
Yukon River at Mountain Village		N.D. to 0.X		2	1958 and 1959	Emission spectrographic			Durum (1960)
Arizona									
Statewide		< D.L.	8	11		Flame spectrophotometry	1		Durum et al. (1971)
		1 to 5	2						
Mineral Creek near Ray	4,500		1						
Colorado River at Yuma		N.D.		2	1958 and 1959	Emission spectrographic			Durum (1960)
Arkansas									
Statewide		< D.L.	3	13		Flame spectrophotometry	1		Durum et al. (1971)
		1 to 5	2						
		6 to 10	3						
White River near Goshen	25		1						
Arkansas River below Ft. Smith	25		1						
Lake Winona on Alum Fork	15		1						
Bayou Meto near Lonoke	20		1						
Hurricane Creek near Sheridan	55		1						
California									
Streams and rivers	-	< D.L.	28	32		Flame spectrophotometry	1		Durum et al. (1971)
		1 to 5	4						
Surface waters		< D.L.	60	65		Direct-reading emission spectrophotometry			Silvey (1967)
		0.74-2.9	4						
	15	-	1						

TABLE IV-21 (continued)

Location	Cobalt Concentration (μg/liter) Mean	Range	No. of Samples	Total No. of Samples	Date of Sampling	Method of Analysis	Detection Limit (μg/liter)	Remarks	Authors
California Lake Miller, Yosemite National Park	 0.9	 < D.L.	 169 1	 170	 1 week of September 1965	 Spectrophotometry	 < 0.3	 Paper does not say whether samples were filtered after collecting.	 Bradford et al. (1968)
Sacramento River at Sacramento		< 1.6 to < 3.3	2	4	1958 and 1959	Emission spectrographic			Durum (1960)
Colorado		< D.L. 1 to 5	18 1	19		Flame spectrophotometry	1		Durum et al. (1971)
Connecticut		< D.L. 1 to 5	11 14	25		Flame spectrophotometry	1		Durum et al. (1971)
Delaware		< D.L. 1 to 5	2 2	4		Flame spectrophotometry	1		Durum et al. (1971)
District of Columbia		1 to 5	1	1		Flame spectrophotometry	1		Durum et al. (1971)
Florida Statewide		 < D.L. 1 to 5	 10 7	 17		 Flame spectrophotometry	 1		 Durum et al. (1971)
Apalachicola River, Blountstown	< 0.75		1	4	1958 and 1959	Emission spectrographic			Durum (1960)
Georgia		< D.L. 1 to 5	9 8	17		Flame spectrophotometry	1		Durum et al. (1971)
Hawaii		< D.L.	8	8		Flame spectrophotometry	1		Durum et al. (1971)
Idaho Statewide South Fork Coeur d'Alene River near Smelterville (Kellogg)	 14	 < D.L.	 7 1	 8		 Flame spectrophotometry	 1		 Durum et al. (1971)

TABLE IV-21 (continued)

Location	Cobalt Concentration (μg/liter) Mean	Range	No. of Samples	Total No. of Samples	Date of Sampling	Method of Analysis	Detection Limit (μg/liter)	Remarks	Authors
Illinois		< D.L. 1 to 5	6 13	19		Flame spectrophotometry	1		Durum et al. (1971)
Indiana		< D.L. 1 to 5	9 12	21		Flame spectrophotometry	1		Durum et al. (1971)
Iowa		< D.L.	11	11		Flame spectrophotometry	1		Durum et al. (1971)
Kansas									
Statewide		< D.L.	12	12		Flame spectrophotometry	1		Durum et al. (1971)
Kansas River at Desoto	 8	< D.L.	8 1	9		Spectrophotometry	20		Kopp and Kroner (1970)
Kansas River	0.02	0 to 0.3				Atomic absorption		Highest Co concentration found during July, August, September 1967	Angino et al. (1969)
Kentucky									
Statewide		< D.L. 1 to 5 6 to 10	2 1 2	8		Flame spectrophotometry	1		Durum et al. (1971)
Ohio River at Louisville	10		1						
Salt River near Louisville	20		1						
Ohio River at Owensboro	16		1						
Louisiana									
Statewide		< D.L. 1 to 5 6 to 10	7 3 2	13		Flame spectrophotometry	1		Durum et al. (1971)
Atchafalaya River at Morgan City	22		1						

TABLE IV-21 (continued)

Location	Cobalt Concentration (μg/liter) Mean	Range	No. of Samples	Total No. of Samples	Date of Sampling	Method of Analysis	Detection Limit (μg/liter)	Remarks	Authors
Louisiana									
Mississippi River at Delta		< D.L.	9	10		Spectrophotometry	20		Kopp and Kroner (1970)
	36		1						
Mississippi River, Baton Rouge		N.D. to 5.8	1	3	1958 and 1959	Emission spectrographic			Durum (1960)
Atchafalaya River, Krotz Springs		N.D. to 5.2	2	4	1958 and 1959	Emission spectrographic			Durum (1960)
Maine		1 to 5	7	7		Flame spectrophotometry	1		Durum et al. (1971)
Maryland									
Statewide		< D.L.	11	13		Flame spectrophotometry	1		Durum et al. (1971)
		1 to 5	1						
North Branch Potomac River at Cumberland Fairgrounds	23		1						
Susquehanna River at Conowingo		N.D. to 0.X	1	3	1958 and 1959	Emission spectrographic			Durum (1960)
Massachusetts									
Statewide		< D.L.	3	14		Flame spectrophotometry	1		Durum et al. (1971)
		1 to 5	10						
Blackstone River at Northbridge	17		1						
Michigan									
Statewide		< D.L.	18	19		Flame spectrophotometry	1		Durum et al. (1971)
		1 to 5	1						
Detroit River at Detroit		< D.L.	8	9		Spectrophotometry	20		Kopp and Kroner (1970)
	21		1						
Minnesota		< D.L.	13	13		Flame spectrophotometry	1		Durum et al. (1971)

TABLE IV-21 (continued)

Location	Cobalt Concentration (μg/liter) Mean	Range	No. of Samples	Total No. of Samples	Date of Sampling	Method of Analysis	Detection Limit (μg/liter)	Remarks	Authors
Mississippi									
Statewide		< D.L. 1 to 5 6 to 10	3 2 2	10		Flame spectrophotometry	1		Durum et al. (1971)
Tallahala Creek at Ellisville	40		1						
Okatibbee Creek at Arundel	12		1						
Pearl River at Byram	40		1						
Missouri									
Statewide		< D.L. 1 to 5 6 to 10	3 1 2	13		Flame spectrophotometry	1		Durum et al. (1971)
Missouri River at Saint Joseph	18		1						
Missouri River at St. Louis	24		1						
Missouri River near St. Louis	15		1						
Meramec River at Paulina Hills	20		1						
McDaniels Lake at Springfield	14		1						
James River near Boaz	14		1						
Turkey Creek near Joplin	30		1						
Missouri River at Kansas City	 8	< D.L.	9 1	10		Spectrophotometry	20		Kopp and Kroner (1970)
Montana		< D.L.	8	8		Flame spectrophotometry	1		Durum et al. (1971)
Nebraska		< D.L.	10	10		Flame spectrophotometry	1		Durum et al. (1971)
Nevada		< D.L.	8	8		Flame spectrophotometry	1		Durum et al. (1971)
New Hampshire		< D.L. 1 to 5	1 3	4		Flame spectrophotometry	1		Durum et al. (1971)
New Jersey		< D.L. 1 to 5	10 8	18		Flame spectrophotometry	1		Durum et al. (1971)

TABLE IV-21 (continued)

Location	Cobalt Concentration (µg/liter) Mean	Range	No. of Samples	Total No. of Samples	Date of Sampling	Method of Analysis	Detection Limit (µg/liter)	Remarks	Authors
New Mexico									
Statewide		< D.L.	9	15		Flame spectrophotometry	1		Durum et al. (1971)
		1 to 5	6						
Animas River at Cedar Hill		< D.L.	19	20		Spectrophotometry	20		Kopp and Kroner (1970)
	10		1						
New York									
Statewide		< D.L.	9	34		Flame spectrophotometry	1		Durum et al. (1971)
		1 to 5	20						
		6 to 10	5						
12 tributary streams of Cayuga Lake				100	March to August 1970	Colorimetric (nitrosocresol method)			Kubota et al. (1974)
8 rural		≤ 0.50 to 0.130							
4 urban		0.13 to 0.20							
Hudson River at Green Island		< 1.1 to < 1.3	2	2	1958 and 1959	Emission spectrographic			Durum (1960)
North Carolina		< D.L.	17	22		Flame spectrophotometry	1		Durum et al. (1971)
		1 to 5	5						
North Dakota		< D.L.	7	7		Flame spectrophotometry	1		Durum et al. (1971)
Ohio									
Statewide		< D.L.	19	24		Flame spectrophotometry	1		Durum et al. (1971)
		1 to 5	5						
Cuyahoga River at Cleveland		< D.L.	21	23		Spectrophotometry	20		Kopp and Kroner (1970)
	20		1						
	46		1						
Oklahoma		< D.L.	11	12		Flame spectrophotometry	1		Durum et al. (1971)
		1 to 5	1						
Oregon									
Statewide		< D.L.	13	13		Flame spectrophotometry	1		Durum et al. (1971)
Columbia River below The Dalles Dam		< 1.1	1	2	1958 and 1959	Emission spectrographic			Durum (1960)

TABLE IV-21 (continued)

Location	Cobalt Concentration (μg/liter) Mean	Range	No. of Samples	Total No. of Samples	Date of Sampling	Method of Analysis	Detection Limit (μg/liter)	Remarks	Authors
Pennsylvania									
Statewide		< D.L.	15	43		Flame	1		Durum et al.
		1 to 5	20			spectrophotometry			(1971)
		6 to 10	3						
Lackawanna River at Old Forge	67		1						
West Branch Susquehanna River at Renovo	63		1						
Lake Altoona near Altoona	37		1						
Kiskiminetas River at Leechburg	110		1						
West Branch Susquehanna River at Lewisburg	11		1						
Allegheny River at		< D.L.	29	37		Spectrophotometry	20		Kopp and
Pittsburgh	18	3 to 48	8						Kroner
Monongahela River at		< D.L.	21						(1970)
Pittsburgh	19	7 to 34	15	36					
Delaware River at		< D.L.	35	36					
Martins Creek	6		1						
Delaware River at		< D.L.	32	33					
Philadelphia	13		1						
Rhode Island		1 to 5	4	4		Flame spectrophotometry	1		Durum et al. (1971)
South Carolina		< D.L.	11	16		Flame	1		Durum et al.
		1 to 5	5			spectrophotometry			(1971)
South Dakota		< D.L.	7	7		Flame spectrophotometry	1		Durum et al. (1971)

TABLE IV-21 (continued)

Location	Cobalt Concentration (μg/liter) Mean	Range	No. of Samples	Total No. of Samples	Date of Sampling	Method of Analysis	Detection Limit (μg/liter)	Remarks	Authors
Tennessee Statewide		< D.L.	4	12		Flame spectrophotometry	1		Durum et al. (1971)
		1 to 5	1						
		6 to 10	2						
Little River above Townsend	35		1						
Clinch River near Lawnville	20		1						
Tennessee River at Chattanooga	34		1						
Buffalo River near Flatwoods	16		1						
Mississippi River near Memphis	25		1						
Texas		< D.L.	5	30		Flame spectrophotometry	1		Durum et al. (1971)
		1 to 5	25						
Utah Statewide		< D.L.	11	11		Flame spectrophotometry	1		Durum et al. (1971)
Green River at Dutch John		< D.L.	8	9		Spectrophotometry	20		Kopp and Kroner (1970)
	11		1						
Vermont		1 to 5	3	3		Flame spectrophotometry	1		Durum et al. (1971)
Virginia		< D.L.	9	11		Flame spectrophotometry	1		Durum et al. (1971)
		1 to 5	2						
Washington		< D.L.	14	14		Flame spectrophotometry	1		Durum et al. (1971)
West Virginia Statewide		< D.L.	4	12		Flame spectrophotometry	1		Durum et al. (1971)
		1 to 5	8						
Ohio River at Huntington		< D.L.	19	20		Spectrophotometry	20		Kopp and Kroner (1970)
	36		1						
Wisconsin		< D.L.	16	16		Flame spectrophotometry	1		Durum et al. (1971)

TABLE IV-21 (concluded)

Location	Cobalt Concentration (μg/liter) Mean	Range	No. of Samples	Total No. of Samples	Date of Sampling	Method of Analysis	Detection Limit (μg/liter)	Remarks	Authors
Wyoming		< D.L.	9	9		Flame spectrophotometry	1		Durum et al. (1971)
Puerto Rico		< D.L. 1 to 5	9 1	10		Flame spectrophotometry	1		Durum et al. (1971)
Canada									
St. Lawrence River, Levis Quebec		< 1.1 to 4.0	3	7	1958 and 1959	Emission spectrographic			Durum (1960)
Fraser River at Mission City, British Columbia		< N.D. to 1.9	2	4	1958 and 1959	Emission spectrographic			Durum (1960)
Mackenzie River at Arctic Red River, Northwest Territory		N.D. to 5.0	1	3	1958 and 1959	Emission spectrographic			Durum (1960)
Nelson River, near Amery, Manitoba		N.D. to 5.1	1	6	1958 and 1959	Emission spectrographic			Durum (1960)
Churchill River, 8 mi. south of Churchill, Manitoba		N.D.		4	1958 and 1959	Emission spectrographic			Durum (1960)

Note: D.L. = Detection limit.
D.F. = Detection frequency.
N.D. = Not detected.

Bogen (1974) reported 1.8 μg Co/liter in rainwater in Heidelberg; Peirson et al. (1973) found 0.25 μg Co/liter in rainwater in Wraymires, England; and a value as high as 3.2 μg Co/liter was found in rainwater at a gas platform in the North Sea (Peirson et al., 1974). Schuytyser et al. (1978) reported 0.79 μg Co/liter in the soluble fraction of a Ghent, Belgium, rainwater sample collected in August 1976; they found 0.433 μg Co/liter in the insoluble fraction.

Miklishanskii et al. (1976) found 8.1 ± 2.2 ng/liter Co in Antarctic snow by neutron activation analysis. In the USSR, Miklishanskii et al. (1977) found 0.18 μg Co/liter in a fresh-fallen snow in Moscow, November 25, 1975. A higher concentration (0.27 μg Co/liter) found in the same snowfall in a spruce-birch forest was attributed to washing of dust from the vegetation. The old snow in Moscow on January 13, 1976, contained 1.4 μg Co/liter.

Hutchinson and Whitby (1974a, 1974b) found sharply diminishing soluble cobalt in rainwater collected from June 23 to July 21, 1970, along transects from the Coniston Smelter in Sudbury, Ontario. (The average amount of cobalt in the stack emissions of the Coniston Smelter for the last quarter of 1971 was 0.6 short tons/28 days.*) The amounts of cobalt collected in the rainwater (in $mg/m^2/28$ days) were 8.5 at 1.6 km from the smelter; 4.2, at 1.9 km; 0.5, at 7.4 km; 0.2, at 13.5 km; and 0.1, at 19.3 km. The pH of the rainwater generally increased (2.85 to 4.34) with distance from the smelter. The low pH was due to the large quantities of sulfur dioxide produced when smelting the nickel-copper sulfide ore.

(ii) Vitamin B_{12} (Cyanocobalamin): Gérôme (1971) reviewed literature data on the concentrations of vitamin B_{12} in natural freshwaters. Benoit had found 0.075 μg/liter at the surface of Linsley Pond (Connecticut) and 0.150 μg/liter at the bottom. Robin found 0.001 to 0.010 μg/liter at the surface of oligotrophic Quebec lakes and 0.001 to 0.070 μg/liter

* Similarly, cobalt emissions from the Copper Cliff Smelter were 3.4 MT/28 days; from the Iron Ore Recovery Plant, 0.04 MT/28 days. Total annual emissions at this rate from these three plants would have been 52.3 MT/year.

at the surface of more eutrophic lakes. Daisley found 0.001 to 0.015 μg/liter in Lake District (Lancashire, England) surface waters with the maximum concentrations being in autumn and winter. Gérôme found that the vitamin B_{12} concentrations in the mesotrophic Lake Geneva (Switzerland) in 1970 increased from 0 to 0.004 μg/liter at the surface to 0.006 to 0.016 μg/liter at the bottom (150 m). (A culture of Asterionella from Lake Geneva was found to grow more rapidly when 20 μg vitamin B_{12} per liter was added to its growing medium.) At Lake Nantua in November 1970, the concentration of B_{12} increased from 0.008 μg/liter at the surface to 0.020 μg/liter from 10 m to 40 m (the bottom).

Mills and Oglesby (1971) followed the fluctuations in the concentrations of vitamin B_{12} and soluble cobalt in the Cayuga Lake Drainage Basin, one of the Finger Lakes in New York State. The basin is 48% agricultural, 31% forest, 2% residential, and 19% other, including low industrial development. Eutrophication in the past 50 years had increased; and in recent periods prior to the report of Mills and Oglesby, the blue-green algae had become more predominant and abundant in the summer. The concentrations of particulate cobalt (lead, zinc, and cadmium) peaked in late March and early April during maximum spring runoff.* Soluble cobalt and cadmium, on the other hand, peaked during a period of intense rainfall in early July, and the concentration of cobalt in vitamin B_{12} reached a maximum in mid-July. The percent of total soluble cobalt represented by cobalt in vitamin B_{12} rose from ~ 0.5 to 0.6% in late June to 11.00% in early August. The data for soluble and B_{12} cobalt are presented in Table IV-22. These data were compared with data of Benoit (1957; cited by Mills and Oglesby, 1971) on Linsley Pond, where in August the B_{12} cobalt concentration (in cellular form) was 2 to 13% of the soluble cobalt concentration. In August, cobalt is converted to vitamin B_{12} more efficiently in the photic zone by the blue-green algae.

---

* The overall particulate concentration of cobalt in Cayuga Lake's tributaries was 0.11 μg/liter compared with 0.04 μg/liter soluble cobalt and 0.009 μg/liter soluble vitamin B_{12}. In Cayuga Lake, the soluble cobalt concentration was 0.02 μg/liter; the B_{12} concentration, 0.005 μg/liter.

TABLE IV-22

SOLUBLE COBALT AND VITAMIN B_{12} COBALT AT THREE DEPTHS AND TWO STATIONS IN THE EUPHOTIC ZONE OF CAYUGA LAKE
(Mills and Oglesby, 1971)

Date and Station	Soluble Cobalt (μg/liter) 0 m	5 m	10 m	Vitamin B_{12} as Cobalt (μg/liter) 0 m	5 m	10 m
June 22, 1970						
A	-	0.006	0.006	-	0.000037	-
B	0.008	0.013	0.015	0.000037	0.000074	-
July 16, 1970						
A	0.038	0.022	0.010	0.0001	0.00022	0.00015
B	0.020	0.030	0.050	0.00044	0.00015	0.00007
August 6, 1970						
A	0.006	0.006	0.003	0.00026	0.00030	0.00015
B	0.003	0.003	0.006	0.00033	0.00022	0.000074
August 17, 1970						
A	0.003	0.005	0.006	0.00019	0.00026	0.000026

(b) Cobalt in freshwater sediments: Except where there is gross heavy metal contamination by industrial or sewage effluents, the concentration of cobalt in freshwater sediments is generally < 20 mg/kg, or about the same range of concentrations as in cobalt-adequate agricultural soils. Of course, freshwater (lake) manganese nodules contain elevated cobalt concentrations (e.g., Green Bay, 160 mg Co/kg; northern Lake Michigan, 234 mg Co/kg; Lake Ontario, 650 mg Co/kg; and Lake Oneida, 70 mg Co/kg). These concentrations are lower than those of deep-sea manganese nodules (average 2,800 mg Co/kg) (Jones and Bowser, 1978). Table IV-23 summarizes many literature data on the occurrence of cobalt in freshwater sediments.

Hutchinson and Fitchko (1974) found that heavy metal contamination of the Great Lakes is higher in the lower Great Lakes due to industrial pollution. Table IV-24 shows that even when Great Lakes' tributaries frequently contain cobalt in their outlet sediments, the frequency of finding samples with cobalt concentrations > 20 mg/kg is usually less than for samples with < 5 mg Co/kg. They found no significant correlation of cobalt concentrations in the sediments with percent clay.

TABLE IV-23

STABLE COBALT IN FRESHWATER SEDIMENTS

Body of Water	Location	Sampling Date	Depth of Sediment	Description, Size of Particles	Cobalt Concentration mg/kg	Analytical Method	Correlations	Sources of Heavy Metals	Other Remarks	Reference
Ottawa and Rideau Rivers	Near Ottawa			4 to 62 μm	15	Atomic absorption (10% precision)	Surface area 0.70 to 0.89		No known mineral deposits in area. Background sediment concentrations 11 to 13 mg/kg.	Oliver (1973)
	Near sewage plant				128	AA		Sewage	Pb, Hg, Zn, Cu, and Ni also high.	
Buffalo River	Boxley (River Mile 130), Arkansas	May to June 1973			7.2 to 15.4	AA; detection limit 0.5 mg Co/kg	Co/100 parts Fe ratio of 150 μ sediments = 0.60.	Contact with shale in upper drainage area. Old Zn, Pb, and Cu mines	Other iron-associated metals also high.	Steele and Wagner (1975)
Buffalo River	Pruitt (River Mile 101), Arkansas	May to June 1973			3.0 to 7.2	AA			Concentration decreased after iron precipitated upon oxidation.	Steele and Wagner (1975)
Redstone River	Near confluence with the Mackenzie River.		Suspended		17	Flameless atomic absorption				Wagemann et al. (1977)
Mackenzie River	East channel		Bottom Suspended		11 0.81	FAA				Wagemann et al. (1977)
Upper Peoria River	Impoundment of the Illinois (River Miles 167 to 172) near Peoria	June 1971		Brownish gray clayey and sandy silts	6 to 17	Neutron activation analysis (± 10% precision)			Slow movement of sediment due to slow water travel time. False bottom resuspended by towboats and temporary currents.	Collinson and Shimp (1972)
Illinois River	River Mile 158 to 167				1.1 to 17.6	NAA				Collinson and Shimp (1972)

TABLE IV-23 (continued)

Body of Water	Location	Sampling Date	Depth of Sediment	Description, Size of Particles	Cobalt Concentration mg/kg	Analytical Method	Correlations	Source of Heavy Metals	Other Remarks	Reference
Nonindustrial streams	Peoria, Illinois area				4.1 to 8.4					Mathis and Cummings (1971; cited by Collinson and Shimp, 1972)
Maumee River Basin	Ohio, see text				0.006 to 83.3 (average 0.90) Detection frequency: 112/113			See text.	See text.	Pettyjohn et al. (1975)
Lake Superior	Harbor of Duluth, Minnesota, and Superior, Wisconsin	1973 to 1975	≤ 15 cm	Total Clay-size (< 2 μm)	7.92, 12.7 ≤ 28	Neutron activation analysis or INAA		Cobalt was assumed to be present from nonanthropogenic sources.	25 mg Co/kg recommended as background concentration in clay fraction. Co and other heavy metals varied < 15% in the clay-size fraction, whose major components are clay minerals, hydrous Al and Fe oxides, and organic matter.	Helmke et al. (1977)
			13 to 33 cm	Clay-size	average 25					
Lake St. Clair	Ontario, between Lake Huron (from which it receives its water) and Lake Erie	1970 and 1974	2 cm		5.5 to 10.1	Flameless atomic absorption		At or near background levels.	Lower values near bluffs.	Thomas et al. (1977)
Lake Huron		1969	2 cm		11.1 to 17.1					Thomas et al. (1977)

TABLE IV-23 (continued)

Body of Water	Location	Sampling Date	Depth of Sediment	Description, Size of Particles	Cobalt Concentration (mg/kg)	Analytical Method	Correlations	Source Heavy Metals	Other Remarks	Reference
Lake Superior, non-contaminated streams flowing into		1972 to 1973			1.2 to 18.0 (mean 8.7)	Flameless atomic absorption				Hutchinson and Fitchko (1974) Fitchko and Hutchinson (1975)
Lake Superior tributaries	Outlet sediments	1972 to 1973			1.2 to 18.0 or 0.5 to 21.4	FAA				Fitchko and Hutchinson (1975)
Lake Superior, Keweenaw Upper Entry		1972 to 1973			17.0	FAA		Copper mining and smelting		Fitchko and Hutchinson (1975)
and Keweenaw Lower Entry		1972 to 1973			6.7	FAA		Copper mining and smelting		Fitchko and Hutchinson (1975)
Lake Michigan tributaries	Outlet sediments; high in Calumet River at South Chicago	1972 to 1973			2.5 to 20.6 or 1.8 to 25.4	FAA				Fitchko and Hutchinson (1975)
Milwaukee River	Milwaukee, Wisconsin	1972 to 1973			18.0	FAA		Tanning, chemical industries, foundries, metalworks.		Hutchinson and Fitchko (1974) Fitchko and Hutchinson (1975)
Saginaw River	Mouth near Aplin Beach	1972 to 1973			68	Atomic absorption			Fitchko and Hutchinson found only 6.5 mg Co/kg, perhaps because they sampled only from shore.	Hesse and Evans (1972)
North Channel and Georgian Bay tributaries	Outlet sediments; highest at French River	1972 to 1973			3.0 to 22.8 or 2.5 to 26.0	FAA		Sudbury smelting operations ~ 100 km north of sampling site for highest value.	Highest Co concentration accompanied by highest Ni, Cu, and Cr and high Zn and Hg concentrations.	Fitchko and Hutchinson (1975) Hutchinson and Fitchko (1974)

TABLE IV-23 (continued)

Body of Water	Location	Sampling Date	Depth of Sediment	Description, Size of Particles	Cobalt Concentration (mg/kg)	Analytical Method	Correlations	Source of Heavy Metals	Other Remarks	Reference
Lake Huron tributaries	Highest in Thunder Bay River	1972 to 1973			2.6 to 16.5 or 2.2 to 20.0	Flameless Atomic Absorption				Fitchko and Hutchinson (1975) Hutchinson and Fitchko (1974)
Lake Huron	North Thunder Bay				80	Atomic Absorption				Hesse and Evans (1972)
St. Clair-Detroit Rivers tributaries					5.4 to 11.8 or 4.6 to 13.1	FAA			Expected more pollution in Rouge River canal, but swift current may have swept it into Detroit River.	Fitchko and Hutchinson (1975) Hutchinson and Fitchko (1974)
Lake Erie tributaries	Highest in Kettle Creek	1972 to 1973			1.8 to 14.7 or 3.9 to 15.3	FAA				
Raisin River	Monroe, Michigan	1972 to 1973			50					Hesse and Evans (1972)
Maumee River	Toledo, Ohio	1972 to 1973			8.6				River very contaminated with heavy metals.	Hutchinson and Fitchko (1974) Fitchko and Hutchinson (1975)
Lake Ontario tributaries	Highest in Moira River at Belleville, Ontario, Canada	1972 to 1973			6.8 to 43.4 or 6.1 to 39.7				High levels of heavy metals. Also found 22.5 mg Co/kg in sediments of Napance River.	Hutchinson and Fitchko (1974) Fitchko and Hutchinson (1975)
Lake Michigan				Manganese nodules	99 to 274	AA		Leaching from Fe deposits in Canadian Shield.	Co, Ni, Cu, and Zn concentrations considerably lower than in deep ocean (e.g., 2,800 mg/kg).	Rossman and Callendar (1969)

TABLE IV-23 (continued)

Body of Water	Location	Sampling Date	Depth of Sediment	Description, Size of Particles	Cobalt Concentration (mg/kg)	Analytical Method	Correlations	Source of Heavy Metals	Other Remarks	Reference
Lakes	Sweden and Finland			Manganese	100					Manheim (1965; cited by Rossman and Callendar, 1969)
Southern Lake Michigan				Cores	0 to 51		None with organic carbon, clay, manganese oxide, or depth.		51 may be a transposition. Compare with values reported by Collinson and Shimp (1972).	Shimp et al. (1971)
Lake Erie	Buffalo Harbor		23 cm		44 (max)		Fe, Co, and Cr related to As, Hg, and Cu. Co loading roughly proportional to increases of all metals.	Concentration depth profiles due to intermittent pollution input.	Very highly polluted. Cobalt enrichment factor 5.3 at surface, compared with background concentration at depth. Average cobalt enrichment factor for Lake Erie was 2.0.	Walters et al. (1974)
Lake Erie	Cleveland Harbor		60 cm		max < 20				General order of increasing enrichment about same as that of increasing ionization potential: Fe < Co < Ni < As < Cr < Zn < Cu < Sb < Cd < Hg.	Walters et al. (1974)
Isar River	Bavaria, Germany		Bottom Suspended		2.1 to 5.4 1 to 10	INAA			Co concentration increased as % fine particles (clay) increased.	Uken et al. (1977)
Rivers	Europe				13 to 43					De Groot (1969; cited by Uken et al., 1977)

TABLE IV-23 (concluded)

Body of Water	Location	Sampling Date	Depth of Sediment	Description, Size of Particles	Cobalt Concentration (mg/kg)	Analytical Method	Correlations	Source of Heavy Metals	Other Remarks	Reference
Rhine River					19 to 31					Patchineelan (1975; cited by Uken et al., 1977)
Freshwater reservoir, pH < 5	Uusikaupunki, southwest coast of Finland	1972	Upper	Mostly gyttja clays	29		Only small-scale industrial input.		High compared to Finnish soils, 12 mg Co/kg in arable soils.	Hinneri (1974)
			Deeper		22		Taken as representing no human cobalt contamination.			
Rhine River	Europe			<16 μm as 100%	24					De Groot et al. (1973)
Ems River	Europe			<16 μm as 100%	22					De Groot et al. (1973)
Chao Phya River	Thailand			<16 μm as 100%	12					De Groot et al. (1973)
Amazon River	Brazil			<16 μm as 100%	13					De Groot et al. (1973)
Danube River and Canal	Near Vienna, Austria			Mud	Range 12.25 to 23.71	Neutron activation analysis		Not especially polluted by heavy metals.	Canal receives runoff from over one million people.	Rehwoldt et al. (1975)
				Dry residue from water	Range 1.30 to 5.00	NAA				Rehwoldt et al. (1975)
Meža River	Yugoslavia, vicinity of a lead-zinc mine and smelter				0.2 to 10.0	Pulse anodic stripping voltammetry			Site receiving flotation discharge had lowest Co concentration.	Kosta et al. (1979)

TABLE IV-24

COBALT IN OUTLET SEDIMENTS OF GREAT LAKES TRIBUTARIES[a/]
(Hutchinson and Fitchko, 1974)

Tributaries of (Study Unit)	Cobalt, mg/kg < 5	> 20	> 25
Lake Superior	6 (26)	2 (9)	-
Lake Michigan	10 (42)	2 (8)	1 (4)
Georgian Bay	2 (13)	2 (13)	1 (7)
Lake Huron	4 (36)	-	-
Detroit--St. Clair River	1 (14)	-	-
Lake Erie	1 (5)	-	-
Lake Ontario	-	2 (13)	1 (7)

a/ Numbers in parentheses represent the percentage of the total number of tributaries for each study unit.

Pettyjohn et al. (1974) reported that "exchangeable" cobalt was detected in 97.4% (110/113) of sediment samples from the Maumee River Basin in Ohio at concentrations ranging from 0.111 to 50.4 mg/kg (average 1.5 mg/kg) and that "extractable"* cobalt was detected in 99.1% (112/113) of samples at 0.006 to 83.3 mg/kg (average 0.90 mg/kg). Despite its high detection frequency in sediments, cobalt had been found in only 26.5% (37/140) of the water samples taken from the Maumee River Basin. Its concentrations ranged from 13 to 49 μg/liter (arithmetic mean when detected 16 μg/liter). In Ohio, anomalous cobalt water concentrations were noted at Van Wert (textiles, metal products); Celina (chemicals, metal products); St. Mary's (rubber and steel products, textiles); Lima (petroleum products, chemicals, aircraft parts); Napoleon (metal products, steel casings); and Findlay (rubber tires and tubes, electronic parts).** High cobalt concentrations were also noted in Indiana at Fort Wayne (electrical equipment, paints), Decatur (cement, electrical machinery), and at several other sites in Ohio, Indiana, and Michigan not readily identifiable from the crude map.

* Adsorbed metal ions were displaced from the sediments and placed in solution by a modification of the extraction techniques described in Perkin-Elmer Analytical Methods (1971; cited by Pettyjohn et al., 1974).

** Industries in each municipality that might discharge cobalt in their wastes are in parentheses.

Williams et al. (1977) found < 0.1 to 0.7 μg Co/liter in the surface waters of six lakes in the Adirondacks region and 0.8 to 1.3 μg Co/liter in Ticonderoga Bay of Lake Champlain. The acid extractable cobalt concentration of the lake and bay sediments was < 0.1 to 1.9 mg/kg. For most samples, there was little difference in the cobalt concentration whether in the surface horizon or the bottom horizon of the core. No trend was noted in the cobalt concentrations in the sediments related to anthropogenic input, but cobalt generally increased with trophic states from oligotrophic to eutrophic.

(3) <u>Seawater and other waters of enriched salinity</u>

(a) <u>Seawater</u>: Reported values of cobalt in seawater range from 0.0097 μg/liter in the North Central Pacific Ocean to 0.045 μg/liter at 100 m depths* in the Northeast Pacific 64 km off the Newport, Oregon coast (Robertson, 1970). Intermediate values are cited by Wedepohl (1974-1978). Bradford (1971) found < 1 μg Co/liter in California oil well brines and brine from the desert Amboy Salt Well.

Lowman and Ting (1970, 1973) reported that in the seawater surrounding Puerto Rico, 35 to 50% of the cobalt occurred as nonionic complexes. The seawater soluble cobalt content was ~ 100 times higher than in open ocean water. Puerto Rican rivers are "loaded with several trace metals and with organic materials from sugar mills, agricultural runoff, and municipal sewage." But they had not determined organic forms of cobalt in freshwater. Of various adsorption, extraction, and precipitation procedures, only activated charcoal removed 100% of the cyanocobalamin from the Puerto Rican seawater.

Vitamin B_{12} is synthesized by 58 species of seven genuses of bacteria: <u>Bacterium</u>, <u>Pseudomonas</u>, <u>Micrococcus</u>, <u>Chromobacterium</u>, <u>Sarcina</u>, <u>Bacillus</u>, and <u>Planococcus</u> as well as blue-green algae and actinomycetes.

---

* Surface values, however, were only 0.0020 μg Co/liter. Presumably, the high cobalt concentrations observed between 100 and 660 m were due to high concentrations of cobalt in detrital material or organisms.

One of the most important sources of enrichment of the seawater of coastal zones with cobalamins appears to be the macrophytes with their relatively high concentrations of vitamin B_{12}. When they die, they release large amounts of cobalamins into the ocean waters. Benzhitskii (1974) tabulated the concentrations of vitamin B_{12} in world oceans (Table IV-25). The values are comparable with those reported for freshwater concentrations.

(b) <u>Marine sediments and interstitial waters</u>

(i) <u>Cobalt in sediments unrelated to anthropogenic cobalt sources</u>: Angino (1966) reported that Antarctic pelagic sediments contain an average 27 mg Co/kg, with an average Co-Al ratio of 0.00037 and an average Ni-Co ratio of 1.4. (Analyses by emission spectrography were acknowledged to be not very accurate or reproducible.) These glacial marine sediments have distinctly different chemical compositions compared with shelf, deep-sea carbonate (Co-Al ratio, 0.00035; Ni-Co ratio, 4.3), and deep-sea clay (Co-Al ratio, 0.00088; Ni-Co, 3.0) sediments. Glacial marine sediments have been largely transported from the land by glacial erosion and discharges.

Duursma (1973) reported on the trace element content of marine sediments from 60 sites throughout the world. Table IV-26 summarizes the concentrations of cobalt found. The concentration of stable cobalt in the sediments was stated to be roughly proportional to the base-exchange capacity of the sediment.

Only up to 3.53% of stable cobalt in marine sediments is exchangeable with ^{60}Co (without carrier*). This exchangeability is due to a mixture of isotopic exchange processes with kinetics similar to those of ion-exchange and of isotopic exchange by precipitation and dilution reactions (Duursma, 1973).

Cobalt belongs to a group of elements (Cd, Co, Pb, and Hg) that have similar background

---

* An excess of an element will cause chemical reactions other than isotopic exchange, e.g., ion-exchange reactions.

TABLE IV-25

CONCENTRATION OF VITAMIN B_{12} IN THE OCEANS

(Benzhitskii, 1974)

Sampling Location	Indicator Organism	B_{12} Concentration, ng/mℓ	Source Cited
Vineyard Sound, Woods Hole, Massachusetts, USA	*Euglena gracilis*	30-200[a]	Provasoli, Pintner, 1953
Northwest Arms, Halifax, Canada	*Stichococcus sp.*	10	Lewing, 1954
Baltic Sea, Sweden	*Escherichia coli*	5	Erickson, Lewis, 1954
Mol in Millpot, Scotland	*Monochrysis luther*	5-10	Droop, 1955
Northeast Part of Atlantic and North Sea, Scotland	*Lactobacillus leichmannii*	0.1-2	Cowey, 1956
Bay of Biscay, France	*Euglena gracilis*	0.3-4.6	Daisley, Fisher, 1958
Northeast Part of Pacific Ocean, Japan	*Euglena gracilis*	0-26.3	Kashiwada et al., 1957a
Kagoshima Bay, Japan	*Euglena gracilis*	0-6.7	Kashiwada et al., 1957b
Phosphorescent Bay, Puerto Rico	*Escherichia coli*	1.3-3.5	Burkholder, Burkholder, 1958
Long Island Sound, USA	*Thraustochrytrium globosum*	4.5-14.6	Vishniac, Riley, 1959
Sargasso Sea, Bermuda Island	*Cyclotella nana*	0-0.22	Menzel, Spaeth, 1962
Sevastopol Bay, Black Sea, USSR	*Escherichia coli*	0.42-3.14	Suprunov, Muravskaya, 1964
Northeast Part of Pacific Ocean, USA	*Cyclotella nana*	0-3.7	Carlucci, Silbernagel, 1966
Catalina Island, California, USA	*Cyclotella nana*	0.67-3.75	Holm-Hanse et al., 1966
Apalachee Bay, Florida USA	*Cyclotella nana* *Euglena gracilis*	0-19.5 1.3-17.6	Stewart et al., 1966 Stewart et al., 1966
Northwest Shore of Florida, USA	*Cyclotella nana* *Euglena gracilis*	0-71.8 5.2-65.8	Stewart et al., 1967 Stewart et al., 1967
Southeast of Yalta, Black Sea, USSR	*Escherichia coli*	0-4.86	Suprunov, Benzhitskii, Bugaeva, 1967
Coastal Zone, LaJolla, California, USA	*Cyclotella nana*	0.4-6.5	Natarajan, 1970
Subarctic Region, Pacific Ocean, Alaska, USA	*Cyclotella nana*	0-3.39	Natarajan, 1970
Berents Sea, USSR	*Escherichia coli*	0-2.16	Propp, 1970

a/ The high concentrations of vitamin B_{12} were caused by lengthy storage of unfiltered water.

concentrations in freshwater, estuarine, and coastal marine sediments. On the other hand, copper, manganese, and zinc show decreasing concentrations with increasing salinity. Taylor (1974; cited by Taylor, 1976) had reported an average 6.6 mg Co/kg in unpolluted coastline sediments (4.6 to 14.0 mg Co/kg). In 1976, Taylor reported on the cobalt concentrations in Solway Estuary, the least polluted major estuary in the United Kingdom, receiving drainage from primarily farmland and lowland hills, with little industry. The estuary sediments contained 3.8 to 7.5 (average 5.7) mg Co/kg (determined by atomic absorption), whereas the freshwater sediments from the river draining into the estuary contained 5.3 to 11.9 (average 7.9) mg Co/kg. Cobalt concentrations correlated well with the nickel contents of the sediments (0.96), zinc (0.84), and copper (0.74); and less with the manganese (0.35) and silt (0.26).

TABLE IV-26

COBALT IN MARINE SEDIMENTS
(Duursma, 1973)

Site (number of samples)	Cobalt Concentration, mg/kg (dry weight)
Atlantic Ocean (12)	0.37-83
Pacific Ocean (7)	38-195
Indian Ocean (5)	12-34
Arctic Ocean (1)	20
Baltic Sea (1)	12
North Sea (1)	8.4
Mediterranean Sea[a/] (5)	6.4-25
Black Sea (1)	21
Red Sea (1) (Discovery Deep)	12
Mean	24.5

a/ A Mediterranean seawater was found to have 0.0023 µg Co/liter by neutron activation analysis.

Similarly, Goldberg et al. (1979) found cobalt and other metals more dilute in estuarine sediments than in sediments from the rivers flowing into it (Wilmington and Savannah). There was no uniform trend for cobalt with depth.

San Francisco Bay does not appear to be significantly contaminated by cobalt. In studies by Bradford et al. (1976), little variation was found in the cobalt concentrations of the 20-μ to 2.0-mm solids fraction of San Francisco Bay sediments (average 3.1 mg Co/kg; range 0.6 to 8.9 mg/kg). The average cobalt concentration was somewhat higher in the < 20-μ fraction (47% of the sediments); it was 8.8 mg/kg. The interstitial water of the sediments (pH 6.42 to 7.71) contained up to 60 μg Co/liter. The study was performed during September to December 1972, when the major source of freshwater was from sewage treatment plants.

Upper Newport Bay, California, receiving runoff from a 376-km^2 watershed whose pollution sources are nonpoint was estimated to receive 17 μg Co/cm^2/year to give a cobalt concentration in the upper sediments of 13 mg/kg. Cobalt did not show clearly increased concentrations in the uppermost strata (deposited since 1955) as did zinc and lead (Christensen et al., 1978).

Red Sea sediments contain 4 to 400 mg Co/kg (semiquantitative analyses by wet chemistry and optical spectrography; cobalt detection limit, 4 mg/kg). The interstitial brines contain up to 4.1 mg Co/liter (Hendricks et al., 1969).

Indian Ocean sediments contain 1 to 83.33 mg Co/kg with highest concentrations being in high-iron (and highest-alumina) sediments. Murty et al. (1977) did not find particularly good correlations between cobalt and other substances in the bulk sediment samples. For example, organic carbon and cobalt had a correlation coefficient of 0.51. However, correlations were better in the nonlithogenous (acid-soluble) fraction of the sediments: Cu or Ni, 0.86; Fe, 0.80; and Mn, 0.74.

(ii) Cobalt in manganese nodules and other iron-rich sediments: Wogman et al. (1973) found 0.2 to 1.1% cobalt in manganese nodules in sea-bed sediments which were analyzed *in situ* by using ^{252}Cf as the neutron source.

Cobalt is concentrated in the iron-rich amorphous oxides of Pacific Ocean manganese nodules. In these oxides, the maximum cobalt concentration reported is 7,400 mg/kg. Copper and nickel are concentrated in the manganese-rich crystalline oxides (birnessite and todorokite). There does not appear to be any linear interelement relationship between concentrations of manganese and iron and of nickel, copper, and cobalt. Concentrations of the same element vary greatly even within individual nodule layers. Banning (1975) explained that any literature reports that there is such a linear relation may be due to mixing of crystalline and amorphous oxides during sampling.

Thin yellow layers (0.5 to 2 cm) of seven sediment cores taken from west and central portions of the Gulf of Mexico were 51% enriched in cobalt, 66% in iron, 33% in manganese, and 28% in nickel. The correlation coefficients for cobalt were Ni, 0.53; Fe, 0.62; and Mn, 0.50. The cobalt concentration in the yellow zones was 60 mg/kg compared with 40 mg/kg in the gray lutites. The iron-rich zones had resulted from reduced sediment accumulation. Watson and Angino (1969) stated that scavenging of cobalt and nickel is probable if the iron and manganese minerals go through a gel state prior to sedimentation, allowing a longer exposure period at the water-sediment interface.

Iron and nickel occurred in higher concentrations in sulfide-rich reducing sediment of a Santa Barbara Basin core than in an oxidizing Santa Cruz Basin core. Presley et al. (1967) had expected that manganese would be the most soluble in the Santa Barbara Basin sediment at a low Eh; but instead, in the interstitial water of cores from oxidizing near shore sediments off southern California, there was a 50-fold enrichment in manganese compared with the sulfide-rich reducing sediments.

Weathering of terrigenous materials is a source of manganese and nickel in coastal sediments. However, the high concentrations of manganese and nickel in deep-sea cores of the east Pacific may originate from shallow vulcanism. All but one of the deep-sea cores sampled from the East Pacific Rise off the coast of Mexico had high dissolved manganese (Mn^{2+} or a soluble complex)* in the interstitial water of the cores.

* e.g., $MnHCO_3^+$, $MnSO_4$, or an ion pair.

In two cores, there was 100-fold enrichment in nickel and cobalt. Apparently, manganese nodules form in deep-sea sediments by diffusion of manganese from the depths of the sediments to their surface (Presley et al., 1967).

The concentration of cobalt in the interstitial waters of the East Pacific cores were significantly higher than its concentration in seawater (see Table IV-27). Generally, the iron concentration was not elevated in the interstitial waters compared with its concentration in oxidizing near shore basins. Soluble manganese is 50 to 3,000 times richer in East Pacific core interstitial water than in seawater.* Nickel does not follow iron as it does in near shore sediments, and cobalt is correlated with nickel. When manganese has diffused from depth to the surface, Mn(II) may become oxidized to Mn_2O_3 so that its colloidal form will accrete with iron oxide and catalyze the precipitation of nickel and cobalt (Presley et al., 1967).

TABLE IV-27

COBALT CONCENTRATION IN INTERSTITIAL WATER OF EAST PACIFIC CORES (μg/liter)
(Presley et al., 1967)

Gray-green mud (no sulfide odor)	<0.3
Black mud with strong sulfide odor	<0.3-2.7
Green-gray clay, 1-cm layer, brown mud at top	5.6-20
Brown clay with light striation	3.6-6.2
Green-gray streaked with black	12 (same 0-66 as in depth
Calcareous ooze, 5 cm red clay at top	2.2-4.2
Red clay	2.6-6.5
Brown clay with light mottling	1.5-2.3

Evrard et al. (1975) found 0.77 to 6.5 μg Co^{2+} per liter in interstitial water of sediments of the Gulf of Gascogne, values of the same magnitude as those reported in Table IV-27.

---

* The behavior of manganese is anomalous with respect to the redox potential. At Eh +400 and pH 7.8, the MnO_2 stability field is entered and the concentration of Mn^{2+} should be < 500 μg/liter.

Recently, massive ore-grade zinc, copper, and iron sulfide deposits were found on the deep ocean floor of the East Pacific Rise near 20°54'N, 109°03'W. (The red clay in the report of Presley et al. (1967) was sampled closest to this site at 20°15'N, 112°42'W.) Although cobalt, cadmium, silver, and lead were detected by x-ray fluorescence spectrometry, manganese was absent even as a minor element (Francheteau et al., 1979).

(iii) Cobalt in industrially polluted sediments: Untreated industrial and domestic wastes from Athens, Greece, are discharged into the Keratsini Bay and Elefsis Bay of the Upper Saronikos Gulf. At Athens in early 1975, the maximum cobalt concentration in the sediments was 18 mg/kg and in Elefsis Bay, 13 mg/kg (the latter receives effluent from at least 25 large industries). Grab samples collected from the upper 5 cm of sediment from the Piraeus Harbor sea floor had up to 140 mg Co/kg (analyses by neutron activation analysis). Cobalt, hafnium, and iron followed the distribution pattern of the rare earths in the sediments. This pattern was similar to those of mercury, arsenic, and antimony (Grimanis et al., 1977).

An electroplating plant operated at Quonset Point, Narragansett Bay, Rhode Island, from 1943 to the spring of 1973. This plant, the Naval Air Rework Facility (NARF) at the Air Station in North Kingston, was the primary reconstruction center for U.S. Navy aircraft in the eastern United States. There are no data for the amounts of metals released. The Bay also received grease, paint stripping, and metal plating wastes through storm sewer runoff. Despite this pollution, the sediments were not greatly enriched in cobalt, even near the NARF. Core depths 0 to 5 cm sampled on October 12 and November 1, 1973, contained 0.2 to 9.6 mg Co/kg. Depths 5 to 15 cm contained 0.5 to 10.1 mg Co/kg, and depths $\geq$ 15 cm contained 2.6 to 9.8 mg Co/kg (analyses by atomic absorption). The overall rank for the elements in the sediments, Zn > Cr > Cu > Pb > Ni > Co > Ag > Cd, was the same as that found in sediments at other polluted sites. For example, sediments at Providence, Rhode Island (about 20 km north of the NARF) contain 7.7 mg Co/kg; at Fix Island, 4.7 mg Co/kg; at Rhode Island Sound dumpsite, 5.0 mg Co/kg (controls,

2.0 mg Co/kg);* the Philadelphia ocean dumpsite, 1.0 mg Co/kg (controls, 0.5 mg Co/kg);** and the Gulf of Mexico shelf, 5.0 mg Co/kg (Eisler et al., 1977).***

Although the mollusks Mercenaria mercenaria are contaminated by cobalt and other heavy metals in the upper polluted part of Narragansett Bay, there was no significant difference between the cobalt concentrations in the populations in contaminated portions and those in uncontaminated portions of the Bay. Perhaps the mollusks regulate cobalt uptake biologically, or the cobalt is fluxed from the sediments in a biologically unavailable form (Phelps and Myers, 1977).

Within 3 weeks of initiation of a continuous nickel refinery discharge to Halifax Bay, North Queensland, Australia, approximately 2 km northwest of Townsville, population 80,000, the concentrations of cobalt in the Bay sediments were increased significantly. In the tailings ponds, the cobalt concentration was 514 mg/kg. An artificial tailings supernatant contained 130 mg Co/liter. (The subsurface sediment adsorbed 9.04 to 9.86% of the cobalt in this artificial tailings supernatant.) At the mouth of Althaus Creek directly adjacent to the refinery, the dried surface sediment contained 27 mg Co/kg, and the subsurface sediment, 37 mg Co/kg. The surface sediments at Queensland Nickel had 11.2 mg Co/kg, and those sediments below 10 cm contained 8.7 mg Co/kg. At Halifax Bay and Cleveland Bay, the surface sediments contained 7.2 and 7.8 mg Co/kg, little different from the depth concentrations. The cobalt and nickel concentrations were linearly related to the iron concentrations in the uncontaminated surface sediments. The Co-Fe correlation in the bays was 0.69 to 0.71, but at Queensland Nickel, the correlation was only 0.13, indicating an artificial elevation in cobalt levels. Cobalt was also correlated with nickel and zinc in the uncontaminated bays (0.51 to 0.80), but not at Queensland Nickel (Knauer, 1977).

---

* Preceding data were from unpublished data of Lapan and Blasco; cited by Eisler et al. (1977).

** Unpublished data of Reynolds; cited by Eisler et al. (1977).

*** Holmes (1973; cited by Eisler et al., 1977).

Loring (1976) reported that the sediments of Saguenay Fjord,* Quebec, contain 5 to 20 (average 14) mg Co/kg. The highest concentrations were in the fine-grained sediments from the upper part of the fjord. The lowest concentrations were in coarse sediments. The metal concentrations in the muds were comparable with those of the St. Lawrence estuary and the Gulf of St. Lawrence. Acetic acid extractions showed that 8 to 25% of the cobalt was contributed by the nondetrital** fraction and may be available to the biota of the fjord. Possibly, the nondetrital cobalt was associated with iron and manganese oxide grain coatings as were nondetrital Cr, V, and Ni. These elements were not enriched in the sediments due to the industrial pollution. The nondetrital contribution of Co, Ni, and Cr in the sediments was < 5 mg/kg.

Most nondetrital cobalt and part of the nondetrital Ni, Cr, and V were adsorbed, held in ion-exchange positions, or situated on broken-edge positions of the ferromagnesian minerals. Adsorption of nondetrital cobalt and nickel onto aluminosilicate minerals controlled their overall dispersal pattern during transport and deposition (Loring, 1976).

Most of the Co, Ni, Cr, and V were in the detrital fraction (71 to 98%) and existed in the sulfide, oxide, and ferromagnesian minerals. Detrital cobalt correlated with water depth (0.68), mud content (0.61), total iron (0.83), magnesium (0.66), manganese (0.59), lithium (0.70), detrital nickel (0.70), detrital chromium (0.74), nondetrital chromium (0.58), and nondetrital vanadium (0.54). There were no correlations with aluminum and negative correlations with sulfur and silicon. Thus, the detrital cobalt was in ferromagnesian and sulfide minerals in the sediments (an average of 20% of the detrital cobalt may have been contributed by sulfide minerals). The detrital Co, Cr, and V were not related significantly to variations in organic matter content nor C-N ratio in the sediments.

---

* Major freshwater input is from the Saguenay River, which receives effluents from aluminum smelters, a chloralkali plant, a fluorspar processing plant, etc.

** Nondetrital cobalt has been removed directly as Co^{2+} from the seawater. Detrital cobalt has been introduced into the sediment in the solid state.

Dispersal of detrital Co, Ni, and Cr was controlled by the sedimentation pattern (Loring, 1976).

Elderfield and Hepworth (1975) studied the heavy metal concentrations in the estuaries of the River Tees (northeast England, classified as grossly polluted) and the River Conway (north Wales; no urban or industrial contamination, but lead-zinc mineralization and tailings dumps contribute heavy metals). The estuarine water of the River Conway contained 0.05 μg Co per liter, and the pore water of the sediments contained 490 μg Co/liter. The lead-zinc mining and concentrating activities contributed more cobalt to the River Conway sediments (6.17 $\mu g/cm^2/year$) than did urban industrial pollution to the River Tees sediments: 0.57 $\mu g/cm^2/year$ at concentrations of 1.7 μg Co/liter in the estuarine water (34 times higher than in the River Conway) but only 46 μg Co/liter in the pore water.

These estuarine sediments were reducing and carried high levels of H_2S, yet the cobalt concentration in the pore water was 130,000 times higher than would be calculated from the solubility of cobalt sulfide. Elderfield and Hepworth (1975) presumed that complexation with organic matter facilitated the mobility of cobalt in the anoxic estuarine sediments. They pointed out that the upward diffusive flux of metals through the sediment, which is followed by oxidation and precipitation at the surface, tends to concentrate metals at the sediment surface. Thus, an enrichment of only 10% at the surface compared with deeper levels cannot prove anthropogenic input.

Kaohsiung Harbor, southwest of Taiwan Island, receives untreated sewage and industrial effluents from a city of population greater than one million with more than 1,000 factories. Cobalt concentrations in sediments sampled from November 1973 to August 1974 ranged from about 10 mg/kg to about 50 mg/kg (atomic absorption analyses). The highest cobalt concentrations were at stations receiving mixed industrial wastes and at stations at the mouth of inflowing rivers, one of which went through the city (Hung et al., 1975).

The Humber Estuary, formed from the confluence of the Ouse and Trent rivers stretches for 40 miles (64 km) before it enters the North Sea. The estuary carries sewage effluent from 20% of the English population as well as industrial effluents, e.g., red mud and synthetic fiber plant discharges. The concentrations of cobalt in the sediments from the relatively shallow depths were: anoxic mud, 30 mg/kg; silty sand, 11 mg/kg; and fine sand, 6 mg/kg. All metal concentrations in the sediments decreased as sampling progressed seaward (Jaffe and Walters, 1975).

The Pamlico River Estuary, North Carolina, receives mining wastewaters and drainage from agricultural and a few suburban areas. The average concentration of cobalt in sediments sampled from the center of the estuary on March 1, 1973, was 10.4 mg/kg (determined by atomic absorption and neutron activation analyses). The average cobalt concentration in samples from the side of the estuary was only 4.4 mg/kg. The highest sediment concentration was in the tributary bearing industrial waste (Harding and Brown, 1975a and 1975b).

Anaerobic conditions in the muds in the center of the Pamlico River Estuary allowed preservation of organic matter there. Although the sediment contents of organic matter and clay were higher in the center of the estuary, the metal concentrations were higher in the sparse clay and organic matter in the sediments just offshore where they were apparently initially scavenged from the incoming water (Harding and Brown, 1975).

Ramondetta and Harris (1978) reported that concentrations of vanadium, cobalt, and, to a lesser extent, nickel correlated with each other very well in Jamaica Bay* sediments in July 1972 and were enriched where petroliferous pollution from fuel terminals and heavy industry was greatest (the southeastern portion). The mean concentration of cobalt in the sediments was 5.98 mg/kg; its peak concentration was 21.4 mg/kg. Average values for nickel, copper, vanadium, chromium, lead, and zinc were 4 to 33 times greater than that of cobalt.

---

* Jamaica Bay is surrounded by Brooklyn and Queens on the north, Kennedy Airport and Nassau County on the east, and Rockaway Beach on the south. On the west, the bay is connected with the Atlantic Ocean.

Sediments in Foundry Cove near Cold Springs Harbor, New York, have been heavily contaminated by discharges from a nickel-cadmium battery plant operating since 1953 (Kneip et al., 1975). The industrial wastewaters contained 3.87 kg Cd/day and 2.22 kg Ni/day (Bower et al., 1978). At one time, the wastes discharged also contained cobalt in amounts representing about 10% of the nickel used. The concentrated nitric acid solutions were neutralized to pH 12 by calcium carbonate before discharge to the sewer and river or to the cove itself. Concentrated acid solutions were probably lost occasionally due to operator errors. Untreated rinse solutions were discharged between 1953 and 1970. After 1965, all liquid waste discharges were directly to the cove. Contaminated sediments contained about 15 to 120 mg Co/kg, dry weight (Kneip et al., 1975).

In the early 1970's, public health environmental agencies ruled that further cadmium discharges to Foundry Cove should be reduced to 0.2 mg/liter. The channel was dredged in the summer and fall of 1972 and again in the spring of 1973 to remove the most contaminated sediments. Dredged sediment was piled on the ground along the channel (Alford, 1973). However, by 1976, sediment cadmium concentrations were very similar to those present before the dredging. A total 20 to 25 MT cadmium was estimated in the sediments in 1970 to 1971. The 1974 estimate was 12.4 MT cadmium. The recontaminated sediment bed was apparently formed by resuspension of the sediments and recycle of dredge spoil water (Kneip and Hazen, 1979). The concentrations of cobalt in the sediments with increasing distance from the outfall are given in Table IV-28.

(4) <u>Cobalt in sewage and sewage sludge and the effect of their disposal into the environment</u>: In 1964, it was estimated that at 11.4 kg sewage sludge solids per year per capita and 130 million people served by sewers, the United States produced 1.5 billion kilograms sewage sludge solids per year (Thompson et al., 1964). A more recent estimate of U.S. dry sewage sludge production is 6 million tons (5 billion kg) (Shen, 1979).

As recently as 1974, 5.676 million tons of U.S. sewage sludge was dumped into the Atlantic Ocean (Anonymous, 1975). Since the ocean already contains most of the heavy metal

TABLE IV-28

CONCENTRATION OF COBALT IN FOUNDRY COVE BOTTOM SEDIMENTS, 1973 (after dredging)
(Kneip and Hazen, 1979)

Distance from Outfall, m	Median Cobalt Concentration mg/kg (dry)
10	224
60	248
175	552
300	24
400	12
480	< 20-41
660	40
690	<52
710	18
760	27
900	19
1,800-2,530	< 6-<17

pollutants present in sewage in vast amounts, Bascom (1974) argued that the ocean is a plausible place for its disposal.

On the other hand, Feldman (1970) pointed out fallacies in "considering the oceans as infinite sinks" for release of heavy metals in wastes to coastal areas:

- We do not have a mass balance for most elements for input, inventory, and removal.
- Detailed information on heavy metal reactions with the biota is lacking.
- Rapid mixing is assumed along with permanent removal to the sediment or to the deep ocean.

Biota may well provide a mechanism for enhancing the availability of the elements from the sediments.

As of 1975, cities like Boston, New York, and Los Angeles were still dumping their sludge a few miles off the coast; some were even pumping raw sewage into the ocean. For example, New York City and New Jersey dumped a billion gallons of sludge annually only 12 miles offshore; 40% of it was raw. Ocean City, Maryland, discharged 1,000 tons raw sewage sludge solids per year at a site 3,800 ft from the coast (McLaughlin, 1975).

Ocean dumping of sewage sludge must cease by the end of 1981 according to a recent EPA regulation (Shen, 1979). About 35% of the sewage sludge is incinerated. Incinerator ashes across the United States contained 11 to 57 mg Co/kg except those at Palo Alto, California, which contained 290 mg Co/kg. The Palo Alto sludge ashes also showed the highest values for copper, 7,300 mg/kg; nickel, 1,500 mg/kg; palladium, 8.6 mg/kg; and silver, 575 mg/kg, probably reflecting the electronics industry there (Gabler and Neylan, 1977, 1978).

In addition to land application and incineration, sewage sludge may one day be used as animal feed ingredients. Experiments using animal excreta and sewage sludge are being conducted at Colorado State University (Capar et al., 1978).

(a) Cobalt in sewage and sewage sludges: Values of 3.4 to 30.5 mg Co/kg were determined in U.S. sewage sludges by colorimetric methods by Clark and Hill (1958) (see Table IV-29). These values are comparable to those determined by more modern instrumental methods (see Table IV-30).

Fortescue (1975) found 13.5 mg/kg cobalt (oven-dry basis), 0.96 mg/kg wet weight, and 20.8 mg/kg ash weight in a mixture of 40 sewage sludges collected on November 27 and 28, 1974, from the Niagara Peninsula, Ontario, Canada. (The coeffi- of variation for oven-dry cobalt determinations by atomic absorption was 25.7%.) Data for nine treatment plants are presented in Table IV-31.

Fortescue (1976) reported 6 to 77 mg/kg (average 25 mg/kg) cobalt in dry sewage sludges collected from six sewage treatment plants in late November 1974. Similar data were reported by Webber and Corneau (1977) in Table IV-32. Canadian soils have similar cobalt concentrations; for example, for six cities, soil with pH 4 to 5 had 8.0 to 40.0 mg/kg cobalt (average about 20 mg/kg).

TABLE IV-29

COBALT IN SEWAGE SLUDGE (mg/kg)
(Clark and Hill, 1958

City	Amount of Cobalt (mg/kg)
Activated sludges	
Chicago, Illinois	7.0, 10.6
Houston, Texas	3.6
McKeesport, Pennsylvania	6.0
Milwaukee, Wisconsin	4.9
Activated and digested sludges	
Des Moines, Iowa	13.6
Hagerstown, Maryland	3.4, 4.2
Los Angeles, California	10.9
Digested sludge	
Beltsville, Maryland	30.5
Beltsville, Maryland, Imhoff tank sludge	4.6
Greenbelt, Maryland	6.8
Washington, D.C.	8.8
Rochester, New York	5.0
Indianapolis, Indiana, lagoon sludge	7.3

TABLE IV-30

COBALT IN U.S. SEWAGE SLUDGE (dry weight)

Location (type sewage sludge)	Cobalt Concentration (dry weight) (mg/kg) Mean	Range	No. of Samples	Method[a]	Detection Limit (ppm)	Remarks	Authors
Keasby, New Jersey (anaerobic digested sludge)	13					The concentration of cobalt in the sewage (20% industrial, 80% domestic) was at least 98 μg/liter at pH 7.1 to 7.2.	Gould and Genetelli (1977)
Six North Carolina cities	7.0	5.4 to 10.8	7	NAA		The high was from a town of fewer than 5,000 population with little or no industry; but one major manufacturer used large amounts of Co and Cr in plating his product.	Weaver et al. (1974)
Ten Indiana cities	16.4	5.85 to 59.1	10	INAA		Highest value is for Kokomo, the location of a large producer of cobalt alloys.	Nadkarni and Morrison (1974)
Binghamton/Johnson City treatment plant, New York	24 (median)					The sewage effluent into the Susquehanna River contained 6 μg Co/liter.	McDuffie (1978)
Northeastern and Central United States				AA		Compilation of composition data from a regional survey of 150 treatment plants in six north central states and two eastern states.	Sommers (1977)
(anaerobic)	7	3 to 18	4				
(other types)	4	1 to 11	9				
(all)	4	1 to 18	13				
Atlanta, Georgia	9.4			NAA			Furr et al. (1976)
Chicago, Illinois	3.7			NAA			Furr et al. (1976)
Denver, Colorado	8.8			NAA			Furr et al. (1976)
Houston, Texas	6.3			NAA			Furr et al. (1976)
Ithaca, New York	10.5			NAA			Furr et al. (1976)
Cayuga Heights, New York	3.9			NAA			Furr et al. (1976)

TABLE IV-30 (concluded

Location (type sewage sludge)	Cobalt Concentration (dry weight) (mg/kg) Mean	Range	No. of Samples	Method[a/]	Detection Limit (ppm)	Remarks	Authors
Los Angeles, California	16.8			NAA			Furr et al. (1976)
Miami, Florida	17.6			NAA			Furr et al. (1976)
Milwaukee, Wisconsin	4.3			NAA			Furr et al. (1976)
New York, New York	6.2			NAA			Furr et al. (1976)
Philadelphia, Pennsylvania	15.6			NAA			Furr et al. (1976)
San Francisco, California	5.3			NAA			Furr et al. (1976)
Schenectady, New York	15.6			NAA			Furr et al. (1976)
Seattle, Washington	9.4			NAA			Furr et al. (1976)
Syracuse, New York	5.1			NAA			Furr et al. (1976)
Washington, D.C.	15.1			NAA			Furr et al. (1976)
Twelve Oklahoma cities	67	$\leq$ 500		Semiquantitative spectrochemical analysis		Detection frequency 8/12. Cobalt contents highest in sludges from sewages treated by primary settling, activated sludge, final settling, and separate sludge digestion (Oklahoma City and Enid). Lower or no amounts were found when the sewage treatment had included a filtration or biosorption step.	Thompson et al. (1964)

Note: NAA = neutron activation analysis.
INAA = instrumental neutron activation analysis.
AA = atomic absorption.

TABLE IV-31

COBALT IN NIAGARA PENINSULA SEWAGE SLUDGES

	City	Primary Sedimentation	Secondary Treatment	Other Treatment	Cobalt (mg/kg, oven-dry weight)
1.	Grimsby	-	X	-	4.0
2.	Fort Erie	X	-	-	4.0
3.	Crystal Beach	No	X	Chlorination	16.0
4.	Port Colbourne (E)	No	X	Recently emptied and cleaned. Not representative.	7.6
5.	Port Colbourne (W)	-	X	Also under startup conditions.	20.0
6.	Welland	X	No	Chlorination	12.0
7.	Niagara Falls	X	X	Chlorination	8.0
8.	Port Dalhousie	X	X	Chlorination	24.0
9.	Port Weller	-	X	-	80.0[a]
	Average	-	-	-	11.95

a/ Also had unusually high Cd and Cr concentrations, indicating industrial contamination that day.

TABLE IV-32

COBALT IN OTHER CANADIAN SEWAGE SLUDGES, AIR-DRIED
(Webber and Carneau, 1977)

City	No Treatment	Lime Treatment	Ferric Chloride, Anaerobic Digestion	Co Concentration (mg/kg)
Ottawa[a/]	X	-	-	12.2
Fergus[b/]	X	-	-	5.3
Newmarket[c/]	-	X	-	ND
Midland[b/]	-	X	-	11.5
North Toronto[c/]	-	-	X	11.5
Guelph[b/]	-	-	X	10.0
Sarnia[b/]	-	-	X	6.9

Note: ND = Not detected.

a/ Dried for several months in a large lagoon.

b/ Suspension obtained directly from anaerobic digesters.

c/ Vacuum-filter cake.

Chattopadhyay (1977) compared the cobalt concentration in raw sludges (2.11 to 7.2 mg/kg) from two industrial Canadian cities with cobalt concentration in chemical sludges* (2.89 to 13.2 mg/kg), sludge-based fertilizers (1.83 to 11.4 mg/kg), and nonsludge fertilizers (1.25 to 6.94 mg/kg). Both instrumental neutron activation analysis and instrumental proton activation analysis were sensitive enough for determining cobalt in sewage sludges. By the former method, 11.1 ± 1 mg/kg cobalt was found in a sewage sludge round-robin test circulated by the Canadian Centre for Inland Waters; by the latter method, 10.2 ± 1.1 mg/kg.

Wiseman and Bedri (1975) detected 22.6 mg/kg cobalt by neutron activation analysis in a sewage sludge from Rye Meade Sewerage Disposal Works, Hoddesdon, Essex, United Kingdom. Egan and Spyrou (1977), however, with a detection limit of 50 mg/kg cobalt for their neutron activation method did not detect cobalt in the United Kingdom sludges they analyzed.

(b) <u>The effect of sewage treatment on contained cobalt</u>: Oliver and Cosgrove (1974) calculated that a model sewage treatment plant released 32 g cobalt daily in 6.5 million gallon of effluent (25 million liters). After atomic absorption analyses of the sewage streams of 10 southern Ontario sewage treatment plants, Oliver and Cosgrove (1975) calculated that < 60 kg cobalt was discharged daily into Lake Ontario. Their data are presented in Table IV-33. They were unable to calculate the cobalt removal efficiency of the activated sludge plant.

Although Matthews and Thomas (1975) found 40 ± 40 μg/liter cobalt in raw sewage in the Glenelg area of Adelaide, Australia (where 17 electroplating plants discharged their wastes) and 0.23 ± 0.06 mg/kg cobalt in the digested sludge, they did not detect cobalt in the primary effluent (after mechanical treatment) nor in the secondary effluent (after aerobic decomposition).

---

* A chemical sludge is produced by treating an anaerobically digested sludge with alum, ferric chloride, or lime.

TABLE IV-33

COBALT IN SEWAGE STREAMS OF SOUTHERN ONTARIO TREATMENT PLANTS

Sewage Stream	Cobalt in 8-hr Composite Samples from a Conventional Activated Sludge Treatment Plant (July-August 1973) (mg/kg) (Oliver and Cosgrove, 1974)	Cobalt in Samples from 10 Sewage Treatment Plants (February-August 1973) (mg/kg) (Oliver and Cosgrove, 1975)
Raw sewage	0.002 (0.001 dissolved)	0.002 to 0.02 (most < 0.03)
Primary effluent	0.001 (total = dissolved) (range < 0.001 to 0.001)	0.001 (100% dissolved at Oakville 50% industrial sewage) to 0.03 at Clarkson (most < 0.03)
Secondary effluent	-	0.001 (100% dissolved) at Oakville (50% industrial sewage) to 0.03 at Clarkson (most < 0.03)
Final effluent	0.001 (total = dissolved) (range < 0.001 to 0.001)	-
Aeration tank contents	0.03 (range 0.02 to 0.08)	-
Raw sludge	-	18 (range 7 to 30; only three studied)[a]
Recycled sludge	0.07 (range 0.03 to 0.13)	-
Secondary digester sludge	1.0 (range 0.3 to 1.8)	-
Digested sludge	-	< 0.8 (only domestic sewage) plus six others: range 8.6 to 35 (average 18)[a]

a/ The values were not given for the same plants.

The effectiveness of wastewater treatment for cobalt removal has been reported in a study by Kostovetskii et al. (1975). The cobalt concentration in the wastewaters (i.e., sewage) of two Ukrainian cities was 5.4 to 13.6 μg/liter. Mechanical purification removed about 50% of the cobalt; further purification in biological treatment ponds permitted a total removal of 83 to 89%.

The fate of cobalt in stabilization or storage ponds designed for land disposal of sewage effluents for the Great Lakes Region was studied by Bulthuis et al. (1974). In contrast to the Ukrainian study, in Michigan stabilization ponds the ambient concentrations of cobalt, cadmium, and nickel were not much reduced by accumulation in plants (anaerobic, facultative, or aerobic) or pond sediments although, due to a high rate of seepage flow and very low output volume, about 85% of the influent cobalt to the last of five ponds was trapped in the sediment. In each pond cobalt in the water was 12 to 17% of the total cobalt stored in the system; and in plants, 0.5 to 1.0% of the total.

McDuffie (1978) presented 1973-1974 results for heavy metal removal from sewage at the Binghamton/ Johnson City, New York, and the Endicott sewage treatment plants. Although the percent removal of heavy metals was quite variable, the concentrations remaining in the effluent were less variable. Cobalt removal efficiency varied from 0 to 85%, but its concentration in the effluents was generally < 10 μg/liter.

Gould and Genetelli (1977) found concentrations of 13 to 28 mg Co/kg in various phases of an anaerobic digested sewage sludge from the municipal wastewater treatment plant at Keasby, New Jersey. The phases considered were: total solids, volatile solids, or manganese in the particulate* (92.8% of the cobalt) and supracolloidal** (7.2% of the cobalt). The plant received 20% industrial and 80% domestic sewage. The association of cobalt in the sludge with iron was 0.505 mg Co/g Fe (0.386 mg/g in the particulate phase and 0.777 mg/g in the supracolloidal phase).

---

* The particulate phase particles were > 100 μm.

** The supracolloidal particles were 0.6 to 100 μm.

During digestion, the high carbonate ion concentration produced by the bacteria indicates that cobalt and other metal carbonates will remain in the solid phase until equilibration with the atmosphere, when the carbonates will dissolve. Sulfide precipitates, which probably control the solubility of heavy metals in digesters until the sulfide is exhausted, are also unstable in an oxidizing environment. The higher concentration of cobalt in the volatile solids perhaps indicate the presence of organic cobalt complexes (Gould and Genetelli, 1977).

Discharge of activated sludge and sewage may represent the largest source of cobalamins in natural waters other than their production by benthic bacteria. Several early studies cited by Beck and Brink (1978) report cobalamin concentrations (determined by bioassays) in sewage sludge of 3.5 to 22.0 mg/kg. Their more recent analyses (by high-pressure liquid chromatography) of activated sludges from the Easterly Marlborough (Massachusetts) Water Pollution Control Facility showed mean values of 3.79, 9.75, and 4.75 mg cobalamin/kg dried sludge (primary stage sludge, secondary stage activated sludge, and tertiary stage nitrification sludge, respectively). Hoover et al. (1951; cited by Beck and Brink (1978) also had found higher cobalamin concentrations in sludges from the secondary aeration stage (albeit 10 times rather than 2 to 3 times higher). The cobalamin productivity in the secondary sludge ranged from 1.87 mg/kg in late January to 27.08 mg/kg in late July. High cobalamin concentrations were associated with elevated environmental temperatures, elevated primary stage effluent biological oxygen demand, and low dissolved oxygen.

(c) Environmental cobalt concentrations related to sewage and sludge disposal: Several workers have reported on the effects of sewage sludge and other waste disposal on cobalt concentrations in marine environments. Ali et al. (1975) reported that about 4.2 million MT waste solids (including sewage sludge from 27 treatment plants) had been dumped annually in the New York Bight from 1964 to 1968. In the apparently uncontaminated sand and gravel in a widely dispersed area south of Long Island, the cobalt concentration was 7 mg/kg. In the fine-grained deposits in the Hudson Shelf receiving sewage, the average cobalt concentration was 19 mg/kg. Here the benthic fauna were reduced. An average concentration of 60 mg/kg cobalt was found in dredge waste from New York Harbor, and an average as high as 70 mg/kg was found in uncharacterized sediments from a few scattered sites (one was opposite Rockaway Beach). Analyses were by semiquantitative optical emission-spectrochemical analysis.*

Schell and Nevissi (1977) found little enrichment of cobalt in surface sediment of Central Puget Sound. The ratio of the average cobalt concentration in surface sediment for the period 1955 to 1975 to that for the period 1900 to 1920 was 1.0 to 1.3 (average 1.1 for 13 sites). However, of the heavy metals studied, only lead showed enrichments greater than 3. Near the outfall of the chlorinated sewage effluent from the Seattle sewage treatment plant at West Point, the enrichments for lead were 3.6; copper, 2.8; zinc, 1.8; mercury, 1.8; nickel, 1.2; and cobalt, 1.2. The highest sediment enrichment value for cobalt was from north of Blake Island (State Park) where the enrichment values for the other heavy metals were 1.0 to 1.6.

Gupta (1972) found 0 to 3.75 μg/liter (ppb) cobalt in Baltic Sea water by using atomic absorption spectrometry. Cobalt concentrations in August 1970 at 10 and 50 m were 0.75 and 0.55 μg/liter, respectively. The highest cobalt concentration was found at 100 m in May 1970 in the Great Karlsö Deep, but it was not detected at this station in January, August, or October of 1970.

Morel et al. (1975) calculated from data in the literature that the ratio of the cobalt concentration in Los Angeles County sewage effluent (10 μg/liter), which is discharged to the ocean,** to its concentration in southern California seawater (0.2 μg/liter) is 50 compared to ratios of 30 to 70 for mercury, nickel, manganese, and silver; ratios of 200 to 300 for copper, zinc, and cadmium; and 1,000 to 6,000 for iron, chromium, and lead. In the outfalls studied, a dilution of 50

---

* The best accuracy to be expected from this method is only ± 30% (Angino, 1979).

** After the sedimented sludge is anaerobically digested and centrifuged, the "centrate" is reblended with the primary effluent; and the blend is discharged to the ocean at 60-m depth.

occurs immediately. About half of the cobalt, manganese, and nickel in the sewage was in the particulate fraction (> 0.45 μm).*

In a sewage modeling experiment, Morel et al. (1975) calculated that in the pε range -4.0 to -3.47, (although most metals are present as insoluble sulfides) the sulfides of cobalt and nickel in the sewage begin to dissolve, with cobalt becoming the more soluble cobalt carbonate.** Cobalt, chromium, nickel, and manganese adsorb strongly on particulates. Upon contact with seawater, adsorbed manganese becomes solubilized, but cobalt does not.

Under conditions of low dilution of the sewage and moderate oxidation, cobalt, zinc, and lead precipitate as the carbonates; but at greater degrees of oxidation, the insoluble sulfides are oxidized to soluble compounds with increasing dilution of the sewage in the ocean. However, reduced sediments around the outfall occur in an area about 16 km long by 1.6 km wide. The metals apparently settle at the same rate as the sewage particulate material. The ratio of cobalt to zinc in the top 4 cm of California Bight sediments*** was 0.22 to 0.32% (± 100% error in core data); the ratio in the sewage particulates was 0.20%. The concentrations in the sediments were about one-third those in the sewage particulates. The calculated rate of

---

* In the final effluent (1-week composites, summer 1971) of all five sewage treatment plants of the Los Angeles County Sanitation District, 42 to 72% of the cobalt (average 54%) was retained by 0.45-μ filters. The cobalt concentrations in these residues were 4 to 15 mg/kg (average 9 mg/kg). In the Hyperion sewage sludge, the cobalt concentration was 23 mg/kg (Young et al., 1973).

** In an inorganic speciation-in-sewage model, cobalt was calculated to be 97% solid CoS and 2% Co^{2+}. With the addition of model organics, the cobalt species were 95% CoS, 2% Co^{2+}, and 2% glutamate complex. With the addition of organics and adsorbing surface, the speciation was 93% CoS, 2% Co^{2+}, 2% glutamate complex, and 2% adsorbed. Experimental data, however, had shown that nickel, cobalt, and manganese were in approximately equal concentrations in the filterable (< 0.45 μm) and nonfilterable fractions.

*** The cobalt concentration in natural near shore sediments is 7.2 mg/kg (Young et al., 1973).

sewage sedimentation was about 6 mg/year/cm^2 compared with a literature value of 9 mg/year/cm^2 for natural sedimentation in this region. Of the total sewage particulate, only 0.5% was postulated to settle in the reduced area.

Young et al. (1973) calculated that, at the rate of discharge of Los Angeles sewage effluent (6 μg Co/liter) and sludge effluent (0.03 μg Co/liter) from the Hyperion Sewage Treatment Plant and the effluent (< 6 μg Co/liter) from the Terminal Island Plant, 3.0 MT cobalt was discharged annually to the southern California Bight.* This value compares with estimates of 5.3 MT cobalt/year added to the Bight from surface runoff (75% particulate), and a total of 14 MT cobalt/year after a consideration of all possible sources for ocean dumping (estimated cobalt concentration in effluents $\leq$ 0.5%). A total amount of 14 MT cobalt contributed to the Bight by anthropogenic sources is dwarfed by an estimate of 8,000 MT cobalt/year brought to the Bight by the California current.

(5) Industrial wastewater and potential spills

(a) Cobalt in industrial wastewaters: Information is sparse on the amounts of cobalt in wastewaters discharged to the environment by industries likely to be using cobalt in their processes. Somewhat more information is available on cobalt as a trace impurity in the major materials being processed and in their wastes.

The Bureau of Mines (George and Cochran, 1970) reported on the cobalt content of wastes from two industrial electroplating companies. The nitric acid rack-stripping stream in one plant accounted for an annual loss of 0.08 short ton (0.07 MT) cobalt. The concentration of cobalt in the wastes was < 0.1 to 6.7 μg/liter. There were no data on the effluent.

In its modeling of jobshop electroplaters based on a wide survey of the industry, Battelle excluded cobalt plating. Cuthbertson and Ponzer (1977) attempted another reasonably ambitious survey of U.S. electroplaters (491 companies),

---

* Values are averages of 24-hr composite samples collected August 24 and September 9, 1970. The analyses were done by atomic absorption.

trying to get information on the amounts of chromium, cobalt, and nickel in their wastes. Only seven replies contained useful information, and none of these listed the cobalt concentration in its waste. They also surveyed hazardous waste management facilities regarding the contents of these metals in industrial wastes. Of 24 wastes described in the replies, only one contained cobalt: a landfilled waste from an unidentified industry contained 0.13 to 1.3% Cr, 0.01 to 0.13% Co, and 0.001 to 0.13% Ni.

Allen et al. (1976) found 0.5 and 3 μg cobalt/liter in final effluents from two aircraft metal finishing plants. To judge from their data for plant streams (see Table IV-34), the various standard waste treatment practices for reducing heavy metals from industrial wastes were only 25 to 50% effective in removing cobalt.

At the Balkash Mining-Metallurgical Combine, Ilialetdninov et al. (1976) found cobalt in the humic-fulvic acid fraction of the secondary waste pond's sludge. Thus, 2% of the cobalt was in the humic acid fraction at 10 μg/liter, and 98% was in the fulvic acid fraction at 490 μg/liter.

Cobalt is a constituent of acid dyes; the average concentration in 300 dyes is 3.2 mg/kg. Its concentration in other types of dyes is < 1 mg/kg. Acid dyes also contain an average of 9 mg Cr/kg, 79 mg Cu/kg, and 37 mg Pb/kg. If only dyes contributed cobalt to dye plant effluents, the cobalt concentration would be < 0.3 μg/liter since the contaminants are diluted approximately 10,000-fold in untreated textile mill effluents (American Dye Manufacturers Institute, 1972). However, Netzer and Beszedits (1975) found < 10 to 90 μg Co/liter in the dye-bath effluent (pH 5 to 7) from an Ontario carpet mill that both spun and dyed synthetic fibers. Massive lime dosage of various dyebath effluents had no apparent effect on removing cobalt at these trace concentrations.

Most municipal incinerators discharge their wastewaters to municipal sewer systems. These wastewaters are heavily loaded with metals because water is used in spray chambers to remove fly ash from the gases and to quench hot residues before disposal. Law (1977) found, on the average, < 0.1 or < 0.2 mg Co/liter in spray and quench waters, respectively, from

TABLE IV-34

COBALT IN THE WATERS FROM TWO AIRCRAFT METAL FINISHING PLANTS, μg/liter
(Allen et al., 1976)

Plant	SSMS[a/]	INAA[b/]	ICPES[c/]
Plant L, May 1973 to September 1974			
1. Tapwater	< 0.5-0.5	2	< 5
2. General Waste	< 0.5-4	4	< 5
3. Holding Pond After Waste Treatment[d/]	< 0.5-4	4	8
4. Discharge to Surface Water (Includes Sanitary Wastes)	< 0.5	3	< 5
Plant B, January 1974			
1. Tapwater	-	< 0.5	-
2. Acid Rinse, pH 3	-	< 0.5	-
3. After Chromium Reduction, Lime, Ferrous Sulfate	-	< 0.5	-
4. Waste From Cyanide Rinse, pH 9	-	1	-
5. Waste From Cyanide Rinse After Treatment with Chlorine and OH^-	-	0.5	-
6. Clarifier Effluent Prior to Lagooning and Discharge	-	0.5	-

a/ SSMS = Spark source mass spectrometry.
b/ INAA = Instrumental neutron activation analysis.
c/ ICPES = Inductively coupled plasma emission spectroscopy.
d/ $FeSO_4$, lime, etc.

incinerators at Alexandria, Virginia, Washington, D.C., and Montgomery County, Maryland, during operation. Recycled incinerator spray-chamber effluents at the Alexandria plant on one occasion were found to contain 0.5 mg Co/liter ± 0.05 (five replicate samples).* On discharge, the effluents at the Washington, D.C. and Montgomery County plants contained, on the average, < 0.1 mg Co/liter.

Metry and Harris (1977) reported < 50 μg Co/liter in taconite suspensions at pH 1-3. Mercury and arsenic were the trace metals detected in leachates from the taconite stockpile.

* This represented an enrichment of 170 over the cobalt concentration found in the city water.

The Southeastern Environmental Research Laboratory of the EPA investigated the metal content of wastes from sugar beet processing plants (Alford, 1975). Cobalt was absent in samples from Kansas and Nebraska, but a Colorado holding pond water contained 10 μg Co/liter compared to 2 μg/liter in the upstream groundwater. Since sugar beet processing wastes would contain contaminants only from added limestone, sulfur dioxide, and possibly fly ash from the coal burned for process heat (Shreve, 1967), it is probable that the source of the cobalt is from the beet wastes themselves.

Bradford (1971) found averages of 4 to 6 μg Co/liter in the wastewater of two California food processing plants and averages of 2 to 11 μg Co/liter in wastewaters from six metal and chemical processing plants.

Barrett et al. (1974) found, among data from Corps of Engineers Discharge Permit Applications, a report of 0.5 g cobalt/0.91 MT of product in the wastes from an inorganic pigment plant. Wastes were neutralized and probably underwent sedimentation before discharge.

As an important industrial catalyst, cobalt may be found in aqueous process streams at 10 to 10,000 mg/liter (Rhodes, 1971).

Deep well injection of wastes (pH 3.8) from terephthalic acid manufactured by oxidation of _p_-xylene by Hercules, Inc., near Wilmington, North Carolina, continued from May 1968 to December 1972. Of the trace metals present in the waste, cobalt had one of the highest concentrations, 1,600 μg/liter (exceeded only by those of aluminum and iron). It was shown that the organic constituents contaminated the groundwater for up to 2,700 ft (823 m) from the injection site within 3 yr of monitoring observation wells (beginning in January 1971). Cobalt was detected at 20 μg/liter in the groundwater of one of the observation wells on June 14, 1973. Here, the cobalt concentration was exceeded by the concentration of barium (not reported for the original waste), chromium, iron, manganese, nickel, selenium, strontium, and zinc. Cobalt was probably adsorbed onto the precipitates of iron hydroxide and iron carbonates that plugged the injection zone adjacent to the wells, which led to the abandonment of the system (Leenheer et al., 1976).

Shaver et al. (1975) showed a flow sheet for titanium dioxide production by the sulfate process* in which 0 to 5% cobalt is said to be a typical ilmenite ore impurity and in which there was 1.3 to 9% cobalt in the digestion sludge. (No reference was given but the high value is clearly erroneous. Cobalt ores do not contain values that high.) Barksdale (1966) listed values of only 50 to 200 mg Co/kg in ilmenite ores.

Additional information on cobalt in industrial aqueous streams was discussed in Sections IV.B.2.b(1) and (2).

The Manufacturing Chemists of America organization has recommended disposal of packaged lots of cobaltous material by diluting, adding a slight excess of soda ash to precipitate Co^{2+}, adding slaked lime to precipitate sulfate, neutralizing with hydrochloric acid, and discharging the supernatant to sewer or stream. The sludge may be added to a landfill (Ottinger et al., 1973).

By the use of algae having stable cobalt concentration factors 76 to 342, Paoletti et al. (1976) removed cobalt from wastewaters with average efficiencies of 15.1 to 53.5%; 90% of the samples had an average cobalt removal efficiency 50 to 54%. Organic sludge and algae were removed continuously to prevent redissolution of adsorbed cobalt. The biofiltering watercourse had to be aerated to prevent eutrophication in hot summer months and to promote adequate cobalt removal in the cold of winter.

Friend et al. (1963) reported that 60% of the ^{60}Co added to the water of an artificial pond system was removed within 1 day by the biota and sediments; $\sim$ 96%, within 4 days. One day after dosing, the ^{60}Co activity in the suspended solids was three times higher than that in the water. Table IV-35 shows that coagulation and filtration removed 40 to 75% of the ^{60}Co, depending on the turbidity. Only 2 to 13% of the ^{60}Co was discharged to wastes.

---

* No cobalt is mentioned on the chloride process flow sheet.

TABLE IV-35

COBALT-60 REMOVAL DURING WATER TREATMENT, %
(Friend, 1963)

	Coagulation and Filtration	Cation Exchanger	Anion Exchanger (Removal of Negatively Charged Silicate Colloids)	To Waste
Low turbidity (~ 60 on Jackson scale) due to quiescence for 50 days	39.9	35.0	12.1	13.1
High turbidity (~ 1,000 on Jackson scale) due to agitation of bottom sediments	74.8	21.8	1.3	2.3

Ion-exchange resins with cobalt ion specificity include m-phenylenediamine, 8-hydroxyquinoline, and insoluble starch xanthate. The latter removed about 99% of the total cobalt present in solutions (Calmon, 1973; Wing and Rayford, 1977 and 1978).

Cadman and Dellinger (1974) reviewed techniques reported in the literature for removing heavy metals, including cobalt from process wastewaters and seawater:

- Chelation by ammonium pyrrolidine dithiocarbamate at pH 2.8 and extraction by methyl isobutyl ketone at pH 4 to 5, or chelation at pH 4.5 to 6 and extraction by chloroform.
- Foam separation involving the use of a surfactant and nitrogen gas.
- Ion exchange by Chelex-100 (a derivative of iminodiacetic acid), Amberlite, Permutit-S1005, and DeAcidite FF.
- Precipitation by hydroxides: Pb, Cu, Cr, Zn, and Co ions will precipitate from solution at pH 9.5 upon addition of 100 mg $Ca(OH)_2$ per liter.

- Dolomitic limestone and $Mg(OH)_2$ are also effective precipitants.

(b) Cobalt in potential spills to water: In a proposed rulemaking notice published December 30, 1975, in the Federal Register (Environmental Protection Agency, 1975), cobaltous materials were listed among the hazardous substances to be regulated in discharges into navigable waters or adjoining shorelines from vessels or facilities. Some of the cobaltous salts that had been listed in the advance notice were deleted because of their low potential for discharge. These were the acetate, chloride, citrate, iodide, nitrate, perchlorate, succinate, and sulfate salts. Co(II) bromide, fluoride, formate, and sulfamate were listed as hazardous materials based on the criterion that 50% of a test population of aquatic animals die within 96 hr if exposed to the chemicals at a concentration $\leq$ 500 mg/liter. The "hazardous quantity unit" chosen for each cobalt salt was 100 lb. These cobaltous salts were assigned to hazard category C, whose compounds had LC_{50}, 10 to $<$ 100 mg/liter. (Category A, the most hazardous, had compounds whose LC_{50} was $<$ 1 mg/liter and a harmful quantity designated as 1.0 lb).

A physical/chemical/dispersal factor reflecting the solubility, density, volatility, and propensity for water dispersal was calculated to be 0.62 for the cobaltous salts. This 0.62 was the proposed rate of penalty in dollars per pound discharged. However, by 1978 (Anonymous, 1978a), among the 270 chemicals that the U.S. EPA designated as hazardous, only three cobaltous salts remained: the bromide, formate, and sulfamate salts. Their penalty rates in dollars per pound discharged were 0.49, 7.50, and 0.75. These penalty rates compare with the highest penalty rates assessed: $880/lb for acrolein and $360/lb for several chlorinated pesticides such as endosulfan and DDT.

Amendments to these proposed rules published in the Federal Register on February 16, 1979 (Environmental Protection Agency, 1979), have substituted "reportable quantity" (to determine when spills must be reported) for the term "harmful quantity." The reportable quantity for each of three cobaltous salts was designated as 1,000 lb (454 kg). Penalty rates were invalidated. Discharges identified under various permits or permit applications are exempt from these regulations.

c. Air: Urban and smelter-related concentrations of cobalt in the ambient air generally average < 4 ng/m^3. Average values around 10 ng/m^3 in the National Air Surveillance Network are definitely unusual. The highest average cobalt concentration reported in ambient air was 48 ng/m^3 at Clydach, Wales, the site of nickel and, presumably, cobalt refining. In the nearby towns of Trebanos and Llansamlet, average values of 15.3 (0.8 to 90) ng Co/m^3 and 15.9 ng Co/m^3 were reported (Peirson et al., 1974; Pattenden, 1971, 1975, and 1977).

Values for cobalt in ambient air reported in the literature have been summarized in Table IV-36. Probably the most detailed information on cobalt concentrations at urban sites relative to their activities was reported by King et al. (1976) for Cleveland, Ohio. Their data appear in Table V-23 in Chapter V. Heavy industry and traffic appear to be related to the highest cobalt maxima reported. The metals industry of Cleveland is discussed in Chapter V with respect to ambient air levels at various monitoring sites.

A large part, if not all, of the cobalt in ambient air is usually attributed to suspended soil particles. Cobalt generally correlates well with elements whose presence in the air is most frequently due to soil, is on soil-size particles, and has a low enrichment factor.

The enrichment factor, EF, is the ratio of normalized atmospheric and crustal concentrations of an element, here cobalt. It may be defined as:

$$EF = \frac{[Co]/[\underline{n}] \text{ atmosphere}}{[Co]/[\underline{n}] \text{ crust}}$$

where $\underline{n}$ is an appropriate normalizing element. Among those that have been used are Al, Fe, Sc, Na, and Eu. When used, these elements are presumed not to have any anthropogenic source.

TABLE IV-36

COBALT IN AMBIENT AIR

Location	Cobalt Concentration Mean ng/m^3	Cobalt Concentration Range ng/m^3	No. of Samples	Method of Analysis[a/]	Remarks	Enrichment Factor (normalized to)	Authors
Northwest Indiana	1.1	0.47-2.6	25 Total	NAA	0.025 Detection limit. Sampling covered one 24-hr period.	3.5 (Fe)	Dams et al. (1971)
Industrial	1.5	0.92-2.6	10 } part of above				
Semirural	0.9	0.55-1.6	11 } part of above				
Background (Niles, Mich.)	0.95		1 } part of above		Analysis was replicated 2-4 times.		
Chicago, Illinois Metropoliton area		0.3-4.8	22	NAA	Cobalt correlated at α 0.01 with Mg, Zn, La, Eu.		Brar et al. (1970)
Cleveland, Ohio					See also Table V-23. Mass median diameters of particulate fractions containing Co were 3.3 to 4.8 μm.		
Urban	2.58 ± 0.52			INAA		1.3 (Al)	King et al. (1976)
Suburban	0.75 ± 0.15						Leibecki et al. (1975)
Boston, Mass. Metropolitan area		0.62-2.3		INAA	Size distribution curves show cobalt mixed over range of 0.5-16 μ.		Gladney et al. (1974)
Chadron, Nebraska							
Summer 1973	3.2	1.6-5.1	17	AAS			Struempler (1975)
1973	3.3	1.6-5.1	52			for 1973 value:	
Summer 1974	2.2	1.2-3.5	17			20 (Al) 22 (Mn)	
San Francisco Bay area	1.0	0.46-1.7	9	NAA	One-day samples. About 33% was calculated as being soil derived.		John et al. (1973)
Kellogg, Idaho (vicinity of Bunker Hill Lead Smelter)	3.21	1.16-7.49		INAA	Collected at center of town; source attributed to natural soils.	14 (Sc)	Ragaini et al. (1977)
Tucson, Arizona (Vicinity of)			3914	AAS	50% of Cu produced in U.S. is from an area 120 miles around Tucson.		Moyers et al. (1977)
Urban	1.9						
Rural	0.7						
Corvallis, Oregon	0.4			INAA			Cheung and Loveland (1979)
Portland, Oregon	4.2						
Airport on Atlantic Seaboard. Aerosol from aircraft engines collected 90 m from runway	< 3.99		1 Anderson Cascade Sampler	PIXE	Only stage 3 (3.3 μ) concentration given; 157 airplanes took off and landed during test.		Jolly et al. (1975)

TABLE IV-36 (continued)

Location	Cobalt Concentration Mean ng/m^3	Cobalt Concentration Range ng/m^3	No. of Samples	Method of Analysis[a]	Remarks	Enrichment Factor (normalized to)	Authors
National Air Surveillance Network, 1958-1965:			3/29		Miller et al. (1974) found 20 ± 5 mg Co/kg in airborne soil dust in Pasadena, a higher value than in the soil itself--8 ± 2 mg Co/kg. This agrees with Calop et al. (1977), who found an average 15 mg/kg in suspended particulates in Lyons, France.		Environmental Protection Agency (1971)
Pasadena, Calif.,	8						
Elizabeth, N.J.,	7						
Loquillo Mountains, Puerto Rico	8						
National Air Surveillance Network 1970-1974. Sites where Co was detected:			270 Stations, 10% nonurban	High-volume emission spectrometry	Detection limit 0.52-0.76 ng/m^3, depending on year. Sites are sampled for 24 hr at least every 6 days (Thompson, 1979).		Aklend (1976)
Rock Island, Ill.							
2nd quarter, 1974	10						
4th quarter, 1974	8						
Ironton, Ohio		< Detection limit to 8.5	6/12				
Johnstown, Penn. 4th quarter, 1970	7						
Reading, Penn.		< Detection limit to 1.4	3/14				
Guayanilla, Puerto Rico, 2nd quarter, 1974	13		1/13				
Nashville, Tenn.		< Detection limit to 10	2/16				
Toronto, Ont., Can:				IPAA	Because of its low enrichment factor and its concentration on larger particles (mass median diameter ~ 4 μm, 19% < 1.1 μm), and its correlation with Al, Mg, Ca, and Ti, Co in air was assumed to be derived from soils.	3.2 (Al)	Jervis et al. (1976)
Near Univ. of Toronto and a major intersection.	0.8						
Industrial sites within 100 m. of secondary metals plants.	1.3		1/4				
City average.	1.0						
Flin Flon, Man., Canada	2			Not given	Site of copper smelter. Data acquired before emission controls.		Robertson (1959)

TABLE IV-36 (continued)

Location	Cobalt Concentration Mean ng/m^3	Cobalt Concentration Range ng/m^3	No. of Samples	Method of Analysis[a]	Remarks	Enrichment Factor (normalized to)	Authors
Ghent, Belgium				NAA	Co normally present in soil-like ratio; 10-fold increase probably due to cool fly ash and dust-rise by automobile.	Industrial: 17 (Al) Residential: 12 (Al)	Demuynck et al. (1976)
Beginning of large stationary anticyclone.		0.33–0.48	2				
During peak period.			3				Heindryckx (1976)
Breakup of anticyclone.	0.85		1		Duration, Sept. 16–24, 1979.		
Belgium, 14 sites		0.47–12	14	NAA	24-hr samples from 3 industrial sites, 7 residential and 4 rural, Oct. 1972. Average concentration in particulates, 28 mg Co/kg. There were anomalously high concentrations of several elements associated with the copper smelter near Antwerp, but not cobalt.	7.4 (Si)	Rahn et al. (1973)
Turin, Italy		0.03–1.22		AAS	Sampled in December 1974.		Natule and Braja (1978)
Midlands, England	2.0			Emission spectrography	Lumped monthly samples.		Turner (1975)
Chilton, Oxfordshire, England	0.4		500 filters grouped in batches of 3 months	NAA	Sampling period 1957–1974; observable seasonal variations, periodic and long-term fluctuations.		Salmon et al. (1977)
Wraymires, England Rural	0.32 ng/kg air			NAA	Weekly or monthly samples from Jan.-Dec. 1971; concentration in rain = 0.250 ng/ℓ; washout factor 780.		Peirson et al. (1973)
3 sites in U.K. 3 in North Sea 1 in The Netherlands		0.051–0.33 ng/kg air		NAA	Sampling between June 1972–May 1973, concentration in rain = 0.190–0.68 μg/ℓ.		Peirson et al. (1974)
Six sites in U.K.		0.066–0.76 ng/kg air		NAA	Jan.-Dec. 1972.	~ 4–7 (Sc)	Peirson et al. (1974)

TABLE IV-36 (continued)

Location	Cobalt Concentration Mean ng/m^3	Cobalt Concentration Range ng/m^3	No. of Samples	Method of Analysis[a/]	Remarks	Enrichment Factor (normalized to)	Authors
Trebanos, Wales, U.K.	8.5 ng/kg air			NAA	Jan.-Dec. 1972. All 3 Welsh sites in the vicinity of nickel refinery in Clydach. Co is a by-product.	~ 90 (Sc)	Pierson et al. (1974)
Trebanos, Wales, U.K.	15.3	0.8-90		NAA		131 (Sc)	Pattenden (1971, 1975, and 1977)
Llansamlet, Wales U.K.	15.9			NAA		114 (Sc)	Pattenden (1971, 1975, and 1977)
Clydach, Wales, U.K.	48			NAA		240 (Sc)	Pattenden (1971, 1975, and 1977)
Karlsruhe, Federal Republic of Germany	1.1		24	NAA	Samples taken by running sampler for 5 min/hr/month from 3/74-2/76.		Vogg and Härtel (1977)
Beer Sheba, Israel Semi-arid zone		4-22.0		NAA	Dead Sea as possible source; 2-day samples over 2 months, Apr.-May 1974.		Shani and Cohen (1977)
Beer Sheba, Israel	1.65			NAA	Soil in desert rich in manganese. May-July 1975.		Shani and Haccoun (1977)
Rural desert area south of city.	4.48						
During day		2.23-5.08					
During night		0.72-3.0					
Ankara, Turkey	1.3			NAA		1.1-1.2 (Fe)	Olmez and Aras (1977)
Budapest, Hungary				NAA	Four sites sampled between Sept. 1973-June 1974; sites were 50 to 500 m. from reactor.		Ördögh and Kālmān (1975)
Site 1	0.09-1.4						
Site 2	0.07-0.16						
Site 3	0.05-0.61						
Site 4 in the vicinity of a nuclear reactor	0.165-3.42						
Osaka, Japan district, winter,	3.67	2.0-5.9		INAA	Data from 4 sampling stations with day/night filters for two months; 2/15/70-3/14/70 and 6/7/70-7/6/70.		Mamuro et al. (1970)
summer.	3.57	2.3-5.8	32				

TABLE IV-36 (continued)

Location	Cobalt Concentration Mean ng/m^3	Cobalt Concentration Range ng/m^3	No. of Samples	Method of Analysis[a]	Remarks	Enrichment Factor (normalized to)	Author
15 areas in Japan				NAA			Mamuro et al. (1973)
A. Unpolluted rural areas		0.08-0.4					
B. Heavy industrial area		1.0-6.0					
C. Small cities with little industry		0.29-3.2					
Japan (indoor air)				AAS	Indoor air has half the suspended particles of outside air.		Tsuji et al. (1973)
Office, day,	10						
night;	2						
with air conditioning, day,	6						
night.	3						
Department store, day.	7						
Outdoor air, day,	7						
night.	7						
Lower Troposphere over World Oceans:	($ng/m^3 \times 10^6$:)			Oxygen/argon spectrographic technique	Cobalt associated with soil-sized particles (~ 0.5-10 μm) and concentrations consistent with crustal abundance.	(Fe)	Chester et al. (1974)
North Atlantic		0.003-0.138				0.3-0.5	
South Atlantic		0.003-0.033				0.3-0.5	
Indian Ocean		0.003-0.053				0.4-1.1	
China Sea		0.002-0.123				0.3-0.4	
Inland Sea, Japan	0.128					0.4	
Sea of Japan	0.042					0.4	
Java Sea	0.021					0.7	
North Atlantic		0.006-0.09		NAA	Collection efficiency better than 95% for particles with radii ≥ 0.1 μm. Samples collected north of 30° N. with a few samples from Bermuda.	2.4 (Al)	Duce et al. (1975)
Bermuda, SW Coast 32° 14' 52" N, 64° 51' 58" W.	0.042	< 0.005-0.50	29	NAA	Sampling in 1973, ~ 60 m. above sea.	1.8 (Al)	Duce et al. (1976)
"Dust" over Ocean	(mg/kg)			NAA	"Dust" refers to insoluble fraction in suspended particulate.		Bressan et al. (1973)
Atlantic	43						
Pacific	18						
Caribbean	16						
Greenland Sea	46						
East Coast U.S.	50						

TABLE IV-36 (concluded)

Location	Cobalt Concentration		No. of Samples	Method of Analysis[a]	Remarks	Enrichment Factor (normalized to)	Author
	Mean ng/m^3	Range ng/m^3					
South Pole	0.00049			INAA	Samples taken 5 km upwind of Amundsen-Scott station	~ 3.5 (Al)	Maenhaut and Zoller (1977)
South Pole	0.00084	0.00036-0.00121	30	NAA	Samples collected upwind of Amundsen-Scott station	4.7 (Al)	Zoller et al. (1974)

a/ AAS = Atomic Absorption Spectrometry
INAA = Instrumental Neutron Activation Analysis
IPAA = Instrumental Proton Activation Analysis
NAA = Neutron Activation Analysis
PIXE = Proton-induced X-ray Emission

Enrichment factors reported for cobalt range from 0.3 to 0.7 (normalized to iron) in the lower troposphere over world oceans (Chester et al., 1974) to 240 (normalized to Sc) at Clydach, Wales (Pattenden, 1975).

In recent years, data gathering from remote areas has been emphasized to place the environmental input of human activity into perspective. Values at the South Pole have been reported at 0.00049 and 0.00084 ng/m^3 by Maenhaut and Zoller (1977) and Zoller et al. (1974), respectively. Samples from wind systems over open ocean have cobalt concentrations in soil-size particles of the same order of magnitude as those of average crustal material (Chester et al., 1974). The major wind systems contribute soil-size particles to the dust veil in the lower troposphere over the oceans and make a major contribution of trace elements to the ocean surface (Chester, 1972; cited in Chester et al., 1974). Open-ocean trade winds such as the southeast trades of the southern Atlantic and the southeast monsoons of the south China Sea have cobalt concentrations on the order of 0.001 to 0.01 ng/m^3. Cobalt concentrations in wind systems blowing off continents and islands such as the northeast trades off northeast Africa and variable winds off Japan into the Inland Sea of Japan have been reported at 0.138 x 10^{-6} and 0.128 x 10^{-6} ng/m^3, respectively (Chester et al., 1974).

Abel, K. H., and L. A. Rancitelli, "Major, Minor, and Trace Element Composition of Coal and Fly Ash, as Determined by Instrumental Neutron Activation Analysis" in Trace Elements in Fuel, Advances in Chemistry Series No. 141, American Chemical Society, Washington, D.C., 1975, pp. 118-138.

Abernethy, R. F., and F. H. Gibson, Rare Elements in Coal, Bureau of Mines Information Circular 8163, U.S. Department of the Interior, Washington, D.C., 1963.

Abernethy, R. F., M. J. Peterson, and F. H. Gibson, Spectrochemical Analyses of Coal Ash for Trace Elements, Report of Investigations 7281, Bureau of Mines, U.S. Department of the Interior, Washington, D.C., 1969.

Agrawal, Y. K., K. P. S. Raj, and M. R. Patel, "Metal Contents in the Drinking Water of Cambay [India]," Water, Air, Soil Pollut., **9**(4), 429-431 (1978).

Akland, G. G., Air Quality Data for Metals 1970 Through 1974 from the National Air Surveillance Networks, EPA-600/4-76-041, PB 260 905, National Technical Information Service, U.S. Department of Commerce, Springfield, Virginia, August 1976.

Alford, A. L., Environmental Application of Advanced Instrumental Analyses: Assistance Projects FY 72, PB 228 147, EPA-660/2-73-013, National Technical Information Service, U.S. Department of Commerce, Springfield, Virginia, September 1973.

Alford, A. L., Environmental Applications of Advanced Instrumental Analyses: Assistance Projects, FY 73, EPA 660/2-74-078, U.S. Government Printing Office, Washington, D.C., August 1974.

Alford, A. L., Environmental Applications of Advanced Instrumental Analyses: Assistance Projects, FY 74, EPA-660/4-75-004, National Technical Information Service, U.S. Department of Commerce, Springfield, Virginia, June 1975.

Ali, S. A., M. G. Gross, and J. R. L. Kishpaugh, "Cluster Analysis of Marine Sediments and Waste Deposits in New York Bight," Environ. Geol., 1(3), 143-148 (1975).

Allen, F. R., A. W. Garrison, and C. E. Taylor, "Chemical Analysis of Metal Finishing Wastewaters" in Preprints of Papers Presented at the 171st National Meeting, Division of Environmental Chemistry, American Chemical Society, New York, New York, April 4-9, 1976.

Allen, W., E. Altherr, R. H. Horning, J. C. King, J. M. Murphy, W. E. Newby, and M. Saltzman," The Contribution of Dyes to the Metal Content of Textile Mill Effluents," Textile Chem. Colorist, 4(12), 275-277 (1972).

Andelman, J. B., "Chapter 18. The Effect of Water Treatment and Distribution on Trace Element Concentrations" in Chemistry of Water Supply Treatement and Disposal, A. J. Rubin, Ed., Ann Arbor, Michigan, 1974, pp. 423-440.

Andelman, J. B., and M. A. Shapiro, "Changes in Trace Element Concentrations in Water Treatment and Distribution Systems" in Trace Substances in Environmental Health VI, D. D. Hemphill, Ed., Proceedings of University of Missouri's 6th Annual Conference on Trace Substances in Environmental Health, Columbia, Missouri, June 13-15, 1972, University of Missouri, Columbia, Missouri, 1973, pp. 87-94.

Andersson, A., "Some Aspects on Significance of Heavy Metals in Sewage Sludge and Related Products Used as Fertilizers," Swed. J. Agric. Res., 7(1), 1-5 (1977).

Angino, E. E., "Geochemistry of Antarctic Pelagic Sediments," Geochim. Cosmochim. Acta, 30, 939-961 (1966).

Angino, E. E., Consulting Geochemist, personal communication to B. L. Carson, July 1979.

Angino, E. E., O. K. Galle, and T. C. Waugh, "Fe, Mn, Ni, Co, Sr, Li, Zn, and SiO_2 in Streams of the Lower Kansas River Basin," Water Resources Res., 5(3), 698-705 (1969).

Anonymous, "Ocean Dumping Rose in 1974, Environmental Protection Agency Reports," Toxic Mater. News, **2**(16), 128 (1975).

Anonymous, "Drinking Water and Health. Recommendations of the National Academy of Sciences," Fed. Reg., **42**(132), 35764-35779 (1977).

Anonymous, "EPA Names Hazardous Water Pollutants, Issues Rules for Their Spill Control," Toxic Mater. News, **5**(10), 60-62 (1978a).

Anonymous, "Power Plant Ash Disposal a Growing Problem," Chem. Eng. News, **56**(45), 26 (1978b).

Ball, J. S., W. J. Wenger, H. J. Hyden, C. A. Horr, and A. T. Myers, "Metal Content of Twenty-Four Petroleums," J. Chem. Eng. Data, **5**(4), 553-557 (1960).

Banning, D. L., "Copper-Nickel-Cobalt Associations in Marine Manganese Nodules" (abstr.), Trans. Am. Geophys. Union, **56**(12), 1000 (1975).

Barksdale, J., Titanium. Its Occurrence, Chemistry, and Technology, 2nd ed., The Ronald Press Company, New York, New York, 1966.

Barrett, W. J., G. A. Morneau, and J. J. Roden, III, Waterborne Wastes of the Paint and Inorganic Pigments Industries, PB-232 019, Southern Research Institute for the Environmental Protection Agency, National Technical Information Service, U.S. Department of Commerce, Springfield, Virginia, 1974.

Barufke, W., "Über die Bedeutung des Kobalts als Spurenelement auf brandenburgischen Grünlandböden" [The Importance of Cobalt as a Micro-Element in Meadow Soils of Brandenberg District], Wiss. Z. Humboldt Univ. Berl. Math.-Naturwiss. Reihe, **11**, 135-144 (1962).

Bascom, W., "The Disposal of Waste in the Ocean," Sci. Am., **231**(2), 16-25 (1974).

Beck, R. A., and J. J. Brink, "Production of Cobalamins During Activated Sewage Sludge Treatment," Environ. Sci. Technol., 12(4), 435-438 (1978).

Bell, K. G., Uranium and Other Trace Elements in Petroleums and Rock Asphalts, U.S. Geol. Survey Professional Paper No. 356-B, U.S. Government Printing Office, Washington, D.C., 1960, pp. 45-65.

Bennett, E. H., Reconnaissance Geology and Geochemistry of the Blackbird Mountain Panther Creek Region, Lemhi County, Idaho, Idaho, Bur. Mines Geol., Pam., 167, Idaho, Bur. Mines Geol., Moscow, Idaho, 1977, pp. 1-108; Chem. Abstr., 88, 92462v (1978).

Benzhitskii, A. G., "Kobal't v Forme Kobalaminov v Vode i Organizmakh Mirovogo Okeana" ["Cobalt in the Form of Cobalamin in the Water and Organisms of the Pacific Ocean"] in Khemoradioekologiya Pelagiali Bentali, G. G. Polikarpov, Ed., Naukova Dumka, Kiev, USSR, 1974, pp. 56ff.

Biesinger, K. E., and G. M. Christensen, "Effects of Various Metals on Survival, Growth, Reproduction, and Metabolism of Daphnia magna," J. Fisheries Res. Board Can., 29(12), 1691-1700 (1972).

Billings, C. E., "Technical Sources of Air Pollution" in Industrial Pollution, N. I. Sax, Ed., Van Nostrand Reinhold Company, New York, New York, 1974, pp. 350-408.

Block, C., and R. Dams, "Concentration - Data of Elements in Liquid Fuel Oils as Obtained by Neutron Activation Analysis," J. Radioanal. Chem., 46, 137-144 (1978).

Bogen, J., "Trace Elements in Precipitation and Cloud Water in the Area of Heidelberg, Measured by Instrumental Neutron Activation Analysis," Atmos. Environ., 8(8), 835-844 (1974); Chem. Abstr., 81, 175307e (1974).

Borneff, J., "Metalle und Metalloide im Wasser (Beurteilung ihrer Bedeutung für den Menschen)" ["Metals and Metalloids in the Water (Their Importance to Humans"], Zentr. Bakteriol. Parasitenk. Abt. I. Orig., 158(6), 524-529 (1974).

Bower, P. M., H. J. Simpson, S. C. Williams, and Y. H. Li, Heavy Metals in the Sediments of Foundry Cove, Cold Spring, New York," Environ. Sci. Technol., **12**(6), 683-687 (1978).

Boyle, R. W., Elemental Associations in Mineral Deposits and Indicator Elements of Interest in Geochemical Prospecting, Paper 68-58, Department of Energy, Mines and Resources, Geological Survey of Canada, 1969, 45 pp.

Bradford, G. R., "Trace Elements in the Water Resources of California," Hilgardia, **41**(3), 45-53 (1971).

Bradford, G. R., F. L. Bair, and V. Hunsaker, "Trace and Major Element Content of 170 High Sierra Lakes in California," Limnol. Oceanog., **13**(3), 526-530 (1968).

Bradford, G. R., F. L. Bair, and V. Hunsaker, "Trace and Major Element Contents of Soil Saturation Extracts," Soil Sci., **112**(4), 225-230 (1971).

Bradford, W. L., Distribution and Movement of Zinc and Other Heavy Metals in South San Francisco Bay, California, PB 251 111, USGS/WRD/WRI-76/014, National Technical Information Service, U.S. Department of Commerce, Springfield, Virginia, February 1976.

Brar, S. S., D. M. Nelson, J. R. Kline, P. F. Gustafson, E. L. Kanabrocki, C. E. Moore, and D. M. Hattori, "Instrumental Analysis for Trace Elements Present in Chicago Area Surface Air," J. Geophys. Res., **75**(15), 2939-2945 (1970).

Bressan, D. J., R. A. Carr, and P. E. Wilkniss, "Geochemical Aspects of Inorganic Aerosols near the Ocean-Atmosphere Interface" in Trace Elements in the Environment, Advances in Chemistry Series No. 123, E. L. Kothny, Ed., American Chemical Society, Washington, D.C., 1973, pp. 17-30.

Brown, H. R., and D. J. Swaine, "Inorganic Constituents of Australian Coals. II. Formation and Composition of Boiler Deposits from Combustion of Australian Coals," J. Inst. Fuel, **37**(285), 434-440 (1964); Chem. Abstr., **62**, 369f (1965).

Bryan, D. W., V. P. Guinn, R. P. Hackleman, and H. R. Lukens, Development of Nuclear Analytical Techniques for Oil Slick Identification (Phase I), GA 9889, Gulf General Atomic, Inc., National Technical Information Service, U.S. Department of Commerce, Springfield, Virginia, 1970.

Bulthuis, D. A., J. R. Craig, and C. D. McNabb, "Metal Dynamics in Municipal Stabilization Ponds" in Trace Substances in Environmental Health - VII, D. D. Hemphill, Ed., Proceedings of University of Missouri's 7th Annual Conference, June 12-14, 1973, Columbia, Missouri, University of Missouri, Columbia, Missouri, 1974, pp. 127-135.

Burnham, C. W., Metallogenic Provinces of the Southwestern United States and Northern Mexico, New Mexico Bureau of Mines and Mineral Resources, Bulletin No. 65, New Mexico Institute of Mining and Technology, Socorro, New Mexico, 1959, pp. 1-76.

Cadman, T. W., and R. W. Dellinger, "Techniques for Removing Metals from Process Wastewater," Chem. Eng., **81**(8), 79-85 (1974).

Calmon, C., "Trace Heavy Metals in Water: Removal Processes by Ion-Exchange" in Traces of Heavy Metals in Water Removal Processes and Monitoring, J. E. Sabadell, Ed., U.S. Environmental Protection Agency, Region II, November 1973, pp. 7-42.

Calop, J., P. Isoard, and R. Fontanges, "Incidences Physio-Pathologiques Possibles de l'Ihalation par l'Homme des Elements Traces Presents dans les Poussieres Atmospheriques Urbaines," ["Physiopathological Conditions Possible from the Inhalation of Trace Elements in Urban Atmospheric Dusts"], Bull. Med. Leg., Urgence Med., Cent. Anti-Poisons, **20**(4), 403-417 (1977).

Cannon, H. L., "Trace Element Excesses and Deficiencies in Some Geochemical Provinces of the United States" in Trace Substances in Environmental Health-III, D. D. Hemphill, Ed., Proceedings of University of Missouri's 3rd Annual Conference on Trace Substances in Environmental Health, Columbia, Missouri, June 24-26, 1969, University of Missouri, Columbia, Missouri, 1970, pp. 21-43.

Capar, S. G., J. T. Tanner, M. H. Friedman, and K. W. Boyer, "Multielement Analysis of Animal Feed, Animal Wastes, and Sewage Sludge," Environ. Sci. Technol., **12**(7), 785-790 (1978).

Carr, M. H., and K. K. Turekian, "The Geochemistry of Cobalt," Geochim. et Cosmochim. Acta, **23**, 9-60 (1961).

Carter, J. A., D. L. Donohue, J. C. Franklin, and R. W. Stelzner, "Environmental and Fuel Materials Analyses by Multi-Element Isotope Dilution Spark-Source Mass Spectrometry" in Trace Substances in Environmental Health - IX, D. D. Hemphill, Ed., Proceedings of University of Missouri's 9th Annual Conference on Trace Substances in Environmental Health, Columbia, Missouri, June 10-12, 1975, University of Missouri, Columbia, Missouri, 1976, pp. 303-310.

Casagran, D. J., and L. D. Erchull, "Metals in Okefenokee Peat-Forming Environments - The Relation to Constituents Found in Coal," Geochim. et Cosmochim. Acta, **40**(4), 387-393 (1976).

Chattopadhyay, A., "Optimal Use of Instrumental Neutron and Photon Activation Analyses for Multielement Determinations in Sewage Sludges," J. Radioanal. Chem., **37**, 785-799 (1977).

Cherry, D. C., and R. K. Guthrie, "Toxic Metals in Surface Waters from Coal Ash," Water Resources Bull., **73**(6), 1227-1236 (1977).

Chester, R., S. R. Aston, J. H. Stoner, and D. Bruty, "Trace Metals in Soil-Sized Particles from the Lower Troposphere Over the World Ocean," J. Rech. Atmos., **8**(3-4), 777-789 (1974).

Cheung, I., and W. D. Loveland, "Indoor/Outdoor Atmospheric Trace Element Abundance Ratios," ACS Div. Environ. Chem. Preprints, **19**(1), 813-814 (1979).

Chichilo, P., and C. W. Whittaker, "Trace Elements in Agricultural Limestones of the United States," Agronomy, **53**, 139-144 (1961).

Christensen, E. R., J. Scherfig, and M. Koide, "Metals from Urban Runoff in Dated Sediments of a Very Shallow Estuary," Environ. Sci. Technol., **12**(10), 1168-1173 (1978).

Chu, T.-Y. J., R. J. Ruane, and G. R. Steiner, "Characteristics of Wastewater Discharges from Coal-Fired Power Plants," Proc. Ind. Waste Conf. 1976, **31**, 690-712 (1977).

Clark, L. J., and W. L. Hill, "Occurrence of Manganese, Copper, Zinc, Molybdenum, and Cobalt in Phosphate Fertilizers and Sewage Sludge," J. Assoc. Off. Anal. Chem., **41**(3), 631-637 (1958).

Clarke, F. W., and H. S. Washington, "The Average Chemical Composition of Igneous Rocks," Proc. Nat. Acad. Sci., **8**, 108-115 (1922).

Coles, D. G., R. C. Ragaini, J. M. Ondov, G. L. Fisher, D. Silberman, and B. A. Prentice, "Chemical Studies of Stack Fly Ash from a Coal-Fired Power Plant," Environ. Sci. Technol., **13**(4), 455-459 (1979).

Collinson, C., and N. F. Shimp, Trace Elements in Bottom Sediments from Upper Peoria Lake, Middle Illinois River--A Pilot Project, Environmental Geology Notes No. 56, Illinois State Geological Survey, Urbana, Illinois, September 1972.

Colombo, U. P., G. Sironi, G. B. Fasolo, and R. Malvano, "Systematic Neutron Activation Technique for the Determination of Trace Metals in Petroleum," Anal. Chem., **36**(4), 802-807 (1964).

Connor, J. J., J. A. Erdman, J. D. Sims, and R. J. Ebens, "Roadside Effects on Trace Element Content of Some Rocks, Soils, and Plants of Missouri" in Trace Substances in Environmental Health-IV, D. D. Hemphell, Ed., Proceedings of University of Missouri's 4th Annual Conference on Trace Substances in Environmental Health, University of Missouri, Columbia, Missouri, June 23-25, 1970, University of Missouri, Columbia, Missouri, 1971.

Costescu, L. M., and T. C. Hutchinson, "The Ecological Consequences of Soil Pollution by Metallic Dust from the Sudbury Smelters," Proc. Inst. Environ. Sci., 18th Tech. Version, 540-545 (1972).

Cralley, L., M. M. Key, D. H. Groth, W. S. Lainhart, and R. M. Ligo, "Fibrous and Mineral Content of Cosmetic Talcum Products," Am. Ind. Hyg. Assoc. J., **29**, 350-354 (1968).

Craun, G. F., and L. J. McCabe, "Problems Associated with Metals in Drinking Water," J. Am. Water Works Assoc., **67** (11), 593-599 (1975).

Crockett, A. B., "Terrestrial Monitoring of Elemental Contaminants Around Geothermal Power Plants" in Geotherm. Environ. Semin., 1976, F. L. Tucker, and M. D. Anderson, Eds., Geothermal Semin., Lakeport, California, 1977, pp. 205-211; Chem. Abstr., **88**, 94195j (1978).

Cuthbertson, G. R., and R. J. Ponzer, Survey of Metal-Containing Wastes for Chromium, Cobalt and Nickel Content, USPM Purchase Order P 4161072, Prepared for the U.S. Department of the Interior, Bureau of Mines, by the Department of Engineering Management, University of Missouri-Rolla, Rolla, Missouri, August 1977.

Dams, R., J. A. Robbins, K. A. Rahn, and J. W. Winchester, "Quantitative Relationships Among the Trace Elements Over Industrialized N.W. Indiana," Nucl. Technol. Environ. Pollut., Proc. Symp., **1970**, 139-157 (1971).

DeGroot, A. J., E. Allersma, N. DeBruin, and J. P. W. Houtman, "Use of Activatable Tracers," Tech. Rep. Ser., I.A.E.A., **145**, 151-166 (1973).

Demuynck, M., K. A. Rahn, M. Janssens, and R. Dams, "Chemical Analysis of Airborne Particulate Matter During a Period of Unusually High Pollution," Atmos. Environ., **10**(1), 21-26 (1976); Chem. Abstr., **84**, 94864u (1976).

Desborough, G. A., J. K. Pitman, and C. Huffman, Jr., "Concentration and Mineralogical Residence of Elements in Rich Oil Shales of the Green River Formation, Piceance Creek Basin, Colorado, and the Uinta Basin, Utah - a Preliminary Report," Chem. Geol., 17(1), 13-26 (1976).

Dreher, G. B., and J. A. Schleicher, "Trace Elements in Coal by Optical Emission Spectroscopy" in Trace Elements in Fuel, Advances in Chemistry Series No. 141, American Chemical Society, Washington, D.C., 1975, pp. 35-47.

Duce, R. A., G. L. Hoffman, and W. H. Zoller, "Atmospheric Trace Metals at Remote Northern and Southern Hemisphere Sites: Pollution or Natural?," Science, 187(4171), 59-61, (1975).

Duce, R. A., G. L. Hoffman, B. J. Ray, I. S. Fletcher, G. T. Wallace, J. L. Fasching, S. R. Pietrowicz, P. R. Walsh, E. J. Hoffman, J. M. Miller, and J. L. Heffter, "Trace Metals in the Marine Atmosphere: Sources and Fluxes" in Marine Pollutant Transfer, Lexington Books, D.C. Heath and Company, Lexington, Massachusetts, 1976, pp. 77-119.

Durfor, C. N., and E. Becker, Public Water Supplies of the 100 Largest Cities in the United States, 1962, U.S. Geological Survey Water - Supply Paper 1812, U.S. Department of the Interior, U.S. Government Printing Office, Washington, D.C., 1962.

Durum, W. H., "Occurrence of Trace Elements in Water" in Proceedings, Conference on Physiological Aspects of Water Quality, Washington, D.C., September 8-9, 1960.

Durum, W. H., J. D. Hem, and S. G. Heidel, Reconnaissance of Selected Minor Elements in Surface Waters of the United States, October 1970, Geological Survey Circular 643, U.S. Geological Survey, U.S. Department of the Interior, Washington, D.C., 1971.

Duursma, E. K., "Specific Activity of Radionuclides Sorbed by Marine Sediments in Relation to the Stable Element Composition" in Radioactive Contamination of the Marine Environment, Symposium Proceedings, Seattle 1972, International Atomic Energy Agency, Vienna, Austria, 1973, pp. 57-71.

Ebens, R. J., and J. J. Connor, "Geochemical Survey of Geologic Units" in Geochemical Survey of Missouri. Plans and Progress for Fourth Six-Month Period (January-June, 1971), Branch of Regional Geochemistry, U.S. Geological Survey, Denver, Colorado, June 1971, pp. 9-22.

Egan, A., and N. M. Spyrou, "Determination of Heavy Metals in Sewage-Based Fertilizers Using Short-Lived Isotopes," J. Radioanal. Chem., **37**, 775-784 (1977).

Eisler, R., R. L. Lapan, G. Telek, E. W. Davey, A. E. Soper, and M. Barry, "Survey of Metals in Sediments Near Quonset Point, Rhode Island," Mar. Pollut. Bull., **8**(11), 260-264 (1977).

Elderfield, H., and A. Hepworth, "Diagenesis, Metals and Pollution in Estuaries," Mar. Pollut. Bull., **6**(6), 85-87 (1975).

Environmental Protection Agency, Air Quality Data for 1967 from the National Air Surveillance Networks, Division of Atmospheric Surveillance, Environmental Protection Agency, Research Triangle Park, North Carolina, August 1971 (revised 1971).

Environmental Protection Agency, Helena Valley, Montana, Area Environmental Pollution Study, AP-91, Research Triangle Park, North Carolina, 1972.

Environmental Protection Agency, "Hazardous Substances. Designation, Removability, Harmful Quantities, Penalty Rates," Federal Register, **40**(250), 59960-60017 (1975).

Environmental Protection Agency, "Environmental Protection Agency, Water Program, Designation of Hazardous Substances, Amendment and Addition of Definitions," Federal Register, **44**(34), 10266-10284 (1979).

Erickson, R. L., A. T. Myers, and C. A. Horr, "Association of Uranium and Other Metals with Crude Oil, Asphalt and Petroliferous Rock," Bull. Am. Assoc. Petrol. Geolog., **38**(10), 2200-2218 (1954).

Evrard, M., G. Michard, and D. Renard, "Dosage des ions manganese, cobalt, nickel et cuivre dans les eaux interstitielles des sediments oceaniques" ["Determination of Manganese, Cobalt, Nickel, and Copper in Oceanic Sediment Pore Waters"], C. R. Hebd. Seances Acad. Sci., Ser. C, **281**(17), 681-682 (1975).

Feldman, M. H. Trace Materials in Wastes Disposed to Coastal Waters: Fates, Mechanisms, and Ecological Guidance and Control, PB 202 346, Working Paper No. 78, Federal Water Quality Administration, Northwest Region, National Technical Information Service, U.S. Department of Commerce, Springfield, Virginia, July 1970.

Filby, R. H., "The Nature of Metals in Petroleum" in The Role of Trace Metals in Petroleum, T. F. Yen, Ed., Ann Arbor Science Publishers, Inc., Ann Arbor, Michigan, 1975, pp. 31-58.

Filby, R. H., and K. R. Shah, "Mode of Occurrence of Trace Elements in Petroleum and Relationship to Oil-Spill Identification Methods" in Nuclear Methods in Environmental Research, Proc. Am. Nuclear Soc. Top. Meet. August 23-24, 1971, J. R. Vogt, T. F. Parkinson, and R. L. Carter, Eds., Office of Conferences and Short Courses, University of Missouri-Columbia, Columbia, Missouri, 1971, pp. 86-96.

Filby, R. H., and K. R. Shah, "Neutron Activation Methods for Trace Elements in Crude Oils" in The Role of Trace Metals in Petroleum, T. F. Yen, Ed., Ann Arbor Science Publishers, Inc., Ann Arbor, Michigan, 1975, pp. 89-110.

Filby, R. H., K. R. Shah, and C. A. Sautter, "Trace Elements in the Solvent Refined Coal Process" in Symposium Proceedings: Environmental Aspects of Fuel Conversion Technology III (September 1977, Hollywood, Florida), PB-282 429, National Technical Information Service, Springfield, Virginia, 1978, pp. 266-282; see also "A Study of Trace Element Distribution in the Solvent Refined Coal (SRC) Process Using Neutron Activation Analysis," J. Radioanal. Chem., **37**, 693-704 (1977).

Fisher, G. L., C. E. Chrisp, and W. G. Jennings, "Physical and Chemical Properties of Mutagens in Coal Fly Ash" in Trace Substances in Environmental Health - XII, D. D. Hemphill, Ed., Proceedings of University of Missouri's 12th Annual Conference on Trace Substances in Environmental Health, Columbia, Missouri, June 6-8, 1978, University of Missouri, Columbia, Missouri, 1979, pp. 293-298.

Fitchko, J., and T. C. Hutchinson, "A Comparative Study of Heavy Metal Concentrations in River Mouth Sediments Around the Great Lakes," J. Great Lakes Res., **1**(1), 46-78 (1975).

Fortescue, J. A., "Studies in Landscape Geochemistry. No. 9. The Content of 13 Elements in 69 Samples of Sewage Sludge Collected from the Niagara Peninsula During Late November 1974," Res. Rep. Ser. - Brock Univ. Dep. Geol. Sci., **19**, 1-38 (1975).

Fortescue, J. A. C., M. D. Silvester, and F. Abercrombie, "Chemical Composition of Sewage Sludges Taken During a Ten Day Period from Six Plants in the Niagara Peninsula" in Trace Substances in Environmental Health - IX, D. D. Hemphill, Ed., Proceedings of University of Missouri's 9th Annual Conference on Trace Substances in Environmental Health, Columbia, Missouri, June 10-12, 1975, University of Missouri, Columbia, Missouri, 1976, pp. 179-188.

Franchetcau, J., H. D. Needham, P. Choukroune, T. Juteau, M. Seguret, R. D. Ballard, P. J. Fox, W. Normark, A. Carranza, D. Cordoba, J. Guerrero, C. Rangin, H. Bougault, P. Cambon, and R. Hekinian, "Massive Deep-Sea Sulphide Ore Deposits Discovered on the East Pacific Rise," Nature, **277**, 523-528 (1979).

Friend, A. G., "The Aqueous Behavior of Strontium-85, Cesium-137, Zinc-65, and Cobalt-60 as Determined by Laboratory Type Studies" in Transport of Radionuclides in Freshwater Systems, Report of Meeting, University of Texas, January 30-February 1, 1963, TIC-7664, U.S. Atomic Energy Commission, Div. Technical Information, 1963, pp. 43-60.

Fruchter, J. S., J. C. Laul, M. R. Peterson, P. W. Ryan, and M. E. Turner, "High Precision Trace Element and Organic Constituent Analysis of Oil Shale and Solvent-Refined Coal Materials" in Analytical Chemistry of Liquid Fuel Sources, Advances in Chemistry Series No. 170, American Chemical Society, Washington, D.C. 1978, pp. 255-281.

Fryklund, V. C., Jr., Ore Deposits of the Coeur d'Alene District Shoshone County, Idaho, Geological Survey Professional Paper 445, U.S. Department of the Interior, U.S. Government Printing Office, Washington, D.C., 1964; Chem. Abstr., **62**, 340e (1965).

Furr, A. K., A. W. Lawrence, S. S. C. Tong, M. C. Grandolfo, R. A. Hofstader, C. A. Bache, W. H. Gutenmann, and D. J. Lisk, "Multielement and Chlorinated Hydrocarbon Analysis of Municipal Sewage Sludges of American Cities," Environ. Sci. Technol., **10**(7), 683-687 (1976).

Gaarenstroom, P. D., and S. P. Perone, "Application of Pattern Recognition and Factor Analysis for Characterization of Atmospheric Particulate Composition in Southwest Desert Atmosphere," Environ. Sci. Technol., **11**(8), 795-800 (1977).

Gabler, R. C., Jr., and D. L. Meylan, "Incinerated Municipal Sewage Sludge as a Secondary Resource for Metals and Phosphorus," Sludge Manage. Disposal Util., Proc. Natl. Conf., 3rd, 1976, Information Transfer, Inc., Rockville, Maryland, 1977, pp. 197-200.

Gabler, R. C., and D. L. Neylan, "Extraction of Metals and Phosphorus from Incinerated Municipal Sewage Sludge" in Proceedings of the 32nd Industrial Waste Conference, May 10, 11, and 12, 1977, Purdue University, Ann Arbor Science Publishers, Inc., Ann Arbor, Michigan, 1978, pp. 39-49.

George, L. C., and A. A. Cochran, Recovery of Metals from Electroplating Wastes by the Waste-Plus-Waste Method, Bureau of Mines Solid Waste Research Program, Technical Progress Report 27, U.S. Department of the Interior, Washington, D.C., 1970.

Gérôme, D., "Etude préliminarie des variations de la vitamine B12 dans les lacs Léman et de Nantua. Consequences de ces variations pour l'evolution biologique des lacs" ["Preliminary Study of the Variations of Vitamin B_{12} in Lakes Léman (Geneva) and Nantua"], Compt. Rend. Ser. D, **272**, 808-811 (1971).

Gladney, E., "Composition and Size Distributions of Atmospheric Particulate Matter in the Boston Area," Environ. Sci. Technol., **8**(6), 551-557 (1974).

Gluskoter, H. J., "Mineral Matter and Trace Elements in Coal" in Trace Elements in Fuel, Advances in Chemistry Series No. 141, American Chemical Society, Washington, D.C., 1975, pp. 1-22.

Goldberg, E. D., J. J. Griffin, V. Hodge, and M. Koide, "Pollution History of the Savannah River Estuary," Environ. Sci. Technol., **13**(5), 588-594 (1979).

Goldschmidt, V. M., "The Principles of Distribution of Chemical Elements in Minerals and Rocks," J. Chem. Soc., 655-673 (1937).

Gould, M. S., and E. J. Genetelli, "Heavy Metal Distribution in Anaerobically Digested Sludges," Proc. Ind. Waste Conf., 1975, **30**, 689-699 (1977).

Greathouse, D. G., and G. F. Craun, "Cardiovascular Disease Study - Occurrence of Inorganics in Household Tap Water and Relationships to Cardiovascular Mortality Rates" in Trace Substances in Environmental Health - XII, D. D. Hemphill, Ed., Proceedings of University of Missouri's 12th Annual Conference on Trace Substances in Environmental Health, Columbia, Missouri, June 6-8, 1978, University of Missouri, Columbia, Missouri, 1979, pp. 31-39.

Grigor'ev, V. M., "Osnovnye Geneticheskie Tipy Zhelezorydnykh Mestorozhdenii i Soderzhashchiesya v Nikh Elements-Primesi" ["Main Genetic Types of Iron Ores Deposits and Accessory Elements Present in Them"], Geol. Mestorozhd. Redk. Elem., Vses. Nauchno-Issled. Inst. Miner. Syr'ya, No. 29, 22-40 (1966).

Grimanis, A. P., M. Vassilaki-Grimani, and G. B. Griggs, "Pollution Studies of Trace Elements in Sediments from the Upper Saronikos Gulf, Greece," J. Radioanal. Chem., **37**, 761-773 (1977).

Gulf Radiation Technology, Validation of Neutron Activation Technique for Trace Element Determination in Petroleum Products, API Publication No. 4188, Committee on Environmental Affairs, American Petroleum Institute, Washington, D.C., August 1973.

Gupta, R. S., "On Some Trace Metals in the Baltic," Ambio, **1** (6), 226-230 (1972).

Guthrie, R. K., and D. S. Cherry, "Pollutant Removal from Coal-Ash Basin Effluent," Water Resour. Bull., **12**(5), 889-902 (1976).

Hamilton, E. I., "The Chemical Elements and Human Morbidity--Water, Air and Places--A Study of Natural Variability," Sci. Total Environ., **3**(1), 3-85 (1974).

Harding, S. C., and H. S. Brown, "Distribution of Selected Trace Metals Elements in Sediments of Pamlico River Estuary, North Carolina," Environ. Geol., **1**, 181-191 (1975).

Harrison, R. M., "Toxic Metals in Street and Household Dusts," Sci. Total Environ., **11**, 89-97 (1979).

Hawkes, H. E., and J. S. Webb, Geochemistry in Mineral Exploration, Harper and Row, Publishers, Inc., New York, New York, 1962.

Heindryckx, R., "Comparison of the Mass-Size Functions of the Elements in the Aerosols of the Ghent Industrial District with Data from Other Areas. Some Physico-Chemical Implications," Atoms. Environ., **10**(1), 65-71 (1976); Chem. Abstr., **84**, 94866w (1976).

Helmke, P. A., R. D. Koons, P. J. Schomberg, and I. K. Iskandar, "Determination of Trace Element Contamination of Sediments by Multielement Analysis of Clay-Size Fraction," Environ. Sci. Technol., **11**(10), 984-989 (1977).

Hemphill, D. D., "Availability of Trace Elements to Plants With Respect to Soil-Plant Interaction" in "Geochemical Environment in Relation to Health and Disease," Ann. N.Y. Acad. Sci., **199**, 46-61 (1972).

Hendricks, R. L., F. B. Reisbick, E. J. Mahaffey, D. B. Roberts, and M. N. A. Peterson, "Chemical Composition of Sediments and Interstitial Brines from the Atlantis II, Discovery and Chain Deeps" in Hot Brines and Recent Heavy Metal Deposits Red Sea, E. T. Degens, Ed., Springer-Verlag New York, Inc., New York, N.Y., 1969, pp. 407-440.

Hesse, J. L., and E. D. Evans, Heavy Metals in Surface Waters, Sediments and Fish in Michigan, Michigan Water Resources Commission, Bureau of Water Management, Department of Natural Resources, State of Michigan, July 1972.

Hildebrand, S. G., R. M. Cashman, and J. A. Carter, "The Potential Toxicity and Bioaccumulation in Aquatic Systems of Trace Elements Present in Aqueous Coal Conversion Effluents" in Trace Substances in Environmental Health - X, D. D. Hemphill, Ed., Proceedings of University of Missouri's 10th Annual Conference on Trace Substances in Environmental Health, Columbia, Missouri, June 8-10, 1976, University of Missouri, Columbia, Missouri, 1977, pp. 305-316.

Hinneri, S., "Enrichment of Elements, Especially Heavy Metals in Recent Sediments of the Fresh Water Reservior of Uusikaupunki Southwest Coastland of Finland," Ann. Univ. Turku., Ser. A2, **56**, 1-30 (1974).

Hitchon, B., R. H. Filby, and K. R. Shah, "Geochemistry of Trace Elements in Crude Oils, Alberta, Canada" in The Role of Trace Metals in Petroleum, T. F. Yen, Ed., Ann Arbor Science Publishers, Inc., Ann Arbor, Michigan, 1975, pp. 111-121.

Hodgson, J. F., "Chemistry of Trace Elements in Soils with Reference to Trace Element Concentration in Plants" in Trace Substances in Environmental Health - III, D. D. Hemphill, Ed., Proceedings of University of Missouri's 3rd Annual Conference on Trace Substances in Environmental Health, Columbia, Missouri, June 24-26, 1972, University of Missouri, Columbia, Missouri, 1973, pp. 45-58.

Hopke, P. K., D. F. S. Natusch, and R. Lamb, "Multielemental Characterization of Urban Street Dust" in ACS Div. Environ. Chem. Prepr., **17**(1), 332-334 (1977).

Horr, C. A., A. T. Myers, and P. J. Dunton, "Methods of Analysis for Uranium and Other Metals in Crude Oils, with Data on Reliability" in Uranium and Other Metals in Crude Oils, U.S. Geological Survey Bulletin 1100, U.S. Department of the Interior, Washington, D.C., 1961, pp. 1-15.

Hung, G. W. C., "Atomic Absorption Spectrophotometric Determination of Some Trace Metals in Biological Systems," Reprint from Proceedings, 4th Joint Conf. on Sensing of Environ. Pollutants, American Chemical Society, Washington, D.C., 1978.

Hung, T. C., Y. H. Li, and D. C. Wu, "The Pollution of Heavy Metals in the Kaohsiung Harbor, Taiwan" in International Conference on Heavy Metals in the Environment October 27-31, 1975, Symposium Proceedings, Vol. 2, Part 2, Institute for Environmental Studies, University of Toronto, Toronto, Ontario, Canada, 1977, pp. 809-819.

Hutchinson, T. C., and J. Fitchko, "Heavy Metal Concentrations and Distributions in River Mouth Sediments Around the Great Lakes," Proceddings International Conference Transp. Persistent Chem. Aquat. Ecosyst., **I**. 69-77 (1974).

Hutchinson, T. C., and L. M. Whitby, "A Study of Airborne Contamination of Vegetation and Soils by Heavy Metals from the Sudbury, Ontario, Copper-Nickel Smelters" in Trace Substances in Environmental Health - VII, D. D. Hemphill, Ed., Proceedings of University of Missouri's 7th Annual Conference on Trace Substances in Environmental Health, Columbia, Missouri, June 12-14, 1973, University of Missouri, Columbia, Missouri, 1974, pp. 179-189.

Hutchinson, T. C., and L. M. Whitby, Environ. Conservation, 1(2), 123-132 (1974).

Hyden, H. J., "Distribution of Uranium and Other Metals in Crude Oils" in Uranium and Other Metals in Crude Oils, U.S. Geological Survey Bulletin 1100, U.S. Department of the Interior, Washington, D.C., 1961, pp. 17-99.

Ilialetdinov, A. N., P. B. Enker, and S. E. Iakubovskii, "Uchastie Geterotrofnykh Mikroorganizmov v Ochistke Stokov ot Ionov Tyazhelykh Metallov" ["Participation of Heterotrophic Microorganisms in the Purification of Drainage Waters from Heavy Metal Ions"], Mikrobiologiia, 45(6), 1092-1099 (1976).

Jackson, N. M., Jr., "Public Water Supplies of North Carolina. I. Northern Piedmont," Public Water Supplies N.C., Pt. 1, 1-277 (1972).

Jackson, N. M., Jr., "Public Water Supplies of North Carolina. II. Southern Piedmont," Public Water Supplies N.C., Pt. 2, 1-225 (1973).

Jacobs, M. L., Evaluation of Spark Source Mass Spectrometry in the Analysis of Biological Samples, Prepared for the National Institute for Occupational Safety and Health, Cincinnati, Ohio, Division of Laboratories and Criteria Development, PB-267 506, National Technical Information Service, U.S. Department of Commerce, Springfield, Virginia, 1975.

Jaffe, D., and J. K. Walters, "Trace Metals in Sediments from the Humber Estuary" in International Conference on Heavy Metals in the Environment, Abstracts of Papers, University of Toronto, Toronto, Ontario, Canada, October 27-31, 1975, pp. C-270 to C-272.

Jervis, R. E., J. J. Paciga, A. Chattopadhyay, "Characterization of Urban Aerosols and Their Hazard Assessment, by Size Sampling Combined with Inter Element Ratios," Meas., Detect. Control Environ. Pollut., Proc. Int. Symp., 125-150 (1976).

Jester, W. A., and E. E. Klaus, "Activation Analysis of High-Purity Lubricants for Trace Elements," Nuclear Appl., **3**, 375-382 (1967).

John, W., R. Kaifer, K. Rahn, and J. J. Wesolowski, "Trace Element Concentrations in Aerosols from the San Francisco Bay Area," Atmos. Environ., **7**(1), 107-118 (1973).

Jolly, R. K., S. K. Gupta, G. Randerspehrson, D. C. Buckle, W. B. Thornton, H. Aceto, J. J. Singh, and D. C. Woods, "Preferential Concentration of Certain Elements in Smaller Aerosols Emitted from Aircraft Engines," J. Appl. Phys., **46**(10), 4590-4594 (1975).

Jones, B. F., and C. J. Bowser, "The Mineralogy and Related Chemistry of Lake Sediments" in Lakes. Chemistry, Geology, Physics, A. Lerman, Ed., Springer-Verlag, New York, New York, 1978, pp. 179-235.

Kessler, T., A. G. Sharkey, and R. A. Friedel, Spark-Source Mass Spectrometer Investigation of Coal Particles and Coal Ash, Bureau of Mines Progress Report No. 42, U.S. Department of the Interior, Washington, D.C., 1971.

King, R. B., J. S. Fordyce, A. C. Antoine, H. F. Leibecki, H. E. Neustadter, and S. M. Sidik, Extensive 1-Year Survey of Trace Elements and Compounds in the Airborne Suspended Particulate Matter in Cleveland, Ohio, NASA-TN D-8110, Washington, D.C., 1976.

Klein, D. H., "Mercury and Other Metals in Urban Soils," Environ. Sci. Technol., **6**(6), 560-562 (1972).

Klemetson, S. L., "Pollution Potentials of Coal Gasification Plants," Proc. Ind. Waste Conf. 1976, **31**, 63-76 (1977).

Knauer, G. A., "Immediate Industrial Effects on Sediment Metals in a Clean Coastal Environment," Mar. Pollut. Bull., **8**(11), 249-254 (1977).

Kneip, T. J., and R. E. Hazen, "Deposit and Mobility of Cadmium in a Marsh-Cove Ecosystem and the Relation to Cadmium Concentration in Biota," Environ. Health Perspect., **28**, 67-73 (1979).

Kneip, T. J., G. Re, and T. Hernandez, "Cadmium in an Aquatic Ecosystem: Distribution and Effects" in Trace Substances in Environmental Health - VIII, D. D. Hemphill, Ed., Proceedings of University of Missouri's 8th Annual Conference on Trace Substances in Environmental Health, Columbia, Missouri, June 11-13, 1974, University of Missouri, Columbia, Missouri, 1975, pp. 173-177.

Kopp, J. F., "The Occurrence of Trace Elements in Water" in Trace Substances in Environmental Health - III, D. D. Hemphill, Ed., Proceedings of University of Missouri's 3rd Annual Conference on Trace Substances in Environmental Health, Columbia, Missouri, June 24-26, 1969, University of Missouri, Columbia, Missouri, 1970, pp. 59-73.

Kopp, J. F., and R. C. Kroner, Trace Metals in Waters of the United States, U.S. Department of the Interior, Federal Water Pollution Control Administration, Cincinnati, Ohio, 1970.

Kosta, L., B. Pihlar, and M. Dermelj, "Interactions of Foreign Substances in the Environment" in Trace Substances in Environmental Health - XII, D. D. Hemphill, Ed., Proceedings of University of Missouri's 12th Annual Conference on Trace Substances in Environmental Health, Columbia, Missouri, June 6-8, 1978, University of Missouri, Columbia, Missouri, 1979, pp. 309-316.

Kostovetskii, Ya I., G. M. Rakhov, and E. I. Shteinberg, "Spetsificheskie Zagryazneniya Gorodskikh Stochnykh Vod i Effektivnost' ikh Ochistki' ["Specific Pollution of Municipal Waste Waters and Effiency of Their Purification"], Vodosnabzh. Sanit. Tekh., No. 10, 34-35 (1975).

Kubota, J., "Distribution of Total and Extractable Forms of Cobalt in Morphologically Different Soils of Eastern United States," Soil Sci., **99**(3), 166-174 (1965).

Kubota, J., "Distribution of Cobalt Deficiency in Grazing Animals in Relation to Soils and Forage Plants of the United States," Soil Sci., **106**, 122-130 (1968).

Kubota, J., 'Sampling of Soils for Trace Element Studies' in "Geochemical Environment in Relation to Health and Disease," Ann. N.Y. Acad. Sci., **199**, 105-117 (1972).

Kubota, J., and V. A. Lazar, "Cobalt Status of Soils of Southeastern United States: II. Ground-Water Podzols and Six Geographically Associated Soil Groups," Soil Sci., **86**(5), 262-268 (1958).

Kubota, J., V. A. Lazar, and K. C. Beeson, "The Study of Cobalt Status of Soils in Arkansas and Louisiana Using the Black Gum as the Indicator Plant," Soil Sci. Soc. Am. Proc., **24**, 527-528 (1960).

Kubota, J. E. L. Mills, and R. T. Oglesby, "Lead, Cadmium, Zinc, Copper, and Cobalt in Streams and Lake Waters of Cayuga Lake Basin, New York," Environ. Sci. Technol., **8**(3), 243-248 (1974).

Lakanen, E., "Outokummun Kaivoksen Ympäristön Hivenaine-Pitoisuuksista" ["Trace Element Levels in the Vicinity of the Outokumpu Copper Mine"], Ann. Agri. Fenniae, **4**(4), 290-298 (1965).

Law, S. L., "Dissolved Metals in Aqueous Effluents from Municipal Incinerators," J. Water Pollut. Control Fed., **49**(12), 2453-2466 (1977).

Lebowitz, H. E., S. S. Tam, G. R. Smithson, Jr., H. Nack, J. H. Oxley, G. Cavanaugh, C. E. Burklin, and J. C. Dickerson, Potentially Hazardous Emissions from the Extraction and Processing of Coal and Oil, U.S. Environmental Protection Agency, National Technical Information Service, U.S. Department of Commerce, Springfield, Virginia, 1975.

Leenheer, J. A., R. L. Malcolm, and W. R. White, "Investigation of the Reactivity and Fate of Certain Organic Components of an Industrial Waste After Deep-Well Injection," Environ. Sci. Technol., **10**(5), 445-451 (1976).

Leibecki, H. F., J. S. Fordyce, R. B. King, and H. E. Neustadter, "Elemental Composition - Size of Air Particulates" in International Conference on Heavy Metals in the Environment, Abstacts of Papers, University of Toronto, Toronto, Ontario, Canada, October 27-31, 1975, pp. D-97 to D-99.

Lieser, K. H., W. Calmano, E. Heuss, and V. Neitzert, "Neutron Activation as a Routine Method for the Determination of Trace Elements in Water," J. Radioanal. Chem., **37**, 717-726 (1977).

Lockwood, R. P., "Geochemistry and Petrology of Some Oklahoma Red-Bed Copper Occurrences," Okla., Geol. Surv., Circ., **77**, 66-68 (1976).

Loran, B. I., and J. B. O'Hara, "A Clean Coal Conversion Technology," Environ. Sci. Technol., **12**(12), 1258-1263 (1978).

Loring, D. H., "Distribution and Partition of Cobalt, Nickel, Chromium, and Vanadium in the Sediments of the Saguenay Fjord," Can. J. Earth Sci., **13**, 1706-1718 (1976).

Lowman, F. G., and R. Y. Ting, The State of Cobalt in Sea Water and Its Uptake by Marine Organisms and Sediments, CONF-720708-8, National Technical Information Service, U.S. Dept. of Commerce, Springfield, Virginia, 1970, 26 pp.

Lowman, F. G., and R. Y. Ting, "The State of Cobalt in Seawater and Its Uptake by Marine Organisms and Sediments" in Radioactive Contamination of the Marine Environment, Proc. Symp. IAEA, Seattle, Washington, July 10-14, 1972, International Atomic Energy Agency, Vienna, Austria, 1973, pp. 369-383.

Lukens, H. R., D. Bryan, N. Z. Hiatt, and H. L. Schlesinger, Development of Nuclear Analytical Techniques for Oil-Slick Identification, Phase IIA, Final Report, GULF-RT-A-10684 Prepared for the U.S. Atomic Energy Commission Division of Isotope Development, National Technical Information Service, Springfield, Virginia, 1971.

Maenhaut, W., and W. H. Zoller, "Determination of the Chemical Composition of the South Pole Aerosol by Instrumental Neutron Activation Analysis," J. Radioanal. Chem., **37**, 637-650 (1977).

Magee, E. M., H. J. Hall, and G. M. Varga, Jr., Potential Pollutants in Fossil Fuels, EPA-R2-73-249, Office of Research and Monitoring, U.S. Environmental Protection Agency, Washington, D.C., 1973.

Mamuro, T., Y. Matsuda, A. Mizohata, T. Takeuchi, and A. Fujita, "Neutron Activation Analysis of Airborne Dust," Ann. Rep. Radiat. Center Osaka Prefect., 11, 1-13 (1970).

Mamuro, T., Y. Matsuda, and A. Mizohata, "Comparative Multielement Analyses of Airborne Particulate Samples Collected in Various Areas," Ann. Rep. Radiat. Center Osaka Prefect., 14, 11-18 (1973); Chem. Abstr., 82, 115,515q (1973).

Matthews, D. B., and G. A. Thomas, "Study of Electroplating Industry in Adelaide in Relation to the Assessment of Environment Impact," Proc. R. Aust. Chem. Inst., 42(12), 323-327 (1975).

Matusevich, V. M., and V. K. Popov, "Mikroelementy v Podzemnykh Vodakh--Pokazateli Neftegasonosnosti" ["Trace Elements in Subsurface Waters as Indexes of Petroleum and Gas Content"], Izv. Vyssh. Uchebn. Zaved., Neft Gaz, 21(8), 3-8 (1978).

Mazur, T., "Copper, Cobalt, Boron, Zinc, and Manganese in Farmyard Manure from Krakow Voivodship," Rocz. Glebozn., 23(2), 305-313 (1972); Chem. Abstr., 78, 134997z (1973).

McCabe, L. J., and J. C. Vaughan, "Trace Metals Content of Drinking Water from a Large System," Presented at the Symposium on Water Quality in Distribution Systems, Division of Water, Air, and Waste Chemistry, American Chemical Society National Meeting, Minneapolis, Minnesota, April 13, 1969.

McCabe, L. J., J. M. Symons, R. D. Lee, and G. G. Robeck, "Survey of Community Water Supply Systems," J. Am. Water Works Assoc., 62(11), 670-687 (1970).

McDuffie, B., "Chemistry and Biochemistry of Toxic Materials" in Toxic Materials in the Environment, A. A. Carlson and R. L. Collin, Eds., Tech. Paper No. 52, New York State Department of Environmental Conservation, Albany, New York, June 1978, pp. 11-35.

McIntyre, S. G., "Noranda's Cobalt Plans," Paper presented at the Cobalt Crisis Conference, April 29 - May 1, 1979, Oak Brook Illinois, Sponsored by Gorham International, Inc., Gorham, Maine.

McLaughlin, E., "Metal Finishers in Phila. Face New Waste Deadlines," Am. Metal Market, 82(38), 26 (1975).

Metry, A. A., and M. E. Harris, "A New Concept for Treating Leachate and Contaminated Runoff Waters from a Taconite Transshipment Facility," Proc. Ind. Waste Conf. 1976, 31, 227-239 (1977).

Miklishanskii, A. Z., Yu. V. Yakovlev, V. Ya. Vyropaev, and B. V. Savel'ev, "Neitronno-Aktivatsionnoe i Rentgenoradiometricheskoe Opredelenie Soderzhaniya Mikroelementov v Snezhnom Pokrove Antarktidy" ["Neutron-Activation and X-Ray Fluorescence Determination of Trace Elements in Antarctic Snow Mantle"], Zh. Anal. Khim., 31(7), 1345-1350 (1976).

Miklishanskii, A. Z., F. I. Pavlotskaya, B. V. Savel'ev, and Yu. V. Yakovlev, ["Content and Forms of Occurrence of Trace Elements in the Surface Air Layer and Atmospheric Precipitation"], Geokhimiya, No. 11, 1673-1682 (1977).

Mills, E. L., and R. T. Oglesby, "Five Trace Elements and Vitamin B_{12} in Cayuga Lake, New York," Proc., Conf. Great Lakes Res., 14th, 256-267 (1971).

Minami, K., and K. Araki, "Distribution of Trace Elements in Arable Soil Affected by Automobile Exhausts," Soil Sci. Plant Nutr. (Tokyo), 21(2), 185-188 (1975); Chem. Abstr., 83, 162922t (1975).

Mitchell, R. L., "Chapter 9. Trace Elements" in Chemistry of the Soil, F. E. Bear, Ed.-in-Chief, Reinhold Publishing Corporation, New York, New York, 1955, pp. 253-285.

Monkman, L. J., "Mineralogical and Chemical Contaminants in Graded Milled Chrysotile," Resumes Commun. - Conf. Int. Phys. Chim. Miner. Amiante, 3rd, 3.13, Univ. Laval, Quebec, Quebec, 1975, 17 pp.

Morel, F. M. M., J. C. Westall, C. R. O'Melia, and J. J. Morgan, "Fate of Trace Metals in Los Angeles County Wastewater Discharge," Environ. Sci. Technol., 9(8), 756-761 (1975).

Morrison, G. H., and R. A. Nadkarni, "Elemental Abundances of Lunar Soils by Neutron Activation Analysis," J. Radioanal. Chem., 18(1), 153-167 (1973).

Mosier, E. L., J. C. Antweiler, and J. M. Nishi, "Spectrochemical Determination of Trace Elements in Galena," J. Res. U.S. Geol. Survey, 3(5), 625-630 (1975).

Murty, P. S. N., Ch. M. Rao, M. Veerayya, and C. V. G. Reddy, "Partition Patterns of Al, Fe, Mn, Ni, Co, and Cu in the Sediments of the Deep Sea Drilling Project Site 219," Ind. J. Mar. Sci., 6(June), 64-71 (1977).

Nadkarni, R. A., and G. H. Morrison, "Multielement Analysis of Sludge Samples by Instrumental Neutron Activation Analysis," Environ. Lett., 6(4), 273-285 (1974).

Natale, P., and M. Braja, "Inquinamento da Metalli nell' Atmosfera Urbana di Torino, nel Quadriennio 1973-'76" ["Metal Pollutants in the Urban Atmosphere of Turin During 1973-1976"], Ig. Mod., 71(5), 665-687 (1978).

Natusch, D. F. S., C. F. Bauer, H. Matusiewicz, C. A. Evans, J. Baker, A. Loh, R. W. Linton, and P. K. Hopke, "Characterization of Trace Elements in Fly Ash" in International Conference on Heavy Metals in the Environment, October 27-31, 1975, Symposium Proceedings, Vol. 2, Part 2, Institute for Environmental Studies, University of Toronto, Toronto, Ontario, Canada, 1977, pp. 553-575.

Netzer, A., and S. Beszedits, "Color and Heavy Metal Removal from Dyebath Effluents by Lime Precipitation," Water Pollut. Res. Can., 10, 151-163 (1975).

Nilsson, R., "Removal of Metals by Chemical Treatment of Municipal Waste Water," Water Res., 5(2), 51-60 (1971).

Oliver, B. G., "Heavy Metal Levels of Ottawa and Rideau River Sediments," Environ. Sci. Technol., 7(2), 135-137 (1973).

Oliver, B. G., and E. G. Cosgrove, "Efficiency of Heavy Metal Removal by a Conventional Activated Sludge Treatment Plant," Water Res., 8, 869-874 (1974).

Oliver, B. G., and E. G. Cosgrove, "Metal Concentrations in Sewage, Effluents, and Sludges of Some Southern Ontario Wastewater Treatment Plants," Environ. Letters, 9(1), 75-90 (1975).

Ölmez, I., and N. K. Aras, "Trace Elements in the Atmosphere Determined by Nuclear Activation Analysis and Their Interpretation," J. Radioanal. Chem., 37(2), 671-677 (1977).

Ördögh, and L. Kálmán, Neutron Activation Analysis of Airborne Inorganic Pollutants, KFKI-73-3, ISBN 963 371 002 2, National Technical Information Service, U.S. Department of Commerce, Springfield, Virginia, 1975.

Ottinger, R. S., J. L. Blumenthal, D. F. Dal Porto, G. I. Gruber, M. J. Santy, and C. C. Shih, Recommended Methods of Reduction, Neutralization, Recovery, or Disposal of Hazardous Waste. Vol. XII. Industrial and Municipal Disposal Candidate Waste Stream Constituent Profile Reports. Inorganic Compounds, TRW Systems Group for the U.S. Environmental Protection Agency, PB-224 591, National Technical Information Service, U.S. Department of Commerce, Springfield, Virginia, 1973.

Parker, L., "Chapter D. Composition of the Earth's Crust" in Data of Geochemistry, 6th edition, M. Fleischer, Tech. Ed., Geol. Survey Professional Paper 440-D, U.S. Government Printing Office, Washington, D.C., 1967, pp. D1-D19.

Paoletti, A., G. Bartoli, and A. Parrella, "Impianti Pilota per L'abbattimento del Co^{58-60} e di Attri Radionuclidi" ["Pilot Treatment Plant for Reduction of Cobalt-58, -59, and -60 and Other Radioisotopes in Nuclear Power Waste Waters"], Ig. Mod., 69(2), 119-149 (1976).

Pattenden, N. J., "The Measurement of Heavy Metals in the Atmosphere and Their Interpretation," National Society for Clean Air. Proceedings of the Annual Conference, **41**(1), 1-22 (1974).

Pattenden, N. J., "Atmospheric Concentrations and Deposition Rates of Some Trace Elements Measured in the Swansea/Neath/Port Talbot Area" in Report of a Collaborative Study on Certain Elements in Air, Soil, Plants, Animals and Humans in the Swansea/Neath/Port Talbot Area, Together with a Report on a Moss Bag Study of Atmospheric Pollution Across South Wales, Welsh Office, Cardiff, Wales, 1975.

Pattenden, N. J., "A Study of Atmospheric Pollution by Trace Elements in the Swansea Area," Clean Air (UK), **7**(27), 15-19 (1977).

Peirson, D. H., P. A. Cawse, L. Salmon, and R. S. Cambray, "Trace Elements in the Atmospheric Environment," Nature, **241**, 252-256 (1973).

Peirson, D. H., P. A. Cawse, and R. S. Cambray, "Chemical Uniformity of Airborne Particulate Material, and a Maritime Effect," Nature, **251**, 675-679 (1974).

Pellizzari, E. D., Identification of Components of Energy-Related Wastes and Effluents, EPA-600/7-78-004, Research Triangle Institute for U.S. Environmental Protection Agency, Athens, Georgia, January 1978.

Persigehl, M., K. Kasperek, H. J. Klein, and L. E. Feinendegen, "Einfluss der Industrialisierung auf den Spurenelementgehalt in menschlichen Lungen," ["Influence of Industrialization on Trace Element Concentration in Human Lungs"], Beitr. Pathol., **157**(3), 260-268 (1976).

Pettyjohn, W. A., L. R. Hayes, and T. R. Schultz, Concentration and Distribution of Selected Trace Elements in the Maumee River Basin, Ohio, Indiana, and Michigan, PB-234 013, National Technical Information Service, U.S. Department of Commerce, Springfield, Virginia, March 1974.

Phelps, D. K., and A. C. Myers, "Ecological Considerations in Site Assessment for Dredging and Spoiling Activities," in Manage. Bottom Sediments Containing Toxic Subst.; Proc. U.S.-Jap. Expert. Meet., 2nd, EPA-600/3-77-083, U.S. Environmental Protection Agency, Office of Research and Development, Washington, D.C., 1977, pp. 266-286.

Presley, B. J., R. R. Brooks, and I. R. Kaplan, "Manganese and Related Elements in the Interstitial Water of Marine Sediments," Science, **158** (3803), 906-910 (1967).

Ragaini, R. C., H. R. Ralston, and N. Roberts, "Environmental Trace Metal Contamination in Kellogg, Idaho, Near a Lead Smelting Complex," Environ. Sci. Technol., **11**(8), 773-781 (1977).

Rahn, K. A., and R. P. Harrison, "The Chemical Composition of Chicago Street Dust," ERDA Symp. Ser., **38** (Atmos.-Surf. Exch. Part. Gaseous Pollut., Proc. Symp., 1974), 557-570 (1976).

Rahn, K. A., M. Demuynck, R. Dams, and J. DeGraeve, "The Chemical Composition of the Aerosol Over Belgium" in Proceedings of the Third International Clean Air Congress Düsseldorf (Federal Republic of Germany), VDI-Verlag GmbH, Düsseldorf, 1973, pp. C81 to C84.

Ramondetta, P. J., and W. H. Harris, "Heavy Metal Distribution in Jamaica Bay Sediments," Environ. Geol., **2**(3), 145-159 (1978).

Rehwoldt, R., D. Karimian-Teherani, and H. Altmann, "Measurement and Distribution of Various Heavy Metals in the Danube River and Danube Canal Aquatic Communities in the Vicinity of Vienna, Austria," Sci. Total Environ., **3**(4), 341-348 (1975).

Reno, H. T., "Cobalt" in Minerals Facts and Problems, 1970 ed., Bureau of Mines Bulletin 650, U.S. Department of the Interior, U.S. Government Printing Office, Washington, D.C., 1970, pp. 263-274

Rhodes, J. R., "Radioisotope Neutron Activation for On-Stream Process Analysis," Neutron Sources Appl., Proc. Am. Nucl. Soc. Natl. Top. Meeting, **3**, IV-1-IV-10 (1971).

Rhodes, W. J., "Potential Pollutants in Coal and the Environmental Impact of Coal Gasification," Presented at the 74th National AIChE Meeting, New Orleans, March 15, 1973.

Rittenhouse, G., R. B. Fulton, III, R. J. Grabowski, and J. L. Bernard, "Minor Elements in Oil-Field Waters," Chem. Geol., 4, 189-209 (1969).

Robertson, D. W., "Distribution of Cobalt in Oceanic Waters," Geochim. Cosmochim. Acta, 34(5), 553-567 (1970).

Robertson, D. J., "Filtration of Copper Smelter Gases at Hudson Bay Mining and Smelting Company, Limited," Can. Mining Metall. Bull., 53, 326-335 (1960).

Rohl, A. N., A. M. Langer, I. J. Selikoff, A. Tordini, R. Klimentidis, D. R. Bowes, and D. L. Skinner, "Consumer Talcums and Powders: Mineral and Chemical Characterization," J. Toxicol. Environ. Health, 2(2), 255-284 (1976).

Rossmann, R., and E. Callender, "Geochemistry of Lake Michigan Manganese Nodules," Proc. 12th Conf. Great Lakes Res. 1969, 306-316 (1969).

Ruch, R. R., H. J. Gluskoter, and N. F. Shimp, Occurrence and Distribution of Potentially Volatile Trace Elements in Coal. An Interim Report, Environmental Geology Notes, No. 61, Illinois State Geol. Survey, Urbana, Illinois, 1973.

Ruch, R. R., H. J. Gluskoter, and N. F. Shimp, Occurrence and Distribution of Potentially Volatile Trace Elements in Coal, Illinois State Geological Survey for U.S. Environmental Protection Agency, PB 238 091, National Technical Information Service, U.S. Department of Commerce, Springfield, Virginia, 1974.

Sallee, E. G., "UV and Other Metal-Decorating Processes" in Environmental Aspects of Chemical Use in Printing Operations, EPA-560/1-75-005, Conference Proceedings, King of Prussia, Pennsylvania, September 22-24, 1975, Office of Toxic Substances, Environmental Protection Agency, PB 251 406, National Technical Information Service, Springfield, Virginia, January 1976, pp. 394-411.

Salmon, L., D. H. F. Atkins, E. M. R. Fischer, and D. V. Law, "Retrospective Analysis of Air Samples in the U.K. 1957-1974," J. Radioanal. Chem., **37**, 867-880 (1977).

Sauchelli, V., Trace Elements in Agriculture, Van Nostrand Reinhold Company, New York, New York, 1969.

Schroeder, H. A., A. P. Nason, and I. H. Tipton, "Essential Trace Metals in Man: Cobalt," J. Chron. Dis., **20**, 869-890 (1967).

Schutyser, P., W. Maenhaut, and R. Dams, "Instrumental Neutron Activation Analysis of Dry Atmospheric Fall-Out and Rainwater," Anal. Chim. Acta, **100**, 75-85 (1978).

Shacklette, H. T., "A U.S. Geological Study Survey of Elements in Soil and Other Surficial Materials in the United States" in Trace Substances in Environmental Health - IV, D. D. Hemphill, Ed., Proceedings of University of Missouri's 4th Annual Conference on Trace Substances in Environmental Health, Columbia, Missouri, June 23-25, 1970, University of Missouri, Columbia Missouri, 1971, pp. 35-45.

Shah, K. R., R. H. Filby, and W. A. Haller, "Determination of Trace Elements in Petroleum by Neutron Activation Analysis. II. Determination of Sc, Cr, Fe, Co, Ni, Zn, As, Se, Sb, Eu, Au, Hg, and U," J. Radioanal. Chem., **6**, 413-422 (1970).

Shani, G., and D. Cohen, "Air Pollution Measurements in a Semi-Arid Zone, Using Neutron Activation Analysis," Tellus, **29**(6), 535-544 (1977).

Shani, G., and A. Haccoun, "Nuclear Methods Used to Compare Air Pollution in a City and a Pollution-Free Area," Meas., Detect. Control Environ. Pollut., Proc. Int. Symp., 89-99 (1976).

Shaver, R. G., L. C. Parker, E. F. Rissman, K. M. Slimak, and R. C. Smith, Assessment of Industrial Hazardous Waste Practices, Inorganic Chemicals Industry, PB 244 832, EPA-SW-104C, National Technical Information Service, U.S. Department of Commerce, Springfield, Virginia, March 1975.

Sheibley, D. W., "Trace Elements Analysis of Coal by Neutron Activation," Am. Chem. Soc., Div. Fuel Chem. Prepr., **18**(4), 59-71 (1973); Chem. Abstr., **83**, 63190w (1975).

Shen, T. T., "Air Pollutants from Sewage Sludge Incineration," J. Environ. Eng. Div. (Am. Soc. Civ. Eng.), **105**(EE1), 61-74 (1979).

Shendrikar, A. D., and G. B. Faudel, "Distribution of Trace Metals During Oil Retorting," Environ. Sci. Technol., **12**(3), 332-334 (1978).

Shimp, N. F., J. A. Schleicher, R. R. Ruch, D. B. Heck, and H. V. Leland, "Studies of Lake Michigan Bottom Sediments-Number Six. Trace Element and Organic Carbon Accumulation in the Most Recent Sediments of Southern Lake Michigan," Environ. Geol. Notes, **41**, 1-25 (1971).

Shreve, R. N., Chemical Process Industries, 3rd ed., McGraw-Hill Book Company, New York, New York, 1967.

Silvey, W. D., Occurrence of Selected Minor Elements in the Waters of California, U.S. Geol. Survey, Water-Supply Paper No. 1535-L, U.S. Government Printing Office, Washington, D.C., 1967, 25 pp.

Sim, P. G., and J. F. Lewin, "Potentially Toxic Metals in New Zealand Coals," N.Z.J. Sci., **18**(4), 635-641 (1975); Chem. Abstr., **84**, 168,932p (1976).

Simon, P. J., B. C. Giessen, and T. R. Copeland, "Categorization of Papers by Trace Metal Content Using Atomic Absorption Spectrometric and Pattern Recognition Techniques," Anal. Chem., **49**(14), 2285-2288 (1977).

Sims, P. K., P. B. Barton, Jr., and P. R. Barnett, "Some Aspects of the Geochemistry of Sphalerite, Central City District, Colorado," Economic Geology, **56**(7), 1211-1237 (1961).

Somerville, M. H., and J. L. Elder, "A Comparison of Trace Element Analyses of North Dakota Lignite Laboratory Ash with Lurgi Gasifier Ash and Their Use in Environmental Analysis" in Symposium Proceedings: Environmental Aspects of Fuel Conversion Technology III (September 1977, Hollywood, Florida), PB-282 429, National Technical Information Service, Springfield, Virginia, 1978, pp. 292-315.

Sommers, L. E., "Chemical Composition of Sewage Sludge and Analysis of Their Potential Use as Fertilizers," J. Environ. Qual., **6**(2), 225-232 (1977).

Sparr, M. C., E. O. Schneider, and L. J. Sullivan, "Micronutrients the 'Fertilizer Shoenails.' Part 1," Reprint from January-February 1968, issue Fertilizer Solutions Magazine, National Fertilizer Association, Peoria, Illinois, 1968, 12 pp.

Steele, K. F., and G. H. Wagner, "Trace Metal Relationships in Bottom Sediments of a Fresh-Water Stream--The Buffalo River, Arkansas," J. Sediment. Petrol., **45**(1), 310-319 (1975).

Stoiber, R. E., "Minor Elements in Sphalerite," Econ. Geol., **35**, 501-519 (1940).

Stojkovska, A., and G. W. Cooke, "Micro-Nutrients in Fertilisers," Chem. Ind., **1958**, 1368 (1958).

Strain, W. H., A. Flynn, E. G. Mansour, F. R. Plecha, W. J. Pories, and O. A. Hill, Jr., "Heavy Metal Content of Household Water" in International Conference on Heavy Metals in the Environment, October 27-31, 1975, Symposium Proceedings, Vol. 2, Part 2, Institute for Environmental Studies, University of Toronto, Toronto, Ontario, Canada, 1977, pp. 1003-1011.

Straughan, I., A. A. Elseewi, and A. L. Page, "Mobilization of Selected Trace Elements in Residues from Coal Combustion with Special Reference to Fly Ash" in Trace Substances in Environmental Health - XII, D. D. Hemphill, Ed., Proceedings of University of Missouri's 12th Annual Conference on Trace Substances in Environmental Health, Columbia, Missouri, June 6-8, 1978, University of Missouri, Columbia, Missouri, 1979, pp. 389-402.

Strock, L. W., "Quantitative Spectrographic Determination of Minor Elements in Zinc Sulphide Ores," Am. Inst. Mining Metall. Eng. Tech. Pub., No. 1866, 1-22 (1945).

Struempler, A. W., "Trace Element Composition in Atmospheric Particulates During 1973 and the Summer of 1974 at Chadron, Nebraska," Environ. Sci. Technol., 9(13), 1164-1168 (1975).

Swaine, D. J., The Trace-Element Content of Fertilizers, Commonwealth Agr. Bureau, Farnham Royal, Bucks, England, 1962.

Swaine, D. J., "Trace Elements in Fly-Ash" in Geochemistry 1977, A. P. W. Hodder, Ed., New Zealand Department of Scientific and Industrial Research, DSIR Bulletin 218, 1977, pp. 127-131.

Swaine, D. J., "Trace Elements in Coal" in Trace Substances in Environmental Health - XI, D. D. Hemphill, Ed., Proceedings of University of Missouri's 11th Annual Conference on Trace Substances in Environmental Health, Columbia, Missouri, June 7-9, 1977, University of Missouri, Columbia, Missouri, 1978, pp. 107-116.

Tarakanova, E. I., "O Raspredelenii Aktsessornykh Elementov v Sovremennykh Torfyanikakh" ["The Distribution of Accessory Elements in Modern Peat Bogs"], Tr. Inst. Geol. Geokhim., Akad. Nauk SSSR, Ural. Filial, No. 90, 56-63 (1971); Chem. Abstr., 76, 143382f (1972).

Taylor, D., "Distribution of Heavy Metals in Sediment of an Unpolluted Estuarine Environment," Sci. Total Environ., 6(3), 259-264 (1976).

Taylor, F. B., "Trace Elements and Compounds in Waters," J. Am. Water Works Assoc., 63(11), 728-733 (1971).

Taylor, S. R., "Abundance of Chemical Elements in the Continental Crust: A New Table," Geochim. Cosmochim. Acta, 28, 1273 (1964).

Thomas, R. L., J.-M. Jaquet, and A. Mudroch, "Sedimentation Processes and Associated Changes in Surface Sediment Trace Metal Concentrations in Lake St. Clair, 1970-1974" in International Conference on Heavy Metals in the Environment, October 27-31, 1975, Symposium Proceedings, Vol. 2, Part 2, Institute for Environmental Studies, University of Toronto, Toronto, Ontario, Canada, 1977, pp. 691-708.

Thompson, R. J., "Collection and Analysis of Airborne Metallic Elements" in Ultratrace Metal Analysis in Biological Sciences and Environment, Advances in Chemistry Series No. 172, T. H. Ribby, Ed., American Chemical Society, 1979, pp. 54-72.

Thompson, R. N., J. E. Zajic, and E. Lichti, "Spectrographic Analysis of Air-Dried Sewage Sludge," J. Water Pollut. Control Fed., **36**(6), 752-759 (1964).

Tsuji, K., Y. Suzuki, M. Tobe, M. Tonomura, Y. Emoto, M. Saito, M. Araki, and M. Muramatsu, "Suspended Particles in Indoor Air. II. Pollution of Indoor Air by Suspended Metal Particles," Eisei Shikenjo Hokoku, No. 90, 27-32 (1973).

Turner, A. C., "Determination of Atmospheric Pollution Around Non-Ferrous Metallurgical Industry," Miner. Environ., Proc. Int. Symp. 1974, 531-541 (1975).

Uken, E.-A., P. Hahn-Weinheimer, and H. Stark, "Distribution of Selected Trace Elements in Samples of Sediment, Suspension and Water Taken Along the Isar River, Bavaria," J. Radioanal. Chem., **37**, 741-750 (1977).

Vhay, J. S., D. A. Brobst, and A. V. Heyl, "Cobalt" in United States Mineral Resources, D. A. Brobst and W. P. Pratt, Eds., Geological Survey Professional Paper 820, U.S. Government Printing Office, Washington, D.C., 1973, pp. 143-155.

Vanderploeg, H. A., D. C. Parzyck, W. H. Wilcox, J. R. Kercher, and S. V. Kaye, Bioaccumulation Factors for Radionuclides in Freshwater Biota, ORNL-5002 UC-11, Oak Ridge National Laboratory, Oak Ridge, Tennessee, 1975.

Vine, J. D., and E. B. Tourtelot, "Geochemistry of Black Shale Deposits. Summary Report," Econ. Geol., **65**(3), 253-272 (1970).

Vinogradov, A. P., The Geochemistry of Rare and Dispersed Chemical Elements in Soils, 2nd ed., translated from the Russian, Consultants Bureau, Inc., New York, New York, 1959.

Vinogradov, A. P., "Average Content of Chemical Elements in the Main Types of Igneous Rocks in the Earth's Crust," Geochemistry (USSR) (English translation), No. 7, 641-664 (1963).

Vogg, H., and R. Härtel, "Experience in the Analysis of Atmospheric Aerosols at the Karlsruhe Nuclear Research Center," J. Radioanal. Chem., **37**, 857-866 (1977).

Vogt, J. R., and D. M. Hashii, "Trace Element Distributions in Man-Made Fibers," J. Radioanal. Chem., **15**(1), 355-366 (1973).

Wagemann, R., G. J. Brunskill, and B. W. Graham, "Composition and Reactivity of Some River Sediments from the Mackenzie Valley, N.W.T., Canada," Environ. Geol., **1**(6), 349-358 (1977).

Walters, L. J., Jr., T. J. Wolery, and R. D. Myser, "Occurrence of As, Cd, Co, Cr, Cu, Fe, Hg, Ni, Sb, and Zn in Lake Erie Sediments," Proc., Conf. Great Lakes Res., **17** (pt. 1), 219-234 (1974).

Watson, J. A., and E. E. Angino, "Iron-Rich Layers in Sediments from Gulf of Mexico," J. Sediment. Petrol., **39**(4), 1412-1419 (1969).

Weaver, J. N., A. Hanson, J. McGaughey, and F. J. Steinkruger, "Determination of Heavy Metals in Municipal Sewage Plant Sludges by Neutron Activation Analysis," Water Air Soil Pollut., **3**, 327-335 (1974).

Webber, M. D., and D. G. M. Corneau, "Metal Extractability from Sludge-Soil Mixtures" in International Conference on Heavy Metals in the Environment, October 27-31, 1975, Symposium Proceedings, Vol. 1, Institute for Environmental Studies, University of Toronto, Toronto, Ontario, Canada, 1977, pp. 205-225.

Wedepohl, K. H., Ed., Handbook of Geochemistry, Element 27, Springer-Verlag, Berlin, Heidelberg, 1974 to 1978, pp. 27-B-1 to 27-O-1.

Williams, J. M., E. M. Wewerka, N. E. Vanderborgh, P. Wagner, P. L. Wanek, and J. D. Olsen, Environmental Pollution by Trace Elements in Coal Preparation Wastes, LA-UR-77-1850, National Technical Information Service, U.S. Department of Commerce, Springfield, Virginia, 1977.

Williams, S. L., D. B. Aulenbach, and N. L. Clesceri, "Distribution of Metals in Lake Sediments of the Adirondack Region of New York State" in Biological Implications of Metals in the Environment, CONF-750929, Proc. 15th Ann. Hanford Life Sci. Symp. Richland, Washington, September 29-October 1, 1975, Technical Information Center, Energy Research and Development Administration, 1977, pp. 153-166.

Wing, R. E., and W. E. Rayford, "Starch-Based Products Effective in Heavy Metal Removal," Proc. Ind. Waste Conf. 1976, **31**, 1068-1079 (1977).

Wing, R. E., and W. E. Rayford, "Heavy Metal Removal Processes for Plating Rinse Waters," Proc. Ind. Waste Conf. 1977, **32**, 838-852 (1978).

Wiseman, B. F. H., and G. M. Bedri, "A Neutron Activation Scheme for the Detection and Determination of Pollutant Heavy Metals in Sewage Based Fertilizers," J. Radioanal. Chem., **24**, 313-320 (1975).

Wogman, N. A., H. G. Rieck, Jr., J. R. Kosorok, and R. W. Perkins, "In Situ Activation Analysis of Marine Sediments with Californium-252," J. Radioanal. Chem., **15**(2), 591-600 (1973); Chem. Abstr., **80**, 90665z (1974).

Yen, T. F., "Chemical Aspects of Metals in Native Petroleum" in The Role of Trace Metals in Petroleum, T. F. Yen, Ed., Ann Arbor Science Publishers, Inc., Ann Arbor, Michigan, 1975, pp. 1-30.

Young, D. R., C.-S. Young, and G. E. Hlavka, "Sources of Trace Metals from Highly Urbanized Southern California to the Adjacent Marine Ecosystem" in Cycling and Control of Metals, Proceedings of an Environmental Researces Conference, Columbus, Ohio, October 31-November 2, 1972, U.S. Environmental Protection Agency, National Environmental Research Center, Cincinnati, Ohio, February 1973, pp. 21-39.

Zoller, W. H., E. S. Gladney, R. A. Duce, and G. L. Hoffman, "Trace Metals in the Anarctic Atmosphere," *Spec. Environ. Rep.-W. M. O.*, **3** (Obs. Meas. Atmos. Pollut., Proc. Tech. Conf., 1973), 380-386 (1974).

Zubovic, P., T. Stadnichenko, and N. B. Sheffey, "Comparative Abundance of the Minor Elements in Coals from Different Parts of the United States" in *Short Papers in the Geological Sciences. Geological Survey Research 1960*, Geological Survey Professional Paper 400-B, U.S. Department of the Interior, U.S. Government Printing Office, Washington, D.C., 1960, pp. B87-B88.

V. ENVIRONMENTAL TRANSPORT

Bonnie L. Carson

Cobalt is held to a greater degree by soils with higher pH and higher contents of clay, natural organics, and hydrous manganese and iron oxides. Conversely, the mobility of cobalt is increased by lowering these factors. Very strong organic-complexing agents, such as EDTA--which is now a widespread contaminant of U.S. waters--greatly enhance cobalt mobility. Weaker organic-complexing agents, such as those present in plant extracts, also increase its extractability. Formation of either weak or strong soluble organic complexes of cobalt decreases its uptake by plants.

Many factors complicate the transport and speciation of cobalt in natural waters and sediments, and few generalities can be made. Anthropogenic pollution appears to enhance the solubility of cobalt in freshwater by complexing it with sewage-derived organics. Predominant cobalt species in unpolluted freshwater are: Co^{2+}, the carbonate, the hydroxide, the sulfate, adsorbed forms, oxide coatings, and crystalline sediments; in seawater: $CoCl^{+}$, Co^{2+}, the carbonate, and the sulfate.

Little has been reported regarding atmospheric transport of cobalt. The data that exist suggest that metallurgical plants can be strong point sources of cobalt if cobalt is an impurity or constituent of the raw material or product of the plant. Metallurgical plants have not been considered to be a major source of airborne cobalt in source inventories, yet certain metal works appear to be the cause of very high ambient air concentrations of cobalt. Usually fossil fuel burning and coking are listed as the major sources of cobalt emitted to ambient air.

The following chapter is divided into discussions of cobalt transport in soils and water plus descriptions of atmospheric emissions of cobalt. Atmospheric transport of cobalt has not been nearly so well studied as its behavior in soils and water. The descriptions of atmospheric emissions and related environmental contamination are, however, necessary preliminaries to understanding the behavior of aerosol cobalt.

A. Soils

1. Cobalt content and fixation related to soil chemistry: Soils developed on acid rocks, as we have noted in the subsection on cobalt-deficient soils, have relatively little cobalt, nickel, or copper. Conversely, these elements are enriched in soils from basic rocks, because the basic rocks themselves are enriched in them (Vinogradov, 1959). (See Table V-1.)

Numerous investigators have reported that elevated concentrations of cobalt in soils increase with increasing amounts of iron, manganese, and clay, increasing pH, and increasing or decreasing organic content. Jenne (1968) has explained most of the reported phenomena in soils and fresh water in terms of sorption and desorption by hydrated iron and manganese oxides.

As was shown in the section on cobalt-deficient soils, cobalt concentrations are high in high-clay soils. Cobalt is bound more strongly by soil clays than are potassium, magnesium, or calcium (Mortvedt and Cunningham, 1971). The data of Webber and Corneau (1977) (See Table V-2) indicate that clay content is more important than organic content in effecting an increase in cobalt concentrations in Ontario soils. Yet Nelson (1972) reported that in Alabama, cobalt is highest in the soils with the highest contents of organic material. (See Table V-3.)

Other studies have provided evidence that high organic content serves to mobilize cobalt and other heavy metals when the soil is not well drained. For example, Shirokov and Panasin (1972) reported that the peat bog soil of polders (maritime and riverine regions close to sea level that are drained by pumping stations) are poor in cobalt, copper, and zinc while well supplied with manganese, molybdenum, and boron. Alluvial bog soils, however, have moderate concentrations of cobalt. These soils also contain more manganese than peat bog polders.

TABLE V-1

AVERAGE COBALT CONCENTRATIONS IN SOILS AND ROCKS, mg/kg
(Vinogradov, 1959)

Soils	8
Sedimentary rocks (clays and shales)	23
Acid rocks (granites, liparites, rhyolites, etc.)	5
Intermediate rocks	20
Basic rocks (basalts, gabbros, norites, diabases, etc.)	45
Ultrabasic rocks (dunites, periodotites, pyroxenites)	200

TABLE V-2

COBALT CONTENT OF SOILS RELATED TO SOIL TEXTURE
AND CONTENTS OF ORGANICS AND CLAY
(Webber and Corneau, 1977)

Ontario Location	pH	Texture	% Clay	Organic Matter, (%)	Cobalt Concentration (mg/kg)
St. Thomas	5.2	Sand	1.3	3.1	8.8
Vaudreuil	4.7	Sandy loam	3.6	10.0	8.0
Grimsby	4.3	Silt loam	7.4	1.0	10.0
Haldimand	4.8	Silty clay loam	36.4	4.0	17.3
Rideau	5.3	Silty clay	45.2	1.7	40.0
Wendover	5.4	Silty clay	45.4	16.2	25.3

TABLE V-3

COBALT CONTENT OF ALABAMA SOILS RELATED TO ORGANIC MATTER
(Nelson, 1972)

Soil	10^{-5} Moles Co/g Soil		
	Total Adsorption	After leaching with KCl (1.0 N)[a]	After leaching with Cu acetate (0.1 N)[b]
Muck	10.85[c]	8.86	6.99
Hoytville	1.77	1.22	0.20
Nebraska	1.80	1.35	1.07
Norfolk	5.06	4.25	3.06
Troup	4.92	4.57	3.58

a/ Exchangeable fraction removed.

b/ Chelated fraction removed.

c/ e.g., when 61.00 mg Co as $CoCl_2$ added/g soil, 29.00 mg Co adsorbed/g soil.

No relation between cobalt concentration and soil depth was observed by either Shirokov and Panasin (1972) in peat bog soils or by Chattopadhay and Jervis (1974), who studied Holland Marsh (Ontario) muck soil, which had been treated in the past with fertilizer, nutrients, heavy metal and organic pesticides, fungicides, herbicides, and insecticides. Their data are shown in Table V-4.

TABLE V-4

COBALT CONCENTRATION IN HOLLAND MARSH (ONTARIO) MUCK SOIL WITH DEPTH (Chattopadhay and Jervis, 1972)[a/]

Depth	Cobalt Concentration mg/kg (dry weight)
Surface	8.52, 8.65
0-7.5 cm	8.24
7.5-15.0 cm	7.66
15.0-22.5 cm	5.48
22.5-30.0 cm	7.32
30.0-37.5 cm	4.37
37.5-45.0 cm	5.28

a/ Analyses by instrumental photon activation analysis.

Jenne (1968) has postulated that the increase in heavy metal mobility and/or plant availability seen in waterlogged soils that are high in organic matter is due principally to the exclusion of atmospheric oxygen, which lowers the redox potential (Eh) of the soil, and to the reduction in the loss of carbon dioxide, which favors an increase in the hydrogen ion concentration (i.e., lowers the pH). These factors favor valence reduction of the iron and manganese oxides, which then dissolve, releasing cobalt and other adsorbed heavy metals. Organic matter also can reduce the iron and manganese oxides directly. Indirect reduction of the iron and manganese oxides is achieved by the electrons released in anaerobic respiration. Acidic conditions favor the removal of organic matter complexed with heavy metals; while at higher soil pH, organic complexation reduces heavy metal availability.

Kubota (1958) reported that the cobalt concentrations of deficient soils of the Atlantic Coastal Plain were related directly to their clay contents. Groundwater podzol soils with up to 15% clay held up to $\sim$ 1.8 mg Co/kg, whereas low humic gley soils with up to $\sim$40% clay held up to $\sim$ 4.5 mg Co/kg. The relations were similar for all horizons of the latter soils and for the B_h, B_2, and A_2' horizons of the groundwater podzols.

Iron content increased exponentially with increasing clay and cobalt contents in all horizons of the low humic gley soils and in the B_h, B_t, and A_2' horizons of the groundwater podzols but not in their A_1 and A_2 horizons. The iron (presumably as free oxides) was at concentrations 2,000 to 10,000 times higher than those of cobalt.

Singh and Singhal (1970) found that the correlation coefficient between cobalt concentration and the cation-exchange capacity (c.e.c.) of illitic soils of India was 0.68. However, Khodzhaeva et al. (1973) reported that when cobalt-60 solutions without carrier were passed through vertical columns of soils, the percent of total cobalt held by the top 3 cm of soil was 95.80% in the sandy loam, 97.2% in the loam, 96.2% in the till soil, 57% in the fine sand, and 72.0% in the coarse sand. The soils apparently exhibited non-ion-exchange character in their sorption of cobalt.

Hodgson (1960) found that two forms of Co^{2+} were bound to montmorillonite. A slowly dissociable form chemisorbed in a monolayer would exchange slowly with transition metal ions, but not with Ca^{2+}, Mg^{2+}, or NH_4^+. A nondissociable form (nonextractable with 2.5% acetic acid) had possibly entered into the crystal lattice or was occluded in the precipitation of another phase. It was not characterized further. In 1962, Tiller and Hodgson reported that cobalt behaved similarly with other layered silicates: vermiculite, muscovite, biotite, etc.

Hodgson et al. (1964) studied the reaction of Co^{2+} with montmorillonite at pH 5.0 to 7.0 in $CaCl_2$ solution with or without competing Mg^{2+}. The results were consistent with adsorption of hydrolyzed ions and inconsistent with the exchange of metal ions for weakly dissociable surface H^+. Since other results had supported the latter hypothesis, perhaps both mechanisms are operative--some surface sites being controlled by hydrolysis, others by ion exchange.

James and Healy (1972) studied the theoretical adsorption of cobalt ions in 6 mg/liter and 0.7 mg/liter solutions by silica. At the higher cobalt concentrations, the adsorption was 0% at pH $\sim$ 5, $>$ 10% at pH $\sim$ 6, $\sim$ 30% at pH $\sim$ 7, $\sim$ 70% at pH $\sim$ 8, and $\sim$ 100% at pH $\sim$ 8.5. At the lower concentration, cobalt was adsorbed more strongly at lower pH's. Thus, at pH $\sim$ 5, $\sim$ 7% was adsorbed; at pH $\sim$ 6, $\sim$ 23%; at pH $\sim$ 7, $\sim$ 88%; and at pH $\sim$ 8, $\sim$ 97%. The greatest increase in adsorption was between pH 6 and 7; yet this is below the range where Co^{2+} changes into $CoOH^+$, which can occupy twice as many sites in the form Si-O-M-OH as when it is doubly charged. Although the coulombic attraction of a divalent metal cation for the negatively charged surface of colloidal silica is greater than that of MOH^+, the effect is partially offset by the opposing attraction for water molecules, which tend to keep M^{2+} in solution (Leeper, 1978).

Hodgson and Tiller (1962) investigated the nature of the clay mineral surfaces involved in selectively bonding low concentrations of cobalt in the presence of high $CaCl_2$ concentrations. They found that edge surfaces of the 2:1 layer silicates, montmorillonite and vermiculite, are not likely involved in cobalt adsorption and that internal basal surfaces of vermiculite had only a limited capacity to sorb cobalt. The principal sites of cobalt sorption appeared to be the external basal surfaces, where chemical weathering and physical abrasion may have introduced defect structures favorable to chemical bonding.

Tiller et al. (1963) compared Co^{2+} sorption of clays of soil origin with that of geologic clays. Although Co^{2+} sorption from very dilute solutions in the presence of 0.1 N $CaCl_2$ by many different minerals varies widely and although Tiller et al. (1963) studied soils with a great range in mineralogical composition, most soils adsorbed about the same amounts of Co^{2+} in the presence of $CaCl_2$. Pachappa soil clay (California) was an exception, showing cobalt sorption similar to that of vermiculite from geological sources. Tiller et al. used subsoils to minimize the influence of organic matter.

Treating the soils with H_2O_2 to destroy organic matter did not significantly affect Co^{2+} sorption except by the Pachappa soil of California and the Nipe soil of Puerto Rico. There was a marked decrease in the first case and an increase in the second. Tiller et al. could not explain the phenomena.

It was thought that the experimental soils had already sorbed substantial amounts of heavy metals, but experiments with EDTA showed that presaturation under natural soil conditions was not operative.

The presence of amorphous materials (silica and alumina) as a factor in cobalt adsorption was also excluded. Free iron oxides were still a possibility, but crystalline iron oxides were ruled out (Tiller et al., 1963).

Hodgson et al. (1969) presented more evidence that it is the clay surface itself (as affected by weathering or formation) that affected cobalt sorption, and not its iron oxide coating nor the silicate crystal structure. Soils were treated by a modification of Jeffries' method for removing iron oxide coatings from minerals without affecting the reactivity of the mineral surface. Heating the soil with a mixture of lithium and sodium oxalates dissolved the free iron oxides without reducing hydrous ferric oxide. Organic matter was removed by peroxide. Although four of the seven soils treated showed increased reactivity after iron removal, the soil clays all behaved similarly in adsorbing cobalt.

These observations with chemical treatments that destroyed ferric and manganese oxides and organic matter do not resolve the question, is it the oxides or the organics that bind more cobalt? Leeper (1978) cited the 1963 report of LeRiche and Weir, who treated soils with oxalate under ultraviolet light. The treatment not only reduced ferric oxide to ferrous but dissolved one-third of the organic matter. As a result, 80% of the soil's manganese, 60% of the iron and cobalt, 50% of the copper and lead, and 20% of the chromium and nickel were released to solution.

Many studies show that adsorption of cobalt by manganese oxides in soils is important.

In soils, manganese minerals occur as nodules, veins, occlusions, and coatings. Cobalt either is present due to precipitation, to sorption, and/or to incorporation in the crystal lattice. When 1.0 mg Co/g Mn oxide was added to a solution, > 0.9 mg cobalt was sorbed after 1.5 days. Desorption with 2.5% acetic acid left ~ 0.4 mg Co/g. However, after aging for 180 days, 86% of the original adsorbed cobalt became nonextractable. This secondary reaction is specific for cobalt.* Cobalt sorption rose sharply as the pH approached neutral in another experiment. At pH 3, manganese minerals dissolved as Mn^{2+} in the presence of H_2O_2. Before removing manganese, the total cobalt sorbed was 0.85 to 1.93 mg Co/g Mn mineral (nonextractable 0.16 to 0.62 mg/g). After manganese removal, total cobalt sorbed was 0.26 to 0.70 mg/g (nonextractable 0.038 to 0.10) (McKenzie, 1967).

Even in the presence of a large excess of Ca^{2+} ions, manganese minerals can remove cobalt completely from solutions at the low concentrations expected in soil solutions.

Thus, it is not surprising that cobalt applied as fertilizer to a manganiferous soil developed on deeply weathered basalt in Tasmania was unavailable for plant uptake and thus did not cure the needs of Co-deficient sheep** (Nicholls and Honeysett, 1964; cited by Leeper).

Taylor and McKenzie (1966, cited by Mortvedt and Cunningham, 1971) found that cobalt in Australian soils is associated with the manganese minerals birnessite and lithophorite, and McKenzie (1967) reported that finely ground manganese nodules isolated from soil samples sorbed an average 20 times more cobalt than did associated soil minerals.

---

* Reversion with time to forms less available to plants has been reported for other heavy metals, notable in calcareous soils (Epstein and Chaney, 1978).

** Cobalt sulfate applied to deficient soils can be extracted by neutral 1 N ammonium acetate soon after application. Later, 0.1 N hydrochloric acid is required to remove it (Banerjee et al., 1953; cited by Mortvedt and Cunningham, 1971).

McKenzie (1967) presented good evidence to show that oxidation of cobalt(II) to cobalt(III) increased binding to MnO_2 (Leeper, 1978). Murray et al. (1968; cited by Loganathan and Burau, 1973) proposed that the high specific sorption potential of cobalt (on manganite) may be due to a surface oxidation of Co^{2+} to Co^{3+} at the oxide-water interface.

Suarez and Langmuir (1976) followed the release of heavy metals from a loamy sand soil on weathered sandy dolomite cored from up to 70 ft (21 m) beneath a municipal waste landfill at State College, Pennsylvania. Most of the metals were bound to amorphous manganese oxides as coatings on quartz grains. Initial mild chemical soil leaching dissolved primarily the manganese oxides and released heavy metals in the order Mn > Fe > Co > Ni > Pb > Zn > Cu > Cd > Ag. A final leaching dissolved primarily the ferric oxides and released the metals in the order Fe > Mn > Ni > Zn > Co > Cu > Pb > Cd > Ag. The chief source of the metals was presumably weathering of sulfides in the dolomite, not leachate from the landfill. If infiltration of the leachate had been the source, readily sorbed cobalt would have been enriched at shallow depths.

δ-MnO_2 is one of the main hydrous manganese oxides stable at ordinary temperature, pressure, and Eh. It is an important form in water, soils, and manganese nodules. Loganathan and Burau (1973) proposed that Co^{2+} may interchange with both Mn^{2+} and Mn^{3+}* in the hydrous δ-MnO_2 crystal structure at pH 4. Mn^{2+} was released to the solution phase during Co^{2+} adsorption. In addition, 2 moles of H^+ were released per mole of Co^{2+} sorbed at bound H^+ sites. McKenzie (1970; cited by Loganathan and Burau) found that Mn^{2+} was released when heavy metals were sorbed. He also suggested that cobalt replaced Mn^{3+} from the δ-MnO_2 structures. Morgan and Stumm (1964; cited by Loganathan and Burau, 1973) suggested that δ-MnO_2 sorbed strongly hydrolyzable metal ions better than weakly hydrolyzable ones.

---

* It has not been proved that Mn^{3+} is part of the crystal structure of δ-MnO_2. Substitution of low-spin Co^{3+} in the octahedral field for Mn^{3+} would achieve a much greater crystal field stabilization energy than if Mn^{4+}, or other suitably sized cations were substituted.

Singhal et al. (1976) reported information that indicated that about 14% of the cobalt in a sodic saline soil at pH 8.8 was held by soil bacteria. The original cobalt concentration of the soil (in pot experiments) was 15.47 mg/kg. After fumigation of the soil with Nemagon (1,2-dibromo-3-chloropropane) or benzene hexachloride, the total content of cobalt increased significantly (e.g., up to 17.90 mg Co/kg after Nemagon treatment) within the first 30 days. The increase was attributed to the cobalt from the protoplasmic constituent and the vitamin B_{12} of the dead soil bacteria. After 30 days, the cobalt concentration in the soil began to decrease as the microorganisms became reestablished and readsorbed or fixed some of the cobalt. The paper does not explain how the cobalt contained in the soil bacteria was excluded in determining the original soil cobalt concentration.

2. <u>Extractability and mobility of cobalt in soils</u>: Before 1955, the prevailing view of the speciation of cobalt in soils was that it existed in an adsorbed state in inorganic and organic ion-exchange complexes, as the phosphate, and as the water extractable sulfate. Mitchell (1955) then suggested that because cobalt extractability by ammonium acetate and acetic acid are so different, either H^+ has a better replacing power for Co^{2+} than NH_4^+ or different forms of cobalt are being extracted.

Table V-5 is a compilation of the experimental results reported by several authors regarding the extractability of cobalt from soils. In general, the extractability of cobalt is enhanced by waterlogging (poor drainage), low pH, the basicity of the parent rock (more cobalt present), the presence of plant extracts and artificial chelating agents, airborne cobalt contamination, strength of the acid extractant, and increasing temperature up to 200 to 400°C. In some cases, the higher the clay content, the less cobalt was extractable. However, since more cobalt is usually present with the higher clay content, the opposite may also occur, i.e., the more clay, the greater the extractability of cobalt. Cobalt is less extractable from sphagnum peat than from other organogenic soils. Organic matter tends to retard cobalt mobility at higher pH. The lower horizons sometimes have more extractable cobalt than the A horizon of poorly drained soils.

TABLE V-5

EXTRACTABILITY OF COBALT FROM VARIOUS SOILS

Soil Composition	Extractant	Extractable Cobalt Content	Comments	Authors
Various German soils, 2.0 to 76.5% clay, containing < 0.3 to 17.0 mg Co/kg	EDTA + ammonium acetate	0 to 0.65 mg/kg[a/]	There was not much difference among the horizons in extractability. Correlation between clay and nonextractable cobalt was 0.823.	Schlichting and Elgala (1975)
Various soils	Water, maximum extractability at 300°C	-	Aiken clay loam and "Soil No. 4FF" did not give any water-extractable cobalt under any treatment.	Nishita and Haug (1974)
Hanford sandy loam and Vina loam	Ammonium acetate	Maximum at 200°C	-	Nishita and Haug (1974)
Aiken clay loam	Ammonium acetate	No extraction at 300-400°C	Cobalt was extractable at temperatures above and below this range.	Nishita and Haug (1974)
Four soils, one with severe iron deficiency	DTPA-$CaCl_2$-sodium acetate	-	Extractability of cobalt from two soils higher at pH 5 than at pH 6-8, but the two soils with a low-labile pool of cobalt did not show much difference with pH (one of those was iron-deficient).	Lopez and Graham (1972)

TABLE V-5 (continued)

Soil Composition	Extractant	Extractable Cobalt Content	Comments	Authors
Ten U.S. soils of various types, containing 25-310 mg Co/kg	Acidic industrial wastes simulated by 1% HCl with $AlCl_3$ and $FeCl_3$, pH $\sim$ 3.0	$\leq$ 7% (six soils), 30-33% (three soils), 84% (one soil)	Analysis by atomic absorption of the six soils having $\leq$ 7% extractable cobalt, only one was a clay soil. The four soils with 30-84% extractable cobalt were all clays. The order of the elements eluted from a soil column was $Mn \gg Co \simeq Ni \approx Zn > Cu \approx Cr > Pb \approx Cd$. Correlation coefficients of extractable cobalt with other soil properties were highest for total metal, total Mn, Fe oxides, and clay content and electrical conductance. The correlation with cation-exchange capacity was very low.	Korte et al. (1975)
Scottish soils developed on serpentinite-derived glacial drift	Acetic acid	5.1% (coarse sand), 9.7% (fine sand), 18.6% (silt), 11.0% (clay)	Cobalt showed no systematic variation in concentration in various sized soil fractions or with differences in drainage.	Wilson and Berrow (1978)
Scottish soils, typically podzolic brown forest soils of low-base status on more basic parent materials	2.5% acetic acid + 1 N ammonium acetate	Higher from soils of basic or ultrabasic origin	There was a gradual fall in extractability of metals with depth regardless of clay content. Cobalt is mobilized in poorly-drained Scottish soils.	Swaine and Mitchell (1960)

TABLE V-5 (continued)

Soil Composition	Extractant	Extractable Cobalt Content	Comments	Authors
Scottish soils	2.5% acetic acid at pH 2.5	< 0.05 mg/ℓ in very deficient sands to 2 mg/ℓ	Level of deficiency, 0.25 mg/kg extractable cobalt (pH 5.6)	Mitchell (1955)
Finnish topsoils	0.5 N ammonium acetate + 0.5 N acetic acid at pH 4.65	-	Order of extractability of metals Mo > V > Co > Cu > Ni > Pb > Zn > P > Sr > Mn > Fe > K > Mg > Ca. Of organogenic soils, the lowest concentrations of soluble elements was in sphagnum peat; the highest in mold soils (15-40% humus).	Sillanpää and Lakanen (1967)
Soddy medium podzol, A_2 horizon 0.7% humus; 1:1 ratio with sand	Lake water Aqueous pine needle extract, linden leaf extract, or aspen leaf extract	62.5% 91.8-94.2%	Most of the ^{60}Co (as $CoCl_2$ without carrier), which had been sorbed by a column of soil, was desorbed within ~ 6 volumes of solution although the experiments continued for 30 volumes of extractant.	Chebotina (1975)
Unspecified soil	0.012 N EDTA Distilled water Extracts from yellow pine needles, yellow birch leaves, green Artemisia, or yellow aspen leaves	100% ~ 20% ~ 13-31%	-	Titlyanova and Timofeeva (1962)
Southeastern U.S. poorly drained soils, A_1 horizon	2.5% acetic acid + dithizone	0.008-0.394 mg/kg	Extractable cobalt invariably less in the A_2 horizon (less organic matter therein). Sometimes B horizon had more extractable Co than A horizon.	Alban and Kubota (1960)

TABLE V-5 (continued)

Soil Composition	Extractant	Extractable Cobalt Content	Comments	Authors
Southeastern U.S. well-drained soils, A_1 horizon	2.5% acetic acid + dithizone	0.06-3.74 mg/kg	-	Alban and Kubota (1960)
Groundwater podzols associated with cobalt deficiency in ruminants	2.5% acetic acid + dithizone	Usually ≤ 0.02 mg/kg	-	Alban and Kubota (1960)
Unspecified soils		Ratio of ^{60}Co concentration in soil solution to that in soil:	-	Chebotina and Kulikov (1973)
	Distilled water	1/948	-	Chebotina and Kulikov (1973)
Unspecified soils	Extracts prepared from macerated herbs and grasses, e.g., wood horsetail, May lily, clover, and meadow pea	1/560 to 1/4 (examples, 1/8-1/4); average 1/56	-	Chebotina and Kulikov (1973)
Soils (≤ 40.0 mg Co/kg) incubated with anaerobically digested Ontario sewage sludges (≤ 12.2 mg Co/kg)	0.005 M DPTA	Extractability increased with increasing organic content of the soils.	-	Webber and Corneau (1978)
Soil incubated with atrazine for 10 days at 20°C and 100% water capacity	Acetate buffer	More than without the atrazine treatment	The amounts of Mn, Cu, and Zn extracted were reduced.	Stancheva (1977)
Lowmoor soils (plants growing thereon had < 0.005-0.08 mg Co/kg)	Acetate	0.09-0.23 mg/kg (< 20%)	-	Barufke (1962)
Highmoor soils, higher pH (plants growing thereon had 0.000-0.07 mg Co/kg; generally plants had more Co than on lowmoor soils)	Acetate	0.11-0.41 (< 30%)	Acetate solubility decreased with increasing Co concentration in a soil.	Barufke (1962)

TABLE V-5 (concluded)

Soil Composition	Extractant	Extractable Cobalt Content	Comments	Authors
Scottish surface soil ($\leq$3 cm) from downwind side of a group of metal refineries	Acetic acid	$\leq$ 1.5 mg/kg	-	Mitchell (1955)
Scottish surface soil ($\leq$3 cm) from upwind side of a group of metal refineries	Acetic acid	0.3 mg/kg	-	Mitchell (1955)
Gleyed sublayers of cultivated Scottish soils on glaciated till of basic igneous origin	Water	Removes much Co, Ni, and Fe^{2+}	Availability increases with impeded drainage	Mitchell (1955)
Well-drained profile, 0-3 cm	2.5% acetic acid	1.3 mg/kg	-	Mitchell (1955)
Well-drained profile, 14-16 cm	2.5% acetic acid	0.15 mg/kg	-	Mitchell (1955)
Very poorly drained profile, 0-3 cm	2.5% acetic acid	1.5 mg/kg	-	Mitchell (1955)
Very poorly drained profile, 14-16 cm	2.5% acetic acid	18 mg/kg	-	Mitchell (1955)
Well-drained profile, 0-3 cm	1 N ammonium acetate	0.04 mg/kg	-	Mitchell (1955)
Well-drained profile, 14-16 cm	1 N ammonium acetate	0.02 mg/kg	-	Mitchell (1955)
Very poorly drained profile, 0-3 cm	1 N ammonium acetate	0.08	-	Mitchell (1955)
Very poorly drained profile, 14-16 cm	1 N ammonium acetate	13	-	Mitchell (1955)

a/ 0 at 2% clay, pH 2.7, or at 55.0% clay, pH 8.1.

Nishita and Haug (1974) attributed the variations in cobalt extractability with heat treatments (see Table V-5) to possible physical-chemical changes: oxidation or reduction of the metal, dehydration, concentration of organic matter, and fusion of fine particulate material.

Several Soviet authors have reported evidence that water soluble substances from decomposing fallen leaves of woody plants enhance the mobility of radioisotopes in the soil but decrease their plant uptake (Chebotina and Kulikov, 1973). These substances are probably fulvic and humic acid complexes. Klechkovskii (1968) studied (by potentiometric titration) the formation of Co^{2+} complexes with fulvic acid fractions separated by adsorption chromatography from dark gray forest soil. Schnitzer and Hansen (1970; cited by Schnitzer 1978) calculated stability constants for cobalt(II)-fulvic acid complexes at pH 3.0 (log K=3.1 to 3.2) and pH 5.0 (log K=4.1 to 4.2). Orlov and Eroshicheva (1967) obtained polarographic evidence for the formation of a cobalt complex with humic acid. Under equilibrium conditions, 1.768 milliequivalents cobalt (104 mg) was bound per gram of humic acid at pH 4.2. Mukherjee et al. (1972) found that about 10% of the cobalt adsorbed by humic acid isolated from a hilly soil could not be leached by treating with 0.1 N sulfuric acid. Dunigan and Francis (1972) found 91 to 99% desorption of cobalt held by humic acids* extracted from the A horizon of Emory silt loam by washing four times with 1 N salt solutions; ferric chloride dislodged the most cobalt.

Singh and Singhal (1977) followed the extractability of cobalt by various salt solutions at pH 8.80 from soils with and without organic matter. The soils were used as adsorbents on thin layer chromatographic plates. Organic matter retarded the mobility of cobalt at this pH with each of four 0.1 N salt solutions (NaCl, $NaHCO_3$, KCl, $CaCl_2$) and enhanced it with two 0.1 N salt solutions ($MgCl_2$ and Na_2SO_4).

---

* At pH 3 to 10, 12 to 28% cobalt from 25 ml 1 N $CoCl_2$ tagged with ^{60}Co was held by 25 to 50 mg humic acid.

Thomas (1975) reported that when ^{60}Co-containing leachate from leaf litter of mockernut hickory (Carya tomentosa) and swamp black gum (Nyssa sylvatica) was applied to the A_1 horizon of a Fullerton silt loam**, 91% of ^{60}Co remained in the upper 9 cm of the soil (the biologically active zones).

Alban and Kubota (1960) found a 0.930 correlation of the concentration of cobalt in swamp black gum and its concentration in the top 0 to 30 cm of poorly drained soils. This correlation was 0.906 at 0 to 15 cm depth (significant at the 1% level) for well-drained soils. To minimize any soil differences that might have resulted from leaf fall, all the soils were sampled from under black gum trees of comparable height (1.8 to 2.1 m tall).

Hodgson et al. (1965) concluded that cobalt was complexed by organic ligands to "only a very limited degree in the soil solution studied" although they calculated that -15 (sic) to 69% of the cobalt in the soil solution of four New York soils existed as organic complexes. The highest calculated cobalt complexing was in the A horizons of hardwood forest Williamson silt loams.

The groundwater and sediments at the Hanford reservation in Washington were monitored from 1944 to 1967 while the chemical processing plants disposed of waste liquids to the vadose zone.** Little movement of the radionuclides, presumably including ^{60}Co, occurred; nearly all were sorbed by the soil and remained in the upper levels of the vadose zone (Rouston, 1973).

Means et al. (1978), however, reported that EDTA, which is commonly used in decontaminating operations at nuclear facilities, is causing low-level migration of ^{60}Co from intermediate level liquid waste disposal pits and trenches at the Oak Ridge National Laboratories (ORNL) burial grounds. Since 1944, solid waste has been buried in shallow trenches at six

---

* The A_1 horizon comprised sand, silt, and clay in the ratio 34:63:3 with 6.5% organic matter at pH 4.5. The cation-exchange capacity of the soil was 5.6 milliequivalents per 100 g.

** Vadose is a term used to denote seepage waters between the surface and the water table.

different burial grounds; and from 1951 to 1965, there was also disposal to seven different seepage pits. ORNL's problem may be because its annual precipitation (> 127 cm) is greater than that at other burial sites in the United States. Groundwater levels at ORNL are high, and fractures are abundant in the rocks. The predominant bedrock is Conasauga shale. Some ^{60}Co were adsorbed by manganese oxides in the shale and soil. The distribution coefficient (Kd) in laboratory experiments for cobalt in Conasauga shale compared with cobalt in solution at pH 6.7 was ~70,000; but in the presence of 0.00001 $\underline{M}$ EDTA, the Kd was reduced to 2.9. The Kd for ^{60}Co in soil and water at the ORNL burial grounds in the vicinity of the most contaminated site was ~ 7 to 70 (average ~ 35).

^{60}Co complexes with EDTA, humic acid, fulvic acid, or other multidentate ligands are so strong that they resist adsorption of Co^{2+} by cation-exchange resins. It had previously been noted that ^{60}Co in groundwater exchanged to the extent of only ~ 5 to 10% with cation-exchange resins. Gel filtration chromatography, which separates solutes according to size, showed that 90 to 95% of the ^{60}Co in the groundwater near the burial location was eluted along with an organic fraction of molecular weights < 700 plus sodium salts of polyvalent anions. The major component of this fraction was EDTA (~ 3.4×10^{-7} $\underline{M}$). Other multidentate ligands such as DTPA had been used sparingly in decontamination. About 5 to 10% of the ^{60}Co was associated with a fraction of molecular weights > 700, identified as humic substances by infrared spectrometry. Weaker complexing agents identified in the trench leachates were palmitic, phthalic, and other mono- and dicarboxylic acids.*

---

* The larger breakdown products of organic materials are formed under anaerobic conditions. These weaker complexing and solubilizing agents represent more complete breakdown of organic materials under aerobic conditions (Patrick and Mikkelsen, 1971).

Use of EDTA and similar compounds in decontamination operations at nuclear facilities is widespread throughout the world. Means et al. (1978) suggested substitution of biodegradable NTA for the environmentally persistent EDTA. Similarly, Orlova et al. (1974) recommended that Trilon B, which forms negatively charged combinations with Co^{2+}, be excluded from detergents used at nuclear power plants because it increases the mobility of ^{60}Co. ^{60}Co was converted to a chemical state not adsorbed by loose rocks.

By radiotracer experiments, Bittel et al. (1969) found that free Co^{2+} (added as $CoCl_2$) was held in the superficial soil profiles of an artificial rice field irrigated with radiocobalt-containing water. When cobalt was complexed with EDTA, its mobility in the soil was greater. Thus, 92.1% of non-chelated cobalt was held in the upper 0 to 2 cm of soil compared with only 61.2% of cobalt chelated by EDTA. Conversely, only 3.1% of the free cobalt was found at 2 to 4 cm depth compared with 26.2% of the Co-EDTA.

The stable cobalt concentration in the field soil was 13 mg/kg (0.4 mg/kg of which was extractable by ammonium acetate). At a concentration of approximately 1 μg Co/liter in the irrigation water, in the absence of EDTA, the distribution of cobalt in the field was 8.29% in the plants, 35.50% sorbed by the soil, and 56.21% remaining in the water. When EDTA was present, the cobalt distribution was 0.38% in the plants, 11.50% in the soil, and 88.12% in the water.

Ammonium hydroxide (0.01 $\underline{N}$) extracts the "mobile fraction" of trace elements--the organically bound components from top soil and oxides from the subsoil. About 80 to 90% of the extract comprises organic substances and aluminum and ferric hydroxides. The order of mobility of trace elements in mineral soil is Zn > Cd = V > Pb > Cu > Co > Be; in organic layers, Zn > V > Cd > Cu > Co > Be > Pb (Keilen et al., 1978).

B. Water

Bradford (1976) pointed out that systems may appear to be unique in their trace metal distributions because "the number of ways in which known [influencing] processes can combine and interact in any given system . . . is astronomical." Other problems leading to a picture of apparent uniqueness may be "inadequate or improper analytical techniques, failure to distinguish between solid and solution phases and inability to determine the chemical speciation of the dissolved trace metal." Bradford continued that models that can successfully describe unique systems as the interactions of a few well-recognized processes do not yet exist.

Bradford (1976) remarked that although the chemical behavior of major inorganic species in natural water systems can be roughly predicted by equilibrium thermodynamics, especially in systems such as the deep ocean where there is a steady state approximating equilibrium, there are a large number of processes involved in the behavior of trace heavy metals, some of which may not yet be identified. The influence of the biota is not readily predictable, although some authors have discussed metal movement through the foodweb. Jenne (1968) has suggested that adsorption by hydrous iron and manganese oxides is one of the most important processes affecting trace metal behavior; and that in high-organic sediments, the organic fraction can be even more important in interacting with trace metals in solution. Many authors have suggested that dissolved organic matter can sequester trace metals in surface waters or interstitial waters of sediments. Some authors propose cation exchange with organic matter as a sediment desorption mechanism. Others emphasize the influence of variable potential environments on trace metal movements and diagenesis.

1. Transport of cobalt in natural waters

a. Dissolved and suspended matter: Although the amount of suspended material in 19 rivers sampled by Turekian and Scott (1967) varied over nearly two orders of magnitude (see Table V-6), the cobalt concentrations in the suspended materials varied by only one order of magnitude except for the Susquehanna River in Pennsylvania. The suspended material in each river, however, carried comparable amounts of cobalt per liter. Most of the

TABLE V-6

COBALT TRANSPORT IN THE SUSPENDED LOAD OF RIVERS
(Turekian and Scott, 1967)

River	Location	Suspended Material, mg/liter	Cobalt Concentration in Suspended Material, mg/kg	Amount of Suspended Cobalt in the Water μg/liter
Brazos	Texas	954	20	19
Colorado	Texas	150	17	2.6
Red	Louisiana	436	7	3.1
Mississippi	Arkansas	185	33	6.1
Tombigbee	Alabama	54	34	1.8
Chattahoochee	Georgia	71	35	2.5
Flint	Georgia	12	39	0.5
Savannah	South Carolina	30	36	1.1
Wateree	South Carolina	37	34	1.3
Pee Dee	South Carolina	188	23	4.3
Cape Fear	North Carolina	61	21	1.3
Neuse	North Carolina	36	30	1.1
Roanoke	North Carolina	33	45	1.5
James	Virginia	41	60	2.5
Rappahannock	Virginia	28	46	1.3
Potomac	Virginia	34	94	3.2
Susquehanna[a/]	Pennsylvania	54	> 500	> 27
Rhone	Avignon, France June 1960	296	29	8.6
Rio Maipo	Puente Alto, Chile south of Santiago September 1966	41	76	3.1

a/ Nearer its source in New York, the insoluble cobalt concentration in 1974 was only 0.2 μg/liter, representing only 9% of the cobalt in the water (El-Barbary et al., 1977).

rivers listed in Table V-6 had concentrations of suspended cobalt varying from 0.5 to 8.6 mg/liter. The Brazos River, with its high suspended load, had 19 μg Co/liter in the suspended state, which was not much different from the > 27 μg Co/liter found in the suspended material of the highly contaminated Susquehanna River. Corrections of the calculations of Turekian and Scott show that the very polluted Susquehanna River transported 150 short tons cobalt annually in its suspended sediments.

Similarly, Windom et al. (1971) found 21 to 34 mg Co/kg in the suspended matter (average 22.4 to 43.5 mg/liter) of three southeastern U.S. rivers: the Altamaha (southeast Georgia), the Ogeechee (eastern Georgia), and the Satilla (south and southeast Georgia). They estimated that southeastern rivers supply the ocean with only 0.011 MT cobalt annually from the suspended load and < 0.013 MT/year in the dissolved load. The average post-glacial accumulation of cobalt in the Atlantic Ocean sediments was estimated to be 46 $\mu g/m^2/1{,}000$ years.

Radosavljević et al. (1973) reported data on dissolved and suspended cobalt concentrations in the Danube River from 1961 to 1970 (see Table V-7). The fluctuations of cobalt in the suspended load (27.4 to 85.9%) were not obviously related to fluctuations in river flow (4,329 m^3/sec in 1961 to 7,304 m^3/sec in 1965), suspended mineral content (22 mg/liter in 1970 to 58 mg/liter in 1968), or percent organics in suspended material (12 to 19%, measured only in 1966 to 1970).

Popp et al. (1979) reported that the Rio Puerco, a tributary of the Rio Grande in New Mexico, carried 200 μg Co/liter in the dissolved state and 1,280 μg Co/liter in suspension (86% of the total load). The concentration of cobalt in the suspended sediment was 23 mg/kg.

In the Susquehanna River discharge, Carpenter et al. (1975) found the highest concentrations of cobalt, manganese, nickel, and zinc in January. To correct for high solids discharge and water flow in February and March, they compared concentrations in the solids themselves. There were secondary peaks for concentrations of cobalt, chromium, nickel, copper, and manganese in July. The authors thought that binding of

TABLE V-7

AVERAGE YEARLY CONCENTRATIONS OF COBALT IN THE DANUBE RIVER AT 1,145 km, 1961-1970
(Radosavljevic et al., 1973)

Year	Cobalt, μg/liter		Percent of Total Cobalt Suspended	Percent of Cobalt Dissolved
	Suspended	Dissolved		
1961	0.578	0.909	38.9	61.1
1962	-	2.65	-	-
1963	2.36	0.64	78.7	21.3
1964	3.38	8.95	27.4	72.6
1965	3.04	4.59	39.8	60.2
1966	5.41	3.14	42.0	58.0
1967	2.71	0.446	85.9[a]	14.1
1968	1.28	0.975	69.1	30.9
1969	0.174	0.422	29.2	70.8
1970	0.127	0.136	48.3	51.7

a/ The cobalt particulate loading appeared especially concentrated that year. In other years, the ratio of cobalt concentration in the particulate to its concentration in the water ranged from ~ 11,000 to 82,000; but in 1967, this ratio was 177,000.

the metals by abundant decaying organic matter at those times contributed to their high concentrations in the suspended solids. At periods of high flow (4 months of the year), the Susquehanna was calculated to be carrying 0.7 MT cobalt per day. During low-flow periods, however, only 0.04 MT cobalt was carried daily. Their total of 90 MT Co/year carried by the Susquehanna compared rather well with the 150 tons/year (136 MT/year) derived from the data of Turekian and Scott.

Leland (1977) found an average 8 to 9 mg Co/kg in the suspended matter of the epilimnion (5 to 7 m) of offshore Lake Michigan waters in July 1972. In the Northern Basin, he found 11 mg Co/kg in suspended sediments 1 meter above the lake floor.

Trefey and Presley (1976) found 18.1 to 22.3 mg Co/kg in the suspended matter of the lower Mississippi River between May 1974 and September 1975. The cobalt concentration in the suspended matter of the Gulf of Mexico was not significantly different: 18.2 to 23.0 mg/kg.

Kopp and Kroner (1968) detected dissolved cobalt in only 3.1% of 1,410 natural water surveillance samples taken across the United States. Only cadmium had a lower detection frequency (2.7%). However, in a survey conducted from October 1965 to December 1966 at seven sites on the Ohio, Delaware, Allegheny, and Kanawha Rivers, cobalt was detected in 12% of the water samples (at 13 to 48 μg Co/liter) and 8% of the suspended solids (at 8 to 13 μg/liter).

Joe Mill Creek is in the Copper Ridge Zinc District in northeast Tennessee. It is $\sim$ 17 km southwest of the Flat Gap Mine, but it does not drain the mine area itself. The stream drains an area of exposed minor zinc mineralization, and the creek's stream bed itself has outcrops of sphalerite-bearing carbonates. Dissolved species account for nearly all (96.8 to 98.6%) of the solids in the water. Thus, by atomic absorption analysis, 4 to 8 μg Co/liter was determined to be in the water. Cobalt represented 20 to 32 mg/kg of the dissolved solids. The cobalt concentration was somewhat higher in the total suspended particulates: 52 to 54 mg/kg (slightly higher in the fine than in the coarse fraction). Even higher cobalt concentrations occurred in the colloidal material (mostly

10 Å illitic material): 167 to 288 mg/kg; but colloidal material represented much less than 1% of the weight of total solids (Perhac and Whelan, 1972).

Perhac (1972) compared the data for Joe Mill Creek, Grainger County, with similar data for Poor Valley Creek, Hawkins County, Tennessee, which flows through sandstone and shales instead of carbonates. The shales and windblown debris from ore trucks may have contributed heavy metals to Poor Valley Creek, which carried 91.8% of its cobalt in the dissolved solids and 8.1% in the coarse particulate. Not only was more cobalt carried by the greater amount of coarse material particulates, compared with Joe Mill Creek, the cobalt was more concentrated in them.

Nix (1970) reported that when cobalt was found in concentrations > 4 μg/liter in impoundments of the Ouachita River between Mount Ida and Malvern, Arkansas, it was almost exclusively water soluble. However, when the concentration was < 4 μg/liter, a greater fraction of the total cobalt was particulate. In the extensively developed Lake Hamilton, on this river, the average cobalt concentrations were 3 μg/liter in the soluble fraction and 2 μg/liter in the particulate. In the less developed Lake Catherine, which received some industrial effluents, the soluble and particulate cobalt concentrations were each 2 μg/liter.

Lieser et al. (1977) (using freeze drying, adsorption on charcoal, or formation of organic complexes and neutron activation analysis) found 34% of the cobalt in the North Sea in the suspended material (0.023 μg/liter versus 0.045 μg/liter in the filtered seawater). This value was not much different from the amount of cobalt in suspended material of the Main River near Frankfurt, Germany: 33.4% of the cobalt was in suspended material in midstream, and 42.2% near the left bank.

Freshwater in Lake Washington (Washington State) was found to have completely soluble cobalt at 0.18 to 0.40 μg/liter, but 11 to 15% of the cobalt of the Strait of Juan de Fuca (between Puget Sound and the Pacific Ocean) was in particulate form (> 0.3 μm) as determined by concentration on the filter/sorption beds of Battelle large volume water samplers (Schell and Nevissi, 1977).

Marchand (1974) found that within 3 hr after ^{60}Co was added as the chloride to natural seawater at pH 7.8, about 89% was still soluble. His results are given in Table V-8.

When particulate concentrations are given, they may comprise not only solid precipitates of an element and amounts adsorbed to solid surfaces but also amounts incorporated into organisms and chemically bound to detritus. "Dissolved" species are usually those that pass through a membrane filter (usually < 0.45 μm) (Sibley and Morgan, 1977).

Because adsorption on the filter and trapping of smaller size particles are apt to cause false results in trying to distinguish "dissolved" from "particulate" elements, Beneš and Steinnes (1975a and 1975b) used dialysis *in situ* to sample the water. For the large, only slightly polluted Glomma River (pH ~ 6.9) at Fetsund, Norway, dialysis *in situ* removed 13% of the cobalt in particulate form from the water (0.23 μg Co/liter in November 1973 and 0.11 μg Co/liter in February 1974). Lake Trehorningen at Baerum, Norway, a small, unpolluted lake, pH ~6.7, is a source of potable water. Dialysis *in situ* removed 66% of the 0.13 μg Co/liter; centrifugation removed 71%; and ion-exchange filtration removed 58%. At least near the surface, most of the cobalt was in particulate form. The Nitelva River at Lillestrom is a small, rather polluted stream receiving industrial and municipal effluents. The cobalt concentration in February 1974 was 0.36 μg/liter, and cationic cobalt was the predominant form. The findings of these studies may have been complicated by adsorption and slow diffusion of low-molecular-weight species, nonattainment of equilibrium within the 24-hr sampling period, and failure to separate organic particulates and small inorganic colloids. During sample storage, Co^{2+} and Mn^{2+} might have been oxidized to the trivalent state, hydrolyzed, and adsorbed to particles at higher pH's; or Co^{2+} may have precipitated as carbonates due to bacterial respiration, a fact which was supported by a pH decrease in the lake water.

TABLE V-8

TRANSPORT OF COBALT IN SEAWATER
(Marchand, 1974)

	Natural Seawater	Seawater Enriched in Marine Algal Organic Matter	
pH	7.8	7.8	4.5
Contact time, hr	3	3	3
Soluble Cobalt	88.8%	89.1%	91.1%
Anionic or neutral[a]	1.0%	2.8%	1.7%
Secondary cationic[b]	5.5	-	5.4
Primary cationic[c]	55.3	57.1	68.3
Remaining activity	27.0	29.2	15.7
Insoluble Cobalt	11.1%	10.5%	8.9%
Soluble in 3N HNO_3 eluant[d]	9.6%	8.7%	8.0%
Remaining on ion-exchange resin	1.5	1.8	0.9

Probable species:

a/ $CoCl_2$, $CoCl_3^-$, $CoCl_4^{2-}$

b/ Co^{2+}

c/ $CoCl^+$

d/ CoOOH with MnO_2 or FeOOH

In the Columbia River, in its course from the Hanford Works in Washington, to the sea, ^{60}Co is predominantly in the particulate form. Robertson et al. (1973a) found about 95 to 98% of the cobalt on the particulate fraction (see Table V-9). Evans and Cutshall (1973) reported that > 80% of the ^{60}Co in the Columbia River at Vancouver, Washington (425 km downstream from Hanford, where it was released) was removed by filtration. The high amounts of cobalt on the particulate material may be due to the unusually high pH of the Columbia River (pH 7.5 to 8.0).* The sediments desorbed little ^{60}Co when in contact with seawater at pH 8. Robertson et al. (1973) found 4.7% desorption at 25°C in seawater, < 1.0% in tapwater, and 25.5% in hydrochloric acid at pH 2.3.

There were eight nuclear reactors in operation at Hanford during much of the 1944 to 1971 period; but after 1971, only small amounts of radionuclides reached the Columbia River from a seepage trench. The sediment profile analyses for ^{60}Co in 1973 showed little variation with depth. The primary entry of radionuclides into the river water in 1973 was by re-suspension of sedimentary particulates with sorbed radionuclides (Robertson, 1973b).

Emel'yanov and Pustel'nikov (1975) reported that 27% of the cobalt in suspended matter in the Baltic Sea was from erosion of the bottom and abrasion of the shores. These processes contributed 345 MT cobalt/year. The remaining 73% of the cobalt in suspended matter was in phytoplankton (934 MT/yr). The concentrations of cobalt in the four different types of suspended matter (terrigenous, transient, organic, and siliceous) were practically identical. The average particulate cobalt concentration in the water at all depths, was 0.012 μg/liter, being somewhat higher (0.014 μg/liter) in the surface water layer. The cobalt concentration in the particulates increased in the open sea due to sorption from the solution by iron hydroxides and, apparently, by organic particles. Cobalt was more strongly associated with organic substances than were iron, aluminum, titanium, manganese, and nickel.

---

* A more likely explanation is that the source of cobalt in the effluents is oxide corrosion products from the reactor coolant water pipes.

TABLE V-9

CONCENTRATION OF COBALT-60 IN VARIOUS PHYSICAL-CHEMICAL FORMS IN COLUMBIA RIVER WATER AFTER THE JANUARY 1971 CLOSURE OF THE LAST OF THE ORIGINAL HANFORD PLUTONIUM PRODUCTION REACTORS[a/]

(Robertson et al., 1974a)

	August 18, 1971		September 13, 1972	
	d/min^3 [b/]	%	d/min^3	%
Particulate	460	97.9	340	94.7
Cationic	4	0.9	12 ± 1	3.3
Anionic	< 1		7 ± 1	2.0
Neutral	6	1.3	< 1	
Total	470	100	359	100

a/ Small amounts of fresh radioactivity being released by a dual-purpose power-Pu production, N reactor.

b/ d = disintegrations

Gibbs (1973) reported that the transport of cobalt in the large, unpolluted Amazon and Yukon Rivers was > 98% in particulate forms (see Table V-10).

Angino and Schneider (1975) found particulate cobalt distributed equally between organic and adsorbed phases in the Kansas (Kaw) River, but they did not find cobalt in the crystalline minerals of the river.

Konovalov et al. (1967) did not find any cobalt in solution in 30 of the largest rivers of the European USSR. It was found in the suspended substances of only a few rivers. Ivanova and Konovalov (1968) reported that the magnitude of transport of cobalt in suspension was less than the amounts of suspended zinc, manganese, vanadium, nickel, or copper. The ratio of the weight of cobalt to that of other suspended trace elements was $\sim 4 \times 10^{-6}$. They calculated that a total of only 3.9 short tons (3.5 MT) cobalt was discharged annually into the seas from the suspended load of these rivers. The major input (2.064 short tons/year; 1.87 MT/year) was into the basin containing the Ural River. The next highest input (0.996 short tons/year; 0.903 MT/year) was to the Kara Sea Basin.

Shuman et al. (1977) used methods patterned after Gibbs to determine cobalt transport phases in two North Carolina rivers. Analyses were by neutron activation analysis. The results of their transport phase study are presented in Table V-11. The average fraction of particulate cobalt in the Haw and New Hope Rivers was 85 to 92%. Very similar results for the Haw River only were published by Shuman et al. (1978). Sampling was done midstream in January through May 1974. They noted in the latter study that only the organically bound fraction of cobalt was significantly higher in polluted streams.

In the Haw and New Hope Rivers of North Carolina, cobalt, iron, and manganese were about twice as concentrated in suspended matter as in bottom sediments. Shuman et al. (1977) found 3.1 to 36.1 mg Co/kg in the sediments of 16 stations. The average for polluted stations was 18.6 mg/kg, not much different from the average for clean stations: 16.4 mg/kg. The highest cobalt concentration was at a station with no known industrial

TABLE V-10

TRANSPORT OF COBALT, %, IN LARGE UNPOLLUTED RIVERS
(Gibbs, 1973)

	Amazon River at Macapa, Brazil	Yukon River Upstream From Alakanuk, Alaska
In solution and organic complexes (remaining after filtration with 0.45 μm filter)	1.6	1.7
Adsorbed (removed by treating filtered residue with $MgCl_2$)	8.0	4.7
Precipitated and coprecipitated in metallic coatings (reduced by Na dithionite and citrate)	27.3	29.2
Inorganic solids (oxidized by NaOCl)	19.3	12.9
In crystalline sediments	43.9	51.4

TABLE V-11

PERCENTAGE OF COBALT TRANSPORTED IN EACH OF FIVE MODES IN THE HAW AND NEW HOPE RIVERS, NORTH CAROLINA, AT CONTROL AND CONTAMINATED STATIONS
(Shuman et al., 1977)

	High Flow		Low Flow		Average	
	Control	Contaminated	Control	Contaminated	Control	Contaminated
Soluble	-	-	10.3	15.3	8.0	12.2
Adsorbed	26.2	36.4	32.8	25.1	31.4	27.4
Oxide coating	26.4	23.3	19.00	18.2	20.6	19.2
Solid organics	11.4	8.4	10.7	16.2	10.7	14.6
Crystalline minerals	36.0[a]	31.9	27.2	25.2	29.1	26.5

a/ May be too large because soluble cobalt could not be measured under high-flow conditions. For most elements, the percentages in the crystalline minerals were the same under both high and low flow conditions.

contamination (Dry Creek). The next highest sediment cobalt concentration, 34.1 mg/kg, was from the New Hope River receiving woodland, urban, and agricultural runoff and sewage treatment plant discharge. These values compare with 15.7 to 34.3 mg Co/kg in suspended particulates from 15 stations. The average polluted particulate material contained 26.3 mg Co/kg; the average particulate material from clean stations contained 22.3 mg Co/kg.

In the filtered water of eight stations, Shuman et al. (1977) found 0.05 to 0.39 μg Co/liter (average 0.31 μg/liter in polluted areas and 0.14 μg/liter in clean areas). The range in unfiltered water for 11 of 12 stations was 0.2 to 7 μg/liter, but the concentration was 24.6 μg/liter for the Haw River station downstream of East Burlington and the Haw River sewage treatment plant plus several points of textile waste discharge. (At this station, the cobalt concentration in the filtered water was 0.38 μg/liter or only 2% dissolved.)

Cobalt was possibly (sic) negatively correlated with the total amount of suspended solids in this study (correlation coefficient -0.40, significant at $\alpha = 0.05$). In the sediments, cobalt was strongly correlated with scandium (0.83) and iron (0.75); in the filterable fraction of the water, with potassium (0.79) and sodium (0.62). The correlations of cobalt with potassium (0.72) and sodium (0.57) were not so strong in the suspended matter (Shuman et al., 1977).

Schutyser et al. (1978) reported that the August 1976 dry deposition* of cobalt in Ghent, Belgium, comprised a 0.82 $\mu g/m^2$/day soluble fraction and a 0.951 $\mu g/m^2$/day insoluble fraction (analyses by instrumental neutron activation analysis). A rainwater sample from the same period contained 0.79 μg Co/liter in a soluble fraction and 0.433 μg/liter in an insoluble fraction. The ratio of wet deposition of cobalt to dry deposition was 21 (the ratios for 15 other elements ranged from 14 to 81). In a November 1976 rainwater sample, the soluble-insoluble concentrations of cobalt were almost equal: 0.17 ± 0.02 μg/liter and 0.179 ± 0.002 μg/liter. If rainwater and minor dry deposition were the only sources of cobalt to a local body of water and no

* Dry deposition is the settling of atmospheric particulates in the absence of precipitation.

physical-chemical changes occurred upon entering the water body, then the expected soluble cobalt content would be 40 to 64% (MRI calculation). However, additional sources of suspended cobalt would reduce this value.

b. Sediments: Pathways of various forms of cobalt into natural water bodies include not only atmospheric loadings, but also drainage and leaching from enriched or contaminated sources (e.g., coal and sulfide ore mines, shales, slag, ore bodies) and mechanical erosion of soils and wastes. When the heavy metal concentrations in sediment cores do not vary appreciably, the sediment history probably includes resuspension, mixing, and redeposition by dredging, shipping activity, and current action (Fitchko and Hutchinson, 1975).

Walters et al. (1974) enumerated other ways that buried pollutants can be transferred back across the sediment water interface: pollutant-adsorbing species in the sediments, such as insoluble organic substances and iron-manganese hydrous oxides, may be chemically degraded via oxidation-reduction reactions. Some elements may be converted to labile forms, e.g., by methylation. They may be physically resuspended by storm actions, burrowing organisms, or by those actions mentioned by Fitchko and Hutchinson.

Bortoli and Gaglione (1970) found that about 3% of the total ^{60}Co released from the Ispra Establishment of the Euratom Joint Nuclear Research Centre via Novellino Brook into Lake Maggiore, Italy, was trapped in the sandy bottom sediments of the brook and lake. The value agrees reasonably well with the 20% ^{60}Co held by the clay and silt sediments of the Clinch River receiving ORNL effluents (Struxness et al., 1967; cited by Bartoli and Gaglione, 1970). The vertical profile of ^{60}Co in the sediments was very different at different points in the Italian lake. This was attributed to mixing during lake storms. The average ^{60}Co activities in the first five layers of 1.5 cm each were approximately the same, and their total represented 90% of the total from 0 to 9.0 cm. Generally, ^{60}Co activity was absent from sediments deeper than 9.0 cm. The ^{60}Co contamination of the lake was confined to an area of about 0.03 km^2.

According to Lowman and Ting (1973), 90 to 95% of an added radionuclide ends up in the marine sediment in the first day. At least two studies have followed the behavior of ^{60}Co in marine sediments.

Windham and Phillips (1973) reported that ^{60}Co levels in the sediments of New London Harbor, Thames River, Connecticut (where nuclear seagoing vessels have been serviced) had decreased by an average factor of 33 between 1966 and 1972. Reducing the amount of radioactivity discharged to the river, loss by radioactive decay, and dilution by uncontaminated sediment effected the reduction. ^{60}Co was not found in fish. Only traces were present in algae growing near the source of discharges. No shift in radioactive sediment sites, however, was noted from 1966 to 1972.

Radioactive wastes have been discharged irregularly to Humboldt Bay from the Pacific Gas and Electric Company 65 Mw nuclear power reactor at Eureka, California. The Bay is a shallow, well-mixed tidal inlet with rapid flushing that removes 72% of nonreacting substances within a day. A total of 100,000 to 400,000 short tons (91,000 to 360,000 MT) sediment is removed annually to keep the channels clear. Between April 1, 1971, and April 1, 1972, a total of 107 mCi ^{60}Co was released to the Bay at an average concentration of 19 pCi per liter. After four releases from June 11, 1971 to September 22, 1971, the water contained an average 167 pCi/m^3; the sediments of the discharge canal, an average 8,185 pCi ^{60}Co/kg. On three of four occasions, the total water of the system contained 2.12 to 3.52% of the total ^{60}Co. However, samples taken on August 19, 1971, showed 62.7% of the cobalt in the sediments and 34.9% in the water (see Table V-12). Even in the discharge canal itself, which was 682 km long, there was an approximately threefold variation in ^{60}Co concentration with canal length in the bottom sediment. Leaching/readsorption and/or scouring/sedimentation may have effected the lateral transport (Heft et al., 1973).

c. Source inventories: Helz (1976) prepared a trace element inventory for the Chesapeake Bay using published literature values. Insufficient data were found for cobalt, but values used included rivers (dissolved plus suspended),

TABLE V-12

COBALT-60 INVENTORY IN HUMBOLDT BAY (%)
(Heft et al., 1973)

	6/11/71	7/8/71	8/19/71	9/22/71
Discharge canal				
Water	0.020	0.001	0.0004	0.006
Sediment	0.323	0.391	0.211	2.035
Algae	-	-	0.144	0.156
South Bay				
Water	3.48	2.01	34.90	1.408
Sediment	26.96	40.70	23.44	35.92
Eel grass	0.435	0.302	0.521	0.352
Center Bay				
Water	0	1.51	0	0.704
Sediment	7.83	25.13	7.81	10.56
North Bay				
Water	0	0	0	0
Sediment	60.87	30.15	31.25	49.30
Eel grass	0	0.015	0.021	0.014
	99.9	100.2	98.3	100.5

4 μg/liter; coastal seawater, 0.5 μg/liter; municipal wastewater, 5 μg/liter; and coastal plain deposits, 3 mg/kg. Table V-13 shows the source inventory prepared for cobalt. It indicates that at least 1% of the sources to the Bay are anthropogenic.

TABLE V-13

CHESAPEAKE BAY SOURCE INVENTORY FOR COBALT, MT/YR
(Helz, 1976)

Source	MT/YR
Northern Chesapeake Bay excluding the Patapsco Estuary	
Rivers	150
Salt water advection	30
Shore erosion	4
Municipal wastewater	0.5
Rain	-
Atmospheric fallout	-
Patapsco Estuary (Baltimore Harbor)	
Rivers and storm drainage	1
Excess sediment	0.7
Municipal wastewater	1
Rain	-
Atmospheric fallout	-
Direct industrial discharge	-
Total	187

Thomas (1979) reported on a full scale sediment survey of the Laurentian Great Lakes by the Canada Centre for Inland Waters. Loadings data for the elements were calculated from the difference between their concentrations at the surface and at depth. Table V-14 gives the estimates for cobalt loadings for Lakes Huron and Superior; no data were given for cobalt for Lakes Ontario and Erie.

TABLE V-14

NATURAL AND ANTHROPOGENIC LOADINGS OF COBALT TO THE DEPOSITIONAL BASINS OF THE GREAT LAKES, MT/YR
(Thomas, 1979)

	Anthropogenic	Natural
Huron	30	95
Superior	35	205

2. Speciation of soluble cobalt in water: The preceding subsection discussed the relative amounts of dissolved and particulate cobalt in natural waters. Table V-15 relates the general types of speciation of metals in water to their solubilities. This subsection discusses primarily the chemical forms of soluble cobalt. The chemical forms of insoluble cobalt will be dealt with more formally in the following subsection on adsorption and desorption of cobalt from sediments and suspended particulates.

Krauskopf (1956) calculated that although a total of 14 mg/liter cobalt had been supplied to the oceans during geologic time, only 0.0007% of this cobalt remained in solution in seawater (i.e., 0.0001 mg Co/liter). He assumed that input and removal processes had long ago achieved a balance that determined the concentration of each element in seawater.

If $CoCO_3$ and $Co(OH)_3$ (solubility product 8×10^{-13}) were the insoluble species controlling the dissolved cobalt concentration in seawater, the latter would be expected to be 0.01 to 0.08 mg/liter in a saturated solution (decreasing with increasing partial pressure of oxygen present). However, the measured value was 25 to 200 mg/liter. Krauskopf assumed that the amount of Co^{2+} was limited by the oxidation potential of its

TABLE V-15

SPECIATION AND SOLUBILITY OF METALS IN WATER
(Stumm and Bilinski, 1973)

Species	True Solution	Dialyzable	Membrane Filtrable	Filtrable	Particulate
Free metal ions	✓	✓	✓	✓	-
Inorganic ion pairs, inorganic complexes, e.g., $CoOH^+$	✓	✓	✓	✓	-
Organic complexes, chelates (≥ 10Å)	✓	✓	✓	✓	-
High-molecular-weight organic complexes (~100°A)	✓ (some)	✓ (some)	✓ (most)	✓	-
Highly dispersed colloids	-	-	-	✓	-
Colloids (~1,000 Å)	-	-	-	✓ (some)	✓ (some)
Precipitates, organic particles, remains of living organisms	-	-	-	-	✓

reaction with oxygen and the solubility of cobaltic hydroxide or by the solubility of cobaltous carbonate. Each process gave the equilibrium concentration.

The calculated cobalt concentration in equilibrium with HS^- and S^{2-} in seawater was 0.003 mg/liter, but the measured cobalt concentration was 0.02 to 0.6 mg/liter. Thus, cobaltous sulfide was also too soluble to explain the solubility of cobalt in seawater. When cobaltous chloride was added to seawater at pH 7.8 to 8.2 at 18 to 23°C until a precipitate formed, the cobaltous chloride concentrations in solution after 4 to 8, 10 to 17, and 21 to 28 days were 153, 218, and 196 mg/liter. When the precipitate was added to fresh seawater, 26 to 33 mg/ Co/liter was found in solution at 4 to 63 days.

According to the inorganic model of Sibley and Morgan (1977)*, the dominant dissolved cobalt species (< 0.45 μm) in seawater is the chloro complex. In this model, adsorption is negligible in seawater for all the metals studied. Free cobalt comprises 15% of the cobalt in seawater. The major dissolved species of cobalt in freshwater are the free ion (~ 45%) and the carbonate (~ 45%); about 8% of the freshwater cobalt of the freshwater model is adsorbed. Table V-16 gives the concentration of cobalt species expected at pH 8.0, pE = 12.0, and $p[Co_2] = 3.5$. The concentrations of ammonia, phosphate, and sorbents used were higher in the freshwater system than in seawater.

The stability constants at infinite dilution for some of the species used in the model are given in Table V-17.

---

* The computer modeling study was first described by Morel and Morgan in 1972 (Sibley and Morgan, 1977).

TABLE V-16

MODELING RESULTS FOR SPECIATION OF COBALT IN SEAWATER AND FRESHWATER
(Sibley and Morgan, 1977)

Ligand	Negative Logarithm of Molar Cobalt Species Concentration: Seawater (rank)		Freshwater (rank)	
Cl^-	9.07	(1)	10.54	(6)
Br^-	12.77	(6)	14.80	(9)
NH_3	15.26	(8)	12.72	(8)
$PO_4{}^{3-}$	16.18	(9)	12.68	(7)
OH^-	11.13	(5)	9.38	(4)
Adsorbed	14.70	(7)	8.88	(3)
$CO_3{}^{2-}$	10.16	(3)	8.19	(2)
Free Co^{2+}	9.62	(2)	8.17	(1)
$SO_4{}^{2-}$	10.32	(4)	9.82	(5)
Total Cobalt	8.87[a/] (0.9 μg/liter)		7.82 (0.08 μg/liter)	

a/ The model uses Goldberg's 1965 seawater data, which, according to recent GEOSECS program data may be an order of magnitude too high. Goldberg gave 0.5 μg Co/liter as a typical seawater concentration. He believed that Co^{2+} and $CoSO_4$ were the major species in seawater.

TABLE V-17

STABILITY OR SOLUBILITY OF AQUEOUS COBALT SPECIES AT 25°C, INFINITE DILUTION
(Sillen and Martell, 1964)

Species	Log of Stability or Solubility Constant
$Co(OH)^+$	-9.65
$Co(OH)_3^-$	-31.5
$CoCl^+$	-2.4
$CoSO_4$	+2.47
$Co(OH)_2$ (s)	-15 (Ksp)

Marchand (1974) found about 89% of the ^{60}Co added to natural seawater as the chloride was still soluble after 3 hr. The probable speciation was 55.3% $CoCl^+$; 5.5% Co^{2+}; 1.0% $CoCl_2$, $CoCl_3^-$, and $CoCl_4^{2-}$; 9.6% solid CoOOH on MnO_2 or FeOOH; 27.0% unassigned soluble cobalt; and 1.5% unassigned insoluble cobalt. (See Table V-8). When the seawater at pH 7.8 was enriched with marine algal organic matter, the most notable differences were that Co^{2+} was not present and that the anionic and neutral soluble species concentration had increased to 2.8%. The greatest speciation difference for organic-enriched seawater at pH 4.5 (compared with natural seawater at pH 7.8) was in the 68.3% $CoCl^+$, the increase being largely at the expense of the unassigned soluble cobalt species.

In the seawater model of Stumm and Brauner, cited by Sholkovitz (1978), the total Co^{2+} concentration is 10^{-8} M. About 10% each $CoCl^+$ and $CoSO_4$ are present at 10^{-9} M. Organic complexes represented by Co-glycine represent only 0.1% of the cobalt at $\sim 10^{-11}$ M.

An aberrant view of cobalt speciation was discussed by Gang and Langmuir (1974), who proposed that Co(III) is the predominant form in oxidized water whereas Co(II) predominates in reduced waters. The pK (-log Ksp) for $Co(OH)_3$ as stainerite is 43.1, and its precipitation would be expected to limit the concentration of Co(III) in many streamwaters. Since all of the groundwaters they analyzed appeared to be supersaturated with a $Co(OH)_3$ of pK 43.1, they proposed that a more soluble form of $Co(OH)_3$ is present or that supersaturation may reflect the presence of $Co(OH)^{2+}$, $Co(OH)_2^{+}$, and $Co(OH)_3$ in solution.

Although many workers have presented evidence for organic-metal complexes in natural waters, Stumm and Bilinski (1973) enumerated several reservations. In particular, they noted that when the concentration of a metal ion is higher than expected from calculated solubility products, rather than attribute formation of soluble organic complexes as the reason, one might consider that solubility calculations are usually off by several orders of magnitude and that "soluble" is usually defined operationally (e.g., membrane filterable) rather than conceptually. The release of metals from colloids upon decrease of the pH or destruction of organic material is not necessarily an indication of release from organic material upon acidification or oxidation. Although biological uptake is affected by organic material or chelate formers, the interpretation is complex. Among effects of organic material are ionic interactions, shifts in adsorption equilibria, and shifts in the extent of colloidal dispersions. For example, Stumm and Bilinski believed that addition of humic acid merely increases the colloidally dispersed metal, that the humic acid complexes are not truly soluble.

Shapiro (1964) reported that the "yellow acid" in yellow-colored lake water is peptized when the pH is raised sufficiently to ionize most carboxyl groups. The iron adsorbed is probably held in the form of a protected colloid. Thus, the metal-holding capacity for copper and iron of yellow acids from Linsley Pond (near New Haven, Connecticut) rose with increasing pH, from pH 5 to 10, and then decreased sharply. Between pH 9 and pH 11 (the only range studied for cobalt), the binding capacity for cobalt decreased from about 14 μmol/mg to about 7 μmol/mg. At pH 10 to 11, the total quantity of cobalt passing through a filter resulted almost entirely from the presence of the yellow acids. At pH 9 to 10, the acids accounted for a considerable fraction of the filter-passing cobalt, but not all.

Sigleo et al. (1979) reported that the colloidally bound fraction of cobalt in the Chesapeake Bay near Howell Point was "a major reservoir" for cobalt (> 10% of the total). Unchlorinated, freeze-dried samples were determined by instrumental neutron activation analysis to have 0.524 μg Co/liter in colloidal form. (A chlorinated duplicate sample contained 0.954 μg Co/liter in colloidal form.)

Humic substances in natural waters and sediments originate from terrestrial matter and partially biodegraded phytoplankton. Humic acid is base soluble but acid insoluble; fulvic acid dissolves in either. The stabilities of humic metal complexes decrease in the order: Fe(III) > Al(III) > Pb(II) > Ca(II) > Fe(II) > Zn(II) > Ni(II) > Co(II) > Mn(II) > Ca(II) > Mg(II). The logarithms of the "apparent stability constants"* reported for Co(II)-fulvic acid complexes at pH 3.0 to 6.0 are 2.7 to 6.6; for the Co(II)-humic acid complex at pH 4.5, log K = 6.0; at pH 6, log K = 8.3 (review by Manahan, 1975). Thus, humic acid forms stronger complexes with Co(II) than does fulvic acid although the latter has a higher capacity for interactions with metal ions (Jackson et al., 1978).

Dement'ev and Nazarova (1977) studied the effect of humic and fulvic acids on the dialyzability of 0.5 to 1 mg Co^{2+} per liter in a 0.05 M sodium bicarbonate solution at pH 9.2 and in a 0.05 M sodium chloride solution at pH 5.0. The dialysis rate for Co^{2+} in the chloride solution was somewhat faster within the first 6 hr than in the bicarbonate solution; but by 24 hr, 97% and 96%, respectively, had dialyzed; and by 100 hr, 100% of each had dialyzed. Complexed Co^{2+} dialyzed more slowly than the free ion. Dialysis was slowed by addition of humic and fulvic acids (20 mg/liter; their molecular weights were assumed to be 300 to 400). Only 90 to 93% of the cobalt dialyzed after 150 hr in the salt solutions when humic acids were added. Fulvic acid

---

* The "apparent stability constant" describes the situation when half of the potentially available complexing sites on the humic substance are bonded to Co^{2+}.

was even more inhibiting in sodium chloride solution, reducing cobalt dialysis after 150 hr to 65%. However, the effect was slight in the bicarbonate solution (96% dialysis after 150 hr).

Sholkovitz (1978) found that as salinity increased, a river water (Water of Luce, Scotland) containing 0.25 μg Co/liter exhibited 11 ± 3% flocculation of the cobalt at high salinities (25 to 30‰). The flocculation of cobalt was explained as due to the flocculation of humic acids and hydrous iron oxides at the high salinities. But Co^{2+} is among the divalent cations having the least affinity for humic and fulvic acids and goethite. Another reason that cobalt flocculated to such a small extent was that its solubility was enhanced by the formation of the very strong chloride complex in seawater.

Analytical techniques involving dehydration, freeze drying, flocculation, chromatography, and extraction alter the conformation and chemistry of organic species found in sediment pore water. Cline and Holland (1975) used absorbance, light scatter, and fluorescence techniques to avoid disrupting an organic rich pore water system while determining the degree of their association with Co^{2+} and Cu^{2+}. Whereas increasing the concentration of Cu^{2+} induced flocculation of the pore water organics, Co^{2+} at $<$ 20 μg/liter apparently did not affect the organic particle size. Either the Ca^{2+}-organic complexes were stable to substitution by Co^{2+}, or the Co^{2+} was forming soluble complexes. According to fluorescence experiments, Co^{2+} reacted with fewer organic molecules at fewer functional sites than did Cu^{2+} and bonded more weakly, possibly by surface or specific adsorption.

Bovard et al. (1975) reported that in the Rhône River in France, organic complexation of heavy metals increased with increasing chemical oxygen demand (COD). Cobalt and cerium were almost 100% complexed when the COD was about 15 mg/liter.

Hanck and Dillard (1977) reported that Co^{2+} was not complexed by organic species in Raleigh, North Carolina, tapwater or in the Tar River at Greenville, North Carolina. They found 0.6 ± 1 μmole ligand per liter in Briery Swamp water (Stokes, North Carolina), 4.1 ± 0.9 μmole/liter in Raleigh sewage effluent,

7.5 $\pm$ 0.1 μmole/liter in aquarium water, and an average 0.7 to 1.6 (range not detected to 2.4) μmole/liter in natural water bodies in Raleigh. They tentatively concluded that the complexing capacity for cobalt was higher in streams receiving waste effluents (an average 1.4 μmole ligand per liter in Raleigh streams receiving moderate industrial or municipal use compared with an average 0.8 μmole/liter in Raleigh streams receiving little industrial or municipal input).

The analytical method used by Hanck and Dillard (1977) involved adding a 10-fold excess of Co^{2+} to the water sample containing unknown complexing ligands. The mixture was treated with hydrogen peroxide to oxidize the Co(II)-complex to the kinetically inert* Co(III)-complex. The free Co(III) formed is rapidly converted back to Co(II) when it oxidizes water. The uncomplexed Co^{2+} was then analyzed by differential pulse polarography in 0.06 <u>M</u> ethylenediamine.

In the discussion of ^{60}Co wastes at ORNL, it was pointed out that Co-EDTA chelates are very environmentally stable. However, at low pH they can be photodegraded, albeit to a much lesser extent than can the Fe(III)-EDTA chelate. Thus at pH 3.5, the Co(II)-EDTA chelate photodegraded to the extent of 2.9% under 4,000 ft candles of artificial sunlight; to the extent of 6.4% at pH 4.5 within 96 hr; and to the extent of 0.4% at pH 6.9. Under these same conditions up to 69.4% of the Fe(III)-EDTA chelate photodegraded. The photodegradation process was followed by evolution of $^{14}CO_2$ (from the EDTA-labeled carboxylate groups) and formaldehyde. Not only is EDTA used in nuclear power plant decontamination procedures, it may be a widespread contaminant of U.S. waters. About 40 million lb (18,000 MT) are produced annually in the United States and are widely used in agriculture, food and drug processing, photography, and textile and paper manufacturing (Lockhart and Blakeley, 1975).

3. <u>Cobalt adsorption from water by suspended particulate and by sediment materials</u>: Most suspended particles in natural waters are negatively charged and attract cations. In addition, trace elements may be adsorbed on ion exchange positions

---

* Co(III) complexes have generally significantly higher formation constants than Co(II) complexes.

of clay minerals and on hydrous iron and manganese oxides. Some elements may also be bound within the crystalline lattice of sediment particles (Troup and Bricker, 1975).

DeGroot et al. (1973) proposed various ways in which cobalt may be fixed by sediments. It may be intercalated as cations between the layers of clay lattices, but only in small amounts because of the low polarizability and relatively small size of Co^{2+}. Cobalt may be held relatively tightly as exchangeable ions in the electrical double layer of clay minerals. Cobalt may be chelated by natural organic matter in sediments (more so in freshwater than in seawater). Cobaltic hydroxide may precipitate on sediment particles after oxidation of coagulated cobaltous hydroxide at pH > 7.7 if there is sufficient calcium carbonate to neutralize the protons released by the reaction.

Murray and Murray (1973) remarked that there are a "bewildering array of possibilities to account for environmental data" since the different mechanisms that have been proposed are difficult to test experimentally. They enumerated factors that affect radionuclide adsorption-desorption equilibria of a sediment:

1. Equilibration time between sediment and trace element in river water.
2. pH of the water body.
3. Concentration of sediment and trace element in river water.
4. Speciation of the trace elements.
5. Mineralogy of the sediment (iron, carbonate, and organic content).
6. Formation of particulate material in the river water and its flocculation by sediment in the estuary and sea.
7. Salinity effects on desorption.

a. Adsorption by inorganic substances

(1) Clays and sands: O'Conner and Kester (1975) found that cobalt concentrations of 50 to 1,800 μg/liter in artificial river water at pH 8 were 95% adsorbed by

1,000 mg/liter illite suspensions (74% of particles < 2 μm). The percent of cobalt that was adsorbed decreased at higher cobalt concentrations. Cobalt at 50 μg/liter was adsorbed less by illite from seawater than from artificial river water at pH < 7. For example, at pH 4, there was 20% adsorption from artificial river water but none from seawater. At pH 4, increasing the Ca^{2+} concentration from 1.5 mg/liter to 15 mg/liter in artificial river water halved the cobalt adsorption. At low pH, cobalt was adsorbed nonspecifically according to its concentration among all solution cations. The variation with pH in adsorption of cobalt at 50 μg/liter by 47 ppt very fine sand from artificial river water and seawater was similar to that by clay.

In laboratory experiments, under conditions similar to those in streams, Kharkar et al. (1968) found that about 90% of the cobalt was adsorbed by montmorillonite and illite. When transferred to seawater, 40 to 70% of the cobalt was released. However, at 1,000 mg/kg kaolinite in the fresh water, cobalt was not adsorbed detectably.

Chester (1965) found that the fraction of cobalt adsorbed by illite suspended in seawater increased with increasing temperature but was relatively constant at one temperature with varying cobalt concentrations. At 26°C and 5 to 200 μg cobalt in 60 ml seawater at pH 8.2, 38 to 48% of the cobalt was adsorbed by 0.2 g of the small size fraction of illite. (At 500 μg cobalt, however, only 22% was adsorbed by the illite.) At 50°C, the amount of total cobalt (5 to 100 μg/60 ml) adsorbed was 59 to 70%; and at 70°C, 84 to 94%. The results support the isotherm model postulated by Kurbatov et al. (1951) for adsorption of trace elements by clay minerals. Presumably, cobalt was taken up as $CoCl^{+}$ or $CoOH^{+}$. The difference in temperature responses might have reflected differences in bonding energy.

In sediments, the cobalt distribution is frequently related to that of fine-grained material, which is usually dominated by clay minerals. Thus, in the Indian Fork (Tennessee) sediment fractions of particle size 0.2 to 5 μm, the cobalt concentration was 46 to 128 mg/kg; and in the New River, the concentrations in these fractions was 79 to 104 mg/kg compared with the overall average sediment concentrations of 26 to 51 mg/kg (Schrader et al., 1977).

Meeussen et al. (1975) found the sorption capacity for $^{60}Co^{2+}$ of 100 to 350 μm particles of various sediments collected from the Scheldt River to be 0.043 to 1.4 mEq/g with clays and silts showing generally higher sorption capacity (see Table V-18). Metal accumulation in the sediment was limited by high salt concentrations. Co^{2+} retention by all sediments was attributed to its presence as the sulfide.

TABLE V-18

COBALT SORPTION BY SCHELDT RIVER SEDIMENTS
(Meeussen et al., 1975)

Sediment Description	Co^{2+} Sorption Capacity, mEq/g
Blue gray mud	1.0
Blue sandy mud	0.65
Rough sand, black gravel, 347 μm	0.043
Fairly fine gray sand, 238 μm	-
Peat	1.4
Sandy clay	0.35
Fine sand with mud, 122 μm	-
Fine sand, 136 μm	-
Plastic clay and clay fragments	-
Fine sand with clay nodules	-
Gray clay, blue mud	-
Brown gray silty clay	-
Fine sand with clay	-
Mud	-
Clay	0.52
Black peat-moor	0.74
Brown peat-moor	0.66
Detached black mud	0.52
Very detached blue mud	0.49

Lachet (1972) found that 15% of the 450 nCi ^{60}Co (as $CoCl_2$ with 0.23 nanomole carrier per liter) in Isere River (France) water at pH 8.1 was taken up by suspended sediments (chiefly fine sand) after 90 min. Cobalt was taken up by the suspended sediments more slowly than were ruthenium, cerium, cesium, and mercury but faster than chromium, iron, and strontium.

From all the data reported for cobalt adsorption by the oxide minerals, α-quartz, manganese dioxides, goethite, rutile, amorphous iron oxide, gibbsite, and montmorillonite, it appears that the systems behave similarly. Adsorption is low at low pH, but when the pH is increased so that hydrolysis becomes significant ($\sim$pH 7 to 10), the percentage adsorption rises to 100%. When the cobalt concentration surface area ratio is high enough, reversal of the zeta potential (the experimentally observed double layer potential) also occurs near this pH (Leckie and James, 1974). The charge reversal from negative to positive that occurs at high pH often corresponds to the isoelectric point of the adsorbing metal ion's pure oxide. Healy et al. (cited by Leckie and James, 1974) interpreted this as meaning that the original substrate oxide has become coated with hydroxides and hydrous oxides of the adsorbing metal ion.

Presumably, lack of adsorption of Co^{2+} by some minerals at low pH is related to a large and unfavorable change in solvation energy associated with adsorption of the unhydrolyzed metal ion on low dielectric solids such as silica and kaolinite (James and Healy, 1972).

(2) <u>Manganese and iron oxides and hydroxides</u>: Vuceta and Morgan (1978) described an inorganic model for an oxic (normally oxygenated) fresh water with $p\epsilon = 12$, $pCo_2 = 10^{-3.5}$, and cobalt concentration 10^{-7} $\underline{M}$ (5.9 μg/liter). Only 0.42% of the total Co^{2+} was adsorbed by silica, $\ll$ 0.001% was adsorbed by ferric hydroxide, and 1.55% was adsorbed by manganese dioxide. Adsorption may well differ in natural waters containing up to 10 times more ferric hydroxide and manganese dioxide and 0.1 as much silica.

Immediately downstream from copper-zinc sulfide mineralization at the Magruder mine (Lincoln County, Georgia), cobalt, nickel, and copper were concentrated more in iron oxides than in the manganese oxides in the stream. However, cobalt and copper were found preferentially in the manganese oxides upstream. The iron and manganese oxides were present in the stream as black coatings on boulders. The results suggested that cobalt was preferentially adsorbed by iron oxides "where metal-rich groundwater percolates into the stream." Quite probably, in the more oxidizing environment of the stream, the iron(II) from the groundwater was oxidized and was freshly precipitated (Carpenter et al., 1978).

Hem (1978) proposed that the scavenging and coprecipitation effects of manganese oxides may be partly attributed to oxidation-reduction processes. As the pH of a water increases, Mn^{2+} precipitates as Mn_3O_4 or MnOOH, followed by disproportionation to MnO_2 and Mn^{2+}. Electron transfers from Mn^{2+} may be diverted to other heavy metal ions such as Co^{2+} and reduce their equilibrium solubilities. Both Co^{2+} and Co^{3+} as well as certain other heavy metal ions with approximately the same physical dimensions as Mn^{2+} and Mn^{3+} in manganese oxides can enter the lattice during disproportionation of Mn_3O_4. Such substitution may be driven by the increasing in the stability of the lattice, e.g.:

$$3Co^{2+} + 2Mn_3O_4 + 4H^+ \rightleftharpoons MnO_2 + Co_3O_4 + 5Mn^{2+} + H_2O$$

With the assumption of unit activity for each species, the equilibrium constant is $10^{18.73}$. The hypothesis also seemed reasonable for the mechanism whereby cobalt and, probably, lead and nickel are incorporated into marine manganese nodules. The Standard Gibbs free energy of formation of cobalt species used for equilibrium calculations are given in Table V-19.

At cobalt concentrations ranging from 7 μg/liter to 194 mg/liter, 15.7 to 100.0% of the cobalt in seawater was removed by hydrous oxides of iron and manganese by varying the pH from 7.40 to 8.23 and Eh from 419 to 510 mv. The maximum cobalt removal was at initial pH 8.12, final pH 8.03, initial

TABLE V-19

STANDARD GIBBS FREE ENERGY OF FORMATION FOR COBALT SPECIES

Species	ΔG°, kcal/mole
Co_3O_4(c)	-185.0
$Co(OH)_2$ (c) pink	-108.8
Co^{2+} (aq)	-13.41
$Co(OH)_2$ (aq)	-101.13
$Co(OH)_3^-$ (aq)	-140.5

Eh 457 mv, final Eh 479 mv. From 78.5 to 100% of the cobalt in the precipitates was in strongly bound (non-ion-exchangeable) positions and could not be removed after 30 days by eluting with 0.03 M $MgCl_2$ and 0.3 M $CaCl_2$ (i.e., at 10 times the concentrations of these metal ions in seawater) (Pronina and Varentsov, 1973). These results are supported by those of Doshi et al. (1973), who recovered 82.1% and 82.6% of ^{58}Co from seawater on manganese dioxide and ferric hydroxide, respectively, precipitated *in situ*. Reagent grade ferric hydroxide showed no detectable cobalt adsorption, but freshly precipitated ferric hydroxide adsorbed 95% of the cobalt and released little of it on contact with seawater. Manganese dioxide adsorbed 20% of the cobalt in freshwater and released none on contact with seawater. They present data for various streams in which the soluble cobalt concentration after contact of the freshwater's suspended particulates with seawater changes from 1 to > 4 times more than in the freshwater (Kharkov et al., 1968).

Lenaers (1971; cited by Jackson et al., 1978) reported that selective reduction of organic matter in Columbia River sediment released ^{54}Mn but little ^{65}Zn, ^{46}Sc, and ^{60}Co. The latter three were thought to be bound to hydrous iron and manganese oxides.

Tessier et al. (1979) found that about 40% of the cobalt in sediment samples from the lower reaches of the Yamaska and Saint Francois rivers collected at Saint Marcel and Pierreville, Quebec, Canada, was bound to the iron and manganese oxides. More than 50% of the cobalt concentration, however, was in a residual fraction comprising detrital silicate minerals, resistant sulfides, and a small amount of organic substances resistant to acidic peroxide treatment.

Particles 10 to 20 μm in size from deep-sea manganese nodules from the Central Pacific Ocean* contained an average 305 mg Co/kg, significantly higher than in smaller (191 to 237 mg Co/kg) or larger particles (188 to 231 mg Co/kg). The concentration of cobalt was positively correlated with concentrations of manganese, nickel, copper, and zinc in these nodules. These metals were bound (15 to 75%) to the micronodules of the hydrogenous** phases of the sediment. Only 15% of the cobalt was bound to the micronodules, which contained an average cobalt concentration of 1,400 mg/kg. About 30% of nonnodular cobalt, nickel, copper, zinc, and lead was bound to the particles > 4 μm, consisting primarily of fragments and tests. (The total organic carbon in the sediment samples was only 0.16 to 0.28%). Metal scavenging by sinking particles, especially planktonic organisms, often after their deaths, and actual incorporation of metals into their skeletons will explain some of the metal content of the nodules. But planktonic material cannot account for all the metals in the 10 to 63 μm particles. No explanation could be offered for the behavior of cobalt (Halback et al., 1979). About 80% of the cobalt in Pacific pelagic clays, on the other hand, is authigenic, only 20% being detrital (Krishnaswami, 1976).

b. Adsorption and desorption by organic substances: Several studies have mentioned the relative roles of organic and inorganic species in keeping cobalt in suspended particulates or sediments. Krauskopf (1956) found that cobalt's adsorption from seawater at initial concentrations of 0.1 to 2.0 mg/liter ranged from 20 to 91% by hydrous ferric oxide loaded at 0.003 to 0.3 g/liter; 93 to 94% by hydrous manganese dioxide loaded at 0.1 g/liter; 15 to 82% by apatite loaded at 0.1 to 0.7 g/liter; 18% by clay loaded at 12 g/liter; and 8 to 18% by "dilute" plankton concentrations. Manganese dioxide was also the best adsorbent for nickel.

---

* The samples were taken within the siliceous ooze province between the Clarion and Clipperton Fracture Zones.

** Hydrogenous material is authigenic. It is formed by precipitation of minerals from seawater or by modification of lithogenic (rock-originated) material.

Krauskopf (1956) concluded that organic reactions in reducing environments must control the concentrations of Co, Ni, V, and W since all other mechanisms studied did not. Nickel and vanadium are most strongly enriched in organic sediments and in marine organisms. Cobalt, too, is concentrated by organisms but not so much nor so consistently in organic sediments.

Using neutron activation analysis, Huljev and Strohal (1972) found a total of 3.9 mg Co/kg bound by chelation (surface sorbed ions already replaced by Na^+ in the isolation process) to humic acids from sediments of the North Adriatic. After hydrolysis of the humic acids (a method that probably did not redistribute the cobalt among the fractions), the cobalt concentrations were 6.4 mg/kg in the amino acids, 0.1 mg/kg in the sugars, and 0.66 mg/kg in the condensed benzene core (none was observed in the phenolic fraction).

Kempf (1976) described the cobalt speciation in sediments of the Lower Rhine: $\sim$ 60% rock dust, $\sim$ 20% carbonate, $\sim$ 10% humate, $\sim$ 7% sorbed, and $\sim$ 3% hydroxide. The best model for cobalt in the sediments was $Co(OH)_3$ at pH 8.2 with a borate buffer.

Cobalt in suspended sediments of the Redstone River (MacKenzie Valley, Canada) were leached faster and to a greater extent (12 to 49%) by 0.1 M hydrochloric acid than by EDTA (2.1 to 17%). Thus, inorganic species seemed to predominate over organic substances in these sediments. Even after 150 hr in EDTA, cobalt had not reached a constant concentration in the leachate. The phenomena were attributed to solid phase dissolution (chiefly of dolomite and calcite) by 0.1 M hydrochloric acid rather than mere desorption. Ammonium acetate leached only 1.0 to 4.5% of the cobalt from the river bottom sediments. There was essentially no uptake or release of cobalt from the sediment with Beaufort seawater as the leachate (Wagemann et al., 1977).

Natural organic compounds in the water frequently cause desorption and solubilization of cobalt from the inorganic fractions of a sediment.

Nissenbaum et al. (1971; cited by Jackson et al., 1978) found large concentrations of cobalt and other heavy metals associated with soluble organic matter in interstitial waters, perhaps as loose ion-colloid assemblages rather than true coordination complexes.

DeGroot et al. (1973) believed that in the Rhine River, cobalt desorption from fine-grained sediments increased as the waters became increasing saline during passage to the sea. Ultimately cobalt mobilization was $\sim$ 60% and was accompanied by the formation of soluble organic cobalt complexes. The intensity of the cobalt mobilization appeared to be dominated by the stability of its soluble organic complexes. Presumably, organic matter in the sediments decomposed in the Rhine delta (the Netherlands).

Duursma (1970) showed how drastically the presence of leucine altered the equilibrium distribution coefficient of cobalt between sediment and solution in freshwater and seawater. For example, without leucine, the distribution coefficient for cobalt between a deep Pacific sediment and Mediterranean seawater was 38,000. Adding 0.00152 to 0.00608 $\underline{M}$ leucine decreased the distribution coefficient to values of 18,000 to 4,000. Similarly, the distribution coefficient for cobalt between a fine clay sediment and water of the Var River (Monaco) decreased from 12,000 in the absence of leucine to 3,400 with 0.00304 $\underline{M}$ leucine. Duursma calculated, however, that the natural leucine concentration in seawater is too low to chelate Co^{2+}.

In an inorganic model for cobalt in an oxic freshwater at a concentration of 10^{-7} $\underline{M}$ (5.9 μg/liter), < 2% of the total cobalt was adsorbed by silica, ferric hydroxide, and manganese dioxide. The organic model of Vuceta and Morgan (1978), showed that concentrations of EDTA greater than 20 times the total cobalt concentration altered the speciation of cobalt; but that when total aspartate was 300 times the total cobalt concentration, cobalt was only 2 to 30% complexed. Cobalt citrate complexes were unimportant. In the model, when silica of low surface area was present at pH 7 with all complexing species, cobalt was predominantly present as the aquated Co^{2+} and the EDTA complex. But at higher surface areas for silica, cobalt adsorption increased up to 30% of the total cobalt.

According to Rashid and Leonard (1973), sediments are commonly comprised of 2 to 3% organic matter. Up to 20 mg organic matter per liter may be in the water.* Humic compounds quite similar to those in soils** are the major fraction of the organic matter. In laboratory experiments at pH 7.0, 1 g humic acid per 500 ml solution solubilized 250.0 mg cobalt from its carbonate precipitate and 90.0 mg from its sulfide. At pH 11, 1 g humic acid hydrolyzate (comprising 17 amino acids) solubilized 280.0 mg cobalt from the carbonate and 217.5 mg from the sulfide.

As a model for humic acid, often present in natural waters in concentrations of a few tenths to several mg/liter, Duursma (1970) used quinoline-2-carboxylic acid at concentrations given in terms of 0.2 to 2.0 mg carbon per liter. At the lowest carbon concentration, 30.5% of the total cobalt (0.9 to 1 μg/liter) existed as the free ion; 0.6%, as CoL^+ (where L = the organic ligand); and 68.9%, as CoL_2. At the highest carbon concentration, the cobalt distribution was 0.44% Co^{2+}, 0.09% CoL^+, and 99.47% CoL_2. The CoL_2 accounted for only 0.01% of the total L present.

Murray and Meinke (1974) showed that increasing the content of soluble organic matter increased the concentration of soluble cobalt in Var River (Monaco) water in experiments where the sediment load was 1.0 g/liter. The $< 2,000$-μm fraction used comprised 45% calcite, 35% quartz, 20% mica, and $< 1\%$ organic matter. When no sewage had been added to the river water at pH 8.1, $> 93\%$ of the cobalt was adsorbed by the sediment. Addition of about 25% sewage decreased adsorption to about 87%; 50% sewage, about 84%; 75% sewage, about 75%; and 100% sewage, about 64%. All solutions were maintained at ≥ 5.8 ml O_2/liter. When shaken for 96 hr with seawater, the Var sediments uniformly

* Stumm and Morgan (1970; cited by Jackson et al., 1978) stated that within natural waters, the concentration of organic carbon is 0.1 to 10 mg/liter.

** Aqueous humic substances are more highly aliphatic, less condensed, poorer in phenolic groups, and richer in carbonyl and carboxyl groups than soil humic substances (Jackson et al., 1978).

desorbed about 20% of the cobalt when from 0 to 100% sewage was present. Cobalt desorption was about 75% at sediment pH 5 to 6.5 but was only 10 to 15% desorbed at a sediment pH of about 9.

In the U.S. rivers, acidic organic matter seldom predominates over inorganic matter except in the Coastal Plain rivers of the Southeast (Beck et al., 1974; cited by Jackson et al., 1978).

c. Adsorption on natural sediments and suspended particulates: Preceding subsections focussed on the roles of inorganic and organic matter in cobalt transport. Most of the following studies are of cobalt added to natural systems--water bodies and their sediments *in situ*, not in the laboratory.

Ijuin et al. (1973) followed the behavior of 200 to 500 μCi ^{60}Co in Rhône River water. The diffusion of ^{60}Co in relation to time seemed to obey Fick's second law of diffusion. After 30 days contact time, only about 10% of the initial ^{60}Co activity remained in solution. After 147 days contact time, abrupt decreases in sediment radioactivity with the square of the depth were noted: 0 cm^2, $\sim$ 10 nCi/g; $\sim$ 2 cm^2, < 1 nCi/g; $\sim$ 17 cm^2, 0.01 nCi/g.

^{60}Co (with carrier) adsorption by 250 mg/liter Var River (Monaco) sediment had reached equilibrium within 3 days; the desorption equilibrium was reached within 1 day. At pH 8 and cobalt concentrations up to 10 μg/liter, 3% particulate formation in distilled water and in seawater was expected according to earlier data of Fukai and Huynh-Ngoc (1966; cited by Ijuin et al., 1973). Murray and Murray (1973) found only 10% cobalt particulate formation in Var water and 2% in seawater. Addition of ferric hydroxide to seawater gave 30% particulate cobalt, but adsorption of cobalt onto ferric hydroxide was not a factor in Var River particulate formation.

In Var River water (at its original pH 6.5) at a sediment load of 250 mg/liter and cobalt concentration (with ^{60}Co) 25 μg/liter, sorption increased from $\sim$ 0% at pH 4 to $\sim$ 30% at pH $\sim$ 7.5. At pH 8, adsorption on the particulate approached 100%.

The weight of cobalt adsorbed per unit weight of sediment was higher for smaller concentrations of sediment, but the total adsorption of cobalt, was, of course, greater with a higher sedimentary load. For example, at pH 8.1 with a sediment load of 500 mg/liter, about 40% (60 μg) of the $\sim$150 μg Co in solution was adsorbed. Adsorption of cobalt was only 17% by 100 mg sediment per liter. Thus, at the higher sediment load of 60 μg Co/500 mg sediment, the cobalt concentration of 120 mg/kg was less than at the lower sediment load: 20 μg/100 mg or 200 mg/kg. In desorption experiments in seawater, the amount of cobalt desorbed was proportionally less as the sediment load increased (Murray and Murray, 1973).

Nelson et al. (1966) followed the uptake of radiocobalt by sediments in the Columbia River after its release from the Hanford Works reactors north of Richland, Washington. Between January 1964 and January 1965, radiocobalt (^{58}Co and ^{60}Co) showed $\geq$ 50% depletion from the river by natural sediment uptake between Pasco and Vancouver except during May through July due to scouring of the river bottoms. The particulate carried only 2 to 3% of the ^{58}Co and $\sim$ 3 to 12% of the ^{60}Co. At McNary Reservoir, the ^{60}Co activity in the fine sediment decreased from 137 to 151 d/min/g (where d = disintegrations) at 0 to 10 cm depth to 30 d/min/g at 25 to 30 cm depth. The ^{60}Co activity in the coarse sediment was only about 10 to 20% that in the coarse plus fine sediments at each depth. Yet these fine particulates comprised only 0.096% of the whole sample. Extraction experiments on the McNary Dam sediments indicated that a large fraction of the ^{60}Co in the fine sediments was associated with organic matter. Thus, bromoethanol extracted 43.9% of the ^{60}Co after acetic acid removed 32.5%.

^{60}Co and ^{106}Ru have been released to the basin of Whiteoak Creek, Tennessee, in high-nitrate solutions (Pickering, 1970). Carrigan (1969) reported that 9% of the ^{60}Co activity released from the ORNL into Whiteoak Creek and then into the Clinch River via Whiteoak Lake was retained in the bottom sediment within the 21-mile reach extending from the mouth of the river (where it flows into the Tennessee River) to the mouth of the creek. The peaks in ^{60}Co concentrations in the sediments along the 21-mile reach had a similar pattern to those of other radionuclides. Highest concentrations were immediately downstream from the mouth of Whiteoak Creek. Secondary peaks occurred at Clinch River miles 17.5 and 14.0. Presumably, diffusion, flow distribution, and turbulence affected the sedimentation pattern. Islands and submerged ridges which influence the flow pattern in the vicinity of Clinch River mile 10.0 produced the largest cross-sectional area of radioactive sediment. Pickering (1969) reviewed the work on sorption of radioions by Whiteoak Creek and Tennessee River sediments. Andrew (1964; cited by Pickering, 1969) had noted a partition in ^{60}Co concentrations between fine and coarse clay fractions. ^{60}Co also correlated with organic matter in the Tennessee River. Together, particle size and organic matter accounted for 50 to 90% of the observed variation in bottom sediment content of radionuclides.

^{60}Co contributed 10 to 13% (average 12%) to the total gamma-ray activity of selected Clinch River sediment cores, indicating little variation with depth from that of ^{137}Cs, the major contributor to radioactivity. The largest concentrations of both radionuclides in sample cores were at the same depths. ^{60}Co correlated better with ^{137}Cs in the cores with increasing depth than with ^{60}Co releases to the Clinch River. Presumably, both were incorporated into Clinch River bottom sediments by sedimentation of suspended solids that had acquired their ^{60}Co before entry into the Clinch River. Pickering also suggested why only a small fraction of the ^{60}Co released had been incorporated into the Clinch River bottom sediments. About 20% of the ^{60}Co released to Whiteoak Creek is associated with the suspended sediment having a diameter $> 7\ \mu$ (Churchill et al., 1965; cited by Pickering, 1969). Rather than being associated with smaller suspended solids, maybe ^{60}Co is present as dissolved ions.

The contents of ^{137}Cs and ^{60}Co correlated with cation-exchange capacity and the contents of free iron and aluminum oxides, adsorbed water, and organic matter in one core but only free iron and aluminum oxides content in another.

Jenne and Wahlberg (1968) had claimed in 1965 that ^{60}Co was 95% scavenged in Whiteoak Creek sediments by manganese and iron oxides. There was more cobalt in manganese oxides although there was 30 times more iron than manganese. In the 1968 study, extraction of the sediment by ammonium acetate-ammonium chloride removed very little of the cobalt from the 0.25- to 2-mm particles; presumably that which dissolved did so upon dissolution of the manganese oxides. Cobalt sorption in these sediments is due to ion exchange with silicate minerals (reversible sorption) as well as to adsorption by manganese and iron oxides.

Meisch et al. (1978) reported a very localized uptake of cobalt and other heavy metals by the sediments of the Saar River (West Germany) in the vicinity of Mettlach. On the right bank, where the sludge had deposited from an industrial wastewater, the cobalt concentrations were 72 mg/kg at 0 to 4 cm, 92 mg/kg at 4 to 8 cm, and 133 mg/kg at 8 to 12 cm. Nickel concentrations were only a little higher, but there was 4 to 5% lead and about 1% zinc in these sediments. In the sediments of the opposite bank, the surface cobalt concentration of 10 mg/kg was not much higher than at nearby points on that side of the river. However, for at least 400 m downstream from the outfall on the right bank, cobalt concentrations were still elevated (surface: 31 to 70 mg/kg).

4. Leaching and/or mechanical mobilization

a. From natural sources: The south central region of Idaho is little disturbed, but Emmett (1975) found cobalt in the Salmon River and tributary streams (pH 7.2 to 8.0) in 40.2% (45/112) of the samples. The concentrations varied from 1 to 4 μg/liter when detected. Cobalt was detected mostly in June and October. The Coeur d'Alene region of Idaho is noted for its lead-zinc-silver (sulfide) ores. Funk et al. (1973; cited by Filby et al., 1974) had reported 27 μg Co/liter in water of the South Fork of the Coeur d'Alene; 0.4 μg/liter, in the unpolluted North Fork, and 3.4 μg/liter, in the Spokane River. In the surface sediments (0 to 3 cm) of the delta of the Coeur d'Alene River, Filby et al., (1974) found 21.5 mg/kg cobalt (dry weight). In the north part of Coeur d'Alene Lake, where lead and zinc concentrations were also much lower, there was only 0.17 mg/kg cobalt in the sediments. The concentrations of cobalt in the lake sediments close to the delta were 5.32 mg/kg at 0 cm depth, 15.9 mg/kg at 10 cm, 22.9 mg/kg at 20 cm, 7.85 to 10.4 mg/kg at 30 to 70 cm, and 0.16 mg/kg at 80 cm.

In the supergene zone of weathered sulfide ores, cobalt is most frequently encountered at concentrations of 0.5 to 3 μg/liter in neutral waters having an oxidizing setting and 0.2 μg/liter in waters of gley settings.* The maximum concentration for cobalt in waters of the supergene zone found by Shvartsev et al. (1975) was 68 μg/liter. The contrast ratio (maximum concentration found divided by minimum concentration) for cobalt, 340, placed it among other elements of low fluctuation: barium (400), strontium (300), selenium (100), germanium (100), nickel (400), vanadium (180), fluorine (230), and mercury (40).

Kroon and DeGrys (1970) did not report any cobalt anomalies in the sediments of streams draining sphalerite, galena, copper-molybdenum porphyry, or disseminated pyrite in the Cordilleras of Central Equador. However, by spectrographic analyses, sediments from streams draining an area of black shales showed anomalies of cobalt (30 mg/kg), copper, zinc, molybdenum, vanadium, nickel, and arsenic.**

b. From anthropogenic sources

(1) Urban runoff: The average cobalt concentration in urban runoff in the Third Fork Creek Drainage Basin*** in Durham, North Carolina, was 0.16 mg/liter and represents an annual yield of 82 g/m^2. The cobalt concentration

---

* Gley is a term for a soil horizon developed under the influence of excessive moisture.

** Hathaway and Galle (1979) showed that cobalt, manganese, nickel, and zinc were increasingly mobilized by solutions of greater salinity from black Heebner shale (Pennsylvanian Age) by cation exchange processes with Na^+ and probably by complexation type reactions with chloride ions.

*** In the north part of Durham, the system drains into the Neuse River; in the south, into the Cape Fear system.

in the samples taken from October 23, 1971, to March 21, 1973, ranged from 0.04 to 0.47 mg/liter. The average and range of concentrations for copper were comparable to those of cobalt. The average concentrations of other heavy metals were: lead, 0.46 mg/liter; copper, 0.15 mg/liter; chromium, 0.23 mg/liter; manganese, 0.67 mg/liter; nickel, 0.15 mg/liter; and zinc, 0.36 mg/liter (Colston, 1975; Tafuri, 1975).

Continental runoff may be the reason the concentrations of cobalt, nickel, cadmium, and zinc are higher in continental shelf waters off the northern coast of Georgia and the southern coast of South Carolina than in shelf waters immediately to the south. Samples collected in late December through early February (and analyzed by atomic absorption spectrophotometry) contained 0.17 μg Co/liter in the north and 0.06 μg Co/liter in the south. Cobalt concentrations in the northern area are similar to those reported for river and coastal waters by Windom et al. (1971) and Kharkar et al. (1968) but significantly higher than values for waters of Atlantic offshore waters such as the Gulf Stream and the Sargasso Sea (Spencer and Brewer, 1969; cited by Windom and Smith, 1972) and the Tropical Northwest Atlantic (Robertson, 1970; cited by Windom and Smith, 1972). The values for cobalt found in the south in this study, however, were similar to those reported for the Atlantic offshore and tropical waters (Windom and Smith, 1972).

(2) Agriculture: Amoozegar-Fard et al. (1975) did not detect cobalt in the leachates from feedlot steer manure containing 8 mg Co/kg (dry weight).

Horvath et al. (1972) found a 42.5-fold difference in cobalt concentrations between water contaminated by agricultural runoff in Collier County, Florida, where cobalt is applied to the soil in fertilizer or to the plants in nutritional sprays. In a developed area of the Barron River Canal System of Chokoloskee Bay, the cobalt concentration in October 1971 was 34.0 μg/liter; in an undeveloped area, 0.8 μg/liter. In the Everglades estuaries, a background concentration for cobalt was 5.8 μg/liter in Lostman's Bay. Estuaries receiving canal effluents had higher cobalt concentrations; thus, in Chokoloskee Bay, the cobalt concentration was 28.0 μg/liter.

(3) Landfills: Carlson and Johansen (1976) found cobalt in the leachate from five of six sanitary landfills in Norway during dry weather in August: 4, 9, 18, 33, and 70 μg/liter. The highest cobalt concentration was associated with by far the highest values for total organic carbon, BOD, COD, and phosphorus; highest concentrations of zinc, iron, and calcium; and lowest pH. The detection limit of 0.98 mg/liter for cobalt by an X-ray spectrographic method was too high for Clark (1975) to detect cobalt in the groundwater polluted by the Little Walnut Creek landfill northeast of Austin, Texas. (The landfill was closed at the time.)

Miller et al. (1977) reported cobalt concentration of 10 to 220 μg/liter in leachates from various land disposal sites. Table V-20 summarizes the more explicit waste identification information in Table V-21.

TABLE V-20

DECREASING COBALT CONCENTRATION mg/liter, IN LEACHATES FROM VARIOUS LAND DISPOSAL SITES
(Miller et al., 1977)

Spills, industrial wastes	0.22
Electronics effluent (treated)	0.17
Municipal and pickling wastes	0.08
Metal finishing wastes	0.05
Pharmaceuticals, chemicals, pesticides plant wastes (two sites)	0.04
Mix of municipal and industrial wastes, fly ash (three sites)	0.01

(4) Mines and mining wastes: Cobalt exists in sulfides associated with coal and ore bodies ($Co_{\underline{m}} X_{\underline{n}}$ where X = S, Se, As, or Te). Weathering of these sulfides produces sulfuric acid and releases ferrous and other heavy metal ions in acid mine drainage. Such drainage also occurs from sulfide mineral mines and tailings. If it cannot be prevented, mine drainage is usually treated by lime neutralization, which reduces the concentration of Co^{2+} and associated heavy metals to

TABLE V-21

COBALT[a] IN MONITORING WELL WATER AT LAND DISPOSAL SITES[b] RECEIVING LARGE VOLUMES OF INDUSTRIAL WASTES

(Miller et al., 1977)

Land Disposal Site	Description of Site	Above-background Concentrations of Leached Cobalt at Monitoring Wells, μg/liter	Sample Description
Massachusetts S-2	North-central Mass. landfill integrated with small lagoon. Landfill operated since 1971. Lagoon receives neutralized acid pickling wastes. Municipal solid refuse plus some wool-processing wastes	80 (Ni 40, As 650, Ba 100)	Nov. 1976. 4.3- to 5.3-m-deep wells with water levels 1.2 to 1.5 m below grade. Well screened in very coarse gravels. 6.2 m from site.
Michigan S-1	Southwestern MI. Untreated wastewater lagoon of metal finishing plant. Since at least 1973.	50 (Cr 10, Cu 2,800, Ni 670, etc.)	Average depth of wells 20 m in sand and gravel. 31 m from site.
Michigan S-4	Central MI. Tile drainage field receiving treated wastewater (~600 gpd) from electronics manufacturing plant. Since 1975. Abandoned seepage ponds nearby.	170 (Cu 20, Ni 80, Se 10)	Wells 3.6 to 4.3 m below land surface. October 1976. 46 m west of filled seepage pond and 61 m south of the tile field 46 m from site.
New Jersey S-1	Southeastern NJ. Dry lake formerly used as lagoon for liquid chemical wastes discharged from tank trucks through a spillway. Former site owners produced pharmaceuticals and pesticides. Located in recharge areas of a major aquifer.	40 (Cr 150, Cu 810, Zn 240,000)	15 m from site.
New Jersey S-8	Landfill receiving chemical processing wastes, much of it in barrels. Some buried containers have corroded and leaked. Overlies recharge area of a major aquifer, which discharges to a nearby river.	40 (Ba 200, Se 10, etc.)	October 1976. 152 m downgradient from the landfill.
New Jersey S-9	Southern NJ near the Delaware River. Storage tanks and lagoons. History of ground surface spills. Situated on outcrop of the principal aquifer in southern NJ.	220 (Cr 10, Cu 40, Ni 70, Zn 400)	February 1977. 92 m from the site.

TABLE V-21 (concluded)

Land Disposal Site	Description of Site	Above-background Concentrations of Leached Cobalt at Monitoring Wells, μg/liter	Sample Description
New York S-4	Southeastern NY west of Hudson River. Operated for ~3 yr. Receives municipal (including sewage sludge) and industrial wastes. Site bounded on three sides by flowing surface water.	10 (Ba 100)	21-24-m-deep wells. Depth to groundwater ~ 4.6 m. 15 to 62 m from site.
Pennsylvania S-1	Southeastern PA. MD border forms one of its boundaries. Used since 1970. About 33% filled by trench method with municipal garbage plus industrial lime, slag, and fly ash. Nearby shoe industry probably contributes wastes such as dyes, leather trimmings, etc.	10 (besides Fe, Ca, and Na, only element above background)	September 1976. Water levels 6.7 to 18 m below grade. 91 m from site.
Wisconsin S-7	West-central Wisconsin. Disposal site 1969-July 1976 for municipal waste and industrial waste said to contain Pb, Zn, and Ni. Trench-method of landfilling. Some open burning prior to covering.	10 (Ni 50, Ba 2,800, Mo 40, Se 590)	Wells at 1.5 to 4.6 m below ground surface, finished in very fine sand (over sandstone bedrock). 15 to 21 m from site.

a/ Emission spectrographic analysis for scanning; atomic absorption for quantification.
b/ Cobalt was found at 9 of 50 industrial sites studied.

< 0.5 mg/liter. Aeration converts Fe(II) to Fe(III), promoting its hydrolysis and precipitation as $Fe(OH)_3$. Other methods that have been studied for treating acid mine drainage are ion exchange, reverse osmosis, cementation, and electrolysis (Hill, 1973).

The sludge or "yellowboy" resulting from neutralization of acid mine water, containing < 10% solids, is currently disposed of in lagoons, abandoned mines, and underground workings. Grady and Akers (1976) discussed potential uses of the neutralization sludge. It might be used as a cement additive or as a soil amendment in revegetation of strip mine soil to increase pH and improve soil texture. They studied recovery of copper by electroplating, ammonia leaching, and organic extraction and recovery of cobalt, manganese, nickel, lead, and zinc by similar processes. The recovery of cobalt from drainage from Easton, West Virginia, was 3 lb/million gallons (0.36 mg/liter) comparable to the recovery of copper. By comparison, 492 lb aluminum and 1,394 lb iron per million gallons drainage were recovered.

Coal preparation wastes are discharged in a damp condition and are wet by rain for long periods before they are covered. Although the cobalt concentration is not high in the wastes, it is highly leachable (Williams et al., 1977).

In Maryland coal mine drainage, Hollyday and McKenzie (1973) found 8 to 1,400 μg Co/liter (median 210 μg Co/liter) at 21 sites. The median value for cobalt was lower than median values (expressed in μg/liter) of aluminum (18,000), manganese (4,000), zinc (1,000), strontium (570), and nickel (540), but considerably higher than those of copper (160), lithium (55), zirconium (46), yttrium (43), boron (34), chromium (22), barium (17), beryllium (12), arsenic (10), and ytterbium (6). Neutralizing a drainage water containing 1,100 μg Co/liter to pH 6.0 removed 75% of the cobalt as a precipitate containing 640 μg Co/kg, leaving a solution containing 270 μg Co/liter.

Several studies have reported on the impact of cobalt in acid mine drainage on the cobalt concentration of receiving streams.

There are major coal beds in the Powder River Basin in Wyoming and Montana. Dettman et al. (1976) found on June 18, 1975, only large concentrations of iron, manganese, and zinc downstream from the Big Horn (Strip) Mine under the prevailing conditions of high dilution of the stream. But on August 24, 1975, the upstream (Goose Creek) cobalt concentration was 2.4 to 3.6 μg/liter. With a mine discharge containing 4.3 to 7.1 μg Co/liter, the downstream (Tongue River) concentration of cobalt was 5.4 μg/liter. On September 27, 1975, the respective cobalt concentrations were 1.2 to 3.0, 1.5 to 4.7, and 3.1 μg/liter.

Schrader et al. (1977) reported on the cobalt concentrations in the sediments of the New River and Indian Fork in eastern Tennessee which are in proximity to the strip- and deep-mined highlands of the Cumberland plateau. Surface sediments ($<$ 1 mm) contained 26 to 51 mg Co/kg; averages were 34 and 38 mg/kg.

Although cobalt was not detected by spectrographic analysis in 75 Ohio River samples, Kopp and Kroner (1967) reported that 38% of streams receiving acid mine drainage (57% of which samples contained approximately 40 μg Co/liter) contained approximately 25 μg Co/liter.

Gang and Langmuir (1974) found an average 0.5 μg Co/liter (0.01, the detection limit, to 4.98 μg/liter) by atomic absorption analysis in surface and ground waters of the Clarion River-Redbank Creek Watershed* receiving acid mine drainage in a bituminous strip-mining region of northwestern Pennsylvania.

Lake Anna, a reservoir created by the impoundment of the North Anna River in Virginia, includes a large part of Contrary Creek, a major tributary of the North Anna River. The creek receives acid drainage from abandoned pyrite mines; and the sediments in this portion of the lake have higher concentrations of cobalt, iron, chromium, and cadmium for a distance of 3 miles below the source of acid mine drainage than does other lake sediment. The cobalt concentrations in the sediments ranged from 8.12 to 34.95 mg/kg (Blood and Reed, 1975).

---

* Six springs, 15 wells (13 flowing abandoned gas and oil wells), and 31 surface waters.

Hawley and Shikaze (1971; cited by Williams, 1975) found that the seepage water (pH 2) from a high-pyrite tailings pile in the Eliot Lake, Ontario, uranium district contained 3.8 mg Co/liter, 3.2 mg Ni/liter, and 3.6 mg Cu/liter. Hawley (1973; cited by Hill, 1973), however, did not report any cobalt concentrations in mine adits in Colorado, Idaho, or Montana.

Moran and Wentz (1974) and Wentz (1974; both cited by Hem, 1978) found 1 to 49 μg Co/liter in Colorado surface streams receiving manganese-enriched mine drainage. Samples were collected from September 1971 to December 1972.

Proctor and Sinha (1979) analyzed 95 water samples, 86 sediment samples, and 42 algae samples from the 200 mi^2 (520 km^2). St Francis River drainage basin in Missouri (see Figure V-1) for cobalt, nickel, and cadmium. The values are reported in Table V-22. The greatest total metal content was found near Fredericktown where copper and associated cobalt and nickel have been intensively mined and milled in the past and where future cobalt recovery is planned. The Doe Run and Big Creek areas are associated with lead-zinc mining. The cobalt concentrations in the surface waters ranged from 1 to 6,190 μg/liter (average 64 μg/liter); in the stream sediments, 3 to 1,744 mg/kg (average 82 mg/kg); and in the stream algae 7 to 1,633 mg/kg (average 162 mg/kg). These ranges are slightly different from those that would be derived from Table V-23, because of disagreements in the text of the original.

Ore-forming processes, erosion and weathering as well as mechanical and chemical mobilization from mining and milling activities account for the presence of the metals in the samples. Several thousand tons of cobalt and nickel (exact figures are not available) have been produced in the Fredericktown vicinity. Major mineral sources were siegenite $(Ni,Co)_3S_4$ and bravoite $(Fe,Ni)S_2$. Surface and underground mining to depths of 400 ft and milling have left large dumps and mill-tailing piles and ponds. These are sources of metals by mechanical and chemical mobilization.

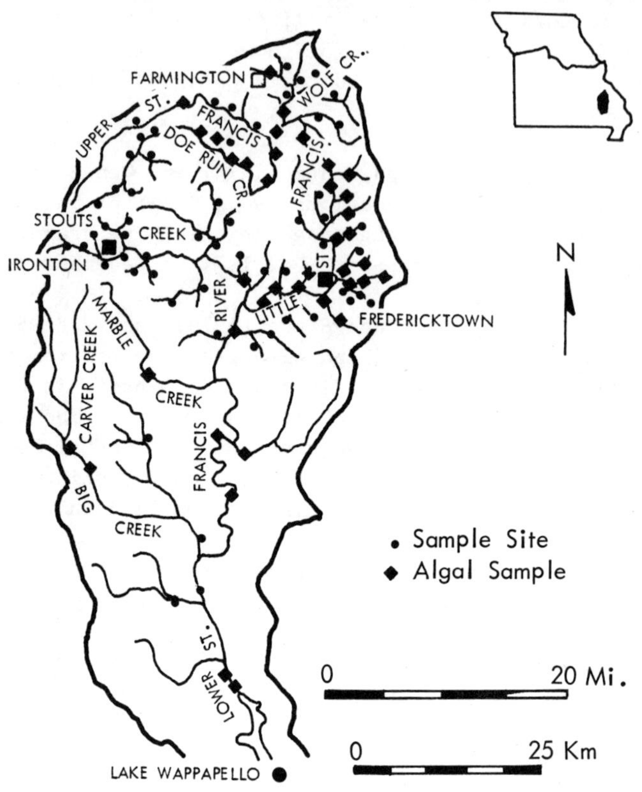

Figure V-1. Index Map and Sample Sites, St. Francis River Drainage Basin, Southeast Missouri (Proctor and Sinha, 1979).

TABLE V-22

COBALT IN STREAM WATERS, SEDIMENTS, AND ALGAE FOR SEVEN SUB-BASINS OF THE ST. FRANCIS RIVER DRAINAGE BASIN, SOUTHEAST MISSOURI

(Proctor and Sinha, 1979)

	Stream water, µg/liter		Stream sediments, mg/kg[a]		Algae, mg/kg	
	Mean	Range	Mean	Range	Mean	Range
Upper St. Francis[b]	1	1	22.9	10 - 56	12.2	10 - 15
Doe Run Creek[c]	1.05	1 - 1.1	38.1	9 - 90	26.3	12 - 40
Wolf Creek[b]	1.51	1 - 4.3	19.0	3 - 35	32.3	18 - 51
Little St. Francis[d]	23.6	1 - 6,500	215.4	6 - 1,744	307.3	10 - 1,663
Stouts Creek	1	1	28.6	13 - 69	-	-
Big Creek[c]	1	1	27.7	14 - 72	22.8	21 - 24
Lower St. Francis[e]	1.03	1 - 1.5	24.9	9 - 72	-	-

a/ Sediments were leached for 24 hr by hot nitric acid. The value reported is for the leachable cobalt.

b/ Sub-basins with little or no former metal mining activity.

c/ Sub-basins associated with lead-zinc mining and milling.

d/ Sub-basin associated with cobalt mining and milling.

e/ Sub-basin that accumulates sediments from all other sub-basins.

By comparing Missouri stream sediment concentrations, it appears that the cobalt enrichment is largely restricted to the Fredericktown area. Doe Run Creek sediments were somewhat elevated at 38 mg Co/kg in an area associated with lead-zinc mining. Sediments from the Upper Meramec River Basin associated with the North New Lead Belt contained comparable concentrations of cobalt (36 mg/kg). In the latter basin, sediments from areas not having former metal mining activities contained 11 and 15 mg Co/kg.

After a major flood in the spring of 1977 and the breaching of a major tailing pond dam, tailings were widely dispersed, to about 10 miles (16 km) downstream in the Little St. Francis. Similar variations in cobalt and nickel contents of the stream sediments indicate that they are probably transported together in closely associated mineral phases (see Figures V-2 and V-3).

Boyle et al. (1966) found $<$ 10 to 700 mg Co/kg in stream sediments in the Bathurst District, New Brunswick, Canada, where the massive lead-zinc-copper sulfide deposits contain 100 to 3,500 mg Co/kg. Apparently, cobalt is released from the sulfide deposits as soluble cobalt sulfate. In the streams, cobalt is adsorbed and/or coprecipitated by the gels associated with the sediments. These gels are both organic and inorganic. Of the gels, manganese hydroxides and oxides appeared to be the most efficient collectors.

C. Atmospheric Emissions

Although most authors who have considered the subject state that soil dust and the burning of coal and petroleum are the major sources of atmospheric cobalt, circumstantial evidence points to the metals industry as an important, if not the most important, localized source of cobalt emissions. Urban refuse burning, on the other hand, does not appear to be a significant source of atmospheric cobalt.

1. Metals industry: No reports were found clearly linking a predominantly cobalt metallurgical source with ambient air concentrations in the United States. The Bunker Hill lead

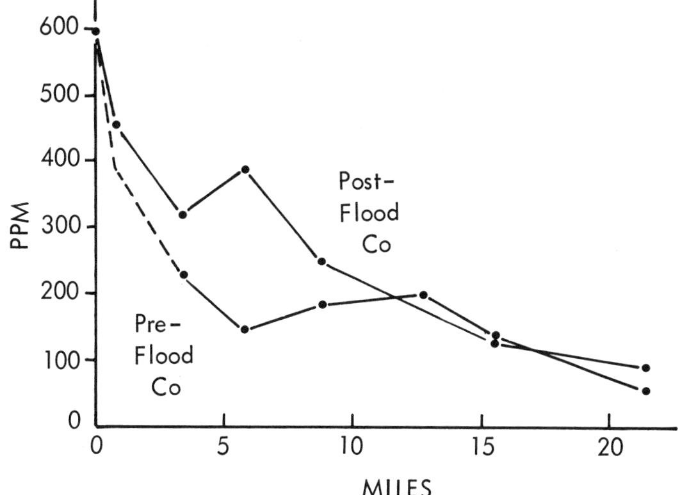

Figure V-2. Preflood and Postflood Cobalt Content (ppm or mg/kg) in Stream Sediments From Point Source in Little St. Francis River Sub-Basin to St. Francis River (Proctor and Sinha, 1979).

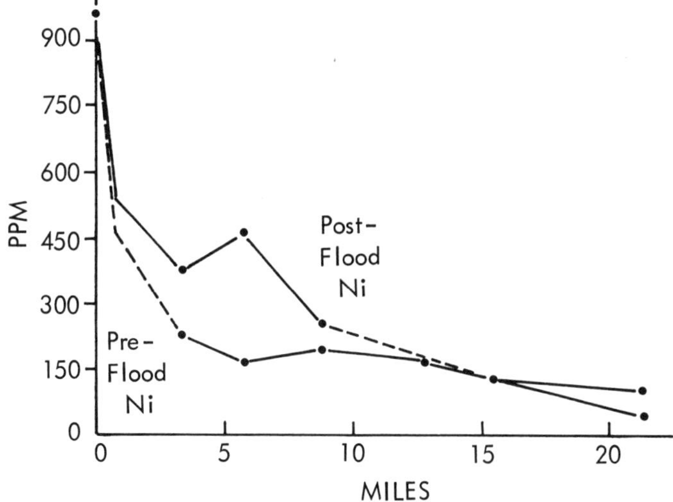

Figure V-3. Preflood and Postflood Nickel Content (ppm or mg/kg) in Stream Sediments From Point Source in Little St. Francis River Sub-Basin to St. Francis River (Proctor and Sinha, 1979).

smelter gave ambient air concentrations of cobalt at Kellogg, Idaho, in the high end of the range of ordinary average urban concentrations (Ragaini et al., 1977). Values near Tucson, Arizona, were about 2 ng/m^3, well within usual urban cobalt concentrations. The clustering methods of Perone et al. (1975) indicated that the presence of cobalt, cesium, and nickel in the air in the vicinity of Tucson was not related to the sources of copper emissions (half of the U.S. copper production is smelted within a 120-mile (193 km) radius of Tucson). These elements apparently had several sources, one of which was crustal.

The cobalt concentrations data for ambient air in Cleveland, Ohio, and vicinity in 1971 to 1972 (see Table V-23 and Figure V-4) are related to sites having variable traffic, industrial, and power plant emission sources. An attempt was made via Table V-24 to relate the metal works in Cleveland to the monitoring sites. The maximum values for cobalt, copper (3,000 ng/m^3) beryllium (1.6 ng/m^3), and arsenic (1,900 ng/m^3) were found at site 21 in downtown Cleveland. Although there are many potential metalworks sources of copper emissions in Cleveland, there is one plant that produces copper-beryllium alloys, which contain cobalt as a grain refiner. The Brush Beryllium Company is east of site 6 and south of site 20. Since the wind rose shows the wind direction 28% of the time was from the northeast, it is not unreasonable that elevated concentrations of metals from the Brush Beryllium Company (and numerous other metal works in the vicinities of sites 6 and 20) could appear downwind at sites 17, 10, 21, 1, 9, and 15. Although the beryllium concentration was not particularly high at site 6, it showed maxima at sites 10, 21, and 15. The maximum cobalt concentrations at site 6 was 46 ng/m^3; site 20, 69 ng/m^3; site 17, 46 ng/m^3; site 10, 15 ng/m^3; site 21, 610 ng/m^3; site 1, 38 ng/m^3; site 9, 23 ng/m^3; and site 15, 35 ng/m^3. The highest mean cobalt concentrations occurred at these downwind sites, except for sites 17 (2.6 μg Co/m^3) and 4 (2.1 μg Co/m^3), plus sites 5 and 13. The range of these highest cobalt means was 3.1 to 4.4 μg/m^3.

TABLE V-23

COBALT IN CLEVELAND, OHIO, AMBIENT AIR
(King et al., 1976)

Site	Description	Average, ng/m^3 [a]	Maximum, ng/m^3
1.	Adjacent to heavy industry on the north, west and south and to an interstate highway on the east.	4.0	38.0
3.	Predominantly upwind from heavy industry and downwind from mixed residential, commercial, and industrial environment.	1.6	11.0
4.	Predominantly residential but heavy rush-hour auto traffic.	2.1	6.9
5.	Mixed industrial and commercial. Predominantly downwind from heavy auto traffic.	3.4	28.0
6.[b]	Mixed residential, commercial, and industrial.	4.4	46.0
7.	Residential with heavy rush-hour traffic on the north and east.	1.5	11.0
8.	Mixed residential, commercial and industrial. Heavy rush-hour traffic on the east and west.	1.4	9.2
9.	Downwind (east) of heavy industry and a major traffic artery. Dirtiest site.	3.5	23.0
10.	Winds off Lake Erie often affect site with electric power plant (2) effluents and major traffic artery. Industry and commerce on the east, south, and north.	3.3	15.0
12.	Commercial and residential with major traffic arteries to the south and west.	1.1	11.0
13.[c]	Major traffic artery. South of steel-rolling mill. Al, Zn, and other chemical and metallurgical operations. Adjacent to Pb fabricating shop.	3.5	25.0
14.	Residential. Major traffic artery with heavy rush-hour traffic.	1.1	9.5

TABLE V-23 (concluded)

Site	Description	Average, ng/m³[a/]	Maximum, ng/m³
15.	Predominantly residential but adjacent to mixed metallurgical processing to the south. Generally upwind of heavy industry to the east.	2.3	35.0
17.	Mixed residential and commercial. Heavy rush-hour traffic on north and east.	2.6	46.0
20.	Shore of Lake Erie to north. Heavy rush-hour traffic to the south. Residential and commercial to the east, south, and northeast.	2.3	69.0
21.	Downtown Cleveland. Heavy railroad traffic to north; commerce to east, south, and west; and heavy industry farther south.	3.1	610.0
91.	Heavy rush-hour traffic. Residential but on flight path of airport, ~ 1.6 km north, Berea, Ohio.	0.82	1.3
92.	Rural. About 0.4 km from nearest medium-traffic-density street, 0.8 km north of major railroad line. Olmstead Township, Ohio.	0.75	1.8
93.	Residential. ~ 0.8 km from major traffic artery.	0.67	1.35
94.	Suburban. Sparsely settled. 0.8 km from major traffic artery, Westlake, Ohio.	0.70	1.35
95.	Medium-density residential, 91.5 m from well-traveled road, Bay Village, Ohio.	0.80	1.13
96.	Low-density residential. Medium-traffic-density streets with heavy rush-hour traffic, Rocky River, Ohio.	0.75	1.55
97.	High-density suburban residential. 0.4 km from a heavily-traveled major highway, Fairview Park, Ohio.	0.75	1.58

a/ Analysis by instrumental neutron activation analysis.

b/ Site 6 was adjacent to an electric lamp filament manufacturing plant.

c/ Site 13 was adjacent to a chemical plant that produces antimony compounds.

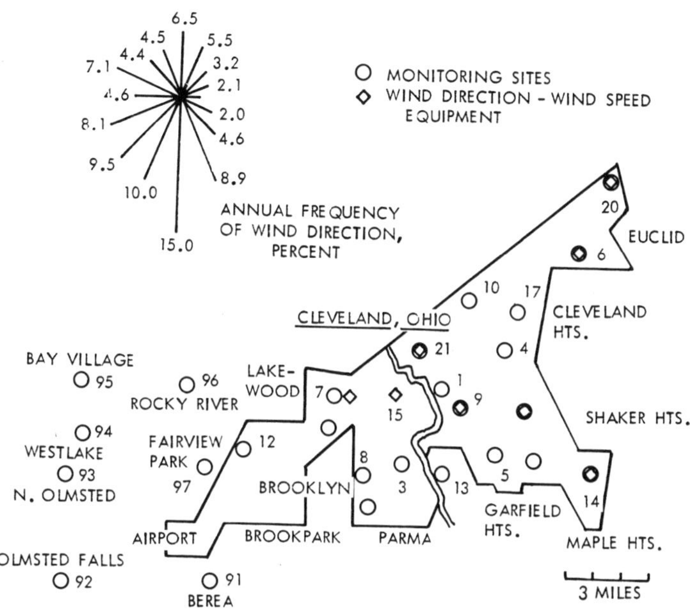

Figure V-4. Cleveland Sampling Sites (King et al., 1976).

TABLE V-24

METAL WORKS (OTHER THAN IRON AND STEEL)
IN CLEVELAND, OHIO, VICINITY IN 1967
(Cordero and Tarring, 1967)

Plant and Address	Nearest Site in Figure V-4	Activity and/or Products
Alloys and Chemicals Corporation 4365 Bradley Road, SW	(? Al max. at 10) 8 ?	Aluminum smelter
Aluminum Smelting and Refining Co., Inc. (Certified Alloys Co. head office) 5463 Dunham Road Maple Heights	14	Secondary aluminum smelter
Brush Beryllium Company 17876 St. Clair Avenue	6 (Be max. 21, 10, 15; rather low at 6)	Beryllium, beryllium-copper, beryllium compounds
Cerro Copper and Brass Company Division of Cerro Corporation 16600 St. Clair Avenue	6	Refrigeration, air conditioning, and heat exchanger tubing in copper and copper alloys
Chase Brass and Copper Co., Inc. Euclid, Ohio	6, 20	Fabrication of nonferrous metals
Cleveland Tungsten Division Chase Brass and Copper Co., Inc. Solon (Head office: 1121 E. 260th Street)	(? W max. avg. at 5, 6, 20)	Reduction of W and Mo ore to metal. Fabrication
General Electric Lamp Metals and Components Dept., 21800 Tungsten Road Cleveland Wire Plant	6	Reduction of W - Mo ores to metal. Fabrication
Master Metals Inc.	(? Sb max. at 13 and 6 chemical plant)	Secondary lead
National Lead Company	(? Pb max. at 10 and 17)	Lead extruding and alloying
The River Smelting and Refining Company 4189 Bradley Road	1 or 13 ? (? Zn and Sn max. avg. at 1)	Brass ingots
I. Schumann and Company 4391 Bradley Road	(Ni max. avg. at 13) 13	Produces nickel and copper alloys from scrap
TRW Inc. 23555 Euclid Avenue 44117	far beyond 6, 20	Cleveland plant activity? All plants: Motor car components, space hardware and electronics, superalloy, semifabrication, aluminum impact forging, titanium and high-temperature metals fabrication, investment castings

Pattenden (1975)* studied air pollution by heavy metals in the Swansea Valley, Wales (see Figure V-5). There was clearly a local pollution source for cobalt at Clydach (48 ng/m^3), Llansamlet (15.9 ng/m^3), and Trebanos (15.3 ng/m^3). The distribution of nickel in the study area was also elevated at these sites. Cobalt appeared to reside mainly in large particles that were more associated with resuspension of soil dust by wind than with the industrial process. The Swansea zinc smelter waste dump was enriched in cobalt; but the dump was discounted as a source since it was much richer in zinc and lead, which were not major air pollutants in the valley, and because of wind direction and element-element correlations. The Mond process nickel refinery in Clydach was the probable source of nickel, cobalt, and copper.

Goodman et al. (1975), who also took part in the collaborative study of the southeast Wales area, pointed out that in an earlier study, air loadings of lead, zinc, and cadmium at Llansamlet were reduced 10- to 20-fold between April and December 1971, a period corresponding to the shutdown of a lead-zinc smelter (mid-May 1971). The nickel works 4.5 km north of Llansamlet kept operating, and nickel, cobalt, and copper levels did not change. Using sphagnum moss bags as monitors of the airborne metals, they found strong intercorrelations among zinc, lead, and cadmium and among nickel, cobalt, and copper (see Table V-25). The wet moss bags in winter appeared to be better collectors around the nickel works than were the dry bags. The correlations for cobalt with nickel and copper as collected in moss bags in April and December 1971 were 0.776 and 0.965, respectively. In addition to the Clydach source, a much weaker source of cobalt was recognized in the Kidwelly-Burryport area. Enrichment above background for cobalt in the Clydach area was about 25; and in Kidwelly and Burryport, ~15 (Goodman and Smith, 1975). Burryport is a site of zinc use and former copper smelting (Goodman and Roberts, 1971).

---

* See Pattenden (1977) for a brief account of this work.

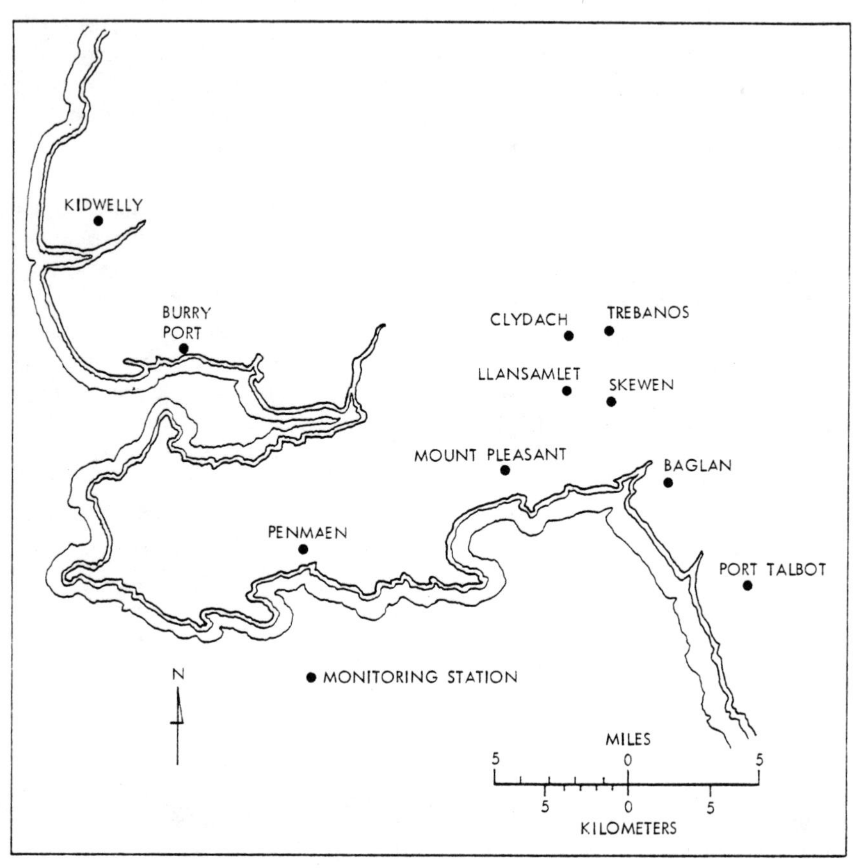

Figure V-5. Distribution of Monitoring Stations, South Wales, (Goodman and Roberts, 1971).

TABLE V-25

COBALT IN AIR COLLECTED BY SPHAGNUM MOSS BAGS IN SWANSEA AREA, WALES
(Goodman et al., 1975)

Area	ng Co/cm^2/day	Ratio to Background
In normal, clean background	0.7	1
"Elevated" background	0.7	1
High Ni, Co, Cu area	50	71
High Cd area	0.3	0.4
High Pb, Zn, Fe, Mn, Ca, Mg area	2	2.9
High Fe, Ca, Mg	1	1.4
Urban background Pb area	0.4	0.6

Table V-26 summarizes the environmental data collected on cobalt in the Swansea-Neath-Port Talbot survey. The enrichment factors for cobalt in the air at Clydach, Trebanos, and Llansamlet were 240, 131, and 114, respectively, with intermediate values of 27 and 31 for cobalt in air at nearby Mt. Pleasant and Skewen. These latter two sites have intermediate values for cobalt in the ambient air, which probably reflects some contribution to the total by transport from Clydach. The percentage of soluble cobalt in the atmospheric particulates was higher at the contaminated sites (65% at Mt. Pleasant; 78 to 86% at the other four sites) than at the three sites farther from Clydach (53 to 60%). Interestingly, the average annual deposition velocity for cobalt was almost as high at remote Kidwelly (1.13 cm/sec) as at Clydach (1.20 cm/sec). The mean annual deposition velocity at Trebanos was 1.68 cm/sec. The mass median diameter (MMD) of the particles at Trebanos was 3.5 μm; 30% were < 2 μm (Pattenden, 1974). Comparable particle size information for the other sites was not found.

Although soil is frequently attributed as a major source of cobalt, it does not appear to be particularly reliable to relate the MMD of the fraction of the particulate on which cobalt resides to the pollution source. Heindryckx (1976) collected the values given in Table V-27. Note that soil-size particles are as common to Trebanos air as to rural sites having no particular large source of cobalt emissions.

TABLE V-26

COBALT IN SWANSEA-NEATH-PORT TALBOT SURVEY IN SOUTHWEST WALES
(Pattenden, 1975; Goodman et al., 1975)

	Baglan	Clydach	Llansamlet	Mt. Pleasant	Port Talbot	Skewen	Trebanos	Kidwelly	Penmaen
Cobalt in air particulate, ng/m^3 (yearly avg. of weekly measurements)	1.37	48	15.9	5.5	0.65	3.7	15.3	0.49	0.88
Cobalt in dry deposition, $\mu g/cm^2$-yr (yearly sum of monthly measurements)	0.033	1.79	0.35	0.13	0.038	0.082	0.81	0.017	0.020
Cobalt in total deposition, $\mu g/cm^2$-yr	0.038	4.5	0.66	0.23	0.10	0.18	2.8	0.028	0.042
Cobalt in total soluble deposition, $\mu g/cm^2$-yr	0.020	3.6	0.54	0.15	0.048	0.14	2.4	0.015	0.025
Percentage of soluble cobalt in total annual deposition	53	80	82	65	48	78	86	54	60
Cobalt in air particulates, ng/m^3 (range of weekly values)	0.4-3	3-300	0.6-70	0.2-30	0.06-2	0.14015	0.8-90	< 0.01-3	0.01-5
Enrichment factor (normalized to Sc) for cobalt in air particulates	8.1	240	114	27	5.9	31	131	5.0	9.3
Average annual deposition velocity for cobalt, cm/sec	0.76	1.20	0.70	0.74	1.87	0.70	1.68	1.13	0.72
Soil cobalt, mg/kg	-	-	-	-	-	-	16.5	10	< 0.8-10
Cobalt in smelter wastes, mg/kg	-	-	< 3-205	-	-	-	-	-	-
Cobalt-element correlations (correlation coefficient/ probability of chance occurrence) ≥ 5 in air particulate	Cr 0.54/0.00	Fe 0.85/0.00 Sc 0.55/0.00 Sb 0.51/0.00	Ni 0.63/0.00 Se 0.52/0.00	Al 0.59/0.00 As 0.72/0.00 Br 0.61/0.00 Cr 0.51/0.01 Sb 0.59/0.00 Sc 0.70/0.00 Se 0.63/0.00	As 0.55/0.00 Cr 0.53/0.00 Sc 0.58/0.00 Se 0.60/0.00	All ≤ 0.26	Br 0.55/0.00 Br 0.55/0.00	Al 0.64/0.00 Fe 0.92/0.00 Mn 0.68/0.00 Cr 0.57/0.00 Sb 0.86/0.00 Sc 0.87/0.00 Se 0.88/0.00 V 0.67/0.00 Zn 0.87/0.00	Al 0.66/0.00 As 0.56/0.00 Fe 0.54/0.00 Mn 0.59/0.00 Ni 0.53/0.00 Pb 0.59/0.00 Sb 0.60/0.00 Sc 0.58/0.00 Se 0.61/0.00 V 0.61/0.00 Zn 0.59/0.00

TABLE V-26 (concluded)

	Baglan	Clydach	Llansamlet	Mt. Pleasant	Port Talbot	Skewan	Trebanos	Kidwelly	Penmaen
Cobalt correlations ≥ 0.5 between dry deposition and air particulate	0.67/0.14	0.72/0.02	0.79/0.02	0.86/0.03	0.85/0.00	-	0.76/0.00	0.67/0.02	-
Cobalt correlations ≥ 0.5 between rain solute and air particulates	-	0.60/0.05	-	0.82/0.04	-	-	0.73/0.01	-	-
Cobalt correlations ≥ 0.5 between dry deposition and rain solute	-	-	0.72/0.01	0.51/0.30	-	0.58/0.05	0.88/0.00	0.63/0.03	-
Cobalt correlations ≥ 0.5 between dry deposition and rain insoluble residue		0.64/0.03	0.80/0.00	0.53/0.28			0.82/0.00		0.68/0.02
Cobalt collected in atmospheric dust and rainfall by moss bags, $\mu g/cm^2$-yr (June 1, 1972 to May 31, 1973)	0.188	6.639	0.354	0.284	0.892	0.336	13.32	0.139	0.070
Cobalt collected in atmospheric dust and rainfall by moss bags, $\mu g/cm^2$-month (April 1971, December 1971)	-	-	0.02, 0.06	-	-	-	0.3, 0.2	-	-
Cobalt collected in atmospheric dust and rainfall by moss bags, $\mu g/cm^2$-month (April 1973, December 1973)	-	-	0.05, 0.0	-	-	-	1.0, 0.4	-	-
Cobalt collected in atmospheric dust and rainfall by moss bags, $\mu g/cm^2$-month (monthly average 1972-1973)	-	-	0.03	-	-	-	1.1	-	-

TABLE V-27

COBALT IN ATMOSPHERIC PARTICULATES AS RELATED TO PARTICLE SIZE
(Heindryckx, 1976)

Site	Mass Median Diameter, μm	Enrichment Factor (Al)
Ghent, Belgium (industrial)	1.9	17
Ghent, Belgium (residential)	0.85	12
Liege, Belgium (industrial)	4	14
Mechelen, Belgium (small town)	2.1	23
Niles, Michigan (rural)	3.3	5.2
Jungfraujoch, Switzerland (3,400 m altitude)	4.3	15
Lakselv, Norway	–	17

The 3-year average ambient air concentration of cobalt was 9 ng/m^3 in the vicinity of an unidentified USSR metallurgical plant whose largest metal emissions were iron and aluminum. The snow cover in March 1973, within 20 to 30 km to the northeast of the plant was least contaminated by cobalt: 0.0002 kg Co/km^2. Average soil cobalt concentrations (7 mg/kg) did not appear to be enriched above world average values (Nazarov et al., 1974).

Total annual emissions of cobalt from the smelters at Sudbury, Ontario, have been about 50 to 70 MT. Decreasing concentrations of cobalt in rainwater and soils with increasing distance from the smelter have been described in Chapter IV, subsection B.4.b.(2) and subsection B.4.a., respectively.

Although not strictly related to a discussion of atmospheric transport of cobalt from metallurgical sites, a recent report by Geladi et al. (1978) implies that many metal-working processes may be sources of airborne cobalt. Data were given for the ratio of the metal concentrations in the particulates of indoor and outdoor air of an antiquated plant producing steel wire products (welding, fencing, and barbed wire) and nuts and bolts in Ghent, Belgium. The concentration of total solid particulate (TSP) in the city air outside the plant was < 100 $\mu g/m^3$ whereas inside the plant the TSP was as high as 14,000 $\mu g/m^3$ at some stations. The elevated cobalt ratios at the bolt factory (10.3), zinc-plating station (1.6), galvanization station (18), copper-plating station (62), and welding-wire-packaging station (4.9) may have been due to cobalt impurities in the steel, zinc, or copper. At the draw benches, the elevated cobalt ratio (107) may have been due both to wear of a cobalt-containing tungsten carbide die and to wear of the steel wire being drawn. (High levels of tungsten and tantalum were noted in the "mechanical workshop" air, but their ratios were not given.)

2. <u>Fossil fuels burning and coking</u>: Bertine and Goldberg (1971) calculated that 8% of the total cobalt in the coal annually consumed worldwide would be mobilized into the atmosphere: 700 MT (based on 1967 statistics and 5 mg Co/kg coal). This was compared with 30 MT from oils (based on 7% mobilization, 0.2 mg Co/kg crude oil, and 1969 statistics); 7,200 MT transported by rivers due to weathering; and 8,000 MT transported by sediments.

Musin (1978) calculated that 450 short tons (408 MT) of fine ash particles were emitted daily to the atmosphere from a USSR power plant burning 14,000 tons brown coal daily. About 38% of the total fly ash emitted annually (85% collection efficiency) was distributed about the plant as follows: within a 6 km radius, 22,600 short tons (20,500 MT); within a 6- to 14-km radius, 32,930 tons (29,870 MT); and within a 14- to 18-km radius, 6,030 tons (5,470 MT). The cobalt concentration in the fly ash was 30 to 200 mg/kg. Since 30 mg/kg is closer to cobalt concentrations in U.S. coal fly ash, we can calculate that 0.68 short ton cobalt/year (0.61 MT/year) may be deposited within 6 km of such a coal-fired plant; 0.99 short ton/year (0.90 MT/year) within a 6- to 14-km radius; and 0.18 short ton/year (0.16 MT/year), within a 14- to 18-km radius. This would be equivalent to an average 0.002 ton (1.8 kg)/km^2/year within the 1,017 km^2 area.

Sponar et al. (1975) reported that up to 9.36 kg Co/km^2 (average 2.96 kg Co/km^2) was deposited annually within a radius of 1.5 km of a coal-fired power plant at Brno, Czechoslovakia.*

Klein and Russell (1973) calculated that 5 short tons (4.5 MT) cobalt should have been emitted from a power plant burning 10^7 short tons (9 Tg) Ohio coal since 1962 if the coal contained 5 mg Co/kg and the collection efficiency was 90%. The power plant, on the eastern shore of Lake Michigan, was only 6 miles (9.7 km) north of Holland, Michigan, which has several metal industries. Soils within a 115 mi^2 (298 km^2) area, 2 cm deep, were enriched in cobalt (4.5 mg Co/kg compared with 2.3 Co mg Co/kg in background soils). They found that the soil contained 34 short tons (31 MT) cobalt from the coal fly ash and that the total discharge of cobalt had been 42 short tons (38 MT). This would be the equivalent to 0.011 ton (10 kg)/km^2/year.

Bolton et al. (1973), (1974) and Lindberg et al. (1975) reported the concentrations of cobalt in the soil for 20 miles (32 km) north and south of the coal-fired Allen Steam Plant in

---

* In the same work, up to 6.0 kg Co/km^2 was found in the dustfall in the vicinity ($\leq$ 2.5 km) of cement works in Mokrá and Maloměřické, Czechoslovakia (averages were 1.5 and 1.0 kg Co/km^2, respectively). Adjacent highways contributed to the heavy metal enrichment of the dustfall.

Memphis, Tennessee. There was little difference in cobalt concentrations at 1 in. (2.5 cm) and 6 in. (15 cm) depths, whether close to or far from the plant. The top 1 in. samples north and south of the plant contained 15.1 and 12.8 mg Co/kg. The highest soil cobalt values were at 5 miles (8 km) north (23.6 mg/kg) and 8 miles (13 km) south (21.0 mg/kg); the minimum values were at 19 miles (30.5 km) north (11.3 mg/kg) and 9 miles (14 km) south (7.6 mg/kg). Analyses were by instrumental neutron activation. None of the other metals determined showed any marked trend with distance. However, copper and zinc showed an enrichment of 5 to 10 times compared to world soil average data. Uranium was four times higher, and the cobalt average was almost twice as high (15.4 mg/kg). The fly ash concentration, 25 to 70 mg Co/kg, ranged up to greater than four times higher than the average soil concentration.

Vaughan et al. (1975) estimated that cobalt emissions from a coal combustion and coal gasification facility would be 0.0038 g/sec. There would be a negligible increase of cobalt in runoff water and soil in the plant vicinity after 40 years of operation. Coal combustion after 40 years would raise the cobalt concentration in green plants by a factor of 1.3; coal gasification, by a factor of 1.1. Values used in their models included a plant concentration factor of 87, only 0.5 mg Co/kg in the coal, < 10% cobalt volatilized during combustion or gasification, and a 0.00087 g/sec emission rate for cobalt during coal gasification. In the region 0 to 20 km from a 1,400 Mw power plant, the expected annual deposition of cobalt from fly ash was 0.586 $\mu g/m^2$ to 0.938 $\mu g/m^2$; in the region 20 to 40 km from the plant, deposition was 0.185 μg Co/m^2/year; and in the region 40 to 50 km from the plant, deposition was 0.0586 $\mu g/m^2$/year. The total deposition in the 0- to 20-km area would be a maximum of 1.2 kg/year, a very low value compared with other estimates.

Cuffe and Gerstle (1967) presented data (see Table V-28) that indicated that the collection efficiency for cobalt in fly ash from a cyclone boiler (64%) was considerably less than that for other types of boilers (< 71 to 93%) in full load tests.

TABLE V-28

COBALT IN COAL-FIRED BOILER EMISSIONS IN FULL-LOAD TESTS[a/]
(Cuffe and Gerstle, 1967)

Type of Boiler	Average Fly Ash Collection Efficiency (%)	Cobalt Content Before Fly-Ash Collector (gr/scf x 10^{-4})	Cobalt Content After Fly Ash Collector (gr/scf x 10^{-4})	Collection Efficiency for Cobalt (%)
Vertical	89	0.48	0.06	87
Corner	94	1.22	0.082	93
Front-wall	86	2.0	0.37	81
Spreader-stoker	47	<0.73	0.21	< 71
Cyclone	69	2.2	0.8	64
Horizontally opposed	84	5.1	0.66	87

a/ Collection efficiency is better at partial loading.

Schwitzgebel et al. (1977) attempted to determine mass balances of numerous elements, including cobalt, at coal-fired power plants. Their data, presented in Table V-29, indicate that when emissions are controlled by either a scrubber or an electrostatic precipitator, cobalt atmospheric emissions comprise only about 2% of the cobalt contained in the coal but that almost 30% of the cobalt is discharged on the fly ashes from a cyclone-fed boiler.* Except for the concentrations of cobalt in the fly ash (60 to 80 mg/kg, MRI calculations), cobalt concentrations in the waste streams are within the range of those found in natural soils and water.

Klein et al. (1975a) reported on an ongoing mass-balance study for elements at the Allen Steam Plant in Memphis, Tennessee. The boiler is cyclone fed. Cobalt was one of the elements that was readily incorporated into the slag. The enrichment factor of cobalt in the slag compared with its concentrations in the coal was at least 6. For example, Lyon et al. (1974) reported cobalt concentrations of 3.0 mg/kg in the coal burned at the Allen Steam Plant, 21 mg/kg in the slag, 41 mg/kg in the inlet fly ash, and 65 mg/kg in the outlet fly ash. The rate of atmospheric discharge of 0.02 g Co/min was less than the rates of other toxic heavy metals: arsenic, 0.2 g/min; barium or chromium, 0.3 g/min; and vanadium, 0.4 g/min.

Andren et al. (1974) calculated the minimum atmospheric discharge of cobalt from the Oak Ridge area's three coal-fired steam plants (including the Allen plant), which burn a total of 9.1 million short tons (8.3 MT) of coal annually, to be 0.12 short ton (99 kg). Fly ash was tentatively deemed the major source of cobalt in the area's ambient air. For example, the air concentration of cobalt at the Walker Branch Watershed, which is within 20 km of the plant, at ground level in July 1974, was 0.34 ng/m^3. Normalized to europium, the enrichment factors for cobalt in the air and in the fly ash were 4.7 and 5.0, respectively.

---

* About 50% of the ash of a cyclone-fed boiler is bottom slag (ash). About 50% is fly ash. For a pulverized coal boiler, $\leq$ 90% of the ash is fly ash (Klein et al., 1975).

TABLE V-29

MASS BALANCE DETERMINATIONS FOR COBALT AT COAL-FIRED POWER PLANTS
(Schwitzgebel et al., 1977)

	Cobalt in Inlet and Outlet Streams					
	Station I Subbituminous coal Venturi scrubber (99.6% efficiency)		Station II Subbituminous coal Electrostatic Precipitator (99.1% efficiency)		Station III North Dakota Lignite Cyclone-Fed (65% efficiency)	
	kg/hr	%	kg/hr	%	kg/hr	%
Coal	0.27 (2.08 mg/kg)	90.0	0.19 (1.6 mg/kg)	~100%		
Ash sluice water	0.0045	1.5				
Inlet sluice water			0.0007			
Cooling tower blowdown	0.005	1.7				
Makeup water	0.015 (8 μg/ℓ)	5.0				
Lime	0.003 (1.3 mg/kg)	1.0				
Total In	0.30 (2.6 MT/yr)	0.04	0.19	0.03		

TABLE V-29 (concluded)

	Cobalt in Inlet and Outlet Streams					
	Station I Subbituminous coal Venturi scrubber (99.6% efficiency)		Station II Subbituminous coal Electrostatic Precipi- tator (99.1% efficiency		Station III North Dakota Lignite Cyclone-Fed (65% efficiency)	
	kg/hr	%	kg/hr	%	kg/hr	%
Bottom ash (to pond)	0.037 (7.1 mg/kg)	14.2				48
Outlet sluice water			<0.0007			
Bottom ash sluice water	0.0029 (5 μg/ℓ)	1.1				
Sluice ash			0.04 (20 mg/kg)			
Scrubber solids (to pond)	0.19 (8.3 mg/kg)	73.1				
Scrubber liquids (to pond)	0.021 (11 μg/ℓ)	8.1				
Precipitator ash			0.19 (27.5 mg/kg)			
Economizer ash	0.0002 (16.7 mg/kg)	0.1				
Flue gas } Fly ash through stack }	0.0059 (78.6 mg/kg if all on fly ash)	2.3	0.004 (63 mg/kg if all on fly ash)	1.5		28.5
Total Out	0.25	0.5	0.23			
Σ Out/Σ In	0.87		1.21			

Mass balance calculations for cobalt at the Allen Steam power plant are presented in Table V-30. The averages for the 1972 and 1973 runs indicate that 0.4 to 0.8% of the cobalt in the coal was emitted to the atmosphere, about 50% was concentrated in the slag, and about 55% was trapped in the electrostatically precipitated fly ash (Klein et al., 1975b). The range of cobalt concentrations in Allen Steam Plant fly ash was 25 to 70 mg/kg (Bolton et al., 1974). These mass balance values are not out of line with those shown above by Schwitzgebel et al. (1977).

Pressurized fluidized-bed combustion of coal in laboratory-scale experiments at 1450 to 1750°F (780 to 950°C) showed mass balances for cobalt, lead, iron, potassium, lanthanum, sodium, and scandium of 100 ± 10%, indicating essentially no losses by volatilization. The lower temperatures as compared with conventional coal burning reduced the preferential concentration of trace elements in the finer particles of ash although some elements, including cobalt, showed a slight tendency to do so (Vogel et al., 1975).

Ragaini and Ondov (1976) calculated that the mass median diameter of 80 ± 9% of the particulate cobalt from an ESP-controlled coal-fired power plant was 1.13 μm. It was puzzling that a significant fraction of cobalt was found on the small particles since most inorganic forms of cobalt likely to be present are nonvolatile at ≤ 1550°C; the typical combustion temperatures are 1500 to 1600°C.* They found that the enrichment factors for cobalt (normalized to scandium) in suspended particulates from two coal-fed power plants with respect to the input coal were 2.2 and 0.3.

At the Huntington Canyon power plant, 80% of the emissions from a stack equipped with an electrostatic precipitator (ESP) were on the large particles (~ 5.2 μm). At the Four Corners plant, only 10% of the total ESP emission was submicron

---

* Davison et al. (1974) remarked that cobalt species boiling or subliming at > 1550°C are Co, CoO, and CoS. These are possibly the forms evolved during coal combustion.

TABLE V-30

COBALT FLOW THROUGH THE COAL-FIRED
ALLEN STEAM PLANT, mg/kg
(Bolton et al., 1973, 1975; Klein et al., 1975)

	Cobalt Concentration, mg/kg					Cobalt Mass Flow, g/min				
	Run A	Run B	Run C[a/]	January 1972 Runs	August 1973 Runs	Run A	Run B	Run C	January 1972 Runs	August 1973 Runs
Coal	3.5	5	3.3		2.9	4.4	6.3	4.1	4.9	4.3
Slag tank solids	15	28	19		20.8	1.5	2.9	2.1	2.7	1.8
Precipitator inlet fly ash	35	51	25		39	2.3	3.5	1.2	2.1	2.9
Precipitator outlet fly ash	-	26	58		65	-	0.048	0.11	-	-
Stack	-	11	-		-	-	0.013	-	0.04 (21 kg/yr)	0.02
Imbalance, %						-14	1.6	-19[b/]	-2	+9

a/ Bolton et al. (1975) compare these values for cobalt obtained by neutron activation analysis (NAA) with those obtained by the less accurate method of spárk source mass spectroscopy (SSMS). Although both methods indicate about 50% of the cobalt remained in the bottom ash, the SSMS measurements indicate 38% of the cobalt (29% by NAA) was in the collected fly ash and 0.9% (2.7% by NAA) in the airborne fly ash.

b/ Cobalt concentrations determined on this run by spark source mass spectroscopy (SSMS) were usually more than twice those found by neutron activation analysis. The imbalance for the SSMS concentrations was -13% (Bolton et al., 1975).

particles,* but for the scrubber-equipped stack, $\geq$ 95% of the mass of the insoluble elements was in submicron particles. Ragaini and Ondov (1977a) calculated that almost twice as much "respirable" cobalt will be emitted from a scrubber as from an ESP and that cobalt's potential alveolar deposition would be four to five times greater for scrubber emissions than for ESP emissions. The enrichment of cobalt in the stack particulates relative to its concentration in coal was 1 to < 1.4.

In a more recent study of the same type (Ondov et al., 1979), data were put forward to predict that twice as much of the aerosol particles from scrubbers will be deposited in the lungs as particles from ESP units. The mass median aerodynamic diameter (MMAD) of cobalt in emissions controlled by an ESP in the stack of a power plant burning coal containing $\sim$ 2 mg Co/kg was 9.1 to 10.9 μm. Aerosol particles from a Venturi scrubber unit contained cobalt in the fraction of MMAD 7.1 to 12 μm. The cobalt emission rates were higher from the ESP unit (median 2.38 pg/J) than from three scrubber unit runs (medians of 0.165 to 0.57 pg/J). Cobalt on intermediate-sized particles passed through the scrubber at 74% the rate that they passed through the ESP. (The ESP was more efficient at collecting the smaller aerosols.)

Natusch et al. (1977) pointed out that 20 to 30% of the mass of all fly ash particles of physical diameter < 5 μm are encapsulated or agglomerated. This natural collecting system tends to reduce the health hazard of small, lung-depositing particles. They found that cobalt predominated on the surface

---

* Gladney et al. (1978) reported that 11 to 15% of the cobalt in the coal combustion particles released from an ESP is < 2 μm in diameter.

layer of particles of bulk fly ash containing 65 mg Co/kg. Only 18% of this cobalt was extracted by a dimethyl sulfoxide water mixture after 48 hr at 25°C.*

Gladney et al. (1978) concluded from their own study of coal combustion particulates at the Chalk Point Generating Station, Southern Prince George County, Maryland, plus literature values that the enrichment factor for cobalt (normalized to aluminum) ranges from < 2 to ~ 5. The enrichment factor for cobalt in the particulates of the ambient air of six U.S. cities overlapped this range, ~ 4 to 9. In an earlier Chalk Point Study, the enrichment factor of cobalt in the particulates compared to its concentration in the coal was only 1.2 (Zoller et al., 1975). The enrichment factor in the stack particulates compared to the crustal abundance was ~ 4. The net plume sample at ground level contained 3.86 ± 0.55 ng Co/m^3, but the average for cobalt in the ambient air was 0.41 ± 0.11 ng Co/m^3.

Under constant operating conditions, the average element collection efficiency of cascade impactors for power plant emissions from an ESP was only 60% that of Nucleopore filters. For example, the cobalt concentration in the stack aerosols as measured by impactors was from 21.9 to 29.7 μg/m^3 compared with 39.6 to 45.2 μg/m^3 as measured by the filters. The variability in the concentrations for Co, Cr, Mn, and Se may have indicated greater inhomogeneities for these elements in the aerosols. Collection substrates were thought to have been contaminated by sulfuric acid mists leaching cobalt, chromium, and manganese from the stainless steel of the impactors and filter holders. Cobalt is a rare addition to stainless steels, however. Because of their greater surface area, impactors would be subject to greater contamination (Ondov et al., 1978).

---

* Fisher et al. (1979) found that $\sim 50\%$ of the cobalt in 50 mg of the fly ash fraction of volume median diameter 2.2 μm was soluble in 8 ml buffer solution (0.1 M 2-amino-2-hydroxymethyl-1,3-propanediol [Tris] and HCl) at pH 7.3 after 5 min contact time with 3 min of vigorous shaking. The fact that several elements dissolved "at physiological pH" was stressed.

Soaps and salts of cobalt may be included as detergent or dispersant additives in lubricating oil formulations (Sargent and Ross, 1970; Anonymous, 1957). Many other metals are used as lubricant additives, and wet scrubbers have been recommended for trapping submicron metal oxides produced during the incineration of oily wastes.

Traces of cobalt may be introduced into certain heating fuels and motor fuels during removal of mercaptans (cobalt is used as the major desulfurization catalyst) and into other fuels when added as a combustion improver to reduce formation of smoke and soot (Hofstader et al., 1976).

The Air Pollution Control Office (Martin et al., 1971) studied the effects of 206 commercial fuel additives (except those for coal and gasoline) on air pollutant emissions from distillate oil-fired furnaces. Probable forms of the 14 metals contained in the additives were organometallics, metallic sulfonates, and metal oxide slurries. Particulate emissions were reduced significantly only by those additives containing iron, manganese, and/or cobalt; particulate emission reductions that were achieved by these additives were 31% to 47%. The names and composition of the three commercial additives containing cobalt are given in Table V-31. At doses of 1:150 Improsoot and 1:200 Formula LSD, the particulate emissions were reduced by 32% and 43%, respectively. It was calculated that 350 lb (159 kg) of Formula LSD would be needed for 10,000 gal. (38,000 liters) fuel oil to reduce particulate emissions by 35%, which was not as economical as additives containing ferrocene or methylcyclopentadienylmanganese tricarbonyl. The commercial use of cobalt compounds in fuel oils could not be verified in 1979.

Levy et al. (1971) studied the emissions from fuel oil combustion for space heating. For two residential units, about 0.9 to 1.0 lb (0.41 to 0.45 kg) cobalt was emitted per million gal. (3.8×10^6 liters) of fuel oil burned. The emissions from three commercial boilers fired with Nos. 4, 5, and 6 residual oils were 17, 1.5, and < 0.56 lb Co/10^6 gal. (7.7, 0.68, and < 0.25 kg/3.8×10^6 liters) fuel oil. The analyses done by emission spectroscopy were only semiquantitative.

TABLE V-31

COBALT-CONTAINING ADDITIVES FOR PETROLEUM FUELS
(Martin et al., 1971)

	Sul-Van-Control	Improsoot	Formula LSD
Manufacturer	Butler-Engineering Association, Hillside, New Jersey	Commercial Chemicals	Commercial Chemicals
Percent metals	0.05 Co, 0.4 Mg, 0.07 Pb	0.3 Co, 0.1 Ca	0.9 Co
Doses at which commonly used	1:440 (tests)	1:8,000 to 1:1,000 to residual fuel oils	1:8,000 to 1:4,000 to all grades of fuel
Doses at which metal emissions were measured from a distillate-oil-fired furnace	-	1:150	1:200
Cobalt emissions (probably as oxide)			
$\mu g/m^3$	-	1,219	2,744
mg/kg fuel	-	20.0	45.0

There is little cobalt in gasoline, and it is doubtful that the addition of cobalt in a gasoline additive would have accounted for the high concentration of cobalt in an automobile exhaust pipe deposit.* The dry carbon removed from the internal zone of the exhaust pipe of a 1,000-cm^3 automobile (presumably of British make) 2 cm from the end of the pipe contained 391.0 mg Co/kg (Hamilton, 1974). This high-cobalt concentration may have reflected attrition of the cobalt-containing hard-facing alloy used on the exhaust valves and valve seats of internal combustion engines or a corrosion product of the cast iron tail pipe.

Winchester and Nifong (1971) calculated that 55 short tons (50 MT) cobalt is emitted annually in the Chicago, Milwaukee, and Northwest Indiana area. The total was comprised of 25 short tons (22.7 MT) from coal burning, 2 short tons (1.8 MT) from coke emissions, and 28 short tons (25.4 MT) from fuel oil. There were assumed to be no cobalt emissions from iron and steel works, cement plants, and transportation. The area was estimated to consume 4% of the U.S. total for coal, fuel oil, and automotive fuel. Coke production was estimated to be 21% of the U.S. total. They assumed 90 mg Co/kg in coal-burning particulate emissions and 1,500 mg/kg in fuel oil particulate emissions. According to their "air pollution inventory," the calculated cobalt concentration in air should have been 3.5 ng/m^3 (the iron concentration was set the same for the inventory as for the ambient concentration). The geometric mean of cobalt in industrialized Northwest Indiana air, however, was 1.5 ng/m^3. They then calculated that the streams feeding Lake Michigan would contribute 6.5 MT cobalt/year (at 0.2 μg Co/liter), and that the atmospheric input of cobalt to Lake Michigan could be significant.

Slater and Hall (1977) estimated the amounts of trace elements emitted nationwide from electricity generation by utilities in 1974, taking into consideration the relative amounts of fossil fuels used, the combustion systems used, control efficiency and application, etc. The typical cobalt concentrations in fuels

---

* Rancitelli (1973; cited by Jungers et al., 1975) examined gasoline additives such as antioxidants, metal deactivators, corrosion inhibitors, anti-icers, and carburetor and valve deposit detergents by neutron activation analysis. Only < 0.0002 to 0.0360 μg Co/ml was found.

consumed in the U.S. were assumed to be (in mg/kg) bituminous coal, 4.0; anthracite coal, 90.00; lignite, 4.0; and residual fuel oil, 2.24. Based on a review of the literature mass balance data, they estimated that 10% of the cobalt present in pulverized coal is emitted from the coal-burning plant. Their detailed estimates for cobalt emissions are given in Table V-32. The 320 short tons (274 MT) total cobalt is emitted from external combustion equipment; none is estimated to be emitted from internal combustion equipment. They calculated that 63.8% of the particulate emissions from conventional stationary combustion sources are from the generation of electricity. Since coal accounts for 95% of the particulate emissions from electrical utilities, the total cobalt emission from coal-burning is 137 MT/0.61 or approximately 225 MT.

Slater and Hall (1977) did not give enough information to scale up the 160 MT cobalt from residual fuel oil combustion for electricity generation to total U.S. fuel oil combustion. However, a value of 50 MT cobalt emitted from power plant fuel oil combustion and a total of 480 MT cobalt from all U.S. fuel oil combustion can be calculated based on recent consumption of 8 billion gallons fuel oil for power plant boilers, 21 billion gallons for industrial and commercial boilers, and 12 billion gallons for domestic boilers (Harper, 1977; and Harper et al., 1977); particulate emission factors* of 8 lb/1,000 gallons, $\sim$20 lb/1,000 gallons, and 10 lb/1,000 gallons for these boiler types, respectively (U.S. Environmental Protection Agency, 1972); and a cobalt emission factor of 1,500 mg/kg fuel oil particulate emissions (Winchester and Nifong, 1971). Note, however, that a maximum cobalt concentration of 0.3 mg/kg** in the fuel oil itself (Table IV-9) would give only about 43 MT cobalt. This would indicate that more than half of the total annual amount of cobalt consumed as salts is added to fuel oils or that the emission factor is too high.

Kowalczyk et al. (1978) resolved the concentration pattern for 27 elements collected from the air at four sites in Washington, D.C., in the summer of 1974 into six source components: coal, oil, and refuse combustion; marine aerosols; soil; and motor vehicle emissions. Their estimates for cobalt

* Unfortunately, the emission factors for burning fuel oil do not distinguish between residual and distillate fuel oils.

** See new information in Part D of the Summary.

TABLE V-32

STACK EMISSIONS OF COBALT FROM THE GENERATION OF ELECTRICITY BY EXTERNAL COMBUSTION, TONS/YR
(Slater and Hall, 1977)

Electric generation	320				
External combustion		320			
Coal			160		
Bituminous				140	
Pulverized dry					100
Pulverized wet					19
Cyclone					19
All stokers					1.0
Anthracite				13	
Pulverized dry					4.6
Pulverized wet					0
Cyclone					0
All stokers					8.5
Lignite				5.5	
Pulverized dry					3.3
Pulverized wet					0.82
Cyclone					0.82
All stokers					0.55
Petroleum			160		
Residual oil				160	
Tangential firing					63
All other					98
Distillate oil				-	
Tangential firing					-
All other					-
Gas				-	
Tangential firing					-
All other					-

TABLE V-33

SOURCES OF COBALT IN WASHINGTON, D.C., ATMOSPHERIC PARTICULATES
(Kolwalczyk et al., 1979)

Source	Relative Cobalt Concentration in Source	Cobalt Contribution From Components, ng/m^3
Coal combustion	0.00048	0.35
Oil combustion	0.007	0.36
Refuse combustion	0.000055	0.007
Marine aerosols	0.000000048	<0.001
Soil	0.00027	0.32
Motor vehicle emissions	-	-
Predicted atmospheric concentration	-	1.0
Observed atmospheric concentration	-	1.1 ± 1.0

in Table V-33 indicate that coal combustion, oil combustion, and soil contributed about equally to atmospheric cobalt concentrations.

3. Urban refuse burning: The average concentration of cobalt in the combustible fraction of urban refuse is 3 or 4 mg/kg (range < 3 to 5 or 7) (Law and Gordon, 1979; Campbell, 1976). Marr et al. (1975) found 4 to 5 mg Co/kg in the heavy combustible fraction of urban refuse (heavy paper and plastics with a significant amount of noncombustibles such as wire and foil) but < 4 mg Co/kg in the light combustible fraction (lighter paper and plastics and $\leq$ 20% dirt and glass). The typical cobalt concentration found in 18 samples of combustible refuse from several locations in the Washington, D.C., metropolitan area was 7 mg/kg (range 2 to 17 mg/kg). These values are compatible with concentrations for cobalt found in paper (1 to 4 mg/kg) and hand-picked plastics from urban refuse (< 8 mg/kg).

There is no cobalt in newsprint, but there is 13 mg Co/kg in the comics sections. Cobalt concentrations in kraft paper and corrugated cardboard are < 2 mg/kg (Campbell, 1976). Table V-34 lists the composition of combustible urban refuse.

TABLE V-34

THE COMPOSITION OF COMBUSTIBLE URBAN REFUSE
(Campbell, 1976)

Material	Percent	
Paper	50.2	
Corrugated cardboard		13.4
Newspapers		9.0
Other		27.8
Yard waste	17.8	
Food	16.8	
Plastics	5.2	
Wood	4.6	
Leather and rubber	3.4	
Textiles	2.0	

Law and Gordon (1979) summarized the reported concentrations of cobalt in municipal incinerator products (see Table V-35). Concentrations of 70, 100, and 12 mg Co/kg in the fine bottom ash, fly ash, and atmospheric particles correspond to enrichment factors (normalized to iron) of 3.4, 3.2, and 1.0, respectively. Table V-36 depicts the mass balance derived by Law and Gordon (1979) for cobalt during municipal incinerator operation. Almost all of the cobalt in the refuse was accounted for by the fine bottom ash and collected fly ash.

Greenberg et al. (1978a) found that 41% of the cobalt in suspended particles from the Alexandria (Virginia) Municipal Incinerator was $\leq 2\ \mu m$ in size and that 34% was $\leq 2\ \mu m$ from the Solid Waste Reduction Center in Washington, D.C. Since cobalt (as well as chromium, vanadium, manganese, iron, and selenium) had mixed size distribution in the suspended particles, it may have had at least two significant sources in the refuse (which was borne out by its presence in both combustibles and non-combustibles). Data for cobalt concentrations in the fly ashes and suspended particles of the two incinerators are given in Table V-37.

TABLE V-35

COBALT IN INCINERATION PRODUCTS FROM A MUNICIPAL INCINERATOR
(Law and Gordon, 1979)

Fine bottom ash, mg/kg[a/]	70 ± 10
Fly ash, mg/kg[b/]	100 ± 30
Atmospheric particles, mg/kg[c/]	12 ± 7
Quench water, mg/liter[d/]	<0.2
Fly ash scrubber water, mg/liter[d/]	<0.2 - 0.5
Spray waters, dissolved solids, kg/week	0.02
Spray waters, undissolved solids, kg/week	0.008
Quench waters, dissolved solids, kg/week	<0.004
Quench waters, undissolved solids, kg/week	0.0007

a/ Bulk metals, glass, and other particles >3 mm diameter were excluded (Law, 1978).
b/ From Law (1978).
c/ Greenberg et al. (1978a).
d/ From Law (1977).

TABLE V-36

COBALT INPUT-OUTPUT DURING ONE WEEK'S OPERATION OF A MUNICIPAL INCINERATOR
(Law and Gordon, 1979)

Input, kg/week	
Combustibles	2 (< 2 to 5)
Not accounted for by combustibles	4
Output, kg/week	
Fine bottom ash (bulk scrap, cans, bottles, etc., excluded)	4 ± 1
Fly ash (collected)	2 ± 0.6
Atmospheric particles	0.05 ± 0.03
Total, kg/week	6 ± 2

TABLE V-37

COBALT IN MUNICIPAL INCINERATOR FLY ASHES AND SUSPENDED PARTICLES, mg/kg
(Greenberg et al., 1978)

	Fly Ash		Suspended Particles	
	Average[a/]	Range	Average[b/]	Range
Alexandria Municipal Incinerator (inefficient pollution control)	35 ± 5	26 - 54	12 ± 7	3.8 - 28
Solid Waste Reduction Center in Washington, D.C. (one of best pollution-control systems)	27 ± 2	25 - 29	5.4 ± 2.5	3.0 - 8.0

a/ Other elements with comparable concentrations: I, 25; Sc, 10.4; As, 40; Ag, 85; Cd, 42; La, 34; Ce, 78; Th, 14; W, 16 mg/kg.

b/ Others: Cs, 3.1; V, 22; Se, 23; In, 6.5; La, 3.8; Ce, 11; W, 17 mg/kg.

Greenberg et al. (1978a) concluded that refuse incineration was not a major source of atmospheric cobalt (or of vanadium, chromium, iron, nickel, selenium, bromine, and iodine).* The enrichment factor for cobalt emitted from the incinerators (normalized to aluminum) was approximately 3 to 7 compared with an enrichment factor for cobalt in Washington, D.C., air of approximately 5 to 8.

Greenberg et al. (1978b) found 1.8 to 3.3 mg Co/kg (average 2.3 $\pm$ 0.5 mg/kg) in the suspended particles emitted from the Nicosia Municipal Incinerator in East Chicago, Indiana. They calculated an average 6.6 $\pm$ 5.0 mg Co/kg in suspended particles for this incinerator serving a highly commercial and industrial area plus the two from the Washington, D.C., residential and lightly commercial areas. They commented that the element variability among these incinerators is much less than among coal-fired power plants despite different types of areas served and different pollution control devices. The emission control system for the Nicosia Incinerator comprises a "spray chamber followed by a three stage, horizontal, plate type scrubbing tower." Disposable filters in the closed loop water system remove the suspended particles that are scrubbed from the gas stream (Greenberg et al., 1978b). The Alexandria emissions pass through a water spray baffle, which partially removes the fly ash. The pollution control devices at the Solid Waste Reduction Center include mechanical separators and electrostatic precipitators (Greenberg et al., 1978a).

---

* Kolwalczyk et al. (1978) estimated the contribution of urban refuse burning to cobalt ambient air concentrations to be 0.7%. See Table V-33 in the preceding subsection.

BIBLIOGRAPHY. CHAPTER V.

Alban, L. A., and J. Kubota, "A Study of Extractable Soil Cobalt in Soils of the Southeastern United States," Soil Sci. Soc. Am. Proc., 24(3), 183-185 (1960).

Amoozegar-Fard, A., W. H. Fuller, and A. W. Warrick, "Migration of Salt from Feedlot Waste as Affected by Moisture Regime and Aggregate Size," J. Environ. Qual., 4(4), 468-472 (1975).

Andren, A. W., B. G. Blaylock, E. A. Bondietti, C. W. Francis, S. G. Hildebrand, J. W. Huckabee, D. R. Jackson, S. E. Lindberg, F. H. Sweeton, R. I. Van Hook, and A. P. Watson, "4. Ecological Research" in Ecology and Analysis of Trace Contaminants. Progress Report October 1973 - September 1974, W. Fulkerson, W. D. Shults, and R. I. Van Hook, Eds., Oak Ridge National Laboratory, Oak Ridge, Tennessee, 1974, pp. 61-104.

Angino, E. E., and H. Schneider, Trace Element, Mineralogical, and Size Distributions in Suspended Material Samples from Selected Rivers in Eastern Kansas, PB-251164, National Technical Information Service, U.S. Department of Commerce, Springfield, Virginia, 1975.

Anonymous, "Additives with a Purpose," Lubrication, 43(3), 25-40 (1957).

Barufke, W., "Über die Bedeutung des Kobalts als Spurenelement auf brandenburgischen Grünlandböden" ["The Importance of Cobalt as a Micro-Element in Meadow Soils of Brandenberg District"], Wiss. Z. Humboldt-Univ. Berlin Math.-Naturw. Reihe, 11, 135-144 (1962).

Beneš, P., and E. Steinnes, "Migration Forms of Trace Elements in Natural Fresh Waters and the Effect of Water Storage," Water Res., 9, 741-749 (1975a).

Beneš, P., and E. Steinnes, "Determination of the State of Trace Elements in Water" in International Conference on Heavy Metals in the Environment, Abstracts of Papers, Institute of Environmental Studies, University of Toronto, Toronto, Ontario, Canada, October 27-31, 1975b, pp. D-8 to D-10.

Bertine, K. K., and E. D. Goldberg, "Fossil Fuel Combustion and the Major Sedimentary Cycle," Science, **173**(3993), 233-235 (1971).

Bittel, R., M. Merlini, C. Myttenaere, and P. Bourdeau, "Etude du comportement du radiocobalt dans un ecosysteme artificiel simple irrigue par submersion" ["Behavior of Radiocobalt in Simple Artificial Ecosystem Irrigated by Submersion"], Commis. Energ. At. [Rep.], CEA-CONF-1431, 1969, 27 pp.; Chem. Abstr., **74**, 121063j (1971).

Blood, E. R., and J. R. Reed, "Heavy Metals in a Lake Affected by Acid Mine Drainage" in International Conference on Heavy Metals in the Environment, Abstracts of Papers, Institute for Environmental Studies, University of Toronto, Toronto, Ontario, Canada, October 27-31, 1975, pp. C-162 to C-164.

Bolton, N. E., R. I. Van Hook, W. Fulkerson, W. S. Lyon, A. W. Andren, J. A. Carter, and J. F. Emery, Trace Element Measurements at the Coal-Fired Allen Steam Plant. Progress Report June 1971 - January 1973, ORNL-NSF-EP-43, Oak Ridge National Laboratory, Oak Ridge, Tennessee, March 1973.

Bolton, N. E., W. Fulkerson, R. I. Van Hook, W. S. Lyon, A. W. Andren, J. A. Carter, J. F. Emery, C. Feldman, L. D. Hulett, H. W. Dunn, C. J. Sparks, Jr., J. C. Ogle, and M. T. Mills, Trace Element Measurements at the Coal-Fired Allen Steam Plant. Progress Report February 1973 - July 1973, ORNL-NSF-EP-62, Oak Ridge National Laboratory, Oak Ridge, Tennessee, June 1974.

Bolton, N. E., J. A. Carter, J. F. Emery, C. Feldman, W. Fulkerson, L. D. Hulett, and W. S. Lyon, "Trace Element Mass Balance Around a Coal-Fired Steam Plant" in Trace Elements in Fuel, Advances in Chemistry Series No. 141, American Chemical Society, Washington, D.C., 1975, pp. 175-187.

Bovard, P., A. Grauby, A. Saas, and R. Schaeffer, "Incidence de la charge polluante des ea ux sur le comportement des radionuclides" ["Effect of the Pollutant Burden of Water on the Behavior of Radionuclides"] in Comb. Eff. Radioact., Chem. Therm. Releases Environ., Proc. Symp., International Atomic Energy Agency, Vienna, Austria, 1975, pp. 179-190.

Boyle, R. W., W. M. Tupper, J. Lynch, G. Friedrich, M. Ziauddin, M. Shafiqullah, M. Carter, and K. Bygrave, Geochemistry of Pb, Zn, Cu, As, Sb, Mo, Sn, W, Ag, Ni, Co, Cr, Ba, and Mn in the Waters and Stream Sediments of the Bathurst-Jacquet River District, New Brunswick, Geol. Survey Canada Paper 65-42, 1966, 49 pp.

Bradford, W. L., Distribution and Movement of Zinc and Other Heavy Metals in South San Francisco Bay California, PB 251 111, National Technical Information Service, U.S. Department of Commerce, Springfield, Virginia, February 1976.

Campbell, W. J., "Metals in the Wastes We Burn?", Environ. Sci. Technol., **10**(5), 436-439 (1976).

Carpenter, J. H., W. L. Bradford, and V. E. Grant, "Processes Affecting the Composition of Estuarine Waters (HCO_3, Fe, Mn, Zn, Cu, Ni, Cr, Co, Cd)" in Estuarine Research, 2nd International Estuarine Research Conference, Vol. I, Chemistry, Biology, and the Estuarine System, L. E. Cronin, Ed., Academic Press, Inc., New York, 1975, pp. 188-214.

Carpenter, R. H., G. D. Robinson, and W. B. Hayes, "Partitioning of Manganese, Iron, Copper, Zinc, Lead, Cobalt, and Nickel in Black Coatings on Stream Boulders in Vicinity of Magruder Mine, Lincoln County, Georgia," J. Geochem. Explor., **10**(1), 75-89 (1978).

Carlson, D. A., and O. J. Johansen, "Aerobic Treatment of Leachates from Sanitary Landfills," Proc. Natl. Congr. Waste Manage. Technol. Resour. Energy Recovery, 4th, 359-382 (1976).

Carrigan, P. H., Jr., Inventory of Radionuclides in Bottom Sediment of the Clinch River Eastern Tennessee, Geol. Surv. Prof. Pap. 433-I, U.S. Government Printing Office, Washington, D.C., 1969.

Chattopadhyay, A., and R. E. Jervis, "Multielement Determination in Market-Garden Soils by Instrumental Photon Activation Analysis," Anal. Chem., **46**(12), 1630-1639 (1974).

Chebotina, M. Ya., and N. V. Kulikov, "Influence of Water-Soluble Decomposition Products of Grassy Plants on the Absorption of Radioisotopes in the Soil," Sov. J. Ecol., **4**(1), 84-85 (1973).

Chebotina, M. Y., "Desorbtsiya Zheleza-59, Kobal'ta-60, Strontsiya-90 iz Pochvy Rastitel'nymi Ekstraktami v Usloviyakh Dinamicheskikh Opytov" ["Desorption of Iron-59, Cobalt-60, and Strontium-90 from Soil by Plant Extracts Under Dynamic Experiment Conditions"], Tr. Inst. Ekol. Rast. Zhivotn., Ural. Nauchn. Tsentr, Akad. Nauk SSR, **95**, 16-20 (1975).

Chester, R., "Adsorption of Zinc and Cobalt on Illite in Seawater," Nature, **206**, 884-886 (1965).

Clark, T. P., "Determination of Trace Element Levels in Landfill Leachate by Ion-Exchange, X-Ray Spectrography" in Trace Substances in Environmental Health - VII, D. D. Hemphill, Ed., Proc. of University of Missouri's 7th Annual Conference on Trace Substances in Environmental Health, June 12-14, 1973, University of Missouri, Columbia, Missouri, 1974, pp. 401-408.

Cline, J. T., and J. F. Holland, "Fluorometric Evidence for the Reaction of Heavy Metals with Pore-Water Organics" in Biological Implications of Metals in the Environment, CONF-750929, Proc. 15th Ann. Hanford Life Sci. Symp., Richland, Washington, September 29 - October 1, 1975, Technical Information Center, Energy Research and Development Administration, 1977, pp. 264-279.

Colston, N. V., Jr., Characterization and Treatment of Urban Land Runoff, EPA-670/2-74-096, PB 240 987/8GI, National Technical Information Service, U.S. Department of Commerce, Springfield, Virginia, 1975.

Cordero, H. G., and T. J. Tarring, Eds., Non-Ferrous Metal Works of the World, 1st ed., Metal Bulletin Books, Ltd., London, 1967.

Cuffe, S. T., and R. W. Gerstle, Emissions from Coal-Fired Power Plants: A Comprehensive Summary, U.S. Department of Health, Education, and Welfare, U.S. Government Printing Office, Washington, D.C., 1967.

Davison, R. L., D. F. S. Natusch, J. R. Wallace, and C. A. Evans, Jr., "Trace Elements in Fly Ash. Dependence of Concentration on Particle Size," Environ. Sci. Technol., **8**(13), 1107-1113 (1974).

DeBortoli, M., and P. Gaglione, "Study of the Dispersion of Radioactive Effluents into a Lake. I. Contamination of the Bottom Sediments," Minerva Fisiconucl., **14**(4), 190-196 (1970).

DeGroot, A. J., E. Allersma, N. DeBruin, and J. P. W. Houtman, "Use of Activatable Tracers," Tech. Rep. Ser., I.A.E.A., **145**, 151-166 (1973).

Dement'ev, V. S., and L. V. Nazarova, "Ob Izuchenie Otnositel'noi Podvizhnosti Metallov v Vode s Pomoshch'yu Dializa" ["Study of the Relative Mobility of Metals in Water Using Dialysis"], Izv. Akad. Nauk Kaz. SSR, Ser. Khim., **27**(2), 74-77 (1977).

Dettmann, E. H., R. D. Olsen, and W. S. Vinikour, "Effects of Coal Strip Mining on Stream Water Quality: Preliminary Results," Pap. Symp. Coal Mine Drain. Res., **6**, 53-63 (1976).

Doshi, G. R., T. M. Krishnamoorthy, V. N. Sastry, and T. P. Sarma, "Sorption Behavior of Trace Nuclides in Sea Water on Manganese Dioxide," Indian J. Chem., **11**(2), 158-161 (1973).

Dunigan, E. P., and C. W. Francis, "Adsorption and Desorption of ^{60}Co, ^{85}Sr, and ^{137}Cs on Soil Humic Acid," Soil Sci., **114**(6), 494-496 (1972).

Duursma, E. K., "Organic Chelation of ^{60}Co and ^{65}Zn by Leucine in Relation to Sorption by Sediments" in Symposium on Organic Matter in Natural Waters, September 2-4, 1968, University of Alaska, D. W. Hood, Ed., Institute of Marine Science, University of Alaska, College, Alaska, 1970, pp. 387-397.

El-Barbary, I., B. McDuffie, R. D. Tiberio, and G. J. Hollod, "Trace Metal Distribution and Speciation in the Susquehanna River Ecosystem in New York," ACS Div. Environ. Chem. Prepr., **17**(1), 285-288 (1977).

Emel'yanov, E. M., and O. S. Pustel'nikov, "Kolichestvo Vzveshennykh Form Elementov (C_{org}, $SiO_{2_{amorf}}$, Fe, Al, Ti, Mn, Ni, Cu, Co) v Vodakh Baltiiskogo Morya" ["Amount of Suspended Forms of Elements (Organocarbon, Amorphous Silica, Iron, Aluminum, Titanium, Manganese, Nickel, Copper, Cobalt) in the Waters of the Baltic Sea"], Geokhimiya, **1975**(7), 1049-1063 (1975).

Emmett, W. W., Channels and Waters of the Upper Salmon River Area, Idaho, U.S. Geol. Surv., Prof. Pap. No. 870-A, U.S. Government Printing Office, Washington, D.C., 1975.

Epstein, E., and R. L. Chaney, "Land Disposal of Toxic Substances and Water-Related Problems," J. Water Pollut. Control Fed., **50**(8), 2037-2042 (1978).

Filby, R. H., K. R. Shah, and W. H. Funk, "Role of Neutron Activation Analysis in the Study of Heavy Metal Pollution of a Lake-River System" in Proc. Int. Conf. Nucl. Methods Environ. Res., 2nd, CONF-740701, J. R. Vogt, and W. Meyer, Eds., National Technical Information Service, U.S. Department of Commerce, Springfield, Virginia, 1974, pp. 10-23.

Fisher, G. L., D. Silberman, B. A. Prentice, R. E. Heft, and J. M. Ondov, "Filtration Studies with Neutron-Activated Coal Fly Ash," Environ. Sci. Technol., **13**(6), 689-693 (1979).

Fitchko, J., and T. C. Hutchinson, "A Comparative Study of Heavy Metal Concentrations in River Mouth Sediments Around the Great Lakes," J. Great Lakes Res., **1**(1), 46-78 (1975).

Gang, M. W., and D. Langmuir, "Controls on Heavy Metals in Surface and Ground Waters Affected by Coal Mine Drainage. Clarion River-Redbank Creek Watershed, Pennsylvania," Pap. Symp. Coal Mine Drain. Res., **5**, 39-69 (1974).

Geladi, P., F. Adams, and W. Eylenbosch, "Evaluation of Air Pollution Hazards in the Work Environment of Some Belgian Factories" in Int. Symp. Control Air Pollut. Work. Environ., [Proc.] 1977, Part 1, Arbetarskyddsfonden, Stockholm, Sweden, 1978, pp. 405-416.

Gibbs, R. J., "Mechanisms of Trace Metal Transport in Rivers," Science, 180, 71-73 (1973).

Gladney, E. S., G. E. Gordon, and W. H. Zoller, "Coal Combustion: Source of Toxic Elements in Urban Air?" J. Environ. Sci. Health, Part A, A13(7), 481-491 (1978).

Goldberg, E. D., "The Oceans as a Chemical System" in The Sea. Ideas and Observations on Progress in the Study of the Seas, Vol. 2, M. N. Hill, E. D. Goldberg, C. O'D. Iselin, and W. H. Munk, Eds., Wiley, New York, New York, 1963, pp. 3-25.

Goodman, G. T., and T. M. Roberts, "Plants and Soils as Indicators of Metals in the Air," Nature, 231, 287-292 (1971).

Goodman, G., and S. Smith, "Relative Burdens of Airborne Metals in South Wales" in Report of a Collaborative Study on Certain Elements in Air, Soil, Plants, Animals and Humans in the Swansea/Neath/Port Talbot Area, Together with a Report on a Moss Bag Study of Atmospheric Pollution Across South Wales, Welsh Office, Cardiff, Wales, 1975, pp. 333-365.

Goodman, G. T., S. Smith, and M. J. Inskip, "Moss Bags as Indicators of Airborne Metals--an Evaluation" in Report of a Collaborative Study on Certain Elements in Air, Soil, Plants, Animals and Humans in the Swansea/Neath/Port Talbot Area, Together with a Report on a Moss Bag Study of Atmospheric Pollution Across South Wales, Welsh Office, Cardiff, Wales, 1975, pp. 267-332.

Grady, W. C., and D. J. Akers, Jr., "Utilization of Acid Mines Drainage Treatment Sludge" in Proc. of the Fifth Mineral Waste Utilization Symposium, E. Aleshin, Ed., Cosponsored by the U.S. Bureau of Mines and IIT Research Institute, Chicago, Illinois, April 13-14, 1976, pp. 114-121.

Greenberg, R. R., W. H. Zoller, and G. E. Gordon, "Composition and Size Distributions of Particles Released in Refuse Incineration," Environ. Sci. Technol., 12(5), 566-573 (1978a).

Greenberg, R. R., G. E. Gordon, W. H. Zoller, R. B. Jacko, D. W. Nuendorf, and K. J. Yost, "Composition of Particles Emitted from the Nicosia Municipal Incinerator," Environ. Sci. Technol., 12, 1329-1332 (1978b).

Halbach, P., E. Rehm, and V. Marchig, "Distribution of Si, Mn, Fe, Ni, Cu, Co, Zn, Pb, Mg, and Ca in Grain-Size Fractions of Sediment Samples from a Manganese Nodule Field in the Central Pacific Ocean," Mar. Geol., 29, 237-252 (1979).

Hamilton, E. I., "The Chemical Elements and Human Morbidity--Water, Air, and Places--A Study of Natural Variability," Sci. Total Environ., 3(1), 3-85 (1974).

Hanck, K. W., and J. W. Dillard, "Determination of the Complexing Capacity of Natural Water by Cobalt(III) Complexation," Anal. Chem., 49(3), 404-409 (1977).

Harper, W. B., "Petroleum" in Commodity Data Summaries 1977, Bureau of Mines, U.S. Department of the Interior, Washington, D.C., 1977, p. 122.

Harper, W. B., B. C. Michalski, and M. R. Parker, "Crude Petroleum and Petroleum Products" in Minerals Yearbook 1975, Vol. I, Metals, Minerals, and Fuels, Bureau of Mines, U.S. Department of the Interior, Washington, D.C., 1977, pp. 1041-1162.

Hathaway, L. R., and O. K. Galle, "Preliminary Investigations of Trace Element Mobilization from the Heebner Shale by Pure Brine Solutions" in Trace Substances in Environmental Health-XII, D. D. Hemphill, Ed., Proc. of University of Missouri's 12th Annual Conference on Trace Substances in Environmental Health, June 6-8, 1978, University of Missouri, Columbia, Missouri, 1979, pp. 99-104.

Heft, R. E., W. A. Phillips, H. R. Ralston, and W. A. Steele, "Radionuclide Transport Studies in the Humboldt Bay Marine Environment" in Radioactive Contamination of the Marine Environment, Proc. Symp. 1972, International Atomic Energy Agency, Vienna, Austria, 1973, pp. 595-614.

Heindryckx, R., "Comparison of the Mass-Size Functions of the Elements in the Aerosols of the Ghent Industrial District with Data from Other Areas. Some Physico-Chemical Implications," Atmos. Environ., 10(1), 65-71 (1976); Chem. Abstr., 84, 94866w (1976).

Helz, G. R., "Trace Element Inventory for the Northern Chesapeake Bay with Emphasis on the Influence of Man," Geochim. Cosmochim. Acta., 40, 573-580 (1976).

Hem, J. D., "Redox Processes at Surfaces of Manganese Oxide and Their Effects on Aqueous Metal Ions," Chem. Geol., 21 (3-4), 199-218 (1978).

Hill, R. D., "Control and Prevention of Mine Drainage" in Cycling and Control of Metals, Proc. of an Environmental Resources Conference, Columbus, Ohio, October 31 - November 2, 1972, U.S. Environmental Protection Agency, National Environmental Research Center, Cincinnati, Ohio, February 1973, pp. 91-94.

Hodgson, J. F., "Cobalt Reactions with Montmorillonite," Soil Sci. Soc. Am. Proc., 24, 165-168 (1960).

Hodgson, J. F., and K. G. Tiller, "The Location of Bound Cobalt on 2:1 Layer Silicates," Clays Clay Minerals, Proc. Natl. Conf. Clays Clay Minerals, 9, 404-411 (1962).

Hodgson, J. F., H. R. Geering, and W. A. Norvell, "Micronutrient Cation Complexes in Soil Solution: Partition Between Complexed and Uncomplexed Forms by Solvent Extraction," Soil Sci. Soc. Am. Proc., 29, 665-669 (1965).

Hodgson, J. F., K. G. Tiller, and M. Fellows, "The Role of Hydrolysis in the Reaction of Heavy Metals with Soil-Forming Materials," Soil Sci. Soc. Am. Proc., 28(1), 42-46 (1964).

Hodgson, J. F., G. Tiller, and M. Fellows, "Effect of Iron Removal on Cobalt Sorption by Clays," Soil Sci., 108(6), 391-396 (1969).

Hoffstader, R. A., O. I. Milner, and J. H. Runnels, Eds., Analysis of Petroleum for Trace Metals, Advances in Chemistry Series No. 156, American Chemical Society, Washington, D.C., 1976.

Hollyday, E. F., and S. W. McKenzie, "Hydrology of the Formation and Neutralization of Acid Waters Draining from Underground Coal Mines of Western Maryland," Maryland, Geol. Surv., Rep. Invest., No. 20, 1-50 (1973).

Horvath, G. J., R. C. Harriss, and H. C. Mattraw, "Land Development and Heavy Metal Distribution in the Florida Everglades," Mar. Pollut. Bull., 3(12), 182-183 (1972).

Huljev, D., and P. Strohal, "Neutron Activation Analysis for Studying the Role of Humic Acids During Transport of Trace Elements in the Marine Biocycle" in Nuclear Activation Techniques in the Life Sciences, from the Symposium on Nuclear Activation Techniques in Life Sciences, Bled, Yugoslavia, April 10, 1972, CONF-720425, International Atomic Energy Agency, Vienna, Austria, 1972, pp. 385-390.

Ijuin, M., P. Picat, A. Saas, and A. Grauby, "Determination of the Diffusion Coefficient of Radioelements in the Rhone Sediments," Health Physics, 24(6), 665-672 (1973).

Ivanova, A. A., and G. S. Konovalov, "Discharge of Dispersed and Rare Elements (Microelements) in Solution and in Suspended Substances from the Territory of the USSR," Transl. from Gidrokhimicheskiye Materialy, 44, 63-82 (1968); Soviet Hydrology: Selected Papers, No. 4, 336-353 (1968).

Jackson, K. S., I. R. Jonasson, and G. B. Skippen, "Nature of Metals-Sediment-Water Interactions in Freshwater Bodies, with Emphasis on Role of Organic Matter," Earth Sci. Rev., 14(2), 97-146 (1978).

James, R. O., and T. W. Healy, "Adsorption of Hydrolyzable Metal Ions at the Oxide-Water Interface. I. Co(II) Adsorption on SiO_2 and TiO_2 as Model Systems," J. Colloid Interface Sci., 40, 42-52 (1972).

Jenne, E. A., "Controls on Mn, Fe, Co, Ni, Cu, and Zn Concentrations in Soils and Water: The Significant Role of Hydrous Mn and Fe Oxides" in Trace Inorganics in Water, R. A. Baker, Ed., Advances in Chemistry Series No. 73, American Chemical Society, Washington, D.C., 1968, pp. 337-381.

Jenne, E. A., and J. S. Wahlberg, Role of Certain Stream - Sediment Components in Radioion Sorption, Geol. Surv. Prof. Pap. 433-F, U.S. Government Printing Office, Washington, D.C., 1968.

Jungers, R. H., R. E. Lee, Jr., and D. J. von Lehmden, "The EPA National Fuels Surveillance Network. 1. Trace Constituents in Gasoline and Commercial Gasoline Fuel Additives," Environ. Health Perspect., 10, 143-150 (1975).

Keilen, K., K. Stahr, and H. W. Zoettl, "'Mobile Fraktionen' von Spurenelementen (Be, Cu, Cd, Co, Pb, V, Zn) in Boden des Barhaldegranitgebietes" ["'Mobile Fractions' of Trace Elements (Be, Cu, Cd, Co, Pb, V, Zn) in Soils of the Barhalde-Granite Area"], Z. Pflanzenernaehr. Bodenkd., 141(5), 583-596 (1978).

Kempf, T., "Zur Frage der Remobilisierung von Kobalt und Chrom" ["Remobilization of Cobalt and Chromium"], Z. Wasser Abwasser Forsch., 9(6), 176-178 (1976).

Kharkar, D. P., K. K. Turekian, and K. K. Bertine, "Stream Supply of Dissolved Silver, Molybdenum, Antimony, Selenium, Chromium, Cobalt, Rubidium, and Cesium to the Oceans," Geochim. Cosmochim. Acta, 32, 285-298 (1968).

Khodzhaeva, L. I., V. N. Kurovich, and M. M. Videnskaya, "K Voprosu o Sorbtsii Izotopa Co^{60} pri Popadanie v Grunt" ["Sorption of Cobalt-60 in the Case of Entry into the Ground"], Gig. Sanit., 38(10), 106-107 (1973).

King, R. B., J. S. Fordyce, A. C. Antoine, H. F. Leibecki, H. E. Neustadter, and S. M. Sidik, Extensive 1-Year Survey of Trace Elements and Compounds in the Airborne Suspended Particulate Matter in Cleveland, Ohio, NASA-TN D-8110, Washington, D.C., 1976.

Klechkovskii, V. M., "O Fraktsiyakh Ful'vokislot i Vzaimodeistvii ikh s Metallami" ["Fraction of Fulvic Acids and Their Interaction with Metals"], Dokl. vses. Akad. Sel'skokhoz. Nauk, No. 8, 22-23 (1968).

Klein, D. H., and P. Russell, "Heavy Metals: Fallout Around a Power Plant," Environ. Sci. Technol., 7(4), 357-358 (1973).

Klein, D. H., A. W. Andren, J. A. Carter, J. F. Emery, C. Feldman, W. Fulkerson, W. S. Lyon, J. C. Ogle, Y. A. Talmi, R. I. Van Hook, and N. E. Bolton, "Trace Element Measurements at the Coal-Fired Allen Steam Plant - Mass Balance and Concentrations in Fly Ash" in Preprints of Papers Presented at the 169th National Meeting before the Division of Environmental Chemistry, American Chemical Society, Philadelphia, Pennsylvania, April 6-10, 1975a, pp. 134-135.

Klein, D. H., A. W. Andren, J. A. Carter, J. F. Emergy, C. Feldman, W. Fulkerson, W. S. Lyon, J. C. Ogle, Y. Talmi, R. I. Van Hook, and N. Bolton, "Pathways of Thirty-Seven Trace Elements Through Coal-Fired Power Plant," Environ. Sci. Technol., **9**(10), 973-979 (1975b).

Konovalov, G. S., A. A. Ivanova, and T. Kh. Kolesnikov, "Trace and Rare Elements Dissolved in Water and Present in Suspended Substances of the Main Rivers of the USSR," Geokhim. Osad. Porod, 72-87 (1968); Chem. Abstr., **71**, 222 (1969) from Ref. Zh., Geol., **V**, Abstr. No. 11V016 (1968).

Kopp, J. F., and R. C. Kroner, "A Comparison of Trace Elements in Natural Waters, Dissolved Versus Suspended," Develop. Appl. Spectrosc., **6**, 339-352 (1968).

Korte, N. E., J. Skopp, E. E. Niebla, and W. H. Fuller, "A Baseline Study on Trace Metal Elution from Diverse Soil Types," Water Air Soil Pollut., **5**, 149-156 (1975).

Kowalczyk, G. S., C. E. Choquette, and G. E. Gordon, "Chemical Element Balances and Identification of Air Pollution Sources in Washington, D.C.," Atmos. Environ., **12**(5), 1143-1153 (1978).

Krauskopf, K. B., "Factors Controlling the Concentrations of Thirteen Rare Metals in Seawater," Geochim. Cosmochim. Acta, **9**, 1-32B (1956).

Krishnaswami, S., "Authigenic Transition Elements in Pacific Pelagic Clays," Geochim. Cosmochim. Acta, **40**, 425-434 (1976).

Kroner, R. C., and J. F. Kopp, "A Spectrographic Study of Trace Metals in Acid Mine Drainage," Anal. Instrum. 1966, **4**, 91-99 (1967).

Kroon, T. P., and A. De Grys, "A Geochemical Drainage Survey in Central Ecuador," Econ. Geol., **65**(5), 557-563 (1970).

Kubota, J., "Cobalt Status of Soils of Southeastern United States. I. Cobalt, Its Distribution and Relationship to Iron and Clay in Five Selected Soils," Soil Sci., **85**(4), 130-140 (1958).

Lachet, B., "Étude cinétique, au laboratoire, de l'épuration d'une eau de riviere (Isère) par des sediments en suspension. Cas des radionucléides ^{51}Cr, ^{59}Fe, ^{60}Co, ^{85}Sr, ^{106}Ru, ^{137}Cs, ^{141}Ce, ^{203}Hg" ["Laboratory Studies of the Purification Kinetics of River Water (the Isére) by Sedimentary Suspensions. Case of ^{61}Cr, ^{59}Fe, ^{60}Co, ^{85}Sr, ^{106}Ru, ^{137}Cs, ^{141}Ce, and ^{203}Hg"], Radioprotection, **7**(3), 143-157 (1972).

Law, S. L., and G. E. Gordon, "Sources of Metals in Municipal Incinerator Emissions," Environ. Sci. Technol., **13**(4), 432-438 (1979).

Leckie, J. O., and R. O. James, "Control Mechanisms for Trace Metals in Natural Waters" in Aqueous-Environmental Chemistry of Metals, A. J. Rubin, Ed., Ann Arbor Science Publishers, Inc., Ann Arbor, Michigan, pp. 1-76.

Leeper, G. W., Managing the Heavy Metals on the Land, Marcel Dekker, Inc., New York, New York, 1978.

Leland, H. V., "Distribution of Solute and Particulate Trace Elements in Southern Lake Michigan" in International Conference on Heavy Metals in the Environment, October 27-31, 1975, Symposium Proceedings, Vol. 2, Part 2, Institute for Environmental Studies, University of Toronto, Toronto, Ontario, Canada, 1977, pp. 709-730.

Levy, A., S. E. Miller, R. E. Barrett, E. J. Schulz, R. H. Melvin, W. H. Axtman, and D. W. Locklin, A Field Investigation of Emissions from Fuel Oil Combustion for Space Heating, American Petroleum Institute Project SS-5, Battelle Columbus Laboratories, Columbus, Ohio, 1971.

Lieser, K. H., W. Calmano, E. Heuss, and V. Neitzert, "Neutron Activation as a Routine Method for the Determination of Trace Elements in Water," J. Radioanal. Chem., 37, 717-726 (1977).

Lindberg, S. E., A. W. Andren, R. J. Raridon, and W. Fulkerson, "Mass Balance of Trace Elements in Walker Branch Watershed: Relation to Coal-Fired Steam Plants," Environ. Health Perspect., 12, 9-18 (1975).

Lockhart, H. B., and R. V. Blakeley, Jr., "Aerobic Photodegradation of X(N) Chelates of (Ethylenedinitrilo)-tetraacetic Acid [EDTA]: Implications for Natural Waters," Environ. Lett., 9(1), 19-31 (1975).

Loganathan, P., and R. G. Burau, "Sorption of Heavy Metal Ions by a Hydrous Manganese Oxide," Geochim. Cosmochim. Acta, 37, 1277-1293 (1973).

Lopez, P. L., and E. R. Graham, "Labile Pool and Plant Uptake of Micronutrients: Determination of Labile Pool of Mn, Fe, Zn, Co, and Cu in Deficient Soils by Isotopic Exchange," Soil Sci., 114(4), 295-299 (1972).

Lowman, F. G., and R. Y. Ting, "The State of Cobalt in Seawater and Its Uptake by Marine Organisms and Sediments" in Radioactive Contamination of the Marine Environment, Proc. Symp. IAEA, Seattle, Washington, July 10-14, 1972, International Atomic Energy Agency, Vienna, Austria, 1973, pp. 369-383.

Lyon, W. S., N. E. Bolton, J. F. Emery, and D. H. Klein, "Instrumental Neutron Activation Analysis of Effluents from Coal Fired Power Plants" in Proc. Int. Conf. Nucl. Methods Environ. Res., 2nd, CONF-740701, J. R. Vogt and W. Meyers, Eds., National Technical Information Service, U.S. Department of Commerce, Springfield, Virginia, 1974, pp. 376-384.

Manahan, S. E., "The Environmental Chemistry of Humic Materials as Related to Trace Substances: Results of a Computerized WRSIC Bibliographic Literature Search," Presented at the 9th Annual Conference on Trace Substances in Environmental Health, University of Missouri, Columbia, Missouri, June 10-12, 1975.

Marchand, M., "Considérations sur les formes physico-chimiques du cobalt, manganèse, zinc, chrome et fer dans une eau de mer enrichie ou non de materière organique" ["Physical-Chemical Forms of Cobalt, Manganese, Zinc, Chromium, and Iron in Seawater Enriched or Non-Enriched in Organic Matter"], J. Conseil. Intern. Exploration Mer., **35**(2), 130-142 (1974).

Marr, H. E., S. L. Law, and D. L. Neylan, "Trace Metals in the Combustible Fraction of Urban Refuse" in International Conference on Environmental Sensing and Assessment, Proceedings of a Conference, Las Vegas, Nevada, September 14-15, 1975, Vol. 1, Paper No. 4-3, Institute of Electrical and Electronics Engineers, Piscataway, New Jersey, 1976, pp. 1-6.

Martin, G. B., D. W. Pershing, and E. E. Berkau, Effects of Fuel Additives on Air Pollutant Emissions from Distillate-Oil-Fired Furnaces, U.S. Environmental Protection Agency, Office of Air Programs, U.S. Government Printing Office, Washington, D.C., 1971.

McKenzie, R. M., "The Sorption of Cobalt by Manganese Minerals in Soils," Aust. J. Soil Res., **5**, 235-246 (1967).

Means, J. L., D. A. Crerar, and J. O. Duguid, "Migration of Radioactive Wastes: Radionuclide Mobilization by Complexing Agents," Science, **200**, 1477-1481 (1978).

Meeussen, M. A., G. J. Willems, and C. J. DeRanter, "Sorption Capacity Studies of Scheldt Sediments for ^{60}Co, ^{85}Sr and ^{137}Cs," J. Pharm. Belg., **30**(1), 82-90 (1975).

Meisch, H. U., W. Reinle, and H. J. Bielig, "Schwermetalle in Sedimenten der Saar" ["Heavy Metals in Saar River Sediments"], Dtsch. Gewaesserkd. Mitt., **22**(1), 2-8 (1978).

Miller, D. W., O. C. Braids, and W. H. Walker, The Prevalence of Subsurface Migration of Hazardous Chemical Substances at Selected Industrial Waste Land Disposal Sites, PB-272973, National Technical Information Service, U.S. Department of Commerce, Springfield, Virginia, 1977, 529 pp.

Mitchell, R. L., "Chapter 9. Trace Elements" in Chemistry of the Soil, F. E. Bear, Ed.-in-Chief, Reinhold Publishing Corporation, New York, New York, 1955, pp. 253-285.

Mortvedt, J. J., and H. G. Cunningham, "Production, Marketing, and Use of Other Secondary and Micronutrient Fertilizers" in Fertilizer Technology and Use, R. A. Olson, Ed.-in-Chief, Soil Science Society of America, Inc., Madison, Wisconsin, 1971, pp. 413-454.

Mukherjee, D. C., A. N. Basu, and M. Adhikari, "Interaction between Humic Acid Fraction of Soil and Trace Element Cations, viz. Mn^{2+}, Co^{2+} and Ni^{2+} with Wider Ratios of Mg^{2+} as a Complementary Ion," Indian J. Appl. Chem., **35**, 1-6 (1973).

Murray, C. N., and S. Meinke, "Influence of Soluble Sewage Materials on Adsorption and Desorption Behaviors of Cadmium, Cobalt, Silver, and Zinc in Sediment-Fresh Water, Sediment-Sea Water Systems," Nippon Kaiyo Gakkai-Shi, **30**(5), 216-221 (1974).

Murray, C. N., and L. Murray, "Adsorption-Desorption Equilibria of Some Radionuclides in Sediment-Fresh-Water and Sediment-Seawater Systems" in Proceedings of Radioactive Contamination of the Marine Environment Symposium, Intern. Atomic Energy Agency, Seattle, Washington, 1973, pp. 105-124.

Musin, G. Kh., "Vliyanie Zol'nykh Otkhodov Teplovoi Elektrostantsii na Sanitarnoe Sostoyanie Pochvy" ["Effect of Ash Wastes from Thermal Power Plants on the Health Status of the Soil"], Gig. Sanit., **10**, 116-117 (1978).

Natusch, D. F. S., C. F. Bauer, H. Matusiewicz, C. A. Evans, J. Baker, A. Loh, R. W. Linton, and P. K. Hopke, "Characterization of Trace Elements in Fly Ash" in International Conference on Heavy Metals in the Environment, October 27-31, 1975, Symposium Proceedings, Vol. 2, Part 2, Institute for Environmental Studies, University of Toronto, Toronto, Ontario, Canada, 1977, pp. 553-575.

Nazarov, I. M., O. S. Renne, Sh. D. Fridman, L. G. Shapovalova, and E. P. Makhon'ko, "Snezhyi Pokrov kak Indikator Zagryazneniya Atmosfery" ["Snow Cover as an Indicator of Atmospheric Pollution"], Fiz. Aspekty Zagryaz. Atmos., Tezisy Dokl., Mezhdunar. Konf., 115-116 (1974).

Nelson, J. L., R. W. Perkins, J. M. Nielson, and W. L. Haushild, "Reactions of Radionuclides from the Hanford Reactors with Columbia River Sediments" in Disposal of Radioactive Wastes, Seas, Oceans, Surface Waters, Proceedings of a Symposium in Vienna, 1966, pp. 139-161.

Nelson, W. E., Final Report on Fate of Trace-Metals (Impurities) in Subsoils as Related to the Quality of Ground Water, OWRR Project B-028 ALA, Carver Research Foundation, Tuskegee Institute, Tuskegee, Alabama, 1972.

Nielsen, J. M., Pacific Northwest Laboratory Annual Report for 1972 to the USAEC Division of Biomedical and Environmental Research. Vol. II. Physical Sciences. Part 2. Radiological Sciences, Battelle Pacific Northwest Laboratories, Richland, Washington, 1973; BNWL-1751 PT2 UC-48, National Technical Information Service, Springfield, Virginia.

Nishita, H., and R. M. Haug, "Water and Ammonium Acetate-Extractable Zn, Mn, Cu, Cr, and Fe in Heated Soils," Soil Sci., **118**(6), 421-424 (1974).

Nix, J., Distribution of Trace Elements in Impoundments, PB-198 089, National Technical Information Service, U.S. Department of the Interior, Springfield, Virginia, January 1970.

O'Connor, T. P., and D. R. Kester, "Adsorption of Copper and Cobalt from Fresh and Marine Systems," Geochem. Cosmochim. Acta, 39(11), 1531-1543 (1975).

Ondov, J. M., R. C. Ragaini, and A. H. Biermann, "Elemental Particle-Size Emissions from Coal-Fired Power Plants; Use of an Inertial Cascade Impactor," Atmos. Environ., 12(5), 1175-1185 (1978).

Ondov, J. M., R. C. Ragaini, and A. H. Biermann, "Elemental Emissions from a Coal-Fired Power Plant. Comparison of a Venturi Wet Scrubber System with a Cold-Side Electrostatic Precipitator," Environ. Sci. Technol., 13(5), 598-607 (1979).

Orlov, D. S., and N. L. Eroshicheva, "K Voprosu o Vzaimodeistrii Guminovykh Kislot s Kationami Nekotorykh Metallov" ["The Interaction of Humic Acids with Cations of Some Metals"], Vost. mosk. gos. Univ. Sor. Biol. Pochv., 1, 98-106 (1967).

Orlova, E. I., V. A. Smirennaya, and R. A. Chelysheva, "Migratsiya Co^{60} v Rykhlykh Gornykh Porodakh pri Zagryaznenii ikh Netekhnologicheskimi Stokami Atomnykh Elektrostantsii (AES)" ["Migration of Cobalt-60 in Loose Rocks Contaminated with Non-technological Effluents of an Atomic Electric Power Station (AES)"], Gig. Sanit., No. 3, 50-53 (1974).

Patrick, W. H., Jr., and D. S. Mikkelsen, "Plant Behavior in Flooded Soil" in Fertilizer Technology and Use, R. A. Olson, Ed.-in-Chief, Soil Science Society of America, Inc., Madison, Wisconsin, 1971, pp. 187-216.

Pattenden, N. J., "The Measurement of Heavy Metals in the Atmosphere and Their Interpretation," National Society for Clean Air. Proceedings of the Annual Conference, 41(1), 1-22 (1974).

Pattenden, N. J., "A Study of Atmospheric Pollution by Trace Elements in the Swansea Area," Clean Air (UK), 7(27), 15-19 (1977).

Pattenden, N. J., "Atmospheric Concentrations and Deposition Rates of Some Trace Elements Measured in the Swansen/Neath/ Port Talbot Area" in Report of a Collaborative Study on Certain Elements in Air, Soil, Plants, Animals and Humans in the Swansea/Neath/Port Talbot Area, Together with a Report on a Moss Bag Study of Atmospheric Pollution Across South Wales, Welsh Office, Cardiff, Wales, 1975.

Perhac, R. M., "Distribution of Cd, Co, Cu, Fe, Mn, Ni, Pb, and Zn in Dissolved and Particulate Solids from Two Streams in Tennessee," J. Hydrol., **15**, 177-186 (1972).

Perhac, R. M., and C. J. Whelan, "A Comparison of Water, Suspended Solid and Bottom Sediment Analyses for Geochemical Prospecting in a Northeast Tennessee Zinc District," J. Geochem. Explor., **1**, 47-53 (1972).

Perone, S. P., M. Pichler, P. Gaarenstroom, and J. L. Moyers, "The Application of Pattern Recognition Techniques to the Characterization of Atmospheric Aerosols" in International Conference on Environmental Sensing and Assessment, Proceedings of a Conference, Las Vegas, Nevada, September 14-15, 1975, Vol. 1, Paper No. 5-4, Institute of Electrical and Electronics Engineers, Piscataway, New Jersey, 1976, pp. 1-4.

Pickering, R. J., Distribution of Radionuclides in Bottom Sediment of the Clinch River Eastern Tennessee, U.S. Geol. Surv. Prof. Pap. 433-H, U.S. Government Printing Office, Washington, D.C., 1969.

Pickering, R. J., Composition of Water in Clinch River, Tennessee River, and Whiteoak Creek as Related to Disposal of Low Level Radioactive Liquid Wastes, U.S. Geol. Surv. Prof. Pap. 433-J, U.S. Government Printing Office, Washington, D.C., 1970.

Popp, C. J., R. W. Smith, and F. Laquer, "Sediment Composition, Trace Metal Transport and Mineral Equilibrium in the Rio Grande in New Mexico," ACS Div. Environ. Chem. Prepr., **19**(1), 719-721 (1979).

Proctor, P. D., and B. Sinha, "Cobalt-Nickel-Cadmium Mobilization, Transportation and Fixation in Surface Waters, Stream Sediments and Selected Aquatic Life in the Fredericktown Co-Ni Metallogenic Province, Southeast Missouri" in Trace Substances in Environmental Health-XII, D. D. Hemphill, Ed., Proceedings of University of Missouri's 12th Annual Conference on Trace Substances in Environmental Health, June 6-8, 1978, University of Missouri, Columbia, Missouri, 1979, pp. 119-128.

Pronina, N. V., and I. M. Varentso, "O Spetsifike Pogloshcheniya Nikelya i Kobal'ta iz Morskoi Vody Prirodnymi Gidrookislami Zheleza i Margantsa" ["Specific Character of Nickel and Cobalt Absorption from Seawater by Natural Iron Hydroxides and Manganese Hydroxides"], Dokl. Akad. Nauk S.S.S.R., **210**(4), 944-947 (1973).

Radosavljević, R., T. Tasovac, R. Drašković, M. Zarić, and V. Marković, "Complex Behaviour of Cobalt in the Danube River," Arch. Hydrobiol., Suppl., **44**(2), 241-248 (1973).

Ragaini, R. C., and J. M. Ondov, "Trace Contaminants from Coal-Fired Power Plants" in International Conference on Environment Sensing and Assessment, Proceedings of a Conference, Las Vegas, Nevada, September 14-15, 1975, Vol. 1, Paper No. 17-2, Institute of Electrical and Electronics Engineers, Piscataway, New Jersey, 1976, pp. 1-4.

Ragaini, R. C., and J. M. Ondov, "Trace-Element Emissions from Western U.S. Coal-Fired Power Plants," J. Radioanal. Chem., **37**, 679-691 (1977).

Ragaini, R. C., H. R. Ralston, and N. Roberts, "Environmental Trace Metal Contamination in Kellogg, Idaho, near a Lead Smelting Complex," Environ. Sci. Technol., **11**(8), 773-781 (1977).

Rashid, M. A., and J. D. Leonard, "Modifications in the Solubility and Precipitation Behaviour of Various Metals as a Result of Their Interaction with Sedimentary Humic Acid," Chem. Geol., **11**, 89-97 (1973).

Robertson, D. E., W. B. Silker, M. R. Petersen, and J. C. Langford, "Radionuclide Decline in Columbia River Water. Transport and Depletion of Radionuclides in the Columbia River" in Pacific Northwest Laboratory Annual Report for 1972 to the USAEC Division of Biomedical and Environmental Research. Vol. II. Physical Sciences. Part 2. Radiological Sciences, Battelle Pacific Northwest Laboratories, Richland, Washington, 1973a; BNWL-1751 PT2 UC-48, National Technical Information Service, Springfield, Virginia.

Robertson, D. E., W. B. Silker, J. C. Langford, M. R. Peterson, and R. W. Perkins, "Transport and Depletion of Radionuclides in the Columbia River" in Radioactive Contamination of the Marine Environment, International Atomic Energy Agency, Vienna, Austria, 1973b.

Routson, R. C., Review of Studies on Soil-Waste Relations on the Hanford Reservation from 1944-1967, BNWL-1464, National Technical Information Service, U.S. Department of Commerce, Springfield, Virginia, 1973, 63 pp.; Chem. Abstr., **79**, 107896e (1973).

Sargent, J. K., and R. D. Ross, "Chapter 7. Disposal of Oily Wastes," Industrial Oily Waste Control, American Petroleum Institute and the American Society of Lubrication Engineers, 1970.

Schell, W. R., and A. Nevissi, "Heavy Metals from Waste Disposal in Central Puget Sound," Environ. Sci. Technol., **11**(9), 887-893 (1977).

Schlicting, E., and A. M. Elgala, "Schwermetallverteilung und Tongehalte in Böden" ["Heavy Metal Distribution and Clay Contents in Soils"], Z. Pflanzenernaehr. Bodenkd., **1975**(6), 563-571 (1975).

Schnitzer, M., "Reactions of Humic Substances with Minerals in the Soil Environment" in Environmental Biogeochemistry and Geomicrobiology. Volume 2. The Terrestrial Environment, W. E. Krumbein, Ed., Ann Arbor Science Publishers, Inc., Ann Arbor, Michigan, 1978, pp. 639-647.

Schrader, E. L., Jr., J. H. Rule, and W. J. Furbish, "Trace Metal Geochemistry of a Fluvial System in Eastern Tennessee Affected by Coal Mining," Southeast. Geol., **18**(3), 157-172 (1977).

Schutyser, P., W. Maenhaut, and R. Dams, "Instrumental Neutron Activation Analysis of Dry Atmospheric Fall-Out and Rainwater," Anal. Chim. Acta, **100**, 75-85 (1978).

Schwitzgebel, K., F. B. Meserole, R. G. Oldham, R. A. Magee, F. G. Mesich, and T. L. Thoem, "Trace Element Discharged from Coal-Fired Power Plants" in International Conference on Heavy Metals in the Environment, October 27-31, 1975, Symposium Proceedings, Vol. 2, Part 2, Institute for Environmental Studies, University of Toronto, Toronto, Ontario, Canada, 1977, pp. 533-551.

Shapiro, J., "Effect of Yellow Organic Acids on Iron and Other Metals in Water," J. Am. Water Works Assoc., **56**(8), 1062-1082 (1964).

Shirokov, V. V., and V. I. Panasin, "Content of Mobile Cu, Zn, Co, B, Mn, and Mo in the Main Soil Groups of Polders in Kaliningrad Oblast," Sov. Soil Sci., **4**(3), 341-344 (1972); translated from Agrokhimiya, No. 5, 136-139 (1972).

Sholkovitz, E. R., "The Flocculation of Dissolved Iron, Manganese, Aluminum, Copper, Nickel, Cobalt, and Cadmium During Estuarine Mixing," Earth Planet. Sci. Lett., **41**(1), 77-86 (1978).

Shuman, M. S., L. A. Smock, and C. L. Haynie, Metals in the Water, Sediments and Biota of the Haw and New Hope Rivers, North Carolina, PB-272 629, National Technical Information Service, U.S. Department of Commerce, Springfield, Virginia, 1977.

Shuman, M. S., C. L. Haynie, and L. A. Smock, "Modes of Metal Transport Above and Below Waste Discharge on the Haw River, North Carolina," Environ. Sci. Technol., **12**(9), 1066-1069 (1978).

Shvartsev, S. L., P. A. Udodov, and N. M. Rasskazov, "Some Features of the Migration of Microcomponents in Neutral Waters of the Supergene Zone," J. Geochem. Explor., **3-4**, 433-439 (1974/75).

Sibley, T. H., and J. J. Morgan, "Equilibrium Speciation of Trace Metals in Freshwater:Seawater Mixtures" in International Conference on Heavy Metals in the Environment, October 27-31, 1975, Symposium Proceedings, Vol. 1, Institute for Environmental Studies, University of Toronto, Toronto, Ontario, Canada, 1977, pp. 319-338.

Sigleo, A. C., G. R. Helz, and W. H. Zoller, "Chlorinated vs. Unchlorinated Natural Macromolecules; Halogen and Trace Metal Comparison by Neutron Activation Analysis," ACS Div. Environ. Chem. Prepr., **19**(1), 228-230 (1979).

Sillanpää, M., and E. Lakanen, "Readily Soluble Trace Elements in Finnish Soils," Ann. Agric. Fenn., **5**, 298-303 (1967).

Sillen, L. G., and A. E. Martell, Stability Constants of Metal-Ion Complexes, Special Publication No. 17, The Chemical Society, London, 1964.

Singh, R., and J. P. Singhal, "Preliminary Studies on the Profile Distribution of Trace Elements in Some Typical Soils of Aligarh District," Indian J. Appl. Chem., **33**(6), 351-356 (1970).

Singhal, J. P., S. U. Khan, and O. P. Bansal, "Influence of Pesticides on the Trace Element Status of Soils. Part I. (Nemagon and BHC on Boron, Cobalt and Copper)," J. Indian Chem. Soc., **53**(3), 313-315 (1976).

Singhal, J. P., and R. P. Singh, "Mobility of Trace Elements in Soils by Thin Layer Chromatography," Colloid Polym. Sci., **255**(5), 488-491 (1977).

Slater, S. M., and R. R. Hall, "Electricity Generation by Utilities: 1974 Nationwide Emissions Estimates" in Dispersion and Control of Atmospheric Emissions: New-Energy-Source Pollution Potential, R. L. Byers, B. B. Crocker, and D. W. Cooper, Eds., AIChE Symposium Series, Vol. 73, No. 165, American Institute of Chemical Engineers, New York, New York, 1977, pp. 291-311; Chem. Abstr., **87**, 156363c (1977).

Sponar, J., M. Kazdoya, and A. Seffova, "Pouziti Atomoyé Absorpce ke Kalitatativnimu Hodnoceni Prašného Spadu" ["Use of Atomic Absorption in Qualitative Determination of Dust Fallout"], Cesk. Hyg., **20**, 324-332 (1975).

Stancheva, P., "Effect of Atrazine on the Mobility of Some Trace Elements in the Soil," Pochvozn. Agrokhim., **12**(1), 53-60 (1977); Chem. Abstr., **88**, 17206m (1978).

Stumm, W., and H. Bilinski, "Trace Metals in Natural Waters; Difficulties of Interpretation Arising from Our Ignorance on Their Speciation" in Advances in Water Pollution Research, Proceedings of the Sixth International Conference held in Jerusalem, June 18-23, 1972, S. H. Jenkins, Ed., Pergamon Press, New York, New York, 1973, pp. 39-52.

Suarez, D. L., and D. Langmuir, "Heavy Metal Relationships in a Pennsylvania Soil," Geochim. Cosmochim. Acta, **40**(6), 589-598 (1976).

Swaine, D. J., and R. L. Mitchell, "Trace-Element Distribution in Soil Profiles," J. Soil Sci., **11**(2), 347-368 (1960).

Tafuri, A. N., "Pollution from Urban Land Runoff," News of Environ. Res. in Cincinnati, April 11, 1-4 (1975).

Tessier, A., P. G. C. Campbell, and M. Bisson, "Sequential Extraction Procedure for the Speciation of Particulate Trace Metals," Anal. Chem., **51**(7), 844-851 (1979).

Thomas, R. L., "Trace Elements in the Sediments of the Laurention Great Lakes," ACS Div. Environ. Chem. Prepr., **19**(1), 440-442 (1979).

Thomas, W. A., "Cerium and Cobalt Movement with Litter Leachate in a Forest Soil," ERDA Symp. Ser., **36**, 625-629 (1975).

Tiller, K. G., and J. F. Hodgson, "The Specific Sorption of Cobalt and Zinc by Layer Silicates," Clays Clay Minerals, Proc. Natl. Conf. Clays Clay Minerals, **9**, 393-403 (1962).

Tiller, K. G., J. F. Hodgson, and M. Peech, "Specific Sorption of Cobalt by Soil Clays," Soil Sci., **95**(6), 392-399 (1963).

Titlyanova, A. A., and N. A. Timofeeva, "Sorbtsiya Radioaktivnykh Izotopov Pochvoi" ["Sorption of Radioactive Isotopes by Soil"], Tr. Inst. Biol. Akad. Nauk SSSR, Ural'sk. Filial, No. 22, 17-29 (1962); Chem. Abstr., **59**, 199g (1963).

Trefry, J. H., and B. J. Presley, "Heavy Metal Transport from the Mississippi River to the Gulf of Mexico" in Marine Pollutant Transfer, H. L. Windom and R. A. Duce, Eds., Lexington Books, D. C. Heath and Company, Lexington, Massachusetts, 1976, pp. 39-76.

Troup, B. N., and O. P. Bricker, "Processes Affecting the Transport of Materials from Continents to Oceans," Presented at the 169th National Meeting, American Chemical Society, Philadelphia, Pennsylvania, April 1975.

Turekian, K. K., and M. R. Scott, "Concentrations of Cr, Ag, Mo, Ni, Co, and Mn in Suspended Material in Streams," Environ. Sci. Technol., **1**, 940-942 (1967).

U.S. Environmental Protection Agency, Compilation of Air Pollutant Emission Factors (Revised), U.S. Environmental Protection Agency, Office of Air Programs, Research Triangle Park, North Carolina, February 1972.

Vaughan, B., K. Abel, D. Cataldo, J. M. Hales, C. E. Hane, L. A. Rancitelli, R. C. Routson, R. E. Wildung, and E. G. Wolf, Review of Potential Impact on Health and Environmental Quality from Metals Entering the Environment as a Result of Coal Utilization, PB 289 658, Battelle Memorial Institute, Columbus, Ohio, August 1975, 87 pp.

Vinogradov, A. P., The Geochemistry of Rare and Dispersed Chemical Elements in Soils, 2nd ed., translated from the Russian, Consultants Bureau, Inc., New York, New York, 1959.

Vogel, G. J., W. M. Swift, J. C. Montagna, J. F. Lenc, and A. A. Jonke, "Application of Pressurized, Fluidized-Bed Combustion to Reduction of Atmospheric Pollution," Inst. Fuel Symp. Ser. (London), No. 1 (Fluidized Combustion), D3/1-D3/11 (1975).

Vuceta, J., and J. J. Morgan, "Chemical Modeling of Trace Metals in Fresh Waters: Role of Complexation and Adsorption," Environ. Sci. Technol., 12(12), 1302-1308 (1978).

Wagemann, R., G. J. Brunskill, and B. W. Graham, "Composition and Reactivity of Some River Sediments from the Mackenzie Valley, N.W.T., Canada," Environ. Geol., 1(6), 349-358 (1977).

Walters, L. J., Jr., T. J. Wolery, and R. D. Myser, "Occurrence of As, Cd, Co, Cr, Cu, Fe, Hg, Ni, Sb, and Zn in Lake Erie Sediments," Proc., Conf. Great Lakes Res., 17(Pt. 1), 219-234 (1974).

Webber, M. D., and D. G. M. Corneau, "Metal Extractability from Sludge-Soil Mixtures" in International Conference on Heavy Metals in the Environment, October 27-31, 1975, Symposium Proceedings, Vol. 1, Institute for Environmental Studies, University of Toronto, Toronto, Ontario, Canada, 1977, pp. 205-225.

Webber, M. D., and D. G. M. Corneau, Canada-Ontario Agreement on Great Lakes Water Quality. Research Program for the Abatement of Municipal Pollution under Provisions of the Canada-Ontario Agreement on Great Lakes Water Quality. Research Report No. 76: Sludge Metal Solubilities in Soils, Environment Canada, Ottawa, Ontario, 1978.

Williams, J. D., E. M. Wewerka, N. E. Vanderbough, P. Wagner, P. L. Wanek, and J. D. Olsen, Environmental Pollution by Trace Elements in Coal Preparation Wastes, LA-UR-77-1850, National Technical Information Service, U.S. Department of Commerce, Springfield, Virginia, 1977, 11 pp.

Williams, R. E., Waste Production and Disposal in Mining, Milling, and Metallurgical Industries, Miller Freeman Publications, Inc., San Francisco, California, 1975.

Wilson, M. J., and M. L. Berrow, "The Mineralogy and Heavy Metal Content of Some Serpentinite Soils in North-East Scotland," Chem. Erde, **37**(3), 181-205 (1978).

Winchester, J. W., and G. D. Nifong, "Water Pollution in Lake Michigan by Trace Elements from Pollution Aerosol Fallout," Water Air Soil Pollut., **1**, 50-64 (1971).

Windom, H. L., K. C. Beck, and R. Smith, "Transport of Trace Metals to the Atlantic Ocean by Three Southeastern Rivers," Southeast. Geol., **12**, 169-181 (1971).

Windom, H. L., and R. G. Smith, "Distribution of Cadmium, Cobalt, Nickel, and Zinc in Southeastern United States Continental Shelf Waters," Deep Sea Res., **19**, 727-730 (1972).

Windham, S. T., and C. R. Phillips, "Radiological Survey of New London Harbor, Thames River, Conn., and Environs," Radiat. Data Rep., **14**(11), 659-666 (1973).

Zoller, W. H., E. S. Gladney, G. E. Gordon, and J. J. Bors, "Emission of Trace Elements from Coal Fired Power Plants" in Trace Substances in Environmental Health - VIII, D. D. Hemphill, Ed., Proceedings of the University of Missouri's 8th Annual Conference on Trace Substances in Environmental Health, June 11-13, 1974, University of Missouri, Columbia, Missouri, 1975, pp. 167-171.

VI. PRIMARY AND SECONDARY PROCESSING OF COBALT-BEARING MATERIALS

Bonnie L. Carson

This chapter describes the recovery of cobalt from materials that have been, are being, or may be processed in the United States. Cobalt recovery in other countries is treated much more cursorily because those processes have little, if anything, to do with environmental losses of cobalt in the United States. Future world production is summarized to show the rank the United States will achieve among producers. What little that has been reported about secondary recycling of cobalt in the United States has been compiled. As a departure from the reports on thallium and indium, a mass balance has not been attempted for cobalt as an impurity in materials processed for base metals. This is because cobalt is not especially enriched in most smelter waste products and because the magnitude of industrial use of cobalt itself is much larger than the magnitude of cobalt in base-metal metallurgy.

A. U.S. Mining, Milling, Smelting, and Refining Processes

1. Materials processed for cobalt recovery in the U.S. (past, present, and potential)

a. Pennsylvania magnetite-pyrite ores: Cobalt has been produced as a by-product of iron ores mined in Pennsylvania by Bethlehem Cornwall Corporation, subsidiary of Bethlehem Steel Corporation. Cobalt recovery was by the Pyrites Company, Inc., a subsidiary of Rio Tinto Zinc Corporation. Cobalt production figures were confidential, but estimates of 320 short tons (290 MT) cobalt in 1966 and 162 short tons (147 MT) cobalt in 1970 were made by Battelle Memorial Institute (Petrick et al., 1973).

Bethlehem produced a flotation concentrate of cobaltous pyrite from magnetite iron ores from the Grace and Cornwall mines. At the sulfuric acid plant in Sparrows Point, near Baltimore, Maryland, the pyrite concentrate was roasted, and the off-gas SO_2 was used to produce sulfuric acid. Water quenching of the roasted concentrate left a residue and produced a copper-bearing solution, which was shipped to the Pyrites Company, Inc., plant in Wilmington, Delaware. The latter company also processed a small amount of petroleum industry cobalt catalyst residues (Petrick et al., 1973).

Figure VI-1 is a simplified flow diagram of the cobalt recovery processes prior to April 1970, when the Pyrites plant switched to a solvent extraction process. Copper was largely removed electrolytically. Precipitation and filtration eliminated iron, manganese, and residual copper. According to this diagram of Petrick et al. (1973), the cobalt hydroxide precipitated from the solution. The solution remaining after filtration was discarded, representing a major loss of cobalt from the system. The hydroxide was calcined to cobalt hydrate or cobalt oxide. Cobalt was also lost in the manganese precipitation. The ratio of manganese to cobalt in the manganese precipitate was 1:1.

No information was found regarding Pyrites Company's solvent extraction process. Apparently, the higher cobalt ores have been exhausted. According to Larsen (1978), the company went out of business due to environmental problems.

b. Missouri cobalt ores: St. Louis Smelting and Refining Company first mined the Fredericktown, Missouri, site for cobalt from 1944 to 1946. The St. Louis Smelting and Refining Division of National Lead Corporation mined the site from 1955 to 1961. The annual capacity in 1955 was $\sim$ 1.4 million lb cobalt (700 short tons, 635 MT) and $\sim$ 1.9 million lb nickel (Sibley, 1979). In 1961, National Lead closed the mine and liquidated the refinery operation (Charles River Associates, 1969).

During the milling operations, the crushed ore was floated with lime to remove iron, and cyanide was added to depress all the sulfides except galena. The tailings from the 78 to 80% lead concentrate were treated to float the copper minerals. The copper concentrate went to the El Paso copper smelter. The iron tails (that were formerly stored in ponds) containing $\sim$ 2% nickel and cobalt were floated to give the cobalt concentrate. The lead concentrate, containing $\sim$ 0.5% Ni and Co, was smelted at Herculaneum, Missouri.

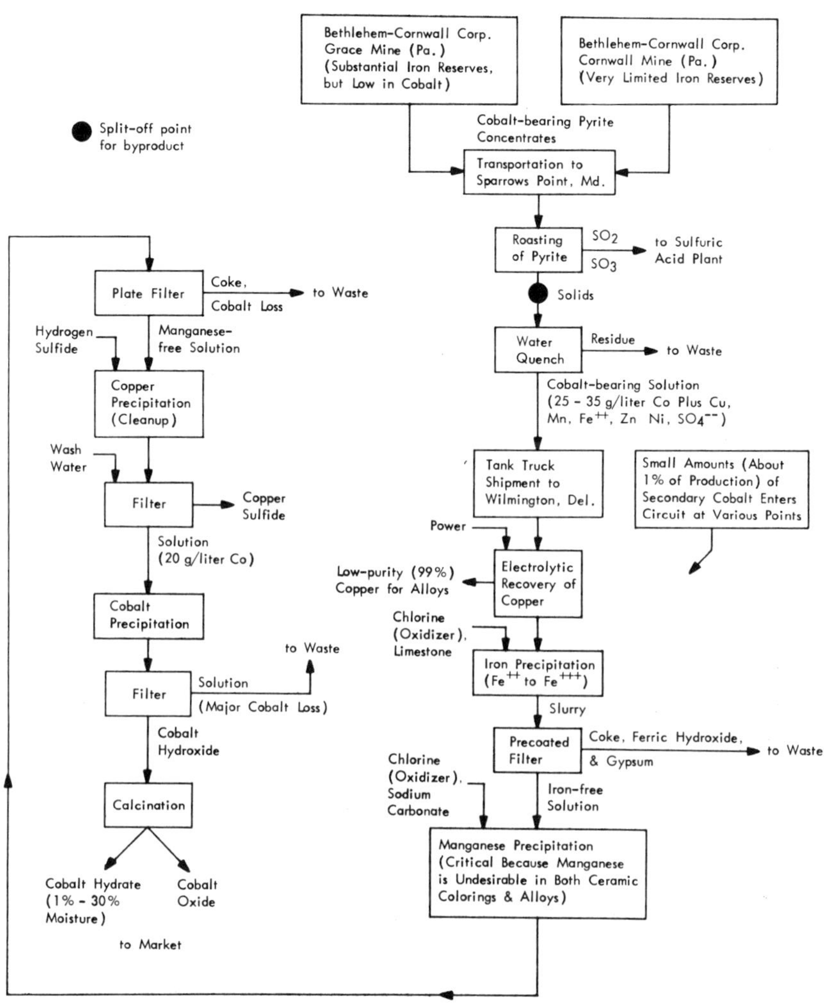

Figure VI-1. Pre-1970 Cobalt Recovery by the Pyrites Company, Inc., Wilmington, Delaware (Petrick et al., 1973).

Pressure leaching was practiced from 1954 to 1960 by the National Lead Company. Partially roasted sulfidic concentrates* were pressure leached in autoclaves at 450°F (230°C) and 560 lb/in.^{2} gauge (3.90 MPa) air. The leaching residue was washed and discarded. It contained most of the iron as a hydrolyzed basic sulfate. The hot solution was treated with hydrogen at 625 lb/in.^{2} gauge (4.30 MPa) air to precipitate 85% of the copper as metal. Residual copper was removed by cementation on iron borings, and iron was removed by oxidation and limestone neutralization to pH 3.5. The solution was clarified, concentrated, and treated with cobalt powder to cement traces of copper. Addition of aqueous ammonia precipitated nickel ammonium sulfate, leaving cobaltic ammine sulfate in solution. The cobaltic compound was reduced to metallic cobalt by hydrogen at 425°F (211°C) and 600 lb/in.^{2} gauge (4.10 MPa). The nickel precipitate was treated to separate cobalt (Mackiw and Benz, 1961; Doane, 1972).

The environmental effects of the mining and milling activity were described in Chapter V. Anschutz Uranium, Inc., plant to reopen the mine in the near future to take advantage of the current high cobalt prices. Planned cobalt production is 1.5 to 2 million lb (680-910 MT) per year (Gorham International, Inc., 1979a).

c. Idaho cobalt-copper ores: Haynes Stellite Company produced 55 tons of gravity cobalt concentrate (17.7% Co) from 4,000 tons ore in the Blackbird District, Lemhi County, Idaho in 1918. The concentrate was processed in Niagara Falls, New York (Wells et al., 1948).

Uncle Sam Mining and Milling Company operated a 75-ton flotation mill from 1938 to 1941, shipping 563 tons of ore and concentrates (containing 1.7% Co) to the Anaconda smelter.

Calera Mining Company began underground development in 1945. A new milling process was developed by 1950, and

---

* The concentrates contained approximately 5.0% Ni, 4.2% Co, 4.8% Cu, 30% Fe, and 40% S (Doane, 1972). The ore contained 2.5 to 2.75% Pb, 0.5 to 0.75% Cu, about 0.25% each Co and Ni, and 5 to 6% Fe (Lutjen, 1953).

the concentrator began operation in July 1951. Production ranged from 591,500 lb (268 MT) cobalt in 1953 to 3,061,000 lb (1,390 MT) cobalt in 1958 (Charles River Associates, 1969).

Wells et al. (1948) described a pilot plant operation for separating copper and cobalt concentrates by flotation and tabling. Cyanide in an alkaline circuit was the only method found to depress cobaltite during copper flotation. A cobalt concentrate contained about 11% cobalt; the copper concentrate contained 0.5% cobalt. About 2% was in the cobalt table middling fraction. The rougher tailing contained 0.045% cobalt. The cobalt distribution in these milling products was cobalt concentrate, 76%; copper concentrate, 6%; cobalt table middling, 7%; and rougher tailing, 11%. Selective flotation of a higher grade ore gave somewhat better recovery of cobalt; only 64% of the cobalt was recovered in the cobalt concentrate when oxidized ores were floated.

The Bureau of Mines, Albany, Oregon, developed a process that could recover cobalt, copper, and nickel from the tailings accumulated during the Calera Mining Company operation. About 4,000 MT of cobalt is contained in the tailings (0.23% cobalt; 80 to 90% of the cobalt is in particles of Tyler mesh screen size minus 200 to plus 400). The tailings treatment process included flotation and reduction roasting steps, an oxidizing ammonia leach, solvent extraction, and electrowinning. In bulk flotation tests, 54 to 89% of the cobalt was found in the bulk concentrate. However, when separate copper and cobalt concentrates were prepared, the cobalt distribution was copper concentrate, 18%; cobalt concentrate, 57%; and tailings, 25%. Eighty-seven percent of the cobalt in the concentrate was recovered through the leaching stage; 80% of the cobalt present was recovered in each cycle through the solvent extraction-electrowinning circuit (Rule and Siemens, 1976).

The Calera Mining Company used sulfuric acid leaching in its cobalt recovery operation from 1953 to 1959. Concentrates containing 17.5% Co, 20.0% Fe, 24.0% As, 29.0% S, 1.0% Ni, and 0.5% Cu were acid leached at $\sim$ 190°C and 500 lb/in^2 gauge (3.45 MPa) air pressure. This step dissolved the cobalt, copper, and nickel, leaving most of the iron and arsenic in the insoluble calcium sulfate leach residue. Ammonia was added to give a 1:2 Co-NH_3 ratio. Cobalt powder precipitated from solution when it was treated with hydrogen at 190°C and 800 lb/in^2 (5.50 MPa) pressure. Sulfur contamination was removed by melting in an arc furnace with a high-lime slag. The product cobalt was 95.6% pure (Doane, 1972; Evans, 1967). Because of problems in the hydrogenation step, an electrolytic scheme giving a 99.8% pure cathode was later adopted (see Figure VI-2) (Evans, 1967).

The 95.6% pure cobalt product met the specifications for the U.S. government stockpile program. Cobalt recovery was no longer economical after U.S. government contracts ended (Evans, 1967).

Mining by Noranda Mines Ltd. of the Blackbird cobalt-copper deposit will be by cut and fill. During the 2-stage differential flotation, the cobaltite and iron sulfides will be depressed to produce the copper concentrate. Noranda has developed new techniques for the complex separation of cobaltite from pyrite and pyrrhotite. Because of the significant amounts of arsenic in the concentrate, the cobalt concentrate will be high-pressure leached (the old Calera Process) with sulfuric acid; and the arsenic will be precipitated as ferric arsenate, which will be buried at the mine. The impure cobalt sulfate recovered will be toll refined until solution purification and electrowinning circuits are constructed. The principal product will be electrolytic cathodes. Correction of current problems due to leaching of old workings and the construction of a new tailings dam are anticipated (McIntyre, 1979).

d. Nickel-bearing materials: Cobalt has been recovered in the United States from Cuban laterites at Port Nickel, Louisiana, and is potentially recoverable from laterites found near the California-Oregon border and from the nickel silicate ores processed near Riddle, Oregon, into ferronickel. Cobalt is also potentially recoverable from the ores in the Duluth gabbro of northeastern Minnesota.

The nickel mineral mined by Hanna Nickel Smelting Company at Nickel Mountain, 4 miles northwest of Riddle, Oregon, is garnierite. This complex nickel magnesium silicate associated with iron, cobalt, chromium, and aluminum is mined by open-cut methods. The soft ore is crushed, screened, and calcined at

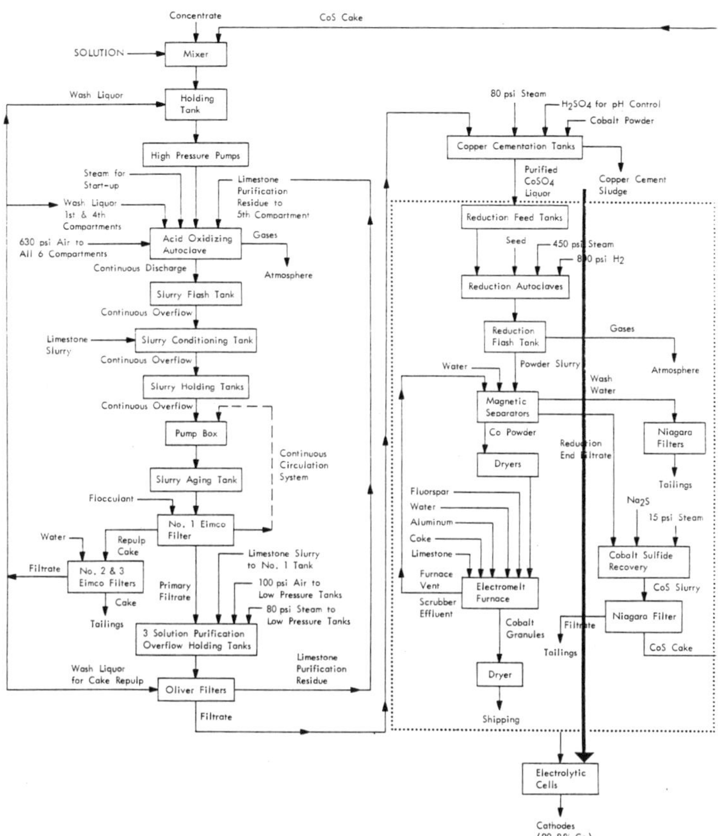

Figure VI-2. Electrolytic Process for Cobalt Recovery Practiced by Howe Sound's Calera Mining Company, Garfield, Utah (Anonymous, 1958).

640 to 700°C to remove water. In the Ugine reduction process used, the molten oxide ore is reduced by melting in electric furnaces and agitating violently in skip mixers with 45% ferrosilicon. At a predetermined weight of ferronickel, a portion is removed for refining (Wedge and Halter, 1971).

The crude ferronickel is refined by dephosphorizing and deoxidizing. Phosphorus is oxidized to the pentoxide, which combines with the high-lime slag to give $Ca_4P_2O_9$; the latter is removed from the furnace by skimming. Lime and iron ore are heated with the dephosphorized ferronickel in the refining furnace in separate slagging operations. Fluorspar and lime are used in the finishing slag operation. Ferrosilicon is added to deoxidize the molten ferronickel, which is poured directly into pigs. The ferronickel product contains 0.5% cobalt. Smelter receptacle "skulls" are reprocessed and returned to the smelter. Dust losses are reduced by bag filters (in 1959, dust losses were only 0.014% of the ore processed) (Coleman and Vedensky, 1961). Figure VI-3 is a flow diagram of the Hanna Nickel Smelting Company smelter.

Although they described the Riddle nickel process in detail, Wedge and Halter (1971) did not mention the fate of cobalt, chromium, or aluminum.

In December 1976, the National Institute for Occupational Safety and Health (NIOSH) observed that cobalt concentrations in the workplace air at this nickel smelter ranged from 0.002 to 0.007 mg Co/m^3 (average 0.002 mg Co/m^3), well below the Occupational Safety and Health Administration (OSHA) standard (0.1 mg Co/m^3). Table VI-1 shows that almost all of the workers were exposed to only 0.002 or 0.003 mg Co/m^3 except the plant cleanup worker, who was exposed to 0.007 mg Co/m^3.

Process water is extensively recycled at the Nickel Mountain mill and smelter in two treatment ponds. The overflow from the first settling pond contains $<$ 0.05 mg Co/liter. The average discharge volume from the second pond at pH 8.7 is 460 m^3/day. Cobalt was not reported in the second pond water (Jarrett and Kirby, 1978b).

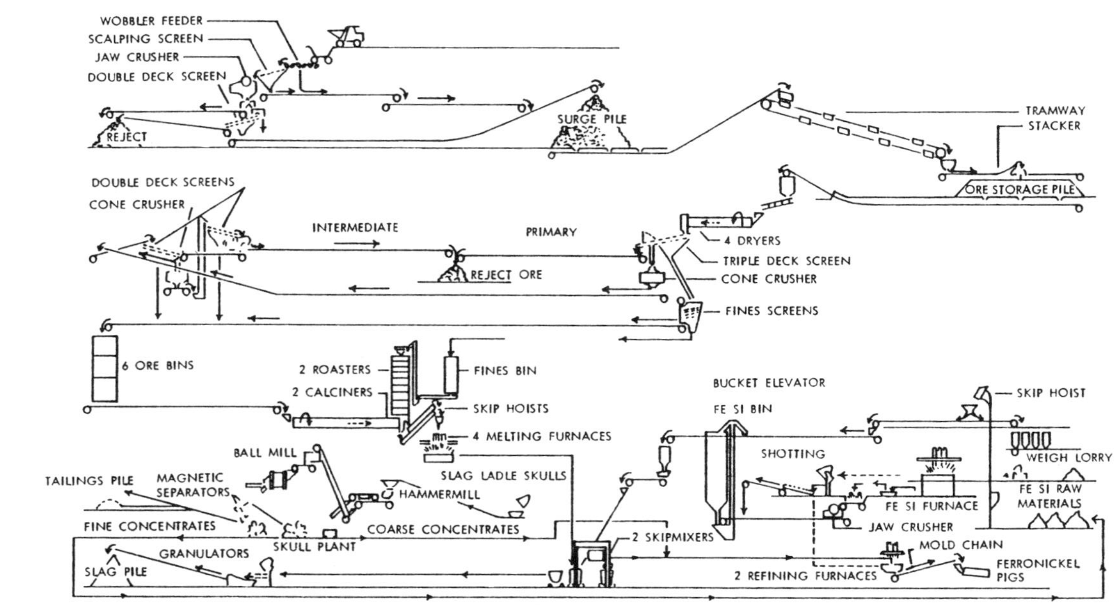

Figure VI-3. Hanna Nickel Smelting Company Smelter Flow Diagram (Donaldson et al., 1978).

TABLE VI-1

COBALT EXPOSURES AT HANNA NICKEL SMELTING PLANT, DECEMBER 1976
(Donaldson et al., 1978)

Operation	Shift	Cobalt (mg/m^3)	Air Volume (m^3)
ORE HANDLING			
Loader operator	2	0.002	0.766
Loader operator	1	0.003	0.690
Crusher operator	1	0.002	0.830
DRYING			
Dryer operator	1	0.002	0.924
Dryer operator	2	0.003	0.782
Dryer helper	2	0.003	0.780
Dryer helper	1	0.002	0.810
CALCINING			
Calciner operator	2	0.002	0.822
Calciner operator	1	0.002	0.840
Calciner helper	2	0.002	0.832
Calciner helper	1	0.002	0.840
SKULL DRILLING			
Skull driller	2	0.002	0.928
Skull driller	2	0.002	0.896
Skull driller	2	0.002	0.898
Skull driller	2	0.002	0.902
Skull driller	2	0.003	0.726
Skull driller	2	0.002	0.708
Skull driller	2	0.002	0.720
Skull driller (helper)	2	0.002	0.898
FERRO SILICON MANUFACTURING			
Stacker operator	2	0.002	0.986
Crusher operator	2	0.003	0.744
Crusher operator	2	0.003	0.754
Loader operator	2	0.003	0.744
Furnace charger	2	0.003	0.738

TABLE VI-1 (continued)

Operation	Shift	Cobalt (mg/m^3)	Air Volume (m^3)
FERROSILICON MANUFACTURING (continued)			
Furnace charger	2	0.003	0.748
Furnace charger	2	0.002	0.882
Furnace feeder	1	0.002	0.820
Furnace feeder	2	0.002	0.796
Furnace tapper	2	0.003	0.746
Furnace tapper	2	0.002	0.868
Furnace tapper (helper)	2	0.002	0.866
Furnace operator	2	0.003	0.750
Furnace operator	1	0.002	0.878
Furnace operator	2	0.003	0.626
MIXING			
Melter smelter helper	2	0.003	0.704
Melter smelter helper	1	0.003	0.700
Melter smelter helper	2	0.002	0.902
Smelter helper (belt)	2	0.003	0.660
Smelter helper (belt)	2	0.002	0.136
Crane operator	1	0.002	0.790
Crane chaser	1	0.003	0.710
Crane chaser	2	0.002	0.900
Slag handler	1	0.003	0.680
Slag handler	2	0.002	0.902
Mixer operator	1	0.002	0.806
Mixer operator	1	0.002	0.820
Mixer operator	2	0.002	0.818
Mixer operator	1	0.002	0.818
Mixer operator	2	0.002	0.796
Mixer operator	1	0.002	0.820
Ladle mixer operator	2	0.002	0.790
REFINING			
Refinery furnace operator	1	0.003	0.700
Refinery furnace operator	1	0.002	0.820
Refinery furnace operator	1	0.003	0.670
Refinery furnace operator	2	0.002	0.926

TABLE VI-1 (concluded)

Operation	Shift	Cobalt (mg/m^3)	Air Volume (m^3)
REFINING (continued)			
Refinery furnace helper	2	0.003	0.703
Refinery furnace helper	2	0.003	0.703
Refinery furnace helper	2	0.002	0.820
Refinery furnace helper	2	0.002	0.930
Refinery furnace helper	2	0.002	0.928
Tapper helper	2	0.003	0.748
HANDLING OF FINISHED PRODUCT			
Pig deck leadman	1	0.002	0.800
Pig deck stacker	1	0.003	0.766
Fork lift operator pig deck	1	0.003	0.790
Fork lift operator pig deck	1	0.002	0.804
Pig deck laborer	1	0.003	0.700
MAINTENANCE			
Welder	1	0.003	0.710
Can welder	1	0.003	0.664
Can welder	1	0.003	0.648
Can welder	1	0.003	0.660
Millwright welder	1	0.003	0.700
Millwright oiler	1	0.002	0.836
Millwright oiler	1	0.002	0.830
Millwright	1	0.003	0.700
Machinist	1	0.003	0.700
MISCELLANEOUS			
Utility man	1	0.002	0.790
Plant cleanup	1	0.007	0.690
Sweeper operator	1	0.002	0.800

Freeport Sulphur Company operated at Port Nickel, Louisiana, on lateritic Cuban ores from Moa Bay for 5 to 6 months in 1959-1960, but was shut down after the Cuban revolution. Initial pressure leaching with sulfuric acid, neutralization, and precipitation of metal values as sulfides were operations performed in Cuba. At Port Nickel, the sulfides were pressure leached in an acid medium at 175°C; the pH of the leach solution was adjusted to pH 5.5 by ammonia addition to precipitate iron, aluminum, and chromium*; and copper, lead, and zinc were precipitated as sulfides at pH 1.5. Treating the solution at 190°C with hydrogen at 650 lb/in^2 gauge (4.48 MPa) in the presence of ammonium sulfate to give pH 2.0 recovered 95% of the nickel as pure metal. The impure sulfates precipitated upon concentration of the solution. Aqueous ammonia removed the remaining nickel as the insoluble ammonium sulfate. The cobaltic ammonium sulfate solution was purified and treated with hydrogen to recover pure cobalt metal (Mackiw and Benz, 1961; Doane, 1972; Lee, 1959).

Prior to the brief commercialization of the process, Freeport Sulphur operated a 10 ton/day pilot plant ca. 1954 at Hoskins Mound, Texas, on 300 tons of Moa Bay ore. Concentrates produced there were treated in a Linden, New Jersey, pilot plant in October 1954, and January 1955, for recovery of nickel and cobalt from solutions (Lee, 1959).

Roorda and Queneau (1973) proposed a process for recovering $\geq$ 90% nickel and cobalt from nickeliferous limonites such as the Moa Bay and Nicaro deposits by aqueous chlorination in seawater. The experimental work was assisted by International Nickel, Ltd., but planned commercial applications are unknown.

Lateritic resources of nickel and cobalt similar to the ores mined in Cuba and the Philippines occur in Oregon and California. Oregon's lateritic deposits are located on Eight Dollar Mountain; in the Woodcock Mountain area of Josephine County; and in the Red Flats area east of Gold Beach, Curry County. The best known district in California is Pine Flat Mountain, northeast of Crescent City; but the largest resource area is in Medocino

---

* The filter cake was redissolved with acid and reprecipitated with ammonia to recover nickel and cobalt (Lee, 1959).

County in the Little Red Mountain District. Private exploration has indicated that Oregon's lateritic resources of nickel are greater than was previously thought to exist. These domestic laterites contain 0.06 to 0.25% cobalt in a manganese oxide wad and 0.5 to 1.2% nickel, principally in goethite (FeOOH) (Siemens and Corrick, 1977).

Large-scale mining of these formerly noncommercial laterites may be started if suitable smelting and refining processes can be developed. Siemens and Good (1976) of the Bureau of Mines' Albany Metallurgical Center in Albany, Oregon, have patented a nickel-cobalt extraction process suitable to the domestic laterites. The process, described more fully by Siemens and Corrick (1977), comprises roasting, selective reduction by carbon monoxide in the presence of pyrite, an oxidizing ammonia-ammonium sulfate leach with solvent extraction, and electrowinning of nickel and cobalt. Figure VI-4 depicts the process flow scheme.

The addition of 1.2% pyrite supplies the sulfur that inhibits the secondary reaction of magnesia with reduced nickel to form inert nickel magnesium silicate. The use of pure carbon monoxide rather than producer gas allows the use of a reduction temperature (525°C) that is 200 to 300°C lower than those used with Nicaro-type processes (Siemens and Corrick, 1977).

The Siemens-Good process allows the recovery of about 80 to 90% of the cobalt from domestic laterites (Siemens and Corrick, 1977).

Universal Oil Products (UOP) Inc.'s Mineral Sciences Division is converting a pilot plant in Tucson, Arizona, to use the Siemens-Good process on domestic laterites. The plant originally leached ore with an ammoniacal carbonate solution. Cracked oil or methane was used for roasting (Mari, 1979; Anonymous, 1978h).

The largest single domestic nickel resource is in the low-grade Duluth gabbro near Ely, Minnesota (Siemens and

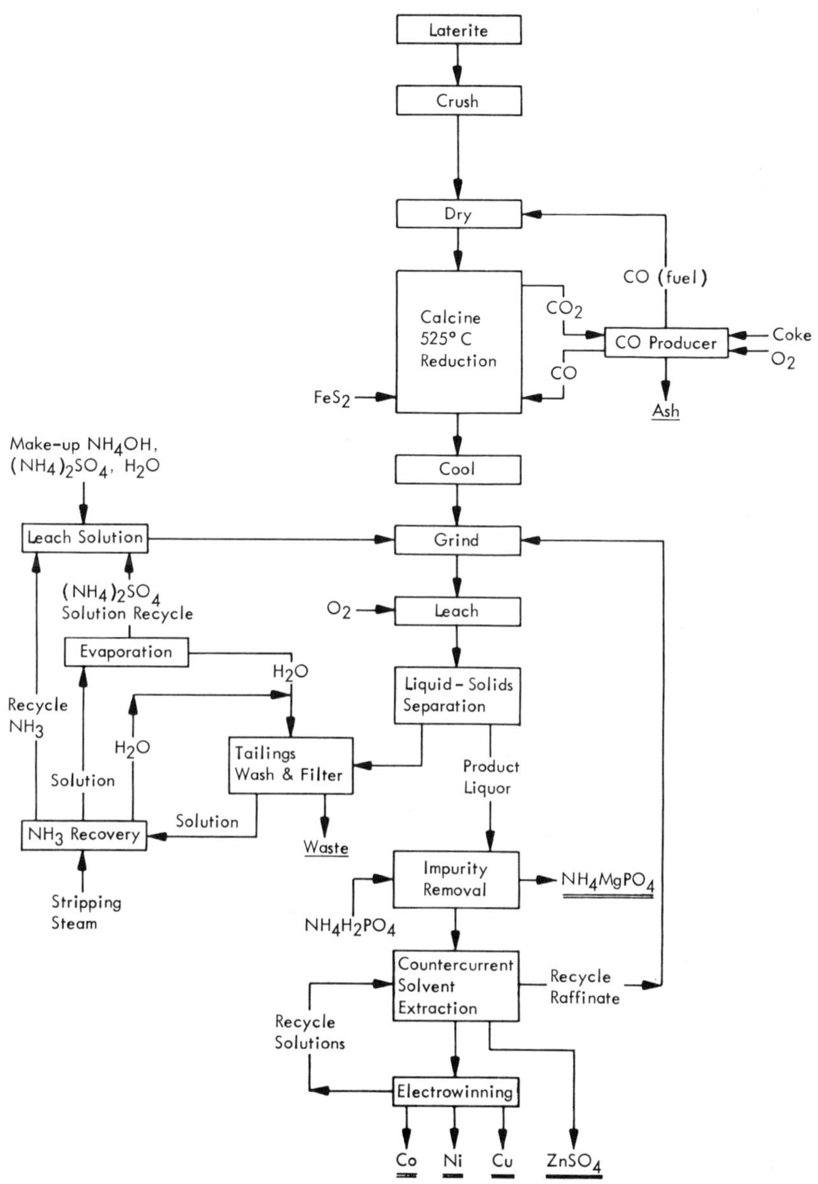

Figure VI-4. U.S. Bureau of Mines Selective Reduction-Oxidizing Ammonia Leach Process for Domestic Laterites (Siemens and Corrick, 1977).

Corrick, 1977). The Nickel Division of Amax Specialty Metals Corporation has made test drillings there. The Bureau of Mines has experimentally produced a copper-nickel concentrate from the ores containing 0.1 to 0.4% cobalt. Gorham International, Inc. (1979a) estimated that 500 to 1,000 short tons (400 to 910 MT) cobalt may eventually be produced annually from the Duluth gabbro complex.

Amax Nickel Division, Amax Inc., has operated the revamped Port Nickel refinery since November 1974. It is currently the only refiner of primary cobalt and pure nickel in the United States. The refinery, situated on the east bank of the Mississippi, 15 miles south of New Orleans, has a capacity of 36,000 MT nickel, 23,000 MT copper, 907 MT cobalt, and 91,000 MT ammonium sulfate. Current feeds are nickel-copper mattes from Botswana ($<$ 0.5% Co), New Caledonia (1.5% Co), Australia (0.6% Co), and South Africa. Apparently the process can be adapted to lateritic and silicate ores. About 600,000 to 850,000 lb cobalt (270 to 385 MT) per year is presently being produced, but the plant aims to reach full capacity by the end of 1979 (Hoppe, 1977; McPhail, 1979; Anonymous, 1979f; Redden, 1979).

A plant flow scheme (Figure VI-5) was presented by Donaldson et al. (1978). Because it is largely a hydrometallurgical process (Hoppe, 1977), most environmental problems have been eliminated.

At the Amax Port Nickel refinery, the atmospheric sulfuric acid leaching step dissolves 50 to 60% of the nickel and cobalt in the matte. Pressure leaching removes most of the remaining cobalt and nickel. The waste residue from the second pressure leaching step is sent to the tailings pond. Spent electrolyte from the copper electrowinning step is recycled to leaching steps. Cobalt in the atmospheric-pressure leachate is precipitated selectively from the sulfate solution as the hydroxide by treating the solution with trivalent nickel hydroxide. Nickel is precipitated upon hydrogenation of the filtrate after cobalt removal. The cobalt hydroxide precipitate is treated in an autoclave with ammonia to dissolve cobalt as the pentammine

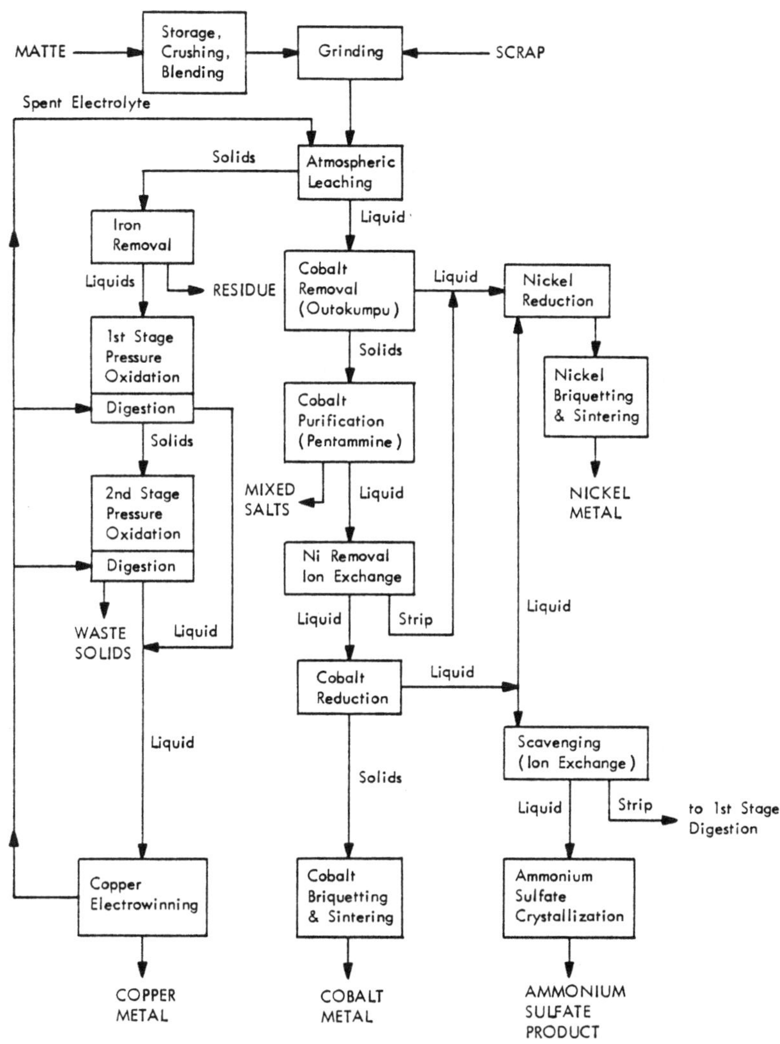

Figure VI-5. Amax Nickel Refinery (Donaldson et al., 1978).

complex and crystallize nickel as a double salt. After ion-exchange removal of residual nickel, cobalt is precipitated by hydrogen gas at elevated temperature and pressure. The by-product ammonium sulfate solutions are sent through an ion-exchange circuit to remove traces of nickel and cobalt prior to crystallization of the ammonium sulfate, which is sold for use as a fertilizer or chemical (Hoppe, 1977).

NIOSH (Donaldson et al., 1978) found an average 0.002 mg Co/m^3 (range < 0.001 to 0.012 mg Co/m^3) in workplace air in October, 1976, at an unidentified U.S. plant in which 33 employees were involved in soluble nickel salt production. The six employees monitored in the operations producing nickel acetate and nitrate were exposed to aerosols containing less cobalt than the detection limit (< 0.001 mg/m^3). However, in the nickel sulfate operation, 3 of 12 workers were exposed to 0.002 to 0.006 mg Co/m^3; and in the nickel chloride operation, 4 of 11 workers were exposed to 0.002 to 0.012 mg Co/m^3 (see Tables VI-2 and VI-3). All values were well below the OSHA standard of 0.1 mg Co/m^3.

e. <u>Manganese nodules</u>: Manganese nodules are abundant at 4- to 5-km depths in all the major ocean basins and at shallow depths in the Baltic Sea, the shelf area of the Soviet Arctic, and some Scottish sea lochs. Surface concentrations of cobalt attained are as high as 35 kg/m^2, but 10 kg/m^2 appears to be an average value for the Pacific Ocean areas explored. Highest cobalt concentrations (1.5%) are found in oceanic seamount concretions. The world average cobalt concentration in manganese nodules is 0.3% (Calvert, 1978).

The mineralization and formation of manganese nodules have been reviewed by Calvert (1978) and Olander (1971).

There are about 200 good-grade nodule mine sites. Initial commercial development of manganese nodules is aimed at an area of the North Pacific Ocean (at depths 5,000 to 6,000 m) midway between Hawaii and Mexico. Good-grade nodules are also found southwest and northwest of Hawaii and in the Indian Ocean (Tinsley, 1976).

TABLE VI-2

COBALT EXPOSURES AT A NICKEL CHEMICAL PLANT, NICKEL CHLORIDE OPERATION, OCTOBER, 1976
(Donaldson et al., 1978)

Job Classification	Shift	Cobalt (mg/m^3)	Air Volume Meters (m^3)
Process helper	1	< 0.001	0.848
Process helper	1	< 0.001	0.844
Process operator	1	< 0.001	0.840
Process operator	1	< 0.001	0.840
Process operator	2	0.012	0.862
Process operator (busts out opens)	2	0.005	0.856
Process operator (makes batches)	2	< 0.001	0.854
Process helper	1	< 0.001	0.840
Process operator	1	< 0.001	0.846
Process helper	2	0.005	0.862
Process helper	2	0.002	0.856

TABLE VI-3

COBALT EXPOSURES AT A NICKEL CHEMICAL PLANT, NICKEL SULFATE OPERATION, OCTOBER, 1976
(Donaldson et al., 1978)

Job Classification	Shift	Cobalt (mg/m^3)	Air Volume Meters (m^3)
Process helper	1	< 0.001	0.866
Process operator	1	0.006	0.852
Chief operator	1	< 0.001	0.858
Process helper	1	< 0.001	0.836
Leadman	1	< 0.001	0.858
Chief operator	2	< 0.001	0.870
Process operator	2	0.002	0.886
Process operator (cleaning tanks)	1	0.002	0.866
Process operator (cleaning tanks)	1	< 0.001	0.842
Bagger	2	0.000	0.864
Process helper (scraping screens)	2	< 0.001	0.858
Operator (mixes Ni and H_2SO_4)	2	< 0.001	0.852

Several consortiums were formed in the 1970's to study the feasibility of mining manganese nodules. Tenneco formed Deepsea Ventures; and Kennecott Copper Company led a five-company consortium with Rio Tinto-Zinc (Great Britain), Consolidated Gold Fields of South Africa, Mitsubishi (Japan), and Noranda Mines (Canada) (Anderson, 1974).

In 1978, Ocean Minerals Company (OMCO) (a consortium that included two Netherlands firms, Lockheed Missiles and Space Co., and Amoco Minerals) leased the Glomar Explorer* from Global Marine Development Inc. to test an ocean mining system. Another consortium-leased ship was exploring and sampling fields between Hawaii and Baja California (Anonymous, 1978c).

Ocean Mining Association** was making feasibility studies at a site 1,200 miles southwest of San Diego, California, in 1978. Samples were collected by 3 miles of pipe with a dredging head. An airlift brought the nodules to the surface. No pilot plant had been built to separate the metal values (30% Mn, 1.2% Ni, 1.3% Cu, 0.25% Co, 0.25% Mo, 3% Al_2O_3, and 4% Fe) (Anonymous, 1978a, 1978d).

In April 1978, Ocean Management Inc.***, Seattle (a consortium of U.S., West German, Japanese, and Canadian companies) recovered 1,800 short tons (1,600 MT) of manganese nodules containing 25% Mn, 1.5% Ni, 1% Cu, and 0.25% Co somewhere between 500 and 1,000 miles (800 and 1,600 km) off Hawaii's southern coast. They were to be stored in Canada until nickel prices improved and until the United Nations Law of the Sea

---

* The Hughes Tool Company's Glomar Explorer sampled manganese nodules as early as 1973 (Anonymous, 1973).

** This consortium comprised U.S. Steel Corp.; the Belgian firm Union Miniere; and the Sun Co., Radnor, Pennsylvania. The consortium was initially financed by Tenneco, Inc. Tenneco and the original five Japanese companies have since dropped out.

*** The association comprised Inco; the Japanese group Domco; a group of German firms organized under the name AMR; and Sedco Inc., a U.S. ocean-engineering company (Anonymous, 1978a).

Conference decides on rights to the deep-sea minerals. One additional cruise of the converted oil drilling ship was planned before mining ceased (Anonymous, 1978).

In 1974, the Office of Ocean Resources, U.S. Department of the Interior (Anderson, 1974), estimated that 15,000 short tons (14,100 MT) could be mined in 1978 by a three-company industry processing 7 million short tons of dry nodules annually. The average metal content was assumed to be 0.22% Co, 1.25% Ni, 1% Cu, and 25% Mn. However, another Office of Ocean Resources estimate based on mining for all four metals was only 2,200 short tons (2,000 MT) cobalt.

J. Daniel Nyhart of Massachusetts Institute of Technology utilized a computer study that took into account all factors relating to ocean mining of manganese nodules and concluded that sea mining will be profitable. The most important factors to be considered included ore grade, production rate, and delays in the initiation of mining. The model estimated 25 yr of productivity for a single mine ship handling 3 million short tons (2.7 million MT) of nodules annually. Commercial production was expected to commence in 5 to 8 yr (Anonymous, 1978f).

Another estimate for first-generation mining production was made in 1979 by Reddy of Charles River Association and Clark of Massachusetts Institute of Technology. One mining operation might recover 6,800 MT cobalt (based on 90% recovery). Forecasting models predict that if four operations come on-stream in 1990 and two more follow in 1995, that cobalt and nickel prices will be affected significantly. Without ocean mining, the model predicts a cobalt price of $10/lb; but six ocean mining ventures would depress the cobalt price to $6.50/lb or less (Reddy and Clark, 1979).

Reddy (as reported by Koflowitz, 1979a) expected that the likelihood of a large supply of cheap cobalt from mining manganese nodules within a decade will discourage producers from investing in plant expansion for cobalt recovery from other sources and will keep the price high in the interim.

Without government guarantees or direct subsidies, McPhail of Inco Metals Company (International Nickel Company), Canada (1979), did not forsee significant deep-sea mining production before 1990. The consortium ventures are presently at the stage equivalent to land-based exploration and ore body definition--about 7 to 10 yr from a producing mine if land-based analogy holds. Capital costs for special ships and shore-based plants will be at least equivalent to land-based mining operations.

Tinsley of Amoco Minerals Company (1976) also pointed out that production forecasts have failed to include adequate plant lead times. He forecast the mining of only 2 or 3 million short tons of nodules in 1985, but as much as 12 million short tons per yr by the year 2000. He estimated that 13% of deep-sea mining gross revenues would be derived from cobalt production. Nodules would provide 8 to 10% of world cobalt supply in 1985 and still less than 25% in 2000.

At least three methods for mining deep-sea nodules have been developed: the continuous bucket-line method; the dragline bucket method; and hydraulic suction dredging (Anonymous, 1975).

The National Academy of Sciences (Panel on Operational Safety in Marine Mining, 1975) concluded that ocean mining would cause few environmental problems. Bottom dwellers would not be seriously affected if heavy metals were leached from resuspended bottom sediment because they are so sparsely distributed.

The National Oceanic and Atmospheric Administration sponsored the Deep Ocean Mining Environmental Study (DOMES), which began in 1975. The mining method studied involved vacuum collection of nodules along with some sediment and small bottom dwellers. A plume of sediment was discharged by the mining ship at the surface. The disturbed material settled rapidly. Particles and chemical differences were not detected after 24 hr (Anonymous, 1978g).

Healy (1976) concluded that the concentrations of Fe, Ni, Co, Mn, and Cu in the water of the DOMES area were between

0.02 to 0.7 mg/liter. He assumed that if the discharge plume from mining operations was diluted by 1,000 and if preliminary experiments (with Zn, Pb, Cu, Ca) were indicative of mixing increases, the increases of heavy metals concentrations in seawater due to mining operations "will be hard to detect."

It is assumed that processing of manganese nodules can be done economically only on shore. Proposed plant sites have not been reported. Hydrometallurgical and pyrometallurgical processes have been developed that are suitable for recovering metal values from deep-sea nodules. These processes do not appear to pose environmental risks any worse than those pyrometallurgical and hydrometallurgical methods presently used for processing nickel, copper, and cobalt ores.

The ammonia leach process, which gives a 60% net recovery of cobalt, involves these steps: crushing, partial drying, grinding, reduction roasting, leaching with ammonia-ammonium carbonate, air sparging and countercurrent decanting to recover soluble ammine metal complexes, solvent extracting of copper and nickel in a fluidized ion-exchange process, electrowinning of nickel and copper, and precipitating of cobalt as the sulfide.

Deepsea Ventures, Inc., obtained three patents in 1975 for metal recovery from manganese nodules. Hydrochloric acid is the leaching agent in all three patents. Recovery operation steps include volatilization of metal chlorides, ion-exchange separation, solvent extraction, and/or electrolysis (Sibley, 1977).

The Kennecott consortium selected an ammonia leach hydrometallurgical process. The Kennecott fluidized ion-exchange process was described in detail by Agarwal et al. (1976). Cobalt concentrations in process streams range up to 200 mg/liter. The aqueous effluent from the ammonia scrubs do not contain cobalt, but the copper strip aqueous effluent contains 12 mg Co/liter, and the nickel strip aqueous effluent (pH 3.0) contains 4.2 mg Co/liter.

A less capital-intensive sulfur oxide roast process gives a 90% cobalt recovery. It involves these steps:

crushing, partial drying, grinding, roasting in an atmosphere of air and sulfur dioxide, water leaching, thickening to reject manganese tails and iron, cementing copper on iron scrap, precipitating cobalt, nickel, and residual copper with hydrogen sulfide, redissolving the precipitate in sulfuric acid, and solvent extracting and electrowinning of cobalt and nickel (Tinsley, 1976, Anonymous, 1978e).

The International Nickel Consortium plans to use a pyrometallurgical process considerably different from the one described by Tinsley (1976). Unground nodules are dried, prereduced with coal, and smelted in an electric furnace to give a matte containing the nickel, copper, and cobalt (5%) and a slag in which the manganese is concentrated. The matte will probably be pressure leached and the metal values recovered by ion exchange and electrowinning (Anonymous, 1978e). The Inco process was described in detail by Sridhar et al. (1976).

Mining and processing technology is ready, but companies are reluctant to proceed with mining deep-sea nodules until legalities are made clear. Although the United Nations Law of the Sea Conference first convened in 1968, an internationally acceptable treaty has yet to be negotiated.

Large profits had been envisioned by the early exploring groups; but terms proposed at the Law of the Sea meeting in 1978 would allow the exploitation of only 10 mine sites per yr, each 40,000 km^2 in area (Anonymous, 1978k).

An international seabed authority would control all deep-sea mining activity under the present draft treaty. The United States, on the other hand, supports a system whereby the authority would have a commercial arm that would grant contracts for harvesting half of the nodules and would mine the other half itself (Abrams, 1978).

The United States administration has decided to push for federal legislation to regulate United States companies engaged in seabed mining (Abrams, 1978). Such a bill was expected in March to be signed into law before the end of 1979. Bills had been introduced in both House and Senate in early 1979. There had been general agreement on similar bills in

1978, but the House-passed bill died in the crush of bills awaiting Senate approval at the close of the legislative session. There will be no provision to protect United States investments if the United States later signs an international treaty preempting the miners' rights. The proposed bills do provide for paying royalties to a fund that would benefit Third World nations (Wesson, 1979).

In April 1979, the U.S. Department of Commerce released a deep-sea mining report as Congress opened hearings on the bills. The report stated that a deep-sea mining policy is crucial to the United States not only for economic but also for security reasons since the critical metals cobalt and manganese are presently almost totally imported from politically unstable or communist countries. This report estimated that a single deep-sea mine could produce 4,000 short tons (3,600 MT) of cobalt annually (11% of world supply) and possibly 750,000 short tons of manganese (6.9%).* The report also stated that exploitation of U.S. land-based cobalt reserves would still not be a significant source of supply for this country (Stern, 1979).

2. <u>Cobalt as an impurity in materials processed in the U.S. for other metals</u>

a. <u>Oxidized iron ores and steel</u>: Kaiser et al. (1954), in their compilation of data on the exhaustive search in U.S. ores and smelter products for rare and critical metals during and shortly after World War II, report that four magnetite ore samples from the Iron Mountain mine (Burke Orebody), Iron Mountain District, Iron County, Utah, contained 200 to 700 mg Co/kg. In addition, the flue dust from a Provo, Utah, steel plant, using ore from the same mine, contained 100 mg Co/kg.

Table IV-5 in Chapter IV gives values of < 4 to 64 mg Co/kg in U.S. iron minerals with 42 mg/kg in hematite from the Mesabi District, Michigan.

---

* Scientists are still not certain that manganese is economically recoverable from the nodules.

Other information on the cobalt content of U.S. iron ores or steel dusts was not found. Probably most of the cobalt in the ore will be carried into the pig iron rather than into the slag (Tolmacheu, 1934; cited by Bogomolov, 1937).

The cobalt source in steels may be both the pig iron or the ferronickel added to produce stainless or low-alloy steels. Banning and Anable (1961) reported that ferronickel produced from Cuban or Philippine Island serpentine containing about 500 mg Co/kg (0.05%)--if the sole source of cobalt--would give standard-grade stainless steels containing $\leq$ 0.3% cobalt. Most stainless steels contain cobalt; some, up to 0.4%.

Reserve Mining Company discharges slurries of taconite tailings to Lake Superior at Silver Bay, Minnesota. These tailings cover the sediment for more than 2,000 km^2. Korda et al. (1977) found similar concentrations of cobalt and other heavy metals in the flesh of the benthic sculpin fish taken from areas of Lake Superior that had high and low concentrations of taconite tailings. Sculpin have relatively small ranges of movement, so that the populations apparently had been restricted to the immediate vicinity where they were sampled. These fish have enhanced exposure to sediment particles and the elements dissolved in the interstitial water because they feed by forcefully expelling water to disturb the sediment, thereby locating their prey. Sediment samples from 0 to 41 cm contained 18.3 to 25.2 mg Co/kg, showing no trend with depth. The taconite tailings were dissolved by acid leaching or fusion with sodium peroxide. Neutron activation analyses of these solutions gave 6.1 and 13.5 mg Co/kg in the taconite tailings, respectively. Apparently, there is less available cobalt in the tailings than in the natural sediment of the lake.

b. Lead-zinc ores: U.S. lead-zinc ores contain up to 100 mg Co/kg. The compilation of Kaiser et al. (1954) was searched for cobalt concentrations of 80 mg/kg or more in zinc concentrates. Since the crustal abundance of cobalt is about 25 mg/kg, this artificial cut-off was chosen to examine concentrations in associated ores, lead concentrates, and tailings (see Table VI-4). Even with this artificial selection, the materials are not strikingly enriched in cobalt.

TABLE VI-4

COBALT IN LEAD-ZINC ORES AND CONCENTRATES
(Kaiser et al., 1954)[a/]

Cobalt-Bearing Material	Cobalt Concentration, mg/kg, in: Ores, Mill Heads	Lead Concentrate	Lead Tailings	Zinc Concentrate	Zinc Tailings	Lead-Zinc Tailings	Cu-Pb-Zn Tailings	Copper Tailings	Other
Lead concentrate, composite mill sample. Ore from Tennessee mine, Wallapai district, Mohave County, Arizona. Tennessee-Schuylkill mill, Chloride, Arizona. Sampled in 1943.		80							
Zinc concentrate, composite mill sample. Ore from Tennessee mine, Wallapai district, Mohave County, Arizona. Tennessee-Schuylkill mill, Chloride, Arizona. Sampled in 1943.				200					
Zinc-lead tailings. Ore from Tennessee mine, Wallapai district, Mohave County, Arizona. Tennessee-Schuylkill mill, Chloride, Arizona. Sampled in 1943.						80			
Lead concentrate. Ore from Duquesne, Bonaza, Estella, Holland, Empire, and Pride of the West mines, Patagonia district, Santa Cruz County, Arizona. Callahan Zinc-Lead Company mill, Nogales, Arizona. Sampled in 1943.		100							
Zinc concentrate. Ore from Duquesne, Bonaza, Estella, Holland, Empire, and Pride of the West mines, Patagonia district, Santa Cruz County, Arizona. Callahan Zinc-Lead Company mill, Nogales, Arizona. Sampled in 1943.				1,000					
Copper-lead-zinc tailings. Ore from Duquesne, Bonaza, Estella, Holland, Empire, and Pride of the West mines, Patagonia district, Santa Cruz County, Arizona. Callahan Zinc-Lead Company mill, Nogales, Arizona. Sampled in 1943.							40		

a/ This is an artificial selection of values appearing in the work by Kaiser et al. Only values of $\geq$ 80 mg Co/kg in the zinc concentrate are included.

TABLE VI-4 (continued)

Cobalt-Bearing Material	Cobalt Concentration, mg/kg, in								
	Ores, Mill Heads	Lead Concentrate	Lead Tailings	Zinc Concentrate	Zinc Tailings	Lead-Zinc Tailings	Cu-Pb-Zn Tailings	Copper Tailings	Other
Zinc concentrate, mill grab sample. Ore from Mammoth and Collins mines, Old Hat district, Pinal County, Arizona. Mammoth-St. Anthony, Ltd., mill, Tiger, Arizona. Sampled in 1942.				800					
Lead concentrate, mill grab sample. Ore from Mammoth and Collins mines, Old Hat district, Pinal County, Arizona. Mammoth-St. Anthony, Ltd., mill, Tiger, Arizona. Sampled in 1942.		600							
Molybdenum-vanadium flotation concentrate, mill grab sample. Ore from Mammoth and Collins mines, Old Hat district, Pinal County, Arizona. Mammoth-St. Anthony, Ltd., mill, Tiger, Arizona. Sampled in 1942.									100
Zinc concentrates (two), mill samples. Copper-zinc deposits, eastern United States. Sampled in 1943.				100					
Zinc-lead heads, mill sample. Ore from Amazon mine, Beaver district, Shoshone County, Idaho. Hercules mill, Wallace, Idaho. Sampled in 1944.	30								
Lead concentrate. Ore from Amazon mine, Beaver district, Shoshone County, Idaho. Hercules mill, Wallace, Idaho. Sampled in 1944.		60							
Zinc concentrate. Ore from Amazon mine, Beaver district, Shoshone County, Idaho. Hercules mill, Wallace, Idaho. Sampled in 1944				400					
Zinc-lead tailings, mill sample. Ore from Amazon mine, Beaver district, Shoshone County, Idaho. Hercules mill, Wallace, Idaho. Sampled in 1944.						0			

TABLE VI-4 (continued)

	Cobalt Concentration, mg/kg, in								
Cobalt-Bearing Material	Ores, Mill Heads	Lead Concentrate	Lead Tailings	Zinc Concentrate	Zinc Tailings	Lead-Zinc Tailings	Cu-Pb-Zn Tailings	Copper Tailings	Other
Zinc concentrate, grab sample. Ore from Morning mine, Hunter district, Shoshone County, Idaho. Morning mill, Mullan, Idaho. Federal Mining and Smelting Company. Sampled in 1943.				400					
Lead-zinc-silver tailings, grab sample. Ore from Morning mine, Hunter district, Shoshone County, Idaho. Morning mill, Mullan, Idaho. Federal Mining and Smelting Company. Sampled in 1943.						30			
Lead heads, mill sample, 48-hr run. Ore from Dayrock mine, Placer Center district, Shoshone County, Idaho. Dayrock mill, Wallace, Idaho. Sampled in 1944.	10								
Lead concentrate, mill sample, 48-hr run. Ore from Dayrock mine, Placer Center district, Shoshone County, Idaho. Dayrock mill, Wallace, Idaho. Sampled in 1944.		100							
Lead tailings, mill sample, 48-hr run. Ore from Dayrock mine, Placer Center district, Shoshone County, Idaho. Dayrock mill, Wallace, Idaho. Sampled in 1944.			0						
Zinc-lead heads, mill sample, 24-hr run. Ore from Parrott mine, Beaver district, Shoshone County, Idaho. Dayrock mill, Wallace, Idaho. Sampled in 1944.	20								
Lead concentrate, mill sample, 24-hr run. Ore from Parrott mine, Beaver district, Shoshone County, Idaho. Dayrock mill, Wallace, Idaho. Sampled in 1944.		20							
Zinc concentrate, mill sample, 24-hr run. Ore from Parrott mine, Beaver district, Shoshone County, Idaho. Dayrock mill, Wallace, Idaho. Sampled in 1944.				200					
Zinc-lead tailings, mill sample, 24-hr run. Ore from Parrott mine, Beaver district, Shoshone County, Idaho. Dayrock mill, Wallace, Idaho. Sampled in 1944.						0			

TABLE VI-4 (continued)

Cobalt-Bearing Material	Cobalt Concentration, mg/kg in								
	Ores, Mill Heads	Lead Concentrate	Lead Tailings	Zinc Concentrate	Zinc Tailings	Lead-Zinc Tailings	Cu-Pb-Zn Tailings	Copper Tailings	Other
Zinc-lead heads. Ore from Silver Tip mine, Beaver district, Shoshone County, Idaho. Dayrock mill, Wallace, Idaho. Sampled in 1944.	10								
Lead concentrate. Ore from Silver Tip mine, Beaver district, Shoshone County, Idaho. Dayrock mill, Wallace, Idaho. Sampled in 1944.		50							
Zinc concentrate. Ore from Silver Tip mine, Beaver district, Shoshone County, Idaho. Dayrock mill, Wallace Idaho. Sampled in 1944				200					
Zinc-lead tailings. Ore from Silver Tip mine, Beaver district, Shoshone County, Idaho. Dayrock mill, Wallace Idaho. Sampled in 1944.						0			
Lead concentrate, grab sample. Ore from Morning mine, Hunter district, Shoshone County, Idaho. Morning mill, Mullan, Idaho. Federal Mining and Smelting Company. Sampled in 1943.		100							
Zinc-lead heads, mill sample. Ore from Constitution mine, Pine Creek district, Shoshone County, Idaho. Spokane-Idaho Mining Company mill, Kellogg, Idaho. Sampled in 1944.	10								
Zinc-lead bulk concentrate. Ore from Constitution mine, Pine Creek district, Shoshone County, Idaho. Spokane-Idaho Mining Company mill, Kellogg, Idaho. Sampled in 1944.		90							
Lead concentrate, mill sample, 24-hr run. Ore from Nabob mine, Pine Creek district, Shoshone County, Idaho. Amy Matchless mill, Pine Creek, Idaho. Sampled in 1944.		80							

TABLE VI-4 (continued)

Cobalt-Bearing Material	Cobalt Concentration, mg/kg, in Ores, Mill Heads	Lead Concentrate	Lead Tailings	Zinc Concentrate	Zinc Tailings	Lead-Zinc Tailings	Cu-Pb-Zn Tailings	Copper Tailings	Other
Zinc concentrates, mill sample, 24-hr run. Ore from Nabob mine, Pine Creek district, Shoshone County, Idaho. Amy Matchless mill, Pine Creek, Idaho. Sampled in 1944.				100					
Zinc-lead tailings, mill sample, 24-hr run. Ore from Nabob mine, Pine Creek district, Shoshone County, Idaho. Amy Matchless mill, Pine Creek, Idaho. Sampled in 1944.						20			
Lead concentrate, grab sample. Ore from Page and Blackhawk mines, Yreka district, Shoshone County, Idaho. Page mill, Page, Idaho. Federal Mining and Smelting Company. Sampled in 1943.		300							
Zinc concentrate, grab sample. Ore from Page and Blackhawk mines, Yreka district, Shoshone County, Idaho. Page mill, Page, Idaho. Federal Mining and Smelting Company. Sampled in 1943.				600					
Zinc-lead-silver tailings, mill grab sample. Ore from Page and Blackhawk mines, Yreka district, Shoshone County, Idaho. Page mill, Page, Idaho. Federal Mining and Smelting Company. Sampled in 1943.						100			
Lead concentrate, grab sample. Ore from upper levels, Frisco mine, Lelande district, Shoshone County, Idaho. Hull Leasing Company mill, Gem, Idaho. Sampled in 1943.		100							
Zinc concentrate, grab sample. Ore from upper levels, Frisco mine, Lelande district, Shoshone County, Idaho. Hull Leasing Company mill, Gem, Idaho. Sampled in 1943.				500					
Zinc-lead-silver tailings, mill grab sample. Ore from upper levels, Frisco mine, Lelande district, Shoshone County, Idaho. Hull Leasing Company mill, Gem, Idaho. Sampled in 1943.						80			

TABLE VI-4 (continued)

Cobalt-Bearing Material	Cobalt Concentration, mg/kg, in: Ores, Mill Heads	Lead Concentrate	Lead Tailings	Zinc Concentrate	Zinc Tailings	Lead-Zinc Tailings	Cu-Pb-Zn Tailings	Copper Tailings	Other
Zinc concentrate from Fresnillo, Zacatecas, Mexico. Granby smelter, E. St. Louis, Illinois. American Zinc Company of Illinois. Sampled in 1943.				400					
Zinc concentrate from Huanchaca, Boliva. Granby smelter, E. St. Louis, Illinois. American Zinc Company of Illinois. Sampled in 1943.				100					
Zinc concentrate from Candalaria mine, Chihauhau, Mexico. Granby smelter, E. St. Louis, Illinois. American Zinc Company of Illinois. Sampled in 1943.				100					
Zinc concentrate from Peru. Granby smelter E. St. Louis, Illinois. American Zinc Company of Illinois. Sampled in 1943.				80					
Zinc-lead mill heads. Ore from High Grade mine, Tri-State district, Jasper County, Missouri. High Grade mill, Chitwood, Missouri. Sampled in 1942.	20								
Zinc concentrate, coarse. Ore from High Grade mine, Tri-State district, Jasper, County, Missouri. High Grade mill, Chitwood, Missouri. Sampled in 1942.				100					
Zinc flotation concentrate. Ore from High Grade mine, Tri-State district, Jasper, County, Missouri. High Grade mill, Chitwood, Missouri. Sampled in 1942.				100					
Lead concentrate, coarse. Ore from High Grade mine, Tri-State district, Jasper County, Missouri. High Grade mill, Chitwood, Missouri. Sampled in 1942.		0							
Zinc-lead tailings, mill sample. Ore from High Grade mine, Tri-State district, Jasper County, Missouri. High Grade mill, Chitwood, Missouri. Sampled in 1942.						10			

TABLE VI-4 (continued)

Cobalt-Bearing Material	Cobalt Concentration, mg/kg, in: Ores, Mill Heads	Lead Concentrate	Lead Tailings	Zinc Concentrate	Zinc Tailings	Lead-Zinc Tailings	Cu-Pb-Zn Tailings	Copper Tailings	Other
Lead-zinc feed, mill sample. Coal removed by picking. Ore from open pit mine, near Versailles, Moniteau County, Missouri. Wemhauer mill, Versailles, Missouri. Sampled in 1942.	40								
Coarse zinc jig concentrate. Ore from open pit mine, near Versailles, Moniteau County, Missouri. Wemhauer mill, Versailles, Missouri. Sampled in 1942.				300					
Coarse lead jig concentrate. Ore from open pit mine, near Versailles, Moniteau County, Missouri. Wemhauer mill, Versailles, Missouri. Sampled in 1942.		10							
Lead-zinc tailings, mill sample. Ore from open pit mine, near Versailles, Moniteau County, Missouri. Wemhauer mill, Versailles, Missouri. Sampled in 1942.						10			
Zinc concentrate, grab sample. Ore from Liberty mine, Troy district, Lincoln County, Montana. Liberty Metals Company mill, Troy, Montana. Sampled in 1944.				90					
Lead-silver-zinc ore from 300-ft level, Liberty mine, Troy district, Lincoln County, Montana. Sampled in 1944.	80								
Lead-silver-zinc ore from 1,100-ft level, Liberty mine, Troy district, Lincoln County, Montana. Sampled in 1944.	60								
Lead-zinc heads, composite mill sample. Ore from Broadwater mine, Montana district, Cascade County, Montana. Klies mill, near Neihart, Montana. Sampled in 1943.	50								
Lead concentrate. Ore from Broadwater mine, Montana district, Cascade County, Montana. Klies mill, near Neihart, Montana. Sampled in 1943.		80							

TABLE VI-4 (continued)

Cobalt-Bearing Material	Cobalt Concentration, mg/kg, in: Ores, Mill Heads	Lead Concentrate	Lead Tailings	Zinc Concentrate	Zinc Tailings	Lead-Zinc Tailings	Cu-Pb-Zn Tailings	Copper Tailings	Other
Zinc concentrate, composite mill sample. Ore from Broadwater mine, Montana district, Cascade County, Montana. Klies mill, near Neihart, Montana. Sampled in 1943.				100					
Lead-zinc tailings, composite mill sample. Ore from Broadwater mine, Montana district, Cascade County, Montana. Klies mill, near Neihart, Montana. Sampled in 1943.						30			
Lead-zinc heads, mill sample. Ore from Lexington, Benton, and Snowdrift mines, Montana district, Cascade County, Montana. Lexington mill, near Neihart, Montana. Sampled in 1943.	30								
Lead-zinc bulk concentrate. Ore from Lexington, Benton, and Snowdrift mines, Montana district, Cascade County, Montana. Lexington mill, near Neihart, Montana. Sampled in 1943.		100							
Lead-zinc tailings, mill sample. Ore from Lexington, Benton, and Snowdrift mines, Montana district, Cascade County, Montana. Lexington mill, near Neihart, Montana. Sampled in 1943.						20			
Lead-zinc heads, mill sample. Ore from Star (Evening Star) and London mines, Montana district, Cascade County, Montana. Star mill, near Neihart, Montana. Sampled in 1943.	40								
Lead-zinc bulk concentrate. Ore from Star (Evening Star) and London mines, Montana district, Cascade County, Montana. Star mill, near Neihart, Montana. Sampled in 1943.		100							
Lead-zinc tailings, mill sample. Ore from Star (Evening Star) and London mines, Montana district, Cascade County, Montana. Star mill, near Neihart, Montana. Sampled in 1943.						20			

TABLE VI-4 (continued)

Cobalt-Bearing Material	Cobalt Concentration, mg/kg, in								
	Ores, Mill Heads	Lead Concentrate	Lead Tailings	Zinc Concentrate	Zinc Tailings	Lead-Zinc Tailings	Cu-Pb-Zn Tailings	Copper Tailings	Other
Lead-zinc ore from Orchard vein, Lyman Township, Grafton County, New Hampshire.	100								
Zinc-lead ore from Ore Hill (Warren) mine, Warren Township, Grafton County, New Hampshire. Sampled in 1943.	100								
Zinc ore, composite sample, from 350-ft level, Oswaldo mine, Central district, Grant County, New Mexico. Kennecott Copper Corporation.	x00								
Zinc-lead ore from Bullfrog and Lutz mines, Central district, Grant County, New Mexico. U.S. Smelting, Refining and Mining Company. Sampled in 1944.	0								
Zinc concentrate. Ore from Bullfrog, Lutz, Slate, and Shingle Canyon mines, Central district, Grant County, New Mexico. Bullfrog mill, Bayard, New Mexico. U.S. Smelting, Refining, and Mining Company. Sampled in 1944.				100					
Lead concentrate. Ore from Bullfrog, Lutz, Slate, and Shingle Canyon mines, Central district, Grant County, New Mexico. Bullfrog mill, Bayard, New Mexico. U.S. Smelting, Refining, and Mining Company. Sampled in 1944.		0							
Zinc-lead tailings, mill sample. Ore from Bullfrog, Lutz, Slate, and Shingle Canyon mines, Central district, Grant County, New Mexico. Bullfrog mill, Bayard, New Mexico. U.S. Smelting, Refining, and Mining Company. Sampled in 1944.						0			
Lead-zinc tailings from tailings pile. Ore from McGhee mine, San Simon district, Hidalgo County, New Mexico. McGhee mill, near Steins, New Mexico. Sampled in 1944.						80			
Lead-zinc ore from Shelburne Lead mine, Shelburne Township, Coos County, New Hampshire. Sampled in 1943.	80								

TABLE VI-4 (continued)

Cobalt-Bearing Material	Cobalt Concentration, mg/kg, in: Ores, Mill Heads	Lead Concentrate	Lead Tailings	Zinc Concentrate	Zinc Tailings	Lead-Zinc Tailings	Cu-Pb-Zn Tailings	Copper Tailings	Other
Coarse lead-zinc feed. Ore from Evans-Wallower No. 7 mine, Tri-State district, Ottawa County, Oklahoma. Evans-Wallower No. 7 mill, Cardin, Oklahoma. Sampled in 1942.	0								
Lead jig concentrate. Ore from Evans-Wallower No. 7 mine, Tri-State district, Ottawa County, Oklahoma. Evans-Wallower No. 7 mill, Cardin, Oklahoma. Sampled in 1942.		0							
Zinc jig concentrate. Ore from Evans-Wallower No. 7 mine, Tri-State district, Ottawa County, Oklahoma. Evans-Wallower No. 7 mill, Cardin, Oklahoma. Sampled in 1942.				80					
Zinc-bearing material. Ruth (old Amalgamated) mine, Little North Santiam River district, Marion County, Oregon. Sampled in 1943.									200
Zinc-bearing material. Ruth (old Amalgamated) mine, Little North Santiam River district, Marion County, Oregon. Sampled in 1943.									300
Zinc-bearing material. Ruth (old Amalgamated) mine, Little North Santiam River district, Marion County, Oregon. Sampled in 1943.									300
Zinc jig concentrate. Ore from Mascot mine, Mascot district, Knox County, Tennessee. Mascot No. 2 mill, Mascot, Tennessee. American Zinc Company of Tennessee. Sampled in 1943.				100					
Zinc flotation concentrate. Ore from Mascot mine, Mascot district, Knox County, Tennessee. Mascot No. 2 mill, Mascot, Tennessee. American Zinc Company of Tennessee. Sampled in 1943.				100					

TABLE VI-4 (continued)

Cobalt-Bearing Material	Cobalt Concentration, mg/kg, in Ores, Mill Heads	Lead Concentrate	Lead Tailings	Zinc Concentrate	Zinc Tailings	Lead-Zinc Tailings	Cu-Pb-Zn Tailings	Copper Tailings	Other
Zinc slime overflow, mill sample. Ore from Mascot mine, Mascot district, Knox County, Tennessee. Mascot No. 2 mill, Mascot, Tennessee. American Zinc Company of Tennessee. Sampled in 1943.					0				
Lead-zinc tailings slimes, mill sample. Ore from Embree mine, Bumpuss cove district, Unicoi and Washington Counties, Tennessee. Embree mill, Embreeville, Tennessee. Sampled in 1943.						500 (only 50 mg/kg in Zn concentrate from carbonate ore)			
Zinc concentrate from El Potosi mine, Santa Eulalia, Mexico. Dumas smelter, Dumas, Moore County, Texas. Peru Mining Company. Sampled in 1943.				100					
Zinc concentrate, probably from Tri-State district, Dumas smelter, Dumas, Moore County, Texas. Peru Mining Company. Sampled in 1943.				700					
Lead-zinc-copper heads, mill sample, 10-day run. Ore from Valzinco mine, Spotsylvania County, Virginia. Valzinco Operation mill, Paytes, Virginia. Sampled in 1943.	60								
Lead-zinc-copper concentrates, 1-mo run. Ore from Valzinco mine, Spotsylvania County, Virginia. Valzinco Operation mill, Paytes, Virginia. Sampled in 1943.									40
Zinc concentrates, 1-mo run. Ore from Valzinco mine, Spotsylvania County, Virginia. Valzinco Operation mill, Paytes, Virginia. Sampled in 1943.				80					
Lead-zinc-copper tailings, mill sample, 10-day run. Ore from Valzinco mine, Spotsylvania County, Virginia. Valzinco Operation mill, Paytes, Virginia. Sampled in 1943.						40			

TABLE VI-4 (concluded)

Cobalt-Bearing Material	Cobalt Concentration, mg/kg, in								
	Ores, Mill Heads	Lead Concentrate	Lead Tailings	Zinc Concentrate	Zinc Tailings	Lead-Zinc Tailings	Cu-Pb-Zn Tailings	Copper Tailings	Other
Lead concentrate. Ore from Grandview and Bella May mines, Metaline district, Pend Oreille County, Washington. Metaline Mining and Leasing Company. Sampled in 1943.				80					
Zinc ore from dump, New England lease, Northport district, Stevens County, Washington. Sampled in 1943.	80								
Zinc flotation tailings from tailings pond. Ores from Wisconsin and northern Illinois. Vinegar Hill Custom mill, Cuba City, Grant County, Wisconsin. Sampled in 1942.					50				
Zinc concentrates. Ores from Wisconsin and northern Illinois. Vinegar Hill Custom mill, Cuba City, Grant County, Wisconsin. Sampled in 1942.				400					

The cobalt concentrations in zinc concentrates range up to 1,000 mg/kg, but the mean for the 80- to 1,000-mg/kg range samples is only 10 times the crustal abundance. The cobalt concentrations in the associated lead concentrates range up to 300 mg/kg. For this artificially chosen group, cobalt concentrations averaged about nine times greater in the zinc concentrate and four times greater in the lead concentrate than in the ore. Vhay et al. (1973) estimated that $\geq$ 100 mg Co/kg is present in some lead ores produced in the new lead belt in Missouri.

For a lead-zinc (massive sulfide) ore associated with pyrite, however, Mukherjee et al. (1973) found that the cobalt was preferentially concentrated in the pyrite concentrate (109 to 167 mg Co/kg) rather than in the zinc (4.6 to 6.5 mg Co/kg) or lead (< 5 to 16.5 mg/kg) concentrates, their tailings (34 to 49 mg Co/kg), or the pyrite tailings (0.9 to 5.5 mg Co/kg).

Other literature data were not found on the cobalt concentrations of U.S. lead-zinc ores. Values as high as 1,000 mg Co/kg may not be unusual. The zinc concentrate used as National Bureau of Standards Standard Reference Material 113a was prepared at the Magmont Mines, Bixby, Missouri. The certificate of analysis (Cali, 1975) gives an uncertified 0.11% as the cobalt concentration.

(1) <u>The lead circuit</u>: Silver-bearing lead concentrates generally have significant amounts of cobalt, nickel, and copper (Schack and Clemmons, 1965). Cobalt was about 20 times more abundant than silver in the ores of the Bonneterre-Flat River areas in the northern part of the old lead belt, which was the major source of all silver produced in Missouri in 1965. At 20 times the 300,000 troy oz silver produced, at least 9 MT cobalt went through the lead circuit that year (Vhay et al., 1973).

Parker (1979) in his flow diagram of the lead circuit (Figure VI-6) showed that the major split-off point for cobalt is the speiss obtained after reverberatory smelting of the copper dross. Drossing (removal of copper dross from the lead melt) of the lead bullion from the blast furnace is done before softening (antimony removal) and desilverizing. Parker

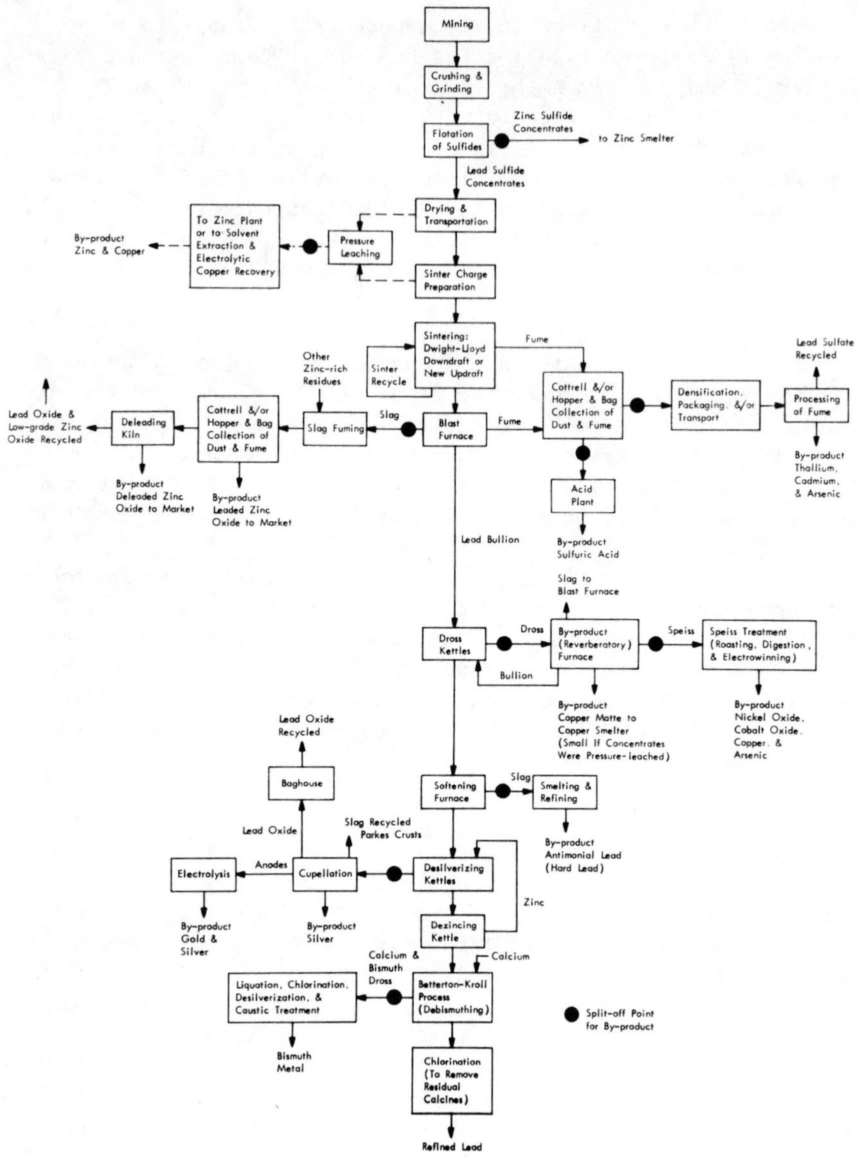

Figure VI-6. Lead System Showing By-Product Metals (Parker, 1979).

(1979) shows that the products of dross reverberatory smelting* are copper matte as well as speiss. Speiss is roasted, digested, and electrowon to give nickel oxide, cobalt oxide, copper, and arsenic. Matte and speiss from copper dross smelting by the three ASARCO Inc. lead-smelters are treated at the Tacoma, Washington, and El Paso, Texas, copper smelters (Cole, 1978; Morris, 1978); however, these materials have been shipped to Belgium for the past few years by Amax Lead and Zinc, Inc., Boss, Missouri; St. Joe Minerals Corporation, Herculaneum, Missouri; and Bunker Hill Company, Kellogg, Idaho (Morris, 1978; Blickensderfer, 1978).

Cobalt may also be found in lead desilverization crusts along with cadmium, copper, nickel, platinum, tellurium, antimony, arsenic, bismuth, silver, and gold. These crusts are processed in silver refineries. Zinc is distilled off, and the gold-silver alloy (doré metal) is cupelled to remove lead and bismuth. The doré is cast into anodes and electrolytically refined (McKay, 1967). None of the many references we have examined on the metal contents of U.S. silver refinery electrolyte or anode slimes mentions cobalt. Presumably, any cobalt in the anode will dissolve in the electrolyte and be removed from the bleed solution by cementation along with copper on scrap iron.

Some cobalt (80 mg/kg) was found in a lead-zinc slag at the Northport, Washington, smelter smelting ores from Coeur d'Alene, Idaho, and the Northport, Washington, district (Kaiser et al., 1954).

(2) The zinc circuit

(a) Pyrometallurgical: Most of the cobalt in the zinc concentrate apparently remains in the calcine (see Table VI-5) and, subsequently, in the sinter. However, if zinc is processed in a blast furnace, cobalt is transferred into

---

* Engineering Science, Incorporated (1973) reported that a dross reverberatory furnace feed may contain reverberatory bullion, copper matte, nickel matte, and lead as well as dross. In the feed examined at one lead plant, the dross contained 0.5% cobalt.

TABLE VI-5

COBALT IN CALCINES FROM ROASTING ZINC-IRON CONCENTRATES
(Kaiser et al., 1954)

Material	Cobalt Concentration, mg/kg
Iron sulfide and oxide magnetic roast (roasted once). Zinc-iron jig concentrates from Wisconsin and northern Illinois. Vinegar Hill Custom mill, Cuba City, Grant Co., Wisc. Sampled in 1943.	100
Iron sulfide and oxide magnetic roast (roasted twice). Zinc-iron jig concentrates from Wisconsin and northern Illinois. Vinegar Hill Custom mill, Cuba City, Grant Co., Wisc. Sampled in 1943.	20
Iron oxide calcines of zinc-iron concentrates from Wisconsin and northern Illinois. Vinegar Hill Custom mill, Cuba City, Grant Co., Wisc. Sampled in 1943.	60
Iron oxide calcines of zinc-iron concentrates from Wisconsin and northern Illinois. Vinegar Hill Custom mill, Cuba City, Grant Co., Wisc. Sampled in 1943.	100
Iron oxide calcines of zinc-iron concentrates from Wisconsin and northern Illinois. Vinegar Hill Custom mill, Cuba City, Grant Co., Wisc. Sampled in 1943.	80
Iron oxide calcines of zinc-iron concentrates from Wisconsin and northern Illinois. Vinegar Hill Custom mill, Cuba City, Grant Co., Wisc. Sampled in 1943.	40

TABLE VI-5 (concluded)

Material	Cobalt Concentration; mg/kg
Iron oxide calcines from flash-roasted flotation concentrate of iron sulfide. Zinc-iron jig concentrates from Wisconsin and northern Illinois. Vinegar Hill Custom mill, Cuba City, Grant Co., Wisc. Sampled in 1943.	100
Iron oxide calcines from flash-roasted flotation concentrate of iron sulfide. Zinc-iron jig concentrates from Wisconsin and northern Illinois. Vinegar Hill Custom mill, Cuba City, Grant Co., Wisc. Sampled in 1943.	100
Zinc calcines (smelter heads) from Ozark Smelting and Mining Company smelter, Coffeyville, Kans. Original zinc concentrate from Tri-State district. American Zinc Company of Illinois smelter, Monsanto, St. Clair Co., Ill. Sampled in 1943.	600
Zinc calcines. Concentrates from Buchans, Newfoundland. Monsanto smelter, Monsanto, St. Clair Co., Ill. American Zinc Company of Illinois. Sampled in 1943.	100
Zinc calcines from Athletic Mining and Smelting Company smelter, Fort Smith, Ark. Concentrates from Fresnillo, Mexico. Monsanto smelter, Monsanto, St. Clair Co., Ill. American Zinc Company of Illinois. Sampled in 1943.	300
Zinc calcines from Dumas, Texas, smelter. Concentrate from Potosi mine, Santa Eulalia, Mexico. Monsanto smelter, Monsanto, St. Clair Co., Ill. American Zinc Company of Illinois. Sampled in 1943.	100
Zinc oxide calcines from American Zinc Lead and Smelting Company oxide plant, Columbus, Ohio. Concentrates from Mascot, Tenn. Granby smelter, E. St. Louis, Ill. American Zinc Company of Illinois. Sampled in 1943.	50

the waste slag. The latter has not been common U.S. pyrometallurgical practice. Zinc sinter has been generally retorted. According to Phillips (1962), the cobalt remains in the retorting residue. Perhaps in the past, retort residues have been returned to a lead circuit (Phillips) or treated by a Waelz process (Parker, 1979); however, neither practice is followed by the only two remaining pyrometallurgical zinc smelters in the United States today. At St. Joe Minerals, Monaca, Pennsylvania, the residue is physically separated into a low-grade ferrosilicon, a zinc-rich fraction that is returned to the sintering operation, and a zinc-lead fraction that is sold as ballast (Lund et al., 1970). Cobalt probably follows the ferrosilicon or the zinc-rich fraction. At the New Jersey Zinc Company plant in Palmerton, Pennsylvania, the residue is sent to the sinter bank (Smith, 1978).

When slab zinc is refined to high-grade zinc by distillation, cobalt becomes enriched in the residue (Gregor'ev and Yaroslavets, 1975).

The concentration of cobalt in other zinc smelter products (slags, dusts, zinc oxide, and clinkers) sampled during World War II ranged from 50 to 500 mg/kg (Kaiser et al., 1954).

(b) Electrolytic: Dennis (1961) described in detail (see Figure VI-7) the common practice for leaching zinc calcines and purifying the leachate (acid extraction and a neutral purification in a two-stage countercurrent continuous process). Copper and cadmium are removed from the neutral thickener overflow by cementation on zinc dust, and cobalt is removed by nitroso-β-naphthol.

Cobalt and the other impurities must be removed from the electrolyte because when present they affect the hydrogen overvoltage, the current efficiency, and degrade the zinc deposit (Andre and Delvaux, 1970; Montague, 1974).

Some information was found in the literature on cobalt concentrations in U.S. electrolytic zinc plant products:

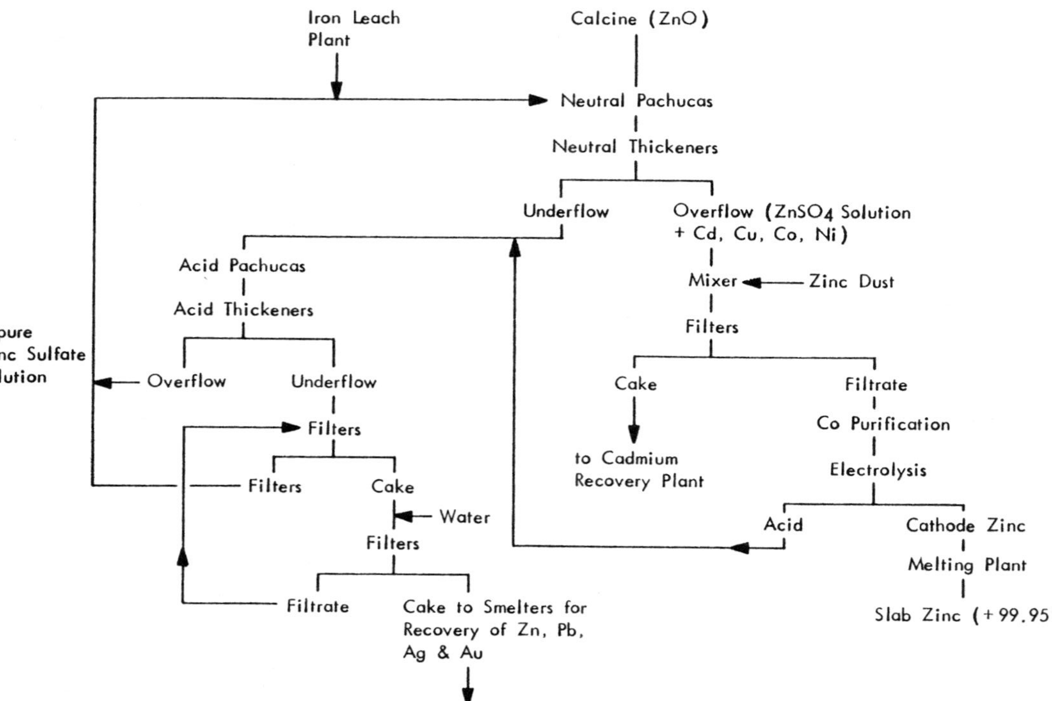

Figure VI-7. Continuous-Leaching Plant Flow Sheet (Dennis, 1961).

At the Sullivan Mining Company electrolytic zinc plant in Kellogg, Idaho, Weimer et al. (1959) reported that cobalt, nickel, and some zinc were removed from the leach solution by a second zinc dust precipitation. The solution after iron removal contained 215 mg Zn/liter, 300 mg Cu/liter, 600 mg Cd/liter, and 30 mg Co/liter. Copper was cemented on zinc dust; and then the solution concentrations of cobalt, antimony, cadmium, and nickel were reduced to < 1 mg/liter by adding more zinc dust (Dennis, 1961). Cobalt was not mentioned as being present at several other U.S. electrolytic zinc plants. In 1965, the plant (then owned by Bunker Hill) recovered 147 tons (131 MT) residue containing 6,703 lb (3.0 MT) cobalt (Lynde, 1966). The fate of these residues was not learned.

The copper cake at the Great Falls (Montana) Reduction Works of Anaconda Copper Mining Company contained 300 mg Co/kg (Kaiser et al., 1954). In 1943, Moore filter cake and copper-cadmium cake at the Monsanto smelter of the American Zinc Company of Illinois contained 100 and 500 mg Co/kg, respectively.

In the first stage cobalt and arsenic removal step at the Sauget, Illinois, plant, when operated by American Zinc Company, the concentration of cobalt in the primary thickener overflow was lowered from 30 mg/liter* to < 0.1 mg/liter after zinc dust and arsenic trioxide were added to the solution at 95°C. Cadmium and cobalt were removed in the second stage purification by adding zinc dust, copper sulfate, and antimony tartrate (Banes et al., 1970).

In 1977, $\sim$ 65% of the Sauget refinery's zinc concentrates were produced by the Boss, Missouri, lead-zinc mine and mill. These concentrates are said to have a "high cobalt content" (Anonymous, 1978b).

Because concentrations for all products at one U.S. plant were not found in the literature, processes at

* The cobalt content represented 200 mg/kg zinc present in the thickener overflow.

some foreign plants were examined to find typical cobalt concentrations. The information is presented in Table VI-6.

Losses of cobalt in iron precipitates from zinc sulfate leach solutions are only about 1% (Scott, 1975). For example, at a concentration of 15 g Co/liter, $\leq$ 1% of the cobalt present in a zinc electrolyte was removed in the goethite (FeOOH) precipitate at pH 2 to 3.5 and 70 to 90°C. The precipitates contained $<$ 0.4% cobalt (Davey and Scott, 1976).

Apparently, a significant fraction of the cobalt is present in the cadmium cake sent to the electrolytic cadmium refinery (see Table VI-6). After leaching the cake with sulfuric acid, the solution is purified in much the same way the zinc electrolyte was purified. Purified cadmium electrolytes at two of three U.S. plants contained 27 to 30 mg Co/liter. Although Lund and Sheppard (1964) surveyed the minor element content of zinc and cadmium refinery intermediates and products, they reported no other cobalt concentrations in U.S. plants.

The purified zinc electrolyte typically contains $<$ 0.1 mg Co/liter (Montague, 1974). Sheka and Karlysheva (1959) studied the distribution of cobalt upon zinc sulfate electrolysis in the presence of manganese dioxide (which collects elements in the anode slime; a common additive). For electrolytes containing 25 to 30 mg Co/liter and 15 to 20 mg Mn/liter, the cobalt distribution was: cathode deposit, 0.1%, anode slime, 0.4 to 2.1%; and spent electrolyte, 97.8 to 99.5%.

Spent zinc electrolyte, which is used for leaching, must be periodically or continuously bled for purification or discharge to waste to remove impurities such as cobalt, fluoride, chloride, magnesium, and sodium that gradually build up in the electrolyte (Schack and Clemmons, 1965).

Thus, minor losses of cobalt from the electrolytic zinc and cadmium circuit occur from discarded spent electrolyte and iron cake and in the cadmium sponge. Larger amounts are transferred via the copper cake to the copper circuit.

TABLE VI-6

COBALT IN ELECTROLYTIC ZINC AND CADMIUM PLANT PRODUCTS

Plant	Process	Cobalt in Product (mg/kg or mg/liter)	Reference
Monteponi e Montevecchio plant, Milan, Italy	Roasting of concentrate	Calcines 120	de Michelis and Gnesotto (1970)
	Leaching	Leachate 11	
	Cementation on Zn powder, nitroso-β-naphthol at pH 2.8 (-Co), activated carbon, aq. $KMnO_4$, $Ca(OH)_2$ (-Fe), Zn dust plus $CuSO_4$ (-Cd)	Electrolyte 0.4[a]	
Vieille Montagne, Balen, Belgium	Neutral leaching	Neutral solution. 5 to 10[b]	Andre and Delvaux (1970)
	Zn scrap (-Cu), Sb dust (-Co and Cu)	Cu-Co cake, 7,500; filtrate, 0.25	
	Zn dust	Cd cake, 3,000	
		Zn electrolyte, 0.00034 to 0.00037	Liessens et al. (1969)
	Leaching of Cd cake with recycled $ZnSO_4$ (50 mg Co/liter), Zn dust, Sb	Cd solution, 600	Andre and Delvaux (1970)
Plant in Spain operating on concentrates from the Cartegena Range	Roasting	Calcines, 11	del Valle Alonso et al. (1970)
	Leaching	Overflow neutral thickener, 12 to 15	
		Overflow acid thickener, 10 to 12	
		Residue, 0	
Outokumpu Oy, Finland	Zn dust plus As_2O_3 to improve Co and Ni removal	Cu cake, 1,000	Fugleberg (1973)
	Zn dust	Cd cake, 50	
	Leaching Cd cake	Cd electrolyte, 500	
	Electrolysis of Cd electrolyte containing 500 mg Co/liter	Cathode Cd, 10 (1 to 5% of Co in the Cd cake)	
	Electrolysis of Cd electrolyte containing 100 mg Co/liter	Cathode Cd, 2.7	

a/ In the absence of dilution, the value would indicate 96% removal of cobalt from the leachate.

b/ Cobalt was not found in the acid solution (from acid leaching of the neutral leaching residue) nor was it in the final residue.

One report of environmental cobalt pollution, presumably from an electrolytic zinc plant, was found. Ayling and Bloom (1976) reported that a residential area in Hobart, Tasmania, was exposed to dust blown over many kilometers from piles of metal concentrates stored by an electrolytic zinc refinery. Cobalt concentrations in the dust fallout remained above 100 mg/kg to within at least 13 km downwind of the refinery. However, at the site where the dust concentrations of zinc, copper, manganese, iron, cadmium, and lead were highest (44.8%, 0.440%, 4.45%, 8.6%, 0.145%, and 3.01%, respectively), the cobalt concentration was only 0.0182% (182 mg/kg). The site where the cobalt dust concentration was highest (1,490 mg/kg) had the second highest concentrations of iron (4.97%), but the other heavy metals were not nearly so high. This would imply that some other metallurgical operation in the vicinity was emitting cobalt.

c. Copper ores: In the survey of minor metals in U.S. ores and smelter products (Kaiser et al., 1954), samples of copper ores (excluding that from the cobalt-copper mine in the Blackbird district in Idaho) and concentrates contained up to hundreds of milligrams cobalt per kilogram. (See Table VI-7.) The cobalt concentration is apparently highly variable. For example, samples from the same mine (the Utah Copper mine in Salt Lake County, Utah) contained none to up to hundreds of milligrams cobalt per kilogram. Copper smelter products, e.g., slags and flue dusts, contained the same range of cobalt concentrations. (See Table VI-8.)

Probably less than 70% of the cobalt in a copper sulfide ore is recovered in the copper concentrate. At the Rokana mine in Zambia, flotation of the copper-cobalt ore recovers 68 to 70% of the cobalt that was in the ore in a single copper-cobalt concentrate. And at Rokana, cobalt recovery has been maximized because of its economic importance to the country.

Figure VI-8 gives a generalized flow sheet for copper processing. The percentages alongside the arrows indicate MRI estimates of the cobalt distribution in that step generalized from the following literature data. The weights beside the arrows represent the maximum transfer of cobalt if it mostly follows the copper until electrolysis.

TABLE VI-7

COBALT IN U.S. COPPER ORES, MILL HEADS, AND CONCENTRATES
(Kaiser et al., 1954)

Cobalt-Bearing Material	Cobalt Concentration, mg/kg
Copper concentrate. Ore from Bagdad mine, Eureka District, Yavapai County, Arizona. Bagdad mill, Bagdad, Arizona. Sampled in 1943.	800
Copper tailings. Ore from United Verde mine, Verde (Jerome) district, Yavapai County, Arizona. United Verde Smelter, Clarkdale, Arizona. Phelps Dodge Corp. Sampled in 1943.	100
Copper concentrate, mill sample. Ore from Castle Dome open pit. Castle Dome mill, near Miami, Gila County, Arizona. Castle Dome Copper Co., Inc. Sampled in 1943.	100
Copper concentrates, composite sample of 32-day run. Ore from Morenci open pit. Phelps Dodge and Defense Plant Corp. mill, Morenci, Greenlee County, Arizona. Phelps Dodge Corp. Sampled in 1945.	80
Copper tailings. Ore from New Cornelia open pit, Ajo. New Cornelia concentrator, Ajo, Pima County, Arizona. Phelps Dodge Corp. Sampled in 1943.	80
Leached capping, porphyry copper, Ray mine, Ray, Pinal County, Arizona. Kennecott Copper Corp. Sampled in 1948.	x00
Schist ore, porphyry copper, Ray mine, Ray, Pinal County, Arizona. Kennecott Copper Corp. Sampled in 1948.	x00
Amphibolite schist ore, porphyry copper. Ray mine, Ray, Pinal County, Arizona. Kennecott Copper Corp. Sampled in 1948.	x00
Copper mill heads. Ore from Ray mine. Hayden mill, Hayden, Pinal County, Arizona. Kennecott Copper Corp. Sampled in 1948.	0
Copper concentrate. Ore from Ray mine. Hayden mill, Hayden, Pinal County, Arizona. Kennecott Copper Corp. Sampled in 1948.	x,000
Copper tails. Ore from Ray mine. Hayden mill, Hayden, Pinal County, Arizona. Kennecott Copper Corp. Sampled in 1948.	x00
Copper concentrate. Ore from Leatherwood mine. Mill of Control Mines, Inc., Oracle, Catalina district, Pima County, Arizona. Sampled in 1942.	1,000
Copper concentrate. Ore from Duquesne, Bonaza, Estella, Holland, Empire, and Pride of the West mines, Patagonia district, Santa Cruz County, Arizona. Callahan Zinc-Lead Company mill, Nogales, Arizona. Sampled in 1943.	300
Copper concentrate, mill sample. Copper-zinc deposits, eastern United States. Sampled in 1943.	800

TABLE VI-7 (continued)

Cobalt-Bearing Material	Cobalt Concentration, mg/kg
Copper concentrate, mill sample. Copper-zinc deposits, eastern United States. Sampled in 1943.	300
Iron sulfide concentrate, mill sample. Copper-zinc deposits, eastern United States. Sampled in 1943.	500
Iron sulfide concentrate, mill sample. Copper-zinc deposits, eastern United States. Sampled in 1943.	400
Iron concentrate, magnetic, mill sample. Copper-zinc deposits, eastern United States. Sampled in 1943.	50
Copper-zinc tailings, mill sample. Copper-zinc deposits, eastern United States. Sampled in 1943.	60
Copper-zinc tailings, mill sample. Copper-zinc deposits, eastern United States. Sampled in 1943.	30
Mill heads. Pyrrhotite deposit, eastern United States. Sampled in 1943.	400
Concentrate from -3 + 8 tables, mill sample. Pyrrhotite deposit, eastern United States. Sampled in 1943.	500
Tails from -3 + 8 tables, mill sample. Pyrrhotite deposit, eastern United States. Sampled in 1943.	300
Concentrate from -8 + 20 tables, mill sample. Pyrrhotite deposit, eastern United States. Sampled in 1943.	600
Tails from -8 + 20 tables, mill sample. Pyrrhotite deposit, eastern United States. Sampled in 1943.	100
Concentrate from -20 + 48 tables, mill sample. Pyrrhotite deposit, eastern United States. Sampled in 1943.	500
Tails from -20 + 48 tables, mill sample. Pyrrhotite deposit, eastern United States. Sampled in 1943.	100
Concentrate from -48 tables, mill sample. Pyrrhotite deposit, eastern United States. Sampled in 1943.	400
Tails from -48 tables, mill sample. Pyrrhotite deposit, eastern United States. Sampled in 1943.	300
Light dust from tables, mill sample. Pyrrhotite deposit, eastern United States. Sampled in 1943.	200
Heavy dust from tables, mill sample. Pyrrhotite deposit, eastern United States. Sampled in 1943.	300
Copper ore, chip sample, from small lens, Uncle Sam mine, Blackbird district, Lemhi County, Idaho. Sampled in 1944.	10,000 (1%)

TABLE VI-7 (continued)

Cobalt-Bearing Material	Cobalt Concentration, mg/kg
Copper ore from the Trade Dollar, Eagle, and Lucky Strike claims, Revais Creek district, Sanders County, Montana. Green Mountain Copper mill, near Dixon, Montana. Sampled in 1943.	400
Copper mill heads. Ore from Chino mine, Central district, Grant County, New Mexico. Hurley mill, Hurley, New Mexico. Kennecott Copper Corp. Sampled in 1948.	$\underline{x}$00
Copper concentrate. Ore from Chino mine, Central district, Grant County, New Mexico. Hurley mill, Hurley, New Mexico. Kennecott Copper Corp. Sampled in 1948.	$\underline{x}$0
Molybdenum concentrate. Ore from Chino mine, Central district, Grant County, New Mexico. Hurley mill, Hurley, New Mexico. Kennecott Copper Corp. Sampled in 1948.	$\underline{x}$00
Copper tailings, sand. Ore from Chino mine, Central district, Grant County, New Mexico. Hurley mill, Hurley, New Mexico. Kennecott Copper Corp. Sampled in 1948.	$\underline{x}$00
Molybdenum tailings, sand. Ore from Chino mine, Central district, Grant County, New Mexico. Hurley mill, Hurley, New Mexico. Kennecott Copper Corp. Sampled in 1948.	$\underline{x}$0
Copper-gold-silver concentrate. Ore from Bonney and Anita No. 1 mines, Lordsburg district, Hidalgo County, New Mexico. Banner mill, Lordsburg, New Mexico. Sampled in 1943.	200
Copper-gold-silver tailings, mill sample. Ore from Bonney and Anita No. 1 mines, Lordsburg district, Hidalgo County, New Mexico. Banner mill, Lordsburg, New Mexico. Sampled in 1943.	60
Copper-gold-silver heads, composite mill sample. Ore from Snake-Opportunity-Litelking group of mines and the Golden Era and Biglow mines, Las Anima's district, Sierra County, New Mexico. Horseshoe mill, (dismantled), 1 mile E of Hillsboro, New Mexico. Black Dome Mining Corp. Sampled in 1944.	80
Copper-gold-silver flotation concentrate. Ore from Snake-Opportunity-Likelking group of mines and the Golden Era and Biglow mines, Las Animas district, Sierra County, New Mexico. Horseshoe mill, (dismantled), 1 mile E of Hillsboro, New Mexico. Black Dome Mining Corp. Sampled in 1944.	800
Copper-gold-silver concentrate. Ore from Snake-Opportunity-Litelking group of mines and the Golden Era and Biglow mines, Las Animas district, Sierra County, New Mexico. Horseshoe mill, (dismantled), 1 mile E of Hillsboro, New Mexico. Black Dome Mining Corp. Sampled in 1944.	700

TABLE VI-7 (continued)

Cobalt-Bearing Material	Cobalt Concentration, mg/kg
Copper-gold-silver pulps, composite mill sample. Ore from Snake-Opportunity-Litelking group of mines and the Golden Era and Biglow mines, Las Animas district, Sierra County, New Mexico. Horseshoe mill, (dismantled), 1 mile E of Hillsboro, New Mexico. Black Dome Mining Corp. Sampled in 1944.	500
Copper-gold-silver tailings, mill sample. Ore from Snake-Opportunity-Litelking group of mines and the Golden Era and Biglow mines, Las Animas district, Sierra County, New Mexico. Horseshoe mill, (dismantled), 1 mile E of Hillsboro, New Mexico. Black Dome Mining Corp. Sampled in 1944.	30
Copper-gold-silver tailings, mill sample. Ore from Snake-Opportunity-Litelking group of mines and the Golden Era and Biglow mines, Las Animas district, Sierra County, New Mexico. Horseshoe mill, (dismantled), 1 mile E of Hillsboro, New Mexico. Black Dome Mining Corp. Sampled in 1944.	80
Copper ore, grab sample from mine. Utah Copper mine, West Mountain district, Salt Lake County, Utah. Kennecott Copper Corp. Sampled in 1948.	0
Copper ore, grab sample from mine. Utah Copper mine, West Mountain district, Salt Lake County, Utah. Kennecott Copper Corp. Sampled in 1948.	x0
Copper ore, grab sample from mine. Utah Copper mine, West Mountain district, Salt Lake County, Utah. Kennecott Copper Corp. Sampled in 1948.	x00
Copper ore, grab sample from mine. Utah Copper mine, West Mountain district, Salt Lake County, Utah. Kennecott Copper Corp. Sampled in 1948.	x0
Copper ore, grab sample from mine. Utah Copper mine, West Mountain district, Salt Lake County, Utah. Kennecott Copper Corp. Sampled in 1948.	x00
Copper ore, grab sample from mine. Utah Copper mine, West Mountain district, Salt Lake County, Utah. Kennecott Copper Corp. Sampled in 1948.	0
Copper ore, composite sample from mine. Utah Copper mine, West Mountain district, Salt Lake County, Utah. Kennecott Copper Corp. Sampled in 1948.	0
Copper ore, composite sample from mine. Utah Copper mine, West Mountain district, Salt Lake County, Utah. Kennecott Copper Corp. Sampled in 1948.	x0
Copper ore, composite sample from mine. Utah Copper mine, West Mountain district, Salt Lake County, Utah. Kennecott Copper Corp. Sampled in 1948.	x0
Copper ore, composite sample from mine. Utah Copper mine, West Mountain district, Salt Lake County, Utah. Kennecott Copper Corp. Sampled in 1948.	0

TABLE VI-7 (continued)

Cobalt-Bearing Material	Cobalt Concentration, mg/kg
Copper ore, composite sample from mine. Utah Copper mine, West Mountain district, Salt Lake County, Utah. Kennecott Copper Corp. Sampled in 1948.	x0
Copper ore, composite sample from mine. Utah Copper mine, West Mountain district, Salt Lake County, Utah. Kennecott Copper Corp. Sampled in 1948.	x00
Mill heads, composite sample. Ore from Utah Copper mine, West Mountain district, Salt Lake County, Utah. Arthur plant, Arthur, Utah. Kennecott Copper Corp. Sampled in 1948.	0
Copper concentrate, composite sample. Ore from Utah Copper mine, West Mountain district, Salt Lake County, Utah. Arthur plant, Arthur, Utah. Kennecott Copper Corp. Sampled in 1948.	x00
Molybdenite concentrate, composite sample. Ore from Utah Copper mine, West Mountain district, Salt Lake County, Utah. Arthur plant, Arthur, Utah. Kennecott Copper Corp. Sampled in 1948.	x0
General tailings, composite sample. Ore from Utah Copper mine, West Mountain district, Salt Lake County, Utah. Arthur plant, Arthur, Utah. Kennecott Copper Corp. Sampled in 1948.	0
Siliceous concentrate from molybdenite plant. Ore from Utah Copper mine, West Mountain district, Salt Lake County, Utah. Arthur plant, Arthur, Utah. Kennecott Copper Corp. Sampled in 1948.	x0
Air wash sediment from molybdenite plant, composite sample. Ore from Utah Copper mine, West Mountain district, Salt Lake County, Utah. Arthur plant, Arthur, Utah. Kennecott Copper Corp. Sampled in 1948.	x00
Unit "C" molybdenite tailings, composite sample. Ore from Utah Copper mine. West Mountain district, Salt Lake County, Utah. Arthur plant, Arthur, Utah. Kennecott Copper Corp. Sampled in 1948.	x00
Filtrate from leach of molybdenite plant, composite sample. Ore from Utah Copper mine, West Mountain district, Salt Lake County, Utah. Arthur plant, Arthur, Utah. Kennecott Copper Corp. Sampled in 1948.	x0
Ash from burlap from concentrate discharge weirs (flat cells), composite sample. Ore from Utah Copper mine, West Mountain district, Salt Lake County, Utah. Arthur plant, Arthur, Utah. Kennecott Copper Corp. Sampled in 1948.	x00
Concentrate from gold launders, composite sample. Ore from Utah Copper mine, West Mountain district, Salt Lake County, Utah. Arthur plant, Arthur, Utah. Kennecott Copper Corp. Sampled in 1948.	x00

TABLE VI-7 (concluded)

Cobalt-Bearing Material	Cobalt Concentration, mg/kg
Copper heads, mill sample. Ore from Elizabeth mine, Orange County, Vermont. Vermont Copper Company mill, South Stafford, Vermont. Sampled in 1943.	500
Copper concentrate. Ore from Elizabeth mine, Orange County, Vermont. Vermont Copper Company mill, South Stafford, Vermont. Sampled in 1943.	1,000
Copper tailings, mill sample. Ore from Elizabeth mine, Orange County, Vermont. Vermont Copper Company mill, South Stafford, Vermont. Sampled in 1943.	300
Copper concentrate, mill sample. Ore from Holden mine, Chelan Lake district, Chelan County, Washington. Howe Sound Company. Sampled in 1943.	100
Zinc concentrate, mill sample. Ore from Holden mine, Chelan Lake district, Chelan County, Washington. Holden mill, Holden, Washington. Howe Sound Company. Sampled in 1943.	100
Copper-zinc tailings, mill sample. Ore from Holden mine, Chelan Lake district, Chelan County, Washington. Howe Sound Company. Sampled in 1943.	80

TABLE VI-8

COBALT IN U.S. COPPER SMELTER PRODUCTS, mg/kg
(Kaiser et al., 1954)

	Slag	Flue Dust	Acid Plant Sludge	Flux	Matte
Copper slag from reverberatory smelter. Ore from Morenci open pit and others (Cu concentration 80 mg/kg Co). Morenci smelter, Morenci, Greenlee County, Arizona. Phelps Dodge Corp. Sampled in 1943.	600				
Copper slag. Ore from United Verde mine, Verde (Jerome) district, Yavapai County, Arizona. United Verde Smelter, Clarkdale, Arizona. Phelps Dodge Corp. Sampled in 1943.	300				
Acid plant sludge from chamber plant. Copper-zinc deposits, eastern United States. Sampled in 1943.			100		
Copper slag from blast furnace. Copper-zinc deposits, eastern United States. Sampled in 1943.	100				
Copper slag from reverberatory furnace. Copper-zinc deposits, eastern United States. Sampled in 1943.	100				
Copper slag from converter. Copper-zinc deposits, eastern United States. Sampled in 1943.	400				
Copper flue dust from blast furnace. Copper-zinc deposits, eastern United States. Sampled in 1943.		100			
Copper flue dust from reverberatory furnace. Copper-zinc deposits, eastern United States. Sampled in 1943.		80			
Copper flue dust from converter. Copper-zinc deposits, eastern United States. Sampled in 1943.		300			
Wet flue dust from iron roaster. Copper-zinc deposits, eastern United States. Sampled in 1943.		200			
Dry flue dust from iron roaster. Copper-zinc deposits, eastern United States. Sampled in 1943.		100			
Copper slag from reverberatory smelter. Ore from Morenci open pit and others. Morenci smelter, Morenci, Greenlee County, Arizona. Phelps Dodge Corp. Sampled in 1943.	600				

TABLE VI-8 (continued)

	Slag	Flue Dust	Acid Plant Sludge	Flux	Matte
Copper slag from reverberatory smelter. Ore from Copper Queen and other mines in vicinity. Douglas smelter, Douglas, Cochise County, Arizona. Phelps Dodge Corp. Sampled in 1943.	100				
Copper slag from dump, grab sample. Ore from mines of the Copper Basin district, Custer County, Idaho. MacKay Standard smelter. Sampled in 1943.	200				
Copper slag from dump. Ore from Ore Knob mine, Ashe County, North Carolina. Sampled in 1943.	500				
Copper ore flux. Beartooth quartzite containing streaks of chalcocite, from Lee Hill, Grant County, New Mexico. Hurley mill, Hurley, New Mexico. Sampled in 1948.				0	
Copper ore flux. Beartooth quartzite containing streaks of chalcocite. From old dump, Hurley mill, Hurley, New Mexico. Sampled in 1948.				$\underline{x}0$	
Copper matte. Hurley mill, Hurley, New Mexico. Sampled in 1948.					$\underline{x}00$
Copper flue dust from waste heat boiler. Hurley mill, Hurley, New Mexico. Sampled in 1948.		$\underline{x}00$			
Heavy copper flue dust from converter. Hurley mill, Hurley, New Mexico. Sampled in 1948.		$\underline{x}00$			
Light copper flue dust from converter. Hurley mill, Hurley, New Mexico. Sampled in 1948.		$\underline{x}00$			
Cottrell flue dust from converter and reverberatory furnace. Hurley mill, Hurley, New Mexico. Sampled in 1948.			$\underline{x}00$		
Copper slag from reverberatory furnace. Hurley mill, Hurley, New Mexico. Sampled in 1948.	$\underline{x}00$				
Copper slag from converter. Hurley mill, Hurley, New Mexico. Sampled in 1948.	$\underline{x}00$				
Copper slag from refining furnace. Hurley mill, Hurley, New Mexico. Sampled in 1948.	$\underline{x}00$				

TABLE VI-8 (concluded)

	Slag	Flue Dust	Acid Plant Sludge	Flux	Matte
Roaster flue dust from molybdenite plant. Ore from Utah Copper mine, West Mountain district, Salt Lake County, Utah. Arthur plant, Arthur, Utah. Kennecott Copper Corp. Sampled in 1948.		x00			
Copper-gold slag from slag pile. Ore from the Rossland district, British Columbia. Northport smelter, Northport, Washington (closed). Sampled in 1944.	80				

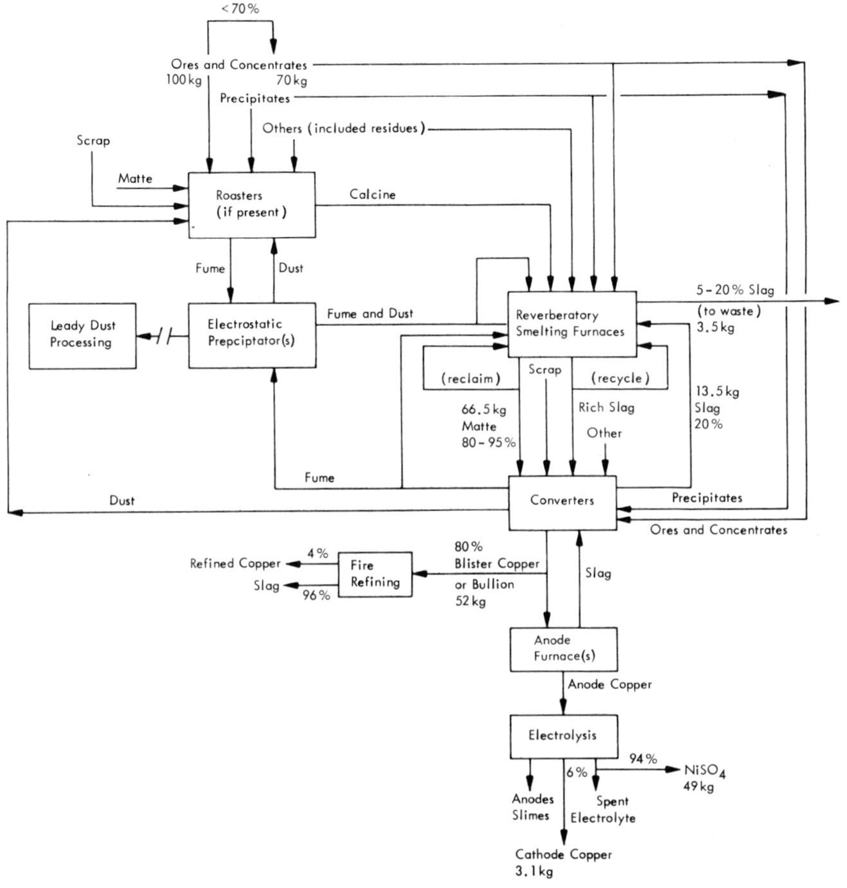

Figure VI-8. Generalized Flow Sheet for Copper Processing (Parker, 1978).

Arbitrarily, 100 kg cobalt was chosen as the ore content (100 kg Co/MT ore equals 100 mg Co/kg, an ore concentration similar to those in Table VI-7).

- Sinev and Govarova (1977) found 83% of the cobalt during reverberatory smelting going into the crude matte and 17% into the waste slag.

- At the Rokana mine in Zambia, the reverberatory furnace matte, and eventually the converter slag, contained 80% of the cobalt in the concentrate. The discarded reverberatory furnace slag contained 20% of the cobalt in the concentrate (Whyte et al., 1977). The converter slag is processed at Rokana for cobalt recovery.

- During reduction (usually by reverberatory smelting) of the concentrate or partially roasted concentrate to produce a matte, the cobalt in the charge is distributed 95% to the matte and 5% to the slag (Biswas and Davenport, 1976). When the matte is converted in a separate step, cobalt primarily remains dissolved in the blister copper (Nelson, 1968). However, in a continuous smelting process most of the cobalt is concentrated in the converter slag (Biswas and Davenport, 1976).

- In a laboratory study of the converting reaction, Kashima et al. (1978) found that the cobalt concentration in the copper after heating with matte and slag at 1300°C and partial SO_2 pressures 0.7 to 10 kPa was 0.014 to 0.083%. The matte contained 0.019 to 0.083% cobalt, and the silica-saturated iron silicate slag contained 0.28 to 1.3% cobalt. The form of cobalt in the slag was presumed to be CoO. Since weights were not given of the various phases, distribution calculations cannot be made.

- Black copper (a somewhat less pure copper than blister copper) that is obtained from smelting copper-nickel sulfide ores contains a significant amount of cobalt, iron, and nickel. An artificial system representative of the constituents present during fire refining (pure copper oxide, copper, nickel, and ^{60}Co-tagged cobalt) was heated under argon at 1300°C.

Approximately 92 to 98% of the cobalt was found in the copper oxide phase, which would be representative of the slag components in actual practice (Burylev et al., 1976).

● Copper containing 0.08% Ni and 0.0244% Co was fire-refined in Morgan crucibles by synthetic slags containing, e.g., 46% FeO, 16% B_2O_3, and 38% SiO_2. Refining at 1300°C for 60 min gave copper containing $\geq$ 0.0150% Ni and $\geq$ 0.0010% Co. If the weight of the copper was not changed significantly in the process, the best cobalt removal by slag was 96% (Komorova, 1973).

Most copper in the United States is refined by electrolysis rather than by fire-refining. The Co, As, Bi, Fe, Ni, and Sb that dissolve* electrochemically in the copper-refining electrolyte must be removed from the electrolyte to avoid their occlusion in the cathode copper. They are removed from a bleed stream of electrolyte in a purification circuit. Commonly, this bleed solution is treated by three stage electrowinning. (Commercial-grade copper may alternatively be removed as copper sulfate by neutralizing the electrolyte with copper shot; this procedure replaces the first two stages). In the third stage, most of the As, Sb, and Bi are deposited on the cathodes. Upon evaporating most of the water from the electrolyte, nickel sulfate precipitates along with iron and cobalt sulfates. After purification, the nickel sulfate is sold for chemical uses including electroplating (Biswas and Davenport, 1976). From MRI calculations, apparently up to 70% of the cobalt in the copper concentrate may be removed from the copper circuit in the nickel sulfate by-product.

Crude nickel sulfate was produced by ASARCO electrolytic copper refineries in Tacoma, Baltimore, and Perth Amboy. The crude nickel sulfate was refined in Perth Amboy starting in 1954 by hot water leaching, air oxidizing, and adjusting the pH with calcium carbonate solution to precipitate iron, precipitating copper and zinc as the sulfides, and crystallizing the purified solution (Busch et al., 1961).

---

* Apparently, a negligible amount of cobalt may remain in the anode slimes. Only one report of cobalt in copper anode slimes was found. Luo et al. (1978) found $\sim$ 20 mg Co/kg in a Chinese copper refinery anode slime.

At SEC Corporation in El Paso, Texas, a plant went on-stream ca. 1972 to recover copper and nickel from the liquor discharged from the final evaporation stage in copper sulfate crystallization from copper-refining electrolytes that were bled for purification. The liquor comes from the Phelps Dodge copper refinery. Before the SEC recovery process, copper was cemented from the mother liquor of the crystallizer and the remainder of the solution, with its nickel content, was discarded. The description of the SEC plant process (Anonymous, 1972; Flett, 1974) does not mention cobalt. At SEC, nickel and copper are extracted by organic solvents; the pH is controlled by ammonia additions. After acid-stripping of the solvent extracts, nickel and copper are recovered by electrowinning.

Data on cobalt in U.S. copper emissions were not found in the literature. For a "worst case," we might examine the smelting of copper-nickel-cobalt ores in Canada. We have calculated from the reports of Hutchinson and Whitby (1974a, 1974b) that the Coniston and Copper Cliff smelters and the Iron Ore Recovery Plant in Sudbury, Ontario, emitted $\sim$ 53 MT/yr cobalt based on emissions of the last quarter of 1971. Calculations based on an internal 1973 Inco Inc. report (International Nickel Company, Inc., 1973) indicate that the average annual cobalt emission from all the Inco smelters was at least 71 MT in the period 1961 to 1971. From these data (presented in Table VI-9), we calculated that the loss of cobalt in smelting emissions is less than 5% of the cobalt actually produced.

Another environmental problem to consider in the copper industry is aqueous wastes. Information was available on cobalt in the U.S. copper industry effluents. Less than 5% of the copper industry must deal with acid mine waters. Some discharge the acid mine waters directly to surface waters (sometimes after lime precipitation and setting), and some use the waters in leach and mill operations. One U.S. mine pumps 27,524.5 m^3 (27.5 million liters) of acid mine water per day containing 0.32 mg Co/liter (8.8 kg Co/day or 3,200 kg/yr). Another pumps 3,274 m^3 acid mine water per day containing 1.9 mg Co/liter or 2,300 kg/yr (Jarrett and Kirby, 1978).

A small amount of copper is recovered from tailings and oxidized ores in the United States by leaching followed

TABLE VI-9

COBALT EMISSIONS FROM THE INCO INC. SMELTERS IN SUDBURY, ONTARIO
(International Nickel Company, Inc., 1973)

Copper Cliff	Total Emissions Within the Period 1961-1971, MT	Average Emissions/Yr, MT	Cobalt Content, %	Annual Cobalt Content, MT
Orford stack	22,501	2,046	0.17	3.5
Converter roof monitors	6,758 (10 yr)	676	?	-
Nickel stack	305,735 (11 yr)	27,794	0.11	30.6
Copper stack	33,833 (11 yr)	3,075	0.24	7.4
Coniston				
Sinter stack	8,943 (7 yr)	1,277	0.19	2.4
Smelter stack	118,751 (11 yr)	10,796	0.20	21.6
Iron ore stack	8,405 (9 yr)	934	0.23	2.1
No. 1 pellet sinter machine	9,093 (9 yr)	1,011	0.022	0.2
No. 2 pellet sinter machine	24,076 (9 yr)	2,675	0.022	0.6
Ducon scrubbers	756 (8 yr)	94	?	-
				70.8

by electrowinning or cementation. Cobalt would be expected to be dissolved by sulfuric acid leaching methods and especially by ammonia solution methods such as the Arbiter process practiced at Anaconda, Montana (Williams, 1975).

Cobalt may build up to high concentrations in sulfuric acid leach solutions (e.g., of six copper mine leach solutions, five contained 3.30 to 72.0 mg Co/liter). Bleed solutions to reduce such buildup are evaporated in lagoons. One barren vat leach acid solution after copper electrowinning contained 51.0 mg Co/liter. Such wastes are evaporated to dryness in an impoundment rather than being discharged (Jarrett and Kirby, 1978).

Although Probert et al. (1975) of Kennecott Copper Corporation mentioned that oxidized cobalt, copper, and other impurities can be selectively removed from a molybdenite concentrate by an ammoniacal leaching solution at 25 to 90°C, the degree of cobalt removal is not illustrated in any of the examples of the patent.

The Bureau of Mines is studying electrowinning of copper from low-grade broken ore at the native copper mine in the Upper Peninsula of Michigan. The ores are leached with ammoniacal ammonium carbonate, and copper is removed by solvent extraction. Copper is recovered by acid stripping the organic solvent and electrowinning with lead-antimony anodes. Small amounts of cobalt sulfate (e.g., 40 to 100 mg Co/liter) were purposely added to the electrolyte to reduce the lead content of the electrowon copper. (Cobalt reduced corrosion at the lead anode.) No cobalt was deposited on the cathode with the copper (Jeffers and Grove, 1977).

Copper mill wastewaters may contain detectable cobalt; however, they are usually recycled fully because of the need for water in the major U.S. copper producing areas. Three of the six operations not in the Southwest use sophisticated lime and settling procedures. The mill wastewater from the same plant producing the barren vat leach solution mentioned above contained only 1.68 mg Co/liter. However, this amounted to 169 kg Co/1,000 MT copper concentrate. Since the annual production of concentrate at this mill was 69,362 MT in a recent year, the cobalt contained in the mill wastes was only 11 MT (Jarrett and Kirby, 1978d).

A pilot plant copper mill wastewater contained 9.8 kg Co/1,000 MT as influent to a polymer coagulant treatment. The effluent discharge contained 0.9 kg/1,000 MT, indicating 90% cobalt removal efficiency (Jarrett and Kirby, 1978b).

d. Other materials: Data on cobalt concentrations in other metal ores from Kaiser et al. (1954) have been compiled in this subsection.

Manganese ores and concentrates produced in the United States make up a very small fraction of the manganese used in domestic metallurgical operations. Therefore, even though the cobalt concentrations in these U.S. materials range up to 0.4% (4,000 mg/kg) (see Table VI-10), they have probably not contributed significant amounts of cobalt to the environment.

Chromite and gold concentrates also are high in cobalt (up to 0.2% Co) but are also of low environmental significance except to the local areas where they are milled (see Tables VI-11 and VI-12). Literature information on environmental pollution by cobalt from gold or chromium mining and milling in the United States was not found. However, Wittmann and Foerstner (1977) reported that South African gold and uranium mines produce acid mine drainage containing up to 2 mg Co/liter. Significant fractions of cobalt are dissolved in the cyanide solutions used for leaching gold. Cobalticyanide is present in uranium solutions and poisons the ion-exchange resin used (Pinkney, 1979).

O'Toole et al. (1973) reported that cobalt was "excessively high" in most samples of soil (1.9 to 50.0 mg/kg), lake water ($\leq$ 63 μg/liter), snow (1.0 to 2.1 μg/liter), vegetation (0.06 to 5.78 mg Co/kg, dry weight), and human hair (mean 0.25 mg/kg) (analyzed by instrumental neutron activation analysis) in the contaminated vicinity ($\leq$ 4.3 km) of the smelters of gold-producing mines at Yellowknife, Northwest Territories, Canada.

TABLE VI-10

COBALT IN MANGANESE ORES AND CONCENTRATES
(Kaiser et al., 1954)

Cobalt-Bearing Material	Cobalt Concentration, mg/kg
Manganese ore, grab sample from high-grade stockpile. Aydelotte mine, Batesville district, Independence County, Arkansas.	500
Manganese ore, grab sample from metals reserve stockpile at Cushman, Arkansas. Aydelotte mine and others, Batesville district, Independence County, Arkansas.	800
Manganese ore, grab sample from metals reserve stockpile at Cushman, Arkansas. Bill Jim, Wildcat, Section 16, Ozark, and other mines, Batesville district, Independence County Arkansas. Sampled in 1942.	800
Manganese ore, grab sample from low-grade metals reserve stockpile at Pfeifer, Arkansas. Bill Jim, Wildcat, Section 16, Ozark, and other mines, Batesville district, Independence County Arkansas. Sampled in 1942.	1,000
Manganese concentrate, shipping grade. Ore from North American mine and custom ores, Glenwood district, Pike County, Arkansas. Black Springs mill, North American Manganese Corp. Sampled in 1942.	3,000
Manganese table tailings. Ore from North American mine and custom ores, Glenwood district, Pike County, Arkansas. Black Springs mill, North American Manganese Corp. Sampled in 1942.	800
Manganese tailings, from log washer. Emerson mine, near Piedmont, Rock Run district, Cherokee County, Alabama. Sampled in 1943.	2,000
Manganese tailings from tailings pile. Ore from Appalachian mine, Cartersville district, Bartow County, Georgia. Appalachian Manganese Corp., Inc. Sampled in 1943.	400
Manganese tailings from tailings pile in gravel pit. Ore from Aubrey mine, Cartersville district, Bartow County, Georgia. Sampled in 1943.	400
Manganese tailings from tailings pile. Ore from Blue Ridge mine, Cartersville district, Bartow County, Georgia. Sampled in 1943.	400
Manganese tailings from tailings pile. Ore from Dobbins mine, Cartersville district, Bartow County, Georgia. Sampled in 1943.	400
Manganese tailings from tailings pile. Ore from Callahan mine, Cave Springs district, Polk County, Georgia. Sampled in 1943.	2,000
Manganese concentrates, metallurgical grade. Manganese ore from Washington and Unicoi Counties, Tennessee. Embree mill, Embreeville, Tennessee. Sampled in 1943.	1,000

TABLE VI-10 (Concluded)

Cobalt-Bearing Material	Cobalt Concentration, mg/kg
Manganese concentrate from dump, low-grade. Manganese ore from Washington and Unicoi Counties, Tennessee. Embree mill, Embreeville, Tennessee. Sampled in 1943.	800
Manganese jig tailings from dump. Manganese ore from Washington and Unicoi Counties, Tennessee. Embree mill, Embreeville, Tennessee. Sampled in 1943.	800
Manganese jig concentrates, Applachian district, Bland County, Virginia. Biggam Manganese Mining Company mill, Bland, Virginia. Sampled in 1942.	2,000
Manganese jig concentrates, Appalachian district, Bland County, Virginia. Conely mill, Hollybrook, Virginia. Sampled in 1942.	2,000
Manganese jig concentrates. Ore from Round Mountain mine, Appalachian district, Bland County, Virginia. Virginia Hardwood Lumber Company mill, Bastian Virginia. Sampled in 1943.	3,000
Coarse manganese concentrate from bins, 1 week's run. Ore from Glade Mountain mine, Smythe County, Virginia Glade Mountain Corp. mill, Marion, Virginia. Sampled in 1942.	4,000
Manganese concentrate (fines), 1 week's run. Ore from Glade Mountain mine, Smythe County, Virginia. Glade Mountain Corp. mill, Marion, Virginia. Sampled in 1942.	1,000
Manganese ore, high-grade, from dump. Ore from mine on W flank of Laramie Mountains, 7 miles south of Marshall, Albany County, Wyoming.	500

TABLE VI-11

COBALT IN CHROMITE ORES AND CONCENTRATES
(Kaiser et al., 1954)

Material	Cobalt Concentration, mg/kg
Chromite heads, composite sample. Ore from Gray Eagle mine, Glenn County, California. Gray Eagle chromite mill, Glenn County, California. Sampled in 1942.	500
Chromite ore, grab sample of mine sample rejects. Ore from Gray Eagle mine, Glenn County, California. Gray Eagle chromite mill, Glenn County, California. Sampled in 1942.	500
Chromite concentrate, composite sample. Ore from Gray Eagle mine, Glenn County, California. Gray Eagle chromite mill, Glenn County, California. Sampled in 1942.	1,000
Chromite concentrate, composite sample. Ore from Gray Eagle mine, Glenn County, California. Gray Eagle chromite mill, Glenn County, California. Sampled in 1942.	1,000
Chromite tailings, composite sample. Ore from Gray Eagle mine, Glenn County, California. Gray Eagle chromite mill, Glenn County, California. Sampled in 1942.	300
Chromite concentrate (fines), grab sample from dewatering tank overflow recovery. Ore from Castro mine, San Luis Obispo district, San Luis Obispo County, California. Castro Chrome Associates mill, San Luis Obispo, California. Sampled in 1943.	1,000
Chromite tailings, grab sample. Ore from Castro mine, San Luis Obispo district, San Luis Obispo County, California. Castro Chrome Associates mill, San Luis Obispo, California. Sampled in 1943.	400
Chromite cleaned concentrate. Ore from raise in No. 2 adit, Benbow mine, Stillwater district, Stillwater County, Montana. Benbow mill, near Nye, Montana. Defense Plant Corp. Sampled in 1943.	800
Chromite heads, mill sample. Ore from Mouat-Sampson mine, Stillwater district, Stillwater County, Montana. Mouat-Sampson Chrome mill, Nye, Montana. Defense Plant Corp. Sampled in 1943.	400
Chromite concentrates, mill sample. Ore from Mouat-Sampson mine, Stillwater district, Stillwater County, Montana. Mouat-Sampson Chrome mill, Nye, Montana. Defense Plant Corp. Sampled in 1943.	800
Chromite cleaned concentrate. Ore from USBM trench G-66, lowest layer of G-zone, Mouat-Sampson mine, Stillwater district, Stillwater County, Montana. Mouat-Sampson Chrome mill, Nye, Montana. Defense Plant Corp. Sampled in 1943.	700
Chromite cleaned concentrate. Disseminated ore from USBM pit H-25, top of H-zone, Mouat-Sampson mine, Stillwater district, Stillwater County, Montana. Mouat-Sampson mill, Nye, Montana. Defense Plant Corp. Sampled in 1943.	800

TABLE VI-11 (continued)

Material	Cobalt Concentration, mg/kg
Chromite cleaned concentrate. Massive ore from USBM pit H-25, base of H-zone, Mouat-Sampson mine, Stillwater district, Stillwater County, Montana. Mouat-Sampson mill, Nye, Montana. Defense Plant Corp. Sampled in 1943.	800
Chromite middlings. Ore from Mouat-Sampson mine, Stillwater district, Stillwater County, Montana. Mouat-Sampson mill, Nye, Montana. Defense Plant Corp. Sampled in 1943.	500
Chromite tailings, mill sample. Ore from Mouat-Sampson mine, Stillwater district, Stillwater County, Montana. Mouat-Sampson mill, Nye, Montana. Defense Plant Corp. Sampled in 1943.	100
Chromite heads, mill sample. Ore from Lagoons gold tailings deposits, Coos district, Coos County, Oregon. Humphreys Gold Corp. mill, Bandon, Oregon. Sampled in 1943.	80
Chromite concentrate. Ore from Lagoons gold tailings deposits, Coos district, Coos County, Oregon. Humphreys Gold Corp. mill, Bandon, Oregon. Sampled in 1943.	200
Chromite tailings, mill sample. Ore from Lagoons gold tailings deposits, Coos district, Coos County, Oregon. Humphreys Gold Corp. mill, Bandon, Oregon. Sampled in 1943.	30
Chromite ore, 4-ft channel sample across chromiferous zone near base of sand. Seven Devils mine, Coos district, Coos County, Oregon. Sampled in 1945.	60
Chromite heads, mill sample. Ore from Seven Devils mine, Coos district, Coos County, Oregon. Krome Corp. rough concentrating plant, Marshfield, Oregon. Sampled in 1943.	100
Chromite concentrate. Ore from Seven Devils mine, Coos district, Coos County, Oregon. Krome Corp. rough concentrating plant, Marshfield, Oregon. Sampled in 1943.	200
Chromite tailings, mill sample. Ore from Seven Devils mine, Coos district, Coos County, Oregon. Krome Corp. rough concentrating plant, Marshfield, Oregon. Sampled in 1943.	40
Chromite mill heads, magnetic, from Krome Corp. and Humphreys Gold Corp. mills, Coos district, Coos County, Oregon. Sampled in 1943.	200
Chromite flotation concentrate. Heads from Krome Corp. and Humphreys Gold Corp. mills, Coos district, Coos County, Oregon. Magnetic separator plant of Defense Plant Corp., Beaverhill, Oregon. Sampled in 1943.	600
Primary chromite concentrate used as heads for secondary concentrator. Primary heads from Seven Devils mine, Coos district, Coos County, Oregon. Magnetic separator plant, Defense Plant Corp., Beaverhill, Oregon. Sampled in 1943.	700

TABLE VI-11 (concluded)

Material	Cobalt Concentration, mg/kg
Chromite heads from Seven Devils mine (Krome Corp.) and Lagoon tailings deposits (Humphreys Gold Corp.), Coos district, Coos County, Oregon, U.S. Plancor Number 1501 plant, 3 miles N of Coquille, Oregon. Sampled in 1945.	50

TABLE VI-12

COBALT IN GOLD-BEARING MATERIALS
(Kaiser et al., 1954)

Material	Cobalt Concentration, mg/kg
Gold-silver tailings. Ore from Golden Belt (Golden Turkey Extension) mine, Black Canyon district, Yavapai County, Arizona. Golden Belt mill, near Cordes, Arizona. Sampled in 1943.	100
Gold cyaniding residue from dump. Ore from Argonaut mine, Mother Lode district, Amador County, California. Argonaut mill, Jackson, California. Sampled in 1943.	200
Gold concentrates. Ore chiefly from mines of Cripple Creek district, Teller County, Colorado. Golden Cycle mill, Colorado Springs, Colorado. Sampled in 1942.	300
Gold mill heads. Ore chiefly from mines of Cripple Creek district, Teller County, Colorado. Golden Cycle mill, Colorado Springs, Colorado. Sampled in 1942.	100
Gold heads, grab sample. Ore from Gold Hill mine, Cedar Creek district, Mineral County, Montana. Gold Mountain mill, S of Superior, Montana. Sampled in 1944.	80
Gold concentrate, grab sample. Ore from Gold Hill mine, Cedar Creek district, Mineral County, Montana. Gold Mountain mill, S of Superior, Montana. Sampled in 1944.	400
Gold cyanide charge (sand). Ore from Homestake mine, Lead district, Lawrence County, South Dakota. South mill, Lead, South Dakota. Homestake Mining Company. Sampled in 1942.	0
Gold cyanide charge (slime). Ore from Homestake mine, Lead district, Lawrence County, South Dakota. South mill, Lead, South Dakota. Homestake Mining Company. Sampled in 1942.	100
Gold tailings (slimes residue), mill sample. Ore from Homestake mine, Lead district, Lawrence County, South Dakota. South mill, Lead, South Dakota. Homestake Mining Company. Sampled in 1942.	0
Gold tailings (sand residue), mill sample. Ore from Homestake mine, Lead district, Lawrence County, South Dakota. South mill, Lead, South Dakota. Homestake Mining Company. Sampled in 1942.	0
Gold tailings (stope filling), mill sample. Ore from Homestake mine, Lead district, Lawrence County, South Dakota. Homestake Mining Company mill, Lead, South Dakota. Sampled in 1943.	10
Gold heads, -4 mesh, grab sample. Ore from Maitland mine, Maitland district, Lawrence County, South Dakota. Maitland mill, Maitland, South Dakota.	200
Gold tailings (70% -200 mesh) from agitation tank discharge to tailings dump. Ore from Maitland mine, Maitland district, Lawrence County, South Dakota. Maitland mill, Maitland, South Dakota.	100

TABLE VI-12 (concluded)

Material	Cobalt Concentration, mg/kg
Gold heads, oxidized ore, from Bald Mountain Mining Company mine, Trojan district, Lawrence County, South Dakota. Trojan mill, Trojan, South Dakota. Sampled in 1942.	200
Gold heads (unoxidized "blue" ore), grab sample. Ore from Bald Mountain Mining Company mine, Trojan district, Lawrence County, South Dakota. Trojan mill, Trojan, South Dakota. Sampled in 1942.	300
Gold tailings from tailings pond. Ore from Bald Mountain Mining Company mine, Trojan district, Lawrence County, South Dakota. Trojan mill, Trojan, South Dakota. Sampled in 1942.	100
Gold-arsenopyrite concentrate. Ore from Bullion Gold mine, Keystone district, Pennington County, South Dakota. Bullion gold mine mill (burned). Sampled in 1942.	800
Gold-arsenopyrite concentrate. Ore from Bullion Gold mine, Keystone district, Pennington County, South Dakota. Bullion Gold mine mill (burned). Sampled in 1942.	800
Gold table concentrate. Ore from Holy Terror mine, Keystone district, Pennington County, South Dakota. Keystone-Holy Terror mill. Sampled in 1942.	1,000
Gold table and jig concentrate. Ore from Holy Terror mine, Keystone district, Pennington County, South Dakota. Keystone-Holy Terror mill. Sampled in 1942.	2,000
Gold refinery roaster dust from Cottrell precipitator. Ore chiefly from mines of Cripple Creek district, Teller County, Colorado. Golden Cycle mill, Colorado Springs, Colorado. Sampled in 1942. Additional analysis: 0.011 Se.	80
Gold refinery by-product from treated slag. Ore chiefly from mines of Cripple Creek district, Teller County, Colorado. Golden Cycle mill, Colorado Springs, Colorado. Sampled in 1942.	80

However, a significant fraction of cobalt in the soil may have been due to natural concentrations. High organic soils contained significantly higher cobalt concentrations than did highly mineralized soils from the same site. Cobalt concentrations in the background lake water 10 to 30 miles (16 to 48 km) from the smelter activity were 1 to 10% the concentrations closest to the smelter.

Little literature information was found on the fate of cobalt in uranium ores. Pitchblende residues containing 2 to 3% cobalt from the Destrehan Street Refinery in St. Louis were leached in a study by the Uranium Division of the Mallinkrodt Chemical Works in Welden Spring, Missouri, for recovery of metal values. In the recommended scheme, there were losses of up to 10% of the cobalt in each of two waste products--the sulfuric acid leaching residue and the iron cake removed at pH 4. Most of the remainder of the cobalt was recovered in a precipitate of nickel and cobalt at pH 8 (Hansen et al., 1966).

Cobalt is apparently not a problem in U.S. uranium milling wastes. One mill's wastewater was reported to contain 1.7 mg Co/liter. The tailing pond decant contained < 0.05 mg/liter. Cobalt was not mentioned in the wastewaters of four other U.S. uranium mills (Jarrett and Kirby, 1978b).

The concentrations of cobalt in tungsten-bearing materials (Table VI-13), titanium-bearing materials (Table VI-14), mercury ores and flue dusts (Table VI-15), and kyanite-garnet (Table VI-16) range up to 500 mg/kg. These materials do not appear to pose any exceptional environmental problems for cobalt beyond those of other heavy metal ores processed in the United States. Probably all areas containing sulfide ore mines of moderate to high rainfall will suffer from acid mine drainage similar to that seen in coal mining areas where cobalt concentrations can range up to 3.8 mg/liter.

The Magnesia Talc Company at Burlington, Vermont, is reported to have produced nickel and cobalt concentrates (0.015% Co and 0.2% Ni) during the froth flotation of talc (Vhay et al., 1973).

TABLE VI-13

COBALT IN TUNGSTEN-BEARING MATERIALS
(Kaiser et al., 1954)

Material	Cobalt Concentration, mg/kg
Tungsten tailings from tailings pile. Ore from Darwin mines, Inyo County, California. Sampled in 1943.	500
Tungsten (ferberite) concentrate. Composite sample from jigs and tables. Ore chiefly from Pride mine, Boulder County, Colorado. Boulder Tungsten Mills, Inc., mill, Boulder Colorado. Sampled in 1942.	100
Tungsten (ferberite) flotation concentrate. Ore chiefly from Pride mine, Boulder County, Colorado. Boulder Tungsten Mills, Inc., mill, Boulder, Colorado. Sampled in 1942.	50
Tungsten (ferberite) flotation tails. Ore chiefly from Pride mine, Boulder County, Colorado. Boulder Tungsten Mills, Inc., mill, Boulder, Colorado. Sampled in 1942.	80
Tungsten (ferberite) concentrates from jigs and table. Ore from several mines of the Wolf Tongue Mining Company, Boulder, Colorado. Wolf Tongue Mining Company mill, Nederland, Boulder County, Colorado. Sampled in 1942.	50

TABLE VI-13 (concluded)

Material	Cobalt Concentration, mg/kg
Tungsten (ferberite) tailings from tailings pond. Ore from several mines owned by Wolf Tongue Mining Company, Boulder, Colorado. Wolf Tongue Mining Company mill, Nederland, Boulder County, Colorado. Sampled in 1942.	100
Tungsten (ferberite) slimes from canvas tables. Ore from several mines owned by Wolf Tongue Mining Company, Boulder, Colorado. Wolf Tongue Mining Company mill, Nederland, Boulder County, Colorado. Sampled in 1942.	500
Scheelite concentrate, composite mill sample. Ore from Tem-Piute-Lincoln and Schofield mines, Tem-Piute district, Lincoln County, Nevada. Lincoln Mines, Inc., mill, Hiko, Nevada. Sampled in 1943.	0
Tungsten tailings (sulfide float tails), mill grab sample. Ore from Tem-Piute mine, Tem-Piute district, Lincoln County, Nevada. Lincoln Mines, Inc., mill, Hiko, Nevada. Sampled in 1943.	100
Tungsten tailings (sulfide float tails), mill grab sample. Ore from Tem-Piute mine, Tem-Piute district, Lincoln County, Nevada. Lincoln Mines, Inc., mill, Hiko, Nevada. Sampled in 1943.	100

TABLE VI-14

COBALT IN TITANIUM-BEARING MATERIALS
(Kaiser et al., 1954)

Material	Cobalt Concentration, mg/kg
Ilmenite heads, grab sample. Ore from open pit mine, Finley, Caldwell County, North Carolina. Yadkin Valley Ilmenite mill, Finley, North Carolina. Sampled in 1944.	100
Ilmenite concentrate, mill sample. Ore from open pit mine, Finley, Caldwell county, North Carolina. Yadkin Valley Ilmenite mill, Finley, North Carolina. Sampled in 1944.	200
Titanium-iron-vanadium tailings, mill sample. Ore from Sanford Hill mine, Sanford Lake district, Essex County, New York. MacIntyre Development mill, Tahawus, New York. National Lead Company. Sampled in 1943.	400

TABLE VI-15

COBALT IN MERCURY ORES AND FLUE DUSTS
(Kaiser et al., 1954)

Material	Cobalt Concentration, mg/kg
Burned mercury ore. Grab sample from dump. Ore from Sulfur Bank mine, Lake County, California. Bradley Mining Company smelter, Clearlake Park (Oaks), California. Sampled in 1942.	80

TABLE VI-15 (concluded)

Material	Cobalt Concentration, mg/kg
Burned mercury ore. Grab sample from dump. Ore from Sulfur Bank mine, Lake County, California. Bradley Mining Company smelter, Clearlake Park (Oaks), California. Sampled in 1942.	70
Flue dust from mercury furnace. Grab sample. Ore from Sulfur Bank mine, Lake County, California. Bradley Mining Company smelter, Clearlake Park (Oaks), California. Sampled in 1942.	100
Flue dust from mercury furnace collected in Sirocco dust collector. Ore from Mount Jackson quicksilver mine, Sonoma County, California. Sonoma Quicksilver Mines, Inc., smelter, Guerneville, California. Sampled in 1942.	200

TABLE VI-16

COBALT IN KYANITE-GARNET ORES AND CONCENTRATES
(Kaiser et al., 1954)

Material	Cobalt Concentration, mg/kg
Crude kyanite ore from Cleo Mountain district, Yancey County, North Carolina. Yancey Cyanite Company mill, Burnsville, North Carolina. Sampled in 1943.	80

TABLE VI-16 (concluded)

Material	Cobalt Concentration, mg/kg
Kyanite concentrate, mill sample. Ore from Cleo Mountain district, Yancey County, North Carolina. Yancey Cyanite Company mill, Burnsville, North Carolina. Sampled in 1943.	0
Coarse garnet concentrate. Ore from Cleo Mountain district, Yancey County, North Carolina. Yancey Cyanite Company mill, Burnsville, North Carolina. Sampled in 1943.	100
Fine garnet concentrate. Ore from Cleo Mountain district, Yancey County, North Carolina. Yancey Cyanite Company mill, Burnsville, North Carolina. Sampled in 1943.	100

Among other U.S. ores that have been studied for cobalt recovery are Arkansas bauxite, which gave a concentrate containing 4.93% Ni and 2.29% Co (Runke and O'Meara, 1944; cited by Bauder, 1957); cobalt (stainerite) ores from Goodsprings, Nevada (Dean, 1936; Shelton, 1946; cited by Bauder, 1957); and low grade Alabama manganese ores (Basore and Eiland, 1936; cited by Bauder, 1957).

3. Processing in other countries: A discussion of cobalt recovery in other countries would not be pertinent to the aims of the present study unless very comparable ores were being processed in quite comparable processes to those used in the United States, or unless something was known about environmental cobalt concentrations in these foreign areas. For these reasons, this subject will not be described in depth.

The Zairian process is depicted by Battelle Columbus Laboratories (1976) in the flow diagram of Figure VI-9. Zaire produces only cobalt metal. Until recently, it was shipped to Belgium for repacking (Chase, 1979). Zairian material is treated in Belgium to produce cobalt powder (Scheidweiler, 1979). Belgium ambient air concentrations of cobalt are reported in Chapter IV. The Belgian process for refining cobalt in Hoboken is depicted in Figure VI-10.

Almost all of Canadian production is refined in Europe. Falconbridge sends a nickel-copper matte to Kristiansand, Norway, to be toll refined. The refinery produces about 1,000 tons (910 MT) cobalt annually and hopes to increase production by 50% (Mari, 1979a). International Nickel Company sends cobalt oxide to the United Kingdom for salts manufacture and matte to the Mond refinery in Clydach, Wales. Sherritt-Gordon, however, produces metallic cobalt from custom feeds. Canadian processes will be altered to improve cobalt recoveries now that higher cobalt prices justify the expense (McPhail, 1979).

The effects of the Sudbury, Ontario, smelters on the environment have been mentioned in Chapter IV. In the preceding subsection, it was noted that up to 78 short tons (71 MT) cobalt are emitted to the atmosphere annually from the smelters of International Nickel Company (Inco), whose copper-nickel smelting process is shown schematically in Figure VI-11.

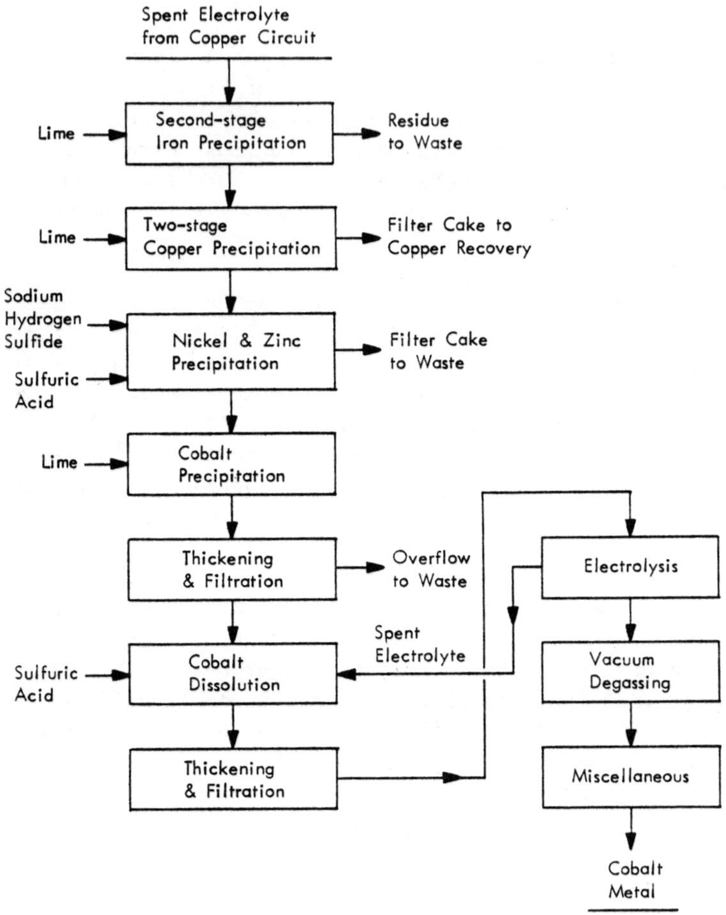

Figure VI-9. Production of Cobalt Metal in Zaire (Battelle Columbus Laboratories, 1976).

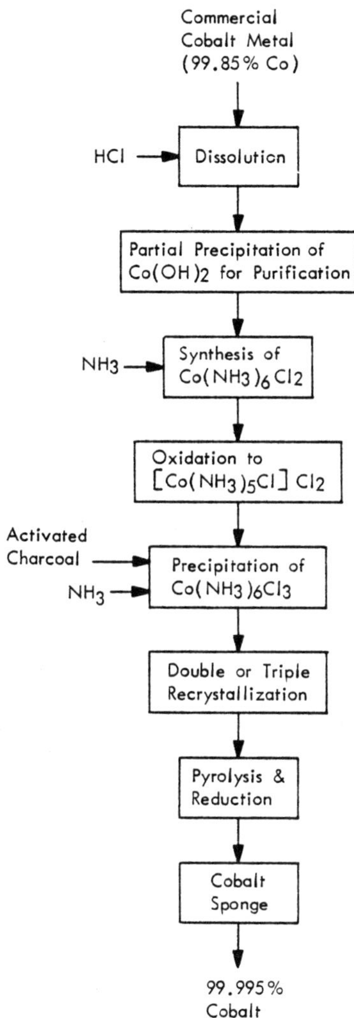

Figure VI-10. Cobalt Refining in Belgium (Tougarihoff, 1966).

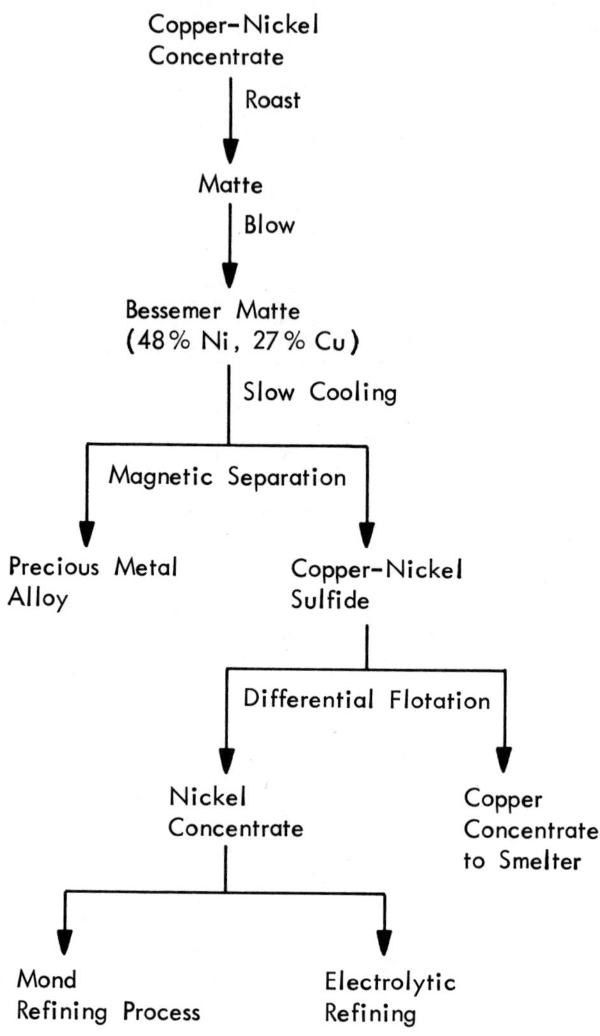

Figure VI-11. Inco Smelting Process for the Copper-Nickel Concentrate (adapted from Beamish, 1961).

According to Pinkney (1979), 75 to 80% of the cobalt in such concentrates is lost primarily in the converter slags. Much of the rest is transferred to the electrolytic nickel unless steps are taken to remove the cobalt.

Although the major fraction of refined nickel has been recovered electrolytically by Inco since 1926, it was not until 1947 that cobalt was recovered commercially from nickel electrolyte (by the Port Colborne refinery of Inco in Ontario). Before that time, commercial nickel cathodes contained up to 1% cobalt.* No process to recover cobalt from the old sulfate-boric acid nickel electrolyte was ever developed. In the early 1960's, at Port Colborne, cobaltic hydroxide was recovered by treating the nickel sulfate electrolyte with hypochlorite to give pH 2.2. Some hydroxide was calcined at 180°F (82°C) to produce the oxide, which was marketed directly. Otherwise, the oxide was reduced to metal, and the metal was cast into anodes for electrolytic refining (Renzoni and Barker, 1961).

Nickel and cobalt are currently refined by Inco at Port Colborne on Lake Erie. Converter matte is cast directly into anodes refined at the Manitoba Division's Thompson refinery to produce nickel metal, cobalt oxide, and elemental sulfur. Nickel refinery anode slimes are converted to secondary anodes to recover nickel, copper, cobalt, and precious metals by procedures similar to those used in a silver refinery. At the older Port Colborne refinery, nickel sulfide anodes are produced from the nickel sulfide concentrate obtained in the matte separation process, but this process is only limited. Generally, the roasted nickel oxide is reduced to nickel before casting into anodes for electrorefining. Cobalt, copper, and other non-noble metal impurities go into solution at normal voltages with the nickel.

A new Copper Cliff nickel refinery heats briquetted nickel smelter and refinery materials in top-blown rotary converters, which directly convert molten nickel sulfide to nickel.

---

* Some Inco nickels still contain as much as 1.3% cobalt. This "utility nickel" is useful for cast nickel applications such as pumps, crushing and pulverizing mill rollers, and pipes for abrasive slurries (Anonymous, 1978j).

Nickel-bearing materials are also treated with carbon monoxide by the Inco Pressure Carbonyl Process at 121 to 177°C and 1,000 lb/in^2 (6.9 MPa). Nickel carbonyl volatilizes whereas copper, cobalt, precious metals, and impurities remain in the residue. The latter is ground and leached to recover the metal content (Anonymous, 1978i).

In years past, little attention was paid to cobalt extraction by Inco Ltd. in Ontario because of low cobalt prices. In 1978, cobalt prices, having climbed from $6 to $20/lb, spurred Inco's plans to double its 1979 cobalt output. The slag handling process was to be modified and other processes were being examined in May of 1978 to increase cobalt recovery. Gorham International, Inc. (1979) stated that the cobalt would be produced as oxide not metal. The 1978-1979 strike, however, has so far kept Inco from its goal.

In the Mond carbonyl practice for nickel recovery from the Inco matte at Clydach, Wales, only the nickel forms a volatile carbonyl at the low temperatures and pressures used (Perelman and Zvorkin, 1975). The residue from the nickel carbonyl volatilization is leached to recover the cobalt (Renzoni and Barker, 1961). The exceptionally high cobalt concentrations in the air of Clydach and its environs noted in Chapters IV and V are hard to explain on the basis of these sketchy facts. To our knowledge, however, a carbonyl process is not used or planned for nickel recovery in the United States.

In West Germany, the Duisburger Kupferhütte Refinery has recovered cobalt from pyrite residues containing up to $\sim$6,000 mg Co/kg from Finland, Germany, Greece, Italy, Norway, Spain, and Sweden. A secondary refinery at Goslar has operated on much scrap from the United States (Charles River Associates, 1969). In Chapter VII, it is noted that the lung tissues of Duisburg residents have been found to contain strikingly more cobalt than the lung tissue of residents of nearby Cologne (Köln) (Persigehl, 1976).

At the Duisburger Kupferhütte, the solution from the chloridizing roasting of pyrites was purified by precipitating ferric iron produced by chlorine oxidation with calcium carbonate and waste lime. The cobaltic ion was precipitated in the next step separately from nickel at pH 3.9 to 4.1 (Teworte, 1969). Information on atmospheric emissions was not found. Annual production fell from 1,100 MT cobalt in 1966 and 600 to 800 MT ca. 1969 to 350 MT in 1976 and 50 MT in 1978 (Klaile, 1979; Teworte, 1969).

The leaching/electrowinning and leaching/hydrogenation processes used in Zambia for cobalt recovery from ores were described in detail by Huxley (1974). Leaching and hydrogenation processes used in Finland and Canada (Sherritt Gordon Mines Ltd.) were described by Evans (1967).

Arsenic-bearing cobalt concentrate from Morocco is processed to cathode cobalt (1,200 MT/year capacity) in Savoie, France, by Metaux Speciaux S.A. Eurotungstene produces fine cobalt power in France from recycled cobalt-containing material (Klaile, 1979).

4. Future world production: World sources of cobalt that will be tapped in the near future include recovery of cobalt sulfate from the nickel electrolyte produced during processing for platinum-group metals at the Rustenberg-Matthey Mine in South Africa. Slag cleaning will probably be practiced, e.g., in Canada, Zaire, and Zambia. In North America, cobalt recovery could improve from 1.5 to 6 million lb/year (680 to 2,700 MT). In Zimbabwe Rhodesia also, more cobalt is being recovered from nickel smelter products. In South Africa, there are 40 million tons (18,000 MT) of gold-uranium tailings that could begin retreatment by Anglo American Company, etc., for cobalt recovery within 18 months of a positive decision. The cobalt content is about 0.1%; and at current prices, the combined copper, cobalt, and nickel content is worth more than the contained gold or uranium. Potentially 1,000 to 1,500 MT Co/year may be recovered. Current gold mining operations themselves could be a source of 1,000 to 3,000 MT Co/year. Up to 10 to 20% of the cobalt contained in concentrates from the platinum- and nickel-bearing Merensky Reef in South Africa reports in the finished electrolytic nickel. One of the platinum refineries has started

producing refined cobalt sulfate; and if all the South African platinum producers were to recover their cobalt, 600 to 1,000 MT Co/year would be available from this source (Klaile, 1979; Pinkney, 1979; Mari, 1979e).

Within the next 5 years, world cobalt production may attain nearly 40,000 MT annually with the United States producing 10% of the total, certainly a far different situation from the 1970's. In the 1950's however, U.S. production was comparably high. Projected U.S. and world production figures are given in Table VI-17.

TABLE VI-17

PROJECTED U.S. AND WORLD COBALT PRODUCTION
(Saville, 1979)

	Projected Level MT/years	Year Attained
Zaire	16,000	1983
Zambia	3,000-3,500	1979
	4,000	1980
	9,500	1984
Blackbird Mine, Idaho	2,000	1981 (year of opening)
Fredericktown cobalt mine, Missouri	~ 800-900	
Outokumpu Oy, Finland	1,500[a]	"within several years"
Morocco	1,000	current[b]

a/ Current production is 1,000 MT/yr. Ore reserves will be exhausted in 15 years. Outokumpu Oy is looking for new feed.

b/ The capacity is 2,200 MT/year; but at the current rate, the 15,000 MT reserves will be exhausted in 15 years.

B. Secondary Recovery

According to the Bureau of Mines data, domestic recycling of cobalt-bearing materials has been a very minor source of cobalt supply. Strauss (1979) of ASARCO Inc. claimed that secondary processing of cobalt in the United States has never exceeded 250 short tons (227 MT) in any year. For example, between 1961 and 1968, the average amount of cobalt in purchased scrap consumed in the United States was < 74 short tons/year (< 67 MT/year) (Charles River Associates, 1969; and in 1974, 135 short tons cobalt (< 127 MT) in scrap was recycled domestically (Sibley, 1976).

Most of the cobalt-bearing scrap that has been recycled in the United States was new (e.g., grindings and turnings) and clean. Off-grade contaminated metal or residues, grindings, and mixed scrap were generally shipped to Japan, West Germany, and Belgium for recycling except during periods of short supply and high prices (Gorham International, Inc., 1979; Charles River Associates, 1969). Charles River Associates (1969) estimated that up to 574 short tons (521 MT) Co/year (average 190 short tons/year or 172 MT/year) was contained in exported scrap, based on the assumption that 30% of the gross weight was cobalt. Most scrap was from high-temperature alloys with lesser amounts from magnet alloys.

Although the government has estimated that only 1% or less of the total cobalt market is recycled, scrap executives at a January 1979, meeting of the Institute of Scrap Iron and Steel (Haflich, 1979) and participants at the Cobalt Crisis Conference in April-May 1979 disputed such a low figure. In fact, an unidentified member of the audience at the latter conference commented that his company alone recycled more cobalt annually than reported by the Bureau of Mines for the entire United States. Sibley, who has compiled these statistics in recent years, remarked from the floor that he had become aware that the domestic cobalt recycling industry is considerably larger than Bureau of Mines figures would indicate.

Whether the recent flurry of advertisements by firms interested in recycling cobalt-bearing materials is indicative of new or continuing recycling activity is a moot point. Gorham

International, Inc. (1979c) commented in June that a concerted effort should be made to survey the extent of recycling of alloys, spent catalysts, hard-facing grindings, cobalt-tungsten [carbide] mixtures, etc.

Table VI-18 is an attempt to compile as many names as possible of companies supplying cobalt metal and oxide products* as well as buying and/or processing secondary cobalt materials. As can be seen from the table, well over half of the companies listed process or merely buy cobalt-bearing scrap materials. Perhaps the trading firms merely sell the scrap to foreign recyclers. Every conceivable type of metal or alloy waste is mentioned in the table. A few firms are also interested in cobalt-bearing catalysts. In most cases, the actual recovery process being used was not found in the literature. (In fact, patents for recycling cobalt-bearing materials are numerous; but since licensing is probably involved in many cases, it did not seem fruitful to pursue these.)

If superalloys have been segregated by alloy type, foundry wastes, rejects, and obsolete parts are clean enough to be reclaimed by melting; but furnace dusts, spills, pickling residue, and grindings cannot be practically segregated for remelting. Wastes from forming superalloys into complicated shapes are contaminated with copper, lead, zinc, oil, moisture, rags, wood, and corundum grinding debris (Brooks et al., 1970). In addition, some superalloy scrap cannot be recycled because of coatings (Mari, 1979c). Sorting and grading are the major problems in economic recycling of high-temperature alloys. The business requires millions of dollars worth of special equipment including pressers, shears, crushers, magnetic sorting devices, and analytical facilities to help sort unidentified materials, such as obsolescent parts from heat-treat furnaces (Anonymous, 1974).

---

* Names of producers of cobalt chemicals can be readily found, e.g., in the current edition of the Products and Services volume of the Thomas Register of American Manufacturers and Thomas Register Catalog.

TABLE VI-18

SUPPLIERS OF COBALT PRODUCTS AND BUYERS AND/OR PROCESSORS OF SECONDARY COBALT-CONTAINING MATERIAL[a/]

Company	Location	Cobalt Product Sold	Materials Processed and Other Comments
ACLI Metal and Ore Company	New York, New York	Metal	
African Metals Corporation	New York, New York	Metal, ore and oxides, and powder	
Allied Metals Corporation	Detroit, Michigan	Metal, Co-Ni alloys	Primary Co and Co-bearing scrap, solids, turnings, grindings, sludges, catalysts, residues.
Alloy Metal Products, Inc.	Davenport, Iowa	Metal, Co-Ni alloys, ore and oxides	Co alloy scrap in all forms: grindings, turnings, spills, and residues. Major Ni recycler.
AMAX Nickel	Greenwich, Connecticut	Metal and powder	
Bache Halsey Stuart, Inc.	New York, New York	Metal	
Belmont Metals, Inc.	Brooklyn, New York	Metal	
Billiton Metals and Ores, U.S., Inc.	New York, New York		Buys and/or processes secondary cobalt
Calabrian International	New York, New York		Co-Mo-Cu catalysts
Colonial Metals Company	Columbia, Pennsylvania	Metal	
Colton Metalex Company[b/]	Los Angeles, California	Metal, Co-Ni alloys	
Continental Metals Corporation	New York, New York	Metal, Co-Ni alloys, ore and oxides, powder, chemicals	
Copalco International Ltd.	New York New York	Metal	
Deka Minerals, Inc.	New York, New York	Metal, Co-Ni alloys, powder	
Diamond Reclamation Company, Inc.[b/]	Niagara Falls, New York		WC blanks, collector dust

a/ Sources: Anderson (1979); Anonymous (1974, 1979a, 1979b, 1979c, 1979g); Alloy Metal Products, Inc.; Diamond Reclamation Company, Inc. (1978); Frankel Company, Inc. (1974); National Nickel Alloy Corporation (1979); S. Wilkoff and Sons Company (1978); Wolverine Metal Company, Inc. (1974); and Samuel J. Zuckerman, Inc. (1978).

b/ Buyer of tungsten scrap (not necessarily cemented tungsten carbine scrap, which would contain significant amounts of cobalt). Other buyers of tungsten scrap include the Chemet Company, Cometal, Inc., Complex Elements Corporation, The Ore and Chemical Corporation, SAMINCORP, Inc. Herman C. Stark, Inc., and Teledyne Wah Chang Huntsville.

TABLE VI-18 (continued)

Company	Location	Cobalt Product Sold	Materials Processed and Other Comments
Dynalloy Industries, Inc.[b/]	Houston, Texas		WC scrap for production of hard metal welding products
Erlanger & Company, Inc.	New York, New York	Metal, ore and oxides	
F&S Alloys and Minerals Corporation		Metal, ore and oxides, powder	
Falconbridge International	Hamilton, Bermuda	Metal	Canadian Co ores in Canada
Frankel Company, Inc.	Detroit, Michigan		Co scraps
F. W. Hempel and Company, Inc.	New York, New York; Bremen and Dusseldorf, West Germany		Broken cathodes and granules
Irland Alloys, Inc.[b/]	Sterling Heights, Michigan		
Irwin Alloys, Inc.[b/]	West Newton, Pennsylvania		WC solids, residues, sweepings, sludges
Kaichen Metal Mart	Paramount, California	Metal, Co-Ni alloys, ore and oxides, powder	Scrap and residue
Samuel G. Keywell Company, Subsidiary of Key International, Inc.[b/]	Detroit, Michigan		High-temperature alloy scrap such as Rene 41 and Stellite® 31
Kolon Trading Company, Inc.	New York, New York	Metal, powder	
Kraft Chemical Company	Chicago, Illinois	Metal, ore and oxides	
Li Tungsten Corporation[b/]	Glen Cove, New York		WC scrap and residues. Producers of W and WC powder.
Lorch Metal Corporation[b/]	Valley Stream, New York		Co/Mo solids, turnings, grindings, and residues; Co and Ni alloys; Stellites®; Co-Fe; W residues; high-speed steel scrap
Mercury Trading, Inc.[b/]	Margate, New Jersey		
Metal World Inc.	East St. Louis, Illinois	Metal, Co-Ni alloys, ores oxides, powder	Mo, Co, W sludges, oxide, ferro grindings catalysts. Will toll refine
Metallurgical Industries, Inc. (formerly Metallurgical International, Inc.)	Tinton Falls, New Jersey	Co-base powders for hardfacing including pre-alloyed WC-Co powders	Scrap WC. Will toll refine.
Metallurgical Products Company[b/]	West Chester, Pennsylvania	Co-Ni alloys	
National Nickel Alloy Corporation	Greenville, Pennsylvania		Solids, turnings, grindings, spills, residues, off-grade and contaminated metal
Northbrook Metals, Inc.	Northbrook, Illinois	Metal, Co-Ni alloys, ore and oxides, powder	
Parkans International, Inc.[b/]	Houston, Texas	Metal, Co-Ni alloys, powder	Scrap, residue, grindings, sludges, wastes in combination with other materials

TABLE VI-18 (concluded)

Company	Location	Cobalt Product Sold	Materials Processed and Other Comments
Penn Iron & Metal Company, Inc.[b/]	Erie, Pennsylvania		WC scrap
The Pesses Company[b/]	Solon, Ohio	Metal, Co-Ni alloys, ore and oxides, powder	Stellites including HS-93, Kovar, Alnico, Remalloy, Remendur, and other Co-containing turnings, grindings, and residues; tool steel scrap
Phillipp Brothers, Division, Engelhard Minerals and Chemical Corporation	New York, New York	Metal, ore and oxides	
Powell Inc.[b/]	Rockford, Illinois	Metal, Co-Ni alloys, ore and oxides, powder	
The Royal Metals Corporation	Stamford, Connecticut	Metal, ore and oxides	All types of cobalt-bearing catalysts and residues
Rushco Alloys[b/]	Chicago, Illinois	Metal, Co-Ni alloys	
SAMINCORP Inc.	New York, New York	Metal	
Schiavone-Bonomo Corporation[b/]	Jersey City, New Jersey	Metal	WC and tool steel scrap
Shir Metal and Steel	Oakland Park, Florida	Metal	
Stainless Processing Company	Chicago, Illinois	Co-Ni alloys	
Steelmet/Klaff Inc.[b/]	See column 4	Metal, Co-Ni alloys, granules, cathodes, powder, oxide, high-temperature base alloys	High-quality superalloy scrap. Production facilities at Port Vue, Pennsylvania; Providence, Rhode Island; E. Providence, Rhode Island; Worcester, Massachusetts; Brooklyn, New York; Baltimore, Maryland; Miami, Florida; Chicago, Illinois; Vicksburg, Mississippi; and Bottrop, West Germany
Strauss International Corporation[b/]	New York, New York		
Suisman and Blumenthal Corporation[b/]	Hartford, Connecticut		Vacuum melts scrap
SUPPO Smelting and Refining Company[b/]	Chicago, Illinois	Metal, Co-Ni alloys, powder	Co and their alloys
Sylvania Chemical and Metallurgical Division, General Telephone and Electronics Corporation	Towanda, Pennsylvania	Fine Co powder	Co-bearing scrap
United Alloys Inc.	Los Angeles, California	Co-Ni alloys	
United Minerals and Chemicals Corporation	New York, New York	Metal, ore and oxides, powder	
S. Wilkoff & Sons Company[b/]	Cleveland, Ohio		W, Mo, and Co scrap
Wolverine Metal Company, Inc.	Detroit, Michigan		High-temperature alloys: tool steel, Co alloys, Stellites
Samuel J. Zuckerman, Inc.	Floral Park, New York		Haynes Stellite-21, Ni and Co-base alloys
Max Zuckerman and Sons, Inc.[b/]	Owings Mill, Maryland	Co-Ni alloys	Stellite, Kovar, Alnico, and high-temperature alloys, grindings, sludges, residues, ores, ferroalloys, other complex materials.

The cobalt in nickel-base superalloys has generally not been recovered. Usually the alloy is sold for its nickel content only (Wein, 1979).

Only a few firms dominate the production of high-temperature alloys for aircraft from scrap; 50 to 67% of the vacuum melting business is done by Colton Metalex, Los Angeles; Frankel Co., Inc., Detroit; Suisman and Blumenthal, Inc., Hartford; Utica Alloys, Inc., Utica; and Vac Air Alloys, Frewsberg, New York (Teplitz, 1976).

Since the current cobalt crisis, superalloy producers are supplementing their supply by having engine makers return their scrap. The scrap supply has been aggravated because the U.S. Air Force is not selling its scrap as it has formerly (Prinzinsky, 1978).*

Pratt and Whitney Aircraft recently gained Air Force approval to have chips from the machining of IN-100 alloy re-melted into ingots from which Homogeneous Metals, Inc., a Pratt and Whitney subsidiary, produces alloy powder by a hydrogen atomization process. The recycling began in the summer of 1978 from a stockpile of chips saved "for several years" (Furst, 1979).

Little information was found in the trade literature on processes being used for reclaiming metal values from used superalloy scrap. The Air Force recently held a conference at Kelly Air Force Base, Texas, for firms interested in sharing technical and economic experience in the feasibility of reclaiming cobalt alloys from condemned aircraft engine parts (Anonymous, 1979h).

---

* The Air Force is encouraging improved superalloy scrap reclamation by military repair depots and by manufacturers, but it intends to retain the reclaimed cobalt for military use (Turk, 1979).

Although the National Materials Advisory Board (1971) stated that superalloys have a 5-year recycle time with a 90% potential recovery of cobalt, it does not appear that cobalt recovery from condemned aircraft engine parts has ever been widely practiced.

Brooks et al. (1970) of the Bureau of Mines developed a process for separately recovering nearly 90% of the nickel, cobalt, and molybdenum in superalloys as the oxides. Figure VI-12 depicts the flowsheet of this hydrometallurgical process for recovering nickel and cobalt oxides from burned superalloy grindings. The process steps include dissolution in a solution of hydrochloric acid and chlorine at 90 to 100°C and separate solvent extractions. After molybdenum and iron removal (by trioctyl phosphate in kerosene and the secondary amine Amberlite LA-1 dissolved in aromatic naphtha), the cobalt chloro complex is extracted by triisooctylamine dissolved in aromatic naphtha. Stripping with water recovers the cobalt chloride. The latter was recrystallized from the aqueous solution or was precipitated with Na_2CO_3. Calcining cobalt carbonate gives the metallurgical grade cobalt oxide (Brooks et al., 1970).

Fletcher (1973) described some other hydrometallurgical processes that have been proposed for recycling superalloy scrap. For example, Metallurgie Hoboken, Belgium, leaches metal scrap with hydrochloric acid plus chlorine, separates nickel and cobalt by solvent extraction with a tertiary amine, and recovers either cobalt oxide by pyrolysis of the cobalt chloride or cobalt powder by hydrogenation of the purified chloride solution.

Cobalt in worn-out cutting tool inserts and wear parts may be reclaimed as pure cobalt powder by three different processes. The zinc embrittlement and Coldstream processes will recover metal in the exact grade contained in the original tool. Chemical extraction will also recover the cobalt. Feedstocks include hard metal scrap, alloy grade or straight tungsten carbide scraps, sludges, slimes, and cutting residues. Certain other materials, if subjected to prior processing, are also suitable (Anderson, 1979).

Leaching and mechanical crushing methods for reclaiming scrap carbides used prior to the zinc embrittlement method were

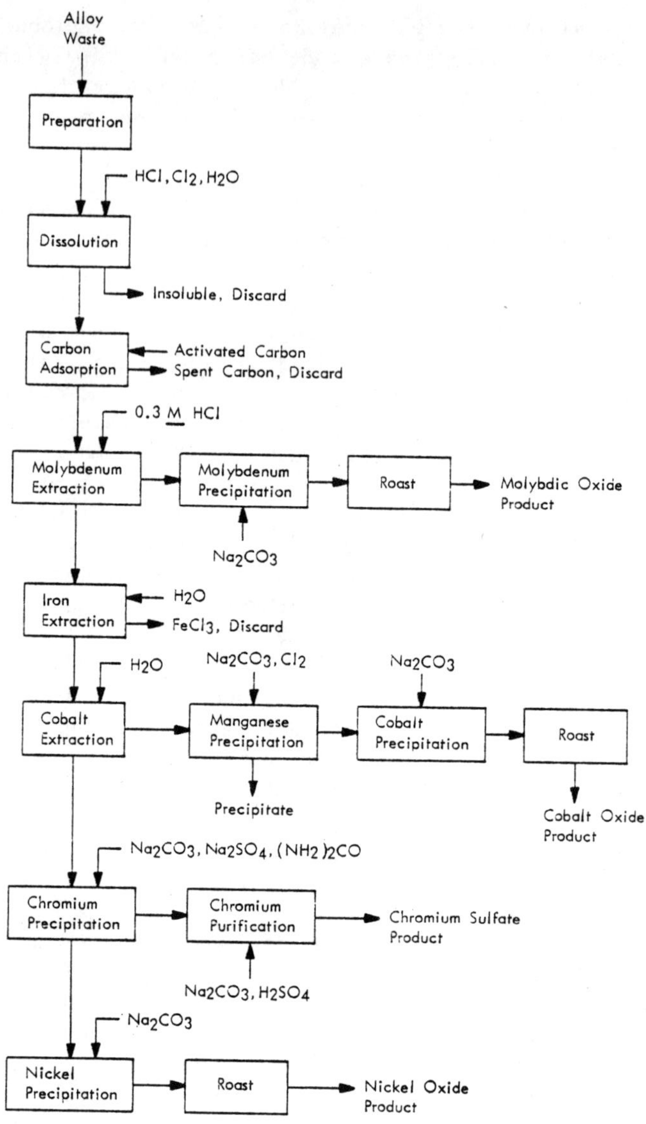

Figure VI-12. Simplified Flow Sheet for Processing Superalloy Grinding (Brooks et al., 1970).

slow or expensive. Metal recoveries were only about 70%. In 1969, the Rolla Metallurgy Research Center of the Bureau of Mines announced a simple and efficient method to recover cobalt binder from the cemented carbide mass by the action of molten zinc at 800°C under hydrogen. The zinc forms an alloy with cobalt, thereby disintegrating the mass. Zinc is recovered from the Zn-Co alloy by distillation (Barnard et al., 1969).

By 1975, there was such a wide difference in the price between virgin cobalt, tungsten, and tantalum and reclaimed pre-alloyed tungsten carbide cobalt that there was strong incentive to use recycled materials for manufacture of cemented carbides. Such recycled material met stringent specifications for critical cutting and mining tools and was no longer suitable merely for tire studs and oil drilling nozzles (Adams, 1975).

Recycling of cobalt, tungsten, and tantalum from cemented carbides is a rapidly growing activity (Mari, 1979b). Currently, about 20% of the carbide powder that is used is recycled (Koflowitz, 1979b). Zinc embrittlement is or soon will be the most common method used (Teplitz, 1979). Zinc embrittlement and/or the patented Coldstream process have been used since 1964 at Tinton Falls, New Jersey, by Metallurgical Industries, Inc., to recycle cemented carbides (Adams, 1975). The Coldstream process, employing a cold, dry atmosphere, pulverizes the zinc-embrittled carbides and allows classification of the particles by grain size (Anonymous, 1979e).

Metallurgical Industries Inc., recently bought a plant in Houston, Texas, to process used drilling bits from major oil companies and independent drillers (Anonymous, 1979d).

Dynalloy Industries currently uses mechanical embrittlement but plans to switch to the zinc embrittlement process within 2 years (Teplitz, 1979).

The fine cobalt powder plant of the Sylvania Chemical and Metallurgical Division came on stream during the first quarter of 1979. To supplement the supply of cobalt-bearing scrap from its customers, the company's worldwide sales network has advertised for additional cobalt-containing scrap. John W. Anderson (1979), President, GTE Precision Materials Group, a part of General Telephone and Electronics Corporation, discussed the current cobalt recycling situation. He remarked on the current sharp interest in cobalt recycling from all types of scrap. Material with a low cobalt content is being partially sintered to increase the cobalt content before being sold. Some attempt has been made to have small companies produce cobalt oxide on a job-lot basis and then sinter the collected batches at one location.

The Sylvania plant's carbide powder will be sold as well as used by GTE's Walmet Cemented Carbides Division, a major producer of carbide cutting tools (Anonymous, 1977).

In addition to these carbide powder plants operated by U.S. firms, Metallurgie Hoboken-Overpelt SA of Belgium plans to construct an extra-fine cobalt powder plant in North Carolina. The planned capacity is about 1 million lb from cobalt-bearing scrap (Nordberg, 1978).

Cobalt in paints, pigments, and ceramics is not recycled (National Materials Advisory Board, 1971). Recycling of spent cobalt-containing catalyst is increasing in the United States. Calabrian International has been recovering cobalt in New York City from Co-Mo-Cu catalysts for 3 to 4 years by a Japanese patented process. Catalysts of cobalt and molybdenum supported on alumina are being collected by the firm, but there is no known recovery process (Cogliandro, 1979). Koflowitz (1979d) reported that in 1978 two chemical companies (Cotter Corporation and Keystone Lenway) brought a plant on stream to recover a mixed nickel cobalt carbonate product from unidentified spent catalysts.

Controlled expansion alloys (17 to 25% Co) can be produced from clean scrap without reverting to recycled or new cobalt. Semihard magnetic materials, which may contain > 50% cobalt, require purer cobalt than clean scrap can provide (Anderson, 1979). The National Materials Advisory Board (1971), however, stated that the recycle time of magnet alloys is 25 years; and that 50% recovery of the cobalt is possible.

The low percentages of cobalt in recycled high-speed steels may have been lost by dilution when mixed with non-cobalt-bearing steels in the past (Wein, 1979). Currently, domestic producers of cobalt-bearing tool steels are making sure that cobalt-bearing scrap is returned to them rather than sold to scrap dealers, who sell it abroad (Thornton, 1979).

Cobalt has been viewed in the past as a contaminant by iron scrap dealers (Haflich, 1979). The National Materials Advisory Board (1971) estimated the cobalt content of ferroalloys in iron and steel scrap consumed annually from 1960 through 1969. The Board's data are presented in Table VI-19. If cobalt had been uniformly diluted in the iron and steel, its average concentration would have been only 1.1 mg/kg. However, certain steels contain significantly higher concentrations ($\leq$ 0.40% Co) because of intentional additions.

TABLE VI-19

FERROALLOY COBALT CONTENT OF IRON AND STEEL SCRAP CONSUMED
(National Materials Advisory Board, 1971)

Year	Cobalt, MT	Iron and Steel, MT	Cobalt Concentrations mg/kg
1960	109	60	1.8
1961	81	59	1.4
1962	94	60	0.6
1963	112	68	1.7
1964	67	77	0.9
1965	40	82	0.5
1966	22	83	0.3
1967	54	78	0.7
1968	66	79	0.8
1969	149	86	1.7

BIBLIOGRAPHY. CHAPTER VI

Abrams, O., "Seabed Mining's Future Contingent on UN Conference," Am. Metal Market, **86**(184), 18 (1978).

Adams, L. A., "See Recycled Tungsten Use Continuing to Grow," Am. Metal Market, **82**(74), 19 (1975).

Agarwal, J. C., N. Beecher, G. L. Hubred, D. L. Natwig, and R. R. Skarbo, "New Fix on Metal Recovery from Sea Nodules," Eng. Mining J., **177**(12), 74-78 (1976).

Alloy Metal Products, Inc., Advertisement, Am. Metal Market, **83**(147), 4 (1976).

Anderson, E. V., "World's Nations Scramble for Sea's Riches," Chem. Eng. News, **52**(10), 18-25 (1974).

Anderson, J. W., "Cobalt Reclamation Attracts Broad Industry Spectrum," Am. Metal Market, **87**(18), 16 (1979).

Andre, J. A., and R. J. Delvaux, "Production of Electrolytic Zinc at the Balen Plant of S. A. Vieille-Montagne, Balen, Belgium" in AIME World Symposium on Mining and Metallurgy of Lead and Zinc, Vol. II, C. H. Cotterill and J. M. Cigan, Eds., Am. Inst. of Mining, Metallurgical, and Petroleum Engineers, New York, N.Y., 1970, pp. 178-197.

Anonymous, "Calera's New Electrolytic Cobalt Refinery," Eng. Mining J., Mining Guidebook, **159**(6a), 72 (1958).

Anonymous, "New Recovery Process Can Yield Both Electrolytic Nickel and Copper," Eng. Mining J., **173**(1) 94-95 (1972).

Anonymous, "The Wealth of Oceans," Newsweek, **82**(12), 85-86 (1973).

Anonymous, "Sorting, Grading Out Problems in the Hi-Temp Scrap Business," Am. Metal Market, **81**(129), 10A (1974).

Anonymous, Photo Caption, Am. Metal Market, **82**(117), 1A (1975).

Anonymous, "GTE Expects to be Nation's First Major Producer of Cobalt Powder," Am. Metal Market, **85**(123),24 (1977).

Anonymous, "Nodules: Treasure of the Deep?," Am. Metal Market, **86**(101), 13A, 21A (1978a).

Anonymous, "Amax Force Majeure on Sauget Raw Materials," Am. Metal Market, **86**(186), 2 (1978b).

Anonymous, "Ocean Minerals Has Ship, System for Seabed Mining," Am. Metal Market, **86**(188), 9 (1978c).

Anonymous, "Test Ship Raises Nodules at 3 Mile Depth," Am. Metal Market, **86**(237), 23 (1978d).

Anonymous, "Processes to Refine Ocean Nodules Assessed," Chem. Eng. News, **56**(12), 22-23 (1978e).

Anonymous, "Sea Mining Profitable Computer Study Finds," Chem. Eng. News, **56**(16), 8 (1978f).

Anonymous, "NOAA Studies Deep Mining Impact," Chem. Eng. News, **56**(24), 18 (1978g).

Anonymous, "Nickel, Cobalt Process to be Tested," Chem. Eng. News, **56**(43), 18 (1978h).

Anonymous, Materials Performance in Mines, Mills and Smelters, The International Nickel Company, Inc., New York, N.Y., 1978i; [reprinted from **7**(3) and **7**(4) of Inco Nickel News (1972)].

Anonymous, "New Alloying Nickel Helps Keep Pioneer Ni-Hard Producer Competitive," Nickel Topics, **31**(4), 4-5 (1978j).

Anonymous (D. S.), "Changing Profile of Deep-Sea Miners," Science, **200**, 1030 (1978k).

Anonymous, Advertisements, Tungsten Supplement, Am. Metal Market, **87**(3), 7-18 (1979a).

Anonymous, "Suppliers of Tungsten Products," Am. Metal Market, **87**(3), 17,18 (1979b).

Anonymous, Advertisements, Cobalt Supplement, Am. Metal Market, **87**(18), 9-20 (1979c).

Anonymous, "Tungsten Carbide Extraction from Worn Drill Bits Slated," Am. Metal Market, **87**(101), 10 (1979d).

Anonymous, "Metallurgical Sees Growth in Prealloys," Am. Metal Market/Metalworking News, **87**(157), 37 (1979e).

Anonymous, "Port Nickel Aims for Full Capacity," Am. Metal Market/Metalworking News, **87**(157), 37 (1979f).

Anonymous, "Suppliers of Cobalt Products," Am. Metal Market, **87**(19), 17, 18 (1979g).

Anonymous, "Reclaiming Cobalt Alloy," Commerce Business Daily, (July 9), 1 (1979h).

Ayling, G. M., and H. Bloom, "Heavy Metal Analyses to Characterize and Estimate Distribution of Heavy Metals in Dust Fallout," Atm. Environ., **10**, 61-64 (1976).

Banes, O. H., R. K. Carpenter, and C. E. Paden, "Electrolytic Zinc Plant of American Zinc Company" in AIME World Symposium on Mining and Metallurgy of Lead and Zinc, Vol. 2, C. H. Cotterill and J. M. Cigan, Eds., Am. Inst. of Mining, Metallurgical, and Petroleum Engineers, New York, N.Y., 1970, pp. 308-328.

Banning, L. H., and W. E. Anable, "Utilization of Nickeliferous Serpentine" in Extractive Metallurgy of Copper, Nickel, and Cobalt, P. Queneau, Ed., Interscience Publishers, Inc., New York, N.Y., 1961, pp. 301-313.

Barnard, P. G., A. G. Starliper, and H. Kenworthy, "Reclaiming Refractory Carbides and Cobalt from Cemented-Carbide Scrap," Secondary Raw Mater., Sept., 19-22 (1969).

Battelle Columbus Laboratories, Energy Use Patterns in Metallurgical and Nonmetallic Mineral Processing (Phase 6--Energy Data and Flow Sheets, Low-Priority Commodities), PB-261 150, prepared for Bureau of Mines, Washington, D.C., National Technical Information Service, U.S. Department of Commerce, Springfield, Va. (1976).

Bauder, R.B., Bibliography on Extractive Metallurgy of Nickel and Cobalt, January 1929-July 1955, U.S. Bureau of Mines Information Circular No. 7805, Bureau of Mines, U.S. Department of the Interior, Washington, D.C., November 1957, 145 pp.

Beamish, F. E., W. A. E. McBryde, and R. R. Barefoot, "The Platinum Metals" in Rare Metals Handbook, 2nd ed., C. A. Hampel, Ed., Reinhold Publishing Corporation, Chapman and Hall, Ltd., London, 1961, pp. 304-334.

Biswas, K., and W. G. Davenport, Extractive Metallurgy of Copper, Pergamon Press, Inc., Elmsford Park, N.Y., 1976.

Blickensderfer, G., Bunker Hill lead plant, Kellogg, Idaho, personal communication [telephone, Carson], April 1978.

Bogomolov, Yu. A., "Redkie Metally v Chernoi Metallurgii" ["Rare Metals in Ferrous Metallurgy"], Teoriya Prakt. Met., **9**(2), 84-88 (1937); Chem. Abstr., 33, 6763[4] (1939).

Brooks, P. T., G. M. Potter, and D. A. Martin, "Processing of Superalloy Scrap," J. Metals, **22**(11), 25-29 (1970).

Burylev, B. P., V. D. Romanov, L. Sh. Tsemekhman, V. V. Mechev, and S. E. Vaisburd, "O Raspredelenii Metallov Gruppy Zheleza Mezhdu Med'yu i ee Zakis'yu" ["Distribution of Iron Group Metals Between Copper and Its Oxides"], Izv. Akad. Nauk SSSR, Met., No. 5, 75-77 (1976).

Busch, D. A., J. R. Stone, and G. Chiszar, "Production of Refined Nickel Sulfate at Asarco's Perth Amboy Plant" in Extractive Metallurgy of Copper, Nickel, and Cobalt, P. Queneau, Ed., Interscience Publishers, Inc., New York, N.Y., 1961, pp. 469-474.

Cali, J. P., National Bureau of Standards Certificate of Analysis. Standard Reference Materials 113a and 329. Zinc Concentrates, Office of Standard Reference Materials, National Bureau of Standards, U.S. Department of Commerce, Washington, D.C., 1975 revision.

Calvert, S. E., "Ferromanganese Nodules on the Deep Sea Floor," Metall. Mater. Technol. (UK), **10**(2), 75-79 (1978).

Charles River Associates, Economic Analysis of the Cobalt Industry, PB-195 821, National Technical Information Service, U.S. Department of Commerce, Springfield, Va., December 1969.

Chase, M., "Zaire Cobalt, Copper--Schwedweiler [sic] Optimistic on Output Boost," Am. Metal Market, **87**(85), 1, 24 (1979).

Cogliandro, C. E., Comments Made from the Floor at the Cobalt Crisis Conference, April 29-May 1, 1979, Oak Brook, Ill., sponsored by Gorham International, Inc., Gorham, Maine.

Cole, E., Bureau of Mines Research Center, Rolla, Mo., personal communication [telephone, Carson], March 1978.

Coleman, E. E., and D. N. Vedensky, "Production of Ferronickel at Riddle, Oregon" in Extractive Metallurgy of Copper, Nickel, and Cobalt, P. Queneau, Ed., Interscience Publishers, Inc., New York, N.Y., 1961, pp. 263-286.

Davey, P. T., and T. R. Scott, "Removal of Iron from Leach Liquors by the 'Goethite' Process," Hydrometallurgy, **2**, 25-33 (1976).

del Valle Alonso, J. L., A. F. de Palencia y Roc, A. Rovira Pereira, J. Moreno Clavel, and R. Guzman Jimenez, "Treatment of the Leaching Residues at the Electrolytic Zinc Plant of Espanola del Zinc, S. A., Cartagena, Spain" in AIME World Symposium on Mining and Metallurgy of Lead and Zinc, Vol. II, C. H. Cotterill and J. M. Cigan, Eds., AIME, New York, N.Y., 1970, pp. 348-388.

De Michelis, T., and F. Gnesotto, "The Porto Marghera Electrolytic Zinc Plant of Monteponi e Montevecchio" in AIME World Symposium on Mining and Metallurgy of Lead and Zinc, Am. Inst. of Mining, Metallurgical, and Petroleum Engineers, New York, N.Y., 1970, pp. 269-307.

Dennis, W. H., Metallurgy of the Nonferrous Metals, 2nd ed., Sir Isaac Pitman and Sons, Ltd., London, 1961.

Diamond Reclamation Company, Inc., Advertisement, Am. Metal Market, **86**(231), 15 (1978).

Doane, R. E., "Treatment of Metal Sulfide Concentrates by Pressure Hydrometallurgy," Colo. Sch. Mines, Miner. Ind. Bull., **15**(6), 1-10 (1972).

Donaldson, H. M., M. Cassady, and J. H. Jones, Environmental Exposure to Airborne Contaminants in the Nickel Industry 1976-1977, PB-287 371, National Technical Information Service, U.S. Department of Commerce, Springfield, Va., August 1978.

Engineering-Science, Incorporated, Field Surveillance and Enforcement Guide for Primary Metallurgical Industries, PB-230 898, Prepared for U.S. Environmental Protection Agency, National Technical Information Service, U.S. Department of Commerce, Springfield, Va., 1973.

Evans, D. J. I., "Production of Metals by Gaseous Reduction from Solution. Processes and Chemistry" in Advan. Extr. Metall., Proc. Symp., London, 1967, Inst. Mining Metall., London, England, 1968, pp. 831-907.

Fletcher, A. W., "Solvent-Extraction in Processes for Metal Recovery from Scrap and Waste," Chem. Ind. (London), **1973**, No. 9, 414-419 (1973).

Flett, D. S., "Solvent Extraction in Copper Hydrometallurgy: A Review," Trans. Inst. Min. Metall., **83**, C 30-C 38 (1974).

Frankel Company, Inc., Advertisement, Scrap Age, **31**(2), 107 (1974).

Fugleberg, S., "Principles of Cadmium Production and Their Application at the Outokumpu Zinc Plant in Kokkola" in International Symposium on Hydrometallurgy, Chicago, Ill., February 25-March 1, 1973, D. J. I. Evans and R. S. Shoemaker, Eds., The American Institute of Mining, Metallurgical, and Petroleum Engineers, New York, N.Y., 1973, pp. 1145-1167.

Furst, A., "Pratt & Whitney to Reduce Cobalt Use 25%," Am. Metal Market/Metalworking News, **87**(89), 10 (1979).

Gorham International, Inc., The Cobalt Communiqué, No. 1, 1-6 (1979a).

Gorham International, Inc., The Cobalt Communiqué, No. 2, 1-5 (1979b).

Gorham International, Inc., The Cobalt Communiqué, No. 3, 1-5 (1979c).

Grigor'ev, V. D., and A. S. Yaroslavets, "Distillyatsionnoe Rafinirovanie Tsinka iz Splavov" ["Distillation Refining of Zinc from Alloys"], Izv. Akad. Nauk SSSR, Met., No. 4, 47-49 (1975).

Haflich, F., "Analyst Predicts Increasing Producer Cobalt Prices," Am. Metal Market/Metalworking News, **87**(15), 2 (1979).

Hampel, C. A., Rare Metals Handbook, Vol. 2, 2nd ed., Reinhold Publishing Corporation, London, England, 1961, pp. 304-334.

Hansen, J. W., E. N. Nelson, T. H. Sublett, and W. J. Robertson, Laboratory Investigation of Recovery Processes for Pitchblende Residues, MCW-1508; Chem. Abstr., **66**, 78526h (1967).

Healy, M. L., Deep Ocean Mining Environmental Study (DOMES). Final Report No. 27. Trace Metal Baselines in the DOMES Study Area, PB-283 243, Prepared for National Oceanic and Atmospheric Administration, National Technical Information Service, U.S. Department of Commerce, Springfield, Va., July 1976.

Hoppe, R. W., "Amax's Port Nickel Refines the Only Pure Nickel in the U.S.," Eng. Mining J., **178**(5), 76-79 (1977).

Hutchinson, T. C., and L. M. Whitby, "A Study of Airborne Contamination of Vegetation and Soils by Heavy Metals from the Sudbury, Ontario, Copper-Nickel Smelters" in Trace Substances in Environmental Health-VII, D. D. Hemphill, Ed., Proceedings of University of Missouri's 7th Annual Conference on Trace Substances in Environmental Health, Columbia, Mo., June 12-14, 1973, University of Missouri, Columbia, Mo., 1974a, pp. 179-189.

Hutchinson, T. C., and L. M. Whitby, "Heavy-Metal Pollution in the Sudbury Mining and Smelting Region of Canada. I. Soil and Vegetation Contamination by Nickel, Copper, and Other Metals," Environ. Conservation, **1**(2), 123-132 (1974b).

Huxley, J. V., "Production of Cobalt," Educ. Chem., **11**(6), 203-205 (1974).

International Nickel Company, Ltd., Internal Report 1973 [on emissions from copper smelters and the Iron Ore Recovery Plant], International Nickel Company, Ltd., Sudbury, Ontario, 1973.

Jarrett, B. M., and R. G. Kirby, Development Document for Effluent Limitations Guidelines for the Ore Mining and Dressing Point Source Category, Vol. I, PB-286 520, EPA 440/1-78/061-d, National Technical Information Service, U.S. Department of Commerce, Springfield, Va., April 1978a.

Jarrett, B. M., and R. G. Kirby, Development Document for Effluent Limitations Guidelines for the Ore Mining and Dressing Point Source Category, Vol. II, PB-286 521, EPA 440/1-78/061-e, National Technical Information Service, U.S. Department of Commerce, Springfield, Va., April 1978b.

Jeffers, T. H., and R. D. Groves, "Electrowinning Copper from Solvent Extraction Acid Strip Solution Using Pb-Sb Anodes," Metall. Trans., B, **8B**(1), 115-119 (1977).

Kaiser, E. P., B. F. Herring, and J. C. Rabbitt, Minor Elements in Some Rocks, Ores, and Mill and Smelter Products, Trace Elements Investigations Report TEI-415, U.S. Atomic Energy Commission, Oak Ridge, Tenn., 1954.

Kashima, M., M. Eguchi, and A. Yazawa, "Distribution of Impurities between Crude Copper, White Metal and Silica-Saturated Slag," Trans. Jap. Inst. Metall., 19(3), 152-158 (1978).

Klaile, B., "Western World Cobalt Production," Paper presented at the Cobalt Crisis Conference, April 29-May 1, 1979, Oak Brook, Illinois, sponsored by Gorham International, Inc., Gorham, Maine.

Koflowitz, L., "Cobalt Export Curbs No Price Help: Reddy," Am. Metal Market, **87**(86), 7 (1979a).

Koflowitz, L., "Friedman: Cobalt Substitution Trend Will Continue," Am. Metal Market/Metalworking News, **87**(89), 31 (1979b).

Komorova, L., "Removal of Nickel and Cobalt in Fire Refining of Copper," Hutn. Listy, **28**(6), 430-435 (1973); Chem. Abstr. 79, 138821f (1973).

Korda, R. J., R. E. Henzler, P. A. Helmke, M. M. Jimenez, L. A. Haskin, and E. M. Larsen, "Trace Elements in Samples of Fish, Sediment and Taconite from Lake Superior," J. Great Lakes Res. **3**(1-2), 148-154 (1977).

Larsen, R., "No Producers, Low-Grade Ore Make U.S. Cobalt Situation Grim," Am. Metal Market, **86**(105), 9 (1978).

Lee, J. A., "Chemical Engineering Know-How Puts New Nickel Process on Stream," Chem. Eng., **66**(18), 145-152 (1959).

Liessens, J. L., R. Dams, and J. Hoste, "Neutron Activation Analysis of Traces in Electrolytic Zinc Sulphate Solution. Part IV. Simultaneous Determination of Cobalt, Cadmium, Iron, and Indium," Anal. Chim. Acta, **45**, 213-218 (1969).

Lund, R. E., and R. E. Sheppard, "Cadmium Purification Practice in Zinc Smelting," J. Met., **16**(9), 724-730 (1964); Chem. Abstr., 61, 15660g (1964).

Lund, R. E., J. F. Winters, B. E. Hoffacker, T. M. Fusco, and D. E. Warnes, "Josephtown Electrothermic Zinc Smelter of St. Joe Minerals Corporation" in AIME World Symposium on Mining and Metallurgy of Lead and Zinc, Vol. II, C. H. Cotterill and J. M. Cigan, Eds., Am. Inst. of Mining, Metallurgical, and Petroleum Engineers, New York, N.Y., 1970, pp. 549-580.

Luo, C. S., M. H. Chang, C. T. Chang, J. C. Chou, H. H. Hsu, and H. T. Tsai, "Quantitative Trace Element Analysis by Proton-Induced X-Rays," Ho Tzu K'o Hsueh, **15**(1), 29-25 (1978); Chem. Abstr., **89**, 225553v (1979).

Lutjen, G. P., "Cobalt at Fredericktown," Eng. Mining J., **154**, 72-76 (1953).

Lynde, H. W., Jr., "Cobalt" in Minerals Yearbook 1965, Vol. I, Bureau of Mines, U.S. Department of the Interior, U.S. Government Printing Office, Washington, D.C., 1966, pp. 329-335.

MacKiw, V. N., and T. W. Benz, "Application of Pressure Hydrometallurgy to the Production of Metallic Cobalt" in Extractive Metallurgy of Copper, Nickel, and Cobalt, P. Queneau, Ed., Interscience Publishers, In., New York, N.Y., 1961, pp. 469-474.

Mari, A., "Analysts Slate World Cobalt Market Study," Am. Metal Market, 87(2), 9 (1979a).

Mari, A., "Consumption Pace Expected to Hold Up," Am. Metal Market, 87(3), 9-11 (1979b).

Mari, A., "Material Problems Squeezing Alloy Producers: Snapp," Am. Metal Market, 87(84), 39 (1979c).

Mari, A., "Lateritic Ore Refining Process Set for Test in '80," Am. Metal Market, 87(125), 8 (1979d).

Mari, A., "Rustenberg, Impala Projects to Boost Cobalt Output," Am. Metal Market, 87(134), 11 (1979e).

McIntyre, S. G., "Noranda's Cobalt Plans," Paper presented at the Cobalt Crisis Conference, April 29-May 1, 1979, Oak Brook, Ill., sponsored by Gorham International, Inc., Gorham, Maine.

McKay, J. E., "Lead" in Kirk-Othmer Encyclopedia of Chemical Technology, Vol. 12, 2nd ed., A. J. Standen, Executive Ed., Interscience Publishers, a Division of John Wiley and Sons, Inc., New York, N.Y., 1967, pp. 207-247.

McPhail, J. A. K., "North American Production and Supply," Paper presented at the Cobalt Crisis Conference, April 29-May 1, 1979, Oak Brook, Illinois, sponsored by Gorham International, Inc., Gorham, Maine.

Montague, H. L., "The Extractive Metallurgy of Zinc. Review of Processes and Projections for the Future," TMS Paper Selection, Paper No. A71-74, The Metallurgical Society of AIME, New York, N.Y., 1974.

Morris, A. E., Professor of Metallurgy, University of Missouri, Rolla, Mo., personal communication [telephone, Carson], March 1978.

Mukherjee, A. D., R. N. Sen, and E. Steinnes, "Distribution of Some Trace Elements in Different Beneficiation Fractions from the Bleikvassli Pyritic Lead-Zinc Ore Body, Nordland," Bull., Nor. Geol. Unders., 20(300), 21-25 (1973).

National Materials Advisory Board, Trends in the Use of Ferroalloys by the Steel Industry of the United States, Publication NMAB-276, National Academy of Sciences--National Academy of Engineering, Washington, D.C., 1971.

National Nickel Alloy Corporation, Advertisement, Am. Metal Market, **86**(230), 11 (1978).

Nelson, K. W., "Nonferrous Metallurgical Operations" in Air Pollution, Vol. 3, 2nd ed. A. C. Stern, Ed., Academic Press, New York, N.Y., 1968, pp. 171-190.

Nordberg, D., "Outokumpu Expanded Cobalt Capacity Goes On-Stream," Am. Metal Market, **86**(152), 3 (1978).

Olander, J. A., "Applications of Chemistry in Deep Ocean Mining" in Marine Chemistry in the Coastal Environment, ACS Symposium Series, No. 18, American Chemical Society, Washington, D.C., 1975, pp. 557-571.

O'Toole, J. J., R. G. Clark, K. L. Malaby, and D. L. Trauger, "Environmental Trace Element Survey at a Heavy Metals Refining Site" in Nucl. Methods Environ. Res., Proc. Am. Nucl. Soc. Top. Meet., J. R. Vogt, Ed., Off. Conf. Short Courses Univ. Mo.-Columbia, Columbia, Mo., 1971, pp. 172-182.

Panel on Operational Safety in Marine Mining, Mining in the Outer Continental Shelf and in the Deep Ocean, National Academy of Sciences, Washington, D.C., 1975.

Parker, J. G., Occurrence and Recovery of Certain Minor Metals in the Processing of Lead and Zinc, Information Circular 8790, Bureau of Mines, U.S. Department of the Interior, Washington, D.C., 1979, 75 pp.

Penn Iron & Metal Company, Inc., Advertisement, Am. Metal Market/Metalworking News, **87**(15), 35 (1979).

Perel'man, F. M., and A. Ya. Zvorykin, Kobal't i Nikel [Cobalt and Nickel (Chemistry)], Nauka, Moscow, USSR, 1975.

Persigehl, M., K. Kasperek, H. J. Klein, and L. E. Feinendegen, "Einfluss der Industrialisierung auf den Spurenelementgehalt in menschlichen Lungen" ["Influence of Industrialization on Trace Element Concentration in Human Lungs"], Beitr. Pathol., **157**(3), 260-268 (1976).

Petrick, A., Jr., H. J. Bennett, E. Starch, and R. C. Weisner, The Economics of Byproduct Metals (in Two Parts). 2. Lead, Zinc, Uranium, Rare-Earth, Iron, Aluminum, Titanium, and Lithium Systems, Bureau of Mines Information Circular 8570, U.S. Department of the Interior, U.S. Government Printing Office Stock No. 2404-01342, Washington, D.C., 1973.

Phillips, A. H., "The World's Most Complex Metallurgy (Copper, Lead and Zinc)," Trans. Metall. Soc. AIME, **224**, 657-668 (1962).

Pinkney, E. T., "Cobalt in South Africa," Paper presented at the Cobalt Crisis Conference, April 29-May 1, 1979, Oak Brook, Illinois, sponsored by Gorham International, Inc., Gorham, Maine.

Prizinsky, D., "Surcharge Deployed as Producers' Insurance Against Rising Costs," Am. Metal Market, **86**(214), 20, 32 (1978).

Probert, T. I., K. J. Richards, and G. E. Entrop, "Purifying Molybdenite Concentrates," U.S. Patent 3,911,076, October 7, 1975.

Redden, J., "Amax Aim: Full Capacity at La. Unit. Port Nickel Refinery Seen Reaching Goal by Year-End," Am. Metal Market, **87**(154), 1, 6 (1979).

Renzoni, L. S., and W. V. Barker, "Production of Electrolytic Cobalt at Inco's Port Colborne Nickel Refinery" in Extractive Metallurgy of Copper, Nickel, and Cobalt, P. Queneau, Ed., Interscience Publishers, Inc., New York, N.Y., 1961, pp. 535-545.

Roorda, H. J., and P. E. Queneau, "Recovery of Nickel and Cobalt from Limonites by Aqueous Chlorination in Sea Water," Institution of Mining and Metallurgy, Transactions, Section C, **82**, 79 C-87 C (1973).

Rule, A. R., and R. E. Siemens, "Recovery of Copper, Cobalt, and Nickel from Waste Mill Tailings" in Proceedings of the Fifth Mineral Waste Utilization Symposium, E. Aleshin, Ed., Cosponsored by the U.S. Bureau of Mines and IIT Research Institute, Chicago, Ill., April 13-14, 1976, pp. 62-67.

Saville, S., "Cobalt Demand Could Belie Expansion Plans," Am. Metal Market, **87**(93), 7 (1979).

Schack, C. H., and B. H. Clemmons, Review and Evaluation of Silver-Production Techniques, U.S. Bureau of Mines Information Circular 8266, U.S. Government Printing Office, Washington, D.C., 1965.

Scheidweiler, P., "Remarks on African and European Cobalt Production," Presented at the Cobalt Crisis Conference, April 29-May 1, 1979, Oak Brook, Illinois, sponsored by Gorham International, Inc., Gorham, Maine.

Scott, T. R., "The Recovery of Nickel and Cobalt from Contaminated Leach Liquors," Inst. Chem. Eng. Symp. Ser., **42** (Hydrometallurgy), 9.1-9.8 (1975).

Sheka, Z. A., and K. F. Karlysheva, "K Voprosu o Roli Margantsa pri Elektrolize Rastvorov Sernokislogo Tsinka" ["Role of Manganese in the Electrolysis of Solutions of Zinc Sulfate"], Ukrain. Khim. Zh., **25**, 668-673 (1959); Chem. Abstr., **59**, 10590b (1963).

Sibley, S. F., "Cobalt" in Minerals Facts and Problems, 1975 ed., Bureau of Mines Bulletin 667, U.S. Department of the Interior, U.S. Government Printing Office, Washington, D.C., 1976, pp. 269-280.

Sibley, S. F., "Cobalt" in Minerals Yearbook 1975, Vol. I, Metals, Minerals, and Fuels, Bureau of Mines, U.S. Department of the Interior, U.S. Government Printing Office, Washington, D.C., 1977, pp. 485-495.

Siemens, R. E., and J. D. Corrick, "Process for Nickel, Cobalt, and Copper from Domestic Laterites," Min. Congress J., **63**(1), 30-34 (1977).

Siemens, R., and P. Good, "Reduction of Laterite Ores," U.S. Patent 3,985,556, October 12, 1976.

Sinev, L. A., and L. K. Govorova, "Metod Rascheta Izvlecheniya Metallov v Tekhnologicheskikh Tsiklakh s Oborotnymi Materialami" ["Method for Calculating the Extraction of Metals in Production Systems with Circulating Materials"], Tsvetn. Met., No. 8, 25-26 (1977).

Smith, J. F., Director of Environmental Technology, G+W Natural Resources Group, Gulf + Western Industries, Bethlehem, Pa., personal communication [telephone, Carson], April 1978.

Sridhar, R., "Extraction of Copper, Nickel and Cobalt from Sea Nodules," J. Met., **28**(4), 32-37 (1976).

Stern, L., "U.S. Seabed Mining Policy Needed for Economic, Security Reasons," Am. Metal Market, **87**(82), 9 (1979).

Strauss, S. D., "Stockpile Strategies Shifting," Am. Metal Market, **87**(18), 15 (1979).

Teplitz, B., "Price Hikes Due on Some Scrap Alloys," Am. Metal Market, **82**(229), 16 (1976).

Teplitz, B., "Recycled Tungsten Finds Ready Market," Am. Metal Market, **87**(3), 15, 16 (1979).

Teworte, W., "Economic Aspects of Recovery of Minerals from Effluents," Chem. Ind. (London), 564-574 (1969).

Thornton, J., "Steel Supplies Worry Midwest Toolmakers," Am. Metal Market, **87**(152), 5A, 7A (1979).

Tinsley, C. R., "The Future Markets for Nodule Metals" in World Mining and Metals Technology, Vol. 1, The Society of Mining Engineers of AIME and The Mining and Metallurgical Institute of Japan, New York, N.Y., 1976, pp. 249-266.

Tougarinoff, B., "Nouveaux métaux et matériaux" ["New Metals and Materials"], Rev. Soc. Belge. Ing. Ind., No. 9-10, 428-439 (1966).

Turk, P., "Cobalt Reduction Efforts Underway at Aerospace Firms," Am. Metal Market, **87**(152), 10A, 18A (1979).

Vhay, J. S., D. A. Brobst, and A. V. Heyl, "Cobalt" in United States Mineral Resources, D. A. Brobst and W. P. Pratt, Eds., U.S. Geological Survey Professional Paper 820, U.S. Government Printing Office, Washington, D.C., 1973, pp. 143-155.

Wedge, H. D., and D. E. Halter, The Evolution of the Riddle Nickel Process, Paper No. A71-65, TMS Paper Selection, The Metallurgical Society of AIME, New York, N.Y., 1965, 8 pp.

Weimer, F. S., G. T. Wever, and R. J. Lapee" "Electrolytic Zinc Processes" in Zinc. The Science and Technology of the Metal, Its Alloys and Compounds, C. H. Mathewson, Ed., No. 142, American Chemical Society Monograph Series, Reinhold Publishing Corporation, New York, N.Y., 1959, pp. 174-225.

Wein, A., "Cobalt: A Study in Incentives for Reclamation," Am. Metal Market, **87**(67), 19A (1979).

Wells, R. R., W. G. Sandell, H. D. Suedden, and T. F. Mitchell, Concentration of Copper-Cobalt Ores from the Blackbird District, Lemhi County, Idaho, Report of Investigations 4279, Bureau of Mines, U.S. Department of the Interior, Washington, D.C., 1948, 21 pp.

Wesson, S., "Seabed Bills in Congress Ready to Sail," Am. Metal Market, **87**(48), 14, 22 (1979).

Whyte, R. M., J. R. Orjans, G. B. Harris, and J. A. Thomas, "Development of a Process for the Recovery of Electrolytic Copper and Cobalt from Rokana Converter Slag" in Adv. Extr. Metall., Int. Symp., 3rd, M. J. Jones, Ed., Institute of Mining and Metallurgy, London, England, 1977, pp. 57-68.

Wilkoff, S., and Sons Company, Advertisement, Am. Metal Market, **86**(231), 15 (1978).

Williams, R. E., Waste Production and Disposal in Mining, Milling, and Metallurgical Industries, Miller Freeman Publications, Inc., San Francisco, Calif., 1975.

Wittmann, G. T. W., and U. Foerstner, "Heavy Metal Enrichment in Mine Drainage: IV. The Orange Free State Goldfield," S. Afr. J. Sci., 73(12), 374-378 (1977).

Wolverine Metal Company, Advertisement, Scrap Age, 31(2), 122 (1974).

Zuckerman, Samuel J., Inc., Advertisement, Am. Metal Market, 86 (230), 15 (1978).

VII. COBALT IN THE FOOD CHAIN

Christopher J. Cole and Bonnie L. Carson

This chapter summarizes available information on cobalt in the aquatic food chain, in the terrestrial food chain, and in the human diet and human tissues. Difficulties in calculating a human body burden are discussed. In general, cobalt concentration factors diminish with increasing trophic levels in any particular food chain. Exceptions are noted in a predator-herbivore-plant-soil chain in a temperate ecosystem and in a reptile/amphibian-insect-plant-soil chain in a tropical moist forest. In contaminated or heavily cobalt-fertilized media, the concentration factors for plants do not increase as rapidly as do the media concentrations.

A. The Aquatic Food Chain

1. Reported concentrations: The concentrations of cobalt measured in marine and freshwater animals and plants are listed in Tables VII-1 through 6. From these studies, the concentration factors (CF's) for cobalt were calculated,

$$CF = \frac{\text{concentration in organism}}{\text{concentration in water}}$$

where the concentrations are given in the same units. Thus, the CF indicates a ratio of the organism's cobalt concentration to the concentration of cobalt in the medium in which it is living. Most of the CF's described below are based on dry weight.* Occasionally, a CF may be calculated on the basis of a sediment for rooted aquatic plants or on the organisms eaten by the organism in question.

---

* For the purposes of generalizations, CF's based on wet weight have been multiplied by 10 since dry weight concentrations should be about 10 times higher than fresh weight concentrations. Amiard-Triquet and Foulquier (1978) estimated that the wet weight of algae was about five times the dry weight in their experiments.

TABLE VII-1

COBALT IN MARINE PLANTS

Organism	Mean Concentration	Unit	Concentration Factor	Analytical Method[a]	Remarks	Reference
Seaweed Brown algae *Fucus vesiculous* *Ascophyllum nodosum* Red algae *Rhodymenia palmata* *Halosaccion ramencetatum*	0.74	mg/kg dry weight	2,800	NAA	34 samples from Barents Sea. *Fucus vesiculous* found in intertidal zone with air bladder. *Ascophyllum nodosum* intertidal. *Rhodymenia palmata* edible.	Vaganov et al. (1978)
Seaweed Brown - *Phaeophyceae* Green - *Chlorophyceae* Red - *Rhodophyceae*	 0.67 0.92 0.52	mg/kg dry weight		Colorimetric spectrophotometry	77 samples. Observed no difference by season or habitat.	Ishibashi et al. (1964)
Brown and green algae	100	mg/kg dry weight			Samples from Sea of Japan.	Gryzhankavo et al. (1973; cited by Leland et al., 1974)
Algae	0.3 to 40	mg/kg dry weight			Compilation based on five literature values.	Phillips (1977)
Algae	0.1	mg/kg fresh weight	2,000	Spectrophotometry		Fukai and Murray (1973)
Seaweed *Ulva* spp. (sea lettuce) *Ecklonia maxima* (kelp) *Porphyra capensis* (purple laver) *Suhria vittata* (red ribbon) *Gigartina rachula* (batters)	 0.014 0.0064 0.047 0.049 0.021	 mg/kg fresh weight mg/kg fresh weight mg/kg fresh weight mg/kg fresh weight mg/kg fresh weight	 600 215 1,600 1,500 600	NAA	Location 25 km north of Cape Town, South Africa.	Van As et al. (1973)
Ulva pertusa	0.17	mg/kg dry weight		AAS		Ishii et al. (1978)
Enteromorpha sp.	0.18	mg/kg dry weight		AAS		Ishii et al. (1978)
Chondrus ocellatus	0.36	mg/kg dry weight		AAS		Ishii et al. (1978)
Undaria pinnatifida	0.16	mg/kg dry weight		AAS		Ishii et al. (1978)
Fucus virsoides (intertidal seaweed	2	mg/kg wet weight		NAA	Samples from North Adriatic Sea.	Lulic and Strohal (1974)
Eisenia bicyclis	0.12	mg/kg dry weight		AAS		Ishii et al. (1978)
Hizikia fusiforme	0.14	mg/kg dry weight		AAS		Ishii et al. (1978)

TABLE VII-1 (concluded)

Organism	Mean Concentration	Unit	Concentration Factor	Analytical Methid	Remarks	Reference
Sargassum spp.		mg/kg dry weight	2,000	AAS		Ishii et al. (1978)
S. thunbergii	0.36					
S. sagamianum	0.17					
S. ringgoldianum	0.23					
S. kjellmanianum	0.43					
S. horneri	0.12					

a/ NAA = neutron activation analysis and AAS = atomic absorption spectrometry.

TABLE VII-2

COBALT IN MARINE INVERTEBRATES

Organism - Organ	Isotope[a]	Mean Concentration	Unit	Concentration Factor	Analytical Method[b]	Remarks	Reference
Mollusks Lamellibranchii and Gastropoda, soft parts		0.100	mg/kg fresh weight	2,000	Spectrophotometry		Fukai and Murray (1973) [7777]
Dunaliellabioculata sp.	^{60}Co	100	pCi/g			Tracer study. Majority lost by 48 hr in seawater.	Fowler (1977)
Acetabularia mediterranae	^{60}Co	20	pCi/g			Tracer study. Majority lost by 48 hr in seawater.	Fowler (1977)
Mytilus edulis	^{60}Co	18 to 96	pCi/g			Tracer study. Majority lost by 48 hr in seawater.	Fowler (1977)
Mollusks Choromytillus meridionalis (mussel) Donax serra (mussel) Haliotus midea (abalone)		 0.037 0.040 0.023	mg/kg fresh weight	 1,250 1,350 800	NAA	Location 25 km north of Cape Town, South Africa. Surf-zone concentration of trace elements in seawater from Cape of Good Hope 0.03 μg/liter.	Van et al. (1972)
Jasus lalandii (lobster)		0.0042	mg/kg fresh weight	140	NAA		Van et al. (1972)
Penaeus japonica, abdomen		0.05	mg/kg dry weight		AAS		Ishii et al. (1978)
Squilla oratoria, abdomen		0.16	mg/kg dry weight		AAS		Ishii et al. (1978)
Sepia esculenta, trunk		0.06	mg/kg dry weight		AAS		Ishii et al. (1978)
Meretrix lamarckii, soft part		0.63	mg/kg dry weight		AAS		Ishii et al. (1978)
Gomphina melanaegis, soft part		0.70	mg/kg dry weight		AAS		Ishii et al. (1978)
Mytilus corscum, soft part		0.64	mg/kg dry weight		AAS		Ishii et al. (1978)
Notohaliotis discus, soft part		0.27	mg/kg dry weight		AAS		Ishii et al. (1978)
Shellfish		-	-	4,000	AAS		Ishii et al. (1978)

TABLE VII-2 (continued)

Organism - Organ	Isotope[a]	Mean Concen-tration	Unit	Concen-tration Factor	Analytical Method[b]	Remarks	Reference
Mya arenaria	^{60}Co	-	-		Gamma well counter	Exposed to discharge of nuclear reactor 1 to 3 pCi/liter. Experimental duration 175 days. Elimination and uptake study.	Harrison (1973)
Peel				2,400			
Body with peel removed				500			
Whole animal				500			
Gills, body fluid, and muscle, each				100			
Stomach, digestive gland				350			
Mytilus edulis	^{60}Co	-	-		Auto-radiography	Experimental values from uptake and elimination study.	Van Weers (1973)
Whole animal				19 to 35			
Soft parts				45 ± 14			
Shell				33 ± 10			
Digestive gland and kidneys							
Mytilus edulis			mg/kg dry weight		NAA	Experimental values from uptake and elimination study.	Van Weers (1973)
Soft part		0.9 to 3.3		2,600			
Lamellibranch mollusk			mg/kg dry weight		AAS	Samples collected from Irish Sea.	Segar et al. (1971)
Modiolus modiolus							
Soft part		0.49 to 5.5					
Shell		0.30					
Glycymeris glycymeris							
Soft part		0.92					
Shell		0.20					
Pecten maximus							
Soft part		8.5					
Shell		0.20 to 0.31					
Mytilus edulis							
Soft part		1.6					
Shell		0.15					
Mercenaria mercenaria							
Soft part		4.3					
Shell		1.2					
Cardium edule							
Soft part		7.1					
Shell		0.16					

TABLE VII-2 (continued)

Organism - Organ	Isotope[a]	Mean Concentration	Unit	Concentration Factor	Analytical Method[b]	Remarks	Reference
Gastropod			mg/kg dry weight		AAS	Samples collected from Irish Sea.	Segar et al. (1971)
Patella vulgata							
Soft part		0.4					
Shell		0.80					
Nucella lapillus							
Soft part		0.30					
Shell		0.25					
Buccinum undatum							
Soft part		0.76					
Shell		0.16					
Crepidula fornicata							
Soft part		17					
Shell		0.15					
Pecten maximus			mg/kg dry weight		AAS	Samples collected from Irish Sea.	Segar et al. (1971)
Whole animal		8.5					
Muscle		0.34					
Gut and digestive gland		0.68					
Mantle and gills, unwashed		0.45					
Mantle and gills, washed		0.58					
Gonad, unwashed		0.44					
Gonad, washed		1.5					
Mixed shell		0.31					
Upper valve shell		0.20					
Lower valve shell		0.20					
Modiolus modiolus			mg/kg dry weight		AAS	Samples collected from Irish Sea.	Segar et al. (1971)
Shell		0.30					
Entire soft parts		0.49 to 5.5					
Mantle and gills		2.9					
Muscle		0.30					
Gonad		0.72					
Gut and digestive gland		1.1					
Mytilus edulis	^{60}Co	-	-	190	AAS	High accumulation on surface.	Shimizu et al. (1971)
Tapes japonica	^{60}Co	-	-	36	AAS	High accumulation on surface.	Shimizu et al. (1971)

TABLE VII-2 (continued)

Organism - Organ	Isotope[a]	Mean Concen-tration	Unit	Concen-tration Factor	Analytical Method[b]	Remarks	Reference
Coral *Solenosmilia* sp., *Desmophyllum* sp., *Caryophyllia* sp., and others		0.35 to 1.0; < 2 to 105	mg/kg dry weight		NAA; emission spectrometry	Coral with ferromanganese stain. Deep open ocean specimens.	Livingston and Thompson (1971)
Coral *Dendrophyllia* sp., *Madracis* sp., *Cladocora* sp., *Anomocora* sp., *Bathycyathus* sp., and others		0.19 to 1.5; < 2 to 3	mg/kg dry weight		NAA; emission spectrometry	Coral with ferromanganese stains. Shallow open ocean specimens.	Livingston and Thompson (1971)
Coral *Meadrina* sp., *Porites* sp., *Madracis* sp., *Montastrea* sp., *Scolmia* sp., and others		0.04 to 0.47; < 2 to 2	mg/kg dry weight		NAA; emission spectrometry	Coral without stain. Shallow coastal coral.	Livingston and Thompson (1971)
Oyster, eastern		0.10	mg/kg wet weight		AAS		Ronk (1971)
Quahaug, northern		0.20	mg/kg wet weight		AAS		Ronk (1971)
Softshell clams		0.10	mg/kg wet weight		AAS		Ronk (1971)
Mytilus galloprovincialis		0.23 to 0.25 ± 0.10	mg/kg dry weight		-	249 samples from the Gulf of Trieste over a 2-year period. Sample concentrations were stable over the 2-year sampling period.	Majori et al. (1978)
Bivalve marine mollusks		0.1 to 8.5	mg/kg dry weight		-	Based on compilation of 14 references.	Phillips (1974)

TABLE VII-2 (continued)

Organism - Organ	Isotope[a]	Mean Concentration	Unit	Concentration Factor	Analytical Method[b]	Remarks	Reference
Donax denticulatus			mg/kg dry weight		Gamma spectrometry	Experimental value and theoretical concentration factor based on 17-day experiment.	Lowman and Ting (1972)
Soft part		1.39					
Soft part	^{57}Co			460			
Shell	^{58}Co			320			
North Adriatic Sea organisms			mg/kg dry weight		NAA		Lulic and Strohal (1974)
Cystoseira abrotanifolia		6.1					
Mytilus galloprovincialis		2.0					
Octopus vulgaris		6.0					
Spheciospongia (sponge)	^{60}Co	11 13 2.8	disintegrations per mg ash weight			First two samples, Port Higuiro, Puerto Rico; third sample, Negro Reef, Puerto Rico. Neither ^{57}Co nor ^{60}Co was detected in seawater.	Lowman et al. (1967)
Echinoderm species			mg/kg dry weight		NAA	Samples from Saronikos Gulf, Greece. The authors suggested S. granularis as an indicater of Co.	Papadopoulou and Kanias (1976)
Ophioderma longicauda		0.20		220			
Echinaster sepositus		0.38		420			
Marthasterias glacialis		0.090		100			
Sphaerechinus granularis		0.66		730			
Paracentrotus lividus		0.28		310			
Arbacia lixula		0.32		350			
Holothuria tubulosa		0.11		120			
Tunicates					NAA		Papadopoulou and Kanias (1977)
Microcosmus sulcatus, whole body		1.9	mg/kg dry weight	340			
		0.31	mg/kg fresh weight				
Ciona intestinalis							
Whole body		0.52	mg/kg dry weight	40			
		0.022	mg/kg fresh weight				
Tunic		0.44	mg/kg dry weight				
		0.017	mg/kg fresh weight				
Rest of body		1.4	mg/kg dry weight				
		0.050	mg/kg fresh weight				
Scrobicularia plana			mg/kg dry weight		AAS	A burrowing bivalve. 47 samples from Tamar estuary, England.	Bryan and Uysal (1978)
Whole soft part		18.6					
Digestive gland		68					

TABLE VII-2 (continued)

Organism - Organ	Isotope[a/]	Mean Concen-tration	Unit	Concen-tration Factor	Analytical Method[b/]	Remarks	Reference
Euphausiid			mg/kg dry weight		NAA	-	Fowler (1977)
Fecal pellets		3.5					
Molts		0.80					
Eggs		0.80					
Whole animal		0.18					
Microplankton Phytoplankton, copepods, chaetognaths		0.87	mg/kg dry weight		NAA		Fowler (1977)
Clam, soft part		0.110 to 0.220	-	2,000	NAA	CF's were calculated on the basis of 0.1 g Co/liter.	Ichikawa and Ohno (1974)
Sea urchin, ovary		0.071 to 0.084	mg/kg fresh weight	800	NAA		Ichikawa and Ohno (1974)
Sea cucumber, whole body		0.0082, 0.019	mg/kg fresh weight	100	NAA		Ichikawa and Ohno (1974)
Shrimp					NAA		Ichikawa and Ohno (1974)
Soft part		0.014,	-	200			
Carapace		0.036		300			
Aplysia (sea hare), whole animal		12 to 29	mg/kg dry weight		AAS	-	Patel et al. (1973)
Copepoda, whole animal		0.200	mg/kg fresh weight	4,000	Spectro-photometry	-	Fukai and Murray (1973)
Decapoda, soft part		0.040	mg/kg fresh weight	800	Spectro-photometry		Fukai and Murray (1973)
Euphausiids, whole animal Amphipods, shrimp, *Acanthephyra palemonetes*		18	mg/kg dry weight		NAA	34% in molt.	Fowler (1977)

TABLE VII-2 (concluded)

Organism - Organ	Isotope[a]	Mean Concentration	Unit	Concentration Factor	Analytical Method[b]	Remarks	Reference
Copepoda	^{57}Co	-	-	30		Experimental value. Eliminated 30% in 12 hr. Copepods attained equilibrium in 20 hr.	Benzhitskii and Sazhina (1975)
Decapoda	^{57}Co	-	-	12			Benzhitskii and Sazhina (1975)
Panulirus argus (juvenile lobster)	^{57}Co	-	-	36	Gamma spectrometry		Lowman and Ting (1972)
	^{58}Co	-	-	100			
Octopus vulgaris	^{60}Co		-			Experimental values at end of 30-day study; water regulated at 5 Ci/liter. Ueda et al. (unpublished data cited by Nakahara et al., 1979) found 9.5 mg Co/kg in the octopus branchial heart.	Nakahara et al. (1979)
branchial heart		-	-	24,000			
gill		-	-	480			
liver		-	-	240			
kidney		-	-	230			
muscle		-	-	15			

a/ Assume stable ^{59}Co if not given.
b/ NAA = neutron activation analysis.
AAS = atomic absorption spectrophotometry.

TABLE VII-3

COBALT IN MARINE FISH

Organism - Organ	Isotope[a]	Concentration Mean	Range	Unit	Concentration Factor[b]	Analytical Method[c]	Remarks	Reference
Line-fish			-	mg/kg fresh weight		NAA	Sample location 25 km north of Capetown, South Africa.	Van As et al. (1973)
Seriola pappei (yellowtail)		0.012			400			
Argyrozona argyrozona (silver fish)		0.0019			90			
Johnius hololepidotus (kabeljou)		0.0029			100			
Pachymentopan grande (Hottentot)		0.0022			60			
Whitefish			-	mg/kg fresh weight		NAA	Sample location 25 km north of Capetown, South Africa.	Van As et al. (1973)
Merluccius capensis (stockfish)		0.0048			150			
Xiphiurus capensis (kingklip)		0.0036			70			
Synaptura marginata (sole)		0.0029			100			
Pelagic fish			-	mg/kg fresh weight		NAA	Sample location 25 km north of Capetown, South Africa.	Van As et al. (1973)
Trachurus trachurus (maasbanker)		0.0081			250			
Fish muscle		-	0.008 to 0.022	mg/kg dry weight	90	AAS	-	Ishii et al. (1978)
Clupea pilchardus (herring)		2.4	-	mg/kg dry weight		NAA	Samples from north Adriatic Sea.	Lulic and Strohal (1974)
Solea solea		7.2	-	mg/kg dry weight		NAA	Samples from north Adriatic Sea.	Lulic and Strohal (1974)
Spheroides niphobles	^{60}Co	-	-	-			15 days after uptake. Not clear if CF's calculated on a net weight basis.	Hiyama and Kahn (1964)
Gills					12			
Digestive tract					11			
Liver and gallbladder					6.6			
Heart					14			
Spleen					15			
Kidney					-			
Muscle					0.54			
Skin					5.1			
Scale and fins					8.7			
Bones					1.2			
Head					-			

TABLE VII-3 (continued)

Organism - Organ	Isotope[a]	Concentration Mean	Concentration Range	Concentration Unit	Concentration Factor[b]	Analytical Method[c]	Remarks	Reference
Albacore, liver, year caught *Thunnus alalunga*	^{60}Co			pCi/kg wet weight	-	-	Reflects atomic testing of 1961 to 1962. Fish collected from North Pacific off San Diego.	Hodge et al. (1973)
1964, 20 fish		145 ± 13	102 to 250					
1965, 28 fish		174 ± 9	80 to 280					
1968, 44 fish		63 ± 3	40 to 110					
1970, 43 fish		45 ± 1	40 to 51					
1971, 39 fish		33 ± 1	26 to 48					
1971		10.8	6.2 to 15.3				Marquesas Islands sample for comparison with above.	
Fish							For continued comparison	
Bigeye, 1970		43.3	35.3 to 51.3				--Hawaiian Islands.	Hodge et al. (1973)
Yellowfin, 1970 to 1971		44.6	24.8 to 67.7					
Yellowfin, 1970		2.16	1.22 to 3.1				--Northwest Africa.	
Yellowfin, 1971		12.15	12.0 to 12.3				--Northwest South America.	
Fish	-		-	mg/kg wet weight		NAA	Samples collected from Japanese fishing boats, native and South American species.	Ichikawa and Ohno (1974)
Skin		0.031			300			
Muscle		0.005			50			
Viscera		0.048			500			
Bone		0.015			200			
Engraulis encrasicholus ponticus (anchovy)	-	-	-	-	< 0.1 to 3	Emission spectrography 0.001% sensitivity	Samples from the Black Sea near Odessa.	Petkevich (1965)
Sprattus sprattus phalericus (sprat)	-	-	-	-	< 0.1 to 4	Emission spectrography 0.001% sensitivity	Samples from the Black Sea near Odessa.	Petkevich (1965)

TABLE VII-3 (concluded)

Organism - Organ	Isotope[a/]	Concentration Mean	Concentration Range	Concentration Unit	Concentration Factor[b/]	Analytical Method[c/]	Remarks	Reference
Fish, muscle		-	0.008 to 0.022	mg/kg dry weight	~90	AAS	-	Ishii et al. (1978)
Bonito				mg/kg wet weight		AAS	8 samples.	Katsuki et al. (1975)
White muscle		2×10^{-5}						
Red muscle		3×10^{-5}						
Pyloric caecum		2×10^{-5}						
Liver		3×10^{-5}						
Spleen		$< 1 \times 10^{-5}$						
Heart		3×10^{-5}						
Kidney		9×10^{-5}						
Digestive gland		3×10^{-5}						
Blood		4×10^{-5}						
Gill		1×10^{-5}						
Spawn		1×10^{-5}						
Spermatic sac		9×10^{-5}						
Stomach content		4×10^{-5}						

a/ Assume stable cobalt unless otherwise stabilized.

b/ To be strictly comparable to concentration factors based on stable cobalt concentration dry weight, the water content of each organism should be known. Wet weight values are roughly comparable to dry weight values by multiplying the former by 10. Concentration factors based on ash weight can be roughly compared to dry weight concentration factors by dividing by 10.

c/ NAA = neutron activation analysis.
AAS = atomic absorption spectrophotometry.

TABLE VII-4

COBALT IN FRESHWATER PLANTS

Organism	Isotope[a/]	Mean Concentration	Unit	Concentration Factor	Analytical Method[b/]	Remarks	Reference
Stream algae in			mg/kg dry weight		AAS	Ni and Co mining on Little St. Francis River, which had the maximum water concentratio, 0.2236 mg Co/liter.	Proctor and Sinha (1978)
Upper St. Francis River		12.2		12,200			
Doe Run Creek		26.3		25,000			
Wolf Creek		32.3		21,400			
Little St. Francis River		307.3		1,374			
Big Creek		22.8		22,800			
Aufwuchs (periphyton)	^{60}Co	1,906	pCi/kg wet weight	4,000		Samples from Po River, Italy, near nuclear power station. Water concentration = 0.467 pCi ^{60}Co/liter. Water concentration = 0.003 mg ^{59}Co/liter.	Smedile and Queirazza (1976)
		134	mg/kg wet weight	45,000			
Plectonema boryanum (blue-green algae) at	^{57}Co	-	-			Samples from streams exposed to outfall of the Savannah River plant (nuclear reactor). CF's were calculated on a dry weight basis.	Harvey (1969)
25°C				6,200			
30°C				4,500			
35°C				3,500			
40°C				2,500			
Algae		8.11 to 12.61	mg/kg dry weight	-	NAA	Samples from the Danube River and Canal near Vienna, Austria.	Rehwoldt et al. (1975)
Algae		3.05	mg/kg ash weight	3,100	NAA	Samples from Danube River.	Radosavljević et al. (1973)
Plankton (unspecified) in			mg/kg wet weight		NAA		Radosavljević et al. (1973)
Danube River		4.65		18,250			
Sava River		3.27		191,000			
Tisa River		4.32		5,500			
Lakes in Switzerland							
Luganer See		2.30		-			
Unter See		0.31		-			
Vierwaldstatter See		2.31		-			
Zuger See		0.54		-			
Algae		0.3 ± 0.5	mg/kg ash weight	-	AAS		Boothe et al. (1972)

TABLE VII-4 (continued)

Organism	Isotope[a]	Mean Concentration	Unit	Concentration Factor	Analytical Method[b]	Remarks	Reference
Algae			mg/kg dry weight		AAS	Samples from Lake St. Clair marshes. CF based on water concentration of < 0.001 mg Co/liter.	Mudroch and Capobianco (1970)
Chara sp. (stonewort)		12.8		> 12,800			
Myriophyllum heterophyllum (water milfoil)		5.6		> 15,600			
Algae	^{60}Co	-				Experimental values reached equilibrium by 15 days; followed for 75 days. Tests were run in a glass aquarium. Trapeznikov and Trapeznikova (1979) calculated much higher CF's of ^{60}Co for *C. demersum* (33,500), *E. canadensis* (21,500), and *Lemna minor* (19,100) growing in cooling ponds of nuclear power plants in the Urals. Possibly, the higher CF's were due to higher water temperatures (24°C).	Timofeeva-Resovskaya (1963)
Myriophyllum spicatum				3,500			
Ceratophyllum demersum				4,665			
Utricularia vulgaris				11,650			
Lemna minor				3,900			
Lemna trisulca				14,000			
Elodea canadensis				3,490; 5,416			
Stratiotes aloides				4,900			
Hydrocharis morsus ranae				5,430			
Scendesmus quadricauda				390			
Scendesmus accuminatus				250			
Cladophora fracta				8,750			
Cladophora glomerata				1,905			
Mougeotia sp.				238,000			
Spirogyra crassa (pondscum)				17,000			
Spirogyra sp. (pondscum)				5,640			
Chara fragilis				7,425			
Carex spp. (sedge)				4,595			Timofeeva-Resovskaya (1963)
Bacteria from cliff lichens				370			
Average for all plants				4,425			
Algae			mg/kg dry weight		AAS	Sample from Lake Eufaula and the Chattahoochee River Reservoir. Cobalt concentration in water ranged from 0.001 to 0.005 mg/liter.	Lawrence (1968)
Chara spp.		16.0		16,000			
Lyngba spp.		18		18,000			
Nitella spp.		4.03		4,030			

TABLE VII-4 (continued)

Organism	Isotope[a]	Mean Concentration	Unit	Concentration Factor	Analytical Method[b]	Remarks	Reference
Lake St. Clair marshland species	-		mg/kg dry weight		AAS	Concentration factors for roots, shoots, and Typha spadix reflect uptake at a maximum stage of growth. CF of water lily based on water concentration of < 0.001 mg Co/liter. Rooted plants CF based on sediment concentration of 7.5 mg Co/kg dry weight.	Mudroch and Capobianco (1978)
Nymphaea odorato (water lily)		1.7		1,700			
Pontederia cordata (pickerelweed)							
Whole plant		1.1		0.1			
Roots				~ 0.1			
Shoots ≤ 40 cm				~ 0.1			
Shoots > 40 cm				~ 0.1			
Typha latifolia (cattail)							
Whole plant		1.3		0.2			
Roots				~ 0.5			
Shoots ≤ 40 cm				~ 0.4			
Shoots > 40 cm				~ 0.1			
Typha spadix				~ 0.3			
Lythrum salicaria (loosestrife)							
Whole plant		1.7		0.23			
Roots				~ 0.2			
Shoots ≤ 40 cm				~ 0.2			
Shoots > 40 cm				~ 0.7			
Carex lacustris (sedge)		1.8		0.24			
Eichornia crassipes (water hyacinths)		-	-	-	-	Study showed uptake of 0.568 mg Co^{2+} per gram (dry weight) per day. Laboratory experiments.	Wolverton and McDonald (1975)
Alternanthera philoxerides (alligator weeds)		-	-	-	-	Study showed uptake of 0.130 mg Co^{2+} per gram (dry weight) per day. Laboratory experiments.	Wolverton et al. (1975)
Limnetic weeds			mg/kg dry weight	-	Colorimetric analysis	-	Ishibashi et al. (1964)
Pleurotaenium nodulosm		2.86					
Myriophyllum spicatum		1.65					
Water lily			mg/kg dry weight		-	Cobalt concentration in water ranged from 0.001 to 0.005 mg/liter.	Lawrence (1968)
White		10.0		10,000			
Banana		0.4		400			

TABLE VII-4 (concluded)

Organism	Isotope[a]	Mean Concen- tration	Unit	Concen- tration Factor	Analytical Method[b]	Remarks	Reference
Limnetic weeds			mg/kg dry weight		-	Cobalt concentration in water ranged from 0.001 to 0.005 mg/liter.	Lawrence (1968)
Bacopa sp.		14.0		14,000			
Hydrocotyl sp.		2.0		2,000			
Patamogeton crispus		8.73		8,730			
Eel grass		3.3 to 4.85		3,300 to 4,850			
Buttonbush		8.0		8,000			
Paspalum fluitans		5.67		5,670			
Spartina roots		2.5 to 32	mg/kg dry weight	0.16 to 0.67	-	Foundry Cove near Cold Springs on the Hudson River. Sediment concentration ~ 16 to ~ 120 mg/kg.	Kneip et al. (1975)
Spartina roots		~ 3 to 50	mg/kg dry weight	-	-	Samples from Danube River and Canal near Vienna, Austria.	Rehwoldt et al. (1975)
Duckweed		1.5	mg/kg wet weight	25[c]	NAA	Area studied is a coal ash settling basin and swamp drainage system of a coal-fired power station. Sediment concentration, 10 mg Co/kg; water concentration, 0.06 mg Co/liter.	Guthrie and Cherry (1979)
Oscillatoria sp.		1.2	mg/kg wet weight	20[c]			
Hydrodictyon sp.		1.4	mg/kg wet weight	23[c]			
Pontederia (pickerelweed)		0.7	mg/kg wet weight	0.07[d]			
Cattail		0.2	mg/kg wet weight	0.02[d]			
Cypress		0.1	mg/kg wet weight	0.01[d]			
Ceratophyllum demersum (hornwort) submerged, nonrooted plant		0.35	mg/kg dry weight		Optical emission spectro- photometry	Samples from Linsley Pond and Cedar Lake in Connecticut.	Cowgill (1973)
Potamogeton praelongus (pondweed) submerged rooted plant		0.30	mg/kg dry weight				
Nuphar advena (yellow water lily) rooted plant with floating leaves		0.08	mg/kg dry weight				
Pontederia cordata (pickerelweed leaves) rooted in sediment		1.7	mg/kg dry weight				

a/ Assume stable cobalt if not given.
b/ AAS = Atomic absorption spectrophotometry.
NAA = Neutron activation analysis.
c/ CF based on water concentration; see text for explanation.
d/ CF based on sediment concentration; see text for explanation.

TABLE VII-10

TERRESTRIAL FOOD CHAIN

Organism/Organ or Ecosystem Compartment	Isotope[a/]	Cobalt Concentration (dry weight)	Concentration Factor	% Distribution in Body	Analytical Method	Remarks	Reference
Rat (*Rattus rattus*)	^{60}Co					Rats at equilibrium after ingestion of five meals. Assimilated 0.60% of uptake.	Amiard (1976)
blood		70 pCi/g		7.2			
liver		93 pCi/g		9.5			
pancreas		12 pCi/g		1.2			
kidneys		131 pCi/g		13.4			
muscles		451 pCi/g		46.2			
digestive gland		175 pCi/g		17.9			
heart		10 pCi/g		1.0			
ovaries		9 pCi/g		0.9			
thyroid and salivary gland		8 pCi/g		0.8			
urinary excretion		3.81 pCi/g		~2.5			
fecal excretion		150.28 pCi/g		~97			
assimilation *in toto*		0.939 pCi/g	0.02	~0.06			
assimilation/g body weight		0.0059 pCi/g		-			
Bobwhite (*Colinus virginianus*)	^{60}Co	8.31 pCi/g	0.011 at 11 days			A 21-day feeding study with an initial (3-day) uptake rate of 87 pCi/g/day. At equilibrium after 11 days with 277 pCi/g body burden. Three percent retention at 21 days.	Anderson et al. (1976)
bone				0			
gut				19 ± 5			
heart				11 ± 3			
liver				0			
muscle				52 ± 15			
Insects	^{60}Co			-		250 Ci ^{60}Co injected into tree; insects feeding on leaf litter around base.	Saas et al. (1971)
Centipedes		9.6 μCi/kg					
Diptera		4.9 μCi/kg					
Minnie's Lake swamp water	-	< 0.06 mg/liter			Atomic absorption	These are the dominant plants from the respective areas of Okefenokee Swamp.	Casagrande and Erchull (1977)
Taxodium sp., needles and twigs		< 6 mg/kg					
Peat		< 8 mg/kg					
Chesser Prairie marsh water		< 0.01 mg/liter					
Nymphea sp., leaves		< 6 mg/kg					
stems		< 6 mg/kg					
rhizomes, roots		< 6 mg/kg					
Oppossum (*Aidelphis* sp.)	^{60}Co	0.5 pCi/g			Gamma spectrometry	Temperate deciduous forest.	Willard (1975)
Raccoon (*Procyon lotor*)		1.0 pCi/g					
White-footed mouse (*Peromyscus*)		50 pCi/g					
Golden mouse		50 pCi/g					
Short-tail shrew (*Blarina* sp.)		80 pCi/g					
Eastern chipmunk (*Tamias* sp.)		20 pCi/g					
Earthworm (*Lumbricus rubellus*)		> 56 pCi/g					

TABLE VII-10 (concluded)

Organism/Organ or Ecosystem Compartment	Isotope[a]	Cobalt Concentration (dry weight)	Concentration Factor	% Distribution in Body	Analytical Method	Remarks	Reference
Animal bone		0.0007 ± 0.000005 mg/kg	-	-	Neutron activation analysis	Run against NBS and IAEA Standards	Ördögh (1978)
Vegetation/soil		-	26	-	-	Tropical moist forest ecosystem in Panama. CF's calculated on dry weight basis.	Golley et al. (1976)
Herbivores/vegetation			9.8				
Reptiles and amphibians/insects			104				
Insects/understory leaves			8.7				
Soil	^{60}Co	4.57 pCi/mg	-	-	-	Herbivore/insect turnover rate, 92 hr. Site, Whiteoak Lake bed, Tennessee.	Crossley (1969)
Vegetation		0.265 pCi/mg	0.058[b] - [c]				
Herbivore		0.114 pCi/mg	0.43 0.025				
Predator		0.122 pCi/mg	1.07 0.026				

a/ Assume stable cobalt unless otherwise stated.
b/ CF is based on ratio of higher trophic level divided by lower; not strictly a CF.
c/ These CF values are related to soil values.

TABLE VII-5

COBALT IN FRESHWATER ANIMALS EXCEPT FISH

Organism - Organ	Isotope[a]	Concentration	Concentration Factor	Analytical Method[b]	Remarks	Reference
Larvae and insects	^{60}Co	-				Timofeeva-Resovskaya (1963)
Culex pipiens, pipiens			4,010			
Halesus interpunctatus			8,500			
Aeschna sp.			635			
Eristalis sp.			390			
Tendipedes			375			
Trichoptera (caddis fly), larvae			9,550			
		(mg/kg ash weight)				
Trichoptera		0.506	52	NAA	Danube River, km 1,148 in 1968.	Radosavljević et al. (1973)
Gastropoda		0.569	58			
Hirudinea		0.103	10			
Lamellibranchiata		0.375	39			
Oligochaeta		3.26	340			
Alburnus alburnus		0.399	8			
Mollusks						
Shell		0.355	365			
Muscle		-	29			
Snail						
Shell		0.252	260			
Flesh		0.420	9			
Stictochironomus annulicrus (bloodworm), larvae	^{60}Co	-	0.11	NaI (Tl) detector with multichannel detector	CF's based on ^{60}Co concentration in the detritus or algae on which the organisms fed.	Wilhm (1970)
Limnodrilus hoffmeisteri (sludgeworm), larvae			0.28			
Procladius sp. (a predaceous midge), larvae			0.15			
Mollusks		(mg/kg dry weight)		AAS	Collected at Saldanha Bay, South Africa. Sediment value, 3 mg Co/kg. No observed difference between sexes or by size, live mass, and shell.	Watling and Watling (1976)
Crassostrea gigas (giant oyster)		1				
Choromytilus meridionalis		2 to 3				
Anadonta sp. (mollusk), soft parts		0.42		AAS	Collected at lake in Aberffraw Anglesey, England.	Segar et al. (1971)
Pond snail	^{60}Co	-	480		Equilibrium at ~90 days with carrier.	Timofeeva-Resovskaya (1963)
Rivulogammarus lacustris (decapoda)			1,100			
Herpobdella (leech)			275			

TABLE VII-5 (concluded)

Organism - Organ	Isotope[a/]	Concentration	Concentration Factor	Analytical Method[b/]	Remarks	Reference
Mollusks	^{60}Co	-			Equilibrium at ∼90 days with carrier.	Timofeeva-Resovskaya (1963)
Anadonta cellensis			1,090			
Limnaea stagnalis			325			
Radix auricularia			925			
Galba palustris			1,160			
Bithynia tentaculata			560			
Alpexa hypnorum			1,380			
Anisus vortex			870			
Planorbis planorbis			800			
Reptiles		-		NAA	Samples from freshwater sites in Panama and Colombia. CF's calculated on wet weight basis.	Templeton et al. (1969)
Turtle, muscle			12.3			
Cayman, muscle			2.74			
Crustaceans						
Shrimp, muscle			341			
Crayfish, muscle			47.6			
Gastropod						
Snail, soft parts			34.2			

a/ Assume stable ^{59}Co if not given.
b/ NAA = Neutron activation analysis.
AAS = Atomic absorption spectrophotometry.

TABLE VII-6

COBALT IN FRESHWATER FISH

Organism - Organ	Isotope[a]	Mean Concentration	Unit	Concentration Factor	Analytical Method[b]	Remarks	Reference
Pimephales promelas (fathead minnow)			mg/kg dry weight		AAS	Fish taken from Lake Eufaula on the Chattahoochee River, Ala., May 1965 to November 1967. Lake concentration ranged from 0.001 to 0.005 mg/liter in unfiltered water. See text for discussion of high values.	Lawrence (1968)
2-in. size group		2.83	mg/kg dry weight	2,830			
3-in. size group		6.25		6,250			
4-in. size group		4.2		4,200			
Dorosoma cepedianum (gizzard shad)		4.37	mg/kg dry weight	4,370			
Notemigonus crysoleucas (golden shiner)		5.0	mg/kg dry weight	5,000			
Ictalurus natalis (yellow bullhead catfish)			mg/kg dry weight				
6-in. size group		3.0		3,000			
6-in. size group		1.5		1,500			
7-in. size group		2.0		2,000			
Lepomis gulosus (warmouth)			mg/kg dry weight				
4-in. size group		10.0		10,000			
5-in. size group		1.5		1,500			
Pomoxis annularis (white crappie)			mg/kg dry weight				
8-in. size group		1.2		1,200			
9-in. size group		12.0		1,200			
Pomoxis nigromaculatus (black crappie)			mg/kg dry weight				
7-in. size group		1.5		1,500			
Cantrarchidae (sunfish)			mg/kg dry weight				
4-in. size group		1.5		1,500			
5-in. size group		1.0		1,000			
5-in. size group		4.5		4,500			

TABLE VII-6 (continued)

Organism - Organ	Isotope[a]	Mean Concentration	Unit	Concentration Factor	Analytical Method[b]	Remarks	Reference
Lepomis macrochirus (bluegills)			mg/kg dry weight		AAS	Same as above.	Lawrence (1968)
2-in. size group		11.0		11,000			
3-in. size group		12.75		12,750			
4-in. size group		6.4		6,400			
5-in. size group		24.0		24,000			
6-in. size group		6.5		6,500			
7-in. size group		13.77		13,770			
8-in. size group		11.5		11,500			
Cyprinus carpio (carp), whole fish, homogenized		0.31	mg/kg dry weight		NAA	Small fish, nonmigratory, in Danube River and canal in vicinity of Vienna, Austria.	Rehwoldt et al. (1975)
Alburnus (whitefish), whole fish, homogenized		0.24	mg/kg dry weight		NAA		Rehwoldt et al. (1975)
Fish, flesh	^{57}Co			100			Hübel and Ruf (1976)
	^{58}Co			100			
	^{60}Co			100			
Esox niger (chain pickerel)		0.13 to 0.32	mg/kg ash weight		SSMS	43 samples from New York lakes as indicated. Cayuga Lake, N.Y., chain pickerel.	Tong et al. (1972)
Micropterus salmoides (largemouth bass)		0.29; 0.95				Cayuga Lake, N.Y.	
Micropterus salmoides (largemouth bass)		0.35				Lake Champlain, N.Y.	
Esox lucius (northern pike)		0.24				Lake Champlain, N.Y.	
Stizostedion vitreum v. (walleye)		0.21				Lake Champlain, N.Y.	
Micropterus dolomieui (smallmouth bass)		1.6				Lake Delta, N.Y.	

TABLE VII-6 (continued)

Organism - Organ	Isotope	Concentration	Unit	Concentration Factor	Analytical Method[a]	Remarks	Reference
Esox lucius (northern pike)		4.7	mg/kg ash weight		SSMS	Lake Delta, N.Y.	Tong et al. (1972)
Stizostedion vitreum v. (walleye)		4.7				Chautauqua Lake, N.Y.	
Salmo gairdneri irideus (rainbow trout)	^{60}Co					Comparative study on uptake and elimination of radionuclides.	Kimura and Honda (1977a)
Advanced fry, whole				11.0			
Fingerling							
Whole				6.5			
Viscera				39.98			
Gills				10.33			
Head				4.79			
Tail				6.03			
Bone				3.77			
Skin				4.18			
Muscle				1.74			
Great Lakes fish			mg/kg dry weight		NAA		Lucas and Edgington (1970)
3 species, 19 whole fish		0.028					
10 species, 40 liver samples		0.040					
Fish, muscle				81.6 (wet weight)	NAA	Representative samples of 121 species taken from freshwater sites of Panama and Colombia.	Templeton et al. (1967)
Barbus barbus		0.02	pCi/g dry weight		-	Studies done on organisms exposed to nuclear power plant wastewater in the Po River, Italy. Only two samples contained ^{60}Co above the detection limit (0.006 pCi/g).	Smedile and Queirazza (1976)
Leuciscus cephalus		0.26					
Salmo gairdneri irideus (rainbow trout), eggs				7.0	Gamma spectrometry	Two component, significant fraction associated with perivitelline fluid. CF calculation based on fresh weight.	Kimura and Honda (1977a)

TABLE VII-6 (concluded)

Organism - Organ	Isotope[a/]	Mean Concentration	Unit	Concentration Factor	Analytical Method[b/]	Remarks	Reference
Black bullheads			mg/kg wet weight		NAA	Living in springwater.	Reed et al. (1968)
Blood		0.305					
Flesh		0.030					
Liver		0.116					
Kidney		1.068					
Bone		0.083					
Black bullheads						Living in White Oak Lake, Tennessee.	
Blood		0.025					
Flesh		0.010					
Liver		0.034					
Kidney		0.510					
Bone		0.026					
Clarias lazera (Nile catfish)	^{60}Co			0.36		Experimental value. Maximum, observed at 1 day. Fresh weight basis.	Ishak et al. (1977)
Sculpin			mg/kg wet weight		NAA	Benthic fish that feed on a burrowing amphipod. Fish concentration values compared over area of taconite dumping; control sediment had higher concentration of Co. No significant difference seen between control and taconite tailings site.	Korda et al. (1977)
Cottus cognatus (slimy)							
Flesh		0.015					
Liver		0.144					
Myoxocephalus quadricornis (four-horned)							
Flesh		0.010					
Liver		0.080					
Largemouth bass			mg/kg dry weight		AAS	Lakes Seminole and Eufaula, Ala. Co concentration in water, 0.001 to 0.005 mg/liter.	Lawrence (1968)
3-in. size group		10.8		10,800			
4-in. size group		7.5		7,500			
7-in. size group		1.5		1,500			
8-in. size group		20.0		20,000			
9-in. size group		6.5		6,500			
10-in. size group		7.5		7,500			
Channel cat		2.17		2,170			

a/ Stable ^{59}Co unless stated otherwise.
b/ AAS = Atomic absorption spectrometry
SSMS = Spark source mass spectrometry
NAA = Neutron activation analysis

a. Marine: In the marine environment, zooplankton CF's for whole animals range from 1,000 to 10,000, with the CF for the soft parts in the hundreds, indicating the sizable contribution to overall accumulation made by adsorption on the relatively large external surface of these invertebrates. The CF's of marine fish range from $\sim$ 100 to 4,000; cobalt is found to accumulate most in the viscera and on the skin. Accumulation in marine fish muscle has been reported at CF's of 5 to 500.

An organism's coming into contact with sediment, either suspended or bed material, can increase the amounts of cobalt adsorbed on its surface. For this reason, the ecological niche of benthic aquatic invertebrates makes them prime targets for cobalt enrichment. For marine shellfish, CF's are generally in the thousands, with CF's for the soft tissues ranging as high as 20,000. The highest level of accumulation is found adsorbed on the peel of lamellibranchs, which is indicative of their feeding mechanisms. Other marine benthic invertebrates such as echinoderms, tunicates, and Crustacea have CF's ranging from 100 to 40,000; the CF for crustacean muscle is reported at approximately 2,000.

Marine algae have reported CF's in the thousands, reflecting the influence on accumulation of the unicellular organism's surface-to-volume ratio. Marine seaweed CF's range from 3,000 to 20,000.

b. Fresh water: Rooted freshwater plants have reportedly accumulated cobalt with CF's of less than 10, which reflects absorption by the roots from sediment enriched with cobalt. Freshwater algae, with CF's ranging from 400 to almost 2 million reflect the effect of their large surface-to-volume ratio on accumulation. Also, submerged and floating leaved plants which obtain nutrients from the water, may attain CF's as high as the algae. The reported values for insects and insect larvae show very high CF's ranging up to 100,000. A possible explanation for this might be adsorption on the chitinous exoskeleton of adult insects and exposure of insect larvae living in the sediment.

Mollusks have the widest CF range of any freshwater invertebrates with CF's of 100 to 14,000 ($\sim$ 1 to 300 for the soft parts), which reflects the potential for accumulation of cobalt when it is present in the sediment. Freshwater oligochaetes do not reflect this sediment contact so clearly, showing CF's of $\sim$ 30. The other invertebrates (hirudinea and Crustacea) have reported CF values from $\sim$ 1 to 11,000; the crustacean muscle CF's are $\sim$ 500 to 3,500. The CF for the muscle of aquatic reptiles has been reported to be $\sim$ 30 to 120.

Most reported freshwater fish and fish muscle CF's range from less than 10 to 1,000, reflecting not only uptake via ingestion but through the gills as well. Lawrence (1968) has reported CF values for freshwater fish of 1,000 to 24,000 in water of Alabama lakes ranging from 0.001 to 0.005 mg Co/liter. A possible explanation for these anomalously high CF values is suggested by the study of Bertine et al. (1978). They reported the increase of cobalt concentration from 60 to 90 mg/kg in the sediment of a remote semiarid desert lake in California after the addition of copper sulfate as an algicide. Cobalt was possibly present as a contaminant in the copper sulfate. Lawrence (1968) also reported that the ponds from which the fish were sampled had been treated with copper sulfate algicide. No mention is made of when or how often the treatment had been done prior to his sampling. Cowgill (1974) suggested that more work needs to be done before the vastly higher concentrations of cobalt in plants growing in Alabama waters as reported by Lawrence (1968) and other discrepancies in the literature can be explained.

2. Trophic relationships: Organisms reflecting habitat and trophic relationships within aquatic ecosystems show plant life obtains the highest accumulation by either a high surface-to-volume ratio as with algae and perhaps some submerged plants, or as rooted plants in the sediment. Benthos organisms have higher CF's than surface organisms, and sessile organisms have higher CF's than free-swimming organisms. Significantly, the higher an organism is in a food chain, the lower the CF. To our knowledge, the one exception is found with the insect carnivores discussed below. This exception may be due to the small size of the insects or it may be related to high adsorption on the chitinous exoskeletons.

Amiard-Triquet and Foulquier (1978) followed ^{60}Co in a laboratory experiment through two food chains and found cobalt concentrated in the macrophytes but decreased in concentration in higher trophic levels (see Table VII-7).

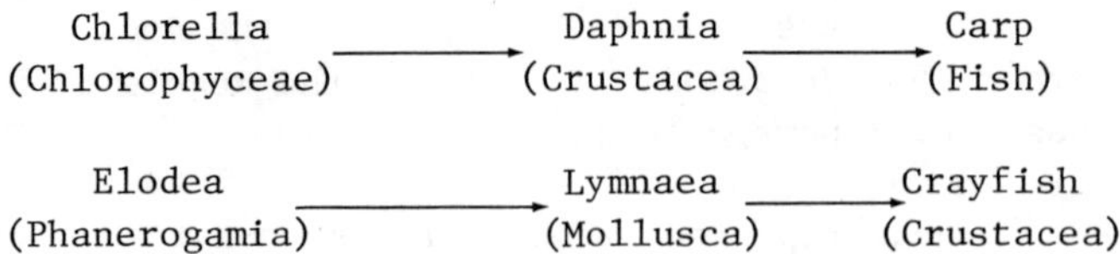

One aspect of being assigned an ecological niche is that an organism's trophic level relationships can appear oversimplified on paper. Osterberg (1968) observed that food webs are complex since animals are opportunistic feeders. Shuman et al. (1977) in their study of the biota of the Haw and New Hope Rivers in North Carolina adopted parts of Leland and McNurney's (1975) grouping of artificial feeding types to develop a better trophic model. These feeding types are: (a) indiscriminate filter feeders which are dependent on the sediment and sediment-associated benthos animals as two groups; (b) filter feeders; (c) discriminate feeders (omnivores); (d) carnivores; and (e) surface feeders which are also predaceous. Type (a) are small insect larvae; type (b), clams; type (c), mayflies, snails, and crayfish; type (d), insects; and type (e), water striders.

Concentration factors based on filterable metal concentrations indicate the highest cobalt accumulators are the filter feeders (CF 55.5) Pelecypoda (clams) and Hydropsychidae (a family of caddisflies) and the carnivores (CF 80.5), the insect classes Megaloptera, Odonata, and Plecoptera. The Andropsychidae are known to ingest fine suspended particles which have the highest metal concentrations. Interestingly, CF's based on sediment metal concentrations indicated an accumulation by the sediment-dependent feeding type that was two times greater than the value of the filter feeders and carnivores. (See Table VII-7 for CF values.) The mayfly (Stenonema) and the crayfish (family Astacidae) had a strongly negative correlation between body length and cobalt concentration. A positive correlation would have suggested a relatively greater correspondence between gut content and internal absorption.

TABLE VII-7

COBALT IN AQUATIC ECOSYSTEMS

Organism Type	Isotope[b]	Mean Concentration (mg/kg)	Concentration Factor	Analytical Method[b]	Remarks	Reference
Fish, marine	^{59}Co and ^{60}Co	-	1 to 10		Experimental value at equilibrium obtained in 1 week.	Hiyama and Khan (1964)
Shellfish, marine, edible part			10 to 100			
Algae, marine			100 to 500			
Biotic components, freshwater		(wet weight)		NAA	Study of the drainage system of an ash basin before and after (in parenthesis) dredging out filled ash basin. Cobalt level in water, 0.06 (0.04) mg/liter.	Cherry and Guthrie (1977)
Plant		1.7	28 (15)			
Invertebrate		1.7	28 (20)			
Vertebrate		0.5	8 (10)			
All studied plant forms, freshwater	^{60}Co	-	4,425	-		Timofeeva-Resovskaya et al. (1961)
Animal forms, freshwater						
All studied animal forms			650			
Large snails			325			
Mollusks			890			
Decapods			1,100			
Insect larvae			2,785			
Algae, marine		-	800	-	Partial review of maximum or average values in the literature.	Jenkins (1975)
Zooplankton, marine			3,200			
Mollusks, marine			6,800			
Fish, marine			2,200			
		(wet weight)				
Enallagma sp., an aquatic insect		2.05	29	NAA	Biota in drainage system receiving coal ash effluent. Co concentration in *Chironimidae* highest at site with lowest Co concentration in water (50 μg/liter).	Cherry et al. (1979)
Libellula, dragonflies		0.79 to 1.47	10 to 25			
Chironomidae, mosquito-like insects		3.50 to 7.40	50 to 148			
Crayfish		0.66 to 1.31	8.3 to 2			
Organism feeding type, freshwater		-		NAA		Shuman et al. (1977)
Discriminate, omnivores			44.4			
Sediment-dependent			35.6			
Filter feeder			55.5			
Carnivore			80.5			
Surface			4.3			
		(ash weight)				
Zooplankton, marine		0 to 0.01	-	Emission spectrometry	-	Petkevich (1965)
Engraulis encrasicholus ponticus, anchovy, muscle		< 0.001 to 0.03	< 0.1 to 3		Plankton-eating fish of the Black Sea. Concentration factor based on accumulation via zooplankton.	

TABLE VII-7 (concluded)

Organism Type	Isotope[a]	Mean Concentration (mg/kg)	Concentration Factor	Analytical Method[b]	Remarks	Reference
Sprattus sprattus phalericus		(ash weight)				
Sprattus sprattus phalericus, sprat, muscle	-	< 0.001 to 0.04	< 0.1 to 4	Emission spectrometry	Plankton-eating fish of the Black Sea. Concentration factor based on accumulation via zooplankton.	Petkevich (1965)
Two food chains, freshwater	^{60}Co	-		Gamma spectro-photometry	Two freshwater food chains studied in laboratory experiments. Daphnia sp. concentration not given. Astracus leptodactylus concentration factor variation probably due to molting.	Amiard-Triquet and Foulquier (1978)
Chlorella sp., algae			565			
Daphnia sp., water flea			-			
Cyprinus carpio, carp			3.5			
Elodea, phanerogamia						
Lymnaea sp., pond snail			4,037 to 4,410			
Astracus leptodactylus, crayfish			401 to 1,038			
Fish, marine, muscle	-	-	90	NAA	CF's calculated on dry weight basis.	Ishii et al. (1978)
Shellfish, marine, soft part			4,000			
Sargassum, marine algae, whole body			10,000			

a/ Stable cobalt unless otherwise indicated.
b/ NAA = Neutron activation analysis.

Furthermore, Shuman et al. (1977) indicate that water quality affected by domestic and industrial wastewater may inhibit cobalt accumulation by organisms because of formation of organic complexes and decreased availability. This was seen as a lower than average cobalt concentration at the sample station with the heaviest load of organic matter from municipal sewage and textile waste discharges.

3. Reported examples of uptake

a. From water: Mudroch and Capobianco (1978) studied metal uptake in marshes on Lake St. Clair, Ontario. Calculation of concentration factors is extremely difficult because the degree to which aquatic plants take up a metal via the sediment and via the water is unresolved. Generally, it is thought that submerged and floating leaved plants are almost completely dependent on the water medium for nutrient supply with the roots serving primarily as anchors. For emergent aquatic plants, the sediment is thought to be the major source of nutrients; and cations are thought to be preferentially absorbed via the roots, especially insoluble cations. (See Mudroch and Capobianco, 1978, for further discussion of this problem.) Therefore, if CF's are based on water, then values of three to four orders of magnitude would indicate accumulation similar to that of algae. On the other hand, CF's based on sediment would indicate significantly more moderate accumulations.

Cowgill (1973) reported that two submerged freshwater plants, the pondweed and hornwort, the former with a creeping rootstock and the latter without roots, had approximately four-fold the concentration as another rooted plant with aerial portions, the yellow water lily (see Table VII-4). This suggests that plant uptake via water is not dependent on roots. However, the cobalt concentration for pickerelweed leaves was the highest value found, 1.7 mg/kg. Pickerelweeds are also rooted aquatic plants with aerial portions.

Kunze et al. (1978) ran a series of tests to determine radionuclide uptake in rainbow trout (*Salmo gairdneri*) eggs from spawning to hatching. A radiotracer with 3% Co was added to normal tapwater. Cobalt concentrations in the eggs were found to be directly proportional to the exposure level in the aquatic medium and probably reversibly bound to the chorion-mucopolysaccharide. Later in the embryonic development, the divalent ^{57}Co was found to desorb from the chorionic membrane. No increased accumulation was noticeable within the hatched larvae. Cobalt concentrations of 1 to 3 ng Co/egg were highest prior to hatching.

Amiard and Amiard-Triquet (1977) showed both assimilation and elimination are affected quantitatively and qualitatively according to whether the pollutant is tranferred through water or food. The authors found that contamination via immersion in water was higher than by ingestion for both the mollusk *Scrobicularia plana* and the crab *Carcinus maenus*. Yet Van Weers (1975) found that uptake via water was not a major source of cobalt for the common shrimp *Crago* sp.; and Harrison (1972) found no significant difference from uptake of ^{60}Co from either food or water when the peel was removed from the marine clam *Mya arenaria*.

The 1972 report on the occurrence of heavy metals in Lake Michigan biota and sediment by the Michigan Water Resources Commission (Hesse and Evans, 1972) lists values for cobalt in bottom feeders (carp, sucker, and redhorse) with a range of 0.1 to 0.2 mg Co/kg wet weight and a mean of 0.12 mg Co/kg wet weight. Predators (rock bass, northern pike, brook trout, crappie, smallmouth bass, and bowfin) also showed a range of 0.1 to 0.2 with a mean of 0.14 mg Co/kg wet weight. In fish living in contaminated areas of the lake where sediment cobalt levels were in excess of the background (approximately 3.0 mg Co/kg), the bottom feeders showed a range of 0.0 to 0.2 mg Co/kg with a mean of 0.13 mg Co/kg wet weight; and the predators, a range of 0.0 to 0.3 mg Co/kg with a mean of 0.15 mg Co/kg wet weight.

b. From sediments: An organism's ecological niche and biological structure can increase its exposure to cobalt; for example, deposit-feeding mollusks filter sediment to select the appropriate size food particle. Especially in the aquatic habitat where an animal feeds, the mechanism by which it eats can make an order of magnitude difference in accumulation of cobalt.

Shuman et al. (1977) found, as mentioned above, that an organism's cobalt concentration (total body burden) closely reflected the sediment's concentration.

Wolfe and Rice (1974) suggested that surface silt layers of high organic content may be a major source of elements for deposit feeders. Surface particles in the sediment layer may arise from flocculation; precipitation of dissolved ionic species; and settling organic matter such as dead organisms, exuviate, feces, and casts. They found that cobalt trapped in deep sedimentary deposits can be re-exposed by deep-rooted plants, e.g., *Spartina* sp., and burrowing animals.

Wolverton and McDonald (1975) showed that water hyacinths (*Eichornia crassipes*) and alligator weeds (*Alternanthera philoxerides*) could remove 0.439 and 0.130 mg, respectively, of divalent cobalt ions per gram of dry plant material daily from water or sediment via roots. Therefore, 1 ha of water hyacinths are capable of removing 340.8 g Co/day.

Bryan and Uysal (1978) found the highest cobalt concentrations in the digestive gland (75%) of the burrowing bivalve *Scrobicularia plana*. This suggests cobalt uptake is mostly by ingestion of sediment. The author recommends this bivalve as a good indicator of cobalt-contaminated sediment.

4. *From contaminated aquatic ecosystems*: Smedile and Queirazza (1976) reported that the cobalt concentration of the aufwuchs (periphyton) community in the Po River (Italy) downstream of a nuclear power station was directly proportional to the concentration in the water. However, the ^{60}Co levels in the herbivorous fish were hardly noticeable. The authors concluded that the aufwuchs accumulated cobalt both by adsorption and absorption and that the radionuclide was not present in a form which could be absorbed through the membrane of the fish's gut. However, Reed (1971) found the black bullhead fish accumulated ^{60}Co rapidly from White Oak Lake water in Tennessee; the gills accumulated 19 to 30% of the whole body activity during an uptake of 3 days.

Templeton (1970) has documented how anthropogenic activities, in this case a nuclear power station, can lead to the inclusion of radiocobalt into a food chain, e.g., oysters to man. Yet, in this case ^{60}Co exposure would under the most unusual situation account for only a small fraction of a consumer's daily exposure rate. Furthermore, "laverbread," which is a daily food made from seaweed in South Wales, was found to contribute over the years 1962 to 1968 only 0.6 rem/year for adults and 0.4 rem/year for children. This is lower than the recommended dose limit of 1.5 rems/year for radiocobalt.

Miles et al. (1974) stated that the principal source of radioactivity from a nuclear power plant of a seagoing ship is the corrosion and wear of the metal surfaces in contact with the reactor cooling water, of which ^{60}Co has the longest half-life.

Lowman (1958) reported after the Redwing Atomic Test in 1956 that ^{60}Co contributed 0.8% of the total radioactivity in the fallout. Thomas et al. (1958) measured marine biological specimens following the Redwing shot and found that during the early months, ^{60}Co concentrations increased slowly by a factor of four, peaking in December 1956, 6 months after the test, at 0.08 pCi/g and then quickly declining.

In the summer of 1964, Lowman et al. (1967) were able to determine 0.08 to 0.40 pCi ^{60}Co/g in sponges (Spechiospongia sp.) at three locations off the coast of Puerto Rico even though cobalt was not detectable in the surrounding water. The researchers assumed that the ^{60}Co was fallout which had been washed from the watersheds on Puerto Rico.

Pesch et al. (1977) collected sea scallops (Placopecten magellanicus) at the Philadelphia and Du Pont sewage disposal sites off the Delaware coast and found a mean concentration for all samples of 0.56 mg/kg dry weight.

Cherry and Guthrie (1977) (see also Guthrie and Cherry, 1976) found cobalt from a coal-burning power plant's ash basin was accumulated by benthic-dwelling midges and tadpoles as well as other flora and fauna. The overall mean cobalt concentration in the drainage system was 2.92 mg/liter with $\sim$ 0.1 mg Co/liter in the water ($\sim$ 1% of the cobalt), 10.6 mg Co/kg in the sediment ($\sim$ 75%), 1.7 mg/kg in the aquatic plants ($\sim$ 10%), 1.7 mg/kg in the invertebrates ($\sim$ 10%), and 0.5 mg/kg in the fish ($\sim$ 5%). Guthrie and Cherry (1979) reported CF's in the twenties for submerged and floating-leaved aquatic plants and < 0.1 for plants rooted in the sediment (see Table VII-4).

Roots of Spartina growing in Foundry Cove, New York, sediments that were contaminated with nickel, cadmium, and cobalt from a nickel-cadmium battery factory showed increased uptake with increasing cobalt content of the bottom sediments: $\sim$ 3 to 50 mg/kg dry weight in the roots at $\sim$ 20 to 400 mg/kg in the sediments (Kneip et al., 1974).

Honda et al. (1973) studied in the laboratory a small scale rice field ecosystem contaminated with ^{60}Co and found approximately 12% of the ^{60}Co was retained within the biomass. Distribution of the ^{60}Co within the rice seedling after 14 days was 40% in the submerged portion and 17% in the emergent portion. Within the mudsnail 88% was in the shell, and within the loach 96% was in tissue other than the visceral mass. A mass balance of ^{60}Co was done at the end of the experiment on Day 21. It appeared that uptake was partially adsorbed on the outside of the roots and stems of the rice plants and on the outer surface of the invertebrates. The mudsnail and earthworm reached maximum concentration in 2 hr and then decreased to their equilibria values. The cobalt concentrations in rice seedlings, loach, and sandy soil reached maximum in 1 day and then decreased to equilibrium. The CF order (based on the water concentration) was rice seedling < mudsnail < sandy soil < earthworm < loach.

5. <u>Discussion of observed variations in uptake and elimination</u>: In studies of cobalt uptake and elimination in an aquatic environment, there are more variables to account for than in the terrestrial environment. The reason for so many variables is that aquatic animals are heterotherms and their metabolisms can function over a wide range of external conditions, whereas, homeotherms must maintain their metabolic rate within a narrow range. The effect of this variability on data collecting is extensive. One direct effect is the unreliability of biological half-life values since rates of metabolic function will vary manyfold depending upon: (a) population size, density of food organisms, and effects of extracellular products; (b) morphology --feeding mechanism and mode of movement; (c) life cycle--ecological niche and average life span; (d) duration of exposure--availability, concentration, and amount of coexisting stable element; (e) migration--seasonal and diurnal; (f) physicochemical state--water temperature, amount of light, and amount of organic material present; and (g) laboratory conditions--closed versus open system recycling, starved versus fed studies and uptake studies, and elimination versus uptake studies.

Some of these variables have been addressed by experiments. In any metabolic experiment, it is essential to run field experiments together with laboratory experiments because there are unknown factors that arise in the field. Also, it is a fact that active animals show increased metabolic rates over their sessile counterparts in an aquarium. For this reason, elimination studies are more dependable than uptake studies since uptake in the field is sporadic and subject to unknown factors. For the above reasons, the data must meet a number of standards before the researcher can look for patterns in the findings.

A major question unanswered at this time is whether cobalt is under homeostatic control or not in higher plants, invertebrates, and fish.

a. Elimination experiments: Harrison (1972) found that ^{60}Co loss was monophasic and that there were two metabolic pools of ^{60}Co present in the body of the marine clam *M. arenaria*, with the major activity residing in the long biological half-life component pools. Moreover, the elimination study showed lower biological loss-rate constants than the accumulation study, which indicates that cobalt follows different metabolic pathways during accumulation and elimination as well as indicates that cobalt may build up faster than it can be removed. She concluded that cobalt regulation may take place at high concentrations. The turnover of ^{60}Co was found to be fastest in organs with a high CF, e.g., viscera, and slowest in organs with a low CF, e.g., body fluid. Biological half-life varied from 100 to 300 days for the various body organs.

Patel et al. (1978) found that the elimination of ^{60}Co in the marine blood clam (*Anadana granose*) was monophasic with a biological half-life ($T_{1/2}$) of 19.4 days.

Harrison (1972) suspended the marine clam *M. arenaria* in cages near a discharge canal of a boiling-water nuclear reactor for 7.5 days. The clams accumulated 1,000 pCi ^{60}Co/kg in soft body tissue, 10,000 to 20,000 pCi ^{60}Co/kg on the peel, and 20,000 to 40,000 pCi ^{60}Co/kg in the settled silt on the floor of the basket.* She calculated a CF of 2,000 for the clam peel, accounting for 65 to 80% of the whole body activity measure. Within the shellfish soft tissue, the digestive gland and stomach had the highest CF (see Table VII-2).

Shimizu et al. (1971) found that ^{58}Co binds to the conchiolin in the periostracum and byssus of the mussel *Mytilus edulis*.

Segar et al. (1971) studied eight species of lamellibranch mollusks in the Irish Sea. They found the highest concentrations of cobalt in the soft tissue rather than the shell. Within the soft tissue, the highest concentrations were in the gut, digestive gland, mantle, and gills. The authors allowed the mollusks to purge themselves before analysis. They did not find a high concentration on the peel or the reason why the values for washed soft gonads, mantle, and gills were higher than for unwashed tissue.

Nakahara and Cross (1978) found the distribution of ^{60}Co in the clam, *Mercenaria mercenaria*, after feeding on *Nitzschia* cells, was initially in the mantle (46.7%), intestine (21.2%), and foot (19.2%) and had moved by 4 hr to the foot (18.5%), intestine (12.4%), body fluid (11.6%), and liver (5.5%), with a large shift at 9 hr to the mantle (2.0%), intestine (0.9%), foot (2.4%), the liver (35.8%), and the body fluid (8.7%). From 9 hr to 9 days after uptake, the significant distribution was a steady decrease in the liver (35.8 to 8.7%), while the body fluid burden rose from 8.7 to 15.3% and the overall whole body (excluding shell) decreased from 51.7 to 30.3% of the initial uptake. The authors also determined that most of the excretions during the first few hours of a clam-feeding experiment with the flagellate phytoplankton *Dunaliella* and *Platymonas* were pseudofeces and that these excretions accounted for more than 90% of the radioactivity lost during that time. Furthermore, their data indicated that retention of cobalt by the clams varied with the species of phytoplankton ingested.

---

* Harrison suggested that silt in the baskets with the clams may have contained an aerobic bacteria synthesizing cobalamin, thus making ^{60}Co available in both organic and inorganic forms.

Deposit-feeding clams _Macoma balthica_ were exposed in separate experiments to four elements bound to six physicochemically different types of sediment. Weak acids, reducing agents, and oxidizing agents removed significant quantities of cobalt nonselectively from the six physicochemically different sediment types. However, extractants appeared to remove a different fraction of cobalt than the digestive process of the clam. Luoma and Jenne (1977) suggested a mild exchange condition and complexation as possible physiological mechanisms for uptake for the digestion and binding of metals to transport sites on biological membranes. This is based on the almost exclusive uptake of cobalt from food and the extractability of cobalt from the sediments by ammonium acetate.

Van Weers (1975) found that 85% of the ^{60}Co activity in the common shrimp _Crangon crangon_ was associated with the exoskeleton when accumulated from seawater. After a single feeding with labeled mussel flesh, a short-lived component with $T_{1/2}$ 1.2 days accounted for 80% of the initial activity; and a long-lived component with $T_{1/2}$ of 10 days for the remainder. After repeated feedings, the short-lived component was totally replaced with the long-lived component. The author found most of the ^{60}Co in the digestive gland and a minimum in the edible abdominal muscles with a whole animal CF of 13. He concluded that ^{60}Co accumulation in the internal organs of the shrimp was of more importance via food than via water.

Reed et al. (1968) reported that the black bullhead fish, _Ictalurus melas_, following 3 days uptake via water lost 28% of the initial activity by excretion and the remainder via three biological components with $T_{1/2}$'s of 4 days, 35 days, and an undetermined number of days. The authors found with single feeding that 95.5% of the initial activity was lost the first day followed by a three-component loss with $T_{1/2}$'s of 1.5 days, 48 days, and an undetermined number of days.

Nelson and Malone (1968) found in their laboratory study of the freshwater snail _Goniobasis clavaeformis_, which fed on ^{60}Co-labeled aufwuchs, that the snails attained equilibrium only when exposed to ^{60}Co for their entire lives. Feeding rates varied by seasonal plant growth and time of day (heaviest in morning), with peak activity in the feces seen 2 hr after intake.

Timofeeva-Resovskaya (1963) followed the concentration factors for cobalt in pond snails for approximately 90 days without their reaching equilibrium, whereas the alga *Elodea canadensis* reached a constant value in 15 days.

Reed et al. (1968) in their study of the biological pathways of cobalt compared separate tissues from lakewater fish and from the experimental fish which had had prior 3-day water uptake of ^{60}Co. The authors found skin and bone concentrations were similar; the blood, liver, and kidney concentrations were greater in the lake fish; and gills, gut, and stomach concentrations were greater in the experimental fish. The kidney was found to retain a higher concentration of stable and radiocobalt than the other tissues. Cobalt uptake by the organs from water was near maximum concentration the first day, reaching equilibria after 3 days. Of the organs, the gills, gut, and stomach took up ^{60}Co the most rapidly and achieved equilibria in 2 days.

Furthermore, Reed et al. (1968) found that ^{60}Co taken up via water dissipated quickly from the gills, gut, and stomach, and from the other tissues within a few days. The kidney and blood ^{60}Co activity remained the longest. The authors suggested that the high cobalt value in the kidney may be due to a cobalt function in kidney tubular metabolism. More probably, the kidney activity reflects the elimination function of the kidney.

Baptist (1970) reported that ^{60}Co, ^{58}Co, and ^{57}Co in the Atlantic croaker *Micropogon undulatus* had a single elimination rate function and a $T_{1/2}$ of 31 days.

Kimura and Honda (1977) found the uptake and accumulation of ^{60}Co in rainbow trout eggs to rapidly attain the maximum CF of 7.0 in 24 hr whereupon they were transferred to an uncontaminated aquarium. The elimination of ^{60}Co consisted of a short component $T_{1/2}$ = 0.3 day) and a long component ($T_{1/2}$ = 35 days) with approximately 33% of the ^{60}Co activity associated with the perivitelline fluid and yolk, 55% with the egg membrane, and 7% in the embryo.

Lucus et al. (1970) reported that cobalt concentrations in Lake Superior bloaters and lake whitefish were lower than those found in the same species in Lake Michigan and concluded cobalt uptake may not be under homeostatic control. Cobalt concentrations in other fish species in the two lakes were not signicantly different, however.

Feldt and Melzer (1978) reported evidence for the homeostatic control of cobalt by freshwater fish. In a study of German rivers, the cobalt concentration in fish flesh was found to be inversely proportional to the cobalt concentration in unfiltered water. Table VII-8 presents the data from a graph in their study (analyses were by flame atomic absorption spectrometry).

TABLE VII-8

COBALT CONCENTRATION FACTORS IN FLESH OF FISH TAKEN FROM GERMAN RIVERS

River Name (Location)	Approximate Cobalt Concentration in Water (μg/liter)	Approximate Concentration Factor
Danube (Grundremmingen)	0.36	7.2
Neckar (Obrigheim)	0.45	6.6
Rhine (Biblis)	0.45	6.6
Neckar (Neckarwestheim)	0.86	4.1
Upper Elbe	1.2	2.8
Lower Elbe	1.7	2.3
Weser	3.7	1.3

In studying the uptake from water of ^{60}Co regulated at 0.5 μ Ci/liter, Nakahara et al. (1979) observed an unusual accumulation by the branchial heart tissue of the Octopus vulgaris. They found the branchial heart tissue accumulated ^{60}Co as the other tissues were eliminating it. At the end of the 30-day uptake and elimination study the percent distribution of ^{60}Co activity was found to be: branchial heart, 58.8; arms and tentacles, 9.7; liver, 8.8; skin, 7.9; gills, 5.8; mantle, 3.1; kidney, 1.9; blood, 1.0; sucker, 1.0; and all others, 2.0.

b. Chronic versus acute uptake: Hiyama and Shimizu (1969) found evidence that the CF values for the marine goby fish obtained under unsteady ^{58}Co concentrations may increase exponentially with time and never reach an equilibrium, which is an important consideration for determining CF's from field data. Therefore, it would seem any "pools" of cobalt along the Japanese or North American coast lines would be evident from periodic sampling of these or other migratory fish.

Amiard and Amiard-Triquet (1977; citing Amiard-Triquet et al., 1976) reported that, with acute ^{60}Co contamination of marine animals, uptake and elimination were high for the clam *Scrobicularia plana* and the crab *Carcinus maenas*. Furthermore, when chronic contamination reached equilibrium, assimilation dropped to zero and elimination was very slow. In these laboratory experiments, the $T_{1/2}$ for the crabs was not observable in 5 months whereas for the clam, it was 66 days. No mention is made of whether the authors were accounting for the adsorption of ^{60}Co on the outside of the shells of the crustaceans and mollusks in this observation. Harrison (1972) found an exponential and monophasic decrease in ^{60}Co concentration in the whole body and in the body with the peel removed. When cobalt carrier was added to aquarium water, the CF was lower over the same time interval than when carrier was not present. The $T_{1/2}$ values ranged over 100 to 300 days during the study.

c. Physicochemical influences: Harrison et al. (1976) reported that the availability of radionuclides to filter-feeding aquatic animals is enhanced by the presence of living microorganisms, organic detritus, and inorganic material. The quantities of these can vary spatially and temporally within the water column, further affecting availability. In oysters, ^{60}Co appears to be gathered primarily from suspended particulate. Concentrations were found to reflect irregular fluctuations in particulate fraction, which probably indicated a relationship with the tidal cycle.

Hiyama and Khan (1964) reported that, with carrier present, cobalt equilibria for fish, algae, and shellfish were obtained in about 1 week. The addition of 0.2 mg Co^{2+}/liter, $>$ 100 times the usual concentration of cobalt in seawater, gave a 50% saturation time of 8.7 days; a stable cobalt concentration of 2 mg/liter dropped the 50% saturation time to 5.8 days; and with no carrier present, there was no tendency to attain equilibrium. Raising the temperature in the seawater aquarium caused a decrease in the 50% saturation time with only a slight rise in CF for the fish. The authors also observed that uptake was lower in artificial seawater.

Harvey (1969) found that the ^{57}Co CF for blue-green algae decreased with a rise in temperature from 25 to 40°C (see Table VII-1).

Grachev (1977) reported that the rate of excretion of ^{60}Co in freshwater fish at various temperatures is inversely proportional to the rate of accumulation whereas the distribution within the organism of ^{60}Co did not depend on temperature.

Patel et al. (1978) reported that uptake by the blood-clam Anadara granosa was independent of the ambient concentrations and temperature, indicating that uptake was at a constant physiological rate and increased with time of exposure. Over a 50-day exposure to wastewater from a boiling-water nuclear reactor, the blood-clam was found to accumulate 60 pCi/g of ^{60}Co.

Vanderploeg et al. (1975) have correlated concentration factors of aquatic biota with the oligotrophic-eutrophic classification of lakes. They cite evidence that CF's for eutrophic lakes would be lower due to the reduced availability of cobalt (see Table VII-9).

d. Age influences: Wilhm (1970) found in his study of the transfer of ^{60}Co from detritus to the benthic invertebrates that only the larvae accumulated ^{60}Co (see Table VII-5). The highest CF was found in the sludge worm larvae, which ingest sediment and have the largest surface-to-volume ratio. The authors advanced no reason for this apparent anomaly. Wilhm speculated that the larvae may undergo a change in skin chemistry, or perhaps feeding habits.

Kunze et al. (1978) found ^{57}Co bound reversibly with the chorion-mucopolysaccharide of rainbow trout eggs Salmo gairdneri. They observed an antagonistic effect during the later embryonic development when ^{57}Co uptake was decreased in the presence of increasing calcium ion. During the embryonic development of rainbow trout, ^{57}Co uptake reached 80% of its eventual maximum concentration during the first 24 hr after spawning when the greatest increase (15%) in weight and volume of the egg took

TABLE VII-9

CONCENTRATION FACTORS FOR COBALT IN AQUATIC BIOTA CORRELATED WITH THE LAKE CLASSIFICATION
(Vanderploeg et al., 1975)[a]

Taxon/Functional Group	Tissue	Environment	Mean Concentration Factors (based on fresh weight)
Fishes	Flesh	Mesotrophic and oligotrophic waters	320
	Whole		440
	Flesh	Eutrophic waters	27
	Whole		44
Algae	Whole	All waters	10,000
Emergent vascular plants	Whole	All waters	1,000
Submerged-leaf macrophytes	Whole	Mesotrophic and oligotrophic waters	10,000
		Eutrophic waters	200
Floating-leaf macrophytes	Whole	Mesotrophic and oligotrophic waters	10,000
		Eutrophic waters	400
Mollusks: Bivalves	Soft	Mesotrophic and olitrophic waters	10,000
		Eutrophic waters	400
Snails	Soft	All waters	10,000
Insect larvae	Whole	All waters	10,000
Tubificid worms	Whole	Eutrophic waters	500
Crustaceans	Whole	All waters	1,000

a/ Concentration factors based on isotope concentration in filtered water.

place. The cobalt concentration in the egg increased with higher cobalt concentrations in the water.

Kulikov and Ozhegov (1975) found the major mechanism of uptake by fish eggs of *Esox lucius* (pike), *Perca fluviatilis* (perch), and *Coregonus lavaretus* (whitefish) to be adsorption on the outer membrane of the egg rather than physiological absorption.

Kimura et al. (1977) reported a rapid uptake in the advanced fry and fingerling of rainbow trout with CF's of 11 and 6.5/day, respectively. The ^{60}Co elimination in the fry and fingerling with $T_{1/2}$'s of 0.8 and 0.4 day, respectively, reflects the high uptake CF 40 in the viscera (see Table VII-6).

e. Influences of body size: Hiyama and Khan (1964) found that various size fish have a significant positive correlation between body size and CF. The highest CF was found in the heart and spleen.

Shuman et al. (1977) found cobalt concentration to be negatively correlated with head capsule width and weight for surface feeders and filter feeders. Similarly, Ivanov (1974) found that the CF decreased with the increasing weight of hydrobionts.

Boyden (1977) found highest concentrations of cobalt in the smallest individuals of the gastropod *Buccinum undatum* and independent of body size in the opisthobranch gastropod *Scraphander lignarius*.

Nakahara and Cross (1978) found that the retention of ^{60}Co in clams (excluding shell) after feeding on labeled phytoplankton varied inversely with size.

Plankton has been found to be a major accumulator of cobalt due to its small size, large surface-to-volume ratio, population density, and ecological niche. Furthermore, plankton appears to be able to contribute to both the concentration of cobalt found in water and turnover of cobalt from the surface to the depths.

Lowman's (1958) observations that phytoplankton have a greater surface area with respect to protoplasmic volume of living material than any other organism holds significance for adsorption of cobalt.

Watling and Watling (1976) in a comparative study of the clam *Choromytilus meridionalis* found no difference between the sexes by live mass, shell dimension, and wet tissue. Geldiay and Uysal (1975) found that the cobalt concentration of the mollusk *Tapes decassutus* (of commercial value) was lower in the summer than fall or winter, inversely proportional to size, and located on the shell and in the body fluids.

Wiser and Nelson (1964) in a comparative study between living and dead crayfish *Cambarus longulus longerostris* measured the ratio of adsorption to absorption at approximately 50:50 with 95% of the body burden located on the exoskeleton and lost at molting. Furthermore, both living and dead large animals were found to accumulate more cobalt than small ones, although the smaller animals had a greater uptake per gram. The integument, hepatopancreas, and gut had the highest levels of cobalt in live animals. The degree of ^{60}Co uptake was proportional to the concentration in the solution.

f. *Population density effects*: Furnica et al. (1974) found that ^{57}Co activity in *Elodea canadensis* increased as the biomass also increased, although the CF declined from 1,479 to 682 over the 30-day study. This was a laboratory study with ^{57}Co activity in the water maintained at 500 pCi ^{57}Co/ml.

Kustin and McLeod (1977) in their review of organism-environment interactions concluded that during algal bloom conditions high growth rates appear to dilute metal levels in the algal mass and that extracellular products would be at their highest concentration during an algal bloom and probably reduce the metal's chemical activity.

Kirchmann et al. (1977) found ^{60}Co to be concentrated 100 times or more at the level of the primary producers, the algae, in seawater.

Nakahara and Cross (1978) reported that the marine clam *M. mercenaria* at high cell densities of phytoplankton increased their pseudofecal elimination and thereby reduced their uptake and retention of cobalt. Furthermore, less retention was seen with increasing size of clams in the presence of higher cell densities of phytoplankton.

Loosanoff et al. (1953; cited by Nakahara and Cross, 1973) reported quantitative and qualitative selectivity in feeding behavior of the clam M. mercenaria and noted that larvae in water containing heavier than optimum food concentrations had less food in their stomachs. Nakahara and Cross (1978) suggested this as another factor in their observation of a decrease in percentage cobalt retention as cell density increased. Also cited in Nakahara and Cross (1978) was the 1958 study by Rice and Smith who had worked with M. mercenaria and found different filtering rates for the clams between a diatom and a small green algae as well as an increase in pseudofeces production with increased density of algae and silt. Nakahara and Cross concluded that it is reasonable to assume that assimilation as well as retention of cobalt is probably influenced more by the species of phytoplankton than by the absorption efficiency of the clam.

g. Diurnal migration: Kuenzler (1967) found that zooplankton migrating diurnally carried the equivalent turnover by vertical eddy diffusion rate through the thermocline. Some of the factors he observed as contributors to the turnover of cobalt by plankton were: losses from molted exoskeletons, gametes, dead organisms, and cobalt-containing organic particles.

Fowler (1977) found the release of zooplankton fecal pellets, molted exoskeletons, and eggs to be an important influence on oceanic residence times of trace metals (see Table VII-2). The molts, containing 34% of the cobalt body burden of the euphausiid, contribute 7 μg/kg euphausiid/day to the cobalt flux rate; and the fecal pellets contribute 130 μg/kg euphausiid/day to the flux rate. Fowler suggested from findings of large numbers of intact pellets in deep water that fecal pellets are a significant factor in trace metal removal from the surface due to high production rate and slow decomposition.

Kuenzler (1969a) noted that plankton have small pools of trace metals which turn over rapidly and are consequently difficult to measure. He suggested that even rinsing them could remove these pools. Furthermore, he suggested that these pools have more of an impact on the occurrence of trace metals in seawater than on the organism itself. As a consequence of these rapid turnovers, he concluded that initial rates and measurements tend to be underestimated in long-term experiments.

Robertson (1970) found unusually high cobalt concentrations in unfiltered seawater samples collected 64 km off the Newport, Oregon, coast. With a surface concentration of 29.0 ng Co/liter, the value at 100-m depth rose sharply to 450 ng Co/liter and slowly dwindled to 110 ng/liter at 660-m depth. The cobalt contribution due to the organisms (living, detritus, and vitamin B_{12}) was calculated to be only 0.9 ng Co/liter if the detritus is assumed to be solely organism-derived and to have concentrations comparable to those of the organisms from whence it came. Detritus, however, might attain higher cobalt concentrations than the organisms by virtue of greater adsorption and complexation, attributable to smaller particle size with greater surface area and exposure of more complexing sites due to fragmentation and decay. Where large numbers of dying-off organisms occur, as at the end of a phytoplankton bloom, a markedly higher cobalt concentration in the water could be related to higher concentrations of detritus.

Kuenzler (1969b) found in his study almost a threefold difference in elimination rates between organisms within the same genus and indicated a need for more extensive elimination studies. He found Chaetognaths with a concentration of 2.5 mg Co/kg who comprised 11% of the mean zooplankton mass and accounted for 92% of the cobalt turnover in the upper 100 m. This may be an effect of population density and migration.

h. Seasonal variation: Patel et al. (1973) measured the occurrence of stable and radiocobalt in the sea hare *Aplysia benedict*. The sea hares accumulated cobalt from the effluent of a boiling water nuclear reactor when they moved shoreward during the winter to spawn. In this case, migration exposed the snail to a higher concentration of cobalt during the time of its reproductive cycle.

Radosavljević et al. (1973) reported that plankton reached a maximum concentration in the late autumn in the Danube and Sava rivers, and in the spring and summer for the Tisa River (all locations within Yugoslavia). No reason was suggested for this difference.

Bochenin and Chebotina (1975) reported that accumulation of ^{60}Co by *E. canadensis* cells in freshwater rivers and lakes varied approximately six- to seven-fold over the year, the peak concentration building from autumn to January with a sharp decrease in the spring.

Ishibashi et al. (1964) using a colorimetric analytical process reported no variation by season or habitat, but they observed that the cobalt concentration decreased in the order, red seaweed > green seaweed > limnetic weeds.

Ishii et al. (1978) observed seasonal variation for cobalt in *Sargassum thumbergii*, with the highest concentration being found in March (0.42 mg/kg dry weight) and lowest in June and January (0.25 and 0.27 mg/kg, respectively). The cobalt concentration increased in distance from the growing tip of the algae *Hizikia fusiforme* and was found in the highest concentration in the lamina of the seaweed *Sargassum* sp.

Katsuki et al. (1975) found low concentrations of cobalt in Bonito tissue with no significant variation due to season, body weight, or the location in which the fish were caught. Hodge et al. (1973) observed that albacore tuna are opportunistic feeders. They are believed to consist of a single population in the North Pacific that migrates to San Diego, California, in the summer months.

Harms (1975) found no difference in cobalt concentration between cod and plaice or whether they were inshore or offshore of the German Bight. The young fish spend the first 1 to 2 years in shallow water nursery grounds and move into deeper water as they grow. Furthermore, Hodge et al. (1973) were able to determine a value for long-term retention of radiocobalt in the upper layers of the Pacific Ocean by yearly sampling of albacore during their annual migration from the Central Pacific to the coast of Southern California. The values showed a decline in activity of ^{60}Co in tuna livers from 1964 to 1971; the cobalt has an apparent half-time of residence of 2.5 years in the upper water layers.

B. Terrestrial Food Chains

1. Temperate versus tropical ecosystems: Cobalt uptake in terrestrial food chains follows several patterns. In temperate regions, a large amount of the available cobalt is present in the soil, whereas in the tropical moist forest, the larger percentage of cobalt is in the biomass and is recycled quickly within the organic system.

The temperate ecosystems generally show a decrease in cobalt assimilation for each succeeding step of the food chain. Cobalt in temperate forest animals appears to accumulate in the muscles, at least as determined by short-term studies. Though data are plentiful for plant concentrations of cobalt, there are few data on stable cobalt concentrations correlated through an entire ecological habitat.

Because the tropical moist forest chain represents the largest natural system studied, as well as a unique environmental situation favoring cobalt accumulation (for that matter, all minerals), it affords a perspective for examining long-term accumulation. In that light, it is interesting to note that tropical forest animals are exposed to higher levels of cobalt in their diet than are their counterparts in temperate forests, and apparently there is a larger accumulation of cobalt in their bones than in their soft tissues.

At White Oak Lake bed in Tennessee, Crossley (1969) found that ^{60}Co was very mobile along plant to arthropod food chains, with the herbaceous and predatory insect concentrations approximately 43% of the level in the plants. The concentration factor increased at each successive transfer to a higher trophic level, soil to plant, CF 0.06; plant to herbivore, 0.43; and herbivore to predator, 1.07.

Willard (1975) studied radiocobalt cycling which had been introduced into a foodweb from a seepage channel at Oak Ridge National Laboratories in Tennessee. Transient animals, raccoon and oppossum, had low ^{60}Co concentrations, whereas permanent residents, the short-tailed shrew, mouse, and chipmunk, had much greater concentrations (see Table VII-10). Tracer studies performed on the earthworms in the seepage channel showed that 70%

TABLE VII-10

TERRESTRIAL FOOD CHAIN

Organism/Organ or Ecosystem Compartment	Isotope[a]	Cobalt Concentration (dry weight)	Concentration Factor	% Distribution in Body	Analytical Method	Remarks	Reference
Rat (*Rattus rattus*)	^{60}Co					Rats at equilibrium after ingestion of five meals. Assimilated 0.60% of uptake.	Amiard (1976)
blood		70 pCi/g		7.2			
liver		93 pCi/g		9.5			
pancreas		12 pCi/g		1.2			
kidneys		131 pCi/g		13.4			
muscles		451 pCi/g		46.2			
digestive gland		175 pCi/g		17.9			
heart		10 pCi/g		1.0			
ovaries		9 pCi/g		0.9			
thyroid and salivary gland		8 pCi/g		0.8			
urinary excretion		3.81 pCi/g		~2.5			
fecal excretion		150.28 pCi/g		~97			
assimilation *in toto*		0.939 pCi/g	0.02	~0.06			
assimilation/g body weight		0.0059 pCi/g		-			
Bobwhite (*Colinus virginianus*)	^{60}Co	8.31 pCi/g	0.011 at 11 days			A 21-day feeding study with an initial (3-day) uptake rate of 87 pCi/g/day. At equilibrium after 11 days with 277 pCi/g body burden. Three percent retention at 21 days.	Anderson et al. (1976)
bone				0			
gut				19 ± 5			
heart				11 ± 3			
liver				0			
muscle				52 ± 15			
Insects	^{60}Co			-		250 Ci ^{60}Co injected into tree; insects feeding on leaf litter around base.	Saas et al. (1971)
Centipedes		9.6 μCi/kg					
Diptera		4.9 μCi/kg					
Minnie's Lake swamp water	-	< 0.06 mg/liter			Atomic absorption	These are the dominant plants from the respective areas of Okefenokee Swamp.	Casagrande and Erchull (1977)
Taxodium sp., needles and twigs		< 6 mg/kg					
Peat		< 8 mg/kg					
Chesser Prairie marsh water		< 0.01 mg/liter					
Nymphea sp., leaves		< 6 mg/kg					
stems		< 6 mg/kg					
rhizomes, roots		< 6 mg/kg					
Oppossum (*Aidelphis* sp.)	^{60}Co	0.5 pCi/g			Gamma spectrometry	Temperate deciduous forest.	Willard (1975)
Raccoon (*Procyon lotor*)		1.0 pCi/g					
White-footed mouse (*Peromyscus*)		50 pCi/g					
Golden mouse		50 pCi/g					
Short-tail shrew (*Blarina* sp.)		80 pCi/g					
Eastern chipmunk (*Tamias* sp.)		20 pCi/g					
Earthworm (*Lumbricus rubellus*)		> 56 pCi/g					

TABLE VII-10 (concluded)

Organism/Organ or Ecosystem Compartment	Isotope[a/]	Cobalt Concentration (dry weight)	Concentration Factor	% Distribution in Body	Analytical Method	Remarks	Reference
Animal bone		0.0007 ± 0.000005 mg/kg	-	-	Neutron activation analysis	Run against NBS and IAEA Standards	Ördögh (1978)
Vegetation/soil		-	26	-	-	Tropical moist forest ecosystem in Panama. CF's calculated on dry weight basis.	Golley et al. (1976)
Herbivores/vegetation			9.8				
Reptiles and amphibians/insects			104				
Insects/understory leaves			8.7				
Soil	^{60}Co	4.57 pCi/mg	-	-	-	Herbivore/insect turnover rate, 92 hr. Site, Whiteoak Lake bed, Tennessee.	Crossley (1969)
Vegetation		0.265 pCi/mg	0.058[b/] - [c/]				
Herbivore		0.114 pCi/mg	0.43 0.025				
Predator		0.122 pCi/mg	1.07 0.026				

a/ Assume stable cobalt unless otherwise stated.

b/ CF is based on ratio of higher trophic level divided by lower; not strictly a CF.

c/ These CF values are related to soil values.

of the initial ^{60}Co activity was eliminated by excretion in the first 24 hr and 30% was retained for the length of the 28-day study. This has significance for the potential exposure to small rodents by polluted groundwaters, especially burrowing animals for whom the earthworm is a source of food.

Amiard (1976) found, during feeding studies with ^{60}Co plus carrier in the rat *Rattus rattus*, a CF of 0.02 for ^{60}Co uptake from food pellets to animal. Approximately 97% of the initial activity of ^{60}Co was never assimilated and had passed through the animal's digestive tract within 48 hr. Of the remainder of the initial activity, 2.5% was eliminated via urinary excretion, and 0.60% was retained (see Table VII-10). Amiard (1976) calculated that after the rats had ingested five meals, the ^{60}Co concentration in the kidneys was 16.0 times higher than in the blood; in the pancreas, 6.7 times higher; in the digestive tract, 4.8 times higher; in the thyroid and salivary glands, 4.0 times higher; and in the liver, heart, and ovaries, 2.3 to 3.1 times higher. On the other hand, the concentration in the muscles was only 0.8 that in the blood.

Heine and Wiechen (1978) found similar concentration factors (0.0091 and 0.0072) for plants growing on heath moor (0.66 mg Co/kg in the soil, 0.006 mg Co/kg fresh weight of plant) and high moor soils (3.6 mg Co/kg in the soil, 0.026 mg Co/kg fresh weight of plant), respectively. The concentrations of cobalt in milk from cows grazing on plants growing in each soil were identical (0.31 μg/kg) as were concentrations of calcium, iron, zinc, and selenium, indicating homeostatic control of essential elements. However, the concentration factors between milk and plant were higher (0.052 vs. 0.012) for the heath moor food chain.

Anderson et al. (1976) found a CF of 0.01 for ^{60}Co from tagged wild birdseed eaten by the bobwhite, *Colinus virginianus*. Tissue absorption of the ^{60}Co ingested was 37% in chronic feeding studies and 25% in acute feeding studies with a 3% whole body retention following a 21-day feeding study (see Table VII-10). The authors suggest that since a small amount of the originally ingested ^{60}Co is retained (3%), potential distribution in the environment is small. However, because 50% of the retained isotope activity is located in the muscle (i.e., 1.5%) there is a potential danger to consumers. The $T_{1/2}$ for the gut component of the 21-day feeding study was 0.17 day; and for the tissue component, 14 days.

Golley et al. (1976) in a comprehensive study of the mineral flow through a tropical moist forest ecosystem found 37% cobalt in the top 30 cm of soil and 59 to 67% in the vegetation depending on the season (see Table VII-11). Here the biomass carries the mineral reserve which prevents the loss of cobalt from the soil.

TABLE VII-11

INVENTORY OF COBALT IN A TROPICAL MOIST FOREST ECOSYSTEM
(Golley et al., 1976)

Compartment	Amount of Cobalt, kg/ha	Percentage of Total
Soil (top 30 cm)	7	37
Vegetation	10-14	67% wet, 59% dry, seasons
Leaves	0.5	3
Stems	10.8[a/]	57
Fruits and flowers	< 0.1	-
Roots	0.2	1
Litter	0.5[b/]	3
Herbivores	0.01	0.05
Carnivores	0.01	0.05
Detritivores	< 0.1	-
Total	19.0	

a/ Overstory stems contribute only 0.05 kg/ha.

b/ The annual vegetation input to the litter was calculated to be 0.3 kg/ha. Perhaps insect debris accounts for the difference.

In this situation cobalt being carried in the biomass helps to minimize loss from leaching. The largest quantity within the biomass is carried in the stems of the vegetation. The highest concentrations, however, are found in reptiles, amphibians, and mammals, especially within the bones (see Table VII-12).

TABLE VII-12

CONCENTRATIONS OF COBALT IN COMPARTMENTS OF A TROPICAL MOIST FOREST ECOSYSTEM (mg/kg dry weight)[a]
(Golley et al., 1976)

Compartment	September	February
Overstory leaves	56	20
Understory leaves	101	26
Overstory stems	35	36
Understory stems	35	38
Overstory fruits and flowers	36	36
Understory fruits and flowers	-	49
Litter	44	33
Roots	35	34
Overall mean	51	31
Soils (to 30 cm)	1.5 to 1.6	
Agouti, soft tissue		354
Agouti, bone		1,371
Crested guan (bird), soft tissue		501
Crested guan (bird), bone		1,713
Iguana, soft tissue		565
Iguana, bone		1,739
Termites		63
Insects on ground vegetation		13
Mammals		568
Birds		1,313
Reptiles and amphibians		1,352
Insects		13
Detritivores (termites)[b]		63
Herbivores (mammals and arthropods)		402
Carnivores (mammals, birds, reptiles, amphibians, and arthropods)[c]		1,175

a/ Analyses by atomic absorption spectrophotometry.

b/ Detritivores comprised 71% of the animal biomass.

c/ Eighty-six percent of the carnivore biomass comprised reptiles and amphibians.

However, it should be kept in mind that only ~ 0.1% of the forest biomass is animal; 71% of the animal biomass is comprised of the detritivores (termites), and the vegetation stems account for 57% of the total biomass.

Cobalt intakes in herbivores consuming leaves and in carnivores were found to be 0.05 and 0.04 kg/ha/year, respectively. The turnover rates for leaves and stems were 2.0 and 15.4 years, respectively.

Additional cobalt concentrations comparing the components of other tropical and semitropical forests to those of the tropical moist forest are presented in Table VII-13.

The concentration factors reflect much higher cobalt accumulation by tropical moist forest organisms than by their niche equivalent cousins in the temperate forest (see Table VII-10).

2. Terrestrial plant uptake: A survey of the literature on cobalt concentrations in plants as related to soils is in Table VII-14. It is difficult to draw patterns from these data as there is such a wide variety of geographical conditions that can affect cobalt availability. Often soil types and parent rock values are not reported along with vegetation concentrations. Generally cobalt concentrations in the plant tissues reflect the soil concentrations and the bioavailability to the plant.

Herbaceous and perennial plants with shallow root structures have a CF range of 0.1 to 10. Woody plants generally have a CF range of 0.1 to 1 and occasionally, greater than 1. This may reflect less uptake because of deep roots in soil horizons of low cobalt content. Cobalt content is highest at the surface in the humus. The cobalt accumulator plants exhibit CF's ranging from 10 to 100, where the parent rock is contributing large amounts of cobalt to the soil.

Cannon and Hopps (1971) surveyed literature values of cobalt concentrations in legumes in the United States. They reported mean values ranging from 0.07 to 0.2 mg Co/kg dry weight (see Figure VII-1).

TABLE VII-13

COBALT CONCENTRATIONS (mg/kg DRY WEIGHT) IN VEGETATION COMPONENTS OF DIFFERENT FORESTS
(Golley et al., 1976)

Compartment	Tropical Forests, Darien, Panama	Premontane Wet Forest a/	Riverine Forest b/	Mangrove Forest c/
Overstory leaves	47	80	54	46
Understory leaves	51	51	72	56
Overstory stems	48	49	91	52
Understory stems	46	36	45	83
Overstory fruits and flowers	d/	31	74	81
Understory fruits and flowers	d/	d/	69	56
Litter	54	39	91	65
Roots	49	44	73	61
Overall mean	49			

a/ 251 kg Co/ha (soil + vegetation)
b/ 119 kg Co/ha (soil + vegetation)
c/ 51 kg Co/ha (soil + vegetation)
d/ Not measured

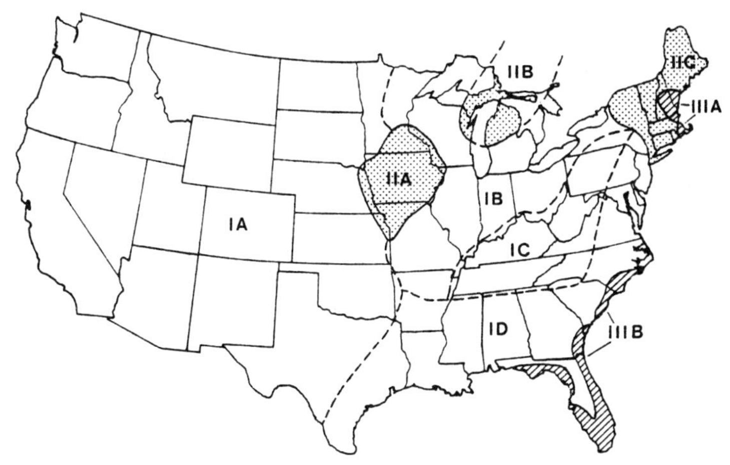

LEGEND

	General Level of Co in Legumes (mg/kg Dry Weight)		Map Symbol	General Region
Areas of Adequate Co	0.1 to 0.2	☐	I A	Western United States and Great Plains States
			I B	North Central States
			I C	Ridge and Valley
			I D	Piedmont and Upper Coastal Plain
Areas of Moderately Low Co	0.1	(dotted)	II A	Central Loess and Drift Plains
			II B	Northern Lake States
			II C	New England and Northern New York
Areas of Low Co	<0.07	(hatched)	III A	New England & Merrimac R. - Saco R. Drainage Basin
			III B	Lower Atlantic Coastal Plain

Figure VII-1. Regional Distribution of Cobalt in Plants in the United States (mg/kg dry weight) (Cannon and Hopps, 1971).

Dobrovol'skii (1969) reported CF's based on plant ash concentration for vegetation in northern regions of the USSR. Most CF's were between 1 and 2. A few of the higher values were CF 12 for forest litter and CF 10 for grassroots in the Murmansk taiga and CF 15 for grass blades in the "superakvalnyi." Furthermore, the high CF of the forest litter would be expected to be due to complexing of the available cobalt by the humus in the soil.

Saas et al. (1971) in a study of the biological cycling of ^{60}Co over several years found that cobalt in decaying plant material, having been converted to organic complexes in the living plant, is more available to soil fauna and more mobile than inorganic cobalt supplements added directly to the soil. In the latter case, the surface layer fine particles of soil adsorb cobalt and reduce cobalt mobility. After injecting 250 μCi ^{60}Co into a holm oak *Quercus ilex*, Saas et al. found 9.6 μCi ^{60}Co/kg in centipedes and 4.9 μCi ^{60}Co/kg in diptera because of their contact with the labeled litter falling from the tree. After direct delivery of comparable ^{60}Co activity to the brown calcareous soil, the same fauna showed negligible activity.

Although Lounamaa (1967) found in Finland, that the cobalt concentration was high in the needles or leaves and twigs of woody plants growing in high-cobalt soils associated with ultrabasic rocks, the CF's for the plants growing on silicic (acid) rocks were somewhat higher (0.38 to 0.81 vs. 0.13 to 0.44, plant ash weight). As expected, plant uptake of cobalt from soils associated with calcareous soils was lowest. *Betula verrucosa* (birch) leaves and twigs contained four to five times as much cobalt as the leaves and twigs of other species growing in soils overlying silicic rock (see Table VII-14).

A compilation by Connor et al. (1975) of United States Geological Survey studies shows CF's for Georgia black gum to be high as expected: stems, 20.3 and leaves, 49. Shagbark hickory from Kentucky had a CF threefold higher than shagbark hickory from Missouri. Most CF's calculated in Table VII-14, are less than 0.5 (based on ash weight) regardless of soil type or species.

Shklyaev et al. (1969) investigated the availability of cobalt under different growing conditions over three successive years.

TABLE VII-14

COBALT CONCENTRATIONS IN PLANTS GROWING IN UNCONTAMINATED SOILS

Plant Species and Part	Soil Type and Location	Cobalt Concentration In Soil (mg/kg)	Cobalt Concentration In Plant (mg/kg) (weight basis) Dry	Cobalt Concentration In Plant (mg/kg) (weight basis) Ash	Concentration Factor (weight basis) Dry	Concentration Factor (weight basis) Ash	Method	Comments	Reference
Helleborus cyclophyllus	Greece	22.0		-		-	Instrumental neutron activation analysis	A medicinal plant	Kanias and Philianos (1978)
Leaves			0.53		0.024				
Petioles			0.29		0.012				
Rhizome			0.66		0.030				
Lichens	Tundra, Khibinsk, USSR	-	-	-	-	1.0	Spectrographic		Dobrovol'skii (1969)
Grasses						1.2			
Saxifraga sp.						1.3			
Berry undergrowth						1.3			
Spruce needles	Taiga, Murmansk, USSR					0.8			
Birch leaves						0.5			
Red bilberry, *Vaccinium nigrum*						1.0			
Crowberry, *Empetrum nigrum*						-			
below-soil part						2.3			
Forest litter						12			
Grasses, above-soil parts						2.2			
below-soil parts						10			
Birch leaves						0.8-3.7			
Willow leaves						0.5			
Grasses, above-soil part						15			
Peat						0.9			
Herbaceous plants of: a birchwood forest	Interfluvial and bottomland of the Okya River, Tarusskii region, USSR								
above-soil mass						1.2			
below-soil mass						1.7			
a bottomland meadow									
above-soil mass						1.1			
below-soil mass						1.5			
Artemisia	"Ustyurt" region, USSR								
above-soil part						1.7			
below-soil part						1.8			
Conifer	Soils from silicic rock, Finland		-	-	-		Spectrographic and colorimetric	Trees on rock outcrops	Lounamaa (1967)
Needles						0.38			
Twigs		16				0.53			
Deciduous Trees and shrubs									
Leaves						0.54			
Twigs						0.81			

TABLE VII-14 (continued)

Plant Species and Part	Soil Type and Location	Cobalt Concentration In Soil (mg/kg)	In Plant (mg/kg) (weight basis) Dry	In Plant Ash	Concentration Factor (weight basis) Dry	Ash	Method	Comments	Reference
Conifer	Soils from ultrabasic parent rock, Finland		-	-	-		Spectrographic and colorimetric	Trees on rock outcrops	Lounamaa (1967)
Needles		140				0.13			
Twigs						0.44			
Deciduous Trees and shrubs									
Leaves						0.41			
Twigs						0.31			
Conifer	Soils from calcareous parent rock	23							
Needles						0.04			
Twigs						0.15			
Deciduous Trees and shrubs									
Leaves						0.06			
Twigs						0.09			
Betula verrucosa	Silicic rock and soil	16	-	-		-	-	-	Wallace et al. (1977)
Leaves					3.75				
Twigs					6.25				
Corn	Missouri		-		-		Atomic absorption spectrometry and emission spectrography	Cultivated soils plow zone	Connor et al. (1975)
	floodplain forest	5.2		<1		<0.19			
	glaciated prairie	8.1		1.0		0.12			
	unglaciated prairie	9.7		0.50		0.05			
	oak-hickory forest	8.7		<1		<0.11			
Soybean	floodplain forest	5.6		1.0		0.18		Cultivated soils plow zone	
	glaciated prairie	10		2.0		0.2			
	unglaciated prairie	10		2.1		0.21			
	oak-hickory forest	7.7		2.1		0.27			
Buckbush	glaciated prairie	11		4.0		0.36		Uncultivated soils	
	unglaciated prairie	14		4.8		0.34			
	cedar glade	9.5		4.4		0.46			
	oak-hickory forest	10		4.7		0.47			
	oak-hickory-pine forest	9.5		5.4		0.57			
Cedar	cedar glade	9.5		1.2		0.13		Uncultivated soils	
	glaciated prairie	11		6.5		0.59			
	oak-hickory forest	10		2.1		0.21			
	oak-hickory-pine forest	9.5		1.9		0.2			
Hickory, shagbark	oak-hickory forest	10		5.9		0.59		Uncultivated soils	
	oak-hickory-pine forest	9.5		7.4		0.78			
Oak, post	cedar glade	9.5		0.95		0.10		Uncultivated soils	
Oak, white	oak-hickory forest	10		2.0		0.20		Uncultivated soils	
	oak-hickory-pine forest	9.5		2.7		0.28			

TABLE VII-14 (concluded)

Plant Species and Part	Soil Type and Location	Cobalt Concentration: In Soil (mg/kg)	Cobalt Concentration: In Plant (mg/kg) (weight basis): Dry	Cobalt Concentration: In Plant (mg/kg) (weight basis): Ash	Concentration Factor (weight basis): Dry	Concentration Factor (weight basis): Ash	Method	Comments	Reference
Sumac, smooth	Missouri		-		-		Atomic absorption spectrometry and emission spectrography	Uncultivated soils	Connor et al. (1975)
	floodplain forest	8.3		0.75		0.09			
	glaciated prairie	11		0.65		0.06			
	unglaciated prairie	14		0.78		0.06			
	cedar glade	9.5		0.71		0.07			
	oak-hickory forest	10		1.1		0.11			
	oak-hickory-pine forest	9.5		0.97		0.10			
Sweet gum	floodplain forest	8.3		1.5		0.18			
Oak, willow	floodplain forest	8.3		0.95		0.11			
Black gum stems	Georgia	5.9		120, 190		20.3, 32.3		Average for each state is average of all reported values given for A, B, and C soil horizons.	
Black gum leaves	Georgia	5.9		290, 400		49, 67.8			
Persimmon stems	Georgia	5.9		2.2		0.37			
Hickory, pignut	Kentucky	7.2		18, 23		2.5, 3.2			
Hickory, shagbark	Kentucky	7.2		21		2.9			

Available cobalt content in the soil arable horizon during the initial pea growth stage ranged from 3.0 to 4.3 mg/kg of dry soil; during the blooming stage, 4.9 to 6.1; and during ripening stage, 4.1 to 5.7 mg/kg of dry soil. It was observed that the available cobalt content in the soil was higher in years with increased rainfall.

Enrichment of the soil with cobalt from natural sources such as serpentine parent rock has led to the identification of some cobalt-accumulating plants as indicators of ore deposit of cobalt and associated metals. Some well-known cobalt accumulators are: *Clethra barbinervis* (pepperbush) of Japan; *Nyssa sylvatica*, the swamp black gum tree of the United States; and *Silene cobalticola* (catchfly) and *Crotalaria cobalticola* (rattle box) of the Katanga copper mining district in Zaire (Peterson, 1971, and Connor et al., 1975). Siegel (1974) also includes white birch, trembling aspen, and white clover as cobalt accumulator plants useful for biogeochemical prospecting.

Connor et al. (1975) in a partial literature survey reported a mean of $\sim$ 400 mg Co/kg for Georgia or swamp black gum *N. sylvatica* and a range of $<$ 7 to 10,000 mg/kg (dry weight). The leaves have a higher concentration than the stems.

Kubota and Lazar (1958) sampled *N. sylvatica* across the southern United States and found the lowest values in plants growing on the cobalt deficient coastal sands of North Carolina (1 to 15 mg/kg dry weight) (see Table VII-15 for the values obtained).

Brooks and Radford (1978) reported that cobalt is concentrated by 11 hyperaccumulators of nickel ($>$ 1,000 mg Ni/kg dry weight) in the genus *Alyssum*. These 11 species showed cobalt concentrations of $<$ 1 to 13.5 mg/kg dry weight in more than 400 samples from European herbarium collections (see Table VII-16).

Okamoto (1978) found the pepperbush *C. barbinervis* accumulated zinc, manganese cobalt, nickel, and cadmium. He recommended it for use as an analytical standard for biological material.

Yamagata and Murakami (1958) reported that the leaves of *C. barbinervis* had a cobalt concentration more than 100 times

TABLE VII-15

COBALT IN SWAMP BLACK GUM LEAVES AND BROOMSEDGE AS RELATED TO SOIL TYPE
(Kubota and Lazar, 1958 and Kubota et al., 1960)

Soil Type	Cobalt in Black Gum Leaves, mg/kg		Cobalt in Broomsedge, mg/kg	
	Average	Range	Average	Range
Gulf and Atlantic Plains States				
Coastal sand[a]	13.2	1-115		0.05-0.12
Coastal sandy loam[b]	47.2	5.3-216		0.11-0.91
Stratified coastal sand, clay, and sandy clay	49.1	10.5-140		0.09-0.76
Coastal clay over marl	199	145-253	0.37	
Loess[b]	67.3	20-187	0.63	
Southern Arkansas and Northern Louisiana				
Coarse-textured soils: Excessively drained Regosols	22.1	6-62	0.10	
Moderately coarse and medium-textured soils: well-drained red-yellow podzolics	51.7	12-91		0.07-0.16
Moderately well-drained red-yellow podzolics	181	53-465		0.10-0.22
Somewhat poorly drained Regosols and Planosols	375	90-845	0.3	0.21-0.38
Fine-textured soils	400	320-480	0.11	

a/ Lowest for groundwater podzol sites even when low humic-gley or humic-gley soils had comparable sand texture.

b/ Lower without fragipan.

TABLE VII-16

COBALT ACCUMULATION BY EUROPEAN SPECIES[a/]
OF THE GENUS ALYSSUM[b/]
(Brooks and Radford, 1978)

Species	Cobalt in Leaves[c/] (mg/kg dry weight) Mean	Range
Alyssum alpestre	0.8	< 1-31
A. argenteum	22	1-135
A. bertolonii	33	2-100
A. corsicum	30	5-38
A. euboeum	2	2-2
A. fallacinum	13	12-13
A. fischerianum	< 1	< 1-10
A. heldreichii	20	2-40
A. markgrafii	5	3-30
A. murale	14	4-34
A. serpyllifolium lusitanicum	5	1-8
A. smolikanum	16	15-17
A. tenium	5	1-8
A. robertianum	40	18-68

a/ Over 400 samples obtained from dried herbarium specimens.
b/ *Alyssum* species are known for hyperaccumulation of nickel.
c/ Analysis by atomic absorption spectrometry.

greater than those of 17 other tree species from 199 locations near a chromite ore mine. The authors recommend this plant as a biogeochemical indicator of chromite (see Table VII-17).

Numerous studies have been made of plant uptake of cobalt from soils contaminated with (or fertilized with) cobalt-bearing materials. The latter include dust fallout from smelters, manure, sewage sludge, and sewage effluent. Many of these studies are summarized in Table VII-17.

TABLE VII-17

COBALT CONCENTRATIONS IN PLANTS GROWING IN SOILS CONTAMINATED WITH STABLE COBALT

Plant Species and Part	Soil Type and Location	Cobalt Concentration in Soil (mg/kg)	Cobalt Concentration in Material Added to Soil (mg/kg)	Cobalt Concentration in Plant (mg/kg)			Cobalt Concentration Factor (weight basis)			Method	Comments	References
				Dry	Wet	Ash	Dry	Wet	Ash			
Leeks, control	Urban garden soil, England Treated with sewage sludge	Not given	Not given	0.17 to 0.19							Soils were treated with sewage sludge from 1942 to 1960; vegetables were grown and analyzed in the following years: Leeks, 1959; carrots, 1967; globe beets, 1960; and potatoes, 1961. This is from the Woburn Market Garden experiment.	LeRiche (1968)
Leeks, treated				0.12 to 0.19								
Globe beets, control				< 0.1								
Globe beets, treated				< 0.1								
Carrots, tops, control				0.06								
Carrots, tops, treated				< 0.05 to 0.11								
Carrots, roots, control				< 0.05								
Carrots, roots, treated				< 0.05								
Potatoes, tops, control				0.30 to 0.45								
Potatoes, roots, control				0.35								
Potatoes, roots, treated				0.02 to 0.03								
Barley grain				0.01 to 0.02								
untreated		0.66	Not given		0.06			0.09			Literature review	Arapov (1968; cited by Yagodin and Tishchenko, 1978).
unspecified treatment with cobalt		1.60			0.37			0.23				
Bean plants (seeds)					13.6						Under intense fertilization. Found reduced germinating capacity above 7.2 mg Co/kg in plant.	Ozolinya and Zarin (1975; cited by Yagodin and Tishchenko, 1978).
Tomato *Lycopersicon esculentum*	Soils from Sudbury, Canada										Smelter emitted over 50 MT Co/yr to the atmosphere.	Hutchinson and Whitby (1974)

TABLE VII-17 (continued)

Plant Species and Part	Soil Type and Location	Cobalt Concentration in Soil (mg/kg)	Cobalt Concentration in Material Added to Soil (mg/kg)	Cobalt Concentration in Plant (mg/kg)			Cobalt Concentration Factor (weight basis)			Method	Comments	References
				Dry	Wet	Ash	Dry	Wet	Ash			
Pasture plants	Treated: irrigated with sewage effluent each year for 16 years at Templeton, Christchurch, New Zealand	3.3 to 3.9 (top 20 cm)	< 0.1 μg/liter	0.24			~ 0.07				Pasture used for grazing sheep. Menzel (1965) listed cobalt among those elements not concentrated by plants when applied to surface soil in water solution. (Relative values fell between 0.1 to 10 based on concentrations in dry weights of plants and soil).	Quin and Syers (1978)
	Control			0.22								
Grass *Calamagrostis canadensis*	Glacial deposit posit and lacustrine silt; Northwest Territory, Canada Distance (km) from smelter									INAA	Gold smelting process.	O'Toole et al. (1971)
	0.5			0.83								
	1.6			5.78								
	2.9			3.29								
	3.7			1.61								
Rose, *Rosa aciculoris*	0.6			1.14								
	1.3			1.09								
	2.9			0.79								
	3.0			0.06								
Alfalfa	Arizo calcareous soil amended with fly ash at:											Straughan et al. (1979)
	0.5 to 2.0%			0.12 to 0.16								
	4.0%			0.36								
	8.0%			0.45								
	Control			0.14								

TABLE VII-17 (continued)

Plant Species and Part	Soil Type and Location	Cobalt Concentration in Soil (mg/kg)	Cobalt Concentration in Material Added to Soil (mg/kg)	Cobalt Concentration in Plant (mg/kg) Dry	Wet	Ash	Cobalt Concentration Factor (weight basis) Dry	Wet	Ash	Method	Comments	References
	Distance (km) from smelter	Depth (cm)										
	0.8	Surface 127			17.9			0.14		AAS	Water extracts from soil contaminated by emissions from Sudbury smelter were used in pot experiments. CF's, however, are based on soil not extract concentrations.	
		5 124			4.9			0.04				
		10 82			8.8			0.11				
	1.5	Surface 72			9.4			0.13				
		5 69			4.7			0.07				
		10 61			-			-				
	1.9	Surface 85			13.9			0.16				
		5 34			9.2			0.27				
		10 31			9.7			0.31				
	3.7	Surface 42			10.4			0.25				
		5 58			2.7			0.05				
		10 43			-			-				
	7.4	Surface 154			3.9			0.02				
		5 98			-			-				
		10 36			-			-				
Radish												
Tops; organic matter added to medium-level-contaminated soil		36		0.35			0.01			INAA	Pot experiment with soil from site of factory in U.K. used for extracting tin from its ores.	Cawse et al. (1975)
Tops; nitrogen added to medium-level-contaminated soil		36		0.4			0.01					
Roots; nitrogen added to medium-level-contaminated soil		36		0.1			< 0.003				Medium and high-level-contamination refer to a mix of metals and minerals not just cobalt.	
Tops; organic matter added to high-level-contaminated soil		30		0.63			0.02					
Roots; organic matter added to high-level-contaminated soil		30		0.23			0.008					

TABLE VII-17 (continued)

Plant Species and Part	Soil Type and Location	Cobalt Concentration in Soil (mg/kg)	Cobalt Concentration in Material Added to Soil (mg/kg)	Cobalt Concentration in Plant (mg/kg)			Cobalt Concentration Factor (weight basis)			Method	Comments	References
				Dry	Wet	Ash	Dry	Wet	Ash			
Grass												
Nitrogen added to medium-level-contaminated soil		36		0.38			0.01					Cawse et al. (1975)
Organic matter added to medium-level-contaminated soil		36		0.63			0.02					
Organic matter added to high-level-contaminated soil		30		0.91			0.03					
Nitrogen added to high-level-contaminated soil		30		1.4			0.05					
Hay												
Deschampsia flexuosa *Agrostis* sp. *Festuca rubra*	Norway	20 to 70			1.0			0.2			Crops growing on soil contaminated by cobalt smelting operations 75 years earlier; samples taken from soil to 0.4 km from smelter. Used 4.5 mg Co/kg as soil concentration for vegetables.	Lag (1978)
Hay												
Deschampsia flexuosa	Norway				1.5			0.3				
Potatoes		3.8 to 8.5			0.2 to 0.3			0.04				
Carrots		3.8 to 8.5			0.2			0.03				
Red Beets		3.8 to 8.5			0.5			0.08				
Moss	Soil enriched with fly ash	15.4	70	7.2			0.47				Soils twice as enriched as world average 32 km north and south of the coal-fired Allen Steam Plant.	Bolton et al. (1974)

TABLE VII-17 (continued)

Plant Species and Part	Soil Type and Location	Cobalt Concentration in Soil (mg/kg)	Cobalt Concentration in Material Added to Soil (mg/kg)	Cobalt Concentration in Plant (mg/kg) Dry	Wet	Ash	Cobalt Concentration Factor (weight basis) Dry	Wet	Ash	Method	Comments	References
Cotton Giza 70 <u>Gossypium</u> <u>barbadense</u>												
Leaf	Yolo soil amended with $CoSO_4$ in nutrient solution	Control		1.5						ES	A pot experiment.	Rehab and Wallace (1978b)
		100		31.9			0.32					
		200		123.7			0.62					
		400		224.0			0.56					
Stem		Control		1.5			-					
		100		23.0			0.23					
		200		130.0			0.65					
		400		200.0			0.5					
Cotton Acala SJ-2 Cultivar <u>Gossypium</u> <u>hirsutum</u>	Yolo soil amended with $CoSO_4$ in a nutrient solution									ES	A pot experiment.	Rehab and Wallace (1978b)
Leaf		Control		1.9			-					
		100		40.0			0.4					
		200		146.3			0.73					
		400		242.5			0.61					
Stem		Control		1.5			-					
		100		59.3			0.59					
		200		173.5			0.87					
		400		200.0			0.50					
<u>Quercus</u> <u>crispula</u> (Oak)			Top soil, 511			2.6			0.0004		Seventeen species of trees were sampled at 199 sites in the vicinity of a chromite ore mine. Used 581 mg Co/kg as average value of top soil and B horizon for calculation of concentration factor.	Yamagata and Murakami (1958)
<u>Deutzia</u> <u>crenata</u>			B horizon, 652			5.3			0.009			
<u>Castanea</u> <u>crenata</u>						3.6			0.006			
<u>Cynoxylon</u> <u>japonica</u>						20.0			0.034			
<u>Lindera</u> <u>umbellata</u>						4.3			0.007			
<u>Clethra</u> <u>barbinervis</u>						1.0 to 22.5			0.02			
Other species						3.9			0.007			

TABLE VII-17 (concluded)

Plant Species and Part	Soil Type and Location	Cobalt Concentration in Soil (mg/kg)	Cobalt Concentration in Material Added to Soil (mg/kg)	Cobalt Concentration in Plant (mg/kg)			Cobalt Concentration Factor (weight basis)			Method	Comments	References
				Dry	Wet	Ash	Dry	Wet	Ash			
Ladino clover	Carp soil (unmanured)	3.4 to 11.7	1.06 (dry weight)		0.15			0.013			After 24 months of manuring, P_2O_5 and K_2O were added to the soil before sowing.	Atkinson et al. (1958)
	Carp soil (manured)	3.3 to 11.0			0.18			0.016				
	Bearbrook soil (manured)	3.3 to 11.0			0.17			0.022				
Clover *Trifolium subterraneum*	Urrbrae, red brown earth;			0.15						AAS		McKenzie (1978)
	Mt. Torrens, lateritic podzilic;			0.53								
	Gumeracha, red podzolic;			0.37								
	Urrbrae	1,000 mg/kg birnessite (MnO_2) added		0.10								
	Mt. Torrens			0.17								
	Gumeracha			0.24								
	Urrbrae	3,000 mg/kg cryptomelane (MnO_2) added		0.07								
	Mt. Torrens			0.09								
	Gumeracha			0.17								
Raphanus sativus var. (radish)												
Leaves		44		18.5			0.42				A total of 48 kg/ha $CoSO_4 \cdot 7H_2O$ had been added to the field over an 11-yr period.	Kloke and Egeis (1976)
Roots				trace			-					
Phaseolus vulgaris (bush bean)												
Leaves		44		4.2			0.10					
Roots				4.0			0.09					

AAS = Atomic absorption spectrophotometry.
INAA = Instrumental neutron activation analysis.
ES = Emission spectrography.

Although the CF's for cobalt growing in contaminated soil are low, they do indicate greater cobalt uptake. For example, Cawse et al. (1975) found one to fourfold increases in cobalt concentrations over those in controls in radish roots and grasses on soils contaminated by unspecified industrial wastes (30 to 36 mg Co/kg in soil). Concentration factors, however, were only 0.003 to 0.05, dry weight basis.

Similarly, Banasova (1978) in Czechoslovakia found the cobalt concentrations of two plants _Meliotus officinalis_ (yellow clover) and _Calamagrostis epigeios_ (reed bent grass) growing on a Dobsina asbestos heap containing 37.5 to 40 mg Co/kg, were 1.5 to 2.0 mg Co/kg but their CF's were only 0.04 to 0.05.

Quin and Syers (1978) did not find any significant increase in cobalt content of plants grown on a pasture treated with a sewage effluent containing < 1 μg Co/liter for 16 years. However, this concentration is far below the irrigation water criterion of 50 μg Co/liter recommended by Chaney (1973). He calculated that only 2.5% of the cobalt added to the soil in such irrigation water applied at the rate of 3 acre-ft/acre/year would be removed by 167 bushels corn per acre. The soil would accumulate 40 lb Co/acre (45 kg/ha) but the corn would remove only 1.0 lb Co/acre.

Sidle et al. (1977) studied the effects of wastewater irrigation on soil content and plant uptake of cobalt. The sewage treatment plant effluent used contained 40 μg Co/liter (57 μg Co/liter when sludge was added to the effluent). The soil was a Hublersburg clay loam, pH 6.3, with a 16.3 mEq./100 g soil cation exchange capacity. The soil (0 to 30 cm) comprised 28.4% sand, 38.7% silt, 32.9% clay, and 3.01% organic matter. In the period 1964 to 1973, three reed canary grass areas were irrigated with the wastewater so that the total cobalt application was 9.9 to 11.6 kg/ha. Of the cobalt in the treated soils at 0 to 30-m depth, 1.615 to 2.562 mg Co/kg was extractable with 0.1 _N_ hydrochloric acid; at 30 to 60 cm, 4.220 to 6.777 mg Co/kg. From 1963 to 1973, corn areas were irrigated with wastewater to give total cobalt applications of 5.8 and 8.5 kg/ha. There were no sustained significant cobalt increases at either soil depth, nor was the 1973 extractable cobalt significantly different from levels found in 1966. Thus, Sidle et al. concluded that there were not any definite accumulation trends for cobalt with time in either the reed canary grass areas or in the cornfield.

A less detailed study by Olson et al. (1979) of the use of secondary effluent for irrigation water (50 μg Co/liter) in California for up to 12 years showed little difference in cobalt concentrations in soil with depth among control sites and irrgated sites.

Grummitt (1976) studied the uptake of ^{60}Co by plants fertilized by sewage sludge. The pH, type of soil, addition of sludge, and ^{60}Co availability had less effect on the degree of ^{60}Co uptake than did the plant species and the part of the plant. Mean CF's, wet weight basis, for radish, potatoes, parsnips, oats, snap beans, carrots, and corn were 0.01 to 0.04. Those plant parts having CF's greater than one upon addition of 300 tons/ha sludge containing 6 to 50 pCi ^{60}Co were potato stems and leaves (1.8 to 2.1) and radish roots and leaves (1.0 and 1.2). In 100% sludge, potato stems and leaves had a CF 4.2 and tomato stems and leaves, 5.2.

Miller et al. (1977) studied the uptake of radiocobalt by vegetation growing along the Columbia River shore in the vicinity of the Hanford Works. Soils contained comparable activities of ^{58}Co (0.016 pCi/g dry weight) and of ^{60}Co (0.022 pCi/g); but the median concentration factors were very different: 0.006 for ^{58}Co but 1.1 for ^{60}Co.

When plants are grown in cobalt salt solutions, their uptake of cobalt can be much higher than when grown in soils. For example, Rehab and Wallace (1978a) grew Gossypium, Acala SJ-2 cultivar, plants in nutrient solutions containing 0.059 to 5.9 mg Co/liter as cobalt sulfate. The cobalt concentrations in the leaf, stem, and root were 6.9 to 180.0, 3.9 to 253.5, and 140.5 to 1,648 mg Co/kg, respectively. The CF's in the leaf, stem, and root were 117, 66.1, and 238.1 at 0.059 mg Co/liter in the solution; 118, 78.3, and 1,881 at 0.59 mg Co/liter; and 30.5, 43.0, and 279 at 5.9 mg Co/liter. (Only the highest solution was toxic to the plant.)

Cobalt can also be taken up from solution directly by the leaves. Shklyaev (1978) applied 5 μCi ^{60}Co/m^2 as the nitrate as a top dressing (leaf application) to young vetch plants. After 2 days' application, $\sim$ 95% of the remaining activity (5.8 to 19.9% of the applied ^{60}Co) was associated with the leaves; but at the time of flowering (within 3 weeks), $\sim$ 47% of the activity was still associated with the leaves, ~37% with the stems, ~12% with the roots, and ~ 4% with the flowers. By the time of pod formation (within 4 weeks), the activity distribution was: leaves, ~ 51%; stems, ~ 32%; pods, ~ 7.5%; roots, ~ 7%; and flowers, ~ 2.5%.

C. Cobalt in the Human Diet and the Human Body Burden

1. Dietary intake: Literature values for daily dietary intakes of cobalt by humans ranged from ~2 to 1,000 μg. The best agreement was found in values determined by neutron activation analysis: 9 μg/day (Italy), 17 μg/day (Germany), 22 and 39 μg/day (Puerto Rico), and > 40 μg/day (United States, teenage male). Five other studies, using other analytical methods, found 10 to 96 μg/day cobalt intakes. Yet the average for the values reported in Table VII-18 is > 150 μg/day. Another study, not tabulated, was that of Belz (1960), who used a colorimetric method to determine that cobalt intake in infants from 1 to 7 months increased from 0.3 to 2 μg/day with age in The Netherlands. These values agree well with those at the low end of the reported range.

As would be expected from the wide range of daily human intakes of cobalt that have been reported, the cobalt concentrations reported within various food groups also vary by as much as three orders of magnitude. The data from several recent and a few older reports are presented in Table VII-19.

A more thorough tabulation of older literature (since 1945) was performed by Schlettwein-Gsell and Mommsen-Straub (1970). Although most foods contained < 0.12 mg/kg (from 11 of 19 citations), there were numerous exceptions. Many of the tabulated values (30 of 68) between 0.13 and 0.49 mg/kg were reported by Schroeder et al. (1967). The data of Ripak et al. (1962; cited by Schlettwein-Gsell and Mommsen-Straub, 1970) represented 16 values in that range (photocolorimetric determination). Schroeder et al. also had reported more values (9/11) in the 0.50 to 0.99 mg/kg range. Although not agreeing in absolute values, there appeared to be agreement in these tabulated reports that beef liver and cocoa had high cobalt concentrations.

TABLE VII-18

COBALT IN TOTAL HUMAN DIETS

Location	Cobalt Dietary Intake (μg/day)	Cobalt Concentration in the Foods (mg/kg)	Remarks	Analytical Method	Remarks on Analytical Procedure	Authors (year)
Canada		<0.02 to 0.14 (fresh weight)	Foods were processed according to normal household practices. Vancouver cereals were high in Co and highest for several other heavy metals. Based on consumption of 1.783 kg food per person per day.	Atomic absorption	Foods were digested in a mixture of HNO_3, H_2SO_4, and $HClO_4$.	Kirkpatrick and Coffin (1974)
Halifax	55					
Vancouver	65					
USA (28 cities)	1,005	0.252 to 0.693 (range of averages); 0.529 (average of averages)	Diets of institutionalized children ages 9-12. Based on average consumption of 1.18 to 2.55 kg food/day (average of averages, 1.90 kg/day). The highest average Co concentration was in foods at Idaho Falls, Idaho; second highest (0.680 mg/kg) was at Tampa, Florida.	Atomic absorption	Foods were wet ashed with HNO_3.	Murthy et al. (1971)
USA	310 ± 40, 470 ± 30		Daily diets of two men for 50 weeks included water and beverages.	Emission spectrography	Standard deviation on replicates of ash 7.7%.	Tipton et al. (1969)
Tarapur Village, India			Diets included coffee and tea. Cereals comprised about 60% of the diets. Seafood comprised about 20% of the fishermen's diet.	Atomic absorption		Bhat (1973)
Fishermen (10 adults)	242.8					
Farmer	233.5 average 254.1 ± 16.5 range 210.0 to 390					
USA	170, 160		Thirty-day average intakes of a man and his wife.	Emission spectrography		Tipton et al. (1966)

TABLE VII-18 (continued)

Location	Cobalt Dietary Intake (μg/day)	Cobalt Concentration in the Foods (mg/kg)	Remarks	Analytical Method	Remarks on Analytical Procedure	Authors (year)
USSR	200 to 300		Calculation based on the cobalt concentration in daily rations.			Kolomiitseva (1966)
Vermont, USA			Water would add only 10 μg to daily Co intake in this area. There were 0.78 to 2.28 μg Co/ℓ in snow and 0.90 to 5.28 μg Co/ℓ in drinking waters.	Atomic absorption	Analytical values thought to be accurate. Co determination in vitamin B_{12} gave 4.29% (calculated value 4.34%).	Schroeder et al. (1967)
Hospital diet	166.41					
Institutional diet	436.21					
USSR			Most diets had highest Co concentrations in the fall and lowest concentrations in the winter.	Not given		Chubukov (1974)
Hospital therapeutic diet	17.3 to 31.9					
in spring	10.4 to 25.3					
in summer	20.0 to 30.8					
in fall	19.8 to 32.6					
in winter	11.2 to 69.9					
Federal Republic of Germany	Average 17, range 3 to 46		One week's diet (food and beverages) of four adult males. Average food and beverage intake 2.40 kg/day.	Neutron activation analysis	Method gave results in good agreement with IAEA Standards (milk powder and potato powder).	Schelenz (1977)
Various	150-600			Various		Crosby (1977) based largely on data from Davidson et al. (1972), WHO (1973), and Monier-Williams (1949)
Puerto Rico				Instrumental neutron activation analysis	Values at 0.05 to 0.5 μg measured with maximum error 2-7%.	Lingappan et al. (1974)
Slum areas	14.7		15% of population			
Rural areas	23.5		42% of population			
Urban areas	23		43% of population			
Average	22					

TABLE VII-18 (continued)

Location	Cobalt Dietary Intake (μg/day)	Cobalt Concentration in the Foods (mg/kg)	Remarks	Analytical Method	Remarks on Analytical Procedure	Authors (year)
Puerto Rico	39 ± 6	Not detected to 0.911 (fresh weight)	Diet included coffee, which had the maximum concentration of Co and contributed 15 μg Co to the daily intake.	Neutron activation analysis		Benítez-Rodríguez (1972)
Japan			A biological half-life for stable cobalt was calculated: 106 days	Spectrophotometric		Yamagata et al. (1963)
Urban adult average	17.9					
Rural adult average	21.1					
Adult average	19.5					
Rural child	2.4 to 14.8					
Two-thirds of adult population	11 to 28					
Tokyo	35.7					
USA, adults	5 to 8		Cereals contributed 12.2 μg/day.			Harp and Scoular (1952; cited by Yamagata et al., 1963)
Switzerland				Not given		Schlettwein-Gsell and Seiler (1972)
Family-type homes	118 to 170					
Germany	5 to 10			Not given		Schormuller (1961; cited by Schlettwein-Gsell and Seiler, 1972)
USSR, children	1.7			Not given		Resetkina (1965 ; cited by Schlettwein-Gsell and Seiler, 1972)
Germany				Not given		Droste and Jekat (1965; cited by Schlettwein-Gsell and Seiler, 1972)
in 1950	360					
in 1962	920					

TABLE VII-18 (continued)

Location	Cobalt Dietary Intake (μg/day)	Cobalt Concentration in the Foods (mg/kg)	Remarks	Analytical Method	Remarks on Analytical Procedure	Authors (year)
Tomsk, USSR, 2- to 6-yr-old children	63.97 to 71.46		Intake was 3.24 μg/kg 6-yr-old child or 6.2 μg/kg 2-yr-old child.	Polarography		Vorob'eva and Osmolovskaya (1970)
USSR, adults	43.5 to 56.4					Parashchak (1961; cited by Vorob'eva and Osmolovskaya, 1970)
USSR, 5- to 6-yr-old children	40 to 42		Amount gives positive Co balance.			Ripak (1961; cited by Vorob'eva and Osmolovskaya, 1970)
USSR, children, recommended	100 to 200					Gabovich and Kul'skaya (1964; cited by Vorob'eva and Osmolovskaya, 1970)
USSR, children, requirement	41.8 to 76.48					Zavertailo (1964; cited by Vorob'eva and Osmolovskaya, 1970)
USA, U.S. Department of Agriculture total diet program	> 40	0.003 to 0.132	Foods were prepared as if for consumption. Data here from 1-yr (1972-1973) collection of market baskets. The diet represents that of a teenage male. Some fruits, oils and fats, sugars, and beverages were not analyzed for Co.	Neutron activation analysis		Tanner and Friedman (1977)
England, 0.25- to 9.83-yr-old children	4.44 ± 1.28/kg (~100/23-kg child)					Alexander et al. (1974)

TABLE VII-18 (concluded)

Location	Cobalt Dietary Intake (μg/day)	Cobalt Concentration in the Foods (mg/kg)	Remarks	Analytical Method	Remarks on Analytical Procedure	Authors (year)
Central Italy, mineralized region, < 10,000 inhabitants	4.4 to 30.5, average 12		Diet sampled over a 3- to 5-day period. Only source of pollution in first central Italy site was a Hg ore processing plant; in second site, automobile major source. NE Italy site had steel and naval industries. Fe, Zn, Se, and Co showed the lowest variability of elements in diet.	Neutron activation analysis	The method showed good agreement for all elements in USA diet except Co and Sb, yet blood concentrations of Co were comparable with literature values. Clemente et al. (1979) compared median daily intake with an airborne intake of only 0.01 g Co/day.	Clemente et al. (1977)
Northeast Italy, well-industrialized area near the sea, ~ 100,000 inhabitants	2.4 to 17.2, average 8					
Northwest Italy, mountainous valley, agricultural area, ~ 50,000 inhabitants	4.2 to 13.2 average 7					
Central Italy, agricultural region, ~ 3 million inhabitants	5.8 to 20.7, average 12					
Crimea, USSR Sanitariums for atherosclerotic patients	96 ± 6			Spectrographic analysis		Koshlyak (1973)
Italy	9		Calculated daily intake from cooked diet, 9; drinking water, 1.6; and inhaled air, 0.008. Average drinking water, 0.8 μg/ℓ.	Neutron activation analysis	Detection limit in diet 0.0005 mg/kg wet weight	Clemente (1976)

TABLE VII-19

COBALT IN HUMAN FOODS[a/]

Location/ Method	Reference/ Units	Beverages	Milk and Dairy Products	Meat, Poultry, Fish	Grain and Cereals	Vegetables	Fruits	Sugars	Oils and Fats
Halifax, Canada Atomic Absorption	Kirkpatrick and Coffin (1974) (μg/person/ day)	(1)	(20)	(3) Poultry	(6)	(6) Potatoes (1) Leafy veg. (1) Legumes (1) Root veg.	(5)	(11)	(1)
Tarapur, India Atomic Absorption	Bhat (1973) (μg/fisherman/ day)		(0.4)	(1.1) Meat (74) Fish	(148)	(19.3) Legumes			
Puerto Rico Neutron Activation Analysis	Benítez Rodríguez (1972) (μg/person/day)	(15) coffee	(6) Milk	(0.4) Fish	(7) Rice	(9) Potatoes (0.1) Leafy Veg. (1) Legumes	(N.D.)/ tomatoes		(0.4) lard
	(mg/kg food)	N.D. to 0.911	0.026 Milk	0.033 Fish	0.055	0.019 Leafy veg. 0.036 Legumes			0.016
Tokyo, Japan Spectrophotometry	Yamagata et al. (1963)		(1.1) 0.0078	(3.1) Marine Products 0.031	(12.2) 0.027	(6.2) 0.031 Leafy veg. (5.9) 0.029 pulses, nuts, fruits (7.2) 0.039 Root veg.			
USA/Neutron Activation Analysis	Tanner and Friedman (1977) (μg/teenage male/ day)		(8)	(8) Poultry	(10)	(10) Potatoes (0) Leafy veg.	(2)		
	(mg/kg food)		0.011	0.030 Poultry	0.029	0.049 Potatoes 0.019 Leafy veg. 0.027 Legumes 0.014 Root veg.	0.018 garden fruits		
USA/Atomic Absorption	Schroeder et al. (1967)	0.01 to 1.71 (high in dry cocoa	0.02 to 0.35, average 0.12 (diary products)	0.04 to 0.94 average 0.22 (meat) 0.07 to 7.21 average 1.56 (sea-food)	0.08 to 1.24 average 0.43	0.01 to 0.69 average 0.14 (fresh vegetables)	0.01 to 0.12 average 0.04 ("canned fruits and vege-tables"); 0.14-0.32 average 0.14 ("fruits"); 0.04 to 0.56 average 0.26 (nuts)	0.01 to 4.62 average 0.52 ("con-diments and spices")	0.11 to 0.92 average 0.37

TABLE VII-19 (continued)

Location/ Method	Reference/ Units	Beverages	Milk and Dairy Products	Meat, Poultry, Fish	Grain and Cereals	Vegetables	Fruits	Sugars	Oils and Fats
USA/Atomic Absorption Using HNO_3/ H_2SO_4/H_2O_2 Digestion and Chelex 100 Resin	Baetz and Kenner (1973)	Cobalt was below the detection limit in all foods analyzed.							
USSR/Atomic Absorption	Andryushchenko and Fertman (1975) (mg/kg food)	0.5 (beer)							
California, USA/ Emission Spectroscopy (direct reading)	Cox et al. (1977) (mg/kg food)	0.3 (wine)							
Ceylon (Sri Lanka) analysis USSR/ Neutron Activation Analysis	Shimkova and Kukula (1974) (mg/kg food)	0.071 to 0.241, average 0.171 (dry tea)							
South Africa/ Atomic Absorption	Horwitz and van der Linden (1974)								
	(mg/kg food)	0.20 (5 dry teas)							
	(mg/kg food)	0.93 (7 dry coffees)							
	(μg/cup)	0.73 to 1.36 (filtered tea infusion)							
	(μg/cup)	2.2 to 10.0 (filtered coffee infusion)							
	(μg/cup)	4.74 to 9.8 (unfiltered coffee effusion)							

TABLE VII-19 (continued)

Location/ Method	Reference/ Units	Beverages	Milk and Dairy Products	Meat, Poultry, Fish	Grain and Cereals	Vegetables	Fruits	Sugars	Oils and Fats
England Atomic Absorption, APDC extraction, no ashing	Thomas et al. (1974)						0.01 to 0.15 (fresh, frozen, and canned fruits and vegetables) 0.02 (average fresh or (canned) 0.03 (average frozen)		
USA/ Colorimetric	Hurwitz and Beeson (1974) (mg/kg food)					1.20 dry weight (highest for spinach and turnip greens)			
USA/Neutron Activation Analysis	Oakes et al. (1978) (mg/kg food)					0.00037 to 0.0056 (fresh vegetables) 0.0056 (turnips)	0.000044 to 0.007 (high in raisins)		
USA/Flame Atomic Absorption	Slavin et al. (1975)			0.24, 0.21 (beef liver) 0.06, 0.09 (turkey) 0.09, 0.08 (beef muscle)					
USA/Neutron Activation Analysis	Zenoble et al. (1977) (mg/kg food)			0.92 to 1.95 average 1.45 (dried cooked turkey breast muscle)					

TABLE VII-19 (continued)

Location/ Method	Reference/ Units	Beverages	Milk and Dairy Products	Meat, Poultry, Fish	Grain and Cereals	Vegetables	Fruits	Sugars	Oils and Fats
Canada/ Atomic Absorption	Kirkpatrick and Coffin (1975) (mg/kg food)			N.D. to 1.27 average 0.06 (cured meats, fresh weight, highest in bologna)					
USA/Neutron Activation Analysis	Furr et al. (1974) (mg/kg dry food)			0.52 to 1.70 (low in fish protein concentrate)					
USA/Spectrochemical	Mitteldorf and Landon (1952) (mg/kg food)			0.002 to 0.01 average 0.002 (raw composite beef rib eye samples)					
Norway/Neutron Activation Analysis	Lunde (1973) (mg/kg food)			0.02 to 0.21 (fish meal lipids)					
Japan (Sea of)/ Flame Atomic Absorption	Morozov et al. (1978) (mg/kg food)			0.06 to 0.26 average 0.13 (fish muscle) (bone > skin > gills > roe > muscle)					

TABLE VII-19 (concluded)

Location/ Method	Reference/ Units	Beverages	Milk and Dairy Products	Meat, Poultry, Fish	Grain and Cereals	Vegetables	Fruits	Sugars	Oils and Fats
Italy (Samples from USSR, Korea, Japan, Madagascar, India, Peru, and Argentina) Neutron Activation Analysis	Rossi et al. (1976) (mg/kg food)			0.0008 to 0.0231, average 0.0053 ± 0.0009 (canned tuna fish)					
USA/California and Colorado Neutron Activation Analysis	Ragaini et al. (1975) (mg/liter)		0.000328 to 0.179 average 0.0126 (milk) (high concentration ÷ low concentration = 546)						
Canada/Atomic Absorption	Meranger and Smith (unpublished) bited by Sommers and Smith (1971)								0.10 (highest concentration in diet)
USA/	Macy et al. (1953, cited by Brink et al. (1969) (mg/liter)		0.0006 (whole milk)						
USSR/Inverse Voltamperometry	Brainina and Kharabadzhi (1973)			0.00353 to 0.00357 (fried elk chuck)					

a/ Values in parentheses are given in units of μg/person/day, whereas those values not in parentheses are given in units of mg Co/kg.

Schroeder et al. (1967) noted other foods that had high cobalt concentrations: smelt, butter, wheat bran, lentil stems and leaves, and sea salt. According to several of the studies cited in Table VII-19, coffee and tea can contribute major amounts of cobalt to the daily intake.

Among cured meats sold in Canada, Kirpatrick and Coffin (1975) found the highest concentrations in partly cooked hams, picnic hams, and bologna (average 0.11 to 0.14 mg/kg fresh weight). Other cured meats contained an average 0.02 to 0.05 mg/kg.

Several authors have noted the effects of food processing on cobalt concentrations.

Seiler et al. (1977) determined (by atomic absorption analysis) 0.00237 to 0.00327 mg Co/kg in fresh veal, beef liver, and pork, compared with 0.00324 to 0.00374 mg Co/kg in the frozen products. They also reported that up to 40% of the cobalt was lost from a fresh food by cooking in boiling water; ~ 10%, by roasting; 0 to 10% by grilling; and ~ 20% by stewing. Brainina and Kharabadzhi (1973) reported 0.03 mg Co/kg in fresh elk breast meat, but only 0.0039 to 0.0055 mg/kg in boiled breast meat (up to 90% loss). Fried elk chuck meat (0.035 mg/kg) showed a slight gain in cobalt content attributed by the authors to the fat used.

Schroeder (1971) claimed that there was a 70.6% loss of cobalt when fresh spinach was canned and a 70% loss in canning fresh carrots. Schroeder et al. (1967) reported that four samples of carrots had 0.03 to 0.16 mg Co/kg (average 0.10 mg Co/kg), but the cooked carrot sample had only 0.01 mg Co/kg. Schroeder et al. (1967) had also noted that metal shavings from a tinned can contained 76.92 mg Co/kg.

Cobalt losses may occur in milling wheat into flour. Kent-Jones and Amos (1957) cited the data of Booth et al. Whereas whole wheat contained 0.002 mg Co/kg, white flour contained only 0.0009 mg/kg and the bran, 0.0001 mg/kg. Schroeder (1971) reported that white bread had 69.4% less cobalt than whole wheat bread. He presented data from the 1967 Millfeed Manual and Czerniejewski et al. (1964) that showed that there was an 88.5%

loss of cobalt in converting wheat to white flour (cobalt concentration, mg/kg): wheat (0.026), mill feeds (0.07 to 0.18), white flour (0.003), and wheat germ (0.017).

Yet Lakhanpal and DaSilva (1973) in Brazil, reported a 12.91% gain in cobalt concentration in a wheat flour (0.00621 mg/kg) compared with whole wheat (0.00550 mg/kg). They noted a concentration of 0.00779 mg Co/kg in manioc flour compared with 0.0000 mg/kg in the manioc. Unprocessed rice contained 0.18 mg Co/kg compared with 0.0971 mg/kg in the hulls and 0.2410 mg/kg in the grains. Where cobalt concentrations are higher in the milled product than in the whole grain, it appears likely that cobalt was present in the materials of the grinding equipment.

Sugar cane contained 0.03 to 0.04 mg Co/kg, but granulated refined sugar contained 0.45 mg Co/kg according to Schroeder et al. (1967). Yet 4 years later, Schroeder (1971) remarked that white sugar has minimum amounts of chromium, cobalt, molybdenum, and zinc compared with raw sugar or sugar cane.

Khanna et al. (1976) found 0.6 to 8.3 mg Co/kg in 36 food colors. The highs were in the permitted lemon yellow (tartrazine plus sunset yellow), 8.3 mg/kg, and the unpermitted acid magenta (CI Acid Violet 19; 42685), 8.0 mg/kg, and Congo red (CI Direct Red 28; 22120), 8.3 mg/kg. Otherwise, the range was 0.6 to 5.5 mg/kg. The maximum per capita intake of synthetic food color, 53.5 mg/day, would lead to a maximum intake of 0.44 μg Co/day from this source. The average intake of synthetic food color was only 15 mg/day.

Malkus and Horacek (1972) studied the extractability of cobalt from glass-fiber-reinforced polyester resins based on isophthalic acid with 1:1 extractant volume per volume of plastic. Water, 50% alcohol, and 8% acetic acid all extracted up to 50 to 70 μg cobalt per dm^2 (a 1-liter bottle has a surface area > 5 dm^2) from certain brands within 14 days. They did not think the extraction of cobalt was unusual but recommended that the plastics not be used for food and beverage containers because of the high extractability of styrene.

Sampaolo et al. (1973) found up to 20.96 mg Co/liter in 1 to 3% acetic acid at 25 to 40°C after 24-hr contact with earthenware and porcelain containers intended for food use that were decorated above the glaze. Where the glaze hindered extraction (by being applied over the decoration, if any), up to 0.05 mg Co/liter was extracted by the acetic acid. The detection of cobalt in the latter cases was only frequent for samples having a colored glaze.

Glazes commonly used for kitchen earthenware may contain 1% cobalt oxide in the transparent base glaze or 2% cobalt oxide in the white slip. Beckman et al. (1974) found that the presence of copper, cobalt, or manganese oxides in the lead-containing glaze was associated with increased extractability of the lead.

Cobalt may be introduced into foods in other ways. Kraght and Marks (1967) of Wallace and Tiernan, Inc., hold a patent for treating raw fruits after harvest with an aqueous wash containing $\sim$ 0.005 to $\sim$ 2% cobalt (as an inorganic compound or as a salt of an organic acid) to inhibit decay during shipment.

Cobaltous chloride, acetate, or sulfate was permitted by the 1965 U.S. Code of Federal Regulations as a foam-stabilizing agent in beer. Such use was associated with outbreaks of cardiac myopathy, and the regulation was revoked on August 12, 1966 (Mahaffey, 1977). The maximum authorized dose added to beer was 1 mg Co/liter (Assenmacher and Stahl, 1972). Because there is apparently a synergistic effect between cobalt and alcohol in producing this disease, we have examined reports on the natural cobalt concentration of beer. Helin and Slaughter (1977) found (by flameless atomic absorption analysis) less than 0.002 mg Co/liter in clarified beer produced from malt containing 0.14 mg Co/kg dry weight and hops with $<$ 0.05 mg Co/kg. The wort (the unfermented infusion of malt) contained 0.01 mg Co/liter. The 80% decrease in cobalt concentration between the wort and the clarified beer was attributed by the authors to uptake by yeast or precipitation.

Andryushchenko and Fertman (1975) performed an elaborate mass balance study for cobalt in the beermaking process. (Analyses were by atomic absorption.) As can be seen from Table VII-20, most of the cobalt in the raw materials originated from

TABLE VII-20

COBALT TRANSFER IN BEERMAKING
(Andryuschenko and Fertman, 1975)

Substance	Cobalt Concentration, mg/kg	Cobalt Content, mg	%
Ingredients			
Malt	0.2 }	2,160.8	73.40
Barley	0.4 }		
Hops	0.1-0.5	26.0	0.88
Sugar	0.5	200.0	6.80
Enzyme preparation	21.8	523.2	17.78
Water	N.D.	-	-
Yeast	0.8[a/]	19.6	0.67
Lactic acid	1.1	12.7	0.43
Corrosion of apparatus		Not observed	
Total		2,942.3	
Products and Losses			
Green beer	0.05	2,112.0	71.80
Malt and hops mash	0.02	377.6	12.83
Losses of wort during transfers	0.05	125.9	4.28
Protein dregs	0.10	50.0	1.70
Spent yeast skimmings	0.12	29.4	1.00
Beer losses during transfers	0.05	126.9	4.31
Lager tank dregs	0.05	18.0	0.61
Beer losses during bottling	0.05	14.9	0.51
Clarification wastes	0.05	44.6	1.52
Unaccounted for losses	-	52.9	1.80

a/ This value is somewhat higher than the average 0.21 mg Co/kg determined by Bergerioux and Žikovsky (1978) in four samples of brewers' yeast including NBS SRM 1569 by neutron activation analysis.

the malt and barley and most of the ingredient cobalt was incorporated into the beer.

The main source of vitamin B_{12} (4.3% Co) in the environment is organisms that fix atmospheric nitrogen (blue-green algae and nodule symbionts of leguminous and nonleguminous plants). Rich food sources (containing more than 0.100 mg B_{12}/kg) are mammalian liver, kidney, and heart and marine bivalves. Sources containing 0.01 to 0.03 mg B_{12}/kg include muscle meat, seafood, and fermenting cheeses. Poorer sources (< 0.01 mg/kg) include milk, cream, and cheddar and cottage cheeses. Fresh fish containing 0.007 to 0.080 mg vitamin B_{12}/kg lost 24 to 96% of the vitamin content on cooking (compared with the content in the boiled samples); goat muscles containing 0.0010 to 0.00126 mg/kg lost 70%; and goat liver (0.040 to 0.100 mg/kg) lost 75% (Tracor Jitco, 1974). These losses of vitamin B_{12} on boiling are comparable to the losses of total cobalt in other boiled foods mentioned above.

2. Tissue concentrations and body burden: Forbes et al. (1954) presented data from colorimetric analyses (*o*-nitrocresol method) of the tissues of a 46-year-old white male, whose death was due to a skull fracture (see Table VII-21). Although not all of the tissues had detectable amounts of cobalt, almost 96% of the man's tissues contained 1 mg total cobalt. This indicates that the average cobalt concentration in this 54-kg man's tissues was about 0.02 mg/kg. The value converts to a body burden of 1.4 mg for a 70-kg individual. The value 0.02 mg/kg was cited by Valkovic (1977) as the concentration of cobalt in humans, based on 1.5 mg total body content (he does not indicate the basis for these values). Kieffer (1973) in a review article listed 3 mg/70 kg as the human body content of cobalt.

The International Commission on Radiological Protection (ICRP) (1974) gave a value of < 1.5 mg cobalt in the total body of Reference Man. Literature references were not cited for the given cobalt contents of the organs and tissues. Values of 0.110 mg in liver, 0.200 mg in muscle, and 0.280 mg in skeleton agree reasonably well with those of Forbes et al. (1954), but values for adipose tissue (0.360 mg) and skin (0.049 mg) do not agree well. In addition, the other major tissue for cobalt accumulation, according to the ICRP, is hair: 0.310 mg cobalt in 20 g

TABLE VII-21

COBALT IN THE TISSUES OF A 46-YEAR-OLD U.S. MALE
(Forbes et al., 1954)

Tissue	Weight, g	Co, mg/kg Fresh Weight	Co Content, mg
Skin	3,372	0.050	0.169
Skeleton	9,223	0.038	0.350
Muscle, striated	20,857	0.006	0.125
Nervous system	1,570	0.011	0.017
Liver	1,225	0.056	0.069
Heart	271	0.032	0.009
Lungs	1,734	0.018	0.031
Kidneys	270	-	-
Gastrointestinal tract	975	0.008	0.008
Adipose tissue	5,963	0.008	0.048
Offal	5,998	0.029	0.174
Blood	310	-	-
Teeth	43	-	-
Spleen	58	-	-
Pancreas	71	-	-
Total	51,940[a/]	> 0.019	> 1.000

a/ Subject's total weight was 53.8 kg.

hair or 16 mg/kg, far higher than any other hair concentration reported except that of a cemented carbide worker (Pillay et al., 1973).

Literature values reported for cobalt concentrations in human tissues are compiled in Tables VII-22, hair; VII-23, bones, teeth, and fingernails; VII-24, blood, blood fractions, saliva, and urine; VII-25, lungs; VII-26, kidneys; VII-27, liver; and VII-28, other human organs (these tables follow the next page of text). The variation in cobalt concentrations in most specific tissues may be due as much to errors arising from analytical methods used as human variability. The extreme variability observed in blood and urine values (0.17 to 262 μg/liter) has been attributed by Bowen (1963) to variations in the time elapsed between sampling and food intake.

Representative values, however, can be selected from these tables for many tissues (mg Co/kg dry weight): lungs, 0.2; hair, 0.5; kidneys, 1; liver, 1; brain, 0.1; heart, 0.2; intestine, 0.8; muscle, 1; spleen, 1; skin, 0.03; and fat, 0.1. These values compare well with those of Forbes et al. (1954; Table VII-21) for lungs, liver, brain, heart, and fat, if it is assumed that fresh weight values can be converted to dry weight by multiplying by 10.

The derived values from the data of Forbes et al. (1954), however, do not agree well with the MRI-selected values for intestine, muscle, and skin. For these tissues, only a few representative values were found in the tables. In addition, there is little or no basis for comparison for hair, spleen, and bone. The ancient skeletal concentrations of cobalt are still $\sim$ 10 times higher than that reported by Forbes et al., when the latter value is adjusted to a maximum 0.06 mg Co/kg dry weight (bone is 33% water). Becker et al. (1968) reported that cobalt was below the detection limit of 5 mg/kg in the ash ($\sim$6% of fresh weight) of tibias or femurs of modern or ancient origin when analyzed by a semiquantitative spectrographic method (arc emission spectroscopy). Such a finding is compatible with the fresh weight value found by Forbes et al. Jones et al. (1972) reported a value of 16.9 mg Co/kg dry weight as the average for tibias of 40 normal humans (determination by atomic absorption spectroscopy); but compared with other human tissue data, and in light of the fact that bone is not considered to be a target organ for radiocobalt isotopes, it would appear that the value of Jones et al. is not accurate.

The critical data compilation by Iyengar et al. (1978) on trace element content of human tissues gives even a larger range for cobalt concentrations in bones: the data of Forbes, the 1975 neutron activation analysis data of Behne et al. (0.01 mg Co/kg), the 1950 neutron activation analysis of McKown et al. (4.6 mg Co/kg wet weight), and the 1965 emission spectrometry data of Nusbaum et al. (43.5 mg Co/kg).

Thus, not enough data have been found to estimate the body burden for cobalt with any confidence. However, the value is probably between 1.5 and 15 mg for a 70-kg Reference Man. If the daily diet of Reference Man contains approximately 0.01 to 0.02 mg Co/kg fresh weight based on an intake of 20 to 40 μg Co/day, it would appear that he does not bioconcentrate cobalt.

TABLE VII-22

COBALT IN HUMAN HAIR

Number of Subjects	Sample Description	Cobalt Concentration in Dry Sample, mg/kg		Analytical Method[a] and Remarks	Other Remarks	Authors	Year
		Mean	Range				
-	"Normal hair values"	-	0.030-0.045	NAA	Data attributed to A. A. Gordus (1973), but this reference does not contain such values	Hopps	1975
	Hair close to scalp			IPAA	-	Chattopadhyay and Jervis	1975
76	Rural population (~ 100 km from industrial areas)	0.41	0.12-1.8			Jervis et al.	1976
45	Urban population, central Toronto, Ontario, Canada	0.48	0.15-2.6				
121	Urban population, Toronto, near two secondary Pb smelters	0.50	0.10-3.3				
	Samples from male and female subjects, most of whom patronized a barber shop. Hair from sides and back of head of males and from ends of hair of females, aged 10 months to 102 years			AAS	No bleached or dyed hair was sampled	Schroeder and Nason	1969
19	Males	0.17 ± 0.025	≤ 0.6				
11	Females	0.28 ± 0.043					
1	Oldest male	< 0.1					
	Males 40-70 years	0.13 ± 0.039					
10	Unspecified German subjects	Detected	-	NAA	Co was detected via X-ray spectra but not γ-spectra after 10-min irradiation	Menke et al.	1973

TABLE VII-22 (continued)

Number of Subjects	Sample Description	Cobalt Concentration in Dry Sample, mg/kg		Analytical Method[a] and Remarks	Other Remarks	Authors	Year
		Mean	Range				
	Scalp hair (from back of head ~1 in. from scalp)	0.04	0.01-0.12	NAA		Gordus et al.	1974
8 male, 1 female	Hair, subjects aged 5 to 19	0.54	-	AAS		Schroeder et al.	1967
1 female	Red hair, 17-year-old subject	0.71		AAS			
1 female	Black hair, 18-year-old subject	0.43		AAS			
1 male	White hair, 102-year-old subject	3.11		AAS			
11	Hair cut close to scalp near back of head, Waika Indians, Amazon Territories, Venezuela	1.70 0.55	0.53-2.83 N.D.[b] to 8.00	NAA Unspecified "previous analyses"	Most elements were found in a narrow range. Co, Na, Mn, Rb, and La ranges were rather broad for these members of two closed population groups. Perhaps much of the metal content is due to ambient contamination rather than metabolic deposition.	Perkons et al.	1977
	Hair, Japanese, collected from 21 barber shops and 7 beauty salons in various unpolluted places			Emission spectrographic		Teraoka and Kobayashi	1977
21 sites	Males	0.10	0.03-0.31		Maximum-minimum ratio 10		
7 sites	Females	0.32	0.16-0.94		Maximum-minimum ratio 5.9		

TABLE VII-22 (continued)

Number of Subjects	Sample Description	Cobalt Concentration in Dry Sample, mg/kg		Analytical Method[a] and Remarks	Other Remarks	Authors	Year
		Mean	Range				
	Washed three times, repeated three times with change of shampoo and water	0.066	-	Emission spectrographic	-	Teraoka and Kobayashi	1977
	Kept in shampoo solution 3 hr after washing procedure above	0.079					
	Kept in shampoo solution 6 hr after washing procedure above	0.069					
	Kept in shampoo solution 15 hr after washing procedure above	0.084					
	Hair, Japanese, collected from areas polluted by industry:			Emission spectrographic	-	Teraoka and Kobayashi	1977
	Aoba Chiba-city						
1	Fe works, male	0.12					
	Annaka Gumma-ken						
1	Zn smelter, male	0.07					
1	Zn smelter, female	0.18					
	Horie Tokyo Cr chemical works,						
1	male	0.13					
1	female	0.16					
	Umachi Nagano-ken						
1	Al refinery, male	0.09					
1	Shiojiri Nagano-ken ferrochrome works, male	0.10					

TABLE VII-22 (continued)

Number of Subjects	Sample Description	Cobalt Concentration in Natural Sample, mg/kg: Mean	Range	Analytical Method[a] and Remarks	Other Remarks	Authors	Year
31	Hair of learning disabled (LD) children, Quebec, Canada, mean age 137.0 months	0.16	-	AAS, level of significance $P < 0.001$	"A discriminant function analysis revealed that by using Cd, Co, Mn, Cr, and Li, all subjects could be classified as LD or normal with 98% accuracy." There was very small positive correlation ($r = 0.25$, $P < 0.05$) between socio-economic index and Co level, which was attributed to chance.	Pihl and Parkes	1977
22	Hair of normal children, Quebec, Canada, mean age 137.5 months	0.23					
1	Worker exposed to WC and WC-graphite mixtures			NAA	The NAA evidence in a homicide-criminal assault case resulted in the worker's conviction. Although not detected in washed samples of fabric from the victim's clothing and bedding, Co and Ta were found in unwashed samples as well as much higher levels of W.	Pillay et al.	1973
	Head hair, unwashed	39					
	Head hair, water washed	50					
	Head hair, ether washed	19					
	Pubic hair, ether washed	N.D.					
	Hair, presumably subjects from the Federal Republic of Germany	1.05	-	NAA	-	Kasperek	1974

TABLE VII-22 (concluded)

Number of Subjects	Sample Description	Cobalt Concentration in Dry Sample, mg/kg		Analytical Method[a/] and Remarks	Other Remarks	Authors	Year
		Mean	Range				
2	Two samples	0.24	0.34	SSMS		Yurachek et al.	1969
	Hair	1.28	-	AAS		Hung	1978
	Hair, median value for Italians	0.066	-	NAA		Clemente et al.	1979
28	Hair of 10-year-old boys from Nováky, Czechoslovakia (area polluted by coal-fired power plant)	0.049	0.02-0.14 + 5 N.D.	Instrumental NAA		Obrusnik and Bencko	1979
29	Hair of 10-year-old boys from Poruba, Czechoslovakia (village, control group)	0.06	0.04-0.11 + 7 N,D.				
25	Hair of adults from nonpolluted area of Prague	0.085	0.0069 -0.78 + 6 N.D.				

a/ NAA = Neutron activation analysis
IPAA - Instrumental proton activation analysis
AAS = Atomic absorption spectroscopy
SSMS = Spark source mass spectrometry

b/ N.D. = Not detected

TABLE VII-23

COBALT IN HUMAN BONES, TEETH, AND FINGERNAILS

Subjects	Sample Description	Cobalt Concentration Dry Weight (mg/kg) Mean	Range	Analytical Method[a] and Remarks	Other Remarks	Author (year)
-	Dental enamel	0.0024 ± 0.00009		NAA		Navia (1970)
-	36 teeth (crown and root) from 3 Scottish locations --Isle of Lewis		3-5	AAS	Pb, Zn, and Co significantly correlated between crown and root concentrations for each metal.	Stack et al. (1974)
69	Ancient skeletons from Europe, Asia, North America, South America, Africa, and Oceania; ages 10 to > 60; 38 male, 31 female:			INAA		Bratter et al. (1977)
66	Iliac crest	0.35			56% of Co in mineral fraction, 44% in collagen fraction	
	One skeleton: Lumbar vertebra	0.6				
	Clavicle and rib	0.28 ± 0.07				
	Pelvis-iliac crest	0.52				
	Humerus	0.86 ± 0.4				
	Ulna and radius	0.06 ± 0.0015				
	Patella (knee cap) and foot bones	0.4 ± 0.05				
	Femur	0.31 ± 0.04				
	Tibia and fibula	0.05 ± 0.02				
40	Fibias of normal humans	16.9		AAS	Value appears to be inaccurate, e.g., bone is not considered a target organ for radiocobalt	Jones et al. (1972)
-	Fingernails of 62-year-old male, U.S., left hand:			NAA Little variation in moisture content among nails: 10.1 to 10.4%		Kanabrocki (1973)
	1--Little finger	0.7				
	2	0.5				
	3	-				
	4	0.3				
	5-Thumb	0.3				
	Right hand: 1--little finger	0.3				
	2	0.2				
	3	1.4				
	4	1.8				
	5-Thumb	1.4				
15	Fingernails and toenails of normal males, aged 12 to 60 years, U.S.	0.04	N.D.[b] to 0.15	NAA Average moisture content--6.7%		
16	Left thumbnails of males hospitalized with myocardial infarct	0.04	0.00 to 0.09	NAA		
9	Left thumbnails of normal females, aged 23 to 60 years, U.S.	0.04	N.D.[b] to 0.15	NAA Average moisture content--1.82%		

a/ NAA = Neutron activation analysis; AAS = atomic absorption spectroscopy; and INAA = instrumental neutron activation analysis
b/ N.D. = Not detected

TABLE VII-24

COBALT IN HUMAN BODY FLUIDS

Sample Description	Mean Concentration Range in:[a/] Whole Blood (μg/ℓ)	Plasma (μg/ℓ)	Serum (μg/ℓ)	Red Blood Cells (μg/ℓ)	Saliva (mg/kg)	Urine (μg/ℓ)	Analytical Method[b/] and Remarks	Other Remarks	Author (year)
Urine and thymol (20 ml) plus nitric acid (5 ml)	-					< 1	SSMS	-	Jacobs (1975)
Blood (10 ml) + 0.2 ml Na heparin ashed in 7.5 ml HNO_3 + 10 ml $HClO_4$ almost to dryness, then dissolved in 25 ml 20% vol/vol HNO_3	30 (25 in Blanks)						SSMS	-	Jacobs (1975)
Blood of 128 male and 61 female English children with pica, or who were suspected of Pb poisoning for other reasons, aged 2 months to 15 yr	3.2 (0-50)						AAS	Because of experimental difficulties, not every element was determined in every sample. Significant positive correlations with Fe and Ni for children, 6-15 yr	Delves et al. (1973a, 1973b)
Blood of 51 male and 31 females of control group of children, with more aged 3-6 yr than above	4.6 (0-21)							Significant positive correlations with Cd in patients 6-15 yr, and with Sr in patients 3-6 yr	
Serum of Co mine workers			380				Colorimetric	Other mammalian serum values compiled were 7.5-260 μg/ℓ. Serum is 57.2% of whole blood. Compiled mammalian whole blood values were < 1-120 μg/ℓ (not given for Co mine workers)	Value cited by Bowen (1963)
Urine of 1 male and 5 female normals, aged 18-24 yr						~ 50-160	AAS	Values given as μg/g were converted to μg/ℓ by multiplying by 1,000/1,000	Schroeder et al. (1967)
Normal urine (first morning void)						< 10 (D.L.[c/])	ICP-AES	-	Haas et al. (1979)

TABLE VII-24 (continued)

Sample Description	Mean Concentration Range in:[a/] Whole Blood (μg/ℓ)	Plasma (μg/ℓ)	Serum (μg/ℓ)	Red Blood Cells (μg/ℓ)	Saliva (mg/kg)	Urine (μg/ℓ)	Analytical Method[b/] and Remarks	Other Remarks	Author (year)
Urine of 2 female psychotics, aged 76 and 80 yr						~ 10-60	AAS	-	Schroeder et al. (1967)
Urine of 1 victim of Wilson's disease, aged 15						~ 20	AAS	-	Schroeder et al. (1967)
Urine, Czech workers in hot-metallurgical shop who drank 2.5 ℓ/day of beer containing 30-54 μg Co/ℓ						262	AAS	-	Linhart et al. (1977)
Urine, Czech, non-beer drinkers						96	AAS	-	Linhart et al. (1977)
Blood, 26 male and 18 female normal Egyptians:							Dithizone method	No significant difference noted due to sex or age. Overall mean calculated to be 7.0 ± 0.25 μg/ℓ	Eisa et al. (1971 and 1972)
Male, newborn to 3 yr	6.6 (6.0-7.0)								
Female, newborn to 3 yr	6.6 (6.2-7.0)								
Male, 3-7 yr	5.6 (5.4-7.5)								
Female, 3-7 yr	7.5 (7.5-7.5)								
Male, 7-15 yr	7.2 (6.5-7.6)								
Female, 7-15 yr	7.1 (6.0-7.7)								
Male, > 15 yr	6.8 (6.0-7.2)								
Female, > 15 yr	7.2 (6.5-7.7)								

TABLE VII-24 (continued)

Sample Description	Mean Concentration Range in:[a/] Whole Blood ($\mu g/\ell$)	Plasma ($\mu g/\ell$)	Serum ($\mu g/\ell$)	Red Blood Cells ($\mu g/\ell$)	Saliva (mg/kg)	Urine ($\mu g/\ell$)	Analytical Method[b/] and Remarks	Other Remarks	Author (year)
Blood	0.5						Not given		Tinker (1973; cited by Gordus et al., 1974)
Blood, presumably subjects from Federal Republic of Germany	0.6						NAA		Kasperek (1974)
Pooled blood	0.4						NAA	Dried animal whole blood (IAEA reference material A-2) contained 0.52 ± 0.08 mg Co/kg	Tjioe et al. (1977)
Blood fractions		43 (3.8-63)		206 (30-508)					Koch et al. (1956; cited by Eisa et al., 1972)
Blood fractions		8.5 (2.8-17.0)		12 (6-23)					Hevrovsky (1952; cited by Eisa et al., 1972)
Dried plasma, United Kingdom		N.D.[d/]					SSMS; D.L.[c/] 0.3 mg/kg		Wolstenholme (1964)
Blood, median value from Italians	1.0						NAA		Clemente et al. (1979)
Freeze-dried parotid saliva (concentrated ~ 100 times):							NAA		Olmez et al. (1979)
Normals (5)					0.008 ± 0.007				
Subjects with loss of taste and/or smell					0.003 to 0.008				

TABLE VII-24 (continued)

Sample Description	Mean Concentration Range in:[a/] Whole Blood (μg/ℓ)	Plasma (μg/ℓ)	Serum (μg/ℓ)	Red Blood Cells (μg/ℓ)	Saliva (mg/kg)	Urine (μg/ℓ)	Analytical Method[b/] and Remarks	Other Remarks	Author (year)
Plasma		0.062, 0.083 μg/kg					Spectrochemical		Thiers et al. (1955)
Red blood cells				0.11, 0.22 μg/kg					
Whole blood (9 samples)	0.083 to 0.38 μg/kg								
Whole blood	58						-	The enormous range for reported mean values of Co in blood and serum may be attributed to variations with time elapsed since food intake. Such variability is also seen with Cr concentrations in blood	Bowen (1974)
Red blood cells				10.5 (0.1-200)					
Serum			2 (0.3-400)						
Range of literature values for whole blood	(0.17-238)		-						
Range of literature values for serum			(0.29-370)						
Serum of 15 normals, 10 patients with nephritis, and 4 patients with essential hypertension from Germany			660 (120-3600)				Emission spectral analysis		Mertz et al. (1968)
Serum of 39 healthy men			(0.0-50)				X-ray spectrochemical	99% confidence limit	Gofmann et al. (1964)

TABLE VII-24 (concluded)

Sample Description	Mean Concentration Range in:[a] Whole Blood (μg/ℓ)	Plasma (μg/ℓ)	Serum (μg/ℓ)	Red Blood Cells (μg/ℓ)	Saliva (mg/kg)	Urine (μg/ℓ)	Analytical Method[b] and Remarks	Other Remarks	Author (year)
Serum, 5 males, 3 females			7.5 (1.7-14.8)				Emission spectrography	-	Wolff (1950)
Plasma (25 samples)		12.2 (3.5-72.8)					NAA	-	Koch et al. (1956)
Red blood cells (17 samples)				206 (30-508)					
Whole blood (12 samples)	43.3 (3.5-63.0)								
Whole blood, 48 normals, Los Angeles area	238		~600				Emission spectrochemical	There was no significant difference among people of different residence times in the Los Angeles area	Butt et al. (1964)
Serum of male, West German blood donors			0.2				NAA	In a companion article, Kasparek et al. (1979) found 1.8 to 14.7 mg Co/kg in plasma-free pure platelets (wet-weight basis; mean 7.5 mg Co/kg) of 7 German donors	Iyengar et al. (1979)

a/ Mean is given first with the range following in parentheses
b/ AAS = Atomic absorption spectrometry; NAA = Neutron activation analysis; SSMS = Spark source mass spectrometry; ICP-AES = Inductively coupled plasma - atomic emission spectroscopy
c/ D.L. = Detection limit
d/ N.D. = Not detected

TABLE VII-25

COBALT IN HUMAN LUNGS

No. of Subjects	Sample Description	Concentration in Ash, (mg/kg)		Concentration, Dry Weight (mg/kg)		Concentration, Fresh Weight (mg/kg)		Analytical Method and Remarks [a/]	Other Remarks	Authors (year)
		Mean	Range	Mean	Range	Mean	Range			
141	Lungs of U.S. accident victims		<2 to 9 (80% range) <2 to 140 (total range)					ES, D.F. 26/141. Ash 1.1% of fresh weight.	-	Tipton and Cook (1963) Tipton and Shafer (1964)
44	Lungs of:								-	Tipton et al. (1965)
	African subjects		<2 to 160					ES, D.F. 32/44		
	Near Eastern subjects		<2 to 67					D.F. 23/34		
	Far Eastern subjects		<2 to 240					D.F. 28/69		
	Swiss subjects		<2 to 240					D.F. 5/7		
27	Lungs of U.S. accident victims, San Francisco		<2 to 18					ES, D.F. 19/27. Ash 1.0% of fresh weight.	Rank correlation coefficient ($P \leq 0.01$) with V 0.22	Tipton and Shafer (1964)
-	Coal miners' lungs		1.5, 2.3					SSMS	-	Jacobs (1975)
	Coal miners' lungs			6.2 (freeze-dried)				SSMS		
	Lightly pigmented lungs		0.85, 0.87					SSMS		
	Heavily pigmented lungs		2.3, 4.5					SSMS		
	Normal lungs		1.7, 2.0					SSMS		
	Lungs, male nonminer	1.2								
2	Lungs, white males ages 38 and 30, Dallas, Texas						0.06 to 0.13	AAS	-	Schroeder et al. (1967)
	Lungs, presumably subjects from Federal Republic of Germany			0.06				NAA	-	Kasperek (1974)

TABLE VII-25 (continued)

No. of Subjects	Sample Description	Concentration in Ash (mg/kg) Mean	Range	Concentration, Dry Weight (mg/kg) Mean	Range	Concentration, Fresh Weight (mg/kg) Mean	Range	Analytical Method and Remarks[a]	Other Remarks	Authors (year)
25	Lungs of patients, 50 to 68 yr old, from Duisburg (site of a large Cu and pyrites smelter)			1.268				NAA	-	Persigehl et al. (1976)
18	Lungs of patients, 0 to 65 yr old, from less industrialized Cologne			0.0957					Cologne-Duisburg concentration ratio was 13.2, indicating a highly significant difference.	
8	Lungs of children from above group			0.0361						
5	Lungs of Cologne adults, 50 to 68 yr old			0.169					Cologne-Duisburg concentration ratio was 7.5, which was also highly significant.	
1	Lung of 55-yr-old male patient with interstitial pneumonia who had worked as a general mechanic, welding an average of 2 hr/day for 18 yr				-	OsO_4 - Stained. Electron dense particles in macrophages comprised of ~35% Fe, 35% Ni, 18% Cr, and 12% Co.		Energy-dispersive X-ray.	-	Siegesmund et al. (1974)
-	Lung, Cologne				-			NAA	-	Persigehl et al. (1977)
	0 to 1 yr old			~0.04						
	20 to 40 yr old			~0.1						
	> 50 yr old			~0.18						
	Lungs of U.S. subjects							NAA	-	Kanabrocki (1973)
4	Myocardial infarct			0.07	0.04 to 0.12					
3	Bronchogenic carcinoma			0.33	0.03 to 0.04					
4	Stillbirth, 5 to 9 months gestation			0.03	0.03 to 0.03					

TABLE VII-25 (concluded)

No. of Subjects	Sample Description	Concentration in Ash (mg/kg) Mean	Range	Concentration, Dry Weight (mg/kg) Mean	Range	Concentration, Fresh Weight (mg/kg) Mean	Range	Analytical Method and Remarks[a]	Other Remarks	Authors (year)
	Lungs, U.S. subjects							NAA	-	Kanabrocki (1973)
4	Myocardial infarct			0.209	N.D. to 0.284[b]					
3	Bronchogenic carcinoma			0.191	0.121 to 0.261					
4	Stillbirth, 5 to 9 months gestation			0.032	N.D. to 0.038[b]					
1	Normal subject, United States							INAA	-	Rancitelli et al. (1969)
	Left lung:									
	Upper lobe				0.165					
	Lower lobe				0.245					
	Right lung:									
	Upper lobe				0.250					
	Middle lobe				0.165					
	Lower lobe				0.170					
	Average				0.200					
1	Subject with heavy element exposure, Great Britain				0.130					
1	Subject with Cu ore exposure				0.092					
1	Subject with U mine exposure				0.076					
1	Subject with U processing exposure				0.079					

a/ ES = emission spectroscopy.
D.F. = detection frequency.
SSMS = spark source mass spectrometry.
AAS = atomic absorption spectroscopy.
NAA = neutron activation analysis.
X-ray = X-ray fluorescence.
INAA = instrumental neutron activation analysis.

b/ N.D. = not detected.

TABLE VII-26

COBALT IN HUMAN KIDNEYS

Number of Samples	Sample Description	Concentration in Ash (mg/kg)		Concentration, Dry Weight (mg/kg)		Concentration, Fresh Weight (mg/kg)		Analytical Method and Remarks[a/]	Other Remarks	Author (year)
		Mean	Range	Mean	Range	Mean	Range			
144	Kidney, U.S. accident victims		< 2 to 2 (80% range)					ES, D.F. 21/144	Ash 1.1% of wet weight	Tipton and Cook (1963)
48	African subjects, ≥ 20 yr		< 2 to 480					ES, D.F. 21/48		Tipton et al. (1965)
31	Near Eastern subjects, ≥ 20 yr		< 2 to 140					D.F. 15/31		Tipton et al. (1965)
66	Far Eastern subjects, ≥ 20 yr		< 2 to > 1,000					D.F. 22/66		Tipton et al. (1965)
9	Swiss, ≥ 20 yr		< 2 to 90 (80% range)					D.F. 4/9		Tipton et al. (1965)
	Noncancerous kidney tissue associated with hypernephroma, U.S.					0.0012		NAA	Same concentration as that in primary tumor	Samsahl et al. (1965)
116	Kidney, normals, Los Angeles, California			1.12				Emission spectrometry, D.F. 11%		Indraprasit et al. (1974)
52	Kidney, victims of acute renal failure, Los Angeles, California			1.03				D.F. 13%		Indraprasit et al. (1974)
52	Kidney, victims of chronic renal failure, Los Angeles, California			1.55				D.F. 15%	Co disproportionately high in kidney of 35-yr-old female with juvenile diabetes in 28th week of pregnancy, dying of viral pneumonia	Indraprasit et al. (1974)
	Kidney, normal Japanese							Flame AAS		Sumino et al. (1975)
8	Males					0.016				
8	Females					0.013				
	Both sexes					0.015 ± 0.0098	0.01 to 0.035			

TABLE VII-26 (concluded)

Number of Samples	Sample Description	Concentration in Ash (mg/kg)		Concentration, Dry Weight (mg/kg)		Concentration, Fresh Weight (mg/kg)		Analytical Method and Remarks[a/]	Other Remarks	Author (year)
		Mean	Range	Mean	Range	Mean	Range			
2	Kidney, accidental death, white males, Dallas, Texas						0.05, 0.09	AAS		Schroeder et al. (1967)
	Kidney, presumably of subjects from Federal Republic of Germany			0.05				NAA		Kasperek (1974)
	Renal medulla, Cologne, Federal Republic of Germany				~0.06 to 0.11			NAA		Persigehl et al. (1977)
	Renal cortex, Cologne, Federal Republic of Germany				~0.07 to 0.11			NAA		Persigehl et al. (1977)
12	Kidney medulla, 10 male, 2 female Pima Indians, Phoenix, Arizona, diabetics and nondiabetics			1.79	0.99 to 6.92			PIXE		Nielson (1975)
12	Kidney cortex, same autopsy subjects as above			0.94	0.43 to 2.04			PIXE		Nielson (1975)

a/ ES = Emission spectroscopy; D.F. = Detection frequency; NAA = Neutron activation analysis; AAS = Atomic absorption spectroscopy; PIXE = Proton-induced x-ray emission.

TABLE VII-27

COBALT IN HUMAN LIVER

Number of Subjects	Sample Description	Concentration in Ash (mg/kg)		Concentration, Dry Weight (mg/kg)		Concentration, Fresh Weight (mg/kg)		Analytical Method and Remarks[a]	Other Remarks	Author (year)
		Mean	Range	Mean	Range	Mean	Range			
149	Liver, U.S. accident victims		< 2 to 10 (80% range)					ES, D.F. 63/149	Ash 1.3% of wet weight	Tipton and Cook (1963)
45	African subjects		< 2 to 300					ES, D.F. 40/45		Tipton et al. (1965)
33	Near Eastern subjects		< 2 to 510					D.F. 27/33		Tipton et al. (1965)
67	Far Eastern subjects		< 2 to 900					D.F. 51/67		Tipton et al. (1965)
9	Swiss subjects		< 2 to 400 (80% range)					D.F. 5/9		Tipton et al. (1965)
	Liver, healthy American males, 5 white, 5 black, accident victims, lived in St. Louis area "for some time," age 26 to 46 yr	8.58						Emission spectrographic, D.L. 1.0 mg Co/kg	The ratio of interhepatic variance to intrahepatic (within the same liver) variance was 23. Best positive correlation with Ag; best negative correlation with Fe and Pb.	Perry et al. (1974) Perry et al. (1978)
2	Liver, normal, accident victims						0.017, 0.035	NAA		Samsahl et al. (1965)
	Postmortem livers from 8 subjects whose deaths were attributed to drug overdose and from 3 subjects whose deaths were not attributed to drugs or poison (New Zealand)			0.59	< 0.5 to 1.0			AAS. Simple HCl/ HNO_3 digestion at 100 ± 20°C.		Johnson (1976)

TABLE VII-27 (continued)

Number of Subjects	Sample Description	Concentration in Ash (mg/kg) Mean	Concentration in Ash (mg/kg) Range	Concentration, Dry Weight (mg/kg) Mean	Concentration, Dry Weight (mg/kg) Range	Concentration, Fresh Weight (mg/kg) Mean	Concentration, Fresh Weight (mg/kg) Range	Analytical Method and Remarks[a]	Other Remarks	Author (year)
	Generalization for normal range				0.3 to 0.5			Not given		Johnson (1976)[b]
116	Liver, normals, Los Angeles, California			1.06				Emission spectrometry, D.F. 11%		Indraprasit et al. (1974)
52	Liver, victims of acute renal failure, Los Angeles, California			1.17				D.F. 14%		Indraprasit et al. (1974)
52	Liver, victims of chronic renal failure, Los Angeles, California			0.97				D.F. 19%		Indraprasit et al. (1974)
2	Liver, accidental death, white males, Dallas, Texas						0.12, 0.28	AAS		Schroeder et al. (1967)
5	Liver, normals, 4 Belgians dying in traffic accidents, 1 from cerebral hemorrhage					0.034	0.023 to 0.039	NAA	Interhepatic variation greater than intrahepatic variation. Dust fallout in laboratory was 0.0036 $ng/cm^2/day$; in a dust-free room, 0.0003 $ng/cm^2/day$.	Lievens et al. (1977)
	Liver, Cologne, Federal Republic of Germany							NAA		Persigehl et al. (1977)
	0 to 1-yr-old subjects			$\sim$ 0.07						
	20 to 40-yr-old subjects			$\sim$ 0.35						
	> 50-yr-old subjects			$\sim$ 0.62						
	Liver, normal Japanese							Flame AAS		Sumino et al. (1975)
14	Males					0.029				
15	Females					0.026				
	Both sexes					0.028 $\pm$ 0.011	0.013 to 0.062			

TABLE VII-27 (concluded)

Number of Subjects	Brief Description	Concentration in Ash (mg/kg)		Concentration, Dry Weight (mg/kg)		Concentration, Fresh Weight (mg/kg)		Analytical Method and Remarks[a/]	Other Remarks	Author (year)
		Mean	Range	Mean	Range	Mean	Range			
15	Livers, male accident victims, mean age 25 yr	3.4				0.05		SSMS		Locke et al. (1979)
10	Liver, 8 male, 2 female Pima Indians, Phoenix, Arizona, diabetics and nondiabetics			0.92	0.34 to 1.76			PIXE		Nielson (1975)

a/ ES = Emission spectroscopy; D.F. = Detection frequency; D.L. = Detection limit; NAA = Neutron activation analysis; AAS = Atomic absorption spectroscopy; PIXE = Proton-induced x-ray emission.

b/ Personal communication from P. van Peteghem, Department of Toxicology, Rijks University, Ghent, Belgium, cited by Johnson (1976).

TABLE VII-28

COBALT IN OTHER HUMAN ORGANS

No. of Subjects	Sample Description[a]	Concentration in Ash, mg/kg Mean	Concentration in Ash, mg/kg Range	Concentration, Dry Weight, mg/kg Mean	Concentration, Dry Weight, mg/kg Range	Concentration, Fresh Weight, mg/kg Mean	Concentration, Fresh Weight, mg/kg Range	Analytical Method[b] and Remarks	Other Remarks	Authors (year)
114	Adrenals, U.S. subjects		< 2-11					ES,DF 6/14	Ash 0.50% of wet weight	Tipton and Cook (1963)
104	Aorta, U.S.		< 2-6					ES, DF 21/104	Ash 1.7% of wet weight	Tipton and Cook (1963)
16	Aorta, African, 20 yr		< 2-700					ES, DF 10/16		Tipton et al. (1965)
15	Aorta, Near Eastern, 20 yr		< 2-29					DF 8/15		Tipton et al. (1965)
65	Aorta, Far Eastern, 20 yr		< 2-1000					DF, 32/65		Tipton et al. (1965)
5	Aorta, Swiss, 20 yr		6-17							Tipton et al. (1965)
1	Aorta, 38-yr-old white male dying of subdural hematoma, Dallas, Texas					0.40		AAS		Schroeder et al. (1967)
	Aorta, Cologne							NAA		Persigehl et al. (1977)
	0 to 1 yr old			~ 0.18						
	20 to 40 yr old			~ 0.26						
	> 50 yr old			~ 0.14						
4	Aorta, U.S.			when detected:				NAA		Kanabrocki (1973)
	myocardial infarct			0.42	N.D.[c]-0.48					
	bronchogenic carcinoma			0.32	N.D.-0.41					
	stillbirth, 5 to 9 months gestation			0.08	N.D.-0.16					
10	Aorta, Pima Indians (9 male, 1 female), Phornix, Arizona			0.93	0.20 to 1.62			PIXE		Nielson (1975)
129	Brain, U.S.	< 2						ES, DF 2/129	Ash 1.7% of wet weight	Tipton and Cook (1963)
17	Brain, African		<2-9					ES, DF 3/17		Tipton et al. (1965)
18	Brain, Near Eastern		ND					DF 0/18		Tipton et al. (1965)
51	Brain, Far Eastern		< 2->1000					DF 16/21		Tipton et al. (1965)
8	Brain, Swiss		< 2-110					DF 4/8		Tipton et al. (1965)

TABLE VII-28 (continued)

No. of Subject	Sample Description[a/]	Concentration in Ash, mg/kg		Concentration, Dry Weight, mg/kg		Concentration, Fresh Weight, mg/kg		Analytical Method[b/] and Remarks	Other Remarks	Authors (year)
		Mean	Range	Mean	Range	Mean	Range			
	Brain							NAA	The subject with by far the highest concentration in cerebral cortex and basal ganglion was a 4-yr-old male with congestive heart failure	Hoch et al. (1975)
	- of six humans aged 5 hr to 74 yr, cerebral cortex				0.0209 to 0.3388					
	- of six humans aged 5 hr to 74 yr, basal ganglia				0.0269 to 0.4049					
	- of seven humans aged 23 to 60 yr, various brain regions				0.0151 to 0.1028					
	Brain, dialysis patients							X-ray (charged particle induced)		Alfrey et al. (1972; cited by Hamilton, 1974)
1	Male					< 0.02				
1	Female					0.005				
3	Brain, two white males, Dallas, Texas; one black male, Miami, Florida					0.05 to 0.37		AAS		Schroeder et al. (1967)
	Brain, Cologne				~0.3 to ~0.4			NAA		Persigehl et al. (1977)
91	Diaphragm, U.S.	< 2						ES, DF 5/91	Ash 0.98% of wet weight	Tipton and Cook (1963)
1	Diaphragm, 30-yr-old male, Dallas, Texas					0.14		AAS		Schroeder et al. (1967)
67	Esophagus, U.S.	< 2						ES, DF 5/67	Ash 0.89% of wet weight	Tipton and Cook (1963)
1	Fat, subject from Hanover, NH					1.24		AAS	Co was found in 19/28 samples of human fat by emission spectroscopy	Schroeder et al. (1967)
4	Fat, abdominal, male Pima Indians, Phoenix, Arizona			0.08	0.06 to 0.12			PIXE		Nielson (1975

TABLE VII-28 (continued)

No. of Subjects	Sample Description[a]	Concentration in Ash, mg/kg Mean	Concentration in Ash, mg/kg Range	Concentration, Dry Weight, mg/kg Mean	Concentration, Dry Weight, mg/kg Range	Concentration, Fresh Weight, mg/kg Mean	Concentration, Fresh Weight, mg/kg Range	Analytical Method[b] and Remarks	Other Remarks	Authors (year)
20	Heart, 15 male, 5 female victims of traumatic accidents, ages 4 months to 65 yr, from Stockholm, Sweden, and surroundings.					0.0121 (medium)	0.001 to 0.018	NAA	Range was wide like that of nonessential elements. Range more like that of essential elements if high value found in the youngest case was excluded (0.001-0,0031)	Wester (1965)
140	Heart, U.S.		< 2 to 5					ES, DF 39/140	Ash 1.1% of wet weight	Tipton and Cook (1963)
43	Heart, African		< 2 to 110					ES, DF 39/43		Tipton et al. (1965)
20	Heart, Near Eastern		< 2 to 60					DF 16/20		
62	Heart, Far Eastern		< 2 to > 1,000					DF 36/62		
8	Heart, Swiss		< 2 to 76					DF 7/8		
3	Heart, two white males, Dallas, Texas; one black male, Miami, Florida						0.16 to 0.23	AAS		Schroeder et al. (1967)
-	Heart, Cologne,							NAA		Persigehl et al. (1977)
	0 to 1 yr old			~0.04						
	20 to 40 yr old			~0.17						
	> 50 yr old			~0.24						
67	Intestine, duodenum, U.S.	< 2						ES, DF 6/67	Ash 0.81% of wet weight	Tipton and Cook (1963)
102	Intestine, jejunum, U.S.	< 2						ES, DF 7/102	Ash 0.92% of wet weight	Tipton and Cook (1963)
3	Intestine, jejunum, U.S. males						0.04 to 0.17			Schroeder et al. (1967)

TABLE VII-28 (continued)

No. of Subjects	Sample Description[a]	Concentration in Ash, mg/kg		Concentration, Dry Weight, mg/kg		Concentration, Fresh Weight, mg/kg		Analytical Method[b] and Remarks	Other Remarks	Authors (year)
		Mean	Range	Mean	Range	Mean	Range			
83	Intestine, ileum, U.S.		< 2 to 10					ES, DF 18/83	Ash 0.75% of wet weight	Tipton and Cook (1963)
1	Intestine, ileum, U.S. white male, Dallas,Tex.					0.08		AAS		Schroeder et al. (1967)
31	Intestine, cecum, U.S.		< 2 to 8					ES, DF 10/31	Ash 0.67% of wet weight	Tipton and Cook (1963)
-	Intestine, large, U.S. associated with adenocarcinoma					0.0002		NAA	Concentration same in primary tumor	Samsahl et al.
108	Intestine, sigmoid colon, U.S.		< 2 to 11					ES, DF 21/108	Ash 0.68% of wet weight	Tipton and Cook (1963)
1	Intestine, sigmoid colon, U.S.					0.01		AAS		Schroeder et al. (1967)
42	Intestine, rectum, U.S.		< 2 to 11					ES, DF 10/42	Ash 0.85% of wet weight	Tipton and Cook (1963)
50	Larynx, U.S.	< 2						ES, DF 3/50	Ash 3.3% of wet weight	Tipton and Cook (1963)
2	Lymph nodes		2.3, 4.5					SSMS		Jacobs (1975)
136	Muscle, U.S.	< 2						ES, DF 6/136	Ash 1.2% of wet weight	Tipton and Cook (1963)
2	Muscle, psoas, white males, Dallas, Texas						0.05, 0.22	AAS		Schroeder et al. (1967)
75	Omentum, U.S.		< 2 to 12					ES, DF 26/75	Ash 0.25% of wet weight	Tipton and Cook (1963)
16	Ovary, U.S.	2 (not detected)						ES, DF 0/16	Ash 1.0% of wet weight	Tipton and Cook (1963)
2 11	Ovary, African Ovary, Far Eastern		< 2 to 9 < 2 to 510					DF 1/2 DF 2/11		Tipton et al. (1965)

TABLE VII-28 (continued)

No. of Subjects	Sample Description[a]	Concentration in Ash, mg/kg		Concentration, Dry Weight, mg/kg		Concentration, Fresh Weight, mg/kg		Analytical Method[b] and Remarks	Other Remarks	Authors (year)
		Mean	Range	Mean	Range	Mean	Range			
2	Pancreas, two white						0.07, 0.12	AAS		Schroeder (1967)
139	Pancreas, U.S.	< 2						ES, DF 13/139	Ash 1.1% of wet weight	Tipton and Cook (1963)
6 36 58 4	Pancreas, African Pancreas, Near East Pancreas, Far East Pancreas, Swiss		< 2 to 12 < 2 to 88 < 2 to > 1,000 < 2 to 1,000					ES, DF 2/6 ES, DF 2/6 DF 25/36 DF 25/58		Tipton et al. (1965)
4	Pancreas, male Pima Indian, Phoenix, Arizona			0.92	0.71 to 1.38			PIXE		Nielson (1975)
14	Pituitary, U.S., myocardial infacrct, carcinoma, and cirrhosis of liver			0.04	0.01 0.09			NAA		Kanabrocki (1973)
50	Prostate, U.S.	< 2						ES, DF 3/50	Ash 1.2% of wet weight	Tipton and Cook (1963)
1	Prostate, Dallas, Texas					0.11		AAS		Schroeder et al. (1967)
22	Skin, U.S.		< 2 to 11					ES, DF 5/22	Ash 0.79% of wet weight	Tipton and Cook (1963)
-	Skin, Cologne			~0.03				NAA	Little variation with age	Persigehl et al. (1977)
116	Spleen, normals, Los Angeles, California			1.08				ES, DF 10%		Indraprasit et al. (1974)
3	Spleen, two white males, Dallas, Texas; one black male, Miami, Florida						0.06 to 0.30	AAS		Schroeder et al. (1967)
143	Spleen, U.S.	< 2						ES, DF 8/143	Ash 1.4% of wet weight	Tipton and Cook (1963)
40 34 62 8	Spleen, African Spleen, Near Eastern Spleen, Far Eastern Spleen, Swiss		< 2 to 300 < 2 to 40 < 2 to 900 < 2 to 100					DF 15/40 DF 7/34 DF 20/62 DF 4/8		Tipton et al. (1965)

TABLE VII-28 (concluded)

No. of Subjects	Sample Description[a]	Concentration in Ash, mg/kg Mean	Concentration in Ash, mg/kg Range	Concentration, Dry Weight, mg/kg Mean	Concentration, Dry Weight, mg/kg Range	Concentration, Fresh Weight, mg/kg Mean	Concentration, Fresh Weight, mg/kg Range	Analytical Method[b] and Remarks	Other Remarks	Authors (year)
11	Spleen, U.S. myocardial infarct, bronchogenic carci-noma, stillbirth				N.D. to 0.08			NAA		Kanabrocki (1973)
7	Spleen, six male, one female, Pima Indians, Phoenix, Arizona			3.69	1.11 to 5.19			PIXE		Nielson (1975)
131	Stomach, U.S.		< 2 to 2					ES, DF 15/131	Ash 0.82% of wet weight	Tipton and Cook (1963)
3	Stomach, U.S. males						0.04 to 0.22	-		Schroeder et al. (1967)
2	Testis, one white male, Dallas, Texas; one black male, Miami, Florida						0.14,	AAS		Schroeder et al. (1967)
72	Testis, U.S.	< 2						ES, DF 6/72	Ash 1.1% of wet weight	Tipton and Cook (1963)
2	Testis, African	N.D.						ES		Tipton et al. (1965)
18	Testis, Near Eastern		< 2 to 12					DF 3/18		
39	Testis, Far Eastern		< 2 to 280					DF 20/39		
4	Testis, Swiss		< 2 to 240					DF 3/4		
21	Thyroid, U.S.		< 2 to 3					ES, DF 3/21	Ash 1.4% of wet weight	Tipton and Cook (1963)
60	Trachea, U.S.		< 2 to 4					ES, DF 13/60	Ash 1.9% of wet weight	Tipton and Cook (1963)
110	Urinary bladder, U.S.		< 2 to 2					ES, DF 12/110	Ash 0.78% of wet weight	Tipton and Cook (1963)
1	Urinary bladder, white male, Dallas, Texas					< 0.06		AAS		Schroeder et al. (1967)
32	Uterus, U.S.	< 2						ES, DF 3/32	Ash 1.0% of wet weight	Tipton and Cook (1963)

a/ All samples of Tipton and Cook (1963) were from "150 adult victims of instantaneous death who had spent their lives in the United States."
b/ ES = Emission spectroscopy; DF - Detection frequency; NAA - Neutron activation analysis; AAS - Atomic absorption spectroscopy; SSMS = Spark source mass spectrometry; and PIXE = Proton-induced x-ray emission.
c/ N.D. = Not detected.

BIBLIOGRAPHY. CHAPTER VII.

Alexander, F. W., B. E. Clayton, and H. T. Delves, "Mineral and Trace-Metal Balances in Children Receiving Normal and Synthetic Diets," Quart. J. Med., **43**(169), 89-111 (1974).

Amiard, J.-C. "Assimilation, Excretion and Distribution of Cobalt-60 in the Rat after Daily Ingestion of Contaminated Marine Food Products," Health Phys., **31**, 371-373 (1976).

Amiard, J.-C., and C. Amiard-Triquet, "Health and Ecological Aspects of Cobalt-60 Transfer in Seawater Food Chain Typical of an Intertidal Mud-Flat," Int. J. Environ. Stud., **10**(2), 113-118 (1977).

Amiard-Triquet, C., and L. Foulquier, "Modalites de la contamination de deux chaines trophiques dulcaquicoles par Le cobalt 60. I. Contamination directe des organismes par l'eau" ["Modes of the Contamination of Two Food Chains by Cobalt-60. I. Direct Contamination of Organisms by Water"], Water, Air, Soil Pollut., **9**(4), 475-489 (1978).

Anderson, S. H., G. J. Dodson, and R. I. Van Hook, "Comparative Retention of ^{60}Co, ^{109}Cd and ^{137}Cs Following Acute and Chronic Feeding in Bobwhite Quail," Spec. Publ. - Ecol. Soc. Am., **1** (Radioecol. Energy Resour.), 321-324 (1976).

Andryushchenko, A. V., and G. I. Fertman, "O Raspredelenii Metallov-Mikroelementov po Etapam Tekhnologii Piva" ["Distribution of Trace Metals in Different Stages of Beermaking"], Izv. Vyssh. Uchebn. Zaved., Pishch. Tekhnol., No. 4, 90-93 (1975).

Assenmacher, C., and A. Stahl, "Etude comparative des methodes de dosage du cobalt - Application à l'analyse des bières" ["Comparative Study of Methods of Determination of Cobalt. Applications to the Analysis of Beers"], Bull. Soc. Pharm. Strasbourg, **15**(2), 177-183 (1972).

Atkinson, H. J., G. R. Giles, and J. G. Desjardins, "Effect of Farmyard Manure on the Trace Element Content of Soils and of Plants Grown Thereon," Plant Soil, **10**, 32-36 (1958).

Baetz, R. A., and C. T. Kenner, "Determination of Heavy Metals in Foods," J. Agric. Food Chem., **21**(3), 436-440 (1973).

Banasova, V., "Gehalt an Ni, Co und Cr in Boden Einigen Pflanzen von Asbesthalden in Dobsina" ["Nickel, Cobalt and Chromium Contents in the Soils and in Some Plants of the Dobsina Asbestos Refuse Piles"], Biologia (Bratislava), **33**(4), 277-282 (1978).

Baptist, J. P., D. E. Hoss, and C. W. Lewis, "Retention of ^{51}Cr, ^{59}Fe, ^{60}Co, ^{65}Zn, ^{85}Sr, ^{95}Nb, ^{141m}In and ^{131}I by the Atlantic croaker (Micropogon undulatus)," Health Phys., **18**(2), 141-148 (1970).

Becker, R. O., J. A. Spadaro, and E. W. Berg, "The Trace Elements of Human Bone," J. Bone Joint Surg., **50**, 326-334 (1968).

Beckman, I., H. Guthenberg, F. Forslund, and E. Sjogren, "Release of Lead from Glazed Earthenware Household Utensils," Var Foeda, **26**(9-10), 236-241 (1974).

Belz, R., "Ijzer, Koper, Mangaan en Kobalt in Germiddelde Diëten van Verschillende Leeftijdsgroepen in Nederland" ["Iron, Copper, Manganese, and Cobalt in Average Diets of Various Age Groups in the Netherlands"], Voeding, **21**, 236-251 (1960); Biol. Abstr., **35**, 71368 (1961).

Benítez Rodríguez, J., Determination of Metal Contamination in Puerto Rican Foods Using Instrumental Neutron Activation Analysis, Report TID-26452, National Technical Information Service, U.S. Department of Commerce, Springfield, Virginia, July 1972.

Benzhitskii, A. G., and L. I. Sazhina, "Nakoplenie Kobal'ta-57 i Zheleza-55 Nekotorymi Zooplanktonnymi Organizmami Sredizemnogo Morya" ["Uptake of Cobalt-57 and Iron-55 by Some Zooplankton of the Mediterranean Sea"], Radiokhemo-Ekol. Issled. Sredizemnom More, 36-40 (1975).

Bergerioux, C., and L. Zikovsky, "Instrumental Neutron Activation Analysis of Brewers' Yeast," J. Radioanal. Chem., **47**, 173-179 (1978).

Bertine, K. K., S. J. Walawender, and M. Koide, "Chronological Strategies and Metal Fluxes in Semiarid Lake Sediments," Geochim. Cosmochim. Acta, **42**(10), 1559-1571 (1978).

Bhat, I. S., "Daily Intake of Cobalt by the Adult Population of Tarapur," Health Phys., **24**, 553-554 (1973).

Bochenin, V. F., and M. Y. Chebotina, "Sezonnaya Dinamika Nakopleniya Co^{60} Elodeei (*Elodea canadensis* Rich.)" ["Seasonal Dynamics of Cobalt-60 Accumulation by Elodea (*Elodea canadensis* Rich.)"], Ekologiya, No. 5, 80-82 (1975).

Bolton, N. E., et al., Trace Element Measurements at the Coal-Fired Allen Steam Plant - Progress Report February 1973 - July 1973, ORNL-NSF-EP-62, Oak Ridge National Laboratory, Oak Ridge, Tennessee, June 1974.

Boothe, P. N., and G. A. Knauer, "The Possible Importance of Fecal Material in the Biological Amplification of Trace and Heavy Metals," Limnol. Oceanogr., **17**(2), 270-274 (1972).

Bowen, H. J. M., The Elementary Composition of Mammalian Blood, U. K. Atomic Energy Authority Research Group Report AERE-R 4196, Isotope Research Division (AERE), Wantage Research Laboratory, Berkshire, 1963.

Bowen, H. J. M., "Problems in the Elementary Analysis of Standard Biological Materials," J. Radioanal. Chem., **19**, 215-226 (1974).

Boyden, C. R., "Effect of Size Upon Metal Content of Shellfish," J. Mar. Biol., **57**(3), 675-714 (1977).

Brainina, Kh. Z., and I. S. Kharabadzhi, "Primenenie Inversionnoi Volt'amperometrii Tverdykh Faz Dlya Opredeleniya Medi, Kobal'ta i Margantsa v Pishchevykh Produktakh" ["Inverse Voltammetry of Solid Phases for Determining Copper, Cobalt, and Manganese in Foods"], Vop. Pitan., No. 6, 56-59 (1973).

Brätter, P., D. Gawlik, J. Lausch, and U. Rösick, "On the Distribution of Trace Elements in Human Skeletons," J. Radioanal. Chem., **37**(1), 393-403 (1977).

Brink, M. F., M. Balsley, and E. W. Speckman, "Nutritional Value of Milk Compared with Filled and Imitation Milks," Am. J. Clin. Nutr., **22**, 168-180 (1969).

Brooks, R. R., and C. C. Radford, "Nickel Accumulation by European Species of the Genus *Alyssum*," Proc. R. Soc. London, Ser. B, **200**(1139), 217-224 (1978).

Bryan, G. W., and H. Uysal, "Heavy Metals in the Burrowing Bivalve *Scrobicularia plana* from the Tamar Estuary in Relation to Environmental Levels," J. Mar. Biol. Assoc. U.K., **58**(1), 89-108 (1978).

Butt, E. M., R. E. Nusbaum, T. C. Gilmour, and S. L. Didio, Arch. Environ. Health, **8**, 52-57 (1964).

Cannon, H. L., and H. C. Hopps, Eds., *Environmental Geochemistry in Health and Disease*, The Geological Society of America, Inc., Boulder, Colorado, 1971.

Casagrande, D. J., and L. D. Erchull, "Metals in Plants and Waters in the Okefenokee Swamp and Their Relationship to Constituents Found in Coal," Geochim. Cosmochim. Acta, **41**(9), 1391-1394 (1977).

Cawse, P. A., I. S. Jones, and G. M. Cox, "Cultivation and Multi-Element Analysis of Vegetables and Grass in Soil Contaminated with Trace Elements," Can. J. Plant Sci., **56**(1), 127-131 (1976).

Chaney, R. L., "Crop and Food Chain Effects of Toxic Elements in Sludges and Effluents," Proceedings of the Joint Conference on Recycling Municipal Sludges and Effluents on Land, July 9-13, 1973, Champaign, Illinois, Environmental Protection Agency, U.S. Department of Agriculture, and the National Association of State Universities and Land-Grant Colleges, Washington, D.C., 1973, pp. 129-141.

Chattopadhyay, A., and R. E. Jervis, "Hair as an Indicator of Multielement Exposure of Population Groups" in Trace Substances in Environmental Health - VIII, D. D. Hemphill, Ed., Proceedings of University of Missouri's 8th Annual Conference on Trace Substances in Environmental Health, Columbia, Missouri, June 11-13, 1974, University of Missouri, Columbia, Missouri, 1975, pp. 31-38.

Cherry, D. S., and R. K. Guthrie, "Toxic Metals in Surface Waters from Coal Ash," Water Resour. Bull., **13**(6), 1227-1236 (1977).

Cherry, D. S., R. K. Guthrie, F. Sherberger, and S. R. Larrick, "The Influence of Coal Ash and Thermal Discharges upon the Distribution and Bioaccumulation of Aquatic Invertebrates," Hydrobiologia, **62**(3), 257-267 (1979).

Chubukov, A. S., "Mikroelementy (Zhelezo, Tsink i Kobal't) v Pishchevykh Ratsionakh" ["Trace Nutrients (Iron, Zinc, and Cobalt) in Food Rations"], Vop. Pitan., No. 2, 68-70 (1974).

Clemente, G. F., "Trace Element Pathways from Environment to Man," J. Radioanal. Chem., **32**, 25-41 (1976).

Clemente, G. F., L. C. Rossi, and G. P. Santaroni, "Trace Element Intake and Excretion in the Italian Population," J. Radioanal. Chem., **37**(2), 549-558 (1977).

Clemente, G. F., L. C. Rossi, and G. P. Santaroni, "Studies on the Distribution and Related Health Effects of the Trace Elements in Italy" in Trace Substances in Environmental Health - XII, D. D. Hemphill, Ed., Proceedings of the University of Missouri's 12th Annual Conference on Trace Substances in Environmental Health, Columbia, Missouri, June 6-8, 1978, University of Missouri, Columbia, Missouri, 1979, pp. 23-30.

Connor, J. J., H. T. Shacklette, et al., Background Geochemistry of Some Rocks, Soils, Plants, and Vegetables in the Conterminous United States, Geological Survey Professional Paper 574-F, U.S. Department of the Interior, U.S. Government Printing Office, Washington, D.C., 1975.

Cowgill, U. M., "The Determination of All Detectable Elements in the Aquatic Plants of Linsley Pond and Cedar Lake (North Branford, Connecticut) by X-Ray Emission and Optical Emission Spectroscopy," Appl. Spectros., **27**, 5-9 (1973).

Cowgill, U. M., "The Hydrogeochemistry of Linsley Pond, North Branford, Connecticut. II. The Chemical Composition of the Aquatic Macrophytes," Arch. Hydrobiol., **45**(1), 1-119 (1974).

Cox, R. J., R. R. Eitenmiller, and J. J. Powers, "Mineral Content of Some California Wines," J. Food Sci., **42**(3), 849-850 (1977).

Crosby, N. T., "Determination of Metals in Foods--Review," Analyst, **102**(1213), 225-268 (1977).

Crossley, D. A., Jr., "Comparative Movement of ^{106}Ru, ^{60}Co, and ^{137}Cs in Arthropod Food Chains," U.S. Atomic Energy Commission, 1967, CONF-670503, National Technical Information Service, Department of Commerce, Springfield, Virginia, 1969, pp. 687-695.

Delves, H. T., B. E. Clayton, and J. Bicknell, "Concentration of Trace Metals in the Blood of Children," Br. J. Prev. Soc. Med., **27**(2), 100-107 (1973a).

Delves, H., J. Bicknell, and B. Clayton, "The Excessive Ingestion of Lead and Other Metals by Children," Proc., Int. Symp., Environ. Health Aspects Lead, 345-356 (1973b).

Dobrovol'skii, V. V., "Akkumulyatsiya Redkikh i Rasseyannykh Khimicheskikh Elementov Rastitel'nost'yu Nekotorykh Zonal'nykh Landshaftov SSSR" ["Accumulation of Rare and Trace Elements in Some USSR Plants"] in Obshch. Teor. Probl. Biol. Prod., O. V. Zalenskii, Ed., Nauka Leningrad (Pub.), Leningrad, USSR, 1969, pp. 51-56; Chem. Abstr., 72, 42336f (1969).

Eisa, E. A., A. E. M. I. Algauhari, S. A. Al-Nagdy, and N. Yanni, "The Effect of Sex and Age on the Blood Content of Zinc, Copper and Cobalt of Normal Egyptian Subjects," Proc. Egypt. Acad. Sci., 24, 49-52 (1971).

Eisa, E. A., A. E. I. Algahari, S. A. Al-Nagdy, and N. Yanni, "Blood Cobalt Content of Egyptian Subjects Under Normal and Pathological Conditions," Ain Shams Med. J., 23(4-5), 499-500 (1972).

Feldt, W., and M. Melzer, "Konzentrationsfaktoren der Elemente Kobalt, Mangan, Eisen, Zink und Silber für Fische" ["Concentration Factors of Elements Cobalt, Manganese, Iron, Zinc and Silver for Fish"], Arch. Fischereiwiss., 29(1-2), 105-112 (1978).

Forbes, R. M., A. R. Cooper, and H. H. Mitchell, "On the Occurrence of Beryllium, Boron, Cobalt, and Mercury in Human Tissues," J. Biol. Chem., 209, 857-865 (1954).

Fowler, S. W., "Trace Elements in Zooplankton Particulate Products," Nature, 269 (5623), 51-53 (1977).

Fukai, R., and C. N. Murray, "Environmental Behavior of Radiocobalt and Radiosilver Released from Nuclear Power Stations into Aquatic Systems" in Environ. Behav. Radionuclides Released Nucl. Ind., Proc. Symp. 1973, International Atomic Energy Agency, Vienna, Austria, 1973, pp. 321-332.

Furnica, G., O. Boldor, E. Dobrescu, S. Bulan, and V. Dobrescu, "Studies on the Concentration Factor of Cobalt-57 and Strontium-85 Radionuclides from the Water in Elodea canadensis," Lucr Gradinii Bot. Bucuresti, 157-163 (1974).

Furr, A. K., F. Kosikowski, C. A. Bache, and D. J. Lisk, "Elemental Analysis of Protein-Containing Food Materials from Various Sources," J. Food Sci., 39(5), 887-891 (1974); Chem. Abstr., **82**, 84760b (1975).

Geldiay, R., and H. Uysal, "Comparative Behavior of Toxic Metals in a Marine Ecosystem" in Origin and Fate of Chemical Residues in Food, Agriculture and Fisheries, International Atomic Energy Agency, Vienna, Austria, 1975, pp. 69-76.

Gofman, J. W., O. F. de Lalla, E. L. Kovich, O. Lowe, W. Martin, D. L. Piluso, R. K. Tandy, and F. Upham, "Chemical Elements in the Blood of Man," Arch. Environ. Health, **8**, 105-109 (1964).

Golley, F. B., J. T. McGinnis, R. Clements, G. I. Child, and M. J. Duever, Mineral Cycling in a Tropical Moist Forest Ecosystem, University of Georgia Press, Athens, Georgia, 1975.

Gordus, A., "Factors Affecting Trace-Metal Content of Human Hair," J. Radioanal. Chem., **15**(1), 229-243 (1973).

Gordus, A. A., C. M. Wysocki, C. C. Maher III, and R. C. Wieland, "Trace-Element Content of Human Scalp Hair" in Proc. Int. Conf. Nucl. Methods Environ. Res., 2nd, CONF-740701, J. R. Vogt and W. Meyer, Eds., National Technical Information Service, U.S. Department of Commerce, Springfield, Virginia, July 1974.

Grachev, M. I., "Vliyanie Temperatury Sredy na Nakoplenie, Raspredelenie i Vyvedenie ^{60}Co u Ryb" ["Effect of Environmental Temperature on the Accumulation, Distribution, and Excretion of Cobalt-60 by Fish"], Radioekol. Zhivotn., Mater. Vses. Konf., 1st, 37-38 (1977).

Grummit, W. E., "Transfer of Cobalt-60 to Plants from Soils Treated with Sewage Sludge" in Radioecology and Energy Resources, C. E. Cushing, Jr., Ed., Dowden, Hutchinson, and Ross, Stroudsburg, Pennsylvania, 1976, pp. 331-335.

Guthrie, R. K., and D. S. Cherry, "Pollutant Removal from Coal-Ash Basin Effluent," Water Resour. Bull., **12**(5), 889-902 (1976).

Guthrie, R. K., and D. S. Cherry, "The Uptake of Chemical Elements from Coal Ash and Settling Basin Effluent by Primary Producers. I. Relative Concentrations in Predominant Plants," Sci. Total Environ., **12**, 217-222 (1979).

Haas, W. J., Jr., V. A. Fassel, F. Grabau IV, R. N. Kniseley, and W. L. Sutherland, "Simultaneous Determination of Trace Elements in Urine by Inductively Coupled Plasma-Atomic Emission Spectrometry" in Ultratrace Metal Analysis in Biological Sciences and Environment, Advances in Chemistry Series No. 172, T. H. Risby, Ed., American Chemical Society, Washington, D.C., 1979, pp. 91-111.

Hamilton, E. I., "The Chemical Elements and Human Morbidity--Water, Air and Places--A Study of Natural Variability," Sci. Total Environ., **3**, 3-85 (1974).

Harms, U., "Levels of Heavy Metals (Mn, Fe, Co, Ni, Cu, Zn, Cd, Pb, Hg) in Fish from Onshore and Offshore Waters of the German Bight," Z. Lebensm. Unters.-Forsch., **157**, 125-132 (1975).

Harrison, F. L., "Accumulation and Loss of Cobalt and Caesium by the Marine Clam, *Mya arenaria*, under Laboratory and Field Conditions" in Radioactive Contamination of the Marine Environment, Proc. Symp. Int. Radioactive Contam. Mar. Environ., Seattle, July 1972, International Atomic Energy Agency, Vienna, Austria, 1973, pp. 453-477.

Harrison, F. L., K. M. Wong, and R. E. Heft, "Role of Solubles and Particulates in Radionuclide Accumulation in the Oyster *Crassostrea gigas* in the Discharge Canal of a Nuclear Power Plant," *Spec. Publ. - Ecol. Soc. Amer.*, **1** (Radioecol. Energy Resour.), 9-20 (1976).

Harvey, R. S., "Effects of Temperature on the Sorption of Radionuclides by a Blue-Green Alga" in *Proc. 2nd Nat. Symp. on Radioecology*, U.S. Atomic Energy Commission Conference 670503, 1969, pp. 266-277.

Heine, K., and A. Wiechen, "Bestimmung der Konzentrationsfaktoren von Elementen in der Nahrungskette Boden-Bewuchs-Milch" ["Determination of Concentration Factors of Elements in the Food Chain Soil-Plant-Milk"], *Milchwissenschaft*, **33**(4), 230-233 (1978).

Helin, T. R. M., and J. C. Slaughter, "Determination of Metals in Brewing Materials by Flameless Atomic Absorption Spectroscopy," *J. Inst. Brew.*, **83**, 15-16 (1977).

Hesse, J. L., and E. D. Evans, *Heavy Metals in Surface Waters, Sediments and Fish in Michigan*, Michigan Water Resources Commission, Bureau of Water Management, Department of Natural Resources, State of Michigan, July 1972.

Hiyama, Y., and J. M. Khan, "On the Concentration Factors of Radioactive I, Co, Fe and Ru in Marine Organisms," *Records Oceanogr. Works Jap.*, **7**(2), 79-106 (1964).

Hiyama, Y., and M. Shimizu, "Uptake of Radioactive Nuclides by Aquatic Organisms: the Application of the Exponential Model" in *Environmental Contamination by Radioactive Materials*, International Atomic Energy Agency, Vienna, Austria, 1969, pp. 463-476.

Höck, A., U. Demmel, H. Schicha, K. Kasperek, and L. E. Feinende, "Trace Element Concentration in Human Brain. Activation Analysis of Cobalt, Iron, Rubidium, Selenium, Zinc, Chromium, Silver, Cesium, Antimony and Scandium," *Brain*, **98**, 49-64 (1975).

Hodge, V. F., R. Folsom, and D. R. Young, "Retention of Fallout Constituents in Upper Layers of the Pacific Ocean as Estimated from Studies of a Tuna Population" in Radioactive Contamination of the Marine Environment. Proceedings of a Symposium, Seattle, Washington, July 10-14, 1972, M. Krippner, Ed., International Atomic Energy Agency, Vienna, Austria, 1973, pp. 263-276.

Honda, Y., T. Koga, and T. Muraoka, "Behavior of ^{60}Co in a Non-Flowing Small-Scale Rice Field Ecosystem," J. Radiat. Res., **14**, 209-218 (1973).

Hopps, H. C., "The Biologic Bases for Using Hair and Nail for Analyses of Trace Elements" in Trace Substances in Environmental Health - VIII, D. D. Hemphill, Ed., Proceedings of the University of Missouri's 8th Annual Conference on Trace Substances in Environmental Health, Columbia, Missouri, June 11-13, 1974, University of Missouri, Columbia, Missouri, 1975, pp. 59-73.

Horwitz, C., and S. E. Van Der Linden, "Cadmium and Cobalt in Tea and Coffee and Their Relationship to Cardiovascular Disease," S. African Med. J., **48**(6), 230-233 (1974).

Hurwitz, C., and K. C. Beeson, "The Cobalt Content of Some Food Plants, Food Res., **9**, 348-357 (1944).

Hutchinson, T. C., and L. M. Whitby, "A Study of Airborne Contamination of Vegetation and Soils by Heavy Metals from the Sudbury, Ontario, Copper-Nickel Smelters" in Trace Substances in Environmental Health - VII, D. D. Hemphill, Ed., Proceedings of the University of Missouri's 7th Annual Conference on Trace Substances in Environmental Health, Columbia, Missouri, June 12-14, 1973, University of Missouri, Columbia, Missouri, 1974, pp. 179-189.

Ichikawa, R., and S. Ohno, "Levels of Cobalt, Cesium and Zinc in Some Marine Organisms in Japan," Bull. Japan. Soc. Sci. Fish., **40**(5), 501-508 (1974).

Indraprasit, S., G. V. Alexander, and H. C. Gonick, "Tissue Composition of Major and Trace Elements in Uremia and Hypertension," J. Chronic Dis., 27(3), 135-161 (1974).

International Commission on Radiological Protection, Report of the Task Group on Reference Man, ICRP Publication 23, Pergamon Press, New York, New York, 1975.

Ishak, M. M., S. R. Khalil, and W. E. Y. Abdelmalik, "Distribution and Tissue Retention of Cesium-134 and Cobalt-60 in the Nile Catfish Clarias lazera (Cuv. & Val.)," Hydrobiologia, 54(1), 41-48 (1977).

Ishibashi, M., T. Yamamoto, and T. Fujita, "Chemical Studies on the Ocean (Part 92). Chemical Studies on the Seaweeds (17). Cobalt Content in Seaweeds," Records Oceanogr. Works Jap., 7(2), 17-24 (1964).

Ishii, T., H. Suzuki, and T. Koyanagi, ["Determination of Trace Elements in Marine Organisms. I. Factors for Variation of Concentration of Trace Element"], Nippon Suisan Gakkaishi, 44(2), 155-162 (1978).

Ivanov, V. N., "Eksperimental'noe Izuchenie Hakopleniya Zooplankton Organizmami Margantsa-54, Zheleza-59, Kobal'ta-60, Tsinka-65 i Ruteniya-106 v Sbyazi s Fiziko-Khimicheskimim Povedeniem Radionuklidov v Morskoi Vode" ["Experimental Study of the Accumulation of Manganese-54, Iron-59, Cobalt-60, Zinc-65, and Ruthenium-106 by Zooplankton in Connection with the Physical-Chemical Behavior of Radionuclides in Seawater"] in Formy Elementov i Radionuklidov v Morskoi Vode [Forms of Elements and Radionuclides in Seawater], Proceedings of the All-Union Symposium Devoted to Problems of the Physical-Chemical State of Elements and Radionuclides in the Ocean, Sevastopol, 1972, Publisher "Nauka," Moscow, 1974, pp. 138-146.

Iyengar, G. V., W. E. Kollmer, and H. J. M. Bowen, The Elemental Composition of Human Tissues and Body Fluids, Verlag Chemie, Weinheim, Federal Republic of Germany, and New York, New York (1978).

Iyengar, G. V., H. Borberg, K. Kasperek, J. Kiem, M. Siegers, L. E. Feinendegen, and R. Gross, "Elemental Composition of Platelets. Part I. Sampling and Sample Preparation of Platelets for Trace Element Analysis," Clin. Chem. (Winston-Salem, North Carolina), 25(5), 699-704 (1979).

Jacobs, M. L., Evaluation of Spark Source Mass Spectrometry in the Analysis of Biological Samples, Prepared for the National Institute for Occupational Safety and Health, Cincinnati, Ohio, Division of Laboratories and Criteria Development, PB-267 506, National Technical Information Service, U.S. Department of Commerce, Springfield, Virginia, 1975.

Jenkins, D. W., "Flow of Toxic Metals in the Environment," Inter. Conf. Environ. Sensing Assess., 1, 1-5 (1976).

Jervis, R. E., J. J. Paciga, and A. Chattopadhyay, "Characterization of Urban Aerosols and Their Hazard Assessment, by Size Sampling Combined with Inter-Element Ratios," Meas., Detect. Control Environ. Pollut., Proc. Int. Symp., 125-150 (1976).

Johnson, C. A., "The Determination of Some Toxic Metals in Human Liver as a Guide to Normal Levels in New Zealand. Part I. Determination of Bi, Cd, Cr, Co, Cu, Pb, Mn, Ni, Ag, Tl and Zn," Anal. Chim. Acta, 81(1), 69-74 (1976).

Jones, J. M., J. T. McCall, and L. M. Elveback, "Trace Metals in Human Osteogenic Sarcoma," Mayo Clin. Proc., 47(7), 476-478 (1972).

Kähäri, J., and H. Nissinen, "Mineral Element Contents of Timothy (*Phleum pratense* L.) in Finland. 1. Elements Calcium, Magnesium, Phosphorus, Potassium, Chromium, Cobalt, Copper, Iron, Manganese, Sodium and Zinc," Act. Agric. Scand., 1978 (Suppl. 20), 26-39 (1978).

Kanabrocki, E. L., "Analysis of Trace Elements in Human Tissue and an Addendum on Chronobiology" in Trace Elements in Relation to Cardiovascular Diseases (WHO/IAEA Joint Research Programme), IAEA-157, Papers Presented at a Research Coordination Meeting Organized by the International Atomic Energy Agency, Vienna, Austria, February 19-23, 1973, pp. 57-69.

Kanias, G. D., and S. M. Philianos, "Determination of Trace Elements in a Medicinal Plant by Neutron Activation Analysis," J. Radioanal. Chem., **46**, 87-93 (1978).

Kasperek, K., "Analytical Techniques for the Determination of 20 Trace Elements in Biological Samples by Means of Instrumental Thermal Neutron Activation Analysis" in Proc. Int. Conf. Nucl. Methods Environ. Res., 2nd, CONF-740701, J. R. Vogt and W. Meyer, Eds., National Technical Information Service, U.S. Department of Commerce, Springfield, Virginia, July 1974.

Kasperek, K., G. V. Iyengar, J. Kiem, H. Borberg, and L. E. Feinendegen, "Elemental Composition of Platelets. Part III. Determination of Silver, Gold, Cadmium, Cobalt, Chromium, Cesium, Molybdenum, Rubidium, Antimony, and Selenium in Normal Human Platelets by Neutron Activation Analysis," Clin. Chem. (Winston-Salem, North Carolina), **25**(5), 711-715 (1979).

Katsuki, Y., K. Yasuda, K. Ueda, and Y. Kimura, "Trace Elements in Marine Fish. II. Distribution of Heavy Metals in Bonito Tissue," Tokyo Toritsu Eisei Kenkyusho Kenkyu Nempo, **26**, 196-199 (1975).

Kent-Jones, D. W., and A. J. Amos, Modern Cereal Chemistry, 5th ed., The Northern Publishing Company, Ltd., Liverpool, England, 1957.

Khanna, S. K., G. B. Singh, and M. Z. Hasan, "Metal Contaminants in Various Food Colors," J. Sci. Food Agric., **27**(2), 170-174 (1976).

Kieffer, F., "Spurenelemente unter den Aspekten der optimalen Versorgung" ["Trace Elements under Aspects of Optimal Maintenance"], Chimia, **27**(11), 596-602 (1973).

Kimura, Y., and Y. Honda, "Uptake and Elimination of Some Radionuclides by Eggs and Fry of Rainbow Trout. I," J. Radiat. Res., **18**, 170-181 (1977a).

Kimura, Y., and Y. Honda, "Uptake and Elimination of Some Radionuclides by Eggs and Fry of Rainbow Trout. II," J. Radiat. Res., **18**, 182-193 (1977b).

Kirchmann, R., S. Bonotto, A. Bossus, G. Nuyts, R. Declerck, and G. Cantillon, "Utilisation d'une chaine trophique experimentale pour l'etude du transfert du ^{60}Co," ["Use of an Experimental Food Chain for Studying Cobalt-60 Transfer"], Rev. Int. Oceanogr. Med., **48**, 117-124 (1977).

Kirkpatrick, D. C., and D. E. Coffin, "The Trace Metal Content of Representative Canadian Diets in 1970 and 1971," J. Inst. Can. Sci. Technol. Aliment., **7**(1), 56-58 (1974).

Kirkpatrick, D. C., and D. E. Coffin. "Trace Metal Content of Various Cured Meats," J. Sci. Food Agric., **26**, 43-46 (1975).

Kloke, A., and W. Egels, "Effect of Excess Fertilization with Boron, Cobalt, Copper, Manganese and Zinc on the Content of These Elements in Soil and Plants," Dokl. Zarub. Uchastnikov-Mezhdunar. Kongr. Miner. Udobr., 8th, **2**(4,5), 115-121 (1976).

Kneip, T. J., G. Re, and T. Hernandez, "Cadmium in an Aquatic Ecosystem: Distribution and Effects" in Trace Substances in Environmental Health - VIII, D. D. Hemphill, Ed., Proceedings of the University of Missouri's 8th Annual Conference on Trace Substances in Environmental Health, Columbia, Missouri, June 11-13, 1974, University of Missouri, Columbia, Missouri, 1975, pp. 173-177.

Koch, H. J., Jr., E. R. Smith, N. F. Shimp, and J. Connor, "Analysis of Trace Elements in Human Tissues," Cancer, **9**, 499-511 (1956).

Kolomiitseva, M. G., ["Trace Element Requirements"], Gig. Pitan., 121-126 (1966); Chem. Abstr., **70**, 17965m (1969).

Korda, R. J., T. E. Henzler, P. A. Helmke, M. M. Jimenez, L. A. Haskin, and E. M. Larsen, "Trace Elements in Samples of Fish, Sediment and Taconite from Lake Superior," J. Great Lakes Res., **3**(1-2), 148-154 (1977).

Koshlyak, T. N., "Soderzhanie Makro- i Mikroelementov v Diete Bol'nykh Ateroskerozom" ["Content of Macronutrients and Trace Elements in a Diet for Patients with Atherosclerosis"], Mikroelem. Med., **4**, 94-96 (1973).

Kraght, A. J., and H. C. Marks, "Cobalt Compounds as Decay Inhibiting Agents for Fruits," U.S. Patent 3,347,683, October 17, 1967.

Kubota, J., and V. A. Lazar, "Cobalt Status of Soils of Southeastern United States. II. Ground-Water Podzols and Six Geographically Associated Soil Groups," Soil Sci., 86(5), 262-268 (1958).

Kubota, J., V. A. Lazar, and K. C. Beeson, "The Study of Cobalt Status of Soils in Arkansas and Louisiana Using the Black Gum as the Indicator Plant," Soil Sci. Soc. Am. Proc., 24, 527-528 (1960).

Kuenzler, E. J., "Elimination of Iodine, Cobalt, Iron, and Zinc by Marine Zooplankton" in Symp. on Radioecology, Proc. 2nd Natl. Symp., May 15-17, 1967, Ann Arbor, Michigan, D. J. Daniel and F. C. Evans, Eds., CONF-670503, National Technical Information Service, U.S. Department of Commerce, Springfield, Virginia, 1969a, pp. 462-473.

Kuenzler, E. J., "Elimination and Transport of Cobalt by Marine Zooplankton" in Symp. on Radioecology, Proc. 2nd Natl. Symp., May 15-17, 1967, Ann Arbor, Michigan, D. J. Nelson and F. C. Evans, Eds., CONF-670503, U.S. Atomic Energy Commission, Washington, D.C., 1969b, pp. 483-492.

Kulikov, N. V., and L. N. Ozhegov, "Nakoplenie Co^{60} Razvivayushcheisya Ikroi Shchuki, Okunya i Siga" ["Accumulation of Cobalt-60 by the Developing Spawn of Pike (Esox lucius), Yellow Perch (Perca fluviatilis), and Whitefish (Coregonus lavaretus)"], Ekologiya, No. 4, 100-102 (1975).

Kunze, J., H. Buehringer, and U. Harms, "Accumulation of Cobalt during Embryonic Development of Rainbow Trout (Salmo gairdneri Rich.)," Aquaculture, 13(1), 61-66 (1978).

Kustin, K., and G. C. McLeod, "Interactions Between Metal Ions and Living Organisms in Seawater," Top. Curr. Chem., 69, 1-37 (1977).

Lag, J., "Arsenic Pollution of Soils at Old Industrial Sites," Acta Agric. Scand., 28(1), 97-100 (1978).

Lakhanpal, R. K., and D. J. Da Silva, "Perda de Elementos Mineralis Essenciais Durante O Processamento dos Alimentos" ["Loss of Essential Mineral Elements during the Processing of Foods"], Seiva, 33(79), 21-26 (1973).

Lawrence, J. M., Dynamics of Chemical and Physical Characteristics of Water, Bottom Muds, and Aquatic Life in a Large Impoundment on a River, Zoology-Entomology Department Series Fisheries No. 6, Agricultural Experiment Station, Auburn University, Alabama, March 1968.

Leland, H. V., E. D. Copenhaver, and L. S. Corrill, "Water Pollution. Heavy Metals and Other Trace Elements," J. Water Pollut. Control Fed., 46(6), 1452-1476 (1974).

Le Riche, H. H., "Metal Contamination of Soil in the Woburn Market-Garden Experiment Resulting from the Application of Sewage Sludge," J. Agric. Sci. Camb., 71, 205-208 (1968).

Lievens, P., J. Versieck, R. Cornelis, and J. Hoste, "The Distribution of Trace Elements in Normal Human Liver Determined by Semi-Automated Radiochemical Neutron Activation Analysis," J. Radioanal. Chem., 37(1), 483-496 (1977).

Lingappan, K., H. Plaza, and K. B. Pedersen, "Mercury, Arsenic, Cadmium and Cobalt Determinations in the Average Puerto Rican Diet Using Instrumental Neutron Activation Analysis" in Proc. Int. Conf. Nucl. Methods Environ. Res., 2nd, J. R. Vogt and W. Meyer, Eds., CONF-740701, National Technical Information Service, U.S. Department of Commerce, Springfield, Virginia, July 1974, pp. 224-233.

Linhart, L., I. Rzymanova, and Z. Jesensky, "Vztah Konzumace Napoju Zaměstnancu Hutniho Prumyslu k Obsanu Kobaltu v Organismu" ["Relation Between Beverage Consumption of Employees in the Metallurgical Industry and the Content of Cobalt in the Body"], Cesk. Hyg., 22(3-4), 158-162 (1977).

Livingston, H. D., and G. Thompson, "Trace Element Concentrations in Some Modern Corals," Limnol. Oceanogr., **16**(5), 786-795 (1971).

Locke, J., D. R. Boase, and K. W. Smalldon, "The Quantitative Multi-Element Analysis of Human Liver Tissue by Spark-Source Mass Spectrometry," Anal. Chim. Acta, **104**(2), 233-244 (1979).

Lounamaa, J., "Trace Elements in Trees and Shrubs Growing on Different Rocks in Finland" in Geochemical Prospecting in Fennoscandia, A. Kvalheim, Ed., Interscience Publishers, New York, New York, 1967, pp. 287-320.

Lowman, F. G., Radionuclides in Plankton Near the Marshall Islands, 1956, U.S. Atomic Energy Commission, UWFL-54, Applied Fisheries Laboratory, University of Washington, Seattle, Washington, 1958; Chem. Abstr., **53**, 12534f (1959).

Lowman, F. G., and R. Y. Ting, "The State of Cobalt in Seawater and Its Uptake by Marine Organisms and Sediments" in Radioactive Contamination of the Marine Environment, Proc. Symp. IAEA, Seattle, Washington, July 10-14, 1972, International Atomic Energy Agency, Vienna, Austria, 1973, pp. 369-383.

Lowman, F. G., R. A. Stevenson, R. McClin Escalera, and S. Lugo Ufret, "The Effects of River Outflows Upon the Distribution Patterns of Fallout Radioisotopes in Marine Organisms" in Radioecological Concentration Processes, Proceedings of an International Symposium, Stockholm, April 25-29, 1966, B. Åberg and F. P. Hungate, Eds., Symposium Publications Division, Pergamon Press, Oxford, England, 1967, pp. 735-752.

Lucas, H. F., Jr., and D. N. Edgington, "Concentrations of Trace Elements in Great Lakes Fishes," J. Fish. Res. Board Can., **27**, 677-684 (1970).

Lulic, S., and P. Strohal, "Application of Neutron Activation Analysis in Studying the Marine Pollution Processes," Rev. Int. Oceanogr. Med., **33**, 119-123 (1974).

Lunde, G., "Trace Metal Content of Fish Meal and of the Lipid Phase Extracted from Fish Meal," J. Sci. Food Agric., **24**(4), 413-419 (1973).

Luoma, S. N., and E. A. Jenne, "Estimating Bioavailability of Sediment-Bound Trace Metals with Chemical Extractants" in Trace Substances in Environmental Health - X, D. D. Hemphill, Ed., Proceedings of the University of Missouri's 10th Annual Conference on Trace Substances in Environmental Health, Columbia, Missouri, June 8-10, 1976, University of Missouri, Columbia, Missouri, 1977, pp. 343-351.

Mahaffey, K. R., "Mineral Concentrations in Animal Tissues: Certain Aspects of FDA's Regulatory Role," J. Anim. Sci., **44**(3), 509-515 (1977).

Majori, L., G. Nedoclan, G. B. Modonutti, and F. Daris, "Study of the Seasonal Variations of Some Trace Elements in the Tissues of *Mytilus galloprovincialis* Taken in the Gulf of Trieste," Rev. Int. Oceanogr. Med., **49**, 37-40 (1978).

Malkus, Z., and J. Horacek, "Zur Migration von Fremdstoffen aus glasfaserverstarkten Polyesterharzen" ["Migration of Residues from Glass Fiber-Reinforced Polyester Resins"], Ernaehrungsforschung, **16**(4), 627-635 (1972).

McKenzie, R. M., "The Effect of Two Manganese Dioxides on Uptake of Lead, Cobalt, Nickel, Copper and Zinc by Subterranean Clover," Aust. J. Soil Res., **16**(2), 209-214 (1978).

Menke, H., C. Leszczynski, and M. Weber, "Comparative Study of X-Ray and Gamma-Ray Spectrometry Applied to Neutron Activation Analysis of Human Head Hair," Radiochem. Radioanal. Lett., **14**(3), 217-226 (1973).

Menzel, R. G., "Soil-Plant Relationships of Radioactive Elements," Health Phys., **11**, 1325-1332 (1965).

Mertz, D. P., R. Koschnick, G. Wilk, and K. Pfeilsticker, "Untersuchungen über den Stoffwechsel von Spurenelementen beim Menschen" ["Metabolism of Trace Elements by Humans"], Z. Klin. Chem., **6**, 171-174 (1968).

Miles, M. E., G. L. Sjoblom, and J. D. Eagles, "Environmental Monitoring and Disposal of Radioactive Wastes from U.S. Naval Nuclear-Powered Ships and Their Support Facilities, 1973," Radiat. Health Data Rep., 15(10), 625-646 (1974).

Miller, M. L., J. J. Fix, and P. E. Bramson, Radiochemical Analyses of Soil and Vegetation Samples Taken from the Hanford Environs, 1971-1976, BNWL-2249, National Technical Information Service, Department of Commerce, Springfield, Virginia, 1977.

Mitteldorf, A. J., and D. O. Landon, "Spectrochemical Determination of the Mineral-Element Content of Beef," Anal. Chem., 24, 469-472 (1952).

Morozov, N. P., S. A. Petukhov, A. S. Romanteeva, A. M. Mel'nikova, and G. S. Borisenko, ["Trace Elements in Fish from the Sea of Japan"], Rybn. Khoz. (Moscow), No. 1, 61-64 (1978); Chem. Abstr., 88, 119540h (1978).

Mudroch, A., and J. Capobianco, "Study of Selected Metals in Marshes on Lake St. Clair, Ontario," Arch. Hydrobiol., 84(1), 87-108 (1978).

Murthy, G. K., U. Rhea, and J. T. Peeler, "Levels of Antimony, Cadmium, Chromium, Cobalt, Manganese, and Zinc in Institutional Total Diets," Environ. Sci. Technol., 5(5), 436-442 (1971).

Nakahara, M., and F. A. Cross, "Transfer of Cobalt-60 from Phytoplankton to the Clam (*Mercenaria mercenaria*)," Nippon Suisan Gakkaishi, 44(5), 419-425 (1978).

Nakahara, M., T. Koyanagi, T. Ueda, and C. Shimizu, "Peculiar Accumulation of Cobalt-60 by the Branchial Heart of Octopus," Nippon Suisan Gakkaishi, 45(4), 539 (1979).

Navia, J. M., "Effect of Minerals on Dental Caries" in Dietary Chemicals vs. Dental Caries, Advances in Chemistry Series No. 94, American Chemical Society, Washington, D.C., 1970, pp. 123-160.

Nelson, D. J., and C. R. Malone, "Feeding Rates of Snails Determined with ^{60}Co" in Health Phys. Div. Ann. Progr. Report, for the Period Ending July 31, 1968, ORNL-4316, National Technical Information Service, U.S. Department of Commerce, Springfield, Virginia, 1968, pp. 122-124.

Nielson, K. K., Development of Proton-Induced X-Ray Emission Techniques with Application to Multielement Analyses of Human Autopsy Tissues and Obsidian Artifacts, Ph.D. Dissertation, Brigham Young University, Provo, Utah, 1975.

Oakes, T. W., K. E. Shank, C. E. Easterly, and L. R. Quintana, "Concentrations of Radionuclides and Selected Stable Elements in Fruits and Vegetables" in Trace Substances in Environmental Health - XI, D. D. Hemphill, Ed., Proceedings of the University of Missouri's 11th Annual Conference on Trace Substances in Environmental Health, Columbia, Missouri, June 7-9, 1977, University of Missouri, Columbia, Missouri, 1978, pp. 123-132.

Obrusnik, I., and V. Bencko, "Study on Trace Elements in Hair of Three Selected Population Groups in Czechoslovakia," Radiochem. Radioanal. Lett., **38**(3), 189-195 (1979).

Okamoto, K., Y. Yamamoto, and K. Fuwa, "Pepperbush Powder, a New Standard Reference Material," Anal. Chem., **50**(13), 1950-1951 (1978).

Olmez, I., M. C. Gulovali, G. E. Gordon, and R. I. Henkin, "Trace Elements in Human Saliva" in Trace Substances in Environmental Health - XII, D. D. Hemphill, Ed., Proceedings of the University of Missouri's 12th Annual Conference on Trace Substances in Environmental Health, Columbia, Missouri, June 6-8, 1978, University of Missouri, Columbia, Missouri, 1979, pp. 231-240.

Olson, B. H., V. P. Guinn, D. C. Hill, and M. Nassiri, "Effects of Land Disposal of Secondary Effluent on the Accumulation of Trace Elements in Terrestrial Ecosystems" in Trace Substances in Environmental Health - XII, D. D. Hemphill, Ed., Proceedings of the University of Missouri's 12th Annual Conference on Trace Substances in Environmental Health, Columbia, Missouri, June 6-8, 1978, University of Missouri, Columbia, Missouri, 1979, pp. 362-376.

Ördögh, M., "A Complex Neutron Activation Method for the Analysis of Biological Materials," J. Radioanal. Chem., **46**, 27-40 (1978).

O'Toole, J. J., R. G. Clark, K. L. Malaby, and D. L. Trauger, "Environmental Trace Element Survey at a Heavy Metal Refining Site," Nucl. Methods Environ. Res., Proc. Am. Nucl. Soc. Top. Meet., 172-182 (1971).

Papadopoulou, C., G. D. Kanias, and E. Moraitopoulou-Kassimati, "Stable Elements of Radioecological Importance in Certain Echinoderm Species," Marine Pollut. Bull., **7**(8), 143-149 (1976).

Papadopoulou, C., and G. D. Kanias, "Tunicate Species as Marine Pollution Indicators," Mar. Pollut. Bull., **8**(1), 229-231 (1977).

Patel, B., P. G. Valanju, C. D. Mulay, M. C. Balani, and S. Patel, "Radioecology of Certain Molluscs in Indian Coastal Waters" in Radioactive Contamination of the Marine Environment, Proc. Symp. Int. Radioactive Contam. Mar. Environ., Seattle, July 1972, International Atomic Energy Agency, Vienna, Austria, 1973, pp. 307-329.

Patel, B., S. Patel, M. C. Balani, and S. Pawar, "Flux of Certain Radionuclides in the Blood-Clam *Anadara granosa* Linneaus under Environmental Conditions," J. Exp. Mar. Biol. Ecol., **35**(2), 177-195 (1978).

Perkons, A. K., J. A. Velandia, and M. Dienes, "Forensic Aspects of Trace Element Variation in the Hair of Isolated Amazonas Indian Tribes," J. Forensic Sci., **22**(1), 95-105 (1977).

Perry, H. M., Jr., E. F. Perry, J. E. Purifoy, and J. N. Erlanger, "A Comparison of Intra- and Interhepatic Variability of Trace Metal Concentrations in Normal Men" in Trace Substances in Environmental Health - VII, D. D. Hemphill, Ed., Proceedings of the University of Missouri's 7th Annual Conference on Trace Substances in Environmental Health, Columbia, Missouri, June 12-14, 1973, University of Missouri, Columbia, Missouri, 1974, pp. 281-288.

Quin, B. F., and J. K. Syers, "Surface Irrigation of Pasture with Treated Sewage Effluent. III. Heavy Metal Content of Sewage Effluent, Sludge, Soil, and Pasture," N. Z. J. Agric. Res., **21**(3), 435-442 (1978).

Radosavljević, R., T. Tasovac, R. Drašković, M. Zarić, and V. Marković, "Complex Behaviour of Cobalt in the Danube River," Arch. Hydrobiol., Suppl., **44**(2), 241-248 (1973).

Ragaini, R. C., A. L. Langhorst, H. R. Ralston, and R. Heft, Instrumental Trace Element Analysis of California Market Milk, UCRL-51859, Lawrence Livermore Laboratory, University of California, Livermore, California, 1975, 27 pp.

Rancitelli, L. A., R. W. Perkins, and A. D. Renzetti, Jr., "The Multielement Analysis of Human Lung Tissue" in Pacific Northwest Laboratory Annual Report for 1968 to the USAEC Div. of Biol. and Med., Vol. II: Physical Sciences. Part 2. Radiological Sciences, D. W. Pearce, Ed., BNWL-1051, Part 2, UC-48, Battelle Memorial Institute, Pacific Northwest Laboratory, Richland, Washington, June 1969.

Reed, J. R., "Uptake and Excretion of ^{60}Co by Black Bullheads Ictalurus melas (Rafinesque)," Health Phys., **21**, 835-844 (1971).

Reed, J. R., Jr., N. A. Griffith, and C. H. Courtney, "Uptake and Excretion of ^{60}Co by Black Bullheads (Ictalurus melas)" in Health Phys. Div. Ann. Progr. Report, for the Period Ending July 31, 1968, ORNL-4316, National Technical Information Service, U.S. Department of Commerce, Springfield, Virginia, 1968, pp. 118-122.

Rehab, F. I., and A. Wallace, "Excess Trace Metal Effects on Cotton. 1. Copper, Zinc, Cobalt and Manganese in Solution Culture," Commun. Soil Sci. Plant Anal., **9**(6), 507-518 (1978a).

Persigehl, M., H. Schicha, K. Kasperek, and L. E. Feinendegen, "Behavior of Trace Element Concentration in Human Organs in Dependence of Age and Environment," J. Radioanal. Chem., 37, 611-615 (1977).

Persigehl, M., K. Kasperek, H. J. Klein, and L. E. Feinendegen, "Einfluss der Industrialisierung auf den Spurenelementgehalt in menschlichen Lungen" ["Influence of Industrialization on Trace Element Concentration in Human Lungs"], Beitr. Pathol., 157(3), 260-268 (1976).

Pesch, G., B. Reynolds, and P. Rogerson, "Trace Metals in Scallops from Within and Around Two Ocean Disposal Sites," Mar. Pollut. Bull., 8(10), 224-228 (1977).

Peterson, P. J., "Unusual Accumulations of Elements by Plants and Animals," Sci. Prog., Oxford, 59, 505-526 (1971).

Petkevich, T. A., "O Khimicheskom Elementarnom Sostave Planktonoyadnykh Ryb Chernogo Morya" ["Mineral Composition of Plankton-Eating Fish of the Black Sea"], Gidrobiol. Zh. Akad. Nauk Ukr. SSR, 1(6), 53-56 (1965); Chem. Abstr., 64, 16334c (1966).

Phillips, D. J., "Use of Biological Indicator Organisms to Monitor Trace Metal Pollution in Marine and Estuarine Environments. Review," Environ. Pollut., 13(4), 281-317 (1977).

Pihl, R. O., and M. Parkes, "Hair Element Content in Learning Disabled Children," Science, 198, 204-206 (1977).

Pillay, K. K. S., C. C. Thomas, Jr., and G. F. Mahoney, "Examination of Evidence Materials for Environmental Contamination Using Activation Analysis," J. Radioanal. Chem., 15(1), 33-39 (1973).

Proctor, P. D., and B. Sinha, "Cobalt-Nickel-Cadmium Mobilization, Transportation and Fixation in Surface Waters, Stream Sediments and Selected Aquatic Life in the Fredericktown Co-Ni Metallogenic Province, Southeast Missouri" in Trace Substances in Environmental Health - XII, D. D. Hemphill, Ed., Proceedings of the University of Missouri'2 12th Annual Conference on Trace Substances in Environmental Health, Columbia, Missouri, June 6-8, 1978, University of Missouri, Columbia, Missouri, 1979, pp. 119-128.

Rehab, F. I., and A. Wallace, "Excess Trace Metal Effects on Cotton: 2. Copper, Zinc, Cobalt and Manganese in Yolo Loam Soil," Commun. Soil Sci. Plant Anal., **9**(6), 519-527 (1978b).

Rehwoldt, R., D. Karimian-Teherani, and H. Altmann, "Measurement and Distribution of Various Heavy Metals in the Danube River and Danube Canal Aquatic Communities in the Vicinity of Vienna, Austria," Sci. Total Environ., **3**(4), 341-348 (1975).

Robertson, D. E., "Distribution of Cobalt in Oceanic Waters," Geochim. Cosmochim. Acta, **34**(5), 553-567 (1970).

Ronk, R. J., "Heavy Metals in United States Fish" in Resource Use and Quality Living, J. B. Trefethen, Ed., Transactions of the 36th North American Wildlife and Natural Resources Conference, March 7-10, 1971, Wildlife Management Institute, 1971, pp. 133-138.

Rossi, L. C., G. G. Mastinu, G. F. Clemente, and G. P. Santaroni, "Contenuto di Mercurio ed in Altri Elementi in Campioni di Tonno in Scatola" ["Content of Mercury and Other Elements in Samples of Canned Tuna Fish"], Riv. Sci. Tecn. Alim. Nutr. Um., **6**(2), 97-100 (1976).

Saas, A., P. Bovard, and A. Grauby, "Biologic Cycle of Manganese-54 and Cobalt-60 in a Soil," Kerntechnik, **13**(5), 230-232 (1971).

Sampaolo, A., L. Rossi, G. Esposito, L. Gramiccioni, S. Di Marzio, M. Piccinini, and S. Bonciani, ["Analytical Inquiry on Migration of Ceramic Material to Be Used in Contact with Food"], Rass. Chim., **25**(3), 249-264 (1973); Chem. Abstr., **79**, 145046k (1973).

Samsahl, K., D. Brune, and P. O. Wester, "Simultaneous Determination of 30 Trace Elements in Cancerous and Noncancerous Human Tissue. Samples by Neutron Activation Analysis," Int. J. Appl. Radiat. Isot., **16**, 273-281 (1965).

Schelenz, R., "Dietary Intake of 25 Elements by Man Estimated by Neutron Activation Analysis," J. Radioanal. Chem., **37**(2), 539-548 (1977).

Schlettwein-Gsell, D., and B. Mommsen-Straub, "Trace Elements in Food. II. Cobalt," Int. Z. Vitaminforsch., **40**, 673-683 (1970).

Schlettwein-Gsell, D., and H. Seiler, "Analysen und Berechnungen des Gehalts der Nahrung an Kalium, Natrium, Calcium, Eisen, Magnesium, Kupfer, Zink, Nickel, Cobalt, Chrom, Mangan und Vanadium in Altersheimen und Familien" ["Analyses and Calculations of Extent of Nutrition of Potassium, Sodium, Calcium, Iron, Magnesium, Copper, Zinc, Nickel, Cobalt, Chromium, Manganese, and Vanadium in Homes for the Aged and Family-Type Homes"], Mitt. Geb. Lebensmittelunters. Hyg., **63**(2), 188-206 (1972).

Schroeder, H. A., "Losses of Vitamins and Trace Minerals Resulting from Processing and Preservation of Foods," Am. J. Clin. Nutr., **24**, 562-573 (1971).

Schroeder, H. A., and A. P. Nason, "Trace Metals in Human Hair," J. Invest. Dermatol., **53**(1), 71-78 (1969).

Schroeder, H. A., A. P. Nason, and I. H. Tipton, "Essential Trace Metals in Man: Cobalt," J. Chron. Dis., **20**, 869-890 (1967).

Segar, D. A., J. D. Collins, and J. P. Riley, "The Distribution of the Major and Some Minor Elements in Marine Animals. Part II. Molluscs," J. Mar. Biol. Assoc. U.K., **51**, 131-136 (1971).

Seiler, H., D. Schlettwein-Gsell, G. Brubacher, and G. Ritzel, "Der Mineralstoffgehalt von Muskelfleisch in Abhangigkeit von der Zubereitungsart" ["Mineral Content of Muscular Substances Depending on the Type of Cooking"], Mitt. Geb. Lebensmittelunters. Hyg., **68**(4), 523-529 (1977).

Shimizu, M., T. Kajihara, I. Suyama, and Y. Hiyama, "Uptake of ^{58}Co by Mussel, Mytilus edulis," J. Radiat. Res., 12(1), 17-28 (1971).

Shklyaev, Yu. N., "Vliyanie Nekornevoi Podkormki Kobal'tom na Raspredelenie ^{60}Co i ^{14}C v Rasteniyakh Viki" ["Effect of Top Dressing with Cobalt-60 and Carbon-14 Distribution in Vetch Plants"], Agrokhimiya, No. 1, 115-118 (1978).

Shklyaev, Yu. N., and V. G. Minibaev, "Izmenenie Podvizhnosti Mikroelementov v Temnoseroi Lesnoi Pochve pod Gorokhim" ["Change in the Availability of Trace Elements in Dark Gray Forest Soil under Pea Plantings"] in Biol. Fiksatsiya Azota Atmos. Gorokhom Nekot. Puti Ee Povysh., S. M. Samosova, Ed., Izd. Kazan. Universitet., Kazan, USSR, 1969, pp. 94-98.

Shuman, M. S., L. A. Smock, and C. L. Haynie, Metals in the Water, Sediments and Biota of the Haw and New Hope Rivers, North Carolina, PB 272 629, National Technical Information Service, U.S. Department of Commerce, Springfield, Virginia, May 1977.

Sidle, R. C., J. E. Hook, and L. T. Kardos, "Accumulation of Heavy Metals in Soils from Extended Wastewater Irrigation," J. Water Pollut. Control Fed., 49(2), 311-318 (1977).

Siegel, F. R., Applied Geochemistry, Wiley-Interscience, a Division of John Wiley and Sons, Inc., New York, New York, 1974.

Siegesmund, K. A., A. Funahashi, and K. Pintar, "Identification of Metals in Lung from a Patient with Interstitial Pneumonia," Arch. Environ. Health, 28(6), 345-349 (1974).

Simkova, M., and F. Kukula, "Opredelenie Sledovykh Kolichestv Elementov v Tseilonskom Chae Metodom Aktivatsionnogo Analiza" ["Trace Element Determination in Ceylon Tea Using Activation Analysis"], Isotopenpraxis, 10(6), 219-223 (1974).

Slavin, S., G. E. Peterson, and P. C. Lindahl, "Determination of Heavy Metals in Meats by Atomic Absorption Spectroscopy," At. Absorpt. Newsl., **14**(3), 57-59 (1975).

Smedile, E., and G. Queirazza, "Uptake of Cobalt-60 and Cesium-137 in Different Components of a River Ecosystem Connected with Discharges of a Nuclear Power Station," Spec. Publ. - Ecol. Soc. Am., **1** (Radioecol. Energy Resour.), 1-8 (1976).

Somers, E., and D. M. Smith, "Source and Occurrence of Environmental Contaminants," Food Cosmet. Toxicol., **9**(2), 185-193 (1971).

Stack, M. V., A. J. Burkitt, and G. Nickless, "Characterization of Teeth by Trace Elements," Int. J. Forensic Dent., **2**(5), 62-65 (1974).

Straughan, I., A. A. Elseewi, and A. L. Page, "Mobilization of Selected Trace Elements in Residues from Coal Combustion with Special Reference to Fly Ash" in Trace Substances in Environmental Health - XII, D. D. Hemphill, Ed., Proceedings of the University of Missouri's 12th Annual Conference on Trace Substances in Environmental Health, Columbia, Missouri, June 6-8, 1978, University of Missouri, Columbia, Missouri, 1979, pp. 389-402.

Sumino, K., K. Hayakawa, T. Shibata, and S. Kitamura, "Heavy Metals in Normal Japanese Tissues. Amounts of 15 Heavy Metals in 30 Subjects," Arch. Environ. Health, **30**(10), 487-494 (1975).

Tanner, J. T., and M. H. Friedman, "Neutron Activation Analysis for Trace Elements in Foods," J. Radioanal. Chem., **37**, 529-538 (1977).

Templeton, W. L., "Radioactive Wastes . . . Disposal of Liquid Wastes into Coastal Waters," Chem. Eng. Prog., **66**(2), 45-50 (1970).

Templeton, W. L., J. M. Dean, D. G. Watson, and L. A. Rancitelli, "Freshwater Ecological Studies in Panama and Colombia," Bioscience, 19(9), 804-808 (1969).

Teraoka, H., and J. Kobayashi, "Chemical Investigation on Minerals in Human Hair. I. Analytical Method for 25 Elements in Hair Samples from Various Parts of Japan," Nippon Eiseigaku Zasshi, 32(4), 574-587 (1977).

Thiers, R. E., J. F. Williams, and J. H. Yoe, "Separation and Determination of Millimicrogram Amounts of Cobalt," Anal. Chem., 27, 1725-1731 (1955).

Thomas, B., J. A. Roughan, and E. D. Watters, "Cobalt, Chromium and Nickel Content of Some Vegetable Foodstuffs," J. Sci. Food Agric., 25(7), 771-776 (1974).

Thomas, C. W., D. L. Reid, and L. F. Lust, Radiochemical Analysis of Marine Biological Samples Following the "Redwing" Shot Series - 1956, U.S. Atomic Energy Commission, HW-58674, Hanford Atomic Products Operation, Richland, Washington, 1958; Chem. Abstr., 55, 5152g (1961).

Timofeeva-Resovskaya, E. A., N. V. Timofeeva-Resovskii, and E. A. Gileva, "O Spetsificheskikh Nakopitelyakh Otdel'nykh Radioizotopov Sredi Presnovodnykh Organizmov" ["Specific Accumulators of Individual Radioisotopes in Fresh-Water Organisms"], Dokl. Akad. Nauk SSSR, 140, 1437-1440 (1961); Chem. Abstr., 56, 7658d (1962).

Timofeeva-Resovskaya, E. A., "Raspredelenie Radioizotopov po Osnovym Komponentam Presnovodnykh Vodoemov" ["Distribution of Radioisotopes in Principal Components of Freshwater Reservoirs"], Tr. Inst. Biol. Akad. Nauk., SSSR Ural. Fil., 30, 3-77 (1963).

Tipton, I. H., and M. J. Cook, "Trace Elements in Human Tissue. Part II. Adult Subjects from the United States," Health Phys., 9, 103-145 (1963).

Tipton, I. H., and J. J. Shafer, "Statistical Analysis of Lung Trace Element Levels," Arch. Environ. Health, **8**, 58-67 (1964).

Tipton, I. H., P. L Stewart, and P. G. Martin, "Trace Elements in Diets and Excreta," Health Phys., **12**(12), 1683-1689 (1966).

Tipton, I. H., H. A. Schroeder, H. M. Perry, Jr., and M. J. Cook, "Trace Elements in Human Tissue. Part III. Subjects from Africa, the Near and Far East and Europe," Health Phys., **11**, 403-451 (1965).

Tipton, I. H., P. L. Stewart, and J. Dickson, "Patterns of Elemental Excretion in Long Term Balance Studies," Health Phys., **16**, 455-469 (1969).

Tjioe, P. S., J. J. M. De Goeij, and J. P. W. Houtman, "Extended Automated Separation Techniques in Destructive Neutron Activation Analysis; Application to Various Biological Materials, Including Human Tissues and Blood," J. Radioanal. Chem., **37**, 511-522 (1977).

Tong, S. C., W. H. Gutenmann, D. J. Lisk, G. E. Burdick, and E. J. Harris, "Trace Metals in New York State Fish," N.Y. Fish Game J., **19**(2), 123-131 (1972).

Tracor-Jitco, Inc., Scientific Literature Reviews on Generally Recognized as Safe (GRAS) Food Ingredients - Vitamin B_{12}, PB-241 966/1GA, National Technical Information Service, U.S. Department of Commerce, Springfield, Virginia, 1974.

Trapeznikov, A. V., and V. N. Trapeznikova, "Cobalt-60 Accumulation by Freshwater Plants under Natural Conditions," Ekologiya (Sverdlovsk), No. 2, 104-106 (1979).

Vaganov, P. A., V. D. Kulikov, and I. V. Shtangeeva, "Biogeokhimicheskie Kharakteristiki Vodoroslei Barentseva Morya (po Rezul'tatam Neitronno-Aktivatsionnogo Analiza)" ["Biogeochemical Characteristics of Barents Sea Algae (Based on Results of Neutron-Activation Analysis)"], Geokhimiya, No. 11, 1740-1745 (1978).

Valkovic, V., Trace Elements in Human Hair, Garland STPM Press, New York, 1977.

Van As, D., H. O. Fourie, and M. Vleggaar, "Accumulation of Certain Trace Elements in Marine Organisms from the Sea Around the Cape of Good Hope" in Proc. Symp. Int. Radioactive Contam. Mar. Environ., Seattle, July 1972, International Atomic Energy Agency, Vienna, Austria, 1973, pp. 615-623.

Vanderploeg, H. A., D. C. Parzyk, W. H. Wilcox, J. R. Kercher, and S. V. Kaye, Bioaccumulation Factors for Radionuclides in Freshwater Biota, ORNL-5002 UC-11, Oak Ridge National Laboratory, Oak Ridge, Tennessee, 1975.

Van Weers, A. W., "Uptake and Loss of ^{65}Zn and ^{60}Co by the Mussel *Mytilus edulis* L." in Radioactive Contamination of the Marine Environment, Proc. Symp. Int. Radioactive Contam. Mar. Environ., Seattle, July 1972, International Atomic Energy Agency, Vienna, Austria, 1973, pp. 385-401.

Van Weers, A. W., "Uptake of Cobalt-60 from Sea Water and from Labelled Food by the Common Shrimp *Crangon crangon* (L.)," Impacts Nucl. Releases Aquat. Environ., Proc. Int. Symp., 349-361 (1975).

Vorob'eva, A. I., and E. V. Osmolovskaya, "O Balanse Molibdena i Kobal'ta v Organizme Detei Doshkol'nogo Vozrasta" ["Molybdenum and Cobalt Balance in Preschool-Age Children"], Gig. Sanit., **35**(11), 108-109 (1970).

Wallace, A., E. M. Romney, and J. Kinnear, "Frequency Distribution of Several Trace Metals in 72 Corn Plants Grown Together in Contaminated Soil in a Greenhouse," Commun. Soil Sci. Plant Anal., **8**(9), 693-697 (1977).

Watling, H. R., and R. J. Watling, "Trace Metals in *Chromytilus meridionalis*," Marine Pollut. Bull. (G.B.), **7**, 91-94 (1976).

Wester, P. O., "Concentration of 24 Trace Elements in Human Heart Tissue Determined by Neutron Activation Analysis," Scand. J. Clin. Lab. Invest., **17**(4), 357-370 (1965).

Wilhm, J. L., "Transfer of Radioisotopes Between Detritus and Benthic Macroinvertebrates in Laboratory Microecosystems," Health Phys., **18**(3), 277-284 (1970).

Willard, W. K., Radiocobalt Cycling in a Small Mammal Food Web, CONF-750503, National Technical Information Service, Department of Commerce, Springfield, Virginia, 1975.

Wiser, C. W., and D. J. Nelson, "Uptake and Elimination of Cobalt-60 by Crayfish," Am. Midland Naturalist, **72**, 181-202 (1964).

Wolfe, D. A., and T. R. Rice, "Cycling of Elements in Estuaries," Fish. Bull., **70**(3), 959-972 (1972).

Wolff, H., "Emissionsspektrographische Untersuchungen über den Kobaltspiegel des Serums" ["Emission Spectrographic Study of Cobalt Level of Serum"], Klin. Wochschr., **28**, 280 (1950).

Wolstenholme, W. A., "Analysis of Dried Blood Plasma by Spark Source Mass Spectrometry," Nature, **203**, 1284-1285 (1964).

Wolverton, B. C., and R. C. McDonald, Water Hyacinths and Alligator Weeds for Removal of Silver, Cobalt, and Strontium from Polluted Waters, TM-X-72727, NASA, National Space Technology Laboratories, Bay St. Louis, Mississippi, May 1975.

Yagodin, B. A., and I. V. Tishchenko, "Soderzhenie Mikroelementov Tsinka i Kobalta v Pochve i Rasteniyakh v Zavisimosti ot Primenyaemykh Udobrenii" ["Content of Zinc and Cobalt Trace Elements in the Soil and in Plants in Relation to Fertilizer Application"], Vestn. S-kh. Nauki (Moscow), No. 3, 42-50. (1978).

Yamagata, N., and Y. Murakami, "A Cobalt-Accumulator Plant, Clethra barbinervis Sieb. et Zucc.," Nature (London), 181, 1808-1809 (1958).

Yamagata, N., W. Kurioka, and T. Shimizu, "Balance of Cobalt in Japanese People and Diet," J. Rad. Res., 4, 8-15 (1963).

Yurachek, J. P., G. G. Clemena, and W. W. Harrison, "Analysis of Human Hair by Spark Source Mass Spectrometry," Anal. Chem., 41, 1666-1668 (1969).

Zenoble, O. C., J. A. Bowers, G. Seaman, and S. D. Howe, "Comparing Methods of Analyses for Selected Trace Elements in Turkey Muscle," J. Food Sci., 42(3), 847-848 (1977).

VIII. PHYSIOLOGICAL EFFECTS

Betty L. Herndon,* Robert A. Jacob,** and Joy McCann***

A. Pharmacokinetics

1. Absorption: Absorption of cobalt and iron is unique in that they are the only trace metals absorbed better by the intestine in an organically bound form (Fletcher, 1975). All the other trace metals can be absorbed better by the intestine in their free ionic state.

a. Humans: Human studies (confirmed by animal measurements) indicate that cobalt absorption is relatively sparce from the gastrointestinal tract at usual levels of intake. Valberg (1972) estimated 44%, and Smith et al. (1972) reported ranges of 5 to 20% and more. The percentage of absorption most frequently cited in the literature is the range of Harp and Scoular (1952; cited by Cass and Lange, 1955) who measured cobalt balance in 23 young adult women on self-selected normal American diets. From 73 to 97% of a daily individual intake of 5.6 to 8.0 μg cobalt was absorbed. The cobalt quantification in this study used a colorimetric method with β-nitroso-α-naphthol.

More recent studies have used neutron activation, atomic absorption, emission spectrography, x-ray fluorescence, and polarographic analysis to quantitate cobalt. These improvements in equipment have often corroborated the older (colorimetric) data. The average dietary daily intake of cobalt in three studies ranged from 170 to 470 μg/day. The average absorption of cobalt, however, was quite similar to the 1952 Harp and Scoular study. On the basis of Tipton and Stewart's 1970 data and a definition of absorption as urinary excretion plus dietary balance divided by intake, cobalt absorption averaged 83%.

The absorption of the divalent metal cation cobalt apparently occurs by the same mechanism by which the body absorbs iron, and absorption of the two is mutually inhibiting (Valberg et al., 1972). The absorption of cobalt was increased in iron-deficient humans, but these studies did not find diminished absorption of cobalt in iron-loaded patients. The rat, on the

---

* Parts A, B, and C.1

** Part C.2

*** Part D

other hand, shows changes in cobalt absorption in both iron-overloaded and iron-depleted situations (Schade et al., 1970).

Absence of a requirement for cobalt in either ionic or loosely bound chemical form contrasts with the body's need for iron, however; and the absorbed cobalt is rapidly lost from the body. There is a direct relationship between the amount of cobalt absorbed following an oral dose and the amount that is excreted in the urine. The cobalt-iron relationship is therefore an index of iron absorption and has been used clinically as a method for detecting iron deficiency.

Absorption of cobalt from normal diets in 11 infants and children was calculated in balance studies by Alexander et al. (1972). With an average daily intake of 4.44 μg/kg, 71% was absorbed. Schroeder et al. (1967) cited the reviews of Howells, which report that an average of 70% of the ingested cobalt appears in urine; therefore, at least this amount has been absorbed.

Paley and Sussman (1963) fed radioactive cobaltous chloride to human subjects and found that 1 to 16% of the dose appeared in urine by 48 hr. More of the $^{60}CoCl_2$ was absorbed by fasting subjects and with higher doses, as well as when albumen was added to the diet.

Smith et al. (1972) gave $^{60}CoCl_2$ either by mouth or intravenously to 24 adult volunteers and made measurements on absorption and retention for as long as 1,018 days. Absorption after an oral dose ranged from 1 to 28%, depending on the size of the dose and the presence or absence of food in the stomach. About 9 to 16% of the absorbed oral or intravenous dose was retained in the body for a prolonged period; elimination rate had a half-life of about 2 years. Techniques for quantitating retention of the $^{60}CoCl_2$ included whole body counting, counts over liver and knee specifically, and blood radioactivity measures. The authors estimated that liver cobalt represented, on the average, one fifth of total body cobalt content.

Absorption of ^{58}Co through the intact skin of the finger, forehead, and scalp area was reported in a scientist who was accidentally contaminated (Suzuki-Yasumoto and Inaba, 1976). Urinary and fecal assays were made daily for 3 weeks, and the half-life of the cobalt in this subject was estimated to be 17.5 days. The activity (no quantities were reported) in the urine decreased exponentially while the fecal cobalt radioactivity decreased by a power function; this indicated to the authors that the nuclide had been incorporated into the body or, in fact, absorbed.

A normal volunteer inhaled a gaseous mixture of several radioelements, and the absorption and excretion of ^{60}Co was measured for a 6-year period (Delpla et al., 1976). ^{60}Co was present in the breathing air at 3.10 $\mu Ci/m^3$; for 16 min, doing work at a level of 70 W (70 J/sec), the subject inhaled air at the rate of 1 m^3/hr. He incorporated 0.89 μCi ^{60}Co, as determined by whole-body counting. Distribution of ^{60}Co, measured 48 hr after inhalation dose, showed the following: head, < 0.001 μCi; abdomen, 0.046 μCi; thorax, 0.370 μCi; and legs, 0.004 μCi.

b. Animals: The species variation in cobalt absorption is well-known, and there has been no single best animal model of human absorption described. Many species and many techniques have been used in cobalt absorption studies; whole body radioisotope counting has been used in the rat (Pollack et al., 1965) and chicken (Suso and Edwards, 1969). Tissue sampling and analysis after dosing is the most common. Forth and Remmel (1971; cited by Salminen et al., 1975) measured absorption in the isolated intestine.

Cattle, sheep, and goats show a cobalt deficiency syndrome which is relieved by oral cobalt salts (Cass and Lange, 1955). Horses in the same cobalt-deficient pasture maintained good health. Comar and Davis (1947) found that absorption of $^{60}CoCl_2$ was slower in ruminants, presumably due to binding by ruminal microflora prior to incorporation into vitamin B_{12}.

A similar binding by cecal flora also occurs in chickens; absorption of ^{60}Co by the oral route was only 13.3%, but was 98.7% when given intramuscularly to Leghorn chicks (Suso and Edwards, 1969). Salminen et al. (1975) measured radioactive cobalt chloride absorption in 58 male and female domestic chickens after both single and multiple oral doses of 35 μg Co/kg. Dissection and counting of organ and tissue activity was done, and the fractional amount of ^{60}Co recovered was taken to indicate the absorption of cobalt from the intestinal tract. There was great individual variation in percent absorbed, but fasting uniformly increased absorption. After the 36-hr fasting, the 24-hr tissue recovery was two- to threefold higher than it was after the 12-hr fasting. The liver received the greatest fraction of the cobalt absorbed.

In experiments on oral, dermal, and injected cobalt salts in guinea pigs and rats, Suzuki-Yasumoto and Inaba (1976) found that absorption varied with the different chemical forms. The order of absorption was $CoCl_2$ > Co tartrate > $Co(OH)_2$.

Merkulova et al. (1969) also found that the form of chemical bond of cobalt in its compounds determined the rate of absorption and incorporation into metal-protein complexes. The absorption of cobalt chloride in the dog was only 6% in 2 hr and that of the glycolate salt about 28% in the Heidenheim stomach model. All investigated cobalt salts were absorbed better in the lower small intestine in this model, which is the main site for metal absorption in most species. Grace (1975) reported, however, that in sheep at least, cobalt is also absorbed in the colon. Without identifying the species, Schroeder et al. (1967) suggests that oral cobalt is absorbed in the jejunum, based on the known presence of an intestine-liver pathway but not intestine-to-pancreas absorption.

Intact skin is a rather good barrier to cobalt absorption in dermally treated guinea pigs. Only 1.72% of a dose of radioactive cobalt appeared in the urine and 0.9% in the liver when the skin was intact; but those values rose to 41.4 and 4.82%, respectively, when the guinea pig skin was abraded before ^{60}Co application (Suzuki-Yasumoto and Inaba, 1976).

Wahlberg (1971) also measured the absorption through unbroken guinea pig skin of cobaltous chloride applied as a patch test and mixed with either petrolatum or with distilled water. A 2% concentration was absorbed better mixed with water than with petrolatum, but there was not much difference in absorption between the two carriers when the concentration was 5.69%. The data are given in Table VIII-1.

TABLE VIII-1

SKIN ABSORPTION OF COBALT BY GUINEA PIGS
(Wahlberg, 1971)

Cobalt Concentration	Carrier	Absorption, μg Co.cm^{-2} hr^{-1}
2.0%, 1 ml	distilled water	3.0-5.0 (10 trials)
2.0%, 1 ml	petrolatum	1.1-3.7 (10 trials)
5.69, 1 ml	distilled water	9.0-12.8 (10 trials)
5.69, 1 ml	petrolatum	8.7-14.5 (10 trials)

Many cobalt absorption studies have been performed on rats. Using urinary excretion as a measure of absorption from gut into the bloodstream, Hatfield (1970) found that inorganic cobalt was poorly absorbed from the gut. Orally dosed inorganic radioactive cobalt appeared in urine at less than 0.01% of the dose, whereas 5% of a soluble salt (not identified) given to rats orally was found in urine.

Pollack et al. (1965) gave Carworth Farms rats iron deficiency by either repeated bleedings or an iron-deficient diet, then dosed them with an isotope-labeled trace metal cation to check for absorption in the iron-deficient state. Approximately 40 rats were used for each cation: 10 for the bleeding experiment and its 10 controls, and 10 for the dietary experiments and 10 controls on a normal diet. The absorption of both cobalt and iron increased highly significantly ($p < 0.001$) in both bled and iron-deficient diet rats when compared to their paired controls. Manganese was also absorbed more in both groups: ($p < 0.05$ in the bled rats, $p < 0.02$ in the iron-deficient diet rats). The authors suggest that these three cations share a transport pathway based on their chemical similarities: similar ionic radii and the tendency to coordinate with six ligands and form octahedral complexes. Cobalt, iron, and manganese absorption in the closed intestinal loop of the rat was also measured by Sheehan (1974). His studies with radiolabeled metals indicate that they are absorbed by a pathway unrelated to that of soluble molecules. Ogawa (1974) noted an important effect of food on the absorption, distribution, and elimination of orally dosed $^{60}CoCl_2$ in rats and mice. Fasted rodents retained all dosed metal; but when the animals were fed, normal elimination occurred. In the iron-deficient rat, cobalt is distributed subcellularly mostly in the soluble fraction of the cell, but additions of cobalt will interfere with iron absorption in the particulate fraction of the cell (Thomson and Valberg, 1972). Schade (1970) also reported that the simultaneous administration of iron and cobalt reduces the absorption of cobalt. In general, it appears that the absorption of iron and cobalt appears to be mediated by a transport system in which two processes operate simultaneously; the first limited largely by the concentration of available metal in the intestine and the second, on a process or mechanism that is active and displays saturation kinetics and competitive inhibition (Sheehan and Finkel, 1972; cited in Thomson and Valberg, 1972).

An overall observation on experimental oral cobalt dosing indicates that the actual percentage absorption depends on the size of the dose, i.e., whether cobalt was given to experimental animals in trace amounts (fractions of micrograms) or carrier amounts (over 100 μg). After oral administration of a carrier dose, blood levels decrease exponentially with a biological half-life of about half a day (Carlberger, 1961, reported 10 hr). With a trace oral dose, the time course for the initial, rapidly dropping phase of the disappearance curve takes only about half as long. The second phase of the disappearance curve is not as dose dependent; both curves become exponential with a biological half-life of about 13 hr (Carlberger's data).

When the cobalt dose is injected instead of given orally, uptake occurs into primarily the liver, kidney, and pancreas, regardless of dose. Liver and kidney were also high in uptake of orally administered cobalt.

c. Vitamin B_{12}-related absorption: All mammals including man require for hemoglobin synthesis a cobalt-containing tetrapyrrolic ring, cyanocobalamin (vitamin B_{12}), which occurs in the normal adult at a level of about 5 mg. This term is a misnomer due to the fact that the linkage of the 5'-deoxyadenosyl group and cobalt is unstable. In the original isolation of vitamin B_{12}, cyanide was used for the linkage to yield a stable product of cyanocobalamin.

A voluminous literature exists on the absorption and metabolism of vitamin B_{12} (cf. Tracor-Jitco, Inc., 1974; FASEB, 1977). Although other forms of this compound exist *in vivo*, the term is used for all the cobalt-containing vitamins found in man (FASEB, 1978). Dietary standards list the recommended daily intake of B_{12} to be 5 μg, which corresponds to 0.2 μg of cobalt.

Vitamin B_{12} is absorbed from the ileum, but its absorption is dependent upon the presence of hydrochloric acid and the intrinsic factor, a mucopolysaccharide secreted by the parietal cells of the gastric mucosa. Free vitamin B_{12} is bound to the intrinsic factor in the proportion of 2 mol cobalamin to 1 mol intrinsic factor dimer (Harper et al., 1977). In the presence of this factor, B_{12} absorption is about 70%; but without it, less than 2%. The intrinsic factor which is necessary for cobalt-containing B_{12} absorption has not been completely characterized, but the glycoprotein may exist as a monomer (molecular weight 55,000 to 60,000) and a dimer (molecular weight 110,000 to 120,000). The intrinsic factor-B_{12} complex is carried by peristalsis to the ileum and normally survives this transit without being digested. Receptors in the epithelial cell surface (microvillous membranes and brush borders) of the ileum are specific for the intrinsic factor-vitamin B_{12} complex (Tracor-Jitco, Inc., 1974).

It is current belief that the intrinsic factor has two receptor sites, one for vitamin B_{12} and the other for ileal intestinal microvilli, the latter requiring calcium ions and a neutral pH. The intestinal-binding receptor site is readily saturated, which results in limiting the absorption of vitamin B_{12} to about 1.5 μg after any single dose. The composition of the receptor for the intrinsic factor-vitamin B_{12} complex is unknown, but the type of attachment has been suggested to resemble that of a mucopolysaccharide blood group antigen and a γ-globulin antibody against that blood group (FASEB, 1978). Goodhart and Shile (1973; cited in FASEB, 1978) have indicated that immature red cells, reticulocytes, in common with the intestine possess "receptor sites" on their surface that are specific for the B_{12}-glycoprotein complex.

The intrinsic factor has not been detected in plasma, so it is assumed that it is released within the intestine, liberating B_{12} to pass into the intestinal mucosal cell. It has also been suggested that the intrinsic factor protects vitamin B_{12} peptide against bacterial attack and that it is in the peptide form that the vitamin is absorbed (Tracor-Jitco, Inc., 1974).

Both very large doses of vitamin B_{12} and intramuscular injection of 10 to 25 μg B_{12} produce similar rises in serum concentration of the vitamin, which demonstrates that the vitamin can be effective without the intrinsic factor. Apparently the only function of the intrinsic factor is to provide for the absorption of the vitamin from the intestine, and then only when it is present in very small amounts, e.g., in foods.

The cobalt absorption mechanism is functional in both animals and man; in fact, there is an indication of species specificity, and humans given animal intrinsic factor usually develop antibodies (Davies, 1972). Guinea pig experiments have shown that the intrinsic factor-vitamin B_{12} complex is split at the brush border membrane, and only the vitamin B_{12} gains access into the mitochondrion (Peters and Hoffbrand, 1970; cited by FASEB, 1978). Tennant et al. (1971; cited by FASEB, 1978) showed that gut bacteria were not involved in B_{12} absorption mechanics in nonruminant animals.

As soon as vitamin B_{12} is passed through the ileum cell wall, it is bound to one of three globulins as it enters the portal system (Hall and Finkler, 1968; cited by Davis, 1979). These binding proteins, transcobalamin I and II (and possibly III) transport or store the cobalt-containing vitamin and are discussed further in subsection VIII.A.2, "Distribution."

Human absorption of cobalt by the intestine ranges from 6 to 72% of the dose as measured by radioactive ^{60}Co incorporated into vitamin B_{12} (Holmes and Bell, 1961, and Weisberg and Glass, 1966; both cited by Snyder et al., 1975). Fish et al. (1973) used ^{60}Co microspheres to measure free and intrinsic-factor bound B_{12} absorption in humans. Thirteen normal subjects were orally given bound B_{12} and free B_{12} attached to different cobalt isotopes. By stool measurements, Fish found an average of 57% free and 66% bound B_{12} had been absorbed.

Within limits, diet appears to have little effect on human B_{12} absorption: Lindenbaum and Lieber (1969) saw a reduction in B_{12} absorption after 2 weeks substituting ethyl alcohol isocalorically for carbohydrate, but protein had little effect.

The absorption, distribution, and excretion of ^{58}Co-cyanocobalamin in rats after both oral and intravenous administration were found to be quite different from the parameters measured for $CoCl_2$ (^{60}Co) (Nishimura et al., 1976). Retention of cyanocobalamin was greater for rats of all ages; the biological half-life of the cobalamin after oral dosage was estimated to be 25 days and that of $CoCl_2$, $\sim$ 10 days. The authors conclude that intestinal absorption (not endogenous excretion) is responsible for the differences.

2. Distribution

a. Humans: Following oral administration of cobalt or its presence in the diet, cobalt is distributed over the body. The metal occurs at a high level in the liver, presumably due to its relationship to the portal circulation; studies have analyzed many human tissues and have found the body burden is highest in that organ (Forbes et al., 1954, and Smith et al., 1972). In the Smith study, employing intravenous administration of radioactive $CoCl_2$, the liver was estimated to hold about one-fifth of the total body burden of cobalt. On the other hand, analysis of human organs in an older study, where the cobalt source was the diet, found the highest cobalt levels in the human pancreas and traces in all other organs except liver, where no cobalt was detected (Dutroit and Zbinden, 1930; cited by Cass and Lange, 1955). Yamagata et al. (1963) used neutron activation analysis and calculated the whole body content of cobalt as 1.1 mg; 43% was stored in muscle, 14% in bone, and the remainder in the soft tissue. The distribution in this study, also, was apparently from normal dietary sources. Bertrand and Macheboeuf (cited by Cass and Lange, 1955) in a 1925 study that analyzed for cobalt in various organs of cattle and man with dimethylglyoxime colorimetric reagent, found human spleen and pancreas contained the highest levels with 0.47 and 0.35 mg Co/kg wet weight, respectively. Human liver and kidney both contained 0.25 mg/kg.

Red cells carry a large proportion of the cobalt in man; Schroeder et al. (1967) reported a range of 0.07 to 0.36 μg Co/liter (mean 0.18) of blood. Clemente et al. (1977) found a range of 0.8 to 11.4 μg Co/liter (mean 2.4) in the blood of 40 male and female Italian adults. They noted the large difference between their work and Schroeder's American population but found no explanation for it.

Kriss et al. (1955) reported cobalt blood levels in a 6-year-old boy receiving 40 to 60 mg $CoCl_2$ orally for anemia. The blood levels during dosing were: 195 μg/100 ml blood; 1 week after discontinuing dosing, 40 μg/100 ml; and 4 weeks after discontinuing dosing, blood cobalt was undetectable.

Tipton and Cook (1963) analyzed several normal human tissues for the presence of cobalt by emission spectrography. The cobalt content, in general, was highest in the gut, which could reflect passage of unabsorbed dietary cobalt. Of tissues that would be receiving absorbed cobalt, the liver and adrenal were high. Their data are summarized in Table VIII-2.

TABLE VIII-2

COBALT DISTRIBUTION IN HUMAN ORGANS
(Tipton and Cook, 1963)

Cobalt Concentration, mg/kg ash weight				
0-2	1-4	3-6	5-8	Over 8
Esophagus	Aorta	Adrenal	Omentum	Cecum
Brain	Heart	Diaphragm		Colon
Larynx	Duodenum	Jejunum		
Ovary	Pancreas	Ileum		
	Prostate	Rectum		
	Spleen	Kidney		
	Stomach	Liver		
	Testis	Lung		
	Thyroid	Muscle		
	Trachea	Skin		
	Uterus	Urinary bladder		

Nagata (1978) measured cobalt transport *in vivo* by quantitating (atomic absorption) the metal content of tissues of humans that had received artificial joints. Cobalt (and other major metal components of these prostheses) are transported through the circulation and accumulate, especially in the spleen, according to this study. It has been noted by Bala et al. (1969) that the cobalt content of the spleen normally decreases with age.

The concentration of wear products in the human body after implantation of cobalt-containing prosthetic devices has been measured by several groups (Coleman et al., 1972; Coleman et al., 1973; Lux et al., 1975). These are discussed later in the review in the subsection VIII.C.1.a.(1) on "Accidental and Intentional Exposure."

Forbes et al. (1954) measured the cobalt content in the organs of a 46-year-old man who died accidentally. The fresh tissue concentrations from their report were given in Table VII-21. Liver and skin contained 0.056 mg Co/kg and 0.050 mg Co/kg for the highest organ concentrations.

Four hospitalized volunteers were administered intramuscular doses of $^{60}CoCl_2$ with a specific activity of 50.6 μCi/μg and a dose range of 415 to 690 μg approximately 1 to 2 days before death. In the two with adequate kidney clearance, the authors describe results "in agreement with animal experiments," and the liver and kidneys were found to contain 42 and 3.2% of the dose, respectively, in one subject and 45 and 4.4% of the dose in the other (Lindgren and Salmi, 1968). Parr and Taylor (1964) measured cobalt concentrations in the tissues of human adults and reported a low of 0.003 μg/g in serum to a high of 0.07 μg/g wet weight of liver.

b. Animals: As a general statement, cobalt given either parenterally or orally does not appear to accumulate in a specific target organ in any species. Higher concentrations are usually found in liver and kidneys, with some of the species and experimental conditions demonstrating cobalt retention in adrenals, thyroid, pancreas, and spleen. After normal dietary intake, the distribution is similar: kidney, liver, pancreas, and thymus were highest in cobalt content; muscle and nervous tissue receives/deposits much lower levels (Masoero, 1972).

(1) Pigs: Some unique studies have quantitated cobalt distribution in animals receiving a normal diet. Ascaris worms were taken from the gut of recently killed pigs and were analyzed for cobalt content. Seven determinations by neutron activation analysis showed 0.39 ppm $\pm$ 0.36 mg Co/kg (Ince, 1976). This is in the range of the gut levels of cobalt in rats fed a normal diet and analyzed by similar methods (Maziere et al., 1977), which indicates the worms are probably not concentrating the trace metal. Distribution of labeled cobalt to the organs of swine is given in Table VIII-3.

(2) Cattle: By far the most studies have dosed animals with cobalt to study distribution of the trace metal to various organs. Comar and Davis (1947) measured the distribution of radioactive cobalt in cattle following both oral dosage and injection. Twice weekly oral doses of 100 μg cobalt for a month produced little tissue distribution, but when the same amount was injected intravenously, cobalt was found in all tissues and organs. Transplacental distribution to fetal calf

TABLE VIII-3

DISTRIBUTION OF LABELED COBALT ADMINISTERED TO SWINE [a/]
(Comar and Davis, 1947)

Tissue	30 μg Radiolabeled Co, Intravenously, Killed After 23.5 Hr	60 μg Radiolabeled Co, Orally, Killed After 4 Days
	% of dose	% of dose
Pituitary	0	0
Thyroid	0.0044	0
Thymus	0.011	0.0043
Adrenals	0.012	0.00067
Reproductive organs	0.096	0.0087
Brain	0.015	0.0022
Eye	0.0033	0.00072
Heart	0.14	0.017
Blood	1.48	0.065
Lung	0.45	0.037
Trachea	0.012	0.0054
Kidney	0.76	0.072
Bladder and urine	0.017	0.0014
Tongue	0.049	0.0034
Esophagus	0.061	0.0064
Stomach (average of six sites)	0.45	0.029
Stomach contents	0.085	0.0095
Small intestine (average of six sites)	1.7	0.070

TABLE VIII-3 (concluded)

Tissue	30 μg Radiolabeled Co, Intravenously, Killed After 23.5 Hr	60 μg Radiolabeled Co, Orally, Killed After 4 Days
	% of dose	% of dose
Small intestine contents	0.35	0.068
Large intestine (three sites)	0.64	0.099
Pancreas	0.20	0.019
Spleen	0.068	0.0067
Liver	3.7	0.60
Gall bladder	0.0052	0
Bile	0.012	0.0039

a/ Two Duroc gilts, average weight 50 kg

liver, kidney, and intestine was demonstrated by dosing the pregnant cow repeatedly with oral cobalt or by repeated injection.

Orally dosed cobalt (42 mg cobalt sulfate) appears in a significant amount in the milk of cattle (Dyatlov, 1972). Although there is good distribution to milk, there is also apparent variability in the pastures where cattle feed, based on the wide range of values in the literature. Values of 0.64 to 60 μg/liter of milk obtained using colorimetric or emission spectroscopy methods were reported in the review of Schroeder et al. (1967).

(3) Dogs: Dogs exposed to ^{60}Co oxide aerosols (Co_3O_4 and CoO) had the following distribution 140 days after exposure: lung highest, followed by liver and kidney (Barnes et al., 1976). The Co_3O_4 remained in the lung to a greater extent than did CoO, with 85% of exposed dose and 10% of exposed dose present at the 8-day measurement, respectively. This makes the biological half-life for the Co_3O_4 about 60 days and that for CoO about 35 days.

The intravenous distribution of radiolabeled cobalt in dogs led Sheline and Chaikoff (1946; cited by Villaume et al., 1976) to conclude that the large amount of cobalt in the feces after oral administration is due largely to unabsorbed cobalt which does not participate in metabolism. They based this on the data from five dogs; an intravenous dose was only returned 0.1 to 0.3% in the pancreatic juice within 3 days and 2 to 5% in bile.

(4) Guinea pigs: The tissue distribution of $CoCl_2$ administered as an aerosol to guinea pigs was reported by Höbel et al. (1969). See Table VIII-4. The method of analysis was not described.

Suzuki-Yasumoto and Inaba (1973) injected or fed ^{58}Co to guinea pigs. Intraperitoneal or subcutaneous injection gave identical distribution/excretion patterns (mostly urinary excretion), but orally dosed cobalt was excreted mainly in feces. After a week, the route of administration was no longer reflected in the excretion pattern.

TABLE VIII-4

DISTRIBUTION OF COBALT IN GUINEA PIGS AFTER INHALATION OF COBALTOUS CHLORIDE
(Höbel et al., 1969)

Organ	Co^{2+} (μMol/g tissue, wet weight)	Co, mg/kg
Larynx	1.19 ± 0.58	0.070
Trachea	1.38 ± 0.40	0.081
Lungs	0.99 ± 0.15	0.058
Carcass	0.014 ± 0.002	0.0083
Blood	0.013 ± 0.003	0.0077
Liver	0.19 ± 0.04	0.011
Kidney	0.08 ± 0.01	0.0047

(5) Sheep: Zharnikov (1967) found the distribution of cobalt in the sheep to be highest in the wall of the abomasum (0.35 mg/kg dry weight) followed by kidney and wall of the third stomach (0.33 mg/kg), lung and liver (0.28 mg/kg), and spleen and wall of the second stomach (0.27 mg/kg). Morrison (1969) measured sheep lung cobalt content using a spark source mass spectrographic analysis and reported 0.25 and 1.4 mg/kg.

(6) Rabbits: In rabbits given 0.25 mg/kg/day cobalt sulfate for 2 months, cobalt accumulated in liver, small intestine, lung, blood, kidney, and stomach (Kichina, 1974). A similar pattern occurred when cobalt was dosed orally or by injection. Comar and Davis (1947) measured the acute distribution of cobalt in 18 Dutch rabbits on a standard diet by oral and intravenous dosage of labeled cobalt. The tissue distribution reported is given in Table VIII-5. The values indicate that ingested cobalt as the chloride was poorly absorbed and almost completely eliminated within the first 24 hr.

Hoeck et al. (1975) measured cobalt in maternal and fetal circulation of iron-dosed rabbits and found a concentration gradient consistent with passive transport to the fetus from maternal circulation. Active transport was indicated for some other elements measured (Fe, Zn, Rb).

TABLE VIII-5

DISTRIBUTION IN RABBITS OF LABELED COBALT AS PERCENT OF ADMINISTERED DOSE (2.4 μg i.v. and 60 μg oral)

(Comar and Davis, 1947)

Tissue	Time After Intravenous Cobalt			Time After Oral Cobalt		
	1 hr	5 hr	159 hr	5 hr	24 hr	144 hr
Thyroid	0	0	0	0	0	0
Thymus	0.13	0.05	0.05	0	0.013	0
Adrenals	0.02	0	0	0	0.0017	0
Reproductive organs	0.52	1.0	0.11	0	0.013	0
Heart	0.39	0.41	0.08	0	0.036	0
Blood	2.4	3.0	0.19	0.0017	0.23	0
Lung	0.8	0.7	0.2	0.0053	0.058	0
Kidney	3.2	2.4	0.4	0.0033	0.16	0.015
Bladder and urine	13.0	44.1	0	0.0744	0.111	0
Stomach and contents	1.58	1.07	1.0	0.031	0.066	0.7839
Small intestine and contents	1.6	2.0	0.49	0.066	0.1536	0.82
Large intestine and contents	2.1	4.8	3.8	26.343	0.179	2.82
Spleen	0.1	0.02	0.04	0	0.0053	0
Liver	16.5	12.7	3.4	0.019	0.47	0.050
Gall bladder and bile	0.05	0.08	0	0	0.0021	0

In rabbits, the distribution pattern of cobalt was reported to vary with the sex of the animal. Babadzhanov (1973) found the cobalt levels of liver, aorta, and kidney higher in male rabbits than in females; and blood and spleen cobalt levels in females were higher. In other organs, the content of the element was about the same in both sexes. By neutron activation analysis, the highest level of cobalt was found in adrenal glands and the lowest in muscle (0.35 mg/kg and 0.04 mg/kg tissue, respectively).

(7) Rats and mice: The rat, like the rabbit showed differences in cobalt distribution patterns when the

radioactive metal was dosed orally versus injection. Greenberg et al. (1943; cited by Cass and Lange, 1955) administered 0.1 mg labeled cobalt to five rats by injection. Although tissue distribution was good, only 5% was retained, and 1% of that by the liver. When rats were orally administered the same 0.1 mg dose, only the liver retained significant amounts (2% of the dose in bile and 3.5% in liver).

Subcutaneously injected $^{60}CoCl_2$ was found chiefly in the liver after 48 hr (Inogamov and Chernomorchenko, 1966); and in liver and kidney when given intraperitoneally to rats (Berlin, 1950). In the Berlin study, injection of 0.125 mg ^{60}Co sulfate gave the following distribution (in mg/kg) after 6 hr: liver, 0.9; kidney, 0.9; spleen, 0.2; lung, 0.2; and blood, 0.2.

Amiard (1976) reported that the female Wistar rat concentrates cobalt in the kidney although other mammals show primary accumulation in the liver. The rats in his study were fed 45.7 to 48.8 nCi ^{60}Co/day in the diet for 5 days, and only 0.6% was retained. The animals reached equilibrium between ingestion and excretion in 48 hr.

Copp and Greenberg (1941; abstracted by Cass and Lange, 1955) reported on the distribution of cobalt after injection or oral dosage to rats. After 96 hr, retention equaled only 4.6% of the injected cobalt and 2.7% of the orally dosed cobalt of a 10-μg dose. The tissues containing the most cobalt were pancreas, liver, spleen, and kidneys.

Maziere et al. (1977) established lifetime baseline levels for distribution of cobalt in the laboratory rat fed a standardized, controlled diet containing 1.3 mg Co/kg dry weight. By neutron activation analysis, the normal cobalt concentrations (mg/kg dry weight) were: kidney, 0.96 ± 0.12; heart 0.30 ± 0.04; liver, 0.37 ± 0.05; muscle, 0.04 ± 0.01; and testes, 0.08 ± 0.02. In general, cobalt concentrations of all tissue were independent of age and were constant for a given organ. For example, cobalt in muscle compared to kidney varied by a factor of 20. This study indicated prime distribution to kidney and liver, but it was difficult to assess how much of the kidney cobalt was tissue accumulation and how much was being excreted.

Kostić et al. (1977) also determined natural cobalt levels in the organs of untreated young adult albino rats using neutron activation analysis. The levels found in mg/kg dry tissue: brain, 4.43; lung, 4.62; heart, 6.75; kidney, 7.45; spleen, 1.21; and liver, 1.11. Cobalt content of the rat chow was not reported; but in the United States, the common brands contain $\sim$ 0.4 mg Co/kg.

Cobalt levels absorbed from food and the cage environment into five mice tissues were determined by thermal neutron activation and high resolution γ-ray spectrometry (Nadkarni and Ehmann, 1972). The food and bedding data are averages of three analyses, and tissue values represent four replicate analyses of freeze dried tissues from 20 ICR/Ha strain female mice. (See Table VIII-6.)

TABLE VIII-6

COBALT IN MOUSE FOOD, BEDDING, AND TISSUE
(Nadkarni and Ehmann, 1972)

Tissue/Material Analyzed		Cobalt Concentration, mg/kg
Mouse food		0.15
Mouse bedding (pine chips)		0.36
Tissues:	skin	0.04
(freeze	brain	0.05
dried)	liver	0.11
	spleen	0.06
	kidney	0.14

(8) Birds: Matsusaka et al. (1972) measured the rate of transfer of fed and injected ^{60}Co and ^{58}Co to the eggs of laying hens and Japanese quail. Both species and both modes of administration yielded cobalt mainly in the egg yolk, with very little in the albumin and shell. None was incorporated before 24 hr. The third egg yolk on an average contained the greatest amount of radioactive cobalt in both species, approximately 0.02% of the dose. The cumulative value of ^{60}Co in the

egg yolk over a period of 2 weeks was about 0.1% of the dose. Whole body retention of ^{58}Co in birds after a single oral dose and a single intraperitoneal injection had similar decay curves, but the intraperitoneal dose was three to four times better retained than the oral. It can be approximated that 1% of an absorbed dose of cobalt is incorporated into bird eggs for about 2 weeks, regardless of mode of administration.

The bone of Antarctic and North American sea birds was found to concentrate cobalt to a high degree compared to seven other heavy metals. The ratio of cobalt concentration (determined by atomic absorption analysis) in breast, liver, and bone was 1:1:12 (Anderlini et al., 1972).

c. Vitamin B_{12}-related distribution: Passage of the cobalt-containing vitamin through the epithelial cell surface of the distal small intestine takes place by poorly described mechanisms, as discussed earlier. Once inside the cell, the absorbed vitamin rapidly binds to a specific transport protein, transcobalamin II, as it enters the portal system (Goodhart and Shile, 1973; cited in FDA, 1974). Transcobalamin II, a β-globulin, has a primary function of transporting the vitamin to the body tissues. A second protein, an α-globulin, transcobalamin I, has some minor transport functions but is mainly a storage depot for B_{12} in the form of methylcobalamin. There is a third transcobalamin with unknown but apparently minor function (FASEB, 1978). The literature contains several references to a variety of protein receptors which bind cobalamin in tissues all over the body, to which it is carried bound to plasma protein fractions.

The plasma level of vitamin B_{12} in normal human subjects ranges from 140 to 750 pg/ml with variations reported in a single day of 80 pg/ml (Tracor-Jitco, Inc., 1974). There is no sex difference in levels; but after 70 years of age, the plasma level tends to decline. The plasma represents, however, only about 0.1% of the total body content of the vitamin, although many reports exist about the plasma-binding fractions.

In plasma, vitamin B_{12} is present as either methylcobalamin, 5'-deoxyadenosylcobalamin, or hydroxocobalamin and is bound to proteins called the transcorrins. Transcorrin I is a strong binder of cobalamin; transcorrin II, a weak binder; and there may be others. A rather large percent of the weakly bound B_{12} is released for excretion in urine.

Vitamin B_{12} is distributed to liver, bone marrow, reticulocytes, and other tissues while the vitamin is bound to the transcobalamins. The liver contains the greatest concentration of B_{12}-bound cobalt, and the bone marrow the least, according to Rappazzo et al. (1970; cited in FDA, 1974). Liver content of B_{12}, probably as 5'-deoxyadenosylcobalamin averages about 1 to 1.5 mg, which is about 50 to 90% of the total body store, estimated to total 2.5 mg by Harper et al. (1977). Goodhart and Shile (1973; cited in FDA, 1974) estimate that turnover of vitamin B_{12} averages up to 2% of the body stores a day (0.1 to 2.0%) with loss mainly via bile. There is enterohepatic recirculation of some of this vitamin.

Glass et al. (1966; cited in FDA, 1974) showed that the blood levels of vitamin B_{12} in humans could be doubled by dosing with hydroxocobalamin (which has an OH on the cobalt) instead of cyanocobalamin (CN on the cobalt). In 7 of 11 patients observed at 15 months, blood levels were still above initial values. Other workers have measured plasma B_{12} in human subjects and have found hydroxo- and methyl-B_{12} to predominate; in fact, Linnell et al. (1969; cited by FDA, 1974) reported that plasma contained cyanocobalamin rarely, except when it was dosed to the subjects. Plasma binding of B_{12} to γ-globulins resembling antibody have been reported but not explained, although there have been suggestions that the binding is high in pernicious anemia patients (FDA, 1974).

Monroe et al. (1951) compared excretion and distribution of radiolabeled inorganic cobalt with cobalt in the form of vitamin B_{12}. Both forms of ^{60}Co-labeled compounds were administered intraperitoneally to 1-week-old New Hampshire chicks, and both forms contained approximately 1 μCi ^{60}Co. Excreta, blood, and tissues were dry ashed, and radioactivity was measured. Table VIII-7, taken from Monroe's data, gives the turnover ratio of inorganic cobalt and B_{12}-cobalt as tissue content in 0.5 hr/ tissue content in 24 hr. The data were expressed as percent of dose per gram of fresh tissue.

TABLE VIII-7

EXCRETION AND DISTRIBUTION OF INORGANIC COBALT AND VITAMIN B_{12} ADMINISTERED INTRAPERITONEALLY TO CHICKS
(Monroe et al., 1952)

Tissue	Turnover Ratio (Tissue Levels at 0.5 hr/24 hr)	
	B_{12}	Inorganic
Skin	2.2	9.1
Liver	1.7	5.6
Spleen	1.5	7.0
Kidney	0.8	7.9
Pancreas	3.5	3.7
Muscle	2.8	6.0
Brain	1.0	2.2
Bone (matrix)	2.3	13.6
Bone (marrow)	1.2	4.2
Blood	1.9	13.3

Monroe's data compare rather well with studies using the same route of injection (intraperitoneal) but different species. The authors noted that the chick liver uptake seemed consistently low compared to other species.

The retention of vitamin B_{12} ^{60}Co was about twice as great as that of the inorganic cobalt in these studies.

3. Excretion

a. Humans

(1) Forms other than vitamin B_{12}: Human cobalt excretion occurs in both the urine and feces. The pattern of excretion or the predominance of one route over the other depends, in experimental studies, primarily on the method of cobalt administration and the amount dosed. Browning (1961) suggests that the nature of the cobalt compound additionally influences the route of excretion. Untersteiner (1931; cited by Kent and

McCance, 1941), for example, found that divalent cobalt was more rapidly eliminated than trivalent cobalt.

In general, the excretion and biological half-life experiments from the 1920's to the present are consistent with the suggestion that orally dosed cobalt, especially at the levels found in the normal diet, is largely excreted by the intestines because of poor absorption. Parenterally administered cobalt, or that which has been absorbed and reaches the tissues, appears to be excreted in large part by the kidneys. This was quantitated by LeGoff (1927), who injected 24 mg $CoCl_2$ intramuscularly into a healthy volunteer and recovered 6.8 mg (28%) in the urine within the next 18 hr.

Exposure of humans to chronic release of cobalt into the body fluids from large orthopedic cobalt-alloy implants was considered by several authors. Onkelinx (1976) prepared a mathematical model of cobalt elimination based on rat data and applied his model to the values of Coleman et al. (1973) and Greene et al. (1975; cited by Onkelinx, 1976). Coleman measured plasma cobalt in the normal human at values approximating 1 μg/liter, and a large implant was suggested by Greene to release cobalt at an upper limit of 1 μg/hr/kg of body weight. Onkelinx suggested that the human plasma values could reach 10 to 20 μg/liter in these cases. The daily "output" of cobalt from such a prosthesis may be comparable to daily dietary intake of cobalt.

In a clinical drug trial, Wester (1973) quantitated cobalt levels in serum of 16 hypertensive patients and 8 normotensive controls, using neutron activation analysis. Baseline (predrug) cobalt serum levels were 0.00052 ± 0.00043 μg/ml for controls and 0.00046 ± 0.00018 μg/ml for hypertensives. These figures were in good agreement with levels by other laboratories (Parr and Taylor, 1964). Cobalt urinary excretion, micrograms per gram creatinine in 24-hr samples, was 0.74 ± 0.48. These values were similar to values determined by Wester (1965).

Four male adult patients of Jacobson and Wester (1977) were measured for blood cobalt (neutron activation analysis) during 5 days of total parenteral nutrition. All four were suffering from conditions or disease states that affected the gastrointestinal tract: gastric resection, pancreatitis, or gastric cancer. The serum cobalt levels of the four subjects averaged 0.42 ng/ml before the start of 5 days on an intravenous diet containing 5.8 μg/day and 0.34 ng/ml after.

Calcium intake was shown not to affect the cobalt balance in four volunteers fed calcium-poor then calcium-rich diets for 5 days each (Wester, 1974).

Bhat et al. (1973) recorded the body elimination of ^{60}Co following inhalation of radionuclide-containing steam by an adult male in Thana, India. Quantitation was by urine monitoring and whole body counting for 50 days. The ^{60}Co body burden was described as a two-compartment model, based on a 7-day fast elimination curve and a 43-day slow elimination curve. The biological half-lives were observed to be smaller than those given by the ICRP, indicating a faster turnover.

Delpla et al. (1976) monitored over an extended period the loss of ^{60}Co from a human volunteer who had inhaled a mixture of radionuclides. The drop in body burden of cobalt, quantitated by whole body counting, showed a gradual decrease plotted by the authors as a single slope. The ^{60}Co radioactivity remaining in the body over a period of 1,641 days is given in Table VIII-8.

TABLE VIII-8

RADIOCOBALT LOSS AFTER ITS INHALATION BY A HUMAN VOLUNTEER
(Delpla et al., 1976)

Time After Inhalation Dose	^{60}Co Activity Remaining, μCi
1 hr	0.890
2 hr	0.700
3 hr	0.600
7 days	0.300
105 days	0.190
378 days	0.100
1,040 days	0.049
1,641 days	0.030

Clinical studies compared mean urinary excretion of cobalt of 20 normal subjects to 17 patients with cirrhosis of the liver (Dean-Guelbenzu et al., 1975). By spark emission spectrography, they determined that the 20 normal patients were not excreting cobalt in their urine, but the 17 cirrhotic patients excreted 0.52 mg/24 hr. Clinical trials in Sweden quantitated control levels of urinary cobalt by neutron activation analysis (six subjects) at 0.94 ± 0.63 μg/24 hr (Wester, 1974).

The cobalt balance for human subjects, gathered from several studies, is shown in Table VIII-9. Some of the numbers were presented in the original studies; others give dosage or diet and excretion data only. The variation is very large, a fact discussed at length by Clemente et al. (1979), who compared only two balance studies: his measurements on an Italian population (Clemente et al., 1977) and the report of the task group on Reference Man (Snyder et al., 1975). The Italian daily balance shows that daily fecal excretion of cobalt (silver and mercury) is much higher than the daily dietary intake; these observed differences greatly exceed the errors due to the analytical method (neutron activation analysis) and the collection and treatment of the samples. Presumably, if food was analyzed as prepared for consumption, it would already be contaminated by the metals from the preparation process. The source of the excess mercury and silver excreted might be dental amalgam.

The study of Kent and McCance (1941) showed that in a single patient (with stomach cancer) the kidney excretion of cobalt from a normal diet was 17%. As tabulated for the Balance Table (Table VIII-9), five intravenous injections of 13 mg cobalt allowed 74% to appear in urine. The total amount excreted after 1 week following the injection was only 2.89 mg of the 13 mg administered, however. Kent and McCance (1941) suggested the gut is the main route of excretion for cobalt in the diet (probably because little is absorbed) and the kidneys for injected cobalt. The authors suggested that, once cobalt has reached the tissues, elimination is very slow. Smith et al. (1972) presented data that are much in agreement with the work of Kent and McCance. When Smith et al. gave intravenous $^{60}CoCl_2$ (1 μCi of ^{60}Co, specific activity 100 μCi/μg) to 24 human subjects, about 22% appeared in the first 24-hr urine collection. Fecal excretion during 2 to 8 days' observation and measurements ranged

TABLE VIII-9

HUMAN COBALT BALANCE

Subjects	Mean Co Intake	Analytical Method[a]	Co Source	Mean Loss: Feces	Mean Loss: Urine	Balance	Reference
Italian adults, 20 male, 20 female	9.75 μg/day	NAA	Diet	26.3 μg/day	1.75 μg/day	-18.3	Clemente et al. (1977)
British children, 11 normal	4.44 μg/kg/day	AAS	Diet	1.32 μg/kg/day	1.12 μg/kg/day	+2.0	Alexander et al. (1974)
Cancer patient, 1 adult male	13 mg/wk	C	I.V. injection	1.78 mg/wk	2.36 mg/wk	+8.86	Kent and McCance (1941)
Two adult males	310 μg/d	ES	Diet	34 μg/day	240 μg/day	+40	Tipton et al. (1969)
	470 μg/day			59 μg/day	230 μg/day	+180	
One adult male	168 μg/day	ES	Diet	42 μg/day	186 μg/day	-60	Tipton et al. (1966)
One adult female	154 μg/day	ES	Diet	74 μg/day	148 μg/day	-68	
		-					
Reference Man	300.1 μg/day	-	Diet and air	90 μg/day[b]	200 μg/day	-3.7	Snyder et al. (1975)
Human: review estimate, overall	306 μg/day	-	Diet and air	23-60 μg/day[c]	120-330 μg/day	0	Schroeder et al. (1967)
Two volunteers	125 mg	-	Experimental oral dose	72 mg/3 days	8.2 mg/3 days	-12.5	Paley et al. (1958; cited by Villaume et al., 1976)
Four male adults	4.3 μg/day	NAA	5-day I.V.	0.4 μg/day	2.0 μg/day	+1.9	Jacobson and Wester (1977)
	7.5 μg/day			0.13 μg/day	-	-	
	6.6 μg/day			0.14 μg/day	7.1 μg/day	-0.86	
	4.7 μg/day			-	5.2 μg/day	-0.5	
Four Russian children, 2 years old	71.46 μg/day	P	Diet	57.12 μg/day	2.29 μg/day	+12.05	Vorob'eva and Osmolovskaya (1970
6 years old	63.97 μg/day	P	Diet	47.06 μg/day	3.65 μg/day	+13.25	
Human: average	41.0 μg/day	-	Diet	38.3 μg/day (92 to 95%)	1.8 μg/day (4 to 5%)	+0.9	Ripak (1961; cited by Elinder and Friberg, 1977)
Males 40 years old	10 ng	W	I.V. injection ($1\ \mu Ci\ ^{60}CoCl_2$)				Smith et al. (1972)
Subject No. 2				10.5%	55.9%	+33.6%	
Subject No. 9				11.9%	28.0%	+66.1%	
Subject No. 8				6.9%	29.1%	+64.0%	
Subject No. 6				3.1%	36.0%	+60.9%	
Subject No. 5				2.0%	35.0%	+63.0%	
Subject No. 11				2.4%	28.2%	+69.4%	
Adult volunteers	9.55 μg/day	NAA	Diet	6.15 μg/day	0.3 μg/day	+3.1	Wester (1974)
	9.15 μg/day			11.5 μg/day	0.45 μg/day	-2.76	
	6.85 μg/day			7.15 μg/day	0.9 μg/day	-1.2	
	5.1 μg/day			3.55 μg/day	0.3 μg/day	+1.25	

a/ NAA = Neutron activation analysis; AAS = Atomic absorption spectrometry; C = Colorimetric analysis; ES = Emission spectrography; P = Polarographic analysis; and W = Whole body gamma counting.

b/ Sweat excretion 4 μg/day, included hair loss 2.4 μg/day.

c/ Also 6 μg/day "other" excretion.

from 5 to 30% (average 15%), and the average ratio of fecal cobalt to urinary cobalt was about 0.2:1. After the first few days, the excretion rate fell sharply and there was a fairly pronounced retention, as shown by reference to the study in the Balance Table (Table VIII-9). The authors reported that 22% of the 10-ng dose of $^{60}CoCl_2$ appeared in the urine the first day. The authors pointed out that a large component of the injected dose (9 to 16%) had a very long biological half-life, on the order of 800 days. They noted a particularly large long-term concentration of cobalt in the liver; about 20% of the total body burden was concentrated there.

In a study by Paley and Sussman (1963) in which $CoCl_2$ was administered to human subjects by mouth, renal excretion was investigated under different conditions. The authors related the excretory amounts to the different degrees of absorption. Absorption was decreased by the following conditions: administration of cobalt after a meal; pretagging cobalt to protein; alkalinization with an antacid preparation orally; and administration of tracer doses. According to the authors, these results suggest that the stomach and upper duodenum in man should be the sites of peak cobalt absorption. Cobalt is more soluble at an acid pH, and the protein-tagging experiments suggested that cobalt was absorbed above the region of protein digestion (Paley and Sussman, 1963).

Valberg et al. (1969) administered 20 μmol of $CoCl_2$ to a human subject and reported 44% absorption of cobalt, calculated from excretion data. Six normal subjects were given intramuscular injection of 10 μmol cobalt tagged with 1 μCi ^{60}Co. Only 4.4% of the injected dose was found in the feces within 5 days.

In 11 healthy children, Alexander et al. (1974) reported that, with a daily dietary intake of 4.44 μg Co/kg, a total of 2.44 μg/day was excreted (see Table VIII-9). The ratio of urinary and fecal excretion in these children on normal diets was not that commonly reported in balance studies on adults. Excretion was 1.12 $\mu g/kg$ in urine and 1.32 $\mu g/kg$ in feces, or an average total of 21 and 24 μg. This nearly equal loss of cobalt by urinary and fecal routes could be related to the subnormal intake, absorption, and retention of iron in these children.

The cobalt blood levels in 17 anemic patients returned to predosage levels 12 weeks after cessation of oral cobaltous chloride, 50 to 200 mg/day (Gardner, 1953; cited by Cass and Lange, 1955). Nuryagdyev (1971) spectrographically analyzed for cobalt in the blood of 238 patients with various types of cancer. Compared to 25 controls, cancer patients had 114 to 200% higher blood trace metal levels; urinary and fecal excretion was subnormal.

Rosen and Sabo (1976) reviewed what is known on the biological half-life of radiocobalt in man. The studies are summarized in Table VIII-10.

TABLE VIII-10

BIOLOGICAL HALF-LIFE OF RADIOCOBALT IN HUMANS
(Rosen and Sabo, 1976)

Route Administered	Isotope	Excretion Data	Half-life Reported	Reference Cited
Inhalation	^{58}Co	exhaled	57 days	Rosen and Sabo (1976)
Inhalation	^{60}Co	exhaled	3.9 years	Rosen and Sabo (1976)
"internal"	^{60}Co	N.R.	9 days	ICRP No. 10 (1972)
Oral (?)	^{60}Co	N.R.	800 days	Smith et al. (1972)
Inhalation	^{60}Co	exhaled	1-17 years	Newton and Rundo (1971)

The whole body counting of 791 persons for routine surveillance scans for body burdens of radionuclides has identified ^{60}Co in "approximately 333 individuals" (Sill et al., 1964). One subject who had inhaled a high level of

^{60}Co apparently more than 2 years earlier had an initial body burden of 1.5 μCi. The authors noted that the body burden dropped to 0.21 μCi 4 days after exposure, after which the elimination rate slowed, and body levels dropped to 0.15 μCi in 11 months for an effective biological half-life of 2.0 years. Sill et al. noted particularly that this inhalation exposure was eliminated nearly completely by the feces. The first two urine samples contained less than 1% of the ^{60}Co lost in feces for a comparable period of time.

Four Indian radiation workers were monitored for ^{60}Co using whole-body counting as well as chest counting, since the chest area had greatest radioactivity (Raghavendran et al., 1978). Decay was followed for 7 to 1,250 days. The retention curve had two components in one subject, with a biological half-life of 68 to 72 days for the short component and 831 to 854 days for the long-lived component. Biological half-life values in the other three cobalt-exposed workers ranged from 539 to 23,243 days (69 years). Raghavendran et al. (1978) suggested that this very long half-life might have been due to pulmonary lymph node uptake. Another of these subjects, Worker "S", had an effective half-life for cobalt of 1,767 days. This is more comparable to the data of Gupton and Brown (1972), who showed a biological half-life of more than 3 years, and to the data of Newton and Rundo (1970), who showed a half-life of about 17 years for activity deposited in the chest region.

Taylor and Marks (1978) cited published values for blood cobalt levels in healthy humans (Table VIII-11). The concentrations vary widely; most analyses cited were performed by atomic absorption spectroscopy.

Weissbecker (1950; cited by Carlberger, 1961) studied the disappearence of cobalt from the blood of humans and showed a curve for plasma cobalt after intravenous injection of 5 mg cobalt. Radioactive cobalt measurements showed that, for the first 4 hr, the average concentration was 1 μg/ml, and this was followed by a rapid decrease. According to the reviewer's observation of the blood or plasma curves for cobalt, the disappearance was an exponential function in semilogarithmic plots.

TABLE VIII-11

COBALT CONCENTRATIONS IN HUMAN BLOOD
(Taylor and Marks, 1978)

Sample	Concentration Range (ng/100 ml)
Whole blood	350-6,300
Whole blood	7-36
Plasma	18
Serum	7,000
Serum	37-100
Serum	6-8
Serum	8-58
Serum	2-6
Serum	4-26

(2) Vitamin B_{12}: The main route of excretion of vitamin B_{12} is by way of the bile; by this pathway about 40 μg (Harper et al., 1977) passes into the jejunum each day; but by an enterohepatic circulation, most of this is reabsorbed in the ileum utilizing the intrinsic factor mechanism. The small amount that remains unabsorbed leaves the body in the feces. This has been estimated to be 3 to 6 μg/day, also counting the amount produced in the colon by bacterial synthesis.

Urinary excretion of vitamin B_{12} is confined to the vitamin that is not protein bound and amounts to 0 to 0.25 μg/day (FASEB, 1977).

The body vitamin B_{12} stores are depleted very slowly. The biologic half-life for the vitamin is estimated to be about 400 days (Harper et al., 1977).

b. Animals: Cook et al. (1956; cited by Salminen et al., 1974) found that organ burdens of ^{60}Co in mice following repeated doses were in agreement with values calculated from single administration data. This experiment supports the validity of retention and body burden calculations made on the basis of single administration dosages.

TABLE VIII-12

ANIMAL BALANCE STUDIES

Species	Dose	Route	Excretion: Urine	Excretion: Fecal	Balance	Reference
Wistar albino and piebald rats						
Adults	1.57 μCi ^{60}Co/wk	Diet	0.268 μCi (17%)	1.22 μCi (78%)	0.087 μCi (-5%)	Cuthbertson et al. (1950)
Weanlings	0.099 μCi ^{60}Co/week	Diet	0.0758 μCi (58%)	0.0361 μCi (36%)	0.005 μCi (-5%)	
Weanlings	1.47 μCi ^{60}Co/week	Diet	0.126 μCi (9%)	1.24 μCi (84%)	0.1035 μCi (-7%)	
Wistar rats, adult females	45.7 to 48.8 nCi ^{60}Co/day	Diet	2.5%	97.3%	+ 0.6% average/ 5 days	Amiard (1976)
Jersey cattle						
Three steers	218 μg radiocobalt	Single oral dose	0.6%	78%	+ 16%	Comar and Davis (1947)
Two steers	218 μg radiocobalt	I.V. injection	62.0%	33%	+ 15%	
Rats						
One adult male	10 μg radiocobalt chloride	I.P.	~ 90%	< 5%	+ 4.6%	Copp and Greenberg (1941, cited by Cass and Lange (1955)
One adult male	10 μg radiocobalt chloride	Oral	~ 37%	60%	+ 2.7%	
Mice and rats	5 to 20 mg/kg	S. C. injection	45%, 1st week	15%, 1st week	+ 40%, 1st week	Hasagawa (1974)

Excretion of cobalt into bile and pancreatic fluid in dogs was measured by Sheline et al. (1946). Dogs, prepared with biliary and pancreatic fistulas and weighing 16 to 19 kg were administered 10 to 26 μg radioactively labeled cobalt intravenously. Biliary excretion over 72 hr averaged about 5% for the animals. Pancreatic fluid contained very little cobalt, 0.3% in 70 hr in one dog and < 0.1% in 48 hr in another.

A few studies have quantitated the cobalt in urine and feces of experimental or domestic animals. While not all studies lend themselves to such presentation, it is extensive enough to show the variability from study to study (Table VIII-12).

An indirect estimate of fecal loss and urinary fraction of ^{60}Co, dosed to chickens orally and by intravenous injection, is shown in Table VIII-13.

TABLE VIII-13

LOSS OF RADIOCOBALT FROM CHICKENS DOSED ORALLY AND INTRAVENOUSLY
(Lee and Wolterink, 1955)

	Intravenously	Oral dose
Calculated fecal loss	13%	51%
Urinary loss and tissue retention	70%	32%

The authors dosed dogs with ^{60}Co intravenously for comparison and reported a similar elimination pattern. LeGoff (1927) noted many years ago that the dosing route influenced the excretory route. A rabbit fed 44.6 mg $CoCl_2$ excreted only 4 mg Co (92 ml urine) in 24 hr by the kidney, but when 31.9 mg cobalt citrate was injected, an average of 11 mg Co and 128 ml urine was excreted.

A similar situation applies to rats. The amount of radiocobalt recovered in feces and urine at 24 hr after oral administration of 0.1 μg are 52 and 29%, respectively; after a 120-μg oral dose to rats, 76 and 9% in feces and urine, respectively. When radiocobalt was injected, excretion was mainly into urine regardless of dose: in 3 hr, 44% of a 0.1-μg dose and 49% of a 120.0-μg dose had been recovered in urine (Carlberger, 1961).

Hasegawa (1974) dosed mice at 5, 10 and 20 mg/kg $CoCl_2$ 6 days/week for 64 days by subcutaneous injection. Accumulation and excretion were determined by atomic absorption analysis of dry-ashed tissue and excreta. Liver and kidney were the main organs where long-term accumulation occurred. After a 40-mg/kg injection to check acute excretion, urine contained 22% of the dose in 24 hr. In 1 week, total feces contained 15% of the dose, and urine, 45% of the dose.

Thomas et al. (1976) hypothesized that bone could be the organ of highest cobalt concentration on a long-term basis. The data supporting this suggestion were from rats dosed by injection of ^{60}Co which had higher bone than tissue levels 400+ days after dosing. At 100 days after dosing, soft tissue had a greater content of ^{60}Co, but tissue levels dropped with time and bone levels remained rather stable.

Elimination of ^{57}Co following a single intravenous injection of $CoCl_2$ was measured in female Wistar rats of three age groups: weanlings (35 days), youth (60 days), and adults (116 days) (Onkelinx, 1976). During the 3 days postinjection, urinary excretion totaled 72.13% of the injected dose and feces contained 7.97% for the combined groups. Expressed as elimination per 100 g of body weight, there are few (or no) differences between age groups. Disappearance of ^{57}Co from the plasma after the intravenous dose occurs in three distinct segments: an 8- to 10-hr rapid clearance, a slower rate between 10 and 75 hr, and still more slowly after that. Onkelinx (1976) described a model for total excretory clearance following intravenous dosage in the rat, based on these data. Clearance is the sum of urinary clearance, fecal clearance, and clearance into a body "sink."

Murdock (1959) reported rapid absorption and clearance of cobalt orally dosed to rats in the form of ^{58}Co-labeled $CoCl_2$ ($\leq$ 1 μCi/dose). At oral dosage levels of 0.25, 1.25, 5, and 10 mg ^{58}Co/kg, first detection was made in urine at 30 min, and 27, 25, 14, and 12% of the respective doses were excreted within 72 hr. The author suggested that there was an inverse relationship between dosage and percent excreted in the urine. Cuthbertson et al. (1950) also noted this inverse relationship and reported retention of 5 to 7% in rats fed a diet containing 0.012 mg Co/kg for 14 days compared to 30 to 40% retention in the work of Houk et al. (1946; cited by Cuthbertson, 1950) who fed rats a diet containing 0.003 mg Co/kg.

Excretion of radiolabeled cobalt was studied for 96 hr in male rats (strain not stated) dosed both orally and by injection (Copp and Greenberg, 1941; cited by Cass and Lange, 1955). During the first 10 hr, 70% of the doses were excreted, but the excretion pattern differed with dosage route; 60% of the oral dose was excreted in the feces, but most of the injected dose was found in urine. After 2 days, 90% of the cobalt was excreted regardless of the route of dosing.

Comar et al. (1946; cited by Carlberger, 1961) fed rats 0.02 or 26.4 μg of radioactive cobalt,* after which feces and urine were collected until radioactivity could not be detected. The rats that received 0.02 μg cobalt excreted 54.1 ± 0.7%, while those receiving 26.4 μg excreted 87.3 ± 4.9% in feces.

---

* The amounts mentioned presumably include a carrier amount of cobalt since the activity of the higher dose would be 30 mCi.

Comar et al. (1946) were not particularly careful in their terminology. The specific activities of the doses were not given in describing their own experiments. Although in their text they state, "injected intravenously with about 120 μg radioactive cobalt," the heading for the tabulated results states "carrier doses (120 μg Co) of radioactive cobalt chloride."

B. Biochemistry

1. Effects on enzymes: The mode of action of cobalt is multilocular and is not fully understood. Most reports and reviews, however, are of the opinion that a large part of the effects of this metal are produced by enzymatic impairment and that this in turn leads to depressed tissue respiration and further depressive effects on energy metabolism. Co^{2+} blocks the Krebs citric acid cycle and cellular respiration. Dingle et al. (1962) showed that cobalt irreversibly chelates lipoic acid, which inactivates the coenzyme required for two vital energy reactions: (1) the oxidative decarboxylation of pyruvate to acetyl coenzyme A and (2) decarboxylation of α-ketoglutarate to succinate.

The depression by cobaltous chloride of the drug-metabolizing enzymes of the liver, particularly the decrease in hepatic cytochrome P-450, is so predictable that $CoCl_2$ dosage is used in animal experimentation to depress enzyme levels for other studies. Because cobalt inhibits the synthesis of P-450, the oxidative metabolism of drugs is decreased (Mitchell et al., 1973; cited by Cooksley, 1978). Allemand et al. (1978), for example, depressed hepatic cytochrome P-450 levels in rats by 69% with two daily doses of 30 mg/kg cobaltous chloride, subcutaneously administered.

Rona and Chappel (1973) produced depression of succinic dehydrogenase, the mitochondrial oxidative enzyme, by dosing rats with 4 to 12 mg/kg/day of cobalt sulfate for 2 weeks.

Dalvi and Robbins (1978) reported on the enzymatic effects of cobalt and some other trace metals on the monooxygenase enzyme levels of mouse liver. While cadmium and lead inhibited the activity of these enzymes, an equimolar (1 m*M*) amount of cobalt enhanced the drug-metabolizing rate. This enzyme-stimulating cobalt effect was inhibited by adding EDTA.

Frieden (1972) emphasized the role of cobalt in the metalloenzymes required for normal metabolic function. Examples of the most important were ribonucleotide reductase (DNA biosynthesis) and glutamate mutase (amino acid metabolism). Cobalt-related enzymes belong to the transition element metalloenzymes, unique in their ability to form strong complexes with protein ligands.

There are other metal-containing proteins which closely resemble the metalloenzymes but lack their catalytic function, such as hemoglobin, hemocyanin (the hemoglobin analog of some invertebrates), metallothionein, and transferrin.

In general, cobalt in its complexes with metalloprotein forms strong bonds, is inactive as a redox catalyst, and has a slow exchange rate. Its most common functions are hydrolysis and stabilization of the complex ternary protein structure (Kench, 1972).

All kinases (phosphotransferases) which catalyze formation of ordinary orthophosphate esters use a high energy phosphate donor (usually ATP) and a divalent cation (usually Mg^{2+}, but Co^{2+} can substitute) (Kench, 1972). In fact, Co^{2+} substitutes for other divalent cations in several enzymes. Zn^{2+} is the main metal in alcohol dehydrogenase, yeast lactate dehydrogenase, carboxypeptidase A, carbonic anhydrase, and alkaline phosphatase of *Escherichia coli*, but Co^{2+} can also activate these enzymes. Co^{3+} is the primary metal ion, of course, in the vitamin B_{12} coenzyme.

It is widely suggested in the literature that the pattern of activity of an enzyme may be altered considerably by metal substitution. Vallee and Williams (1968; cited by Kench, 1972) suggested that cobalt and zinc heighten catalytic function of enzymes by their ability to enter low-symmetry sites readily. Coleman and Vallee (1961; cited by Eichhorn, 1974) showed that, when the zinc of carboxypeptidase was replaced by cobalt, the activity of the enzyme was increased 160%. On the other hand, Zn^{2+} replacement by Co^{2+} in carbonic anhydrase B almost halved the activity of the enzyme. The mechanism of the frequent increase in activity of an enzyme with metal substitution was discussed by Kench (1974) with an example:

The cation draws into its vicinity a group of bases as ligands. In this divergent group, there are several geometric requisites which cannot be fully met, and one group binds weakly under considerable strain. When substrate is added, the cobalt ion-protein catalytic center readily releases acids and base groups since these are weakly bound and under strain anyway.

The above postulated action of cobalt at the subcellular level is consistent with other results and with x-ray crystallographic data (Kench, 1972; Blank and Britten, 1975).

Cobalt apparently has little gross effect on the liver as a whole. In the review of Browning (1969), for example, structural changes were not found by several reports which included light microscopy after cobalt doses. Despite the absence of clinical or histological evidence of hepatic damage following cobalt exposure, cobalt used experimentally in high doses has been found to affect various functions of the liver.

In experimental studies in pigs fed cobalt in their diets, elevations in the serum concentrations of lactate dehydrogenase, serum glutamic oxaloacetic transaminase (SGOT), and serum glutamic pyruvic transaminase (SGPT) were found (Burch et al., 1973).

The effects of cobalt on other liver enzyme systems have been studied. In contrast to the reduction in drug-oxidizing activity, cobalt increases heme oxidation markedly (Cooksley, 1978). The mechanism for this increase appeared to be enzyme induction by the heavy metal since there was an increase in hepatic microsomal protein and an absence of effects *in vitro*. The effect is specific for liver; cobalt appears to have no effect on the heme oxygenase system in the spleen. Cooksley (1978) also reported that at least nine other heavy metals induce heme oxygenase but none as efficiently as cobalt. There are also effects on hepatic catalase (a hemoprotein) and on the mitochondrial enzyme aminolevulinic acid synthetase.

It is evident that cobalt has complex actions on the enzymes and intermediates related to heme metabolism. It is noted, however, that in most studies on the biochemistry of cobalt, doses of cobaltous chloride ranging from 30 to 60 mg/kg daily have been administered to rats, so the relationship of the findings to cobalt-related disease in man is probably not significant.

Although cobalt ion is not essential for any biological function, as a divalent cation, it is able to activate a number of enzymes. These enzymes lack metal specificity, though, and other divalent cations may be equally effective, according to Cooksley (1978). O'Dell and Campbell (1971) pointed out three coenzyme B_{12}-dependent enzyme systems which are of significance either to man or higher animals. Methylmalonyl-CoA mutase plays a key role in the metabolism of propionate; coenzyme B_{12} is the only cobamide that serves as a cofactor for the mammalian mutase. N^5-methyltetrahydrofolate-homocysteine cobalamin methyltransferase is a coenzyme B_{12}-dependent enzyme involved in formation of methionine from methylation of homocysteine. A ribonucleotide reductase which also requires coenzyme B_{12} is important to some bacteria and may be shown important to the higher animals.

Niebroj (1967) described cobalt effects on nucleic acids and proteins. The author administered 1 mg Co per kilogram (dosed as $CoCl_2$ intraperitoneally) to Swiss mice. A significant rise in the number and activity of gonocytes in the mice was observed with enhanced spermatogenesis. Phosphate metabolism was enhanced with a rise in activity of several enzymes involved. Cobalt does not influence spermatozoa motility directly.

The important action of cobalt on increasing hemoglobin and the apparent synergism or sparing effect on iron may also be due to the enzymatic effects of cobalt. A suggested mechanism of this effect was made recently by Maines and Kappas (1977).

An important mitochondrial enzyme, δ-aminolevulinic acid (ALA) synthetase, is rate-limiting for the sequence of enzymatic steps to heme production. Cobalt appears to regulate (by repression) the synthesis of ALA synthetase, thus controlling cellular synthesis of heme. In addition, the rate-limiting enzyme for heme degradation, heme oxygenase, is regulated by cobalt in a negative feedback manner analogous to the control exerted on this enzyme by heme itself. Recent studies (Maines and Kappas, 1977) suggest that the specific site of the metal's action on hemoglobin is in the displacement by the divalent form (Co^{2+}) of two protons from two pyrrole rings allowing the spatial positioning necessary for the formation of heme methene bridges. The action of cobalt, therefore, is profound. Cobalt reduced total microsomal heme, inhibited heme synthesis, and induced the enzyme heme oxygenase.

After 15 mg/kg cobaltous chloride was administered subcutaneously to rats, 80% of the δ-ALA synthetase activity was lost within 2 hr, and this low level remained for 5 hr (De Matteis and Gibbs, 1977). This depression was followed by a dramatic rise of as much as sevenfold in the enzyme activity. Taylor and Marks (1978) noted that this enzymatic increase was inhibited by standard inhibitors of protein synthesis, which indicates that much of the increased enzymatic activity is newly synthesized enzyme.

Currently published research has not sorted out these conflicting data. The decrease synthesis of heme produced by cobalt and the erythropoietic response to cobalt are opposing effects which need much further clarification.

Enzymatic changes elicited by cobalt were addressed (with other heavy metals) by de Bruin (1976). The data he presented for cobalt are given in Table VIII-14.

The function of vitamin B_{12} as a coenzyme in metabolism was proved when three B_{12}-containing coenzymes, the cobamides, were isolated from bacterial sources. These coenzymes contain an adenine nucleoside linked to the cobalt as in Figure VIII-1 instead of a cyano group attached to the cobalt, as does cyanocobalamin. In bacteria, these coenzymes play a vital role in the cobamide-dependent ribonucleotide reductase reaction whereby the ribose moiety in a ribonucleotide is converted to deoxyribose when DNA is to be formed.

The ruminal microflora of multigastric animals have the capability of converting dietary cobalt to vitamin B_{12}. Man and the other monogastric species are dependent on these animals and bacteria for B_{12}, as humans cannot incorporate dietary cobalt into the vitamin (WHO, 1973).

TABLE VIII-14

ENZYME CHANGES ELICITED BY COBALT
(de Bruin, 1976)

Site	Enzyme	Change
Pancreas	LDH, G-6-PD	Selective injurious on α-pancreatic cells accompanied by histochemically detectable enzyme inhibition
Liver	Transaminases, LDH	Marked to small decreases in liver enzyme levels along with serum enzyme increase. Liver damage established.
Heart	Transaminase, ALD, CPK, LDH	All levels rise; associated with myocardial injury
Kidney	Cytochrome-C oxidase, Succinate dehydrogenase	Repeated cobalt chloride dosage produced loss of these enzymes in proximal tubule epithelium with cellular mitochondrial depression
Adrenals	Ascorbic acid biosynthetic enzymes	Loss of vitamin C from adrenals, increased urinary loss, increases in ascorbate biosynthesis

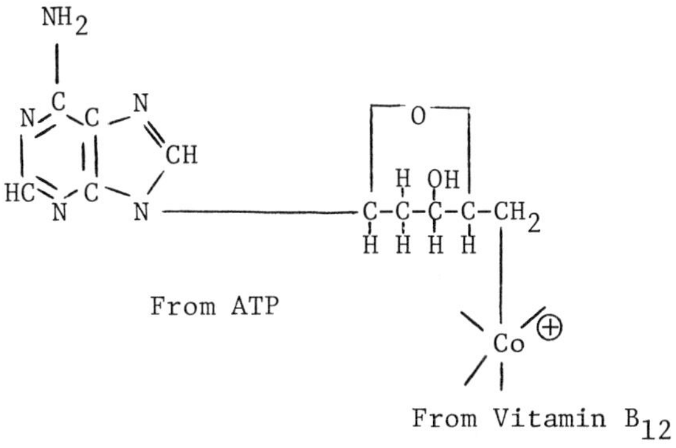

From ATP

From Vitamin B_{12}

Figure VIII-1. B_{12} Coenzyme.

2. Metabolism

a. Cobalt metal and salts: Cobalt, by reacting with the thiol groups of amino acids, proteins, and other coenzymes or cofactors, inactivates several biochemical pathways, especially those pathways associated with the production of cellular energy. Dihydrolipoic acid is required for the entry of pyruvate into the tricarboxylic acid (Krebs) cycle as well as for the conversion of α-oxoglutaric acid to succinyl coenzyme A. According to several studies (Goldwasser et al., 1958; Dingle et al., 1962; Wiberg et al., 1969; and Taylor and Marks, 1978), cobalt inactivates this enzyme cofactor, thus limiting the activity of the Krebs cycle, the production of ATP, and energy metabolism in general.

There are several enzymes that have been suggested as being involved in cobalt metabolism. It has been suggested by some studies (Dingle et al., 1962; Wiberg et al., 1967) that cobalt interferes with ketoacid metabolism, thus blocking the Krebs cycle and aerobic cellular respiration. Dingle et al., for example, achieved a 25% depression in respiration of rat liver at an injection dose of 1.25 mg cobalt as $CoCl_2$.

They suggested that this effect was due to the formation of a complex between the ion and the dithiol form of lipoic acid, which is a coenzyme of ketoacid dehydrogenation. Wiberg et al. (1967) described how the addition of α-lipoic acid to an *in vitro* system of cobalt-treated tissues greatly enhanced the mitochondrial metabolism of pyruvate.

The "ketoacid metabolism" theory of cobalt action is not held by all authors, however. Strickland and Goucher (1963) reported a stimulant action on succinate oxidation rather than a depressive action of cobalt on enzymes. Fishchenko (1971) injected subcutaneously 1 mg Co/kg as the chloride into guinea pigs and observed an increased respiration rate of liver, spleen, lymph nodes, adrenals, heart, and brain tissue. He cited earlier studies by his laboratory which had shown stimulating effects of cobalt injection on immunogenesis.

Wiberg et al. (1969) studied the concept of cobalt's influencing the metabolism of body proteins and amino acids. In a study of the cardiotoxic potential of cobalt on rats, they reported that cysteine, histidine, glutamic acid, and glycine afford protection against a lethal dose of cobalt and decrease the cardiac uptake of cobalt ions. The direct action of cobalt on protein and amino acid metabolism and protective effect of some amino acids on cobalt toxicity can be interpreted as another sign of impairment of enzymes which also influence protein metabolism.

In a study designed to measure thyroid effects of cobalt, Caplan and Block (1963; cited by Villaume et al., 1976) found that cobalt action on basic metabolic functions included elevated blood lipid levels. A rabbit given 8 mg $CoCl_2$/kg intraperitoneally for 11 days showed total cholesterol levels increasing from 186 to 560 mg/100 ml, phospholipids increasing from 161 to 810 mg/100 ml, and triglycerides increasing from 257 to 3,452 mg/100 ml. Triglycerides were so high there was visible clouding of the serum; both α-2 and β-globulins increased with the lipid changes. One week after withdrawing cobalt dosage, both protein and lipid levels returned to normal. Lempert and Levina (1974) showed very similar results. Rabbits fed 36 daily doses of 10 mg $CoCl_2$/kg increased serum cholesterol and β-lipoprotein levels by day 12. The authors suggested there is toxic activity of cobalt on the body organs, causing the organs to release β-lipoproteins into the blood. Taylor and Marks (1978) reported concentrations of fasting plasma lipids in anephric patients treated with cobalt for anemia. Before cobalt treatment (dose not given), the triglyceride levels were 1.7 ± 0.25 mmol/liter and the cholesterol 6.57 ± 0.45 mmol/liter. At completion of cobalt dosage, the levels were significantly higher; triglycerides were 3.5 ± 0.57 mmol/liter, and cholesterol, 7.64 ± 0.79 mmol/liter. Taylor and Marks suggested that the mechanism of the lipemic response to cobalt is not understood.

Histologically, Taylor and Marks (1978) reported pancreatic α-cell degranulation and destruction following administration of cobalt. They reported that the enzyme responsible for the catabolism of circulating lipoproteins (lipoprotein lipase) is inhibited both *in vivo* and *in vitro* by cobalt. This inhibition could explain the hypertriglyceridemia of cobalt-treated rats, and perhaps man.

Metabolic changes were reported by Klucik et al. (1967) in 51 workers exposed to 0.063 to 1.605 μg Co/liter of workplace air. In comparison with nonexposed subjects, serum protein increased 5.5%, sialic acid increased 21.9%, and hexosamine levels increased 23.2%. The cholesterol levels in this population averaged 248 ± 37 mg (high normal), and the albumin: globulin ratio was 0.83.

The relationship which can exist between the biochemical metabolism of trace elements and pathological processes in human medicine has been the subject of several studies during the past few years. Table VIII-15 gives the variations in cobalt levels occurring in kidney stones or urine of patients with various pathologic processes. These levels are compared with those of controls when the author has quantitated such levels for comparison.

TABLE VIII-15

COBALT LEVELS IN URINE OR IN KIDNEY STONES IN VARIOUS DISEASE STATES

Population (No.)	Disease State	Cobalt Level	Material or Fluid Analyzed	Analytical Method[a]	Cobalt Level, Control Population	Reference
Spanish adults (26) (average age 59 years)	Maturity-onset diabetes	0.530 mg/24 hr	Urine	AAS	0.0 mg/24 hr	Culebras-Poza et al. (1974)
Spanish adults (12) (average age 24 years)	Juvenile-onset diabetes	0.670 mg/24 hr	Urine	AAS	0.0 mg/24 hr	Culebras-Poza et al. (1974)
Spanish adults (17) (average age 59 years)	Hepatic cirrhosis	0.520 mg/24 hr	Urine	AAS	0.0 mg/24 hr	Culebras-Poza et al. (1974)
Swedish adults (4)	Hyperten-sion	0.73 μg/24 hr	Urine	NAA	0.2 to 1.0 g/24 hr	Jacobson and Wester (1977)
Filipino adults (47)	Kidney stones	7.94 mg/kg	Kidney stones	NAA	None	Lawas et al. (1973)

a/ AAS = Atomic absorption spectrometry
NAA = Neutron activation analysis

b. Functions of vitamin B_{12}: Vitamin B_{12} coenzyme catalyzes the isomerase reaction where methylmalonyl CoA is converted to succinyl-CoA, a reaction studied in mammalian liver. It also has been shown to catalyze the enzymatic conversion in bacterial systems of glutamate to β-methylaspartate (Harper et al., 1977). The dependence of the isomerase reaction in man upon the B_{12} coenzyme is reflected by the fact that methylmalonic acid excretion in the urine of healthy humans is less than 2 mg/day but is significantly higher in B_{12} deficiency such as occurs in patients with pernicious anemia. As such, the methylmalonic acid in the urine appears to be a sensitive index of the body stores of vitamin B_{12}.

The most characteristic sign in humans of a deficiency of vitamin B_{12} is development of a macrocytic anemia or characteristic lesions of the nervous system. Neurologic symptoms may also occur in B_{12} deficiency states without the prior development of anemia. In general, when the intake of B_{12} is low, demand for this vitamin in the body for the function of hemopoiesis exceeds that for any other clinically recognizable physiologic function; therefore, macrocytosis is a sensitive indicator of a vitamin B_{12} deficiency (Harper et al., 1977).

The most vital influence of B_{12} is its effect on nucleic acid formation in its role as a component of the various coenzymes needed in synthesis. According to the recent review by the Life Sciences Research Office of FASEB (1978), the metabolic roles of vitamin B_{12} and folic acid are intimately combined. B_{12} action cycles 5-methyltetrahydrofolate back into the folate pool; this is exemplified by the reaction whereby homocysteine is converted to methionine. For this purpose, the single carbon unit on $\underline{N}^5$-methyltetrahydrofolic acid is transferred to vitamin B_{12} coenzyme to form methylcobalamin, which subsequently transfers the methyl group to homocysteine, thus forming methionine. The demethylated tetrahydrofolic acid can then return to the folate pool for use in the many other reactions that use the one-carbon moiety, such as purine synthesis and the methylation of uracil. It is possible that the functional impairment of the hematopoietic system resulting from B_{12} deficiency is a nonspecific result of this nucleic acid inhibition. Harper et al. (1977) suggested that the isomerization of methylmalonate to succinate (catalyzed by the cobamide coenzyme) was the rate-limiting reaction for the production of lipoprotein in myelin sheaths. If so, it would provide a specific biochemical explanation for the effects of B_{12} deficiency on the peripheral nervous system.

3. Antagonistic/synergistic effects

a. Polycythemic action of cobalt: Antagonism of the polycythemic effect has suggested to some researchers that this blood effect may not be a normal physiologic stimulus of red cell synthesis, but rather may be a toxic manifestation of cobalt activity. For example, Saikkonen (1959, cited by Villaume et al., 1976) reported that at least eight common agents such as ascorbic acid, EDTA, and choline counteract cobalt's polycythemic action. In a study using a daily 10 mg/kg oral dose of cobalt as cobalt chloride administered 5 days/week to 30 adult male rats, RBC count, hematocrit, and hemoglobin concentration rose at a rapid rate for 60 days and at a slower rate to 180 days. The mean corpuscular volume and hemoglobin concentration per cell, however, remained unchanged. This indicates that the effect of the cobalt was only to produce more red blood cells (Murdock, 1959).

Taylor and Marks (1978) suggested that the stimulus for the polycythemic effect of cobalt in animals and man is an increase in circulating erythropoietin, a glycoprotein hormone which stimulates bone marrow stem cells to produce mature erythrocytes. The probable stimulus to the secretion of erythropoietin is tissue hypoxia resulting from cobalt inhibition of tissue respiration and oxidative phosphorylation.

b. Synergisms in cardiomyopathy: An apparent synergistic effect of cobalt and another component--commonly suggested to be alcohol--existed in the syndrome, "beer-drinkers' cardiomyopathy," discussed in detail in subsection VIII.C.a.(1)(a)(i), on human toxicity of cobalt. Synergistic effects of cobalt are indicated because the toxicity of the metal was far greater than would have been predicted from the dose. In 50 cases in one location, 20 deaths occurred. Cardiomyopathy, polycythemia, and thyroid lesions were reported. The beer, however, contained less than 1.36 mg/liter cobalt sulfate. Beer-drinkers never consumed more than 10 mg/day. Compared to the

25 to 50 mg cobalt/day often used in treating anemia, that was a very small dose (Taylor and Marks, 1968; Alexander, 1972). Morin and Daniel (1967; cited by Alexander, 1972) suggested that alcohol and cobalt were synergists; also, malnutrition in the heavy beer drinkers probably contributed to other biochemical synergistic effects with cobalt.

Many animal experiments have been conducted to look at the mechanism of action of cardiomyopathy and the probable synergisms involved.

Cobalt, 20 mg/kg/day fed for up to 5 weeks to guinea pigs along with 2 g ethanol/day, did not produce additional effects beyond those seen with the cobalt alone (Mohiuddin et al., 1970; cited by Alexander, 1972). Addition of 6% ethanol to rat drinking water had no effect on either acute toxicity or histological change produced by cobalt (Wiberg et al., 1969). Wiberg also fed young rats 1.2 and 3.6 mg Co/liter beer. Effects included minimal thyroid toxicity at the high dose. The LD_{50} for cobalt in rats given ethanol in the drinking water did not change. It is noted that synergistic effects between cobalt and ethanol in guinea pigs and rats did not occur during simultaneous administration of the materials. However, pretreatment of rats for 36 days with beer, then dosing with cobalt, did produce a cardiomyopathy similar to the human cobalt-beer cardiomyopathy (Wiberg et al., 1969).

In Nutrition Reviews (Anonymous, 1971), several elegant experiments by Derr et al. (1970) were cited which presented some experimental evidence in animals that synergism exists between alcohol and cobalt. The design consisted of four groups given (a) water only, (b) water plus 10% ethanol, (c) 1 mg Co (as cobaltous chloride) per 10 ml water, and (d) the cobalt plus ethanol. The animals were given these solutions as drinking water for 35 days. Cobalt plus ethanol depressed the growth rate of the young male albino rats to a significantly greater degree than did either cobalt or ethanol alone. Wet and dry heart weights of rats given cobalt plus ethanol were significantly less than in all other groups. The zinc content of the hearts of these animals was also higher than in the other groups.

Consumption studies showed that the combination group consumed less ethanol and cobalt, thus eliminating the possibility that the effects were due to increased consumption.

Derr et al. (1970; cited by Villaume et al., 1976) noted that 35-day growth rates in male albino rats were significantly more depressed with cobalt plus ethanol than with cobalt dosed alone. They interpreted this finding as an indication of a synergistic effect between cobalt and ethanol. Hematocrit measurements increased with cobalt dosage demostrating a typical polycythemia. The heart wet weight was significantly less in the cobalt-plus-ethanol group than in the cobalt-dosed group (6.2 g and 8.4 g, respectively).

c. Interactions with other elements: Gabbedy (1970; cited in Anonymous, 1972) noted that cobalt deficiency in sheep has been said to render them more susceptible to selenium toxicity. A report that cobalt deficiency enhanced kidney levels of selenium was cited as a possible mechanism of this effect. In other work cited in this review, the addition of cobalt to selenium-supplemented rat diets decreased selenium retention in heart, skeletal muscle, liver, and kidneys.

Myocardial toxicity induced in swine with 100 mg cobalt sulfate/kg was effectively counteracted with 0.25 mg selenite and 17 international units (IU) of vitamin E/kg body weight. The cardiotoxicity produced was mainly atrial, with initial mitochondrial damage and a distinctive accumulation of granular material in the mitochondrial matrix, shown by microanalytical analysis to be deposits of calcium and phosphorus (Van Vleet et al., 1977).

Kichina (1974) reported a specific cobalt-titanium ratio for each of six different tissues in rabbits. Tissue accumulation occurred with cobalt sulfate doses of 0.25 mg/kg/day for two months; dosage by oral route increased titanium levels in liver, small intestine, and lungs. Subcutaneous cobalt dosage increased titanium levels in blood, kidneys, and stomach.

Chelating or sequestering agents remove metallic ions from solution, and the resulting nonionic organometallic complex is often less toxic and more rapidly excreted than the metallic ions. There is no effective chelating agent or other antidotal therapy against cobalt, however (Meyers et al., 1974). For example, an investigational new drug, dithizon (diphenylthiocarbazone), is highly effective at increasing excretion of nine different heavy metals in animal trials but is totally ineffective at increasing the amount of cobalt excreted (Polakowski, 1974). The lack of cobalt poisoning antidotes may reflect the magnitude of the problem and indicate that such toxicity is rare. Cobalt is itself an antidote for cyanide poisoning (Davison, 1969). A British chemical company recommends cobalt-EDTA, 300 g in 20 ml intravenously over 1 min. The acute toxic symptoms of cobalt poisoning--fall in blood pressure, rise in pulse rate, and sometimes nausea--occur in cyanide exposure victims so treated.

Albert (1958) extensively reviewed metal-binding agents and mentioned the antagonism by cobalt of the bacteriostat oxine (8-hydroxyquinoline). In large amounts, several metals inhibited oxine; cadmium, zinc, and nickel all protected bacteria from the effects of oxine but only in large excess. Cobalt was unique in that the bacteriocidal effects were inhibited by as little as 0.00004 M cobaltous sulfate (1.5 μg Co/liter). The mechanism was suggested to be the prevention of a chain-reaction that occurs in oxine, an oxidation by atmospheric oxygen producing hydrogen peroxide, which in turn oxidizes more substrate. Albert suggested that, in some reactions of this kind, traces of cobalt could act as an efficient chainbreaker which would greatly moderate the destruction.

Kaplun and Mezentseva (1967) dosed white rats intratracheally with 10 to 50 mg mixtures of tungsten carbide and cobalt. The minimum lethal dose was 10 mg (1.5 mg Co) for the 15:85 mixture and 15 mg (1.2 mg Co) for the 8:92 mixture. The authors noted that the minimum and absolute lethal doses of pure metallic cobalt were 5 and 10 mg, respectively; however, the absolute lethal dose of the 15% cobalt-tungsten dust mixture contained only 2.25 mg of cobalt. They concluded that the dust mixtures were more toxic than the separate dust components and that the toxicity of cobalt was increased in the presence of tungsten and titanium carbides. The larger volume of the toxic doses was not considered.

The effects of cobalt on cadmium toxicity were reviewed by Fox (1974). In general, 5 mg cadmium/kg diet is the lowest level of cadmium that will produce adverse physiological effects in the rat. At acute, high-level dosage, testicular necrosis is seen with cadmium; and Kar and Kamboj

(1965; cited by Fox, 1974) reported that dosage with cobalt provided protection against this effect. Fox found no synergisms on blood effects, however, with cobalt and cadmium when Japanese quail were fed both elements in longer term toxicity studies.

Sunfish, Lepomis gibbosus, were exposed to 0.04 mg $CdSO_4$/liter in water for 2 weeks before being fed ^{58}Co-vitamin B_{12}. The cadmium-exposed fish had statistically less vitamin in their livers and more in the excretory pathway (gills, intestine) than did control sunfish. Merlini (1978) suggested that since man accumulates cadmium as he ages, the effects of this metal on hepatic storage of vitamin B_{12} may have interesting toxicity relevance to man.

d. Other synergisms/antagonisms: Myakisheva and Shirobokova (1976) showed that cobalt deficiency increased the sensitivity of young children (aged 2 months to 2 years) to vitamin D, and toxicity resulted during moderate dosing of vitamin D supplements to this group. Savitskii (1971) noted that an increase in the environmental temperature above 28 to 29°C potentiated the toxicity of cobalt and six other heavy metals in rats, rabbits, and guinea pigs. Potential mechanisms were suggested: enzymatic disorders, protein metabolic disturbances, or change in oxidation rate or in immunobiological activity.

Cobalt ion was toxic to Vibrio cholerae and Vibrio eltor organisms. It inhibited growth by 50% at 15 mg/liter and 10 mg/liter, respectively, and destroyed their metabolic activity at 25 mg/liter. Fe^{3+} supplementation of the medium reversed the effects in both strains; Mg^{2+} reversed Co^{2+} effects to a considerable extent in both organisms (Nagesha et al., 1977).

Florkin and Stotz (1971) described an unexplained interaction between cobalt and a constituent of the perennial grass, Phalaris tuberosa. When cattle and sheep consume this grass, a disease called Phalaris staggers develops, which is characterized by muscular tremors, incoordination, and tachycardia acutely, and demyelination of lower brain and spinal cord chronically. In many aspects, the disease also resembles pernicious anemia, but vitamin B_{12} is ineffective; cobalt supplements are effective in the acute phase.

Hill (1975) reported that ascorbic acid had no effect on absorption or retention of orally dosed cobalt, 0.02% in the diet in chicks. In unspecified species, Griffith et al. (1942; cited by Browning, 1961) found that cysteine had a detoxifying effect in cobalt-poisoned animals.

Underwood (1976) has described a syndrome of anemia that is common in animals having a very high intake of cobalt (4 mg/kg or more). It is suggested that iron uptake is depressed in the animals by the very high intakes of cobalt, and he cited Forth and Rummel (1971; cited by Underwood, 1975), who showed that the absorption of ^{59}Fe from jejunal loops of the rat was reduced by almost two-thirds in the presence of a 10-fold higher cobalt concentration. A 100-fold excess of cobalt suppressed the absorption of ^{59}Fe nearly completely.

4. Physiological requirements

a. Human nutrition: The requirements for human levels of vitamin B_{12} according to the World Health Organization (WHO, 1970) have been estimated on the basis of three different types of studies: amounts of B_{12} necessary to prevent megaloblastic anemia; amounts of B_{12} in serum and tissue of healthy versus deficient subjects; and amount of body stores and turnover rate of B_{12}. The lower limit of 200 pg/ml (0.2 μg/liter) of serum B_{12} unequivocally excludes any signs of clinical, hematological, or biochemical deficiencies.

(1) Vitamin B_{12} deficiency: In the United States, the effect of income on levels of vitamin B_{12} intake is illustrated by the finding that 31, 16, and 2.7 μg/day are average levels for persons eating adequate high cost, adequate low cost, and poor diets, respectively (WHO, 1970). Megaloblastic anemia, the most serious sign of B_{12} deficiency, rarely results from dietary inadequacy; usually there are mild deficiency symptoms. Malabsorption is the primary cause of the B_{12} deficiency anemias.

A classical example of megaloblastic anemia from vitamin B_{12} deficiency is that of pernicious anemia, as outlined by WHO (1970). It appears to be genetic and affects a high number of Scandinavian or North European subjects. Partial gastrectomy for peptic ulcers has led to vitamin B_{12} deficiency. In developing countries or in communities where the intake of animal protein is low for economic or religious reasons, dietary deficiencies rather than absorption problems are contributing factors for megaloblastic anemias.

The fish tapeworm, *Diphyllobothrium latum*, absorbs vitamin B_{12} contained in food and may contribute to some human deficiencies; the contribution of other helminths to B_{12} deficiency is not so clear. According to the World Health Organization (WHO, 1970), vitamin B_{12} deficiency occurs frequently in infants fed on goat's milk instead of human or cow's milk.

In man, vitamin B_{12} deficiency produces a defect in utilization of 5-methyltetrahydrofolate, which requires the B_{12} for its conversion back into the folate pool. Deficiency of either vitamin B_{12} or folate, therefore, results in defective synthesis of deoxyribonucleic acid. The results of the defective DNA synthesis are seen early in hemopoietic tissue where megaloblastic changes are seen in developing red and white cells and in the megakaryocytes. Vitamin B_{12} deficiency (but not folate deficiency) causes demyelinating neurological lesions of unknown biochemical etiology.

Cuthbertson (1973) reviewed the human requirements for trace elements and noted that conclusive evidence of the existence of a dietary deficiency of cobalt has not been produced, even in areas where cobalt deficiency in ruminants is severe. Several recent epidemiological studies, however, have hinted that a correlation exists between some human clinical signs and cobalt deficiency. Anan'ev (1977) correlated cobalt deficiency with dental caries in USSR children; Konstantinov et al. (1975; cited by Masironi, 1977) reported endemic osteogenesis in areas where soil and water are low in cobalt compared to other minerals. Two studies failed to correlate cobalt in water or soil to goiter incidence (Chigrina et al., 1974; Grekalova, 1973).

Cuthbertson (1973) noted that the richest dietary sources of cobalt are green, leafy vegetables, when grown on normal soil. Since the only positive need of man for cobalt is for the integral part of vitamin B_{12}, all ordinary diets supply much more cobalt than can be accounted for as vitamin B_{12}.

Engel et al. (1967; cited by Salminen et al., 1975) performed metabolic balance studies and suggested that humans must absorb 8 μg/day to keep humans in cobalt equilibrium.

Mertz (1975) has summarized the problems with analysis and interpretation of physiological values in trace element nutrition. Cobalt data interpretation is seriously influenced by the fraction analyzed since, to perform its only known physiological function, cobalt is incorporated into vitamin B_{12}. If serum concentrations of cobalt present as part of vitamin B_{12} and serum concentrations as total cobalt are compared, they approximate 10 and 8,500 ng/liter, respectively, according to Mertz (1975). This means that attempts to use any data on total cobalt concentration as a basis of an assessment of vitamin B_{12} nutriture is futile; no existing analytical method would be able to detect the change in cobalt concentration that would be caused by even drastic declines in vitamin B_{12} concentrations.

Mertz (1970) made several generalizations about trace metals that are relevant to the nutrition of cobalt. Regarding age differences in tissue burden, he stressed that the demonstrated trend of most trace metals to accumulate in tissues with age has not been correlated with functional aspects except at excessive concentrations and the significance of such accumulation is not clear. The review also suggested that the concentration of trace elements in grossly pathological tissues are very difficult to interpret, but values obtained from mildly abnormal tissue can be of value if a strong hypothesis backed up by animal experiments allows a cause-effect relationship to be established.

Vitamin B_{12} absorption was significantly decreased by ethanol intake in the presence of adequate protein and vitamins in four healthy male adult volunteers (Lindenbaum and Lieber, 1969). In these trials, protein levels were set at 12.5% of diet in two subjects and 25% of diet in the other two; the level of protein did not influence the B_{12} absorption. Ethanol was substituted isocalorically for carbohydrate for 3 to 8 weeks, with a maximal daily dose of 158 to 253 g ethanol. Excretion in urine and feces was monitored after oral administration of 0.75 μg ^{57}Co-cyanocobalamin with intrinsic factor concentrate. Urinary excretion was used as a sign of absorption. One subject excreted 42% in urine pre-ethanol, 27% after 13 days on ethanol, and 11.9% after 26 days on ethanol. Post ethanol urinary excretion of labeled B_{12} was back up to 40.7%. Impairment of ileal function was suggested as the mechanism of this poor absorption.

Mills (1974) examined the methods of diagnosing cobalt deficiency in man and farm animals. His brief review stressed the importance of identifying all the physiological roles of the trace elements in addition to determining their active forms (cyanocobalamin for cobalt) in plasma so that the form responsible for a syndrome is identified in blood rather than total cobalt.

(2) Effects of excess vitamin B_{12}: Acute toxicity studies on vitamin B_{12} are scarce; both human and animal work is cited here. In 1950, Traina (cited by FASEB, 1978) reported that a commercial vitamin B_{12} preparation injected intraperitoneally at a level of 1.5 mg/kg killed 2 of 10 mice, and 3 mg/kg was an LD_{100} in mice by both the subcutaneous and the intraperitoneal routes. The same review cited the acute toxicity studies by Winter and Mushett (1950), who also tested crystalline vitamin B_{12} in solution. Up to 1,600 mg/kg injected intraperitoneally and intravenously into mice produced no deaths; 100 mg/kg injected intraperitoneally into rats and guinea pigs produced no death or toxic effect, with 48-hr to 1-week observation. Since they observed no deaths with 500 times the dose that Traina reported, Winter and Mushett concluded that the preparation used by Traina was contaminated in some manner.

Berlin et al. (1968; cited by FASEB, 1978) gave 100 mg vitamin B_{12} in a single oral dose to human subjects and reported no evidence of discomfort or toxicity. This author performed an extensive study on oral vitamin B_{12} treatment of pernicious anemia (low secretion of intrinsic factor). His subjects received 500 to 1,000 μg daily for periods of 10 to 70 months without unfavorable reaction from the long-term therapy.

The reviews of vitamin B_{12} contain large numbers of short-term studies in which massive doses of the vitamin were administered to experimental animals. The main conclusions that can be drawn from these studies are that the vitamin is not toxic to the animal nor to the offspring of pregnant dosed females when given at levels far in excess of nutritional requirements.

b. Animal nutrition: Dreyfus (1967) reviewed the literature on cobalt and summarized that the minimum daily dose for most animals lies between 1 and 10 μg except for ruminants, whose daily dose would be 100 μg. This will suffice, for the only use of cobalt in living organisms is its participation in the structure of vitamin B_{12}.

Cobalt deficiency in soil, which makes low cobalt content in herbage, occurs widely. Young (1948) has discussed the signs of cobalt deficiency in cattle and sheep in various countries including the United States, Canada, New Zealand, Great Britain, and Kenya. The wasting diseases are known by several terms: bush sickness, pining, salt sickness, Grand Traverse disease, nakuriutis, Mairoa dopiness, and Morton Mains disease. Young mentioned the use of cobalt chloride, cobalt sulfate, and cobalt nitrate as treatment, either applied to fertilizer and used for top dressing of pastures to increase the cobalt concentration of the herbage, or concentrated in a salt lick.

The estimated dietary requirement of vitamin B_{12} for pigs and chicks is 0.01 to 0.02 mg/kg of diet (Hays and Swenson, 1977). Since 4.5% of the molecular weight of B_{12} is contributed by elemental cobalt, the dietary requirement for pigs and chicks would represent only 0.0005 to 0.0010 mg Co/kg.

The cobalt requirement for ruminants (whose intestinal bacteria synthesize their own B_{12}) is 0.07 to 0.10 mg/kg of the diet, or several times that included in the required B_{12} for nonruminants.

Marston's review (1952) reported the minimum quantity of cobalt necessary to ensure an adequate nutritional status in sheep. A dietary intake of 0.07 to 0.08 mg Co/day was just sufficient to provide requirements for normal health during 3 years of observation. The sheep had received a daily supplement of cobalt and had grazed on deficient pasture of known cobalt concentrations.

Neathery (1976) reported the maximum safe dietary level of cobalt that can be fed without causing adverse effects on health or performance to several domestic animals. The author suggested that, in ration formulation, the following cobalt maxima exist:

Cattle: 20 mg/kg
Sheep: 50 mg/kg
Poultry: 4 mg/kg

Miller (1972) noted, without giving concentrations, that 23% of 228 common beef cattle feeds and ingredients of feeds list cobalt among their trace minerals.

The nutrition literature has consistently held that cobalt has one essential function in mammals and that is in the B_{12} molecule. Roginski and Mertz (1977) found a biphasic response to cobalt at very low levels in rats on highly purified diets. Hemoglobin concentration, hematocrits, and thyroid function were lowest in rats receiving 1 μg Co/g, and increased with lower and higher concentrations of cobalt. One plausible hypothesis is that cobalt *per se* has an essential function in the rat at low concentrations, although the authors suggested there were alternative explanations such as synergisms with other dietary constituents (Roginski and Mertz, 1977). Both epidemiologic and laboratory studies from the USSR have suggested a possible role for cobalt in addition to its B_{12} function, however. Novikova (1963) suggested that the rat showed an interaction between physiologic doses of cobalt and iodine and that cobalt is required for the optimal utilization of low doses of iodine by the thyroid. Epidemiologic data have indicated negative correlations between cobalt in the environment and the incidence of goiter in the USSR (Kovalsky and Blokhina, 1963; cited by Roginski and Mertz, 1977).

Hays and Swenson reviewed the nutritional physiology of cobalt (Dukes, 1977) in domestic animals. They noted that the ruminant appears not to utilize preformed vitamin B_{12} and is completely dependent on the microflora in the rumen for its synthesis. A dietary source of cobalt, however, is necessary for this synthesis. They stressed that propionate metabolism is more important to energy metabolism in the ruminant than it is in nonruminants, and B_{12} is required for the reaction. This likely accounts for the apparent high requirement for cobalt or B_{12} by the ruminant. In these animals, practical control of B_{12} deficiency is met by supplementing the diet with cobalt. Hays and Swenson (cited by Dukes, 1977) noted that in ruminants nutritional deficiency of cobalt has been met by the use of cobalt oxide pellets, which remain in the reticulum or rumen to yield a relatively steady supply of cobalt to the rumen fluid.

Becker and Smith (1951) fed yearling sheep a complete ration with 10 to 500 mg cobalt chloride (as elemental cobalt) added per centiweight (100 lb) per day, for periods up to 8 weeks. Their data are given in Table VIII-16.

TABLE VIII-16

TOXIC EFFECTS OF COBALT IN SHEEP FEED
(Becker and Smith, 1951)

Cobalt Dose/Cwt/Day	Time (weeks)	Effect
160 mg	8	No-effect level
200 mg	1	Weight loss
200 mg	6	1/5 died with fatty liver, lung edema
500 mg	6	3/5 died with liver, lung, thyroid, and heart pathologies. Intestinal hemorrhages in survivor.

Symptoms of cobalt deficiency in sheep were reviewed by Marston (1952). Bowstead et al. (1941-1942; cited by Marston, 1952) and others had reported a syndrome of a progressively wasting disease with features of anemia and anorexia. Blood volume and protein concentration are reduced, and the oxygen-carrying capacity of the blood falls to 30% of normal, with a profound macrocytic anemia. Sheep show no nervous system signs of cobalt deficiency according to Marston (1952), but necropsy almost invariably shows a fatty liver and hemosiderosed spleen. The growth rate of the young animals is retarded, and the adults lose weight with no obvious symptoms.

Comar (1948) made some generalizations on cobalt toxicity based on his extensive work with nutrition and trace elements in many species:

Ruminants: physiological cobalt dose:
0.1 mg/100 lb

Rats: polycythemic dose: 40 mg/kg oral
2.5 mg/kg subcutaneous

Mammals in general: minimum lethal dose:
50 to 75 mg/kg injected

Cattle: toxicity: 50 to 60 mg/100 lb calves
1 to 5 g total dose, dairy cattle

Cobalt deficiency in ruminants was reviewed by Marston (1952). The author noted that seasonal changes in marginally deficient grazing areas could produce deficiency symptoms in the grazing animal. He noted that, in general, grasses contain less cobalt than other fodder plants, so the botanical composition of a pasture is related to the nutritional state of ruminants in marginally deficient terrain.

Burch et al. (1973) studied the effects of cobalt, thiamine deficiency, beer, and combinations of the three in 32 pigs and attempted to clarify the mechanisms of beer-drinkers' cardiomyopathy. They found that cobalt supplements were associated with an increased ratio of heart weight to body weight, and also that cobalt was associated with elevated serum enzymes. Thiamine deficiency and cobalt excess were not additive or synergistic in their effects on serum enzymes. Many earlier studies in several laboratories had been unsuccessful also in producing a model of beer-drinkers' cardiomyopathy with addition of cobalt alone to the diet of experimental animals.

The major enzymes requiring vitamin B_{12} (therefore cobalt) are (Pier, 1975):

methylmalonyl-coenzyme A mutase
methyl tetrahydrofolate oxido reductase
homocysteine methyl transferase
ribonucleotide reductase

Pope (1971) noted that cobalt must be supplied to sheep very frequently for the best effectiveness. Daily supplements are superior to intermittent dosing or feeding according to four studies described in Pope's review. In areas of cobalt-deficient soil, adding cobalt to salt, to the soil itself, or induction of cobalt pellets into the rumen were all reported to be effective (Pope, 1971). Addition of cobalt to animal feeds is quite common even if less effective for sheep, however. Sax (1975) reported that five different cobalt compounds are added to animal feeds: acetate, carbonate, chloride, oxide, and sulfate.

Huck (1975) reported that the pig can tolerate up to 200 mg Co/kg of a corn-soy diet (cobalt as $CoCl_2 \cdot 6H_2O$). The addition of $\geq$ 400 mg Co/kg was toxic. The toxicity was completely alleviated by addition of either 0.5 or 1.0% methionine to the diet, even when the pigs received 600 mg/kg added cobalt.

Several case reports of overdosing cattle with cobalt were presented (Dickson and Bond, 1974). Two of 100 cattle died after drinking for 3 days water to which 100 mg of cobalt sulfate/liter had been added; the liver of one contained 69 mg Co/kg. Cattle receiving unknown amounts of cobalt by salt licks and feed supplements also died. Cobalt levels in the livers were 62 and 59 mg/kg dry weight. Dickson and Bond (1974) reported that 14 calves that had been given "a half fingertip full" of cobalt sulfate daily died with diarrhea and heart changes resembling nutritional myopathy. The cobalt level in the liver of one was 58 mg/kg. On one Australian farm, a liter dipper of cobalt sulfate was added per ton of cereal ration which was given to 25 animals each week. The result was an approximate daily intake of 1 g cobalt per animal. The cattle showed chronic diarrhea, stiffness of gait, and poor coat color. No deaths occurred; however, on slaughter one animal had a liver cobalt level of 20 mg/kg.

Dickson and Bond noted that in their laboratory, livers from healthy cattle contained 0.06 to 0.28 mg Co/kg, with a mean of 0.15 mg/kg in young cattle.

Keener et al. (1949) determined the limits of cobalt tolerance in growing Holstein dairy cattle. Symptoms of cobalt excess toxicity were anorexia, decreased water consumption, rough hair coat, and lack of muscular coordination. There is an increase in hemoglobin and packed red cell volume. Keener's group (1949) reported that the chronic no-effect level for these dairy cattle was 50 mg/100 lb (1.1 mg/kg).

C. Effects on Man and Animals

1. Toxicity

a. Human

(1) Accidental/intentional exposure: A large number of current literature reports exist on the signs and symptoms of human exposure to cobalt, either accidentally or medically. The older literature from the mid-1950's and before also contains a large number of studies and reports, some very good and some poorly controlled. These older studies are summarized in Table VIII-17. Only rarely will this toxicity review refer to the older literature in case reports.

(a) Case histories and clinical studies

(i) Heart. Beer cardiomyopathies: When cobalt was added to beer at a level of 1 mg/liter to stabilize the foam, the first human demonstration was made that cobalt was related to cardiovascular disease. Epidemics of severe congestive heart failure due to cobalt were documented for heavy beer drinkers in Louvain, Belgium; Omaha, Nebraska; Minneapolis, Minnesota; and Quebec City, Canada. Since the etiology of the syndrome was first described by the Quebec investigators, the term Quebec beer-drinkers' cardiomyopathy has been used to designate the pathology.

The actual levels of cobalt in beer samples were measured by Stone (1965; cited by Elinder and Friberg, 1977). Beer with foam stabilizer agent was reported to contain 1,100 μg Co/liter maximum, and beer without stabilizer measured under 100 μg/liter. Kesteloot (cited by Alexander, 1972), however, stated that some of the beer in Belgium contained as much as 5.5 mg Co/liter.

Kraybill (1976) stated that $CoCl_2$, $CoSO_4$, and Co acetate were all used as foam stabilizers in fermented malt beverages. The reports of the Quebec cases, however, stated that the brand of beer that was apparently responsible for the outbreak of cardiomyopathy "contained ten times more cobalt sulfate" than a beer of the same brand in another location without the appearance of toxic reports (Morin et al., 1969).

TABLE VIII-17

HUMAN TOXIC EFFECTS FROM OLDER COBALT STUDIES
(summarized from the review of Cass and Lange, 1955)

Subject (No.)	Cobalt Dose, Route, Repetitions	Effect	Authors Cited
Patients, male (9)	$CoCl_2$ 4.62 g max., 20 to 60 mg/day, oral, 14 to 77 days	Blue skin discoloration, increased RBC, Hb.	Robinson et al. (1949)
Anemic children (23)	$CoCl_2$ 29.7 mg/day, oral, time not given	Acute mild nausea, anorexia chronic, increased RBC and Hb.	Rohn and Bond (1953)
Cancer patients (23)	$CoCl_2$ 60 to 240 mg/day, oral, time not given	Acute: nausea, epigastric pain, vomiting.	Shen and Hamberger (1951)
Humans, anemic	$CoCl_2$ 100 to 200 mg/day, oral, 20 to 40 mg/day/ intravenous	Effective anemia therapy; nausea, vomiting in (subacute) treatment.	Weissbecker (1950)
Humans, normal	$CoCl_2$ 500 mg/day, oral, time not given	Toxic effects: anorexia, nausea, vomiting, diarrhea, flushing of face; Therapeutic effects: increased RBC, Hb, hematocrit.	Weissbecker and Maurer (1947)
Patients, renal disease and anemia (17)	$CoCl_2$ 31 to 125 mg Co/day, oral, 28 to 308 days	Anorexia, then nausea, vomiting for doses > 100 mg/day; 4/17 = tinnitus, 1/17 = reversible deafness.	Gardner (1953)
Adult volunteers (4)	$CoSO_4$ 0.1 ml repeated intracutaneous injections	2/4 = skin response to many Co salts, 4/4 = skin response to $CoSO_4$.	Haxthausen (1936)
Normal adults (7)	Co metal powder 250 mg/day, oral, 3 to 4 days	Mild G.I. disturbance.	Penati and Ruata (1940)
Pottery factory workers (436)	Dermal exposure to 0.5 to 2.5% $Co(NO_3)_2$ solution, 1 wk to 35 years work exposure	4.4% has positive skin test and dermatitis.	Pirila (1953)
Infants	Powdered Co metal 0.1 mg, oral, 1 X/day		Waltner (1930)
Children	Powdered Co metal 0.1 mg, oral, 2 to 3 X/day	No-effect level, 1 month max. dosage period.	
Adults	Powdered Co metal 0.6 to 1.0 mg, oral, 1 X/ day		
Anemic adults (8)	630 mg $CoCl_2$, oral, over 26 days	6/8 = G.I. disturbance, 2/8 = skin rash, 1/8 = red cell increase.	Baxter (1939)
Adults (8)	3,416 mg $CoCl_2$, oral, over 84 days	7/8 = red cell, hemoglobin, hematocrit increase.	Berk et al. (1949)
Anemic Adult (1)	800 mg $CoCl_2$ over 37 days (I.V. 50 mg 250 ml, 16 X)	50 mg I.V. dose = no effect, 75 mg I.V. = acute G.I. symptoms.	
Adult (1)	3,150 mg $CoCl_2$, oral, 37 days	Severe G.I. symptoms.	
Adult (1)	11,550 mg $CoCl_2$ over 42 days	Some anorexia.	
Anemic adults (32)	75 mg $CoCl_2$/day for "4 wk to several months"	31/32 = G.I. distress, 11/32 = red cell, hemoglobin increases.	
Premature infants (31)	120 mg/day Co sulfate, oral, 12 days	No toxic symptoms or effects on weight gain.	Coles and James (1954)
	560 to 1,120 mg Co sulfate, oral, max. 56 days	No toxicity.	
	560 to 1,120 mg Co sulfate + 3.8 to 10.14 g ferrous sulfate, oral, max. 56 days	No toxicity.	

TABLE VIII-17 (concluded)

Subject (No.)	Cobalt Dose, Route, Repetitions	Effect	Authors Cited
3-year-old child (1)	150 mg $CoCl_2$, oral, 5 days	No toxicity.	Cronin (1939)
Female adult (1)	300 mg $CoCl_2$, oral, 10 days	No toxicity.	
Male adults	3 mg to 79 mg Co/m^3, 2 to 6 years inhalation in work place	Skin sensitivity to Co by patch test.	Fairhall et al. (1947)
20-year-old male	25 mg $CoCl_2$, I.M.	Flushing of face and ears occurred immediately after injection and persisted for 10 min.	LeGoff (1930)
-	40 to 50 mg $CoCl_2$, I.M.	Face, neck, and ears very flushed for 15 min; Nausea and G.I. pain also occurred acutely.	
Males, 25 to 64-years old (4)	I.M. injections of 25 to 40 mg $CoCl_2$, citrate, and salicylate, over a period of 3 years	Redness of face and ears, warmth of skin and neck; Pulse rate increases and blood pressure falls 30 min after injection.[a]	

a/ Meyer and Meyer (1952; cited by Cass and Lange, 1955) used the cat to study the blood pressure drop with cobalt injection. The threshold, with a 10 to 20 mm Hg drop, is 10 μg Co/kg body weight; greater reductions are produced by increased dose. The blood pressure drop is rapidly reversed up to a dose of 500 μg/kg after which recovery is slow.

The background of the cobalt-beer cardiomyopathy was reviewed by Alexander (1972), who traced to a Danish Chemical firm the process of adding 1 to 1.5 mg Co/liter to beer as a stabilizer of beer foam. This patented process became desirable when detergent film on beer glasses destroyed the foam, and certain local breweries in Quebec City, Canada; Omaha, Nebraska; and Minneapolis, Minnesota apparently added cobalt to the beer. It has been claimed that none of the large national brewers added cobalt to beer but perhaps 20 to 25% of all beer sold in the United States between 1964 and 1966 contained added cobalt. Kerr (1967; cited by Alexander, 1972) described the syndrome in 14 beer drinkers who obtained their drink from a local brewery in upper New York State. Kesteloot et al. (1968; cited in Louria et al., 1972) reported on 24 cases in Louvain (Leuven), Belgium, with one fatality. The cobalt levels in Belgian beer were as high as 5.5 mg/liter. Beer containing cobalt may also have been consumed in London (Alexander, 1972).

Bonenfant et al. (1969) reported diffuse vacuolar degeneration and myofibrilysis in the heart in addition to thyroid changes at autopsy of 20 deaths out of 50 cases. The authors noted bizarre-shaped mitochondria; severely damaged sarcoplasmic reticulum; and, in a few cells, large electron-dense bodies suggestive of lipid. Thus, the histological findings were quite similar to those found in animal experiments with cobalt alone. Morin et al. (1971 and 1969) described the symptoms related to the cardiovascular system. In most cases, a cyanosis limited to the face and trunk was diagnostic. Progressive signs of heart failure appeared: tachycardia, low blood pressure, and small volume arterial pulses. Serum enzyme levels showed striking elevations: serum glutamic oxaloacetic transaminase was above 2,000 units in seven cases. There was important elevation of serum globulins.

The electrocardiographic recordings showed typical patterns (Morin et al., 1969): tachycardia (but, strikingly, not arrythmias); elevations of the ST-waves in the right precordial leads; and nonspecific T-wave anomalies, with low voltage of the QRS-segment in the limb leads. After a time, T-waves became negative (ischemic type) with high voltage QRS for 2 to 3 months. After 6 to 9 months, 17 of 22 cardiograms were considered normal in the Canadian patients; Kesteloot, in Belgium, found normalization in only 3 of 18. Alexander (1972) reported residual electrocardiographic changes in 11 patients over a 3-year follow-up (about half of the Minneapolis cases), which is a similar level of improvement to that reported by Sullivan for electrocardiographic normalization in the Omaha patients (Sullivan et al., 1968; cited by Alexander, 1972).

Alexander's report on the Minneapolis incidence of cobalt-beer cardiomyopathy (1972) included clinical signs that were strikingly similar to the Quebec reports. The mortality for the acute illness in Minneapolis was 18%, but later deaths accounted for a total mortality of 43% in 28 patients.

Several mechanisms for the beer-drinkers' pathology have been suggested. According to Shanoff (1972), a long-standing high level intake of alcohol appears to be associated in a casual way with myocardial disease. Since beer drinkers are particularly disposed, it may be that the toxicity of alcohol is enhanced by other substances found in beer (as was the case in cobalt-beer cardiomyopathy). Further, because only a few chronic heavy drinkers develop cardiomyopathy, individual susceptibility must be important. Alcoholic cardiomyopathy is a main example of congestive cardiomyopathy. The mechanism of action on the heart muscle by alcohol appears to be a deleterious effect on the membrane or on myofibrillary function with accumulation of lipid.

In patients with chronic alcoholism, the cardiac muscle releases intramitochondrial enzymes isocitric and malic dehydrogenases both at rest and during exercise, even in those patients without clinical evidence of heart disease (Shanoff, 1972). Thus, the biochemical status of the myocardium may not be normal in alcoholics at the outset. The addition of even low levels of cobalt, with suspected enzymatic toxicities and recognized cardiac pathologies of its own, could cause the major effects cited in the reports.

To summarize, beer containing approximately 1 mg Co/liter produced the syndrome after about 1 month, and no cases appeared later than 1 month after discontinuation of cobalt. A mortality of more than 40% in 100 patients in four localities occurred with striking differences in the different studies.

Symptoms of toxicity: patients were admitted to the hospital with congestive heart failure of recent origin. Specific complaints included: dyspnea, weakness, abdominal pain, nausea and vomiting, cyanosis, hepatomegaly, massive cardiac enlargement, venous distention, peripheral edema, tachycardia, gallop rhythm, hypotension. Serum transaminase, immunoglobulins, and C-reactive proteins were usually elevated. There were always prominent electrocardiographic changes: low QRS voltage, S-T and T-wave anomalies, and displaced transition zones.

Cellular postmortem findings included excess glycogen, degeneration of myofibrils, dilation of sarcoplasmic reticulum, vacuolization of the cytoplasm, and mitochondrial shrinking. The cardiac muscle contained 0.48 mg Co/kg (Sullivan et al., 1968), which is 10 times the normal level.

Several workers suggested that some biochemical effects of chronic alcohol intake followed by cobalt intake have a specific synergistic toxicity on the myocardial function at the subcellular level. Other reviews suggested that cobalt is more toxic to heavy beer drinkers, who have nutritional deficiencies frequently, and that these cobalt effects are nonspecific. Animal or human data are not adequate to back up either hypothesis at this time.

Myocardial toxicity due to industrial cobalt exposure is not common in the literature. Barborik and Dusek (1972) and Barborik et al. (1974) reported a fatal case, a metal worker 41 years of age who had been exposed to cobalt in the hard metal industry for 6 years. The autopsy findings showed a picture similar to the symptoms of the beer-drinkers' cardiomyopathy, with fragmented myocardial fibers, vacuolar change, and the absence of inflammatory reaction. There was diffused thickening of the endocardium. Cobalt concentration in the heart of the victim was 1.4 mg/kg dry weight, as compared to 0.10 to 0.20 mg/kg dry weight of controls, as determined by a colorimetric method. Also like the beer-drinkers' syndrome, the thyroid gland showed focal signs of activation; there was no apparent history of heavy alcohol intake in this case history, however.

Five cases of a typical myocarditis resulting from inhalation of cobalt fumes were reported (Kucharin and Sinitsin, 1976). The workers were engaged in cobalt smelting and were under identical working conditions. The character of the clinical pathology, electrocardiographic signs, and fluorographic data was identical in all five patients, which provided to the authors the diagnostic clue to the etiological factor responsible.

It was determined in an epidemiologic survey that serum cobalt levels were apparently unrelated to blood pressures in measurements on 520 subjects (Bierenbaum et al., 1975). Both systolic and diastolic blood pressures were significantly higher in the 260 subjects in Kansas City, Kansas, where the cobalt blood levels were undetectable, than in Kansas City, Missouri, where the cobalt blood levels were 0.146 mg/liter.

To summarize, the toxicity of cobalt to the organism in general appears to be by enzyme inhibition, most probably by competitive inhibition. A large literature has documented interference by cobalt with the breakdown of metabolites in the Krebs cycle, with pyruvate metabolism, with fatty acids, and with the terminal electron transport mechanism. The effect of cobalt on the myocardium, however, seems to be more complex, since the enzymatic inhibition may be further worsened by partial replacement of calcium on sites where it is necessary for muscular contraction. Such a situation would lead to a particular type of cardiac insufficiency characterized by defective utilization of high energy phosphates. Clinically, this interference of cobalt with cardiac metabolism produces dilation of the heart and secondary thrombosis. Such a myocardial picture is also seen in vitamin B_1 (thiamine) deficiency heart disease.

(ii) Thyroid toxicity to man: Thyroid function tests including ^{131}I uptake were depressed from 20% to over 90% in various studies of patients receiving 0.17 to 3.9 mg Co/kg/day for a duration of 6 days to 8 months (Villaume et al., 1976). The uptake patterns cited in this review were found to be typical of disease states producing an inhibition of the protein binding of inorganic iodide. In some of the case histories reviewed for this document, goiters and classic signs of hypothyroidism occurred as early as 6 weeks after initiation of cobalt treatment (usually for anemia). Several human cases are reviewed below.

Paley et al. (1958) demonstrated that 2 weeks' treatment with cobaltous chloride by mouth to healthy adult volunteers caused the thyroid ^{131}I uptake to drop near zero; it returned to normal when cobalt treatment was discontinued.

In one experiment, four euthyroid and two hyperthyroid patients received 12.5 mg cobalt as $CoCl_2$, three times a day for 6 to 25 days. In two of four euthyroid and one of two hyperthyroid patients, the cobalt depressed thyroid function as determined by the 15-min ^{131}I uptake. In another experiment, the $CoCl_2$ was administered intravenously to six euthyroid and three hyperthyroid patients, 0.08 mg Co/min for up to 30 min with similar results.

Kriss et al. (1955) administered cobaltous chloride to five patients with chronic anemia at dose levels of 2.9 to 3.9 mg/kg/day cobalt for 3 to 7.5 months. Three of the five developed bilateral goiter; and the other two, a microscopic hyperplasia. The thyroid enlargements were observed 7 months, 6 weeks, and 9 weeks after initiation of cobalt treatment. The two patients with microscopic hyperplasia died of unrelated causes. Histologic sections of thyroids showed hypertrophy of the lining epithelium and low colloid levels, which is generally consistent with hyperplasia, not carcinoma.

Schirrmacher (1967) reported on a 35-year-old female patient with chronic anemia and nephritis who received oral cobaltous chloride (approximately 0.4 to 0.8 mg Co/kg/day) to stimulate hemopoiesis. After 5 months' treatment, the patient showed unsteadiness in walking with absent ankle jerks and no vibration sense in the legs. She exhibited bilateral nerve deafness and had a diffuse goiter without signs of hypothyroidism. She complained of nausea and vomiting. A glycosuria was present. Improvement or reversal of all signs occurred after cessation of cobalt medication. Schirrmacher (1967) concluded that this patient's history indicated that targets of cobalt toxicity in humans includes the thyroid, central nervous system, and peripheral nervous system.

Gardner (1974) noted a special susceptibility of thyroid glands of infants or pediatric patients who are given cobalt (the tonic "Roncovite" with unstated cobalt concentration was blamed). It is particularly interesting that Gardner suggested that unknown environmental or genetic factors could be at work, since there was variability in the response of the infants' thyroids to cobalt, which was apparently unrelated to the dose received.

Sederholm et al. (1968) reported a case of severe goiter and hypothyroidism as well as polycythemia in an 11-year-old boy with renal disease who had received 10 mg $CoCl_2$/day for about 3 months in addition to 44 mg Fe/day. In addition, the serum cholesterol was 1,520 mg/100 ml. After cessation of cobalt and giving thyroxin, the thyroid returned to normal size and the serum cholesterol dropped to 640 mg/100 ml by 3 weeks.

Sederholm (1968) summarized reports on the clinical effects of cobalt. The most common symptons were anorexia, nausea, vomiting, and diarrhea; the most serious symptoms were tinnitus and neurogenic deafness (Gardner, 1953, was cited); and other symptons were erythema and hot sensations, skin rashes, and substernal aches. By 1970, due to the goitrogenic effects of cobalt, all antianemia preparations containing cobalt had been removed from the U.S. market (Anonymous, 1970).

Toxicity of cobalt to a 3.5-year-old child being treated for low hemoglobin levels was reported by Washburn and Kaplan (1964; cited by Villaume et al., 1976). Administered orally and daily, 12 mg cobaltous chloride plus 7.5 mg iron produced in 3 months an enlarged thyroid and in 8 months laboratory confirmation of a depressed thyroid function. The symptoms largely disappeared after treatment was discontinued, but it was noted that 2 years later thyroid enlargement remained in this child.

The toxicity of cobalt to the thyroid has been reviewed also by Nguyen-Phu-Lich (1971), as summarized in Table VIII-18.

TABLE VIII-18

TOXICITY OF COBALT TO THE THYROID
(Nguyen-Phu-Lich, 1971)

Cobalt Dose	Effect
3 to 4 mg for anemia	Goitrogenic effects
Undescribed cobalt dose to 10 anemic infants	Goiter
4 mg/kg $CoCl_2$, 17-month-old	Thyroid enlargement
150 mg $CoCl_2$, single dose, adult	Diminution of ^{131}I uptake test 1 week later

Human toxicity to the thyroid was noted at autopsy in 11 of 14 thyroid glands examined among the Quebec beer-drinker cases (Bonenfant et al., 1969). The authors noted that there had been no clinical signs of thyroid dysfunction, but histological changes showed a significant reduction of the colloid contents of the thyroid gland.

Davies (1972) reported on the suppression of thyroid activity in humans receiving high levels of cobalt salts, as well as goiter in humans who live in areas where the soil contains the wrong proportion of cobalt to iodine. Davies suggested that cobalt is necessary for the first stage of thyroid hormone production; the uptake of iodine by the thyroid may be the stage for which cobalt is required. The goitrogenic action of excess cobalt may be due to interference with uptake and transport of iodine by the thyroid gland.

This synergistic action of cobalt and iodine was also suggested by the work of Novikova (1963), who carried out experiments on 120 male rats fed with different quantities of iodine and cobalt for 3 months. Novikova's data indicate that either excessive or deficient intake of cobalt renders a decrease in absolute content and concentration of iodine in the thyroid gland. The greatest decrease in the iodine-concentrating power of the thyroid was observed when a simultaneous deficiency of cobalt and iodine existed in the diet. It appeared that a combination of cobalt and iodine deficiency was worse than deficiency of iodine alone as regards to the functional activity of the thyroid. Shtenberg et al. (1963) also performed animal studies on cobalt-iodine-low protein effects on the thyroid gland. Although cobalt depresses the thyroid gland, increased environmental concentrations of cobalt have not been related to the disease rate in areas of endemic goiter in the USSR.

(iii) Erythropoietic effects: One of the most commonly described effects of cobalt is erythropoietic action; it has been known for over 40 years that the metal produces polycythemia in man and other animals. The mechanism seems sensitive, as long-term experiments have rendered this effect in the absence of other toxic effects.

There are many reports of human clinical studies on cobalt administration for treatment of anemia. Gardner (1953; cited by Cass and Lange, 1955) dosed 17 patients with 50 to 200 mg cobaltous chloride/day for 4 to 44 weeks and produced a significant increase in the erythrocyte count within a month; maximum values were reached within 2 to 3 months. During a 12-week period after cessation of cobalt therapy, the blood values of all patients returned to pretreatment levels.

Increases in hemoglobin, erythrocyte count, and hematocrit were produced in nine adult male patients with chronic suppurative infections treated with 20 to 60 mg cobaltous chloride daily for 2 to 11 weeks (Robinson et al., 1949; cited by Taylor and Marks, 1978). Many reports exist of treatment of adults and children with cobalt for anemia, particularly in the clinical case reports of the 1950's.

A recent study (Duckham and Lee, 1976) reported on a group of anephric, refractory anemic dialysis patients who were given cobaltous chloride orally at doses of 25 to 50 mg/day for 12 to 32 weeks. The authors reported a decrease in the need for blood transfusions and an increase of 46% in average hemoglobin concentration.

Curtis et al. (1976) described the effect as a toxic response. Several animal studies have looked at the mechanisms involved.

There are several studies that indicate the erythropoietic effect of cobalt is a toxic effect rather than a beneficial tool for increasing hemoglobin levels in anemic patients, a common older medical use.

The work of Saikkonen (1959; cited by Villaume et al., 1976) presented three animal experiments which indicate that cobalt polycythemia is a toxic effect. First, the action of cobalt on red blood cell (RBC) synthesis in dogs, rabbits and rats was inhibited by several disparate materials, "especially nutrients" such as ascorbic acid, choline, cysteine, and nicotinamide. This finding indicated to the author that the action of cobalt on the blood-forming tissue was not physiologic but toxicologic. The second experiments dosed rats daily with 4 and 0.25 mg $CoCl_2$ for 4 weeks and found a dose-response agreement in the size of the adrenal. These data indicated to the author that cobalt had produced a generalized stress response in the dosed animals. The third experiment measured the hemoglobin precursor coproporphyrin in the urine of cobalt-dosed rats, which increased from 0.34 μg/rat/24 hr at the baseline to over 0.8 μg by day 28. Saikkonen (1959; cited by Villaume et al., 1976) observed similar coproporphyrin increases in urine after heavy metal poisoning, and such a finding is indicative of toxicity to the blood-forming tissues.

Murdock (1959) suggested that the polycythemic effect of cobalt is a blood-borne factor that is transferrable, such as erythropoietin might be. In Murdock's bioassay, donor rats were dosed with 100 mg $CoCl_2$/kg orally, and their plasma was injected into recipient rats (2 ml/day for 2 days). The recipients also received ^{59}Fe to check on blood synthesis. The rats receiving cobalt-treated donor plasma incorporated 19.3% iron, whereas (uninjected?) controls increased iron incorporation only 8.3%.

It has been suggested by Schirrmacher (1967) that cobalt's erythropoietic effects, used to treat anemia in many medical studies, have no role in modern medicine.

(iv) Nervous system effects: There are only brief references to nervous system toxicity in man resulting from cobalt; a few references are to ill-defined paresthesias, numbness, and other peripheral signs that coincided with more pronounced symptoms of cobalt effects.

A 35-year-old woman treated for anemia with doses of 100 mg $CoCl_2$/day developed an apparent central nervous system toxicity after 4 months' treatment (Schirrmacher, 1967). The author described signs of auditory nerve symptoms with both vestibular and cochlear involvement. The patient had paresthesia in the limbs, muscle tenderness in the legs, and no tendon reflex. The syndrome was diagnosed as a peripheral neuritis. The author reported all effects were reversed at withdrawal of the cobalt. It was noted that this patient had a diagnosis of chronic nephritis predating the anemia treatment, and, according to Schirrmacher, evidence of tubular damage. The contribution of nephritis or poor renal function to cobalt toxicity could be important in this case history.

Alfrey et al. (1972) reported that cobalt levels were less than 25 mg/kg ash weight in postmortem brain samples of both normal (3) and chronically dialyzed patients (5), as quantitated by charged particle-induced x-ray pattern technique.

Cobalt metal implants have been shown to be highly toxic to brain tissue, but Chusid and Kopeloff (1962) found no toxic signs after implanting a monkey motor cortex with cobalt metal and leaving it 21 months. Both postmortem histopathology and clinical studies were performed; the latter included serial electrocardiograms, synergism studies with seizure-producing drugs, and clinical observation.

Kalekin and Brichenko (1972) reported that 118 workers showed acute industrial cobalt carbonyl vapor intoxication. The acute intoxication, which was brought about by use of cobalt carbonyl catalyst for the oxo synthesis of butanol from propylene, produced only transient symptoms, according to these authors. The symptoms related to the central nervous system included headaches, weakness and irritability, changes in the electrical activity of the brain, and changes in reflexes, especially visual. Skin disorders were also reported.

Licht et al. (1972) reported on a case of cobalt toxicity in a 32-year-old male who was given a total of 73 g Co over a 2.5-year period in four dosage courses. The fourth course, with a total dose of 30 g $CoCl_2$, produced visior symptoms. The earlier dosages had been discontinued due to nausea and vomiting in this patient. The eye examination showed pallor of the optic disks, and small hemorrhages were seen on fundal examination. The patient was diagnosed as having optic nerve atrophy, and Licht et al. (1972) attributed the findings to the cobalt dosage. These authors cited another case history of optic nerve damage attributed to cobalt chloride (Alagna and D'Aguino, 1956; cited by Licht et al., 1972). In the Italian study, pathological changes were reported in a patient's lens, retina, choroid, and optic nerve following the administration of a total dose of 900 mg $CoCl_2$ for 30 days. Nerve damage with fatty degeneration of retinal and ganglion cells was seen. Alagna and D'Aguino report that, in their patient, the "fibers of the optic nerve were swollen and tortuous with fragmentation in the myelin sheath."

(v) Renal effects: Patients in chronic renal failure are commonly anemic; several clinical studies have been performed with cobalt as the treatment for this anemia (Anonymous, 1976a). Other workers have noted that the mechanism of the hematopoietic effects are not known and that cobalt may be toxic to the patients (Wardle, 1976). Several reports are summarized here.

Duckham and Lee (1976) concluded that cobaltous chloride has a definite place in the treatment of the refractory anemia of chronic renal failure after their study of 12 anephric patients on maintenance hemodialysis. These patients received enteric-coated tablets of cobaltous chloride, 25 to 50 mg daily for 12 weeks. Eight patients completed the 12-week, 50 mg/day course, and six of them had a significant rise in hemoglobin (from 26 to 70%). When cobalt was withdrawn, hemoglobin fell to near the original level. The hemoglobin rose again in four patients who had a second course and in one patient who had a third course. After about 2 months at 50 mg/day, serum cobalt levels stabilized at 40 to 100 μg/100 ml. Two of 12 of the patients reported toxic symptoms; one, nausea and constipation, and the other, high-tone deafness.

Fourteen patients on maintenance hemodialysis were given cobaltous chloride orally twice a day in a total daily dose of 25 mg for the first 4 weeks and then 50 mg for a further 4 weeks (Bowie and Hurley, 1975). The hematocrit increased very little during the first 4 weeks, but the 11 patients who finished 8 weeks of cobalt had a 23% increase in hematocrit. Toxic signs included one patient with nausea, and the authors also reported three of the patients had a hearing loss confirmed by audiometry. No liver or thyroid dysfunction was seen.

The cobalt serum levels reported by Bowie and Hurley (1975) are of interest in that three of the patients showed increases in serum cobalt after undergoing 6-hr hemodialysis. Hamilton (1974) recorded concentrations of cobalt in water from a soft water area in the United Kingdom used in a hospital renal dialysis unit (RDU).

TABLE VIII-19

COBALT CONCENTRATIONS IN WATER USED IN A HOSPITAL RENAL DIALYSIS UNIT
(Hamilton, 1974)

Source	Cobalt, μg/liter
Filtered (0.45 μ) natural water	0.5
Unfiltered acidified natural water	0.9
Local tapwater (new house, plastic and copper plumbing)	1.0
RDU raw feed	1.0
RDU coarsely filtered water	1.0
RDU post-course filtered water	2.0
RDU post-main filter and sintered bronze filter	1.0
RDU post-deionizer	< 1

As can be seen in Table VIII-19, the post-course filtered water of the RDU had more cobalt than the raw feeds, but the deionized RDU water had less.

Curtis et al. (1976) gave 50 mg/day enteric-coated cobaltous chloride for 3 months to 23 patients on maintenance hemodialysis. The side effects were mainly nausea and vomiting, which appeared early (Edwards and Curtis, 1971). One patient, however, died of resistant congestive cardiac failure 3 months after receiving the cobalt dosage. Myocardial cobalt concentration by neutron activation analysis was 1.65 mg/kg wet weight, which was 25 to 80 times greater than the cobalt levels in tissue of two dialysis patients who never received cobalt (average 0.03 mg/kg) and in two other patients who had not had renal failure (average 0.06 mg/kg). Curtis et al. (1976) also measured with neutron activation analysis the whole blood cobalt concentrations in dialysis patients who had been treated 13 to 20 months previously with cobalt. The metal levels were significantly higher in those patients than in dialysis patients who had not received cobalt. The authors concluded that the prolonged retention of cobalt in the blood of patients with severe renal failure, as well as in normal patients, could lead to toxicity. The possibility that one patient had a cobalt-induced cardiomyopathy and the "limited therapeutic gains" by cobalt treatment led to the recommendation that the metal not be used.

The mechanism of the hemoglobin increase by cobalt is still open to speculation, but much of the literature has indicated that the kidney and erythropoietin are involved (Goldwasser et al., 1958). Duckham and Lee's patients, however, were anephric (1976), and Curtis et al. (1976) reported no difference in response to cobalt between patients who lacked kidneys and those with some kidney function. Necas and Neuwirt (1971; cited by Anonymous, 1976b) suggested that cobalt acts by producing local ischemia of kidney receptors controlling erythropoietin production. This is done by suppression of cellular energy metabolism by inhibition of sulfhydryl groups of some oxidative enzymes, resulting in diminished oxygen utilization by the tissues. It would have to be hypothesized, however, that in the anephric patient, cobalt acts by extrarenal sources of erythropoietin.

(vi) Allergy and dermal hypersensitivity

(a') Inhalation exposure: The immunologic status of four subjects with pulmonary fibrosis related to inhalation of cobalt and tungsten was measured (Baudouin et al., 1974). They were all employees of the hard metal industry. Two had developed respiratory signs after only 2 months' exposure, and the other two developed a bronchiogenic irritation after about 9 years' exposure. Their ages varied from 22 to 65 years. The immunologic studies performed included eight measures of humoral and cellular responses. One of the four had a greatly elevated IgE (975 IU); two of the four had antinuclear antibodies and positive tuberculin reactions. None of the four reacted positively to intradermal challenge with cobalt or the other metals. The lymphoblast transformation, DNCB challenge, PHA stimulation of cultured lymphocytes, and some *in vitro* autoimmune tests were all within normal ranges. The authors concluded, however, that the tests in general indicate a cellular immunologic deficit but could not say whether this was preexistent or due to the dust exposure (Baudouin et al., 1974).

(b') Dermal exposure: In a small percentage of exposed persons, a hypersensitivity to cobalt may develop. An NIOSH criteria document on cemented tungsten carbide (1977) cited six studies which gave numbers of workers as well as those reporting dermal hypersensitivity or respiratory allergic symptoms. There were over 4,200 workers included in these reports, and the incidence of allergic symptoms in the populations ranged from 1.7 to 9.4%, an average 4.8% of the workers studied. Symptoms on exposure included an erythematous, papular type of dermatitis, and positive patch tests to metallic cobalt powder were often reported. The distribution of skin eruptions in industrial workers exposed to cobalt dusts was most marked in friction areas.

Cobalt concentrations of 200 mg/kg, which are found in the cutting fluids used in grinding hard metal alloys containing cobalt, were reported to produce eczema when the human skin was exposed (Einarsson et al., 1979). Using atomic absorption spectrometry (AAS), the authors quantitated the dissolved cobalt in a Swedish factory's circulating coolant liquid to see if the levels posed a hazard to workers' skin. The alloys contained 5 to 10% cobalt; the cutting fluid, 99% water and 1% oil. After 5 days, eight different formulations of cutting fluid contained an average of 196.8 mg Co/liter in solution, with a range of 7 to 552 mg Co/liter. Air samples contributed little to the workers' hazard from cobalt; breathing zone samples contained up to 0.02 mg Co/m^3.

According to Hueper (1974), human contact with cobalt may occur by use or work with alloys, plating, glass pigment, ceramics, paint, rubber, and drugs. Although it does not occur as a cosmetic ingredient in the United States, a recent Japanese patent for a bluegreen cosmetic coloring included cobalt lithium titanium oxide (Suzuki et al., 1976) which, without giving detailed data, was said to be non-irritating. The Polish literature contained two references on contact allergy to the "chemical preparations used in modern agriculture," and one of the causative agents was said to be cobalt (Szarmach and Poniecka, 1975). In skin tests, 18 of 200 farm workers with allergic-toxic dermatitis showed sensitivity to cobalt (Szarmach and Poniecka, 1973). In western Scotland, over 9% of 4,500 dermal allergy patients tested in a large clinic had a positive reaction to a patch test of 1% cobaltous chloride in paraffin oil (Husain, 1977).

Fregert et al. (1972) reported finding 9 and 7 mg Co/kg in samples from different stages of a sulfate pulp process (where the metal was not added). The cobalt was sought when a worker given a routine patch test was found to have developed a skin sensitivity to cobalt.

Calnan et al. (1970) cited a group of 281 European domestic workers who had positive patch tests and contact dermatitis. Eight percent gave a positive reaction to 2% cobaltous chloride. A domestic-use product, detergent, was recently analyzed for cobalt by Wahlberg et al. (1977). Sixteen of 19 samples of Swedish detergent were found very low in cobalt; they contained less than 1 mg/kg and in no case did a sample exceed 4 mg/kg (atomic absorption spectrophotometry).*

There is a high incidence of dermal sensitivity to cobalt among cement workers; Wahlberg (1969) reported a positive reaction to 2% $CoCl_2$ in 5 of 47 common laborers, and Förström and Pirilä (1975) reported 291 cases of cobalt reactivity in 4,529 cases of occupational dermatosis in Finland. Patch tests on 200 French cement workers using a 2% cobalt sulfate challenge dose gave 20% positive reactions (Robin and Brunetière, 1969). Robin and Brunetière cited two other French studies that had used cobalt chloride for skin testing and had shown dermal reactivity in 51 and 38.3% of the cases out of populations of undescribed size.

Cobalt was present in a series of eight brands of Swedish cement at concentrations of 5 to 16 mg/kg as determined by atomic absorption spectrophotometry by Wahlberg and co-workers (1977). A brand of Swedish cement containing 41 mg Co/kg (atomic absorption spectrometric determination) was used to study allergic contact dermatitis by Fregert and Gruvberger (1978). These workers reported that free cobalt oxides occur in cement that are good skin sensitizers and poorly water soluble. These cobalt oxides are more soluble in the presence of amino acids (such as are found in eczematous skin) with which the cobalt forms complexes.

Christie et al. (1976) found that cobalt poorly penetrates normal skin. Cobalt-containing ointment was rubbed on the skin, removed, and the skin content of cobalt studied using spark source mass spectrographic analysis. Cobalt quantities measured in mg/kg: 1,250, immediately after application; 6.61 to 9.55, 1 week later; 2.41 to 4.11, 2 weeks later; 0.18 to 0.31, 3 weeks later.

---

* Detergents might be expected to contain cobalt since cobalt catalysts are used in the synthesis of the long-chain fatty alcohols that are used to manufacture synthetic detergents.

Wahlberg (1973) looked at skin penetration of cobalt in another study. He showed that in 60 patients with dermal allergy to cobalt or cobalt plus other metals that 32% had a lower threshold of skin sensitivity when the challenge dose of $CoCl_2$ was mixed in water and 25% had a lower threshold when the challenge dose was mixed in petrolatum.

Several rather unusual case histories of allergic dermatitis due to cobalt were noted in the literature. An 11-year-old boy had a 4-year history of dermatitis around the mouth, eyes, and wrist. The hypersensitivity to cobalt was discovered, and the contact points of his polyester spectacle frames, his nickel-plated wristwatch band, and his sucking on polyester ballpoint and fountain pens were diagnosed as causing the tact allergic reaction (Grimm, 1971). Cobalt is present in polyester as a trace impurity due to its use as a catalyst, and it is a common impurity in nickel. The cobalt oxide in a high-energy magnetic tape was found to produce a cobalt eczema in a Swedish TV worker (Krook et al., 1977), and Malten (1975) described a patient with contact eczema to cobalt chloride who also gave a bullous flare reaction on her hands and feet after an injection of vitamin B_{12}.

A 23-year-old male with a long-standing dermal response to cobalt gave a positive response to intradermal challenge with cobalt nitrate 1:100,000 dilution. He also gave a positive response to vitamin B_{12} (Rostenberg and Perkins, 1945; cited by Cass and Lange, 1955).

As noted above, some patients with dermal hypersensitivity to cobalt are reactive to very low levels of cobalt in patch tests. Wahlberg (1973), using serial dilutions of cobalt chloride, determined that the mean threshold for 60 cobalt-allergic patients was 0.27% in distilled water and 0.31% in petrolatum. He also noted, however, that 14 of the subjects were exquisitely sensitive to the cobalt challenge and at no dilution was the threshold of sensitivity obtained.

Many nonindustrial cases of dermal hypersensitivity to cobalt have been reported. A large number of orthopedic studies on the rejection of prosthetic implants have suggested that a cobalt allergy contributes to this rejection (Marinoni and Torelli, 1976). Herrmann, on the other hand (1977), suggests that the chromium-cobalt alloys used in dental materials are rarely immunogenic with that use.

An allergic etiology for the rejection of metal prosthetic implants was suggested by Munro-Ashman and Miller (1976), who tested 35 patients who had severe reactions to implants made of 60 to 65% cobalt, 27% chromium, and 2.5% nickel plus traces of five other metals. Sixteen of the 35 gave a positive skin test to metals, and 13 of those were sensitive to the cobalt in the alloy. One of the patients had a localized dermal reactivity to cobalt, a widespread scattered circular erythematous lesion suggestive, according to the authors, of a generalized allergic vasculitis.

Halpin (1975) reported a hypersensitivity reaction to a Vitallium implant which occurred 7 years after inserting the plate over fractures. Patch tests showed sensitivity to cobalt. In the localized area, there was massive fibrosis and patchy necrosis of the muscle.

(c') <u>Allergic reactions to vitamin B_{12}</u>: Sensitivity to vitamin B_{12} as an allergic reaction was described by Nalivko and Fedorovich (1974; cited by Tracor-Jitco, 1974). A 47-year-old woman welder with allergic dermatitis had a positive patch test for cobaltous chloride, and the dermatitis was thought connected to exposure to metal dust and fumes with which she had been in contact for 12 years. She received vitamin B_{12} injection as part of her treatment, 3.3 μg/kg or a total dose of 200 μg. Symptoms which occurred within 40 min of injection included headache; nausea; increased body temperature; intense itching and blistering; and edema of subcutaneous tissue on the buttocks, back, abdomen, extremities, and breasts. Since 4.5% of vitamin B_{12} is cobalt, the suggestion of the authors that the allergic response was due to vitamin B_{12} is reasonable.

Thirteen patients 13 to 67 years of age, receiving vitamin B_{12} either orally or by injection, developed acneiform exanthemata or had a worsening of preexisting acne (Braun-Falco and Lincke, 1976; cited in FDA, 1977). The acne improved when the vitamin was discontinued.

(vii) <u>Other toxic effects of exposure</u>: Takki et al. (1972) reported a fatality due to intentional ingestion of 150 ml of a plant nutrient substance that contained 0.1% $CoSO_4$ as well as copper and nickel salts in an ammonium base. In this case, the lungs were severely affected with congestion, edema, and hemorrhage on gross and microscopic examination. The authors, in experiments on rabbits, duplicated the effects and suggested that the toxicity of the plant nutrient substance might be due to the combined effect of the NH_4^+ with Cu^{2+}, Co^{2+}, and Ni^{2+}. All the heavy metal salts were present at the 0.1% level.

Penati and Ruata (1940; cited by Cass and Lange, 1955) fed seven normal adults 3 to 6 mg Co/kg/day as finely powdered metallic cobalt. Mild gastrointestinal disturbances were the only effects noted.

After 3 to 4 days, the subjects excreted in the urine 1.13 to 6.52% of the total dose; the maximum was excreted from the second to fourth day (Penati and Ruata, 1940; cited by Cass and Lange, 1955).

Jacobziner and Raybin (1961) reported the death of a 19-month-old male child who drank about 1 fl oz of a $CoCl_2$ solution. (The $CoCl_2$ was to be used as a "weather indicator" and was probably a saturated aqueous solution.) Although the child was induced to vomit 4 to 5 times and was hospitalized, he died 6.5 hr after the incident with necrosis of the gastric mucosa and cerebral edema.

(b) <u>Therapeutic use of cobalt in man and other mammals</u>: Both cobalt chloride and cobalt sulfate, 1% added to salt as part of their feed, improved the weight gain and disease resistance of sheep in a study reported from central Kazakhstan (USSR). Zhilin (1969) reported that the cobalt-dosed sheep also had a lower incidence of cestode infestation than did controls.

Dikov et al. (1973) showed the effectiveness of $CoCl_2$ in treating experimental pisiform cysticercosis in rabbits. Daily doses of 0.1, 0.25 or 0.5 mg/kg for 2 months was 100% effective against cysticercosis, and the metal was reported as producing no adverse effects on the blood composition of the rabbits.

Young (1948) reported that the role of cobalt in animal deficiency diseases (such as pining disease of cattle, grazed on cobalt-deficient soil) has been recognized since 1935. Elvehjem's group, however, was uniformly unsuccessful in several early attempts to establish cobalt deficiency disease experimentally in rats (Underwood and Elvehjem, 1938).

Experimental dysentery in swine was successfully prevented and treated with cobalt arsanilate 0.05% added to feed (Olson and Rodabaugh, 1973). Survival among medicated exposed swine groups fed the 0.05% concentration was 63%, and among nonmedicated exposed groups, 40% survived. Clinical signs of cobalt toxicity were observed in animals dosed with 0.2% of the feed weight. Based on the feed consumption data and swine weights furnished by the authors, the effective dose (0.05% of feed weight) was an average total dose of 1.46 g/km body weight given over 5 days. Toxicity appeared at a total cobalt arsanilate dose of 12.8 g/kg given over 5 days.

In 1955, Holly (cited by Villaume et al., 1976) treated 94 pregnant women with 60 to 100 mg cobalt daily as cobaltous chloride for at least the third trimester of pregnancy. Holly concluded that cobalt did not alter blood indicators when dosed alone and produced only a slight improvement over iron alone when dosed with the iron. (See Table VIII-20.) No mention was made of toxic effects to either mother or fetus.

TABLE VIII-20

HEMOGLOBIN AND RBC VOLUME AT DELIVERY AFTER DOSING PREGNANT WOMEN WITH COBALTOUS CHLORIDE
(Holly, 1955; cited by Villaume et al., 1976)

Population Group	Gram Hemoglobin/100 ml	% RBC Volume
Untreated controls	11.4	35.5
60 to 100 mg Co/day	11.4	35.5
0.8 to 1.0 mg iron/day	12.9	39.5
Cobalt + iron	13.1	41.5

El-Masry et al. (1975) demonstrated the effectiveness of cobalt nitrate, cobalt acetate, and cobalt-EDTA as antidotes for cyanide poisoning in albino mice. See Table VIII-21.

TABLE VIII-21

COBALT SALTS AS ANTIDOTE TO CYANIDE POISONING IN MICE
(El-Masry et al., 1975)

Treatment Group	Percent Survival, 5 mg/kg Cyanide Dose: Antidote given with cyanide	Antidote 3 min after cyanide	Antidote 5 min after cyanide
Control	(no antidote, 0 survivors of 20 in 10 min)		
Cobalt-EDTA (60 mg/kg)	100%	95%	85%
Cobalt nitrate (40 mg/kg)	100%	90%	80%
Cobalt acetate (60 mg/kg)	95%	85%	70%
Na nitrite (usual antidote)	80%	55%	40%

The toxicity of cyanide is due to its very rapid mucous membrane diffusion and penetration, stopping cellular respiration by complexing with the trivalent ferric ion of the cytochrome oxidase in the mitochondria, and, with the iron of hemoglogin, forming cyanomethemoglobin. Cyanide is metabolized or counteracted by three pathways: high concentrations of methemoglobin allow minimal formation of cytochrome oxidase-cyanide complexes; rhodanase catalyzes the conversion of cyanide to thiocyanate; and a minor pathway combines cystine with cyanide to form 2-aminothiazolidine-4-carboxylic acid, which is oxidized to carbon dioxide and formate with the formation of cyanocobalamin (Williams, 1959; cited by El-Masry et al., 1975). It may be enhancement of the third pathway by which the cobalt compounds work.

Nagler et al. (1978) gave three case reports of workers treated with cobalt-EDTA for exposure to hydrogen cyanide. The mechanism of the effect was in lowering the CN^- level in plasma. The toxicity of the CN^- radical is essentially due to its ability to form complexes with metal ions, specifically trivalent iron in enzymes such as cytochrome oxidase. Cobalt-EDTA in an effective dose of two ampules of 300 mg cobalt-EDTA, given intravenously, was effective and lifesaving in one case in which about 275 g of HCN was liberated; 50 ppm in air can be fatal. Nagler et al. (1978) mentioned, however, that the treatment had toxic effects of its own. The intravenous cobalt-EDTA produced profuse sweating, angina, intense nausea, vomiting, and ectopic heart beats. Naughton (1974) was cited (Nagler, 1978) as finding sudden development of atrial fibrillation with ventricular ectopic beats after injection of cobalt-EDTA. Nagler's three patients showed major ventricular rhythm disturbances.

Bryson (1978) in a letter to a major medical journal, also noted the toxicity of the cobalt-EDTA when used as a cyanide antagonist. He suggested that since cyanide works very rapidly, the severe toxicity of the antidote should be reserved for acute cases of cyanide exposure that are truly life-threatening.

(c) Prosthetics containing cobalt: There are three principal metallic implant materials for medical and orthopedic use: stainless steel, cobalt alloys, and titanium.

Cobalt-base alloys have been used for orthopedic devices for about 40 years. Their superior corrosion resistance and strength as compared with stainless steel alloys have made them widely used (Harth, 1974). The cobalt-base alloys are marketed for orthopedic devices under trade names of Zimaloy, Vitallium, and others. All of these are currently of two basic types of alloy. One type is for cast parts (ASTM F75-67) and is 60 to 65% cobalt with chromium, molybdenum, and nickel. The other is for wrought products (ASTM F90-68) 50 to 55% cobalt with chromium, tungsten, nickel, and iron (Wagner et al., DMIC report 171, 1962; cited in Harth, 1974).

Smith (1974) reviewed the literature on the materials used for prosthetic implants, but no physiological data on the cobalt-containing alloys were included. It was noted that Vitallium alloy has been used in dentistry since 1936; and it currently consists of a slightly different alloy than surgical Vitallium, higher chromium and carbon and lower silicon and manganese (Fitzpatrick, 1968; cited by Smith, 1974). The cobalt content of both surgical casting alloy and dental alloy are apparently similar: about 62%. Since the saliva tends to be a more acidic medium than the tissue fluids around a prosthetic joint, it is possible that the loss of cobalt may be greater from dental than from bone prostheses, but no studies on this were found.

Herrmann (1977) reviewed the topic of tissue reactions to dental materials including cobalt-chromium alloys used to cast prostheses. Adverse reactions of an allergic nature were said to be rarer than generally assumed; and, according to the review, evidence shows that the nickel component of the alloys (which is 1 to 20% of the total metal) is most often responsible for tissue reactivity.

Although the cobalt-based alloys have been in use some years, Parsons and Ruff (1973) report finding little specific long-term toxicity data on implants (greater than 10 years' continuous implant time).

Laing et al. (1967; cited by Parsons and Ruff, 1973) performed a 6-month toxicity study, implanting both cast and wrought cobalt-containing alloys in rabbits. He found the tissue reaction greater for cobalt alloys than for titanium alloys and reported "significant" (no numbers) element release into the surrounding tissue in the 6-month period. The cobalt alloys released fewer elements into the tissues than did the titanium alloys.

Heath et al. (1971) prepared wear debris from a machine-operated prosthesis and injected 28 mg mixed in 0.4 ml horse serum into muscle tissue of female hooded rats. The particles contained 66.5% cobalt, 26.0% chromium, 6.65% molybdenum and 1.12% manganese, and some were as small as 0.1 μm in diameter. Seven fibrosarcomas of 14 which appeared over the next 15 months were histologically analyzed and were attributed to the wear debris.

A letter in the same medical journal that published Heath's work later emphasized that the rat was quite prone to fibrosarcoma production, and that in hundreds of human patients receiving prosthetic implants made of cobalt-chromium-molybdenum alloy for over 14 years, no fibrosarcomas had been seen clinically (McKee, 1971).

Jones et al. (1975) described the symptoms of seven patients who had received prosthetic implants made of cobalt alloy, six of which were cobalt-positive (but nickel- and chromium-negative) on patch testing. Four of the patients had high urinary cobalt levels, averaging 33.5 ± 12.04 μg/liter. All reported had no evidence of infection but had repeated sterile abscesses around the transplants and necrosis of the bone and muscle in the area of the transplant. The cobalt content in the fluid aspirated from the joint in one patient averaged 250 μg/liter. Jones et al. (1975) stressed the importance of hypersensitivity to cobalt in these results. Thirty good prosthetic implant results were made in patients with no reaction to cobalt, and six of the seven patients with poor results had positive reactions to the metal.

No reference was made to an allergic etiology of metallosis by Lux et al. (1975). This group used neutron activation analysis to measure several metals in tissue samples of 30 to 100 mg removed from the area of prosthetic implants in experimental animals (species not given). Lux et al. suggested that the causative factors of the inflammatory change near the metal implants involved the chemical supercession of essential trace elements from their biological functions by the components of the implants which are released.

Ungethuem (1973) measured the levels of cobalt and chromium in autopsy samples taken from the capsule of a Vitallium prosthesis, implanted 3 years before. Chromium was found at expected levels, but there was less cobalt than expected. This finding suggested to the author that cobalt diffuses from the joint capsule around a Vitallium prosthesis to other parts of the body. The low level of cobalt around the implant in this patient may also reflect an inflammatory state, since Jones et al. (1975) reported very high cobalt levels around the bone in his patients with sterile abscesses. Transport of cobalt into the circulation was quantitated by Coleman et al. (1972), who measured cobalt in hair, urine and blood of patients with cobalt-chrome alloy hip joints. In blood, cobalt averaged 0.6 μg/liter before surgery to 6.5 μg/liter after a few months.

Marinoni and Torelli (1976) measured the reactivity to several implant components including cobalt and a nickel-cobalt combination in 42 patients who had surgically received implants of 31 metal and 11 plastic prostheses. Both patch test and *in vitro* cellular reactivity to the metals were tested. Cobalt reactivity occurred in 6.4% of the patients; fewer were sensitive to plastic (1%) and 9.7% had a sensitivity to chromium.

Ludwigson in 1964 discussed two promising cobalt-containing alloys which were apparently never widely used as prosthetic materials. Elgiloy was the name of the complex cobalt-base alloy used in watch springs and heart valves. No reports of toxicity studies on Elgiloy were found (Smith, 1974), but Ludwigson said it was highly resistant to corrosive attack by the cardiac tissue. The second alloy, Ticonium, contained almost equal amounts of nickel, cobalt, and chromium, plus about 6% molybdenum. Without giving data, Ludwigson (1964) reported that Ticonium gave "favorable animal experiments."

Younkin (1974) described a new material for metallic implants that has highly satisfactory mechanical properties and corrosion resistance, with years of satisfactory *in vitro* use. The material, MP35N, contains 20% Cr, 35% Ni, 35% Co, and 10% Mo. The toxicology information on the metal included a 1-month muscle irritation study using albino rabbits (Hazleton Laboratories, Dr. L. D. Shott). In this study, the tissue from the rabbits containing the MP35N implants had lower incidences of inflammatory changes than tissue from Vitallium or stainless steel-implanted rabbits. A long-term toxicity study being conducted by the Orthopedic University Clinic in Frankfurt, Germany was incomplete at the time, but early indications showed MP35N to be no different in tissue reactions from Vitallium or stainless implants.

(d) Exposure through tobacco products: Cogbill and Hobbs (1957; cited by Stahly, 1973) failed to identify cobalt in analysis of tobacco and smoke from five brands of cigarettes. Zarrin cigarette tobacco, wrapping paper, and its smoke (Iranian brand) were analyzed for cobalt using neutron activation analysis (Abedinzadeh et al., 1977). Tobacco contained 0.21 ± 0.01 mg/kg; cigarette wrapping paper, 12.9 ± 0.8 mg/kg; and cigarette smoke condensate, 0.14 ± 0.01 mg/kg. The authors calculated that only 0.68% of the cobalt present in the paper and tobacco transferred into the ~ 90 μg/cigarette smoke condensate. This percentage transference varies with the metal studied.

Four brands of cigarettes and eight samples of pipe tobaccos (all probably of U.S. origin) were burned by intermittent "puffing." The paper, tobacco, smoke, and cigarette filters were analyzed for cobalt and cobalt carbonyls using atomic absorption, emission spectroscopy, and colorimetry with EDTA reagent. Results for a single brand of cigarette were (μg Co/cigarette): tobacco, 1.0; filter (before smoking), 0.25; paper, 0.05; ashes, 0.5; butts, 0.2; smoke, 0.1; and filter (after smoking), 0.3. Qualitative data also showed the presence of cobalt carbonyls in tobacco smoke formed between puffs at temperatures between 50 and 140°C (Stahly, 1973).

Nadkarni and Ehmann (1969) prepared a reference cigarette by blending the four major tobacco types (flue-cured, Burley, Oriental, and Maryland), using glycerol and sugar as humectant, and wrapping in low porosity phosphate paper. The reference cigarette is 85 mm long, 25 mm in circumference, and is nonfilter. These cigarettes were distributed for work on tobacco and health research. Neutron activation analysis showed the reference cigarette contained 0.50 mg Co/kg, which agrees with values in a popular commercial brand of cigarette.

Nadkarni et al. (1970) determined (by neutron activation analysis) cobalt concentrations (mg/kg) in filter (and nonfilter) cigarette tobacco, 0.49 (0.44); smoke condensate, 1.26 (0.69); and ash, 4.2 (3.9). The percent of transference of cobalt from tobacco to the smoke condensate was 4.2% for the unfiltered cigarettes and only 0.87% for filtered. The authors noted significant differences in the elements found in smoke condensates depending on which method was used for preparing smoke condensate.

Mass spectrographic analysis of mainstream smoke from cigarettes (average of two or more) showed 0.1 ng cobalt per cigarette (Allen and Vickroy, 1976).

Wyttenbach et al. (1976) reported that the cobalt content of 20 different raw tobaccos was ~ 0.3 mg/kg with a coefficient of variation of 4% on four replicates. Neutron activation analysis was used for the analysis.

(2) Occupational exposure

(a) Symptoms: Kaplun (1955) described the symptoms of industrial workers exposed to cobalt compounds (probably in the tungsten carbide industry). General complaints and symptoms are enumerated below.

- Effect on sense of smell suggested as an early symptom of cobalt metal toxicity; 25 of 28 cobalt workers had decreased sense of smell and 3 of 28 had anosmia.
- Nausea was common; occasional vomiting occurred. Workers had stomach pains and lowered appetite.
- Four of 28 cobalt workers showed a decrease in hemoglobin and RBC count.
- Four of 28 had enlarged spleen.
- Four of 28 had dermatitis.
- Eight of 28 had respiratory changes: four had x-ray evidence of early pneumosclerosis. Some complained of a chronic productive cough.

Dorsit (1970) described the first symptoms of toxicity from chronic inhalation of cobalt dust as weight loss and dyspnea that is progressive. Lung sounds included rales at the bases of the lungs. There was no increase in the sedimentation rate, but there was increased red cell count and elevated hemoglobin. Dorsit, although commenting that cobalt fibrosis is known to associate with pulmonary tuberculosis in the French clinical cases, found little association with cancer. Bech et al. (1962; cited by Dorsit, 1970) described the one associated case. A man, age 59, with 13 years' exposure to industrial dusts in the tungsten carbide industry was seen with lung fibrotic changes. He worked in the industry four more years and then died of bronchiogenic cancer occurring in the same area of the respiratory tract where the fibrotic change was first seen 8 years earlier.

Scherrer et al. (1970) reported three cases of hard metal lung disease; and in all three cases, the symptoms were different. The first patient had a rapid onset and slow healing; the second patient, a rapid onset and development of an irreversible form of pulmonary fibrosis; and the third patient, a slow onset and irreversible form of pulmonary fibrosis. In all cases, exposure to cobalt or tungsten dusts was strictly controlled after diagnosis.

Inhalation of cobalt dust in the hard metal industry and signs of resultant toxic effects were summarized by Tolot et al. (1970). Lung signs included specific radiographic images of fibrotic change. Dyspnea and weight loss occurred consistently in the described populations.

The symptoms most commonly reported after occupational exposure to cobalt (and related metals) in the cemented tungsten carbide industry are (NIOSH, 1977): upper respiratory tract irritation, exertional dyspnea, coughing, weight loss, extrinsic asthma, diffuse interstitial pneumonitis, fibrosis, pneumoconiosis (or some combinations of the above).

As a general observation, about 5% of the complaints were allergic in nature (data of NIOSH, 1977).

(b) Case histories: The 1977 review of the occupational exposure by workers in the cemented tungsten carbide industry for the National Institute for Occupational Safety and Health (NIOSH, 1977) covers a large number of case histories and group exposure reports with the level of cobalt quantitated. These have been summarized in Table VIII-22.

Lichtenstein et al. (1975) monitored workplace air and 22 workers 31 to 60 years of age who had worked at grinding tungsten carbide tools for 1 to 30 years (average of 11 years). Red cell counts on the workers were normal, and the frequent complaints of nonproductive coughs were not correlated with length of employment when the population was divided by smokers and nonsmokers.

The workplace air was monitored, and breathing zone exposure to cobalt in the coolant mist aerosols ranged from 0.03 to 0.56 mg/m^3, with a mean time-weighted average of 0.24 mg/m^3. Corrective actions including well-engineered local exhausts reduced exposure considerably in this workplace.

McDermott (1971) estimated that 3,000 workers may be at risk of pneumoconiosis from cobalt plus tungsten carbide in Michigan's hard metal industrial plants or other plants where carbide tools are sharpened. The author obtained 173 samples from seven plants and concluded that the cobalt levels in 121 of the 173 samples were below the 0.1 mg/m^3 TLV in effect at the time of the study.

X-ray examination revealed "granulated conglomerate" markings in the lungs of 24%* of the white male workers in a tungsten carbide industrial study, where conjunctiviti and upper respiratory tract infection also occurred (Fairhall et al., 1947; cited by Browning, 1969). Although the workers had been exposed to a variety of lung-reactive toxicants, the authors reported that cobalt was thought to be the most likely cause of the effects. They cite another study wherein three cases improved and seven had fewer lung shadows on x-ray after cobalt was removed from their environment.

---

* This was a lower incidence than in coal mine, metal mine, and smelter workers.

TABLE VIII-22

OCCUPATIONAL EXPOSURE TO COBALT FROM CEMENTED TUNGSTEN CARBIDES, CASE HISTORIES
(NIOSH, 1977)

Cobalt Content of Cemented Tungsten Carbide[a]	Exposure		No. Exposed	No. Showing Effects					Authors Cited
	Level[a]	Duration[a]		Skin, eye Irritation	Respiratory Toxicity	Cobalt Allergy Shown[b]	General Effects (Liver, Fever, Cardiovascular Toxicity)[b]	Death	
6% Co	195 to 1,230 particles/ml	NR	255		21				Bech et al. (1962)
5 to 25% Co	13 to 100 mg/m^3	NR	163		91				Barborik (1966)
NR	0.27 to 1.75 mg Co/m^3	NR	247		185				Kaplun and Mezentseva (1959)
2.4 to 4.1% Co	2.1 to 3.5 mg/m^3	6 years	1,802	X	36	X			Fairhall et al. (1947)
NR	0.6 to 3.2 mg Co/m^3	Up to 3 years	178		88		X		Vengerskaya and Salikhodzhaev (1962)
NR	0.04 to 0.93 mg Co/m^3	1 to 30 years	22		11				Lichtenstein et al. (1975)
									(1974)
NR	2.3 to 62.3 mg/m^3	Up to 17 years	29		10	3			Dorsit et al. (1970)
"With Co metal"	NR	NR	100		15		5		Baudouin et al. (1975)
3 to 25% Co	NR	12 to 28 years	12		12		8		Coates and Watson (1971)
7.6 to 21.2% Co	277 to 4,064 particles/m^3	NR	208		X	X			Heuer (1962)
NR	NR	1 month, plus	1,200	20		6			Schwartz et al. (1945)
5 to 25% Co	NR	NR	361	34		14			Skog (1963)
6% Co	195 to 505 particles/ml	~ 2 to 22 years	6		6			1	Bech et al. (1974)

a/ NR = Not reported
b/ X = Reported no No.

Pulmonary complaints by Swedish workers in the tungsten carbide industry occurred when the cobalt content of the atmospheric dust was only 2.8% (Lindgren and Ohman, 1954; cited by Browning, 1969). The Fairhall report, above, stated that the cobalt dust in some parts of their operation (calcination of cobalt nitrate) reached 79 mg/m^3, but averaged only 3 to 4 μg Co/m^3 overall.

Brakhnova (1975) reviewed the morbidity of workers exposed to cobalt dust in Russian industry, citing cases of skin lesions in workers exposed to cobalt. Without giving exposure levels, the following symptoms were cited: "acute dermatitis in the form of numerous red papules and nodules, and also edema of the hands and exposed portions of the body; sometimes surficial (sic) ulcerations, papules and nodules." Similar effects were reported for dicobalt octacarbonyl, $Co_2(CO)_8$, which in man produces depilation and dry scabs, which are said to heal despite continued contact with the compound.

In 1975, it was noted that diffuse interstitial fibrosis related to production of hard metal was not recognized as an industrial disease in France. Baudouin et al. (1975) gave a case history of 100 workers engaged in tungsten carbide processing, 15 of which showed signs of lung irritation, and 5 of which had progressed to functional clinical, radiological pulmonary fibrosis.

Long-term exposure (12.6 years average) to industrial dusts containing less than 0.1 mg Co/m^3 produced diffuse, interstitial lung disease with early symptoms of a cough and dyspnea on exertion, with a 50% vital capacity. In the 12 cases reported by Coates and Watson (1971; cited by NIOSH, 1977), pulmonary function changes preceded x-ray evidence of non-specific diffuse interstitial fibrosis. There is an allergy-like symptom pattern to the respiratory involvement, with improvement of symptoms on weekends or leave from the work environment and a return of symptoms on re-exposure. Bruckner (1967; cited in NIOSH, 1977) noted that asthmatic symptoms returned to one of hi hypersensitive workers as soon as 1 to 3 min after beginning work.

Kaplun (1963; translated 1967) reported on 247 workers exposed to cobalt in the USSR hard metal industry. They had all received medical examinations at regular intervals. Respiratory protection was not described. Twenty-five of the 247 workers complained of digestive disturbances such as nausea, "heartburn," and epigastric pain. Complaints of breathing difficulties and coughing were "frequent." Medical examination of this group produced the results shown in Table VIII-23.

TABLE VIII-23

MEDICAL ABNORMALITIES AMONG 247 SOVIET HARD METAL WORKERS
(Kaplun, 1967)

Effect	Numbers of Cases Among Long-Service Workers: 3 to 6 years	Numbers of Cases Among Long-Service Workers: over 6 years
Blood changes		
reticulocytosis	4	13
anemia	6	19
Inflammation, upper respiratory tract	27	85
Chronic bronchitis	12	23
Signs, early pulmonary fibrosis	14	19
Gastritis	5	12
Hypotension	15	33

Linhart et al. (1977) quantitated the cobalt in urine of a large number of employees in the hard metal industry (eastern European), dividing the groups into beer drinkers and non-beer drinkers. The authors hypothesized additional risk to the cardiovascular system from heavy beer drinking in these industrially exposed workers, although the beer did not exceed the (Czech) maximum permissible value for cobalt. Data backing the hypothesis were not given.

Dorsit et al. (1970) reviewed the literature on pulmonary disorders in the hard metal industry, presenting in detail the medical findings on three (French) industrial workers. They reported several generalizations about cobalt-tungsten pulmonary fibrosis noted in their clinical and literature work on the subject relative to duration of exposure, age of disease onset, sex of primary occurrence, and genetic tendency.

Maximum delay from work inhalation was an average of 13 years. Other studies have reported an average of 6 years' exposure; inhalation for 5, 11, and 16 years in the workplace have been reported.

Dorsit et al. noted that the age of pulmonary fibrosis onset in industrial workers was 30 to 50; the sex of primary occurrence is male, but of course the industrial workers are predominantly male. A Russian study (Kochtekova, 1960; cited by Dorsit et al., 1970) reported one case of cobalt pulmonary fibrosis in a female.

Dorsit et al. cited work that suggested a genetic tendency to pulmonary fibrosis; a man of 48 years and his son aged 28 had worked 7 and 4 years, respectively, in the hard metal industry; both had pulmonary fibrosis diagnosed.

Aerosols and their characteristics were discussed at length by Roshchin (1974) as they relate to the toxicity of metals and prophylaxis of occupational poisoning. Most inhalation in alloy production is from powders--aerosols of disintegration. They occur during condensation of metal vapors formed during pyrometallurgical treatment. These metal dusts may be in the form of the metal, its alloys, its salts, but most often, its oxides, both in the biosphere and during production.

The aerodynamics of metal powders to which workers are industrially exposed were described by Roshchin (1974). His data indicate that 80 to 90% of the metal powder was less than 2 μm diameter, which may penetrate to the alveoli, and 10 to 20% were 2 to 5 μm. Ferron (1977) calculated that the aerodynamic diameter of soluble salts having a diameter of 1 μm increased linearly with residence time in the lung. $CoCl_2.6H_2O$ behaved almost ideally, increasing the aerodynamic diameter of its particles from 1 to 2.8 μm in 1.0 sec. Particles of 0.4 to 6 μm showed less deposition in the nasopharyngeal region and more in the tracheobronchial region according to Ferron's model.

Brakhnova (1975) cited the 1963 studies of Kaplun and other workers detailing the toxicity of cobalt oxides. Co_2O_3, 25 to 50 mg, produced no deaths in animals (species not reported). This dose range was an LD_{50} for CoO and was an LD_{100} for cobalt metal. (Kaplun's 1963 report stated that 10 mg was a minimum lethal dose of cobalt oxide given intratracheally but the absolute lethal dose for metallic cobalt.) Kaplun (1967) measured the air in USSR industry and found that, during the production of cobalt oxide from wastes, there was 2 to 5 mg Co/m^3 in the air; around the reduction furnace, cobalt levels were 2 to 6 mg/m^3. At the vibrating sieve, where cobalt oxide was sifted, he noted cobalt in the air at hundreds of mg per cubic meter (no actual numbers given).

Experimental studies by Rabotnikova (1971) demonstrated a linear relationship between the threshold doses of 24 different metal oxides given as a single injection to mice and the official maximum allowable concentration (MAC) in workplace air in the USSR. The threshold doses of CoO and Co_2O_3 were 19.3 and 242 mg/kg, respectively, based on effects in liver, nervous system, spleen, blood, and body weight of the experimental animals. Rabotnikova prepared an equation to calculate an approximate MAC based on the mouse threshold dose of metal oxides; it appeared to fit her data base.

Lazarev (1963; cited by Brakhnova, 1975) detailed a case of cobaltous acetate dust inhalation at undescribed levels of exposure. The following symptons occurred about an hour after the end of work: irritation of the mucosa of the fauces, bloody vomit, marked hypersensitivity of the abdomen, intestinal colic, high temperature, and weakness in the legs. The review hinted that this case may have been (allergic) hypersensitivity, as it was noted that cobalt dust inhalation occasionally produces symptoms of bronchial asthma.

Cobalt acetate powder inhaled acutely during the workday produced vomiting and severe gastric pain in a worker. Marked weakness and pain in limbs occurred acutely, and after 3 days, some vomiting of blood and bloody stools occurred. Recovery was complete in 3 weeks (Hagen, 1940; cited by Browning, 1961). A second case of gastric symptoms and hematuria after acute, high-level cobalt exposure was briefly cited.

Fifty-one workers in a Czech plant processing an iron-nickel ore containing 600 mg Co/kg were exposed to 0.063 to 1.605 mg Co/m^3 of air in the form of aerosols. Urinary-serum cobalt levels were followed for approximately a week. In three selected workers, blood retention values were calculated: 162 μg, 175 μg, and 233 μg. For the entire workforce, the cobalt in urine averaged 694 μg Co/liter. The forms of cobalt in the aerosols included CoNiS, $Co(NH_3)_6CO_3$, $[Co(NH_3)_5SO]_2$-$SO_3 \cdot 2H_2O$, $Co(OH)_2$, $CoSO_4$, and Co metal (Klucik et al., 1967). Klucik et al. (1973) measured several blood parameters in workers exposed to average concentrations of 0.434 to 1.605 mg Co/m^3 air, presumably at this same plant now described as a hydrometallurgical cobalt plant. Erythropoietin, red cell count, hematocrit, hemoglobin, iron, and transferrin appeared to have no statistically significant relationship to cobalt exposure relative to unexposed controls, but the albumin and lactic dehydrogenase isoenzyme (LDH) were highly significantly greater in exposed than in control groups ($p < 0.001$).

Verhamme (1973) performed a survey of a cobalt plant (Hoboken-Overpelt) in Belgium and observed the health of 120 workers with an average length of employment of 12.5 years: 50%, 8 to 20 years; 25%, 20 to 25 years; and 15%, over 25 years. The survey included radiographic and hematological examinations. The radiographic examinations showed that one of the 120 workers had bronchiectasis, four had old tuberculosis lesions, and three had pneumoconioses. General findings without numbers of subjects showing the effects were also reported. Verhamme (1973) noted that some cases of "allergic cutaneous affections" have been observed. This allergic reaction was described as consisting of two types. The first was a papular, slightly pruriginous erythematous eruption confined mainly to the hands, wrists, and forearms. This reaction occurred primarily in summer and most often affected persons who had not been in contact with cobalt salts very long. The second type of skin reaction to cobalt was an eczema with slight hyperkeratosis and skin fissuring which was seen after several years' exposure.

"Several" workers who for 15 to 20 years had been in contact with cobalt oxides where the atmosphere was the most dust laden (no data were given) showed chronic bronchitis with productive coughs and occasionally dyspnea, emphysema, and chronic respiratory insufficiency. Verhamme noted that removal from the dusty environment produced improvement "and in some cases," remission of symptoms.

(c) Accepted levels of exposure: Industrial exposure sources have been reviewed elsewhere in this document and are covered briefly here only as such exposure concerns the toxicity of cobalt. The cobalt content in cemented tungsten carbide is considered to be the major health hazard of the hard metal when the cobalt content exceeds 2%. Cobalt levels in tungsten carbide are usually under 10% but may be as high as 25% (NIOSH, 1977).

Recommendations for the safe handling of cobalt are promulgated by state and federal agencies. The recommendations include air levels, use of environmental control measures such as ventilators and respirators, personnel protective equipment, and medical control.

In the United States, the threshold limit value (TLV) for cobalt in airborne metal fumes and dust for the workplace is 0.1 mg/m^3 (OSHA, 1972; OSHA, 1974; Winell, 1975). The value refers to a time-weighted average for a conventional 7- or 8-hr workday, and the term "threshold limit value" refers to airborne concentrations of substances and refer to conditions under which it is believed nearly all workers may be repeatedly exposed, day after day, without adverse effect, according to the International Labor Organization (ILO, 1970; cited in Schubert, 1973). Neither ACGIH or the Occupational Safety and Health Administration has adopted the tentative TLV of 0.05 mg/m^3 revised downward from the current standard of 0.1 mg/m^3 by the American Conference of Governmental Industrial Hygienists ca. 1976 (ACGIH, 1977 printing of 1971 edition). The short-term excursion level is still recommended at 0.1 mg/m^3 (Anonymous, 1979).

The standards for the work environment levels of cobalt are also 0.1 mg/m^3 in the German Democratic Republic, Sweden, and Czechoslovakia. The occupational exposure standards in the USSR (1972) and the Federal Republic of Germany (1974) were 0.5 mg/m^3. In Argentina, Great Britain, Norway, and Peru, the U.S. standards are applied (Winell, 1975). The higher level of cobalt metal dust in workroom air that is allowed in the USSR was noted by Popov (1976); although higher than the U.S. standards, the level was still considerably below the MAC allowed for tungsten carbide dusts in the USSR. Threshold limit values on some other cobalt compounds in the USSR are: cobalt oxide, 0.5 mg Co/m^3; dicobalt octacarbonyl, 0.01 mg Co/m^3; and cobalt hydrocarbonyl and its decomposition products, 0.01 mg Co/m^3 (Institut National de Recherche et de Securité, 1979).

In the United States, occupational exposure to dust of cemented tungsten carbide which contains more than 2% cobalt should be controlled so that workers may not be exposed to more than the 0.1 mg/m^3 determined as a time-weighted-average concentration for up to a 10-hr work shift in a 40-hr workweek. Positive-pressure, self-contained breathing apparatus with a supplied-air respirator is required if the cobalt concentration range is greater than 200 mg/m^3 in the air.

These recommended standards are based on data which indicate that the cobalt in the cemented tungsten carbide may cause lung damage. There is also the skin irritation and hypersensitivity to cobalt that may occur, as noted in the case histories discussed earlier in this chapter in part VIII.C. 1.a.(vi). The workplace exposure controls for skin protection are designed to minimize mechanical irritation from dust but may not eliminate the toxic reaction of individuals who are allergic to cobalt and who may have adverse reaction to extremely small amounts of the metal or dusts containing it.

Maximum workplace concentrations (MAC) of tungsten carbide dusts (containing unspecified levels of cobalt) were 6 mg/m^3 in 1975 in the USSR and Romania; the German Democratic Republic had an MAC of 6 mg/m^3 with a 30-min excursion limit of 12 mg/m^3 for tungsten carbide in 1973 (NIOSH, 1977). Again, no level of cobalt was specified for the dust mixture.

The maximum permissible concentrations of ^{60}Co as insoluble particles in air for occupational exposure is 9×10^{-9} $\mu Ci/cm^3$ for a 40-hr week. This value applies to the lung, the most "critical organ" according to the regulations from the HEW Division of Radiological Health (USHEW, 1960).

The limit for cobalt in water reservoirs in the USSR set in 1959 at 1.0 mg/liter was substantiated by acute and chronic animal toxicity tests reported by Krasovskii and Fridlyand (1971). The experimental data included animal studies and _in vitro_ measurements:

- Chronic exposure of rats to 2.5 and 0.5 mg Co^{2+}/kg/day, dosed as cobaltous chloride, produced a decline of immunological reactivity (phagocyte function) after 6 months of dosing.

- Chronic exposure (4 months) of rats to 2.5 mg/kg/day also produced a hemopoietic shift and disturbed the conditioned reflexes. A dosage of 0.5 mg/kg/day produced in the rats a mild polycythemia at 6 to 7 months with no effect on conditioned reflexes.

- When added to water, cobalt salts (chloride, nitrate, and sulfate) in concentrations of 10 and 5 mg/liter "considerably inhibited the biochemical oxidation of organic substances."

- The threshold concentration of investigated cobalt compounds with respect to taste when added to water was 16 to 18 mg/liter.

(3) _Epidemiology_: Human epidemiology in this section includes summaries of studies relating nonindustrial cobalt exposure, that is, cobalt in the environment, with health effects. Large animal studies that relate geographical cobalt levels to a toxic effect are also included.

A contact dermatitis investigative unit in western Scotland patch-tested 4,500 persons between 1970 and 1977. The routine test series included 1% $CoCl_2$ in paraffin oil. Cobalt was the fifth most common sensitizer, with over 9% of the group showing positive reactions (approximately 5.5% male). In the specific series reported (Husain, 1977), 77 males and 44 females showed a positive reaction. The author suggested that the frequency may not be representative, since this clinic tested only patients referred with dermal problems.

Agrawal et al. (1978) has reported human toxicity and has analyzed the water for metals by atomic absorption in Cambay, a coastal Indian city. There are four wells that supply the water to the city of 75,000 persons, and the authors suggest that the "high incidence of mental deformation and defective speech" among the inhabitants of Cambay could be attributable to poor water quality. Twelve metals including cobalt were measured from each of the four wells, with 25 determinations for each sample. Cobalt levels were 2.00 mg/liter in all the wells. The health hazard of chronic intake of 4 mg Co/day (calculating 2 liters of drinking water) was not discussed by the authors. Water levels of lead and cadmium averaged eight and four times the maximum allowable limits, and the sodium concentration was 80 to 1,000 mg/liter. The hazards of the latter three elements were stressed.

Grazing on pasture with a topsoil high in cobalt (also nickel, chromium, copper, and titanium) was suggested as a possible cause of a common periodontal disease in sheep in New Zealand (Steele and Henderson, 1977). In a large epidemiological study, a total of 4,128 sheep were examined, and the syndrome was seen to have no correlation with breed, age, exercise, or fertilizer used on the soil. In an analysis on 138 large sheep farms, periodontal disease was related to the soil type, occurring when the major soil type was derived from a parent material of igneous origin.

Several studies have attempted correlation in humans between the trace element composition including cobalt, of soil, water, and food products and the presence of dental caries. Two studies have found no correlation with cobalt levels; one has found a positive correlation. Nastase et al. (1974; cited by

Masironi, 1977) measured the dental caries in school children aged 6 to 15 years in Romania and found no relationship to cobalt in soil, water, or food. Ludwig et al. (1970) found no relationship between dental caries in school children aged 12 to 14 years and cobalt levels in water supplies of 19 towns along the East Coast of the United States. Anan'ev (1977) studied 28,841 children 7 to 17 years of age in the USSR and found that the concentration of cobalt (and several other elements) in drinking water varied inversely with occurrence of dental caries.

Two human epidemiological studies related bone disease to localized geobiochemical characteristics including cobalt levels. Egiazarian (1971; cited by Masironi, 1977) found that osteochrondopathy of the femur head in children in the area of Krasnodar, east of the Black Sea, was twice as prevalent in a mineral-rich area as it was in a mineral-poor area. Cobalt and five other trace elements were high in soil and water in the mineral-rich area. Altered osteogenesis was also reported to be endemic in the Urov region in southeastern Siberia where the soil and water have "imbalanced ratios" of cobalt (and nine other elements) compared to other trace elements (Konstantinov et al., 1975; as cited by Masironi, 1977).

Ilynski and Babadzhanov (1974) measured, using neutron activation analysis, the postmortem cobalt levels in whole blood, aorta wall, and liver in 94 men ranging in age from 30 to over 70 years. Roughly half the group had died in accidents and half died of coronary atherosclerosis. The population was divided by age into decades.

In controls, cobalt concentrations in blood and aorta walls in 30- to 39-year-olds were approximately 1.5 times the levels seen in the over-70 group. In patients that had died of atherosclerosis, the cobalt concentration of blood and aorta was even further decreased with age. Liver cobalt levels remained generally stable in all groups at all ages. Estimates of these values (from plotted data) are given in Table VIII-24.

Cobalt balance studies on patients (numbers not given) with atherosclerosis showed a negative balance compared to controls. The authors suggest the microelements studied (five besides cobalt) participate in vital enzymatic reactions and those elements that decrease with age such as cobalt may contribute to development of metabolic disturbances and thereby to disease processes.

TABLE VIII-24

COBALT CONCENTRATIONS IN BLOOD AND AORTA OF SUBJECTS DYING OF ATHEROSCLEROSIS AND ACCIDENTAL CAUSES
(Ilynski and Babadzhanov, 1974)

Subjects	Co μg% Blood	Co μg% Aorta
30 to 39-year-old controls	~ 12.8	~ 14.2
30 to 39-year-old atherosclerotics	~ 12.8	~ 12.0
70-year-old controls	~ 8.8	~ 10.5
70-year-old atherosclerotics	~ 7.2	~ 5.5

Chigrina et al. (1974) analyzed the drinking water supply sources in four areas of the northern Kazakhstan USSR for levels of cobalt. The water concentration was from 0 to 18.0 μg/liter with considerable geographical variation. The authors evaluated the occurrence of endemic pathological goiter in the population, and found no correlation between the cobalt content of the water and increased goiter occurrence.

Vezhnevets et al. (1975; cited by Masironi, 1977) analyzed for iodine and cobalt in soil, water, and food in the Uzbek SSR in an attempt to clarify the cause of goiter which is endemic there. Water iodine was related to prevalence, but low cobalt concentrations in the food consumed plus low iodine levels seem more related to high goiter prevalence.

The Kuba-Khachmas zone of the Azerbaidzhan SSR has a high level of endemic goiter in the local population. Grekalova (1973) measured trace elements in water resources, soil of the area, and also in food products grown or raised locally. The intake of cobalt by the population was said to be "less than 1% of the daily requirement." The author cited V. A. Leonov et al. (1957), who suggested that children in the first 3 years of life require 3 to 10 mg Co/day. Soil, foods, animal fodder, and drinking water did not contain unusually low cobalt concentrations as the author maintained. In fact, the average 50.7 μg Co in the diet of school children is in very good agreement with the U.S. market basket data (see Chapter VII). The cobalt content of the drinking water was very low (average 0.52 μg/liter). No correlations with goiter occurrence and specific trace elements were presented in this paper.

Masironi's review (1977) of the Russian literature on geochemistry and health has included three epidemiological studies which relate cancer mortality rates and cobalt levels. Cobalt-deficient soil was correlated with a three- to fourfold excess in mortality from esophageal cancer (Dulganov, 1973; cited by Masironi, 1977), but the high cobalt levels in soil and water were correlated with the prevalence and geographic distribution of stomach and lung cancer in another study (Babenko et al., 1970; cited by Masironi, 1977). Dalgat et al. (1974; cited by Masironi, 1977) also found stomach cancer in an area near the Caspian Sea to be associated with high soil and vegetation content of cobalt. At least three trace elements in addition to cobalt were analyzed in each study; there is no pattern of synergism or antagonism of health effects with any of these and cobalt.

In a study sponsored by Consolidated Edison Company of New York, Kneip et al. (1973) reported that trace element tissue burdens in humans were used to identify trends in environmental exposures. Samples obtained at autopsy were analyzed in conjunction with an air sampling program for determining lung and hilar lymph node metal concentrations for persons exposed to New York City air. At the time of the most recent report available, cobalt lymph node-lung concentration ratios indicated that cobalt was distributed in the body in a pattern similar to aluminum and silicon, which would indicate that the cobalt-bearing particulates inhaled were derived from soil dusts.

Twelve percent of the 166 persons dying of anthracosilocosis in the Wyoming-Lackawanna Valley of Pennsylvania between 1951 and 1965 also had bronchogenic carcinoma. El-Ashry and Getzoff (1974) analyzed 70 shale samples from this area for cobalt and four other trace elements (chromium, nickel, iron, and manganese) using atomic absorption spectrophotometry. The pattern of cobalt concentration in the shale of the valley is highest in the south, and the number of lung cancer cases reported among coal miners was also higher in the south. Cobalt concentrations ranged from 13 to 122 mg/kg, with the largest number of values occurring within the 30 to 60 mg/kg range. Trends in distribution patterns of iron and manganese were also positively correlated with tumor occurrence.

There was no correlation between median soil concentrations of cobalt and the occurrence of cancer in an epidemiological survey done for four election districts of Washington County, Maryland (Cannon, 1977).

Bierenbaum et al. (1975) measured water composition and trace metal levels in serum of groups of 260 adults from each of two matched cities (Kansas City, Kansas, and Kansas City, Missouri) who received their water from the same source, but one of which had softened the water (Kansas). The combined cardiovascular/renal mortality data from the two cities over a 3-year period was compared as a part of the study, which was designed to seek cardiovascular disease-related toxic water factors. Cobalt levels in the water supplies of both cities were below detectable limits (1 μg/liter, atomic absorption spectroscopy). Serum trace metals were measured in the 260 adults in each city; cobalt levels were undetectable in the serum of the Kansas population (soft water) although the systolic and diastolic blood pressures were significantly higher in this city ($p < 0.01$). Cobalt serum levels in the Missouri population were 0.146 ± 0.026 mg/liter.

Sharrett and Feinleib (1975) reviewed significant inverse correlations between the hardness of drinking water and local cardiovascular mortality rates in major epidemiologic studies in the United States, Canada, England, and Wales. Ir United States, they reported positive correlations to hardness greater than 0.90 are found for cobalt, cadmium, chromium, molybdenum, nickel, and vanadium in untreated river water. The concentrations of all these metals in raw water correlated negatively with cardiovascular disease death rates by state.

Neri et al. (1975) also sought correlations between the minerals in water and cardiovascular mortality rates; water analysis data came from 575 Canadian communities. Their summaries eliminated cobalt, since cobalt incidence and levels have insignificant correlations with either cardiovascular or total mortality in their population.

Schroeder (1974) reviewed several studies on the role of trace elements in cardiovascular disease. He noted that death rates from coronary heart disease in white males aged 45 to 64 years in 42 states were inversely correlated with cobalt (and nine other trace elements) levels in water. Schroeder estimated that a person who drinks 2 liters of water a day will receive an average cobalt dose of 6 μg from water but will get 300 μg in the diet.* There seems to be no reason to believe cobalt levels in the water supply are causal of hypertensive heart disease in these studies. Schroeder points out, without average levels, that cobalt increases in the heart muscle in atherosclerosis but decreases in injured heart tissue after myocardial infarction.

An epidemiologic study on trace metals in water in Evans County, Georgia (Anonymous, 1976b) noted that heart patients in the study had a higher cobalt intake than did patients without the disease. Shacklette et al. (1970) measured trace elements in Georgia soil and vegetables grown in the soil in counties with high and low cardiovascular mortality rates. No conclusions could be made for cobalt; the metal was frequently present in soils at levels below detection limits (0.0003%) using the method of semiquantitative spectrographic analysis of ashed soils.

---

* Chapter VII points out wide variations in the reported content of cobalt in the daily diet. The value 300 μg/day is about an order of magnitude greater than several studies in which cobalt was determined by neutron activation analysis.

b. Animals

(1) Mammalian studies

(a) Acute, LD_{50}, and lethal dose determinations: Acute toxicity data for laboratory mammals are given in Tables VIII-25 (LD_{100}'s) and VIII-26 (chiefly LD_{50}'s). Table VIII-27 summarizes subacute toxicity data for mammals.

TABLE VIII-25

APPROXIMATE LD_{100} OF COBALT COMPOUNDS TO SEVERAL MAMMALIAN SPECIES
(Stokinger, 1963)

Form of Cobalt	LD_{100} dose, mg/kg	Route	Species
Co	1,500	oral	rat
Co	250 - 500	IP	rat
Co albuminate	9	SC	mouse
$CoCO_3$	1,000	IP	rat
Co lactate	10	IP	rat
Co	700	oral	rabbit
Co	100	IV	rabbit
CoO	70	oral	dog
$CoCl_2.6H_2O$	6.3	IV	rabbit
$CoCl_2.6H_2O$	7.5	IV	dog
$CoSO_4$	3 - 4	IV	dog

Bienvenu et al. (1963) found a 30-day LD_{50} of 21 mg Co/kg as the sulfate dosed intraperitoneally in white male Swiss mice. Co^{2+} ranked 19th among 42 metal ions so tested being 127.2 times more toxic than Na^+. For comparison, Pb^{2+} and Hg^{2+} were 120.3 and 2,283.0 times more toxic than Na^+, respectively.

TABLE VIII-26

ACUTE TOXICITY TO LABORATORY MAMMALS (LETHAL DOSE LEVELS)

Species, Strain	Cobalt compound	Dose	Route	Comments	Reference
Piebald rat	Co (metal)	25 mg/kg	I.P.	100% survival with respiratory symptoms	Harding (1950)
Rat, white	Co, powdered metal	100 to 200 mg/kg	I.P.	LD_{50}; dose is age dependent	Frederick and Bradley (1946)
Rat	Co (metal)	100 mg/kg	I.V.	Lowest lethal dose for species and route	Fairchild et al. (1977)
Rabbit	Co (metal)	100 mg/kg	I.V.	Lowest lethal dose for species and route	Fairchild et al. (1977)
Rat	Co (metal)	25 mg/kg	Intratracheal	Lowest lethal dose for species and route	Fairchild et al. (1977)
Piebald rat	Co (metal)	12.5 mg/kg	Intratracheal	Lethargy and death to all in 15 min to 6 hr	Harding (1950)
Rat	Co tetracarbonyl	753.8 mg/kg	Intragastric	LD_{50} dose range: 675 to 832.6 mg/kg	Spiridonova and Shabalina (1973)
Mouse	Co tetracarbonyl	377.7 mg/kg	Intragastric	LD_{50} dose range: 281.7 to 473.7 mg/kg; hemodynamic and dystrophic changes of internal organs; local damage to gastric mucosa	Spiridonova and Shabalina (1973)
Rat, Carworth-Wistar	CoO	1,700 mg/kg	Oral	LD_{50}: 1,070 to 2,828 mg/kg = 95% confidence limits	Smyth et al. (1969)
Mouse, white	CoO	19.3 mg/kg	"Injection"	"Threshold dose" based on six organ effects	Rabolnikova (1971)
Rat, white	Co_2O_3	5,000 mg/kg	I.P.	LD_{50}	Frederick and Bradley (1946)
Mouse, white	Co_2O_3	242 mg/kg	"Injection"	"Threshold dose"	Rabolnikova (1971)
Rat	$CoCO_3$	2,002 mg/kg	I.P.	Lowest lethal dose for species and route	Fairchild et al. (1977)
Rat, male	$CoSO_4$	597 mg Co/kg	Oral	LD_{50}	Wiberg et al. (1969)
Rat	$CoSO_4$	13 mg Co/kg	I.P.	LD_{50}	Wiberg et al. (1969)
Mouse, white Swiss male	$CoSO_4$	21 mg Co/kg	I.P.	LD_{50} (30-day)	Bienvenu et al. (1963)
Dog	$CoSO_4 \cdot 7H_2O$	16 mg/kg	I.V.	Lowest lethal dose for species and route	Fairchild et al. (1977)
Rat, spontaneously hypertensive	"Common soluble salt	13.8 mg/kg	I.P.	7-day LD_{50}	Lewis (1975)
Rabbit	$Co(NO_3)_2$	400 mg/kg	Oral	Lowest lethal dose for species and route	Fairchild et al. (1977)
Rat	Cobalt stearate	2,390 mg/kg	Oral	LD_{50}	Schmidt et al. (1975)
Rat	Tris(monoethanolamine) cobalt	408.0 mg/kg	S.C.	LD_{50}	Chernov et al. (1973)
Rat, young	$CoCl_2 \cdot 6H_2O$	50.0 mg Co/kg of diet	Oral	25% mortality (corn starch, skim milk diet)	Huck (1975)
Rat, young	$CoCl_2 \cdot 6H_2O$	400.0 mg Co/kg of diet	Oral	100% mortality (corn starch, skim milk diet)	Huck (1975)
Rat	$CoCl_2$	80 mg Co/kg	Oral	LD_{50}	Krasovskii and Fridlyand (1971)
Rat	$CoCl_2$	144 mg Co/kg	Oral	LD_{50}	Murdock and Klotz (1959)
Rat, Wistar	$CoCl_2$	860 mg/kg	Oral	24 hr LD_{50} dose range: 750 to 980 mg/kg	Puget et al. (1975)
Mouse, Swiss	$CoCl_2$	620 mg/kg	Oral	24 hr LD_{50} dose range: 570 to 690 mg/kg	Puget et al. (1975)
Mouse	$CoCl_2$	80 mg Co/kg	Oral	LD_{50}	Krasovskii and Fridlyand (1971)
Guinea pig	$CoCl_2$	55 mg Co/kg	Oral	LD_{50}	Krasovskii and Fridlyand (1971)

TABLE VIII-26 (concluded)

Species, Strain	Cobalt compound	Dose	Route	Comments	Reference
Rat, Wistar	$CoCl_2$	140 mg/kg	S.C.	24 hr LD_{50} dose range: 110 to 170 mg/kg	Puget et al. (1975)
Mouse, Swiss	$CoCl_2$	160 mg/kg	S.C.	24 hr LD_{50} dose range: 140 to 180 mg/kg	Puget et al. (1975)
Mouse, female	$CoCl_2$	40.8 mg/kg	S.C.	LD_{50}	Hasegawa (1974)
Guinea pig	$CoCl_2$	2.0 ml of 0.239 <u>M</u> solution (30 mg)	P.C.	$\sim LD_{50}$: 11/20 dead in 3 weeks	Wahlberg (1965)
Rat, Wistar	$CoCl_2$	50 mg/kg	I.P.	24 hr LD_{50} dose range: 40 to 60 mg/kg	Puget et al. (1975)
Rat	$CoCl_2$	7.9 mg Co/kg	I.P.	LD_{50}	Murdock and Klotz (1959)
Mouse, male	$CoCl_2$	36.7 mg/kg	I.P.	LD_{50}	Hasegawa (1974)
Mouse, Swiss	$CoCl_2$	130 mg/kg	I.P.	24 hr LD_{50} dose range: 110 to 150 mg/kg	Puget et al. (1975)
Guinea pig	$CoCl_2$	2.0 ml of 0.239 <u>M</u> solution (30 mg)	I.P.	$\sim LD_{50}$: 6/11 dead in 24 hr	Wahlberg (1965)
Rat	$CoCl_2$	20 mg/kg	I.V.	LD_{50}	Fairchild et al. (1977)
Rat, white	Co lactate	7.5 mg/kg	I.P.	LD_{50}	Frederick and Bradley (1946)
Rat	Co niacinamide	134 mg/kg	Oral	LD_{50}	Murdock and Klotz (1959)
Dog	$[Co(NH_3)_6]$ nicotinate	45 mg/kg range	I.V.	LD_{50}; hypotension and tachycardia noted with this trivalent Co salt	Polonovski et al. (1954)
Rat	Na_2[Co-EDTA]	> 1,000 mg Co/kg	Oral		Murdock and Klotz (1959)
Rat	Na_2[Co-EDTA]	> 800 mg Co/kg	I.P.		Murdock and Klotz (1959)

TABLE VIII-27

SUBACUTE, MAXIMUM TOLERATED DOSES OF COBALT IN MAMMALS

Species	Dose, mg Co/kg/day	Route	Effect	Reference
Rat	2.5 as $CoSO_4$	Oral	Polycythemia only symptom	Smith (1962)
Rat	600	In diet	Weight loss	Smith (1962)
Dogs	10.0	Oral	Polycythemia only symptom	Smith (1962)
Sheep	3.0	Not given	Tolerated level (no toxic symptoms)	Smith (1962)
Cattle	1.0	Not given	Tolerated level	Smith (1962)
Young cattle	0.5 as $CoSO_4$	Oral	Tolerated level	Keener et al. (1949; cited by Smith, 1962)
Young cattle	> 0.5	Oral	Toxic symptoms: increased hemoglobin and number of red cells, weight loss, anorexia, loss of muscular coordination	Keener et al. (1949; cited by Smith, 1962)
Heifers	50	Oral	No polycythemia	Smith (1962)
Humans	96 as $CoCl_2$	Oral	Polycythemia. Five times as much produced toxic symptoms	Smith (1962)

(b) Cardiovascular and blood effects

(i) Myocardial and circulatory: After the 1965-1966 "epidemic" of cardiac failure in heavy beer drinkers in Quebec was reported and attributed to cobalt added to the beer to stabilize the foam, many studies of the myocardial toxicity of cobalt have been undertaken on several species (Grice et al., 1969; Hall and Smith, 1968; Lin and Duffy, 1970; Huy et al., 1972; Wiberg et al., 1967).

Rats were used in several studies. The heart tissue of old female breeder rats with atherosclerotic changes contained 0.34 mg Co/kg compared to undetectably low levels in young control rats, according to Strain et al. (1971), who used neutron activation analysis to determine the metal.

Lin and Duffy (1970) dosed young (50 to 80 g) rats intraperitoneally with 5 mg Co/kg as the nitrate and studied the effects on the heart after one single dose and after three or more daily injections. Electron microscopy showed some myocardial degeneration after the acute dose, but the authors reported that the single dose did not cause serious toxicity. Histopathology on the repeated doses showed focal irregular patches of myocardial degeneration characterized by muscle damage, densely acidophilic cytoplasm, and intracytoplasmic vacuoles. The mitochondria gradually exhibited generalized enlargement with denser matrices. Lin and Duffy (1970) considered these cobalt-induced cardiac changes to be nonspecific, and similar to the alcoholic cardiopathy, thiamine deficiency, and other chemical toxicity.

Wiberg et al. (1967) suggested that protein deficiency or thiamine deficiency was involved in the cardiac toxicity seen with cobalt dosage. They demonstrated that mitochondria isolated from cobalt-dosed animals did not use pyruvate and the fatty acids stearate and octanoate, which are norm substrates in the normal energy pathways in the heart of their model system.

Treating thiamine-deficient animals with cobalt did not accentuate this effect in Wiberg's study but did enhance the cobalt toxicity in Grice's experiments (discussed below). It appears, however, that deficiency of thiamine produced the same metabolic toxicity as did cobalt; they both interfered with thiamine-requiring pyruvate and α-ketoglutarate oxidation.

The mechanism of cobalt toxicity to the rat heart, according to Grice et al. (1969) is based on a cobalt-induced reduction in the oxidation of pyruvate. Rat myocardium depends primarily on long-chain fatty acids for energy requirements; but in their absence, pyruvate becomes the preferred substrate. In this study, 26 of 30 rats given daily oral doses of cobalt sulfate, 26 mg/kg, for 8 weeks had myocardial histological changes. The pathology was more pronounced in the group with a diet-induced thiamine deficiency.

Long-term administration of cobalt acetate at 5 mg Co/kg/day for 5 months subcutaneously to rats induced the monoamine oxidase activity in the myocardium. Secondary to the increased enzyme, levels of norepinephrine decreased. A monoamine oxidase (MAO) inhibitor, nialamide, 20 mg/kg administered subcutaneously, prevented the cobalt-induced effect (Chekunova, 1976).

There is a flushing or vasodilation that occurs after injection of cobaltous ions in man and experimental animals; the mechanism of this effect was investigated in the rat and cat by Vetterlein (1972). Intra-arterial cobalt gave an earlier, a stronger, and a more localized effect than intravenous dosing, indicating that the site of action is peripheral. Histamine, serotonin, and epinephrine were investigated as potential mediators of this vasodilation effect, but there was no evidence of their involvement.

Huy et al. (1972) studied the acute effects of cobalt on cardiac metabolism and mechanical performance by administration of 0.06 and 0.10 m*M* cobalt chloride to the guinea pig heart. The findings corroborated *in vivo* studies that suggest cobalt slows myocardial energy metabolism and, as a result, substrate uptake. Cobalt additionally increased coronary vascular tonus and, as a result, coronary resistance. Another study by this group (Mohiuddin et al., 1970; cited by Alexander et al., 1972) induced myocardial changes in guinea pigs by feeding the animals 20 mg Co/kg/day as cobalt sulfate. Myocardial changes of fragmented myofibrils, intracytoplasmic vacuoles, dilated sarcoplasmic reticulum, mitochondrial swelling, and increase in lipid droplets were reported. There was also endocardial and pericardial edema and cellular infiltration with progressive loss of QRS voltage and S-T segment elevation in the electrocardiogram.

Subacute cobalt dosage to pigs produced myocardial toxicity and death at levels far below those used in rats. Male weanling Yorkshire-Hampshire pigs developed cardiomyopathy, and two of four died within 5 days after oral doses of 100 mg Co/kg/day as the sulfate. When given the same dose, four of six pigs that had been bred for stress susceptibility died within 5 days (Van Vleet et al., 1977).

Tauberger and Klimmer (1962) studied the toxicity to the cardiovascular system of acute intravenous $CoCl_2$ to cats in 0.016 mg Co/kg, 1.6 mg Co/kg, and 8 mg Co/kg doses. The blood pressure following small doses showed a short decrease followed by an increase. Higher doses also produced a prolonged decrease as a third phase. The animals on high dose also exhibited a tachypnea as high as 300 breaths/min. The mechanisms were defined using adrenalectomized and denervated animals. The authors attributed the blood pressure rise following cobalt dosage to stimulation of the medullary circulatory center. The drop after high doses of the metal was said to be a direct action on the blood vessels.

Beer-drinkers' cardiomyopathy was reproduced in dogs by Greenberg et al. (1977). Cobalt sulfate injections by three routes, 12.5 mg/kg/day for 3 weeks, produced myocardial changes in dogs which histochemically resembled human disease. The injurious effect of cobalt occurred both in protein- and thiamine-deprived dogs and in those receiving protein supplements. The lung and liver congestion seen in the dogs was ascribed by the authors as secondary to heart toxicity.

To determine cobalt effects on the heart of rats, Chenkunova (1978) dosed the rodents acutely with injection of 20 mg/kg or one-fifth the LD_{50} of cobalt metal. Symptoms of toxicity observed were rapid changes in catecholamine metabolism, decreased heart norepinephrine, and increased blood and urine monoamine levels. Permeability of the membranes of the heart was changed as were the following biochemical parameters: pyruvate concentrations, phosphorylase activity, and free lysosomal enzyme activity all increased; glycogen levels and lactate dehydrogenase activity decreased.

(ii) Serum protein effects: Paulov (1971) added $CoCl_2 \cdot 6H_2O$ to the feed of young ducklings for 20 days at levels of 0.02, 0.1, and 0.2% of the diet. Growth of the birds was slowed and the blood proteins were abnormal with an increase of β-globulins and a decrease of albumin. The decrease in albumin was also a toxic symptom noted in chronic beer drinkers when cobalt was used as a foam-stabilizing agent.

Cobalt chloride, 0.01 M (590 μg Co/liter) was added to mouse drinking water, and the mice drank an average of 6.5 ml/mouse/day. Short-term exposure of 4, 24, and 48 hr to the dose was instituted after withholding water for 12 hr, so the cobalt dosage is not estimable. Three-week and 3-month exposures to the 0.01 M cobalt chloride solution were also made. Blood samples were taken from the mice by cardiac puncture and separated by electrophoresis. Bryan and Bright (1973) reported an initial depression of several haptoglobins, transferrin, and ceruloplasmin by the cobalt, all of which reappeared after 48 hr in somewhat shifted positions. After 3 months' exposure, the sera were essentially identical to controls, which suggested to the authors that adaptive processes had occurred in the cobalt-dosed mice.

(iii) Erythropoietic effects: In a study by Stanley et al. (1946), male rats, 4 to 5 months old, were given $CoCl_2 \cdot 6H_2O$ subcutaneously six times weekly for 8 months as a 0.4% solution in doses of 2.5, 5.0, and 10.0 mg/kg body weight. The average increases in blood volume and total erythrocyte mass as compared to controls were 80 and 192%, respectively. The increases were due to increased cell size and cell number, since plasma volume decreased 16% on the average.

Hemoglobin concentration in the blood remained essentially unchanged in these studies; Hb concentration per cell decreased. No generalized toxicity was noted. Brewer (1940; cited by Cass and Lange, 1955) administered 5 to 30 mg Co/kg/day as cobaltous chloride for 4 weeks to five adult female dogs. Two dogs received 50 mg/kg/day for 2 weeks. An increase in the erythrocyte number was caused by daily doses above 10 mg. The hemoglobin concentration increased when doses above 5 mg were given, and the percentage volume of packed red cells increased at daily doses exceeding 15 mg/kg.

Goldwasser et al. (1958), in experiments on rats, suggested that the mechanism of the erythropoietic effect of cobalt is a stimulating effect on the bone marrow. He discussed the possibility of the effect being a hormonal stimulant factor identical with erythropoietin.

Krasovskii and Fridlyand (1971) administered $CoCl_2$ in water solution to rats 6 days/week for 7 months. The 2.5 mg/kg dose and, to a lesser degree, the 0.5 mg/kg dose stimulated erythrocyte and hemoglobin formation and produced larger diameter erythrocytes. These doses also suppressed phagocytic activity of white cells by the sixth to seventh month of testing. The authors recommended 1 mg Co/liter as the maximum allowable concentration in public drinking water supplies. If the average weight of the rats was 0.350 kg and the average drinking water consumption was 0.035 liter/day, the lower dose corresponded to $\sim$ 2.3 mg Co/liter.

The mechanism of cobalt's hemopoietic effect on white rats was investigated by Popov (1976). Two weeks' round-the-clock exposure to cobalt metal aerosols of 0.5 and 0.05 mg/m^3 lowered erythrocyte and hemoglobin levels, but the levels rose after 2.5 months in the inhalation chambers with continuous exposure. Animals exposed to 0.001 mg/m^3 for 3 months showed no significant change in erythrocyte count or hemoglobin levels. Those exposed to 0.005 mg/m^3 showed an increase only in hemoglobin. The mechanisms appear to involve early decomposition of both old and newly formed erythrocytes at higher levels of cobalt exposure, which elicits the hemopoietic effect on bone marrow. Neither cobalt metal particle size nor method of aerosol generation was described.

Kaplun (1955), in chronic trials on dogs, gave 5 mg $CoCl_2$ orally per day (0.2 to 0.4 mg/kg dose) for 6 to 7 months and recorded hematologic parameters. At the beginning of dosing, average red blood cell count was 4.7 to 5.7 million with 56 to 63% hemoglobin; and at the second month of dosing, the erythrocyte count was 8.7 to 8.9 million with 80 to 85% hemoglobin.

Sprague-Dawley rats showed marked polycythemia and characteristic changes in bone marrow and spleen following 32 weeks' subcutaneous injection of 2 mg Co/kg/day, 6 days a week, as a 0.4% solution of cobaltous chloride (Hopps et al., 1958). The authors noted that bone marrow blood vessels were completely normal in dosed animals; therefore, the theory that the pathogenesis of cobalt-induced polycythemia was due to marrow hypoxia was contradicted by histological evidence. Hopps et al. histopathologically examined heart, lungs, liver, kidney, pancreas, salivary glands, testicle, and epididymis and found no evidence of cobalt toxicity in these organs in the dosed rats.

Hopps et al. (1954) injected 44 Sprague-Dawley rats with cobaltous chloride at doses of 2.5 to 10 mg/kg for 6 days/week for 32 weeks, then examined their blood parameters and several tissues microscopically. The dosage, approximately 10 times the recommended human dose for anemia, produced hyperplasia of bone marrow particularly in the erythropoietic elements, as well as some spleen cell loss. Cobalt action, according to the authors' analysis of the data, is by inhibition of the enzymes that deal with oxygen transport.

(c) <u>Pancreas, thyroid, and endocrine effects</u>

(i) <u>Pancreas</u>: Van Campenhout and Cornelis (1951; cited by Telib, 1972) reported that cobalt chloride produced severe toxicity to the α cells of the pancreas in guinea pigs. The intravenous injection of $CoCl_2$ at a dose of approximately 10 mg/kg to dogs produced a complete loss of α cells in many of the pancreatic islets 24 hr following the dosage (Elinder and Friberg, 1977). Lazarus et al. (1953; cited by Cass and Lange, 1955) showed that an intravenous injection of 200 mg cobalt as cobalt chloride (75 mg/kg) in the dog caused a selective autolysis of the pancreatic α cells. Regeneration of the cells began after 5 days. These authors also produced a transient hyperglycemia in the pancreatectomized animal by injecting cobalt. This finding appears to support the postulate of some authors that hyperglycemia following cobalt was not due to cobalt's toxic action on the α cells but perhaps to its action on the exocrine part of the pancreas.

Telib (1972) reported on the histological and electron microscopal changes in the α and the β cells of the pancreas in rabbits dosed intravenously with cobalt chloride at 35 mg/kg as a 1% solution. The changes observed were:

- 1 to 4 hours: Destructive changes in α cells (pancreas). Degranulation in β cells 1 hr after injection, reaching maximum 2 hr after dose. β cells increased number of mitochondria (pancreas). The glycogen disappeared almost entirely from the liver.

- 1 to 4 days: Glycogen reappeared in liver. α cells began regenerating in pancreas; β cells showed signs of protein synthesis in pancreas.

- 1 to 3 weeks: Increase in total number of pancreatic α cells; "signs of activity" in alpha cells. Normal β cells.

The hyperglycemic action of cobalt is effective in animals other than mammals. Khanna and Gill (1975) noted that, although several mammalian species had been studied for cobalt action on the pancreatic α cells of the pancreatic islets of Langerhans, little work had been done on the influence of cobalt chloride in lower vertebrates. Cobalt chloride 0.5% and cobalt nitrate 1% were injected intramuscularly into freshwater teleost fish, _Channa punctatus_, at a dose of 100 mg/kg. Blood glucose in the control fish was 33 mg %; it peaked at 220 mg % 75 hr after a dose of 50 mg/kg; the death rate was high. Both cobalt chloride and cobalt nitrate elicited a rise in the blood glucose level, which became manifest at 0.5 hr;

but a further increment occurred after 24 hr. Cobalt chloride was more effective in producing the hyperglycemic effect than was cobalt nitrate. Histological alterations in the principal islets of C. punctatus consisted initially of a mild degranulation of both the α and β cells, but there was widespread necrosis of β cells in the later stages. The degenerative changes were more pronounced in the $CoCl_2$-treated fish, probably because of sustained hyperglycemia (Khanna and Gill, 1975).

Taylor and Marks (1978) report histological evidence of pancreatic α-cell degranulation and destruction following administration of cobalt and in their review cite Eaton (1972), who found that rats treated with cobalt (no dosage data) had an increased sensitivity to insulin. Khanna and Gill (1975) noted in their review, however, that rats were less susceptible to the pancreatic effects of cobalt than some other species such as rabbits and guinea pigs.

Bieger et al. (1975) cultured pancreas tissue with 0.00001 to 0.001 M $CoCl_2$ solutions and noted that newly synthesized zymogen granules, formed after addition of cobalt, would not secrete or discharge protein. This inhibition of the secretory process in the pancreas was shown to be unrelated to cellular respiration, but other mechanisms are unclear.

Pigeons (Columba livia) were given $CoCl_2$, 1 mg/kg intraperitoneally, and their pancreases were checked for toxicity. Within 1 day after dosage, there was inhibition of DNA synthesis and decrease in acinar cell mitosis and edema. Cobalt damaged both the endocrine α cells of the pancreas and the exocrine pancreatic epithelium in the pigeon as determined by histochemical evaluation (Mikheeva, 1976).

Biochemical investigations of the damage to pancreas by cobalt were done using guinea pigs (Izmirov et al., 1972). $CoCl_2$ as a 0.5% solution was injected subcutaneously in male and female adult guinea pigs at a dose of 25 mg/kg. One hour later, animals were killed and the pancreatic lipids subjected to chromatography and x-ray spectral analysis. In the neutral lipid fraction, cobalt was found to localize with the free fatty acid fraction. With phospholipids, cobalt was localized in the lysocephalins, lecithins, and cephalins. Izmirov et al. (1972) concluded that cobalt accumulates primarily in the exocrine pancreas, in the zymogen granule lipids.

Cobalt chloride, dosed orally to rats for 7 months at 2.5 mg/kg, produced no change in liver function as determined by paper electrophoresis of liver proteins, by cholinesterase activity, or by excretion of a load of bromosulfophthalein (BSP) (Krasovskii and Fridlyand, 1971).

It is common knowledge to physiologists that calcium is essential for insulin release from the pancreas. Henquin and Lambert (1975) investigated the effects of cobalt on pancreatic insulin release in the rat, as this release deals with the calcium flux. They based their work on previous investigations (Kohlhardt et al., 1973; cited by Henquin and Lambert, 1975) which demonstrated that cobalt inhibits calcium flux in neurons and heart muscle.

These studies demonstrated that cobalt adversely affects the pancreas secretory system: *in vitro* incubation of isolated islets of Langerhans in the presence of cobalt ions results in an inhibition of insulin release. This toxicity was also shown by Kern and Kern (1969; cited by Bieger et al., 1975), who saw degranulation in exocrine pancreatic cells after *in vivo* dosage of guinea pigs with cobalt.

The mechanisms through which cobalt salts induce changes were discussed by most authors reviewed. Khanna and Gill (1975) cited Graig (1962) whose data indicated that cobalt stabilizes insulin in the β cells; and the β cells, in addition, hold and concentrate the cobalt from the blood. The selectively absorbed metal produces toxicity to these cells, leading to metabolic derangements. These authors suggest that the initial hyperglycemia was due to some extrapancreatic factor (epinephrine?) while subsequent increment in the blood glucose level was ascribed to the degranulation of the β cells. They reported a direct action of cobalt on the β cells of the pancreas.

(ii) Thyroid: Thyroid effects in experimental animals have not been as clear as the effects in humans. Antila et al. (1955; cited by Sederholm, 1968) produced goiter in guinea pigs, citing histological activation of the thyroid and a decrease in protein-bound iodine (PBI) after cobalt administration. Kriss et al. (1955) produced thyroid effects in chicks; but, according to Sederholm (1968), thyroid effects in rats, mice, and rabbits have been unsatisfactorily shown with cobalt dosage.

(d) Central nervous system toxicity: Bekh (1972) reported that in rabbits intravenous administration of cobalt chloride (10 μg/kg every other day for 10 to 12 days) produced a generalized reduction of excitability of the neuromuscular system. Transmission of nerve impulses to muscle was abnormal and impaired.

By far the most extensive work relating nervous system toxicity to cobalt has been the use of implantation of cobalt on the cortex of the brain in experimental animals as a model of epilepsy. This has been done since Kopeloff, in 1960, implanted mouse brain cortex and observed seizures (Kopeloff, 1960; cited by Clark, 1977). Clark (1977) has reviewed the literature on the different methods used in the epilepsy model. She has additionally shown that $CoCl_2$ would produce seizures acutely when applied to the brain of amphibia (adult bullfrogs, *Rana catesbeiana*). Cobalt implants into numerous specific subcortical structures have been shown to induce seizures. Epileptic activity has been induced by metallic cobalt and cobalt-impregnated gelatin strips applied to the cortex of rats, cats, and monkeys.

The epileptogenic etiology of cobalt is unknown, although severe local effects have been noted by the several researchers using this model. Extensive neuronal death, abundant glial proliferation, neovascularity, and an inflammatory mononuclear response in the leptomeninges have been observed. Dow et al. (1962; cited by Clark, 1977) ascribed epileptic activity to the chemical effects of cobalt rather than the scarring effects due to the implant. It was noted that cobalt interferes with transport ATPase activity in whole brain, and ionized cobalt has been shown to compete with calcium ions in neurons of several lower life forms.

The time course to the development of epilepsy from this model following cobalt application depends upon the form in which the cobalt is applied. Clark (1977) reported that the slowest inducer of experimental epilepsy is a cobalt powder-gelatin mixture which may take weeks to produce an effect. Metallic cobalt powder or cobalt wire was effective within several days. Soluble cobalt, on the other hand, acts within an hour or two to produce an effective epileptic response.

One of the more complete reviews of the pathological features involved in cobalt-induced experimental epilepsy was made by Emson (1976). He described the features of cobalt epilepsy in cat, monkey, and rat that particularly resembled cortical epilepsy in man. These features included the presence of abnormal, high amplitude spikes in the electroencephalogram (EEG) and the fact that seizures were more frequent during the day (as opposed to during sleep) and also occurred more frequently during periods of stress. Biochemically, cobalt produces selective pyramidal cell loss, with particular damage to the neurons in layers II, III, and V. In human epilepsy, there is usually no definite focus; but when there is, the histochemistry resembles cobalt epilepsy, with abnormal neurons that have few dendrites. In both human and cobalt-induced epilepsy, there are reduced levels of transmitter amino acids and impaired ability to synthesize acetylcholine. In human epilepsy, if a focal area can be outlined, the clinical case shows reduced levels of Na^+- and K^+-ATPase activity. The enzymatic levels are reduced in cobalt epilepsy, also.

Emson (1976) included his own earlier work on the levels of biogenic amines and associated enzymes in brain cortex focal areas in cobalt epilepsy (cat and rodent). Clayton and Emson (1975; cited by Emson, 1976) report the following changes from normal in cobalt-epileptic brain:

Increased	Decreased
Norepinephrine metabolites	Tyrosine hydroxylase
Dopamine metabolites	Monoamine oxidase
(Specifically: 4-hydroxy-3-methoxyphenylethylene glycol and homovanillic acid)	Catechol-o-methyl transferase
	Serotonin, 5-HIAA, Tryptophan } decreased only after acute spiking has occurred

Histochemical examination of the epileptic foci in chronic experimental cobalt epilepsy shows that the initial cobalt lesion becomes surrounded by a glial capsule which ultimately calcifies. The pyramidal tract neurons (the cells most severely damaged in cobalt-induced epilepsy) have been shown by histochemical staining to contain high levels of heavy metals (the stain is apparently not specific enough to identify cobalt) (Emson, 1976).

Chronic tests on rats showed behavioral effects elicited by cobalt dosage (Krasovskii and Fridlyand, 1971). Rats were given $CoCl_2$ in water 6 days/week for 7 months at dose levels of 0.05, 0.5 and 2.5 mg $CoCl_2$/kg,* and the conditioned reflex activity was measured by the "motor-food method" in a Kotlyarovskii chamber. At the 2.5 mg/kg dose of cobalt, there was a significant neurological effect on the appearance of reflex activity, and reinforcement was never obtained in this group. Effects of cobalt on the formation of reflex learning were seen after 4 months of dosing; the 0.5 mg/kg dose lengthened the latent period at the 6- to 7-month period significantly and increased the number of extinctions of conditioned reflexes. Rats receiving the 0.05 mg/kg dose showed no behavioral differences from those of the controls (Krasovskii and Fridlyand, 1971).

An interesting effect on thermoregulatory control by cobaltous chloride injection has been noted in rodents (Burke and Brooks, 1979). Intraperitoneal cobaltous chloride, 25 mg/kg, will drop the body temperature of mice and rats as much as 4°C in 30 min, apparently through a centrally mediated decrease in heat production.

---

* These doses roughly correspond to 11.5, 2.3, and 0.23 mg Co/liter.

(e) <u>Lung toxicity; cobalt effects by inhalation/intratracheal dosing</u>: Several papers by Popov's laboratory have appeared in the USSR literature on the effects in rats of metallic cobalt aerosols. Popov (1977) exposed rats for 3 months to metallic cobalt aerosols in concentrations of 0.5, 0.05, 0.005 and 0.001 mg/m^3. Blood phospholipids, cholesterol, and β-lipoproteins were lowered in the three highest exposure groups compared to control rats. Concentrations of 0.001 mg Co/m^3 were inactive. Popov and Markina (1977) determined the cobalt (unstated method) in the rats exposed by inhalation. Control rats showed 0.5 $\pm$ 0.03 mg Co/kg in liver and 0.7 $\pm$ 0.1 mg Co/kg in spleen on a dry weight basis. The high-exposure group, receiving 0.5 mg/m^3 metallic cobalt aerosol, increased the cobalt content 12 times in thyroid, 2 times in kidney, and 1.2 times in liver during the period from 1.5 months' exposure to 3 months; actual data were not given.

Popov (1977b) also reported that exposing rats to cobalt aerosol for 3 months resulted in the activation of the hemopoietic system; disturbances in protein and carbohydrate metabolism and enzyme systems; and pathomorphological changes in respiratory organs, liver, kidney, spleen, thyroid, and brain. Concentrations of 0.5, 0.05 and 0.005 mg/m^3 all produced the effects. Cobalt aerosol concentration of 0.001 mg/m^3 had only insignificant effects, and Popov suggested this level be used for maximum permissible average daily concentration of cobalt in air in populated areas. Popov et al. (1977) also noted a dose-response relationship in cobalt accumulation and distribution in the bodies of the cobalt aerosol-exposed rats. Cobalt accumulated in lungs, thyroid, liver, and kidneys of rats exposed to all concentrations (0.001 through 0.5 mg/m^3), except in the lungs at 0.001 mg/m^3. The degree of metal accumulation was proportional to the applied concentration and to the duration of exposure. Toxic effects such as decreased activity of lymphatic tissue in spleen, decreased reactivity of bronchial cell membranes and lymphatic tissue in lungs, and protein dystrophy in kidney were noted. These effects, as well, were more severe at higher levels of exposure and greater duration of exposure.

Miniature swine from the Hormel Institute were exposed to cobalt metal powder by inhalation of 0.1 mg/m^3 (five), or 1.0 mg/m^3 (five) for 6 hr/day 5 days a week. Exposure was for 1 week (a "sensitizing dose") followed by a 10-day lapse period, then 3 months' exposure. Five swine were used as controls (Kerfoot, 1973; and Kerfoot et al., 1975).

Results showed that lung compliance decreased in a dose-response curve after three months' exposure. Controls showed 35.5 $cm^3/cm\ H_2O$; low cobalt (0.1 mg/m^3) gave 23.3 $cm^3/cm\ H_2O$ and high cobalt (1.0 mg/m^3) gave 19.8 $cm^3/cm\ H_2O$. The compliance in the exposed groups returned to control levels 1 month after cessation of cobalt exposure. ECG changes showed a loss in strength of ventricular contraction and aberrant repolarization in cobalt-exposed pigs when compared to controls. Radiology revealed no differences between cobalt-exposed and control pigs. Hematology also revealed no persistent changes between groups in hemoglobin, hematocrit, red blood cell, white blood cell, white blood cell counts, and white blood cell differential counts. Skin tests revealed no allergic reactions in groups exposed to cobalt. Blood chemistry studies revealed an increase in α-, β-, and γ-globulins over those of controls; a net increase in total protein in exposed groups; and inversion of the albumin/globulin ratio. These changes may be interpreted as early indicators of lung cell damage. Urinalysis indicated the intake of cobalt: controls excreted 18 μg/liter average; low-cobalt swine, 29 μg/liter; and high-cobalt swine, 220 μg Co/liter. Pathology reports from gross and light microscopic examination showed no effects. Electron microscopy showed that alveolar septa in cobalt-exposed swine were markedly thickened with collagen in a manner resembling human cases of interstitial fibrosis.

The 1977 printing of the 1971 edition, Documentation of Threshold Limit Values, gives a tentative TLV of cobalt metal dust and fume as 0.05 mg/m^3. Cited is the above work of Kerfoot et al. (1975), who showed effects of a serious nature with relatively short exposure.

Harding (1950) exposed 12 rats and undisclosed numbers of hamsters, rabbits, and mice to cobalt dust in high concentration for 6 hr. All were moribund (or dead) at the end of the exposure period; and when necropsied, the lungs were edematous and hemorrhagic. Six anesthetized piebald rats were administered 1 ml of a 5% suspension of cobalt metal in physiological saline intratracheally. Acutely, rats showed lethargy, and death occurred within 15 min to 6 hr. Grossly, lungs were hemorrhagic and edematous; microscopically, some cobalt was visible in the bronchi and atria. Intraperitoneal injection of 2 ml of the same solution given to six rats caused lethargy and respiratory effects, but all six survived.

Georgiadi and Elkind (1978) studied the respiratory tract mucosa in rats that were exposed 4 months to cobalt metal powder at 0.48 mg/m^3 concentration. Circulatory and dystrophic changes were seen in the mucosal tissue, with destruction of epithelium cellular elements and some epithelial atrophy. No normalization occurred in respiratory tract mucosal structure by 1 month after cessation of exposure in these animals.

Wehner and Craig (1972) administered cobaltous oxide to hamsters by inhalation at levels of 2 to 160 mg/m^3. Less than 1% CoO was retained in the lungs acutely and none was found after 6 days. All hamsters exposed to CoO at 106 mg/m^3 died by the fourth of the 3-hr exposures. The animals did survive the 3-month subacute toxicity study at 6.47 to 17.2 mg/m^3 and the 3-week subacute toxicity study at 8.4 to 22.9 mg/m^3 exposure. The authors noted that the percentage of hamster mortality is a function not only of total dose but of dose rate. There is evidence that the hamsters that survive the first few exposures develop some tolerance for the inhaled CoO. Wehner and Craig (1972) compared retention of cobalt monoxide to nickel monoxide and found the latter was retained by lungs much more than the cobalt compound.

Barnes et al. (1976) exposed seven beagles to Co_3O_4 (3) and CoO (4) aerosols, both labeled with ^{60}Co for a lung burden of 2.8 to 4.6 μCi.* Dogs were sacrificed at 8, 64, and 128 days to check retention and distribution of the isotopically labeled aerosols. CoO left the body with a shorter effective biological half-life than Co_3O_4. Cobalt detected in the blood was at least an order of magnitude higher in the dogs exposed to ^{60}CoO than in the dogs exposed to $^{60}Co_3O_4$. CoO

* The Co_3O_4 and CoO aerosols were generated by heating $Co(NO_3)_2$ to 850 and 1400°C, respectively.

was also accumulated to higher concentrations in the organs than Co_3O_4. Thus CoO was more soluble in vivo than Co_3O_4, which remained longer in the lungs. Cobalt from the inhalation doses translocated from the lung and accumulated predominantly in kidney, liver, skeleton, and cartilagenous structures (like trachea).

Acute exposure to cobalt carbonyl (14 mg/m^3 for 2 hr) by inhalation decreased the enzyme levels in liver, kidney, and lung in the rats so exposed. Succinate dehydrogenase, lactate dehydrogenase, alkaline phosphatase, and cytochrome oxidase were all depressed but returned to normal levels in 2 days (Sedov and Kalekin, 1971).

A 2-month inhalation exposure to much lower levels of cobalt dust produced a similar drop in enzyme levels. Georgiadi (1978) administered 0.48 mg Co dust/m^3 to rats for 2 months, and there was a drop in all respiratory mucosal enzymes examined. Further exposure produced varied results, with some enzyme activity returning. Higher dose levels (4.4 mg Co/m^3) also affected the enzyme levels of rats, in some cases producing a sex-related change in enzyme activity.

Cobalt oxide dust, 50 mg total dose, was administered intratracheally to guinea pigs in three weekly doses (10% suspension) (Schepers, 1955a). Observations were made at 1, 4, 8, and 12 months after dosing. An acute response occurred, but within 1 month the lungs were normal. In contrast, 12 months after a single intratracheal injection to rats of 50 mg Co in the form of metal particles, Schepers (1955b) reported diffuse central fibrocellular infiltration in the lungs. The author noted that cobalt has an effect on animal tissue when introduced into lung tissue as a metal that is different from its effect when introduced as an oxide.

Höbel et al. (1972) administered aerosols of $CoCl_2$, Co[Co-EDTA], and Na_2[Co-EDTA] to guinea pigs and rats for 3 hr. The authors noted that $CoCl_2$ produced higher concentrations of cobalt in lungs and liver than after inhalation of Na_2-[Co-EDTA]. Edema and congestion of the lungs occurred and there was decreased activity of the animals after inhalation of the ionized aerosols $CoCl_2$ and Co[Co-EDTA] containing free Co^{2+}; but no effect on either lungs or motility followed inhalation of Na_2[Co-EDTA].

Delahant (1955) performed a study somewhat similar to that of Schepers, described above. Cobalt metal dust at levels of 10, 25, and 50 mg was administered intratracheally to six guinea pigs at each dose level, and the animals were observed for up to 360 days after the injection. In Delahant's experiments, only one animal survived the 10-mg dose for a year, one survived the 25-mg dose a year, and two survived the high, 50-mg dose for a year. At the high level, obliterative pleuritis and firm circumscribed dust lesions were seen. No gross reaction was seen at 360 days when the low-dose guinea pig was killed and observed.

The inhalation toxicity of cobalt carbonyl was studied in rats by Lazarev (1963; cited by Brakhnova, 1975). $Co(CO)_4H$ has a 30-min LD_{50} of 165 mg/m^3. Inhalation experiments with this compound at unspecified dose levels produced some deaths in 3 months, with an increase in hemoglobin. Cellular dust nodules with fibrotic phenomena and inflammation were seen in the dead rats' lungs; as much as 0.31 mg cobalt was found in the lung tissue. Lazarev (1963; cited by Brakhnova, 1975) noted that about 12 to 15% of the inhaled cobalt carbonyl hydride was excreted daily in the urine in his chronic inhalation studies in rats.

(f) Allergy, dermal effects: Cobalt metal fragments of less than 1.0 mm were placed in the eyes of two Dutch Belted rabbits by sterile procedures, and the eyes were observed for pathologic changes. The four eyes involved in the cobalt series were enucleated at 4 months; they showed abscesses involving the lens, ciliary body, vitreous humor, and retina. An eosinophilic reaction occurred, suggesting this was an allergic reaction. Eleven other industrial metals were tested; besides cobalt, only nickel and cadmium produced pathologic changes in the rabbit eye, during a 1-year observation (Lauring and Wergeland, 1970).

Rats, fed cobalt chloride at 100 mg/kg for 4 weeks, had a significant reduction in thymus size and body weights, along with marked decreases in cell-mediated immune response. Chetty et al. (1978) suggested that cobalt was exerting its toxicity by inhibiting the ATPase system.

Female Sprague-Dawley rats, when injected either intraperitoneally or intravenously with 5 mg cobalt sulfate, exhibited erythema and edema involving ears, paws, and snout. These tissues are rich in mast cells, and this phenomenon has been ascribed to histamine release (Jasmin, 1974). The reaction also developed after bilateral nephrectomy indicating that it cannot be ascribed to the release of a renal erythropoietic factor. Nickel salts also produced the reaction.

(2) Aquatic species: The toxicity of cobalt and other heavy metals to aquatic wildlife has been reviewed by Doudoroff and Katz (1953), McKee and Wolf (1963), Battelle Columbus Laboratories (1971), Skidmore (1964; cited by Buikema et al., 1974), and Jones (1964; cited by Buikema et al., 1974). There is a considerable base of information on the lethality and dose effect of cobalt to fish and other aquatic organisms. These studies have been summarized in Table VIII-28, "Toxicity of Cobalt to Aquatic Species." The quantitation of stable cobalt levels in aquatic organisms and comparison of these body burdens with the environmental levels have been less well published. Some values may be found in Tables VII-1 through VII-6.

Most of the current literature on toxicity of heavy metals to aquatic organisms recognize that there is a considerable difference in the effective concentration depending on water hardness. Cobalt is not unique; the metal, presented in soft water, shows more toxicity than when added to hard water for several species, including fish (Tabata, 1969), shrimp (Boutet and Chaisemartin, 1973), rotifers (Buikema et al., 1974), and microbial species (Nagesha et al., 1977).

Tabata (1969) measured the threshold of lethality for Co^{2+} and other heavy metals in water that was soft and artificially hardened with addition of $CaCl_2$ and $MgSO_4$.

Cobalt was $\sim$ 2 times more toxic in soft water than it was in water with hardness increased fourfold with salts described above. This relationship held generally with all test species: Japanese killifish, carp, rainbow trout, or *Daphnia*. When both sodium ion in the water and the hardness increased, the cobalt threshold of toxicity increased 2.2- to 2.5-fold. When potassium ion was increased, there was no change in toxicity (Tabata, 1969). In another study on complexing agents, this laboratory continued their studies on ways to decrease the toxicity to aquatic animals of heavy metals in water (Nishikawa and Tabata, 1969). Addition of EDTA, although decreasing the toxicity of some heavy metals, increased the toxicity of Co^{2+} and Ni^{2+} to carp and other (unspecified) aquatic animals. Since the toxicities of Co^{2+} and Ni^{2+} were low, it was suggested the increase in toxicity observed might be attributed to the toxicity of EDTA.

Freshwater zooplankton of three species were used by Baudouin and Scoppa (1974) to test the 48-hr lethal concentrations (LC_{50}) of cobalt and several other metals in water (see Table VIII-28 for values). The linear relation between metal concentration and median effective killing time for *Daphnia* as well as the high sensitivity of this organism suggests that it may be a useful test organism for water pollution. Baudouin and Scoppa (1974) also found toxicity correlated with metal electronegativity and equilibrium constant for the metal-ATP complex. Toxicity curves showed that, for each species of zooplankton, there was a straight-line relation between median effective time to death and cobalt concentration, without any evidence of a lethal threshold. The authors noted that the mechanisms of this toxic action on zooplankton are unknown.

Good models for the toxicity of cobalt to aquatic life were suggested to be the mayfly, *Ephemerella* (Warnick and Bell, 1969); the rotifer, *Philodina acuticornis* (Buikema et al., 1974); the freshwater plankton, *Daphnia hyalina* (Baudouin and Scoppa; 1974); and the ciliated protozoan, *Tetrahymena pyriformis* (Carter and Cameron, 1973).

Doudoroff and Katz (1953) cited Bandt (1946), who reported that the toxicities to fish of cobalt sulfate and nickel sulfate solutions were merely additive, unlike copper and

TABLE VIII-28

TOXICITY OF COBALT TO AQUATIC ANIMALS

Aquatic Species Affected	Co Form	Dose to Produce Effect, mg/liter			Effect/Comment	Reference
		Concentration of Salt	Concentration of Co^{2+}	Time		
Tetrahymena (Protozoan ciliate)	Cobalt ion		0.008		Optimum growth response	Slater (1952; cited by McKee and Wolf, 1963)
Tetrahymena (Protozoan ciliate)	Cobalt ion		0.0005		Definite growth response	Slater (1952; cited by McKee and Wolf, 1963)
Crustacea, worms, insect larvae	Cobalt ion		1.0			Schweiger (1957; cited by McKee and Wolf, 1973)
Daphnia magna	$CoCl_2$	< 3.1	< 1.4	64 hr	Threshold for immobilization in lake water	Anderson (1948; cited by McKee and Wolf, 1963)
Daphnia magna	$CoCl_2$	2.8	1.3		Threshold for immobilization	Report No 3 to Ohio River Water Sanitation Committee (1950; cited by McKee and Wolf, 1963)
Daphnia magna	$CoCl_2 \cdot 6H_2O$		5		Median threshold effect level in river water	Bringmann and Kuhn (1959; cited by McKee and Wolf, 1963)
Mummichog *Fundulus heteroclitis*	$CoCl_2$	200.0	91		No effect level in sea-water	Doudoroff and Katz (1953)
A protozoan, genus *Microregma*	$CoCl_2 \cdot 6H_2O$		0.5		Median threshold effect level in river water	Bringmann and Kuhn (1959; cited by McKee and Wolf, 1963)
Flatworm *Polycelis nigra*	$Co(NO_3)_2$		83	48 hr	Threshold toxic effect	Jones (1940; cited by McKee and Wolf, 1963)
Fish: 1-yr-old tench, carp, rainbow trout, char	Cobalt ion	1.0	1.0		Nontoxic	Schweiger (1957; cited by McKee and Wolf, 1963)
Carassius auratus (Goldfish)	$CoCl_2$	1,000	454	30–32 hr	Death. Hard water pH 7.8	Ellis (1937; cited by Doudoroff and Katz, 1953)
Carassius auratus	$CoCl_2$	100	45.4	168 hr (7 days)	Death. Soft water pH 6.2	Ellis (1937; cited by Doudoroff and Katz, 1953)
Stickleback *Gasterosteus aculeatus*	$CoCl_2$	22	10		Death. Tap water	Anderson (1948; cited by McKee and Wolf, 1963)
Fish: *Fundulus* spp.	$CoCl_2$	16	7.3	5 days	Death	Doudoroff and Katz (1953)
Fish	$CoCl_2$	4,050	1,839	24 hr	Death	Doudoroff and Katz (1953)
Fish: *Orizias* spp.	$CoCl_2$	1,840	835	24 hr	Death. Used 20 ml volume for test solution	Iwao (1936; cited by Doudoroff 1953)
Fish	Cobalt ion		10		Toxicity	Schweiger (1957; cited by Warmick and Bell, 1969)
Stickleback fish *Gasterosteus aculeatus*	$Co(NO_3)_2$		10	10 days	Lethal concentration limit (i.e. survival same as that of unfed controls)	Data of Jones; cited by Doudoroff and Katz (1953)
Stickleback fish *Gasterosteus aculeatus*	$Co(NO_3)_2$		15	7 days	Average survival time	Data of Jones; cited by Doudoroff and Katz (1953)

TABLE VIII-28 (continued)

Aquatic Species Affected	Co Form	Dose to Produce Effect, mg/liter: Concentration of Salt	Concentration of Co^{2+}	Time	Effect/Comment	Reference
Stickleback fish *Gasterosteus aculeatus*	$Co(NO_3)_2$		20	4 days	Average survival time	Data of Jones; cited by Doudoroff and Katz (1953)
Stickleback fish *Gasterosteus aculeatus*	$Co(NO_3)_2$		50	2 days	Average survival time	Data of Jones; cited by Doudoroff and Katz (1953)
Stickleback fish *Gasterosteus aculeatus*	$Co(NO_3)_2$		150	1 day	Average survival time	Data of Jones; cited by Doudoroff and Katz (1953)
Freshwater protozoa:						
Paramecium	$CoCl_2.6H_2O$	> 2,500	> 613	10 min	Death of 100%	Ruthven and Cairns, Jr. (1973)
Paramecium	$CoCl_2.6H_2O$	1,000	248	3 hr	Some viable	Ruthven and Cairns, Jr. (1973)
Trichophorum	$CoCl_2.6H_2O$	> 5,000	> 1240	10 min	Death of 100%	Ruthven and Cairns, Jr. (1973)
Trichophorum	$CoCl_2.6H_2O$	2,500	613	3 hr	Some viable	Ruthven and Cairns, Jr. (1973)
Aquatic beetle						
Tropistermus lateralis nimbatus	$CoCl_2$	130	59	2 week	Toxicity to gut, gonads	Wooldridge and Wooldridge (1969)
Freshwater zooplankton:						
Cyclops abyssorum prealpinus	$CoCl_2.6H_2O$	15.5	3.84	48 hr	LC_{50}	Baudouin and Scoppa (1974)
Eudiaptomus padanus padanus	$CoCl_2.6H_2O$	4.0	0.99	48 hr	LC_{50}	Baudouin and Scoppa (1974)
Daphnia hyalina	$CoCl_2.6H_2O$	1.32	0.33	48 hr	LC_{50}	Baudouin and Scoppa (1974)
Shrimps						
Austropotamobius pallipes pallipes	$CoCl_2.6H_2O$		0.79	30 day	LC_{50} for all 4 conditions: water O_2 saturated, pH 7.0 ± 0.3 and 16°C	Boutet and Chaisemartin (1973)
Austropotamobius pallipes pallipes	$CoCl_2.6H_2O$		8.8	96 hr		Boutet and Chaisemartin (1973)
Orconectes limosus	$CoCl_2.6H_2O$		0.88	30 day		Boutet and Chaisemartin (1973)
Orconectes limosus	$CoCl_2.6H_2O$		10.2	96 hr		Boutet and Chaisemartin (1973)
Rotifers						
Philodina acuticornis	$CoCl_2$	32	14.5	24 hr	EC_{50} (dead or immobilized) Soft water	Buikema et al. (1974)
Philodina acuticornis	$CoCl_2$	183	83.1 (est.)	48 hr		Buikema et al. (1974)
Snail						
Taphius glabratus	$CoSO_4.7H_2O$		30	24 hr	ED_{100} = muscle toxicity, relaxation	Harry and Aldrich (1963)
Insects (flies)						
Ephemerella	$CoSO_4.7H_2O$		16	96 hr	ED_{50} (0.03 mg Co absorbed)	Warnick and Bell (1969)
Acroneuria	$CoSO_4.7H_2O$		32	8 days	LD_{50} (0.01 mg Co absorbed)	Warnick and Bell (1969)
Hydropsyche	$CoSO_4.7H_2O$		32	7 days	LD_{50} (0.01 mg Co absorbed)	Warnick and Bell (1969)
Ciliated protozoan						
Tetrahymena pyriformis	$CoSO_4$	4.08	1.55	96 hr	LC_{50}, distilled water	Carter and Cameron (1973)
Crustacean larvae:						
Carcinus maenas	$CoCl_2$	10	4.5	1 day	Speed of movement of larvae toward light adversely affected	Amiard (1976)
Carcinus maenas	$CoCl_2$	0.001	.045	2-3 days		Amiard (1976)

TABLE VIII-28 (concluded)

Aquatic Species Affected	Co Form	Dose to Produce Effect, mg/liter			Effect/Comment	Reference
		Concentration of Salt	Concentration of Co^{2+}	Time		
Fish						
Species not given	$CoCl_2$		7-15		Toxic when dissolved in fresh water	Thomas (1915; cited by Doudoroff and Katz, 1953)
Rainbow trout	Cobalt ion		1.0	56 hr	No effect level in tapwater	Ebeling (1928; cited by Doudoroff and Katz, 1953)
Fish, *Colisa fascialatus*	$CoCl_2$	225	102	96 hr	LC_{50}	Srivastiva and Agrawal (1979)
Fish, *Colisa fascialatus*	$CoCl_2$	195	88.5	90 hr	LC_{20}. Lymphocytopenia	Srivastiva and Agrawal (1979)

zinc for example, where mixed solutions of the salts showed considerable synergistic toxicity. In general, their review indicated that cobaltous ions in water had relatively low toxicity to fish, usually not being fatal to fish in water at concentrations below 50 mg/liter. Doudoroff and Katz classified the toxicity of cobaltous ions to fish with that of potassium, lithium, barium, and manganous ions, particularly in highly mineralized waters because of precipitation of insoluble compounds.

Shabalina (1969) found that cobalt at 0.1 mg/kg/day depressed the growth of fingerling fish (*Salmo irideus*) but stimulated the growth of adult, 2- to 3-year-old fish by 10%.

2. Special toxicities

a. Cellular toxicity

(1) Mutagenicity: Heath (1954a) injected 0.028 g pure cobalt metal powder intramuscularly into 30 rats, and after 5 to 12 months observed rhabdomyofibrosarcomata as discussed in subsection b, Carcinogenicity. A histological study of the tumors showed a great number of abnormal cells and cells in mitosis. They specifically showed multipolarity, aberrant chromosomes, and failure of the chromosomes to separate in anaphase and polyploidy.

Cobalt effects on the root tips of *Vicia fabais* were described by Herich (1965). An increase in the "stickiness" of chromosomes was reported, with a decrease in cell division.

Levan (1945) measured the effects of some inorganic salts, including cobalt (probably nitrate) in a dilution series of 10 to 16 concentrations. The effects seen on the root tips of *Allium cepa* included: a tendency to spindle disturbances such as are typical of colchicine mitosis; sticky chromosomes manifested mainly by the formation of anaphase bridges; and an aberrant tendency of the chromosomes to stain, manifested by a differential staining of the heterochromatin at metaphase-anaphase.

Prazmo et al. (1975) produced *Saccharomyces cerevisiae* (yeast) mutations with additions of 2 m*M* $CoCl_2$. Both dividing and nondividing cells were induced to form the cytoplasmically inherited respiratory-deficient trait. The authors also showed that, at a given concentration, cobalt can be either strongly mutagenic or nonmutagenic, depending on cell density.

Cobalt was not included in the tabulation of "mutagenic effects of heavy metals" in an extensive review (Schubert, 1973) although the toxicity was detailed.

Bacillus subtilis strains H 17 Rec^+ and M 45 Rec^- were used by Nishioka (1975) to screen several metals including cobalt for mutagenic activity. $CoCl_2$, 0.05 ml of a 0.05-*M* solution, was used to saturate a paper overlaying agar cultures of the *B. subtilis* recombination-deficient strains. After incubation 24 hr at 37°C, inhibition by the metal was quantitated by measuring the distance in mm between the cobalt-impregnated paper disk and the growing culture. In three of three assays, there was no difference between cobalt inhibition of Rec^+ and Rec^- cells. This "no difference" result suggests that the cobalt chloride solution was not DNA-damaging.

Paton and Allison (1972) measured the effects of cobalt nitrate on cultures of human leukocytes and diploid fibroblasts. Preliminary toxicity tests demonstrated that 8.0×10^{-8} *M* cobalt nitrate (4.6 μg Co/liter) caused fibroblast culture cells to round up and detach from the glass surface after 24 hr exposure. For leukocyte cultures, 8.0×10^{-7} *M* cobalt nitrate markedly reduced the mitotic index as a toxic dose level. This is rather remarkable, since 46 μg Co/liter is not an unusual serum concentration in humans, but free Co^{2+} is probably not present in the blood.

Subtoxic doses of cobalt nitrate were added to leukocyte cultures and fibroblast cells at various times between 2 and 24 hr before fixation; no chromosome aberrations were seen.

To investigate long-term effects of metal treatment, human fibroblast cultures were treated with subtoxic doses of cobalt nitrate or were grown in medium supplemented with metal-treated serum, and cultured several weeks. Paton and Allison (1972) noted that the chromosomes were often over-contracted but otherwise did not differ from control cells.

Some chronic dosing studies (Gilman and Ruckerbauer, 1962; and Gilman, 1962) failed to give evidence of mutagenic tendency of cobalt.

Von Rosen (1964) reported that the EDTA salt of cobalt produced the same mutation spectrum as that from ionizing radiation when injected into the buds of *Pisum* species (the common pea plant). The author found that cobalt broke chromosomes in *Pisum* rootlets (1954) and that the nitrate salt produced abnormalities in *Pisum* seeds and young plants (1957).

Von Rosen (1964) measured the effect of bivalent metal ions including cobalt on young *Pisum* plants and reported a "low but significant increase in cell disturbances." Some other metals form complex compounds with proteins and consequently induce chromosome disturbances; the cobalt effects on the plant buds were additive to effects of other heavy metals and to x-ray effects.

Corbett et al. (1970) reported that cobalt ions (Co^{2+}) were toxic but not mutagenic to bacteriophage T4 strains.

Cobalt toxicity but not mutagenicity was also reported in a study of some cobalt-poisoned cattle. Leonard et al. (1974) analyzed the leukocytes of nine surviving animals that had received high levels of contamination in the feed with lead, chromium, cadmium, zinc, copper as well as cobalt. No chromatid changes nor chromosome aberrations were found in the survivors when compared to control cattle cells.

Cultured human lymphocytes exposed to cobalt acetate, 0.06 to 0.6 μg/liter, increased the frequency of diploidy formation (Voroshilin et al., 1978).

The ions Co^{2+}, Pb^{2+}, Cd^{2+}, Cu^{2+}, and Mn^{2+}, known to exhibit mutagenic or carcinogenic activity, stimulated the rate of initiation of RNA synthesis *in vitro* at concentrations that diminish overall RNA synthesis 10 to 15%; whereas nonmutagenic and noncarcinogenic metal ions inhibited the rate of initiation of RNA synthesis at such concentrations. At 44 mg Co^{2+}/liter, Co^{2+} ranked third below lead and cadmium (Hoffman and Niyogi, 1977).

Lück and Souci (1957) showed no mutagenic activity for $CoCl_2$ in an *E. coli* system.

(2) Organ, tissue, or *in vitro* effects: The biochemical effects of cobalt at the cellular level are those related to the tissue changes of hypoxia. Dosed at toxic levels, it produces histological anoxia, it is a mitochondrial toxin, and it inactivates cytochrome oxidase and a variety of sulfhydryl enzymes. Experiments and case histories detailing these toxicities have been discussed in earlier sections. It appears, however, that at the molecular level, cobalt binds sulfhydryl groups, and the toxic manifestation depends on the target tissue of that binding. Has the reaction occurred predominantly in myocardial tissue? Cobalt cardiomyopathy may result. Have liver enzymes been involved? Oxidative enzymatic changes may be the toxic manifestation. Protein deficiency appears to contribute to or even be necessary for a large amount of cobalt toxicity, in animal experiments at least. The amino and sulfhydryl groups of amino acids can combine with excesses of cobalt and prevent its reaction with vital enzymes and coenzymes, according to suggestions of Wiberg et al. (1969). Protein deficiency disallows this protective mechanism. Optimal protection in rats to cardiac effects were observed, for example, when physiologically required levels of the sulfur-containing amino acids, DL-methionine and L-cysteine were added to a protein-deficient, high cobalt diet (Rona and Chappel, 1973).

Toxicity to a cell culture of fetal mouse fibroblasts was produced in a dose-dependent pattern by addition of cobalt to the medium. Brambilla et al. (1975) showed a proliferation inhibition with 1 to 64 mg $CoCl_2 \cdot 6H_2O$/liter of medium, when the time of exposure was 24 to 96 hr. They also noted that the colony-forming ability of the cultured cells was stimulated at the lower levels of cobalt, rather than inhibited; but the mechanism of this effect was undefined.

b. Carcinogenicity

(1) Cobalt metal and cobalt salts

(a) Animal studies: Animal studies indicate that the carcinogenicity of cobalt metal and its salts is low. Most of the tumors observed are sarcomas at the injection sites. Studies pertaining to the carcinogenic potential of cobalt metal and its salts are summarized in Table VIII-29. Species variability was demonstrated by Gilman (Gilman, 1962), since mice were found to be more tumor-resistant to cobalt compounds than rats.

A study by Stoner et al. (1976) (details in Table VIII-29) did report tumors remote from the site of injection of a cobalt compound, but the incidence was statistically nonsignificant.

Heath (cited by Gilman, 1962) studied the carcinogenicity of cobalt and the intracellular distribution of tumor-inducing cobalt (Heath and Webb, 1967; cited by Webb, 1972). Weinzierl and Webb (1972) discussed the interaction of carcinogenic metals, including that of cobalt, with tissue and body fluids. The intranuclear distribution of cobalt, which induced rhabdomyosarcomas in rats, has also been reported (Webb et al., 1972).

Animal studies regarding the potential of cobalt compounds to alter the effect of other carcinogens have been reported. Cobalt in doses of 10 μg (as $CoCl_2.6H_2O$) given intraperitoneally twice a week for 8 weeks potentiated the development of 3-methylcholanthrene-induced skin tumors in mice, but doses of 100 μg had no appreciable effect (Finogenova, 1973; and Furst and Haro, 1969).

A preparation of tris(monoethanolamine) cobalt and ascorbic acid, when injected intraperitoneally after oral administration of urethane, decreased the number of animals with adenomas induced by urethane. However, the preparation had no significant effect on the growth of Guerin carcinoma or sarcoma 45 (Chernov et al., 1973). The number of respiratory tract tumors induced by subcutaneous diethylnitrosamine injections in hamsters was increased by post-injection intratracheal doses of cobalt oxide particulates.

TABLE VIII-29

ANIMAL CARCINOGENICITY STUDIES OF COBALT METAL AND COBALT SALTS

Species, Strain, Sex (if given)	Compound Tested, Dosage, Route, and Duration[a/]	Tumors	Reference
Rabbits	Cobalt metal, intraosseous injections.	Sarcomas	Vollmann (1940; cited by Sunderman, 1971)
Rabbits	Cobalt metal, intraosseous injections.	Sarcomas	Schinz and Uehlinger (1942; cited by Sunderman, 1971)
Rats, Hooded, male and female	Cobalt powder, 400 mesh, 0.028 g/0.4 ml fowl serum, one I.M. injection, > 7 months (20).	Five of 20; 3 rhabdomyosarcomas	Heath (1954a)
Guinea pigs	Cobalt dust, 10% in saline, one 25 mg metal dose injected I.T.,[b/] 360 days (6).	No tumors	Delahant (1955)
	2.5 mg metal dust doses, in saline, injected I.T., a week apart, 360 days (6).	No tumors	
	One 50 mg dose of metal dust in saline, injected I.T., 1 year (6).	No tumors	
	Cobaltic oxide, 150 mg dust, in saline, I.T., at 50 mg/week for 3 weeks, 360 days (6).	No tumors	
Guinea pigs	Cobalt dust, one 50 mg dose, in saline, I.T., ≥ 12 months.	No tumors	Schepers (1955b)
	Cobalt dust, two 5 mg doses, 1 week apart, in saline, I.T., 12 months.	No tumors	
	Cobalt dust, one 25 mg dose, in saline, I.T., ≥ 8 months.	No tumors	
	Cobaltic oxide, 150 mg dust in saline, I.T., at 50 mg/week for 3 weeks. ≥ 12 months.	No tumors	
Rats, Hooded, male and female	Cobalt powder, 0.028 g/0.4 ml fowl serum, I.M., 122 weeks (20).	Nine of 20	Heath (1956)
Rats, Hooded, female	Cobalt powder, 0.028 g/0.4 ml fowl serum, I.M., 105 weeks (10).	Eight of 10	
Rats	Cobalt dust, inhalation of 200 mg/m^3 air, 12 hr every other day for 4 months, ≥ 4 months.	No tumors	Kaplun (1957; cited by Shubik and Hartwell, 1969)
	Cobalt metal, 3 to 10 mg in saline, I.T., ≥ 8 months.	No tumors	
Rabbits, Albino, male	Cobalt fumes, 1.5 mg/m^3 6 hr/day, every third week for 24 weeks, inhalation, 24 weeks (12).	No tumors; lung pathology only	Stokinger and Wagner (1957; cited by Shubik and Hartwell, 1969)
Rats, Hooded, male	Cobalt powder, 0.028 g/0.4 ml fowl serum, I.M., 20 weeks.	Sarcomas at injection site	Heath (1960; cited by Shubik and Hartwell, 1969)
Rats, Hooded, female	Cobalt powder, 0.028 g/0.4 ml fowl serum injected in diaphragm (10) and intercostal space (10), to 28 months.	Intrathoracic tumors	Heath and Daniel (1962)
Rats, Hooded	Cobalt powder, 0.028 g/0.4 ml fowl serum, I.M., 16 to 20 weeks.	Sarcomas	Daniel et al. (1967)
Rats, Albino	Hard metal dust mixture containing 8% and 15% cobalt, I.T., 10 to 50 mg of dust mixture, 1 day to 8 months.	Pre-carcinogenic papillomata at the bronchi	Kaplun and Mezentseva (1967)
Rabbits, New Zealand White	Cobalt prosthesis alloy specimens, muscle implants, 6 months.	No tumors	Laing et al. (1967)
Rats, Hooded, female	28 mg of Co/Cr prostheses wear debris in 0.4 ml of horse serum injected I.M., 17 months (74).	14 of 74, sarcomas	Heath et al. (1971)
Rats	Cobalt/chromium prostheses wear debris, I.M.	Sarcomas	Swanson et al. (1973; cited by Sunderman, 1977)

TABLE VIII-29 (continued)

Species, Strain, Sex (if given)	Compound Tested, Dosage, Route, and Duration[a/]	Tumors	Reference
Swine, Miniature	Cobalt powder inhalation at 0.1 mg/m^3 (5), and 1.0 mg/m^3 (5), 6 hr/day, 5 days/week, for 3 months.	No tumors in any organ system	Kerfoot (1973)
Rats, Sprague-Dawley, female	Cobalt metal (18) and cobalt sulfide (20), 5 mg in glycerin once intrarenally, 12 months.	No tumors	Jasmin and Riopelle (1976)
Rats, Albino, male	Cobalt sulfate, 0.5 mg/day in aqueous solution, S.C., 10 weeks (10).	No tumors	Orten and Bucciero (1948, cited by Shubik and Hartwell, 1957)
	Cobalt sulfate, 0.5 g/kg of diet, 20 weeks (40).	No tumors	
Rats, Albino, male	Cobalt sulfate, 2.0 or 6.0 g/kg of diet, 20 weeks (36).	No tumors	Bucciero and Orten (1948; cited by Shubik and Hartwell, 1957)
Guinea pigs	Cobalt chloride, 1 ml of 0.5% aqueous solution, S.C.: every 3rd day $\leq$ 60 days (10); on alternate days $\leq$ 60 days (10); 2 ml daily, 30 days (10).	No tumors	Cajano (1951; cited by Shubik and Hartwell, 1957)
Mice, C3H, CF, C57, dba	Cobalt chloride, 58 mg/day in the diet, 10 weeks.	No pathology discussed	Mirone and Wade (1953; cited by Shubik and Hartwell, 1957)
Rats, Albino	0.01 to 0.1% CoF_2 in the diet, 35 days (60).	No tumors	Hodge and Maynard (1953; cited by Shubik and Hartwell, 1957)
Rabbits, male and female	Cobalt nitrate, 10 ml of solution containing 1 mg metal, S.C., daily, 5 days.	Liposarcomas	Thomas and Thiery (1953; cited by Shubik and Hartwell, 1957)
Rabbits, female	Cobalt nitrate, 10 ml of solution containing 1 mg metal, S.C., daily, 5 days, 22 to 37 days (2).	2 liposarcomas	
Rats	Cobalt nitrate, 2 ml of solution containing 0.2 mg metal, S.C., daily, 5 days (12).	No tumors	
Mice	Cobalt nitrate, 1 ml of solution containing 0.1 mg metal, S.C., daily, 5 days (6).	No tumors	
Rats, Sprague-Dawley	Cobalt chloride, 0.4% aqueous solution six times weekly, receiving 2.5 to 10 mg/kg body weight, S.C., $\geq$ 32 weeks.	No tumors	Hopps et al. (1954)
Rats, Wistar, male adult	Cobalt chloride, 2 mg/day, I.P., 90 days.	No tumors	Guillet (1957; cited by Shubik and Hartwell, 1969)
Rats, Albino	Cobalt chloride, orally, dose equivalent to elemental cobalt of 4 mg/kg (10), 218 days, and 10 mg/kg (10), 240 days.	No tumors	Murdock and Klotz (1959)
	Metal dust containing 1.0% cobalt oxide, single I.M. injection of 10% aqueous suspension into one or both thigh muscles. Dosage was:		Gilman and Ruckerbauer (1962)
Mice, Swiss	10 mg into both thighs, 553 days (40);	56% of animals;	
Rats, Wistar	30 mg into one thigh, 525 days (20); and	40% of animals;	
Rats, Hooded	20 mg into both thighs, 388 days (66).	71% of animals	
	Cobalt oxide, single I.M. injection of 10% aqueous suspension into thigh muscles. Dosage was:		
Mice, Swiss	10 mg into both thighs, 751 days (50);	2 pulmonary adenomas;	
Rats, Wistar	30 mg into one thigh, 489 days (10).	5 sarcomas at sites	

TABLE VIII-29 (concluded)

Species, Strain, Sex (if given)	Compound Tested, Dosage, Route, and Duration[a]	Tumors	Reference
Rats, Wistar	Cobalt oxide, single I.M. injection of 10% aqueous suspension into thigh muscles, 20 mg, 342 days (32).	12 of 24	Gilman (1962)
Mice, ICR and CBA/J agouti	Cobalt chloride, single intradermal injection of 0.02 ml of 0.003 M $CoCl_2$ in saline solution, 60 days, 150 days (10).	No tumors	Shelley (1973)
Mice, A/Strong, male and female	Cobalt(III) acetate, 19 I.P. injections, three times weekly, in saline, 30 weeks:	Lung tumors[c]	Stoner et al. (1976)
	maximum tolerated dose, 475 mg/kg (20);	10 of 17;	
	1/2 maximum tolerated dose, 237 mg/kg (20);	8 of 20;	
	1/5 maximum tolerated dose, 95 mg/kg (20).	8 of 20	
Rats, Wistar albino, male	4 mg $CoCl_2$/100 g body weight, saline solution injected S.C., daily for two 5-day courses separated by 9-day intervals, 8 months (20), and 12 months (20).	14 of 27 survivors, fibrosarcomas	Shabaan et al. (1977)
Hamster, Syrian golden, ENG: ELA	Cobalt oxide aerosol, inhalation of 10 g/liter, 7 hr/day, 5 days/week, 17 months (51).	No carcinogenic effects observed	Wehner et al. (1977)

a/ Number of animals started on test in parenthesis.
b/ I.T. = Intratracheal.
c/ Since the control mice developed lung tumors spontaneously, the incidence of lung tumors here was not statistically significant.

The cobalt oxide dust appeared to act not only as a carrier, but also as a synergist in the diethylnitrosamine carcinogenesis of the respiratory tract (Farrell and Davis, 1974). Cobaltous sulfate (0.01 M) administered to mice in drinking water induced splenomegaly after injection with Rauscher leukemia virus (RLV), whereas untreated virus-control mice resisted RLV infection (Gainer, 1973). The carcinogenic effects of ethylnitrosourea in rats were not altered by administration of cobalt chloride (Ivankovic et al., 1972).

A number of reports consider the possible role of trace metals, including cobalt, in asbestos-induced carcinoma (Dixon et al., 1970; Cralley, 1971; Holmes et al., 1971; Gibbs and Lachance, 1972; Morgan and Cralley, 1972; Roy-Chowdhury et al., 1973; and Shettigara and Morgan, 1975).

Although increased concentrations of cobalt appeared in the liver, blood, and lungs of rats exposed to welding fumes, no pulmonary tumors were observed (Hewitt and Hicks, 1973). There was no significant change in the cobalt concentrations of mouse tissues and organs due to application of tobacco smoke condensate (Nadkarni and Ehmann, 1972). Mangal and Verma (1979) compared cobalt levels by neutron activation analysis in normal and cancerous mouse skin. Nine other trace elements were also measured in the 20-methylcholanthrene-induced tissue. Although some elements were elevated over controls as much as 52%, and some were depressed as much as 36% from control tissue, the levels of cobalt in normal and cancerous skin were nearly identical.

In vitro culture techniques have been utilized to assess the carcinogenic potential of cobalt compounds. Cobalt chloride (590 μg Co/liter) induced 3.7% transformed colonies of fetal hamster cells, indicating carcinogenic character (Costa, 1979). Cobalt metal powder after dissolution in horse serum (e.g., 208 to 222 μg Co^{2+}/ml at 28 days) produced cytological changes in cultured rat myoblasts similar to those seen in cobalt-powder-induced rhabdomyosarcomas *in vivo*.

Apparently, a cobalt-protein complex rather than Co^{2+} was the carcinogenic agent *in vivo* (Heath et al., 1969). The effect of cobalt on mitosis of chick cells in tissue culture suggested malignant changes (Heath, 1954a; 1954b).

Reviews of metal carcinogenicity which include cobalt have been published by Furst and Haro (1969), Sunderman (1971), Elinder and Friberg (1977), Taylor and Marks (1978), Roe and Lancaster (1964), and Friberg (1978). Cobalt chloride has been found to inhibit overall RNA synthesis *in vitro* (Hoffman and Niyogi, 1977). A review of metal ion-DNA interactions, including cobalt ion, has been published (Sissoëff et al., 1976).

Certain cobalt compounds have been tested for anticancer activity. A cobalt chloride-imidazole complex was found to be inactive against viral-induced tumors (Kirschner et al., 1966) and the growth of egg-cultivated tumors was inhibited more than 70% by cobalt chloride (Taylor and Carmichael, 1953).

Chernov et al. (1973) reported that the incidence of adenomas in urethane-dosed mice (20 mg/mouse) was significantly reduced by intraperitoneal injection of tris-(monoethanolamine)cobalt in doses of 2 mg/mouse. The cobalt preparation, however, had no inhibitory effect on the growth of Guerin carcinoma or sarcoma 45.

(b) Human studies: Dvizhkov (1967) reviewed the literature on blastomogenic properties (tumor-causing properties) of industrial metals; he suggested that cobalt was a potentially hazardous compound in this regard. Elinder and Friberg (1977), however, in their review of cobalt toxicity did not note any carcinogenic effects of cobalt metal or cobalt salts in humans.

Although present evidence is not conclusive, it appears that cobalt metal and cobalt salts are not significant causative agents of human cancer. Industrial cases where malignancy has been associated with metals have usually involved exposure to a mixture, thereby preventing assignment of causation to one specific metal such as cobalt.

Pertinent epidemiological data have been reported. Pulmonary carcinoma was more frequent in cobalt shop workers who were exposed to cobalt dust along with nickel compounds and sometimes arsenic (Saknyn and Shabynina, 1973). A high incidence of pulmonary cancer has been reported in English cobalt miners the etiology being unknown (an unattributed statement in the review of occupational skin diseases by Schwartz et al. 1947). Payne (1977), however, stated that epidemiologic studies in Zaire, Canada, the United States, and other countries where cobalt is mined do not support an association between cobalt and neoplasm.

No statistically significant correlations were found in U.S. water supplies between total cancer or categorized cancers and cobalt (Berg and Burbank, 1972; cited by Berg, 1975). Study of the cobalt content of human tissues showed no correlation with cause of death. No common chronic disease was associated with either high or low tissue cobalt groups (Schroeder et al., 1967).

The toxic effects of hard metal dust (tungsten or titanium carbide and cobalt cementing metal) have been well studied. Coates and Watson (1971; cited by Louria) found considerable lung disease, including alveolar cell metaplasia in tungsten carbide workers.

Bech et al. (1961; cited by NIOSH, 1977) reported anaplastic adenocarcinoma of the right bronchial lobe in a 63-year-old male exposed to hard metal dusts for 17 years. Ten years after beginning this work, he developed a persistent cough, chest pains, and breathlessness. Radiologic changes became progressively worse. Since the worker had been exposed to a number of industrial dusts of unknown composition and levels, the tumor cannot be attributed to exposure to any specific metal. Kaplun and Mezentseva (1967) analyzed the findings of routine medical examinations of workers exposed to hard metal dust and found upper respiratory conditions, bronchitis, and early fibrosis.

Kaplun concluded that lung pathology of workers exposed to industrial dusts containing cobalt involved diffuse fibrosis of the peribronchial and perivascular tissue

(Kaplun, 1957; cited by Kaplun and Mezentseva, 1967). Payne (1977) stated that it is generally accepted that cobalt is the responsible factor for hard metal respiratory disease. Presenting elements of the disease include asthma, dyspnea, cough, lung edema, radiologic shadows, and stages of lung fibrosis.

The blood concentrations of cobalt and four other metals in 238 cancer patients were found to be at least 114% higher than in 25 healthy controls. Blood metal concentrations tended to normalize after anti-tumor therapy (Nuryagdyev, 1971).

As a staff physician for the Olen, Belgium, cobalt plant for more than 20 years, Verhamme (1973) summarized his observations on cobalt toxicity. No carcinogenic action of cobalt was observed. Exposure to fine cobalt metal dust resulted in respiratory irritation and bronchitis (reversible) but not lung fibrosis as observed in cases of persons exposed to complex hard metal dusts. Cobalt salts caused benign dermatoses. Taylor and Marks (1978) state that tumor formation in anemic patients treated with cobalt chloride has not been reported except by Berk et al. (1971; cited by Taylor and Marks, 1978), who observed one patient with a reticulum cell sarcoma and another with a giant follicular lymphoma.

Although nickel carbonyl is carcinogenic, no studies have reported carcinogenic effects of various cobalt carbonyl compounds which arise in certain industrial environments (Sova and Khaustova, 1969; Brief et al., 1971; and separate studies by Stokinger, Fanney, and Palmers et al., cited by Brief et al., 1971).

(2) Organocobalt compounds: The organocobalt class of cobalamins are physiologically active, vitamin B_{12} being an essential nutrient for both animals and humans. Vitamin B_{12} is Generally Recognized As Safe (GRAS) as a food additive, and no reports of direct carcinogenic effects due to its exposure are known. The vitamin has been shown, however, to enhance hepatoma formation in rats fed *p*-dimethylaminoazobenzene (Day et al., 1950; cited by Tracor-Jitco, 1974) and accelerate the development of methylcholanthrene-induced sarcomas (FASEB, 1978). The data may indicate an involvement for the vitamin in tumor growth. A

number of studies have shown that tumors take up and accumulate relatively large amounts of the vitamin; however, reports contrary to this have also appeared. Studies of the effect of vitamin B_{12} on malignant cell growth have been reviewed (FASEB, 1978; Tracor-Jitco, 1974).

No carcinogenic effects were noted in mice and rabbits dosed with hexamethylenetetraminecobalt complex (Nikitina and Adzhikulov, 1973).

No gross or microscopic effects on tissues and organs were observed in rats after prolonged oral dosing with cobalt polyglucopyranose or cobalt disodium ethylenediaminetetraacetate. Experimental details are in Table VIII-29 (Murdock and Klotz, 1959).

c. Reproductive and teratogenic effects

(1) Animal studies: Animal studies pertaining to the teratogenic or reproductive effects of cobalt can be grouped into whole animal studies, cobalt deficiency effects, and fetal/placental transport and distribution data.

Cobalt, as well as other metal intrauterine wires, was found to inhibit pregnancy in rats and hamsters at the time the fertilized egg reacted the uterus, but before implantation (Chang et al., 1970). Exposure to cobalt chloride and sulfate salts at concentrations of $\leq$ 0.1 mg/liter had negligible effects on the reproduction of *Aeolosoma headleyi* worms (Newman and Buikema, 1975). Grosch studied the effect of chemical agents on fecundity and fertility of wasps. There was a slight increase in malformed adult offspring from mothers fed a one time dose of 0.35 μl 0.013 *M* cobalt sulfate (770 mg Co/liter) solution. Cobalt sulfate lowered total egg production and decreased hatchability (Grosch, 1974).

Cobaltous chloride and nitrite salt solutions induced fetal cleft palates when injected alone into mouse dams but inhibited cleft formation caused by cortisone injections (Kasirsky, 1969, and Kasirsky, 1967; cited by Kasirsky, 1969). Pretreatment of pregnant mice with cobalt chloride before injection of radioactive cortisone decreased the radioactivity found in the fetuses (Waddell, 1971). Mitala et al. (1978) studied the influence of cobalt, cobalamins, and cobalt nitrate and chloride salts on mouse teratogenesis induced by the anticonvulsant phenytoin and cortisone. Cobalt prevented cleft palate caused by phenytoin challenge, yet the cobalt compounds had no effect on the cortisone-induced anomalies.

The nutritive effects of trace elements, including cobalt, on intrauterine calf development (300 mg Co/day) (Gireeva, 1968) and on trout larvae (100 μg Co/liter) during resorption of the yolk sac (Borisov, 1969) have been reported. No adverse effects were mentioned.

Evidence that cobalt deficiency has adverse effects on animal reproduction has been reported. Chow et al. found that B_{12}-depleted rats lost their fertility and recovered it after B_{12} injections. An interaction effect of B_{12} and thyroid on rat fertility was suggested (Chow et al., 1961; cited by Tracor-Jitco, 1974). A high incidence of abortion was found in sows fed B_{12}-deficient diets, while piglets fed B_{12}-deficient milk died within 14 days. The data indicated that B_{12} affected reproduction and neonatal survival in pigs (Frederick and Brisson, 1961; cited by Tracor-Jitco, 1974). Rats and mice, on an "unsupplemented synthetic diet" of unspecified cobalt content, received three types of supplements: 0.02 mg Co/day as $CoCl_2$ to drinking water, 4 μg % B_{12} in the diet, or 0.4 mg Co as $CoCl_2$ in the diet. Litters of dams fed each of the three supplements showed increased survival in both rats and mice, and an average weight gain significantly above nonsupplemented controls (Jaffe, 1952; cited by Cass and Lange, 1955). A survey of 40,000 hog producers from 1967 to 1970 showed that cobalt deficiencies in Missouri soils were related to birth defects in swine (Selby and Tidball, 1976). Wojdala and Niebroj (1968) claimed that cobalt plays an essential role in animal fertility and that administration of cobalt in early pregnancy increases trophoblast activity and enhances mother-fetus exchange. Ewes with high fertility showed blood cobalt levels 57% greater than did animals with low fertility (Annamuradov, 1974).

Radiocobalt was used to study transmission of the element from mother to fetus in cows. The fraction of orally dosed cobalt which transferred across the placenta to the fetus was small, about 0.01% of a total dose administered over a period of the last 2 months of pregnancy. It was speculated that the amount of cobalt so transfered would suffice for an internal function but not for a later life rumen function (Comar, 1948). Dzenite and Lieldiens (1972) found that the concentration of cobalt in fetal cow blood increased from 0.01 to 0.30 mg/kg from 3 to 9 months of gestation.

Flodh (1970) reported the placental and fetal accumulation of radioactive cobalt chloride and vitamin B_{12} in mice. After injection of labeled vitamin B_{12} into dams, there was a rapid and abundant uptake of the compound into the placenta, followed by a slow further transfer to the fetus. Fetuses had much higher post-injection tissue concentrations of cobalt than did the dams. A relative decrease in fetal uptake of vitamin B_{12} with increasing B_{12} dose was found.

Placental transfer of $^{60}CoCl_2$ from mother to fetus was found to increase with the time of gestation in rats (Zylicz et al., 1976), but fetal uptake of an intravenous injection of $^{58}CoCl_2$ into pregnant mice was found to decrease with time (Matsusaka, 1976). Cobalt administered to rats as a chelate complex of diethylenetriaminepentaacetic acid (DTPA) did not cross the rat placental barrier in significant amounts, and yet the DTPA was effective in reducing radionuclide content in the fetuses (Zylicz, 1975).

(2) In vitro studies: Chick embryos have been utilized to study the teratogenic effects of cobalt. Heath (1954b) exposed embryonic chick tissue cultures to cobalt chloride and found a level ($\sim$ 10 mg Co/liter) above which growth was completely inhibited. Cobaltous chloride (0.75 mg/kg embryo) was seen to modify development of the endocrine pancreas in the chick embryo (Ghiani, 1957). Chick embryo DNA incorporated $^{60}CoCl_2$ injected into the yolk up to a saturation limit of 10 μg Co/egg (Menke and Sarif-Sarban, 1966). Treatment of chick embryos with 0.5% cobalt chloride resulted in delayed organization patterns, over-development of neural and mesodermal structures, and neural tube and notochord position abnormalities (Adhikari, 1967). Kury

and Crosby (1968) found a teratogenicity rate of 2.8% for injection of 0.4 to 0.5 mg/egg of $CoCl_2$ into 4-day incubated chicken eggs. Gross abnormalities were confined to the eyes and lower extremities while histological examination revealed hyperplasia, necrosis, and edema. Teratogenic effects of cobalt and other metals on mammalian embryos (Ferm, 1972) and the chick embryo (Ridgeway and Karnofsky, 1952) have been reviewed.

Collagen synthesis was considerably reduced in embryonic rat calvaria (skullcaps) maintained in tissue culture due to 24-hr treatment with 10^{-4} M $CoCl_2$ (5.9 mg Co/liter) (Srivastava et al., 1976). Morrill (1963) found development abnormalities of shells, feet, and gastrulae in freshwater snails (*Limnaea stagnalis* and *L. palustris*) exposed to cobaltous chloride (0.16 to 8.1 μg Co/liter). Exposure of snail (*Helisoma*) eggs to cobalt chloride solutions from 1 to 100 mg/liter resulted in decreased survival and deformities of the shell and gut (Jaroensastraraks and McLaughlin, 1974). High concentrations (10^{-3} M; 59 mg Co/liter) of $CoCl_2$ in artificial seawater had a general toxic effect on sea urchin embryos, whereas lower concentrations (10^{-5} to 10^{-4} M; 0.59 to 5.9 mg/liter) affected mainly the skeletal structure (Barrett and Timourian, 1970). Morisawa and Mohri (1974) studied the effect of metal cations, including Co^{2+} on isolated sea urchin sperm flagella. Cobalt ion at 29.5 mg/liter had only a moderate effect in increasing the turbidity of the suspensions of flagella.

A cobalt-resistant strain of *Neurospora crassa* was isolated by training the mold to grow on media containing high concentrations of Co^{2+}. The cobalt-resistant strain was genetically stable and resulted from a resistance mutation induced by Co^{2+} (Venkateswerlu and Sastry, 1973). A concentration of 10 μg Co^{2+}/liter was found to cause 16% reproductive impairment in the zooplankton *Daphnia magna* (water flea) (Biesinger and Christensen, 1972). Cobalt was found to be one of the most inhibitory metals in germination tests with *Nicotiana tabacum* seeds. The 10 and 50% inhibitory concentrations of Co^{2+} were 0.50 ± 0.04 and 9.6 ± 0.7 μM, respectively, in 100-hr incubation tests (~ 30 and 570 μg/liter) (Siegel, 1977). Numerous other reports of seed treatments with cobalt salt solutions are reviewed in subsection VIII.D.3. In general, they increase germination and plant yield.

(3) Human studies: Cobalt has not been shown to cause significant teratogenic or reproductive sequelae in humans. Brill et al. (1974) found no differences in metal levels, including cobalt, in blood and/or tissues between normal and congenitally abnormal fetuses and infants, and their mothers. Twenty pregnant women delivered normal children after receiving 75 to 100 mg $CoCl_2$/day for treatment of anemia. Another group received cobalt-iron mixture (dose or number of subjects not reported); also, all delivered normal children. No toxicity was reported in any women receiving the cobalt either alone or in combination (Jacobziner and Raybin, 1961). There have been no reports of human teratogenic or reproductive effects of cemented tungsten carbide dust (which contains about 25% cobalt), although there is a large workplace population at risk (NIOSH, 1977).

Data on the content and distribution of cobalt in maternal and fetal blood and tissues are available. Levels of cobalt were determined in human placenta, maternal and fetal blood, and hair. Placental cobalt levels correlated with cobalt concentrations in maternal and fetal blood (Baglan et al., 1974a; 1974b). Alexiou et al. (1976) determined concentrations of cobalt in sera of maternal and cord blood. Cobalt in maternal sera was significantly lower than in non-pregnant controls. Cobalt concentrations in human placenta at delivery were significantly less than values in the liver of newborn at birth (Alexiou et al., 1977). Höck et al. (1975) found a concentration gradient of cobalt between mother and fetus and passive transport of cobalt across the placenta. Thieme et al. (1975) determined placental cobalt by neutron activation analysis and explained some differences by environmental exposure. The placental concentrations in Munich averaged 0.064 mg Co/kg compared to 0.031 mg Co/kg in the Bavarian forest. Baglan et al. (1974a; 1974b) also studied the use of placental metal levels, including cobalt, as an index of environmental exposure. Placentae contained 0.023 mg Co/kg dry weight. Fetal blood contained 0.003 mg Co/kg dry weight and maternal blood contained 0.004 mg Co/kg. Placental levels correlated against those in blood.

(4) Vitamin B_{12}: Effects of cobalt and vitamin B_{12} deficiency on animal fertility have been covered in the previous section on animal studies.

There is no evidence that excess vitamin B_{12} is of any significant teratogenic or reproductive hazard in either animals or humans. Supplementation of female rats on synthetic diets with 10 to 1,000 μg vitamin B_{12} per kg of diet showed no adverse effects on offspring over five generations (Richardson and Brock, 1956; cited by FASEB, 1978). Similarly, Newberne found no abnormalities in offspring of rats supplemented with vitamin B_{12} (Newberne, 1962; cited by FASEB, 1978). Offspring of female rats fed excess B_{12} (200 μg/kg diet) showed no signs of toxicity from the mothers' milk (Daniel et al., 1953; cited by FASEB, 1978).

Study of maternal/fetal cobalamin transport in rats shows the transfer occurring against a concentration gradient because cobalt levels in maternal plasma are lower than those in fetal plasma in late pregnancy (Lowenstein et al., 1960; cited by FASEB, 1978). The amount of cobalamin transported to the rat fetus increases 10-fold from day 10 to day 19 of pregnancy (Graber et al., 1971; cited by FASEB, 1978).

D. Effects of Cobalt on Higher Plants and Microflora

At this time, cobalt has not been shown to be an essential element for green plants, but it is known to be essential for some blue-green algae. A few studies have indicated a cobalt requirement by certain green plants, but the evidence is nonconclusive. However, plants with nodules containing nitrogen-fixing bacteria (legumes) are exceptions. The nitrogen-fixing bacteria require cobalt in order to fix atmospheric nitrogen for use by the plants. A few studies have indicated that legumes themselves and other green plants are benefited by small amounts of cobalt, but whether cobalt is essential has not been established (Epstein, 1972; Bonner and Varner, 1976; Eyster, 1964).

No uncertainty exists, however, about cobalt's potential effect on plants. Cobalt above certain levels, is toxic to plants, but the level varies depending on species and conditions. Negative responses began appearing when cobalt was applied above 1 mg/liter in various nonleguminous plants; in legumes, above 0.050 mg/liter; and in algae, above 5 mg/liter.

Cobalt at very low levels may in some way stimulate plant growth, but cobalt solutions as a seed treatment especially affects seed germination and subsequent plant growth. There is evidence of this in studies conducted mainly in the USSR.

1. Effects on nonleguminous plants: Studies have shown that applying cobalt to nonleguminous plants in the growing medium has slight beneficial effects for some plants at very low levels (< 1 mg Co/liter) and increasing detrimental effects at levels $\geq$ 1 mg Co/liter (see Table VIII-30). Favorable effects were reported for tomatoes and sweet peppers at 0.05 mg Co/liter. Increased growth in rubber plant seedlings was observed at 0.005 and 0.25 mg Co/liter, and increased fresh weight and lower nicotine levels were seen in tobacco plants when 1 mg Co/liter was added to their nutrient solution. Wilt resistance in cotton was increased with 1 mg Co/liter. With tomatoes, 2 mg Co/liter severely retarded root elongation. Negative effects on growth were shown at 2 mg Co/liter in oats, 0.1 mg Co/liter in timothy, 0.05 mg Co/liter in cabbage, 1 mg Co/liter in tomatoes and sweet peppers, and 5.9 mg Co/liter in cotton.

Despite these reports of its positive effects on some plants, cobalt has not been demonstrated to be essential to nonleguminous plants; but in one study, Wilson and Nicholas (1967) reported that wheat (*Triticum durum*) grown in a nutrient solution with nitrogen but containing no cobalt developed leaf chlorosis, possibly a cobalt deficiency symptom.

More frequently, leaf chlorosis has been reported as one of the visible toxic symptoms of excess cobalt. Oats grown in cobalt-treated soil with 92 mg Co/liter developed moderate chlorosis; 778 mg Co/liter caused very severe chlorosis (Anderson et al., 1973). Chino and Nitsui (1968a) reported that absorption and transport of cobalt and other heavy metals in plants interfered with iron absorption and transport, thereby inducing chlorosis. The same investigators also found that excess cobalt

TABLE VIII-30

EFFECTS OF COBALT ON NONLEGUMINOUS PLANTS

Plant	Growing Medium	Cobalt Concentration (mg/kg for soils; mg/liter for solution)	Form	Effect	Reference
Oats, field grown	Soil	< 20		No detectable chlorosis	Anderson et al. (1973)
		92		Moderate chlorosis	
		778		Very severe chlorosis	
Oats	In sand with nutrient solution	2		Seven percent reduction in height	
		5		Twelve percent reduction in height	
Timothy	Soil. N, P, and K fertilizer added	0		Dry weight of plants 0.5 g	Young (1935)
		0.1		Dry weight of plants 0.4 g	
		10		Dry weight of plants 0.2 g	
		100		Dry weight of plants 0.1 g	
		500		No growth	
		2,000		No growth	
Tobacco (*Nicotiana tabacum*) cv. Connecticut Broad Leaf	Nutrient solution	1	$CoCl_2$	0.2% increase in fresh wt. over control 36% decrease in nicotine over control	Tso et al. (1973)
Cabbage var. Haruhikari No. 1	Nutrient solution	0	$Co(NO_3)_2 \cdot 6H_2O$	Dry weight 62.9 g/plant, 26.9 g inner leaves	Hara et al. (1976)
		0.05		62.0 g/plant, 23.3 g inner leaves	
		0.50		36.8 g/plant, 6.8 g inner leaves	
		5.0		6.7 g/plant	
Chrysanthemum morifolium Ramat. cv. Bright Golden Anne	1/2 strength nutrient solution	0.06 (10^{-6} *M* Co)	sulfate and chelate	Growth 90% of control	Patel et al. (1976)
		0.59 (10^{-5} *M* Co)		Growth 90% of control	
		5.9 (10^{-4} *M* Co)		Growth 56% of control	
Strawberries		56.8 (0.25% solution; 50 ml/plant; two applications/season)	$CoCl_2$	9% increase in vitamin C content of berries	Chikalova (1968)

TABLE VIII-30 (concluded)

Plant	Growing Medium	Cobalt Concentration (mg/kg for soils; mg/liter for solution	Form	Effect	Reference
Tomato (*Lycopersicum esculentum*) var. Bonny Best	Solution	2		Root-elongation 60% of controls	Whitby and Hutchinson (1974)
		15		Elongation inhibited	
Tomato and Sweet Peppers	Nutrient solution	0.05		Favorable effect	Pais et al. (1969)
		0.2		Favorable effect	
		1		Toxic to plants	
Cotton	Sand	1	$CoCl_3$ [sic in abstract]	Susceptible to 62% fewer strains of *Fusarium* wilt, effect of wilt milder	Poletaeva (1969)
Cotton var. Acala SJ-2	Nutrient solution	5.9 (10^{-4} *M*)	$CoSO_4$	82% reduction in leaf yield	Rehab and Wallace (1978a)
var. Giza 70		5.9 (10^{-4} *M*)		87% reduction in leaf yield	
Cotton var. Acala SJ-2	Soil-pots	100	$CoSO_4$	62% reduction in leaf yield	Rehab and Wallace (1978b)
		400	$CoSO_4$	92% reduction in leaf yield	
var. Giza		100		44% reduction in leaf yield	
		400		91% reduction in leaf yield	
Rubber plant (*Hevea brasiliensis*)	Nutrient solution	0.005		Seedlings treated 1 yr, 68% increase in root wt., 25% increase in girth, 14% increase in height.	Bolle-Jones and Mallikarjuneswara (1957)
		0.25		Increased growth, less than 0.005 treatment level (normal nutrient concentration)	
Oat	Sand with nutrient solution	5	$CoSO_4$	Reduced height 13%, low chorosis of leaves	Hunter and Vagnano (1953)
		10		Reduced height 25%, medium chorosis of leaves	
		15		Reduced height 53%, high chorosis of leaves	

caused leaf chlorosis in alfalfa but that iron interrelations were not the main cause of the chlorosis (Chino and Nitsui, 1968b). Lipskaya (1970a) found changes in tissues of cucumber leaves such as a decrease in the number of chloroplasts per cell and per unit leaf area with increasing cobalt concentrations. Lipskaya (1970b) reported that 0.02 mg Co/liter applied to sugar beet plants increased the chlorophyll and carotinoid content per leaf surface unit.

Other reported cytologic effects of excess cobalt were: inhibited mitosis and chromosome damage in onion root tips (Gori and Zucconi, 1957) and damage to the endoplasmic reticulum of horse bean root tips (Herich and Bobak, 1976).

Treatment of plants with cobalt is reported to affect the water content and transpiration of potato leaves. Leaves grown from tubers treated with 0.05% (237 mg Co/liter) to 0.25% (1,185 mg Co/liter) solutions of $Co(NO_3)_2$ had 1.5 to 2.9% higher water content. Plants had increased ability to retain water, and maximum transpiration was in the morning rather than later in the day as it was for controls (Gorid'ko, 1967). Semina (1970) reported the same findings for buckwheat.

Spraying solutions of cobalt on sugar beet tops was reported beneficial to seed yield; it increased the amount of monogerm seed balls and sugar yield. Repeated spraying lowered seed germination (Shmilliar, 1968).

2. Effects on leguminous plants: The major effect of cobalt on leguminous plants is on their capacity to fix nitrogen from air under nitrogen-free or nitrogen-restricted conditions. In studies with soybeans inoculated with nitrogen-fixing bacteria and grown in nitrogen-free or nitrogen-restricted solution, 0.1 μg Co/liter greatly increased growth. Normal growth was reported at 1.0 μg Co/liter. Levels of 50 μg Co/liter showed no additional benefit, but neither were they harmful. Detrimental effects in soybeans occurred at 60 to 300 μg Co/liter. Under nitrogen-free growing conditions, alfalfa growth and nitrogen in plants was increased 100% by 5.9 μg Co/liter (see Table VIII-31).

TABLE VIII-31

EFFECTS OF COBALT ON LEGUMES

Plant	Growing Medium	Cobalt Concentration (μg/liter)	Form	Plant Weight				Effect	Reference
Alfalfa (*Medicago sativa*) inoculated with two *Rhizobia* strains	N-restricted nutrient solution	5.9 (0.1 μM Co)						Increase in dry wt of plant, 195%; of root nodules, 278%; of N fixed, 697%; of N in plants, 218%.	Reisenaur (1960)
				Dry Weight in Grams					
				Shoots	Roots	Root Nodules	Seed		
Soybean (*Glycine max*) var. Roaroke inoculated with *Rhizobium japonicum*	N-free nutrient solution	No Co		4.4	4.0	0.7	0.4		Ahmed and Evans (1961)
		0.1	$CoCl_2$	63.3	15.0	4.4	31.0	Increase in N content: roots, 7 x; seed, 73 x; total plant, 12 x. Significant increases in chlorophyll content of leaves and hemoglobin and vitamin B_{12} in root nodules.	
		1.0		55.3	13.0	5.1	33.1	Increase in N content: roots, 6.5 x; seed, 77 x; total plant, 12 x. Significant increases in chlorophyll content of leaves and hemoglobin and vitamin B_{12} in root nodules.	
No inoculation	Nutrient solution with NH_4^+	No Co 0.1 1.0		17.3 19.1 15.5				No significant differences in N content or chlorophyll content of three treatments.	
				Mean Dry Weight (g/pot)					
Soybean (*Glycine max*) var. Roaroke inoculated with *Rhizobium japonicum*	N-free nutrient solution	No Co		16.6				Nitrogen deficiency symptoms in foliage.	Ahmed and Evans (1959)
		1	Vitamin B_{12}	22.1				Normal foliage.	
		1	$CoCl_2$	25.3				Normal foliage, vigorous growth	
		50	$CoCl_2$	24.6				Normal foliage, vigorous growth	

TABLE VIII-31 (concluded)

Plant	Growing Medium	Cobalt Concentration (μg/liter)	Form	Plant Weight			Effect	Reference
				Shoot Dry Weight, % of Control				
				Experiment No. I	Experiment No. II	Experiment No.III		
Soybean (*Glycine max*) var. Roanoke inoculated with *Rhizobium japonicum*	N-free nutrient solution	0.1	$CoCl_2$		169	147 to 427	N and chlorophyll content of shoots and vitamin B_{12} in nodules increased in all three experiments and at the four different treatments.	Evans and Ahmed (1960)
		1.0		150 to 152	188	127 to 386		
		50		138 to 148				
		1	Vitamin B_{12}	123 to 133				
Soybean (*Glycine max* L.) var. Biloxi	Nutrient solution	6 to 60	$Co(NO_3)_2$				Increase in fresh and dry weight, slight increase in % moisture, slight decrease in relative turgidity.	Pushpalata (1967)
		60 to 300					Sharp drop in thickness of leaves, % moisture, and relative turgidity, slight decrease in fresh and dry weight.	
Pea		2,000					Reduced dry weight of plant 70%.	Tishchenko (1970; cited Yagodin and Tishchenko)
Subterranean clover (*Trifolium subterranean*) (non-nodulated)	Nutrient solution with N	No Co	$CoSO_4$				Leaf chlorosis.	Wilson and Nicholas (1967)
		0.1					No chlorosis; dry weight of plants increased 29%.	
				Plant Weight, g/plant (plant Co content, mg/kg)				
				Leaves	Stems	Roots		
Bush bean (*Phaseolus vulgaris* L. cv. Improved Tendergreen)	Nutrient solution	0	$CoSO_4$	1.87 (< 1.5)	0.48 (< 0.15)	0.405 (3)		Wallace et al. (1977)
		59 (10^{-6} M)		1.45 (3.6)	0.39 (43.0)	0.314 (4.4)		
		590 (10^{-5} M)		1.39 (43.0	0.39 (56.1)	0.344 (496)	Leaf chlorosis.	
		5,900 (10^{-4} M)		0.47 (142.0)	0.19 (353.0)	0.150 (2,720)	Severe leaf chlorosis.	

Danilova and Demkina (1967) reported that $CoSO_4$ at 0.1 mg/liter (38 μg Co/liter) applied to uninoculated peas tended to minimize the effect of nitrogen deficiency on early growth; and in later stages of growth, the nitrogen content of the plants increased 100%. In field experiments on pea plants, Bereznitskaya et al. (1969) found that cobalt accelerated the intake of nitrogen, phosphorus, and potassium by the plant and increased protein accumulation in the pea plants.

Yagodin et al. (1970) reported that cobalt stimulates the nitrate reduction activity in the leaves and nodules of legumes but that amounts toxic enough to cause chlorosis decrease enzymic activity. I'lina (1967), after studying several nitrogen-fixing bacteria, concluded that cobalt is a cofactor stimulating nitrogen fixation under certain conditions.

3. Cobalt as a seed treatment and its effect on plants: A number of studies showed cobalt to be a plant stimulant when the seeds were soaked up to 24 hr in 0.001 to 0.1% cobalt chloride (3.8 to 450 mg Co/liter) solution prior to planting. Increased yield from seed treatment was reported for a number of plants: an 11% increase for oats, 13% for sugar beets, and 18 to 51% for buckwheat. Other positive effects reported were increased germination, plant growth, protein content, chlorophyll content and activity, disease resistance, and enzyme activity. Toxic effects on oak seeds and Serdste field beans resulted from seed treatment in 0.1% Co solution (see Table VIII-32).

Godner and Leshina (1967) reported that soaking pea seed in a 0.1% solution of $CoCl_2$ (450 mg Co/liter) increased the height and weight of stems and roots; the surface area, number, and weight of leaves; and the dry matter in the plants by 1 to 3%. The positive effects carried over into the second generation.

A possible mechanism for one of the effects of the seed treatment was reported by Krause and Winowski (1966), who found that certain metabolic products of wheat seeds in the presence of Co^{2+} ions promoted seed peroxidase activity. Increased peroxidase-promoting activity increased the germinating capacity of wheat.

TABLE VIII-32

SEED TREATMENT BY COBALT SOLUTIONS

Plant	Treatment Method	Growing Medium	Cobalt Concentration (mg/liter)	Form	Effect	Reference
Oat var. Condor	Soaked for 24 hr	Soil-field plots	3.8 (0.001% solution)	$CoSO_4$	Increased tillering rate, panicle length, plant height, weight of grain. No effect on germination. Yield increased 11%.	Saric and Saciragic (1969)
Barley (*Hordeum sativum*) cv. Slovensky Dunajsky'trh	Soaked for 24 hr	Nutrient solution	190 (0.05% solution)	$CoSO_4$	Stimulated amylase activity in seedlings, slight stimulation of urease activity, and depressed glutamate oxalacetate transaminase (GOT) activity in seedlings.	Pavel and Zakova (1967)
Buckwheat	Sprinkled seed with 400 mg/kg of solution	Soil-field plots	129 (0.04% solution) 193 (0.06% solution) 258 (0.08% solution)	$Co(NO_3)_2$ $Co(NO_3)_2$ $Co(NO_3)_2$	All three levels increased chlorophyll content, intensity of photosynthesis, and yield; 0.04% level, gave greatest increase.	Elagin (1970)
Buckwheat var. Bogatyr	Seeds soaked	Soil-field plots	97 (0.03% solution) 161 (0.05% solution)	$Co(NO_3)_2$ $Co(NO_3)_2$	Yield increased 18.6% Yield increased 26.7%	Semina (1967)
Buckwheat	Seed soaked	Soil-pots	97 (0.03% solution)	$Co(NO_3)_2$	Increased water content of leaves, decreased transpiration in hot daytime periods	Semina (1970)
		Soil-field plots	Not given	Not given	Over 4-yr test, 51% increase in yield in hot, dry year.	
Sunflower		Soil at 60% of field water capacity	64 (0.02% solution)	$Co(NO_3)_2$	Increased RNA and DNA content, decreased RNase activity.	Bozhenko (1968)
Sunflower	Soaked 10 hr	Soil-field plots	? (0.01% solution)	Not given	70% fewer plants infected with *Scterotinia*	Polyakov (1971)
Sugar beets		Not given	? (0.01% solution)	Not given	13.0 to 13.3% increase in yield. Beets contained more sugar than controls.	Guseva et al. (1968)
Corn		Not given	45 (0.01% solution Co)	Not given	Increased yield and vitamin C in plants.	Guseva et al. (1968)

TABLE VIII-32 (concluded)

Plant	Treatment Method	Growing Medium	Cobalt Concentration (mg/liter)	Form	Effect	Reference
Corn	Soaked for 24 hr	Soil-field plots	360 (0.08% solution)	$CoCl_2$	Increased protein in plant and grain.	Gavrilova (1965)
Oak	Acorns soaked for 12 hr		4.5 (0.001% solution)	$CoCl_2$	Accelerated shoot appearances 7 to 9 days.	Khonin (1965)
			450 (0.1% solution)		Retarded germination 4 to 5 days.	
Pine and alder	Soaked for 24 hr		10-100		Increased germination 5 to 7%, dry matter 40 to 65%, height 10 to 15%.	Kaposts (1968)
	Soaked for 24 hr		500		"Deleterious" effects.	
Pea and bean	Soaked for 24 hr		450 (0.1% solution)	$CoCl_2$	Increased yield, protein content, activities of peroxidase and catalase, respiration intensity, chloroplast size, and amount of chlorophyll.	Leshina (1969)
Serdste field beans (*Vicia faba*)	Seed soaked	Field plots	380 (0.1% solution)	$CoSO_4$	20% of plants became chlorotic with 40% reduction in chlorophyll, decrease in high mol. wt. proteins, and increase in low mol. wt. proteins.	Vagodin and Zhiznevskaya (1969)
Barley			380-3800 (0.1-1% solution)	$CoSO_4$	Decreased levels of chlorophyll a and b, no effect on Hill reaction in 7 to 10-day-old seedlings.	Lipskaya and Malyash (1974)
Barley	Soaked for 24 hr		45 (0.01% solution)	$CoCl_2$	Decreased leaf area; increased leaf and chlorenchyma thickness, no. of chloroplasts, and chloroplasts width, surface, and volume in 7-day-old seedlings.	Lipskaya (1974)
Rice	Seed soaked		? (0.05% solution)		Increased germination 40 to 50%.	

4. Effects on algae: At very low levels (> 0.5 mg/liter), cobalt added to the growing media of algae inhibited growth. Little or no growth was reported for levels of > 20 mg Co/liter except for one species (Nitzschia closterium). It was growing in a chelating nutrient solution and, at 20 mg Co/liter, the algal growth was 50% that of the control. Evidently, the chelating amino acids had reduced the toxic effects of the cobalt. Algal growth was much more sensitive to cobalt than algal photosynthesis. In two species, cobalt levels of 100 mg/liter did not adversely affect photosynthesis. Other species did experience reduced photosynthesis at this level. One study reported that very high levels (589 mg/liter) of cobalt did not inhibit the life cycle of Chlorella vulgaris (see Table VIII-33).

Holm-Hansen et al. (1954) found that cobalt met the requirement for an element essential to the healthy growth of several blue-green algae. The cobalt requirement of the algae was low; 2 to 40 μg Co/liter was sufficient for optimum growth.

Mills and Oglesby (1971) reported that the lack of sufficient cobalt as well as copper and zinc may be the limiting factor in phytoplankton growth during late summer in Cayuga Lake, New York.

5. Effects on microflora: The effect of cobalt varies from one microorganism to another. Slight positive effects were noted at very low levels (< 5 mg Co/liter) in some microorganisms. Several negative effects at higher levels were observed. Some of these were reduced growth, reduced respiration, reduced glucose conversion, and longer generation times. Yeast strains were killed at levels of 227 mg Co/liter added to their nutrient solution (see Table VIII-34).

Peciulis (1969) reported that cobalt stimulated accumulation of proteins in yeast and that addition of chromium along with molybdenum, boron, and zinc increased the amount of thiamine, riboflavin, and folic acid in yeast.

One microorganism was reported to develop resistance to high levels of cobalt. Neurospora crassa was cultured on media with successively higher concentrations of cobalt until a strain was tolerant to 2,000 mg Co/liter.

TABLE VIII-33

EFFECT OF COBALT ON ALGAE

Species	Growing Medium	Cobalt Concentration (mg/liter)	Form	Effect	Reference
Chlorella	Nutrient solution	0.02		Growth greater than control.	Young (1935)
		0.2		Growth greater than control.	
		2.0		No growth.	
		20		No growth.	
		200		No growth.	
		0.002		Growth equal to check.	
Crucigina	Nutrient solution	0.02		Growth equal to control.	
		0.2		Growth greater than control.	
		2.0		Growth less than control.	
		20		No growth.	
		200		No growth.	
Chlorella vulgaris	Arnon culture medium	58.9	$Co(NO_3)_2$	Little inhibition of life cycle.	Ishizaka et al. (1966)
		Co^{2+} 5.9	$CoCl_2$	No inhibition of life cycle.	
		Co^{2+} 5.9	$Co(O_2CCH_3)_2$	Slight inhibition of life cycle.	
		Co^{2+} 58.9	$CoCL_2$	Slight inhibition of life cycle.	
		Co^{2+} 58.9	$Co(O_2CCH_3)_2$	No inhibition of life cycle.	
Chlorella pyrenoidosa (Chick)	Nutrient medium	2,480 (1% by wt)	$CoCL_2 \cdot 6H_2O$	Apparent photosynthesis, 80% of control after 3 hr, cells killed within 36 hr.	Phelps and Schlichting (1968)
Chlorella	Nutrient solution	0.5		Growth 50% of control.	Hutchinson (1973)
		1.0		Growth 7% of control.	
Chlamydomonas	Nutrient solution	0.5		No growth.	Hutchinson (1973)
Haematococcus	Nutrient solution	0.5		Growth 78% of control.	Hutchinson (1973)
		1.0		Growth 24% of control.	
		5.0		Growth 20% of control.	
Dunaliella	Nutrient solution	25	$CoSO_4$	No change in photosynthesis.	Mills and Colwell (1977)
		100		No change in photosynthesis.	
Chlorella	Nutrient solution	25		No change in photosynthesis.	Mills and Colwell (1977)
		100		No change in photosynthesis.	

TABLE VIII-33 (concluded)

Species	Growing Medium	Cobalt Concentration (mg/liter)	Form	Effect	Reference
Nitzschia closterium	Nutrient solution (nonchelating)	5.6	$CoSO_4$	Growth 70.1% of control.	Rosko and Rachlin (1975)
		7.5		Growth 61.8% of control.	
		10.0		Growth 50.4% of control.	
		18.0		Growth 29.9% of control.	
	Nutrient solution (chelating)	18.0	$CoSO_4$	Growth 62.5% of control.	
		24.0		Growth 49.7% of control.	
		32.0		Growth 35.0% of control.	
		42.0		Growth 23.6% of control.	
		56.0		Growth 14.5% of control.	
Microcystis aeruginosa		0.03	$Co(NO_3)_2$	Did not affect algae but decreased number of saprophytes present.	Migal (1970)
Chlorella fusca		0.20-0.35	$Co(NO_3)_2$	Significantly lower respiration rate.	Sandstroem and Soedergren (1974)
Chesapeake Bay water nannoplankton	Nutrient solution	25	$CoSO_4$	7% reduction in photosynthesis.	Mills and Colwell (1977)
		100		40% reduction in photosynthesis.	

TABLE VIII-34

EFFECTS OF COBALT ON MICROFLORA

Organism	Medium	Cobalt Concentration (mg/liter in solution; mg/kg in soils)	Form	Effect	Reference
Yeast	Nutrient solution	55 115	Absolute salt	Reduced growth 50%. Completely inhibited growth.	White and Munns (1951)
				% Respiratory Deficient Colonies	
Yeast (5 strains)	Nutrient solution	59 (0.013% solution) 113 (0.025% solution) 172 (0.038% solution) 227 (0.050% solution)	$CoCl_2$	< 1 to 33 < 1 to 100 5 to 100 all five strains killed	Lindegren et al. (1958)
				% Conversion Glucose to L-malate	
Schizophyllum commune	Nutrient solution	0 0.6 (10^{-5} M) 5.9 (10^{-4} M) 59 (10^{-3} M)	$CoSO_4$	63.3 63.6 48.9 31.8	Tachibana et al. (1972)
Phytophthora cinnamomi and Phytophthora drechsleri		0.6 (10^{-5} M Co^{2+}) 0.06 (10^{-6} M Co^{2+}) 0.006 (10^{-7} M Co^{2+})	$Co(NO_3)_2$	No inhibition of zoosporangial. No inhibition of zoosporangial. No inhibition of zoosporangial.	Halsall (1977)
				% Change in Generation Times	
Leuconostoc citrovorum (4 strains)	Skim milk - 22°C	4.5 (0.001% solution) 45 (0.01% solution) 450 (0.1% solution)	$CoCl_2$	-16.3 to +41.2 +21.1 to +271.1 +904.1 to 3,390.3	Goel and Marth (1972)
	Skim milk - 30°C	4.5 (0.001% solution) 45 (0.01% solution) 450 (0.1% solution)		-2.7 to +29.8 +16.0 to +212.6 +464.1 to 1,445.6	
Azotobacter chroococcum I	Nutrient agar	5.6 (10^{-4} M)	$CoCl_2$	Produced inhibitory zone, 60 mm diameter.	Den Dooren De Jong (1971)
Azotobacter chroococcum IV	Nutrient agar	5.9 (10^{-4} M)	$CoCl_2$	Produced inhibitory zone, 65 mm diameter.	
Azotobacter vinelanddii	Nutrient agar	5.9 (10^{-4} M)	$CoCl_2$	Produced inhibitory zone, 65 mm diameter.	
Ritzville silt loam soil microorganisms	Soil	100	Not given	Changes in species distribution in population, but not in CO_2 evolution.	Fujihara et al. (1973)

TABLE VIII-34 (concluded)

Organism	Medium	Cobalt Concentration (mg/liter in solution; mg/kg in soils)	Form	Effect	Reference
Actinomycetes	Solution	1 to 5	$CoCl_2$	Produced "conspicuous growth."	Uesata et al. (1953)
		1 to 5	$CoSO_4$	Produced "conspicuous growth."	
Sclerotinia sclerotiorum	Solution	0.3	$Co(NO_3)_2$ $CoCl_2$ or $CoSO_4$	Inhibited pectolytic activity 20-30%. Decreased pectinesterase activity 50%.	Astapovich and Grel (1975)
Aspergillis awamori	Solution	2	$CoSO_4$	Increased pectolytic activity 170%.	

The cobalt resistance appeared to be a permanent mutation in *N. crassa* (Venkateswerlw and Sastry, 1973).

Studies of the effect of Co(III) complexes on *Escherichia coli* indicate Co(III) causes specific changes in cell metabolism whereas Co(II) salts such as $CoSO_4$ cause more general inhibition of cell activity. The Co(III) complexes were found to inhibit production of protein (Crawford et al., 1974; Talburt et al., 1974; Johnson and Talburt, 1974).

McKee and Wolf (1963) cite studies that 25 to 29 mg Co^{2+}/liter in sewage caused 50% reduction by the sewage microorganisms in oxygen consumption. Burger et al. (1977) found that even 0.5 mg Co^{2+}/liter significantly slowed the oxidative degradation of organic matter by microbial species after incubation for 12 hr. This negative effect was enhanced when the water was acidic or when turbulence was low. The toxic effect was reduced in very hard water.

BIBLIOGRAPHY. CHAPTER VIII

Abedinzadeh, Z., M. Razeghi, and B. Parsa, "Neutron Activation Analysis of an Iranian Cigarette and Its Smoke," J. Radioanal. Chem., **35**, 373-376 (1977).

American Conference of Governmental Industrial Hygienists, Documentation of the Threshold Limit Values for Substances in Workroom Air with Supplements for Those Substances Added or Changed Since 1971, 3rd ed., American Conference of Governmental Industrial Hygienists, Cincinnati, Ohio, 1971 (4th printing, 1977).

Adhikari, S., "Effects of Cobalt Chloride on Chick Embryos," Anat. Anz. Bd., **120**, 75-83 (1967).

Agrawal, Y. K., K. P. S. Raj, and M. R. Patel, "Metal Contents in the Drinking Water of Cambay," Water, Air, Soil Pollut., **9**, 429-431 (1978).

Ahmed, S., and H. J. Evans, "Effect of Cobalt on the Growth of Soybeans in the Absence of Supplied Nitrogen," Biochem. Biophys. Res. Commun., **1**(5), 271-274 (1959).

Ahmed, S., and H. J. Evans, "The Essentiality of Cobalt for Soybean Plants grown under Symbiotic Conditions," Proc. Natl. Acad. Sci., **47**, 24-36 (1961).

Albert, A., "Metal-Binding Agents in Chemotherapy: The Activation of Metals by Chelation," Symp. Soc. Gen. Microbiol., 8th, Strategy of Chemotherapy, 112-138 (1958).

Alexander, C. S., "Cobalt-Beer Cardiomyopathy," Am. J. Med., **53**, 395-417 (1972).

Alexiou, D., A. P. Grimanis, M. Grimani, G. Papaevangelou, and C. Papadatos, "Concentrations of Zinc, Cobalt, Bromine, Rubidium and Gold in Maternal and Cord Blood Serum," Biol. Neonate, **28**(3-4), 191-195 (1976).

Alexiou, D., A. P. Grimanis, M. Grimani, G. Papaevangelou, E. Koumantakis, and C. Papadatos, "Trace Elements (Zinc, Cobalt, Selenium, Rubidium, Bromine, Gold) in Human Placenta and Newborn Liver at Birth," Pediat. Res., **11**, 646-648 (1977).

Alfrey, A. C., J. M. Mishell, J. Burks, S. R. Contiguglia, H. Rudolph, E. Lewin, and J. H. Holmes, "Syndrome of Dyspraxia and Multifocal Seizures Associated with Chronic Hemodialysis," Trans. Am. Soc. Artif. Int. Organs, **18**, 257-261 (1972).

Allen, R. E., and D. G. Vickroy, "The Characterization of Cigarette Smoke from Cytrel® Smoking Products and Its Comparison to Smoke from Flue-Cured Tobacco," Beitr. Tabakforsch., **8**(7), 430-437 (1976).

Amiard, J. C., "Assimilation, Excretion, and Distribution of Cobalt-60 in the Rat after Daily Ingestion of Contaminated Marine Food Products," Health Phys., **31**(4), 371-373 (1976).

Anan'ev, N. I., "Vliyanie Mikro- i Makroelementov Pit'evoi Vody na Rasprostranennost' i Intensivnost' Kariesa Zubov" ["Effect of Trace and Major Elements of Drinking Water on the Distribution and Intensity of Dental Caries"], Gig. Sanit., No. 3, 86-87 (1977).

Anderlini, V. C., P. G. Connors, R. W. Risebrough, and J. H. Martin, "Concentrations of Heavy Metals in Some Antaractic and North American Sea Birds," Proc. Colloq. Conserv. Probl. Antarct., 49-61 (1972).

Anderson, A. J., D. R. Meyer, and F. K. Meyer, "Heavy Metal Toxicities: Levels of Nickel, Cobalt, and Chromium in the Soil and Plants Associated with Visual Symptoms and Variation in Growth of an Oat Crop," Aust. J. Agric. Res., **24**(4), 557-571 (1973).

Annamuradov, A., "Effect of Trace Elements on Ewe Fertility," Zhivotnovodstov, No. 1, 83 (1974); Chem. Abstr. **81**, 48084e (1974).

Anonymous, "Drug-Induced Goiters in the Fetus and in Children and Adults," Med. Lett. Drugs Ther., 12(15), 61-62 (1970).

Anonymous, "Synergism of Cobalt and Ethanol," Nutr. Rev., 29, 43-45 (1971).

Anonymous, "The Selenium Paradox," Food Cosmet. Toxicol., 10(6), 867-873 (1972).

Anonymous, "Heart Disease, Cancer Linked to Trace Metals," Chem. Eng. News, 54(19), 24-28 (1976a).

Anonymous, "Cobalt in Severe Renal Failure," Lancet, 2(7975), 26-27 (1976b).

Anonymous, "Adopted by ACGIH for 1978. Threshold Limit Values for Chemical Substances in Workroom Air," National Safety News, (January), 65-73 (1979).

Aprod, A. I., "Improvement in the Quality of Rice Seeds," Zernovoe Khoz., No. 9, 42 (1977); Chem. Abstr. 88, 84439c (1978).

Astapovich, N. I., and M. V. Grel, "Effect of Various Cobalt Salts and Their Concentrations on the Activity of Pectolytic Enzymes Synthesized by Microscopic Fungi," Biol. Akt. Veshchestva Mikroorg., 36-39 (1975); Chem. Abstr., 85, 057445h (1976).

Babadzhanov, S. N., "Distribution of Some Trace Nutrients in Tissues and Organs of Healthy Rabbits," Uzb. Biol. Zh., 17(2), 71-72 (1973); Chem. Abstr., 80, 118571s (1974).

Baglan, R. J., A. B. Brill, D. Wilson, W. Schaffner, A. Schulert, K. Larsen, J. Davies, and L. Hoffman, "Utility of Placental Tissue as an Indicator of Environmental Exposure," Proc. Ann. NSF Trace Contam. Conf., 1st; Conf. 730802, 505-509 (1974a).

Baglan, R. J., A. B. Brill, A. Schulert, D. Wilson, K. Larsen, N. Dyer, M. Mansour, W. Schaffner, L. Hoffman, and J. Davies, "Utility of Placental Tissue as an Indicator of Trace Element Exposure to Adult and Fetus," Environ. Res., 8(1), 64-70 (1974b).

Bala, Yu., V. M. Lifshits, S. A. Plotko, G. I. Aksenov, and L. M. Kopylova, "Age Level of Trace Elements in the Human Body," Tr. Voronezh. Gos. Med. Inst., **64**, 37-44 (1969); Chem. Abstr., **13**, 96263u (1970).

Barborik, M., and J. Dusek, "Cardiomyopathy Accompanying Industrial Cobalt Exposure," Br. Heart J., **34**(1), 113-116 (1972).

Barborik, M., J. Dusek, and J. Jelinkova, "Organ Concentration of Cobalt and Its Excretion in Hard Metal Workers," Acta Univ. Palacki. Olomuc. Fac. Med., No. 70, 321-330 (1974).

Barnes, J. E., G. M. Kanapilly, and G. J. Newton, "Cobalt-60 Oxide Aerosols: Methods of Production and Short-Term Retention and Distribution Kinetics in the Beagle Dog," Health Phys., **30**, 391-398 (1976).

Barrett, M. M., and H. Timourian, Effect of Cobalt on Sea Urchin Morphogenesis, UCRL-50717, U.S. AEC Univ. Calif. Radiat. Lab., Livermore, California, 1969, pp. 1-6

Battelle Columbus Laboratories, Water Quality Criteria Data Book - Vol. 3. Effects of Chemicals on Aquatic Life. Selected Data from the Literature through 1968, Prepared for the Environmental Protection Agency, Project No. 18050 GWV, Contract No. 68-01-0007, U.S. Government Printing Office, Washington, D.C., May 1971.

Baudouin, M. F., and P. Scoppa, "Acute Toxicity of Various Metals to Fresh Water Zooplankton," Bull. Environ. Contam. Toxicol., **12**, 745-751 (1974).

Baudouin, J., C. Thevenol, G. Dezile, M. Lavandier, J. P. Homasson, and A. Roullier, "Fibrose pulmonaire aux metaux durs. Etude fonctionnelle et immunologique de 4 cas" ["Pulmonary Fibrosis by Hard Metals. Functional and Immunological Study of 4 Cases"], Rev. Inst. Hyg. Mines (Hasselt), **29**(3), 125-129 (1974).

Baudouin, J., P. Jobard, J. Moline, M. Lavandier, A. Roullier, and J. P. Homasson, "Fibrose pulmonaire interstitielle diffuse" ["Diffuse Interstitial Pulmonary Fibrosis. Responsibility of Hard Metals"], Nouv. Presse Med., **4**(18), 1353-1355 (1975).

Becker, D. E., and S. E. Smith, "The Level of Cobalt Tolerance in Yearling Sheep," J. Animal Sci., 10, 266-271 (1951).

Bekh, K. I., "Vliyanie Ionov Kobal'ta na Funktsional'noe Sostoyanie Nervno-Myshechnogo Apparata" ["The Effect of Cobalt Ions on Functional State of the Neuromuscular Apparatus"], Vrach. Delo, 10, 122-124 (1972).

Bereznitskaya, N. I., P. P. Levenets, and L. I. Sulimova, "Uptake of Nitrogen, Phosphorus, and Potassium into the Pea Plant under Various Nutritive Conditions," Tr. Khar'kov. Sel'skokhoz. Inst., 78, 30-40 (1969); Chem. Abstr., 72, 110378y (1970).

Berg, J. W., "Diet" in Persons at High Risk of Cancer. An Approach to Cancer Etiology and Control, Proceedings of a Conference, Key Biscayne, Florida, U.S.A., December 10-12, 1974, Academic Press, New York, New York, 1975, pp. 201-224.

Berlin, N. I., "The Distribution of Cobalt in Polycythemic Rats," J. Biol. Chem., 187, 41-45 (1950).

Bhat, I. S., A. G. Hedge, S. Chandramouli, and R. S. Iyer, "Evaluation of Internal Exposure to Radionuclides of I, Cs, and Co during Maintenance Operations on Primary Steam Leak in a Nuclear Power Station," Health Phys., 25(Aug.), 135-139 (1973).

Bieger, W., J. Seybold, and H. F. Kern, "Studies on Intracellular Transport of Secretory Proteins in the Rat Exocrine Pancreas," Virchows Arch. A., Pathol. Anat. Histol., 368(4), 329-345 (1975).

Bienvenu, P., C. Nofre, and A. Cier, "Toxicité générale compareé des ions métálliques. Relation avec la classification périodique" ["General Comparative Toxicity of the Metallic Ions. Relation with the Periodic Classification"], Compt. Rend., 256, 1043-1044 (1963).

Bierenbaum, M. L., A. I. Fleischman, J. Dunn, and J. Arnold, "Possible Toxic Water Factor in Coronary Heart Disease," Lancet, 1(7914), 1008-1010 (1975).

Biesinger, K. E., and G. M. Christensen, "Effects of Various Metals on Survival, Growth, Reproduction, and Metabolism of Daphnia magna," J. Fish. Res. Board, 29(12), 1691-1700 (1972).

Blank, M., and J. S. Britten, "Effects of Cations on Biologically Active Surfaces," Adv. Chem. Ser., **144** (Monolayers, Men. Symp., 1974), 231-238 (1975).

Bolle-Jones, E. W., and V. R. Mallikarjuneswara, "A Beneficial Effect of Cobalt on the Growth of the Rubber Plant (*Hevea brasiliensis*)," Nature, **179**, 738-739 (1957).

Bonenfant, J.-L., C. Auger, G. Miller, J. Chenard, and P.-E. Roy, "Québec Beer-Drinkers' Myocardosis: Pathological Aspects," Ann. N.Y. Acad. Sci., **156**, 577-582 (1969).

Bonner, J., and J. E. Varner, Eds., *Plant Biochemistry*, Academic Press, New York, New York, 1976.

Boutet, C., and C. Chaisemartin, "Propriétés toxiques spécifiques des sels métalliques chez Austropotamobius pallipes pallipes et Orconectes limosus" ["Specific Toxic Properties of Metal Salts on *Austro-Potamobius pallipes pallipes* and *Orconectes limosus*"], C. R. Soc. Biol. (Paris), **167**(12), 1933-1938 (1973).

Bowie, E. A., and P. J. Hurley, "Cobalt Chloride in the Treatment of Refractory Anaemia in Patients Undergoing Long-Term Hemodialysis," Aust. N. Z. J. Med., **5**, 306-314 (1975).

Bozhenko, V. P., "Effect of Aluminum and Cobalt on the Nucleic Acid Content and Ribonuclease Activity in the Growth Areas of the Sunflower Plant under Water Deficiency Conditions," Fiziol. Rast., **15**(1), 116-122 (1968); Chem. Abstr., **68**, 94925j (1968).

Brambilla, G., M. C. Boaretto, and C. Bolognesi, "Effects of the Cobalt Ion on the *In Vitro* Growth of Mouse Fetus Fibroblasts," Tumori, **61**(4), 327-332 (1975).

Brief, R. S., J. W. Blanchard, R. A. Scala, and J. H. Blacker, "Metal Carbonyls in the Petroleum Industry," Arch. Environ. Health, **23**(5), 373-384 (1971).

Brill, A. B., R. J. Baglan, W. Fleet, W. Schaffner, and A. Schulert, "Network to Determine Causality Between Abnormal Trace Element Levels in Congenital Defects" in *Proceedings of the First Annual NSF Trace Contaminants Conference*, W. Fulkerson, W. D. Shults, and R. I. Van Hook, Eds., Conf. 730802, Oak Ridge National Laboratory, Oak Ridge, Tennessee, March 1974.

Browning, E., Toxicity of Industrial Metals, Butterworths, London, England, 1961.

Browning, E., Toxicity of Industrial Metals, 2nd ed., Butterworths, London, England, 1969.

Bryan, S. E., and J. E. Bright, "Serum Protein Responses Elicited by Iron, Cobalt, and Mercury," Toxicol. Appl. Pharm., **26**, 109-117 (1973).

Bryson, D. D., "Cyanide Poisoning," Lancet, **1**, 92 (1978).

Buikema, A. L., Jr., J. Cairns, Jr., and G. W. Sullivan, "Evaluation of *Philodina acuticornis* (Rotifera) as a Bioassay Organism for Heavy Metals," Water Res. Bull., **10**(4), 648-661 (1974).

Burch, R. E., R. V. Williams, and J. F. Sullivan, "Effect of Cobalt, Beer, and Thiamine-Deficient Diets in Pigs," Am. J. Clin. Nutr., **26**(4), 403-408 (1973).

Burger, G., P. Pfeiffer, V. Risse, and G. Weise, "Modelluntersuchung zur Toxizität von Zink-, Kobalt-, Kupfer- and Nickelionen in Gewässern" ["Model Studies on Toxicity of Zinc, Cobalt, and Nickel Ions in Water"], Acta Hydrochim. Hydrobiol., **5**(4), 351-361 (1977).

Burke, D. H., and J. C. Brooks, "Cobaltous Chloride-Induced Hypothermia II: Pretreatment with Sympathoplegics, Antihistamines, and Narcotic Antagonists," J. Pharm. Sci., **68**(6), 693-696 (1979).

Calnan, C. D., H. J. Bandmann, E. Cronin, S. Fregert, N. Hjorth, B. Magnusson, K. Malten, C. L. Meneghini, V. Pirilä, and D. S. Wilkinson, "Hand Dermatitis in Housewives," Br. J. Dermatol., **82**, 543-548 (1970).

Cannon, H. L., "The Geochemical Environment as Related to Cancer in Washington County, Maryland," in International Conference on Heavy Metals in the Environment, Toronto, Ontario, Canada, October 27-31, 1975, Vol. II, Part 1, T. C. Hutchinson, Ed., Institute for Environmental Studies, University of Toronto, Toronto, Ontario, Canada, 1977, pp. 1-16.

Carlberger, G., "Kinetics and Distribution of Radioactive Cobalt Administered to the Mammalian Body," Acta Radiol., Suppl. 205, 72-126 (1961).

Carter, J. W., and I. L. Cameron, "Toxicity Bioassay of Heavy Metals in Water Using Tetrahymena pyriformis," Water Res., 7(7), 951-961 (1973).

Cass, J. S., and L. B. Lange, Report on the Physiological Effects of Cobalt on Man, The Kettering Laboratory in the Department of Preventive Medicine and Industrial Health, University of Cincinnati, Cincinnati, Ohio, April 1955.

Chang, C. C., H. J. Tatum, and F. A. Kincl, "Effect of Intrauterine Copper and Other Metals on Implantation in Rats and Hamsters," Fertil. Steril., 21, 274-278 (1970).

Chekunova, M. P., "K Mekhanizmu Snizheniya Soderzhaniya Noradrenalina v Miokarde pri Deistvii Nekotorykh Promyshlennykh Yadov" ["On the Mechanism of Decrease in Noradrenaline Content in Myocardium under the Effect of Some Industrial Poisons"], Vopr. Med. Khim., 22(1), 55-58 (1976).

Chekunova, M. P., "Mekhanizm Kardiotoksicheskogo Deistviya Nekotorykh Promyshlennykh Yadov (Metallov)"["Mechanism of the Cardiotoxic Effect of Certain Industrial Poisons (Metals)"], Kardiologiya, 18(5), 54-61 (1978).

Chernov, O. V., I. I. Khitsenko, and V. N. Evreev, "Toksicheskie i Antiblastomogennye Svoistva Trismonoetanolaminkobal'ta" ["Toxicity and Antiblastomogenic Properties of Tris(monoethanolamine)-cobalt"], Farmakol. Toksikol. (Kiev), 8, 86-88 (1973).

Chetty, K. N., D. S. V. Subba Rao, and D. Desaiah, "Immuno-Biochemical Changes in Cobalt Chloride Fed Rats," Fed. Proc., 37(3), 604 (1978).

Chigrina, T. A., N. D. Makushinskaya, and V. D. Shumaev, "Biologicheskaya Rol' Kobal'ta i Kharakteristika ego Soderzhaniya v Istochnikakh Pit'evogo Vodosnabzheniya Nekotorykh Oblastei Severnogo Kazakhstana" ["Biological Role of Cobalt and Characteristics of Its Content in Drinking Water Supply Sources in Some Regions of Northern Kazakhstan"], Tr. Nauchno-Issled. Inst. Kraev. Patol., Alma-Ata, 26, 66-69 (1974).

Chikalova, E. A., "Change in the Vitamin C Content of Strawberries During Spray Nutrition with Manganese and Cobalt Salts," Tr. Vses. Semin. Biol. Aktiv. (Lech). Veshchestvam Plodov Yagod, 3rd, 1966, 160-162 (1968); Chem. Abstr., 73, 119781c (1970).

Chino, M., and S. Mitsui, "Occurrence of Heavy Metal Induced Iron Chlorosis in Plants. II. Effects of Heavy Metal Excess on Absorption, Translocation, and Distribution of ^{59}Fe in Rice Plants," Nippon Dojo-Hiryogaku Zasshi, 38(7), 255-259 (1967); Chem. Abstr., 68, 21251b (1968a).

Chino, M., and S. Mitsui, "Occurrence of Heavy Metal-Induced Iron Chlorosis in Plants. III. Distribution of ^{59}Fe, ^{60}Co, and ^{56}Mn in Alfalfa," Nippon Dojo-Hiryogaku Zasshi, 38(8), 280-286 (1967); Chem. Abstr., 68, 27570b (1968b).

Christie, O. H. J., Ng. Dinh-Nguyen, J. Vincent, L. Hellgren, and W. Pimlott, "Spark Source Mass Spectrographic Study of Metal Allergenic Substances on the Skin," J. Invest. Dermatol., 67 (5, Part 1), 587-590 (1976).

Chusid, J. G., and L. M. Kopeloff, "Epileptogenic Effects of Pure Metals Implanted in Motor Cortex of Monkeys," J. Appl. Physiol., 17, 697-700 (1962).

Clark, M. B., Cobaltous Chloride-Induced Epilepsy in Bullfrogs, Ph.D. dissertation, Tulane University, New Orleans, Louisiana, 1977.

Clemente, G. F., L. Cigna Rossi, and G. P. Santaroni, "Trace Element Intake and Excretion in the Italian Population," J. Radioanal. Chem., 37, 549-558 (1977).

Clemente, G. F., L. C. Rossi, and G. P. Santaroni, "Studies on the Distribution and Related Health Effects of the Trace Elements in Italy" in Trace Substances in Environmental Health - XII, D. D. Hemphill, Ed., Proceedings of University of Missouri's 12th Annual Conference on Trace Substances in Environmental Health, Columbia, Missouri, June 6-8, 1978, University of Missouri, Columbia, Missouri, 1979, pp. 23-30.

Coleman, R. F., J. Herrington, and J. T. Scales, "Concentration of Wear Products in Hair, Blood, and Urine after Total Hip Replacement," Br. Med. J., 1, 527-529 (1973).

Coleman, R. F., J. Herrington, and J. T. Scales, "The Concentration of Wear Products in the Body Following Total Joint Replacement," Phys. Med. Biol., 17(5), 744 (1972).

Comar, C. L., "Radioisotopes in Nutritional Trace-Element Studies. II," Nucleonics, 3(4), 30-42 (1948).

Comar, C. L., and G. K. Davis, "Cobalt Metabolism Studies. IV. Tissue Distribution of Radioactive Cobalt Administered to Rabbits, Swine, and Young Calves," J. Biol. Chem., 170, 379-389 (1947).

Cooksley, W. G. E., "Cobalt" in Metals and the Liver, L. W. Powell, Ed., Marcel Dekker, New York, New York, 1978, pp. 363-391.

Corbett, T. H., W. F. Dove, and C. Heidelberger, "Attempts to Correlate Carcinogenic with Mutagenic Activity Using Bacteriophage T4," Int. Cancer Congr. Abstr., 10, 61-62 (1970).

Costa, M., "Preliminary Report on Nickel-Induced Transformation in Tissue Culture" in Ultratrace Metal Analysis in Biological Sciences and Environment, T. H. Risby, Ed., Advances in Chemistry Series No. 172, American Chemical Society, Washington, D.C., 1979.

Cralley, L. J., "Electromotive Phenomenon in Metal and Mineral Particulate Exposures: Relevance to Exposure to Asbestos and Occurrence of Cancer," Am. Ind. Hyg. Assoc. J., 32(10), 653-661 (1971).

Crawford, B. J., D. E. Talburt, and D. A. Johnson, "Effects of Cobalt III Complexes on Growth and Metabolism of Escherichia coli," Bioinorg. Chem., 3, 121-133 (1974).

Cresta, M., M. Allegrini, E. Casadei, M. Gallorini, E. Lanzola, and G. B. Panatta, "Benin: Nutritional Considerations on Trace Elements in the Diet," Food Nutr. (Roma), 2(2), 8-18 (1976).

Culebras-Poza, J. M., M. Dean-Guelbenzu, M. Santiago-Corchado, and A. Santos-Ruiz, "Normal and Pathological Excretion of Trace Elements by Urine in Man" in Trace Element Metabolism in Animals, No. 2. Proceedings of the Second International Symposium, Madison, Wisconsin, June 18-22, 1973, W. G. Hoekstra et al., Eds., University Park Press, Baltimore, Maryland, 1974, pp. 473-475.

Curtis, J. R., G. C. Goode, J. Herrington, and L. E. Urdaneta, "Possible Cobalt Toxicity in the Maintenance of Hemodialysis Patients after Treatment with Cobaltous Chloride - A Study of Blood and Tissue Cobalt Concentrations in Normal Subjects and in Patients with Terminal Renal Failure," Clin. Nephrol., 5(2), 61-65 (1976).

Cuthbertson, D. P., "Human Requirements for Minerals and Trace Elements," Indian J. Nutr. Dietet., 10(1), 31-49 (1973).

Cuthbertson, W. F. J., A. A. Free, and D. M. Thornton, "Distribution of Radioactive Cobalt in the Rat," Br. J. Nutr., 4, 42-48 (1950).

Dalvi, R. R., and T. J. Robbins, "Comparative Studies on the Effect of Cadmium, Cobalt, Lead, and Selenium on Hepatic Microsomal Monooxygenase Enzymes and Glutathione Levels in Mice," J. Environ. Pathol. Toxicol., 1(5), 601-607 (1978).

Daniel, M. R., J. C. Heath, and M. Webb, "Respiration of Metal Induced Rhabdomyosarcomata," Br. J. Cancer, 780-786 (1967).

Danilova, T. A., and E. N. Demkina, "The Role of Cobalt in Nitrogen Accumulation by Leguminosae," Dokl. Akad. Nauk SSSR, 172(2), 487-490 (1967); Chem. Abstr., 66, 75320a (1967).

Davies, I. J. T., "Chapter VII. Cobalt" in The Clinical Significance of the Essential Biological Metals, William Heinemann Medical Books, Ltd., London, England, 1972, pp. 94-103.

Davison, V., "Cyanide Poisoning: Kelocyanor--A New Treatment," Occup. Health (London), 21(6), 306-308 (1969).

Dean-Guelbenzu, M., J. M. Culebras-Poza, and M. Santiago-Corchado, "Acción de la desferrioxamina sobre la oligoelementuria" ["Action of the Despherrioxamine in the Urinary Excretion of Oligoelements"], An. R. Acad. Farm., 41 (Apr-May-Jun), 299-307 (1975).

De Bruin, A., Biochemical Toxicology of Environmental Agents, Elsevier/North-Holland Biomedical Press, New York, 1976.

Delahant, A. B., "An Experimental Study of the Effects of Rare Metals on Animal Lungs," AMA Arch. Ind. Health, 12, 116-120 (1955).

Delpla, M., J.-P. Moroni, P. Bournay, R. Schaeffer, R. Gaulard, G. Rocquet, J.-M. Chinardet, and H. Létard, "Évolution sur six ans d'une contamination radioactive humaine expérimentale" ["A Six Year Evolution of Human Experimental Radioactive Contamination"], C. R. Hebd. Seances Acad. Sci., Ser. D, 283(12), 1465-1468 (1976).

De Matteis, F., and A. H. Gibbs, "Inhibition of Haem Synthesis Caused by Cobalt in Rat Liver," Biochem. J., 162, 213-216 (1977).

Den Dooren De Jong, L. E., "Tolerance of Azotobacter for Metallic and Non-Metallic Ions," Antonie van Leeuwenhoek, 37, 119-124 (1971).

Dickson, J., and M. P. Bond, "Cobalt Toxicity in Cattle," Australian Vet. J., 50(5), 236 (1974).

Dikov, G. I., E. N. Ermolova, and L. B. Katsova, "Protivopatogeniticheskoe i Etiotropnoe Deistvie Khloristogo Kobal'ta, Monoetanolamina i Sul'fena pri Eksperimental'nom Tsistitserkoze Krolikov" ["Antipathogenetic and Etiotropic Effect of Cobalt Chloride, Monoethanolamine, and Sulfene during Experimental Cysticercosis of Rabbits"], Tr. Kaz. Nauchno-Issled. Vet. Inst., 15, 33-34 (1973).

Dingle, J. T., J. C. Health, M. Webb, and M. Daniel, "The Biological Action of Cobalt and Other Metals. II. The Mechanism of the Respiratory Inhibition Produced by Cobalt in Mammalian Tissues," Biochim. Biophys. Acta, 65, 34-46 (1962).

Dixon, J. R., D. B. Lowe, D. E. Richards, L. J. Cralley, and H. E. Stokinger, "The Role of Trace Metals in Chemical Carcinogenesis: Asbestos Cancers," Cancer Res., 30(4), 1068-1074 (1970).

Dorsit, G., R. Girard, H. Rousset, J. Brune, T. Wiesendanger, F. Tolot, J. Bourret, and P. Galy, "Pulmonary Fibrosis in 3 Workers in the Same Factory Exposed to Cobalt and Tungsten Carbide Dust. Pulmonary Disorders in the Hard Metal Industry. Apropos of an Occupational Survey," Sem. Hop. Paris, 46(51), 3363-3376 (1970).

Doudoroff, P., and M. Katz, "Critical Review of Literature on the Toxicity of Industrial Wastes and Their Components to Fish. II. The Metals, as Salts," Sewage Ind. Wastes, **25**, 802-839 (1953).

Dreyfus, J. C., "Metabolic Cycles of Elements," Traite Biochem. Gen., **3**, 574-577 (1967); Chem. Abstr., **67**, 51592k (1967).

Duckham, J. M., and H. A. Lee, "Treatment of Refractory Anemia of Chronic Renal Failure with Cobalt Chloride," Quart. J. Med., **45** (April), 277-294 (1976).

Dvizhkov, P. P., "O Blastomogennykh Svoistvakh Promyshlennykh Metallov i ikh Soedinenii" ["Blastomogenic Properties of Industrial Metals and Their Compounds"], Arkhiv Patologii, **29**(3), 3-11 (1967).

Dyatlov, M. K., "Dynamika Medzyi, Kobal'tu yi Tsynku yi Malodzyive yi Malatseh Karow pry Roznym Uzrowni Gehtykh Elementaw u Ratsyene" ["Dynamics of Copper, Cobalt, and Zinc in the Colostrum and Milk of Cows When These Elements Are Present in the Ration in Different Amounts"], Vestsi Akad. Navuk Belorus. SSR, Ser. Sel'skagaspad. Navuk, No. 2, 90-92 (1972).

Dzenite, A., and R. Lieldiens, "Dinamika Soderzhaniya Margantsa, Kobal'ta, Medi i Tsinka v Krovi Plodov Korov v Fetal'nyi Period" ["Dynamics of Manganese, Cobalt, Copper, and Zinc Levels in the Blood of Cow Fetuses"], Nauka-Zhivotnovod., No. 12, 135-138 (1972).

Edwards, M. S., and J. R. Curtis, "Use of Cobaltous Chloride in Anaemia of Maintenance Hemodialysis Patients," Lancet, **2**, 582-583 (1971).

Eichhorn, G. L., "Active Sites of Biological Macromolecules and Their Interaction with Heavy Metals" in Ecological Toxicology Research. Effects of Heavy Metal and Organohalogen Compounds, Vol. 7, Environmental Science Research Series, A. D. McIntyre and C. F. Mills, Eds., Plenum Publishing Corporation, New York, New York, 1975, pp. 123-142.

Einarrson, O., E. Eriksson, G. Lindstedt, and J. E. Wahlberg, "Dissolution of Cobalt from Hard Metal Alloys by Cutting Fluids," Contact Dermatitis, **5**(3), 129-132 (1979).

Elagin, I. N., "Effect of Cobalt on the Chlorophyll Level, Intensity of Photosynthesis and Yield of Buckwheat," Dokl. Vses. Akad. Sel'skokhoz. Nauk, No. 7, 22-23 (1970); Chem. Abstr., 73, 97986u (1970).

El-Ashry, M. T., and R. Getzoff, "Geochemistry of Pennsylvanian Shales and Environmental Health in Northern Pennsylvania," Proc. Pa. Acad. Sci., 48, 24-27 (1974).

Elinder, C.-G., and L. Friberg, "Cobalt" in Toxicology of Metals. Vol. II, Permanent Commission and International Association of Occupational Health, PB 268 324, National Technical Information Service, U.S. Department of Commerce, Springfield, Virginia, May 1977, pp. 188-205.

El-Masry, Z., M. Abdel-Latif, A. El-Habashy, and N. Swedan, "Assessment of the Antidotal Potential of Some Cobalt Compounds in Cyanide Poisoning," Ain Shams Med. J., 26(5-6), 701-707 (1975).

Emson, P. C., "Metal Implants as Models of Epilepsy," Chapter 14 in Biochemistry and Neurology, H. F. Bradford and C. D. Marsden, Eds., Academic Press, London, 1976.

Epstein, E., Mineral Nutrition of Plants: Principles and Perspectives, John Wiley and Sons, New York, New York, 1972.

Evans, H. J., and S. Ahmed, "Cobalt: A Micronutrient Element for the Growth of Soybean Plants under Symbiotic Conditions," Soil Sci., 90, 205-210 (1960).

Eyster, C., "Micronutrient Requirements for Green Plants, Especially Algae" in Algae and Man, D. F. Jackson, Ed., Plenum Press, New York, New York, 1964, pp. 86-119.

Fairchild, E. J., R. J. Lewis, Sr., and R. L. Tatken, Registry of Toxic Effects of Chemical Substances, National Institute for Occupational Safety and Health, Cincinnati, Ohio, September 1977.

Farrell, R. L., and G. W. Davis, "Effects of Particulates on Respiratory Carcinogenesis by Diethylnitrosamine," Exp. Lung Cancer: Carcinog. Bioassays, Int. Symp., 219-233 (1974).

(FASEB), Federation of American Societies for Experimental Biology, Review of Health Aspects of Vitamin B12 as a Food Ingredient, Bureau of Foods, Food and Drug Administration, Department of Health, Education, and Welfare, PB-275 755, National Technical Information Service, Department of Commerce, Springfield, Virginia, 1977.

(FASEB), Federation of American Societies for Experimental Biology, Evaluation of the Health Aspects of Vitamin B12 as a Food Ingredient, PB-289 922, National Technical Information Service, Department of Commerce, Springfield, Virginia, 1978.

Ferm, V. H., "The Teratogenic Effects of Metals on Mammalian Embryos Zinc, Manganese, Cobalt, Copper, Molybdenum, Lead, Cadmium, Arsenic, Mercury, Indium, Selenium, Nickel, Lithium," Adv. in Teratol., **5**, 51-75 (1972).

Ferron, G. A., "The Size of Soluble Aerosol Particles as a Function of the Humidity of the Air. Application to the Human Respiratory Tract," J. Aerosol Sci., **8**(4), 251-267 (1977).

Finogenova, M. A., "Effect of Cobalt on Induced Carcinogenesis of the Skin," Bull. Exp. Biol. Med., **75**(2), 168-169 (1973); Translated from Byull. Eksp. Biol. Med., **75**(2), 73-75 (1973).

Fish, M., M. Pollycove, R. Wallerstein, K. Cheng, and M. Tono, "Simultaneous Measurement of Free and Intrinsic Factor (IF) Bound Vitamin B_{12} Absorption: Absolute Quantitation with Incomplete Stool Collection and Rapid Relative Measurement Using Plasma B_{12}(IF): B_{12} Absorption Ratio," J. Nucl. Med., **14**(8), 568-575 (1973).

Fishchenko, L. Ya., "Effect of Cobalt, Copper, and Zinc on Tissue Respiration and Permeability of the Histohematic Barriers of Organs of the Reticuloendothelial System," Mikroelem. Med., No. 2, 56-59 (1971); Chem. Abstr., **77**, 579k (1972).

Fletcher, D. C., "Absorption of Enzymes and Minerals in the Digestive Process," J. Am. Med. Assoc., **232**(6), 657 (1975).

Flodh, H., "Distribution and Kinetics of Cobalt-60 Chloride and Labeled Vitamin B_{12} Using Autoradiography and Impulse Counting," Trace Elem. Metab. Anim., Proc. WAAP (World Assoc. Anim. Prod.)/ISP (Int. Biol. Progr.) Int. Symp., 1969, 67-69 (1970).

Forbes, R. M., A. R. Cooper, and H. H. Mitchell, "On the Occurrence of Beryllium, Boron, Cobalt, and Mercury in Human Tissues," J. Biol. Chem., **209**, 857-865 (1954).

Förström, L., and V. Pirilä, "27 Years of Occupational Dermatology in Finland," Berufsdermatosen, **23**(6), 207-213 (1975).

Fox, M. R. S., "Effect of Essential Minerals on Cadmium Toxicity. Review," J. Food Sci., **39**(2), 321-324 (1974).

Frederick, W., and W. Bradley, "Toxicity of Some Materials Used in the Manufacture of Cemented Tungsten Carbide Tools," Paper presented at the 7th Annual Meeting of the American Industrial Hygiene Association, Chicago, 1946, Summarized in Ind. Med., **15**(8), 482-483 (1946).

Fregert, S., and B. Gruvberger, "Solubility of Cobalt in Cement," Contact Dermatitis, **4**(1), 14-18 (1978).

Fregert, S., B. Gruvberger, and A. Heijer, "Sensitization to Chromium and Cobalt in Processing of Sulphate Pulp," Acta. Derm. Venereol. (Stockh.), **52**(3), 221-224 (1972).

Friberg, L., Toxicology of Metals, Vol. III, PB 280 115, National Technical Information Service, U.S. Department of Commerce, Springfield, Virginia, February 1978.

Frieden, E., "The Chemical Elements of Life," Sci. Am., **229**(1), 52-60 (1972).

Fujihara, M. P., T. R. Garland, R. E. Wildung, and H. Drucken, "Response of Microbiota to the Presence of Heavy Metals in Soil," Abstr. Ann. Meet. Am. Soc. Microbiol., **73**, 32 (1973).

Furst, A., and R. T. Haro, "A Survey of Metal Carcinogenesis," Prog. Exp. Tumor Res., **12**, 102-133 (1969).

Gainer, J. H., "Activation of the Rauscher Leukemia Virus by Metals," J. Natl. Cancer Inst., **51**(2), 609-613 (1973).

Gardner, L. I., "Special Susceptability of the Host to Chemical Agents: Endocrinologic Considerations," Pediatrics, **53**(5, Part 2), 813-815 (1974).

Gavrilova, P. N., "Effect of Trace Elements on Accumulation of Protein in Corn," Tr. Saratov. Zootekh.-Vet. Inst., **13**, 56-66 (1965); Chem. Abstr., **66**, 94359m (1967).

Georgiadi, G. A., "Izmenenie Aktivnosti Nekotorykh Degidrogenaz i Nespetsificheskikh Fermentov v Slizistoi Obolochke Respiratornogo Trakta Krys pod Vozdeistviem Pyli Metallicheskogo Kobal'ta v Khronicheskom Eksperimente" ["Change in the Activity of the Dehydrogenases and Nonspecific Enzymes in the Respiratory Tract Mucosa of Rats Exposed to Metallic Cobalt Dust in a Chronic Experiment"], Zh. Ushn. Nos. Gorl. Bolezn., No. 1, 63-67 (1978).

Georgiadi, B. A., and L. A. El'kind, "Morfologicheskie Izmeniya Slizistoi Obolochki Dykhatel'nykh Putei pod Vliyaniem Aerozolya Metallicheskogo Kobal'ta v Khronicheskom Eksperimente" ["Morphological Changes in the Respiratory Trace Mucosa under the Influence of a Metallic Cobalt Aerosol in a Chronic Experiment"], Zh. Ushn. Nos. Gorl. Bolezn., No. 3, 41-45 (1978).

Ghiani, P., "Aspetti di fisiomorfolologia cellulare del pancreas endocrino. Ricerche con cloruro di cobalto in embrione di pollo" ["Aspects of Cellular Physio-Morphology of the Endocrine Pancreas. Investigations with Cobalt Chloride in Chick Embryo"], Atti. Accad. Ligure Sci. Lett., **13**, 175-178 (1957).

Gibbs, G. W., and M. Lachance, "Dust Exposure in the Chrysotile Asbestos Mines and Mills of Quebec," Arch. Environ. Health, **24**(3), 189-197 (1972).

Gilman, J. P. W., "Metal Carcinogenesis. II. A Study on the Carcinogenic Activity of Cobalt, Copper, Iron, and Nickel Compounds," Cancer Res., **22**, 158-162 (1962).

Gilman, J. P. W., and G. M. Ruckerbauer, "Metal Carcinogenesis. I. Observations on the Carcinogenicity of a Refinery Dust, Cobalt Oxide, and Colloidal Thorium Dioxide," Cancer Res., **22**, 152-156 (1962).

Gireeva, T. M., "Vliyanie Mikroelementov Medi, Kobal'ta, Zheleza i Tsinkana Telyat vo Vremya ikh Vnutriutrobnogo Razvitiya" ["Effect of the Trace Elements Copper, Cobalt, Iron, and Zinc on Calves during Their Intrauterine Development"], Sb. Nauch Rab., Dagestan. Nauch-Issled Vet. Inst., **2**, 338-344 (1968).

Godnev, T. N., and A. V. Leshina, "After Effects of Molybdenum and Cobalt on Peas," Dokl. Akad. Nauk Beloruss. SSR, **11**(4), 359-361 (1967); Chem. Abstr., **67**, 63363z (1967).

Goel, M. C., and E. H. Marth, "Growth of *Leuconostoc citrovorum* in Skim Milk Fortified with Cobalt, Manganese, and Iron Compounds," J. Milk Food Technol., **35**(5), 269-275 (1972).

Goldwasser, E., L. O. Jacobson, W. Fried, and L. Plzak, "Mechanism of the Erythropoietic Effect of Cobalt," Science, **125**, 1085-1086 (1958).

Gori, C., and L. Zucconi, "L'Azione Citologica Indotta da un Gruppo di Composti Inorganici su *Allium cepa*" ["Cytologic Effects Induced in *Allium cepa* by a Group of Inorganic Compounds"], Caryologia, **10**, 29-45 (1957).

Gorid'ko, I. V., ["The Effect of Cobalt on Water Content, Water Retention, and Transpiration of Potato Leaves"], Nauch. Dokl. Vyssh. Skh., Biol. Nauki, No. 3, 84-87 (1967); Chem. Abstr., **67**, 10807r (1967).

Grace, N. D., "Flow of Zinc, Cobalt, Copper, and Manganese Along the Digestive Tract of Sheep Given Fresh Perennial Ryegrass, or White or Red Clover," Br. J. Nutr., **34**(1), 73-82 (1975).

Greenberg, S. R., S. Burman, T. Casale, E. Millhouse, J. Kamarryt, and W. Schumer, "The Pathology of Experimental Cobalt Cardiomyopathy," Bull. Soc. Pharmacol. Environ. Pathol., **5**(3), 5-7 (1977).

Grekalova, T. V., "Soderzhanie Mikroelementov (Ioda, Ftora, Kobal'ta, Medi, Tsinka, Margantsa i Molibdena) v Ob"ektakh Vneshnei Sredy i Pishchevykh Ratsionakh Zobnogo Ochaga Kuba-Khachmasskoi Zony Azerbaidzhanskoi SSR" ["Concentration of Trace Elements (Iodine, Fluorine, Cobalt, Copper, Zinc, Manganese, and Molybdenum) in the External Environment and Food in an Endemic Goitrogenic Region of the Kuba-Khachmas Zone of Azerbaidzhan SSR"], Azerb. Med. Zh., **50**(5), 25-29 (1973).

Grice, H. C., T. Goodman, I. C. Munro, G. S. Wiberg, and A. B. Morrison, "Myocardial Toxicity of Cobalt in the Rat," Ann. N.Y. Acad. Sci., **156**(1), 189-194 (1969).

Grimm, I., "Ungewöhnliche Form einer Kontaktdermatitis durch Kobalt bei einem 11 jährigen Kind," Unusual form of a Contact Dermatitis due to Cobalt in an 11-Year-Old Child," Berufsdermatosen, 19(1), 39-42 (1971).

Grosch, D. S., "Combined Effects of Radiation and Chemical Agents in Altering the Fecundity and Fertility of a Braconid Wasp," Proc. Symp. Steril. Princ. Insect Control, 1974, 243-259 (1975).

Gupton, E. D., and P. E. Brown, "Chest Clearance of Inhaled Cobalt-60 Oxide," Health Phys., 23, 767-769 (1972).

Guseva, V. A., V. A. Valutina, E. I. Noskova, and O. P. Ganicheva, "Effect of the Presowing Treatment of Seeds with Trace Elements on the Metabolism and Productivity of Corn and Sugar Beets," Uch. Zap. Gor's, Gos. Univ., No. 84, 32-36 (1968), Chem. Abstr., 74, 111002z (1971).

Hall, J. L., and E. B. Smith, "Cobalt Heart Disease. An Electron Microscopic and Histochemical Study in the Rabbit," Arch. Pathol., 86, 403-412 (1968).

Halpin, D. S., "An Unusual Reaction in Muscle in Association with a Vitallium Plate: a Report of Possible Metal Hypersensitivity," J. Bone Joint Surg., 57-B(4), 451-453 (1975)

Halsall, D. M., "Effects of Certain Cations on the Formation and Infectivity of Phytophthora Zoospores. 2. Effects of Copper Boron, Cobalt, Manganese, Molybdenum, and Zinc Ions," Can. J. Microbiol., 23(8), 1002-1010 (1977).

Hamilton, E. I., "The Chemical Elements and Human Morbidity--Water, Air and Places--A Study of Natural Variability," Sci. Total Environ., 3(1), 3-85 (1974).

Hara, T., Y. Sonoda, and I. Iwai, "Growth Response of Cabbage Plants to Transition Elements Under Water Culture Conditions. II. Cobalt, Nickel, Copper, Zinc, and Molybdenum," Soil Sci. Plant Nutr., 22(3), 317-325 (1976).

Harding, H. E., "Notes on the Toxicology of Cobalt Metal," Br. J. Ind. Med., 7, 76-78 (1950).

Harper, H. A., V. W. Rodwell, and P. A. Mayes, Review of Physiological Chemistry, 16th ed., LANGE Medical Publication, Los Altos, California, 1977.

Harry, H. W., and D. V. Aldrich, "The Distress Syndrome in Taphius glabratus (Say) as a Reaction to Toxic Concentrations of Inorganic Ions," Malacologica, **1**, 283-289 (1963).

Harth, G. H., Metal Implants for Orthopedic and Dental Surgery, MCIC-74-18, Metals and Ceramics Information Center, Battelle Columbus, Ohio, 1974.

Hasegawa, T., "Tissue Cobalt Content and Excretion in Mice and Rats Following Injection of Cobalt Chloride," Nippon Eiseigaku Zasshi, **29**(2), 289-299 (1974).

Hatfield, M., "A Possible Hazard from Nuclear Submarine Corrosion Products," J. R. Nav. Med. Serv., **56**(1), 183-185 (1970).

Hays, V. W., and M. J. Swenson, "Minerals, Bones and Joints," Chapter 33 in Dukes' Physiology of Domestic Animals, 9th ed., M. J. Swenson, Ed., Cornell University Press, Ithaca, New York, 1977, pp. 395-412.

Heath, J. C., "Cobalt as a Carcinogen," Nature, **173**(4409), 822-823 (1954a).

Heath, J. C., The Effect of Cobalt on Mitosis in Tissue Culture," Exp. Cell. Res., **6**, 311-320 (1954b).

Heath, J. C., "The Production of Malignant Tumours by Cobalt in the Rat," Br. J. Cancer, **10**, 668-673 (1956).

Heath, J. C., and M. R. Daniel, "The Production of Malignant Tumours by Cobalt in the Rat: Intrathoracic Tumours," Br. J. Cancer, **16**, 473-478 (1962).

Heath, J. C., M. Webb, and M. Caffrey, "The Interaction of Carcinogenic Metals with Tissues and Body Fluids. Cobalt and Horse Serum," Br. J. Cancer, **23**(1), 153-166 (1969).

Heath, J. C., M. A. Freeman, and S. A. Swanson, "Carcinogenic Properties of Wear Particles from Prostheses made in Cobalt-Chromium Alloy," Lancet, **1** (7699), 564-566 (1971).

Henquin, J. C., and A. E. Lambert, "Cobalt Inhibition of Insulin Secretion and Calcium Uptake by Isolated Rat Islets," Am. J. Physiol., 228(6), 1669-1677 (1975).

Herich, R., "The Effect of Cobalt on the Structure of Chromosomes and on the Mitosis," Chromosoma, 17, 194-198 (1965).

Herich, R., and M. Bobak, "The Influence of Cobalt on the Endoplasmatic Reticulum of the Horse Bean (Vicia faba L.)," Experientia, 32(5), 570-571 (1976).

Hermann, D., "Allergische Reaktionen durch zahnarztliche Werkstoffe," ["Allergic Reactions to Dental Materials"] , Munch. Med. Wochenschr., 119(8), 265-270 (1977).

Hewitt, P. J., and R. Hicks, "Effects of Inhaled Welding Fume in the Rat," Ann. Occup. Hyg., 16(3), 213-221 (1973).

Hill, C. H., "Reduction of Metal Toxicities by Ascorbic Acid in the Chick" (abstr.), Fed. Proc., 34(3), 926 (1975).

Höbel, V. M., K. Wegener, and C. Glanzmann, "Über die toxische Wirkung von Co^{++},[Co[Co-EDTA] bzw. Na_2[Co-EDTA] enthaltenden Aerosolen auf die Rattenlunge und die Verteilung von Co^{++} sowie [Co-EDTA] -- in Organendes Meerschweinchens," ["The Toxic Action of Aerosols Containing Co^{++}, Co[Co-EDTA] and Na_2[Co-EDTA] on the Rat Lung and Studies of the Distribution of Co^{++} and [Co-EDTA] -- in Organs of Guinea Pig"], Naunyn Schmiedebergs Arch. Pharmacol., 263(1), 270-271 (1969).

Höck, A., K. Kasperek, H. R. Scholz, and I. E. Feinendegen, ["Concentrations of Zinc, Selenium, Rubidium, Cobalt and Iron in Maternal Serun As Well As in the Arteria and Vena Umbilicals, Especially with Regard to Iron Circulation Following Oral Administration of Enriched Iron-58"], Spurenelem. Entwickl. Mensch Tier Vernachlaessigte Elem. Saeuglingsernaehr., Symp., 1974, 89-95 (1975), Chem. Abstr., 86, 169613p (1977).

Hoffman, D. J., and S. K. Niyogi, "Metal Mutagens and Carcinogens Affect RNA Synthesis Rates in a Distinct Manner," Science, 198 (4316), 513-514 (1977).

Holmes, A., A. Morgan, and J. Sandalls, "Determination of Iron, Chromium, Cobalt, Nickel, and Scandium in Asbestos by Neutron Activation Analysis," Am. Ind. Hyg. Assoc. J., 32(5), 281-286 (1971).

Holm-Hansen, O., G. C. Gerloff, and F. Skoog, "Cobalt as an Essential Element for Blue-Green Algae," Physiol. Plantarum, 7, 665-675 (1954).

Hopps, H. C., A. J. Stanley, and A. M. Shideler, "Polycythemia Induced by Cobalt. III. Histologic Studies with Evaluation of Toxicity of Cobaltous Chloride," Am. J. Clin. Pathol., 24, 1374 (1954).

Hopps, H. C., A. J. Stanley, and A. M. Shideler, "Polycythemia Induced by Cobalt. III. Histologic Studies with Evaluation of Toxicity of Cobaltous Chloride," Am. J. Clin. Path., 24, 1374-1380 (1958).

Huck, D. W., "The Study of Cobalt Toxicity in Pigs and Rats," Diss. Abstr. Int. B, 37(1), 159 (1975).

Hueper, W. C., "Environmental Cancer Risks in an Industrialized Economy" in Industrial Pollution, N. I. Sax, Ed., Van Nostrand Reinhold Company, New York, New York, 1974, pp. 118-149.

Hunter, J. G., and O. Vergnano, "Trace-Element Toxicities in Oat Plants," Ann. Appl. Biol., 40, 761-777 (1953).

Husain, S. L., "Contact Dermatitis in the West of Scotland," Contact Dermatitis, 3(6), 327-332 (1977).

Hutchinson, T. C. "Comparative Studies of the Toxicity of Heavy Metals to Phytoplankton and Their Synergistic Interactions, Water Pollut. Res. Can., 8, 68-90 (1973).

Huy, N. D., P. J. Morin, S. M. Mohiuddin, and Y. Morin "Acute Effects of Cobalt on Cardiac Metabolism and Mechanical Performance," Can. J. Physiol. Pharmacol., 51(1), 46-51 (1973)

Il'ina, T. K., "Effect of Cobalt on the Growth of Nitrogen-Fixing Mycobacteria Utilizing Molecular and Bound Nitrogen," Mikrobiologiya, 36(4), 626-631 (1967); Chem. Abstr., 67, 106159g (1967).

Ilynsky, B. V., and S. N. Babadzhanov, "On the Contents of Some Microelements in the Blood, Aorta and Liver in Atherosclerosis," Epatologia, 20(4), 195-206 (1974).

Ince, A. J., "Some Elements and Their Relationships in Ascaris suum," Int. J. Parasitol, 6(2), 127-128 (1976); Chem. Abstr., 84, 176974u (1976).

Inogamov, A. M., and S. G. Chernomorchenko, "Raspredelenie Radioaktivnogo Kobal'ta i Koamida v Organizme Krys" ["Distribution of Radioactive Cobalt and Coamide in the Rat"], Tr. Tashkent. Farmatsevt. Inst. 4, 681-684 (1966).

Institut National de Recherche et de Securite, "Valeurs limites de concentration des substances toxiques dans l'air" ["Concentration Limit Values of Toxic Substances in the Air"], Cah. Notes Doc., 94, 73-103 (1979)

Ishizaka, O., Y. Ito, and S. Yamanoto, "Studies on Toxicity of Food Additives by Disturbance Test of Chorella Life Cycle I. Fundamental Research on Culture and Test Method," Nagoya Shiritsu Daigatu Yakugakubu Kenkyu Nempo, 14, 66-78 (1966).

Ivankovic, S., W. J. Zeller, and D. Schmähl, "Steigerung der carcinogenen Wirkung von Athyl-nitrosoharnstoff durch Schwermetalle," ["Increasing the Carcinogenic Effects of Ethyl-Nitroso-Urea using Heavy Metals"], Naturwissenschaften, 59(8), 369 (1972).

Izmirov, I., R. Galabowa, I. Kolev, and P. Petkov, "La détection du cobalt dans les lipides du pancréas des cobayes traités par $CoCl_2$ (Analysis chromatographique et spectrale par RX)," ["Cobalt Detection in Pancreas Lipids of Guinea Pigs Treated with $CoCl_2$ (Chromatographic Analysis and by X-Ray Spectrum)"] 17(4), 261-266 (1972).

Jacobson, S., and P.-O. Wester, "Balance Study of Twenty Trace Elements During Total Parenteral Nutrition in Man," Br. J. Nutr., 37, 107-126 (1977).

Jacobziner, H., and H. W. Raybin, "Poison Control--Accidental Cobalt Poisoning," Arch. Pediat., 78, 200-205 (1961).

Jaroensastraraks, P., and E. McLaughlin, "The Effects of Different Metallic Compounds on the Embryonic Development of the Fresh Water Snail Helisoma," J. Ala. Acad. Sci., 45(3), 231 (1974).

Jasmin, G., "Anaphylactoid Edema Induced in Rats by Nickel and Cobalt Salts," Proc. Soc. Exp. Biol. Med., 147(1), 289-292 (1974).

Jasmin, G., and J. L. Riopelle, "Renal Carcinomas and Erythrocytosis in Rats following Intrarenal Injection of Nickel Subsulfide," *Lab. Invest.*, **35**(1), 71-78 (1976).

Johnson, D. A., and D. E. Talburt, "Effects of Cobalt (III) Complexes on the Endogenous Metabolism of *Escherichia coli*," *Chem.-Biol. Interact.*, **9**(1), 71-74 (1974).

Jones, D. A., H. K. Lucas, M. O'Driscoll, C. H. G. Price, and B. Wibberley, "Cobalt Toxicity After McKee Hip Arthroplasty," *J. Bone Joint Surg.-Brit. Vol.*, **57**(3), 289-296 (1975).

Kaplun, Z. S., "Proizvodstvennye i Eksperimentalnye Issledovaniya po Gigienicheskoi Karakteristike Soedinenii Kobal'ta" ["Industrial and Experimental Studies of Hygienic Characteristics of Cobalt Compounds"], *Gig. Sanit.*, **5**, 26-31 (1955).

Kaplun, Z. S., "Cobalt" in *Toxicology of the Rare Metals*, Z. I. Izrael'son, Ed., 1963, Translated from the Russian by the Israel Program for Scientific Translations, Jerusalem, 1967, pp. 110-118.

Kaplun, Z. S., and N. V. Mezentseva, "Industrial Dusts Encountered in Powder Metallurgy (Hard Alloys)" in *Toxicology of the Rare Metals*, Z. I. Izrael'son, Ed., 1963, Translated from the Russian by the Israel Program for Scientific Translations, Jerusalem, 1967, pp. 155-163.

Kaposts, V., "Importance of Trace Elements for Tree Seed Germination," *Povysh. Prod. Lesa*, 121-129 (1968); *Chem. Abstr.*, **71**, 90370p (1969).

Kasirsky, G., W. T. Sherman, R. F. Gautieri, and D. E. Mann, Jr., "Cobalt-Cortisone Interrelationships in the Induction and Inhibition of Cleft Palate in Mice," *J. Pharm. Sci.*, **58**(6) 766-767 (1969).

Keener, H. A., G. P. Percival, K. S. Morrow, and G. H. Ellis, "Cobalt Tolerance in Young Dairy Cattle," *J. Dairy Sci.* **32**, 527-533 (1949).

Kench, J. E., "Metals and Proteins," *Trans. Roy. Soc. S. Afr.*, **40**(4), 209-238 (1972).

Kent, N. L., and R. A. McCance, "The Absorption and Excretion of 'Minor' Elements by Man. 2. Cobalt, Nickel, Tin and Manganese," *Biochem. J.*, **35**, 877-883 (1941).

Kerfoot, E. J., Chronic Animal Inhalation Toxicity to Cobalt, PB 232-247, National Technical Information Service, U.S. Dept. of Commerce, Springfield, Virginia, 1973.

Kerfoot, E. J., W. G. Fredrick, and E. Domeier, "Cobalt Metal Inhalation Studies on Miniature Swine," J. Am. Ind. Hyg. Assoc., 36(1), 17-25 (1975).

Khanna, S. S., and T. S. Gill, "Effect of Cobalt Salts on the Glycemia and Islet Histology of *Channa punctatus* (Bloch)," Acta. Anat., 92(2), 194-201 (1975).

Khonin, P. N., "Effect of Trace Elements on the Germination of Acorns, and the Acclimatization and Growth of Annual Crops of Oak and Pine," Tr. Saratov. Selskokhoz. Inst., 15(3), 47-49 (1965); Chem. Abstr., 66, 27915b (1967).

Kichina, M. M., "Soderzhanie Kobal'ta i Titana v Organizme Zhivotnykh pod Vliyaniem Sul'fata Kobal'ta" ["Cobalt and Titanium Levels in Animals Under the Influence of Cobalt Sulfate"], Sb. Rab., Leningr. Vet. Inst., 38, 83-87 (1974).

Kirschner, S., Y.-K. Wei, D. Francis, and J. G. Bergman, "Anticancer and Potential Antiviral Activity of Complex Inorganic Compounds," J. Med. Chem., 9(3), 369-372 (1966).

Klucik, I., R. Kemka, and M. Vladar, "O Metabolických Zmenách po Zaťažení Kobaltom," ["Metabolic Changes After Cobalt Load"], Bratislav. Lek. Listy, 47(6), 355-365 (1967).

Klucik, I., M. Palkovicova, and R. Fabianova, "Hladina Erytropoetinu v Sere Zamestnancov Hydrometallurgickej Vyroby Kobaltu" ["Level of Erythropoietin in the Serum of Employees in the Hydrometallurgical Production of Cobalt"], Bratislav. Lek. Listy, 50(4), 445-455 (1973).

Kneip, T. J., "Environmental Pollution and Ecology" (Research Project Descriptions) (undated).

Kostić, K., R. J. Drašković, M. Ratković, D. Kostić, and R. S. Drašković, "Determination of Some Trace Elements in Different Organs of Normal Rats," J. Radioanal. Chem., 37(1), 405-413 (1977).

Krasovskii, G. N., and S, A., Fridlyand, "Ekperimental'nye Dannye k Obosnovaniyu Predel'no Dopustimoi Kontsentratsii Kobal'ta v Vodoemykh" ["Experimental Data for Substantiating the Maximum Permissible Concentration of Water Reservoirs"], Gig. Sanit., 36(2), 95-96 (1971); Hyg. Sanit. 36(3), 277-279 (1971).

Krause, A., and Z. Winowski, "Germination Capacity of Wheat and its Estimation by Catalytic Means Using Fe^{+++} or Co^{++} as Promoting Ions", Landwirt. Forsch., 19(3-4), 243-245 (1966); Chem. Abstr., 67, 90124d (1967).

Kraybill, H. F., "Food Chemicals and Food Additives" in Trace Substances and Health. A Handbook. Part I., P. M. Newberne, Ed., Marcel Dekker, Inc., New York, New York, 1976, pp. 245-318.

Kriss, J. P., W. H. Carnes, and R. T. Gross, "Hypothyroidism and Thyroid Hyperplasia in Patients Treated with Cobalt," J. Am. Med. Assoc., 157(2), 117-121 (1955).

Krook, G., S. Fregert, and B. Gruvberger, "Chromate and Cobalt Eczema due to Magnetic Tapes," Contact Dermatitis, 3(1), 60-61 (1977).

Kucharin, G. M., and V. F. Sinitsin, "Pyat' Sluchaev Allergicheskogo Miokardita u Rabochikh Kobal'tovogo Otdeleniya" ["Five Cases of Allergic Myocarditis in Workers of the Cobalt Industry"], Gig. Tr. Prof. Zabol., No. 12, 40-41 (1976).

Kury, G., and R. J. Crosby, "Studies on the Development of Chicken Embryos Exposed to Cobaltous Chloride," Toxicol. Appl. Pharmacol., 13, 199-206 (1968).

Laing, P. G., A. B. Ferguson, Jr., and E. S. Hodge, "Tissue Reaction in Rabbit Muscle Exposed to Metallic Implants," J. Biomed. Mater. Res., 1, 135-149 (1967).

Lauring, L., and E. L. Wergeland, Jr., "Ocular Toxicity of Newer Industrial Metals," Milit. Med., 135(12), 1171-1174 (1970).

Lawas, I. C., A. del Callar, and J. A. Feria, "Neutron Activation Analysis of Urinary Stones," J. Radioanal. Chem., 13, 75-86 (1973).

Lee, C.-C., and L. F. Wolterink, "Metabolism of Cobalt 60 in Chickens," Poultry Sci., 34, 764-776 (1955).

LeGoff, J.-M., "Elimination du cobalt par le rein" ["Cobalt Elimination by the Kidney"], Compt. rend. soc. biol., **96**, 455-456 (1927).

Lempert, B. L., and L. V. Levina, "Vliyanie Khronicheskogo Kormeleniya Toksicheskimi Dozami Kobal'ta na Hekotorye Pokazateli Zhiro-Lipoidnogo Obmena y Krolikov" ["Effect of the Chronic Feeding of Toxic Doses of Cobalt Upon Some Indexes of the Fat Lipoid Metabolism in Rabbits"], Farmakol. Toksikol (Moscow), **37**, 466-469 (1974).

Leonard, A., G. Deknudt, and M. Debackere, "Cytogenetic Investigations on Leukocytes of Cattle Intoxicated with Heavy Metals," Toxicology, **2**(3), 269-273 (1974).

Leshina, A. V., "Effect of the Presowing Treatment of Seeds with Cobalt and Molybdenum Salts on Some Physiological and Biochemical Indexes of Leguminous Crops," Botanika (Minsk), No. 11, 179-183 (1969); Chem. Abstr., **71**, 90331b (1969).

Levan, A., "Cytological Reactions Induced by Inorganic Salt Solutions," Nature, **156**(3973), 751-752 (1945).

Lewis, S. C., "Acute Lethality of Selected Heavy Metals in Spontaneously Hypertensive Rats," Toxicol. Appl. Pharmacol., **33**(1), 128 (1975).

Licht, A., M. Oliver, and E. A. Rachmilewitz, "Optic Atrophy Following Treatment with Cobalt Chloride in a Patient with Pancytopenia and Hypercellular Marrow," Isr. J. Med. Sci., **8**(1), 61-66 (1972).

Lichtenstein, M. E., F. Bartl, and R. T. Pierce, "Control of Cobalt Exposures During Wet Process Tungsten Carbide Grinding," J. Am. Ind. Hyg. Assoc., **36**(12), 879-885 (1975).

Lin, J. H., and J. L. Duffy, "Cobalt-Induced Myocardial Lesions in Rats," Lab. Invest., **23**, 158-162 (1970).

Lindegren, C. C., S. Nagai, and H. Nagai, "Induction of Respiratory Deficiency in Yeast by Manganese, Copper, Cobalt and Nickel," Nature, **182**, 446-448 (1958).

Lindgren, I., and H. A. Salmi, "Cobalt in the Human Organism," Ann. Med. Exp. Fenn., **46**, 1-7 (1968).

Lindenbaum, J., and C. S. Lieber, "Alcohol-Induced Malabsorption of Vitamin B_{12} in Man," Nature, 224(221), 806 (1969).

Linhart, L., I. Rzymanova, and Z. Jesensky, "Relation Between Beverage Consumption of Employees in the Metallurgical Industry and the Content of Cobalt in the Body," Cesk. Hyg., 22(3-4), 158-162 (1977); Chem. Abstr., 87, 128240s (1977).

Lipskaya, G. A., "Anatomo-Cytological Features of Cucumber Leaves in the Presence of Cobalt and Manganese in the Nutrient Mixture," Fiziol. Rast., 17(5), 997-1003 (1970a); Chem. Abstr., 74, 122223r (1971).

Lipskaya, G. A., "Accumulation of Chlorophyll in Sugar Beet Chloroplasts Under the Influence of Cobalt Applied Separately and Together with Boron, Manganese, Copper, Zinc, and Molybdenum," Agrokhimiya, No. 2, 105-110 (1970b); Chem. Abstr., 73, 34282r (1970).

Lipskaya, G. A., "Effect of Cobalt and Heteroauxin on the Morphology and Structure of a Barley Leaf," Vestsi Akad. Navuk Belaruski SSR, Ser. Biyal. Navuk, No. 5, 121-123 (1974); Chem. Abstr., 82, 69137k (1975).

Lipskaya, G. A., and M. K. Malyush, "Accumulation of Different Forms of Chlorophylls a and b and Activity of the Hill Reaction During the Action of High Cobalt Concentrations," Dokl. Akad. Nauk Beloruss. SSR, 18(2), 160-163 (1974); Chem. Abstr., 81, 21664g (1974).

Louria, D. B., M. M. Joselow, and A. A. Browder, "The Human Toxicity of Certain Trace Elements," Ann. Int. Med., 76, 307-319 (1972).

Luck, H., and S. W. Souci, "Lebensmittel-Zusatzstoffe und Mutagene Wirkung. II. Mutagene Stoffe in unserer Nahrung," ["Food Additives and Mutagenic Action. II. Mutagenic Materials in Our Foods"], Z. Lebensm.-Untersuch. u.-Forsch., 107, 236-256 (1958); Chem. Abstr., 52, 11302c (1958).

Ludwig, T. G., B. L. Adkins, and F. L. Losee, "Relationship of Concentrations of Eleven Elements in Public Water Supplies to Caries Prevalence in American Schoolchildren," Dent. J., 16, 126-132 (1970)

Ludwigson, D. C., "Today's Prosthetic Metals," J. Metals, 16(3), 226-229 (1964).

Lux, F., R. Zeisler, and J. Schuster, Activation Analysis of the Metabolic Behavior of Corrosion Products and Biological Tracer Elements in Tissue Affected by Metallosis, AED-Conf-75-404-027, Inst. Radiochem., Tech. Univ. Muenchen, Munich, Germany, 1975; Chem. Abstr., 85, 187302 (1976).

Maines, M. D., and A. Kappas, "Metals as Regulators of Heme Metabolism," Science, 198, 1215-1221 (1977).

Malten, K. E., "Flare Reaction Due to Vitamin B_{12} in a Patient with Psoriasis and Contact Eczema," Contact Dermatitis, 1(5), 325-326 (1975).

Marinoni, I., and L. Torelli, "Sensibilizzazione ai Metalli in Pazienti Sottoposti ad Artroplastica," ["Sensitization to Metals in Patients Submitted to Arthroplasty"], Boll. Inst. Sieroter Milan, 55(5), 447-450 (1976).

Marston, H. R., "Cobalt, Copper and Molybdenum in the Nutrition of Animals and Plants" Physiol. Rev., 32, 66 (1952).

Masironi, R., "Geochemistry and Health in Eastern Europe," Interface, 6(2), 13-18 (1977).

Matsusaka, N. "Relation Between the Fetal Uptake of Radioactive Cobalt and the Gestation Period in Mice," Igaku To Seibutsugaku, 92(6), 457-461 (1976).

Matsusaka, N., Y. Nishimura, and R. Ichikawa, "Incorporation of Radioactive Cobalt into the Eggs of Domestic Fowls," J. Radiat. Res., 13, 156-162 (1972).

Mazière, B., C. Loc'h, O. Stulzaft, A. Gaudry, and D. Comar, "Application of Neutron Activation Analysis to the Study of the Variations of the Concentration of Trace Elements in Various Organs of Rat as a Function of Age," J. Radioanal. Chem., 37, 617-629 (1977).

McDermott, F. T., "Dust in the Cemented Carbide Industry," Am. Ind. Hyg. Assoc. J., 32(3), 188-193 (1971).

McKee, G. K., "Carcinogenic Properties of Wear Particles from Prostheses made in Cobalt-Chromium Alloy," Lancet, 1, 750 (1971).

McKee, J. E., and H. W. Wolf, Eds., Water Quality Criteria, 2nd ed., Publication 3-A, California State Water Resources Control Board, 1963 (Reprint January 1973).

Menke, K. H., and M. Sarid-Sarban, "Incorporation of Cobalt into Nucleic Acids of the Chick Embryo," Nature, 212, 821-822 (1966).

Merkulova, I. S., V. R. Soroka, and N. Z. Rudenko, "Vliyanie Formy Khimicheskoi Svyazi Kobal'ta v Soedineniyakh na ego Vsasyvanie v Zheludochno-Kishechnon Trakte i Vklyuchenie v Metalloproteidnye Kompleksy Pecheni" ["Effect of the Form of Cobalt Chemical Bonding in Compounds on Its Capacity to Become Absorbed in the Gastrointestinal Tract and Incorporated into Metal-Protein Liver Complexes"], Vop. Pitan., 28(6), 48-51 (1969).

Merlini, M., "Hepatic Storage Alteration of Vitamin B_{12} by Cadmium in a Freshwater Fish," Bull. Environ. Contam. Toxicol., 19(6), 767-771 (1978).

Mertz, D. P., G. Wilk, and R. Koschnick, "Renale Ausscheidungsbedingungen von Kobalt beim Menschen," ["Conditions for the Renal Excretion of Cobalt by Humans"], Z. Klin. Chem. Klin. Biochem., 8(1), 30-32 (1970).

Mertz, W., "Some Aspects of Nutritional Trace Element Research," Fed. Proc., 29, 1482-1488 (1970).

Mertz, W., "Trace-Element Nutrition in Health and Disease: Contributions and Problems of Analysis," Clin. Chem., 21(4), 468-475 (1975).

Meyers, F. H., E. Jawetz, and A. Goldfien, Eds., Review of Medical Pharmacology, 4th ed., Lan Medical Publications, Los Altos, California, 1974, pp. 656-663.

Migal, O. K., "Effect of Some Trace Nutrients on the Number of Saprophytes in a Culture of Microcystis aeruginosa," Ukr. Bot. Zh., 27(5), 658-661 (1970); Chem. Abstr., 80, 10775X (1974).

Mikheeva, E. A., "Reaktivnye Izmeneniya Epiteliya Podzheludochnoi Zhelezy Ptits v Otvet na Vvedenie Khloristogo Kobal'ta," ["Change in the Pancreatic Epithelium of Birds in Response to Cobalt Chloride Administration"], Tr. Leningr. Sanit.-Gig. Med. Inst., **112**, 87-92 (1976).

Miller, W. J., "Use of Mineral Data Presents Unusual Problems in Feed Formulation," Feedstuffs, **45**, 23-24 (1972).

Mills, A. L., and R. R. Colwell, "Microbiological Effects of Metal Ions in Chesapeake Bay Water and Sediment," Bull. Environ. Contam. Toxicol., **18**(1), 99-103 (1977).

Mills, C. F., "The Detection of Trace Element Deficiency and Excess in Man and Farm Animals," Proc. Nutr. Soc., **33**(3) 267-274 (1974).

Mills, E. L., and R. T. Oglesby, "Five Trace Elements and Vitamin B_{12} in Cayuga Lake," Proc., 14th Conf. Great Lakes Res., 14th, 256-267 (1971).

Mitala, J. J., D. E. Mann, R. F. Gautieri, "Influence of Cobalt (Dietary), Cobalamins, and Inorganic Cobalt Salts on Phenytoin- and Cortisone-Induced Teratogenesis in Mice," J. Pharm. Sci., **67**(3), 377-380 (1978).

Monroe, R. A., H. E. Sauberlich, C. L. Comar, and S. L. Hood, "Vitamin B_{12} Biosynthesis after Oral and Intravenous Administration of Inorganic Co^{60} and Vitamin B_{12} Labeled with Co^{60}," Poultry Sci., **31**, 79-84 (1952).

Morgan, A., and L. J. Cralley, "Chemical Characteristics of Asbestos and Associated Trace Elements" in Biological Effects of Asbestos. Proceedings of a Working Conference, Lyon, France, October 2-6, 1972, IARC Scientific Publication No. 8, World Health Organization, Geneva, Switzerland, 1973, pp. 113-118.

Morin, Y., A. Têtu, and G. Mercier, "Quebec Beer-Drinkers' Cardiomyopathy: Clinical and Hemodynamic Aspects," Ann. N. Y. Acad. Sci., **156**, 566-577 (1969).

Morin, Y., A. Têtu, and G. Mercier, "Cobalt Cardiomyopathy: Clinical Aspects," Br. Heart J., Suppl., **33**, 175-178 (1971).

Morisawa, M., and H. Mohri, "Heavy Metals and Spermatozoan Motility. II. Turbidity Changes Induced by Divalent Cations and Adenosine Triphosphate in Sea Urchin Sperm Flagella," Exp. Cell Res., 83(1), 87-94 (1974).

Morrison, G. H., "Trace Element Survey Analysis of Biological Materials by Spark Source Mass Spectroscopy" in Trace Substances in Environmental Health-II, D. D. Hemphill, Ed., Proceedings of University of Missouri's 2nd Annual Conference on Trace Substances in Environmental Health, Columbia, Missouri, July 16-18, 1968, University of Missouri, Columbia, Missouri, 1969, pp. 307-317.

Munro-Ashman, D., and A. J. Miller, "Rejection of Metal Prosthesis and Skin Sensitivity to Cobalt," Contact Dermatitis, 2(2), 65-67 (1976).

Murdock, H. R., Jr., "Studies on the Pharmacology of Cobalt Chloride," J. Am. Pharm. Assoc., 48, 140-142 (1959).

Murdock, H. R., and L. J. Klotz, "A Pharmacologic Evaluation of Certain Cobalt-Containing Hemopoietic Agents," Am. Pharm. Assoc. J., 48(3), 143-148 (1959).

Myakisheva, L. S., and L. G. Shirobokova, "D-Vitaminnaya Intoksikatsiya i Defitsit Mikroelementov u Detei Rannego Vozrasta" ["Vitamin D Poisoning and Trace Element Deficiency in Young Children"], Kazan. Med. Zh., 57(9), 249-250 (1976).

Nadkarni, R. A., and W. D. Ehmann, "Determination of Trace Elements in the Reference Cigaret Tobacco by Neutron Activation Analysis," Radiochem. Radioanal. Lett., 2(3), 161-168 (1969), Chem. Abstr., 72, 87 (1970).

Nadkarni, R. A., and W. D. Ehmann, "Investigations of the Relative Transference of Trace Elements from Tobacco Smoke Condensate into Mice Tissues," Radiochem. Radioanal. Lett., 11(1), 45-57 (1972).

Nadkarni, R. A., W. D. Ehmann, and D. Burdick, "Investigations on the Relative Transference of Trace Elements from Cigaret Tobacco into Smoke Condensate," Tobacco, 170(11), 25-27 (1970).

Nagata, A., "Systemic Transport of Cobalt, Chromium, and Nickel Ions after a Total Joint Replacement," Nippon Seikeigeka Gakkai Zasshi, 52(7), 931-941 (1978); Chem. Abstr., 89, 177647e (1978).

Nagesha, C. N., B. P. Lalithamma, S. N. Rao, N. C. Ananthakrishna, and I. K. Sagar, "Cobalt Toxicity and its Reversal by Iron and Magnesium in Ogawa Serotypes of Vibrio cholerae and Vibrio eltor," Curr. Sci., 46(22), 770-773 (1977).

Nagler, J., R. A. Provost, and G. Parizel, "Hydrogen Cyanide Poisoning: Treatment with Cobalt EDTA," J. Occup. Med., 20(6), 414-416 (1978).

Neathery, M. W., "Tolerance Levels of Essential Elements for Livestock and Poultry," J. Anim. Sci., 43(1), 328-329 (1976).

Newman, J. P. Jr., and A. L. Buikema, Jr., "The Effect of Heavy Metals on the Reproduction of Aeolosoma headley, Beddard (Annelida)," ASB (Assoc. Southeast. Biol.) Bull., 22(2), 71 (1975).

Newton, D., and J. Rundo, "The Long-Term Retention of Inhaled Cobalt 60," Health Phys., 21, 377-384 (1971).

Nguyen Phu Lich, "Effects toxiques de certains oligo-élements," ["Toxic Effects of Trace Elements"], Aliment. Vie, 59(2), 103-153 (1971).

Niebroj, T. K., "Influence of Cobalt on the Histopathology of Mouse Testis," Endokrynol. Pol., 18(1), 1-13 (1967).

Nikitina, T. S., and E. A. Adzhikulov, "O Toksichnosti Novogo Kompleksnogo Soedineniya Kobal'ta s Geksametilentetraminom" ["Toxicity of a New Complex of Cobalt with Hexamethylenetetramine"], Sov. Zdravookhr. Kirg., No. 3, 26-29 (1973).

(NIOSH) National Institute of Occupational Safety and Health, Criteria for a Recommended Standard. Occupational Exposure to Tungsten and Cemented Tungsten Carbide, PB-275 594, National Tehcnical Information Service, U.S. Dept. of Commerce, Springfield, Virginia, September 1977.

Nishikawa, K., and K. Tabata, "Studies on the Toxicity of Heavy Metals to Aquatic Animals and the Factors to Decrease the Toxicity-III. On the Low Toxicity of Some Heavy Metal Complexes to Aquatic Animals," Bull. Tokai Reg. Fish. Res. Lab., No. 58, 233-241 (1969).

Nishimura, Y., J. Inaba, and R. Ichikawa, "Whole-Body Retention of $^{60}CoCl_2$ and ^{58}Co-Cyanocobalamin in Young and Adult Rats," J. Radiat. Res., 17, 240-246 (1976).

Nishioka, H., "Mutagenic Activities of Metal Compounds in Bacteria," Mutat. Res., 31, 185-189 (1975).

Novikova, E. P., "Effect of Different Amounts of Dietary Cobalt on Iodine Content of Rat Thyroid Gland," Vopr. Pitan. 22(2), 45 (1963); FASEB Transl. Suppl., 23, T459-T460 (1964).

Nuryagdyev, S. K. "Soderzhanie Mikroelementov u Bol'nykh Rakom do i posle Lecheniya" ["Levels of Trace Elements in Patients with Cancer Prior to and After Treatment"], Vop. Onkol., 17(5), 7-12 (1971).

O'Dell, B. L., and B. J. Campbell, "Chapter II. Trace Elements: Metabolism and Metabolite Function" in Comprehensive Biochemistry, Vol. 21, Metabolism of Vitamins and Trace Elements, M. Florkin, and E. H. Stotz, Eds., Elsevier Publishing Co., New York, New York, 1971, pp. 179-266.

Ogawa, E., "Heavy Metal Poisoning and Antidotes," Jap. J. Pharmacol., 24(5), 17 (1974).

Olson, L. D., and D. E. Rodabaugh, "Evaluation of Cobalt Arsanilate for Prevention and Treatment of Swine Dysentery," Am. J. Vet. Res., 34(7), 903-907 (1973).

Onkelinx, C., "Compartment Analysis of Cobalt(II) Metabolism in Rats of Various Ages," Toxicol. Appl. Pharm., 38, 425-438 (1976).

(OSHA) Occupational Safety and Health Administration, "Occupational Safety and Health Standards. Rules and Regulations," Fed. Reg., 37(202) Part II, 22140-22142 (1972).

(OSHA) Occupational Safety and Health Administration, "Occupational Safety and Health Standards," Fed. Reg., 39(125), 23508 (1974).

Pais, I., A. Somos, L. Duda, F. Tarjanyi, and F. Nagymihaly, "Trace Elements Tests on Tomato and Sweet Pepper," Kertesz. Kozlem, 33, 63-81 (1969); Chem. Abstr., 73, 13611g (1970).

Paley, K. R., and E. S. Sussman, "Absorption of Radioactive Cobaltous Chloride in Human Subjects," Metab. Clin. Exp., 12, 975-982 (1963).

Paley, K. R., E. S. Sobel, and R. S. Yalou, "Effect of Oral and Intravenous Cobaltous Chloride on Thyroid Function," J. Clin. Endocrinol. Metab., 18, 825-833 (1958).

Parr, R. M., and D. M. Taylor, "The Concentrations of Cobalt, Copper, Iron and Zinc in Some Normal Human Tissues as Determined by Neutron Activation Analysis," Biochem. J., 91, 424-431 (1964).

Parsons, J. R., and A. W. Ruff, Survey on Metallic Implant Materials, COM-74-11092, National Technical Information Service, December 1973.

Patel, P. M., A. Wallace, and R. T. Mueller, "Some Effects of Copper, Cobalt, Cadmium, Zinc, Nickel, and Chromium on Growth and Mineral Element Concentration in Chrysanthemum," J. Am. Soc. Hort. Sci., 101(5), 553-556 (1976).

Paton, G., and A. C. Allison, "Chromosome Damage in Human Cell Cultures Induced by Metal Salts," Mutat. Res., 16, 332-336 (1972).

Paulov, S., Changes of Growth and of Serum Proteins in Ducklings Intoxicated with Cobalt," Nutr. Metab, 13(1), 66-70 (1971).

Pavel, J., and J. Zakova, "The Effect of Various Microelements on Changes in the Activity of Certain Enzymes in Hydroponically Cultivated Barley During the First Period of Growth," Biol. Plant. (Praha), 9(5), 383-391 (1967).

Payne, L. R., "The Hazards of Cobalt," J. Soc. Occup. Med., 27(1), 20-25 (1977).

Peciulis, J., E. Augustaitiene, K. Pakarskyte, and J. Valavicius, "The Role of Microelements on the Accumulation of Proteins and Some Vitamins in Yeast Cells," Antonie van Leeuwenhoek, Supplement: Yeast Symposium, 35, G13 (1969).

Phelps, R. A., and H. E. Schlichting, Jr., "The Use of a Warburg Apparatus to Test Algicidal Compounds," Advan. Frontiers Plant Sci., 19, 181-194 (1968).

Poletaeva, V. F., "Effect of Cobalt on Fusarium Cotton Wilt," Izv. Akad. Nauk Turkm. SSR, Ser. Biol Nauk, No. 3, 73-74 (1969); Chem. Abstr., 71, 109881e (1969).

Pollack, S., J. N. George, R. C. Reba, R. M. Kaufman, and W. H. Crosby, "The Absorption of Nonferrous Metals in Iron Deficiency," J. Clin. Invest., 44(9), 1470-1472 (1965).

Polonovski, M., J. Gicquel, H. Schmitt, and M. L. Binet, "Etude pharmacodynamique d'un sel trivalent de cobalt," ["A Pharmacodynamic Study of a Trivalent Salt of Cobalt"], Compt. Rend., 239, 1711-1712 (1954).

Pier, S. M., "The Role of Heavy Metals in Human Health," Tex. Rep. Biol Med., 33(1), 85-106 (1975).

Polakowski, P., "Pharmacological Properties of Dithizone and its Influence on Excretion and Distribution of Various Metals in Tissues after Experimental Poisoning," Pol. Tyg. Lek., 29(5), 199-200 (1974).

Polyakov, P. V., "Effect of Trace Elements on the Resistance of the Sunflower to Storage Rot," Khim. Sel. Khoz., 9(2), 109-111 (1971); Chem. Abstr., 74, 124265n (1971).

Pope, A. L., "A Review of Recent Mineral Research with Sheep," J. Anim. Sci., 33(6), 1332-1343 (1971).

Popov, L. N., "Izuchenie Nekotorykh Pokazatelei Krovetvornoi Funktsii pri Gigienicheskoi Otsenke Deistviya Aerozolya Metallicheskogo Kobal' ["Study of Some Indexes of Hemopoietic Functions During the Hygienic Evaluation of the Effect of a Metallic Cobalt Aerosol on Experimental Animals"], Aktual'n. Vopr. Gigieny Okruzhayushchei Sredy, 138-142 (1976); Chem. Abstr., 87, 063698a (1977).

Popov, L. N., "Gigienicheskay Otsenka Deistvii Malykh Kontsentratsii Aerozolya Metallicheskogo Kobal'ta na Serdechno-Sosudistuyu Sistemu Belykh Krys v Eksperimente" ["Hygienic Evaluation of the Effect of Small Concentrations of a Metallic Cobalt Aerosol on the Cardiovascular System of White Rats under Experimental Conditions"], Gigienicheskie Aspekty Okhrany Zdorov'ya Naseleniya., 19-20 (1977a).

Popov, L. N. "Izuchenie Vliyaniya Malykh Kontsentratsii Aerozolya Metallicheskogo Kobal'ta na Organizm Zhivotnykh v Gigienicheskom Eksperimente" ["Study on the Effect of Low Concentrations of Metallic Cobalt Aerosol on Experimental Animals"], Gig. Sanit., No.4, 97-98 (1977b).

Popov, L. N., and N. A. Markina, "Nakoplenie Kobal'ta pri Ingalyatsionn Deistvii na Eksperimental'nykh Zhivotnykh" ["Cobalt Accumulation in Experimental Animals During Inhalation"], Gigienicheskie Aspekty Okhrany Zdorov'ya Naseleniya, 20-21 (1977).

Popov, L. N., T. A. Kochetkova, M. I. Gusev, N. A. Markina, E. V. Elfimova, and M. A. Timonov, "Nakoplenie, Raspredelenie i Morfologicheskie Izmeneniya v Organizme pri Ingalytsionnom Vozdeistvii," Accumulation, Distribution, and Morphological Changes in the Body Due to Inhalation of Metallic Cobalt Aerosol, Gig. Sanit., No. 6, 12-15 (1977).

Prazmo, W., E. Balbin, H. Baranowska, A. Ejchart, and A. Putrament, "Manganese Mutagenesis in Yeast," Genet. Res., Camb., **26**, 21-29 (1975).

Puget A., H. Vergnes, and C. Gouarderes, "Sensibilite de l'Ochotone Afgan (Ochotonna refescens rufescens) au chlorure de cobalt," ["The sensitivity of the Red Pika (Ochotona rufescens rufescens) to Cobalt Chloride"], Zentralbl. Veterinaermed. [A], **22**(7), 583-596 (1975).

Pushpalata, "Effect of Cobalt on Biloxi Soybean Leaf," Indian J. Exp. Biol., **5**(1), 56-57 (1967).

Rabotnikova, L. V., "Issledovanie Sravitel'noi Biologicheskoi Aktivnosti Okislov Metallov na Urovne Porogovykh Doz pri Odnokratnom Vvedenii" ["Comparative Biological Activity of Metal Oxides at the Level of their Threshold Doses Introduced Singly"], Gig. Tr. Prof. Zabol., **15**(8), 33-36 (1971).

Rehab, F. I., and A. Wallace, "Excess Trace Metal Effects on Cotton. 1. Copper, Zinc, Cobalt and Manganese in Solution Culture," Commun. Soil Sci. Plant Anal., **9**(6), 507-518 (1978a), Chem. Abstr., **89**, 123993w (1978a).

Rehab, F. I., and A. Wallace, "Excess Trace Metal Effects on Cotton. 2. Copper, Zinc Cobalt and Manganese in Yolo Loam Soil," Commun. Soil Sci. Plant Anal., **9**(6), 519-527 (1978b).

Reisenaur, H. M., "Cobalt in Nitrogen Fixation by a Legume," Nature, **186**, 375-376 (1960).

Ridgway, L. P., and D. A. Karnofsky, "The Effects of Metals on the Chick Embryo: Toxicity and Production of Abnormalities in Development," Ann. N. Y. Acad. Sci., **55**, 203-215 (1952).

Robin, J., and R. A. Brunetière, "Bilan de l'enquete allergologigue chez 200 cimentiers atteints de dermite," ["Record of an Allergologic Survey of 200 Cement Workers with Dermatitis"], Rev. Fr. Allergol., **9**(2), 97-102 (1969).

Roe, F. J. C., and M. C. Lancaster, "Natural, Metallic and other Substances, as Carcinogens," Br. Med. Bull., 20, 127-133 (1964).

Roginski, E. E., and W. Mertz, "A Biphasic Response of Rats to Cobalt," J. Nutr., 107(8), 1537-1542 (1977).

Rona, G., and C. I. Chappel, "Pathogenesis and Pathology of Cobalt Cardiomyopathy" in Recent Advances in Studies on Cardiac Structure and Metabolism. Vol. 2. Cardiomyopathies Symposium, E. Bajusz, and G. Rona, Eds., University Park Press, Baltimore, Maryland, 1973, pp. 407-422.

Rosen, J. C., and A. T. Sabo, "Differences in the Biological Half Lives of Inhaled Cobalt 58 and Cobalt 60," Oper. Health Phys., Proc. Midyear Top. Symp. Health Phys. Soc., 9th, 621-626 (1976).

Roshchin, A. V., "Toksikologiya Metallov i Profilaktika Professional'nykh Otravlenii" ["Toxicology of Metals and Prophylaxis of Occupational Poisonings"], Zh. Vses. Khim. O-va im. D. I. Mendeleeva, 19(2), 186-192 (1974).

Rosko, J. J., and J. W. Rachlin, "Effect of Copper, Zinc, Cobalt, and Manganese on the Growth of the Marine Diatom Nitzschia closterium," Bull. Torrey Bot. Club, 102(3), 100-106 (1975).

Roy-Chowdhury, A. K., T. F. Mooney, Jr., and A. L. Reeves, "Trace Metals in Asbestos Carcinogenesis," Arch. Environ. Health, 26, 253-255 (1973).

Ruthven, J. A., and J. Cairns, Jr., "Response of Fresh-Water Protozoan Artificial Communities to Metals," J. Protozool., 20(1), 127-135 (1973).

Saknyn, A. V., and N. K. Shabynina, "Epidemiologiy Zlokachestvennykh Novoobrazovanii na Nikelevykh Predpriyatiyakh" ["Epidemiology of Malignant Neoplasms in Nickel Smelters"], Gig. Tr. Prof. Zabol., 17(9), 25-29 (1973).

Salminen, K., O.-P. Obermeier, and W. Kreuzer, "Metabolism of ^{60}Co in Chickens. II. Absorption After Single and Repeated Peroral Administration," Comp. Biochem. Physiol., 51B, 267-272 (1975).

Salminen, K., O.-P. Obermeier and W. Kreuzer, "Radioactive Contamination of Poultry Meat with ^{60}Co and Its Assessment," Poult. Sci., 53(6), 2065-2069 (1974).

Sandstroem, O., and S. Soedergren, "Dialysis Cultures as Indicators of Water Pollution. Effect of Cobalt on the Respiration of Green Algae," Vatten, 30(2), 229-234 (1974); Chem. Abstr., 83, 72863v (1975).

Saric, T., and B. Saciragic, "Effect of Oat Seed Treatment with Microelements," Plant Soil, 31(1), 185-187 (1969).

Savitskii, I. V., "Znachenie Teplovogo Faktora pri Otpravleniyakh, Vyzvannykh Professional'nymi Yadami iz Gruppy Tyazhelykh Metallov" ["Significance of the Heat Factor in Poisonings Induced by Industrial Poisons from the Heavy Metal Group"], Vrach Delo, 3, 138-142 (1971).

Sax, N. I., Dangerous Properties of Industrial Materials, 4th ed., Litton Educational Publishing, Inc., Van Nostrand Reinhold Company, New York, New York, 1975.

Schade, S. G., B. F. Felsher, G. M. Bernier, and M. E. Conrad, "Interrelationship of Cobalt and Iron Absorption," J. Lab. Clin. Med., 75(3), 435-441 (1970).

Schepers, G. W. H., "The Biological Action of Cobaltic Oxide," A. M. A. Arch. Ind. Health, 12, 124-126 (1955a).

Schepers, G. W. H., "The Biological Action of Tungsten Carbide and Cobalt," A. M. A. Arch. Ind. Health, 12, 140-146 (1955b).

Scherrer, M., A. Parambadathumalil, H. Bürki, A. Senn, and R. Zürcher, "Drei Fälle von Hartmetallstaublunge," "3 Cases of Hard Metal Dust Lung Disease" , Schweiz. Med. Wochenschr., 100(52), 2251-2255 (1970).

Schirrmacher, U. O. E., "Case of Cobalt Poisoning," Br. Med. J., 1, 544-545 (1967).

Schmidt, P., D. Burck, G. Fox, R. Gohlke, G. M. Müller, and L. Sabiers, "Tierexperimentalle Untersuchungen zur Toxizität des Managan- and des Kobaltstearats," ["Animal Experiment Studies on the Toxicity of Manganese and Cobalt Stearates"], Z. Gesamte Hyg., 21(1), 13-18 (1975).

Schroeder, H. A., "Role of Trace Elements in Cardiovascular Diseases," Med. Clin. N. Am., 58(2), 381-396 (1974).

Schroeder, H. A., A. P. Nason, and I. H. Tipton, "Essential Trace Metals in Man: Cobalt," J. Chronic Dis., **20**, 869-890 (1967).

Schubert, J., Heavy Metals--Toxicity and Environmental Pollution," Adv. Exp. Med. Biol., **40**, 239-297 (1973).

Schwartz, L., L. Tulipan, and S. M. Peck, Occupational Diseases of the Skin, Lea and Febiger, Philadelphia, Pennsylvania, 1947.

Sederholm, T., K. Kouvalainen, and B. A. Lamberg, "Cobalt-Induced Hypothyroidism and Polycythemia in Lipoid Nephrosis," Acta Med. Scand., **184**, 301-306 (1968).

Selby, L. A., and R. A. Tidball, "The Environmental Geochemistry of Agricultural Soils in Missouri and Its Association with Swine Birth Defects" in Trace Substances in Environmental Health-X D. D. Hemphill. Ed., Proceedings of Univerisity of Missouri's 10th Annual Conference on Trace Substances in Environmental Health. Columbia, Missouri, June 3 - 10, 1976, University of Missouri, Columbia, Missouri, 1977, pp. 63-70.

Semina, R. M., "Effect of Cobalt on the Water Regime of Buckwheat," Biol. Nauki, No. 6, 69-72 (1970); Chem. Abstr., **74**, 52540y (1971).

Semina, R. M., "The Action of Cobalt on Chlorophyll Accumulation, the Intensity of Photosynthesis, and the Seed Crop of Buckwheat," Nauch. Dokl. Vyssh. Shk., Biol. Nauki, No. 3, 80-83 (1967); Chem. Abstr., **67**, 10806q (1967).

Shabaan, A. A., V. Marks, M. C. Lancaster, and G. N. Dufeu, "Fibrosarcomas Induced by Cobalt Chloride ($CoCl_2$) in Rats," Lab. Anim., **11**, 43-46 (1977).

Shabalina, A. A., "Vliyanie Khloristogo Kobal'ta na Rost i Fiziologicheskie Pokazateli Raduzhnoi Foreli" ["Effect of Cobaltous Chloride on the Growth and Physiological Indexes of *Salmo irideus*"], Izv. Gos. Nauch.-Issled. Inst. Ozer. Rechn. Ryb. Khoz., **68**, 110-118 (1969).

Shacklette, H. T., H. I. Sauer, and A. T. Miesch, Geochemical Environments and Cardiovascular Mortality Rates in Georgia, Geological Survey Professional Paper 574-C, U.S. Department of the Interior, U.S. Government Printing Office, Washington, D. C., 1970.

Shanoff, H. M., "Alcoholic Cardiomyopathy: An Introductory Review," Can. Med. Assoc. J., **106**(1), 55-62 (1972).

Sharrett, A. R., and M. Feinleib, "Water Constituents and Trace Elements in Relation to Cardiovascular Disease," Prev. Med., **4**, 20-36 (1975).

Sheehan, R. G., "Interrelationships of Iron and Cobalt Absorption: Mucosal Distribution of Cobalt During Absorption," Proc. Soc. Exp. Biol. Med., **146**(4), 993-996 (1974).

Sheline, G. E., I. L. Chaikoff, and M. L. Montgomery, "The Elimination of Administered Cobalt in Pancreatic Juice and Bile of the Dog, as Measured with Its Radioactive Isotopes," Am. J. Physiol., **145**, 285-290 (1946).

Shelley, W. B., "Chondral Dysplasia Induced by Zirconium and Hafnium," Cancer Res., **33**(2), 287-292 (1973).

Shettigara, P. T., and R. W. Morgan, "Asbestos, Smoking, and Laryngeal Carcinoma," Arch. Environ. Health, **30**(10), 517-519 (1975).

Shmilliar, M., "Fertilization by Spraying the Foliage of Sugar Beets with Solutions Containing Microelements," Novenynemesitesi Novenytermesztesi Kut. Intez., Sopronhorpacs, Kozlem., **4**, 75-95 (1968); Chem. Abstr., **73**, 3115j (1970).

Shtenberg, A. I., I. A. Kusevitskii, and E. E. Abolyn, "Effect on Thyroid Gland of Cobalt in Low Protein Diets with Varying Iodine Contents," FASEB Transl. Suppl., **23**, T763-T766 (1964).

Siegel, S. M., "The Cytotoxic Response of 'Nicotiana' Protoplasts to Metal Ions: A Survey of the Chemical Elements," Water, Air, Soil Pollut., **8**, 293-304 (1977).

Sill, C. W., J. I. Anderson, and D. R. Percival, Comparison of Excretion Analysis with Whole-Body Counting for Assessment of Internal Radioactive Contaminants, AEC Accession No. 43318, Report No. IDO-12038, U.S. Atomic Energy Commission, Health and Safety Division, Idaho Falls, Idaho.

Sissoëff, I., J. Grisvard, and E. Guillé, "Studies on Metal Ions-DNA Interactions: Specific Behaviour of Reiterative DNA Sequences," Prog. Biophys. Mol. Biol., **31**(2), 165-199 (1976).

Smith, D. C., "Materials Used for Construction and Fixation of Implants," Oral Sci. Rev., 5, 23-55 (1974).

Smith, E. L., "Cobalt" in Mineral Metabolism. An Advanced Treatise Vol. II, The Elements, Part B. Comar, C. L., and F. Bronner, Eds. Academic Press, New York, New York, 1962, pp. 349-370.

Smith, T., C. J. Edmonds, and C. F. Barnaby, "Absorption and Retention of Cobalt in Man by Whole-Body Counting," Health Phys., 22, 359-367 (1972).

Smyth, H. F. Jr., C. P. Carpenter, C. S. Weil, U. C. Pozzani, J. A. Striegel, and J. S. Nycum, "Range-Finding Toxicity Data: List VII," Am. Ind. Hyg. J., 30, 470-476 (1969).

Spiridonova, V. S., and L. P. Shabalina, "Eksperimental'noe Issledovanie Toksichnosti Tetrakarbonila Kobal'ta" ["Experimental Study of the Toxicity of Cobalt Tetracarbonyl"], Gig. Sanit., 73(1), 97-99 (1973).

Srivastava, A. K., and S. J. Agrawal, "Hematological Anomalies in a Fresh Water Teleost, Colisa fasciatus, on Acute Exposure to Cobalt," Acta Pharmacol. Toxicol., 44(3), 197-199 (1979).

Srivastava, R., N. Lefebvre, and C. Onkelinx, "Effects of Metal Salts on Collagen Synthesis in Embryonic Rat Calvaria," Toxicol. Appl. Pharmacol., 37, 229-235 (1976).

Stahly, E. E., "Some Considerations of Metal Carbonyls in Tobacco Smoke," Chem. Ind. (London), No. 13, 620-623 (1973).

Stanley, A. J., H. C. Hopps, and A. A. Hellbaum, "Observations on Cobalt Polycythemia. I. Studies on the Peripheral Blood of Rats Proc. Soc. Exp. Biol. Med., 61, 130-133 (1946).

Steele, K. W., and H. V. Henderson, "Occurrence of Periodontal Disease in Sheep in the Mangonui, Whangaroa, Hokianga and Bay of Islands Counties," N. Z. J. Agric. Res., 20, 301-308 (1977).

Stokinger, H. E., "The Metals (Excluding Lead)" in Industrial Hygiene and Toxicology, 2nd revised edition, F. A. Patty, Ed., Vol. II, Toxicology, D. W. Fassett and D. D. Irish, Eds., Interscience Publishers, a division of John Wiley and Sons, Inc., New York, New York, 1963 pp. 987-1194.

Stoner, G. D., M. B. Shimkin, M. C. Troxell, T. L. Thompson, and L. S. Terry, "Test for Carcinogenicity of Metallic Compounds by the Pulmonary Tumor Response in Strain A Mice," Cancer Res., 36(5), 1744-1747 (1976).

Strain, W. H., W. J. Pories, and O. A. Hill, Jr., "Correlation of Results of Activation Analysis of Rat Aortas with Radiocobalt Retention Determinations," Neutron Sources Appl., Proc. Am. Nucl. Soc. Nat. Top. Meet., 1(5), 33-38 (1971).

Strickland, E. H., and C. R. Goucher, "Effects of Cobalt on Mitochondrial Respiration," Nature, 198, 790-791 (1963).

Sullivan, J. F., M. Parker, and S. B. Carson, "Tissue Cobalt Content in 'Beer Drinkers' Myocardiopathy," J. Lab. Clin. Med., 71, 893 (1968).

Sunderman, F. W., Jr., "Metal Carcinogenesis in Experimental Animals," Food Cosmet. Toxicol, 9, 105-120 (1971).

Suso, F. A., and H. M. Edwards, Jr., "Whole Body Counter Studies on the Absorption of ^{60}Co, ^{59}Fe, ^{54}Mn and ^{65}Zn by Chicks, as Affected by Their Dietary Levels and Other Supplemental Divalent Elements," Poultry Sci., 48(3), 933-938 (1969).

Suzuki, F., S. Fukushima, and N. Tsujita, "Cobalt Lithium Titanium Oxide as Cosmetic Colors," Japanese Patent No. 76 130538, November 12, 1976; Chem. Abstr., 86, 195070d (1977).

Suzuki-Yasumoto, M., and J. Inaba, "Absorption and Metabolism of Radioactive Cobalt Compounds Through Normal and Wounded Skins," Diagn. Treat. Inc. Radionuclides, Proc. Int. Semin., 119-136 (1976).

Szarmach, H., and H. Poniecka, "Contact Allergy Caused by Contact with Farm Animal Feed," Przegl. Dermatol., 62(2), 193-197 (1975).

Tabata, K., "Studies on the Toxicity of Heavy Metals to Aquatic Animals and the Factors to Decrease Toxicity. II. The Antagonistic Action of Hardness Components in Water on the Toxicity of Heavy Metal Ions," Bull. Tokai Reg. Fish. Res. Lab., No. 58, 215-232 (1969).

Tachibana, S., T. Murakami, and S. Sawada, "Studies on CO_2-Fixing Fermentation. XXIII. Effects of Trace Elements on L-Malate Fermentation Using Schizophyllum commune," J. Ferment. Technol., 50(3), 171-177 (1972).

Takki, S., J. Heinonen, and E. Taskinen, "Poisoning Caused by a Mixture of Plant-Nutrient Substrates," Arch. Toxikol., 28(4), 270-278 (1972).

Talburt, D. E., D. A. Johnson, and P. S. Sheridan, "Influence of Cobalt(III) Chloro Complexes on the Cellular Chemistry of Escherichia coli," J. Inorg. Nucl. Chem., 37(2), 582-584 (1974).

Tanaka, A., T. Tadano, and Y. Ebine, "Comparison of Adaptability to Heavy Metals Among Crop Plants. III. Adaptability to Nickel and Cobalt," Nippon Dojo Hiryogaku Zasshi, 49(4), 314-320 (1978); Chem. Abstr., 90, 150916a (1979).

Tauberger, G., and O. R. Klimmer, "Tierexperimentelle Untersuchunge einiger Kobaltverbindungen nach intravenöser Injektion" ["Animal Experimental Study of Several Cobalt Compounds after Intravenous Injection"], Arch. int. Pharmacodyn., 143(1-2), 219-238 (1963).

Taylor, A., and N. Carmichael, "The Effect of Metallic Chlorides on the Growth of Tumor and Nontumor Tissue," Univ. Texas Publ., No. Biochem. Inst. Studies 5, Cancer Studies 2, 36-79 (1953); Chem. Abstr., 48, 3565i (1954).

Taylor, A. and V. Marks, "Cobalt: A Review," J. Human Nutr., 32, 165-177 (1978).

Telib, M., "Effects of Cobaltous Chloride in Laboratory Animals. I. The Histological and Electron Microscopical Changes in the Islets of Rabbits," Endokrinologie, 60(1), 81-102 (1972).

Thieme, R., P. Schramel, B.-J. Klose, and E. Waidl, "Spurenelemente in der menschlichen Plazenta," Geburtsh. u. Frauenheilk., 35, 349-353 (1975).

Thomas, R. G., J. E. Furchner, J. E. London, G. A. Drake, J. S. Wilson, and C. R. Richmond, "Comparative Metabolism of Radionucli in Mammals. X. Retention of Tracer-Level Cobalt in the Mouse, Rat, Monkey and Dog," Health Phys., 31(4) 323-333 (1976).

Thomson, A. B. R. and L. S. Valberg "Intestinal Uptake of Iron, Cobalt, and Manganese in the Iron-Deficient Rat," Am. J. Physiol. 223(6), 1327-1329 (1972); Chem. Abstr., 78, 27041m (1973).

Tipton, I. H., and M. J. Cook, "Trace Elements in Human Tissue. Part II. Adult Subjects from the United States," Health Phys., 9, 103-145 (1963).

Tipton, I. H., and P. L. Stewart, "Analytical Methods for the Determination of Trace Elements-Standard Man Studies" in Trace Substances in Environmental Health-III, Proceedings of University of Missouri's 3rd Annual Conference on Trace Substances in Environmental Health, Columbia, Missouri, June 24-26, 1969, University of Missouri, Columbia, Missouri, 1970, pp. 305-330.

Tipton, I. H., P. L. Stewart, and J. Dickson, "Patterns of Elemental Excretion in Long Term Balance Studies," Health Phys., **16**, 455-469 (1969).

Tolot, F., R. Girard, G. Dortit, G. Tabourin, P. Galy, and J. Bourret, "Manifestations pulmonaires des 'métaux durs': troubles irritatifs et fibrose (Enquete et observations cliniques)," ["Pulmonary Manifestations of Hard Metals: Irritative Disorders and Fibrosis (Survey and Clinical Observations)"], Arch. Mal. Prof., **31**(9), 453-470 (1970).

Tracor-Jitco, Inc., Scientific Literature Reviews on Generally Recognized as Safe (GRAS) Food Ingredients-Vitamin B_{12}, PB-241 966/1GA, National Technical Information Service, U.S. Department of Commerce, Springfield, Virginia, 1974.

Tso, T. C., T. P. Sorokir, and M. E. Engelhaupt, "Effects of Some Rare Elements on Nicotine Content of the Tobacco Plant," Plant Physiol., **51**, 805-806 (1973).

Uesata, S., H. Takahasi, and T. Y. Chang, "Importance of Trace Elements in Farm Animal Feeding. II. Effect of Addition of Iron, Cobalt, and Fluorine on Propagation of Actinomycetes," Bull. Res. Inst. Food Sci., Kyoto Univ., **12**, 15 (1953); cited in Water Quality Criteria, 2nd ed., J. E. McKee and H. W. Wolf, Eds., The Resources Agency of California, State Water Resources Control Board, Sacramento, California, 1963, p. 167.

Underwood, E. J., "Cobalt," Nutr. Rev., **33**, 65-69 (1975).

Underwood, E. J., "Cobalt" in Present Knowledge in Nutrition, 4th ed., Nutrition Foundation Inc., Washington, D.C., 1976, pp. 317-324.

Underwood, E. J., and C. A. Elvehjem, "Is Cobalt of any Significance in the Treatment of Mild Anemia with Iron and Copper?" J. Biol. Chem., **124**, 419-424 (1938).

Ungethuem, M., "Material Problems of Wear of Artificial Hip Joints," Fachber.-Jahrestag. Dtsch. Ges. Biomed. Tech., 31-34 (1973); Chem. Abstr., 84, 22064k (1976).

(U.S. HEW) Division of Radiological Health, Department of Health, Education, and Welfare, Ed., Radiologic Health Handbook, Revised ed., PB 121 784R, Department of Health, Education, and Welfare, Washington, D.C., 1960.

Valberg, L. S., J. Sorbie, W. E. N. Corbett, and J. Ludwig, "Cobalt Test for The Detection of Iron Deficiency Anemia," Ann. Int. Med., 77, 181-187 (1972).

Van Vleet, J. F., A. H. Rebar, and V. J. Ferrans, "Acute Cobalt and Isoproterenol Cardiotoxicity in Swine: Protection by Selenium-Vitamin E Supplementation and Potentiation by Stress-Susceptible Phenotype," Am. J. Vet. Res., 38(7), 991-1002 (1977).

Venkateswerlu, G., and K. S. Sastry, "Interrelationships in Trace-Element Metabolism in Metal Toxicities in a Cobalt-Resistant Strain of Neurospora crassa," Biochem. J., 132(4), 673-680 (1973).

Verhamme, E. N., "Contribution to the Evaluation of the Toxicity of Cobalt," Cobalt, 1973, 29-32 (1973).

Vetterlein, F., "Untersuchungen zur vasodilatatorischen Wirkung von Kobaltionen" ["The Vasodilator Effects of Cobalt Ions"], Arch. Int. Pharmacodyn. Ther., 199(1), 53-66 (1972).

Villaume, J., R. Levine, H. Schwartz, A. Craigmill, and G. Miller, Heavy Metals Used as Driers: Cerium, Cobalt, Manganese, Vanadium, Zinc and Zirconium. A Monograph, PB-280 473, National Technical Information Service, U.S. Department of Commerce, Springfield, Virginia, 1976.

Von Rosen, G., "Breaking of Chromosomes by the Action of Elements of the Periodical System and by Some Other Principles," Hereditas, 40, 258-263 (1954).

Von Rosen, G., "Mutations Induced by the Action of Metal Ions in Pisum," Hereditas, 43, 644-664 (1957).

Von Rosen, G., "Mutations Induced by the Action of Metal Ions in Pisum. II. Further Investigations on the Mutagenic Action of Metal Ions and Comparison with the Activity of Ionizing Radiation," Hereditas, 51, 89-134 (1964).

Vorob'eva, A. I., and E. V. Osmolovskaya, "O Balanse Molibdena i Kobal'ta v Organizme Detei Doshkol'nogo Vozrasta" ["Molybdenum and Cobalt Balance in Preschool-Age Children"], Gig Sanit., 35(11), 108-109 (1970).

Voroshilin, S. T., E. G. Plotko, T. V. Fink, and V. Y. Nikiforova, "Tsitogeneticheskoe Deistvie Neorganicheskikh Soedinenii Vol'frama, Tsinka, Kadmiya i Kobal'ta na Somaticheskie Kletki Cheloveka i Zhivotnykh" ["Cytogenetic Effect of Inorganic and Acetate Compounds of Tungsten, Zinc, Cadmium, and Cobalt on Animal and Human Somatic Cells"], Tsitol. Genet., 12(3), 241-243 (1978).

Waddell, W. J., "The Distribution of Cortisone-^{14}C in Pregnant Mice," Teratology, 4(3) 355-365 (1971).

Wahlberg, J. E., "Percutaneous Toxicity of Metal Compounds," Arch. Environ. Health, 11, 201-204 (1965).

Wahlberg, J. E., "Health-Screening for Occupational Skin Diseases in Building Workers," Berufsdermatosen, 17(4), 184-198 (1969).

Wahlberg, J. E., "Vehicle Role of Petrolatum. Absorption Studies with Metallic Test Compounds in Guinea Pigs," Acta Dermato-Venereol., 51(2), 129-134 (1971); Chem. Abstr., 74, 97434h (1971).

Wahlberg, J. E., "Thresholds of Sensitivity in Metal Contact Allergy. 1. Isolated and Simultaneous Allergy to Chromium, Cobalt, Mercury and/or Nickel," Berufsdermatosen, 21(1), 22-33 (1973a).

Wahlberg, J. E., "Thresholds of Sensitivity in Metal Contact Allergy. 2. The Value of Percutaneous Absorption Studies for Selection of the Most Suitable Vehicle," Berufsdermatosen, 21(4), 151-158 (1973b).

Wahlberg, J. E. G. Lindstedt, and O. Einarsson, "Chromium, Cobalt and Nickel in Swedish Cement, Detergents, Mold and Cutting Oils," Berufsdermatosen, 25(6), 220-228 (1977).

Wallace, A., G. V. Alexander, and F. M. Chaudhry, "Phytotoxicity of Cobalt, Vanadium, Titanium, Silver, and Chromium," Commun. Soil Sci. Plant Anal., 8(9), 751-756 (1977).

Wardle, E. N., "Cobalt, Cardiomyopathy, and Dialysis," Lancet, 2 (7976), 106 (1976).

Warnick, S. L., and H. L. Bell, "The Acute Toxicity of Some Heavy Metals to Different Species of Aquatic Insects," J. Water Pollut. Control Fed., 41, 280-284 (1969).

Webb, M., J. C. Heath, and T. Hopkins, "Intranuclear Distribution of the Inducing Metal in Primary Rhabdomyosarcomata Induced in the Rat by Nickel, Cobalt, and Cadmium," Br. J. Cancer, 26, 274-278 (1972).

Wehner, A. P., and D. K. Craig, "Toxicology of Inhaled NiO and CoO in Syrian Golden Hamsters," Am. Ind. Hyg. Assoc. J., 33(3), 146-155 (1972).

Wehner, A. P., R. H. Busch, R. J. Olson, and D. K. Craig, "Chronic Inhalation of Cobalt Oxide and Cigarette Smoke by Hamsters," Am. Ind. Hyg. Assoc. J., 38(7), 338-346 (1977).

Weinzierl, S. M., and M. Webb, "Interaction of Carcinogenic Metals with Tissue and Body Fluids," Br. J. Cancer, 26, 279-291 (1972).

Wester, P. O., "Concentration of 24 Trace Elements in Human Heart Tissue Determined by Neutron Activation Analysis," Scand. J. Clin. Lab. Invest., 17(4), 357-370 (1965).

Wester, P. O., "Trace Elements in Serum and Urine from Hypertensive Patients Before and During Treatment with Chlorthalidone," Acta Med. Scand., 194, 505-512 (1973).

Wester, P. O., "Trace Elements in Serum and Urine from Hypertensive Patients Treated for Six Months with Chlorthalidone," Acta Med. Scand., 196(6), 489-494 (1974).

Wester, P. O., "Trace Element Balances in Relation to Variations in Calcium Intake," Atherosclerosis, 20(2), 207-215 (1974).

Whitby, L. M., and T. C. Hutchinson, "Heavy-metal Pollution in the Sudbury Mining and Smelting Region of Canada. II. Soil Toxicity Tests," Environ. Conserv., 1(3), 191-200 (1974).

White, J., and D. J. Munns, "Inhibitory Effect of Common Elements Towards Yeast Growth," J. Inst. Brew., 54, 175-179 (1951).

(WHO) World Health Organization, Joint FAO/WHO Expert Group on Requirements of Ascorbic Acid, Vitamin D, Vitamin B_{12}, Folate and Iron, WHO Tech. Rep. Series No. 452, 1970.

(WHO) Expert Committee, Trace Elements in Human Nutrition, World Health Organ. Tech. Rep. Ser. No. 532, World Health Organization, Geneva, Switzerland, 1973.

Wiberg, G. S., I. G. Munro, and A. B. Morrison, "Effect of Cobalt Ions on Myocardial Metabolism," Can. J. Biochem., **45**, 1219-1223 (1967).

Wiberg, G. S., I. C. Munro, J. C. Meranger, A. B. Morrison, and H. C. Grice, "Factors Affecting the Cardiotoxic Potential of Cobalt," Clin. Toxicol., **2**(3), 257-271 (1969).

Wilson, S. B., and D. J. D. Nicholas, "A Cobalt Requirement for Non-Nodulated Legumes and for Wheat," Phytochem., **6**, 1057-1066 (1967).

Winell, M., "An International Comparison of Hygienic Standards for Chemicals in the Work Environment," Ambio, **4**(1), 34-36 (1975).

Wojdala, Z., and T. Niebroj, "Role of Trace Elements in Obstetrics," Ginekol. Pol., **39**(2), 247-251 (1968); Chem. Abstr., **69**, 25643Z (1968).

Wooldridge, C. R., and D. P. Wooldridge, "Internal Damage in an Aquatic Beetle Exposed to Sublethal Concentrations of Inorganic Ions," Ann. Entomol. Soc. Am., **62**(4), 921-922 (1969).

Wyttenback, A., S. Bajo, and A. Haekkinen, "Determination of 16 Elements in Tobacco by Neutron Activation Analysis," Beitr. Tabakforsch., **8**(5), 247-249 (1976).

Yagodin, B. A., and I. V. Tischenko, "Soderzhanie Mikroelementov Tsinka i Kobal'ta v Pochve i Rasteniyakh v Zavisimosti ot Primenyaemykh Udobrenii" ["Content of Zinc and Cobalt Trace Elements in the Soil and in Plants in Relation to Fertilizer Application"], Vestn. S-kh. Nauki (Moscow), No. 3, 42-50 (1978).

Yagodin, B. A., and G. Ya. Zhiznevskaya, "Variations in Protein Composition of *Vicia faba* Leaves During Chlorosis Induced by Excessive Cobalt," Fiziol Rast., **16**(3), 505-511 (1969); Chem. Abstr., **71**, 69767p (1969).

Yagodin, B. A., G. A. Ovcharenko, V. Yu. Vasil'eva, and M. A. Ivanova, ["Effect of Cobalt on Nitrate Reductase Activity in Legumes"], Sel'skokhoz. Biol., **5**(1), 134-136 (1970); Chem. Abstr., **73**, 34329m (1970).

Young, R. S., Certain Rarer Elements in Soils and Fertilizers and Their Role in Plant Growth, Agricultural Experiment Station Memoi 174, Cornell University, Ithaca, New York, 1935.

Young, R. S., "Biological and Biochemical Relationships of Cobalt" in Cobalt, Am. Chem. Soc. monograph No. 108, Reinhold Publishing Company, New York, New York, 1948, pp. 137-149.

Younkin, C. N., "Multiphase MP35N Alloy for Medical Implants," Biomed. Mater. Symp., 5(1), 219-226 (1974).

Zharnikov, I. I., "Topografiya Mikroelementov v Organizme Ovets Buryatskii ASSR" ["Distribution of Trace Elements in Sheep of the Buryat ASSR"], Mikroelem. Sib., No. 5, 91-94 (1967).

Zhilin, A. D., "Opyt Kompleksnoi Khimioprofilaktiki Zheludochno-Kishechnykh Gel'mintozov Ovets v Tsentral'nom Kazakhstane" ["Complex Chemoprophylaxis of Gastrointestinal Helminthiases of Sheep in Central Kazakhstan"], Tr. Kaz. Nauchno-Issled. Vet. Inst., 13, 193-197 (1969).

Zylicz, E., R. Zablotna, J. Geisler, and Z. Szot, "Effects of DTPA on the Deposition of ^{65}Zn, ^{60}Co and ^{144}Ce in Pregnant Rat and in Foetoplacental Unit," Int. J. Radiat. Biol., 28(2), 125-136 (1975).

Zylicz, E., R. Zablotna, and Z. Szot, "Placental Transfer of ^{60}Co as a Function of Gestation Age," Nukleonika, 21(10), 1203-1210 (1976).

IX. ENVIRONMENTAL COBALT LOSSES AND ASSESSMENT OF HEALTH HAZARDS TO HUMANS AND OTHER LIFE FORMS*

Bonnie L. Carson

A. Chemical Forms, Relative Magnitude, and Ultimate Fate of Environmental Losses and Human Exposure to Them

Table IX-1 assesses the relative pollution potential of current and past human activities, particularly in the United States. The probable chemical forms, their distribution to the environment, and some of the most likely human populations exposed are suggested in this table. A summary of potential cobalt losses by new technologies is given in Table IX-2.

Coal is one of the major identified sources of environmental losses of cobalt: an estimated 240 MT Co/year are emitted to the atmosphere from coal burning and coking and possibly as much as 2,500 MT/year are discharged to land in collected fly ash and bottom slags. Scrubber waters, acid mine drainage and leachates from refuse piles may contain up to 3,800 μg Co/liter. Only cobalt mining in Missouri and copper-molybdenum mining near Ray, Arizona, are known to produce higher cobalt levels in environmental waters.

Burning fuel oils is estimated to release 100 MT Co/year to the atmosphere. Although no estimates of the absolute amounts of cobalt from manufacture or in-service attrition or corrosion have been made for superalloys, hard-facing alloys, and cemented tungsten carbides, reports of environmental contamination by cobalt are frequently linked to proximity to jet airports, automobile or truck traffic, or metallurgical industries. Because of their long service lives and the large amounts of cobalt used annually in these materials, losses of cobalt as a very small percentage of all cobalt in service could contribute cobalt to the atmosphere in amounts possibly rivaling those from burning fossil fuels.

---

* Only information not cited in the appropriate chapters and in the Summary are cited in this chapter.

TABLE IX-1

COBALT RELEASES TO THE ENVIRONMENT RELATED TO CURRENT AND PAST HUMAN ACTIVITIES

Initial Material	Process	Estimated Cobalt Content in Initial Material: mg/kg or mg/liter	Estimated Cobalt Content in Initial Material: MT/yr	Probable Initial Forms	Probable Forms Released or Transferred	Probable Ultimate Form in the Environment	Initial Concentration in Release	Ultimate Concentration in the Environment Due to Release	Primary Mode of Distribution to the Environment[a]: Air	Water	Land	Human Population Exposure and Other Remarks[b]
Coal	Burning and coking	5	2,700	Sulfides, organics?	CoO	CoO	Collected ash, ~10 to 30 mg/kg; net plume at ground level, ~4 ng/m^3; flue gas scrubber water, 100 to 700 μg/liter	0.35 ng/m^3 air	240 MT/yr		~2,500 MT/yr	1 to 10 kg Co/km^2/yr deposited within a radius of ≤ 18 km from coal-burning plants. Most authors estimate that cobalt contributions to the atmosphere from coal burning are equivalent to contributions from natural soil dust. Exposure will increase with increased reliance on coal.
	Mine and refuse pile drainage				$CoSO_4$	See salts Co^{2+}, $CoCO_3$, $Co(OH)_2$, adsorbed	≤ 3,800 μg/liter; 30 mg/kg in coal cleaning wastes	≤ 40 μg/liter		XXX	X	Concentrations of cobalt in rivers somewhat higher due to coal mine drainage than to urban discharges.
Residual Fuel oils	Burning	0.7 residual No. 6	100	Incorporated into asphaltene sheet structure; not associated with porphyrins	CoO	CoO	0.15 mg/kg in partiulate emissions[c]	0.36 ng/m^3[c]	100 MT/yr			Authors who attempt the estimates generally derive almost equivalent cobalt emissions from burning fuel oil and coal.
Urban Refuse	Burning	4	45	Impurity in paper, plastics; Ingredient of paints, putty, can coatings	CoO	CoO	Particulates, 7 to 12 mg/kg; Collected ash, 70 to 100 mg/kg; process waters, < 200 μg/liter	Air, 0.007 ng/m^3	< 0.5 MT/yr	X	< 45 MT/yr	Not considered a major source of atmospheric cobalt.
	Leaching at landfill disposal sites	4	400				4 to 70 μg/liter			X		Although observed in Norway in dry summer weather, it is a likely U.S. occurrence.

TABLE IX-1 (continued)

Initial Material	Process	Estimated Cobalt Content in Initial Material: mg/kg or mg/liter	Estimated Cobalt Content in Initial Material: MT/yr	Probable Initial Forms	Probable Forms Released or Transferred	Probable Ultimate Form in the Environment	Initial Concentration in Release	Ultimate Concentration in the Environment Due to Release	Primary Mode of Distribution to the Environment[a]: Air	Water	Land	Human Population Exposure and Other Remarks[b]
Municipal Sewage	Effluent discharge to water	0.002 to 0.040		93% CoS, 2% Co^{2+}, 2% Co-organics and 2% adsorbed	$CoCO_3$, Co organic complexes including cobalamins	Sediment	1 to 10 μg/liter	50-fold dilution in water, ≤ 70 mg/kg in sediments		XX		Near points of sewage discharge, cobalt is usually slightly more water soluble than in uncontaminated waters.
	Sludge disposal		15 to 60		CoO (air); CoS, $CoCO_3$, Co-organics (land)	Soil-CoOH complex	Avg. 10 mg/kg in dry sludge; 11 to 290 mg/kg in incinerator ash (highest due to electronics manufacture)	< 10 mg/kg	X	X	X	Little effect (if any) on plant uptake when applied to land. One of highest sewage sludge values was reported from Kokomo, Indiana, site of a major producer of cobalt alloys.
Miscellaneous Manufacturing Wastes	Discharge of effluents from:											Urban centers tend to elevate cobalt concentrations in water bodies (≲ 10 μg/liter). Urban runoff alone contains 160 μg Co/liter.
	Nickel-cadmium battery manufacture			$Co(OH)_2$		Sediment	?	< 6 to 552 mg/kg, Foundry Cove, N.Y. sediments		X		
	Aircraft finishing	≤ 0.003						≤ 10 mg/kg Narragansett Bay		X		
	Carpet dyeing	< 0.01 to 0.09		Acid dyes						X		
	Food processing	0.004 to 0.006		Trace in foods	Co EDTA, cobalamins					X		
	Metal and chemical processing	0.002 to 0.011			Various					X		
	Terephthalic acid manufacture	1.6					20 μg/liter in contaminated groundwater			X		Wilmington, North Carolina, deep well waste disposal.

TABLE IX-1 (continued)

Initial Material	Process	Estimated Cobalt Content in Initial Material: mg/kg or mg/liter	Estimated Cobalt Content in Initial Material: MT/yr	Probable Initial Forms	Probable Forms Released or Transferred	Probable Ultimate Form in the Environment	Initial Concentration in Release	Ultimate Concentration in the Environment Due to Release	Primary Mode of Distribution to the Environment[a/]: Air	Water	Land	Human Population Exposure and Other Remarks[b/]
	Leaching of solid wastes			CoP			10 to 220 µg/liter in contaminated groundwater			X		Electronics wastes generally one of highest cobalt sources in industrial wastes category.
Superalloys, High-temperature alloys, and steels	Attrition or corrosion in service	1 to 74%	~1,500	Metallic Co	Co_3O_4, CoO, $Co(OH)_2$	Co_3O_4, CoO, $Co(OH)_2$; more water extractable than natural soil cobalt (dusts also included Co from automobile traffic)	?	Airport vicinity soils and sediments about three times higher than in adjacent areas	X	X		Some evidence correlates higher environmental cobalt concentrations with closeness to jet airports.
Hard-facing alloys	Attrition or corrosion in service	1.3 to 67%	~300	Metallic Co	Co, CoO	CoO			X			Automotive valve use will decline; but for past 10 years, urban populations near automobile and truck traffic have probably been exposed to slightly higher ambient air concentrations of cobalt due to use of cobalt hard-facing alloys for valves. Coal miners and chain saw users handle hard-faced tools. Cutting fluids contain up to 550 mg Co/liter after 5 days' use, but the bulk of the Co settles in the noncirculating sludge (Einarsson et al., 1979).
Cemented tungsten carbides	Attrition in tool use	3 to 25%	~300	Co-WC	Co, CoO, $CoWO_4$	CoO, $CoWO_4$	?	?	X			Auto workers and miners are exposed
	Manufacture			Co, CoC_2	Co, CoC_2	Usually < 0.1 mg/m^3						
Cobalt salts	Electroplating and electroless coating			$Co(SONH_2)_2$, also $CoCl_2$, $Co(NH_4)_2(SO_4)_2$, $CoSO_4$	Co^{2+}, $Co(NH_3)_6{}^{2+}$	Co^{2+}, $CoCO_3$, $Co(OH)_2$, adsorbed	?	?		X		Electronics manufacturers may be one of largest point sources. Potential for substitution for chromium plating.

TABLE IX-1 (continued)

Initial Material	Process	Estimated Cobalt Content Initial Material mg/kg or mg/liter	Estimated Cobalt Content Initial Material MT/yr	Probable Initial Forms	Probable Forms Released or Transferred	Probable Ultimate Form in the Environment	Initial Concentration in Release	Ultimate Concentration in the Environment Due to Release	Primary Mode of Distribution to the Environment[a/] Air	Water	Land	Human Population Exposure and Other Remarks[b/]
	Manufacture		~ 1,200	Various	Co^{2+}, $CoCO_3$, soluble Co complexes	Co^{2+}, $CoCO_3$ $Co(OH)_2$, adsorbed	?	?		X		There are four producers of cobalt salts in Cleveland, Ohio, whose addresses are upwind of sites 4, 15, and 20 in Figure V-4. The mean cobalt concentrations in 1971-1972 at these sites were near the city-wide average (2.6 ± 0.5 ng/m^3), but maxima for these sites were somewhat elevated (6.9 to 69.0 ng/m^3).
?	Miscellaneous metal-working process	?	?					610 ng/m^3	XX			Highest ambient air concentration in Cleveland, Ohio, observed at site near, but not closest to, a plant producing beryllium-copper alloys (cobalt is an ingredient).
Cobalt ores	Mining, milling, and natural mobilization	200 to 6,000	~ 4,000 (1950's, mid-1980's)	CoAsS, CoS_2, CoS, etc.	Ore dust $CoSO_4$ to water	Same as released forms?	Tailings concentrations lower than in those in ores	Soils, up to > 100 mg/kg; water, ≤ 6,500 μg/liter		XXX	X	Lebanon and Berks counties, Pennsylvania; Fredericktown, Missouri (population < 4,000; Cobalt, Idaho (population 35).
	Roasting, smelting, and/or refining	4.2% in Missouri concentrates; 0.06 to 0.25% Oregon-California laterites							?		X	Pennsylvania materials roasted formerly in Sparrows Point (near Baltimore), Maryland. Fredericktown, Missouri, old process involved partial roasting. Precipitation residue probably discarded to land in the past in Wilmington, Delaware.
Nickel-cobalt mattes	Refining in U.S.	< 0.5 to 1.5% in imported mattes	300 to 900	CoNi sulfides	CoNiS $Co(NH_3)_6$ salts, $Co(OH)_2$, $CoSO_4$, Co metal	Same as released forms, CoO	?	?		X		Port Nickel, Louisiana (15 miles from New Orleans). Hydrometallurgical process.

TABLE IX-1 (continued)

Initial Material	Process	Estimated Cobalt Content Initial Material mg/kg or mg/liter	Estimated Cobalt Content Initial Material MT/yr	Probable Initial Forms	Probable Forms Released or Transferred	Probable Ultimate Form in the Environment	Initial Concentration in Release	Ultimate Concentration in the Environment Due to Release	Primary Mode of Distribution to the Environment[a/] Air	Water	Land	Human Population Exposure and Other Remarks[b/]
	Nickel refining by carbonyl process (non U.S.)	?	?	Co sulfides	CoO (if from roasting)	CoO	?	Avg. 48 ng/m^3	XXX		X ?	No effects attributed to cobalt in study of exposed population and environment in Swansea Valley, Wales. If cobalt is released during a roasting process, this would imply U.S. roasting operations may release cobalt to the atmosphere. A carbonyl refining process is also used at Copper Cliff, Ontario.
Nickel-cobalt laterites	Mining and milling	500		Co Mg silicate	Unchanged	Unchanged	?	?		Nil	X	
	Ferronickel production				Co	CoO	Avg. 0.002 mg/m^3 in smelter air; not observed in pond effluent		X			Both mine and smelter about 4 miles from Riddle, Oregon (population 1,042) ferronickel product contains 0.5% cobalt
Nickel salts	Manufacture of nickel salts			?	Probably mists containing Co in the same chemical form as the Ni salt being produced.		Avg. 0.002 mg/m^3 in workplace air		X	?		See copper refining comments, below.
Lead-zinc ores	Mining and milling	≤ 100		CoS	$CoSO_4$ (water) CoS (air)	Co^{2+}, $CoCO_3$, $Co(OH)_2$, adsorbed (water)	?	1 - 4.3 μg/liter	X	X		See Kellogg, Idaho, below. Values given for Missouri rivers not contaminated by Fredericktown activity.
	Roasting zinc concentrates	Concentrates avg. 250 ?		CoS	CoO	CoO	?	3 ng/m^3	X			Ambient air concentrations given for Kellogg, Idaho (population < 4,000). Electrolytic refinery at Sauget, Illinois, operating on Boss, Missouri, concentrates might emit cobalt to air.

TABLE IX-1 (continued)

Initial Material	Process	Estimated Cobalt Content Initial Material mg/kg or mg/liter	MT/yr	Probable Initial Forms	Probable Forms Released or Transferred	Probable Ultimate Form in the Environment	Initial Concentration in Release	Ultimate Concentration in the Environment Due to Release	Primary Mode of Distribution to the Environment[a/] Air	Water	Land	Human Population Exposure and Other Remarks[b/]
	Sintering zinc calcines, retorting, etc.	Calcines, avg. 250?			CoO (air); retorting residue form? (recycled, sold, or discharged)	Same as initial	?		?		X	Only two remaining pyrometallurgical zinc plants in U.S. (Palmerton and Monaca, Pennsylvania). The latter is slated to be closed in the near future (Anonymous, 1979).
	Electrolytic zinc and cadmium production			$CoSO_4$	$CoSO_4$ bled from electrolyte; copper cake to copper circuit (major); 1 to 5% of Co in Cd cake transfers to cathode Cd; $Co(OH)_3$ in iron cake (minor)	$CoSO_4$ (water)	< 100 µg/liter	?		X	X	Cadmium contains ≤ 10 mg Co/kg. A by-product containing cobalt was isolated for many years at Kellogg, Idaho.
	Sintering lead concentrates and zinc plant residues			CoS, Co	Slag, dusts, clinkers, ZnO (CoO in latter)				?		X	
	Lead refining		≤ 9 (Missouri)	CoS ?	Speiss and matte (Co sulfides); Lead desilverization crusts (Co ?)		See remarks	-	-	-	-	Speiss is roasted in Tacoma, Wash. (population 411,000) and El Paso, Texas (population 359,000). Other lead smelters send speiss to Belgium. No information on cobalt at silver refineries.
Copper ores	Mining and milling	0 to x00	2,500	Co sulfides	Co sulfides in tailings, $CoSO_4$ in water	Co sulfides, $CoSO_4$	?	≤ 4,500 µg/liter		XX	700 MT/year	Acid mine waters from < 5% of copper industry. Maximum water ultimate concentration found near Ray, Arizona. Process waters recycled in Southwest, lime precipitation practiced elsewhere.

TABLE IX-1 (continued)

Initial Material	Process	Estimated Cobalt Content Initial Material: mg/kg or mg/liter	MT/yr	Probable Initial Forms	Probable Forms Released or Transferred	Probable Ultimate Form in the Environment	Initial Concentration in Release	Ultimate Concentration in the Environment Due to Release	Primary Mode of Distribution to the Environment[a/]: Air	Water	Land	Human Population Exposure and Other Remarks[b/]
	Roasting, smelting, converting	x00 to x000	1,740	-	340 MT in recycled converter slag	Co silicate (slags), CoO (dusts)	x00 in dusts and slags	?	?	?	90 MT	See Lead refining above. No specific information found on cobalt emissions from roasting speiss or copper concentrates. Not much difference in cobalt concentration of eastern and western dusts and slags.
	Electrolytic refining				1,230 MT $CoSO_4$ in $NiSO_4$ by-product	-	-	-	-	-	-	Cathode copper ~ 80 MT (~ 60 mg/kg).
Iron ores	Mining and milling	< 4 to 64 (iron minerals); 200 to 700 (iron ores, Utah)	2,000 ?	Co_3O_4, Co_2O_3, CoO(OH), $CoCO_3$	Unchanged	Unchanged in sediments				X	?	No reports of elevated cobalt in environment due to processing of major iron ores, e.g., taconite.
	Smelting	?	?						?		?	Most of cobalt tranfers into pig iron?
Scrap iron and steel	Steel-making	Avg. ~ 1	~ 60 to 90 in 1960's	Co								No information found on emissions or wastes.
Uranium ores	Mining and milling			CoO			Not detected to < 50 μg/liter	?		X		Detection frequency 1 of 5 mills.
	Refining				$CoSO_4$ and Co hydroxides in leaching residues and iron cake			?			X	No environmental cobalt information related to processing uranium ores was found.

TABLE IX-1 (continued)

Initial Material	Process	Estimated Cobalt Content Initial Material mg/kg or mg/liter	MT/yr	Probable Initial Forms	Probable Forms Released or Transferred	Probable Ultimate Form in the Environment	Initial Concentration in Release	Ultimate Concentration in the Environment Due to Release	Primary Mode of Distribution to the Environment[a/] Air	Water	Land	Human Population Exposure and Other Remarks[b/]
	Roasting, smelting, converting	x00 to x000	1,740	-	340 MT in recycled converter slag	Co silicate (slags), CoO (dusts)	x00 in dusts and slags	?	?	?	90 MT	**See Lead refining above. No specific information found on cobalt emissions from roasting speiss or copper concentrates. Not much difference in cobalt concentration of eastern and western dusts and slags.**
	Electrolytic refining				1,230 MT $CoSO_4$ in $NiSO_4$ by-product	-	-	-	-	-	-	**Cathode copper ~ 80 MT (~ 60 mg/kg).**
Iron ores	Mining and milling	< 4 to 64 (iron minerals); 200 to 700 (iron ores, Utah)	2,000 ?	Co_3O_4, Co_2O_3, CoO(OH), $CoCO_3$	Unchanged	Unchanged in sediments				X	?	**No reports of elevated cobalt in environment due to processing of major iron ores, e.g., taconite.**
	Smelting	?	?						?		?	**Most of cobalt tranfers into pig iron?**
Scrap iron and steel	Steelmaking	Avg. ~ 1	~ 60 to 90 in 1960's	Co								**No information found on emissions or wastes.**
Uranium ores	Mining and milling			CoO			Not detected to < 50 μg/liter	?		X		**Detection frequency 1 of 5 mills.**
	Refining				$CoSO_4$ and Co hydroxides in leaching residues and iron cake			?			X	**No environmental cobalt information related to processing uranium ores was found.**

TABLE IX-1 (concluded)

Initial Material	Process	Estimated Cobalt Content Initial Material mr/kg or mg/liter	Estimated Cobalt Content Initial Material MT/yr	Probable Initial Forms	Probable Forms Released or Transferred	Probable Ultimate Form in the Environment	Initial Concentration in Release	Ultimate Concentration in the Environment Due to Release	Primary Mode of Distribution to the Environment[a/] Air	Water	Land	Human Population Exposure and Other Remarks[b/]
Gold ores	Mining and milling	≤ 2,000 in concentrates	?	Co sulfides	$Co^{II}(CN)_5^{3-}$, $Co^{II}(CN)_6^{4-}$, $CoSO_4$		≤ 2 mg/liter in South Africa acid mine drainage, 100 to 200 mg/kg in tailings	?		X	X	No reports of U.S. environmental contamination from gold processing. Gold recovered by cyanidation at Lead, South Dakota; Carlin, Nevada; Cortez, Nevada; Republic, Washington; and San Manuel, Arizona (copper ore) (McQuiston and Shoemaker, 1975).
	Smelting	?	?	Co sulfides	CoO		80 mg/kg in dusts	?	X			Environment within 4.3 km around gold smelter at Yellowknife, Northwest Territories, Canada is contaminated by cobalt. Forty-three percent of the gold produced at amalgamation and cyanidation mills is smelted (presumably roasted).
Agriculture	Runoff due to use of fertilizers and nutritional sprays	< 10	?					34.0 μg/liter				Observed in Collier County, Florida. Concentration > 40 times higher than in a water body of an undeveloped area.

a/ Where a quantitative estimate cannot be made, X's have been entered into the appropriate column: X for a minor release, XX for a moderate release, and XXX for the most important releases.

b/ Population figures from Hammond Inc. (1977).

c/ Values from the literature, which may be substatially lower according to MRI calculations.

d/ Cobalt content in ores estimated in two ways: (1) MT Co/69,400 MT concentrate for mill wastes at one plant and ~ 30% of total cobalt in the ore transfers into the mill wastes; (2) average cobalt concentration in ores assumed to be 10 mg/kg.

TABLE IX-2

POTENTIAL ENVIRONMENTAL LOSSES OF COBALT BY NEW TECHNOLOGIES

Process	Potential Environmental Loss, Etc.
Coal liquefaction Sulfur recovery Reactor off-gas Fixed bed catalyst regeneration	Cobalt carbonyl possibly emitted--most likely during shut-down operations.
Coal gasification Quenching and cooling of gasifier off-gas Sulfur recovery	Cobalt carbonyl possibly emitted--most likely during shut-down operations Estimated 5,450 MT Co/year emissions from gasifying 1 billion short tons coal. Process waters ≤ 10 μg Co/liter. Aqueous effluents 2 μg Co/liter.
Solvent refining of coal	Almost 97% of cobalt in the original coal is transferred to the residue from evaporating the solvent. Fate of residue? If discarded on land, leaching would be a problem.
Oil shale retorting Average 10 mg Co/kg	Over 99% of the cobalt in the raw shale remains in the spent shale. Process waters contain 2 to 30 μg Co/liter.
Processing domestic laterites 0.06 to 0.25% Co	Roasting possible source of atmospheric cobalt emissions. Pilot plant in Tucson, Arizona. Total losses, 10 to 20% of cobalt.
Processing low-grade nickel ores in Duluth gabbro near Ely, Minnesota	Commercialization not imminent. Possible annual cobalt production 410 to 910 MT.
Processing deep-sea manganese nodules, average 0.3% Co	Collection and concentration at sea expected to cause minimal marine disturbance. Land-based recovery methods or sites not yet chosen. Forecasts of cobalt production 2,000 to 14,000 MT/year.

Copper mining and milling is seldom implicated as a source of environmental cobalt in the United States. The cobalt in lead concentrates, however, ultimately is transferred to speiss that is roasted at the copper smelters in Tacoma, Washington, and El Paso, Texas, or sent to Belgium. Roasting operations at U.S. copper and zinc plants are likely emitters of cobalt to the atmosphere, but few U.S. data were found to substantiate this supposition. Mining, milling, and/or smelting of gold and uranium ores may be minor contributors of cobalt to environmental waters and the atmosphere. Large amounts of cobalt might be discharged to water in taconite tailings, but increased environmental cobalt concentrations have not been observed.

Electrolytic zinc and cadmium plants and electronics plants discharge aqueous and solid wastes containing cobalt. These could be major point sources of environmental cobalt because of either direct discharge to water or leaching from waste disposal sites. Disposal of electronics wastes and other industrial wastes such as those containing cobalt catalysts has been reported to contaminate groundwater.

Urban refuse burning is not thought to be a large contributor of cobalt to air, but cobalt concentrations in landfill leachates are probably fairly high. An urban runoff concentration of 160 μg Co/liter has been reported, so it is not surprising that water bodies receiving urban runoff as well as sewage effluents are more highly contaminated with cobalt than water bodies near less populated areas. Although the concentrations of cobalt in sewage effluents are not high, sewage effluents are probably a major source of cobalt in U.S. waters.

Although cobalt is a known impurity in metallic nickel and salts and has been detected in the workplace air in nickel plants, environmental contamination by cobalt from this source has not been reported. At Port Colborne, Ontario, where nickel is electrolytically refined, nickel is very high in sewage sludges ($\leq$ 700 mg/kg) (Atkins and Hawley, 1977), but cobalt is not.

In areas where cobalt is added to cobalt-deficient soils (i.e., via a superphosphate fertilizer or by direct spraying on pastures), agricultural runoff is probably a significant source of cobalt to ambient waters.

In areas where mining, milling, and/or recovery of cobalt has been renewed, additional losses of cobalt to the environment will probably not attain the levels of the 1950's because of increased environmental awareness. The former activity in Missouri left the highest environmental water concentrations of cobalt reported in the 1960's and 1970's. Pending legislation (the Idaho Wilderness Bill) may close the Blackbird Mine area to future developments (Thornton, 1979).

Of the new technologies mentioned in Table IX-2, coal gasification appears to have the greatest pollution potential. The loss, however, was estimated on the basis of gasifying almost twice as much coal as is currently burned and coked in the United States.

Levels of cobalt in urban ambient air and dust may well decline since the use of cobalt for hard-facing alloys for automotive engine valves is apparently being phased out.

The major anthropogenic losses of cobalt are as the species CoO in air emissions and as various soluble forms in effluents, including the sulfate from acid mine drainage. Cobalt probably assumes the forms clay-CoOH and adsorbed forms on manganese and iron oxides in soils and sediments and Co^{2+}, $CoCO_3$, $Co(OH)_2$ or adsorbed forms in fresh water. The relative amounts of cobalt species are highly variable among different water bodies, and can vary widely with time in the same body.

B. Assessment of the Human Health Hazard

Anemic patients have been treated with cobalt salts at oral dose levels of 0.17 to 3.9 mg Co/kg. For the smallest dose, a 70-kg adult would ingest 12 mg/day. Higher doses were associated with goiter (inhibition of iodine uptake), nerve deafness, and gastrointestinal symptoms. Even the lowest doses were higher than the amount of cobalt that could have been ingested when cobalt salts were added to beers to stabilize their foams. The cardiomyopathies reported in victims drinking beer containing 1 mg Co/liter were apparently due to a synergistic action with ethanol and associated nutritional deficiencies of heavy beer-drinkers. Cobalt has since been banned as a foam stabilizer. The natural cobalt content of beers is about 2 to 50 μg/liter.

Interestingly, chronic drinking water studies with rats at 2 mg Co/liter have been observed to have an erythropoietic effect, to cause immunosuppression, and to inhibit reflex learning. Concentrations of 200 μg/liter exhibited no such effects. In the USSR, a limit of 1 mg Co/liter in drinking water has been suggested. In light of the beer-drinkers' cardiomyopathies, surely such a concentration is ill advised for heavy consumers of alcoholic beverages.

For about 75% of people allergic to cobalt, the mean threshold concentration in water is 2.7 mg/liter. The rest are still sensitive at much lower cobalt concentrations.

Long-term exposure to cobalt aerosols at the present TLV (0.1 mg/m^3) produces diffuse lung disease and decreased vital capacity. The American Conference of Governmental Industrial Hygienists recently suggested lowering the TLV to 0.05 mg Co/m^3 because of results of an experiment wherein pigs exposed to 0.1 mg Co/m^3 for 3 months showed abnormalities in lung compliance and increased septal collagen. Rats exposed to 0.48 mg Co/m^3 for 4 months exhibited epithelial destruction. Even concentrations of 0.005 mg Co/m^3 produced pathological changes in the lungs and other organs of rats. The authors of this latter study suggest that the ambient air limit should be 0.001 mg Co/m^3. This level corresponds to 1,000 ng/m^3 or more than 20 times greater than the highest average ambient air levels of cobalt reported.

It appears that the ambient concentrations of cobalt in drinking water (up to 107 μg/liter) and in air are 1 to 5% of those concentrations known to produce adverse effects in experimental animals.

Some epidemiological studies have correlated cobalt with cancer of the lung and stomach; other studies were unable to make any correlation with cobalt. Malignancies seen in people occupationally exposed to cobalt cannot be ascribed to cobalt because the workers have been simultaneously exposed to other metals. No association of cobalt with neoplasms has been made in countries where cobalt is mined. In animal experiments, the low carcinogenicity of cobalt metal, salts, or oxides is usually manifested by tumors at the injection site. However, cobalt potentiates the skin and lung cancers induced by 20-methylcholanthrene and diethylnitrosamine, respectively.

Cobalt salts have proved mutagenic to some higher plants, yeast, rat tumor cells, and human lymphocytes but not so in tests with *Bacillus subtilis*, leukocyte cultures, *E. coli*, and bacteriophage T_4 strains. Various teratogenic effects have been observed *in vivo* or *in vitro* in wasps, mice, rats, chicks, snails, and sea urchins; but none were found in the children borne of mothers who were dosed with cobalt salts at 75 to 100 mg $CoCl_2$/day for anemia during their pregnancies. Although transfer of cobalt to the fetus is low, human placental concentrations of cobalt reflect environmental contamination.

There is considerable disagreement in the results of human diet and tissue studies as to the cobalt concentrations. The best values appear to be about 50 μg Co/day in the diet and tissue concentrations that lead to the calculation of a human body burden of 1.5 mg. The wide cobalt concentrations reported by many studies may be due: to analytical problems; or to contamination by knives, enamel ware, grinders, glazed earthenware, and polyester containers; or to cooking losses, especially if the preparation method involved water.

It is not likely that cobalt could attain dietary concentrations that would produce any adverse effects. Mollusks and fish may bioconcentrate cobalt from contaminated waters by a factor of several thousand; but the highest cobalt concentrations reported for terrestrial organisms were those of mammals, birds, reptiles, and amphibians of the tropical moist forest ecosystem. These animals showed concentration factors of about 100 based on an insect diet. In the terrestrial food chain, CF's are generally < 1. Plants and animals in the most highly cobalt-contaminated soils and water bodies tend to show CF's that do not rise proportionally with increasing contamination.

Normal concentrations of cobalt in drinking water (2 μg/liter) and in coffee (5 to 10 μg/cup) may also contribute to the daily intake. Smoking two packs of cigarettes per day at 0.1 μg Co in the smoke condensate of each cigarette might contribute as much as 4 μg/day to the intake. Except for people who are occupationally exposed, inhalation of cobalt from the air is usually not an important fraction of the daily intake. Yet strikingly elevated concentrations of cobalt were found in the lungs of long-time residents of the city of Duisburg, West Germany, where cobalt-bearing pyrites have been smelted.

C. Hazard for Other Life Forms

Table IX-3 summarizes effects on aquatic life forms and hydroponically grown rooted plants of concentrations up to 50 mg Co/liter. At concentrations most commonly found in ambient waters (< 200 μg/liter), snails show developmental abnormalities, *Daphnia* show reproductive impairment, soybeans show reduced turgidity and growth, and the alga *Chlorella fusca* shows inhibited respiration. At high concentrations, fish are generally not adversely affected below 7.3 mg Co/liter, a highly unlikely environmental level. Concentrations of 200 to 250 μg Co/liter are generally beneficial to the growth of crop plants and algae, but the adverse effects seen in soybeans at about 60 μg/liter suggest that caution be exercised in the use of cobalt-contaminated irrigation waters. The recommended limit for cobalt in irrigation

TABLE IX-3

EFFECTS OF COBALT ION ON AQUATIC ANIMALS AND ON PLANTS[a/]

Co^{2+} Concentration (μg/liter)	Effect	Species
0.16 to 8.1	Developmental abnormalities of shells, feet, and gastrulae	Limnaea stagnalis and L. palustris (snails)
0.5	Definite growth response	Tetrahymena pyriformis (protozoa)
2 to 6	No effect on growth	Chlorella alga, Phytophora (microbe)
5 (1 year)	Increased growth	Hevea brasiliensis (rubber plant)
6 to 60	Reduction in turgidity	Glycine max (soybean)
8	Optimum growth response	T. pyriformis
10	16% reproductive impairment	Daphnia magna
10 to 100	Favorable for growth	Tomato, sweet pepper, legumes, and Chlorella
45 (2 to 3 days)	Retardation of larval movement toward light	Carcinus maenas (crab)
60 to 300	Slight reduction in growth and turgidity	Glycine max
100	Growth accelerated during resorption of the yolk sac	Trout larvae
200 to 250	Favorable for growth	Tomato, sweet pepper, H. brasiliensis, Chlorella, and the alga Crucigina
200 to 350	Respiration rate reduced	Chlorella fusca
330	48-hr LC_{50}	Daphnia hyalina
≤ 450	Generally beneficial to growth	Seeds of a variety of plants
500 to 590	Growth reduced 10 to 20%	Chrysanthemum, Chlorella, Haematococcus (alga)
590 to 5,900	Teratogenic effect, especially on skeletal structure (in artificial seawater)	Sea urchin
600	Ability to convert glucose to L-malate reduced by 36%	Microflora
790	30-day LC_{50}	Austropotamobius pallipes pallipes (snail)
880	30-day LC_{50}	Orconectes limosus (snail)
990	48-hr LC_{50}	Eudiaptomus padanus padanus (fresh water zooplankton)
1,000	Growth inhibited 93%	Chlorella
1,000	Nitrogen uptake from air 12 times higher than in	Soybeans inoculated with Rhizobium
1,000	Nontoxic	One-year-old tench, carp, and char (fish) and the organisms on which they feed (crustacea, worms, insect larvae)
1,000 (56 hr)	No-effect level in tapwater	Rainbow trout
1,000 to 5,000	Growth highly inhibited	Algae and microflora
1,000 to 5,000	Favorable to growth	Oats and oak shoots
1,300	Threshold concentration for immobilization	D. magna
< 1,400 (64 hr)	Immobilization in lake water	D. magna
1,550	96-hr LC_{50} in distilled water	T. pyriformis
3,840	48-hr LC_{50}	Cyclops abyssorum prealpinus (fresh water zooplankton)
4,500 (1 day)	Retardation of larval movement toward light	C. maenas
5,000	Median threshold effect level in river water	D. magna
5,900 to 10,000	Moderate to severe growth retardation	Most kinds of plants
7,300	Death in 5 days	Fundulus spp. (fish)
8,800	96-hr LC_{50}	A. pallipes pallipes
10,000	Average survival in tapwater at this concentration no worse than that of the controls	G. aculeatus
10,200	96-hr LC_{50}	O. limosus
14,500	EC_{50} (death or immobilization) in soft water	Philodina acuticornis (rotifer)
15,000	7-day LC_{50}	G. aculeatus
20,000	96-hr LC_{50}	G. aculeatus
30,000	24-hr ED_{100} (muscle toxicity, relaxation)	Taphius glabratus (snail)
45,400	Death, softwater, pH 6.2 (ten times higher concentration required in hard water at pH 7.8)	Goldfish
50,000	48-hr LC_{50}	G. aculeatus

a/ Ordinarily terrestrial plants are included if they were grown hydroponically.

waters is 50 μg/liter; but because of adsorption by soil minerals, the effective concentration experienced by the exposed plant is considerably less.

Because plants growing on cobalt-contaminated soils do not concentrate cobalt to very high levels, most domestic animals would probably be able to tolerate such forage. For example, poultry tolerates 4 mg Co/kg feed, and ruminants, 200 mg Co/kg.

D. Conclusions and Recommendations

Generally, when cobalt is found in contaminated environments, the enrichment factor is seldom above 20. Although ambient cobalt concentrations are sometimes high enough to be detrimental to aquatic species and hydroponically grown rooted plants, cobalt is almost always associated with higher concentrations of other heavy metals. Thus it would be difficult to attribute adverse effects solely to cobalt. Cobalt at its present levels in the environment of nonoccupationally exposed individuals poses no recognized health hazard.

Because of its physiological effects, including mutagenicity, teratogenicity, and cancer-promoting activity in some species, multielement surveys of the environment, diet, and human tissue levels should continue to include cobalt. The role of methylcobalamin in methylating certain heavy metals in the environment, rendering some of them more toxic to higher life forms, should be further elucidated as well as the effect on the methylation of increasing environmental availability of noncobalamin cobalt. Because of changing consumption patterns of cobalt, compilers of future inventories of environmental cobalt may come to very different conclusions. At present, less exposure from reduced consumption might balance increased exposure from expanded U.S. production.

BIBLIOGRAPHY. CHAPTER IX

Anonymous, "St. Joe Notifies Employees, Unions of Monaca Closure. Engages SNC/Dravo in Refinery Venture," Am. Metal Market/Metalworking News, **87**(226), 5, 24 (1979).

Atkins, E. D., and J. R. Hawley, Sources of Metals and Metal Levels in Municipal Wastewaters, Canada-Ontario Agreement Research Report No. 80, Technical Committee of the Canada-Ontario Agreement on Great Lakes Water Quality, February 1977.

Einarrson, O., E. Eriksson, G. Lindstedt, and J. E. Wahlberg, "Dissolution of Cobalt from Hard Metal Alloys by Cutting Fluids," Contact Dermatitis, **5**(3), 129-132 (1979).

Hammond Inc., Hammond Collector's Edition World Atlas, Hammond Inc., Maplewood, New Jersey, 1977.

McQuiston, F. W., Jr., and R. S. Shoemaker, Gold and Silver Cyanidation Plant Practice, The American Institute of Mining, Metallurgical, and Petroleum Engineers, Inc., New York, New York, 1975.

Thornton, J., "Bill Would Kill Cobalt Self Sufficiency Hope," Am. Metal Market/Metalworking News, **87**(226) 5, 24 (1979).

INDEX